Lecture Notes in Computer Science 5757

Advanced Research in Computing and Software Science

Subline of Lectures Notes in Computer Science

Amos Fiat   Peter Sanders (Eds.)

Algorithms - ESA 2009

17th Annual European Symposium
Copenhagen, Denmark, September 7-9, 2009
Proceedings

Springer

Volume Editors

Amos Fiat
Tel Aviv University, School of Computer Science
Tel Aviv, Israel
E-mail: fiat@tau.ac.il

Peter Sanders
Universität Karlsruhe (TH), Fakultät für Informatik
Am Fasanengarten 5, 76131 Karlsruhe, Germany
E-mail: sanders@ira.uka.de

Library of Congress Control Number: 2009933188

CR Subject Classification (1998): F.2, G.1, G.2, I.1, E.5, E.1

LNCS Sublibrary: SL 1 – Theoretical Computer Science and General Issues

ISSN 0302-9743
ISBN-10 3-642-04127-2 Springer Berlin Heidelberg New York
ISBN-13 978-3-642-04127-3 Springer Berlin Heidelberg New York

springer.com

Printed in Germany

Typesetting: Camera-ready by author, data conversion by Scientific Publishing Services, Chennai, India
Printed on acid-free paper SPIN: 12747795 06/3180 5 4 3 2 1 0

Preface

This volume contains the papers presented at ESA 2009: The 17th Annual European Symposium on Algorithms, September 7–9, 2009. ESA has been held annually since 1993, and seeks to cover both theoretical and engineering aspects of algorithms. The authors were asked to classify their paper under one or more categories as described in Fig. 1.

Since 2001, ESA has been the core of the larger ALGO conference, which typically includes several satellite conferences. ALGO 2009 was held at the IT University of Copenhagen, Denmark. The five members of the ALGO 2009 Organizing Committee were chaired by Thore Husfeldt.

The ESA submission deadline was April 12, Easter Sunday. This was clearly an error and we offer profuse apologies for this mistake. Albeit no excuse, the hard constraints we faced were *(a)* ICALP notification, April 6, and *(b)* ESA in Copenhagen, September 7. Between these two endpoints we needed to design a schedule that allowed modifying ICALP rejections for resubmission (1 week), Program Committee deliberations (7 weeks), preparing final versions (4 weeks), and, to prepare, publish, and transport the proceedings (9 weeks).

ESA 2009 had 272 submissions of which 14 were withdrawn over time. Of the remaining 222 submissions to Track A (*Design and Analysis*), 56 were accepted. Of the remaining 36 submissions to Track B (*Engineering and Applications*), 10 were accepted. This gives an acceptance rate of slightly under 25%.

Authors were affiliated with institutions in 41 countries, to wit: Algeria, Argentina, Australia, Austria, Bangladesh, Belgium, Brazil, Bulgaria, Canada, Chile, China, Czech Republic, Denmark, Finland, France, Germany, Greece, Hong Kong, Hungary, Iceland, India, Ireland, Israel, Italy, Japan, Republic of Korea, The Netherlands, Norway, Poland, Romania, the Russian Federation, Singapore, Slovakia, Slovenia, Spain, Sweden, Switzerland, Taiwan – Province of China, Turkey, the United Kingdom, and the United States. Most successful were authors affiliated with institutions from Chile, Iceland, and Turkey (100 % acceptance rate).

The program also included three invited talks, "Some Open Questions Related to Cuckoo Hashing" by Michael Mitzenmacher, "Algorithms Meet Art, Puzzles, and Magic" by Eric D. Demaine, and "Google's Auction for TV Ads" by Noam Nisan.

Following the lead of Claire Mathieu at SODA 2009, ESA 2009 required that full proofs be given in the appendix. This proved very useful for the review process.

The Program Committee for Track A had 19 members, the Track B Program Committee had 14 members. Every submission not withdrawn had at least 3 reviews by members of the Program Committee, aided by 344 external reviewers (this may include duplicates), overall — 818 individual reviews.

Category	Submitted	Accepted	% acc.
Machine learning	1	0	0 %
Parallel algorithms	1	1	100 %
Quantum computing	2	0	0 %
Hierarchical memories	3	0	0 %
Databases and information retrieval	4	0	0 %
Computational biology	6	1	17 %
Streaming algorithms	7	3	43 %
Distributed computing	8	2	25 %
Pattern matching	9	0	0 %
Data compression	9	3	33 %
Mathematical programming	10	3	30 %
Randomized algorithms	20	6	30 %
Algorithmic game theory	22	6	27 %
On-line algorithms	23	5	22 %
Algorithmic aspects of networks	27	9	33 %
Data structures	30	4	13 %
Parameterized complexity	31	9	29 %
Computational geometry	37	10	27 %
Approximation algorithms	57	17	30 %
Combinatorial optimization	73	16	22 %
Graph algorithms	89	27	30 %

Fig. 1. Self characterization of the 272 submissions, a single paper may belong to multiple categories

We stand amazed at the great work and outstanding volunteer spirit of the members of both the Program Committees and the many external reviewers. Infinite thanks are due to them. Six weeks to review 40 papers seems impossible yet the reviews clearly show that very well thought out reviews can be produced in such circumstances (with the right Program Committee).

In striking contrast, the great difficulty of getting timely reviews for journal publication may be strong evidence that a major paradigm shift is in order. Given that many conference submissions now require full proofs, perhaps this suggests new directions for scientific publication.

The European association for theoretical computer science, EATCS, sponsors a best student paper award at ESA and a best paper award at ESA. A submission was deemed eligible for the best student paper award if all authors were students; there were 19 such submissions.

The best student paper award went to Heidi Gebauer for her submission "Disproof of the Neighborhood Conjecture with Implications to SAT."

The best paper award went to Christoph Dürr, Flavio Guiñez and Martín Matamala for their submission "Reconstructing 3-Colored Grids from Horizontal and Vertical Projections Is NP-Hard."

```
\documentstle{lncs}
\addtolength{\textwidth}{2cm}
\addtolength{\oddsidemargin}{-1cm}
\addtolength{\evensidemargin}{-1cm}
\addtolength{\textheight}{1.5cm}
\addtolength{\topmargin}{-0.5cm}
```

Fig. 2. Random samples of Latex source from camera-ready ESA 2009 papers

Many thanks to Thore and the Organizing Committee, who were extremely helpful throughout.

The process of producing a program and the proceedings would have been infinitely more difficult without the EasyChair system and we greatly appreciate the very commendable efforts by the system developers and supporters.

We humbly suggest that future versions of EasyChair include antivirus defenses. This feature may be useful to counteract the great ingenuity, perseverance, and skill of the authors in getting a 30-page paper to fit in 12 pages of LNCS style (See Fig. 2).

June 2009

Amos Fiat
Peter Sanders

Conference Organization

Program Committee

Design and Analysis Track

Avrim Blum	Carnegie Mellon University
Ioannis Caragiannis	University of Patras
Debora Donato	Yahoo! Research Barcelona
Uriel Feige	The Weizmann Institute of Science
Michal Feldman	The Hebrew University of Jerusalem
Amos Fiat (Chair)	Tel Aviv University
Pierre Fraigniaud	CNRS and University Paris Diderot
Klaus Jansen	University of Kiel
Rohit Khandekar	IBM T.J. Watson Research Center
Alberto Marchetti-Spaccamela	Sapienza University of Rome
Adam Meyerson	UCLA
Seffi Naor	Technion
Ely Porat	Bar Ilan University
Piotr Sankowski	Sapienza University of Rome and University of Warsaw
Jiří Sgall	Charles University, Prague
Martin Skutella	TU Berlin
Angelika Steger	ETH Zürich
Uli Wagner	ETH Zürich
Gerhard Woeginger	TU Eindhoven

Engineering and Applications Track

Tetsuo Asano	JAIST, Kanazawa
David Bader	Georgia Institute of Technology
Hannah Bast	MPI for Informatics, Saarbrücken *and* Google, Zürich
Siavash Daneshmand	University of Mannheim
Paolo Ferragina	University of Pisa
Giuseppe F. Italiano	University of Rome, Tor Vergata
Jyrki Katajainen	University of Copenhagen
Juha Kärkkäinen	University of Helsinki
Rolf Möhring	TU Berlin
Tomasz Radzik	King's College, London
Abhiram Ranade	IIT Bombay
Knut Reinert	FU Berlin
Kunihiko Sadakane	National Institute of Informatics, Tokyo
Peter Sanders (Chair)	University of Karlsruhe

Local Organization

Philip Bille	Technical University of Denmark, Copenhagen
Thore Husfeldt (Chair)	IT University of Copenhagen and Lund University
Bengt J. Nilsson	Malmö University
Rasmus Pagh	IT University of Copenhagen
Nhi Quyen Le	IT University of Copenhagen

External Reviewers

Scott Aaronson
Eyal Ackerman
Peyman Afshani
Nir Ailon
Ali Akhavi
Susanne Albers
Noga Alon
Ernst Althaus
Amihood Amir
Fabrizio d'Amore
Aris Anagnostopoulos
Alexandr Andoni
Eric Angel
Itai Ashlagi
Konstantin Avrachenkov
Yossi Azar
Yoram Bachrach
Evripidis Bampis
Nikhil Bansal
Gill Barequet
Luca Becchetti
Wolfgang Bein
Pietro Belotti
Andre Berger
Nicla Bernasconi
Nadja Betzler
Vittorio Bilò
Henrik Björklund
Andreas Bley
Johannes Blömer
Paolo Boldi
Vincenzo Bonifaci
Ilaria Bordino
Endre Boros
Glencora Borradaile
René Brandenberg
Ulrik Brandes
Andreas Brandstädt
Gunnar Brinkmann
Yves Brise
Gerth Stølting Brodal
Hajo Broersma
Niv Buchbinder
Chris Calabro
Saverio Caminiti
Alberto Caprara
Hamish Carr
Marjan Celikik
Ho-Leung Chan
Jessica Chang
Shuchi Chawla
Frederic Chazal
Panagiotis Cheilaris
Steve Chien
Flavio Chierichetti
Markus Chimani
Janka Chlebikova
Tobias Christ
George Christodoulou
Marek Chrobak
Raphael Clifford
Hagai Cohen
Vincent Conitzer
Jacomo Corbo
Graham Cormode
Derek Corneil
Jose Correa
Aaron Cote
David Coudert
Bruno Courcelle
Maxime Crochemore
Marek Cygan
Peter Damaschke
Mayur Datar
Raghavan Dhandapani
Florian Diedrich
Martin Dietzfelbinger
Shahar Dobzinski
Frederic Dorn
Daniel Dressler
Anne-Katrin Emde
Yuval Emek
Matthias Englert
David Eppstein
Amir Epstein
Leah Epstein
Thomas Erlebach
Bruno Escoffier
Angelo Fanelli
Martin Farach-Colton
Dan Feldman
Michael Fellows
Henning Fernau
Jiri Fiala
Felix Fischer
Lisa Fleischer
Rudolf Fleischer
Fedor Fomin
Dimitris Fotakis
Paolo Franciosa
Satoru Fujishige
Stefan Funke

Bernd Gärtner
Iftah Gamzu
Jie Gao
Naveen Garg
Leszek Gasieniec
Michel Goemans
Shayan Oveis Gharan
Arpita Ghosh
Raffaele Giancarlo
Joachim Giesen
Aristides Gionis
Robert Görke
Paul Goldberg
Elazar Goldenberg
Daniel Golovin
Fabrizio Grandoni
Clemens Gröpl
Roberto Grossi
Luca Gugelmann
Antonio Gullí
Jiong Guo
Mohammad Taghi Hajiaghayi
Magnus M. Halldorsson
Danny Halperin
Dan Halperin
Sariel Har-Peled
Tobias Harks
Avinatan Hassidim
Dan Hefetz
Danny Hermelin
Kirsten Hildrum
Wiebke Höhn
Michael Hoffmann
Andreas Holmsen
Stefan Hougardy
Benot Hudson
Falk Hüffner
Thore Husfeldt
Nicole Immorlica
Gabor Ivanyos
Satoru Iwata
Martin Jaggi
Vit Jelinek
Charanjit Jutla
P. Kanellopoulos
Haim Kaplan
Alexis Kaporis
George Karakostas
Nikos Karanikolas
Srinivas Kashyap
Gjergji Kasneci
Matthew J. Katz
Michael Kaufmann
Telikepalli Kavitha
Hans Kellerer
Iordanis Kerenidis
Samir Khuller
Christian Knauer
Sigrid Knust
Stephen Kobourov
Stavros Kolliopoulos
Petr Kolman
Jochen Könemann
Spyros Kontogiannis
Tsvi Kopelowitz
Guy Kortsarz
Ioannis Koutis
Miroslaw Kowaluk
Daniel Kral
Jan Kratochvil
Dieter Kratsch
Stefan Kratsch
Robert Krauthgamer
Sven Krumke
Ariel Kulik
Maria Kyropoulou
Stefan Langerman
Luigi Laura
Van Bang Le
Stefano Leonardi
Asaf Levin
Meital Levy
Moshe Lewenstein
Katrina Ligett
Andrzej Lingas
Nelly Litvak
Daniel Lokshtanov
Shachar Lovett
Hannes Luz
Marco Lübbecke
Veli Mäkinen
Yury Makarychev
Kazuhisa Makino
David F. Manlove
Martin Mares
Evangelos Markakis
Daniel Marx
Ajith Mascarenhas
Jiří Matoušek
Ross McConnell
Andrew McGregor
Aranyak Mehta
Dimitrios Michail
Pauli Miettinen
Vahab Mirrokni
Gianpiero Monaco
Luca Moscardelli
Robin Moser
Hannes Moser
Ahuva Mu'alem
M. Müller-Hannemann
S. Muthukrishnan
Torsten Mütze
Uri Nadav
Viswanath Nagarajan
Moni Naor
Saketh Nath
Gonzalo Navarro
Guyslain Naves
Hani Neuvirth
Kobbi Nissim
Steve Noble
Igor Nor
Simeon Ntafos
Zeev Nutov
Yoshio Okamoto
Svetlana Olonetsky
Hirotaka Ono
Lorenzo Orecchia
Christina Otte
Steve Oudot
Sang-il Oum
Martin Pal
Alessandro Panconesi

Vinayaka Pandit
Ondrej Pangrac
Gyula Pap
Evi Papaioannou
Srinivasan Parthasarathy
Matthias Peinhardt
Eelko Penninkx
Ulrich Pferschy
Todd Phillips
Marcin Pilipczuk
Sylvain Pion
Benny Porat
Roberto Posenato
Lars Prädel
Ariel Procaccia
Kirk Pruhs
Mathieu Raffinot
Deepak Rajan
Christoforos Raptopoulos
Pasi Rastas
Tobias Rausch
Andreas Razen
Igor Razgon
Oded Regev
Liam Roditty
Adi Rosen
Günter Rote
Amir Rothschild
Thomas Rothvoß
Alan Roytman
Natan Rubin
Srinivasa Rao Satti
Petr Savicky
Gabriel Scalosub
Guido Schäfer
Michael Schapira
Dominik Scheder
Dennis Schieferdecker
Christiane Schmidt
Warren Schudy
Roy Schwartz
Ulrich Schwarz
Ariel Shiftan
Gennady Shmonin
Anastasios Sidiropoulos
René Sitters
Shakhar Smorodinsky
Roberto Solis-Oba
Mauro Sozio
Bettina Speckmann
Frits Spieksma
Paul Spirakis
Reto Spöhel
Milind Sohoni
Anand Srivastav
Rob van Stee
Damien Stehle
Sebastian Stiller
Leen Stougie
Quentin F. Stout
Martin Strauss
S. Sudarshan
Marek Sulovsky
Jukka Suomela
Maxim Sviridenko
Zoya Svitkina
John Talbot
Xuehou Tan
Martin Tancer
Kanat Tangwongsan
Orestis Telelis
Evimaria Terzi
Jukka Teuhola
Torsten Tholey
Henning Thomas
Mikkel Thorup
Jarkko Toivonen
Patrick Traxler
Panayiotis Tsaparas
Dekel Tsur
Torsten Ueckerdt
Antti Ukkonen
Pavel Valtr
Suresh Venkatasubramanian
Rossano Venturini
Éric Colin de Verdière
José Verschae
Stefan Vigerske
Jan Vondrak
Tjark Vredeveld
Niko Vuokko
David Weese
Matthias Weller
Matthias Westermann
Ryan Williams
Pawel Winter
Mingyu Xiao
Neal Young
Raphael Yuster
Aviv Zohar
Philipp Zumstein
Uri Zwick

Table of Contents

Invited Talk

Trees

Geometry I

Mathematical Programming

Geometry II

Algorithmic Game Theory I

Geometry III

Algorithmic Game Theory II

Navigation and Routing

Invited Talk

Graphs and Point Sets

Bioinformatics

Wireless Communications

Flows, Matrices, Compression

Scheduling

Streaming

Online Algorithms

Bluetooth and Dial a Ride

Invited Talk

Decomposition and Covering

Algorithm Engineering

Parameterized Algorithms I

Data Structures

Parameterized Algorithms II

Hashing and Lowest Common Ancestor

Best Paper Awards

Some Open Questions Related to Cuckoo Hashing

Michael Mitzenmacher*

School of Engineering and Applied Sciences
Harvard University, Cambridge, MA 02138
michaelm@eecs.harvard.edu

Abstract. The purpose of this brief note is to describe recent work in the area of cuckoo hashing, including a clear description of several open problems, with the hope of spurring further research.

1 Introduction

Hash-based data structures and algorithms are currently a booming industry in the Internet, particularly for applications related to measurement, monitoring, and security. Hash tables and related structures, such as Bloom filters, dictionaries, and their derivatives, are used billions of times a day, and new uses keep proliferating. Indeed, one of the most remarkable trends of the last five years has been the growing prevalence of hash-based algorithms and data structures in networking and other areas. At the same time, the field of hashing, which has enjoyed a long and rich history in computer science (see e.g., [28]), has also enjoyed something of a theoretical renaissance. Arguably, this burst of activity began with the demonstration of the power of multiple choices: by giving each item multiple *possible* hash locations, and storing it in the least loaded, remarkably balanced loads can be obtained, yielding quite efficient lookup schemes [4,7,23,30,38]. An extension of this idea, cuckoo hashing, further allows items to be moved among its multiple choices to better avoid collisions, improving memory utilization even further.

In this brief note I plan to describe some recent work in the area of cuckoo hashing, providing some focus on several remaining open problems, with the hope of spurring further research. The presentation may admittedly be somewhat biased, focusing on my own recent research in the area; this is hopefully excused by the fact that this note is written in conjunction with an invited talk for the 2009 ESA conference in Denmark. The topic seems apropos; the paper introducing cuckoo hashing by Pagh and Rodler appeared in the 2001 ESA conference, also held in Denmark! [34,35] Also for this reason, the focus here will be primarily on *theoretical* results and problems. There is of course also recently a great deal of interesting work in hashing combining theory and practice, as detailed for example in the survey [27].

* Supported in part by NSF grants CNS-0721491 and research grants from the Cisco University Research Program, Yahoo! University Research Program, and Google University Research Program.

A. Fiat and P. Sanders (Eds.): ESA 2009, LNCS 5757, pp. 1–10, 2009.

2 Background : Multiple-Choice Hashing and Cuckoo Hashing

The key result behind multiple choice hashing was presented in a seminal work by Azar, Broder, Karlin, and Upfal [4], who showed the following: suppose that n items[1] are hashed sequentially into n buckets by hashing each item d times to obtain d choices of a bucket for each item, and placing each item in the choice with the smallest current number of items (or *load*). When $d = 1$, which is standard hashing, then the maximum load grows like $(1 + o(1))(\log n/\log\log n)$ with high probability [22]; when $d \geq 2$, the maximum load grows like $\log\log n/\log d + O(1)$ with high probability, which even for 2 choices gives a maximum load of 6 in most practical scenarios. The upshot is that by giving items just a small amount of choice in where they are placed, the maximum load can be greatly reduced; the cost is that now d locations have to be checked when trying to look up the item, which is usually a small price to pay in systems where the d locations can be looked up in parallel. A variant later introduced by Vöcking [38], that we refer to as d-left hashing, both gives slightly improved performance and is particularly amenable to parallelization. The hash table is split into d equal-sized subtables; when inserting an item, one bucket is chosen uniformly and independently from each subtable as a possible location; the item is placed in the least loaded bucket, breaking ties to the left. This combination of splitting and tie-breaking reduces the maximum load to $\log\log n/d\phi_d + O(1)$, where ϕ_d is the asymptotic growth rate of the dth order Fibonacci numbers [38].

In practice, the $\log\log n$ terms are so small in the analysis above that one can generally assume that a suitably sized bucket will never overflow. As noted for example in [7], this effectively means that d-left hash tables can provide an "almost perfect" hash table in many settings, which can then be used to bootstrap further data structures. The hash table is only almost perfect in that technically there is some probability of failure, and of course it is not minimal in terms of size.

Cuckoo hashing [35] is a further variation on multiple choice hashing schemes. In the original description, an item can be placed in one of two possible buckets. But if on insertion there is no room for an item at any of its two choices, instead of this causing an overflow, we consider *moving the item* in one of those buckets to the other location consistent with its set of two choices. Such a move may require the move of yet another element in another bucket to prevent overflow, and so on until an empty spot for the current item is found (or until sufficiently many attempts have been made to declare a failure). An excellent picture and description is available on Wikipedia's entry for cuckoo hashing, and I encourage everyone who has not already read this entry to do so now. The name cuckoo hashing comes from the cuckoo bird in nature, which kick other birds out of their nest, much like the hashing scheme recursively kicks items out of their location as needed. Successfully placing an element corresponds to finding an augmenting

[1] We use the term *item* for the objects to be hashed, which are generally keys or key-value pairs; we assume throughout that items are a fixed size.

path in the underlying graph where buckets are nodes and elements correspond to edges between nodes. When there are n items to be placed in $2(1+\epsilon)n$ buckets, that is when the load of the table is less than 1/2, all such augmenting paths are $O(\log n)$ in length with high probability. A failure occurs if an item can't be placed after $c \log n$ steps for an appropriately chosen constant c.

Although cuckoo hashing was originally introduced with just two choices per items and buckets of unit capacity, it was naturally generalized to situations with more than two choices per bucket and more than one item per bucket [17,19]. These variations share the properties that they require checking only $O(1)$ memory locations even in the worst case. Hence, in general, we refer to the entire range of variations as cuckoo hashing, and clarify in context when necessary. For cuckoo hashing the case of $d = 2$ choices with one item per bucket is now well understood [29,35], the cases with more choices and more items per bucket have left many remaining open questions [17,19]. The case of $d = 2$ is so well understood because there is a direct correspondence to random graphs. We can think of buckets as vertices, and items as edges, where the edge for an item connects the two vertices corresponding to its two buckets. The choice of a bucket by an item corresponds naturally to an orientation of a directed edge. For $d > 2$, there is a correspondence to random hypergraphs, which are more technically challenging, and for more than one item in a bucket, the edge orientation problems become more technically challenging. The questions that remain for these variations are both theoretically interesting and potentially important practically, as these cuckoo hashing variants can allow very high memory utilizations, significantly higher than previous multiple choice hashing schemes.

3 Insertion Times for Random Walk Cuckoo Hashing

Let us consider the *online setting* for cuckoo hashing, where new items may arrive to be inserted and old items may be deleted. Note that this is in contrast to the *offline setting*, where all items are initially present and one merely wants to make a lookup table, without updates to the underlying set. When there are $d > 2$ choices or more than one item per bucket, the question of what to do when inserting a new item is more subtle than in the case with two choices. One approach is to do a breadth first search to find an augmenting path in the underlying graph structure, looking at all paths that require one move, then two moves, and so on. For constant d in both settings it is known that an insertion only takes constant expected time, although high probability bounds on the insertion time are generally very weak [17,19]. Moreover, both because of memory and time requirements, this approach does not suitable for many practical implementations.

Let us describe an alternative approach generally much more amenable to practical implementation, is to at each step kick out a random item. Specifically let us consider the case of one item per bucket and $d > 2$ choices; in this case, we randomly kick out of the d choices the first time, and of the $d-1$ "other choices" after the first time. This avoids the storage required for the breadth first search and is usually much faster. This approach gives a random walk on items being kicked out of their location, until an item that has an empty bucket to be placed in is found. Intuition suggests that this approach should also find an augmenting path in $O(\log n)$ steps with high probability, since at each step there seems to be a constant probability of finding an open space. While simulations suggest good, possibly logarithmic performance, the intuition is quite speculative, as it ignores dependencies in the placement of items that are troublesome for analysis.

Until recently, there was no proof of even polylogarithmic performance for the random walk cuckoo hashing approach. A current result of Frieze, Melsted, and Mitzenmacher shows that in fact with high probability over the choices of the cuckoo hashing algorithm any insertion will, with high probability, take polylogarithmic time under suitable loads for large enough numbers of choices d [21]. The argument breaks into a pair of steps: first, most buckets have an augmenting path of length at most $O(\log \log n)$ to an empty bucket; and second, the graph representing the cuckoo hashing process expands sufficiently so that, regardless of the starting point, the random walk cuckoo hashing process will find itself at one of these buckets with an augmenting path of length at most $O(\log \log n)$ to an empty bucket after only $O(\log n)$ steps. While this represents a significant step forward, the picture for random walk cucko hashing remains incomplete.

Open Question 1: Find tight bounds on the performance of random-walk cuckoo hashing in the online setting, for $d > 3$ choices and possibly more than one item per bucket.

4 Threshold Loads for Cuckoo Hashing

Cuckoo hashing schemes appear to have natural load thresholds. As the number of items approaches some constant c times the number of buckets (where c depends on the variant of cuckoo hashing), the time to find an augmenting path increases, and as one reaches the threshold collisions become unavoidable. Given the connection to random graphs, this behavior in unsurprising. Indeed, when $d = 2$ and there is just one item per bucket, it is known that cuckoo hash tables with load less than $1/2$ succeed with high probability, but fail when the load is larger than $1/2$. See [29] for more detailed analysis. There is a large jump in moving to $d = 3$ choices, where the threshold appears to be around a 91% load based on experiments.

When $d = 2$ and there is more than one choice per bucket, results are well understood for the offline case. Again thinking of buckets as vertices and items as edges, the problem in the offline case becomes how to orient each edge so that no vertex has degree more than k. Hence the problem corresponds to the threshold for k-orientability on random graphs, which provides a framework for finding the threshold [8,18]. Because in the offline case there is no moving of items needed, as items are simply placed, whether these loads can be achieved by a natural cuckoo hashing variant in the online setting remains open. Specifically, it would be intereseting to determine if the threshold is the same for random walk cuckoo hashing, or for a different scheme with constant average time and logarithmic time with high probability per insertion and deletion.

When $d > 2$ choices (and one item per bucket), the threshold for the offline case is also nearly settled. Upper bounds on the theshold can found by again viewing the problem as an orientation problem on random hypergraphs, and while some additional considerations are needed, an upper bound can be calculated [5]. Lower bounds have been achieved, based on a new approach for designing dictionary and retrieval structures, based on matrix techniques [15]. (See also [36].) These techniques are quite interesting and highly recommended but a full description is beyond the scope of this short note; essentially, one utilizes a full-rank matrix with at most d ones per column derived from a hash function on the set of keys, and solves for a vector such that the multiplication of the matrix times the vector yields the value associated with each key. Storing the vector is then sufficient to generate the value associated with each key, and further requires just d lookups into the vector. As a specific example, for the important case of $d = 3$, there is an upper bound of 0.9183 for the threshold load [5], and a lower bound of 0.8894 [15]. Again, however, the question of bounds for efficient algorithms in the online setting remains more open.

Open Question 2: Tighten the bounds on the thresholds on the load capacity of cuckoo hashing schemes for $d > 2$ choices and 1 item per bucket for the offline setting.

Open Question 3: Prove bounds on thresholds for other settings, such as for cuckoo hashing with $d > 2$ choices and more than 1 item per bucket (offline or online), or for specific or general online schemes.

5 Using Stashes and Queues with Cuckoo Hashing

The failure rate of cuckoo hashing is surprisingly high. With standard cuckoo hashing using $d = 2$ choices, if n items are placed into $2(1 + \epsilon)$ buckets, the probability of a failure – that some item can't be placed or takes too long to place – is $\Theta(1/n)$, with the constant factor in the asymptotic notation depending on ϵ [29]. In theoretical papers the standard suggested response is to rehash everything in case of such a failure; this does not change the important fact that the expected average insertion time per item is constant. Rehashing, however, is unsuitable for many applications. The failure rate is smaller with more choices of items [19] or more items per bucket [17], but the high failure probability still remains a potential problem.

In [26] we show that one needs only a small, constant-sized *stash* to greatly reduce the probability of a failure. A stash should be thought of as a small, fully-associative memory, that allows an arbitrary lookup in a single time step. In hardware, this can be implemented as a content-addressable memory (CAM), as long as the size of the stash is small, since CAMs are expensive. In software, this can be implemented with a small number of dedicated cache lines. We show a stash of constant size s reduces the probability of any failure to fall from $\Theta(1/n)$ to $\Theta(1/n^{s+1})$ for the case of $d = 2$ choices. Similar results hold for other variants, in that the failure probability provably falls linearly by a factor of the stash size s in the exponent. Such a reduction is key for scaling to applications with millions of users. The original motivation was for potential applications to routers, and applications of this result to devices using history-independent hash tables have also been suggested [33].

This idea of allowing a *small* amount of additional space to handle collisions seems quite powerful, although it is not commonly studied in theoretical work. (Interestingly, though, one can think of the seminal work on perfect hashing of Fredman, Komlós, and Szemerédi [20] in this context.) The issue of the right scale of the additional space seems to be an interesting question. For example, in other work, we have alternatively suggested using a CAM as a *queue* for pending move operations in a cuckoo hash table [24]. The advantage of this approach is it gives an effective de-amortization of cuckoo hash inserts: by queueing operations, we can arrange for inserts to have worst-case constant time (corresponding to the average time for an insert in standard cuckoo hashing). This technique appears potentially useful as an approach for deamortizing other algorithms or data structures in hardware. We conjectured in this setting that the CAM size is required to scale like $O(\log n)$, corresponding to a maximum size achieved by a queue over $O(n)$ steps. For the case of $d = 2$, this conjecture has recently been proven in [3] (see also the similar [12]).

Finally, in other work, we have considered variants that allow only one move of an item in a hash table on each insertion [25]. The motivation for this work was to consider the benefits of making the *minimum possible change* to multiple-choice hashing, which is already being used in some hardware solutions, in order to convince builders of devices to consider trying systems that allow items to move within the hash table. Besides showing significant gains, we were able to analyze several schemes using a fluid limit/differential equations analysis. Here, we require using a CAM that scales linearly in n. That is, we find such schemes require a CAM of size ϵn for a very small ϵ chosen by the designer (e.g., 0.2%). So now we have examples where the natural choice of a stash size is constant, logarithmic, and linear, depending on our overall goal.

Open Question 4: Extend the de-amortization analysis for cuckoo hashing to other variants, including the case of $d > 2$ choices. Can this de-amortization technique be applied to other related problems as well?

Open Question 5: Develop a more general theory of the power of stashes and appropriate scalings in the setting of hash tables.

6 Limited Randomness and Cuckoo Hashing

Even from the inception of cuckoo hashing, the question of how much randomness is required was considered a worthwhile question. While assuming perfectly random hash functions is useful for analysis, it is both theoretically and practically unappealing, since perfectly random hash functions are not readily available. From the connection with random graphs in the case of $d = 2$ choices, it is apparent that if each hash function is independently chosen from a $c \log n$-wise independent family for an appropriate constant c, the analysis showing expected constant time per operation continues to hold. Pagh and Rodler [35] in fact showed that a hash function family derived from the work of Siegel [37] with limited independence suffices for cuckoo hashing in the case where $d = 2$. However, these hash functions still appear to be too complex to be utilized in practice. They also experimented with weaker hash functions.

Recent advances in the area include the work of [3], where a result by Braverman [6] is used to show that the analysis of cuckoo hashing with a queue holds even with only polylogarithmically-wise independent hash functions. Cohen and Kane [11] demonstrate that 5-independence (which is slightly different than but close to 5-wise independence) is insufficient for constant amortized cost per operation for cuckoo hashing with $d = 2$ choices, but also show that only one of the two hash functions needs to be $c \log n$-wise independent to obtain constant expected time per operation.

An alternative direction, taken by Mitzenmacher and Vadhan, started with the question of why simple hash functions work so well in practice [32]. As mentioned, when analyzing hash-related data structures such as cuckoo hashing, one commonly assumes that the underlying hash functions are completely random, even though this is unrealistic. But in practice, such analysis generally turns out to be accurate, even when weak hash functions, such as pairwise independent (or universal) hash functions [9], are used.

The proposed resolution was to model the data as coming from a random source, where the i'th item X_i has at least some k bits of entropy (specifically, Renyi entropy) conditioned on the previous items $X_1, \ldots, X_{i-1}$. Then results from the theory of randomness extraction imply that when a hash function H is chosen from even a pairwise independent family, the sequence $(H(X_1), \ldots, H(X_T))$ has small statistical difference from the distribution obtained if H were a perfect hash function. That is, a weak hash function is good enough, as long as there is sufficient randomness in the data. The implications of this model apply to cuckoo hashing as well as other hashing-based algorithms and data structures. Improvements on the bounds of [32] are developed in [10].

As shown by Dietzfelbinger and Schellbach, however, one cannot use this insight blindly. They demonstrate that natural families of universal hash functions, namely multiplicative hash functions and standard linear hash functions over a prime field, fail even for fully random key sets, when the key set is sufficiently dense over the universe of keys [16]. In such cases, there is *not* sufficient entropy for the results of [32] to hold, so there is no contradiction. The implications of these results to practical settings certainly appear to be a worthy of further study.

Open Question 6: Determine better bounds on the amount of randomness needed for cuckoo hashing to be effective, either in terms of the requirements of the underlying family of hash functions, the amount of randomness in the data, or both.

7 Parallelized Variations of Cuckoo Hashing

As a final area for future work, there appears to be renewed interest in parallel algorithms for constructing hash tables and related data structures, inspired by the development of multi-core processors and other mainstream hardware that allows parallelization, such as graphics processor units (GPUs). In [2], we design a practical parallel scheme for constructing hash tables on GPUs motivated in part by cuckoo hashing techniques. The setting is offline, with all items available. Essentially, items perform the random walk cuckoo hashing approach in parallel: each item tries to place itself in its first choice; each item that fails to capture its first choice location tries to place itself it its second choice, and then it third choice. (Three choices per item were used in this implementation.) Any unplaced item then tries to kick out the item placed at its first choice, and then its second choice, and so on. In order to ensure quick convergence, a two-level scheme is used, where items are first partitioned using a separate hash function, in order to give with high probability a bounded number of items (in this case 512) per partition. The parallel cuckoo hashing approach is then run in parallel on each partition. This random partitioning trades additional space for efficiency. For details, see [2].

While there is a fair amount of historical work on parallel hashing and load balancing schemes (see, for example, [1,13,14,23,31,30]), the significant advances made in the last decade in terms of analysis and understanding of the power to move items suggests that we can obtain both stronger results and tighter analyses in theory for such parallel hashing schemes. Moreover, there may be significant opportunities for the design of efficient parallel hash table construction schemes for real hardware systems. Given the inherent potential for parallelization with multiple-choice hash tables in general and cuckoo hashing in particular, this appears to be an interesting area for future growth.

Open Question 7: Design and analyze efficient schemes for constructing and maintaining hash tables in parallel architectures, particularly modern multicore architectures.

8 Conclusion

This note provides a smattering of open questions related to the theme of cuckoo hashing. There are certainly others, and more waiting to found. Indeed, at this very conference, there are a number of papers specifically on the theme of cuckoo hashing and on the more general themes of dictionary data structures and hash-based data structures. There remains plenty of interesting work to do in this area, which offers both rich theory and practical payoff.

Acknowledgments

I thank Martin Dietzfelbinger and Rasmus Pagh for helpful discussions, assistance with references, and comments on an earlier draft of this work.

References

1. Adler, M., Chakrabarti, S., Mitzenmacher, M., Rasmussen, L.: Parallel randomized load balancing. In: Proceedings of the 27th Annual ACM Symposium on the Theory of Computing, pp. 238–247 (1995)
2. Alcantara, D., Sharf, A., Abbasinejad, F., Amenta, N., Mitzenmacher, M., Owens, J., Sengupta, S.: Real-time parallel hashing on the GPU (submitted)
3. Arbitman, Y., Naor, M., Segev, G.: De-amortized cuckoo hashing: provable worst-case performance and experimental results. In: Albers, S., Marchetti-Spaccamela, A., Matias, Y., Nikoletseas, S., Thomas, W. (eds.) ICALP 2009. Part I. LNCS, vol. 5555, pp. 107–118. Springer, Heidelberg (2009)
4. Azar, Y., Broder, A., Karlin, A., Upfal, E.: Balanced allocations. SIAM Journal of Computing 29(1), 180–200 (1999)
5. Batu, T., Berenbrink, P., Cooper, C.: Balanced allocations: Balls-into-bins revisited and chains-into-bins. CDAM Research Report LSE-CDAM-2007-34
6. Braverman, M.: Poly-logarithmic independence fools AC^0 circuits. To appear in Proceedings of the 24th Annual IEEE Conference on Computational Complexity (2009)
7. Broder, A., Mitzenmacher, M.: Using multiple hash functions to improve IP Lookups. In: Proceedings of IEEE INFOCOM, pp. 1454–1463 (2001)
8. Cain, J., Sanders, P., Wormald, N.: The random graph threshold for k-orientability and a fast algorithm for optimal multiple-choice allocation. In: Proceedings of the Eighteenth Annual ACM-SIAM Symposium on Discrete Algorithms, pp. 469–476 (2007)
9. Carter, J.L., Wegman, M.N.: Universal classes of hash functions. Journal of Computer and System Sciences 18(2), 143–154 (1979)
10. Chung, K.M., Vadhan, S.: Tight bounds for hashing block sources. In: Goel, A., Jansen, K., Rolim, J.D.P., Rubinfeld, R. (eds.) APPROX and RANDOM 2008. LNCS, vol. 5171, pp. 357–370. Springer, Heidelberg (2008)
11. Cohen, J., Kane, D.: Bounds on the independence required for cuckoo hashing (preprint)
12. Dalal, K., Devroye, L., Malalla, E., McLeish, E.: Two-way chaining with reassignment. SIAM Journal on Computing 35, 327–340 (2006)
13. Dietzfelbinger, M., Meyer auf der Heide, F.: An optimal parallel dictionary. Information and Computation 102(2), 196–217 (1993)
14. Dietzfelbinger, M., Meyer auf der Heide, F.: Simple, efficient shared memory simulations. In: Proceedings of the Fifth Annual ACM Symposium on Parallel Algorithms and Architectures, pp. 110–119 (1993)
15. Dietzfelbinger, M., Pagh, R.: Succinct data structures for retrieval and approximate membership (Extended abstract). In: Aceto, L., Damgård, I., Goldberg, L.A., Halldórsson, M.M., Ingólfsdóttir, A., Walukiewicz, I. (eds.) ICALP 2008, Part I. LNCS, vol. 5125, pp. 385–396. Springer, Heidelberg (2008)
16. Dietzfelbinger, M., Schellbach, U.: On risks of using cuckoo hashing with simple universal hash classes. In: Proceedings of the 20th Annual ACM-SIAM Symposium on Discrete Algorithms, pp. 795–804 (2009)
17. Dietzfelbinger, M., Weidling, C.: Balanced allocation and dictionaries with tightly packed constant size bins. Theoretical Computer Science 380(1-2), 47–68 (2007)
18. Fernholz, D., Ramachandran, V.: The k-orientability thresholds for $G_{n,p}$. In: Proceedings of the Eighteenth Annual ACM-SIAM Symposium on Discrete Algorithms, pp. 459–468 (2007)
19. Fotakis, D., Pagh, R., Sanders, P., Spirakis, P.: Space efficient hash tables with worst case constant access time. Theory of Computing Systems 38(2), 229–248 (2005)

20. Fredman, M., Komlós, J., Szemerédi, E.: Stoaring a sparse table with $O(1)$ worst case access time. Journal of the Association of Computing Machinery 31(3), 538–544 (1984)
21. Frieze, A., Melsted, P., Mitzenmacher, M.: An analysis of random-walk cuckoo hashing. In: Dinur, I., Jansen, K., Naor, J., Rolim, J. (eds.) APPROX 2009 and RANDOM 2009. LNCS, vol. 5687, pp. 490–503. Springer, Heidelberg (2009)
22. Gonnet, G.: Expected length of the longest probe sequence in hash code searching. Journal of the Association for Computing Machinery 28(2), 289–304 (1981)
23. Karp, R., Luby, M., Meyer, F., Meyer auf der Heide, F.: Efficient PRAM simulation on a distributed memory machine. Algorithmica 16(4), 517–542 (1996)
24. Kirsch, A., Mitzenmacher, M.: Using a queue to de-amortize cuckoo hashing in hardware. In: Proceedings of the Forty-Fifth Annual Allerton Conference on Communication, Control, and Computing (2007)
25. Kirsch, A., Mitzenmacher, M.: The power of one move: hashing schemes for hardware. In: Proceedings of the 27th IEEE International Conference on Computer Communications (INFOCOM), pp. 106–110 (2008)
26. Kirsch, A., Mitzenmacher, M., Wieder, U.: More robust hashing: Cuckoo hashing with a stash. In: Halperin, D., Mehlhorn, K. (eds.) ESA 2008. LNCS, vol. 5193, pp. 611–622. Springer, Heidelberg (2008)
27. Kirsch, A., Mitzenmacher, M., Varghese, G.: Hash-based techniques for high-speed packet processing. Preprint, to appear in Algorithms for Next Generation Networks. Springer, Heidelberg (2009)
28. Knuth, D.: The Art of Computer Programming, Sorting and Searching, vol. 3. Addison-Wesley, Reading (1973)
29. Kutzelnigg, R.: Bipartite random graphs and cuckoo hashing. In: Proceedings of the Fourth Colloquium on Mathematics and Computer Science (2006)
30. Mitzenmacher, M., Richa, A., Sitaraman, R.: The power of two choices: a survey of techniques and results. In: Pardalos, P., Rajasekaran, S., Reif, J., Rolim, J. (eds.) Handbook of Randomized Computing, pp. 255–312. Kluwer Academic Publishers, Norwell (2001)
31. MacKenzie, P., Plaxton, C.G., Rajaraman, R.: On contention resolution protocols and associated probabilistic phenomena. Journal of the ACM 45(2), 324–378 (1998)
32. Mitzenmacher, M., Vadhan, S.: Why simple hash functions work: exploiting the entropy in a data stream. In: Proceedings of the Nineteenth Annual ACM-SIAM Symposium on Discrete Algorithms (SODA), pp. 746–755 (2008)
33. Naor, M., Segev, G., Wieder, U.: History-Independent Cuckoo Hashing. In: Aceto, L., Damgård, I., Goldberg, L.A., Halldórsson, M.M., Ingólfsdóttir, A., Walukiewicz, I. (eds.) ICALP 2008, Part II. LNCS, vol. 5126, pp. 631–642. Springer, Heidelberg (2008)
34. Pagh, A., Rodler, F.: Cuckoo hashing. In: Meyer auf der Heide, F. (ed.) ESA 2001. LNCS, vol. 2161, pp. 121–133. Springer, Heidelberg (2001)
35. Pagh, A., Rodler, F.: Cuckoo hashing. Journal of Algorithms 51(2), 122–144 (2004)
36. Porat, E.: An optimal Bloom filter replacement based on matrix solving. Technical report, arxiv:0804.1845v1 [cs.DS] (April 2008)
37. Siegel, A.: On universal classes of fast high performance hash functions, their time-space tradeoff, and their applications. In: Proceedings of the 30th Annual Symposium on Foundations of Computer Science, pp. 20–25 (1989)
38. Vöcking, B.: How asymmetry helps load balancing. Journal of the ACM 50(4), 568–589 (2003)

Efficient Computation of the Characteristic Polynomial of a Tree and Related Tasks

Martin Fürer*

Department of Computer Science and Engineering
Pennsylvania State University
University Park, PA 16802, USA
Visiting: ALGO EPFL
1015 Lausanne
Switzerland
and
Institut für Mathemtik
Universität Zürich
CH-8057 Zrich
Switzerland
furer@cse.psu.edu
http://cse.psu.edu/~furer

Abstract. An $O(n \log^2 n)$ algorithm is presented to compute the characteristic polynomial of a tree on n vertices improving on the previously best quadratic time. With the same running time, the algorithm can be generalized in two directions. The algoritm is a counting algorithm, and the same ideas can be used to count other objects. For example, one can count the number of independent sets of all possible sizes simultaneously with the same running time. These counting algorithms not only work for trees, but can be extended to arbitrary graphs of bounded tree-width.

Keywords: characteristic polynomial, counting matchings, counting independent sets, bounded tree-width, efficient algorithms.

1 Introduction

It is easy to find a maximum independent set or a maximum matching in a tree in linear time. The size of the latter determines the rank of the adjacency matrix and therefore the number of trailing zero coefficients of the characteristic polynomial. Still in linear time, one can also compute the number of the maximum matchings in a tree, and therefore determine the lowest non-trivial coefficient of the characteristic polynomial. But if the goal is to compute all coefficients of the characteristic polynomial or count the number of independent sets of size r for all possible values of r simultaneously, it has been believed that the necessary time would increase by a factor of n. We show that an increase by a factor of only $O(\log^2 n)$ is sufficient for such tasks.

* Research supported in part by NSF Grant CCF-0728921.

A. Fiat and P. Sanders (Eds.): ESA 2009, LNCS 5757, pp. 11–22, 2009.

For any graph with adjacency matrix A, elementary considerations of the characteristic polynomial

$$\chi(A;\lambda) = \det(\lambda I - A) = \sum_{i=0}^{n} c_i \lambda^{n-i}$$

show the well known result that

$$c_r = \sum_{\sigma} \operatorname{sgn}(\sigma)$$

where σ ranges over all directed cycle packings covering r vertices, i.e., permutations with with exactly $n-r$ fixed points and $A_{i\sigma(i)} = 1$ whenever i is not a fixed point of σ.

We consider undirected graphs without self-loops. Every single edge $\{u,v\}$ represents one directed cycle $(u,v),(v,u)$, while every undirected cycle represents two directed cycles (one in each direction). Naturally, in a tree the only directed cycles are those corresponding to single edges. Thus the number of cycle packings covering $2r$ vertices in a tree is the number of matchings of size r. We call them *r-matchings*.

For a tree or forest, the previous observation implies $c_{2r+1} = 0$ and

$$c_{2r} = (-1)^r \#r\text{-matchings}$$

(see, e.g., [1, p. 49]).

An early algorithm [2] for the characteristic polynomial of a tree runs in time $O(n^3)$. More complicated algorithms are needed for general graphs, but the time can even be improved. Computing the characteristic polynomial of an arbitrary real matrix has actually the same algebraic complexity as matrix multiplications [3] (see [4, Chap. 16]). Thus, with the fastest known algorithm, it can be computed in time $O(n^{2.376})$ [5]. All running times are based on the algebraic complexity measure where every arithmetic operation counts as one step.

As adjacency matrices of trees are sparse and have special structural properties, one could hope for faster algorithms. Indeed, there are algorithms to compute the determinant of the adjacency matrix of a tree in linear time [6] and the characteristic polynomial in time $O(n^2)$ [7] (also later rediscovered [8]).

A main result of this paper is to improve the running time for the computation of the characteristic polynomial of a tree to $O(n \log^2 n)$ using a novel divide and conquer approach. Computing the characteristic polynomial of a tree is equivalent to counting the number r-matchings simultaneously for all r. Thus, it is not astonishing that our new method can be applied to a wider class of simultaneous counting problems.

Many computational approaches use self-reduction. A problem is solved by solving a set of smaller problems of the same type. Quite often these smaller problems are not completely independent, but actually have a fair amount of common substructures. Our novel divide and conquer approach aims at using this similarity and solving the collection of smaller problems together with significant savings.

Our simultaneous counting method is not restricted to trees, but extends in a natural way to graphs of bounded tree-width k. For constant k, we obtain several $O(n \log^2 n)$ time simultaneous counting algorithms even for problems that are NP hard without a bound on the tree-width. The time improvement is always a factor of $\Omega(n/\log^2 n)$ compared to algorithms based on traditional techniques.

The area of algorithms has a significant branch dealing with parameterized complexity. The complexity of problems is not just studied depending on the size n of an instance, but together with an additional parameter k. The idea is that even for large n an instance of a difficult problem might still be easy if its parameter k is small. A problem is fixed-parameter tractable if it can be solved in time $O(f(k)n^c)$ for an arbitrary function f and a constant c. For example, the NP-complete Independent Set problem is solvable in time $O(2^k n)$ for graphs of tree-width k. Our result implies that with just a factor of $O(\log^2 n)$ more time, we can simultaneously count the number of independent sets of every size r in graphs of tree-width k.

2 Computing the Characteristic Polynomial

When we count objects like r-matchings (i.e., matchings of size r) it is convenient to encode them by a generating polynomial.

Definition 1. *With a_r being the number of r-matchings*

$$f_M(G;x) = \sum_{r=0}^{\lceil n/2 \rceil} a_r x^r$$

is the matching generating polynomial (see e.g. [9]).

This greatly simplifies the description of the algorithms, as the polynomial multiplication is actually an important computational step.

Similarly, we could define the generating polynomials for independent sets, vertex covers, dominating sets, and so on. These polynomials are defined for all graphs, and some might well be worth studying for their structural properties.

Definition 2. *With b_r being the number of independent sets of size r*

$$f_I(G;x) = \sum_{r=0}^{n} b_r x^r$$

is the independent set generating polynomial.

For trees (and forests), but not for general graphs, there is a well known strong relationship between the matching generating polynomial

$$f_M(G;x) = \sum_{r=0}^{\lceil n/2 \rceil} a_r x^r$$

Algorithm Characteristic-Polynomial:

Input: A tree $T = (V, E)$ with $|V| = n$.
Output: The coefficients $c_0, c_1, \ldots, c_n$ of the characteristic polynomial $\chi(A; \lambda) = \det(\lambda I - A) = \sum_{i=1}^{n} c_i \lambda^{n-i}$, where A is the adjacency matrix of T.
Comment: Use the fact that for trees $c_{2r+1} = 0$ and $c_{2r} = (-1)^r \#$ r-matchings.

$(a_0, \ldots, a_{\lfloor n/2 \rfloor}) =$ Matchings(T)
for $r = 1$ **to** $\lceil n/2 \rceil$ **do**
  $c_{2r-1} = 0$
for $r = 0$ **to** $\lfloor n/2 \rfloor$ **do**
  $c_{2r} = (-1)^r a_r$
Return $(c_0, \ldots, c_n)$

Fig. 1. The algorithm Characteristic-Polynomial

Algorithm Matchings:

Input: A tree $T = (V, E)$ with $|V| = n$.
Output: The vector $(a_0, a_1, \ldots, a_{\lfloor n/2 \rfloor})$ where a_r is the number of r-matchings in T.

$(a_0 + a_1 x + \cdots + a_{\lfloor n/2 \rfloor} x^{\lfloor n/2 \rfloor}) =$ Restricted-Matchings($T, \emptyset$)
Return $(a_0, \ldots, a_{\lfloor n/2 \rfloor})$

Fig. 2. The algorithm Matchings

and the characteristic polynomial $\chi(G; \lambda)$, namely

$$\chi(G; \lambda) = \lambda^n f_M(G; -\lambda^{-2})$$

This is a direct consequence of the characterization of the coefficients c_r of the characteristic polynomial for forests. $c_{2r+1} = 0$ and $c_{2r} = (-1)^r \#r$-matchings (see, e.g., [1, p. 49]).

Thus, we could actually have used the characteristic polynomial directly in our algorithms. But besides the waste of half the coefficients (being 0), the use of the matching generating polynomial is more natural. It also emphasizes that no hidden algebraic properties of the characteristic polynomial are used, and algorithms immediately generalize to counting other things like independent sets.

We describe the algorithm to compute the characteristic polynomial in detail using pseudo-code. The algorithm Characteristic-Polynomial (Figure 1) inputs a tree T and just outputs the coefficients of the characteristic polynomial after receiving the coefficients of the matching generating polynomial $f_M(T, x)$ from the algorithm Matching. The algorithm Matching itself (Figure 2) inputs the tree T and outputs the coefficients of the matching generating polynomial $f_M(T, x)$, after calling the recursive procedure Restricted-Matchings.

The actual work is done in the recursive procedure Restricted-Matchings (Figure 3). Besides the tree T, it receives a small subset U of the vertices as input. Its task is not only to compute the matching generating polynomial for T, but for the subgraphs of $T = (V, E)$ induced by $V \setminus W$ for all $W \subseteq U$. Naturally, these subgraphs of T are forests.

A minor feature of the procedure Restricted-Matchings (Figure 3) is the use of approximate sizes of graphs. The approximate size of a graph with n vertices is defined to be $2^{\lfloor \lg n \rfloor}$ where lg is the logarithm to the base 2. The procedure Restricted-Matchings repeatedly selects a pair of approximately smallest trees, i.e., trees of minimal approximate size. Approximately smallest trees are as good a smallest trees, but there is no need to sort the trees by size. A bucket for each approximate size is sufficient.

Procedure Restricted-Matchings:

Input: A tree $T = (V, E)$ and a subset $U \subseteq V$.

Output: The function $\mathbf{f}$ from the powerset of U into the polynomials $\mathbb{Z}[x]$ where for every subset $W \subseteq U$, $\mathbf{f}(W) = a_0^W + a_1^W x \ldots, a_{\lfloor n/2 \rfloor}^W x^{\lfloor n/2 \rfloor}$ with a_r^W being the number of r-matchings in $T \setminus W$ (the subtree of T induced by $V \setminus W$).

Comment: This procedure is only called for some sets U whose size is bounded by a constant.

$n = |V|$
if $n = 1$ **then** $\mathbf{f}(\emptyset) = \mathbf{f}(U) = 1$ // In this case U is either $\emptyset$ or V.
Return $\mathbf{f}$
else if $n = 2$ **then** $\mathbf{f}(\emptyset) = 1 + x$
for all non-empty sets W **do** $\mathbf{f}(W) = 1$
Return $\mathbf{f}$
$v =$ Select-Root(T, U)
Consider v to be the root of T, and let d be the degree of v.
Let $v_1, \ldots, v_d$ be the neighbors of the root v.
For $i = 1, \ldots, d$, let $T_i = (V_i, E_i)$ be the subgraph of T induced by all the vertices reachable from v_i without going through v as an intermediate vertex.
// Thus the sets E_i form a partition of E, the sets $V_i \setminus \{v\}$ form a partition
// of $V \setminus \{v\}$, and $v \in V_i$ for all i.
$U = U \cup \{v\}$
for $i = 1$ **to** d **do**
$\mathbf{f}_i =$ Restricted-Matchings$(T_i, V_i \cap U)$
$S = \{T_1, \ldots, T_d\}$
while $|S| > 1$ **do**
Let T_i and T_j be two approximately smallest trees in S of sizes n_i and n_j respectively.
// Replace T_i and T_j by their union. Call it T_k.
$n_k = n_i + n_j - 1$
$S = S \setminus \{T_i, T_j\} \cup \{T_k\}$
for all $W \subseteq U$ **do**
if $v \in W$ **then**
$\mathbf{f}_k(W) = \mathbf{f}_i(W)\mathbf{f}_j(W)$
else
$\mathbf{f}_k(W) = \mathbf{f}_i(W)\mathbf{f}_j(W) - (\mathbf{f}_i(W) - \mathbf{f}_i(W \cup \{v\}))(\mathbf{f}_j(W) - \mathbf{f}_j(W \cup \{v\}))$
Now S is a singleton $\{T_k\}$ with $T_k = T$.
Return $\mathbf{f}_k$

Fig. 3. The procedure Restricted-Matchings

The algorithms are natural, easy to understand and yet efficient. Their correctness immediately follows from the following principles.

- The well known relationship between numbers of matchings and the coefficients of the characteristic polynomial of a tree.
- The characteristic polynomial $\chi(G;\lambda)$ of a union $G = (V_1 \cup V_2, E_1 \cup E_2)$ of disjoint graphs $G_1 = (V_1, E_1)$ and $G_2 = (V_2, E_2)$ (with $V_1 \cap V_2 = \emptyset$) is the product of the characteristic polynomials $\chi(G_1;\lambda)$ and $\chi(G_2;\lambda)$. This is seen immediately from the block structure of the adjacency matrix in the definition of $\chi(G;\lambda)$ as a determinant.
- Under the same conditions the matching generating polynomials are multiplicative too. $f_M(G;x) = f_M(G_1;x)\, f_M(G_2;x)$. This follows from the fact that each matching in G_1 can be combined with each matching in G_2.
- For every vertex v with the set of neighbors $\{v_1, \ldots, v_k\}$ in any graph G, the number $\mathrm{match}(G, r)$ of r-matchings in G is decomposed as follows.

$$\mathrm{match}(G,r) = \mathrm{match}(G \setminus \{v\}, r) + \sum_{r=1}^{k} \mathrm{match}(G \setminus \{v, v_r\}, r-1)$$

- In a tree, every internal vertex v is an articulation point, meaning that almost all these graphs obtained by deleting vertices, decompose into connected components for which the product rule holds.

All these simple properties could be used to design a straightforward algorithm computing the matching generating polynomial by a simple tree traversal, computing the polynomial for the tree rooted at v recursively from the polynomials of the subtrees rooted at the children of v. The problem is that this natural algorithm runs in quadratic time.

On the positive side, this design immediately results in a linear time algorithm to count the number of maximal independent sets and maximum independent sets.

When computing the whole matching generating polynomial, we overcome the quadratic time problem by a cleverer selection of the articulation points v. The simple idea of splitting as evenly as possible is not enough. We also have to deal with the vertices of U. We don't solve just one matching problem, but one for each possible restriction on the vertices of U. We would also like to split the set U evenly. The time analysis shows that we can just switch back and forth as needed between the two objectives of splitting V and splitting U nicely.

In both cases, we want to select a point v that according to the current criterion is located in the center of the graph.

Definition 3. *For $T = (V, E)$, $U \subseteq V$, and $|U| \geq 2$, let* $\mathrm{Center}(T, U)$ *be one of the nodes $v \in V$ such that every tree in $T \setminus \{v\}$ (the subgraph of T induced by $V \setminus \{v\}$) contains at most $|U|/2$ points of U.*

For $U = V$, there are either one or two vertices v with this property. In the latter case, the procedure $\mathrm{Center}(T, U)$ in Figure 4 picks an arbitrary one of them. For

Procedure Select-Root:

Input: A tree $T = (V, E)$ and a subset $U \subseteq V$.
Output: A vertex v of T which will be viewed as the root of T.
Comment: Let $n_0 \geq 3$ be an integer constant. n_0 is an upper bound on the size of U. The choice of n_0 only affects the running time by a constant factor. $n_0 = 5$ might be the optimal choice.

```
if |U| ≥ n0 then
    Return Center(T, U)
else
    Return Center(T, V)
```

Fig. 4. The procedure Select-Root

$|E| \geq 2$ and U a set of leaves, the set of vertices with this property consists of the vertices of a path. If this path has positive length, then the procedure Center(T, U) picks any vertex of this path. A simple traversal of the tree T (with counting the number of vertices of U in the subtree of v on post-visiting v) finds a center in linear time.

An alternative approach is to keep $|U| \leq 2$. As before, for $|U| \leq 1$ Center(T, V) is called to pick a vertex minimizing the size of the largest tree in $V \setminus \{v\}$. Otherwise for $U = \{u_1, u_2\}$, Center(T, U) picks a vertex v on the path from u_1 to u_2, still minimizing the size of the largest tree in $V \setminus \{v\}$. This approach seems somewhat more efficient (by a constant factor), but is not analyzed here.

3 Time Complexity

As the previously cited papers, this paper uses the customary algebraic computation model. All arithmetic operations, including multiplications are counted as one step. This is not a serious problem, as all our numbers have at most a linear length in binary.

Analysing the procedure Restricted-Matchings, we first note that the size of U is under control as long as it is initially bounded by n_0. In the recursive call for T_i, the set $U_i = V_i \cap U \cup \{v\}$ plays the role of U.

Lemma 1. *Let $n_0 \geq 3$ be the constant used in the procedure Select-Root. If the procedure Restricted-Matchings is called with $|U| \leq n_0$ then all the recursive calls are with $|U_i| \leq n_0$. Furthermore, if $|U| = n_0$, then all $|U_i| < n_0$.*

Proof. For $|U| < n_0$ the set U_i satisfies $|U_i| \leq |U| + 1$, while for $|U| = n_0$ the algorithm is designed to split U evenly, resulting in the inequality $|U_i| \leq \lfloor |U|/2 \rfloor + 1 \leq \lfloor n_0/2 \rfloor + 1 < n_0$ for $n_0 \geq 3$. □

Therefore, as $U = \emptyset$ at the beginning, the size of the set U will stay bounded by $n_0 \geq 3$, and U does not even reach the bound n_0 twice in a row.

Let $m = n - 1$ be the number of edges in T. Assume $m \geq 1$, as the one vertex case is trivial and does not show up during recursive calls.

Lemma 2. *For $m \geq 1$ and suitable constants c, c', and c'', the running time of the procedure Restricted-Matchings is at most $c\,m \lg^2 m + c''m$ for $|U| < n_0$ and at most $c\,m \lg^2 m + c'\,m \lg m + c''m$ for $|U| = n_0$.*

Proof. The lemma trivially holds for $m = 1$. Let $m \geq 2$ and assume the lemma is true for all trees with less than m edges.

Recall that the procedure Restricted-Matchings partitions the tree T edge-wise into trees $T_1, \ldots, T_d$ with $T_i = (V_i, E_i)$, $|E_i| = m_i$, and for $|U| < n_0$, the sizes m_i are bounded by $m/2$ for all i. After the recursive calls for these trees T_i with common root v, repeatedly pairs of approximately smallest trees are merged into single trees until there is just one tree left, i.e., T has been reassembled. Obviously, there is at most one tree T' among the trees T_i with $|U_i| = n_0$. Let m' be the number of edges of T'. Let $m_{\min} = \min_i m_i$.

Claim: If after some sequence of merges of pairs of trees, we have the trees $T_1, \ldots, T_{d'}$, then the time spent for the recursive calls and the merges together has been at most

$$t(d') = c \sum_{i=1}^{d'} m_i \lg^2 m_i + c'\, m_{\min} \lg m_{\min} + b\, m' \lg m' + c''m \qquad (1)$$

where c, c' and c'' are from the lemma and b is defined by $b = c'$ if the tree T' (with $|U_i| = n_0$) exists and $b = 0$ otherwise.

For the total time, until all merges have been done, i.e., for $d' = 1$, we will show a different bound t' later.

The proof of the claim is by induction on the number of merges. The base case (just before any merges) follows immediately from the inductive hypothesis of the lemma, without any need for the second term $c'\, m_{\min} \lg m_{\min}$. For the inductive step, we look at the difference $t(d') - t(d'+1)$ of the allowed time after and before the merge of two trees T_i and T_j into T_k. We show in each case that this time difference is enough to perform the merge, i.e., to compute the polynomial for T_k from the polynomials for T_i and T_j.

First note that none of the four terms in $t(d')$ decreases during a merge (i.e., as d' decreases by 1). The first term always increases by

$$c\,(m_i + m_j) \lg^2(m_i + m_j) - c\,m_i \lg^2 m_i - c\,m_j \lg^2 m_j > 0$$

The second term increases when $m_{\min}$ increases. The last two terms clearly don't decrease.

We consider two kind of merges depending on whether the merged trees are of similar size or not. W.l.o.g., we assume $m_i \leq m_j$.

Case "not similar": Assume T_i and T_j are merged with $1 \leq m_{\min} = m_i < m_j/4$, and $\{T_i, T_j\}$ are approximately minimal, i.e., $m_\ell > m_j/2$ for all $\ell \neq i$. Here we do not assume $m_j \leq m/2$. For $|U| = n_0$, it is possible to have a large tree with m_j very close to m. Now the second term in $t(d') - t(d'+1)$ increases by at least

$$\begin{aligned}
& c' \frac{m_j}{2} \lg \frac{m_j}{2} - c' \, m_{\min} \lg m_{\min} \\
& > c' \frac{m_j}{2} \lg \frac{m_j}{2} - c' \frac{m_j}{4} \lg \frac{m_j}{4} \\
& > c' \frac{m_j}{4} \lg \frac{m_j}{2} \\
& > c' \frac{m_j}{8} \lg m_j \quad (\text{as } m_j > 4) \\
& > C' \, m_j \lg m_j
\end{aligned}$$

First C' is chosen large enough to make it possible to do the last merge in time $C' \, m_j \lg m_j$, i.e., to do the multiplications of $O(1)$ pairs of polynomials of degree m_i and m_j respectively using the fast Fourier transformation (FFT). Then we make sure c' is chosen sufficiently large that the last inequality holds.

Case "similar": Assume T_i and T_j are merged with $m_i \leq m_j \leq 4m_i$. Now the first term in $t(d') - t(d'+1)$ increases by

$$\begin{aligned}
& c\,(m_i + m_j) \lg^2 (m_i + m_j) - c\, m_i \lg^2 m_i - c\, m_j \lg^2 m_j \\
& \geq c\,(m_i + m_j) \lg^2 (\tfrac{5}{4} m_j) - c\, m_i \lg^2 m_j - c\, m_j \lg^2 m_j \\
& = c\,(m_i + m_j)((\lg \tfrac{5}{4} + \lg m_j)^2 - \lg^2 m_j) \\
& = c\,(m_i + m_j)(2 \lg \tfrac{5}{4} \, \lg m_j + \lg^2 \tfrac{5}{4}) \\
& > C \, m_j \lg m_j
\end{aligned}$$

Again C is first chosen large enough to make it possible to do the last merge in time $C \, m_j \lg m_j$, i.e., to do the multiplications of $O(1)$ pairs of polynomials of degree m_i and m_j respectively using FFT. Then we make sure c is chosen sufficiently large that the last inequality holds. This proves the claim.

At this point, we should notice that Claim (1) is not always strong enough to show the inductive step in the induction proof of the lemma. Indeed we can do better during the last merge. We claim a different bound t' instead of just $t(1)$ to hold after the last merge, when we have just one tree $T_k = T$.

$$t' = c\, m \lg^2 m + a\, m \lg m + c'' m \tag{2}$$

where $a = c'$ if $|U| = n_0$ (where U is the set associated with the tree T), and $a = 0$ otherwise.

The case with $|U| = n_0$ and therefore $|U_i| < n_0$ for all i causes no problem. Then $b = 0$, $m_{\min} = m$, and the first time bound (1) implies the second (2).

In the case $|U| < n_0$, the last merge has to be handled separately. We show that this merge is always balanced and therefore significantly cheaper. We are left with two trees with m_i and m_j edges to be merged into a tree of $m = m_k = m_i + m_j$ edges. We assume $m_i \leq m_j$. We claim $m_j < \frac{4}{5}m$. Otherwise, the large tree with more than $\frac{4}{5}m$ edges would have been produced by a merge involving a tree of size at least $\frac{2}{5}m$, omitting a tree of size at most $\frac{1}{5}m$, contradicting the rule of always merging approximately smallest trees.

Now the difference of bounds is

$$\begin{aligned} t' - t(2) &= cm \lg^2 m - c(m_i \lg^2 m_i + m_j \lg^2 m_j) - c'\, m_{\min} \lg m_{\min} - b\, m' \lg m' \\ &> cm_i(\lg^2 m - \lg^2 m_i) + cm_j(\lg^2 m - \lg^2 m_j) - 2c'm \lg m \\ &\geq cm(\lg^2 m - \lg^2 m_j) - 2c'm \lg m \\ &> cm(\lg m + \lg m_j)(\lg m - \lg m_j) - 2c'm \lg m \\ &= cm \lg(mm_i) \lg(m/m_j) - 2c'm \lg m \\ &> cm \lg m \lg \tfrac{5}{4} - 2c'm \lg m \\ &> Cm \lg m \end{aligned}$$

Once more, C has been chosen large enough to make it possible to do the last merge in time $Cm \lg m$, i.e., to do the multiplications of $O(1)$ pairs of polynomials of degree m_i and m_j respectively. Then we make sure c is chosen sufficiently large that the last inequality holds. □

Lemma 2 immediately implies the desired complexity result for the procedure Restricted-Matchings and therefore also for the algorithm Matchings.

Theorem 1. *For $|U| \leq n_0$, the running time of the algorithm Matchings is $O(n \log^2 n)$.*

4 Other Problems

The algorithms and their analysis easily transfer to many other counting problems, like computing the independent set generating polynomial, the vertex cover generating polynomial and so on. Another example is computing the number of 3-colorings which color exactly r vertices being colored red, simultaneously for every r. All these problems can be solved in time $O(n \log^2 n)$ for trees with basically the same algorithm. All we need is that these are all local properties. If a set is not independent, then you can put your finger on an edge where both incident vertices are selected.

Indeed the algorithms for these problems are even slightly easier than computing the matching generating polynomial, because the counted objects are sets of vertices not sets of edges. Instead of $\mathbf{f}(W)$ whose r-th coefficient is the number of matchings not involving the vertices of W, we would use $\mathbf{f}(U, W)$ whose r-th coefficient is the number of independent sets including all the vertices of $(U \setminus W) \cap V$ and excluding all the vertices of $W \cap V$, when counting independent sets. This would change the "if" clause of the procedure Restricted-Matchings near the end of the procedure to

$$\mathbf{f}_k(U, W) = \mathbf{f}_i(U, W)\, \mathbf{f}_j(U, W)$$

and the corresponding "else" clause to

$$\mathbf{f}_k(U, W) = \mathbf{f}_i(U, W)\, \mathbf{f}_j(U, W)/x$$

Note that now in both cases, the number of independent sets is just the product of the the number of independent sets in the two subtrees. In the second case, the vertex v is double counted, as it is in the independent sets of both subtrees. This is corrected by the division by x.

This change would be exactly the same, if we wanted to count other locally testable sets of vertices, like the numbers of vertex covers. But there are additional changes of the initialization. These changes are different for different kinds of polynomials. It would be a bit tedious to describe the complete initialization, because there are so many cases. For example, for $n = 2$ with one vertex $u \in U \setminus W$ (i.e., u is required in the set), and the other vertex $v \in V \setminus U$ (i.e., v is allowed by not required in the set), $\mathbf{f}(U, W) = x + x^2$ for Vertex Cover, but $\mathbf{f}(U, W) = x$ for Independent Set, as in both cases $\{u\}$ is the only allowed singleton set, while $\{u, v\}$ is a vertex cover but not an independent set.

5 Graphs of Bounded Tree-Width

Things get much more tedious, but the method clearly carries through graphs of bounded tree-width. Instead of the recursive calls having to deal with all ways of handling one additional vertex v of the tree, now recursive calls have the additional task of dealing with all possible ways of handling all the graph vertices assigned to the same tree vertex v. Naturally, the running time is exponential in the tree-width, but that is still just a constant. The dependence on n remains $O(n \log^2 n)$.

There is an extensive literature on graph polynomials for graphs of bounded tree-width. A main focus is on parametrized complexity of counting and evaluation problems on graphs definable in Monadic Second Order Logic [10,11,12]. For bounded tree-width these problems are solvable in polynomial time. The resulting running time is $O(f(k)n^4)$ with the dependence on the parameter $f(k)$ double exponential. When our algorithm is applied to graph polynomials for graphs of bounded tree width, then $f(k)$ is singly exponential.

There is some confusion caused by a linear time algorithm for the interlace polynomial for graphs of bounded tree-width [13]. This is a more complicated multivariate polynomial. It is important to notice that these authors have a different notion of computing a polynomial. While they compute one value of a polynomial in linear time, we compute all the coefficients of our polynomials in almost linear time.

6 Final Remark

We have presented a simple algorithmic paradigm with very wide applicability for counting problems in graphs of bounded tree-width. It always saves a factor of order $n/\log^2 n$ over previously known methods. A detailed description in general terms will be quite tedious, even though, there is no principle hurdle.

We have decided to present the method with detailed descriptions and proofs for the special, but interesting case of computing the characteristic polynomial of a tree. This problem has been investigated before, and it is not astonishing that previous progress has stopped at the "natural" bound of $O(n^2)$.

Only a new multitasking divide and conquer method has allowed to obtain a significantly more efficient algorithm. The method allows to divide according to two different natural strategies, taking turns when needed, and basically reaching both goals.

References

1. Biggs, N.: Algebraic graph theory, 2nd edn. Cambridge Mathematical Library. Cambridge University Press, Cambridge (1993)
2. Tinhofer, G., Schreck, H.: Computing the characteristic polynomial of a tree. Computing 35(2), 113–125 (1985)
3. Keller-Gehrig, W.: Fast algorithms for the characteristic polynomial. Theor. Comput. Sci. 36(2,3), 309–317 (1985)
4. Bürgisser, P., Clausen, M., Shokrollahi, M.A.: Algebraic complexity theory. Grundlehren der Mathematischen Wissenschaften or Fundamental Principles of Mathematical Sciences, vol. 315. Springer, Berlin (1997); With the collaboration of Thomas Lickteig
5. Coppersmith, D., Winograd, S.: Matrix multiplication via arithmetic progressions. Journal of Symbolic Computation 9(3), 251–280 (1990)
6. Fricke, G.H., Hedetniemi, S., Jacobs, D.P., Trevisan, V.: Reducing the adjacency matrix of a tree. Electron. J. Linear Algebra 1, 34–43 (1996) (electronic)
7. Mohar, B.: Computing the characteristic polynomial of a tree. J. Math. Chem. 3(4), 403–406 (1989)
8. Jacobs, D.P., Machado, C.M.S., Trevison, V.: An $O(n^2)$ algorithm for the characteristic polynomial of a tree. J. Combin. Math. Combin. Comput. 54, 213–221 (2005)
9. Ellis-Monaghan, J., Merino, C.: Graph polynomials and their applications ii: Interrelations and interpretations (2008)
10. Courcelle, B., Makowsky, J.A., Rotics, U.: On the fixed parameter complexity of graph enumeration problems definable in monadic second-order logic. Discrete Appl. Math. 108(1-2), 23–52 (2001)
11. Makowsky, J., Marino, J.: Farrell polynomials on graphs of bounded tree width. Advances in Applied Mathematics 30, 160–176 (2003)
12. Makowsky, J.A.: From a zoo to a zoology: Descriptive complexity for graph polynomials. In: Beckmann, A., Berger, U., Löwe, B., Tucker, J.V. (eds.) CiE 2006. LNCS, vol. 3988, pp. 330–341. Springer, Heidelberg (2006)
13. Bläser, M., Hoffmann, C.: Fast computation of interlace polynomials on graphs of bounded treewidth. CoRR abs/0902.1693 (2009); 35 pages informal publication

Improved Approximation Algorithms for Label Cover Problems

Moses Charikar[1,⋆], MohammadTaghi Hajiaghayi[2], and Howard Karloff[2]

[1] Department of Computer Science, Princeton University, Princeton, NJ 08540, USA
moses@cs.princeton.edu

[2] AT&T Labs — Research, 180 Park Ave., Florham Park, NJ 07932, USA
{hajiagha,howard}@research.att.com

Abstract. In this paper we consider both the maximization variant Max Rep and the minimization variant Min Rep of the famous Label Cover problem, for which, till now, the best approximation ratios known were $O(\sqrt{n})$. In fact, several recent papers reduced Label Cover to other problems, arguing that if better approximation algorithms for their problems existed, then a $o(\sqrt{n})$-approximation algorithm for Label Cover would exist.

We show, in fact, that there are a $O(n^{1/3})$-approximation algorithm for Max Rep and a $O(n^{1/3}\log^{2/3} n)$-approximation algorithm for Min Rep. In addition, we also exhibit a randomized reduction from Densest k-Subgraph to Max Rep, showing that any approximation factor for Max Rep implies the same factor (up to a constant) for Densest k-Subgraph.

1 Introduction

Label Cover was first introduced in Arora et al. [2] and is a canonical problem used to show strong hardness results for many NP-hard problems [12]. It is known that for Label Cover, there is no approximation algorithm achieving a ratio $2^{\log^{1-\varepsilon} n}$, for any $0 < \varepsilon < 1$, unless $\mathbf{NP} \subseteq \mathbf{DTIME}(n^{\mathrm{polylog}(n)})$ [2,12]. Label Cover has both maximization and minimization variants for both of which the above hardness holds. Kortsarz [14] introduced slight variants of these two problems called Max Rep and Min Rep. (See the end of this section for formal definitions of both problems.) Indeed Max Rep is equivalent to the maximization version of Label Cover, but Min Rep is slightly different from the minimization version of Label Cover. Kortsarz [14] showed that for both Max Rep and Min Rep, there is the same hardness of $2^{\log^{1-\varepsilon} n}$, for $0 < \varepsilon < 1$, unless $\mathbf{NP} \subseteq \mathbf{DTIME}(n^{\mathrm{polylog}(n)})$. The simpler definitions of Max Rep and Min Rep make them particularly attractive for use in hardness reductions.

⋆ Supported by NSF ITR grant CCF-0426582, NSF CAREER award CCF-0237113, MSPA-MCS award 0528414, and NSF expeditions award 0832797.

A. Fiat and P. Sanders (Eds.): ESA 2009, LNCS 5757, pp. 23–34, 2009.

For the upper bound, it is known that both MAX REP and MIN REP admit relatively simple $O(\sqrt{n})$ approximation algorithms [6,16]. Recently some authors suggested the possibility that $O(\sqrt{n})$ is the best approximation factor for these two problems. See, e.g., [8], in which the authors write, *"This ratio [$O(\sqrt{n})$] seems hard to improve and better ratio algorithms for LABEL-COVER$_{\max}$ are not known even for very simple versions of the problem (e.g., when the structure of the graph obeys the rules of the Unique Game Conjecture...). If LABEL-COVER$_{\max}$ is indeed $\Omega(\sqrt{n})$ hard to approximate, then so is DSF [Directed Steiner Forest].* Indeed several recent papers reduced MIN REP/MAX REP to other problems in order to obtain hardness results; therefore studying the approximability of MIN REP/MAX REP is an important goal. See [8] for DIRECTED STEINER FOREST, [16] for RED-BLUE SET COVER, [4,11] for SET COVER WITH PAIRS, [3] for SPARSEST k-TRANSITIVE-CLOSURE-SPANNER, [10] for MIN-POWER k-EDGE-DISJOINT PATHS, [1] for ℓ-ROUND POWER DOMINATING SET, [5] for TARGET SET SELECTION, [15] for VERTEX CONNECTIVITY SURVIVABLE NETWORK DESIGN, and [9] for STOCHASTIC STEINER TREE WITH NON-UNIFORM INFLATION.

In this paper, we refute the possibility of $\Omega(\sqrt{n})$ hardness for both MAX REP and MIN REP by developing a $O(n^{1/3})$-approximation algorithm for MAX REP and a $O(n^{1/3}\log^{2/3} n)$-approximation algorithm for MIN REP. Our result for MIN REP (see Section 2) uses a natural LP relaxation for the problem. We round this LP based on an interesting generalization of the birthday paradox. Our result for MAX REP (see Section 3) uses a direct combinatorial approach. Indeed, we show that for MAX REP the integrality ratio for a natural LP relaxation is $\Omega(\frac{\sqrt{n}}{\ln n})$ (in contrast to MIN REP for which the integrality ratio is $\Omega(n^{1/3-\varepsilon})$, for all $\varepsilon > 0$.)

Our $O(n^{1/3})$- and $O(n^{1/3}\log^{2/3} n)$-approximation algorithms for MaxRep and MinRep might suggest a connection between these problems and the related well-studied problem DENSEST k-SUBGRAPH, for which the best approximation factor so far is $O(n^{1/3-\delta})$ [7], for some small *fixed* $\delta > 0$. The current best inapproximability result only rules out a polynomial time approximation scheme (PTAS) under the assumption that $\mathbf{NP} \not\subseteq \mathbf{BPTIME}(2^{n^\varepsilon})$ [13]. We show indeed that there is a randomized reduction from DENSEST k-SUBGRAPH to MAX REP, which preserves the approximation factor up to a constant factor (see Section 4).

We end this section with exact definitions of MAX REP and MIN REP.

Definition 1. (MAX REP)
Instance: *A bipartite graph $G = (A, B, E)$, where $|A| = |B| = n$, and an equitable partition $\mathcal{A}$ of A and $\mathcal{B}$ of B into k sets of same size $q = \frac{n}{k}$ each (assuming that $n \bmod k = 0$).*
Objective: *Choose $A' \subseteq A$ and $B' \subseteq B$ with $|A' \cap A_i| = |B' \cap B_j| = 1$ for each $i, j = 1, \ldots, k$ such that the subgraph induced by $A' \cup B'$ has the maximum number of edges.*

In Definition 1 the bipartite graph and the partition of A and B induce a "supergraph" $\mathcal{H}$ in the following way: The vertices of $\mathcal{H}$ are the sets A_i and B_j.

Two sets A_i and B_j are adjacent by a "superedge" in $\mathcal{H}$ if and only if there exist $a_i \in A_i$ and $b_j \in B_j$ which are adjacent in G. In this case, we say pair (a_i, b_j) *covers* the superedge (A_i, B_j). In the MAX REP problem the goal is to select one element, called a *representative*, from each A_i and each B_j such that the number of covered superedges in $\mathcal{H}$ is maximized. Another natural objective function considered in the literature is as follows:

Definition 2. (MIN REP)
Instance: *A bipartite graph* $G = (A, B, E)$, *where* $|A| = |B| = n$, *and equitable partitions* $\mathcal{A}$ *of* A *and* $\mathcal{B}$ *of* B *into* k *sets of same size* $q = \frac{n}{k}$.
Objective: *Choose* $A' \subseteq A$ *and* $B' \subseteq B$ *such that pairs* (a, b), $a \in A'$ *and* $b \in B'$, *cover all the superedges of* $\mathcal{H}$, *while minimizing* $|A'| + |B'|$.

2 $O(n^{1/3} \log^{2/3} n)$-Approximation Algorithm for MIN REP

There is a trivial k-approximation algorithm for MIN REP, namely, select both vertices of one edge corresponding to each superedge. (The optimum selects at least one vertex of each A_i (B_j) to which there is a superedge attached and we choose at most k since there are at most k superedges attached to each A_i (B_j).) In this section, we present a $O(\sqrt{q} \log k)$ approximation algorithm for the MIN REP problem using a natural LP relaxation and a rounding scheme whose analysis is based on a generalization of the birthday paradox. By using the better of these two algorithms, and remembering that $q = n/k$, we obtain an $O(n^{1/3} \log^{2/3} n)$-approximation algorithm.

First, we start with an LP relaxation as follows:

$$\text{OPT} = \text{minimize} \quad \sum_{u \in A} p_u + \sum_{v \in B} p_v \tag{1}$$

subject to

$$\sum_{u \in A_i, v \in B_j \text{ s.t. } (u,v) \in E(G)} f_{uv} = 1 \quad \forall i, j : (A_i, B_j) \text{ is a superedge}$$

$$\sum_{v \in B_j \text{ s.t. } (u,v) \in E(G)} f_{uv} \leq p_u \quad \forall 1 \leq i, j \leq k, \forall u \in A_i$$

$$\sum_{u \in A_i \text{ s.t. } (u,v) \in E(G)} f_{uv} \leq p_v \quad \forall 1 \leq i, j \leq k, \forall v \in B_j$$

$$f_{uv} \geq 0 \quad \forall u \in A, v \in B \text{ s.t. } (u, v) \in E(G).$$

In the IP corresponding to LP 1, p_x for $x \in A \cup B$ is a binary variable which specifies whether vertex x has been chosen or not in our integral solution. (In the LP, intuitively it specifies the fraction of vertex x that is chosen.) In the IP, for all i, j such that (A_i, B_j) is a superedge, choose $u \in A_i, v \in B_j$ such that u, v are both chosen and set $f_{uv} = 1$; set $f_{u'v'} = 0$ for all other $u' \in A_i, v' \in B_j$. (In the LP, f specifies the "flow" from u to v and satisfies capacity constraint p_x on each *vertex* $x \in A \cup B$.)

Our algorithm, called MinRepAlg, for rounding LP 1 is relatively simple, though its proof is involved and is based on an interesting generalization of the birthday paradox. The algorithm is as follows.

1. Find an optimal solution f^*, p^* to LP 1.
2. For each $x \in A \cup B$, let $p^1_x = \min\{1, \sqrt{q}p_x\}$.
3. Let $S_1 = \emptyset$ be the current set of selected elements.
4. Repeat the following $O(\log k)$ times: for each vertex $x \in (A \cup B) - S_1$, flip an independent biased coin and put x into S_1 with probability p^1_x.

Since in MinRepAlg, we amplify each (probability) variable p by a factor $\sqrt{q}$, the objective function would be at most $\sqrt{q}$ times the optimum solution to LP 1. Next, we show that, for each currently-uncovered superedge, the probability that one iteration will cover that superedge is boundedly away from 0. Since there are at most k^2 such pairs, with high probability after $O(\log k)$ iterations of the while loop, with total cost $O(\sqrt{q}\log k)z^*_{LP}$, we cover all superedges in supergraph $\mathcal{H}$.

Theorem 1. *If we choose each vertex $x \in A \cup B$ with probability p^1_x, any single superedge (A_i, B_j) is covered with constant probability.*

The proof of the above theorem uses the following lemma.

Lemma 2. *Consider a superedge (A_i, B_j) for which LP 1 routes one unit of flow f from vertices $u \in A_i$ to $v \in B_j$ and satisfies capacity constraints with respect to the p variables. Then there exists a flow $\hat{f}$ from vertices $u \in A_i$ to vertices $v \in B_j$ that*

1. *has value at least $\frac{1}{3}$ and at most 1,*
2. *that satisfies the capacity constraint p_x on each vertex $x \in A_i \cup B_j$, and*
3. *such that every nonzero $\hat{f}_{uv}$ is at least $1/(6q)$.*

Proof. We start with a flow $f' = f$ initially and decrease it in iterations until its flow from A_i to B_j becomes at most $\frac{1}{3}$. We also maintain node capacities $p' = p$ initially and reduce them in each iteration maintaining the property that the modified flow f' is a feasible flow for the modified node capacities p'. We build a new flow $\hat{f} = 0$ initially and increase it iteratively such that in each iteration, we increase flow $\hat{f}$ by at least some α on one edge from A_i to B_j and simultaneously, we decrease flow f' by at most 2α; we will ensure that $\alpha \geq 1/(6q)$. We do this increasing of $\hat{f}$ and decreasing of f' in such a manner that flow $f' + \hat{f}$ always satisfies capacity constraints p. Thus when the flow f' becomes less than $\frac{1}{3}$, the flow $\hat{f}$ is at least $\frac{1}{3}$ and we are done.

Now consider f' whose flow is at least $\frac{1}{3}$ during the process. Let $A'_i = \{x \in A_i, p'_x < \frac{1}{6q}\}$ and $B'_j = \{x \in B_j, p'_x < \frac{1}{6q}\}$. First, we show that there is an edge $(u, v) \in E(G)$ with $u \in A_i - A'_i$ and $v \in B_j - B'_j$. If it is not the case, all flow of f' should pass through either a vertex of A'_i or a vertex B'_j and thus its flow is less than $2q\frac{1}{6q} = \frac{1}{3}$, a contradiction. Let $\alpha = \min\{p'_u, p'_v\} \geq \frac{1}{6q}$. We now add a flow α from u to v in $\hat{f}$ and reduce the flow f' as follows: Assume without loss of generality that $p'_u \leq p'_v$. Then we reduce the flow in f' along all edges incident on u to zero. Note that the total flow reduction in this step is at most α. We also reduce flow arbitrarily along edges incident to v other than (u, v) such that the total flow on these edges is at most $p'_v - \alpha$. The total flow reduction in this step is also bounded by α. Finally, we reduce p'_u and p'_v by α. This maintains the property that flow f' is feasible for capacities p', and that $f' + \hat{f}$ is feasible for the original capacities p. In this way, the flow of f' is decreased by at most 2α, and the flow of $\hat{f}$ on any one edge has been increased by at least $\frac{1}{6q}$. □

We are now ready to prove Theorem 1.

Proof. [**of Theorem 1**] Fix i, j such that (A_i, B_j) is a superedge. First by Lemma 2, we obtain a flow $\hat{f}$ from the flow f in LP 1 and use the properties of $\hat{f}$ instead of f in the statement of the lemma in the rest of the proof. Let $\hat{p}_x \leq p_x$, for each vertex $x \in A_i \cup B_j$, be the total flow of $\hat{f}$ passing through vertex x.

If $\hat{p}_x \geq \frac{1}{\sqrt{q}}$, for $x \in A_i \cup B_j$, then since $\sqrt{q}p_x \geq \sqrt{q}\hat{p}_x \geq 1$, vertex x is chosen in our random selection with probability 1. Without loss of generality, assume that $x \in A_i$. Let N_x be the set of all vertices $y \in B_j$ for which x has positive flow $\hat{f}_{xy}$ to y. If there is a y with $\hat{p}_y \geq \frac{1}{\sqrt{q}}$, then vertex y is also chosen in our random selection with probability 1. Thus in this case we will satisfy the superedge (A_i, B_j) with probability 1 and we are done. If it is not the case, then each vertex $y \in N_x$ will be selected in our random process with probability at least $\sqrt{q}\hat{f}_{xy}$. This means that the probability that we do not select in our random process any vertices in N_x is at most $\Pi_{y \in N_x}(1 - \sqrt{q}\hat{f}_{xy}) \leq e^{-\sqrt{q}\sum_{y \in N_x} \hat{f}_{xy}} \leq e^{-\sqrt{q}\frac{1}{\sqrt{q}}} = \frac{1}{e}$ (since $\sum_{y \in N_x} \hat{f}_{xy} = \hat{p}_x \geq \frac{1}{\sqrt{q}}$). Thus with probability at least $1 - \frac{1}{e}$, we select a vertex in N_x and thus satisfy the superedge (A_i, B_j).

In the rest of the proof, we assume $\hat{p}_x \leq \frac{1}{\sqrt{q}}$, for $x \in A_i \cup B_j$, and thus $\sqrt{q}\hat{f}_{uv} \leq 1$ for $u \in A_i$ and $v \in B_j$.

The outline of the rest of the proof is as follows. Instead of directly analyzing the probability that the randomized rounding chooses both endpoints of some edge in $G[A_i \cup B_j]$, for a general bipartite graph between A_i and B_j with $q = |A_i| = |B_j|$, we first transform the bipartite graph, in a natural way, into a perfect matching graph. We do this by replacing a vertex v of degree $d(v)$ by $d(v)$ "clones," associating a different edge incident to v with each clone, and keeping the flow values on edges unchanged. We then choose each clone with probability $\sqrt{q}$ times the flow on the incident edge. (Note that the scaling factor is the square root of q, not the square root of the number of boys or girls in the perfect matching.) We argue that with at least positive constant probability, there is an edge e of $G[A_i \cup B_j]$ with at least one clone of each endpoint of e chosen. However, this is not what algorithm MinRepAlg does, in fact (it doesn't detour through a perfect matching graph), so we then argue that the probability that algorithm MinRepAlg chooses both endpoints of some edge of $G[A_i \cup B_j]$ is at least as high as it is in the perfect matching, and hence at least a positive constant.

First, we construct a bipartite graph $M = (A', B', E')$, for the given i, j, in which for each vertex $a \in A_i$ (resp., $b \in B_j$), we put r vertices $a^1, a^2, \ldots, a^r$ (resp., $b^1, b^2, \ldots, b^r$) in A' (resp., B') called *clones* of vertex a (resp., b), where r is the number of edges incident to a (resp., b) in $G[A_i \cup B_j]$ that carry nonzero flow (and thus a flow of at least $\frac{1}{6q}$) in $\hat{f}$. We associate each clone a^i of a (resp., b^j of b) with a different edge of G incident to a (resp., b). We put edges between vertices (clones) in E' corresponding to edges in $G[A_i \cup B_j]$ that carry a nonzero flow in $\hat{f}$ (and we put this flow as the flow of the new edge). Since each edge carrying positive flow in the bipartite graph between A_i and B_j gives rise to one edge in M whose endpoints have degree 1, M is a bipartite perfect matching, with $|A'| = |B'|$, which is at most the number of edges in $G[A_i \cup B_j]$.

We now consider a random process in which we build a set S by selecting each vertex (clone) c in M independently with probability $\sqrt{q}$ times the $\hat{f}$ flow of the unique edge incident to c in M. Let the subset of $A_i \cup B_j$ chosen by MinRepAlg be called S_1. We will prove two things: (1) first, that the chance that, in the perfect matching graph M, S contains an edge, is at least $1 - e^{-1/54} > 0$, and (2) second, the chance that S_1 contains an edge in G is at least as large as the chance that S contains an edge in the perfect matching graph M.

Now we prove (1), that both endpoints of some edge in M are chosen with constant probability. Consider one fixed edge $d = (c_A, c_B)$ carrying a flow $\hat{f}_d$. We select both c_A and c_B with probability $(\sqrt{q}\hat{f}_d)^2 = q\hat{f}_d^2$. Thus with probability $1 - q\hat{f}_d^2$, edge d will be not selected. The probability that no edges are selected then is at most $\Pi_{d \in E'}(1 - q\hat{f}_d^2)$. (We have independence because the graph is a perfect matching.) Since each edge carries a flow of at least $\frac{1}{6q}$ and the total flow is at most one (by Lemma 2), $|E'| \leq 6q$. Hence, since flow of $\hat{f}$ is at least $\frac{1}{3}$ by Lemma 2, $\Pi_{d \in E'}(1 - q\hat{f}_d^2) \leq e^{-q\sum_{d \in E'} \hat{f}_d^2} \leq e^{-q\frac{(\sum_{d \in E'} \hat{f}_d)^2}{|E'|}} \leq e^{-q\frac{(\frac{1}{3})^2}{6q}} = e^{-\frac{1}{54}}$. Thus with constant probability $1 - e^{-\frac{1}{54}} > 0$ we satisfy any one given superedge. This completes the proof of (1).

Now we prove (2). Build a new probabilistic process as follows. Define p_x^2, for $x \in A_i \cup B_j$, to be the probability that at least one of the clones of x is chosen to be in S. This is, of course, at most the sum of the probabilities that each individual clone is chosen to be in S, which is itself at most the probability that $x \in S_1$ (since the flow values add). Build a set S_2 by choosing each node $x \in A_i \cup B_j$ independently with probability p_x^2. The algorithm, on the other hand, builds S_1 using probabilities $p_x^1 \geq p_x^2$. It is a fairly obvious fact that, since $p_x^1 \geq p_x^2$, the chance that S_1 contains an edge is no smaller than the chance that S_2 contains an edge, but we prove it anyway.

Lemma 3. *Suppose we are given an r-node graph H and two probabilities $p_x^1 \geq p_x^2$ for each vertex x. Consider experiment E_ℓ, for $\ell = 1, 2$, with probability measure P_ℓ, in which we build set S_ℓ by putting each vertex x into S_ℓ with probability p_x^ℓ, independently. Then $P_1[S_1$ contains an edge$] \geq P_2[S_2$ contains an edge$]$.*

Proof. We can pick *one* sequence of r independent random reals ξ_x in $[0,1]$ and put x into S_ℓ if $\xi_x \leq p_x^\ell$. As $S_2 \subseteq S_1$ always, in every run in which S_2 contains an edge, so does S_1. □

But now we can view the construction of S_2 as putting a node x into S_2 if and only if at least one of its clones is chosen for S. It is clear that S contains an edge in M implies that S_2 contains an edge in G (but not the converse), so that the chance that S contains an edge is dominated by the chance that S_2 contains an edge, which itself is dominated by the chance that S_1 contains an edge, and we are done with the proof of Theorem 1. □

2.1 The Integrality Ratio of Min Rep

Next, we show that the integrality ratio of LP 1 is indeed $\Omega(n^{\frac{1}{3}-\varepsilon})$ for all $\varepsilon > 0$ and thus our algorithm in this section is essentially the best that we can hope for using the LP.

Theorem 4. *The integrality ratio of LP 1 for* Min Rep *is* $\Omega(n^{1/3-\varepsilon})$ *for any* $\varepsilon > 0$*, for all large enough* n.

Proof. Consider an instance of Min Rep with $k = n/q$ groups of q boys each and $k = n/q$ groups of q girls each. Between the ith group A_i of boys and the jth group B_j of girls there is a random perfect matching. It is clear that one can assign $f_e = 1/q$ for any edge e and $p_u = 1/q$ for any vertex u. This implies that $z_{LP}^* \leq 2n/q$. To study the integrality ratio, we look at the smallest feasible set S (i.e., the smallest set of vertices such that for all i, j, there is at least one edge between A_i and B_j both of whose endpoints are in S). Let S be a feasible set, $s = |S|$. The size s of S is the sum of $2n/q$ terms, one for each A_i and B_j. Let $a = s/(2n/q) = sq/(2n)$, the average size of the intersection of S with some A_i or B_j. Of the $2n/q$ terms, whose sum is s, fewer than $1/4$ of them (i.e., $(1/2)n/q$) can exceed $4a = 2sq/n$, and hence at least $(3/2)n/q$ of them are at most $4a$. Since at most n/q of them can be intersections with the n/q A_i's, at least $(1/2)n/q$ of them are intersections with n/q B_j's. Similarly, at least $(1/2)n/q$ of them are intersections with the n/q A_i's. Hence there are sets $I \subseteq \{1, 2, ..., n/q\}$ and $J \subseteq \{1, 2, ..., n/q\}$, $|I|, |J| = (1/2)n/q$ (provided that n/q is even), such that $|S \cap A_i| \leq 4a = 2sq/n$ for all $i \in I$ and $|S \cap B_j| \leq 4a = 2sq/n$ for all $j \in J$, and such that there is an edge between $S \cap A_i$ and $T \cap B_j$. Let S_i be any subset of A_i which contains $S \cap A_i$ and which has size exactly $4a$. Analogously, let T_j be any subset of B_j which contains $T \cap B_j$ and which has size exactly $4a$. Clearly there is an edge between S_i and T_j.

Fix an s and let $a = sq/(2n)$. As just shown, the existence of an S, of size s, for graph G implies the existence of $I, J \subseteq \{1, 2, ..., n/q\}$, $|I| = |J| = (1/2)n/q$, $S_i \subseteq A_i$ for all $i \in I$, and $T_j \subseteq B_j$ for all $j \in J$, $|S_i| = |T_j| = 4a$, such that for all $i \in I, j \in J$, the perfect matching in G between A_i and B_j contains an edge between S_i and T_j. (S_i, T_j have size exactly $4a$.) Hence the probability that there is an S is at most the probability that there exist $I, J \subseteq \{1, 2, ..., n/q\}$, both of size $(1/2)n/q$, and $S_i \subseteq A_i, T_j \subseteq B_j$ for all $i \in I, j \in J$, with $|S_i| = |T_j| = 4a$, such that for all $i \in I, j \in J$, the perfect matching in G between A_i and B_j contains an edge between S_i and T_j.

Given fixed $I, J, (S_i), (T_j)$, what is the probability that the random graph contains, for each $i \in I, j \in J$, an edge whose left endpoint is in S_i and whose right one is in T_j? The chance that the random graph does *not* contain both endpoints of some edge in the random perfect matching between A_i and B_j is the chance that all the edges in the (i, j) perfect matching (the one between A_i and B_j) emanating from S_i end outside T_j. There are exactly $4a$ such edges. We will prove a *lower* bound on the probability that a random perfect matching does *not* contain both endpoints of some edge whose left endpoint is in S_i and whose right one is in T_j. In order for the mate of each vertex in S_i to lie outside of T_j, the mate of the first one must be chosen to be one of $q - 4a$ nodes not in T_j among the q vertices in B_j, the mate of the second must be chosen to be one of the remaining $q - 4a - 1$ nodes not in T_j among the remaining $q - 1$ vertices in B_j, etc. Hence the probability is exactly $\frac{q-4a}{q}\frac{q-4a-1}{q-1}\cdots\frac{q-4a-(4a-1)}{q-(4a-1)} \geq \left(\frac{q-8a}{q}\right)^{4a} = \left(1 - \frac{8a}{q}\right)^{4a}$. Hence the chance that the (i, j) perfect matching *does* contain an edge whose left endpoint is in S_i and whose right one is in T_j is at most $1 - (1 - 8a/q)^{4a}$. The chance that the random matching works for all $(n/q)^2/4$ pairs (i, j) with $i \in I, j \in J$ is at most $[1 - (1 - 8a/q)^{4a}]^{(n/q)^2/4}$. Let $A = \binom{n/q}{n/(2q)}^2 \binom{q}{4a}^{n/q} \left[1 - \left(1 - \frac{8a}{q}\right)^{4a}\right]^{(n/q)^2/4}$, the first binomial coefficient representing the choices of I and J, the second representing the subsets S_i of A_i and T_j of B_j. If $A < 1$, then there is a fixed graph for which no set S of size s is good. $A \leq 2^{2n/q} q^{4an/q} \left[1 - \left(1 - \frac{8a}{q}\right)^{4a}\right]^{(n/q)^2/4}$. We choose $a = \sqrt{q/32}$ so that $q/(8a) = 4a$. Note that $(1 - 8a/q)^{4a} = (1 - 1/(q/(8a)))^{q/(8a)} \geq 1/4$ for $q/(8a) = 4a$ sufficiently large. So $[1-(1-8a/q)^{4a}]^{(n/q)^2/4} \leq [1-(1/4)]^{(n/q)^2/4}$. Letting $q = n^\delta$ for a fixed δ, we have $A \leq 2^{2n^{1-\delta}} (n^\delta)^{4an/q} (3/4)^{(n/q)^2/4}$. Since $4an/q = 4n^{1-\delta}\sqrt{q/32} = (1/\sqrt{2})n^{1-\delta/2}$, we have $A \leq (3/4)^{(1/4)n^{2(1-\delta)}} 2^{2n^{1-\delta}} n^{\delta(1/\sqrt{2})n^{1-\delta/2}}$. We have $A \leq 2^{-0.01n^{2(1-\delta)} + 2n^{1-\delta} + (\lg n)(\delta/\sqrt{2})n^{1-\delta/2}}$. Since obviously $2(1 - \delta) > 1 - \delta$, we will have $A < 1$, in fact, $A \to 0$, if $2(1 - \delta) > 1 - \delta/2$, i.e., $2 - 2\delta > 1 - \delta/2$, i.e., $1 > (3/2)\delta$, i.e., $\delta < 2/3$. Hence if $\delta < 2/3$, then for $q = n^\delta$ and $a = \sqrt{q/32}$, as specified above, there is an instance for which no set S of size $a(2n/q)$ is feasible. For this instance, $z^*_{IP} > \sqrt{q/32}(2n/q)$. Since $z^*_{LP} \leq 2n/q$, the integrality ratio exceeds $\sqrt{q/32}$, which is $\Omega(n^{\delta/2})$. Since $\delta < 2/3$ is arbitrary, for all $\varepsilon > 0$ the integrality ratio is $\Omega(n^{1/3-\varepsilon})$. □

3 $O(n^{\frac{1}{3}})$-Approximation Algorithm for Max Rep

In this section, we provide an $O(n^{\frac{1}{3}})$-approximation algorithm for Max Rep. However, in contrast to Section 2, in which we use a natural LP for the problem, we can show that the integrality gap of a natural LP for Max Rep is $\Omega(\sqrt{n})$. This forces us to use a combinatorial approach to obtain a non-trivial approximation factor $O(n^{\frac{1}{3}})$.

We consider the best of three algorithms:

1. *Matching:* Find a maximal matching in the supergraph $\mathcal{H}$. For each edge (A_i, B_j) in this matching, pick $a_i \in A_i$ and $b_j \in B_j$ such that $(a_i, b_j) \in E(G)$.
2. *Random-Choice:* For each B_j, pick $b_j \in B_j$ at random. For each A_i, pick $a_i \in A_i$ that has the maximum number of edges to the set of all selected b_j vertices. Repeat, flipping the roles of A and B.
3. *Random-Neighbor:* For each $a \in A$, construct a solution in the following fashion and eventually pick the best such solution: For each B_j, pick $b_j \in B_j$ at random from amongst those vertices that are neighbors of a (if there is no neighbor of a in B_j, pick an arbitrary $b_j \in B_j$). For each A_i, pick $a_i \in A_i$ that has the maximum number of edges to the selected b_j vertices over all j. Repeat, flipping the roles of A and B.

Theorem 5. *The best of these three algorithms is a* $2(2n)^{1/3}$*-approximation algorithm.*

Proof. Suppose that the maximal matching in $\mathcal{H}$ has size ℓ. Renumber the A_i's and B_j's such that the matching has edges $(A_i, B_i), i = 1, \ldots, \ell$. There are no edges between A_i and B_j for $i, j > \ell$. Let $A' = \cup_{i=1}^{\ell} A_i$, and $B' = \cup_{j=1}^{\ell} B_j$. The edges in the optimal solution can be decomposed into two groups: those that go between A' and B and those that go between A and B'. (Edges between A' and B' appear in both.) Hence the optimal solution restricted to one of these two groups much contain at least half the number of edges in the optimal solution. Without loss of generality, assume that the optimal solution restricted to edges between A and B' contains at least half the number of edges in the optimal solution.

We introduce some notation to facilitate the analysis. Let $X_{ij} = 1$ if there is an edge in the optimal solution from A_i to B_j (and 0 otherwise). Let N_{ij} be the number of edges from the optimal vertex a_i^* in A_i to the remaining vertices in B_j, called "nonoptimal" since they're not in the optimal solution.

Define p and r as follows:

$$\sum_{i=1}^{k} \sum_{j=1}^{\ell} X_{ij} = p(k\ell) \tag{2}$$

$$\sum_{i=1}^{k} \sum_{j=1}^{\ell} N_{ij} = r(\ell n). \tag{3}$$

Thus $OPT \leq 2\sum_i^k \sum_{j=1}^{\ell} X_{ij} = 2pk\ell$ and algorithm *Matching* gives a $2pk$ approximation.

Next, we analyze algorithm *Random-Choice.* The algorithm picks random vertices in B and picks the best vertices in A for the chosen vertices in B. In order to obtain a lower bound on the number of superedges covered, we compute the expected number of superedges covered if we pick random vertices in $B_j, j = 1, \ldots, \ell$, and instead of the best vertex in A_i, we use the vertex $a_i^* \in A_i$ which is in the optimal solution.

For a superedge (A_i, B_j), $i = 1, \ldots, k$ and $j = 1, \ldots, \ell$, the probability that this edge is covered by *Random-Choice* is $(X_{ij} + N_{ij})/(n/k)$. Hence the expected number of superedges covered is at least $\frac{k}{n}\sum_{i=1}^{k}\sum_{j=1}^{\ell}(X_{ij} + N_{ij}) = \frac{k}{n}(pk\ell + r\ell n)$. Hence the approximation ratio of algorithm *Random-Choice* is at most $\frac{2pk\ell}{\frac{k}{n}(pk\ell + r\ell n)} \leq \min\left\{\frac{2n}{k}, \frac{2p}{r}\right\}$.

Finally, we analyze algorithm *Random-Neighbor*. Suppose the vertex a chosen by the algorithm in the first step is in, say, A_h, and also is in the optimal solution. Consider set B_j and the vertex $b_j^* \in B_j$ in the optimal solution. The number of edges from a to B_j is $X_{hj} + N_{hj}$. The algorithm picks a random neighbor of a in B_j. Thus the probability that b_j^* is chosen is $\frac{X_{hj}}{X_{hj}+N_{hj}}$. As before, instead of picking the best choice of vertices in A for the chosen vertices in B, we lower bound the expected number of superedges covered by replacing the vertex a_i by the vertex $a_i^* \in A_i$ in the optimal solution. If $b_j^* \in B$ is chosen, the number of edges from the set of a_i^*'s is $\sum_{i=1}^{k} X_{ij}$. Thus the expected number of superedges covered is at least $\sum_{j=1}^{\ell}\frac{X_{hj}}{X_{hj}+N_{hj}}(\sum_{i=1}^{k} X_{ij})$. In this calculation, we assumed that $a = a_h^*$ was chosen in the first step. We average over $h = 1, \ldots, k$. Thus the expected number of covered edges is at least $\frac{1}{k}\sum_{h=1}^{k}\sum_{j=1}^{\ell}\frac{X_{hj}}{X_{hj}+N_{hj}}(\sum_{i=1}^{k} X_{ij})$. Let $C_j = \sum_{i=1}^{k} X_{ij}$ and let $N_j = \sum_{i=1}^{k} N_{ij}$. Then the previous expression is $\frac{1}{k}\sum_{j=1}^{\ell} C_j \sum_{h=1}^{k}\frac{X_{hj}}{X_{hj}+N_{hj}}$.Note that $\sum_{h=1}^{k}\frac{X_{hj}}{X_{hj}+N_{hj}} \geq C_j\left(\frac{1}{1+\frac{N_j}{C_j}}\right)$ by the arithmetic-geometric-harmonic means inequality. In order to obtain a lower bound for this expression, consider the minimum value of $\displaystyle\sum_j \frac{C_j^3}{C_j + N_j}$ over all choices of C_j and N_j subject to the constraint that $\sum_j C_j$ and $\sum_j N_j$ are fixed. Now let C_j, N_j be the respective values that minimize this expression. Then for any indices $f \neq g$, the function (of x) $\displaystyle\frac{{C_f}^3}{C_f + (N_f - x)} + \frac{{C_g}^3}{C_g + (N_g + x)}$ must be minimized for $x = 0$. Thus, the derivative of this function at $x = 0$ must be zero. Hence ${C_f}^3/(C_f + N_f)^2 = {C_g}^3/(C_g + N_g)^2$. Hence there is a constant α such that for all indices f, $(C_f + N_f)^2 = \alpha {C_f}^3$. Hence $\alpha\sum_j C_j^{3/2} = \sum_j C_j + \sum_j N_j = pk\ell + r\ell n$. Thus the expected number of superedges covered is at least $\frac{1}{k}\sum_{j=1}^{\ell}\frac{C_j^3}{\alpha C_j^{3/2}} = \frac{1}{k}\sum_{j=1}^{\ell}\frac{C_j^{3/2}}{\alpha} \geq \frac{1}{k}\frac{(\sum_{j=1}^{\ell} C_j^{3/2})^2}{(pk\ell + r\ell n)}$. Convexity of $f(x) = x^{3/2}$ shows that this expression is minimized when all C_j are equal. Hence a lower bound on the expected number of superedges covered is given by $\frac{1}{k}\frac{(\ell(pk)^{3/2})^2}{(pk\ell+r\ell n)} = \frac{\ell^2 p^3 k^2}{pk\ell + r\ell n}$. Thus the approximation ratio of this procedure is at most $\frac{2pk\ell(pk\ell+r\ell n)}{\ell^2 p^3 k^2} = \frac{2(1+(\frac{r}{p})\frac{n}{k})}{p}$.

Thus we have the following upper bounds on the approximation ratio of the algorithm: $2pk, 2\frac{n}{k}, 2\frac{p}{r}, 2\frac{1+(\frac{r}{p})\frac{n}{k}}{p}$. We consider two cases:

Case 1: $(\frac{r}{p})\frac{n}{k} \geq 1$. In this case, the fourth bound is at most $\frac{4(\frac{r}{p})\frac{n}{k}}{p}$. The product of the first, third and fourth bounds is $16pk \times \frac{p}{r} \times \frac{(\frac{r}{p})\frac{n}{k}}{p} = 16n$. Hence at least one upper bound is at most $2(2n)^{1/3}$.

Case 2: $(\frac{r}{p})\frac{n}{k} < 1$. In this case, the fourth bound is at most $\frac{4}{p}$. Now the product of the first, second and fourth bound is $2pk \times \frac{2n}{k} \times \frac{4}{p} = 16n$. Hence at least one upper bound is at most $2(2n)^{1/3}$. □

4 Reduction from Densest *k*-Subgraph to Max Rep

In this section, we consider the Densest k-Subgraph (DkS) problem, in which the goal is to find an induced subgraph of order k of a given graph with the maximum number of edges.

Theorem 6. *An $f(n)$-approximation algorithm for* Max Rep *implies the existence of a randomized $O(f(n))$-approximation algorithm for DkS.*

Proof. From an instance of DkS, we produce an instance of Max Rep by randomly dividing vertices of the given graph for DkS into k groups of equal size $s = \lfloor \frac{n}{k} \rfloor$, e.g., by using a random permutation of all vertices, and disregard the rest of vertices. Next we place $\lfloor k/2 \rfloor$ groups on one side (call this L) and the other $\lceil k/2 \rceil$ groups on the other side (call this R) of the instance for Max Rep. Any feasible solution to the Max Rep instance obtained directly gives a solution to the original DkS instance of the same value. Both instances have the same number n of vertices.

We claim that the expected value of the optimal solution to the Max Rep instance obtained thus is at least a constant times the optimal value for DkS. Consider the optimal solution S of size k to the DkS instance. We produce a solution to the Max Rep instance as follows. For every group in the instance, if the group contains a unique vertex of S, then this unique vertex is picked as the group representative. If there are zero or at least 2 vertices from S then an arbitrary vertex is picked as the group representative (and we don't count edges incident to that vertex). We show that the expected value of this solution is at least a constant times the value of the DkS optimal solution. For any vertex $v \in S$, with constant probability v is placed alone in its group. Furthermore, for two distinct vertices $u, v \in S$, the probability that u and v are both alone in their groups, u is in the L side and v is on the R side, is bounded below by a constant greater than 0. Hence $E[z^*_{MaxRep}] \geq c z^*_{DkS}$, for a positive constant c.

Now the reduction is apparent. Given an $f(n)$-approximation algorithm A for Max Rep, take an n-node instance I of DkS, randomly convert it as above into an n-node instance I' of Max Rep, use A to generate a solution $A(I')$ of value at least $f(n) z^*_{MaxRep}$, and report $A(I')$ as a feasible solution to the DkS instance. That $E[z^*_{MaxRep}] \geq c z^*_{DkS}$ implies that the expected size of the DkS solution returned is at least $cf(n) \cdot z^*_{DkS}$. □

5 Conclusion

Obtaining improvements over the approximation guarantees in this paper would be instructive. Given the reduction demonstrated in Section 4, possibly one can use ideas from the DENSEST k-SUBGRAPH algorithm to build an $n^{1/3-\delta}$-approximation algorithms for some fixed $\delta > 0$. However, the main remaining open problem is whether, for MAX REP or MIN REP, there is a $O(n^{\varepsilon})$-approximation algorithm for each $\varepsilon > 0$.

References

1. Aazami, A., Stilp, M.D.: Approximation algorithms and hardness for domination with propagation. In: Charikar, M., Jansen, K., Reingold, O., Rolim, J.D.P. (eds.) RANDOM 2007 and APPROX 2007. LNCS, vol. 4627, pp. 1–15. Springer, Heidelberg (2007)
2. Arora, S., Babai, L., Stern, J., Sweedyk, Z.: The hardness of approximate optima in lattices, codes, and systems of linear equations. J. Comput. System Sci. 54, 317–331 (1997)
3. Bhattacharyya, A., Grigorescu, E., Jung, K., Raskhodnikova, S., Woodruff, D.P.: Transitive-Closure Spanners, ArXiv e-prints (2008)
4. Breslau, L., Diakonikolas, I., Duffield, N., Gu, Y., Hajiaghayi, M., Johnson, D., Karloff, H., Resende, M., Sen, S.: Optimal Node Placement For Path-Disjoint Network Monitoring (manuscript) (2008)
5. Chen, N.: On the approximability of influence in social networks. In: SODA, pp. 1029–1037 (2008)
6. Elkin, M., Peleg, D.: The hardness of approximating spanner problems. Theory Comput. Syst. 41, 691–729 (2007)
7. Feige, U., Kortsarz, G., Peleg, D.: The dense k-subgraph problem. Algorithmica 29, 410–421 (2001)
8. Feldman, M., Kortsarz, G., Nutov, Z.: Improved approximation for the directed steiner forest problem. In: SODA 2009, pp. 922–931 (2009)
9. Gupta, A., Hajiaghayi, M.T., Kumar, A.: Stochastic steiner tree with non-uniform inflation. In: Charikar, M., Jansen, K., Reingold, O., Rolim, J.D.P. (eds.) RANDOM 2007 and APPROX 2007. LNCS, vol. 4627, pp. 134–148. Springer, Heidelberg (2007)
10. Hajiaghayi, M.T., Kortsarz, G., Mirrokni, V.S., Nutov, Z.: Power optimization for connectivity problems. Math. Program. 110, 195–208 (2007)
11. Hassin, R., Segev, D.: The set cover with pairs problem. In: Sarukkai, S., Sen, S. (eds.) FSTTCS 2005. LNCS, vol. 3821, pp. 164–176. Springer, Heidelberg (2005)
12. Hochbaum, D.S. (ed.): Approximation algorithms for NP-hard problems. PWS Publishing Co., Boston (1997); see the section written by Arora and Lund
13. Khot, S.: Ruling out PTAS for graph min-bisection, dense k-subgraph, and bipartite clique. SIAM J. Comput. 36, 1025–1071 (2006)
14. Kortsarz, G.: On the hardness of approximating spanners. Algorithmica 30, 432–450 (2001)
15. Kortsarz, G., Krauthgamer, R., Lee, J.R.: Hardness of approximation for vertex-connectivity network design problems. SIAM Journal on Computing 33, 185–199 (2004)
16. Peleg, D.: Approximation algorithms for the Label-Cover$_{\mathrm{MAX}}$ and Red-Blue Set Cover problems. J. Discrete Algorithms 5, 55–64 (2007)

A Linear Time Algorithm for $L(2, 1)$-Labeling of Trees

Toru Hasunuma[1], Toshimasa Ishii[2], Hirotaka Ono[3], and Yushi Uno[4]

[1] Department of Mathematical and Natural Sciences,
The University of Tokushima, Tokushima 770-8502, Japan
hasunuma@ias.tokushima-u.ac.jp

[2] Department of Information and Management Science,
Otaru University of Commerce, Otaru 047-8501, Japan
ishii@res.otaru-uc.ac.jp

[3] Department of Computer Science and Communication Engineering,
Kyushu University, Fukuoka 812-8581, Japan
ono@csce.kyushu-u.ac.jp

[4] Department of Mathematics and Information Sciences,
Graduate School of Science, Osaka Prefecture University,
Sakai 599-8531, Japan
uno@mi.s.osakafu-u.ac.jp

Abstract. An $L(2, 1)$-labeling of a graph G is an assignment f from the vertex set $V(G)$ to the set of nonnegative integers such that $|f(x) - f(y)| \geq 2$ if x and y are adjacent and $|f(x) - f(y)| \geq 1$ if x and y are at distance 2, for all x and y in $V(G)$. A k-$L(2, 1)$-labeling is an $L(2, 1)$-labeling $f : V(G) \to \{0, \ldots, k\}$, and the $L(2, 1)$-labeling problem asks the minimum k, which we denote by $\lambda(G)$, among all possible assignments. It is known that this problem is NP-hard even for graphs of treewidth 2, and tree is one of very few classes for which the problem is polynomially solvable. The running time of the best known algorithm for trees had been $O(\Delta^{4.5}n)$ for more than a decade, and an $O(\min\{n^{1.75}, \Delta^{1.5}n\})$-time algorithm has appeared recently, where Δ is the maximum degree of T and $n = |V(T)|$, however, it has been open if it is solvable in linear time. In this paper, we finally settle this problem for $L(2, 1)$-labeling of trees by establishing a linear time algorithm.

1 Introduction

Let G be an undirected graph. An $L(2, 1)$-*labeling* of a graph G is an assignment f from the vertex set $V(G)$ to the set of nonnegative integers such that $|f(x) - f(y)| \geq 2$ if x and y are adjacent and $|f(x) - f(y)| \geq 1$ if x and y are at distance 2, for all x and y in $V(G)$. A k-$L(2, 1)$-labeling is an $L(2, 1)$-labeling $f : V(G) \to \{0, \ldots, k\}$, and the $L(2, 1)$-*labeling problem* asks the minimum k among all possible assignments. We call this invariant, the minimum value k, the $L(2, 1)$-*labeling number* and is denoted by $\lambda(G)$. Notice that we can use $k + 1$ different labels when $\lambda(G) = k$ since we can use 0 as a label for conventional reasons.

The original notion of $L(2, 1)$-labeling can be seen in the context of frequency assignment, where 'close' transmitters must receive different frequencies and 'very close' transmitters must receive frequencies that are at least two frequencies apart so that they can avoid interference. Due to its practical importance, the $L(2, 1)$-labeling problem has

A. Fiat and P. Sanders (Eds.): ESA 2009, LNCS 5757, pp. 35–46, 2009.

been widely studied. From the graph theoretical point of view, since this is a kind of vertex coloring problem, it has attracted a lot of interest [4,10,13,16]. In this context, $L(2,1)$-labeling is generalized into $L(p,q)$-labeling for arbitrary nonnegative integers p and q, and in fact, we can see that $L(1,0)$-labeling ($L(p,0)$-labeling, actually) is equivalent to the classical vertex coloring. We can find a lot of related results on $L(p,q)$-labelings in comprehensive surveys by Calamoneri [2] and by Yeh [17].

Related Work: There are also a number of studies on the $L(2,1)$-labeling problem from the algorithmic point of view [1,8,15]. It is known to be NP-hard for general graphs [10], and it still remains NP-hard for some restricted classes of graphs, such as planar graphs, bipartite graphs, chordal graphs [1], and it turned out to be NP-hard even for graphs of treewidth 2 [5]. In contrast, only a few graph classes are known to have polynomial time algorithms for this problem, e.g., we can determine the $L(2,1)$-labeling number of paths, cycles, wheels within polynomial time [10].

As for trees, Griggs and Yeh [10] showed that $\lambda(T)$ is either $\Delta+1$ or $\Delta+2$ for any tree T, and also conjectured that determining $\lambda(T)$ is NP-hard, however, Chang and Kuo [4] disproved this by presenting a polynomial time algorithm for computing $\lambda(T)$. Their algorithm exploits the fact that $\lambda(T)$ is either $\Delta+1$ or $\Delta+2$ for any tree T. Its running time is $O(\Delta^{4.5}n)$, where Δ is the maximum degree of a tree T and $n = |V(T)|$. This result has a great importance because it initiates to cultivate polynomially solvable classes of graphs for the $L(2,1)$-labeling problem and related problems. For example, Fiala et al. showed that $L(2,1)$-labeling of t-almost trees can be solved in $O(\lambda^{2t+4.5}n)$ time for λ given as an input, where a t-almost tree is a graph that can be a tree by eliminating t edges [8]. Also, it was shown that the $L(p,1)$-labeling problem for trees can be solved in $O((p+\Delta)^{5.5}n) = O(\lambda^{5.5}n)$ time [3]. Both results are based on Chang and Kuo's algorithm, which is called as a subroutine in the algorithms. Moreover, the polynomially solvable result for trees holds for more general settings. The notion of $L(p,1)$-labeling is generalized as $H(p,1)$-labeling, in which graph H defines the metric space of distances between two labels, whereas labels in $L(p,1)$-labeling (that is, in $L(p,q)$-labeling) take nonnegative integers; i.e., it is a special case that H is a path graph. In [6], it has been shown that the $H(p,1)$-labeling problem of trees for arbitrary graph H can be solved in polynomial time, which is also based on Chang and Kuo's idea. In passing, these results are unfortunately not applicable for $L(p,q)$-labeling problems for general p and q. Recently, Fiala et al. [7] showed that the $L(p,q)$-labeling problem for trees is NP-hard if q is not a divisor of p, which is contrasting to the positive results mentioned above.

As for $L(2,1)$-labeling of trees again, Chang and Kuo's $O(\Delta^{4.5}n)$ algorithm is the first polynomial time one. It is based on dynamic programming (DP) approach, and it checks whether $(\Delta+1)$-$L(2,1)$-labeling is possible or not from leaf vertices to a root vertex in the original tree structure. The principle of optimality requires to solve at each vertex of the tree the assignments of labels to subtrees, and the assignments are formulated as the maximum matching in a certain bipartite graph. Recently, an $O(\min\{n^{1.75}, \Delta^{1.5}n\})$ time algorithm has been proposed [11]. It is based on the similar DP framework to Chang and Kuo's algorithm, but achieves its efficiency by reducing heavy computation of bipartite matching in Chang and Kuo's and by using an amortized analysis. We give a concise review of these two algorithms in Subsection 2.2.

Our Contributions: Although there have been a few polynomial time algorithms for $L(2, 1)$-labeling of trees, it has been open if it can be improved to linear time [2]. In this paper, we present a linear time algorithm for $L(2, 1)$-labeling of trees, which finally settles this problem. It is based on the similar DP approach to the preceding two polynomial time algorithms [4,11]. In our new algorithm, besides using their ideas, we introduce the notion of "label compatibility", which indicates how we flexibly change labels with preserving its $(\Delta + 1)$-$L(2, 1)$-labeling. Interestingly, we can show that only $O(\log_\Delta n)$ labels are essential for $L(2, 1)$-labeling in any input tree by using this notion. By utilizing this fact, we can replace the bipartite matching of graphs with the maximum flow of much smaller networks as an engine to find the assignments. Consequently, our algorithm finally achieves its linear running time.

Organization of this Paper: The rest of this paper is organized as follows. Section 2 gives basic definitions and introduces as a warm-up the ideas of Chang and Kuo's $O(\Delta^{4.5}n)$ time algorithm and its improvement into $O(n^{1.75})$ time. Section 3 introduces the crucial notion of label compatibility that can bundle a set of compatible vertices and reduce the size of the graph constructed for computing bipartite matchings. Moreover, this allows to use maximum-flow based computation for them. In Section 4, we give precise analyses to achieve linear running time. Some parts of the detailed analyses are omitted due to space limitation. Interested readers can find them in the technical report version of this paper [12].

2 Preliminaries

2.1 Definitions and Notations

A graph G is an ordered pair of its vertex set $V(G)$ and edge set $E(G)$ and is denoted by $G = (V(G), E(G))$. We assume throughout this paper that all graphs are undirected, simple and connected, unless otherwise stated. Therefore, an edge $e \in E(G)$ is an unordered pair of vertices u and v, which are *end vertices* of e, and we often denote it by $e = (u, v)$. Two vertices u and v are *adjacent* if $(u, v) \in E(G)$. A graph $G = (V(G), E(G))$ is called *bipartite* if the vertex set $V(G)$ can be divided into two disjoint sets V_1 and V_2 such that every edge in $E(G)$ connects a vertex in V_1 and one in V_2; such G is denoted by (V_1, V_2, E).

For a graph G, the *(open) neighborhood* of a vertex $v \in V(G)$ is the set $N_G(v) = \{u \in V(G) \mid (u, v) \in E(G)\}$, and the *closed neighborhood* of v is the set $N_G[v] = N_G(v) \cup \{v\}$. The *degree* of a vertex v is $|N_G(v)|$, and is denoted by $d_G(v)$. We use $\Delta(G)$ to denote the maximum degree of a graph G. A vertex whose degree is $\Delta(G)$ is called *major*. We often drop G in these notations if there are no confusions. A vertex whose degree is 1 is called a *leaf vertex*, or simply a *leaf*.

When we describe algorithms, it is convenient to regard the input tree to be rooted at a leaf vertex r. Then we can define the parent-child relationship on vertices in the usual way. For a rooted tree, its *height* is the length of the longest path from the root to a leaf. For any vertex v, the set of its children is denoted by $C(v)$. For a vertex v, define $d'(v) = |C(v)|$.

2.2 Chang and Kuo's Algorithm and Its Improvement

Before explaining algorithms, we give some significant properties on $L(2,1)$-labeling of graphs or trees that have been used so far for designing $L(2,1)$-labeling algorithms. We can see that $\lambda(G) \geq \Delta + 1$ holds for any graph G. Griggs and Yeh [10] observed that any major vertex in G must be labeled 0 or $\Delta + 1$ when $\lambda(G) = \Delta + 1$, and that if $\lambda(G) = \Delta + 1$, then $N_G[v]$ contains at most two major vertices for any $v \in V(G)$. Furthermore, they showed that $\lambda(T)$ is either $\Delta + 1$ or $\Delta + 2$ for any tree T. By using this fact, Chang and Kuo [4] presented an $O(\Delta^{4.5}n)$ time algorithm for computing $\lambda(T)$.

Chang and Kuo's Algorithm. Now, we first review the idea of Chang and Kuo's dynamic programming algorithm (CK algorithm) for the $L(2,1)$-labeling problem of trees, since our linear time algorithm also depends on the same formula of the principle of optimality. The algorithm determines if $\lambda(T) = \Delta + 1$, and if so, we can easily construct the labeling with $\lambda(T) = \Delta + 1$.

To describe the idea, we introduce some notations. We assume for explanation that T is rooted at some leaf vertex r. Given a vertex v, we denote the subtree of T rooted at v by $T(v)$. Let $T(u,v)$ be a tree rooted at u that forms $T(u,v) = (\{u\} \cup V(T(v)), \{(u,v)\} \cup E(T(v)))$. Note that this u is just a virtual vertex for explanation and $T(u,v)$ is uniquely determined by $T(v)$. For $T(u,v)$, we define

$$\delta((u,v),(a,b)) = \begin{cases} 1, & \text{if } \lambda(T(u,v) \mid f(u) = a, f(v) = b) \leq \Delta + 1, \\ 0, & \text{otherwise,} \end{cases}$$

where $\lambda(T(u,v) \mid f(u) = a, f(v) = b)$ denotes the $L(2,1)$-labeling number on $T(u,v)$ under the condition that $f(u) = a$ and $f(v) = b$, i.e., the minimum k of k-$L(2,1)$-labelings on $T(u,v)$ satisfying $f(u) = a$ and $f(v) = b$. This δ function satisfies the following formula:

$$\delta((u,v),(a,b)) = \begin{cases} 1, & \text{if there is an injective assignment } g\colon C(v) \to \{0,1,\ldots,\Delta+1\} - \{a, \\ & \quad b-1, b, b+1\} \text{ such that } \delta((v,w),(b,g(w))) = 1 \text{ for each } w \in C(v), \\ 0, & \text{otherwise.} \end{cases}$$

The existence of such an injective assignment g is formalized as the maximum matching problem: For a bipartite graph $G(u,v,a,b) = (C(v), X, E(u,v,a,b))$, where $X = \{0, 1, \ldots, \Delta, \Delta + 1\}$ and $E(u,v,a,b) = \{(w,c) \mid \delta((v,w),(b,c)) = 1, c \in X - \{a\}, w \in C(v)\}$, we can see that there is an injective assignment $g\colon C(v) \to \{0,1,\ldots,\Delta+1\} - \{a, b-1, b, b+1\}$ if there exists a matching of size $d'(v)$ in $G(u,v,a,b)$. Namely, for $T(u,v)$ and two labels a and b, we can easily (i.e., in polynomial time) determine the value of $\delta((u,v),(a,b))$ if the values of δ function for $T(v,w)$, $w \in C(v)$ and any two pairs of labels are given. Now let $t(v)$ be the time for calculating $\delta((u,v),(*,*))$ for vertex v. CK algorithm solves the bipartite matching problems of $O(\Delta)$ vertices and $O(\Delta^2)$ edges $O(\Delta^2)$ times for each v, in order to obtain δ-values for all combinations of labels a and b. This amounts $t(v) = O(\Delta^{2.5}) \times O(\Delta^2) = O(\Delta^{4.5})$, where the first $O(\Delta^{2.5})$ is the time complexity of the bipartite matching problem [14]. Thus the total running time is $\sum_{v \in V} t(v) = O(\Delta^{4.5}n)$.

An $O(n^{1.75})$-time Algorithm. Next, we review the $O(n^{1.75})$-time algorithm proposed in [11]. The running time $O(n^{1.75})$ is roughly achieved by two strategies. One is that the problem can be solved by a simple linear time algorithm if $\Delta = \Omega(\sqrt{n})$, and the other is that it can be solved in $O(\Delta^{1.5}n)$ time for any input tree.

The first idea of the speedup is that for computing $\delta((u, v), (*, b))$, the algorithm does not solve the bipartite matching problems every time from scratch, but reuse the obtained matching structure. More precisely, the bipartite matching problem is solved for $G(u, v, -, b) = (C(v), X, E(u, v, -, b))$ instead of $G(u, v, a, b)$ for a specific a, where $E(u, v, -, b) = \{(w, c) \mid \delta((v, w), (b, c)) = 1, c \in X, w \in C(v)\}$. A maximum matching of $G(u, v, -, b)$ is observed to satisfy the following properties:

Property 1. If $G(u, v, -, b)$ has no matching of size $d'(v)$, then $\delta((u, v), (i, b)) = 0$ for any label i. □

Property 2. $\delta((u, v), (i, b)) = 1$ if and only if vertex i can be reached by an M-alternating path from some vertex in X unmatched by M in $G(u, v, -, b)$, where M denotes a maximum matching of $G(u, v, -, b)$ (of size $d'(v)$). □

From these properties, $\delta((u, v), (*, b))$ can be computed by a single bipartite matching and a single graph search, and its total running time is $O(\Delta^{1.5}d'(v)) + O(\Delta d'(v)) = O(\Delta^{1.5}d'(v))$ (for solving the bipartite matching of $G(u, v, -, b)$, which has $O(\Delta)$ vertices and $O(\Delta d'(v))$ edges, and for a single graph search). Since this calculation is done for all b, we have $t(v) = O(\Delta^{2.5}d'(v))$.

The other technique of the speedup introduced in [11] is based on preprocessing operations for amortized analysis. By some preprocessing operations, the shape of input trees can be restricted while preserving $L(2,1)$-labeling number, and the input trees can be assumed to satisfy the following two properties.

Property 3. All vertices connected to a leaf vertex are major vertices. □

Property 4. The size of any path component of T is at most 3. □

Here, a sequence of vertices $v_1, v_2, \ldots, v_\ell$ is called a *path component* if $(v_i, v_{i+1}) \in E$ for all $i = 1, 2, \ldots, \ell - 1$ and $d(v_i) = 2$ for all $i = 1, 2, \ldots, \ell$, and ℓ is called the *size* of the path component.

Furthermore, this preprocessing operations enable the following amortized analysis. Let V_L and V_Q be the set of leaf vertices and the set of major vertices whose children are all leaf vertices, respectively. Also, let $d''(v) = |C(v) - V_L|$ for $v \in V$. (Note that $d''(v) = 0$ for $v \in V_L \cup V_Q$.)

By Property 3, if we go down the resulting tree from a root, then we reach a major vertex in V_Q. Then, the following facts are observed: (i) for $v \in V_Q$ $\delta((u, v), (a, b)) = 1$ if and only if $b = 0$ or $\Delta + 1$ and $|a - b| \geq 2$, (ii) $|V_Q| \leq n/\Delta$. Note that (i) implies that it is not required to solve the bipartite matching to obtain δ-values. Also (ii) and Property 4 imply that $|V - V_Q - V_L| = O(n/\Delta)$ (this can be obtained by pruning leaf vertices and regarding V_Q vertices as new leaves). Since it is not necessary to compute bipartite matchings for $v \in V_L \cup V_Q$, and this implies that the total time to obtain δ-values for all v's is $\sum_{v \in V} t(v) = O(\sum_{v \in V - V_L - V_Q} t(v))$, which turned out to be $O(\Delta^{2.5} \sum_{v \in V - V_L - V_Q} d''(v))$. Since $\sum_{v \in V - V_L - V_Q} d''(v) = |V - V_L - V_Q| + |V_Q| - 1 = O(n/\Delta)$, we obtain $\sum_{v \in V - V_L - V_Q} t(v) = O(\Delta^{1.5}n)$. Since we have a linear time algorithm if $\Delta = \Omega(\sqrt{n})$ as mentioned above, we can solve the problem in $O(n^{1.75})$ time in total.

3 Label Compatibility and Flow-Based Computation of δ

As reviewed in Subsection 2.2, one of keys of an efficient computation of δ-values is reusing the matching structures. In this section, for a further speedup of the computation of δ-values, we introduce a new novel notion, which we call 'label compatibility', that enables to treat several labels equivalently under the computation of δ-values. Then, the faster computation of δ-values is achieved on a maximum flow algorithm instead of a maximum matching algorithm. Seemingly, this sounds a bit strange, because the time complexity of the maximum flow problem is larger than the one of the bipartite matching problem. The trick is that the new flow-based computation uses a smaller network (graph) by this notion than the graph $G(u, v, -, b)$ used in the bipartite matching.

3.1 Label Compatibility and Neck/Head Levels

Let $L_h = \{h, h+1, \ldots, \Delta - h, \Delta - h + 1\}$. Let T be a tree rooted at v, and $u \notin V(T)$. We say that T is *head-L_h-compatible* if $\delta((u, v), (a, b)) = \delta((u, v), (a', b))$ for all $a, a' \in L_h$ and $b \in L_0$ with $|a - b| \geq 2$ and $|a' - b| \geq 2$. Analogously, we say that T is neck-L_h-compatible if $\delta((u, v), (a, b)) = \delta((u, v), (a, b'))$ for all $a \in L_0$ and $b, b' \in L_h$ with $|a - b| \geq 2$ and $|a - b'| \geq 2$. The neck and head levels of T are defined as follows:

Definition 1. *Let T be a tree rooted at v, and $u \notin V(T)$.*
(i) *The* neck level (*resp.,* head level) *of T is 0 if T is neck-L_0-compatible (resp., head-L_0-compatible).* (ii) *The* neck level (*resp.,* head level) *of T is h (≥ 1) if T is not neck-L_{h-1}-compatible (resp., head-L_{h-1}-compatible) but neck-L_h-compatible (resp., head-L_h-compatible).*

An intuitive explanation of neck-L_h-compatibility (resp., head-L_h-compatibility) of T is that if for $T(u, v)$, a label in L_h is assigned to v (resp., u) under $(\Delta + 1)$-$L(2, 1)$-labeling of $T(u, v)$, the label can be replaced with another label in L_h without violating a proper $(\Delta + 1)$-$L(2, 1)$-labeling; labels in L_h are compatible. The neck and head levels of T represent the bounds of L_h-compatibility of T. Thus, a trivial bound on neck and head levels is $(\Delta + 1)/2$.

For the relationship between the neck/head levels and the tree size, we can show the following lemma, whose proof can be found in the technical report version [12]:

Lemma 1. *Let T' be a subtree of T. If $|V(T')| \leq (\Delta - 3 - 2h)^{h/2} - 1$ and $\Delta - 2h \geq 10$, then the head level and neck level of T' are both at most h.*

By this lemma, we obtain the following theorem:

Theorem 1. For a tree T, both the head and neck levels of T are $O(\log |V(T)| / \log \Delta)$.

3.2 Flow-Based Computation of δ

We are ready to explain the faster computation of δ-values. Recall that $\delta((u, v), (a, b)) = 1$ holds if there exists a matching of $G(u, v, a, b)$ in which all $C(v)$ vertices are just matched; which vertex is matched to a vertex in X does not matter. From this fact, we can treat vertices in X corresponding to L_h equally in computing δ, if T is neck- and head-L_h-compatible. The idea of the fast computation of δ-values is that, by bundling compatible vertices in X of G, we reduce the size of a graph (or a network) to compute the assignments of labels, which is no longer the maximum matching; the maximum flow.

The algorithm introduced in Subsection 2.2 computes δ-values not by solving the maximum matchings of $G(u,v,a,b)$ for all pairs of a and b but by finding a maximum matching M of $G(u,v,-,b)$ once and then searching M-alternating paths. In the new flow-based computation, we adopt the same strategy; for a tree $T(v)$ whose head and neck levels are at most $h(v)$, we do not prepare a network for a specific pair (a,b), say $\mathcal{N}(u,v,a,b)$, but a general network $\mathcal{N}(u,v,-,b) = (\{s,t\} \cup C(v) \cup X_{h(v)}, E(v) \cup E_X \cup E_\delta, cap)$, where $X_{h(v)} = (L_0 - L_{h(v)}) \cup \{h(v)\}$, $E(v) = \{(s,w) \mid w \in C(v)\}$, $E_X = \{(c,t) \mid c \in X_{h(v)}\}$, $E_\delta = \{(w,c) \mid w \in C(v), c \in X_{h(v)}\}$, and $cap(e)$ function is defined as follows: $\forall e \in E(v), cap(e) = 1$, for $e = (w,c) \in E_\delta$, $cap(e) = 1$ if $\delta((v,w),(b,c)) = 1$, $cap(e) = 0$ otherwise, and for $e = (c,t) \in E_X$, $cap(e) = 1$ if $c \neq h(v)$, $cap(e) = |L_{h(v)} - \{b, b+1, b-1\}|$ if $c = h(v)$.

For a maximum flow $\psi : e \to R^+$, we define X' as $\{c \in X_h \mid cap((c,t)) - \psi((c,t)) \geq 1\}$. By the flow integrality and arguments similarly to Properties 1 and 2, we can obtain the following properties:

Lemma 2. *If $\mathcal{N}(u,v,-,b)$ has no flow of size $d'(v)$, then $\delta((u,v),(i,b)) = 0$ for any label i.* □

Lemma 3. *$\delta((u,v),(i,b)) = 1$ if and only if vertex i can be reached by a ψ-alternating path from some vertex in X' in $\mathcal{N}(u,v,-,b)$.* □

Here, a ψ-alternating path is defined as follows: Given a flow ψ, a path in E_δ is called *ψ-alternating* if its edges alternately satisfy $cap(e) - \psi(e) \geq 1$ and $\psi(e) \geq 1$. By these lemmas, we can obtain $\delta((u,v),(*,b))$-values for b by solving the maximum flow of $\mathcal{N}(u,v,-,b)$ once and then applying a single graph search.

The current fastest maximum flow algorithm runs in $O(\min\{m^{1/2}, n^{2/3}\}\, m \log(n^2/m) \log U) = O(n^{2/3} m \log n \log U)$ time, where U, n and m are the maximum capacity of edges, the number of vertices and edges, respectively [9]. Thus the running time of calculating $\delta((u,v),(a,b))$ for a pair (a,b) is

$$O((h(v) + d''(v))^{2/3}(h(v)d''(v)) \log(h(v) + d''(v)) \log \Delta) = O(\Delta^{2/3}(h(v)d''(v)) \log^2 \Delta),$$

since $h(v) \leq \Delta$ and $d''(v) \leq \Delta$ (recall that $d''(v) = |C(v) - V_L|$). By using a similar technique of updating matching structures introduced in [11], we can obtain $\delta((u,v),(*,b))$ in $O(\Delta^{2/3}(h(v)d''(v)) \log^2 \Delta) + O(h(v)d''(v)) = O(\Delta^{2/3}(h(v)d''(v)) \log^2 \Delta)$ time. Since the number of candidates for b is also bounded by $h(v)$ from the neck/head level property, we have the following lemma.

Lemma 4. *$\delta((u,v),(*,*))$ can be computed in $O(\Delta^{2/3}(h(v))^2 d''(v) \log^2 \Delta)$ time, that is, $t(v) = O(\Delta^{2/3}(h(v))^2\, d''(v) \log^2 \Delta)$.* □

Combining this with $\sum_{v \in V - V_L - V_Q} d''(v) = O(n/\Delta)$ shown in Subsection 2.2, we can show the total running time for the $L(2,1)$-labeling is $O(n(\max\{h(v)\})^2(\Delta^{-1/3} \log^2 \Delta))$. By applying Theorem 1, we have the following theorem:

Theorem 2. *For trees, the $L(2,1)$-labeling problem can be solved in $O(\min\{n \log^2 n, \Delta^{1.5} n\})$ time. Furthermore, if $n = O(\Delta^{poly(\log \Delta)})$, it can be solved in $O(n)$ time.* □

Corollary 1. *For a vertex v in a tree T, we have $\sum_{w \in V(T(v))} t(w) = O(|T(v)|)$ if $|T(v)| = O(\Delta^{poly(\log \Delta)})$.* □

Only by directly applying Theorem 1 (actually Lemma 1), we obtain much faster running time than the previous one. In the following section, we present a linear time algorithm, in which Lemma 1 is used in a different way.

4 Proof of Linear Running Time

As mentioned in Subsection 2.2, one of keys for achieving the running time $O(\Delta^{1.5}n) = O(n^{1.75})$ is equation $\sum_{v \in V_\delta} d''(v) = O(n/\Delta)$, where V_δ is the set of vertices in which δ-values should be computed via the matching-based algorithm; since the computation of δ-values for each v is done in $O(\Delta^{2.5}d''(v))$ time, it takes $\sum_{v \in V_\delta} O(\Delta^{2.5}d''(v)) = O(\Delta^{1.5}n)$ time in total. This equation is derived from the fact that in leaf vertices we do not need to solve the matching to compute δ-values, and any vertex with height 1 has $\Delta - 1$ leaves as its children after the preprocessing operation.

In our new algorithm, we generalize this idea: By replacing leaf vertices with subtrees with size at least Δ^4 in the above argument, we can obtain $\sum_{v \in V_\delta} d''(v) = O(n/\Delta^4)$, and in total, the running time $\sum_{v \in V_\delta} O(\Delta^{2.5}d''(v)) = O(n)$ is roughly achieved. Actually, this argument contains a cheating, because a subtree with size at most Δ^4 is not always connected to a major vertex, whereas a leaf is, which is well utilized to obtain $\sum_{v \in V_\delta} d''(v) = O(n/\Delta)$. Also, whereas we can neglect leaves to compute δ-values, we cannot neglect such subtrees. We resolve these problems by best utilizing the properties of neck/head levels and the maximum flow techniques introduced in Section 3.

4.1 Efficient Assignment of Labels for Computing δ

In this section, by compiling observations and techniques for assigning labels in the computation of $\delta((u, v), (*, *))$ for $v \in V$, given in Sections 2 and 3, we will design an algorithm to run in linear time within the DP framework. Throughout this section, we assume that an input tree T satisfies Properties 3 and 4. Below, we first partition the vertex set V into five types of subsets defined later, and give a linear time algorithm for computing the value of δ functions, specified for each type.

We here start with defining such five types of subsets V_i ($i = 1, \ldots, 5$). Throughout this section, for a tree T', we may denote $|V(T')|$ simply by $|T'|$. Let V_M be the set of vertices $v \in V$ such that $T(v)$ is a "maximal" subtree of T with $|T(v)| \leq \Delta^5$; i.e., for the parent u of v, $|T(u)| > \Delta^5$. Divide V_M into two sets $V_M^{(1)} := \{v \in V_M \mid |T(v)| \geq (\Delta - 19)^4\}$ and $V_M^{(2)} := \{v \in V_M \mid |T(v)| < (\Delta - 19)^4\}$ (notice that $V_L \subseteq \cup_{v \in V_M} V(T(v))$). Define $\tilde{d}(v) := |C(v) - V_M^{(2)}|$ $(= d'(v) - |C(v) \cap V_M^{(2)}|)$. Let

$$
\begin{aligned}
V_1 &:= \cup_{v \in V_M} V(T(v)),\\
V_2 &:= \{v \in V - V_1 \mid \tilde{d}(v) \geq 2\},\\
V_3 &:= \{v \in V - V_1 \mid \tilde{d}(v) = 1, C(v) \cap (V_M^{(2)} - V_L) = \emptyset\},\\
V_4 &:= \{v \in V - (V_1 \cup V_3) \mid \tilde{d}(v) = 1, \textstyle\sum_{w \in C(v) \cap (V_M^{(2)} - V_L)} |T(w)| \leq \Delta(\Delta - 19)\},\\
V_5 &:= \{v \in V - (V_1 \cup V_3) \mid \tilde{d}(v) = 1, \textstyle\sum_{w \in C(v) \cap (V_M^{(2)} - V_L)} |T(w)| > \Delta(\Delta - 19)\}.
\end{aligned}
$$

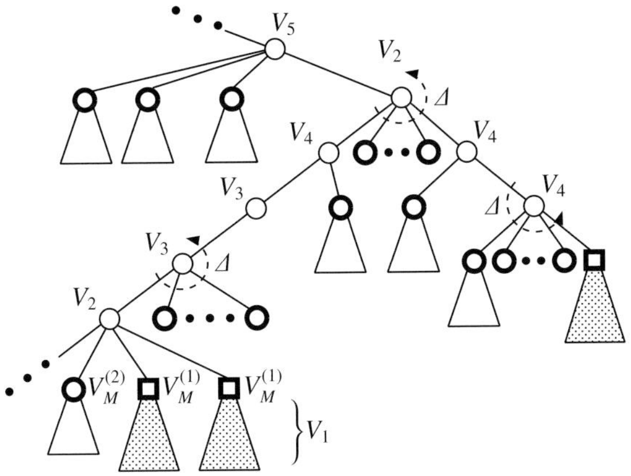

Fig. 1. Partition of V into V_i's ($i = 1, \ldots, 5$). Bold circles are leaves (V_L) or pseudo-leaves ($V_M^{(2)} - V_L$) with their subtrees, while bold squares are vertices in $V_M^{(1)}$ with their subtrees.

Notice that $V = V_1 \cup V_2 \cup V_3 \cup V_4 \cup V_5$, and $V_i \cap V_j = \emptyset$ for each i, j with $i \neq j$ (see Figure 1).

Here we describe an outline of the algorithm for computing $\delta((u, v), (*, *))$, $v \in V$, named COMPUTE-$\delta(v)$ (Algorithm 1), which can be regarded as a subroutine of the DP framework. Below, we show that for each V_i, $\delta((u, v), (*, *))$, $v \in V_i$ can be computed in linear time in total; i.e., $O(\sum_{v \in V_i} t(v)) = O(n)$. Namely, we have the following theorem.

Theorem 3. *For trees, the $L(2, 1)$-labeling problem can be solved in linear time.*

Algorithm 1. COMPUTE-$\delta(v)$

1: /** Assume that the head and neck levels of $T(v)$ are at most h. **/
2: If $v \in V_1 \cup V_2$, then for each $b \in (L_0 - L_h) \cup \{h\}$, compute $\delta((u, v), (*, b))$ by the max-flow computation in the network $\mathcal{N}(u, v, -, b)$ defined in Subsection 3.2.
3: If $v \in V_3$, execute the following procedure for each $b \in L_0$ in the case of $C(v) \cap V_L = \emptyset$, and for each $b \in \{0, \Delta + 1\}$ in the case of $C(v) \cap V_L \neq \emptyset$.
/** Let w^* denote the unique child of v not in $V_M^{(2)}$. **/
3-1: If $|\{c \mid \delta((v, w^*), (b, c)) = 1\}| \geq 2$, then let $\delta((u, v), (*, b)) := 1$.
3-2: If $\{c \mid \delta((v, w^*), (b, c)) = 1\} = \{c^*\}$, then let $\delta((u, v), (c^*, b)) := 0$ and $\delta((u, v), (a, b)) := 1$ for all other labels $a \notin \{b - 1, b, b + 1\}$.
3-3: If $|\{c \mid \delta((v, w^*), (b, c)) = 1\}| = 0$, then let $\delta((u, v), (*, b)) := 0$.
4: If $v \in V_4 \cup V_5$, then similarly to the case of $v \in V_1 \cup V_2$, compute $\delta((u, v), (*, *))$ by the max-flow computation in a network such as $\mathcal{N}(u, v, -, b)$ specified for this case (details will be described in Subsection 4.3).

We first show $O(\sum_{v \in V_1} t(v)) = O(|V_1|)$. For each $v \in V_M$, we have $O(\sum_{w \in V(T(v))} t(w)) = O(|T(v)|)$, by Corollary 1 and $|T(v)| = O(\Delta^5)$. Hence, we have $O(\sum_{v \in V_1} t(v)) = O(\sum_{v \in V_M} \sum_{w \in V(T(v))} t(w)) = O(\sum_{v \in V_M} |T(v)|) = O(|V_1|)$.

The sketch of proofs for V_2, V_3, V_4 and V_5 are given in the subsequent subsections, where some proofs of lemmas are omitted. See [12] for details.

4.2 Computation of δ-Value for V_2

By Lemma 4, we can see that $\sum_{v\in V_2} t(v) = \mathrm{O}(\sum_{v\in V_2} \Delta^{2/3} d'(v) h^2 \log^2 \Delta) = \mathrm{O}(\Delta^{8/3} \log^2 \Delta \sum_{v\in V_2} d'(v))$ (note that $h \leq \Delta$ and $d''(v) \leq d'(v)$). Now, we have $d'(v) \leq \tilde{d}(v) + \Delta \leq \Delta\tilde{d}(v)$. It follows that $\sum_{v\in V_2} t(v) = \mathrm{O}(\Delta^{11/3} \log^2 \Delta \sum_{v\in V_2} \tilde{d}(v))$. Below, in order to show that $\sum_{v\in V_2} t(v) = \mathrm{O}(n)$, we prove that $\sum_{v\in V_2} \tilde{d}(v) = \mathrm{O}(n/\Delta^4)$.

By definition, there is no vertex whose all children are vertices in $V_M^{(2)}$, since if there is such a vertex v, then for each $w \in C(v)$, we have $|T(w)| < (\Delta - 19)^4$ and hence $|T(v)| < \Delta^5$, which contradicts the maximality of $T(w)$. It follows that in the tree T' obtained from T by deleting all vertices in $V_1 - V_M^{(1)}$, each leaf vertex belongs to $V_M^{(1)}$ (note that $V(T') = V_M^{(1)} \cup V_2 \cup V_3 \cup V_4 \cup V_5$). Hence,

$$\begin{aligned} |V(T')| - 1 = |E(T')| &= \tfrac{1}{2} \textstyle\sum_{v\in V(T')} d_{T'}(v) \\ &= \tfrac{1}{2}(|V_M^{(1)}| + \textstyle\sum_{v\in V_2\cup V_3\cup V_4\cup V_5} (\tilde{d}(v) + 1) - 1) \\ &= \tfrac{1}{2}(|V_M^{(1)}| + \textstyle\sum_{v\in V_2} (\tilde{d}(v) + 1) + 2|V_3| + 2|V_4| + 2|V_5| - 1) \\ &\geq \tfrac{1}{2}|V_M^{(1)}| + \tfrac{3}{2}|V_2| + |V_3| + |V_4| + |V_5| - \tfrac{1}{2} \end{aligned}$$

(the last inequality follows from $\tilde{d}(v) \geq 2$ for all $v \in V_2$). Thus, $|V_M^{(1)}| - 1 \geq |V_2|$. Therefore, we can observe that $\sum_{v\in V_2} \tilde{d}(v) = |E(T')| - |V_3| - |V_4| - |V_5| = |V_M^{(1)}| + |V_2| - 1 \leq 2|V_M^{(1)}| - 2$ (the first equality follows from $|E(T')| = \sum_{v\in V_2\cup V_3\cup V_4\cup V_5} \tilde{d}(v) = \sum_{v\in V_2} \tilde{d}(v) + |V_3| + |V_4| + |V_5|$ and the second equality follows from $|E(T')| = |V(T')| - 1 = |V_M^{(1)}| + |V_2| + |V_3| + |V_4| + |V_5| - 1$). It follows by $|V_M^{(1)}| = \mathrm{O}(n/\Delta^4)$ that $\sum_{v\in V_2} \tilde{d}(v) = \mathrm{O}(n/\Delta^4)$.

4.3 Computation of δ-Value for V_3, V_4, and V_5

We sketch proofs for V_3, V_4, and V_5. Since Property 3 indicates that $|T(w)| \geq \Delta$ for each $w \in V_M - V_L$ (resp., $\sum_{w\in C(v)\cap(V_M^{(2)} - V_L)} |T(w)| > \Delta(\Delta - 19)$), we have $|V_4| = \mathrm{O}(n/\Delta)$ (resp., $|V_5| = \mathrm{O}(n/\Delta^2)$). By Property 4, we can observe that $|V_3| = \mathrm{O}(n/\Delta)$. Hence, it suffices to show that for each $v \in V_3 \cup V_4$ (resp., V_5), $\delta((u, v), (*, *))$ can be computed in $\mathrm{O}(\Delta)$ (resp., $\mathrm{O}(\Delta^2)$) time. Now,

$$\text{the head and neck levels of } T(w) \text{ are at most 8 for each } w \in V_M^{(2)} \tag{1}$$

by Lemma 1 and $|T(w)| < (\Delta - 19)^4$ (note that we assume that $\Delta \geq 26$, since otherwise the original CK algorithm is already a linear time algorithm). Let w^* be the unique child of v in $C(v) - V_M^{(2)}$.

First consider the case where $v \in V_3$ (i.e., Step 3 in algorithm Compute-$\delta(v)$). Let b be a label such that $b \in L_0$ if $v \in V_3^{(1)} := \{v \in V_3 \mid C(v) \cap V_L = \emptyset\}$, and $b \in \{0, \Delta + 1\}$ if $v \in V_3^{(2)} := V_3 - V_3^{(1)}$. Notice that if $v \in V_3^{(2)}$ (i.e., $C(v) \cap V_L \neq \emptyset$), then by Property 3, v is major and hence $\delta((u, v), (a, b)) = 1$, $a \in L_0$ indicates that $b = 0$ or $b = \Delta + 1$. Observe that if there is a label $c \in L_0 - \{b-1, b, b+1\}$ such that $\delta((v, w^*), (b, c)) = 1$, then for all $a \in L_0 - \{b-1, b, b+1, c\}$, we have $\delta((u, v), (a, b)) = 1$. It is not difficult to see that this shows the correctness of the procedure in this case. Obviously, for each $v \in V_3$, we can check which case of 3-1, 3-2, or 3-3 in algorithm Compute-$\delta(v)$ holds, and determine the values of $\delta((u, v), (*, b))$, in $\mathrm{O}(1)$ time. Therefore, the values of $\delta((u, v), (*, *))$ can be determined in $\mathrm{O}(\Delta)$ time.

Next consider the case where $v \in V_4$. For a label b, we divide $C(v) \cap (V_M^{(2)} - V_L)$ into two subsets $C_1(b) := \{w \in C(v) \cap (V_M^{(2)} - V_L) \mid \delta((v, w), (b, c)) = 1$ for all $c \in L_8 - \{b-1, b, b+1\}\}$ and $C_2(b) := \{w \in C(v) \cap (V_M^{(2)} - V_L) \mid \delta((v, w), (b, c)) = 0$ for all $c \in L_8 - \{b-1, b, b+1\}\}$. By the following property, we only have to consider the assignments for $\{w^*\} \cup C_2(b)$.

Lemma 5. *Let $v \in V_4$ and a and b be labels with $|b - a| \geq 2$ such that $b \in L_0$ if $C(v) \cap V_L = \emptyset$ and $b \in \{0, \Delta + 1\}$ otherwise. Then, $\delta((u, v), (a, b)) = 1$ if and only if there exists an injective assignment $g : \{w^*\} \cup C_2(b) \to L_0 - \{a, b-1, b, b+1\}$ such that $\delta((v, w), (b, g(w))) = 1$ for each $w \in \{w^*\} \cup C_2(b)$.*

Below, we will show how to compute $\delta((u, v), (*, b))$ in O(1) time for a fixed b, where $b \in L_0$ if $C(v) \cap V_L = \emptyset$ and $b \in \{0, \Delta + 1\}$ otherwise. If $|C_2(b)| \geq 17$, then $\delta((u, v), (*, b)) = 0$ because in this case, there exists some $w \in C_2(b)$ to which no label in $L_0 - L_8$ can be assigned since $|L_0 - L_8| = 16$. Assume that $|C_2(b)| \leq 16$. There are the following three possible cases: (Case-1) $\delta((v, w^*), (b, c_i)) = 1$ for at least two labels $c_1, c_2 \in L_8$, (Case-2) $\delta((v, w^*), (b, c_1)) = 1$, for exactly one label $c_1 \in L_8$, and (Case-3) otherwise.

(Case-1) By assumption, for any a, $\delta((v, w^*), (b, c)) = 1$ for some $c \in L_8 - \{a\}$. By Lemma 5, we only have to check whether there exists an injective assignment $g : C_2(b) \to L_0 - L_8 - \{a, b-1, b, b+1\}$ such that $\delta((v, w), (b, g(w))) = 1$ for each $w \in C_2(b)$. According to Subsection 3.2, this can be done by utilizing the maximum flow computation on the subgraph $\mathcal{N}'$ of $\mathcal{N}(u, v, -, b)$ induced by $\{s, t\} \cup C_2(b) \cup X'$ where $X' = \{0, 1, \ldots, 7, \Delta-6, \Delta-5, \ldots, \Delta+1\}$. Obviously, the size of $\mathcal{N}'$ is O(1) and it follows that its time complexity is O(1).

(Case-2) For all $a \neq c_1$, the value of $\delta((u, v), (a, b))$ can be computed similarly to Case-1. Consider the case where $a = c_1$. In this case, if $\delta((v, w^*), (b, c)) = 1$ holds, then it turns out that $c \in L_0 - L_8$. Hence, by Lemma 5, it suffices to check whether there exists an injective assignment $g : \{w^*\} \cup C_2(b) \to L_0 - L_8 - \{b-1, b, b+1\}$ such that $\delta((v, w), (b, g(w))) = 1$ for each $w \in \{w^*\} \cup C_2(b)$. Similarly to Case-1, this can be done in O(1) time, by utilizing the subgraph $\mathcal{N}''$ of $\mathcal{N}(u, v, -, b)$ induced by $\{s, t\} \cup (C_2(b) \cup \{w^*\}) \cup X'$.

(Case-3) By assumption, if $\delta((v, w^*), (b, c)) = 1$ holds, then it turns out that $c \in L_0 - L_8$. Similarly to the case of $a = c_1$ in Case-2, by using $\mathcal{N}''$, we can compute the values of $\delta((u, v), (*, b))$ in O(1) time.

We analyze the time complexity for computing $\delta((u, v), (*, *))$. It is dominated by that for computing $C_1(b)$, $C_2(b)$, and $\delta((u, v), (*, b))$ for each $b \in L_0$. By (1), we have $C_i(b) = C_i(b')$ for all $b, b' \in L_8$ and $i = 1, 2$. It follows that the computation of $C_1(b)$ and $C_2(b)$, $b \in L_0$ can be done in $O(|C(v) \cap (V_M^{(2)} - V_L)|)$ time. On the other hand, the values of $\delta((u, v), (*, b))$ can be computed in constant time in each case of Cases-1, 2 and 3 for a fixed b. Thus, $\delta((u, v), (*, *))$ can be computed in $O(\Delta)$ time.

Finally, we consider the case where $v \in V_5$. We will prove that the values of $\delta((u, v), (*, b))$ can be computed in $O(\Delta)$ time for a fixed b. A key is that the children $w \in C(v) \cap V_M^{(2)}$ of v can be classified into 2^{17} (= O(1)) types, depending on its δ-values $(\delta((v, w), (b, i)) \mid i \in (L_0 - L_8) \cup \{\tilde{c}_8\})$ where $\tilde{c}_8$ is some label in $L_8 - \{b-1, b, b+1\}$, since by (1), $\delta((v, w), (b, c)) = \delta((v, w), (b, \tilde{c}_8))$ for any $c \in L_8 - \{b-1, b, b+1\}$. Then, we can construct in $O(d'(v))$ time a network $\mathcal{N}'(u, v, a, b)$ with O(1) vertices, O(1) edges, and $O(\Delta)$ units of capacity from $\mathcal{N}(u, v, a, b)$ by letting $X_h := X_8$ and replacing $C(v)$ with a set of 2^{17} vertices corresponding to types of vertices in $C(v) \cap V_M^{(2)}$, and compute in $O(\log \Delta)$ time the values of $\delta((u, v), (a, b))$ by applying the maximum flow techniques to $\mathcal{N}'(u, v, a, b)$ (see [12] for the details about $\mathcal{N}'(u, v, a, b)$). Furthermore, by the following lemma, we can see that $\delta((u, v), (*, b))$ can be obtained by checking $\delta((u, v), (a, b))$ for $O(1)$ candidates of a; $\delta((u, v), (*, b))$ can be obtained in $O(\Delta)$ time.

Lemma 6. *If* $\delta((u, v), (a_1, b)) \neq \delta((u, v), (a_2, b))$ *for some* $a_1, a_2 \in L_8 - \{b - 1, b, b + 1\}$ *(say,* $\delta((u, v), (a_1, b)) = 1$*), then we have* $\delta((v, w^*), (b, a_2)) = 1$ *and* $\delta((v, w^*), (b, a)) = 0$ *for all* $a \in L_8 - \{a_2, b - 1, b, b + 1\}$*, and moreover,* $\delta((u, v), (a, b)) = 1$ *for all* $a \in L_8 - \{a_2, b - 1, b, b + 1\}$.

References

1. Bodlaender, H.L., Kloks, T., Tan, R.B., van Leeuwen, J.: Approximations for λ-coloring of graphs. The Computer Journal 47, 193–204 (2004)
2. Calamoneri, T.: The L(h,k)-labelling problem: A survey and annotated bibliography. The Computer Journal 49, 585–608 (2006), `http://www.dsi.uniroma1.it/~calamo/PDF-FILES/survey.pdf` (January 13, 2009)
3. Chang, G.J., Ke, W.-T., Kuo, D., Liu, D.D.-F., Yeh, R.K.: On $L(d, 1)$-labeling of graphs. Discr. Math. 220, 57–66 (2000)
4. Chang, G.J., Kuo, D.: The L(2,1)-labeling problem on graphs. SIAM J. Discr. Math. 9, 309–316 (1996)
5. Fiala, J., Golovach, P.A., Kratochvíl, J.: Distance constrained labelings of graphs of bounded treewidth. In: Caires, L., Italiano, G.F., Monteiro, L., Palamidessi, C., Yung, M. (eds.) ICALP 2005. LNCS, vol. 3580, pp. 360–372. Springer, Heidelberg (2005)
6. Fiala, J., Golovach, P.A., Kratochvíl, J.: Distance constrained labelings of trees. In: Agrawal, M., Du, D.-Z., Duan, Z., Li, A. (eds.) TAMC 2008. LNCS, vol. 4978, pp. 125–135. Springer, Heidelberg (2008)
7. Fiala, J., Golovach, P.A., Kratochvíl, J.: Computational complexity of the distance constrained labeling problem for trees (Extended abstract). In: Aceto, L., Damgård, I., Goldberg, L.A., Halldórsson, M.M., Ingólfsdóttir, A., Walukiewicz, I. (eds.) ICALP 2008, Part I. LNCS, vol. 5125, pp. 294–305. Springer, Heidelberg (2008)
8. Fiala, J., Kloks, T., Kratochvíl, J.: Fixed-parameter complexity of λ-labelings. Discr. Appl. Math. 113, 59–72 (2001)
9. Goldberg, A.V., Rao, S.: Beyond the flow decomposition barrier. J. ACM 45, 783–797 (1998)
10. Griggs, J.R., Yeh, R.K.: Labelling graphs with a condition at distance 2. SIAM J. Disc. Math. 5, 586–595 (1992)
11. Hasunuma, T., Ishii, T., Ono, H., Uno, Y.: An $O(n^{1.75})$ algorithm for $L(2, 1)$-labeling of trees. In: Gudmundsson, J. (ed.) SWAT 2008. LNCS, vol. 5124, pp. 185–197. Springer, Heidelberg (2008); Journal version to appear in Theoretical Comp. Sci., doi:10.1016/j.tcs.2009.04.025
12. Hasunuma, T., Ishii, T., Ono, H., Uno, Y.: A linear time algorithm for L(2,1)-labeling of trees. CoRR abs/0810.0906 (2008)
13. Havet, F., Reed, B., Sereni, J.-S.: L(2,1)-labelling of graphs. In: Proc. 19th SIAM-SODA, pp. 621–630 (2008)
14. Hopcroft, J.E., Karp, R.M.: An $n^{5/2}$ algorithm for maximum matchings in bipartite graphs. SIAM J. Comput. 2, 225–231 (1973)
15. Kratochvíl, J., Kratsch, D., Liedloff, M.: Exact algorithms for $L(2, 1)$-labeling of graphs. In: Kučera, L., Kučera, A. (eds.) MFCS 2007. LNCS, vol. 4708, pp. 513–524. Springer, Heidelberg (2007)
16. Wang, W.-F.: The L(2,1)-labelling of trees. Discr. Appl. Math. 154, 598–603 (2006)
17. Yeh, R.K.: A survey on labeling graphs with a condition at distance two. Discr. Math. 306, 1217–1231 (2006)

On Inducing Polygons and Related Problems*

Eyal Ackerman[1], Rom Pinchasi[2], Ludmila Scharf[1], and Marc Scherfenberg[1]

[1] Institute of Computer Science, Freie Universität Berlin, Takustr. 9,
14195 Berlin, Germany
{eyal,scharf,scherfen}@mi.fu-berlin.de

[2] Mathematics Department, Technion—Israel Institute of Technology,
Haifa 32000, Israel
room@math.technion.ac.il

Abstract. Bose et al. [1] asked whether for every simple arrangement $\mathcal{A}$ of n lines in the plane there exists a simple n-gon P that *induces* $\mathcal{A}$ by extending every edge of P into a line. We prove that such a polygon always exists and can be found in $O(n \log n)$ time. In fact, we show that every finite family of curves $\mathcal{C}$ such that every two curves intersect at least once and finitely many times and no three curves intersect at a single point possesses the following Hamiltonian-type property: the union of the curves in $\mathcal{C}$ contains a simple cycle that visits every curve in $\mathcal{C}$ exactly once.

1 Introduction

Arrangements of lines in the plane are among the most studied structures in Combinatorial and Computational Geometry (see, e.g., [4,5]). Every set of straight-line segments S naturally *induces* an arrangement of lines, simply by extending every segment in S into a line. Bose et al. [1] asked the following natural question.

Problem 1. Does every *simple* arrangement $\mathcal{A}$ of n lines contain a simple n-gon that induces $\mathcal{A}$?

An arrangement of lines is *simple* if every pair of lines intersects, and no three lines intersect at a single point. A polygon (resp., curve) is *simple* if it is non-self-intersecting. Fig. 1(a) shows a simple arrangement of six lines and a simple hexagon that induces this arrangement.

Problem 1 remained open until now, though a few partial results were obtained. In [1] it was shown that a simple arrangement $\mathcal{A}$ of n lines contains a subarrangement of $m \geq \sqrt{n-1}+1$ lines that has an inducing simple m-gon, and that $\mathcal{A}$ always has an inducing simple n-path (a polygonal chain consisting of n line segments), which can be constructed in $O(n^2)$ time. Recently, the third and fourth authors [8] showed that an inducing n-path can be constructed in $O(n \log n)$ time, and that there always exists an inducing simple $O(n)$-gon, which can be found in $O(n^2)$ time.

Our main result is an affirmative answer to Problem 1.

* Research by Eyal Ackerman was supported by a fellowship from the Alexander von Humboldt Foundation. Research by Rom Pinchasi was supported by the Israeli Science Foundation (grant No. 938/06).

A. Fiat and P. Sanders (Eds.): ESA 2009, LNCS 5757, pp. 47–58, 2009.

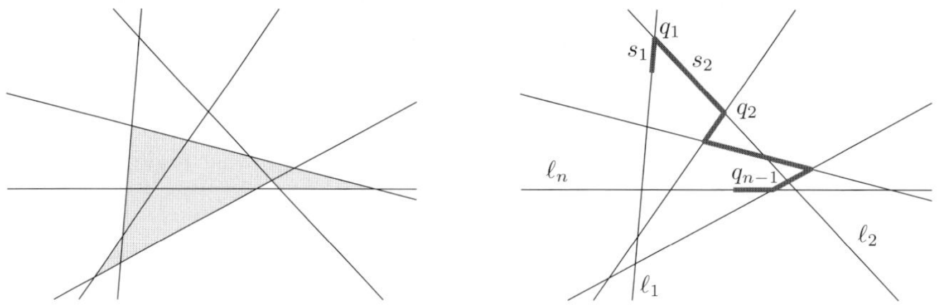

(a) (b)

Fig. 1. An inducing simple n-gon and n-path

Theorem 1. *For every simple arrangement $\mathcal{A}$ of $n > 2$ lines in the plane there is a simple n-gon that induces $\mathcal{A}$. Given the set of n lines that form $\mathcal{A}$, such a polygon can be constructed in $O(n \log n)$ time.*

We give two different constructive proofs for the existence of an inducing simple n-gon. The first proof is short and elegant and yields a non-optimal but polynomial-time algorithm for finding such a polygon. The second proof yields an $O(n \log n)$-time algorithm. It is based on a simple idea, however, it involves several case distinctions and is, thus, quite technical.

During our quest for a solution to Problem 1, we proved the following interesting fact.

Theorem 2. *For every simple arrangement $\mathcal{A}$ of n non-vertical lines in the plane there is an x-monotone n-path that induces $\mathcal{A}$.*

Note that the first part of Theorem 1 can also be phrased as follows: Every arrangement of lines contains a simple cycle (i.e., a closed curve) that visits every line exactly once. To be more precise, we say that a curve x *visits* another curve y if their intersection contains a point in which they neither cross nor touch. A simple curve visits y *exactly once* if it visits y and their intersection is connected. The first part of Theorem 1 is then equivalent to saying that every simple line arrangement contains a simple (polygonal) cycle that visits every line exactly once. We also have the following generalization of Theorem 1.

Theorem 3. *Let $\mathcal{C}$ be a finite family of $n > 2$ simple curves in $\mathbb{R}^3$, such that every pair of curves in $\mathcal{C}$ intersects at least once and at most finitely many times, and no three curves intersect at the same point. Then $\bigcup_{C \in \mathcal{C}} C$ contains a simple cycle that visits every curve exactly once.*

The rest of this paper is organized as follows. A first proof for the existence of an inducing simple n-gon is given in Section 2. This proof is then extended in Section 3 to a proof of Theorem 3. In Section 4 we describe a different and more efficient way of finding an inducing simple n-gon. Due to space limitations, we only sketch the idea of the proof and omit most of the details, which can be found in the full version of this paper. Theorem 2 is proved in Section 5, while Section 6 contains some concluding remarks.

2 First Proof of the Existence of an Inducing Simple n-gon

Let $\mathcal{A}$ be a simple arrangement of n lines in the plane. We begin by constructing a simple path that visits every line in $\mathcal{A}$ exactly once. This is done in a way similar to the construction of an inducing path in [1]. Consider an arbitrary intersection point of two lines, denote these lines by ℓ_1 and ℓ_2. Walk a short distance on ℓ_1 toward its intersection point with ℓ_2. Remove ℓ_1, and walk on ℓ_2 in a direction that contains at least one intersection point, until reaching the first intersection point. Let ℓ_3 be the other line that determines this intersection point. Remove ℓ_2 and repeat the same process for ℓ_3 and so on and so forth, until reaching a line that has no additional intersection points. Finally, walk a short distance on this line in some direction. See Fig. 1(b) for an example.

Since every pair of lines intersects, no line is missed and this process results in a path that induces every line in $\mathcal{A}$. Denote this path by Q. The lines in $\mathcal{A}$ are denoted by $\ell_1, \ell_2, \dots, \ell_n$ according to the order they are visited by Q. Denote by s_j the segment of ℓ_j on Q. Assume to the contrary that Q is self-intersecting. Then there are two intersecting segments, s_i and s_j, such that $i \leq j-2$. But this is a contradiction to the definition of ℓ_{i+1} as the first line, different from $\ell_1, \dots, \ell_i$, that we encounter while walking along ℓ_i.

Observe that Q lies in one of the two half-planes determined by ℓ_n. Indeed, otherwise ℓ_n would have crossed Q, contradicting the definition of ℓ_n as the last line in $\mathcal{A}$ we encounter while creating the path Q. We assume without loss of generality that ℓ_n coincides with the x-axis and that Q lies in the half-plane above ℓ_n. For convenience, denote the line ℓ_n by ℓ.

We call a simple inducing n-path of $\mathcal{A}$ *rooted above* ℓ if it lies in the closed half-plane above ℓ and ℓ includes an extreme segment of the path. As we have just seen, there is at least one simple inducing n-path rooted above ℓ, namely Q. For such a path W we denote by $q_1(W), \dots, q_{n-1}(W)$ the $n-1$ internal vertices of the path starting from $q_1(W)$ on ℓ. We denote by $\ell(W)$ the line through $q_{n-1}(W)$ that includes the other extreme segment of W. Denote by $\ell^-(W)$ the half-line of $\ell(W)$ that consists of all points with y-coordinates smaller than the y-coordinate of $q_{n-1}(W)$. We denote by $q_n(W)$ the topmost (also first) intersection point of $\ell^-(W)$ with $\ell \cup [q_1(W)q_2(W)] \cup \ldots \cup [q_{n-2}(W)q_{n-1}(W)]$. (Here, $[ab]$ denotes the line segment connecting point a to point b.)

Let $z_1, \dots, z_m$ denote all the intersection points in $\mathcal{A}$ indexed in any way such that $i < j$ if the y-coordinate of z_i is smaller than the y-coordinate of z_j. We then define for every j, $Y(z_j) = j$.

For a simple inducing n-path W rooted above ℓ, let $Y(W) = \sum_{i=1}^{n} Y(q_i(W))$. Consider the simple inducing n-path W rooted above ℓ such that $Y(W)$ is minimum. If $q_n(W)$ lies on ℓ, then observe that the vertices $q_1(W), \dots, q_n(W)$ define

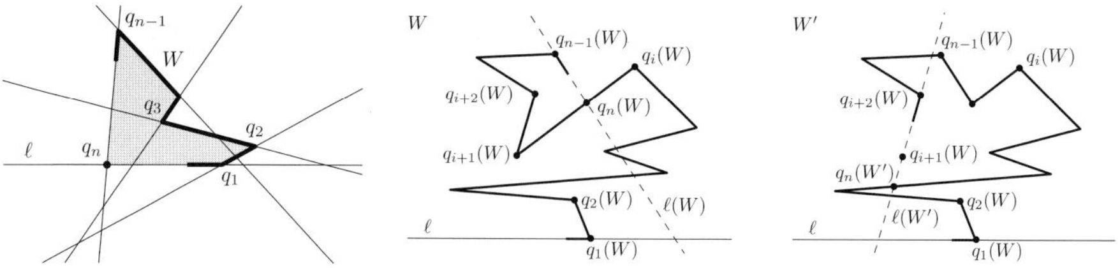

(a) q_n lies on ℓ (b) q_n lies on W (c) W'

Fig. 2. The paths W and W'

a simple inducing closed n-path of $\mathcal{A}$ (see Fig. 2(a)). Assume therefore that $q_n(W)$ is the intersection point of $\ell(W)$ with the segment $[q_i(W)q_{i+1}(W)]$ for some $1 \leq i \leq n-2$ (see Fig. 2(b)). Then we define W' as the path whose internal vertices are

$$q_1(W), \ldots, q_i(W), q_n(W), q_{n-1}(W) \ldots, q_{i+2}(W),$$

and hence $\ell(W')$ is the line through $q_{i+1}(W)$ and $q_{i+2}(W)$. Observe that W' is a simple inducing n-path rooted above ℓ. We have $Y(W') < Y(W)$ because $q_n(W')$ has a smaller y-coordinate than the y-coordinate of $q_{i+1}(W)$ (see Fig. 2(c)). We have thus reached a contradiction to the minimality of W. □

Remark. The proof of Theorem 1, presented above, yields an algorithm with running time polynomial in n. This is because $Y(W)$ is always smaller than n^3 and this gives a bound on the number of iterations going from W to W' required to find a simple inducing closed n-path for $\mathcal{A}$.

3 Proof of Theorem 3

Let $\mathcal{C}$ be a family of n simple curves in $\mathbb{R}^3$, such that every pair of curves in $\mathcal{C}$ intersects at least once and at most finitely many times, and no three of the curves meet at a point. We will show that $\bigcup_{C \in \mathcal{C}} C$ contains a simple closed path that visits every curve in $\mathcal{C}$ exactly once.

The proof is a modification of the argument in the proof of Theorem 1. We first find a simple path Q that visits every curve in $\mathcal{C}$ exactly once, exactly in the same way that was described in Section 2, applied this time to $\mathcal{C}$. Let c be a curve in $\mathcal{C}$ containing the last segment of Q thus constructed. As we observed in the case of lines, c does not meet Q at any point outside the segment of Q contained in c.

A simple (oriented) path W that visits every curve in $\mathcal{C}$ exactly once will be called *rooted in* c if c is the first curve visited by W. Clearly, Q is an example for such a path.

For a path W, as above, we denote by $q_1(W), \ldots, q_{n-1}(W)$ the $n-1$ internal vertices of the path starting from $q_1(W)$ on c. For $i = 1, \ldots, n-2$ we denote by $s_i(W)$ the segment of W whose vertices are $q_i(W)$ and $q_{i+1}(W)$, these will be called the *internal segments* of W. We denote by $c(W)$ the curve in $\mathcal{C}$ that passes through $q_{n-1}(W)$ and contains the last segment of W.

Let s be a portion of a curve in $\mathcal{C}$. We define $|s|$ as the number of intersection points of pairs of curves in $\mathcal{C}$ that lie on s. Finally, we define

$$Y(W) = f(|s_1(W)|, \dots, |s_{n-2}(W)|),$$

where $f(x_1, \dots, x_{n-2})$ is a strictly monotone increasing function of the lexicographic order of $(x_1, \dots, x_{n-2})$.[1]

Consider the simple path W that is rooted in c and visits every curve in $\mathcal{C}$ exactly once, such that $Y(W)$ is minimum. Let p be an intersection point of $c(W)$ and c. Let $q_n(W)$ be the intersection point of $c \cup s_1(W) \cup \dots \cup s_{n-2}(W)$ and the portion of $c(W)$ between $q_{n-1}(W)$ and p that is closest to $q_{n-1}(W)$ along the curve $c(W)$.

If $q_n(W)$ lies on c, then observe that the vertices $q_1(W), \dots, q_n(W)$ define a simple closed path that visits every curve in $\mathcal{C}$ exactly once. Assume therefore that $q_n(W)$ is an intersection point of $c(W)$ with $s_i(W)$ for some $1 \le i \le n-2$. Let s' denote the portion of $s_i(W)$ delimited by $q_i(W)$ and $q_n(W)$. Let s'' denote the portion of $c(W)$ delimited by $q_n(W)$ and $q_{n-1}(W)$. Then we define W' as the path rooted in c whose internal segments are

$$s_1(W), \dots, s_{i-1}(W), s', s'', s_{n-2}(W), s_{n-3}(W) \dots, s_{i+2}(W),$$

and $c(W')$ is the curve containing the segment $s_{i+1}(W)$.

Observe that W' is a simple path rooted on c that visits every curve in $\mathcal{C}$ exactly once. It immediately follows that $Y(W') < Y(W)$, because $s_j(W') = s_j(W)$ for $j = 1, \dots, i-1$ while it is easy to see that $|s_i(W')| < |s_i(W)|$ as $s_i(W') = s' \subset s_i(W)$ and $q_{i+1}(W)$ is an intersection point in $s_i(W) \setminus s_i(W')$. We have thus reached a contradiction to the minimality of W. □

Remarks. (1) Because Theorem 3 is stated in $\mathbb{R}^3$, geometry actually does not play any role here. We may conclude the same result for "combinatorial curves" that "intersect" finitely many times, as long as there is a total order on the set of intersection points in each curve.
(2) The result in Theorem 3 is valid also if the curves in $\mathcal{C}$ are not simple and have self-crossings. In this case we repeat the proof and ignore self-intersections of curves. Finally, when obtaining the resulting closed path we observe that self-intersections of the closed path result only from loops in the path. These loops can easily be canceled.

4 Finding an Inducing Simple n-gon Efficiently

Let $\mathcal{A}$ be a simple arrangement of n lines in the plane. We incrementally construct a polygon inducing $\mathcal{A}$ by starting with the boundary of a cell of $\mathcal{A}$. In every

[1] For example, $f(11, 0, 6, \dots) > f(6, 9, 5, \dots) > f(6, 9, 4, \dots)$.

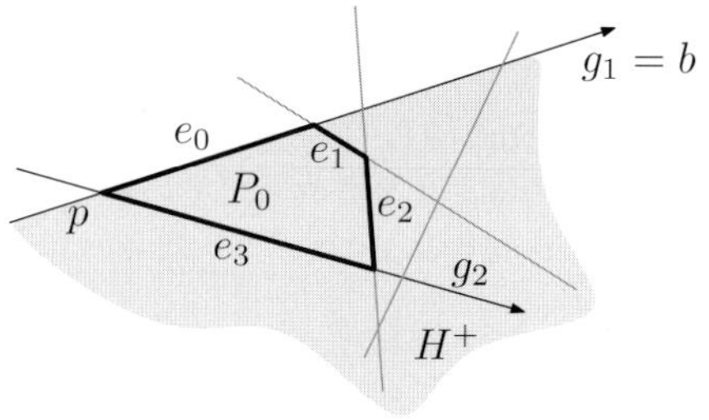

Fig. 3. Initialize P_0 to be the boundary of the bounded face of $\mathcal{A}$ incident to a critical point p

construction step the polygon is extended using a part of the boundary of the cell containing it. We assume that $n > 4$, since, combinatorially, there is only one arrangement of size three and one of size four and their inducing polygons can be easily found.

We start with a so-called critical point p, i.e., p is the first intersection point on both lines g_1 and g_2 containing it. The initial polygon P_0 is then the boundary of the only bounded face incident to p, see Fig. 3.

Let P_i denote the polygon constructed in step i, and $|P_i|$ its number of edges. Denote by $\mathcal{A}_i$ the arrangement of all the lines except the ones induced by P_i. We maintain the following invariants throughout the construction of the polygons P_i.

Property 1.

1. P_i is a simple polygon;
2. P_i induces $|P_i|$ lines of the arrangement $\mathcal{A}$; and
3. P_i is contained in an unbounded face of the arrangement $\mathcal{A}_i$.

The unbounded face of $\mathcal{A}_i$ containing P_i is denoted by $C^{(i)}$ and its by $R^{(i)}$. Define the orientation of the two initial lines g_1 and g_2 in direction from p towards the remaining intersection points. Without loss of generality we can assume that all intersection points of g_2 lie in the positive half-plane of g_1, denoted by $H^+(g_1)$, as in Fig. 3.

For every construction step we maintain a so-called *base line* $b^{(i)}$. Intuitively, the base line will be the line that determines the direction in which P_i is extended. For P_0 the base line is $b^{(0)} = g_1$. The edges of P_i are labeled in the following way: the edge contained in the base line $b^{(i)}$ is the edge $e_0^{(i)}$. In counter-clockwise order we enumerate with negative indices the edges contained in the previous base lines $e_{-1}^{(i)}, \ldots, e_{-m}^{(i)}$, where $e_{-m}^{(i)}$ is contained in the first base line $b^{(0)}$. These edges are referred to as *base edges*. It can be shown that the base edges form a connected concave chain in P_i. The remaining *non-base edges* are enumerated in clockwise order with positive indices $e_1^{(i)}, \ldots, e_k^{(i)}$, where $e_1^{(i)}$ is incident to $e_0^{(i)}$ and $e_k^{(i)}$ is incident to $e_{-m}^{(i)}$.

A line containing an edge $e_j^{(i)}$ is denoted by $l_j^{(i)}$ and the intersection point of two lines $l_j^{(i)}, l_m^{(i)}$ by $x_{j,m}^{(i)}$. We define the orientation of base edges in clockwise direction and the orientation of non-base edges in counter clockwise direction with respect to the polygon P_i. For each line $l_j^{(i)}$ its orientation is defined by the orientation of the edge $e_j^{(i)}$. The part of $l_j^{(i)} \setminus e_j^{(i)}$ oriented in positive (negative) direction of $l_j^{(i)}$ is called positive (negative) half-line and is denoted by $l_j^{(i)+}$ ($l_j^{(i)-}$), respectively. For simplicity we will omit the index (i) if all identifiers refer to the same step i, and will use the index in order to distinguish between different steps.

We maintain the following properties for base lines, and non-base lines, respectively.[2]

Property 2. All intersection points of a base line l_j, $j \leq 0$, with $\mathcal{A}_i$ lie in the positive half-line, i.e., $l_j^- \cap \mathcal{A}_i = \emptyset$ for $j \leq 0$. The base edges $e_0, e_{-1}, \ldots, e_{-m}$ form a concave chain in P_i, and every non-base edge is contained in the union of the positive half-planes (i.e., half-planes to the right of the oriented line) of the base lines $l_0, l_{-1}, \ldots, l_{-m}$.

Property 3. The intersection of a non-base line l_j with a non-base edge e_k is empty, for $k > j + 1$.

For the line l_1 it would be helpful to have an even stronger property:

Property 4. The intersection of l_1 with P_i is exactly the edge e_1. That is, l_1 supports P_i.

The idea of the extension step is to extend the polygon P_i in direction of the base line by modifying the edges e_0 up to at most e_3 and adding a part of the boundary $R^{(i)}$ to the new polygon P_{i+1}. In every extension step we remove a chain of edges from the polygon P_i, and attach a simple polygonal chain to the open ends. Thus, if the added chain does not intersect the unchanged part of P_i, the polygon P_{i+1} is simple.

Depending on the combinatorial configuration of the lines l_1, l_2, l_3, the chain of base edges, and the boundary R, one of several extension construction steps is taken, until all lines of $\mathcal{A}$ are induced by P_j, for some j. The inducing polygon for $\mathcal{A}$ is then $P = P_j$.

The first case distinction is whether the negative half-line of l_1 intersects the boundary R. If it does, Case 1 applies.

Case 1 [$l_1^- \cap R \neq \emptyset$]: The edge e_1 is replaced by the part of l_1^- from $x_{1,2}$ to its intersection with R. The edge e_0 is extended until the intersection of b and R. Finally, we add the segment of R between these two intersection points, see Fig. 4(a). The base line for P_{i+1} remains unchanged $b^{(i+1)} = b^{(i)}$.

The next distinction is whether l_2^+ intersects R:

Case 2 [$l_1^- \cap R = \emptyset$ and $l_2^+ \cap R \neq \emptyset$]: In this case e_1 is replaced by the part of l_1^+ from $x_{0,1}$ to its intersection with R. The edge e_2 is extended following the

[2] Property 3 can be violated in a special case that is considered in the full version of the paper.

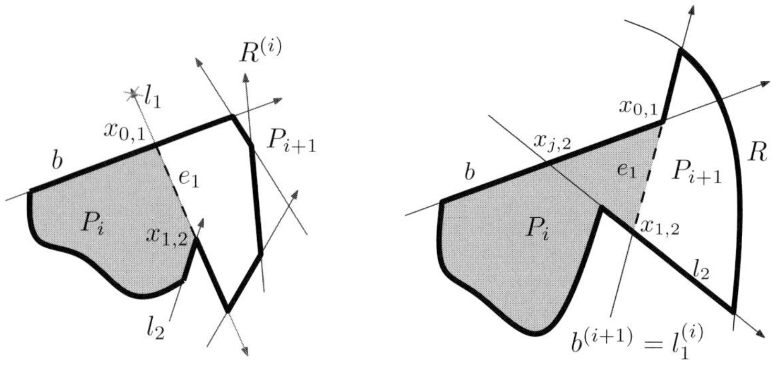

Fig. 4. Case 1 and 2: The polygon P_i is the shaded area. The identifiers refer to P_i, and the new polygon P_{i+1} is outlined by the bold black line.

orientation of l_2 until the intersection of l_2^+ and R. Finally, we add the segment of R between these two intersection points, see Fig. 4(b). The new base line for the polygon P_{i+1} is now $b^{(i+1)} = l_1^{(i)}$.

It is easy to verify that all the above-mentioned properties are maintained when applying Cases 1 or 2. Due to space limitations we do not include the remaining and more complicated cases in this extended abstract, and refer the reader to the full version of the paper for those missing details.

Running Time. In the initialization step we need to find an intersection point of the arrangement that is the last point on both lines intersecting in it. Ching and Lee [3] showed that such points are a subset of the intersection points between two neighboring lines sorted by slope. Thus, the initialization can be performed in $O(n \log n)$ time by sorting the lines by slope, computing the intersection points of the neighboring lines and selecting the point with the maximum or minimum x-coordinate.

For the extension steps we consider the dual points of the lines of the arrangement, where the dual space π^* is defined as in [2]: The dual of a point $p : (a, b)$ in the primal space is the line $p^* : f(x) = ax - b$ in π^*; the dual of a line $l : f(x) = ax + b$ in the primal space is the point $l^* : (a, -b) \in \pi^*$.

Let $\mathcal{A}^*$ denote the set of points in π^* dual to the lines of the arrangement $\mathcal{A}$. We will utilize the following property of the dual points: the points of the lower/upper convex hull of $\mathcal{A}^*$ are the duals of the lines in $\mathcal{A}$ that form the boundary of the upper/lower unbounded face of the arrangement.

For that purpose we can rotate the arrangement $\mathcal{A}$ such that the initial two lines g_1 and g_2 have the maximal and the minimal slope, the initial point $p = g_1 \cap g_2$ is a vertex of the lower unbounded face, and no line of $\mathcal{A}$ is vertical. Observe that p must be the only vertex of the lower unbounded face.

When the lines g_1 and g_2 are removed from $\mathcal{A}$ the point p is contained in the new lower unbounded face. Similarly, after every extension step the constructed polygon is contained in the lower unbounded face of the arrangement of the remaining lines.

In every extension step we need to determine the intersection points of a constant number of lines with the boundary of the lower unbounded face of the arrangement of the remaining lines and to update the boundary of the lower unbounded face after deleting some lines. Updating the boundary of the lower unbounded face corresponds to updating the upper convex hull of the dual point set. Using the dynamic convex hull data structure by Hershberger and Suri [6] updates of the upper convex hull of the point set can be performed in $O(\log n)$ time, that is $O(n \log n)$ time in total.

Intersection points of a line l with the boundary of the lower unbounded face correspond in dual space to lines through l^* that are tangent to the upper convex hull of the remaining points. These tangent lines can be found in $O(\log n)$ time.

Thus the total time complexity of the construction algorithm is $O(n \log n)$.

5 x-Monotone Inducing n-Path: Proof of Theorem 2

In this section we show that every simple arrangement of n non-vertical lines, contains an inducing x-monotone n-path. Since the path is x-monotone, it is clearly simple. Suppose first that n is an even number. We sort the lines according to their slopes, and denote by A the set of the first $n/2$ lines in this order, and by B the rest of the lines. Initially, all the lines are unmarked. Pick the leftmost intersection point of two unmarked lines, one from A and one from B, then mark these lines. Continue to pick a total of $n/2$ points $p_1, p_2, \ldots, p_{n/2}$ this way. We will construct an x-monotone n-path through $p_1, p_2, \ldots, p_{n/2}$.

Denote the lines that intersect at p_i by $a_i \in A$ and $b_i \in B$, $i = 1, 2, \ldots, n/2$. First, pick arbitrarily one of the lines that intersect at p_1, say a_1, walk a short distance on a_1 from a point left of p_1 to p_1, then walk a short distance on b_1 rightwards. Assume that we have built an x-monotone $2i$-path that goes a short distance rightwards beyond p_i and induces the lines $a_1, \ldots, a_i$ and $b_1, \ldots, b_i$. We will show how to extend it into an x-monotone $2(i+1)$-path that goes a short distance rightwards beyond p_{i+1} and induces the lines $a_1, \ldots, a_{i+1}$ and $b_1, \ldots, b_{i+1}$.

Observation 4 *The intersection of a_{i+1} (resp., b_{i+1}) and b_i (resp., a_i) is to the right of p_i.*

Proof. Otherwise this intersection point would be picked instead of p_i.

Consider the triangle with a vertex at p_i, an edge on the vertical line through p_{i+1}, an edge e_a on a_i, and an edge e_b on b_i (see Fig. 5).

Observation 5 *e_a (resp., e_b) is crossed by a_{i+1} or b_{i+1}.*

Proof. We consider two cases based on whether p_{i+1} is inside the wedge determined by a_i and b_i. Suppose that it is. Then a_{i+1} (resp., b_{i+1}) must cross either

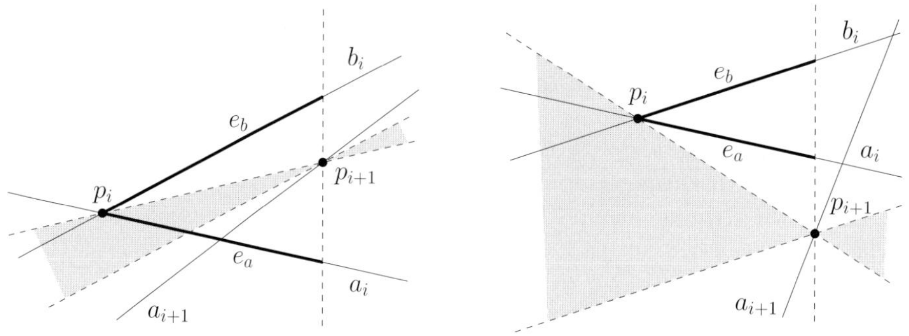

(a) p_{i+1} is inside the wedge determined by a_i and b_i.

(b) p_{i+1} is outside the wedge determined by a_i and b_i.

Fig. 5. An illustration for the proof of Observation 5. a_{i+1} cannot be in the shaded region.

e_a and e_b. Suppose, w.l.o.g., that they both cross e_a (otherwise, we can reflect everything with respect to the x-axis). See Fig. 5(a). Then a_{i+1} must have a larger slope than b_i, otherwise it will cross b_i to the left of p_i, contradicting Observation 4. This is of course impossible.

Suppose that p_{i+1} is outside the wedge determined by a_i and b_i. We can assume, w.l.o.g., that it is below the wedge, for otherwise we can reflect everything with respect to the x-axis. If a_{i+1} does not cross both e_a and e_b, then it must have a larger slope than b_i, or cross b_i to the left of p_i, which is impossible. See Fig. 5(b) for an illustration.

Now, suppose that the path built so far goes a short distance rightwards beyond p_i on e_a (resp., e_b). Then by Observation 5 there is a line $\ell \in \{a_{i+1}, b_{i+1}\}$ that crosses e_a (resp., e_b). Walk on e_a (resp., e_b) until the intersection point with ℓ, then walk on ℓ until p_{i+1}, and finally walk a short distance rightwards on the other line in $\{a_{i+1}, b_{i+1}\}$. The new path is an x-monotone $2(i+1)$-path that goes a short distance rightwards beyond p_{i+1} and induces the lines $a_1, \ldots, a_{i+1}$ and $b_1, \ldots, b_{i+1}$.

It remains to consider the case that n is an odd number. Let ℓ be the line with the median slope. Create a new line ℓ' that is a slightly rotated copy of ℓ such that its slope is slightly smaller than the slope of ℓ, and their intersection point is the leftmost intersection point in the arrangement $\mathcal{A} \cup \{\ell'\}$. Now continue as before, while choosing ℓ' as the first induced line. Finally, remove the segment of ℓ' from the constructed path.

Time complexity. An inducing x-monotone n-path can be found in $O(n^2)$ time as follows. First we construct the arrangement of lines. This can be done in $O(n^2)$ time [2]. Then we find the sets A and B in $O(n \log n)$ time. For every line in A we find its leftmost intersection point with a line from B. The first vertex

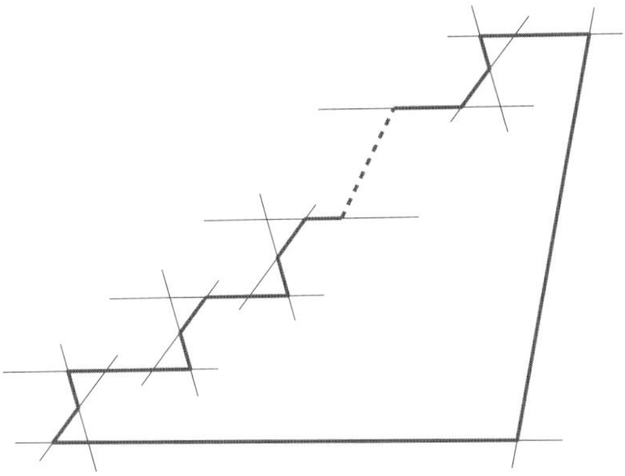

Fig. 6. n lines with exponentially many inducing simple n-gons. At every "step" of the "stairs" one can "climb" either from left or from right.

of the path is the leftmost point among these points. The two lines defining this minimum point are removed from the arrangement while updating the minimum leftmost points for the other lines. This can be done in $O(n)$ time. The process of finding the next leftmost intersection point between a line from A and a line from B (among the remaining lines), removing the corresponding lines, and making appropriate updates is then repeated $O(n)$ times.

6 Concluding Remarks

We proved in two different ways that every simple arrangement of n lines contains an inducing simple n-gon. The proof given in Section 2 actually works also for *pseudoline* arrangements. A pseudoline arrangement consists of a finite set of x-monotone curves, unbounded in both directions, such that every two curves intersect at exactly one point where they properly cross each other. It is enough to show that there is a partial order of the intersection points that lie above the pseudoline ℓ_n. Such an order can be derived from orienting every pseudoline toward its intersection point with ℓ_n. The proof then shows that there is a simple cycle that visits every pseudoline exactly once, and that such a cycle can be found in polynomial time. In fact, the proof also works for *pseudo-parabolas* (pseudo-parabolas are defined similarly to pseudolines, except that two curves cross exactly twice). Here, a partial order of the intersection points can be defined as in [7].

The second proof, given in Section 4, yields an $O(n \log n)$-time algorithm for finding an inducing simple polygon. We believe that this time complexity is the best possible, but leave it as an open question.

An inducing simple polygon need not be unique. It would be interesting to determine the maximum and minimum number of inducing simple n-gons of an arrangement of n lines. Fig. 6 shows an arrangement with exponentially many inducing simple n-gons.

References

1. Bose, P., Everett, H., Wismath, S.: Properties of arrangement graphs. Int. J. Comput. Geom. Appl. 13, 447–462 (2003)
2. de Berg, M., Cheong, O., van Kreveld, M., Overmars, M.: Computational Geometry, 3rd edn. Springer, Berlin (2008)
3. Ching, Y.T., Lee, D.T.: Finding the diameter of a set of lines. Pattern Recognition 18(3–4), 249–255 (1985)
4. Felsner, S.: Geometric Graphs and Arrangements. Some Chapters from Combinatorial Geometry. Advanced Lectures in Mathematics. Friedr. Vieweg & Sohn, Wiesbaden (2004)
5. Grünbaum, B.: Arrangements and spreads. In: Conference Board of the Mathematical Sciences Regional Conference Series in Mathematics, vol. 10. American Mathematical Society, Providence (1972)
6. Hershberger, J., Suri, S.: Applications of a semi-dynamic convex hull algorithm. BIT 32(2), 249–267 (1992)
7. Perlstein, A.: Problems in Combinatorial Geometry, Ph.D. Thesis, Mathematics Department, Technion—Israel Institute of Technology (2008)
8. Scharf, L., Scherfenberg, M.: Inducing polygons of line arrangements. In: Hong, S.-H., Nagamochi, H., Fukunaga, T. (eds.) ISAAC 2008. LNCS, vol. 5369, pp. 507–519. Springer, Heidelberg (2008)

Computing 3D Periodic Triangulations*

Manuel Caroli and Monique Teillaud

INRIA Sophia Antipolis – Méditerranée
{Manuel.Caroli,Monique.Teillaud}@sophia.inria.fr

Abstract. This work is motivated by the need for software computing 3D periodic triangulations in numerous domains including astronomy, material engineering, biomedical computing, fluid dynamics etc. We design an algorithmic test to check whether a partition of the 3D flat torus into tetrahedra forms a triangulation (which subsumes that it is a simplicial complex). We propose an incremental algorithm that computes the Delaunay triangulation of a set of points in the 3D flat torus without duplicating any point, whenever possible; our algorithmic test detects when such a duplication can be avoided, which is usually possible in practical situations. Even in cases where point duplication is necessary, our algorithm always computes a triangulation that is homeomorpic to the flat torus. To the best of our knowledge, this is the first algorithm of this kind whose output is provably correct. The implementation will be released in CGAL [7].

1 Introduction

Computing Delaunay triangulations of point sets is a well-studied problem in Computational Geometry. Several algorithms as well as implementations [31,26,19,38,25,21] are available. However, these algorithms are mainly restricted to triangulations in $\mathbb{R}^d$. In this paper, we take interest in triangulations of a *periodic* space, represented as the so-called *flat torus* [35].

This research was originally motivated by the needs of astronomers who study the evolution of the large scale mass distribution in our universe by running dynamical simulations on periodic 3D data. In fact there are numerous application fields that need robust software for geometric problems in periodic spaces. A small sample of these needs, in fields like astronomy, material engineering for prostheses, mechanics of granular materials, was presented at the CGAL Prospective Workshop on Geometric Computing in Periodic Spaces.[1] Many other diverse application fields could be mentioned, for instance biomedical computing [36], solid-state chemistry [29], physics of condensed matter [15], fluid dynamics [10], this list being far from exhaustive.

* This work was partially supported by the ANR *(Agence Nationale de la Recherche)* under the "Triangles" project of the *Programme blanc* (No BLAN07-2_194137) `http://www-sop.inria.fr/geometrica/collaborations/triangles/`.

[1] http://www.cgal.org/Events/PeriodicSpacesWorkshop/

A. Fiat and P. Sanders (Eds.): ESA 2009, LNCS 5757, pp. 59–70, 2009.

So far we are not aware of any robust and efficient algorithm for computing Delaunay triangulations from a given point set $\mathcal{S}$ in a periodic space. In the literature, proved algorithms usually need to compute with 9 copies of each input point in the planar case [23,17], or with 27 copies in 3D [14], which obviously leads to a huge slow-down. Additionally, their output is a triangulation in $\mathbb{R}^d, d = 2, 3$, of the copies of the points, whereas our approach always outputs triangulations of the flat torus.

In the engineering community, an implementation for computing a periodic Delaunay *"tessellation"* was proposed, avoiding duplications of points [34]. However, the tessellation is not necessarily a simplicial complex. Moreover, the algorithm heavily relies on the assumption that input points are well distributed.

In fact, as shown in Section 4, using copies of the input points may actually be necessary: in some cases, the flat torus may be partitioned into tetrahedra having the points as vertices and satisfying the Delaunay property, but such a partition does not always form a simplicial complex. Figure 1 shows a simple partition of the 2D torus that is not a triangulation. However, in practice, input data sets are likely to admit a Delaunay triangulation.

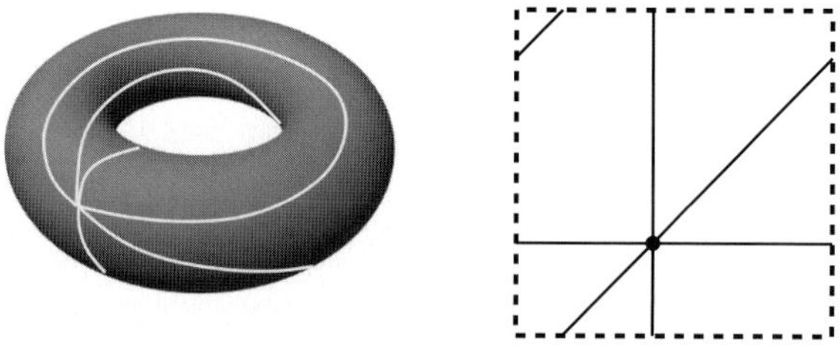

Fig. 1. The partition of the torus (left) and the flat torus (right) is not a triangulation: All simplices have a unique vertex

Let us insist here on the fact that computing a "true" triangulation, i.e. a *simplicial complex*, is important for several reasons. First, a triangulation is defined as a simplicial complex in the literature [2,9,16,20,22,33,39]. Moreover, designing a data structure to efficiently store tetrahedral tessellations that are non-simplicial complexes (e.g. Δ-complexes [18]) would be quite involved. The CGAL 3D triangulation data structure, that we reuse in our implementation, assumes the structure to be a simplicial complex [24]. Even more importantly, algorithms using a triangulation as input are heavily relying on the fact that the triangulation is a simplicial complex; this is the case for instance for meshing algorithms [27,28], as well as algorithms to compute α-shapes, which are actually needed in the periodic case by several applications mentioned at the beginning of this introduction. We are planning to use the 3D periodic triangulation as the fundamental ingredient for computing these structures in the future.

Contributions of the paper

We prove conditions ensuring that the Delaunay triangulation can be computed without duplicating the input points. To this aim, we design an algorithmic test for checking whether a set $\mathcal{K}$ of simplices in the flat torus forms a simplicial complex.

We present an adaptation of the well-known incremental algorithm in $\mathbb{R}^3$ [3] that allows to compute three-dimensional Delaunay triangulations in the flat torus. We focus on the incremental algorithm for several reasons: Its practical efficiency has been proved in particular by the fully dynamic implementation in CGAL [25]; moreover, a dynamic algorithm, allowing to freely insert (and remove) points, is a necessary ingredient for all meshing algorithms and software based on Delaunay refinement methods (see for instance [32,27,8]).

For sets of points that cannot be triangulated in the flat torus, our algorithm outputs a triangulation of an h-sheeted covering space, where h depends on some parameters of the flat torus, i.e. a triangulation that is still homeomorphic to the flat torus and containing $h > 1$ explicit copies of the input point set. However, as soon as the above mentioned conditions are fulfilled, the algorithm switches to a 1-sheeted covering and so does not duplicate points. In this way, the algorithm always computes a triangulation and is provably correct. It has optimal randomized worst case complexity.

Our implementation of the algorithm has been accepted for version 3.5 of the CGAL library [7]. We presented a video demonstration of the software [5].

The paper is organized as follows. In Section 2 we review some general notions about triangulations and simplicial complexes. In the next section, we adapt the definition of simplicial complexes to the flat torus. In Section 4 we give a criterion to decide whether a point set has a triangulation in the flat torus. We give a second criterion that is based on the same idea but can be verified easily by the algorithm that is presented in Section 5. We show the correctness of the algorithm and finish with its complexity analysis and experimental observations. Proofs are omitted in this paper due to lack of space. They can be found in [6].

2 Triangulations

Before talking about triangulations we need to recapitulate the well-known notions of simplices and simplicial complexes. A *k-simplex* σ in $\mathbb{R}^3$ ($k \leq 3$) is the convex hull of $k+1$ affinely independent points $\mathcal{P}_\sigma = \{p_0, p_1, \ldots, p_k\}$. A simplex τ defined by $\mathcal{P}_\tau \subseteq \mathcal{P}_\sigma$ is a *face* of σ and has σ as a *coface*. This is denoted by $\sigma \geq \tau$ and $\tau \leq \sigma$. Note that $\sigma \geq \sigma$ and $\sigma \leq \sigma$.

The following definitions are completely combinatorial. With an appropriate definition of a simplex, they will remain valid in any topological space $\mathbb{X}$.

There exist several definitions of simplicial complexes in the literature. Often they restrict to a finite number of simplices [39,30]. In the sequel, we deal with infinite simplicial complexes, so, we use the definition given in [22]:

Definition 1 (Simplicial complex). *A* simplicial complex *is a set $\mathcal{K}$ of simplices such that:*

(i). $\sigma \in \mathcal{K}, \tau \leq \sigma \Rightarrow \tau \in \mathcal{K}$
(ii). $\sigma, \sigma' \in \mathcal{K} \Rightarrow \sigma \cap \sigma' \leq \sigma, \sigma'$
(iii). Every point in a simplex of $\mathcal{K}$ has a neighborhood that intersects at most finitely many simplices in $\mathcal{K}$ (local finiteness).

Note that if $\mathcal{K}$ is finite, then the third condition is always fulfilled.

A *triangulation* of a topological space $\mathbb{X}$ is a simplicial complex $\mathcal{K}$ such that $|\mathcal{K}| = \bigcup_{\sigma \in \mathcal{K}} \sigma$ is homeomorphic to $\mathbb{X}$. A *triangulation of a point set* $\mathcal{S}$ is a triangulation such that the set of vertices of the triangulation is identical to $\mathcal{S}$.

Some more definitions are needed for the following discussion: Let $\mathcal{K}$ be a simplicial complex. If a subset of $\mathcal{K}$ is a simplicial complex as well, we call it *subcomplex* of $\mathcal{K}$. The *star* of a subcomplex $\mathcal{L}$ of $\mathcal{K}$ consists of the cofaces of simplices in $\mathcal{L}$: $\mathrm{St}(\mathcal{L}) = \{\sigma \in \mathcal{K} | \sigma \geq \tau \in \mathcal{L}\}$. In the following sections, we will be interested in the union of simplices in the star of a set $\mathcal{L}$ of simplices, denoted as $|\mathrm{St}(\mathcal{L})|$.

3 The Flat Torus $\mathbb{T}^3_{\boldsymbol{c}}$

At first we give a precise definition of the space of study $\mathbb{T}^3_{\boldsymbol{c}}$. Then we review some of its well-known properties and establish the notations used in the following discussion. Finally, we give a definition of simplices in $\mathbb{T}^3_{\boldsymbol{c}}$.

Definition 2 ($\mathbb{T}^3$)**.** *Let* $\boldsymbol{c} := (c_x, c_y, c_z) \in (\mathbb{R} \setminus \{0\})^3$ *and* G *be the group* $(\boldsymbol{c} * \mathbb{Z}^3, +)$, *where* $*$ *denotes coordinate-wise multiplication*[2]. *The quotient space* $\mathbb{T}^3_{\boldsymbol{c}} = \mathbb{R}^3/G$ *is called flat torus [35]. We denote the quotient map by* $\boxed{\pi : \mathbb{R}^3 \to \mathbb{T}^3_{\boldsymbol{c}}}$.

The elements of $\mathbb{T}^3_{\boldsymbol{c}}$ are the equivalence classes under the equivalence relation $p_1 \sim p_2 \Leftrightarrow p_1 - p_2 \in \boldsymbol{c} * \mathbb{Z}^3$, for $p_1, p_2 \in \mathbb{R}^3$. Hence, these equivalence classes are isomorphic to $\mathbb{Z}^3$ and $\mathbb{T}^3_{\boldsymbol{c}} \times \mathbb{Z}^3$ is isomorphic to $\mathbb{R}^3$. We also call the points of $\mathbb{T}^3_{\boldsymbol{c}}$ *orbits* and refer to their elements as *representatives.* $\mathbb{T}^3_{\boldsymbol{c}}$ is a metric space with $d_{\mathbb{T}}(\pi(p), \pi(q)) := \min d_{\mathbb{R}}(p', q')$ for $p' \sim p, q' \sim q$. Note that π is continuous.

The space $\mathbb{T}^3_{\boldsymbol{c}}$ is homeomorphic to the hypersurface of a 4-dimensional torus. Consider the closed cuboid $[u, u + c_x] \times [v, v + c_y] \times [w, w + c_z]$. Identifying the pairs of opposite sides results in a space homeomorphic to $\mathbb{T}^3_{\boldsymbol{c}}$. Such a cuboid is usually called a *fundamental domain* or a *fundamental region.* A fundamental domain contains at least one representative of each orbit. The half-open cuboid $\boxed{\mathcal{D}_{\boldsymbol{c}} = [0, c_x) \times [0, c_y) \times [0, c_z)}$ contains exactly one representative for each element of $\mathbb{T}^3_{\boldsymbol{c}}$. We call it the *original domain.* The map

$$\begin{aligned} \varphi_{\boldsymbol{c}} : \mathcal{D}_{\boldsymbol{c}} \times \mathbb{Z}^3 &\to \mathbb{R}^3 \\ (p, \zeta) &\mapsto p + \boldsymbol{c} * \zeta \end{aligned}$$

is bijective. The longest diagonal of $\mathcal{D}_{\boldsymbol{c}}$ has length $\|\boldsymbol{c}\|$, which denotes the L_2-norm of $\boldsymbol{c}$. We say that two points $p_1, p_2 \in \mathbb{R}^3$ are *periodic copies* of each other if they both lie in the same orbit, or equivalently if there is a point $p \in \mathcal{D}_{\boldsymbol{c}}$ such that $p_1, p_2 \in \varphi_{\boldsymbol{c}}(\{p\} \times \mathbb{Z}^3)$.

---

[2] Coordinate-wise multiplication: $(a_x, a_y, a_z) * (b_x, b_y, b_z) := (a_x b_x, a_y b_y, a_z b_z)$.

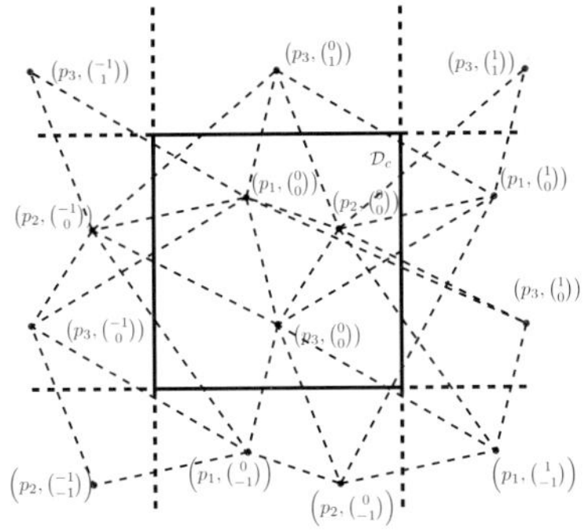

Fig. 2. (2D case) The three points p_1, p_2, and p_3 do not uniquely define a triangle. Intuitively, the offset allows to know which way the triangle "wraps around" the torus.

Now we turn towards the definition of simplices in $\mathbb{T}^3_{\boldsymbol{c}}$. There is no meaningful definition of a convex hull in $\mathbb{T}^3_{\boldsymbol{c}}$ and a tetrahedron is not uniquely defined by four points. We attach with each vertex an integer vector, named *offset*, that specifies one representative out of an orbit (see Figure 2). In the above definition of $\varphi_{\boldsymbol{c}}$, the offsets are the numbers $\zeta \in \mathbb{Z}^3$. We can adapt the definition of a simplex in $\mathbb{R}^3$ in the following way to $\mathbb{T}^3_{\boldsymbol{c}}$ [37]:

Definition 3 (simplex). *Let $\mathcal{P}$ be a set of $k+1$ ($k \leq 3$) point offset pairs (p_i, ζ_i) in $\mathcal{D}_{\boldsymbol{c}} \times \mathbb{Z}^3$, $0 \leq i \leq k$. Let* $\mathrm{Ch}(\mathcal{P})$ *denote the convex hull of* $\varphi_{\boldsymbol{c}}(\mathcal{P}) = \{p_i + \boldsymbol{c} * \zeta_i \mid 0 \leq i \leq k\}$ *in* $\mathbb{R}^3$. *If the restriction* $\pi|_{\mathrm{Ch}(\mathcal{P})}$ *of* π *to the convex hull of* $\mathcal{P}$ *is a homeomorphism, the image of* $\mathrm{Ch}(\mathcal{P})$ *by* π *is called a* k-simplex *in* $\mathbb{T}^3_{\boldsymbol{c}}$.

In other words, the image under π of a simplex in $\mathbb{R}^3$ is a simplex in $\mathbb{T}^3_{\boldsymbol{c}}$ only if it does not self-intersect or touch. Figure 3 shows the convex hulls A and B of three point-offset pairs in $[0,1)^2 \times \mathbb{Z}^2$; $(p_1, \binom{0}{2})$ is a representative of the equivalence class of a vertex of A that lies *inside* A.

There are infinitely many sets of point-offset pairs specifying the same simplex. The definition of face and coface is adapted accordingly: Let σ be a k-simplex defined by a set $\mathcal{P}_\sigma \subseteq \mathcal{D}_{\boldsymbol{c}} \times \mathbb{Z}^3$. A simplex τ defined by a set $\mathcal{P}_\tau \subseteq \mathcal{D}_{\boldsymbol{c}} \times \mathbb{Z}^3$ is a *face* of σ and has σ as a *coface* if and only if there is some $\zeta \in \mathbb{Z}^3$ such that $\{(p_i, \zeta_i + \zeta) \mid (p_i, \zeta_i) \in \mathcal{P}_\tau\} \subseteq \mathcal{P}_\sigma$.

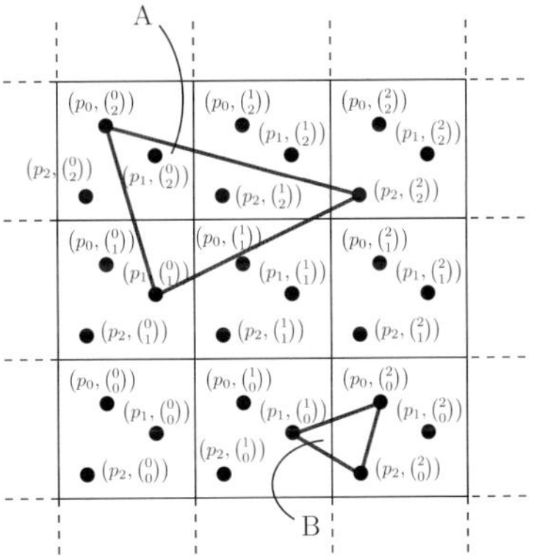

Fig. 3. (2D case) $\pi(A)$ is not a simplex; however, $\pi(B)$ is a simplex

4 Delaunay Triangulation in $\mathbb{T}^3_{\boldsymbol{c}}$

This section is organized as follows: At first we give a definition of the Delaunay triangulation in $\mathbb{T}^3_{\boldsymbol{c}}$. We observe that there are point sets in $\mathbb{T}^3_{\boldsymbol{c}}$ whose Delaunay triangulation is in fact not defined. The second part elaborates on this question, finally giving a criterion to decide whether or not a point set has a Delaunay triangulation in $\mathbb{T}_c^3$. In the last part we discuss how to deal with point sets that do not have a Delaunay triangulation in $\mathbb{T}_c^3$.

Let us recall that a triangulation of a point set $\mathcal{S}$ in $\mathbb{R}^3$ is a *Delaunay* triangulation *iff* each tetrahedron satisfies the Delaunay property, i.e. its circumscribing ball does not contain any point of $\mathcal{S}$ in its interior. If the point set is not degenerate, i.e. if no five points of $\mathcal{S}$ are cospherical, then its Delaunay triangulation is uniquely defined. Still, even for degenerate point sets, it is possible to specify a unique Delaunay triangulation, using a symbolic perturbation [13]. In the sequel we always assume Delaunay triangulations in $\mathbb{R}^3$ to be uniquely defined in that way (see Lemma 2). Let $\mathcal{S}$ now denote a finite point set in $\mathcal{D}_c$. We want to define the Delaunay triangulation of $\pi(\mathcal{S})$ in $\mathbb{T}_c^3$. The idea is to use the projection under π of a Delaunay triangulation of the infinite periodic point set $\boxed{\mathcal{S}^c := \varphi_c(\mathcal{S} \times \mathbb{Z}^3)}$ in $\mathbb{R}^3$.

Lemma 1. *For any finite point set* $\mathcal{S} \subset \mathcal{D}_c$*, a set of simplices* $\mathcal{K}$ *in* $\mathbb{R}^3$ *that fulfills (i) and (ii) in Definition 1 as well as the Delaunay property with respect to* $\mathcal{S}^c$ *is a simplicial complex in* $\mathbb{R}^3$*.*

Since $\mathcal{S}^c$ contains points on an infinite grid, any point $p \in \mathbb{R}^3$ is contained in some simplex defined by points in $\mathcal{S}^c$. Together with Lemma 1, this implies that the set of all simplices with points of $\mathcal{S}^c$ as vertices and respecting the Delaunay property is a Delaunay triangulation of $\mathbb{R}^3$ and we denote it by $DT_{\mathbb{R}}(\mathcal{S}^c)$. Since $|DT_{\mathbb{R}}(\mathcal{S}^c)|$ is homeomorphic to $\mathbb{R}^3$ and π is surjective, then $\pi(|DT_{\mathbb{R}}(\mathcal{S}^c)|)$ is homeomorphic to $\mathbb{T}_c^3$. So, if $\pi(DT_{\mathbb{R}}(\mathcal{S}^c))$ is a simplicial complex, it is also a triangulation of $\mathbb{T}_c^3$. We can now define a Delaunay triangulation in $\mathbb{T}_c^3$:

Definition 4. *Let* $DT_{\mathbb{R}}(\mathcal{S}^c)$ *be the Delaunay triangulation of the point set* $\mathcal{S}^c$ *in* $\mathbb{R}^3$*. If* $\pi(DT_{\mathbb{R}}(\mathcal{S}^c))$ *is a simplicial complex in* $\mathbb{T}_c^3$*, then we call it the* Delaunay triangulation *of* $\mathcal{S}$ *in* $\mathbb{T}_c^3$ *and denote it by* $DT_{\mathbb{T}}(\mathcal{S})$*.*

We show now that Definition 4 actually makes sense: Lemma 2 is used to prove Theorem 1, which gives a sufficient condition for $\pi(DT_{\mathbb{R}}(\mathcal{S}^c))$ to be a simplicial complex.

Lemma 2. *If the restriction of* π *to any simplex in* $DT_{\mathbb{R}}(\mathcal{S}^c)$ *is a homeomorphism, then conditions (i) and (iii) in Definition 1 are fulfilled.*

Theorem 1. *If for all vertices* v *of* $DT_{\mathbb{R}}(\mathcal{S}^c)$ *the restriction of the quotient map* $\pi|_{|\mathrm{St}(v)|}$ *is a homeomorphism, then* $\pi(DT_{\mathbb{R}}(\mathcal{S}^c))$ *forms a simplicial complex.*

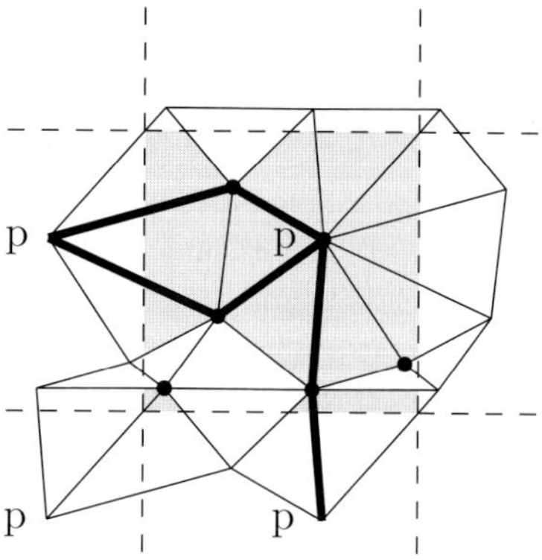

Fig. 4. (2D case) The shaded region is $\varphi_c(\mathrm{St}(p) \times \mathbb{Z}^3) \cap \mathcal{D}_c$. There are several cycles of length two originating from p.

In the following theorem we give another criterion that is algorithmically easier to check. Let us recall that the *1-skeleton* of a simplicial complex is the subcomplex that consists of all edges and vertices.

Theorem 2. *Assume the restriction of* π *to any simplex in* $DT_{\mathbb{R}}(\mathcal{S}^c)$ *is a homeomorphism. If the 1-skeleton of* $\pi(DT_{\mathbb{R}}(\mathcal{S}^c))$ *does not contain any cycle of length less than or equal to two, then* $\pi(DT_{\mathbb{R}}(\mathcal{S}^c))$ *forms a simplicial complex.*

See Figure 4 for an illustration of Theorems 1 and 2.

In the remaining part of this section, we explain how we can give a finite representation of the periodic triangulation $DT_{\mathbb{R}}(\mathcal{S}^c)$ that is a simplicial complex, even if $\pi(DT_{\mathbb{R}}(\mathcal{S}^c))$ is not a simplicial complex.

Definition 5. *[2] Let* $\mathbb{X}$ *be a topological space. A map* $\rho : \widetilde{\mathbb{X}} \to \mathbb{X}$ *is called a* covering map *and* $\widetilde{\mathbb{X}}$ *is said to be a* covering space *of* $\mathbb{X}$ *if the following condition holds: For each point* $x \in \mathbb{X}$ *there is an open neighborhood* V*, and a decomposition of* $\rho^{-1}(V)$ *as a family* $\{U_\alpha\}$ *of pairwise disjoint open subsets of* $\widetilde{\mathbb{X}}$*, in such a way that* $\rho|_{U_\alpha}$ *is a homeomorphism for each* α*. Let* h_x *denote the cardinality of the family* $\{U_\alpha\}$ *corresponding to some* $x \in \mathbb{X}$*. If the maximum* $h := \max_{x\in\mathbb{X}} h_x$ *is finite, then* $\widetilde{\mathbb{X}}$ *is called an* h-sheeted covering space.

$\mathbb{R}^3$ with the quotient map π as covering map is a *universal covering* of $\mathbb{T}^3_c$, which means that it is a covering space for all covering spaces of $\mathbb{T}^3_c$ [2].

Let $\boldsymbol{h} = (h_x, h_y, h_z) \in \mathbb{N}^3$. $\mathbb{T}^3_{\boldsymbol{h}*\boldsymbol{c}}$ is a covering space of $\mathbb{T}^3_{\boldsymbol{c}}$ together with the covering map $\rho_{\boldsymbol{h}} := \pi \circ \pi_{\boldsymbol{h}}^{-1}$, where $\pi_{\boldsymbol{h}} : \mathbb{R}^3 \to \mathbb{T}^3_{\boldsymbol{h}*\boldsymbol{c}}$ denotes the quotient map of $\mathbb{T}^3_{\boldsymbol{h}*\boldsymbol{c}}$. As $\rho_{\boldsymbol{h}}^{-1}(p)$ for any $p \in \mathbb{T}^3_{\boldsymbol{c}}$ consists of $h_x \cdot h_y \cdot h_z$ different points, $\mathbb{T}^3_{\boldsymbol{h}*\boldsymbol{c}}$ is a $h_x \cdot h_y \cdot h_z$-sheeted covering space. The original domain is $D_{\boldsymbol{h}*\boldsymbol{c}} = [0, h_x c_x) \times [0, h_y c_y) \times [0, h_z c_z)$. If $h_x = h_y = h_z$ we use the notation $\pi_h := \pi_{\boldsymbol{h}}$ with $h := h_x \cdot h_y \cdot h_z$, like for π_{27} in Theorem 3 below.

Dolbilin and Huson [14] showed that only the points of $\mathcal{S}^c$ contained in $\mathcal{D}_{\boldsymbol{c}}$ and the 26 copies that surround it can have an influence on the Delaunay property for simplices that are completely contained in $\mathcal{D}_{\boldsymbol{c}}$. The ideas of their proof can be used to show the following:

Theorem 3. $\pi_{27}(DT_{\mathbb{R}}(\mathcal{S}^c))$ *is always a simplicial complex.*

We prefer to use the framework of covering spaces, rather than just talk about copies of the points as in [14], for several reasons: A major part of the code can be reused for any finite covering space. Also, the simplicial complex we compute is actually homeomorphic to $\mathbb{T}^3_c$. So we do not have any artificial boundaries in the data structure and we get all adjacency relations between simplices.

The algorithm we use to compute triangulations of $\mathbb{T}^3_{\boldsymbol{c}}$ requires a slightly stronger result, which we present in the next section.

5 Algorithm

As mentioned in the introduction, there is a strong motivation for reusing the standard incremental algorithm [3] to compute a periodic Delaunay triangulation.

We propose the following algorithm:

- We start computing in some finitely-sheeted covering space $\mathbb{T}^3_{\boldsymbol{h}*\boldsymbol{c}}$ of $\mathbb{T}^3_{\boldsymbol{c}}$, with $\boldsymbol{h}$ chosen such that $\pi_{\boldsymbol{h}}(DT_{\mathbb{R}}(\mathcal{S}^{\boldsymbol{c}}))$ is guaranteed to be a triangulation.
- If the point set is large and reasonably well distributed, it is likely that after having inserted all the points of a subset $\mathcal{S}' \subset \mathcal{S}$, all the subsequent $\pi(DT_{\mathbb{R}}(\mathcal{S}''^{\boldsymbol{c}}))$ for $\mathcal{S}' \subset \mathcal{S}'' \subset \mathcal{S}$ are simplicial complexes in $\mathbb{T}^3_{\boldsymbol{c}}$. In this case, we discard all periodic copies of simplices of $\pi_{\boldsymbol{h}}(DT_{\mathbb{R}}(\mathcal{S}'^{\boldsymbol{c}}))$ and switch to computing $\pi(DT_{\mathbb{R}}(\mathcal{S}^{\boldsymbol{c}}))$ in $\mathbb{T}^3_{\boldsymbol{c}}$ by adding all the points left in $\mathcal{S} \setminus \mathcal{S}'$.

In this way, unlike [14], we avoid duplicating points as soon as this is possible. However, if $\mathcal{S}$ is a small and/or badly distributed point set, the algorithm never enters the second phase and returns $\pi_{\boldsymbol{h}}(DT_{\mathbb{R}}(\mathcal{S}^{\boldsymbol{c}}))$. Note that, before switching to computing in $\mathbb{T}^3_{\boldsymbol{c}}$, it is not sufficient to test whether $\pi(DT_{\mathbb{R}}(\mathcal{S}'^{\boldsymbol{c}}))$ is a simplicial complex. Indeed, adding a point could create a cycle of length two (see Figure 5). So, a stronger condition is needed before the switch.

See Algorithm 1 for a pseudo-code listing of the algorithm.

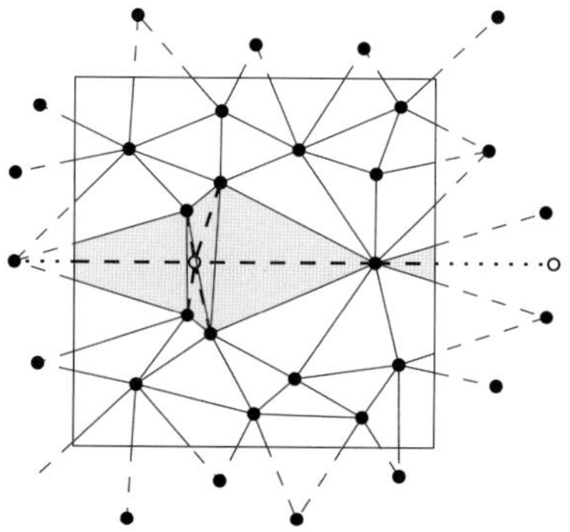

Fig. 5. (2D case) Adding a point in a simplicial complex can create a cycle of length two

---

Algorithm 1. Compute Delaunay triangulation of a point set in $\mathbb{T}^3_{\boldsymbol{c}}$

Input: Set $\mathcal{S}$ of points in $\mathcal{D}_{\boldsymbol{c}}$, $\boldsymbol{c}$ such that $\mathcal{D}_{\boldsymbol{c}}$ is a cube with edge length $c \in \mathbb{R}^3 \setminus \{0\}$.
Output: $DT_{\mathbb{T}}(\mathcal{S})$ if possible, otherwise $\pi_{27}(DT_{\mathbb{R}}(\mathcal{S}^{\boldsymbol{c}}))$

$\mathcal{S}' \Leftarrow \mathcal{S}$
Pop p from $\mathcal{S}'$
$\mathcal{S} \Leftarrow \{p\}$
$TR_{27} \Leftarrow \pi_{27}(DT_{\mathbb{R}}(\varphi_{\boldsymbol{c}}(\{p\} \times \mathbb{Z}^3)))$ `// can be precomputed`
while the longest edge in TR_{27} is longer than $\frac{1}{\sqrt{6}}c$ **do**
  Pop p from $\mathcal{S}'$; $\mathcal{S} \Leftarrow \mathcal{S} \cup \{p\}$
  for all $p' \in \{p + \boldsymbol{c} * \zeta \mid \zeta \in \{0,1,2\}^3\}$ **do**
    Insert p' into TR_{27}
  end for `//` $TR_{27} = \pi_{27}(DT_{\mathbb{R}}(\mathcal{S}^{\boldsymbol{c}}))$
  if $\mathcal{S}' = \emptyset$ **then return** $TR_{27} = \pi_{27}(DT_{\mathbb{R}}(\mathcal{S}^{\boldsymbol{c}}))$ `// non-triangulable point set`
end while
Compute $DT_{\mathbb{T}}(\mathcal{S})$ from TR_{27} `// switch to` $\mathbb{T}^3_{\boldsymbol{c}}$
Insert all points remaining in $\mathcal{S}'$ into $DT_{\mathbb{T}}(\mathcal{S})$ one by one
return $DT_{\mathbb{T}}(\mathcal{S})$

Two central points must be established to show the correctness of the algorithm:

1. After each insertion, TR_{27} is a Delaunay triangulation in $\mathbb{T}^3_{3\boldsymbol{c}}$. Let us emphasize on the fact that Theorem 3 cannot be used here because in the inner loop (step 8), the set of points present in TR_{27} does not contain all the periodic copies of p. Let p be a point in $\mathcal{D}_{\boldsymbol{c}}$ and $\mathcal{T}_p \subseteq \varphi_{\boldsymbol{c}}(\{p\} \times \mathbb{Z}^3) \cap \mathcal{D}_{3\boldsymbol{c}}$, i.e. $\mathcal{T}_p$ is a subset of the grid of 27 copies of p that lie within $\mathcal{D}_{3\boldsymbol{c}}$. Then TR_{27} is always of the form $\pi_{27}(DT_{\mathbb{R}}(\mathcal{S}^{\boldsymbol{c}} \cup \mathcal{T}_p^{3\boldsymbol{c}}))$ with $\mathcal{T}_p^{3\boldsymbol{c}} = \varphi_{3\boldsymbol{c}}(\mathcal{T}_p \times \mathbb{Z}^3)$. Lemma 3 shows that this is a triangulation.
2. If all edges in $\pi_{27}(DT_{\mathbb{R}}(\mathcal{S}^{\boldsymbol{c}}))$ are shorter than $\frac{1}{\sqrt{6}}\, c$, then we can switch to computing in $\mathbb{T}^3_{\boldsymbol{c}}$.

Lemma 3. *Let $\mathcal{S} \subset \mathcal{D}_{\boldsymbol{c}}$ be a finite point set and $p \in \mathcal{D}_{\boldsymbol{c}}$ a point. If $\mathcal{D}_{\boldsymbol{c}}$ is a cube, then $\pi_{27}(DT_{\mathbb{R}}(\mathcal{S}^{\boldsymbol{c}} \cup \mathcal{T}_p^{3\boldsymbol{c}}))$ is a triangulation.*

Lemma 4 gives a criterion to decide whether $\pi(DT_{\mathbb{R}}(\mathcal{S}^{\boldsymbol{c}}))$ is a simplicial complex and thus a triangulation in $\mathbb{T}^3_{\boldsymbol{c}}$.

Lemma 4. *If the 1-skeleton of $DT_{\mathbb{R}}(\mathcal{S}^{\boldsymbol{c}})$ contains only edges shorter than $\frac{1}{\sqrt{6}}\, c$, where c is the edge length of $\mathcal{D}_{\boldsymbol{c}}$, then $\pi(DT_{\mathbb{R}}(\mathcal{T}^{\boldsymbol{c}}))$ is a simplicial complex for any finite $\mathcal{T} \subset \mathcal{D}_{\boldsymbol{c}}$ with $\mathcal{S} \subseteq \mathcal{T}$.*

Note that the criterion in Lemma 4 is only sufficient: There are triangulations without cycles of length two that have edges longer than $\frac{1}{\sqrt{6}}\, c$.

Lemmas 3 and 4 prove the correctness of Algorithm 1 in the case of a cubic domain. The above discussion still remains valid if the original domain $\mathcal{D}_{\boldsymbol{c}}$ is a general cuboid, i.e. $\boldsymbol{c} = (c_x, c_y, c_z)$. Only the constants, like the number of sheets of the covering space to start with and the edge length threshold need to be adapted. Analogously, the algorithm can be extended to weighted Delaunay triangulations. For a more detailed discussion see [6].

6 Theoretical and Practical Analysis

Complexity analysis. Let us first discuss the following two points: (1) How to test for the length of the longest edge and (2) how to switch from the triangulation in $\mathbb{T}^3_{\boldsymbol{h}*\boldsymbol{c}}$ to the triangulation in $\mathbb{T}^3_{\boldsymbol{c}}$.

(1) We maintain an unsorted data structure $\mathcal{E}$ that references all edges that are longer than the threshold $\frac{1}{\sqrt{6}}\, c_{\min}$. As soon as $\mathcal{E}$ is empty, we know that the longest edge is smaller than the threshold. The total number of edges that are inserted and removed in $\mathcal{E}$ is proportional to the total number of simplices that are created and destroyed during the algorithm. We can have direct access from the simplices to their edges in $\mathcal{E}$. Hence, the maintenance of $\mathcal{E}$ does not change the algorithm complexity.

(2) To convert the triangulation in $\mathbb{T}^3_{\boldsymbol{h}*\boldsymbol{c}}$ to $DT_{\mathbb{T}}(\mathcal{S})$ when we switch to $\mathbb{T}^3_{\boldsymbol{c}}$, we need to iterate over all cells and all vertices to delete all periodic copies, keeping only one; furthermore, we need to update the incidence relations of

Table 1. Current running times in seconds on a 2.33 GHz Intel Core 2 Duo processor

No. of points	$\mathbb{T}^3$	$\mathbb{R}^3$	factor
1000	0.032	0.012	2.65
10000	0.230	0.128	1.79
100000	2.24	1.36	1.65
1000000	23.0	14.2	1.62

those tetrahedra whose neighbors have been deleted. This is linear in the size of the triangulation and thus dominated by the main loop.

The overall algorithm is incremental and using the Delaunay hierarchy [12] the following result can be shown:

The randomized complexity of Algorithm 1 is the same as the complexity of [12], and thus it has randomized worst-case optimal complexity $O(n^2)$.

Experimental observations. Algorithm 1 has been implemented in CGAL, so, it benefits from some of the optimizations that are already available in the CGAL Delaunay triangulations in $\mathbb{R}^3$ [25], such as the spatial sorting [11].

We tested the implementation on real data from research in cosmology. The input sets consist of up to several hundreds of thousands of points, and they are sufficiently well distributed to have triangulations in $\mathbb{T}^3_c$. This property holds for most of the applications mentioned in the introduction. With these real data, usually less than 400 points are needed for Algorithm 1 to reach the threshold on the edge length and switch to computing in $\mathbb{T}^3_c$.

We compared the running time of our implementation for computing Delaunay triangulations in $\mathbb{T}^3_c$ with the running time of computing the Delaunay triangulation in $\mathbb{R}^3$ with the CGAL package [25]. Table 1 shows for large random point sets a factor of about 1.6 between the running time of our current implementation, using the above optimization, and the CGAL implementation for $\mathbb{R}^3$. The timings have been measured for the unit cube $\mathcal{D}_c = [0,1)^3$ using specialized predicates; if we allow $\mathcal{D}_c$ to be any cube, we currently lose about 12%. More experiments can be found in [6].

7 Conclusion and Future Work

We proposed an algorithm to compute 3D periodic Delaunay triangulations. The algorithm is guaranteed to produce a correct finite representation of the periodic triangulation for any given point set. We avoid duplications of points whenever possible, and if there is no triangulation for some point set in the flat torus $\mathbb{T}^3_c$, we output a triangulation in a covering space that is homeomorphic to $\mathbb{T}^3_c$. The algorithm has optimal randomized worst case complexity. Note that the main parts of the discussion are not bound to three-dimensional space and will still hold for higher dimensions. The constants in the geometric criteria and the complexity of the underlying algorithm for computing the Delaunay triangulation will have to be adapted.

Future work will mainly concentrate on two topics: (1) Extend in a similar way some meshing and α-shape algorithms based on Delaunay triangulations so that they can handle periodic data. (2) Extend this work to more general orbifolds. There is ongoing work to unify our results with the results of [1].

Acknowledgments

We are very grateful to Nico Kruithof, who initiated this work and made important contributions when he visited INRIA for a short post-doctoral stay in 2006 [4]. We thank Mridul Aanjaneya for fruitful discussions on Section 4 and Olivier Devillers for discussions on his Delaunay hierarchy.

Also, we wish to thank Rien van de Weijgaert for providing us with data sets from cosmology research projects to test our implementation.

References

1. Aanjaneya, M., Teillaud, M.: Triangulating the real projective plane. In: Mathematical Aspects of Computer and Information Sciences (2007)
2. Armstrong, M.A.: Basic Topology. Springer, Heidelberg (1982)
3. Bowyer, A.: Computing Dirichlet tessellations. The Computer Journal 24, 162–166 (1981)
4. Caroli, M., Kruithof, N., Teillaud, M.: Decoupling the CGAL 3D triangulations from the underlying space. In: Workshop on Algorithm Engineering and Experiments, pp. 101–108 (2008)
5. Caroli, M., Teillaud, M.: Video: On the computation of 3D periodic triangulations. In: Proceedings of the twenty-fourth Annual Symposium on Computational Geometry, pp. 222–223 (2008)
6. Caroli, M., Teillaud, M.: Computing 3D periodic triangulations. Research Report 6823, INRIA (2009), `http://hal.inria.fr/inria-00356871`
7. Cgal, Computational Geometry Algorithms Library, `http://www.cgal.org`
8. Cheng, S.-W., Dey, T.K., Levine, J.A.: A practical Delaunay meshing algorithm for a large class of domains. In: Proceedings of the sixteenth International Meshing Roundtable, pp. 477–494 (2007)
9. Daverman, R.J., Sher, R.B. (eds.): Handbook of Geometric Topology. Elsevier, Amsterdam (2002)
10. de Fabritiis, G., Coveney, P.V.: Dynamical geometry for multiscale dissipative particle dynamics (2003), `http://xxx.lanl.gov/abs/cond-mat/0301378v1`
11. Delage, C.: Spatial sorting. In: CGAL editorial Board (eds.) CGAL User and Reference Manual, 3.4 edn. (2008)
12. Devillers, O.: The Delaunay hierarchy. International Journal of Foundations of Computer Science 13, 163–180 (2002)
13. Devillers, O., Teillaud, M.: Perturbations and vertex removal in a 3D Delaunay triangulation. In: Proceedings of the fourteenth ACM-SIAM Symposium on Discrete Algorithms, pp. 313–319 (2003)
14. Dolbilin, N.P., Huson, D.H.: Periodic Delone tilings. Periodica Mathematica Hungarica 34(1-2), 57–64 (1997)
15. Campayo, D.D.: Sklogwiki - Boundary conditions, `http://www.sklogwiki.org/SklogWiki/index.php/Boundary_conditions`
16. Graham, R.L., Grötschel, M., Lovász, L. (eds.): Handbook of Combinatorics. Elsevier, Amsterdam (1995)

17. Grima, C.I., Márquez, A.: Computational Geometry on Surfaces. Kluwer Academic Publishers, Dordrecht (2001)
18. Hatcher, A.: Algebraic Topology. Cambridge University Press, Cambridge (2002)
19. Held, M.: Vroni: An engineering approach to the reliable and efficient computation of Voronoi diagrams of points and line segments. Computational Geometry: Theory and Applications 18, 95–123 (2001)
20. Henle, M.: A Combinatorial Introduction to Topology. Dover publication, New York (1979)
21. Hert, S., Seel, M.: dD convex hulls and Delaunay triangulations. In: CGAL Editorial Board (eds.) CGAL User and Reference Manual, 3.4 edn. (2008)
22. Lee, J.M.: Introduction to Topological Manifolds. Springer, New York (2000)
23. Mazón, M., Recio, T.: Voronoi diagrams on orbifolds. Computational Geometry: Theory and Applications 8, 219–230 (1997)
24. Pion, S., Teillaud, M.: 3D triangulation data structure. In: CGAL Editorial Board (eds.), CGAL User and Reference Manual. 3.4 edn. (2008)
25. Pion, S., Teillaud, M.: 3D triangulations. In: CGAL Editorial Board (eds.) CGAL User and Reference Manual, 3.4 edn. (2008)
26. Qhull, `http://www.qhull.org`
27. Rineau, L., Yvinec, M.: Meshing 3D domains bounded by piecewise smooth surfaces. In: Proceedings of the sixteenth International Meshing Roundtable, pp. 443–460 (2007)
28. Rineau, L., Yvinec, M.: 3D surface mesh generation. In: CGAL Editorial Board (eds.) CGAL User and Reference Manual, 3.4 edn. (2008)
29. Robins, V.: Betti number signatures of homogeneous Poisson point processes. Physical Review E 74(061107) (2006)
30. Rote, G., Vegter, G.: Computational topology: An introduction. In: Boissonnat, J.-D., Teillaud, M. (eds.) Effective Computational Geometry for Curves and Surfaces. Mathematics and Visualization, pp. 277–312. Springer, Heidelberg (2006)
31. Shewchuk, J.R.: Triangle: Engineering a 2d quality mesh generator and Delaunay triangulator. In: First Workshop on Applied Computational Geometry, May 1996. Association for Computing Machinery (1996)
32. Shewchuk, J.R.: Tetrahedral mesh generation by Delaunay refinement. In: Proceedings of the fourteenth Annual Symposium on Computational Geometry, pp. 86–95. ACM Press, New York (1998)
33. Spanier, E.H.: Algebraic Topology. Springer, New York (1966)
34. Thompson, K.E.: Fast and robust Delaunay tessellation in periodic domains. International Journal for Numerical Methods in Engineering 55, 1345–1366 (2002)
35. Thurston, W.P.: Three-Dimensional Geometry and Topology. Princeton University Press, Princeton (1997)
36. Weiss, D.: How hydrophobic Buckminsterfullerene affects surrounding water structure. INRIA Geometrica Seminar (March 2008), `http://www-sop.inria.fr/geometrica`
37. Wilson, P.M.H.: Curved Spaces. Cambridge University Press, Cambridge (2008)
38. Yvinec, M.: 2D triangulations. In: CGAL Editorial Board (eds), CGAL User and Reference Manual. 3.4 edn. (2008)
39. Zomorodian, A.: Topology for Computing. Cambridge University Press, Cambridge (2005)

Cauchy's Theorem for Orthogonal Polyhedra of Genus 0

Therese Biedl[1,⋆] and Burkay Genc[2,⋆⋆]

[1] David R. Cheriton School of Computer Science,
University of Waterloo, Waterloo, Ontario N2L 3G1, Canada
biedl@uwaterloo.ca

[2] Izmir University of Economics, Faculty of Computer Science,
Sakarya Cad. No:156, Balcova, Izmir, Turkey
burkaygenc@gmail.com

Abstract. A famous theorem by Cauchy states that the dihedral angles of a convex polyhedron are determined by the incidence structure and face-polygons alone. In this paper, we prove the same for orthogonal polyhedra of genus 0 as long as no face has a hole. Our proof yields a linear-time algorithm to find the dihedral angles.

1 Introduction

A famous theorem by Cauchy states that for a convex polyhedron, the incidence structure and the face-polygons determine the polyhedron uniquely. Put differently, if we are given a graph with a fixed order of edges around each vertex, and we are given the angles at every vertex-face incidence and edge lengths, then there can be at most one set of dihedral angles such that graph, facial angles, edge lengths and dihedral angles are those of a convex polyhedron. Many books and monographs give this theorem, proofs, and related results [1,10].

Cauchy's theorem does not hold for polyhedra that are not convex. An easy example is a polyhedron where one face has a rectangular "hole" where a small box can be popped to the "inside" or "outside". But in fact, there are even so-called *flexible* polyhedra where the dihedral angles change continuously (see e.g. [10].)

We show in this paper that Cauchy's theorem *does* hold for orthogonal polyhedra of genus 0, as long as we exclude holes in faces. (Rather than defining holes, we will express this by saying that the graph of the polyhedron must be connected; see Section 2 for precise definitions.) Thus, while a big cube with a small cube attached on one face has two possible realizations, this is in essence the only way in which multiple realizations are possible.

Our proof is algorithmic and yields a linear-time algorithm to find the only possible set of dihedral angles of a realizing orthogonal polyhedron. This is in

⋆ Supported by NSERC.

⋆⋆ This work was done while the author was a student at University of Waterloo and supported by NSERC.

A. Fiat and P. Sanders (Eds.): ESA 2009, LNCS 5757, pp. 71–82, 2009.

contrast to convex polyhedra, where determining whether a working set of dihedral angles exist is not even known to be in NP; see [4,9] and the references therein for some recent progress on this tantalizing problem.

1.1 Roadmap

We first briefly outline the approach of this paper. Rather than proving uniqueness and then deriving an algorithm from the proof, we provide an algorithm that reconstructs an orthogonal polyhedron. There will never be any choice in the assignment of dihedral angles, except at one moment when we can choose one dihedral angle. Hence we obtain two sets of dihedral angles, and can argue that only one of them could possibly do; this then proves uniqueness.

Our algorithm proceeds in three steps. In the first step in Section 3, we only identify which dihedral angles must be flat, i.e., have value 180°. We do this by determining the orientation of each face; the algorithm to do so is simple, but proving its correctness is not.[1] Two adjacent faces with the same orientation must have a flat dihedral angle between them, so this determines all flat dihedral angles.

The problem hence reduces to reconstructing an orthogonal polyhedron where all dihedral angles are non-flat. In Section 4, we show that there are only 7 possible configurations of vertices for such a polyhedron. Moreover, if we fix one dihedral angle and know all facial angles, this determines all other dihedral angles at a vertex, and hence with a simple propagation scheme, all dihedral angles can be computed as long as the graph is connected.

Finally, we study in Section 5 which of the two resulting sets of dihedral angles can possibly be the correct set of dihedral angles. This is the only part of the algorithm that uses edge lengths. We conclude with remarks in Section 6.

2 Definitions

A *polygonal curve* is a simple closed curve in the plane that consists of a finite number of line segments. A *polygon* is a set in a plane whose boundary is one polygonal curve. A *polygonal region* is an interior connected set in a plane that is a finite union of polygons. A *polyhedral surface* is a connected 2-manifold that is a finite union of polygonal regions. A *polyhedron* is a set in 3D whose boundary is a polyhedral surface. Its *genus* is the genus of the surface that bounds it.

A *face* of a polyhedron is a maximal polygonal region on the boundary of the polyhedron. Note that a face need not be a polygon, because its boundary may be disconnected and/or touch itself and hence not be simple. A *vertex* is a point that belongs to at least three faces. An *edge* is a maximal line segment that belongs to two faces and contains no vertex other than its endpoints. A *facial angle* is the interior angle of a face at a vertex. A *dihedral angle* is the interior angle at an edge between two adjacent faces.

[1] A preliminary version of this algorithm appeared in 2004 [2], but its correctness was shown only for orthogonally convex polyhedra for which all faces are rectangles.

The incidences between vertices and edges of a polyhedron determine a graph called the *graph of the polyhedron.* Looking at the polyhedron from the outside fixes a cyclic order of edges around each vertex; this is called the *induced embedding of the graph.*

Every polyhedral surface S bounds a polyhedron P, but the polygonal regions that define S need not be the same as the faces of the polyhedron P: the faces of P may have been subdivided. For a polyhedral surface S, we can also define a graph by using as faces of the graph the polygonal regions that defined S, and then carry over all other definitions (vertex, edge, graph, facial angles, dihedral angles). The main difference is that in a polyhedral surface, some dihedral angles may be *flat*, i.e., have value 180°.

We usually assume that we are given an *embedded graph*, i.e. a graph with a fixed cyclic order of edges around each vertex. (Note that "embedded" does not imply a mapping to coordinates; the embedding is given combinatorially only.) From the order of edges around vertices we can determine the *faces of the graph*, which are the cycles obtained by always taking the next edge in cyclic order. We also assume that we are given *facial angles of the graph*, which are values at each incidence between a vertex and a face of the graph.

Given an embedded graph and facial angles (and sometimes also the lengths of the edges), we say that a polyhedral surface S *realizes* the input if its graph (with the induced embedding) is the given embedded graph, and its facial angles (and edge lengths, if given) are as prescribed in the input. For a connected graph, the polyhedral surface has genus 0 if and only if its graph is *planar*, i.e. it can be drawn in the plane without crossing.

We will almost only study orthogonal polyhedra of genus 0 in this paper. A polyhedral surface is *orthogonal* if all its faces are perpendicular to a coordinate axis. This implies that all facial angles and all dihedral angles are multiples of 90°, and all edges are parallel to a coordinate axis.

We use the term *orientation* $o \in \{x, y, z\}$ for any 1-dimensional object that is parallel to a coordinate axis. Thus an edge of an orthogonal polyhedron is an o-edge if it is parallel to the o-axis, and a face is said to be an *o-face* if the normal of the plane that contains the face has orientation o (for $o \in \{x, y, z\}$.)

3 Flat Dihedral Angles

In this section, we present an algorithm that, given a connected embedded planar graph and facial angles that are multiples of 90°, determines which of the edges of the graph must have a flat dihedral angle in any realization. Since the graph is planar, any realization must have genus 0. Since the facial angles are multiples of 90°, any face is the union of rectangles. As proved (independently of each other) in [7] and [8], this implies that any realization must have all dihedral angles that are multiples of 90°, i.e., it is an orthogonal polyhedral surface after a suitable rotation.

3.1 Algorithm

For each face of the input graph, the facial angles determine relative orientations of edges within the face. We write $e \parallel e'$ if e and e' are edges on one face and have the same orientation within that face. We can extend $\parallel$ into an equivalence relation $\sim$ by defining that $e \sim e'$ if there exists a set of edges $e = e_1, \ldots, e_k = e'$ with $e_i \parallel e_{i+1}$ for $1 \leq i < k$. We define an *edge-bundle* to be an equivalence class under equivalence relation $\sim$. The edge-bundles can easily be computed in linear time from the embedded graph and the facial angles. Directly from the definition of $\parallel$ and $\sim$, the following holds:

Observation 1. *All edges in an edge-bundle must be parallel in any realization.*

It hence makes sense to say that an edge-bundle has *orientation* o (for $o \in \{x, y, z\}$.) Two edge-bundles $\mathcal{B}_1 \neq \mathcal{B}_2$ are said to *cross* if there exists a face in the graph that contains edges from both. Since $\mathcal{B}_1$ and $\mathcal{B}_2$ were equivalence classes, their edges are not parallel to each other, so if $\mathcal{B}_1$ had orientation o in some realization then $\mathcal{B}_2$ cannot have orientation o.

This gives rise to a simple greedy-propagation algorithm to determine orientations of edge-bundles. Initally pick two edge-bundles $\mathcal{B}_1$ and $\mathcal{B}_2$ that cross and set $T(\mathcal{B}_1) = \{y\}$ and $T(\mathcal{B}_2) = \{z\}$. Here, $T(\mathcal{B}_i)$ is a set of possible orientations of edge-bundle $\mathcal{B}_i$; we hence arbitrarily fix one rotation of a realization (which must be orthogonal as discussed earlier.) For all other bundles, initialize $T(\mathcal{B}_i) = \{x, y, z\}$. Now propagate orientations by picking any edge-bundle $\mathcal{B}$ with $|T(\mathcal{B})| = 1$ that has not been identified yet (initially they are all unidentified.) Mark $\mathcal{B}$ as identified and let o be the unique orientation left in $T(\mathcal{B})$. For all edge-bundles $\mathcal{B}'$ that cross $\mathcal{B}$, remove o from $T(\mathcal{B}')$, because o cannot possibly be the orientation of $\mathcal{B}'$.

This algorithm, which we refer to as algorithm BundleOrientation, stops if either all edge-bundles are identified, or if there are unidentified edge-bundles, but none of them has $|T(\mathcal{B})| = 1$.

Algorithm BundleOrientation can be implemented in linear time if we precompute an auxiliary graph H of edge-bundles, which has a vertex for every edge-bundle and an edge between two edge-bundles if and only if the edge-bundles cross. By storing edge-bundles in buckets by the size of $|T(\mathcal{B})|$, we can then in $O(1)$ time find the next edge-bundle $\mathcal{B}$ to be identified, and in $O(\deg_H(\mathcal{B}))$ time update all the edge-bundles that $\mathcal{B}$ crosses; this is $O(m + n)$ time overall since H has size $O(m + n)$.

We will show the following result in the next subsection:

Lemma 1. *If an embedded connected planar graph with facial angles has a realization S, then algorithm* BundleOrientation *identifies all edge-bundles. Moreover, after applying a suitable rotation of S, for all edge-bundles $\mathcal{B}$ the unique value left in $T(\mathcal{B})$ at termination is the orientation of $\mathcal{B}$ in S.*

Thus algorithm BundleOrientation determines all edge orientations, which in turn determines the face orientations. We cannot determine dihedral angles directly from this (because there are still two possible directions for each face normal), but we can determine flat dihedral angles, since for any two adjacent co-planar faces, the dihedral angles at the edges shared by them must be 180°.

Theorem 1. *Given an embedded connected planar graph with facial angles that are multiples of* 90°*, we can in* $O(m+n)$ *time*

- *report that no polyhedral surface can realize this graph and facial angles, OR*
- *report all edges of the graph for which the dihedral angle must be 180° in any polyhedral surface that realizes this graph and facial angles.*

3.2 Correctness

In this section, we prove correctness of algorithm BUNDLEORIENTATION, i.e., we prove Lemma 1. We assume throughout this section that some orthogonal polyhedral surface S exists that realizes the given graph and facial angles. We furthermore assume that S has been rotated such that all edges in edge-bundle $\mathcal{B}_1$ (the first edge-bundle picked by algorithm BUNDLEORIENTATION) are parallel to the y-axis, and all edges in edge-bundle $\mathcal{B}_2$ are parallel to the z-axis. Since we only eliminate orientations that cannot possibly be used in an edge-bundle, the following observation has a straightforward proof by induction:

Lemma 2. *At any time during algorithm* BUNDLEORIENTATION, *for any edge-bundle* $\mathcal{B}$ *the orientation of* $\mathcal{B}$ *in* S *remains in* $T(\mathcal{B})$.

So no edge-bundle $\mathcal{B}$ will ever have $|T(\mathcal{B})| = 0$ during algorithm BUNDLEORIENTATION. So if not all edge-bundles are identified, then the algorithm must stop when some edge-bundles have two or three possible orientations left. We claim that this cannot happen if the realizing polyhedral surface S has genus 0.

To show that this is non-trivial, observe that it is not true for higher genus. Fig. 1 shows a polyhedral surface of genus 1, where algorithm BUNDLEORIENTATION, for this embedded graph, facial angles and choice of initial edge-bundles $\mathcal{B}_1$ and $\mathcal{B}_2$, does not identify any edge-bundles since none of them crosses both $\mathcal{B}_1$ and $\mathcal{B}_2$.

From now on, assume that the input graph is planar and connected and has a realization S. We will prove the following:

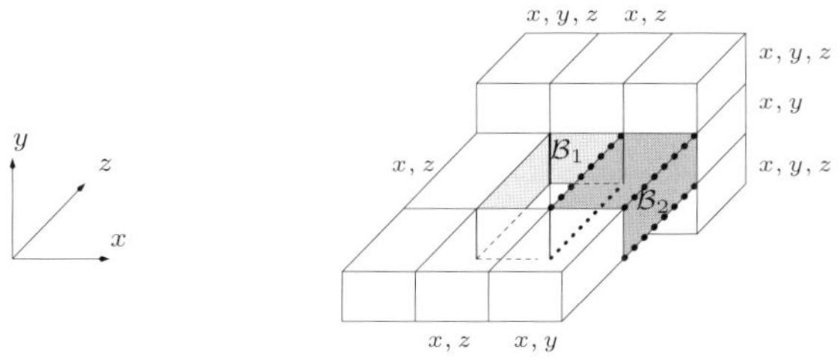

Fig. 1. An example of genus 1 where algorithm BUNDLEORIENTATION fails. We inidicate at selected edges the possible orientations that remain.

Lemma 3. *Let $\mathcal{B}^*$ be an edge-bundle that has been identified by algorithm* BundleOrientation. *Then all edge-bundles that cross $\mathcal{B}^*$ will also be identified by algorithm* BundleOrientation.

We first argue why it suffices to prove this lemma. When algorithm BundleOrientation stops, then there are three possible types of faces: those where 0, 1 or 2 of the two edge-bundles meeting the face have been identified. At least one face has type 2 (the one where the two initial edge-bundles cross.) No face has type 1 if Lemma 3 holds. If any face had type 0, then there would be two adjacent faces where one has type 0 and the other type 2, since the surface is connected. But then the edge common to the two faces is in an edge-bundle that is both identified and unidentified, a contradiction. So all faces have type 2 and hence all edge-bundles are identified, which together with Lemma 2 implies Lemma 1.

Proof. (of Lemma 3) We first give an outline of the proof, which is by contradiction. Assume there exists faces which are incident to some of the edges in $\mathcal{B}^*$ where the crossing edge-bundle has not been identified. There also exists at least one edge-bundle that crosses $\mathcal{B}^*$ and whose orientation has been identified. This holds if $\mathcal{B}^*$ is one of the initial two edge-bundles, because they cross each other, and also holds if $\mathcal{B}^*$ was identified later, because then $|T(\mathcal{B}^*)|$ became 1 due to some crossing identified edge-bundle.

We can argue that among these identified and unidentified edge-bundles that cross $\mathcal{B}^*$, there exists an identified one $\mathcal{B}$ and an unidentified one $\mathcal{B}'$ that "interleave" in the sense that the faces where they cross $\mathcal{B}^*$ alternate. By genus 0, interleaving edge-bundles must cross. So there are three edge-bundles $\mathcal{B}^*, \mathcal{B}, \mathcal{B}'$ that pairwise cross and two have been identified. This means that algorithm BundleOrientation will also identify the third, a contradiction.

The difficulty of the proof lies in clarifying what "interleave" means. This is much easier if every face of the graph has 4 vertices (and hence the realizing surface S is *quadrangulated* and has rectangular faces.) Thus, we will prove Lemma 3 first for quadrangulated surfaces, and discuss later why it holds in general.

If S is quadrangulated, edge-bundles (which were defined as a set of edges) naturally become a (cyclic) sequence of edges, since every face contains only two parallel edges, and every edge belongs to two faces. See Fig. 2. It now also makes sense to speak of a sequence of faces of an edge-bundle $\mathcal{B}$; we call this sequence of faces a *band* B (similarly as in [6]). Bands and edge-bundles of a quadrangulated surface are in 1-to-1 correspondence, and so all terms defined for one (such as "crossing" and "identified") will also be applied to the other.

We need some definitions. For the edge-bundle $\mathcal{B}^*$, let $c(\mathcal{B}^*)$ be the cycle on surface S obtained by connecting the midpoints of consecutive edges of $\mathcal{B}^*$. Because S has genus 0, $c(\mathcal{B}^*)$ splits S into two connected regions; arbitrarily pick one of them and call it the *interior* of $c(\mathcal{B}^*)$. Let B^* be the band corresponding to edge-bundle $\mathcal{B}^*$. For any band B that crosses B^*, a *chord* of B is a subsequence $f_1, \dots, f_k$ of the faces of B where f_1 and f_k are on B^* and $f_2, \dots, f_{k-1}$ are not on B^* and in the interior of $c(B^*)$. See also Fig. 2. Faces f_1 and f_k of a chord

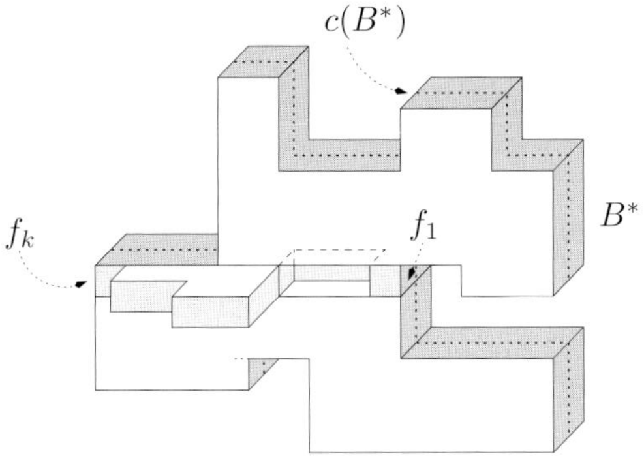

Fig. 2. Band B^* (dark-gray) and one of its chords (light-gray). We do not show all edges of the quadrangulated polyhedron.

are called the *chord-anchors* and belong to both B^* and B. Since all faces of a band span the same range in one of the coordinate axis, the two faces that are chord-anchors span the same range in two of the coordinate axes.

Two chords are said to *interleave* if their chord-anchors appear alternatingly in the cyclic sequence of faces B^*. By planarity, if two chords interleave, then the bands that defined the chords must cross (i.e., have a face in common); see also Fig. 3.

We now consider a restricted version of algorithm BUNDLEORIENTATION, where we only propagate orientations along interleaving chords; this alone is enough to identify all bands and hence all edge-bundles. We argued earlier that there exists an edge-bundle $\mathcal{B}_0$ that crosses $\mathcal{B}^*$ and was identified. Pick one face common to their bands B_0 and B^*, and let C_0 be the chord that is a subsequence of B_0 starting at this face. Mark all chords that have chord-anchors on B^* and can be reached from C_0 via interleaving chords, i.e., mark all chords C^1 that interleave C^0, then in turn mark all chords that interleave C^1, and so on until no more chords can be marked. One can easily see that for any marked chord, the edge-bundle that defined it was identified by algorithm BUNDLEORIENTATION, since interleaving

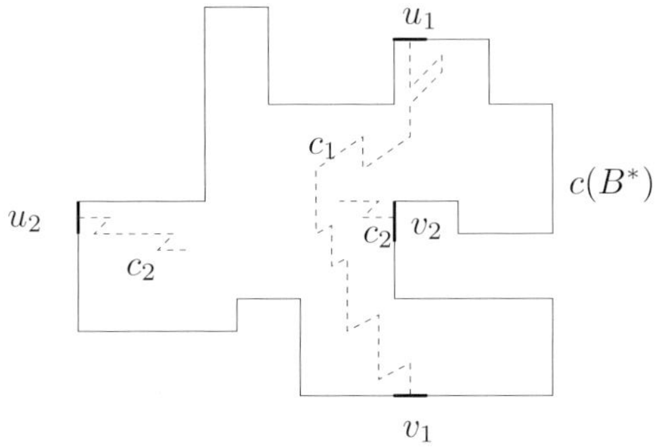

Fig. 3. Chord c_1 with anchors $\{u_1, v_1\}$ interleaves chord c_2 with anchors $\{u_2, v_2\}$

chords means that their edge-bundles cross, and all chords belong to edge-bundles that cross $\mathcal{B}^*$.

Every face f of B^* is an anchor of a chord, since there is a band crossing B^* at f, and the part of the band that enters the interior of $c(B^*)$ forms a chord. On the other hand, every face of B^* belongs to only one chord, since it belongs to band B^* and only one other band. Call a face of B^* *marked* if and only if the unique chord that contains it is marked. If all faces of B^* are marked, then all edge-bundles that cross $\mathcal{B}^*$ are identified and Lemma 3 holds as desired. So assume not all faces of B^* are marked, and let U be a maximal contiguous set of faces of B^* that is not marked.

Claim. There exists a chord with anchors $\{f, f'\}$ such that $f \in U$ and $f' \notin U$.

Proof. The intersection of U with $c(B^*)$ forms an open curve in the plane that contains $c(B^*)$. By considering the region between the two endpoints of that open curve, we can find a line ℓ that is parallel to a coordinate axis, does not intersect an edge, and intersects U an odd number of times. See Fig. 4.

Recall that for every face f on B^*, there is another face f' on B^* such that $\{f, f'\}$ are the anchords of a chord. If f is intersected by ℓ, then so is f' since f and f' span the same range in two coordinate directions. But ℓ intersects U an odd number of times, so for at least one face f in $U \subset B^*$, the face f' with which it forms a chord-anchor cannot also be in U. □

Going in order of faces along band B^*, we hence encounter: faces in U (including f); a set of faces S_1 which may be marked or not, but the first of them is marked by definition of U; face f'; and another set of faces S_2 which may be marked or not, but the last of them is marked by definition of U. Recall that faces were marked during propagation along interleaving chords, and consider the first time when both S_1 and S_2 contained marked faces. The chord that caused this to happen hence had one anchor in S_1 and the other in S_2. But then this chord interleaves with the chord anchored at $\{f, f'\}$, which means that f and f' should have been marked as well, a contradiction.

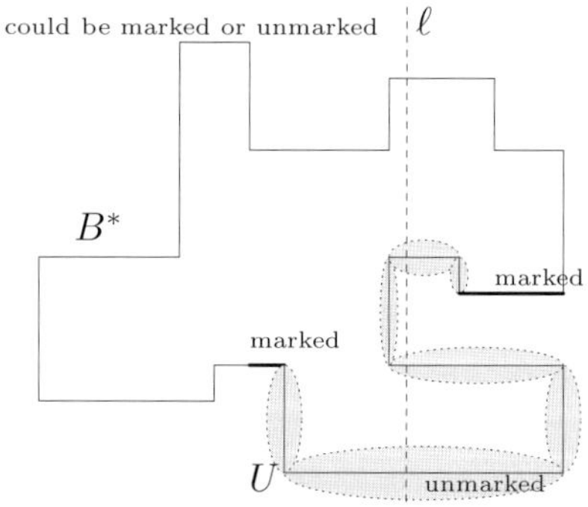

Fig. 4. There must be a horizontal or vertical line ℓ that intersects U an odd number of times. The picture shows the cross-section with the plane that contains $c(\mathcal{B}^*)$.

This finishes the proof of Lemma 3 for a quadrangulated surface. Now we briefly discuss the case when the input is not quadrangulated. A simple, but inefficient, approach would be to subdivide faces until all of them are rectangles. This is undesirable for two reasons: it uses edge lengths and it increases the time complexity to quadratic. A better approach is to only use the existence of such a quadrangulation.

Thus, assume the input can be realized by some polyhedral surface S, let $\mathcal{B}^*$ be the edge-bundle of Lemma 3, and $\mathcal{B}$ any other edge-bundle that crosses it, say at face f. Let S' be a quadrangulation of surface S. Let B^* and B be bands in S' that contain edges (or parts of edges) of f. By the above proof for the quadrangulated surface S', if B^* is identified by BundleOrientation, then so is B. But this implies Lemma 3 for S as well: If $\mathcal{B}^*$ is identified, then so is B^*, and so are all other bands that are identified from B^* in S'. So in particular, B (and hence $\mathcal{B}$) are identified.

Note that the quadrangulated surface S' need not actually be computed; its existence is used to show that $\mathcal{B}$ is reached from $\mathcal{B}^*$, but the sequence of edge-bundles to reach $\mathcal{B}$ will be found without knowing S' by algorithm BundleOrientation. □

4 Non-flat Dihedral Angles

If we know all flat dihedral angles, we can delete the corresponding edges in the graph, and then delete the resulting isolated vertices and contract the resulting vertices of degree 2 into their neighbours. Doing this merges co-planar faces of a polyhedral surface S until they become faces of the polyhedron bounded by S, and the resulting graph is the graph of the polyhedron. In this section, we are interested in determining the remaining dihedral angles, and we can thus assume that we are given the graph of the polyhedron.

Let v be a vertex of an orthogonal polyhedron. The incident 8 octants of v may or may not be occupied by the polyhedron within a small neighbourhood of v, yielding 2^8 possible configurations at vertex v. Of those, many cannot occur in an orthogonal polyhedron, since the resulting surface is not a 2-manifold. Some more have a flat dihedral angle. Eliminating all these cases and omitting rotational symmetries, we are left with only 7 vertex configurations, which are given in Fig. 5.

Each vertex of an orthogonal polyhedron can have three, four or six incident edges (so it has degree 3, 4 or 6 in the graph.) In Fig. 5, we give the vertex configurations together with the facial angles and dihedral angles in the graph. The reader should at this point start to forget the geometry and view this as an embedded graph with facial angles and labels on all edges.

We group the 7 configurations into four groups; configurations in different groups have different degrees or different facial angles. Within each group, any mapping from one configuration to the other that preserves order and facial angles maps every dihedral angle α to its opposite $360° - \alpha$. Since $\alpha \neq 180°$, this implies the following:

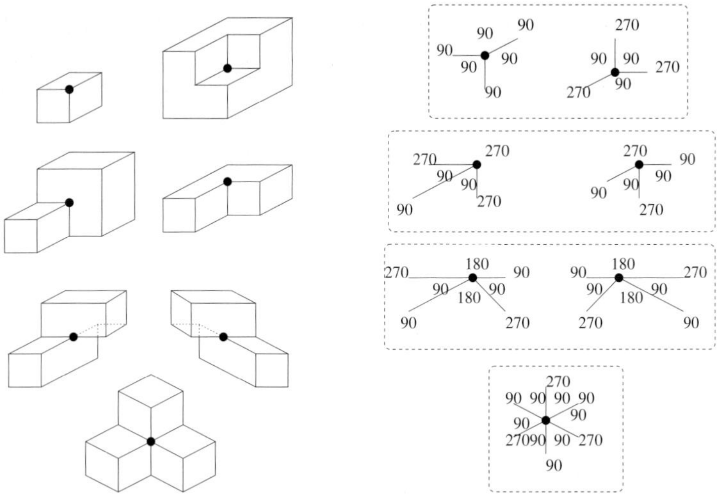

Fig. 5. The vertex configurations with facial and dihedral angles

Observation 2. *All dihedral angles at a vertex v are determined by the degree of v, the facial angles at v, and one dihedral angle of an edge incident to v.*

If the graph is connected, all dihedral angles can hence be computed if one initial dihedral angle is fixed, by propagating the information along the edges of the graph and updating dihedral angles at the other endpoint according to the appropriate vertex configuration. The running time for this is $O(m+n)$ time.

5 Selecting among Two Sets

At this point, we have computed two possible sets of dihedral angles $\{d_1(e)\}$ and $\{d_2(e)\}$ (depending on how we fixed the initial dihedral angle), and we now need to determine which of them is the correct one.

These two sets are in fact opposite to each other, i.e., $d_1(e) = 360° - d_2(e)$ for all edges e. This clearly holds for the initial edge, and by induction also for the other edges, since the two configurations within a group in Fig. 5 have opposite dihedral angles. So if the set $\{d_1(.)\}$ is realized by an orthogonal polyhedron P, then $d_2(e)$ is the outside angle between the faces adjacent to e in P; we could thus call $\{d_2(.)\}$ the *outside dihedral angles*.

To determine which of the two sets are the inside and which the outside dihedral angles, we use edge lengths and reconstruct the coordinates of all vertices. To be precise, pick some vertex of degree 3, assign it to be located at the origin, and arbitrarily assign three orientations and directions to its three incident edges. Using the facial angles, edge lengths, and the dihedral angles from $\{d_1(e)\}$, we can then easily compute all coordinates of all vertices in $O(m+n)$ time. (If this assigns two different coordinates to the same vertex, output an error message; the edge lengths cannot have been correct.)

Now find a vertex v with maximal x-coordinate (breaking ties arbitrarily), and let f be a face adjacent to v and perpendicular to the x-axis. Since there are no flat dihedral angles, the edges incident to f must have dihedral angle 90°, otherwise there would be a vertex with even larger x-coordinate. This decides which of $\{d_1(.)\}$ and $\{d_2(.)\}$ was correct, and only one of them can be correct.

Putting all three algorithms together, we hence obtain the following:

Theorem 2. *Given an embedded planar graph with facial angles and edge lengths, we can in $O(m+n)$ time*

- *find the dihedral angles of any orthogonal polyhedral surface that has this graph, facial angles and edge lengths, OR*
- *report that this graph and facial angles can only belong to an orthogonal polyhedral surface for which the polyhedron bounded by it has a disconnected graph, OR*
- *report that no orthogonal polyhedral surface can realize this graph, facial angles and edge lengths.*

Moreover, if a realizing orthogonal polyhedral surface exists, then it is unique.

6 Remarks

We assumed that we are given a graph, facial angles and edge lengths, and that the reconstructed orthogonal polyhedron has a connected graph and genus 0. We now briefly discuss these assumptions.

- Inspection of the proof of Cauchy's theorem shows that it does not use edge lengths, so for a convex polyhedron the graph and facial angles determine the dihedral angles. Our proof also does not use edge lengths, except at the very last step where we determine which of two possible sets of dihedral angles is the correct one.
  It seems exceedingly likely that this step could be done without using edge lengths. In particular, in the corresponding 2D problem (given a set of angles, can this be the set of angles of an orthogonal polygon?) there is a simple solution: the set of n angles can be realized if and only if it adds up to $180°(n+2)$. If any edge-bundle happens to have only two parallel edges on each face, then the cycle of the edge-bundle (as defined in Section 3) lies within a plane, and studying the dihedral angles at this cycle tells us which set is correct. But in general the incidence structure of faces used by edge-bundles is more complicated. Can we use it somehow to determine the correct set of dihedral angles without using edge lengths?
- We demanded that the graph of the polyhedron is connected, i.e., no face has holes. If this condition is dropped, then testing whether a realizing polyhedral surface exists becomes NP-hard. In fact, the problem is NP-hard in the strong sense, and holds even for polyhedral surface where every face is a unit rectangle [3].

- We demanded that the orthogonal polyhedral surface has genus 0, which was used frequently throughout the proof of correctness of algorithm Bundle-Orientation. The example in Fig. 1 shows that this algorithm can fail for higher genus. Is there some embedded graph of higher genus (with given facial angles) where different edge-bundle orientations are in fact possible? We suspect that this is not true, but this remains open.
  The other algorithms work without modification for surfaces of higher genus, so Cauchy's theorem holds for higher genus orthogonal polyhedra (not polyhedral surfaces, i.e., no flat dihedral angles are allowed), as long as they have a connected graph.
- Our algorithm computes the set of dihedral angles, and as a by-product also vertex coordinates, but it does not check whether the resulting surface is indeed a 2-manifold (i.e., that edges are incident to exactly two faces, etc.) This can be done in polynomial time (see [5]).

References

1. Aigner, M., Ziegler, G.: Proofs from THE BOOK, 1st edn. Springer, Heidelberg (1998); 3rd edn. (2004)
2. Biedl, T., Genc, B.: When can a graph form an orthogonal polyhedron. In: Canadian Conference on Computational Geometry (CCCG 2004), August 2004, pp. 53–56 (2004)
3. Biedl, T., Genc, B.: Cauchy's theorem for orthogonal polyhedra of genus 0. Technical Report CS-2008-26, University of Waterloo, School of Computer Science (2008)
4. Biedl, T.C., Lubiw, A., Spriggs, M.J.: Cauchy's theorem and edge lengths of convex polyhedra. In: Dehne, F., Sack, J.-R., Zeh, N. (eds.) WADS 2007. LNCS, vol. 4619, pp. 398–409. Springer, Heidelberg (2007)
5. Biedl, T., Lubiw, A., Sun, J.: When can a net fold to a polyhedron? Computational Geometry: Theory and Applications 31(3), 207–218 (2005)
6. Damian, M., Flatland, R., O'Rourke, J.: Unfolding Manhattan towers. Computational Geometry: Theory and Applications 40, 102–114 (2008)
7. Dolbilin, N.P., Shtan'ko, M.A., Shtogrin, M.T.: Rigidity of a quadrillage of a torus by squares. Russian Math. Surveys 54(4), 839–840 (1999)
8. Donoso, M., O'Rourke, J.: Nonorthogonal polyhedra built from rectangles. In: 14th Canadian Conference on Computational Geometry (CCCG 2002), pp. 97–100 (2002)
9. O'Rourke, J.: Computational geometry column 49. ACM SIGACT News 38(2) (2007)
10. Pak, I.: Lectures on Discrete and Polyhedral Geometry. Cambridge University Press, Cambridge (in print, 2009), `http://www.math.umn.edu/~pak/book.htm` (last accessed, April 2009)

Approximability of Sparse Integer Programs

David Pritchard

Department of Combinatorics & Optimization, University of Waterloo
dagpritchard@math.uwaterloo.edu

Abstract. The main focus of this paper is a pair of new approximation algorithms for sparse integer programs. First, for covering integer programs $\{\min cx : Ax \geq b, \mathbf{0} \leq x \leq d\}$ where A has at most k nonzeroes per row, we give a k-approximation algorithm. (We assume A, b, c, d are nonnegative.) For any $k \geq 2$ and $\epsilon > 0$, if $\mathsf{P} \neq \mathsf{NP}$ this ratio cannot be improved to $k - 1 - \epsilon$, and under the unique games conjecture this ratio cannot be improved to $k - \epsilon$. One key idea is to replace individual constraints by others that have better rounding properties but the same nonnegative integral solutions; another critical ingredient is knapsack-cover inequalities. Second, for packing integer programs $\{\max cx : Ax \leq b, \mathbf{0} \leq x \leq d\}$ where A has at most k nonzeroes per column, we give a $2^k k^2$-approximation algorithm. This is the first polynomial-time approximation algorithm for this problem with approximation ratio depending only on k, for any $k > 1$. Our approach starts from iterated LP relaxation, and then uses probabilistic and greedy methods to recover a feasible solution.

1 Introduction and Prior Work

In this paper we investigate the following problem: what is the best possible approximation ratio for integer programs where the constraint matrix is sparse? To put this in context we recall a famous result of Lenstra [1]: integer programs with a constant number of variables or a constant number of constraints can be solved in polynomial time. Our investigations analogously ask what is possible if the constraints each involve at most k variables, or if the variables each appear in at most k constraints.

Rather than consider the full class of all integer programs, we consider only packing and covering problems. One sensible reason for this is that *every* integer program can be rewritten (possibly with additional variables) in such a way that each constraint contains at most 3 variables and each variable appears in at most 3 constraints, if mixed positive and negative coefficients are allowed. Aside from this, packing programs and covering programs represent a substantial portion of the literature on integer programs; and sparse programs of this type are interesting in their own right as "multiple-knapsack" problems where each item affects a bounded number of knapsacks, or each knapsack is affected by a bounded number of items.

We use CIP (resp. PIP) as short for *covering* (resp. *packing*) *integer program*, which is any integer program of the form $\{\min cx : Ax \geq b, \mathbf{0} \leq x \leq d\}$ (resp.

A. Fiat and P. Sanders (Eds.): ESA 2009, LNCS 5757, pp. 83–94, 2009.

$\{\max cx : Ax \geq b, \mathbf{0} \leq x \leq d\}$) with A, b, c, d nonnegative and rational. Note that CIPs are sometimes called *multiset multicover* when A and b are integral. We call constraints $x \leq d$ *multiplicity constraints* (also known as *capacity constraints*). We allow for entries of d to be infinite, and without loss of generality, all finite entries of d are integral. An integer program with constraint matrix A is *k-row-sparse*, or *k-RS*, if each row of A has at most k entries; we define *k-column-sparse (k-CS)* similarly. As a rule of thumb we ignore the case $k = 1$, since such problems trivially admit fully polynomial-time approximation schemes (FPTAS's) or poly-time algorithms. The symbol $\mathbf{0}$ denotes the all-zero vector, and similarly for $\mathbf{1}$. For covering problems an *α-approximation algorithm* is one that always returns a solution with objective value at most α times optimal; for packing, the objective value is at least $1/\alpha$ times optimal. We use n to denote the number of variables and m the number of constraints (i.e. the number of rows of A).

1.1 k-Row-Sparse Covering IPs: Previous and New Results

The special case of 2-RS CIP where A, b, c, d are 0-1 is the same as Min Vertex Cover, which is APX-hard. More generally, 0-1 k-RS CIP is the same as k-Bounded Hypergraph Min Vertex Cover (a.k.a. Set Cover with maximum frequency k) which is not approximable to $k - 1 - \epsilon$ for any fixed $\epsilon > 0$ unless P=NP [2] ($k-\epsilon$ under the unique games conjecture [3]). This special case is known to admit a matching positive result: set cover with maximum frequency k can be k-approximated by direct rounding of the naive LP [4] or local ratio/primal-dual methods [5].

The following results are known for other special cases of k-RS CIP with multiplicity constraints: Hochbaum [6] gave a k-approximation in the special case that A is 0-1; Hochbaum et al. [7] and Bar-Yehuda & Rawitz [8] gave pseudopolynomial 2-approximation algorithms for the case that $k = 2$ and d is finite. For the special case $d = \mathbf{1}$, Carr et al. [9, §2.6] gave a k-approximation, and Fujito & Yabuta [10] gave a primal-dual k-approximation. Moreover [9,10] claim a k-approximation for general d, but there seems to have been some oversights as the papers do not provide full proofs and their methods alone seem to be insufficient for general d. Our first main result, given in Section 2, is a simple and correct proof of the same claim.

Theorem 1. *There is a polynomial time k-approximation algorithm for k-RS CIPs with multiplicity constraints.*

Our approach is to first consider the special case that there are no multiplicity constraints (i.e. $d_j = +\infty$ for all j); we then extend to the case of finite d via *knapsack-cover inequalities*, using linear programming (LP) techniques from Carr et al. [9]. A $(k + 1)$-approximation algorithm is relatively easy to obtain using LP rounding; in order get the tighter ratio k, we replace constraints by other "$\mathbb{Z}_+$-equivalent" constraints (see Definition 5) with better rounding properties. The algorithm requires a polynomial-time linear programming subroutine.

Independently of our work, a recent paper of Koufogiannakis & Young [11] also gives a full and correct proof of Theorem 1. Their primal-iterative approach works for a broad generalization of k-RS CIPs and runs in low-degree strongly polynomial time. Our approach has the generic advantage of giving new ideas that can be used in conjunction with other LP-based methods, and the specific advantage of giving integrality gap bounds. See the full version [12] for details.

1.2 k-Column-Sparse Packing IPs: Previous and New Results

So far, no constant-factor approximation is known for k-CS PIPs, except in special cases. If every entry of b is $\Omega(\log m)$ then randomized rounding provides a constant-factor approximation. *Demand matching* is the special case of 2-CS PIP where (i) in each column of A all nonzero values in that column are equal to one another and (ii) no two columns have their nonzeroes in the same two rows. Shepherd & Vetta [13] showed demand matching is APX-hard but admits a $(\frac{11}{2} - \sqrt{5})$-approximation algorithm when $d = \mathbf{1}$; their approach also gives a $\frac{7}{2}$-approximation for 2-CS PIP instances satisfying (i). Results of Chekuri et al. [14] yield a $11.542k$-approximation algorithm for k-CS PIP instances satisfying (i) and such that the maximum entry of A is less than the minimum entry of b.

The special case of k-CS PIP where A, b are 0-1 is the same as *min-weight k-set packing*, *hypergraph matching with edges of size $\leq k$*, and *strong independent sets in hypergraphs with degree at most k*. The best approximation ratio known for this problem is $(k+1)/2 + \epsilon$ [15] for general weights, and $k/2 + \epsilon$ when $c = \mathbf{1}$ [16]. The best lower bound is due to Hazan et al. [17], who showed $\Omega(k/\ln k)$-inapproximability unless P=NP, even for $c = \mathbf{1}$.

Our second main result, given in Section 3, is the following result.

Theorem 2. *There is a polynomial time $2^k(k^2 - 2k + 1) + 1$-approximation algorithm for k-CS PIPs with multiplicity constraints.*

Our methodology begins by using *iterated LP relaxation* [18] to find an integral solution with super-optimal value, but violating some constraints in an additively-bounded way. Then we use a combination of probabilistic and greedy methods to recover a high-weight feasible solution. An extension of this methodology gives improved results in two special cases: we get a 4-approximation when $k = 2$, and we get a $(W + k)/(W - k)$-approximation when the program's *width*, defined as $W := \min_{i,j:A_{ij} \neq 0} \frac{b_i}{A_{ij}}$ satisfies $W > k$. These results also require a polynomial-time linear programming subroutine.

Subsequent to the initial release of this paper on the arXiv [12], C. Chekuri, A. Ene and N. Korula (personal communication) have obtained results for k-CS PIPs: a fairly simple $O(k2^k)$-approximation algorithm without iterated rounding, and a $O(k^2)$-approximation (and integrality gap bound) that builds on our iterated rounding ideas.

1.3 Other Related Work

Srinivasan [19,20] showed that k-CS CIPs admit a $O(\log k)$-approximation. Kolliopoulos and Young [21] extended this result to handle multiplicity constraints. There is a matching hardness result: it is NP-hard to approximate k-Set Cover,

Table 1. The landscape of approximability of sparse integer programs. Our main results are in boldface.

	k-Column-Sparse		k-Row-Sparse	
	lower bound	upper bound	lower bound	upper bound
Packing	$\Omega(k/\ln k)$	$\mathbf{(k^2-2k+1)2^k+1}$	$n^{1-o(1)}$	ϵn
Covering	$\ln k - O(\ln\ln k)$	$O(\ln k)$	$k-\epsilon$	$\mathbf{k}$

which is the special case where A, b, c are 0-1, better than $\ln k - O(\ln\ln k)$ for any $k \geq 3$ [22]. Hence for k-CS CIP the best possible approximation ratio is $\Theta(\log k)$. A $(k+\epsilon)$-approximation algorithm can be obtained by separately applying an approximation scheme to the knapsack problem corresponding to each constraint. Hochbaum [23] showed 2-CS CIPs are NP-hard to optimize and gave a bicriteria approximation algorithm. Although 0-1 2-CS CIP is Edge Cover which lies in P, 2-CS CIP in general is NP-hard to $(17/16-\epsilon)$-approximate, due to methods from [24], even if A has 2 *equal* nonzeroes per column and d is 0-1 or d is all-$+\infty$. See the full version [12] for details.

The special case of 2-RS PIP where A, b, c are 0-1 is the same as Max Independent Set, which is not approximable within $n/2^{\log^{3/4+\epsilon} n}$ unless $\mathsf{NP} \subset \mathsf{BPTIME}(2^{\log^{O(1)} n})$ [25]. On the other hand, n-approximation of any packing problem is easy to accomplish by looking at the best singleton-support solution. A slightly better n/t-approximation, for any fixed t, can be accomplished by exhaustively guessing the t most profitable variables in the optimal solution, and then solving the resulting t-dimensional integer program to optimality via Lenstra's result [1].

We remark that integer CIPs and PIPs where A has at most k rows are known as k-*dimensional knapsack* problems, and for any fixed $k \geq 2$ they have a PTAS and pseudopolynomial-time algorithm, but no FPTAS unless P=NP — see [26, §9.4] for references. In particular, to clarify Lenstra's result [1], it is NP-hard to get an FPTAS for PIPs with 2 constraints plus a nonnegativity constraint for each variable [27].

1.4 Summary

We summarize the existing and new results in Table 1. Note that in all four cases, the strongest known lower bounds are obtained even in the special case that A, b, c, d are 0-1.

2 k-Approximation for k-Row-Sparse CIPs

By scaling rows and clipping coefficients that are too high, there is no loss of generality in the following definition.

Definition 1. *A* k-RS CIP *is an integer program* $\{\min cx : Ax \geq \mathbf{1}, \mathbf{0} \leq x \leq d\}$ *where* A *is* k*-RS and all entries of* A *are at most 1.*

To begin with, we focus on the case $d_j = +\infty$ for all j, which we will call *unbounded k-RS CIP*, since it already illustrates the essence of our new technique. Motivated by LP rounding methods, we make the following definition, in which x is a vector-valued variable and α is a vector of real coefficients. Throughout, we assume coefficients are nonnegative. When we apply $\lfloor\cdot\rfloor$ to vectors we mean the component-wise floor.

Definition 2. *A constraint $\alpha x \geq 1$ is* ρ-roundable *for some $\rho > 1$ if for all nonnegative real x, $(\alpha x \geq 1)$ implies $(\alpha\lfloor \rho x\rfloor \geq 1)$.*

Note that ρ-roundability implies ρ'-roundability for $\rho' > \rho$. The relevance of this property is explained by the following proposition.

Proposition 3. *If every constraint in an unbounded covering integer program is ρ-roundable, then there is a ρ-approximation algorithm for the program.*

Proof. Let x^* be an optimal solution to the program's linear relaxation. Then cx^* is a lower bound on the cost of any optimal solution. Thus, $\lfloor \rho x^*\rfloor$ is a feasible solution with cost at most ρ times optimal. □

Another simple observation helps us get started.

Proposition 4. *The constraint $\alpha x \geq 1$ is $(1 + \sum_i \alpha_i)$-roundable.*

Proof. Let $\rho = (1+\sum_i \alpha_i)$. Since $\lfloor t\rfloor > t-1$ for any t, if $\alpha x \geq 1$ for a nonnegative x, then

$$\alpha\lfloor \rho x\rfloor \geq \sum_i \alpha_i(\rho x_i - 1) = \rho \sum_i \alpha_i x_i - \sum_i \alpha_i \geq \rho \cdot 1 - (\rho - 1) = 1,$$

as needed. □

Now consider an unbounded k-RS CIP. Since each constraint has at most k coefficients, each less than 1, it follows from Proposition 4 that every constraint in these programs is $(k+1)$-roundable, and so such programs admit a $(k+1)$-approximation algorithm by Proposition 3. It is also clear that we can tighten the approximation ratio to k for programs where the sum of the coefficients in every constraint (row) is at most $k-1$. What we will now do is show that rows with sum in $(k-1, k]$ can be replaced by other rows which are k-roundable.

Definition 5. *Two constraints $\alpha x \geq 1$ and $\alpha' x \geq 1$ are* $\mathbb{Z}_+$-equivalent *if for all nonnegative integral x, $(\alpha x \geq 1) \Leftrightarrow (\alpha' x \geq 1)$.*

In other words, $\alpha x \geq 1$ and $\alpha' x \geq 1$ are $\mathbb{Z}_+$-equivalent if $\alpha x \geq 1$ is valid for $\{x : x \geq 0, \alpha' x \geq 1\}$ *and* $\alpha' x \geq 1$ is valid for $\{x : x \geq 0, \alpha x \geq 1\}$.

Proposition 6. *Every constraint $\alpha x \geq 1$ with at most k nonzero coefficients is $\mathbb{Z}_+$-equivalent to a k-roundable constraint.*

Before proving Proposition 6, let us illustrate its use.

Theorem 3. *There is a polynomial time k-approximation algorithm for unbounded k-RS CIPs.*

Proof. Using Proposition 6 we replace each constraint with a $\mathbb{Z}_+$-equivalent k-roundable one. The resulting IP has the same set of feasible solutions and the same objective function. Therefore, Proposition 3 yields a k-approximately optimal solution. □

With the framework set up, we begin the technical part: a lemma, then the proof of Proposition 6.

Lemma 7. *For any positive integers k and v, the constraint $\sum_{i=1}^{k-1} x_i + \frac{1}{v} x_k \geq 1$ is k-roundable.*

Proof. Let $\alpha x \geq 1$ denote the constraint. If x satisfies the constraint, then the maximum of $x_1, x_2, \ldots, x_{k-1}$ and $\frac{1}{v} x_k$ must be at least $1/k$. If $x_i \geq 1/k$ for some $i \neq k$ then $\lfloor k x_i \rfloor \geq 1$ and so $\alpha \lfloor kx \rfloor \geq 1$ as needed. Otherwise x_k must be at least v/k and so $\lfloor k x_k \rfloor \geq v$ which implies $\alpha \lfloor kx \rfloor \geq 1$ as needed. □

Proof of Proposition 6. If the sum of coefficients in the constraint is $k-1$ or less, we are done by Proposition 4, hence we assume the sum is at greater than $k-1$. Without loss of generality (by renaming) such a constraint is of the form

$$\sum_{i=1}^{k} x_i \alpha_i \geq 1 \tag{1}$$

where $\mathbf{0} < \alpha \leq \mathbf{1}$, $k - 1 < \sum_i \alpha_i \leq k$, and the α_i's are nonincreasing in i.

Define the *support* of x to be $\text{supp}(x) := \{i \mid x_i > 0\}$. Now $\alpha_{k-1} + \alpha_k > 1$ since $k - 1 < \sum_{i<k-1} \alpha_i + \alpha_{k-1} + \alpha_k \leq \alpha_{k-1} + \alpha_k + (k-2)$. Since the α_i are nonincreasing, $\alpha_i + \alpha_j > 1$ for any $i < k, j \leq k$; more generally, any integral $x \geq 0$ with $|\text{supp}(x)| \geq 2$ must satisfy $\alpha x \geq 1$. To express the set of *all* feasible integral solutions, let $t = \max\{0\} \cup \{i \mid \alpha_i = 1\}$, let e_i denote the ith unit basis vector, and let $v = \lceil 1/\alpha_k \rceil$. Then it is not hard to see that the nonnegative integral solution set to constraint (1) is the disjoint union

$$\begin{aligned} &\{x \mid x \geq 0, |\text{supp}(x)| \geq 2\} \uplus \{z e_i \mid 1 \leq i \leq t, z \geq 1\} \\ \uplus &\{z e_i \mid t < i < k, z \geq 2\} \uplus \{z e_k \mid z \geq v\}. \end{aligned} \tag{2}$$

The special case $t = k$ (i.e. $\alpha_1 = \alpha_2 = \cdots = \alpha_k = 1$) is already k-roundable by Lemma 7, so assume $t < k$. Consider the constraint

$$\sum_{i=1}^{t} x_i + \sum_{i=t+1}^{k-1} \frac{v-1}{v} x_i + \frac{1}{v} x_k \geq 1. \tag{3}$$

Every integral $x \geq 0$ with $|\text{supp}(x)| \geq 2$ satisfies constraint (3). By also considering the cases $|\text{supp}(x)| \in \{0, 1\}$, it is easy to check that constraint (3) has precisely Equation (2) as its set of feasible solutions, i.e. constraint (3) is $\mathbb{Z}_+$-equivalent to $\alpha x \geq 1$. If $t < k - 1$, the sum of the coefficients of constraint (3) is $k - 1$ or less, so it is k-roundable by Proposition 4. If $t = k - 1$, constraint (3) is k-roundable by Lemma 7. Thus in either case we have what we wanted. □

2.1 Multiplicity Constraints

We next obtain approximation guarantee k even with multiplicity constraints $x \leq d$. For this we use *knapsack-cover inequalities*. These inequalities represent residual covering problems when a set of variables is taken at maximum multiplicity. Wolsey [28] studied inequalities like this for 0-1 problems to get a primal-dual approximation algorithm for submodular set cover. The LP we use is most like what appears in Carr et al. [9] and Kolliopoulos & Young [21], but we first replace each row with a k-roundable one.

Specifically, given a CIP $\{\min cx \mid Ax \geq \mathbf{1}, \mathbf{0} \leq x \leq d\}$ with A, d nonnegative, we now define the knapsack cover LP. Note that we allow d to contain some entries equal to $+\infty$. For a subset F of $\text{supp}(A_i)$ such that $\sum_{j \in F} A_{ij} d_j < 1$, define $A_{ij}^{(F)} = \min\{A_{ij}, 1 - \sum_{j \in F} A_{ij} d_j\}$. Following [9,21] we define the *knapsack cover LP* for our problem to be

$$\text{KC-LP} = \Big\{\min cx : \mathbf{0} \leq x \leq d;$$
$$\forall i, \forall F \subset \text{supp}(A_i) \text{ s.t. } \sum_{j \in F} A_{ij} d_j < 1 : \sum_{j \notin F} A_{ij}^{(F)} x_j \geq 1 - \sum_{j \in F} A_{ij} d_j \Big\}.$$

Theorem 1. *There is a polynomial time k-approximation algorithm for k-RS CIPs.*

Proof. Using Proposition 6, we assume all rows of A are k-roundable. Let x^* be the optimal solution to KC-LP. Define $\widehat{x} = \min\{d, \lfloor kx^* \rfloor\}$, where min denotes the component-wise minimum. We claim that $\widehat{x}$ is a feasible solution to the CIP, which will complete the proof. In other words, we want to show for each row i that $A_i \widehat{x} \geq 1$.

Fix any row i and define $F = \{j \in \text{supp}(A_i) \mid x_j^* \geq d_j/k\}$, i.e. F is those variables in the constraint that were rounded to their maximum multiplicity. If $F = \varnothing$ then, by the k-roundability of $A_i x \geq 1$, we have that $A_i \widehat{x} = A_i \lfloor kx^* \rfloor \geq 1$ as needed. So assume $F \neq \varnothing$.

If $\sum_{j \in F} A_{ij} d_j \geq 1$ then the constraint $A_i \widehat{x} \geq 1$ is satisfied; consider otherwise. Since $\lfloor kx_j^* \rfloor > kx_j^* - 1$ for $j \notin F$, since x^* satisfies the knapsack cover constraint for i and F, and since $A_{ij}^{(F)} \leq 1 - \sum_{j \in F} A_{ij} d_j$ for each j, we have

$$\sum_{j \notin F} A_{ij}^{(F)} \widehat{x}_j \geq k \sum_{j \notin F} A_{ij}^{(F)} x_j^* - \sum_{j \notin F} A_{ij}^{(F)}$$
$$\geq k\Big(1 - \sum_{j \in F} A_{ij} d_j\Big) - \Big|\{j : j \in \text{supp}(A_i) \backslash F\}\Big| \Big(1 - \sum_{j \in F} A_{ij} d_j\Big).$$

Since $F \neq \varnothing$ and $|\text{supp}(A_i)| \leq k$, this gives $\sum_{j \notin F} A_{ij}^{(F)} \widehat{x}_j \geq 1 - \sum_{j \in F} A_{ij} d_j$. Rearranging, and using the facts $(\forall j : A_{ij} \geq A_{ij}^{(F)})$ and $(\forall j \in F : d_j = \widehat{x}_j)$, we deduce $A_i \widehat{x} \geq 1$, as needed.

For fixed k, we may solve KC-LP explicitly, since it has polynomially many constraints. For general k, we follow the ellipsoid algorithm-based approach of

[9,21]: rather than solve KC-LP in polynomial time, we obtain a solution x^* which is optimal for a modified KC-LP having not all knapsack-cover constraints, but at least all those for the the m specific (i, F) pairs (depending on x^*) used in our proof; thus we still get a k-approximation in polynomial time. □

3 Column-Sparse Packing Integer Programs

In this section we give an approximation algorithm for k-column-sparse packing integer programs with approximation ratio $2^k(k^2-2k+1)+1$, and better results for $k=2$. The results hold even in the presence of multiplicity constraints $x \leq d$. Broadly speaking, our approach is rooted in the demand matching algorithm of Shepherd & Vetta [13]; their path-augmenting algorithm can be viewed as a restricted form of *iterated relaxation*, which is the main tool in our new approach. Iterated relaxation yields a superoptimal solution that violates some constraints, and with probabilistic rounding and greedy ideas we are able to obtain a feasible solution while retaining at least a constant fraction of the weight.

By scaling rows and eliminating variables whose coefficients are too high, there is no loss of generality in the following definition.

Definition 8. *A* k-CS PIP *is an integer program* $\{\max cx : Ax \leq \mathbf{1}, \mathbf{0} \leq x \leq d\}$ *where* A *is* k*-CS and all entries of* A *are at most 1.*

We begin this section by explaining a simpler version of our new mechanism; this simpler version gives a $2^k(k^2-k+1)$-approximation algorithm for k-CS PIP in the special case $d = \mathbf{1}$.

By analogy with the demand matching problem and hypergraphic matching, it is natural to think of the rows of the constraint matrix A as indexed by a *vertex set* V and the columns as indexed by a *hyperedge set* E. Specifically, define a vertex for each row, let A_{ve} denote the entry of A at row v and column e, and for each column define its corresponding hyperedge e to be $\{v \mid A_{ve} > 0\}$; the resulting hypergraph may not be simple. We define the term *endpoint* to mean a pair (v, e) such that $A_{ve} > 0$.

The following intermediate result of iterated rounding is key for our approach. For a k-CS PIP $\mathcal{P}$ let $\mathcal{L}(\mathcal{P})$ denote its linear relaxation $\{\max cx \mid Ax \leq \mathbf{1}, \mathbf{0} \leq x \leq d\}$. Our iterated rounding algorithm computes a set S of *special* endpoints; for such a set we let $A_{S\to 0}$ denote the matrix obtained from A by zeroing out the entries corresponding to each special endpoint.

Lemma 9. *Given a* k*-CS PIP* $\mathcal{P}$ *with* $d = \mathbf{1}$*, we can in polynomial time find* $y \in \{0,1\}^E$ *and* S *such that*

(a) $cy \geq \mathrm{OPT}(\mathcal{L}(\mathcal{P}))$
(b) $\forall v \in V$*, we have* $|\{e : (v,e) \in S\}| \leq k$
(c) $A_{S\to 0}\, y \leq \mathbf{1}$.

Proof of Lemma 9. First, we give a sketch. Since $\mathcal{P}$ is k-column sparse, every hyperedge has size at most k. Let x^* be an extreme optimal solution to $\mathcal{L}(\mathcal{P})$.

The crux of our approach deals with the case that x^* has no integral values: then x^* is a *basic feasible solution* all of whose tight constraints correspond to vertices, so the number of vertices is greater than or equal to the number of hyperedges. Thus by double-counting the average vertex degree is at most k, so some vertex has degree at most k. In other words there is some constraint which contains at most k nonzero variables, which allows iterated rounding to take place.

Since y is a 0-1 vector we can alternatively view it as a subset Y of E. With this convention, we now give pseudocode for our iterated rounding algorithm, ITERATEDSOLVER.

ITERATEDSOLVER(A, c)
1: Set $S = Y = N = \varnothing, V' = V, E' = E$
2: **loop**
3: Let x^* be an extreme optimum of

$$\{\max cx \mid x \in [0,1]^E; A_{S\to 0}x \leq \mathbf{1}; \forall e \in Y : x_e = 1; \forall e \in N : x_e = 0\}$$

4: For each $e \in E'$ with $x^*_e = 0$, add e to N, delete e from E'
5: For each $e \in E'$ with $x^*_e = 1$, add e to Y, delete e from E'
6: If $E' = \varnothing$, terminate and return S and y, the characteristic vector of Y
7: **for** each vertex $v \in V'$ with degree less than or equal to k in (V', E') **do**
8: Mark each endpoint $\{(v, e) \mid e \in E'\}$ special, delete v from V'

Now we explain the pseudocode. The sets Y, N are disjoint subsets of E, and $E' = E \backslash Y \backslash N$. When e leaves E', the value of x_e is fixed at 0 or 1. After deleting a vertex v from V', it will not be possible to later violate the constraint corresponding to v. Hence the linear program effectively only has variables for E' and constraints for V'. As remarked previously, since x^* is a basic feasible solution the average vertex degree is at most k each time Step 7 is reached, so $|V'|$ decreases in each iteration, and the algorithm has polynomial running time. (In fact, it is not hard to show that there are at most $O(k \log |V|)$ iterations.)

The algorithm has the property that cx^* does not decrease from one iteration to the next; since $x^* = y$ at termination, property (a) holds. Properties (b) and (c) can be seen immediately from the definition of the algorithm. □

Next, we show the kind of rounding which takes the output of ITERATEDSOLVER to a feasible solution.

Theorem 4. *There is a polynomial time $2^k(k^2 - k + 1)$-approximation algorithm for k-CS PIPs with $d = \mathbf{1}$.*

Proof. After running ITERATEDSOLVER, suppose we find a subset Z of Y with the property that if any $e, f \in Z$ intersect at a vertex v, neither (v, e) nor (v, f) is special. Then from Lemma 9(c) and the fact that entries of A are at most 1, it follows that Z is a feasible solution to $\mathcal{P}$ (the original k-CS PIP). In the rest of the proof, we show there exists such a set with at least a constant fraction of Y's profit.

To accomplish this we have each vertex $v \in V$ independently declare "special" or "non-special," each with probability 1/2. We say that the eth column is *accepted* if (1) for every endpoint $(v, e) \in S$ the vertex v declares special and (2) for every endpoint $(v, e) \notin S$ the vertex v declares non-special. It follows from the k-column-sparseness of A that each column is accepted with probability at least $1/2^k$.

Let $Y^a \subset Y$ denote the set of accepted columns, and $V^s \subset V$ denote the set of vertices that declared special. So $E[c(Y^a)] \geq c(Y)/2^k$. We need the following claim, whose easy proof is in the full version [12].

Claim 10. *If $Z \subset Y^a$ has the property that every two hyperedges e_1, e_2 in Z satisfy $e_1 \cap e_2 \cap V_s = \varnothing$, then Z is a feasible solution to $\mathcal{P}$.*

Another way of stating Claim 10 is that whenever Z is a matching on the induced subhypergraph $(V, Y^a)[V^s]$, Z is a feasible solution to $\mathcal{P}$. (Note, we do not discard "empty" hyperedges in this view — hyperedges disjoint from V_s can be freely added to any matching.) Consider the greedy algorithm for finding such a matching: we iteratively select the maximum-weight hyperedge that does not intersect any previously selected hyperedges on V^s. Since the subhypergraph has hyperedges of size at most k and degree at most k, it is easy to see that each chosen hyperedge precludes at most $k(k-1)$ other hyperedges for future selection. Thus the greedy algorithm outputs a set Z with cost at least $c(Y^a)/(k(k-1)+1)$, which is at least $c(Y)/2^k(k^2-k+1)$ in expectation. Using the fact that $c(Y) \geq \text{OPT}(\mathcal{L}(\mathcal{P})) \geq \text{OPT}(\mathcal{P})$, we are done. □

3.1 Strongest Results

Our strongest results are obtained by using a somewhat more refined iterated rounding algorithm. The details appear are deferred to the full version [12] due to lack of space; we simply state the results here.

Theorem 2. *There is a polynomial time $2^k(k^2 - 2k + 1) + 1$-approximation algorithm for k-CS PIPs.*

Theorem 5. *There is a deterministic polynomial time 4-approximation algorithm for 2-CS PIPs, and a randomized $6-\sqrt{5} \approx 3.764$-approximation algorithm when $d = \mathbf{1}$ and no two columns have the same support.*

The *width* W of an integer program is $1/\left(\max_{ij} A_{ij}/b_i\right)$. Note that without loss of generality, $W \geq 1$. If we normalize $b = \mathbf{1}$ by row scaling as in the rest of this paper, then a program has width $\geq W$ iff every entry of A is at most $1/W$.

In many settings better approximation can be obtained as W increases. For example in k-RS CIPs with $b = \mathbf{1}$, the sum of the entries in each row is at most k/W, so Propositions 3 and 4 give a $(1 + k/W)$-approximation algorithm. Srinivasan [19,20] gave a $(1+\ln(1+k)/W)$-approximation algorithm for unbounded k-CS CIPs. Using *grouping and scaling* techniques introduced by Kolliopoulos and Stein [29], Chekuri et al. [14] showed that no-bottleneck demand multicommodity flow in a tree admits a $(1 + O(1/\sqrt{W}))$-approximation algorithm, and gave general sufficient conditions for a problem to admit a $(1+O(1/\sqrt{W}))$-approximation algorithm. Along the same vein, using iterated rounding, Könemann et al. [30] obtained a $(1 + O(1/W))$-approximation algorithm for ordinary multicommodity flow in a tree, and general sufficient conditions for a problem to admit a $(1 + O(1/W))$-approximation algorithm [30]. Using a new technique (iteratively using an LP to reduce an additively-violating solution to a feasible solution), we get the following — again see [12] for details.

Theorem 6. *There is a polynomial time* $(1 + k/W)/(1 - k/W)$*-approximation algorithm to solve* k*-column-sparse PIPs with* $k/W < 1$.

Acknowledgement. We would like to thank Glencora Borradaile, Christina Boucher, Deeparnab Chakrabarty, Stephane Durocher, Jochen Könemann and Christos Koufogiannakis for helpful discussions, and the ESA referees for useful feedback.

References

1. Lenstra, H.: Integer programming with a fixed number of variables. Math. Oper. Res. 8, 538–548 (1983)
2. Dinur, I., Guruswami, V., Khot, S., Regev, O.: A new multilayered PCP and the hardness of hypergraph vertex cover. SIAM J. Comput. 34(5), 1129–1146 (2005); Preliminary version appeared in Proc. 35th STOC, pp. 595–601 (2003)
3. Khot, S., Regev, O.: Vertex cover might be hard to approximate to within $2-\epsilon$. J. Comput. Syst. Sci. 74(3), 335–349 (2008); Preliminary version appeared in Proc. 18th CCC, pp. 379–386 (2003)
4. Hochbaum, D.S.: Approximation algorithms for set covering and vertex cover problems. SIAM J. Comput. 11, 555–556 (1982)
5. Bar-Yehuda, R., Even, S.: A linear time approximation algorithm for the weighted vertex cover problem. J. Algorithms 2, 198–203 (1981)
6. Hall, N.G., Hochbaum, D.S.: A fast approximation algorithm for the multicovering problem. Discrete Appl. Math. 15(1), 35–40 (1986)
7. Hochbaum, D.S., Megiddo, N., Naor, J.S., Tamir, A.: Tight bounds and 2-approximation algorithms for integer programs with two variables per inequality. Math. Program. 62(1), 69–83 (1993)
8. Bar-Yehuda, R., Rawitz, D.: Efficient algorithms for integer programs with two variables per constraint. Algorithmica 29(4), 595–609 (2001)
9. Carr, R.D., Fleischer, L., Leung, V.J., Phillips, C.A.: Strengthening integrality gaps for capacitated network design and covering problems. In: Proc. 11th SODA, pp. 106–115 (2000)
10. Fujito, T., Yabuta, T.: Submodular integer cover and its application to production planning. In: Persiano, G., Solis-Oba, R. (eds.) WAOA 2004. LNCS, vol. 3351, pp. 154–166. Springer, Heidelberg (2005)
11. Koufogiannakis, C., Young, N.E.: Greedy degree-approximation algorithm for covering with arbitrary constraints and submodular cost. In: Albers, S., Marchetti-Spaccamela, A., Matias, Y., Nikoletseas, S., Thomas, W. (eds.) ICALP 2009. Part I. LNCS, vol. 5555, pp. 634–652. Springer, Heidelberg (2009) arXiv:0807.0644
12. Pritchard, D.: Approximability of sparse integer programs (2009) arXiv:0904.0859

13. Shepherd, F.B., Vetta, A.: The demand-matching problem. Mathematics of Operations Research 32(3), 563–578 (2007); Preliminary version appeared in Proc. 9th IPCO, pp. 457–474, (2002)
14. Chekuri, C., Mydlarz, M., Shepherd, F.B.: Multicommodity demand flow in a tree and packing integer programs. ACM Trans. Algorithms 3(3), 27 (2007); Preliminary version appeared in Proc. 30th ICALP, pp. 410–425 (2003)
15. Berman, P.: A d/2 approximation for maximum weight independent set in d-claw free graphs. Nordic J. of Computing 7(3), 178–184 (2000); Preliminary version appeared in Proc. 7th SWAT, pp. 214–219 (2000)
16. Hurkens, C.A.J., Schrijver, A.: On the size of systems of sets every t of which have an SDR, with an application to the worst-case ratio of heuristics for packing problems. SIAM J. Discret. Math. 2(1), 68–72 (1989)
17. Hazan, E., Safra, S., Schwartz, O.: On the complexity of approximating k-set packing. Comput. Complex. 15(1), 20–39 (2006); Preliminary versions appeared in Proc. 6th APPROX, pp. 83–97 (2003); ECCC-TR03-020 (2003)
18. Singh, M.: Iterative Methods in Combinatorial Optimization. PhD thesis, Carnegie Mellon University (2008)
19. Srinivasan, A.: Improved approximation guarantees for packing and covering integer programs. SIAM J. Comput. 29(2), 648–670 (1999); Preliminary version appeared in Proc. 27th STOC, pp. 268–276 (1995)
20. Srinivasan, A.: An extension of the Lovász Local Lemma, and its applications to integer programming. SIAM J. Comput. 36(3), 609–634 (2006); Preliminary version appeared in Proc. 7th SODA, pp. 6–15 (1996)
21. Kolliopoulos, S.G., Young, N.E.: Approximation algorithms for covering/packing integer programs. J. Comput. Syst. Sci. 71(4), 495–505 (2005)
22. Trevisan, L.: Non-approximability results for optimization problems on bounded degree instances. In: Proc. 33rd STOC, pp. 453–461 (2001)
23. Hochbaum, D.S.: Monotonizing linear programs with up to two nonzeroes per column. Oper. Res. Lett. 32(1), 49–58 (2004)
24. Chakrabarty, D., Goel, G.: On the approximability of budgeted allocations and improved lower bounds for submodular welfare maximization and GAP. In: Proc. 49th FOCS, pp. 687–696 (2008)
25. Khot, S., Ponnuswami, A.K.: Better inapproximability results for maxClique, chromatic number and min-3Lin-deletion. In: Bugliesi, M., Preneel, B., Sassone, V., Wegener, I. (eds.) ICALP 2006. LNCS, vol. 4051, pp. 226–237. Springer, Heidelberg (2006)
26. Kellerer, H., Pferschy, U., Pisinger, D.: Knapsack Problems. Springer, Heidelberg (2004)
27. Magazine, M.J., Chern, M.S.: A note on approximation schemes for multidimensional knapsack problems. Math. of Oper. Research 9(2), 244–247 (1984)
28. Wolsey, L.: An analysis of the greedy algorithm for the submodular set covering problem. Combinatorica 2(4), 385–393 (1982)
29. Kolliopoulos, S.G., Stein, C.: Improved approximation algorithms for unsplittable flow problems. In: Proc. 38th FOCS, pp. 426–436 (1997)
30. Könemann, J., Parekh, O., Pritchard, D.: Max-weight integral multicommodity flow in spiders and high-capacity trees. In: Bampis, E., Skutella, M. (eds.) WAOA 2008. LNCS, vol. 5426, pp. 1–14. Springer, Heidelberg (2009)

Iterative Rounding for Multi-Objective Optimization Problems

Fabrizio Grandoni[1], R. Ravi[2], and Mohit Singh[3]

[1] Department of Computer Science, Systems and Production, University of Rome Tor Vergata
grandoni@disp.uniroma2.it

[2] Tepper School of Business, Carnegie Mellon University, Pittsburgh, USA
Supported in part by NSF grant CCF-0728841
ravi@cmu.edu

[3] Microsoft Research, New England, Cambridge, USA
mohsingh@microsoft.com

Abstract. In this paper we show that *iterative rounding* is a powerful and flexible tool in the design of approximation algorithms for multi-objective optimization problems. We illustrate that by considering the multi-objective versions of three basic optimization problems: spanning tree, matroid basis and matching in bipartite graphs. Here, besides the standard weight function, we are given k length functions with corresponding budgets. The goal is finding a feasible solution of maximum weight and such that, for all i, the ith length of the solution does not exceed the ith budget. For these problems we present polynomial-time approximation schemes that, for any constant $\epsilon > 0$ and $k \geq 1$, compute a solution violating each budget constraint at most by a factor $(1+\epsilon)$. The weight of the solution is optimal for the first two problems, and $(1-\epsilon)$-approximate for the last one.

1 Introduction

Most real-life optimization problems involve finding a feasible solution trading off many mutually conflicting goals. This is a rich area of study in Operations Research, Economics and Computer Science in the broad area of *Multi-objective Optimization* [10,14,26]. A variety of approaches have been employed to formulate such problems including Goal Programming [4], Pareto-Optimality [9], and Multi-objective Approximation Algorithms [26]. We adopt the latter approach and cast one of the goals as the objective function, and the others as *budget constraints*. More precisely, we are given a (finite) set $\mathcal{F}$ of feasible solutions for the problem; we are also given a weight function $\mathrm{w} : \mathcal{F} \rightarrow \mathbb{R}_+$ and a set of k length functions $\ell^i : \mathcal{F} \rightarrow \mathbb{R}_+$, $1 \leq i \leq k$, that assign a weight $\mathrm{w}(S)$ and k lengths $\ell^i(S)$, $1 \leq i \leq k$, to every feasible solution $S \in \mathcal{F}$. For each length function ℓ^i, we are also given a non-negative budget $\mathrm{L}_i \in \mathbb{R}_+$. The *multi-objective optimization problem* can then be formulated as follows[1].

[1] With a slight notation abuse, we will use OPT also to denote the actual optimal solution (besides its weight).

A. Fiat and P. Sanders (Eds.): ESA 2009, LNCS 5757, pp. 95–106, 2009.

$$\text{OPT} := \text{maximize } \text{w}(S) \;\; \text{subject to} \;\; S \in \mathcal{F},\ \ell^i(S) \leq \text{L}_i,\ 1 \leq i \leq k. \tag{1}$$

In this paper we study the multi-objective version of three fundamental maximization problems, namely spanning tree, matroid basis, and matching in bipartite graphs.

1. In the MULTI-OBJECTIVE SPANNING TREE problem, we are given an n-node undirected graph $G = (V, E)$ with edge weights $\text{w} : E \to \mathbb{R}_+$, k edge lengths $\ell^i : E \to \mathbb{R}_+$, $1 \leq i \leq k$, and positive budgets $\text{L}_1, \ldots, \text{L}_k$. The set of all feasible solutions $\mathcal{F}$ is given by the spanning trees of G. Define the weight of $T \in \mathcal{F}$ as $\text{w}(T) := \sum_{e \in T} \text{w}(e)$, and its ith-length as $\ell^i(T) := \sum_{e \in T} \ell^i(e)$. The goal is finding $T \in \mathcal{F}$ of maximum weight $\text{w}(T)$ such that $\ell^i(T) \leq \text{L}_i$ for each $1 \leq i \leq k$.

2. The MULTI-OBJECTIVE BIPARTITE MATCHING problem is defined analogously. Here the goal is finding a matching M in a bipartite graph of maximum-weight $\text{w}(M)$ such that $\ell^i(M) \leq \text{L}_i$ for all $1 \leq i \leq k$.

3. In the MULTI-OBJECTIVE MATROID BASIS problem, we are given a matroid $\mathcal{M} = (E, \mathcal{E})$, $\mathcal{E} \subseteq 2^E$, on the ground set E, $m = |E|$ (for basic definitions and results on matroids, see e.g. [27]). Moreover, we are given element weights $\text{w} : E \to \mathbb{R}_+$, element lengths $\ell^i : E \to \mathbb{R}_+$ and budgets $\text{L}_i \in \mathbb{R}_+$, $1 \leq i \leq k$. The set of all feasible solutions $\mathcal{F}$ is given by then bases of $\mathcal{M}$. The weight of a basis $B \in \mathcal{E}$ is defined as $\text{w}(B) := \sum_{e \in B} \text{w}(e)$, while its ith-length is $\ell^i(B) := \sum_{e \in B} \ell^i(e)$. The goal is computing a basis $X \in \mathcal{F}$ of maximum weight satisfying all the budget constraints. This naturally generalizes the multi-objective spanning tree problem which results when we consider a graphic matroid.

All three problems are polynomial-time solvable in their unbudgeted version ($k = 0$), but become NP-hard [1,6] even for a single budget constraint ($k = 1$).

Our Results. We give a PTAS for multi-objective spanning trees and generalize it to multi-objective matroid basis and also give a PTAS for multi-objective matchings in bipartite graphs. Our results however require that the number of budget constraints k is fixed.

Theorem 1. *For any $\epsilon > 0$, there exists an algorithm for* MULTI-OBJECTIVE SPANNING TREE *with $k \geq 1$ budget constraints which returns a spanning tree T of optimal weight and $\ell^i(M) \leq (1 + \epsilon)\text{L}_i$ for each $1 \leq i \leq k$. The running time of the algorithm is $O(n^{O(k^2/\epsilon)})$.*

Theorem 1, proved in Section 2, generalizes the result of Ravi and Goemans [25] who gave the same guarantees for the special case of a single budget constraint($k = 1$), and improves on the (much more involved) algorithm of Papadimitriou and Yannakakis [22] which returns a suboptimal ($(1 - \epsilon)$-approximate) solution with a similar (i.e., $1 + \epsilon$) violation of the budget constraints. The latter result of [22] also holds for our case of many different but fixed number of objectives, and even in this case, we improve on the approximation factor in the main objective (with the same violation in the budgets).

Theorem 2. *For any $\epsilon > 0$, there exists an algorithm for* MULTI-OBJECTIVE MATROID BASIS *with $k \geq 1$ budget constraints which returns a basis B of optimal weight and $\ell^i(B) \leq (1+\epsilon)\mathrm{L}_i$ for each $1 \leq i \leq k$. The running time of the algorithm is $O(m^{O(k^2/\epsilon)})$.*

Theorem 2 is discussed in Section 2.1, and generalizes a similar result for the $k = 1$ case as above in [25].

Theorem 3. . *For any $\epsilon > 0$, there exists a deterministic algorithm for* MULTI-OBJECTIVE BIPARTITE MATCHING *with $k \geq 1$ budget constraints which returns a matching M of weight $\mathrm{w}(M) \geq (1-\epsilon)\mathrm{OPT}$ and length $\ell^i(M) \leq (1+\epsilon)\mathrm{L}_i$ for each $1 \leq i \leq k$. The running time of the algorithm is $O(n^{O(k^2\sqrt{k \log k}/\epsilon^2)})$.*

Theorem 3 is proved in Section 3. A similar approximation guarantee was known earlier via the work of [22]. However, their result implies a fully polynomial RNC scheme rather than a PTAS, and thus Theorem 3 provides the first deterministic approximation scheme for MULTI-OBJECTIVE BIPARTITE MATCHING. A PTAS, based on a completely different approach, was known earlier only for the case of one budget constraint, i.e. $k = 1$ [6].

Our Techniques. Perhaps even more importantly than our specific results, our main contribution is to demonstrate that the general framework of *iterative techniques* can be used to obtain approximation algorithms for various multi-objective optimization problems. This technique was introduced by Jain [15] for approximating survivable network design problems. The basic idea in iterative *rounding* for covering problems is as follows: Consider the optimal (fractional) vertex (or extreme point or basic feasible) solution to a linear programming relaxation to the problem, and show that there is a variable with high fractional value (e.g. at least 0.5) which can be rounded up to an integer without losing too much (e.g. 2) in the approximation. The method includes this rounded variable in the integral solution and iterates. Since the basis iterative rounding loses a constant factor in approximation, we refine the method by replacing the rounding step by the following: relax (remove) a constraint that can be ignored without losing too much in the feasibility and iterate on the residual problem. The resulting iterative *relaxation* method has been very successful for approximating degree-constrained network design problems [16,17,29].

We now outline how the iterative technique is applied to our problems. The algorithm for MULTI-OBJECTIVE SPANNING TREE is rather simple; a vertex solution for the natural LP relaxation of the problem is already sparse: it has about k edges more than a spanning tree in its support due to the well-known laminarity of an independent set of tight spanning tree constraints [27]. We remove all edges corresponding to variables of value zero, relax (remove) all the budget constraints, and solve optimally the residual problem (which is a standard spanning tree problem). A preliminary guessing phase ensures that the k edges not used in the tree do not add much to the approximation bound for any of the budgets. This approach also gives a very simple proof of the earlier result for the case $k = 1$ [25]. An identical approach works also for the more general MULTI-OBJECTIVE MATROID BASIS problem.

Our algorithm for MULTI-OBJECTIVE BIPARTITE MATCHING is more involved: after an initial preprocessing phase, where the algorithm removes all edges with large weight and large length, there is a decomposition phase. In that phase, we run an iterative relaxation algorithm which uses the optimal solution of the natural LP formulation to obtain a modified LP solution. The iterative algorithm ensures that the support of the modified solution is a collection of $h \leq k$ vertex disjoint paths. Moreover, each of these paths has small weight and length. In the final combination phase, we combine the solutions on these paths to return one feasible matching. Each path can be decomposed in two matchings. The algorithm picks one matching from each of the paths. While the algorithm is a brute force enumeration over all choices (which are $2^h \leq 2^k$ in number), a probabilistic argument is used to show that there exists a choice of a matching from each path which provides a solution with the desired guarantee.

Related Work. Multi-objective optimization has been studied extensively in Operations Research, Microeconomics and Computer Science. We refer the reader to more general sources [4,9,10,14], and restrict our attention to work which closely relates to our problems. There are many examples of single-budget versions of polynomial-time solvable optimization problems addressed in the literature. In the *constrained shortest path problem* the goal is finding a minimum-weight path in a directed graph between two nodes s and t such that the length of the path does not exceed a budget L [5]. In the *constrained minimum arborescence problem* we are given a directed graphs with edge weights and lengths. The aim is computing an *arborescence* of minimum weight whose length is below the input budget [13]. Previous work on budgeted optimization problems also includes results on budgeted scheduling [18,28] and bicriteria results for several budgeted network design problems [19].

Jain [15] introduced the iterative rounding framework and applied it to approximating general network design problems. Subsequently, it was applied to various other network design problems [8,12,20]. The iterative relaxation technique has recently been successfully applied to degree constrained network design problems [2,16,17,29].

There are few general tools for designing approximation algorithms for budgeted problems. One is the Lagrangian relaxation method. The basic idea is relaxing the budget constraint, and lifting it into the objective function weighting it by a Lagrangian multiplier. Solving the relaxed problem, one obtains two or more optimal solutions, which are then patched together to get a good solution for the original problem. Demonstrating this method, Goemans and Ravi [25] gave the first PTAS for MULTI-OBJECTIVE SPANNING TREE with a single budget constraint. Using the same approach, but a more involving patching step, Berger, Bonifaci, Grandoni, and Schäfer [6] obtained a PTAS for the single-budget version of the matching problem. This approach does not seem to generalize to the case of multiple budget constraints.

A second general tool, due to Papadimitrou and Yannakakis [22], is based on the construction of succinct approximation of Pareto curves. In order to efficiently construct such ϵ-approximate Pareto curves, a sufficient condition is the existence of a pseudo-polynomial-time algorithm for the *exact* version of the problem considered. The task in the exact version of the problem is to return a feasible solution of *exactly* some pre-specified value. The existence of such pseudo-polynomial-time algorithm for the spanning tree problem [3] implies a polynomial-time algorithm which returns a $(1-\epsilon)$-approximate solution violating all the budget constraints by a factor of $(1+\epsilon)$ for the corresponding multi-objective version. Unfortunately, it is not known whether such an algorithm exists for matchings in bipartite graphs, while the famous randomized algorithm of Mulmuley, Vazirani and Vazirani [21] can be used to obtain a polynomial-time randomized approximation scheme for MULTI-OBJECTIVE BIPARTITE MATCHING[2]. Their method, however, only approximates the objective while our algorithm matches the value of the objective function with the optimal for two out of the three problems addressed here, while for the third we obtain a deterministic rather than an RNC algorithm.

A third approach is based on parametric search and is advocated in [19]; their results imply that a ρ-approximation algorithm for the single objective problem gives a $(k \cdot \rho)$-approximation for *each* of the budget violations as well as for the objective in the corresponding k-objective problem. This only gives a much weaker k-approximation for each objective for the problems considered here.

Other general tools for multi-objective problems such as Matching-Based Augmentation [24] advocates building the solution iteratively using one (path) matching at a time controlling the various objectives, and Randomized Rounding of fraction LP solutions while bounding all objectives simultaneously [7,23]. While these techniques are useful in handling more than one type of objective, their performance ratios tend to be in the higher logarithmic range.

In the context of these methods, our paper shows that iterative rounding is a powerful and flexible tool for approximating multi-objective optimization problems giving even better results than all of the above methods. This was already the case for degree-constrained spanning trees and survivable network design problems [16,29] and directed network design problems [2], and our results extend these to some more multi-objective problems.

2 Multi-Objective Spanning Tree and Matroid Basis

We formulate the following linear programming relaxation for MULTI-OBJECTIVE SPANNING TREE which is a standard extension of the linear program for the maximum spanning tree problem. There is a variable x_e for each edge $e \in E$. For a subset $F \subseteq E$ of edges, we denote $x(F) = \sum_{e \in F} x_e$ and for a subset $S \subseteq V$, we denote $E(S) = \{e : |e \cap S| = 2\}$ to be the set of edges with both endpoints in S.

[2] The same algorithm works for general graphs also.

$$
\begin{aligned}
\text{(LP-ST)} \qquad & \text{maximize} && \sum_{e \in E} \mathrm{w}(e)\, x_e && \\
& \text{subject to} && x(E(V)) = |V| - 1, && \\
& && x(E(S)) \le |S| - 1, && \forall S \subset V \\
& && \sum_{e \in E} \ell^i(e) x_e \le \mathrm{L}_i, && \forall 1 \le i \le k \\
& && x_e \ge 0, && \forall e \in E.
\end{aligned}
$$

The following characterization of any vertex solution of (LP-ST) follows directly from the uncrossing technique (see [27]).

Lemma 1. *Let x be a vertex solution of the linear program (LP-ST) such that $x_e > 0$ for each edge e and let $\mathcal{T} = \{S \subseteq V : x(E(S)) = |S| - 1\}$ be the set of all tight subset constraints. Then there exists a laminar family $\mathcal{L} \subseteq \mathcal{T}$ and a subset $J \subseteq \{1 \le i \le k : \sum_{e \in E} \ell^i(e) x_e = \mathrm{L}_i\}$ of tight length constraints such that*

1. *The vectors $\{\chi(E(S)) : S \in \mathcal{L}\}$ are linearly independent.*
2. *$span(\mathcal{L})$=$span(\mathcal{T})$*
3. *$|\mathcal{L}| + |J| = |E|$*

Algorithm for Multi-Objective Spanning Tree
1. Guess all edges in the optimal solution such that $\ell^i(e) \ge \frac{\epsilon}{k}\mathrm{L}_i$. Include these edges in the solution and contract them. Delete all other edges with $\ell^i(e) \ge \frac{\epsilon}{k}\mathrm{L}_i$ from G. Update L_i.
2. Find a vertex solution x of (LP-ST) for the residual problem and remove every edge e with $x_e = 0$.
3. Pick any maximum-weight spanning tree in the support.

The algorithm for MULTI-OBJECTIVE SPANNING TREE above proceeds in two phases. The first phase is the pruning step which we describe below. Observe that no feasible solution can include an edge whose ith-length is more than L_i. We extend this step further and *guess* all edges in the solution whose ith-length is at most $\frac{\epsilon}{k}\mathrm{L}_i$. For any i there can be at most $\frac{k}{\epsilon}$ such edges in the optimal solution. Hence, trying all such possibilities for inclusion in a partial initial solution takes time $O(m^{k/\epsilon})$ where m is the number of edges in G. There are k length function to try which amounts to the total number of choices being at most $O(m^{k^2/\epsilon})$. After guessing these edges correctly, we throw away all other edges which have ℓ^i length more than $\epsilon \mathrm{L}_i$ and contract the guessed edges in the optimal solution. Clearly, the rest of the edges in the optimal solution form a spanning tree in the contracted graph. Also, now we have an instance where $\ell^i(e) \le \frac{\epsilon}{k}\mathrm{L}_i$ for each e and i. We also update the bound L_i by subtracting the lengths of the selected edges. Let L'_i denote the residual bounds. We solve the linear program (LP-ST) with updated bounds L'_i. Step (3) can be interpreted as removing all the k constraints bounding the length under the length functions $l^1 \ldots, l^k$. Removing these constraints gives us the linear program for the spanning tree problem which is integral and its optimal solution is a maximum weight spanning tree.

Proof. (Theorem 1) First observe that the support of (LP-ST) on a graph with n vertices has at most $n+k-1$ edges. In fact, from Lemma 1, we have $|E| = |\mathcal{L}|+|J|$. But $|\mathcal{L}| \le n-1$ since $\mathcal{L}$ is a laminar family without singletons and $|J| \le k$ proving the claim.

Observe that the weight of the tree returned by the algorithm is at most the weight of the LP-solution and hence is optimal for the correct guess of *heavy* edges. Now, we show that the ith-length is at most $L'_i + \epsilon \mathrm{L}_i$. Observe that any tree must contain $n-1$ edges out of the $n+k-1$ edges in the support. Hence, the maximum ith-length tree has length no more than $k \cdot \frac{\epsilon}{k}\mathrm{L}_i = \epsilon \mathrm{L}_i$ more than the minimum ith-length tree. In turn, the tree of minimum ith-length has ith-length no larger than the ith-length of the optimal fractional solution, which is at most L'_i by feasibility. Altogether, the maximum ith-length of the solution returned is no more than $\mathrm{L}'_i + \epsilon \mathrm{L}_i$. Adding the length of edges guessed in the first step we obtain that the tree returned by the algorithm has ith-length at most $\mathrm{L}'_i + \epsilon \mathrm{L}_i + \mathrm{L}_i - \mathrm{L}'_i = (1+\epsilon)\mathrm{L}_i$.

2.1 Multi-Objective Matroid Basis

The results on Multi-objective spanning tree can be naturally generalized to the case of Multi-objective matroid basis. Consider the following linear programming relaxation (LP-MB) for the problem. There is a variable x_e for each element $e \in E$. For any subset $S \subseteq E$, we denote $x(S) = \sum_{e \in S} x_e$. Here r denotes the rank function of the matroid $\mathcal{M}$.

$$
\begin{array}{llll}
\text{(LP-MB)} & \text{maximize} & \sum_{e \in E} \mathrm{w}(e)\, x_e & \\
& \text{subject to} & x(E) = r(E), & \\
& & x(S) \le r(S), & \forall\, S \subseteq E \\
& & \sum_{e \in E} \ell^i(e) x_e \le \mathrm{L}_i, & \forall\, 1 \le i \le k \\
& & x_e \ge 0, & \forall\, e \in E.
\end{array}
$$

The polynomial time solvability of the linear program (LP-MB) follows from the polynomial time separation of the rank constraints [11]. Our algorithm for Multi-objective matroid basis is described below. Its analysis follows along the same line as in the case of Multi-objective spanning tree and is omitted due to space restrictions.

Algorithm for Multi-Objective Matroid Basis
1. Guess all elements in the optimal solution such that $\ell^i(e) \ge \frac{\epsilon}{k}\mathrm{L}_i$. Include all such elements in the solution and update the matroid by contracting these elements in the matroid. Delete all other heavy elements e with $\ell^i(e) \ge \frac{\epsilon}{k}\mathrm{L}_i$ for any i from $\mathcal{M}$. Update L_i.
2. Find a vertex solution x of (LP-MB) for the residual problem and remove every element e with $x_e = 0$.
3. Pick any maximum weight basis in the support.

3 Multi-Objective Bipartite Matching

In this section we present a polynomial-time approximation scheme for Multi-Objective Bipartite Matching and prove Theorem 3.

We formulate the following linear programming relaxation (LP-BM) for the problem. We use $\delta(v)$ to denote the set of edges incident to $v \in V$.

$$\text{(LP-BM)} \qquad \text{maximize} \quad \sum_{e \in E} \mathrm{w}(e)\, x_e$$

$$\text{subject to} \quad \sum_{e \in \delta(v)} x_e \le 1, \qquad \forall v \in V$$

$$\sum_{e \in E} \ell^i(e) x_e \le \mathrm{L}_i, \qquad \forall 1 \le i \le k$$

$$x_e \ge 0, \qquad \forall e \in E.$$

Algorithm for Multi-Objective Bipartite Matching

Preprocessing

(a) Let $\delta = \epsilon^2 / 36k\sqrt{2k\ln(k+2)}$. Guess all the edges e in OPT such that $\mathrm{w}(e) \ge \delta\, OPT$ or $\ell^i(e) \ge \delta\, \mathrm{L}_i$ for some i, and add them to the solution. Reduce the problem consequently.

Decomposition

(b) Compute the optimal fractional vertex solution x^b to LP-BM for the reduced problem. As long as there is an integral variable, reduce the problem appropriately and iterate.

(c) Remove all the nodes of degree zero and of degree at least 3, and all the edges incident to the removed nodes. Compute an optimal fractional vertex solution x^c to the problem LP-BM in the remaining graph. As long as there is an integral variable, reduce the problem appropriately and iterate. Finally, remove one edge from each remaining cycle.

(d) Compute an optimal fractional vertex solution x^d to the problem LP-BM in the remaining graph. As long as there is an integral variable, reduce the problem appropriately and iterate.

(e) Let $\gamma = \epsilon / 2\sqrt{2k\ln(k+2)}$. As long as there is a path $P = (e_1, e_2, \ldots, e_t)$ induced by x^d such that $\mathrm{w}(P) > \gamma\, \mathrm{w}(x^d)$ or $\ell^i(P) > \gamma\, \ell^i(x^d)$ for some i, find a minimal subpath $P' = (e_1, e_2, \ldots, e_{t'})$ of P satisfying the condition above and remove $e_{t'}$ from the graph.

(f) Compute an optimal fractional vertex solution x^f to the problem LP-BM in the remaining graph. As long as there is an integral variable, reduce the problem appropriately and iterate.

(g) Let $P_1, P_2, \ldots, P_q$ be the set of paths induced by x^f. Return the subpaths $S_1, S_2, \ldots, S_h$ formed after deleting the internal nodes whose matching constraints are not tight with respect to x^f. Return the solution x^g which is x^f induced on the edges in S_i for each $1 \le i \le h$.

Combination

(h) Let M_j and $\bar{M}_j$ be the two matchings partitioning S_j. Return the matching M' satisfying the following properties: (i) For each S_j, $M' \cap S_j \in \{M_j, \bar{M}_j\}$; (ii) $\mathrm{w}(M') \ge (1-\epsilon/2)\mathrm{w}(x^g)$ and $\ell^i(M') \le (1+\epsilon/2)\ell^i(x^g)$ for all i.

The algorithm for Multi-Objective Bipartite Matching above works in three phases.

In the *Preprocessing Phase*, the algorithm guesses all the edges in OPT of weight at least $\delta\, OPT$ or ith-length at least $\delta \mathrm{L}_i$ for some i. Here δ is a proper function of ϵ and k. This guessing can be performed in time polynomial in n (but exponential in δ). The algorithm then includes all the guessed edges in the solution, and deletes the remaining heavy edges. It also reduces the L_i's accordingly. After this phase $\mathrm{w}(e) \leq \delta\, OPT$ and $\ell^i(e) \leq \delta L_i$ for each edge e.

In the *Decomposition Phase* our algorithm computes over a series of pruning and iterative steps, a solution to the multi-objective matching problem on a reduced graph that is eventually a collection of paths. In Step (c), we discard nodes of degree 0 or of degree 3 or higher so as to leave only paths and cycles; Finally, one edge from each cycle is removed in this step. In Step (e), we further break each path into subpaths of bounded total weight and length. This pruning is useful in the later Combination Phase when we choose one of the two matchings in each path: the bounded difference ensures that one such combination is near optimal. The use of vertex solutions in all the residual problems ensures that the total number of edges thrown away in all the above stages is roughly of the order of the extra budget constraints in the problem which is $O(k/\gamma)$ for a parameter $\gamma \simeq O(\epsilon/\sqrt{k})$. Finally, we output a feasible fractional vertex solution x^g to the LP with the following properties.

(1) The support of x^g is a collection of vertex disjoint paths $S_1, \ldots, S_h$ where $h \leq k$.
(2) x^g is a $(1+\epsilon/4)$-approximate solution.
(3) For each S_i, the degree constraints of the vertices of S_i are tight except for its endpoints.
(4) For each S_i, $w \cdot x^g(S_i) \leq \gamma OPT$ and $\ell_i \cdot x^g(S_j) \leq \gamma L_i$ for each $1 \leq i \leq k$ and $1 \leq j \leq h$ where $\gamma = \epsilon/2\sqrt{2k\ln(k+2)}$.

In the final *Combination Phase*, the paths $S_1, \ldots, S_h$ are used to compute an approximate feasible (integral) solution. The algorithm enumerates over all the 2^h matchings which are obtained by taking, for each S_i, one of the two matchings which partition S_i. This enumeration takes polynomial time since $h \leq k = O(1)$. A probabilistic argument is used to show that one of these matchings satisfies the claimed approximation guarantee of the algorithm.

Analysis. We now analyze the three phases of the algorithm, bounding the corresponding approximation guarantee and running time. Consider first the *Preprocessing Phase.* In order to implement Step (a), we have to consider all the possible choices, and run the algorithm for each choice. Observe that there are at most $(k+1)/\delta$ such heavy edges in the optimal solution, and hence the number of possibilities is $O(m^{(k+1)/\delta}) = O(m^{O(k^2\sqrt{k\log k}/\epsilon^2)})$. The algorithm generates a different subproblem for each possible guess of the edges. In the following we will focus on the run of the algorithm where the guessed edges correspond to an optimal solution to the multi-objective problem.

Consider now the *Decomposition Phase.* We prove that the output of this phase satisfies the four properties stated above. Observe that by construction the algorithm returns a collection of edge disjoint paths whose interior vertices have tight degree constraints. Properties (3) and (4) follow by construction. We now argue that the number of paths is bounded by k, proving Property (1).

Lemma 2. *The number h of subpaths in Step (g) is upper bounded by k.*

Proof. Consider the solution x^f. The number of variables $|E| = \sum_{i=1}^{q} |P_i|$ is upper bounded by the number of tight constraints. Let q' be the number of internal nodes whose matching constraint is not tight in x^f. Note that the matching constraints at the endpoints of each path are not tight. Hence the number of tight constraints is at most $\sum_{i=1}^{q}(|P_i|-1)-q'+k = |E|-q-q'+k \geq |E|$, from which $q+q' \leq k$. Observe that, by definition, the number h of subpaths is exactly $q+q'$ (we start with q subpaths, and create a new subpath for each internal node whose matching constraint is not tight). The claim follows.

Clearly, solution x^g satisfies all the constraints. We next argue that the weight of x^g is nearly optimal. In Steps (c), (e) and (g) we remove a subset of edges whose optimal fractional value is larger than zero in the step considered. In the following lemma, whose proof is omitted for lack of space, we bound the number of edges removed. Due to the Preprocessing Phase, the weight of these edges is negligible, which implies that the consequent worsening of the approximation factor is sufficiently small. This proves Property (2).

Lemma 3. *The algorithm removes at most $7k$, $(k+1)/\gamma$, and $2k$ edges in Steps (c), (e), and (g), respectively.*

Each of the steps (b) to (g) is run polynomially many times and takes polynomial time. Hence the overall running time of the Decomposition Phase is polynomial.

Consider eventually the *Combination Phase.* As described earlier, the running time of this phase is bounded by $O(2^k n^{O(1)})$. The following lemma, which is the heart of our analysis, shows that a subset M' satisfying Properties (i) and (ii) always exists. Henceforth the algorithm always returns a solution. Although we use a randomized argument to prove the lemma, the algorithm is completely deterministic and enumerates over all solutions. Recall that M_j and $\bar{M}_j$ are the two matchings which partition subpath S_j.

Lemma 4. *In Step (h) there is always a set of edges M' satisfying Properties (i) and (ii).*

Proof. Consider the following packing problem

$$(PACK) \qquad \text{maximize} \quad \sum_{j=1}^{h}(y_j\, \mathrm{w}(M_j) + (1-y_j)\, \mathrm{w}(\bar{M}_j))$$

$$\text{subject to} \quad \sum_{j=1}^{h}(y_j\, \ell^i(M_j) + (1-y_j)\, \ell^i(\bar{M}_j)) \leq \mathrm{L}_i, \qquad \forall 1 \leq i \leq k$$

$$y_j \in \{0,1\}, \qquad \forall 1 \leq j \leq h.$$

We can interpret the variables y_j in the following way: $M' \cap S_j = M_j$ if $y_j = 1$, and $M' \cap S_j = \bar{M}_j$ otherwise. Given a (possibly fractional and infeasible) solution y to PACK, we use $\mathrm{w}(y)$ and $\ell^i(y)$ as shortcuts for $\sum_{j=1}^{h}(y_j\,\mathrm{w}(M_j) + (1-y_j)\,\mathrm{w}(\bar{M}_j))$ and $\sum_{j=1}^{h}(y_j\,\ell^i(M_j) + (1-y_j)\,\ell^i(\bar{M}_j))$, respectively.

Observe that x^g induces a feasible fractional solution y^g to the linear relaxation of PACK. In fact, consider each subpath S_j. By definition, each matching constraint at an internal node of S_j is tight. This implies that all the edges e of M_j (resp., $\bar{M}_j$) have the same value $x^g_e =: y^g$ (resp., $x^g_e =: 1 - y^g$). Thus, we have $\mathrm{w}(y^g) = \mathrm{w}(x^g)$. Now, we construct a (near) feasible integral solution y' to PACK which satisfies (i) and (ii). Independently, for each path S_i, select M_i with probability y^g_i and $\bar{M}_i$ with probability $1 - y^g_i$. Note that $E[\mathrm{w}(y')] = \mathrm{w}(y^g)$ and $E[\ell^i(y')] = \ell^i(y^g) \leq \mathrm{L}_i$ for all i.

By Step (e), switching one variable of y' from 1 to 0 or vice versa can change the cost and ith-length of y' at most by $\gamma\,\mathrm{w}(x^g)$ and $\gamma\,\ell^i(x^g)$, respectively. The proof of the lemma now follows directly from the following proposition, which derives from Chernoff's bounds.

Proposition 1. *With positive probability, $w(y') \geq (1-\epsilon/2)w(x^g)$ and $l^i(y') \leq (1+\epsilon/2)l^i(x^g)$ for all i.*

Proof. (Theorem 3) It is easy to see that the solution returned is a matching. Moreover a solution is always returned by Lemma 4. The approximation guarantee of the algorithm follow from the properties of the Decomposition step and Lemma 4. The running time of each step is polynomial (for fixed k and ϵ) thus proving Theorem 3.

Acknowledgement. We thank an anonymous referee for suggesting improvements in Section 3 in an earlier draft of the paper.

References

1. Aggarwal, V., Aneja, Y.P., Nair, K.P.K.: Minimal spanning tree subject to a side constraint. Computers & Operations Research 9, 287–296 (1982)
2. Bansal, N., Khandekar, R., Nagarajan, V.: Additive Guarantees for Degree Bounded Directed Network Design. In: STOC, pp. 769–778 (2008)
3. Barahona, F., Pulleyblank, W.R.: Exact arborescences, matchings and cycles. Discrete Applied Mathematics 16(2), 91–99 (1987)
4. Barichard, V., Ehrgott, M., Gandibleux, X., T'Kindt, V. (eds.): Multiobjective Programming and Goal Programming: Theoretical Results and Practical Applications. Lecture Notes in Economics and Mathematical Systems, vol. 618. Springer, Heidelberg (2009)
5. Beasley, J.E., Christofides, N.: An algorithm for the resource constrained shortest path problem. Networks 19, 379–394 (1989)
6. Berger, A., Bonifaci, V., Grandoni, F., Schäfer, G.: Budgeted matching and budgeted matroid intersection via the gasoline puzzle. In: Lodi, A., Panconesi, A., Rinaldi, G. (eds.) IPCO 2008. LNCS, vol. 5035, pp. 273–287. Springer, Heidelberg (2008)
7. Bilò, V., Goyal, V., Ravi, R., Singh, M.: On the Crossing Spanning Tree Problem. In: Jansen, K., Khanna, S., Rolim, J.D.P., Ron, D. (eds.) RANDOM 2004 and APPROX 2004. LNCS, vol. 3122, pp. 51–64. Springer, Heidelberg (2004)

8. Cheriyan, J., Vempala, S., Vetta, A.: Network design via iterative rounding of setpair relaxations. Combinatorica 26(3), 255–275 (2006)
9. Chinchuluun, A., Pardalos, P.M., Migdalas, A., Pitsoulis, L. (eds.): Pareto Optimality, Game Theory and Equilibria. Optimization and Its Applications, vol. 17 (2008)
10. Climacao, J.: Multicriteria Analysis. Springer, Heidelberg (1997)
11. Cunningham, W.H.: Testing Membership in Matroid Polyhedra. Journal of Combinatorial Theory B 36(2), 161–188 (1984)
12. Fleischer, L., Jain, K., Williamson, D.P.: Iterative rounding 2-approximation algorithms for minimum-cost vertex connectivity problems. Journal of Computer and System Sciences 72(5), 838–867 (2006)
13. Guignard, M., Rosenwein, M.B.: An application of Lagrangean decomposition to the resource-constrained minimum weighted arborescence problem. Networks 20, 345–359 (1990)
14. Hartley, R.: Survey of Algorithms for Vector Optimization Problems. In: French, S., Hartley, R., Thomas, L.C., White, D.J. (eds.) Multiobjective Decision Making, pp. 1–34. Academic Press, London (1983)
15. Jain, K.: A factor 2 approximation algorithm for the generalized steiner network problem. Combinatorica 21, 39–60 (2001)
16. Lau, L.C., Naor, S., Salavatipour, M., Singh, M.: Survivable network design with degree or order constraints. In: STOC, pp. 651–660 (2007)
17. Lau, L.C., Singh, M.: Additive approximation for bounded degree survivable network design. In: STOC, pp. 759–768 (2008)
18. Levin, A., Woeginger, G.J.: The constrained minimum weight sum of job completion times. Mathematical Programming 108, 115–126 (2006)
19. Marathe, M.V., Ravi, R., Sundaram, R., Ravi, S.S., Rosenkrantz, D.J., Hunt III, H.B.: Bicriteria network design problems. In: Fülöp, Z., Gecseg, F. (eds.) ICALP 1995. LNCS, vol. 944, pp. 487–498. Springer, Heidelberg (1995)
20. Melkonian, V., Tardos, E.: Algorithms for a Network Design Problem with Crossing Supermodular Demands. Networks 43, 4 (2004)
21. Mulmuley, K., Vazirani, U., Vazirani, V.: Matching is as Easy as Matrix Inversion. Combinatorica 7(1), 101–104 (1987)
22. Papadimitriou, C.H., Yannakakis, M.: On the approximability of trade-offs and optimal access of Web sources. In: FOCS, pp. 86–92 (2000)
23. Ravi, R.: Rapid rumor ramification: Approximating the minimum broadcast time. In: FOCS, pp. 202–213 (1994)
24. Ravi, R.: Matching Based Augmentations for Approximating Connectivity Problems. In: Correa, J.R., Hevia, A., Kiwi, M. (eds.) LATIN 2006. LNCS, vol. 3887, pp. 13–24. Springer, Heidelberg (2006)
25. Ravi, R., Goemans, M.X.: The constrained minimum spanning tree problem (extended abstract). In: Karlsson, R., Lingas, A. (eds.) SWAT 1996. LNCS, vol. 1097, pp. 66–75. Springer, Heidelberg (1996)
26. Ravi, R., Marathe, M.V., Ravi, S.S., Rosenkrantz, D.J., Hunt, H.B.: Many Birds with One Stone: Multi-objective Approximation Algorithms. In: STOC, pp. 438–447 (1993)
27. Schrijver, A.: Combinatorial Optimization: Polyhedra and Efficiency. Algorithms and Combinatorics, vol. 24. Springer, Berlin (2003)
28. Shmoys, D.B., Tardos, É.: Scheduling unrelated machines with costs. In: SODA, pp. 448–454 (1993)
29. Singh, M., Lau, L.C.: Approximating minimum bounded degree spanning trees to within one of optimal. In: STOC, pp. 661–670 (2007)

A Global-Optimization Algorithm for Mixed-Integer Nonlinear Programs Having Separable Non-convexity

Claudia D'Ambrosio[1], Jon Lee[2], and Andreas Wächter[2]

[1] Dept. of ECSS, University of Bologna, Italy
c.dambrosio@unibo.it
[2] IBM T.J. Watson Research Center, NY, U.S.A.
{jonlee,andreasw}@us.ibm.com

Abstract. We present a global optimization algorithm for MINLPs (mixed-integer nonlinear programs) where any non-convexity is manifested as sums of non-convex univariate functions. The algorithm is implemented at the level of a modeling language, and we have had substantial success in our preliminary computational experiments.

1 Introduction

The global solution of practical instances of Mixed-Integer NonLinear Programming (MINLP) problems has been considered for some decades. Over a considerable period of time, technology for the global optimization of convex MINLP (i.e., the continuous relaxation of the problem is a convex program) had matured (see, for example, [7,13,8,3]), and recently there has been considerable success in the realm of global optimization of non-convex MINLP (see, for example, [14,12,11,2]).

Global optimization algorithms, e.g., spatial branch-and-bound approaches like those implemented in codes like `BARON` [14] and `COUENNE` [2], have had substantial success in tackling complicated, but generally small scale, non-convex MINLPs (i.e., mixed-integer nonlinear programs having non-convex continuous relaxations). Because they are aimed at a rather general class of problems, the possibility remains that larger instances from a simpler class may be amenable to a simpler approach.

We focus on MINLPs for which the non-convexity in the objective and constraint functions is manifested as the sum of non-convex univariate functions. There are many problems that are already in such a form, or can be brought into such a form via some simple substitutions. In fact, the first step in spatial branch-and-bound is to bring problems into nearly such a form. For our purposes, we shift that burden back to the modeler. We have developed a simple algorithm, implemented at the level of a modeling language (in our case `AMPL`, see [9]), to attack such separable problems. First, we identify subintervals of convexity and concavity for the univariate functions using external calls to `MATLAB`. With such an identification at hand, we develop a convex MINLP relaxation of the problem.

A. Fiat and P. Sanders (Eds.): ESA 2009, LNCS 5757, pp. 107–118, 2009.

Our convex MINLP relaxation differs from those typically employed in spatial branch-and-bound; rather than relaxing the graph of a univariate function on an interval to an enclosing polygon, we work on each subinterval of convexity and concavity separately, using linear relaxation on only the "concave side" of each function on the subintervals. The subintervals are glued together using binary variables. Next, we employ ideas of spatial branch-and-bound, but rather than branching, we repeatedly refine our convex MINLP relaxation by modifying it at the modeling level. We attack our convex MINLP relaxation, to get lower bounds on the global minimum, using the code BONMIN [3,4] as a black-box convex MINLP solver. Next, by fixing the integer variables in the original non-convex MINLP, and then locally solving the associated non-convex NLP restriction, we get an upper bound on the global minimum, using the code IPOPT [15]. We use the solutions found by BONMIN and IPOPT to guide our choice of further refinements.

We implemented our framework using the modeling language AMPL. In order to obtain all of the information necessary for the execution of the algorithm, external software, specifically the tool for high-level computational analysis MATLAB, the convex MINLP solver BONMIN, and the NLP solver IPOPT, are called directly from the AMPL environment. A detailed description of the entire algorithmic framework, together with a statement of its convergence properties, is provided in §2.

We present computational results in §3. Some of the instances arise from specific applications; in particular, Uncapacitated Facility Location problems, Hydro Unit Commitment and Scheduling problems, and Nonlinear Continuous Knapsack problems. We made further computational tests on selected instances of GLOBALLib and MINLPLib and those results can be found in an extended version of this paper. We have had significant success in our preliminary computational experiments. In particular, we see very few major iterations occurring, with most of the time being spent in the solution of a small number of convex MINLPs. As we had hoped, our method does particularly well on problems for which the non-convexity is naturally separable. An advantage of our approach is that it can be implemented easily using existing software components and that further advances in technology for convex MINLP will immediately give us a proportional benefit.

2 Our Algorithmic Framework

We focus now on MINLPs, where the non-convexity in the objective and constraint functions is manifested as the sum of non-convex univariate functions. Without loss of generality, we take them to be of the form

$$\begin{array}{l} \min \sum_{j \in N} C_j x_j \\ \quad \text{subject to} \\ f(x) \leq 0\,; \\ r_i(x) + \sum_{k \in H(i)} g_{ik}(x_k) \leq 0\,,\ \forall i \in M\,; \\ L_j < x_j \leq U_j\,,\ \forall j \in N\,; \\ x_j \text{ integer},\ \forall j \in I\,, \end{array} \qquad (\mathcal{P})$$

where $N := \{1, 2, \ldots, n\}$, $f : \mathbb{R}^n \to \mathbb{R}^p$ and $r_i : \mathbb{R}^n \to \mathbb{R}\ \forall i \in M$, are convex functions, $H(i) \subseteq N\ \forall i \in M$, the $g_{ik} : \mathbb{R} \to \mathbb{R}$ are non-convex univariate function $\forall i \in M$, and $I \subseteq N$. Letting $H := \cup_{i \in M} H(i)$, we can take each L_j and U_j to be finite or infinite for $j \in N \setminus H$, but for $j \in H$ we assume that these are finite bounds.

We assume that the problem functions are sufficiently smooth (e.g., twice continuously differentiable) with the exception that we allow the univariate g_{ik} to be continuous functions defined piecewise by sufficiently smooth functions over a finite set of subintervals of $[L_k, U_k]$. Without loss of generality, we have taken the objective function as linear and all of the constraints to be inequalities, and further of the less-then-or-equal variety. Linear equality constraints could be included directly in this formulation, while we assume that nonlinear equalities have been split into two inequality constraints.

Our approach is an iterative technique based on three fundamental ingredients:

- A reformulation method with which we obtain a convex MINLP relaxation $\mathcal{Q}$ of the original problem $\mathcal{P}$. Solving the convex MINLP relaxation $\mathcal{Q}$, we obtain a lower bound of our original problem $\mathcal{P}$;
- A non-convex NLP restriction $\mathcal{R}$ of the original MINLP problem $\mathcal{P}$ obtained by fixing the variables within the set $\{x_j : j \in I\}$. Locally solving the non-convex NLP restriction $\mathcal{R}$, we obtain an upper bound of our original problem $\mathcal{P}$;
- A refinement technique aimed at improving, at each iteration, the quality of the lower bound obtained by solving the convex MINLP relaxation $\mathcal{Q}$.

The main idea of our algorithmic framework is to iteratively solve a lower-bounding relaxation $\mathcal{Q}$ and an upper-bounding restriction $\mathcal{R}$ so that, in case the value of the upper and the lower bound are the same, the global optimality of the solution found is proven; otherwise we make a refinement to the lower-bounding relaxation $\mathcal{Q}$. At each iteration, we seek to decrease the gap between the lower and the upper bound, and hopefully, before too long, the gap will be within a tolerance value, or the lower bounding solution is deemed to be sufficiently feasible for the original problem. In this case, or in the case a time/iteration limit is reached, the algorithm stops. If the gap is closed, we have found a global optimum, otherwise we have a heuristic solution (provided that the upper bound is not $+\infty$). The lower-bounding relaxation $\mathcal{Q}$ is a convex relaxation of the original non-convex MINLP problem, obtained by approximating the concave part of the non-convex univariate functions using piecewise linear approximation. The novelty in this part of the algorithmic framework is the new formulation of the convex relaxation: The function is approximated only where it is concave, while the convex parts of the functions are not approximated, but taken as they are. The convex relaxation proposed is described in details in §2.1. The upper-bounding restriction $\mathcal{R}$, described in §2.2, is obtained simply by fixing the variables with integrality constraints. The refinement technique consists of adding one or more breakpoints where needed, i.e., where the approximation of the non-convex function is bad and the solution of the lower-bounding problem lies. Refinement strategies are described in §2.3, and once the ingredients of the algorithmic framework are described in detail, we give a pseudo-code description of our algorithmic framework (see §2.4). Here, we also discuss some considerations about the general framework and the similarities and differences with popular global optimization methods. Theoretical convergence guarantees are discussed in §2.5. In §3, computational experiments are presented, detailing the performance of the algorithm and comparing the approach to other methods.

2.1 The Lower-Bounding Convex MINLP Relaxation $\mathcal{Q}$

To obtain our convex MINLP relaxation $\mathcal{Q}$ of the MINLP problem $\mathcal{P}$, we need to locate the subintervals of the domain of each univariate function g_i for which the function is uniformly convex or concave. For simplicity of notation, rather than refer to the constraint $r_i(x) + \sum_{k \in H(i)} g_{ik}(x_k) \leq 0$, we consider a term of the form $g(x_k) := g_{ik}(x_k)$, where $g : \mathbb{R} \to \mathbb{R}$ is a univariate non-convex function of x_k , for some k $(1 \leq k \leq n)$.

We want to explicitly view each such g as a piecewise-defined function, where on each piece the function is either convex or concave. This feature also allows us to handle functions that are already piecewise defined by the modeler. In practice, for each non-convex function g , we compute the points at which the convexity/concavity may change, i.e., the zeros of the second derivative of g , using `MATLAB`. In case a function g is naturally piecewise defined, we are essentially refining the piecewise definition of it in such a way that the convexity/concavity is uniform on each piece.

Now, on each concave piece we can use a secant approximation to give a piecewise-convex lower approximation of g . We can obtain a better lower bound by refining the piecewise-linear lower approximation on the concave pieces. We let

$$L_k =: P_0 < P_1 < \cdots < P_{\overline{p}} := U_k$$

be the ordered breakpoints at which the convexity/concavity of g changes, including, in the case of piecewise definition of g, the points at which the definition g changes. We define:

$[P_{p-1}, P_p]$:= the p-th subinterval of the domain of g $(p \in \{1 \ldots \overline{p}\})$;
$\check{H}$:= the set of indices of subintervals on which g is convex;
$\hat{H}$:= the set of indices of subintervals on which g is concave.

On the concave intervals, we will allow further breakpoints. We let B_p be the ordered set of breakpoints for the concave interval indexed by $p \in \hat{H}$. We denote these breakpoints as

$$P_{p-1} =: X_{p,1} < X_{p,2} < \cdots < X_{p,|B_p|} := P_p \,,$$

and in our relaxation we will view g as lower bounded by the piecewise-linear function that has value $g(X_{p,j})$ at the breakpoints $X_{p,j}$, and is otherwise linear between these breakpoints.

Next, we define further variables to manage our convexification of g on its domain:

$z_p :=$ a binary variable indicating if $x_k \geq P_p$ $(p = 1, \ldots, \overline{p} - 1)$;

$\delta_p :=$ a continuous variable assuming a positive value iff $x_k \geq P_{p-1}$ $(p = 1, \ldots, \overline{p})$;

$\alpha_{p,b} :=$ weight of breakpoint b in the piecewise-linear approximation of the interval indexed by p $(p \in \hat{H},\, b \in B_p)$.

In the convex relaxation of the original MINLP $\mathcal{P}$, we substitute each univariate non-convex term $g(x_k)$ with

$$\textstyle\sum_{p\in\check{H}} g(P_{p-1}+\delta_p) + \sum_{p\in\hat{H}} \sum_{b\in B_p} g(X_{p,b})\, \alpha_{p,b} - \sum_{p=1}^{\overline{p}-1} g(P_p)\,, \tag{1}$$

and we include the following set of new constraints:

$$P_0 + \textstyle\sum_{p=1}^{\overline{p}} \delta_p - x_k = 0\,; \tag{2}$$

$$\delta_p - (P_p - P_{p-1}) z_p \geq 0\,,\ \forall p \in \check{H} \cup \hat{H}\,; \tag{3}$$

$$\delta_p - (P_p - P_{p-1}) z_{p-1} \leq 0\,,\ \forall p \in \check{H} \cup \hat{H}\,; \tag{4}$$

$$P_{p-1} + \delta_p - \textstyle\sum_{b\in B_p} X_{p,b}\, \alpha_{p,b} = 0\,,\ \forall p \in \hat{H}\,; \tag{5}$$

$$\textstyle\sum_{b\in B_p} \alpha_{p,b} = 1\,,\ \forall p \in \hat{H}\,; \tag{6}$$

$$\{\alpha_{p,b} : b \in B_p\} := \text{SOS2}\,, \qquad \forall p \in \hat{H}\,; \tag{7}$$

with two dummy variables $z_0 := 1$ and $z_{\overline{p}} := 0$.

Constraints (2–4) together with the integrality of the z variables ensure that, given an x_k value, say $x_k^* \in [P_{p^*-1}, P_{p^*}]$:

$$\delta_p = \begin{cases} P_p - P_{p-1}\,, & \text{if } 1 \leq p \leq p^* - 1\,; \\ x_k^* - P_{p-1}\,, & \text{if } p = p^*\,; \\ 0\,, & \text{otherwise.} \end{cases}$$

Constraints (5–7) ensure that, for each concave interval, the convex combination of the breakpoints is correctly computed. Finally, (1) approximates the original non-convex univariate function $g(x_k)$.

Constraints (7) define $|\hat{H}|$ Special Ordered Sets of Type 2 (SOS2), i.e., ordered sets of positive variables among which at most 2 can assume a non-zero value, and, in this case, they must be consecutive (Beale and Tomlin [1]). Unfortunately, at the moment, convex MINLP solvers do not typically handle SOS2 like most MILP solvers do (also defining special-purpose branching strategies). For this reason, we substitute constraints (7), $\forall p \in \hat{H}$, with new binary variables $y_{p,b}$, with $b \in \{1, \ldots, |B_p| - 1\}$, and constraints:

$$\alpha_{p,b} \leq y_{p,b-1} + y_{p,b}\ \forall b \in B_p\,; \tag{7.a}$$

$$\sum_{b=1}^{|B_p|-1} y_{p,b} = 1\,, \tag{7.b}$$

with dummy values $y_{p,0} = y_{p,|B_p|} = 0$. In the future, when convex MINLP solvers will handle the definition of SOS2, variables y and constraints (7.a–b) would be not necessary.

It is important to note that if we utilized a very large number of breakpoints at the start, solving the resulting convex MINLP $\mathcal{Q}$ would mean essentially solving globally the original MINLP $\mathcal{P}$ up to some pre-determined tolerance related to the density of the breakpoints. But of course such a convex MINLP $\mathcal{Q}$ would be too hard to be solved in practice. With our algorithmic framework, we dynamically seek a significantly smaller convex MINLP $\mathcal{Q}$, thus generally more easily solvable, which we can use to guide the non-convex NLP restriction $\mathcal{R}$ to a good local solution, eventually settling on and proving global optimality of such a solution to the original MINLP $\mathcal{P}$.

2.2 The Upper-Bounding Non-convex NLP Restriction $\mathcal{R}$

Given a solution $\underline{x}$ of the convex MINLP relaxation $\mathcal{Q}$, the upper-bounding restriction $\mathcal{R}$ is defined as the non-convex NLP:

$$\begin{aligned}
&\min \textstyle\sum_{j\in N} C_j x_j \\
&\quad \text{subject to} \\
&f(x) \le 0\,; \\
&r_i(x) + \textstyle\sum_{k\in H(i)} g_{ik}(x_k) \le 0\,,\ \forall i \in M\,; \\
&L_j \le x_j \le U_j\,,\ \forall j \in N\,; \\
&x_j = \underline{x}_j,\ \forall j \in I\,.
\end{aligned} \tag{$\mathcal{R}$}$$

A solution of this non-convex NLP $\mathcal{R}$ is a heuristic solution of the non-convex MINLP problem $\mathcal{P}$ for two reasons: (i) the integer variables x_j, $j \in I$, might not be fixed to globally optimal values; (ii) the NLP $\mathcal{R}$ is non-convex, and so even if the integer variables x_j, $j \in I$, are fixed to globally optimal values, the NLP solver may only find a local optimum of the non-convex NLP $\mathcal{R}$ or even fail to find a feasible point. This consideration emphasizes the importance of the lower-bounding relaxation $\mathcal{Q}$ for the guarantee of global optimality. The upper-bounding problem resolution could be seen as a "verification phase" in which a solution of the convex MINLP relaxation $\mathcal{Q}$ is tested to be really feasible for the non-convex MINLP $\mathcal{P}$. To emphasis this, the NLP solver for $\mathcal{R}$ is given the solution of the convex MINLP relaxation as starting point.

2.3 The Refinement Technique

At the end of each iteration, we have two solutions: $\underline{x}$, the solution of the lower-bounding convex MINLP relaxation $\mathcal{Q}$, and $\overline{x}$, the solution of the upper-bounding non-convex NLP restriction $\mathcal{R}$; in case we cannot find a solution of $\mathcal{R}$, e.g., if $\mathcal{R}$ is infeasible, then no $\overline{x}$ is available. If $\sum_{j\in N} C_j \underline{x}_j = \sum_{j\in N} C_j \overline{x}_j$ within a certain tolerance, or if $\underline{x}$ is sufficiently feasible for the original constraints, we return to the user as solution the point $\overline{x}$ or $\underline{x}$, respectively. Otherwise, in order to continue, we want to refine the approximation of the lower-bounding convex MINLP relaxation $\mathcal{Q}$ by adding further breakpoints. We employed two strategies:

- *Based on the lower-bounding problem solution $\underline{x}$: For each $i \in M$ and $k \in H(i)$, if $\underline{x}_k$ lies in a concave interval of g_{ik}, add $\underline{x}_k$ as a breakpoint for the relaxation of g_{ik}.*
  This procedure drives the convergence of the overall method since it makes sure that the lower bounding problem becomes eventually a sufficiently accurate approximation of the original problem in the neighborhood of the global solution. Since adding a breakpoint increases the size of the convex MINLP relaxation, in practice we do not add such a new breakpoint if it would be within some small tolerance of an existing breakpoint for g_{ik}.
- *Based on the upper-bounding problem solution $\overline{x}$: For each $i \in M$ and $k \in H(i)$, if $\overline{x}_k$ lies in a concave interval of g_{ik}, add $\overline{x}_k$ as a breakpoint for the relaxation of g_{ik}.*
  The motivation behind this option is to accelerate the convergence of the method. If the solution found by the upper-bounding problem is indeed the global solution, the relaxation should eventually be exact at this point to prove its optimality. Again, to keep the size of the relaxation MINLP manageable, breakpoints are only added if they are not too close to existing ones.

We found that these strategies work well together. Hence, at each major iteration, we add a breakpoint in each concave interval where $\underline{x}$ lies in order to converge and one where $\overline{x}$ lies to speed up the convergence.

2.4 The Algorithmic Framework

Algorithm 1 details our `SC-MINLP` (Sequential Convex MINLP) Algorithm.

At each iteration, the lower-bounding MINLP relaxation $\mathcal{Q}$ and the upper-bounding NLP restriction $\mathcal{R}$ are redefined: What changes in $\mathcal{Q}$ are the sets of breakpoints that refine the piecewise-linear approximation of concave parts of the non-convex functions. At each iteration, the number of breakpoints used increases, and so does the accuracy of the approximation. What may change in $\mathcal{R}$ are the values of the fixed integer variables $\underline{x}_j$, $j \in I$. Moreover, what changes is the starting point given to the NLP solver, derived from an optimal solution of the lower-bounding MINLP relaxation $\mathcal{Q}$.

Our algorithmic framework bears comparison with spatial branch-and-bound, a successful technique in global optimization. In particular:

- during the refining phase, the parts in which the approximation is bad are discovered and the approximation is improved there, but we do it by adding one or more breakpoints instead of branching on a continuous variable as in spatial branch-and-bound;
- like spatial branch-and-bound, our approach is a rigorous global-optimization algorithm rather than a heuristic;

Algorithm 1. SC-MINLP (Sequential Convex MINLP) Algorithm

```
Choose tolerances ε, ε_feas > 0; initialize LB := −∞; UB := +∞;
Find P^i_p, Ĥ^i, Ȟ^i, X^i_pb (∀i ∈ M, p ∈ {1 . . . p̄^i}, b ∈ B^i_p).
repeat
   Solve the convex MINLP relaxation Q of the original problem P to obtain x̲;
   if (val(Q) > LB) then
      LB := val(Q);
      if (x̲ is feasible for the original problem P (within tolerance ε_feas)) then
         return x̲
      end if
   end if
   Solve the non-convex NLP restriction R of the original problem P to obtain x̄;
   if (solution x̄ could be computed and val(R) < UB) then
      UB := val(R); x_UB := x̄
   end if
   if (UB − LB > ε) then
      Update B^i_p , X^i_pb ;
   end if
until ((UB − LB ≤ ε) or (time or iteration limited exceeded))
return the current best solution x_UB
```

- unlike spatial branch-and-bound, our approach does not utilize an expression tree; it works directly on the broad class of separable non-convex MINLPs of the form $\mathcal{P}$, and of course problems that can be put in such a form;
- unlike standard implementations of spatial branch-and-bound methods, we can directly keep multivariate convex functions in our relaxation instead of using linear approximations;
- unlike spatial branch-and-bound, our method can be effectively implemented at the modeling-language level.

2.5 Convergence Analysis

For the theoretical convergence analysis of Algorithm 1, we make the following assumptions, denoting by l the iteration counter for the **repeat** loop.

A1. The functions $f(x)$ and $r_i(x)$ are continuous, and the univariate functions g_{ik} in ($\mathcal{P}$) are uniformly Lipschitz-continuous with a bounded Lipschitz constant L_g.

A2. The problem $\mathcal{P}$ has a feasible point. Hence, for each l, the relaxation $\mathcal{Q}^l$ is feasible, and we assume its (globally) optimal solution $\underline{x}^l$ is computed.

A3. The refinement technique described in Section 2.3 adds a breakpoint for every lower-bounding problem solution $\underline{x}^l$, even if it is very close to an existing breakpoint.

A4. The feasibility tolerance $\varepsilon_{\text{feas}}$ and the optimality gap tolerance ε are both chosen to be zero, and no iteration limit is set.

We have the following result (the proof will appear in an extended version of this paper).

Theorem 1. *Under assumptions A1-A4, Algorithm 1 either terminates at a global solution of the original problem $\mathcal{P}$, or each limit point of the sequence $\{\underline{x}^l\}_{l=1}^{\infty}$ is a global solution of $\mathcal{P}$.*

3 Computational Results

We implemented our algorithmic framework as an `AMPL` script, and we used `MATLAB` as a tool for numerical convexity analysis, `BONMIN` as our convex MINLP solver, and `IPOPT` as our NLP solver.

We used `MATLAB` to detect the subintervals of convexity and concavity for the non-convex univariate functions in the model. In particular, `MATLAB` reads a text file generated by the `AMPL` script, containing the constraints with univariate non-convex functions, together with the names and bounds of the independent variables. With this information, using the Symbolic Math Toolbox, `MATLAB` first computes the formula for the second derivative of each univariate non-convex function, and then computes its zeros to split the function into subintervals of convexity and concavity. The zeros are computed in the following manner. On points of a rather fine uniform discretization, we evaluate the second derivative. Then, between pairs of adjacent points for which the second derivative changes sign, we compute precisely the associated zero using the `MATLAB` function "fzero". For each univariate non-convex function, we use `MATLAB` to return the number of subintervals, the breakpoints, and associated function values in a text file which is read by the `AMPL` script.

In this section we present computational results for three problem categories. Details of the problem categories and test instances are presented in the extended version of this paper. The tests were executed on a single processor of an Intel Core2 CPU 6600, 2.40 GHz with 1.94 GB of RAM, using a time limit of 2 hours per instance. The relative optimality gap and feasibility tolerance used for all the experiments is 10^{-4}, and we do not add a breakpoint if it would be within 10^{-5} of an existing breakpoint.

Two tables with computational results exhibit the behavior of our algorithm on some instances of each problem class. Table 1 presents the iterations of our `SC-MINLP` Algorithm, with the columns labeled as follows:

- instance: the instance name;
- var/int/cons: the total number of variables, the number of integer variables, and the number of constraints in the convex relaxation $\mathcal{Q}$;
- iter #: the iteration count;
- LB: the value of the lower bound;
- UB: the value of the upper bound;
- int change: indicated whether the integer variables in the lower bounding solution $\underline{x}$ are different compared to the previous iteration;
- time MINLP: the CPU time needed to solve the convex MINLP relaxation $\mathcal{Q}$ to optimality (in seconds);
- # br added: the number of breakpoints added at the end of the previous iteration.

Table 2 presents comparisons of our SC-MINLP Algorithm with COUENNE and BONMIN. COUENNE is an open-source Branch-and-Bound algorithm aimed at the global solution of MINLP problems [2,6]. It is an exact method for the problems we address in this paper. BONMIN is an open-source code for solving general MINLP problems [3,4], but it is an exact method only for convex MINLPs. Here, BONMIN's nonlinear branch-and-bound option was chosen. When used for solving non-convex MINLPs, the solution returned is not guaranteed to be a global optimum. However, a few heuristic options are available in BONMIN, specifically designed to treat non-convex MINLPs. Here, we use the option that allows solving the root node with a user-specified number of different randomly-chosen starting points, continuing with the best solution found. This heuristic use of BONMIN is in contrast to its use in SC-MINLP, where BONMIN is employed only for the solution of the *convex* MINLP relaxation $\mathcal{Q}$.

The columns in Table 2 have the following meaning:

- instance: the instance name;
- var/int/cons: the total number of variables, the number of integer variables, and the number of constraints;
- for each approach, in particular SC-MINLP, COUENNE, BONMIN 1, BONMIN 50, we report:
  - time (LB): the CPU time (or the value of the lower bound (in parentheses) if the time limit is reached);
  - UB: the value of the upper bound.

BONMIN 1 and BONMIN 50 both refer to the use of BONMIN, but they differ in the number of multiple solutions of the root node; in the first case, the root node is solved just once, while in the second case, 50 randomly-generated starting points are given to the root-node NLP solver. If BONMIN reached the time limit, we do not report the lower bound because BONMIN cannot determine a valid lower bound for a non-convex problem.

The first category of problems are Uncapacitated Facility Location (UFL) problems (see [10]). In the first section of Table 1 the performance of SC-MINLP is shown. For the first instance, the global optimum is found at the first iteration, but 4 more iteration are needed to prove global optimality. In the second instance, only one iteration is needed. In the third instance, the first feasible solution found is not the global optimum which is found at the third (and last) iteration. In the first section of Table 2 demonstrates good performance of SC-MINLP. In particular, instance ufl_1 is solved in about 117 seconds compared to 530 seconds needed by COUENNE, instance ufl_2 in less than 18 seconds compared to 233 seconds. In instance ufl_3, COUENNE performs better than SC-MINLP, but this instance is really quite easy for both algorithms. BONMIN 1 finds solutions to all three instances very quickly, and these solutions turn out to be globally optimal (but note that BONMIN 1 is a heuristic with no guarantee of global optimality). BONMIN 50 also finds the global optima, but in non-negligible time.

The second set of test instances are Hydro Unit Commitment and Scheduling problems (see [5]). Univariate non-convexity in the model arises due to the dependence of the power produced by each turbine on the water flow passing

Table 1. Behavior of SC-MINLP

instance	var/int/cons	iter #	LB	UB	int change	time MINLP	# br added
ufl_1	153/39/228	1	4,122.000	4,330.400	-	1.35	-
	...	2	4,324.780	4,330.400	no	11.84	11
	...	3	4,327.724	4,330.400	no	19.17	5
	...	4	4,328.993	4,330.400	no	30.75	5
	205/65/254	5	4,330.070	4,330.400	no	45.42	5
ufl_2	189/57/264	1	27,516.600	27,516.569	-	4.47	-
ufl_3	79/21/101	1	1,947.883	2,756.890	-	2.25	-
	...	2	2,064.267	2,756.890	no	2.75	2
	87/25/105	3	2,292.743	2,292.777	no	3.06	2
hydro_1	324/142/445	1	-10,231.039	-10,140.763	-	18.02	-
	332/146/449	2	-10,140.760	-10,140.763	no	23.62	4
hydro_2	324/142/445	1	-3,950.697	-3,891.224	-	21.73	-
	...	2	-3,950.583	-3,891.224	no	21.34	2
	...	3	-3,950.583	-3,891.224	no	27.86	2
	336/148/451	4	-3,932.182	-3,932.182	no	38.20	2
hydro_3	324/142/445	1	-4,753.849	-4,634.409	-	59.33	-
	...	2	-4,719.927	-4,660.189	no	96.93	4
	336/148/451	3	-4,710.734	-4,710.734	yes	101.57	2
nck_20_100	144/32/205	1	-162.444	-159.444	-	0.49	-
	146/33/206	2	-159.444	-159.444	-	0.94	1
nck_20_200	144/32/205	1	-244.015	-238.053	-	0.67	-
	...	2	-241.805	-238.053	-	0.83	1
	...	3	-241.348	-238.053	-	1.16	1
	...	4	-240.518	-238.053	-	1.35	1
	...	5	-239.865	-238.053	-	1.56	1
	...	6	-239.744	-238.053	-	1.68	1
	156/38/211	7	-239.125	-239.125	-	1.81	1
nck_20_450	144/32/205	1	-391.499	-391.337	-	0.79	-
	146/32/206	2	-391.364	-391.337	-	0.87	1
nck_50_400	356/78/507	1	-518.121	-516.947	-	4.51	-
	...	2	-518.057	-516.947	-	14.94	2
	...	3	-517.837	-516.947	-	23.75	2
	...	4	-517.054	-516.947	-	25.07	2
	372/86/515	5	-516.947	-516.947	-	31.73	2
nck_100_35	734/167/1035	1	-83.580	-79.060	-	3.72	-
	...	2	-82.126	-81.638	-	21.70	2
	...	3	-82.077	-81.638	-	6.45	2
	744/172/1040	4	-81.638	-81.638	-	11.19	1
nck_100_80	734/167/1035	1	-174.841	-171.024	-	6.25	-
	...	2	-173.586	-172.631	-	24.71	2
	742/171/1039	3	-172.632	-172.632	-	12.85	2

Table 2. Comparison of solvers

		SC-MINLP		COUENNE		BONMIN 1		BONMIN 50	
instance	var/int/cons original	time (LB)	UB	time (LB)	UB	time	UB	time	UB
ufl_1	45/3/48	116.47	4,330.400	529.49	4,330.400	0.32	4,330.400	369.85	4,330.39
ufl_2	45/3/48	17.83	27,516.569	232.85	27,516.569	0.97	27,516.569	144.06	27,516.569
ufl_3	32/2/36	8.44	2,292.777	0.73	2,292.775	3.08	2,292.777	3.13	2,292.775
hydro_1	124/62/165	107.77	-10,140.763	(-11,229.80)	-10,140.763	5.03	-10,140.763	5.75	-7,620.435
hydro_2	124/62/165	211.79	-3,932.182	(-12,104.40)	-2,910.910	4.63	-3,928.139	7.02	-3,201.780
hydro_3	124/62/165	337.77	-4,710.734	(-12,104.40)	-3,703.070	5.12	-4,131.095	13.76	-3,951.199
nck_20_100	40/0/21	15.76	-159.444	3.29	-159.444	0.02	-159.444	1.10	-159.444
nck_20_200	40/0/21	23.76	-239.125	(-352.86)	-238.053	0.03	-238.053	0.97	-239.125
nck_20_450	40/0/21	15.52	-391.337	(-474.606)	-383.149	0.07	-348.460	0.84	-385.546
nck_50_400	100/0/51	134.25	-516.947	(-1020.73)	-497.665	0.08	-438.664	2.49	-512.442
nck_100_35	200/0/101	110.25	-81.638	90.32	-81.638	0.04	-79.060	16.37	-79.060
nck_100_80	200/0/101	109.22	-172.632	(-450.779)	-172.632	0.04	-159.462	15.97	-171.024

through the turbine. Our computational results are reported in the second sections of Tables 1 and 2 We observe good performance of SC-MINLP. It is able to find the global optimum of the three instances within the time limit, but COUENNE does not solve to global optimality any of the instances. Also, BONMIN 1 and BONMIN 50 show good performance. In particular, often a good solution is found in few seconds, and BONMIN 1 finds the global optimum in one case.

Our third set of test instances are Nonlinear (purely) Continuous Knapsack problems. Our computational results are reported in the third sections of Tables 1 and 2. SC-MINLP finds the global optimum for all the 6 instances in less than

3 minutes. COUENNE is able to close the gap for only 2 instances within the time limit. BONMIN 1 and BONMIN 50 terminate quickly, but the global optimum is found only for 1 instance for BONMIN 1 and 2 instances for BONMIN 50.

Overall, we have had substantial success in our preliminary computational experiments. In particular, we see very few major iterations occurring, so most of the time is spent in the solution of a small number of convex MINLPs.

References

1. Beale, E., Tomlin, J.: Special facilities in a general mathematical programming system for non-convex problems using ordered sets of variables. In: Lawrence, J. (ed.) Proc. of the 5th Int. Conf. on Operations Research, pp. 447–454 (1970)
2. Belotti, P., Lee, J., Liberti, L., Margot, F., Wächter, A.: Branching and bounds tightening techniques for non-convex MINLP, IBM Research Report RC24620 (2008); to appear in: Optimization Methods and Software
3. Bonami, P., Biegler, L., Conn, A., Cornuéjols, G., Grossmann, I., Laird, C., Lee, J., Lodi, A., Margot, F., Sawaya, N., Wächter, A.: An algorithmic framework for convex mixed integer nonlinear programs. Discrete Optimization 5, 186–204 (2008)
4. BONMIN. projects.coin-or.org/Bonmin (v. 1.0)
5. Borghetti, A., D'Ambrosio, C., Lodi, A., Martello, S.: An MILP approach for short-term hydro scheduling and unit commitment with head-dependent reservoir. IEEE Transactions on Power Systems 23(3), 1115–1124 (2008)
6. COUENNE. projects.coin-or.org/Couenne (v. 0.1)
7. Duran, M., Grossmann, I.: An outer-approximation algorithm for a class of mixed-integer nonlinear programs. Mathematical Programming 36(3), 307–339 (1986)
8. Fletcher, R., Leyffer, S.: Solving mixed integer nonlinear programs by outer approximation. Mathematical Programming 66(1), 327–349 (1994)
9. Fourer, R., Gay, D., Kernighan, B.: AMPL: A Modeling Language for Mathematical Programming. 2nd edn. Duxbury Press/Brooks/Cole Publishing Co. (2003)
10. Günlük, O., Lee, J., Weismantel, R.: MINLP strengthening for separable convex quadratic transportation-cost UFL, IBM Research Report RC24213 (2007)
11. Liberti, L.: Writing global optimization software. In: Liberti, L., Maculan, N. (eds.) Global Optimization: From Theory to Implementation, pp. 211–262. Springer, Heidelberg (2006)
12. Nowak, I., Alperin, H., Vigerske, S.: LaGO – an object oriented library for solving MINLPs. In: Bliek, C., Jermann, C., Neumaier, A. (eds.) COCOS 2002. LNCS, vol. 2861, pp. 32–42. Springer, Heidelberg (2003)
13. Quesada, I., Grossmann, I.: An LP/NLP based branch and bound algorithm for convex MINLP optimization problems. Comp. & Chem. Eng. 16, 937–947 (1992)
14. Sahinidis, N.: BARON: A general purpose global optimization software package. J. Global Opt. 8, 201–205 (1996)
15. Wächter, A., Biegler, L.T.: On the implementation of a primal-dual interior point filter line search algorithm for large-scale nonlinear programming. Mathematical Programming 106(1), 25–57 (2006)

Constructing Delaunay Triangulations along Space-Filling Curves*

Kevin Buchin

Department of Mathematics and Computer Science, TU Eindhoven
k.a.buchin@tue.nl

Abstract. Incremental construction con BRIO using a space-filling curve order for insertion is a popular algorithm for constructing Delaunay triangulations. So far, it has only been analyzed for the case that a worst-case optimal point location data structure is used which is often avoided in implementations. In this paper, we analyze its running time for the more typical case that points are located by walking. We show that in the worst-case the algorithm needs quadratic time, but that this can only happen in degenerate cases. We show that the algorithm runs in $O(n \log n)$ time under realistic assumptions. Furthermore, we show that it runs in expected linear time for many random point distributions.

1 Introduction

Delaunay triangulations (DTs) and their dual Voronoi diagrams are frequently used in many application areas, such as surface reconstruction, molecular modeling, and geographical information systems. They have been extensively studied in computational geometry and many different construction algorithms have been devised. Since its introduction in 2003 *Incremental Construction con BRIO* (biased randomized insertion order) [1] has been one of the favorite algorithms for constructing DTs. Points are inserted in rounds of increasing size which avoids full randomization. In a round the insertion order can be chosen, for which mostly space-filling curve (SFC) orders are used (see Fig. 1(d) for such an order). Already considered in the original article [1] (see also [20]), these orders have been popularized by Liu and Snoeyink [15] who used them in their program for constructing DTs of finite-precision input points. A variant of the algorithm is available as package in the *Computational Geometry Algorithms Library*[1] (CGAL) [7].

In the incremental construction, to insert a point it first has to be located in the current DT. When inserting points along a SFC order, this is typically done without an additional point location data structure. Using the spatial coherence of the order, a new point is located by *walking* from the previous point

* This research was supported by the Deutsche Forschungsgemeinschaft within the European graduate program 'Combinatorics, Geometry, and Computation' (No. GRK 588/2) and by the Netherlands' Organisation for Scientific Research (NWO) under BRICKS/FOCUS grant number 642.065.503 and project no. 639.022.707.

[1] *http://www.cgal.org/*

A. Fiat and P. Sanders (Eds.): ESA 2009, LNCS 5757, pp. 119–130, 2009.

in the order, i.e., by traversing the triangulation data structure starting at this point. Incremental construction con BRIO with SFC orders have been tested thoroughly, and their running time on surface, protein, terrain and random data is linear or near-linear in experiments [1,4,15,20]. The algorithm can be made asymptotically optimal by using a point location data structure like the conflict graph, but this not only requires additional space but also does not make use of the spatial coherence of the order. Without such a data structure non-trivial bounds on the running time were not known so far.

Most commonly the running time of DT algorithms is analyzed with respect to the worst-case point distribution. The drawback of such an analysis is that worst-case point sets might be degenerate and that the worst-case bound might not represent the running time on typical points well. On the other extreme, some DT algorithms have been analyzed with respect to the average-case running time on points drawn independently and uniformly from a unit square, or in higher dimensions from the unit d-cube [3,9,10,13,16,19]. This is an insightful alternative to the worst-case analysis, although such an input might be rather unlikely. Such an analysis can be strengthened by extending it to further random point distributions. However, except for trivial extensions to nearly uniform points, this has not been done for DT algorithms.

An alternative to the traditional worst- and average-case analysis are *realistic input models*. A global parameter for point sets which can often be bounded (in the size of the set) is its *spread*, i.e., the quotient between the largest and the smallest point to point distance. Frequently bounds on the spread result from minimum separation distances between the points and limited precision. In many cases the spread can be assumed to be polynomially bounded in the number of points. A further reason why the spread can be expected to be bounded, in particular when the points come from measurements, is noise in the data. *Smoothed analysis* [6,18] models this by allowing arbitrary input point sets, but by performing an average-case analysis with the points perturbed by random noise. In the case of surface reconstruction a realistic assumption is that the surface is *well-sampled*, i.e., every surface point has one (but not too many) sample point close to it. So far, realistic input models have to the best of our knowledge not been used explicitly for analyzing DT algorithms. They have been used to bound the complexity of DTs in $\mathbb{R}^3$. Although most three-dimensional point sets occurring seem to have DTs of linear size, their worst-case complexity is quadratic. Point sets in R^3 with spread Φ may induce DTs with complexity $O(\Phi^3)$ [11], but for many suitably sampled surfaces the complexity is linear or near-linear (see e.g. [2]).

Our results. Our aim is to give a theoretical explanation for the linear or near-linear running time in experiments of incremental construction con BRIO with SFC orders. As we will show, the worst-case running time is quadratic. We therefore turn to realistic and probabilistic models. We prove that the running time is $O(N \log \Phi)$ for an N-point set in the plane with spread Φ. This bound is tight for worst-case point sets as long as the spread is at least $w(\sqrt{N})$ (and at most exponential). Thus, if the spread is polynomially bounded then the running time is in $O(N \log N)$. This directly implies a similar bound for the smoothed complexity and can be easily extended to a bound for well-sampled surfaces. The bound also holds for points drawn from any typical random distribution, but for this case we even show a stronger bound. For independent identically distributed points we provide a condition on the distribution function, under which a variant of the algorithm runs in linear expected time after computing a SFC order (which can be computed in linear expected time in a suitable model of computation). We give the explicit analysis for uniformly and for normally distributed points. Our results extend to higher dimensions, in which case the running time depends on the structural change of the incremental construction.

This is the first analysis of a DT algorithm for realistic input models. It is also the first probabilistic analysis of a DT algorithm that goes beyond uniformly distributed points. Besides the analysis of incremental search [10] it is the only probabilistic analysis for DT of points in higher dimensions. Proving linear expected running time for this algorithm also solves an open problem posed in the original con BRIO paper [1]. It is especially surprising that the algorithm achieves these running times without a point location data structure. So far, the fastest incremental construction algorithm without point location data structure was the *jump & walk* algorithm [9,16] which runs in $O(n^{3/2})$ expected time for uniformly distributed points in a square (and close to $O(n^{4/3})$ expected time in a 3-cube), with no worst-case guarantee except for the straightforward $O(n^2)$ bound. In addition to analyzing incremental construction con BRIO with SFC orders, we also present a generalized analysis of incremental constructions con BRIO which also applies to settings other than DTs.

2 Algorithm

Consider the following construction of a map from the unit interval to the unit square: Divide the unit interval into four intervals, divide the unit square into four squares, and assign each interval to one of the squares (see Fig. 1(a)). This process can be continued recursively and furthermore it can be done in such a way that neighboring intervals are assigned to neighboring squares. The first three steps of this construction are shown in Fig. 1(a-c). In the limit this yields a surjective, continuous map from the unit interval to the unit square. By its recursive construction the Hilbert curve maps an interval to a region with an area equal to the length of the interval. This is a property shared by many space-filling curves referred to as *bi-measure-preserving property*. Another property shared by many space-filling curves including the Hilbert curve is that they are *Hölder-$1/d$ continuous*. This means for a space-filling curve $\psi\colon [0,1] \to [0,1]^d$ that there is a constant C such that $\|\psi(s) - \psi(t)\| \le C|s-t|^{1/d}$ for all $s, t \in [0,1]$.

The algorithm combines a *biased randomized insertion order (BRIO)* [1] and *space-filling curve (SFC) orders* (see Algorithm 1). In a BRIO points are grouped into rounds. Each point is independently assigned to the last round with probability 1/2. Points not assigned to the last round are assigned to the next to last round with the probability of 1/2 and so on [1]. After a logarithmic number

Algorithm 1. Incremental Construction along Space-Filling Curves

Input: Point set in $\mathbb{R}^d$
Output: Delaunay triangulation of the point set

1 Compute BRIO with SFC in rounds:
1.1 Sample points to rounds (using coin flips with sampling ratio 1/2),
1.2 Order points in a round using a space-filling curve order, for every other round use reversed order (see Remark 1).
2 Incrementally construct Delaunay triangulation using order from Step 1: In each step do
2.1 Locate new point from the previously inserted point by walking,
2.2 Update Delaunay triangulation.

of rounds an expected constant number of points remain, and we can therefore stop the sampling and assign the remaining points to the first round. If $p \in S_i$ denotes that the point p is inserted in round i or before ($i \geq 1$) then the assignment can be described in terms of probabilities as $\mathrm{P}\left[p \in S_i \,|\, p \in S_{i+1}\right] = \frac{1}{2}$ for $1 \leq i < \lceil \log_2 N \rceil + 1$ and $\mathrm{P}\left[p \in S_{\leq \lceil \log_2 N \rceil + 1}\right] = 1$. In the analysis we will use the fact that the expected structural change, i.e., the total number of simplices created and deleted, using a BRIO is asymptotically bounded by the expected structural change using a randomized order (see Sect. 3.1).

Within a round we sort points along a space-filling curve. A SFC maps a 1-dimensional space onto a higher-dimensional space, e.g., the unit interval onto the unit square. We will use SFCs in the form of the SFC heuristic for the Euclidean traveling salesperson problem [17]. We demonstrate the SFC heuristic for this task by the example of the two-dimensional Hilbert curve [12].

For our purposes it suffices to repeat the subdivision process until there is only one point per square of the subdivision. In this example, for one of the squares one more subdivision step is necessary. Fig. 1(d) shows the resulting order. We call this order of the points a *space-filling curve order*. We will call the graph obtained by connecting the points in this order *space-filling curve tour*. To efficiently compute the SFC order of a point set in $\mathbb{R}^d$, we do $\log_{2^d} N$ subdivision steps at once. This results in $\Theta(N)$ cells in the subdivision. The order of the cells and the orientation of the curve in a cell can be stored in a

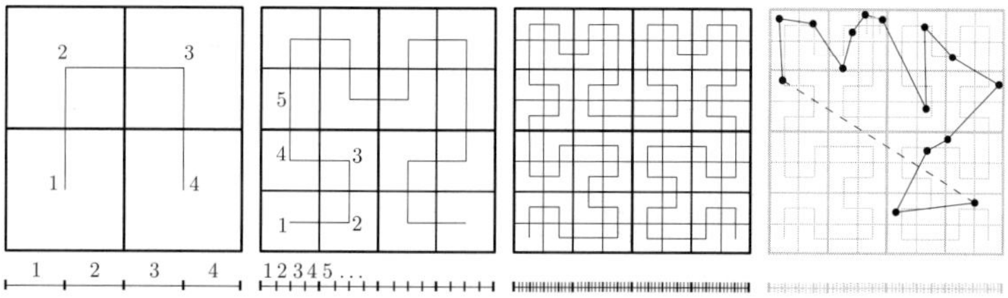

(a) (b) (c) (d)

Fig. 1. Hilbert curve and order

look-up table and the sorting can then be done efficiently with radix sort. If the spread of a point set is polynomially bounded, a constant number of such rounds suffices.

Remark 1. Using a floor function restricted to $\log N$ bits, a SFC order of a point set of size N with polynomially bounded spread can be computed in linear time. In particular this yields a linear expected time for the random distributions we consider. Restricting the floor function avoids issues about creating an unreasonably powerful model of computation. Without the restricted floor function we get an additional factor $\log N$. Since we are interested in the point location cost of the DT construction, we will assume in the rest of the paper that the points are already given in a SFC order.

After computing the insertion order, we incrementally construct the Delaunay triangulation (DT) using the order. A point is located by a *straight line walk* from the previously inserted point, i.e, we trace the line segment from the previous point to the new point in the DT data structure. In our experiments (see [4, Section 4.5]) the computation of the SFC order made up about 10% of the running time in two dimensions and less in three dimensions. The experiments also confirm that a sampling ratio smaller than 1/2 (in two dimensions between 1/10 and 1/4) speeds up the algorithm as has been already observed in earlier experiments [15,20]. Our analysis easily generalizes to other sampling ratios.

3 General Analysis

3.1 Incremental Construction con BRIO Revisited

Amenta, Choi, and Rote [1] introduced biased randomized insertion orders in the context of Delaunay triangulations of points sampled from a surface in $\mathbb{R}^3$. They consider point sets for which the expected complexity of the DT of a random sample of the point set is linear in the size of the sample. They prove for this case that the expected total update and point location cost with a history are for BRIOs asymptotically the same as for random orders. For our analysis we need to generalize their result to points in any dimension. We simplify their analysis by directly linking the costs for a construction with biased randomized insertion order to the costs for a randomized construction.

For a d-simplex with vertices in P and with s conflicting points in P let $p_B(s)$ and $p_R(s)$ denote the probabilities that the simplex occurs in an incremental construction with biased randomized insertion order and with randomized insertion order, respectively. For simplicity we assume that the sampling to rounds is not stopped after $\log(n)$ steps, but when no points remain (see [4, Proposition 3.7] for an analysis without this assumption). We bound $p_B(s)$ in terms of $p_R(s)$. This directly yields bounds for the costs determining the expected run-time of the construction, i.e., the expected *structural change* $\sum_{k=0}^{n} k_s p_B(s)$ and the expected *conflict change* $\sum_{k=0}^{n} s k_s p_B(s)$, where k_s denotes the total number of d-simplices with s conflicts. Note that the following lemma directly generalizes to arbitrary degree bounded configuration spaces and sampling ratios $1/\alpha$ by replacing (in the lemma and its proof) $d+1$ by the degree bound and 2 by α.

Lemma 1. *For $d \geq 1$ it holds in $\mathbb{R}^d$ that $p_B(s) \leq 2^{d+1} p_R(s)$.* [4, Lemma 3.5]

3.2 Counting Intersections

In the following we develop a general scheme to count the number of intersections of a space-filling curve tour with a possibly changing Delaunay triangulation. Viewing this number as a double sum over the simplices of the DT and the line segments of the tour, there are two natural ways to count the intersections. In this section we will count for each simplex the number of line segments it intersects. More specifically, we will bound for each vertex of the DT the number of line segments of the SFC tour that might be intersected by a simplex with this vertex as a corner. This analysis allows us to focus on the structure of the tour. Alternatively, we could count for each line segment of the tour the number of simplices it intersects, which shifts the focus of the analysis to the structure of the DT. We will follow this alternative approach later (Theorem 5).

Setup. Let $x_1, \ldots, x_n$ and $y_1, \ldots, y_m$ be points in $\mathbb{R}^d$. Assume that we want to insert $y_1, \ldots, y_m$ into the Delaunay triangulation $\mathrm{DT}(x_1, \ldots, x_n)$ of the points $x_1, \ldots, x_n$. We insert $y_1, \ldots, y_m$ along a space-filling curve tour denoted by $\mathrm{T}(y_1, \ldots, y_m)$ which is given by a permutation $\pi\colon \{1, \ldots, m\} \to \{1, \ldots, m\}$. Let $f(x, \mathrm{DT})$ denote the number of d-dimensional faces incident to x in the Delaunay triangulation DT, e.g., in the plane the number of triangles incident to x. Let $F(\mathrm{DT})$ denote the total number of d-dimensional faces of DT and $C(\mathrm{DT}, \mathrm{T})$ the structural change when inserting the points of the tour T into DT in the order given by the tour. Let B_{y_i,y_j} denote the ball with the line segment (y_i, y_j) as a diameter. Furthermore, let $b(x, \mathrm{T}(y_1, \ldots, y_m)) := \sum_{i=1}^{m-1} \mathbf{1}_{B_{y_i,y_{i+1}}}(x)$, i.e., the number of balls around tour segments in which x lies. In a probabilistic setting we denote the random variables corresponding to $x_1, \ldots, x_n$ and $y_1, \ldots, y_m$ as $X_1, \ldots, X_n$ and $Y_1, \ldots, Y_m$, respectively.

Counting Scheme. For points in general position the faces of the DT intersected by tour segments are $(d-1)$-dimensional or d-dimensional with these two cases alternating along the tour segment. Of these, we will count the d-dimensional faces.

Let I be the number of intersections between d-simplices of the current DT and line segments of the SFC tour. We will consider two scenarios: In the first, we directly insert a new point after we located it. This corresponds to the situation in Algorithm 1. For the line segment $y_{\pi(i)} y_{\pi(i+1)}$ we count the number of intersections with $\mathrm{DT}(x_1, \ldots, x_n, y_{\pi(1)}, \cdots y_{\pi(i)})$ $(1 \leq i < m)$. In a second scenario, we will simply count the number of intersections between $\mathrm{DT}(x_1, \ldots, x_n)$ and $\mathrm{T}(y_1, \ldots, y_m)$. Most of the analysis will handle both scenarios simultaneously.

We split the number of intersections into $I = I_1 + I_2$ where

- I_1 is the number of intersections where the d-simplex is in conflict with one of the endpoints of the tour segment,
- I_2 is the number of intersections where the d-simplex is not in conflict with the endpoints of the tour segment.

Bounding I_1. A Delaunay face in conflict with a vertex of the tour needs to be counted at most once for each tour segment adjacent to the vertex, i.e., at most twice for the vertex. In the first scenario it is actually only counted once, since it is no longer in the DT after the insertion of the vertex. We can bound the cost induced by these faces by the structural change, i.e., $I_1 \in O(C(\mathrm{DT}(x_1,\dots,x_n),\mathrm{T}(y_1,\dots,y_m)))$.

Bounding I_2. Consider a fixed line segment $(y_{\pi(i)}, y_{\pi(i+1)})$ on the SFC tour. By the following lemma any d-face of the DT intersecting this segment and not in conflict with one of the endpoints of the tour segment must have one vertex in the ball with the tour segment as diameter.

Lemma 2. *Let Δ be a d-simplex and s a line segment intersecting Δ. If the endpoints of s lie outside of the circumsphere of Δ then the ball with s as diameter contains a vertex of Δ.* [4, Lemma 4.1]

Thus, for any intersection counted in I_2 the corresponding Delaunay simplex has a vertex in the ball with the corresponding tour segment as a diameter. We bound I_2 by counting for each vertex of the DT in a ball of a tour segment the total number of d-simplices at this vertex. In the first scenario, i.e., if we insert points while traversing the tour, we have

$$\begin{aligned} I_2 \le & \sum_{i=1}^{m-1}\sum_{j=1}^{n} \mathbf{1}_{B_{y_{\pi(i)},y_{\pi(i+1)}}}(x_j) f(x_j, DT(x_1,\dots,x_n,y_{\pi(1)},\dots,y_{\pi(i)})) \\ & + \sum_{i=1}^{m-1}\sum_{j=1}^{i-1} \mathbf{1}_{B_{y_{\pi(i)},y_{\pi(i+1)}}}(y_{\pi(j)}) f(x_j, DT(x_1,\dots,x_n,y_{\pi(1)},\dots,y_{\pi(i)})) \ . \end{aligned}$$

In the second scenario this bound is simply

$$I_2 \le \sum_{i=1}^{m-1}\sum_{j=1}^{n} \mathbf{1}_{B_{y_{\pi(i)},y_{\pi(i+1)}}}(x_j) f(x_j, DT(x_1,\dots,x_n)).$$

Proposition 1. *If points are inserted directly (scenario 1) then*

$$I_2 \le b_m(d+1)(F(DT(x_1,\dots,x_n)) + C(DT(x_1,\dots,x_n),T(y_1,\dots,y_m))) \ ,$$

where $b_m := \max_{z\in P} b(z, T(y_1,\dots,y_m))$.

Proof. Any vertex is covered by at most b_m balls of the tour. Counting a simplex b_m times for each incident vertex counts it $b_m(d+1)$ times. Thus, we can bound I_2 by $b_m(d+1)$ times the total number of simplices occurring. The number of simplices is bounded by $F(DT(x_1,\dots,x_n)) + C(DT(x_1,\dots,x_n),T(y_1,\dots,y_m))$. □

Proposition 1 gives a worst-case bound on I_2. The straightforward generalization of the proposition to a probabilistic setting, would replace b_m by the *expected maximum coverage.* In the following we show that if we turn to the second scenario, i.e., do not insert points directly, we can replace expected maximum coverage by the typically smaller *maximum expected coverage* instead.

Proposition 2. *Let $X_1, \ldots, X_n, Y_1, \ldots, Y_m \in D \subseteq \mathbb{R}^d$ be independent random variables. If points are located without directly inserting them (scenario 2) then $E[I_2] \leq (d+1)\widehat{b_m} F_n$, where $\widehat{b_m} := \sup_{x \in D} E[b(x, T(Y_1, \ldots, Y_m))]$ and $F_n := E[F(DT(X_1, \ldots, X_n))]$.* [4, Proposition 4.2]

4 Analysis for Bounded Spread

4.1 Lower Bound

In this section we will focus on DTs in the plane. In two dimensions the running time of Algorithm 1 is trivially in $O(n^2)$, since the time needed to locate one point is at most linear. Unfortunately, for worst-case point sets the algorithm indeed needs quadratic time, as we show next. We construct a point set for the Hilbert curve. Fig. 2(a) shows the point set for $N = 9$. The first point is placed at $(0, 0)$. All further points are placed on the line $y = 2/3 - x$. Note that by adding a small offset to the points, they can be placed in strictly convex position instead. The x-coordinates of these $N - 1 = 2K$ points are $1/8, 1/4 + 1/(4 \cdot 8), \ldots, \sum_{i=1}^{K-1} 1/4^i + 1/(4^{K-1} \cdot 8)$ and $2/3 - 1/8, 2/3 - (1/4 + 1/(4 \cdot 8)), \ldots, 2/3 - (\sum_{i=1}^{K-1} 1/4^i + 1/(4^{K-1} \cdot 8))$. The points are chosen such that the SFC tour first traverses the points closest to the diagonal $(0, 0), (1, 1)$, going outward from there. This can be seen from the self-similar structure of the point set, i.e., the situation in a sub-square is essentially the same as in the original square (with two points less). Now we pair up the points on the line by their distance to the diagonal. Any such pair has probability $1/4$ to be inserted in the last round, and

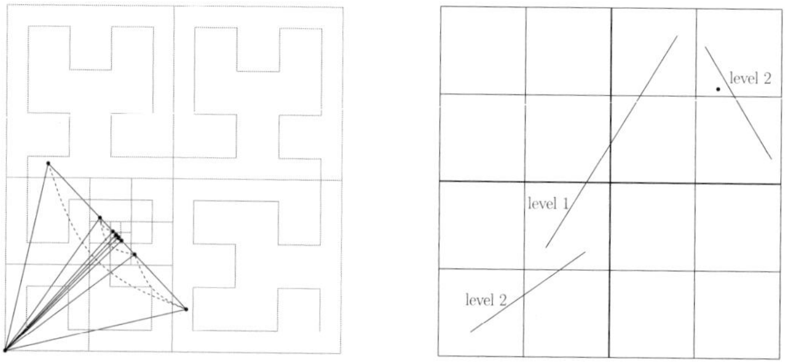

(a) Points with $\Omega(n^2)$ running time

(b) Levels of tour segments

Fig. 2. Lower and upper bound constructions

the ith pair intersects $2(i-1)$ lines. Thus the expected number of intersections is $\Omega(n^2)$ which dominates the running time.

Our worst-case example is highly degenerate. Most notably it has exponential spread. We therefore study how the running time parameterizes in terms of the spread. Adapting the worst-case above yields the following bound.

Theorem 1. *For $\Phi(N) \in \omega(\sqrt{N}) \cap 2^{O(N)}$ there are point sets of size N for which the spread is at most $\Phi(N)$ and the running time of the* incremental construction along space-filling curves *(Algorithm 1) is in $\Omega(N \log \Phi(N))$.*

Proof. We place instances of the construction above with $k = \log(\Phi(N)\sqrt{N})$ points on a $\sqrt{N/k} \times \sqrt{N/k}$ grid. The total number of intersections occuring in the last round is in $\Omega(Nk) = \Omega(N \log \Phi(N))$, since $\Phi(N) \in \omega(\sqrt{N})$. □

While Theorem 1 shows that Algorithm 1 needs super-linear time on certain inputs, it does not show that the DT cannot be computed in linear time from a SFC order. Indeed the DT can be computed in linear time from a quadtree (which is closely related to SFCs) [5].

4.2 Upper Bound

In the following we show an upper bound matching the lower bound.

Theorem 2. *The* incremental construction along space-filling curves *runs in $O(|P| \log \Phi(P))$ time in the plane.*

Proof. We assign a *level* to each edge of $\mathrm{T}(y_1, \ldots, y_m)$ according to the highest subdivision level (counting from coarse to fine) for which the edge is still contained in a single cell (see Fig. 2(b) for an example). Any point can be in the ball of at most a constant number of edges per level, for instance the point in the upper right of Fig. 2(b) cannot be in a ball corresponding to a level-2-edge with vertices in the lower left square. Further, the number of levels is in $O(\log \Phi(N))$, which yields the claimed running time using Proposition 1. □

Many point sets have polynomially bounded spread, in which case our bound implies that the algorithm runs in $O(N \log N)$ time. In higher dimensions the complexity of the DT is not necessarily linear, so we get as bound on the running time $O(C(P) \log \Phi(P))$, where $C(P)$ denotes the structural change. In a smoothed analysis the noise added (as long as it is not exponentially small) will bound the expected smallest point-to-point distance and therefore the expected spread, if the largest point-to-point distance is bounded. Thus, we again obtain a $O(N \log N)$ running time. For well-sampled domains we can typically restrict the number of levels we need to consider.

5 Average-Case Analysis

5.1 Structure of Random Space-Filling Curves

In the previous section we obtained a running time of $O(N \log N)$ for typical inputs. To prove even stronger bounds we turn to an average-case analysis. For this we will use Proposition 2, which only holds if we do not insert points directly. We therefore consider the following variant of Algorithm 1: The point location of a round is done in two steps. First, points are located in the DT of the points of the previous rounds by a walk along the SFC order. Second, points are located from the location found by the walk using the history. Note that for this we only need to maintain the history of the current round. We have the choice of inserting the points in a random order or in the order given by the space-filling curve. In the first case we directly obtain an expected constant point location cost [8], but the same can be obtained in the second case [4, Theorem 3.9, Corollary 3.10].

Let B_T be a ball chosen uniformly at random from the balls along the tour with m vertices. To prove $\mathrm{E}[I_2] \in O(n)$ it suffices to prove that for all $x \in D$ it holds that $\mathrm{P}[x \in B_T] \in O(1/m)$, where D is the domain from which the X_i are drawn ($1 \leq i \leq n$). Now, $\mathrm{P}[x \in B_T]$ does not depend on properties of the DT, thus we have reduced the problem to a problem on properties of the tour. To bound $\mathrm{P}[x \in B_T]$ we will now use the Hölder-continuity of space-filling curves. First it is important to consider how the SFC was computed. If the points come from a certain region we can simply compute the space-filling curve based on a subdivision of this region. But for points from an unbounded region, like in the case of the normal distribution, the bounding cube for the SFC depends on the actual points. For simplicity we will assume that the bounding cube is chosen as $[-u, u]^d$ where u is the largest occurring coordinate, i.e., the largest L_∞-norm of a point. For a space-filling curve $\psi\colon [0,1] \to [0,1]^d$ we denote by $\hat{\psi}\colon [0,1] \to [-u,u]^d$ the scaled space-filling curve. The mapping $\hat{\psi}$ is Hölder continuous with exponent $1/d$ and Hölder constant $c_{\hat{\psi}} = 2u \cdot c_\psi$, i.e., for $t_1, t_2 \in [0,1]$ we have $\left\|\hat{\psi}(t_1) - \hat{\psi}(t_2)\right\| \leq c_{\hat{\psi}} \left\|t_1 - t_2\right\|^{1/d}$. We denote by $\hat{\psi}^*\colon [-u,u]^d \to [0,1]$ the selection of preimages according to ψ^*. The following lemma provides a bound on the length of a tour edge in this setting.

Lemma 3. *Let $Y_1, \ldots, Y_m$ be independent identically distributed random variables in $\mathbb{R}^d$ with Lebesgue density function g_{Y_1}. Let $\psi : [0,1] \to [0,1]^d$ be a Hölder continuous and bi-measure preserving space-filling curve with Hölder constant c_ψ. Let L be a random tour segment of a space-filling curve tour through $Y_1, \ldots, Y_m$ based on $\hat{\psi}$. Then for all $\ell > 0$, $P[|L| > \ell]$ is at most*

$$\int_{\mathbb{R}^d} g_{Y_1}(y) \left(1 - \frac{d}{c_\psi^d} \int_{[0,\ell]} s^{d-1} \min\left\{g_{Y_1}(y') \mid \|y - y'\| < s\right\} ds\right)^{m-1} d\lambda^d(y) \; .$$

[4, Lemma 4.3]

Using Lemma 3 we can bound $\mathrm{P}[x \in B_T]$ by

$$\int_{\mathbb{R}^d} g_{Y_1}(y) \left(1 - \frac{d}{c_\psi^d} \int_{[0,\|x-y\|]} s^{d-1} \min\left\{g_{Y_1}(y') \mid \|y - y'\| < s\right\} ds\right)^{m-1} d\lambda^d(y) \; .$$

Two examples for distributions handled by Lemma 3 are uniformly distributed points in $[0,1]^d$ and normally distributed points in the plane. For normally distributed points we apply the lemma to all points except a few (an expected logarithmic number of points) far away from the center of the distribution. To handle the remaining points we assume in the analysis that an additional point location structure like Kirkpatrick's point location hierarchy [14] is used.

Theorem 3. *The* incremental construction along space-filling curves *(using the history of a round) computes the Delaunay triangulation of points drawn independently and uniformly from a d-cube in linear expected time.* [4, Theorem 4.4]

Theorem 4. *The* incremental construction along space-filling curves *(using the history of a round and an $O(\log n)$ point location data structure) computes the Delaunay triangulation of independent, identically normally distributed points in the plane in linear expected time.* [4, Theorem 4.5]

For points drawn independently and uniformly from a bounded convex region in the plane, we next give an alternative analysis, which yields a linear bound on the expected running time, even for the case that points are inserted directly during the walk. It suffices to analyze the run-time of the last round. We assume that at the beginning of the last round n points have already been inserted into the DT, while the m points from the last round are to be inserted. The points are located by traversing the DT along a SFC order. Therefore, the time for locating the points is proportional to the number of intersections between the order and the DT. Let L be a line segment that is not too close (no closer than $c\sqrt{\log n/n}$ for a suitable constant c) to the boundary of the bounded convex region and that is independent from the points of the DT. Then the expected number of intersections between L and the DT is in $O(1 + \sqrt{n}|L|)$ [9]. Now, the SFC order of m points in a bounded region in the plane yields a walk through the points of length $O(\sqrt{m})$ [17]. This yields an expected number of intersections in $O(m+\sqrt{nm})$, but there are two pieces missing in this argument. First, points close to the boundary are not handled. Second, points are inserted during the walk. Therefore the DT changes and depends on the points to be inserted and their insertion order. Both of these problems can be overcome [4, Sect. 3.4].

Theorem 5. *The* incremental construction along space-filling curves *(with no additional point location data structure) computes the Delaunay triangulation independent, uniformly distributed points in a bounded convex region in linear expected time.* [4, Theorem 3.27]

Acknowledgments. The author would like to thank Günter Rote and Scot Drysdale and Maike Buchin for many helpful ideas.

References

1. Amenta, N., Choi, S., Rote, G.: Incremental constructions con BRIO. In: Proc. 19th Annu. ACM Sympos. Comput. Geom., pp. 211–219. ACM Press, New York (2003)
2. Attali, D., Boissonnat, J.-D.: A linear bound on the complexity of the Delaunay triangulation of points on polyhedral surfaces. Discrete Comput. Geom. 31(3), 369–384 (2004)

3. Bentley, J.L., Weide, B.W., Yao, A.C.: Optimal expected-time algorithms for closest-point problems. ACM Trans. Math. Softw. 6, 563–580 (1980)
4. Buchin, K.: Organizing Point Sets: Space-Filling Curves, Delaunay Tessellations of Random Point Sets, and Flow Complexes. PhD thesis, Free University Berlin (2007), http://www.diss.fu-berlin.de/diss/receive/FUDISS_thesis_000000003494
5. Buchin, K., Mulzer, W.: Delaunay triangulations in $O(\mathrm{sort}(n))$ and other transdichotomous and hereditary algorithms in computational geometry. In: Proc. 50th Annu. IEEE Sympos. Found. Comput. Sci. (to appear, 2009)
6. Damerow, V., Meyer auf der Heide, F., Räcke, H., Scheideler, C., Sohler, C.: Smoothed motion complexity. In: Di Battista, G., Zwick, U. (eds.) ESA 2003. LNCS, vol. 2832, pp. 161–171. Springer, Heidelberg (2003)
7. Delage, C.: Spatial sorting. In: CGAL Editorial Board (eds), CGAL User and Reference Manual (2007)
8. Devillers, O.: Randomization yields simple O(n $\log^*$ n) algorithms for difficult Omega(n) problems. Int. J. Comput. Geometry Appl. 2(1), 97–111 (1992)
9. Devroye, L., Mücke, E., Zhu, B.: A note on point location in Delaunay triangulations of random points. Algorithmica 22, 477–482 (1998)
10. Dwyer, R.A.: Higher-dimensional Voronoi diagrams in linear expected time. Discrete Comput. Geom. 6(4), 343–367 (1991)
11. Erickson, J.: Dense point sets have sparse Delaunay triangulations: or but not too nasty. In: Proc. 13th Annu. ACM-SIAM Sympos. Discrete Algorithms, pp. 125–134 (2002)
12. Hilbert, D.: Ueber die stetige Abbildung einer Linie auf ein Flächenstück. Math. Ann. 38, 459–460 (1891)
13. Katajainen, J., Koppinen, M.: Constructing Delaunay triangulations by merging buckets in quadtree order. Fundam. Inform. 11, 275–288 (1988)
14. Kirkpatrick, D.G.: Optimal search in planar subdivisions. SIAM J. Comput. 12(1), 28–35 (1983)
15. Liu, Y., Snoeyink, J.: A comparison of five implementations of 3d Delaunay tesselation. In: Goodman, J.E., Pach, J., Welzl, E. (eds.) Combinatorial and Computational Geometry. MSRI Publications, vol. 52, pp. 439–458. Cambridge University Press, Cambridge (2005)
16. Mücke, E.P., Saias, I., Zhu, B.: Fast randomized point location without preprocessing in two- and three-dimensional Delaunay triangulations. Comput. Geom. Theory Appl. 12(1-2), 63–83 (1999)
17. Platzman, L.K., Bartholdi III, J.J.: Spacefilling curves and the planar travelling salesman problem. J. ACM 36(4), 719–737 (1989)
18. Spielman, D.A., Teng, S.-H.: Smoothed analysis of algorithms: Why the simplex algorithm usually takes polynomial time. J. ACM 51(3), 385–463 (2004)
19. Su, P., Drysdale, R.: A comparison of sequential Delaunay triangulation algorithms. Comput. Geom. Theory Appl. 7, 361–386 (1997)
20. Zhou, S., Jones, C.B.: HCPO: an efficient insertion order for incremental Delaunay triangulation. Inf. Process. Lett. 93(1), 37–42 (2005)

Piercing Translates and Homothets of a Convex Body

Adrian Dumitrescu[1,⋆] and Minghui Jiang[2,⋆⋆]

[1] Department of Computer Science, University of Wisconsin-Milwaukee, WI 53201-0784, USA
ad@cs.uwm.edu

[2] Department of Computer Science, Utah State University, Logan, UT 84322-4205, USA
mjiang@cc.usu.edu

Abstract. According to a classical result of Grünbaum, the transversal number $\tau(\mathcal{F})$ of any family $\mathcal{F}$ of pairwise-intersecting translates or homothets of a convex body C in $\mathbb{R}^d$ is bounded by a function of d. Denote by $\alpha(C)$ (resp. $\beta(C)$) the supremum of the ratio of the transversal number $\tau(\mathcal{F})$ to the packing number $\nu(\mathcal{F})$ over all families $\mathcal{F}$ of translates (resp. homothets) of a convex body C in $\mathbb{R}^d$. Kim et al. recently showed that $\alpha(C)$ is bounded by a function of d for any convex body C in $\mathbb{R}^d$, and gave the first bounds on $\alpha(C)$ for convex bodies C in $\mathbb{R}^d$ and on $\beta(C)$ for convex bodies C in the plane. In this paper, we show that $\beta(C)$ is also bounded by a function of d for any convex body C in $\mathbb{R}^d$, and present new or improved bounds on both $\alpha(C)$ and $\beta(C)$ for various convex bodies C in $\mathbb{R}^d$ for all dimensions d. Our techniques explore interesting inequalities linking the covering and packing densities of a convex body. Our methods for obtaining upper bounds are constructive and lead to efficient constant-factor approximation algorithms for finding a minimum-cardinality point set that pierces a set of translates or homothets of a convex body.

1 Introduction

A *convex body* is a compact convex set in $\mathbb{R}^d$ with nonempty interior. Let $\mathcal{F}$ be a family of convex bodies. The *packing number* $\nu(\mathcal{F})$ is the maximum cardinality of a set of pairwise-disjoint convex bodies in $\mathcal{F}$, and the *transversal number* $\tau(\mathcal{F})$ is the minimum cardinality of a set of points that intersects every convex body in $\mathcal{F}$.

Let G be the *intersection graph* of $\mathcal{F}$ with one vertex for each convex body in $\mathcal{F}$ and with an edge between two vertices if and only if the two corresponding convex bodies intersect. The *independence number* $\alpha(G)$ is the maximum cardinality of an independent set in G. The *clique partition number* $\vartheta(G)$ is the minimum number of classes in a partition of the vertices of G into cliques. Since a set of pairwise-disjoint convex bodies in $\mathcal{F}$ corresponds to an independent set in G, we have $\nu(\mathcal{F}) = \alpha(G)$. Also, since any subset of convex bodies in $\mathcal{F}$ that share a common point corresponds to a clique in G, we have $\tau(\mathcal{F}) \geq \vartheta(G)$. For the special case that $\mathcal{F}$ is a family of axis-parallel boxes in $\mathbb{R}^d$, we indeed have $\tau(\mathcal{F}) = \vartheta(G)$ since any subset of pairwise-intersecting boxes share a common point. In general, we clearly have the inequality $\vartheta(G) \geq \alpha(G)$, thus also $\tau(\mathcal{F}) \geq \nu(\mathcal{F})$. But what else can be said about the relation between $\tau(\mathcal{F})$ and $\nu(\mathcal{F})$?

⋆ Supported in part by NSF CAREER grant CCF-0444188.
⋆⋆ Supported in part by NSF grant DBI-0743670.

A. Fiat and P. Sanders (Eds.): ESA 2009, LNCS 5757, pp. 131–142, 2009.

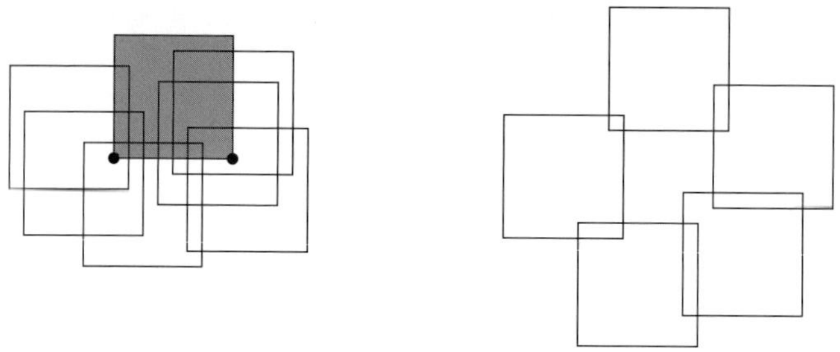

Fig. 1. Piercing a family $\mathcal{F}$ of axis-parallel unit squares. Left: all squares that intersect the highest (shaded) square contain one of its two lower vertices. Right: five squares form a 5-cycle.

For example, let $\mathcal{F}$ be any family of axis-parallel unit squares in the plane, and refer to Figure 1. One can obtain a subset of pairwise-disjoint squares by repeatedly selecting the highest square that does not intersect the previously selected squares. Then $\mathcal{F}$ is pierced by the set of points consisting of the two lower vertices of each square in the subset. This implies that $\tau(\mathcal{F}) \leq 2 \cdot \nu(\mathcal{F})$. The factor of 2 cannot be improved below $\frac{3}{2}$ since $\tau(\mathcal{F}) = 3$ and $\nu(\mathcal{F}) = 2$ for a family $\mathcal{F}$ of five squares arranged into a 5-cycle [11].

For a convex body C in $\mathbb{R}^d$, $d \geq 2$, define

$$\alpha(C) = \sup_{\mathcal{F}_t} \frac{\tau(\mathcal{F}_t)}{\nu(\mathcal{F}_t)} \quad \text{and} \quad \beta(C) = \sup_{\mathcal{F}_h} \frac{\tau(\mathcal{F}_h)}{\nu(\mathcal{F}_h)},$$

where $\mathcal{F}_t$ ranges over all families of translates of C, and $\mathcal{F}_h$ ranges over all families of (positive) homothets of C. In the definitions of α and β, both the convexity of C and the homothety of $\mathcal{F}_t$ and $\mathcal{F}_h$ are necessary for the values $\alpha(C)$ and $\beta(C)$ to be bounded. Our previous discussion (Figure 1) yields the bounds $\frac{3}{2} \leq \alpha(C) \leq 2$ for any square C.

Define $\alpha_1(C)$ (resp. $\beta_1(C)$) as the smallest number k such that for any family $\mathcal{F}$ of *pairwise-intersecting* translates (resp. homothets) of a convex body C, there exists a set of k points that intersects every member of $\mathcal{F}$. Note that α and β generalize α_1 and β_1. For any convex body C, the four numbers $\alpha(C)$, $\beta(C)$, $\alpha_1(C)$, and $\beta_1(C)$ are invariant under any non-singular affine transformation of C, and we have the four inequalities $\alpha_1(C) \leq \alpha(C)$, $\beta_1(C) \leq \beta(C)$, $\alpha_1(C) \leq \beta_1(C)$, and $\alpha(C) \leq \beta(C)$.

Grünbaum [10] showed that, for any convex body C in $\mathbb{R}^d$, both $\alpha_1(C)$ and $\beta_1(C)$ are bounded by functions of d. Deriving bounds on $\alpha_1(C)$ and $\beta_1(C)$ for various types of convex bodies C in $\mathbb{R}^d$ is typical of classic Gallai-type problems [7,17], and has been extensively studied. For example, a result by Karasev [12] states that $\alpha_1(C) \leq 3$ for any convex body C in the plane, i.e., for any family of pairwise-intersecting translates of a convex body in the plane, there always exists a set of three points that intersects every member of the family. It is folklore that $\alpha_1(C) = \beta_1(C) = 1$ for any parallelogram C; see [10] and the references therein. Also, $\alpha_1(C) = 2$ for any affinely regular hexagon C [10], $\alpha_1(C) = \beta_1(C) = 3$ for any triangle C [4], $\alpha_1(C) = 3 < 4 = \beta_1(C)$ for any (circular) disk C [10,6], and $\beta_1(C) \leq 7$ for any centrally symmetric convex body C in the plane [10]. Perhaps the most celebrated recent result on point transversals of convex sets is Alon and Kleitman's solution to the Hadwiger-Debrunner (p, q)-problem [1]. We refer to the two surveys [7, pp. 142–150] and [17, pp. 77–78] for more related results.

The two numbers $\alpha_1(C)$ and $\beta_1(C)$ bound the values of $\tau(\mathcal{F})$ for special families $\mathcal{F}$ of translates and homothets of a convex body C with $\nu(\mathcal{F}) = 1$. It is thus natural to study the general case $\nu(\mathcal{F}) \geq 1$, and to obtain estimates on $\alpha(C)$ and $\beta(C)$. Despite the many previous bounds on $\alpha_1(C)$ and $\beta_1(C)$ [7,17], first estimates on $\alpha(C)$ and $\beta(C)$ have been only obtained recently by Kim et al. [14], who showed that $\alpha(C)$ is bounded by a function of d for any convex body C in $\mathbb{R}^d$, and gave the first bounds on $\alpha(C)$ for convex bodies C in $\mathbb{R}^d$ and on $\beta(C)$ for convex bodies C in the plane. In this paper, we show that $\beta(C)$ is also bounded by a function of d for any convex body C in $\mathbb{R}^d$, and present new or improved bounds on both $\alpha(C)$ and $\beta(C)$ for various types of convex bodies C in $\mathbb{R}^d$ for all dimensions d.

Definitions. For a convex body C in $\mathbb{R}^d$, denote by $|C|$ the Lebesgue measure of C, i.e., the area in the plane, or the volume in d-space for $d \geq 3$. For a family $\mathcal{F}$ of convex bodies in $\mathbb{R}^d$, denote by $|\mathcal{F}|$ the Lebesgue measure of the union of the convex bodies in $\mathcal{F}$, i.e., $|\bigcup_{C \in \mathcal{F}} C|$.

For two convex bodies A and B in $\mathbb{R}^d$, denote by $A+B = \{a+b \mid a \in A, b \in B\}$ the Minkowski sum of A and B. For a convex body C in $\mathbb{R}^d$, denote by $\lambda C = \{\lambda c \mid c \in C\}$ the *scaled copy* of C by a factor of $\lambda \in \mathbb{R}$, denote by $-C = \{-c \mid c \in C\}$ the *reflexion* of C about the origin, and denote by $C + a = \{c + a \mid c \in C\}$ the *translate* of C by the vector from the origin to a. Write $C - C$ for $C + (-C)$.

For two parallelepipeds P and Q in $\mathbb{R}^d$ that are parallel to each other (but are not necessarily axis-parallel), denote by $\lambda_i(P, Q)$, $1 \leq i \leq d$, the length ratios of the edges of Q to the corresponding parallel edges of P. Then, for a convex body C in $\mathbb{R}^d$, define

$$\gamma(C) = \min_{P,Q} \left(\lceil \lambda_d(P, Q) \rceil \prod_{i=1}^{d-1} \lceil \lambda_i(P, Q) + 1 \rceil \right),$$

where P and Q range over all pairs of parallelepipeds in $\mathbb{R}^d$ that are parallel to each other, such that $P \subseteq C \subseteq Q$. Note that in this case $\lambda_i(P, Q) \geq 1$ for $1 \leq i \leq d$.

For a convex body C in $\mathbb{R}^d$, denote by $\delta(C)$, $\delta_T(C)$, and $\delta_L(C)$, respectively, the packing density, the translative packing density, and the lattice packing density of C, that is, the maximum densities of a packing in $\mathbb{R}^d$ with congruent copies of C, translates of C, and translates of C by vectors of a lattice, respectively. Similarly, denote by $\theta(C)$, $\theta_T(C)$, and $\theta_L(C)$, respectively, the covering density, the translative covering density, and the lattice covering density of C. See [3, Chapter 1]. Note that the four densities $\theta_T(C)$, $\theta_L(C)$, $\delta_T(C)$, and $\delta_L(C)$ are invariant under any non-singular affine transformation of C. For any convex body C in $\mathbb{R}^d$, we have the inequalities $\delta_L(C) \leq \delta_T(C) \leq \delta(C) \leq 1 \leq \theta(C) \leq \theta_T(C) \leq \theta_L(C)$.

For two convex bodies A and B in $\mathbb{R}^d$, denote by $\kappa(A, B)$ the smallest number κ such that A can be covered by κ translates of B.

Main results. Kim et al. [14] proved that, for any family $\mathcal{F}$ of translates of a convex body in $\mathbb{R}^d$, $\tau(\mathcal{F}) \leq 2^{d-1}d^d \cdot \nu(\mathcal{F})$, in particular $\tau(\mathcal{F}) \leq 108 \cdot \nu(\mathcal{F})$ when $d = 3$, and moreover $\tau(\mathcal{F}) \leq 8 \cdot \nu(\mathcal{F}) - 5$ when $d = 2$. We improve these bounds for all dimensions d in the following theorem:

Theorem 1. *For any family $\mathcal{F}$ of translates of a convex body C in $\mathbb{R}^d$,*

$$\tau(\mathcal{F}) \leq \gamma(C) \cdot \nu(\mathcal{F}), \quad \text{where } \gamma(C) \leq d(d+1)^{d-1}. \tag{1}$$

In particular, $\tau(\mathcal{F}) \leq 48 \cdot \nu(\mathcal{F})$ when $d = 3$, and $\tau(\mathcal{F}) \leq 6 \cdot \nu(\mathcal{F})$ when $d = 2$.

For any parallelepiped C in $\mathbb{R}^d$, we can choose two parallelepipeds P and Q such that $P = Q = C$ hence $P \subseteq C \subseteq Q$. Then $\lambda_i(P, Q) = 1$ for $1 \leq i \leq d$, and $\gamma(C) = 2^{d-1}$. This implies the following corollary:

Corollary 1. *For any family $\mathcal{F}$ of translates of a parallelepiped in $\mathbb{R}^d$, $\tau(\mathcal{F}) \leq 2^{d-1} \cdot \nu(\mathcal{F})$.*

In contrast, for a family $\mathcal{F}$ of (not necessarily congruent or similar) axis-parallel parallelepipeds (boxes) in $\mathbb{R}^d$, the current best upper bound [8] (see also [13]) is

$$\tau(\mathcal{F}) \leq \nu(\mathcal{F}) \log^{d-2} \nu(\mathcal{F})(\log \nu(\mathcal{F}) - 1/2) + d.$$

Kim et al. [14] also proved that, for any family $\mathcal{F}$ of translates of a centrally symmetric convex body in the plane, $\tau(\mathcal{F}) \leq 6 \cdot \nu(\mathcal{F}) - 3$. The following theorem gives a general bound for any centrally symmetric convex body in $\mathbb{R}^d$ and an improved bound for any centrally symmetric convex body in the plane:

Theorem 2. *For any family $\mathcal{F}$ of translates of a centrally symmetric convex body S in $\mathbb{R}^d$,*

$$\tau(\mathcal{F}) \leq 2^d \cdot \frac{\theta_L(S)}{\delta_L(S)} \cdot \nu(\mathcal{F}). \tag{2}$$

Moreover, $\tau(\mathcal{F}) \leq 24 \cdot \nu(\mathcal{F})$ when $d = 3$, and $\tau(\mathcal{F}) \leq \frac{16}{3} \cdot \nu(\mathcal{F})$ when $d = 2$.

For special types of convex bodies in the plane, the following theorem gives sharper bounds than the bounds implied by Theorem 1 and Theorem 2. Also, as we will show later, (3) may give a better asymptotic bound than (1) and (2) for high dimensions.

Theorem 3. *Let $\mathcal{F}$ be a family of translates of a convex body C in $\mathbb{R}^d$. Then*

$$\tau(\mathcal{F}) \leq \min_L \kappa((C - C) \cap L, C) \cdot \nu(\mathcal{F}), \tag{3}$$

where L ranges over all closed half spaces bounded by hyperplanes through the center of $C - C$. In particular, $\tau(\mathcal{F}) \leq 5 \cdot \nu(\mathcal{F})$ if C is a centrally symmetric convex body in the plane. Moreover, (i) *if C is a square, then $\tau(\mathcal{F}) \leq 2 \cdot \nu(\mathcal{F}) - 1$,* (ii) *if C is a triangle, then $\tau(\mathcal{F}) \leq 5 \cdot \nu(\mathcal{F}) - 2$,* (iii) *if C is a disk, then $\tau(\mathcal{F}) \leq 5 \cdot \nu(\mathcal{F}) - 2$.*

Having presented our bounds for families of translates, we now turn to families of homothets. Kim et al. [14] proved that, for any family $\mathcal{F}$ of homothets of a convex body C in the plane, $\tau(\mathcal{F}) \leq 16 \cdot \nu(\mathcal{F})$ and, if C is centrally symmetric, $\tau(\mathcal{F}) \leq 9 \cdot \nu(\mathcal{F})$. The following theorem gives a general bound for any convex body in $\mathbb{R}^d$, an improved bound for any centrally symmetric convex body in the plane, and additional bounds for special types of convex bodies in the plane:

Theorem 4. *Let $\mathcal{F}$ be a family of homothets of a convex body C in $\mathbb{R}^d$. Then*

$$\tau(\mathcal{F}) \le \kappa(C - C, C) \cdot \nu(\mathcal{F}). \tag{4}$$

In particular, $\tau(\mathcal{F}) \le 7 \cdot \nu(\mathcal{F})$ if C is a centrally symmetric convex body in the plane. Moreover, (i) *if C is a square, then $\tau(\mathcal{F}) \le 4 \cdot \nu(\mathcal{F}) - 3$,* (ii) *if C is a triangle, then $\tau(\mathcal{F}) \le 12 \cdot \nu(\mathcal{F}) - 9$,* (iii) *if C is a disk, then $\tau(\mathcal{F}) \le 7 \cdot \nu(\mathcal{F}) - 3$.*

For any parallelepiped C in $\mathbb{R}^d$, $C - C$ is a translate of $2C$ and can be covered by 2^d translates of C, thus $\kappa(C - C, C) \le 2^d$. This implies the following corollary:

Corollary 2. *For any family $\mathcal{F}$ of homothets of a parallelepiped in $\mathbb{R}^d$, $\tau(\mathcal{F}) \le 2^d \cdot \nu(\mathcal{F})$.*

Both Theorem 3 and Theorem 4 are obtained by a simple greedy method, used also previously by Kim et al. [14]. Although we have improved their bounds using new techniques in Theorem 1 and Theorem 2, we show that a refined analysis of the simple greedy method yields even better asymptotic bounds for high dimensions in Theorem 3 and Theorem 4. We will use the following lemma by Chakerian and Stein [4] in our analysis:

Lemma 1. (Chakerian and Stein [4]). *For every convex body C in $\mathbb{R}^d$ there exist two parallelepipeds P and Q such that $P \subseteq C \subseteq Q$, where P and Q are homothetic with ratio at most d.*

For any convex body C in $\mathbb{R}^d$, let P and Q be the two parallelepipeds in Lemma 1. Since $C - C \subseteq Q - Q$ and $P \subseteq C$, it follows that $\kappa(C - C, C) \le \kappa(Q - Q, P) = \kappa(2Q, P) \le (2d)^d$; see also [14, Lemma 4]. The classic survey by Danzer, Grünbaum, and Klee [7, pp. 146–147] lists several other upper bounds due to Rogers and Danzer: (i) $\kappa(C - C, C) \le \frac{2^d}{d+1} 3^{d+1} \theta_T(C)$ for any convex body C in $\mathbb{R}^d$, (ii) $\kappa(C - C, C) \le 5^d$ and $\kappa(C - C, C) \le 3^d \theta_T(C)$ for any centrally symmetric convex body C in $\mathbb{R}^d$. Note that $\theta_T(C) < d \ln d + d \ln \ln d + 5d = O(d \log d)$ for any convex body C in $\mathbb{R}^d$, according to a result of Rogers [15]. The following lemma summarizes the upper bounds on $\kappa(C - C, C)$:

Lemma 2. *For any convex body C in $\mathbb{R}^d$,*

$$\kappa(C - C, C) \le \min\left\{(2d)^d,\ \tfrac{2^d}{d+1} 3^{d+1} \theta_T(C)\right\} = O(6^d \log d).$$

Moreover, if C is centrally symmetric, then

$$\kappa(C - C, C) \le \min\left\{5^d,\ 3^d \theta_T(C)\right\} = O(3^d d \log d).$$

From Lemma 2 and Theorem 4, it follows that $\beta(C)$ is bounded by a function of d, namely by $O(6^d \log d)$, for any convex body C in $\mathbb{R}^d$. Since $\min_L \kappa((C - C) \cap L, C) \le \kappa(C - C, C)$, Lemma 2 also provides upper bounds on $\min_L \kappa((C - C) \cap L, C)$ in Theorem 3. As a result, (3) implies an upper bound $\tau(\mathcal{F}) \le O(6^d \log d) \cdot \nu(\mathcal{F})$ for any family $\mathcal{F}$ of translates of a convex body in $\mathbb{R}^d$, which is better than the upper bound

Table 1. Upper bounds on $\alpha(C)$ and $\beta(C)$ for a convex body C in $\mathbb{R}^d$. †By Theorem 4 and Lemma 2: for $d = 3$, $(2d)^d = 216$ and $5^d = 125$.

Convex body C in $\mathbb{R}^d$		$\alpha(C)$ *upper*	
arbitrary	$d = 2$	6	T1
centr. symm.	$d = 2$	5	T3
arbitrary	$d = 3$	48	T1
centr. symm.	$d = 3$	24	T2
arbitrary	$d > 3$	$\min\{d(d+1)^{d-1}, \frac{2^d}{d+1}3^{d+1}\theta_T(C)\}$	T1 T4-L2
centr. symm.	$d > 3$	$\min\{d(d+1)^{d-1}, 2^d\frac{\theta_L(C)}{\delta_L(C)}, 5^d, 3^d\theta_T(C)\}$	T1 T2 T4-L2
parallelepiped	$d \geq 2$	2^{d-1}	C1
Convex body C in $\mathbb{R}^d$		$\beta(C)$ *upper*	
arbitrary	$d = 2$	16	[14]
centr. symm.	$d = 2$	7	T4
arbitrary	$d = 3$	216	†T4-L2
centr. symm.	$d = 3$	125	†T4-L2
arbitrary	$d > 3$	$\min\{(2d)^d, \frac{2^d}{d+1}3^{d+1}\theta_T(C)\}$	T4-L2
centr. symm.	$d > 3$	$\min\{5^d, 3^d\theta_T(C)\}$	T4-L2
parallelepiped	$d \geq 2$	2^d	C2

$\tau(\mathcal{F}) \leq d(d+1)^{d-1} \cdot \nu(\mathcal{F})$ in (1) when d is sufficiently large. Also, (3) implies an upper bound $\tau(\mathcal{F}) \leq 3^d\theta_T(S) \cdot \nu(\mathcal{F})$ for any family $\mathcal{F}$ of translates of a centrally symmetric convex body S in $\mathbb{R}^d$. Schmidt [16] showed that, for any centrally symmetric convex body S, $\delta_L(S) = \Omega(d/2^d)$; hence (2) implies the bound $\tau(\mathcal{F}) \leq O(4^d/d)\theta_L(S) \cdot \nu(\mathcal{F})$. Note that $\theta_T(S) \leq \theta_L(S)$. So (3) may be also better than (2) for high dimensions. Table 1 summarizes the current best upper bounds on $\alpha(C)$ and $\beta(C)$ (obtained by us and by others) for various types of convex bodies C in $\mathbb{R}^d$:

A natural question is whether $\alpha(C)$ or $\beta(C)$ need to be exponential in d. The following theorem gives a positive answer:

Theorem 5. *For any convex body C in $\mathbb{R}^d$, $\beta(C) \geq \alpha(C) \geq \frac{\theta_T(C)}{\delta_T(C)}$. In particular, if C is the unit ball B^d in $\mathbb{R}^d$, then $\beta(C) \geq \alpha(C) \geq 2^{(0.599 \pm o(1))d}$ as $d \to \infty$.*

Kim et al. [14] asked whether the upper bound $\tau(\mathcal{F}) \leq 3 \cdot \nu(\mathcal{F})$ holds for any family $\mathcal{F}$ of translates of a centrally symmetric convex body in the plane. This upper bound, if true, is best possible because there exists a family $\mathcal{F}$ of congruent disks (i.e., translates of a disk) such that $\tau(\mathcal{F}) = 3 \cdot \nu(\mathcal{F})$ for any $\nu(\mathcal{F}) \geq 1$ [10]; see also [14, Example 10]. On the other hand, Karasev [12] proved that $\tau(\mathcal{F}) \leq 3 \cdot \nu(\mathcal{F}) = 3$ for any family $\mathcal{F}$ of pairwise-intersecting translates of a convex body in the plane. Also, for any family $\mathcal{F}$ of congruent disks such that $\nu(\mathcal{F}) = 2$, Kim et al. [14] confirmed that $\tau(\mathcal{F}) \leq 3 \cdot \nu(\mathcal{F}) = 6$. Our Corollary 1 confirms that $\tau(\mathcal{F}) \leq 2 \cdot \nu(\mathcal{F})$ for any family $\mathcal{F}$ of translates of a parallelogram. The following theorem confirms the upper bound $\tau(\mathcal{F}) \leq 3 \cdot \nu(\mathcal{F})$ for another special case:

Theorem 6. *For any family $\mathcal{F}$ of translates of a centrally symmetric convex hexagon, $\tau(\mathcal{F}) \leq 3 \cdot \nu(\mathcal{F})$. Moreover, if $\nu(\mathcal{F}) = 1$, then $\tau(\mathcal{F}) \leq 2$.*

Table 2. Lower and upper bounds on $\alpha(C)$ and $\beta(C)$ for special convex bodies C in the plane

Special convex body C in the plane	$\alpha(C)$ *lower*		$\alpha(C)$ *upper*		$\beta(C)$ *lower*		$\beta(C)$ *upper*	
centrally symmetric convex hexagon	2	[10]	3	T6	2	[10]	7	T4
square	$\frac{3}{2}$	[11]	2	T3	$\frac{3}{2}$	[11]	4	T4
triangle	3	[4]	5	T3	3	[4]	12	T4
disk	3	[10]	5	T3	4	[10]	7	T4

Grünbaum [10] showed that $\alpha_1(C) = 2$ for any affinely regular hexagon C. Theorem 6 implies a stronger and more general result that $2 = \alpha_1(C) \leq \alpha(C) \leq 3$ for any centrally symmetric convex hexagon C. The example in Figure 1 gives the bound $\frac{3}{2} \leq \alpha(C) \leq 2$ for any square C. For any triangle or disk C, it follows by Theorem 3 (ii) and (iii) that $\alpha(C) \leq 5$, and we have the lower bound $\alpha(C) \geq \alpha_1(C) = 3$ [4,10]. Theorem 4 (i), (ii), and (iii) imply that $\beta(C) \leq 4$ for any square C, $\beta(C) \leq 12$ for any triangle C, and $\beta(C) \leq 7$ for any disk C. We also have the lower bounds $\beta(C) \geq \alpha(C) \geq \frac{3}{2}$ for any square C, $\beta(C) \geq \beta_1(C) = 3$ for any triangle C [4], and $\beta(C) \geq \beta_1(C) = 4$ for any disk C [10]. Table 2 summarizes the current best bounds on $\alpha(C)$ and $\beta(C)$ for some special convex bodies C in the plane:

2 Upper Bound for Translates of an Arbitrary Convex Body in $\mathbb{R}^d$

In this section we prove Theorem 1. Let $\mathcal{F}$ be a family of translates of a convex body C in $\mathbb{R}^d$. Let P and Q be any two parallelepipeds in $\mathbb{R}^d$ that are parallel to each other, such that $P \subseteq C \subseteq Q$. Since the two values $\tau(\mathcal{F})$ and $\nu(\mathcal{F})$ are invariant under any non-singular affine transformation of C, we can assume that P and Q are axis-parallel and have edge lengths 1 and e_i, respectively, along the axis x_i, $1 \leq i \leq d$.

We first show that $\tau(\mathcal{T}) \leq \lceil e_d \rceil \cdot \nu(\mathcal{T})$ for any family $\mathcal{T}$ of C-translates whose corresponding P-translates intersect a common line ℓ parallel to the axis x_d. Define the *x_d coordinate* of a C-translate as the smallest x_d-coordinate of a point in the corresponding P-translate. Set $\mathcal{T}_1 = \mathcal{T}$, let C_1 be the C-translate in $\mathcal{T}_1$ with the smallest x_d-coordinate, and let $\mathcal{S}_1$ be the subfamily of C-translates in $\mathcal{T}_1$ that intersect C_1 ($\mathcal{S}_1$ includes C_1 itself). Then, for increasing values of i, while $\mathcal{T}_i = \mathcal{T} \setminus \bigcup_{j=1}^{i-1} \mathcal{S}_j$ is not empty, let C_i be the C-translate in $\mathcal{T}_i$ with the smallest x_d-coordinate, and let $\mathcal{S}_i$ be the subfamily of C-translates in $\mathcal{T}_i$ that intersect C_i. The iterative process ends with a partition $\mathcal{T} = \bigcup_{i=1}^{m} \mathcal{S}_i$, where $m \leq \nu(\mathcal{T})$.

Denote by c_i the x_d-coordinate of C_i. Then each C-translate in the subfamily $\mathcal{S}_i$, which is contained in a Q-translate of edge length e_d along the axis x_d, has an x_d-coordinate of at least c_i and at most $c_i + e_d$, and the corresponding P-translate, whose edge length along the axis x_d is 1, contains at least one of the $\lceil e_d \rceil$ points on ℓ with x_d-coordinates $c_i + 1, \ldots, c_i + \lceil e_d \rceil$. These $\lceil e_d \rceil$ points form a piercing set for $\mathcal{S}_i$, hence $\tau(\mathcal{S}_i) \leq \lceil e_d \rceil$. It follows that

$$\tau(\mathcal{T}) \leq \sum_{i=1}^{m} \tau(\mathcal{S}_i) \leq \lceil e_d \rceil \cdot m \leq \lceil e_d \rceil \cdot \nu(\mathcal{T}). \tag{5}$$

For $(a_1, \ldots, a_{d-1}) \in \mathbb{R}^{d-1}$, denote by $\ell(a_1, \ldots, a_{d-1})$ the following line in $\mathbb{R}^d$ that is parallel to the axis x_d:

$$\{ (x_1, \ldots, x_d) \mid (x_1, \ldots, x_{d-1}) = (a_1, \ldots, a_{d-1}) \}.$$

Now consider the following (infinite) set $\mathcal{L}$ of parallel lines:

$$\{ \ell(j_1 + b_1, \ldots, j_{d-1} + b_{d-1}) \mid (j_1, \ldots, j_{d-1}) \in \mathbb{Z}^{d-1} \},$$

where $(b_1, \ldots, b_{d-1}) \in \mathbb{R}^{d-1}$ is chosen such that no line in $\mathcal{L}$ is tangent to the P-translate of any C-translate in $\mathcal{F}$. Recall that P and Q are axis-parallel and have edge lengths 1 and e_i, respectively, along the axis x_i, $1 \le i \le d$. So we have the following two properties:

1. For any C-translate in $\mathcal{F}$, the corresponding P-translate intersects exactly one line in $\mathcal{L}$.
2. For any two C-translates in $\mathcal{F}$, if the two corresponding P-translates intersect two different lines in $\mathcal{L}$ of distance at least $e_i + 1$ along some axis x_i, $1 \le i \le d-1$, then the two C-translates are disjoint.

Partition $\mathcal{F}$ into subfamilies $\mathcal{F}(j_1, \ldots, j_{d-1})$ of C-translates whose corresponding P-translates intersect a common line $\ell(j_1 + b_1, \ldots, j_{d-1} + b_{d-1})$. Let $\mathcal{F}'(k_1, \ldots, k_{d-1})$ be the union of the families $\mathcal{F}(j_1, \ldots, j_{d-1})$ such that $j_i \bmod \lceil e_i + 1 \rceil = k_i$ for $1 \le i \le d-1$. It follows from (5) that the transversal number of each subfamily $\mathcal{F}'(k_1, \ldots, k_{d-1})$ is at most $\lceil e_d \rceil$ times its packing number. Therefore we have

$$\begin{aligned} \tau(\mathcal{F}) \le \sum_{(k_1,\ldots,k_{d-1})} \tau\left(\mathcal{F}'(k_1, \ldots, k_{d-1})\right) &\le \lceil e_d \rceil \sum_{(k_1,\ldots,k_{d-1})} \nu\left(\mathcal{F}'(k_1, \ldots, k_{d-1})\right) \\ &\le \left(\lceil e_d \rceil \prod_{i=1}^{d-1} \lceil e_i + 1 \rceil \right) \cdot \nu(\mathcal{F}). \end{aligned} \tag{6}$$

Since (6) holds for any pair of parallelepipeds P and Q in $\mathbb{R}^d$ that are parallel to each other and satisfy $P \subseteq C \subseteq Q$, it follows by the definition of $\gamma(C)$ that $\tau(\mathcal{F}) \le \gamma(C) \cdot \nu(\mathcal{F})$. By Lemma 1, there indeed exist two such parallelepipeds P and Q with length ratios $\lambda_i(P, Q) = d$ for $1 \le i \le d$. It then follows that $\gamma(C) \le d(d+1)^{d-1}$ for any convex body C in $\mathbb{R}^d$. This completes the proof of Theorem 1.

3 Upper Bound for Translates of a Centrally Symmetric Convex Body in $\mathbb{R}^d$

In this section we prove Theorem 2. Recall that $|C|$ is the Lebesgue measure of a convex body C in $\mathbb{R}^d$, and that $|\mathcal{F}|$ is the Lebesgue measure of the union of a family $\mathcal{F}$ of convex bodies in $\mathbb{R}^d$. To establish the desired bound on $\tau(\mathcal{F})$ in terms of $\nu(\mathcal{F})$ for any family $\mathcal{F}$ of translates of a centrally symmetric convex body S in $\mathbb{R}^d$, we link both $\tau(\mathcal{F})$ and $\nu(\mathcal{F})$ to the ratio $|\mathcal{F}|/|S|$. We first prove a lemma that links the transversal number $\tau(\mathcal{F})$ to the ratio $|\mathcal{F}|/|S|$ via the lattice covering density of S:

Lemma 3. *Let $\mathcal{F}$ be a family of translates of a centrally symmetric convex body S in $\mathbb{R}^d$. If there is a lattice covering of $\mathbb{R}^d$ with translates of S whose covering density is θ, $\theta \geq 1$, then $\tau(\mathcal{F}) \leq \theta \cdot |\mathcal{F}|/|S|$.*

Proof. Denote by S_p a translate of the convex body S centered at a point p. Since S is centrally symmetric, for any two points p and q, p intersects S_q if and only if q intersects S_p. Given a lattice covering of $\mathbb{R}^d$ with translates of S, every point $p \in \mathbb{R}^d$ is contained in some translate S_q in the lattice covering, hence every translate S_p contains some lattice point q.

Let Λ be a lattice such that the corresponding lattice covering with translates of S has a covering density of θ. Divide the union of the convex bodies in $\mathcal{F}$ into pieces by the cells of the lattice Λ, then translate all cells (and the pieces) to a particular cell, say σ. By the pigeonhole principle, there exists a point in σ, say p, that is covered at most $\lfloor |\mathcal{F}|/|\sigma| \rfloor$ times by the overlapping pieces of the union. Let k be the number of times that p is covered by the pieces. Now fix $\mathcal{F}$ but translate the lattice Λ to Λ' until p becomes a lattice point of Λ'. Then exactly k lattice points of Λ' are covered by the S-translates in $\mathcal{F}$. Since every S-translate in $\mathcal{F}$ contains some lattice point of Λ', we have obtained a transversal of $\mathcal{F}$ consisting of $k \leq \lfloor |\mathcal{F}|/|\sigma| \rfloor$ lattice points of Λ'. Note that $\theta = |S|/|\sigma|$, and the proof is complete. □

The following lemma[1] is a dual of the previous lemma, and links the packing number $\nu(\mathcal{F})$ to the ratio $|\mathcal{F}|/|S|$ via the lattice packing density of S:

Lemma 4. *Let $\mathcal{F}$ be a family of translates of a centrally symmetric convex body S in $\mathbb{R}^d$. If there is a lattice packing in $\mathbb{R}^d$ with translates of S whose packing density is δ, $\delta \leq 1$, then $\nu(\mathcal{F}) \geq \frac{\delta}{2^d} \cdot |\mathcal{F}|/|S|$.*

Proof. Let S' be a homothet of S scaled up by a factor of 2. Since S is centrally symmetric, an S-translate is contained by an S'-translate if and only if the S-translate contains the center of the S'-translate. Given a lattice packing in $\mathbb{R}^d$ with translates of S', two S'-translates centered at two different lattice points are disjoint, hence two S-translates containing two different lattice points are disjoint.

Let Λ be a lattice such that the corresponding lattice packing with translates of S' has a packing density of δ (such a lattice exists because S' is homothetic to S). Divide the union of the convex bodies in $\mathcal{F}$ into pieces by the cells of the lattice Λ, then translate all cells (and the pieces) to a particular cell, say σ. By the pigeonhole principle, there exists a point in σ, say p, that is covered at least $\lceil |\mathcal{F}|/|\sigma| \rceil$ times by the overlapping pieces of the union. Let k be the number of times that p is covered by the pieces. Now fix $\mathcal{F}$ but translate the lattice Λ to Λ' until p becomes a lattice point of Λ'. Then exactly k lattice points of Λ' are covered by the S-translates in $\mathcal{F}$. Choose k translates in $\mathcal{F}$, each containing a distinct lattice point of Λ'. Since any two S-translates containing two different lattice points of Λ' are disjoint, we have obtained a subset of $k \geq \lceil |\mathcal{F}|/|\sigma| \rceil$ pairwise-disjoint S-translates in $\mathcal{F}$. Note that $\delta = |S'|/|\sigma| = 2^d |S|/|\sigma|$, and the proof is complete. □

Lemma 3 and Lemma 4 are then connected by the following "sandwich" lemma:

[1] The planar case of this lemma is implied by a recent result [2, Theorem 5].

Lemma 5. *Let $\mathcal{F}$ be a family of translates of a (not necessarily centrally symmetric) convex body C in $\mathbb{R}^d$. Let A and B be two centrally symmetric convex bodies in $\mathbb{R}^d$ such that $A \subseteq C \subseteq B$. Then*

$$\tau(\mathcal{F}) \leq 2^d \cdot \frac{|B|}{|A|} \cdot \frac{\theta_L(A)}{\delta_L(B)} \cdot \nu(\mathcal{F}).$$

Proof. Since $A \subseteq C$, it follows by Lemma 3 that

$$\tau(\mathcal{F}) \leq \theta_L(A) \cdot \frac{|\mathcal{F}|}{|A|}.$$

Since $C \subseteq B$, it follows by Lemma 4 that

$$\nu(\mathcal{F}) \geq \frac{\delta_L(B)}{2^d} \cdot \frac{|\mathcal{F}|}{|B|}.$$

Putting these together yields

$$\tau(\mathcal{F}) \leq \theta_L(A) \cdot \frac{|\mathcal{F}|}{|A|} = 2^d \cdot \frac{|B|}{|A|} \cdot \frac{\theta_L(A)}{\delta_L(B)} \cdot \frac{\delta_L(B)}{2^d} \cdot \frac{|\mathcal{F}|}{|B|} \leq 2^d \cdot \frac{|B|}{|A|} \cdot \frac{\theta_L(A)}{\delta_L(B)} \cdot \nu(\mathcal{F}).$$

□

By setting $A = B = C$ in Lemma 5, we obtain (2) in Theorem 2. For the planar case, the following lemma is now folklore [3, Theorems 2.5 and 2.8]:

Lemma 6. *For any centrally symmetric convex body S in the plane, there are two centrally symmetric convex hexagons H and H' such that $H \subseteq S \subseteq H'$ and $|H|/|H'| \geq 3/4$.*

Note that $\theta_L(H) = \delta_L(H) = 1$ for a centrally symmetric convex hexagon H. Set $A = H$, $B = H'$, and $C = S$ in the previous two lemmas, and we have, for any family $\mathcal{F}$ of translates of a centrally symmetric convex body in the plane,

$$\tau(\mathcal{F}) \leq 2^2 \cdot \frac{4}{3} \cdot \frac{1}{1} \cdot \nu(\mathcal{F}) = \frac{16}{3} \cdot \nu(\mathcal{F}).$$

This completes the proof of Theorem 2.

4 Upper Bounds by Greedy Decomposition

In this section we sketch the proofs of Theorems 3 and 4 (the discussion of the special cases is omitted). First let $\mathcal{F}$ be a family of translates of a convex body C in $\mathbb{R}^d$. Without loss of generality, assume that $\kappa((C - C) \cap L, C)$ is minimized when $L = \{(x_1, \ldots, x_d) \mid x_d \geq 0\}$. Perform a *greedy decomposition* as follows. For $i = 1, 2, \ldots$, while $\mathcal{T}_i = \mathcal{F} \setminus \bigcup_{j=1}^{i-1} \mathcal{S}_j$ is not empty, let C_i be the translate of C in $\mathcal{T}_i$ that contains a point of the largest x_d-coordinate, and let $\mathcal{S}_i$ be the subfamily of translates in $\mathcal{T}_i$ that intersect C_i ($\mathcal{S}_i$ includes C_i itself). The iterative process ends with a partition $\mathcal{F} = \bigcup_{i=1}^{m} \mathcal{S}_i$, where $m \leq \nu(\mathcal{F})$. We next show that $\tau(\mathcal{S}_i) \leq \kappa((C - C) \cap L, C)$.

Choose any point in C as a reference point. We have the following lemma:

Lemma 7. *Let A and B be two translates of C with reference points a and b, respectively.* (i) *A contains b if and only if $-(B-b)+b$ contains a.* (ii) *If A intersects B, then a is contained in a translate of $C-C$ centered at b.*

By Lemma 7 (ii), the reference point of each translate of C in $\mathcal{S}_i$ is contained in a translate of $C-C$ centered at the reference point of C_i. Since the translate of $C-C$ is covered by $\kappa(C-C,-C)$ translates of $-C$, it follows by Lemma 7 (i) that each translate of C in $\mathcal{S}_i$ contains one of the $\kappa(C-C,-C)$ corresponding reference points. Therefore,

$$\tau(\mathcal{S}_i) \leq \kappa(C-C,-C) = \kappa(C-C,C). \tag{7}$$

The stronger bound $\tau(\mathcal{S}_i) \leq \kappa((C-C)\cap L,C)$ follows by our choice of C_i. We have

$$\tau(\mathcal{F}) \leq \sum_{i=1}^{m} \tau(\mathcal{S}_i) \leq \kappa((C-C)\cap L,C)\cdot m \leq \kappa((C-C)\cap L,C)\cdot \nu(\mathcal{F}).$$

In the special case that C is a centrally symmetric convex body in the plane, $C-C$ is a translate of $2C$. Assume without loss of generality that C is centered at the origin. Then $C-C=2C$. We have the following lemma on covering $2C$ with translates of C, which is implicit in a result by Grünbaum [10, Theorem 4]:

Lemma 8. (Grünbaum [10]). *Let C be a centrally symmetric convex body in the plane. Then $2C$ can be covered by seven translates of C, including one translate concentric with $2C$ and six others centered at the six vertices, respectively, of an affinely regular hexagon H_C concentric with $2C$.*

Choose the halfplane L through the center of $2C$ and any two opposite vertices of the hexagon H_C in Lemma 8. Then $\kappa((C-C)\cap L,C)\leq 5$. It follows that $\tau(\mathcal{F})\leq 5\cdot\nu(\mathcal{F})$ for any family $\mathcal{F}$ of translates of a centrally symmetric convex body in the plane.

Next let $\mathcal{F}$ be a family of homothets of a convex body C in $\mathbb{R}^d$. We again use greedy decomposition. The only difference in the algorithm is that C_i is now chosen as the smallest homothet of C in $\mathcal{T}_i$. By our choice of C_i, each homothet in $\mathcal{S}_i$ contains a translate of C_i that intersects C_i. Hence the bound $\tau(\mathcal{S}_i)\leq\kappa(C-C,C)$ follows in a similar way as the derivation of (7). It then follows that $\tau(\mathcal{F})\leq\kappa(C-C,C)\cdot\nu(\mathcal{F})$. By Lemma 8, $\kappa(C-C,C)\leq 7$ if C is a centrally symmetric convex body in the plane.

5 Concluding Remarks

A computational problem related to the results in this paper is finding a minimum-cardinality point set that pierces a given set of geometric objects. This problem is NP-hard even for the special case of axis-parallel unit squares in the plane [9], and it admits a polynomial-time approximation scheme for the general case of fat objects in $\mathbb{R}^d$ [5]. The approximation scheme has a very high time complexity of $n^{O(1/\epsilon^d)}$, and hence is impractical. Our methods for obtaining the upper bounds in Theorems 1, 2, 3, and 4 are constructive and lead to efficient constant-factor approximation algorithms for piercing a set of translates or homothets of a convex body. The approximation factors, which depend on the dimension d, are the multiplicative factors in the respective bounds on $\tau(\mathcal{F})$ in terms of $\nu(\mathcal{F})$ in the theorems, see also Table 1 and Table 2. For instance, Theorem 1 yields a factor-6 approximation algorithm for piercing translates of a convex body in the plane, and Theorem 4 yields a factor-216 approximation algorithm for piercing homothets of a convex body in 3-space.

References

1. Alon, N., Kleitman, D.J.: Piercing convex sets and the Hadwiger-Debrunner (p, q)-problem. Advances in Mathematics 96, 103–112 (1992)
2. Bereg, S., Dumitrescu, A., Jiang, M.: On covering problems of Rado. Algorithmica, doi:10.1007/s00453-009-9298-z (to appear); A preliminary version in: Proceedings of the 11th Scandinavian Workshop on Algorithm Theory, pp. 294–305 (2008)
3. Braß, P., Moser, W., Pach, J.: Research Problems in Discrete Geometry. Springer, New York (2005)
4. Chakerian, G.D., Stein, S.K.: Some intersection properties of convex bodies. Proceedings of the American Mathematical Society 18, 109–112 (1967)
5. Chan, T.: Polynomial-time approximation schemes for packing and piercing fat objects. Journal of Algorithms 46, 178–189 (2003)
6. Danzer, L.: Zur Lösung des Gallaischen Problems über Kreisscheiben in der Euklidischen Ebene. Studia Scientiarum Mathematicarum Hungarica 21, 111–134 (1986)
7. Danzer, L., Grünbaum, B., Klee, V.: Helly's theorem and its relatives. In: Proceedings of Symposia in Pure Mathematics., vol. 7, pp. 101–181. American Mathematical Society (1963)
8. Fon-Der-Flaass, D.G., Kostochka, A.V.: Covering boxes by points. Discrete Mathematics 120, 269–275 (1993)
9. Fowler, R.J., Paterson, M.S., Tanimoto, S.L.: Optimal packing and covering in the plane are NP-complete. Information Processing Letters 12, 133–137 (1981)
10. Grünbaum, B.: On intersections of similar sets. Portugaliae Mathematica 18, 155–164 (1959)
11. Gyárfás, A., Lehel, J.: Covering and coloring problems for relatives of intervals. Discrete Mathematics 55, 167–180 (1985)
12. Karasev, R.N.: Transversals for families of translates of a two-dimensional convex compact set. Discrete and Computational Geometry 24, 345–353 (2000)
13. Károlyi, G.: On point covers of parallel rectangles. Periodica Mathematica Hungarica 23, 105–107 (1991)
14. Kim, S.-J., Nakprasit, K., Pelsmajer, M.J., Skokan, J.: Transversal numbers of translates of a convex body. Discrete Mathematics 306, 2166–2173 (2006)
15. Rogers, C.A.: A note on coverings. Mathematika 4, 1–6 (1957)
16. Schmidt, W.M.: On the Minkowski-Hlawka theorem. Illinois Journal of Mathematics 7, 18–23 (1963)
17. Wenger, R.: Helly-type theorems and geometric transversals. In: Handbook of Discrete and Computational Geometry, 2nd edn., pp. 73–96. CRC Press, Boca Raton (2004)

Output-Sensitive Algorithms for Enumerating Minimal Transversals for Some Geometric Hypergraphs

Khaled Elbassioni[1], Kazuhisa Makino[2], and Imran Rauf[1]

[1] Max-Planck-Institut für Informatik, Saarbrücken, Germany
{elbassio,irauf}@mpi-inf.mpg.de
[2] Graduate School of Information Science and Technology, University of Tokyo, Tokyo, Japan
makino@mist.i.u-tokyo.ac.jp

Abstract. We give a general framework for the problem of finding all minimal hitting sets of a family of objects in $\mathbb{R}^d$ by another. We apply this framework to the following problems: (i) hitting hyper-rectangles by points in $\mathbb{R}^d$; (ii) stabbing connected objects by axis-parallel hyperplanes in $\mathbb{R}^d$; and (iii) hitting half-planes by points. For both the covering and hitting set versions, we obtain incremental polynomial-time algorithms, provided that the dimension d is fixed.

1 Introduction

Let $\mathcal{V}$ and $\mathcal{F}$ be two finite sets of geometric objects in $\mathbb{R}^d$. A subset of objects $\mathcal{X} \subseteq \mathcal{V}$ is said to be a *hitting set* (or *transversal* or *cover*) for $\mathcal{F}$ if for every $O \in \mathcal{F}$, there exists an $O' \in \mathcal{X}$ such that $O \cap O' \neq \emptyset$. A hitting set is *minimal* if none of its proper subsets is also a hitting set.

In this paper, we are interested in finding all minimal hitting sets of one family of objects by another. For such generation problems, we measure the time complexity in terms of both input and output length. An algorithm is said to run in *incremental polynomial-time*, if the time required to find k minimal transversals is polynomial in $|\mathcal{V}|$, $|\mathcal{F}|$, and k.

When $\mathcal{V}$ is a finite set of points and each object in $\mathcal{F}$ is an arbitrary finite subset of $\mathcal{V}$, we obtain the well-known *hypergraph transversal* or *dualization* problem [2], which calls for finding all minimal hitting sets for a given hypergraph $\mathcal{G} \subseteq 2^V$, defined on a finite set of vertices V. Denote by $\mathrm{Tr}(\mathcal{G})$ the set of all minimal hitting sets of $\mathcal{G}$, also known as the *transversal hypergraph* of $\mathcal{G}$. The problem of finding $\mathrm{Tr}(\mathcal{G})$ has received considerable attention in the literature (see, e.g., [3,12,13,19,29,31]), since it is known to be polynomially or quasi-polynomially equivalent with many problems in various areas, such as artificial intelligence (e.g., [12,24]), database theory (e.g., [30]), distributed systems (e.g., [23]), machine learning and data mining (e.g., [1,7,20]), mathematical programming (e.g., [5,25]), matroid theory (e.g., [26]), and reliability theory (e.g., [9]).

A. Fiat and P. Sanders (Eds.): ESA 2009, LNCS 5757, pp. 143–154, 2009.

The currently fastest known algorithm [17] for solving the hypergraph transversal problem runs in quasi-polynomial time $|V|N^{o(\log N)}$, where N is the combined input and output size $N = |\mathcal{G}| + |\operatorname{Tr}(\mathcal{G})|$. Several quasi-polynomial time algorithms with some other desirable properties also exist [8,16,18,33]. While it is still open whether the problem can be solved in polynomial time for arbitrary hypergraphs, polynomial time algorithms exist for several classes of hypergraphs, e.g. hypergraphs of bounded edge-size [4,12], of bounded-degree [11,13], of bounded-edge intersections [4], of bounded conformality [4], of bounded treewidth [13], and read-once (exact) hypergraphs [15].

Almost all previously known polynomial-time algorithms for the the hypergraph transversal problem assume that at least one of the hypergraphs $\mathcal{G}$ or $\operatorname{Tr}(\mathcal{G})$ either (i) has bounded size $\min\{|\mathcal{G}|, |\operatorname{Tr}(\mathcal{G})|\} \leq k$, (ii) is k-conformal[1], or (iii) is k-degenerate[2], for a constant k. One can verify that all the special classes mentioned above belong to one of these categories.

In this paper, we shall extend these polynomially solvable classes to include hypergraphs arising in geometry. More precisely, we consider the following problems $\text{HIT}(\mathcal{V}, \mathcal{F})$:

- **Hitting hyper-rectangles by points:** Given a finite set of points $\mathcal{V} \subseteq \mathbb{R}^d$ and a finite collection $\mathcal{F}$ of axis-parallel hyper-rectangles (also called orthotopes or boxes) in $\mathbb{R}^d$, find all minimal sets of points from $\mathcal{V}$ that hit every hyper-rectangle in $\mathcal{F}$;
- **Hitting (Stabbing) connected objects by axis-parallel hyperplanes:** Given a finite set of axis-parallel hyperplanes $\mathcal{V} \subseteq \mathbb{R}^d$ and a finite collection $\mathcal{F}$ of connected objects in $\mathbb{R}^d$, find all minimal sets of hyperplanes from $\mathcal{V}$ that stab every object in $\mathcal{F}$;
- **Hitting half-spaces by points:** Given a finite set of points $\mathcal{V} \subset \mathbb{R}^d$ and a finite collection $\mathcal{F}$ of half-spaces in $\mathbb{R}^d$, find all minimal sets of points from $\mathcal{V}$ that hit every half-space in $\mathcal{F}$.

We show that the first two problems can be solved in incremental polynomial time, if the dimension d of the underlying space is bounded, and that the last problem can be solved in incremental polynomial time, if $d = 2$.

To construct efficient algorithms for the above problems, we first propose a general framework to solve the hypergraph transversal problem, which can be regarded as a generalization of the algorithms given in [13,28], and apply it to the above problems. We remark that when we apply the framework to the problem of hitting half-planes by points, we need to run a backtracking algorithm at the base level of the recursion in the framework. While such an algorithm is inefficient in general, as it requires solving an NP-hard problem as a subroutine, we exploit the geometry to show that it can be made to work in the case of hitting half-planes by points.

---

[1] A hypergraph is said to be k-conformal [2] if any set $X \subseteq V$ is contained in a hyperedge of $\mathcal{G}$ whenever each subset of X of cardinality at most k is contained in a hyperedge of $\mathcal{G}$.

[2] A hypergraph $\mathcal{G}$ is said to be k-degenerate [13] if for every set $X \subseteq V$, the minimum degree of a vertex in the induced hypergraph $\mathcal{G}_X$ on X is at most k.

We also consider the covering versions $\text{COVER}(\mathcal{V}, \mathcal{F})$ $(= \text{HIT}(\mathcal{F}, \mathcal{V}))$ of the above problems. For example, we consider the problem of finding all minimal sets of hyper-rectangles from $\mathcal{F}$ that hit all points in $\mathcal{V}$. We propose incremental polynomial-time algorithms for finding all minimal covers for the first two problems, by exploiting the fact that the geometric hypergraphs arising in the first two problems have the bounded *Helly property* [2], and show that minimal covers for the last problem can be generated in incremental polynomial time, by using *geometric duality* from the corresponding result for minimal hitting sets.

The enumeration of minimal geometric hitting sets, as the ones described above, arises in various areas such as computational geometry, machine learning, and data mining [14]. Moreover, our efficient enumeration algorithm might be useful in developing exact algorithms, fixed-parameter tractable algorithms, and polynomial-time approximation schemes for the corresponding optimization problems (see, e.g., [22]).

The rest of this paper is organized as follows. In the next section, we give a general framework for finding all minimal hitting sets for a given hypergraph. In Sections 3, 4 and 5, we apply this framework to hitting hyper-rectangles by points, stabbing connected objects by axis-parallel hyperplanes, and hitting half-planes by points.

2 A Framework for Computing Transversal Hypergraphs

Let $\mathcal{G} \subseteq 2^V$ be a hypergraph, and σ be an ordering of vertices in V. For $i = 1, \ldots, n$, let $\mathcal{G}_i$ be the sub-hypergraphs of $\mathcal{G}$ defined as $\mathcal{G}_i = \{G \in \mathcal{G} : G \subseteq \{v_{\sigma(1)}, \ldots, v_{\sigma(i)}\}\}$. Let us denote the set of hyperedges in $\mathcal{G}_i$ which are not contained in $\mathcal{G}_{i-1}$ as Δ_i, i.e., $\Delta_i = \mathcal{G}_i \setminus \mathcal{G}_{i-1}$ and for a set $X \subseteq V$, let $\Delta_i[X] = \{G \in \Delta_i : G \cap X = \emptyset\}$. Given a hypergraph $\mathcal{G}$, Eiter et.al. [13] describe an algorithm to generate $\text{Tr}(\mathcal{G})$, the hypergraph consisting of all minimal transversals of $\mathcal{G}$. The algorithm proceeds inductively, for $i = 1, \ldots, n$, by extending each minimal transversal X in $\text{Tr}(\mathcal{G}_{i-1})$ to a set in $\text{Tr}(\mathcal{G}_i)$ by finding $\text{Tr}(\Delta_i[X])$, each set of which is combined with X to obtain a minimal transversal of $\mathcal{G}_i$. In Figure 1, we present a generic algorithm which recursively reduces the problem of finding $\text{Tr}(\mathcal{G}_i)$ into the smaller subproblems of computing $\text{Tr}(\Delta_i[X])$ for $i \in [n]$ and $X \in \text{Tr}(\mathcal{G}_{i-1})$.

The algorithm uses a sequence of permutations $\Sigma = \sigma_1 \ldots \sigma_k$ as a part of an input. When called initially as DUALIZE-INC$(\mathcal{G}, \Sigma, 1)$, it dualizes $\mathcal{G}$ by using the above mentioned approach where σ_j is used for partitioning in the j-th level of the recursion. The operator minimal$(\mathcal{H})$ in Step 9 returns the hypergraph obtained from a given hypergarph $\mathcal{H}$ by removing the non-minimal edges.

After k levels of recursion, procedure DUALIZE-SIMPLE() is used directly to solve the problem. As we will see in the later sections, for several classes of geometric hypergraphs, the subproblem after k levels can be solved easily, where k depends only on the dimension of the geometric space under consideration.

As an illustration, consider the problem of dualizing an *interval* hypergraph: Let $v_1, v_2, \ldots, v_n$ be a set of points on the line ordered from left to right, and

Procedure DUALIZE-INC($\mathcal{G}, \Sigma, j$)**:**
Input: A hypergraph $\mathcal{G}$ over $n = |V(\mathcal{G})|$ vertices, an index j ($\leq k$)
and a sequence $\Sigma = (\sigma_1, \ldots, \sigma_k)$ of permutations of vertices in $V(\mathcal{G})$.
Output: The hypergraph $\mathcal{G}^d$.
$\mathcal{G}_0 \leftarrow \emptyset$, $\mathcal{X}_0 \leftarrow \{\emptyset\}$, $\mathcal{X}_i \leftarrow \emptyset$ $\forall i = 1, \ldots, n$
for $i = 1, \ldots, n$ **do**
    Let $\mathcal{G}_i \leftarrow \{G \in \mathcal{G} \ : \ G \subseteq \{v_{\sigma_j(1)}, \ldots, v_{\sigma_j(i)}\}\}$
    for each $X \in \mathcal{X}_{i-1}$ **do**
        Let $\Delta_i[X] = \{G \in \mathcal{G}_i \setminus \mathcal{G}_{i-1} \ : \ G \cap X = \emptyset\}$
        if $j \geq k$ or $|\Delta_i[X]| \leq 1$ **then**
            $\mathcal{A} \leftarrow$ **DUALIZE-SIMPLE**($\Delta_i[X]$)
        else $\mathcal{A} \leftarrow$ **DUALIZE-INC**($\Delta_i[X], \Sigma, j+1$)
        $\mathcal{X}_i \leftarrow \text{minimal}\,(\mathcal{X}_i \bigcup \{X \cup Y \ : \ Y \in \mathcal{A}\})$
return $\mathcal{X}_n$

Fig. 1. A generic sequential method for finding minimal transversals

let $\mathcal{G} \subseteq 2^V$ be a *Sperner*[3] hypergraph, in which each edge $G \in \mathcal{G}$ consists of consecutive points from V. Denote by σ the left-to-right ordering of the vertices, and consider the execution of the algorithm when called as DUALIZE-INC($\mathcal{G}, \Sigma, 1$) with $\Sigma = \sigma$. The algorithm incrementally dualizes the hypergraphs $\mathcal{G}_i = \{G \in \mathcal{G} : G \subseteq \{v_1, \ldots, v_i\}\}$ for $i = 1 \ldots n$. Note that the subproblem $\Delta_i = \mathcal{G}_i \setminus \mathcal{G}_{i-1}$ contains at most one edge because of our assumption that $\mathcal{G}$ is Sperner and thus can be solved trivially.

The correctness of the procedure follows from the following statement.

Proposition 1 ([13]). *For* $i = 1, \ldots, n$, $\text{Tr}(\mathcal{G}_i) = \textit{minimal}(\{X \cup Y \ : \ X \in \text{Tr}(\mathcal{G}_{i-1}),\ Y \in \text{Tr}(\Delta_i[X])\})$.

It is shown in [13] that the intermediate hypergraphs obtained in the algorithm never get too large, more specifically, $|\text{Tr}(\Delta_i[X])| \leq |\text{Tr}(\mathcal{G}_i)| \leq |\text{Tr}(\mathcal{G})|$. Consequently, we get the following bound on the worst-case running time.

Theorem 1. *Let* $\mathcal{G} \subseteq 2^V$ *be a hypergraph over vertex set* V *and* $\Sigma = (\sigma_1, \ldots, \sigma_k)$ *be a sequence of permutation functions of vertices of* $\mathcal{G}$. *Then the procedure DUALIZE-INC*($\mathcal{G}, \Sigma, 1$) *computes* $\text{Tr}(\mathcal{G})$ *in* $\mathcal{O}((nm')^k(mm' + T))$ *time, where* $n = |V|$, $m = |\mathcal{G}|$, $m' = |\text{Tr}(\mathcal{G})|$ *and* T *is time required by DUALIZE-SIMPLE in each k-th level recursion of the procedure.*

3 Points and Hyper-rectangles in $\mathbb{R}^d$

Let $\mathcal{V}$ be a set of points and $\mathcal{F}$ be a collection of axis-parallel hyper-rectangles in $\mathbb{R}^d$. In this section, we consider the problem of enumerating all minimal hitting sets for $\mathcal{F}$ from $\mathcal{V}$ as well as the related problem of enumerating all minimal

[3] A hypergraph $\mathcal{H}$ is said to be *Sperner* if no hyperedge of $\mathcal{H}$ contains another.

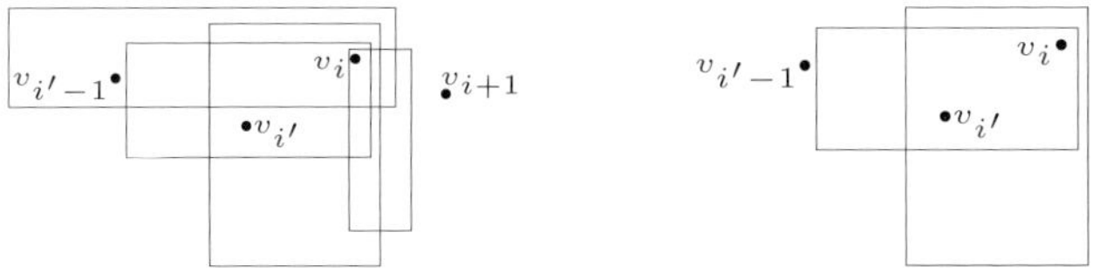

Fig. 2. An example of points and rectangles in $\mathbb{R}^2$. Left: The set Δ_i consists of all rectangles that contain v_i and other points only from the subset $\{v_1, \ldots, v_i\}$. Right: The subproblem in the recursive call considers all rectangles which contain both v_i and $v_{i'}$ and no points from the strict left of $v_{i'}$ nor from the strict right of v_i.

covers of $\mathcal{V}$ by $\mathcal{F}$. Let $\mathcal{G} \subseteq 2^V$ be the hypergraph defined by $V = \mathcal{V}$ and $\mathcal{G} = \mathcal{G}(\mathcal{V}, \mathcal{F}) \stackrel{\text{def}}{=} \{\{v \in \mathcal{V} : v \cap F \neq \emptyset\} : F \in \mathcal{F}\}$. Then the transposed hypergraph $\mathcal{G}^T \subseteq 2^{\mathcal{F}}$ is defined as $\mathcal{G}^T = \mathcal{G}(\mathcal{F}, \mathcal{V})$. Clearly, a minimal set of points from $\mathcal{V}$ hitting every hyper-rectangle in $\mathcal{F}$ corresponds to a minimal hitting set of $\mathcal{G}$, while a minimal set of hyper-rectangles from $\mathcal{F}$ covering every point in $\mathcal{V}$ corresponds to a minimal hitting set for $\mathcal{G}^T$. For a hypergraph $\mathcal{G}$, we will denote by $V(\mathcal{G})$ the vertex set of $\mathcal{G}$.

3.1 Minimal Hitting Sets

To illustrate the idea, let us first consider the problem in $\mathbb{R}^2$. The algorithm is based on the framework presented in Figure 1. We order the points in $\mathcal{V}$ from *left* to *right* and if their x-coordinates are equal, we sort them from *bottom* to *top*. Let $v_1, v_2, \ldots, v_n$ be the corresponding ordering of the vertices of the hypergraph $\mathcal{G} \subseteq 2^{\mathcal{V}}$, defined above. Note that because of our ordering of the vertices, no rectangle in the hypergraph $\mathcal{G}_i$ contains any point strictly to the right of v_i and by definition, every rectangle in $\Delta_i \subseteq \mathcal{G}_i$ contains v_i.

Consider the subproblem of dualizing $\Delta_i[X]$ for each $i \in [n]$ and $X \in \mathrm{Tr}(\mathcal{G}_{i-1})$ and let the primed variables denote the corresponding variables in the recursive call of the algorithm. We order the vertices of $\mathcal{G}' = \Delta_i[X]$ in the *reverse* order i.e., from *right* to *left*, breaking ties by sorting them from *top* to *bottom*.

Consider now a further recursive call of the procedure on a hypergraph $\mathcal{G}'' = \Delta'_{i'}[X']$, where $i' \in \{1, \ldots, |V(\mathcal{G}')|\}$ and $X' \in \mathrm{Tr}(\mathcal{G}'_{i'-1})$. The crucial observation is that each rectangle corresponding to an edge in the hypergraph $\Delta'_{i'}[X']$ contains both the points v_i and $v_{i'}$, and because of our ordering, no rectangle in the subproblem contains a point from the left of $v_{i'}$ nor from the right of v_i (see Figure 2). Hence, as can be easily seen, only the y-coordinates matter when deciding whether a given point $v \in \{v_{i'} \ldots v_i\}$ intersects a rectangle from $\Delta'_{i'}[X']$. So we can project the subproblem $\Delta'_{i'}[X']$ on the y-axis and reduce it to a problem of dualizing an interval hypergraph, which can be solved in polynomial time, as seen in Section 2.

The above algorithm can be extended to higher dimensions in an obvious manner. In dimension d, we use the orderings $\Sigma = (\sigma_1, \ldots, \sigma_{2d-2})$, where σ_i is the *lexicographical ordering* of the points using their last $d - \lfloor \frac{i-1}{2} \rfloor$ coordinates. Moreover, we define the ordering σ_i to be *increasing* when i is odd and *decreasing* otherwise. To generate all minimal transversals of $\mathcal{G}$, we call DUALIZE-INC $(\mathcal{G}, \Sigma, 1)$, and use the dualization procedure for interval hypergraphs in place of DUALIZE-SIMPLE(). After the second recursive call the subproblems we obtain contain all hyper-rectangles that intersect two given points, say $v_{i'}$ and v_i with $v_{i'}$ being lexicographically smaller than v_i, and have the property that they contain no point that is lexicographically smaller then $v_{i'}$ or lexicographically greater than v_i. Hence after two levels of recursion, the first coordinate of the points can be ignored and thus the problem reduces to $d-1$-dimensional subproblems.

3.2 Minimal Covers

As mentioned above, to generate all minimal covers of the given points $\mathcal{V}$ by hyper-rectangles from $\mathcal{F}$, we consider the *transposed* hypergraph $\mathcal{G}^T(\mathcal{V}, \mathcal{F})$ and compute its transversal hypergraph.

Halman [21] showed that any hypergraph $\mathcal{G} = \mathcal{G}(\mathcal{V}, \mathcal{F})$ defined by a set of hyper-rectangles $\mathcal{F}$ and a set of points $\mathcal{V}$ in $\mathbb{R}^d$ is $2d$*-Helly*, that is, for any $\mathcal{G}' \subseteq \mathcal{G}$, the following holds: if every $2d$ edges in $\mathcal{G}'$ have a non-empty intersection then all edges in $\mathcal{G}'$ have a non-empty intersection. Thus the transposed hypergraph $\mathcal{G}^T$ is $2d$-conformal (cf. [2]), and hence can be dualized by the algorithm of Khachiyan et al. [27] in incremental polynomial time.

Theorem 2. *Both problems* $\text{Hit}(\mathcal{V}, \mathcal{F})$ *and* $\text{Cover}(\mathcal{V}, \mathcal{F})$ *can be solved in time* $O((|\mathcal{V}||\mathcal{F}|k)^{O(d)})$*, when* $\mathcal{V}$ *and* $\mathcal{F}$ *are, respectively, a set of points and a set of axis-parallel hyper-rectangles in* $\mathbb{R}^d$*, and* k *is the size of the output.*

4 Stabbing Connected Objects in $\mathbb{R}^d$

4.1 Minimal Stabbing Sets

Given a collection of connected objects $\mathcal{F}$ and a set of axis-parallel hyperplanes $\mathcal{V}$, both in $\mathbb{R}^d$, we are interested in the problem of finding all minimal sets of hyperplanes from $\mathcal{V}$ such that every object in $\mathcal{F}$ is stabbed by at least one of the hyperplanes in the set. Let $\mathcal{G} = \mathcal{G}(\mathcal{V}, \mathcal{F})$ be the corresponding hypergraph with vertex set $\mathcal{V}$ and each object $F \in \mathcal{F}$ defining an edge consisting of all hyperplanes from $\mathcal{V}$ which intersect F.

We consider the simple case first, that is, when all hyperplanes in $\mathcal{V}$ are parallel to each other. This turns out to be equivalent to the interval hypergraph case since we can project the problem on any line L that is perpendicular to the hyperplanes in $\mathcal{V}$. The projection maps hyperplanes in $\mathcal{V}$ into points on L, and because of the connectivity, the objects in $\mathcal{F}$ are mapped to intervals on L.

More generally, for $d > 1$, assuming there is at least one hyperplane perpendicular to every principal axis, there are exactly d groups of axis-parallel hyperplanes in $\mathbb{R}^d$ such that every group contains hyperplanes that are parallel to one principal axis. This assumption can be made without loss of generality, since if there is no hyperplane along a particular principal axis, say z, then we

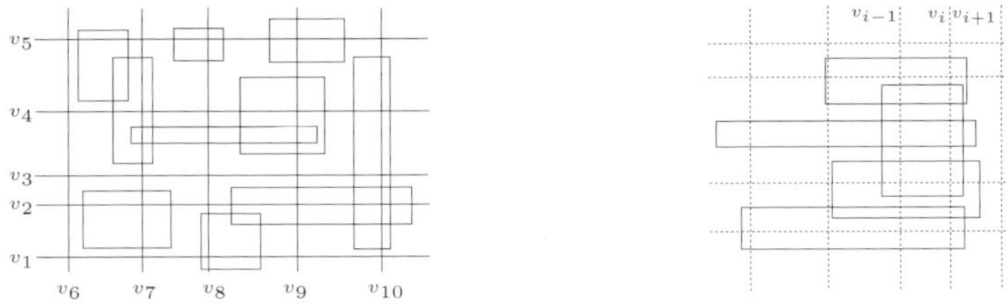

Fig. 3. Left: An example of lines and objects in $\mathbb{R}^2$ and a valid ordering of lines. Right: The set $\Delta_i[X]$ consists of new objects that intersect v_i and only lines from the subset $\{v_1, \dots, v_{i-1}\}$.

can orthogonally project all other hyperplanes and objects on the hyperplane $z = 0$ and reduce the dimension of the problem by one.

The dual of $\mathcal{G}$ can be found incrementally by following the algorithm of Figure 1. Fixing the order of principal axes, we order the hyperplanes sequentially: starting with hyperplanes perpendicular to the first principal axis, sorted in increasing order, we continue with the hyperplanes perpendicular to the second principal axis and so on. For an example in $\mathbb{R}^2$, the set of lines $\{x = 1, y = -1, x = 0, y = 1\}$ would be ordered as $x = 0, x = 1, y = -1, y = 1$ assuming that x-axis comes before y-axis in our fixed ordering of principal axes. Let σ be an ordering of hyperplanes in $\mathcal{G}$ as defined above and let $j_0^{\mathcal{G}} < j_1^{\mathcal{G}} < \dots < j_d^{\mathcal{G}}$ be indices with $j_0^{\mathcal{G}} = 0$ and $j_d^{\mathcal{G}} = n$, such that the hyperplanes in the group $\sigma(j_{r-1}^{\mathcal{G}} + 1), \dots, \sigma(j_r^{\mathcal{G}})$ are parallel to each other and perpendicular to r-th principal axis for $r \in [d]$.

Consider the subproblem of dualizing $\mathcal{G}_i$ for $i = 1, \dots, n$ as defined in the algorithm, where $\mathcal{G}_i$ contains only those edges which form subsets of vertices from $v_{\sigma(1)}, \dots, v_{\sigma(i)}$. As discussed above, the problem reduces to dualizing an interval hypergraph when $1 \le i \le j_1^{\mathcal{G}}$. Now consider the case when $j_{r-1}^{\mathcal{G}} < i \le j_r^{\mathcal{G}}$ for $r \in [d]$. Consider the subproblem of dualizing $\mathcal{G}' = \Delta_i[X]$ for $X \in \mathrm{Tr}(\mathcal{G}_{i-1})$, and let the primed variables denote the corresponding variables in the recursive call of the algorithm. Note that the subproblem for $\Delta_i[X]$ contains all objects that do not intersect any hyperplane "above" $v_{\sigma(i)}$. Let σ' be an ordering of vertices of $\mathcal{G}'$ defined similarly as σ for the hyperplanes perpendicular to first $r-1$ principal axes except the r-th group of hyperplanes which are sorted in decreasing order. As before, let $j_0^{\mathcal{G}'} < \dots < j_r^{\mathcal{G}'}$ be indices with $j_0^{\mathcal{G}'} = 0$ and $j_r^{\mathcal{G}'} = n'$, where $n' = |V(\mathcal{G}')|$ and the hyperplanes in the group $\sigma'(j_{r'-1}^{\mathcal{G}'} + 1), \dots, \sigma'(j_{r'}^{\mathcal{G}'})$ are parallel to each other and perpendicular to r'-th principal axis for $r' \in [r]$.

In the recursive call, we use σ' as our ordering and dualize $\mathcal{G}'$ incrementally by considering $\mathcal{G}'_{i'}$ for $1 < i' \le i$. Note that for $i' \le j_{r-1}^{\mathcal{G}'}$, $V(\mathcal{G}'_{i'}) \subseteq V(\mathcal{G}_{i-1})$ and hence the subproblem $\mathcal{G}'_{i'}$ is already taken care of when $\mathrm{Tr}(\mathcal{G}_{i-1})$ is computed[4]. Alternatively, when $j_{r-1}^{\mathcal{G}'} < i' \le i$ then similar to the case in Section 3.1, the subproblems $\Delta'_{i'}$ we get contain all objects that intersect both v_i and $v_{i'}$ with

[4] Note that the set of all minimal transversals of the sub-hypergraph $\mathcal{H}_S$ induced by vertices in S is equivalent to the set $\mathrm{Tr}(\mathcal{H}_S) = \mathrm{minimal}(\{H \cap S : H \in \mathrm{Tr}(\mathcal{H})\})$.

the property that no hyperplane above v_i or below $v_{i'}$ stabs any of them. Note that both $\{v_i\}$ and $\{v_{i'}\}$ as well as all hyperplanes between them are trivially hitting sets for the subproblems for hypergraphs of the form $\Delta'_{i'}[\cdot]$. The other hitting sets can be found recursively by observing that they do not involve any of the hyperplanes parallel to v_i and $v_{i'}$. Thus we are able to reduce the dimension of the problem by 1 after two levels of recursion.

In summary, to generate $\mathrm{Tr}(\mathcal{G})$, we call DUALIZE-INC($\mathcal{G}, \Sigma, 1$) with $\Sigma = (\sigma_1, \sigma_2, \ldots, \sigma_{2d-1})$ and a trivial dualization procedure for DUALIZE-SIMPLE(). For odd r, $1 \leq r < 2d$, the ordering σ_r sorts each group of parallel hyperplanes in increasing order (of the points of intersection with the common orthogonal line), whereas for even r, $1 < r < 2d$, σ_r is defined by sorting each group of parallel hyperplanes in increasing order except the last group, which is sorted in decreasing order (of the points of intersection with the common orthogonal line). As we noted above, at every r-th level of recursion for even r, the dual hypergraphs $\mathrm{Tr}(\mathcal{G}_i)$ such that $v_{\sigma(i)}$ does not belong to the last group of hyperplanes, can be easily computed from the corresponding dual hypergraph in the recursion level $r-1$. This observation can be used to avoid redundant computations by solving those subproblems directly instead of following the algorithm of Figure 1.

4.2 Minimal Covers

We use again the algorithm of [27] for dualizing conformal hypergraphs.

Lemma 1. *Let $\mathcal{F}$ be the set of connected objects and $\mathcal{V}$ be set of axis-parallel hyperplanes in $\mathbb{R}^d$. Then the corresponding hypergraph $\mathcal{G}(\mathcal{F}, \mathcal{V})$ is 2d-Helly.*

Theorem 3. *Both problems* HIT($\mathcal{V}, \mathcal{F}$) *and* COVER($\mathcal{V}, \mathcal{F}$) *can be solved in time $O((|\mathcal{V}||\mathcal{F}|k)^{O(d)})$, when $\mathcal{V}$ and $\mathcal{F}$ are, respectively, a set of axis-parallel hyperplanes and a set of connected objects in $\mathbb{R}^d$, and k is the size of the output.*

5 Hitting and Covering with Half-Planes

In this section we consider the case when the hypergraph $\mathcal{G} = \mathcal{G}(\mathcal{V}, \mathcal{F})$ is defined by a finite set of points $\mathcal{V} \subseteq \mathbb{R}^2$ and a set of half-planes $\mathcal{F}$. In the following, we will refer to the elements of $\mathcal{V}$ as points or vertices interchangeably.

5.1 Minimal Hitting Sets

Let $\Sigma = (\sigma'_1, \sigma''_1, \sigma'_2, \sigma''_2)$, where for $i \in \{1, 2\}$, σ'_i and σ''_i are respectively the ascending and descending orderings of the points in $\mathcal{V}$ along the ith coordinate. We call the procedure DUALIZE-INC($\mathcal{G}, \Sigma, 1$).

When the procedure reaches the base level, we end-up with an instance on a hypergraph $\mathcal{G}' = \mathcal{G}(\mathcal{V}', \mathcal{F}')$ in which all points $\mathcal{V}$ lie inside a rectangle R, the boundary of which contains four (not necessarily distinct) points $p_1, p_2, p_3, p_4 \in \mathcal{V}$, one on each side of R, and such that for each $F \in \mathcal{F}'$, $F \supseteq \{p_1, p_2, p_3, p_4\}$. Consider the polygon connecting these four points: it partitions the rectangle R

Fig. 4. Left: All hyperplanes in the subproblem contain the points $\{p_1, p_2, p_3, p_4\}$. Right: One of the subproblems consists of all points in the right-angle triangle and all halfplanes containing the longest side of it.

into 5 regions (some possibly empty); see Figure 4, Left. Let $B \subseteq \mathcal{V}'$ be the set of points inside the polygon and A_1, A_2, A_3, A_4 be the sets of points inside the other four regions. Note that, without loss of generality, we can assume that any line defining a half-plane in $\mathcal{F}'$ intersects *exactly one* of the 4 regions (otherwise, there is only one hyperplane and $\mathrm{Tr}(\mathcal{G}') = \{\{v\} : v \in \mathcal{V}'\}$). For $i = 1, 2, 3, 4$, let $\mathcal{F}_i$ be the half-planes that hit the ith region, and $\mathcal{H}_i = \mathcal{G}(A_i, \mathcal{F}_i)$ be the corresponding hypergraph.

Lemma 2. *$\mathrm{Tr}(\mathcal{G}') = \text{minimal}\ (\mathcal{T}' \cup \mathcal{T}'' \cup \bigcup_{i=1}^{4} \mathrm{Tr}(\mathcal{H}_i))$ where $\mathcal{T}' = \{\{v\} : v \in B\}$ and $\mathcal{T}'' = \{\{v, v'\} : v \in A_i,\ v' \in A_j,\ i \neq j$ and $\{v, v'\}$ is transversal to $\mathcal{G}'\}$.*

5.2 Backtracking Method

Given a hypergraph $\mathcal{H} \subseteq 2^V$ and a subset $S \subseteq V$ of vertices, [6] gave a criterion to decide if S is a *sub-transversal* of $\mathcal{G}$, i.e., if there is a minimal transversal $T \in \mathrm{Tr}(\mathcal{G})$ such that $T \supseteq S$. In general, this is an NP-hard problem even if $\mathcal{G}$ is a graph (see [4]). However, if $|S|$ is bounded by a constant, or if the hypergraph is read-once [15], then such a check can be done in polynomial time (see [6]). Here we give another instance in which such a check can also be performed in polynomial time.

For a subset $S \subseteq V$, and a vertex $v \in S$, let $\mathcal{H}_v(S) = \{H \in \mathcal{H} \mid H \cap S = \{v\}\}$. A selection of $|S|$ hyperedges $\{H_v \in \mathcal{H}_v(S) \mid v \in S\}$ is called *covering* if there exists a hyperedge $H \in \mathcal{H}_0 = \{H \in \mathcal{H} : H \cap S = \emptyset\}$ such that $H \subseteq \bigcup_{v \in S} H_v$.

Proposition 2 ([6]). *A non-empty subset $S \subseteq V$ is a sub-transversal for $\mathcal{H} \subseteq 2^V$ if and only if there is a non-covering selection $\{H_v \in \mathcal{H}_v(S) \mid v \in S\}$ for S.*

The algorithm is given in Figure 5, and is based on the standard backtracking technique for enumeration (see e.g. [32,15]). The procedure is called initially with $S_1 = S_2 = \emptyset$ and $i = 1$. It is easy to verify that the algorithm outputs all elements of the transversal hypergraph $\mathrm{Tr}(\mathcal{H})$, without repetition (and in lexicographic ordering according to the input permutation σ). Since the algorithm essentially builds a backtracking tree whose leaves are the minimal transversals of $\mathcal{G}$, the time required to produce each new minimal transversal is bounded by the depth of the tree (at most $\min\{|V|, |\mathcal{H}|\}$) times the maximum time required at each node. The efficiency of such procedure depends on being able to perform the test in Step 3, which is addressed in the next subsection.

Procedure DUALIZE-BT$(\mathcal{H}, \sigma, i, S)$**:**

Input: A hypergraph $\mathcal{H} \subseteq 2^V$, an ordering σ on V,
an integer $i \in \{1, \ldots, |V|\}$ and a subset $S \subseteq \sigma([i-1]) \stackrel{\text{def}}{=} \{\sigma(j) : j \in [i-1]\}$
Output: The set $\{T \in \text{Tr}(\mathcal{H}) : T \supseteq S, T \cap (\sigma([i-1]) \setminus S) = \emptyset\}$

1. **if** $S \in \text{Tr}(\mathcal{H})$ **then**
2. **output** S and **return**
3. **if** $\exists T \in \text{Tr}(\mathcal{H})$ s.t. $S \cup \{\sigma(i)\} \subseteq T$ and $(\sigma([i-1]) \setminus S) \cap T = \emptyset$ **then**
4. **DUALIZE-BT**$(\mathcal{H}, \sigma, i+1, S \cup \{\sigma(i)\})$
5. **DUALIZE-BT**$(\mathcal{H}, \sigma, i+1, S)$

Fig. 5. The backtracking method for finding minimal transversals

5.3 Solving the Special Instance

Without loss of generality we concentrate on finding $\text{Tr}(\mathcal{H}_1)$, where $\mathcal{H}_1$ is the hypergraph defined in Section 5.1. The other 3 sets of transversals can be found similarly. In other words, we may assume that all points lie inside a right-angle triangle of which the hyperplanes contain the longest side (see Figure 4, Right).

We use the backtracking method with σ being the following lexicographic order of the points: if $p = (p_1, p_2)$ and $q = (q_1, q_2)$ then $p \prec_\sigma q$ if and only if $p_1 < q_1$ or $p_1 = q_1$ and $p_2 < q_2$. Without loss of generality, we assume $V(\mathcal{H}_1) = \{1, \ldots, n\}$ and reorder the points such that they are numbered from 1 to n according to σ, i.e., assume that σ is the identity permutation.

Now we show that the sub-transversal test in Step 3 of the backtracking procedure in Figure 5 can be performed in polynomial time. Given $i \in [n]$ and $S \subseteq [i-1]$, we would like to check if

$$\exists T \in \text{Tr}(\mathcal{H}_1) \text{ such that } S \cup \{i\} \subseteq T \text{ and } ([i-1] \setminus S) \cap T = \emptyset. \tag{1}$$

Lemma 3. *Fix $i \in [n]$ and let $S \subseteq [i-1]$ be a sub-transversal of $\mathcal{H}_1$. Suppose that $F, F' \in \mathcal{H}_1$ satisfy: $F \cap (S \cup \{i\}) = \{j\}$ for $j \neq i$ and $F' \cap (S \cup \{i\}) = \{i\}$. Then $F \setminus [i-1] \subseteq H'$.*

Proof. If there is a point with index $k \in F \setminus ([i-1] \cup F')$ then k comes before i with respect to the first coordinate (see Figure 4, Right), in contradiction to the fact that we process the points according to the order imposed by σ. □

Now the sub-transversal criterion of Proposition 2 reduces to a simple check.

Lemma 4. *Given $i \in [n]$ and $S \subseteq [i-1]$. Then (1) holds if and only if there exists $F \in \mathcal{H}_1$ such that $F \cap (S \cup \{i\}) = \{i\}$ and for all $F' \in \mathcal{H}_1$ for which $F' \cap (S \cup \{i\}) = \emptyset$, we have $(F' \setminus [i]) \setminus (F \setminus [i]) \neq \emptyset$.*

Proof. We apply the sub-transversal criterion for $S \cup \{i\}$ in the restricted hypergraph $\mathcal{H}_1' = \{H \setminus ([i-1] \setminus S) : H \in \mathcal{H}_1\}$. By Proposition 2, (1) holds if and only if there exists $F_1, \ldots, F_i \in \mathcal{H}_1$ such that $F_j \cap (S \cup \{i\}) = \{j\}$, for $j = 1, \ldots, i$, and $(F' \setminus [i-1]) \not\subseteq \bigcup_{j=1}^{i} (F_j \setminus [i-1])$ for all $F' \in \mathcal{H}_1$ such that $F' \cap (S \cup \{i\}) = \emptyset$. By Lemma 3, the union $\bigcup_{j=1}^{i} (F_j \setminus [i-1])$ is equal to $F_i \setminus [i-1]$. □

5.4 Minimal Covers

By *geometric* duality (see e.g. [10], Chapter 8), the problem of finding minimal set covers reduces to finding all minimal hitting sets. Indeed, we may assume by rotating the given instance if necessary that all points are in general position. If we use the mapping $(p_1, p_2) \mapsto y = p_1 x + p_2$ that maps the point $p = (p_1, p_2)$ in the original space to the line $\{(x, y) : y = p_1 x + p_2\}$, then one can easily see that minimal set covers in one space correspond to minimal hitting sets in the other space and vice versa.

Theorem 4. *Both problems* $\textsc{Hit}(\mathcal{V}, \mathcal{F})$ *and* $\textsc{Cover}(\mathcal{V}, \mathcal{F})$ *can be solved in time* $O((|\mathcal{V}||\mathcal{F}|k)^{O(1)})$*, when* $\mathcal{V}$ *and* $\mathcal{F}$ *are, respectively, a set of points and a set of half-planes in* $\mathbb{R}^2$*, and* k *is the size of the output.*

References

1. Anthony, M., Biggs, N.: Computational learning theory: an introduction. Cambridge University Press, New York (1992)
2. Berge, C.: Hypergraphs. Elsevier, Amsterdam (1989)
3. Bioch, J.C., Ibaraki, T.: Complexity of identification and dualization of positive boolean functions. Information and Computation 123(1), 50–63 (1995)
4. Boros, E., Elbassioni, K., Gurvich, V., Khachiyan, L.: Generating maximal independent sets for hypergraphs with bounded edge-intersections. In: Farach-Colton, M. (ed.) LATIN 2004. LNCS, vol. 2976, pp. 488–498. Springer, Heidelberg (2004)
5. Boros, E., Elbassioni, K., Gurvich, V., Khachiyan, L., Makino, K.: Dual-bounded generating problems: All minimal integer solutions for a monotone system of linear inequalities. SIAM J. Computing 31(5), 1624–1643 (2002)
6. Boros, E., Gurvich, V., Hammer, P.L.: Dual subimplicants of positive boolean functions. Optim. Methods Softw. 10, 147–156 (1998)
7. Boros, E., Gurvich, V., Khachiyan, L., Makino, K.: On the complexity of generating maximal frequent and minimal infrequent sets. In: Alt, H., Ferreira, A. (eds.) STACS 2002. LNCS, vol. 2285, pp. 133–141. Springer, Heidelberg (2002)
8. Boros, E., Makino, K.: A fast and simple parallel algorithm for the monotone duality problem. In: Albers, S., Marchetti-Spaccamela, A., Matias, Y., Nikoletseas, S., Thomas, W. (eds.) ICALP 2009. Part I. LNCS, vol. 5555, pp. 183–194. Springer, Heidelberg (2009)
9. Colbourn, C.J.: The Combinatorics of Network Reliability. Oxford University Press, NY (1987)
10. de Berg, M., van Kreveld, M., Overmars, M., Schwarzkopf, O.: Computational Geometry, Algorithms and Applications. Springer, Heidelberg (1997)
11. Domingo, C., Mishra, N., Pitt, L.: Efficient read-restricted monotone cnf/dnf dualization by learning with membership queries. Machine Learning 37(1), 89–110 (1999)
12. Eiter, T., Gottlob, G.: Identifying the minimal transversals of a hypergraph and related problems. SIAM J. Computing 24(6), 1278–1304 (1995)
13. Eiter, T., Gottlob, G., Makino, K.: New results on monotone dualization and generating hypergraph transversals. SIAM J. Computing 32(2), 514–537 (2003)
14. Eiter, T., Makino, K., Gottlob, G.: Computational aspects of monotone dualization: A brief survey. Discrete Applied Mathematics 156(11), 2035–2049 (2008)

15. Eiter, T.: Exact transversal hypergraphs and application to Boolean μ-functions. Journal of Symbolic Computation 17(3), 215–225 (1994)
16. Elbassioni, K.M.: On the complexity of the multiplication method for monotone CNF/DNF dualization. In: Azar, Y., Erlebach, T. (eds.) ESA 2006. LNCS, vol. 4168, pp. 340–351. Springer, Heidelberg (2006)
17. Fredman, M.L., Khachiyan, L.: On the complexity of dualization of monotone disjunctive normal forms. Journal of Algorithms 21, 618–628 (1996)
18. Gaur, D.R., Krishnamurti, R.: Average case self-duality of monotone boolean functions. In: Canadian AI 2004, pp. 322–338 (2004)
19. Gottlob, G.: Hypergraph transversals. In: Seipel, D., Turull-Torres, J.M.a. (eds.) FoIKS 2004. LNCS, vol. 2942, pp. 1–5. Springer, Heidelberg (2004)
20. Gunopulos, D., Mannila, H., Khardon, R., Toivonen, H.: Data mining, hypergraph transversals, and machine learning (extended abstract). In: PODS 1997 (1997)
21. Halman, N.: On the power of discrete and of lexicographic Helly-type theorems. In: FOCS 2004, pp. 463–472 (2004)
22. Hüffner, F., Niedermeier, R., Wernicke, S.: Techniques for practical fixed-parameter algorithms. Comput. J. 51(1), 7–25 (2008)
23. Ibaraki, T., Kameda, T.: A theory of coteries: Mutual exclusion in distributed systems. IEEE Trans. on Parallel and Distributed Systems 4(7), 779–794 (1993)
24. Kavvadias, D.J., Papadimitriou, C.H., Sideri, M.: On horn envelopes and hypergraph transversals. In: Ng, K.W., Balasubramanian, N.V., Raghavan, P., Chin, F.Y.L. (eds.) ISAAC 1993. LNCS, vol. 762, pp. 399–405. Springer, Heidelberg (1993)
25. Khachiyan, L.: Transversal hypergraphs and families of polyhedral cones. In: Advances in Convex Analysis and Global Optimization, honoring the memory of K. Carathéodory, pp. 105–118. Kluwer, Dordrecht (2000)
26. Khachiyan, L., Boros, E., Borys, K., Elbassioni, K., Gurvich, V., Makino, K.: Enumerating spanning and connected subsets in graphs and matroids. In: Azar, Y., Erlebach, T. (eds.) ESA 2006. LNCS, vol. 4168, pp. 444–455. Springer, Heidelberg (2006)
27. Khachiyan, L., Boros, E., Elbassioni, K., Gurvich, V.: A global parallel algorithm for the hypergraph transversal problem. Information Processing Letters 101(4), 148–155 (2007)
28. Lawler, E., Lenstra, J.K., Rinnooy Kan, A.H.G.: Generating all maximal independent sets: NP-hardness and polynomial-time algorithms. SIAM J. Computing 9, 558–565 (1980)
29. Lovász, L.: Combinatorial optimization: some problems and trends. DIMACS Technical Report 92-53, Rutgers University (1992)
30. Mannila, H., Räihä, K.J.: Design by example: An application of armstrong relations. Journal of Computer and System Sciences 33(2), 126–141 (1986)
31. Papadimitriou, C.: NP-completeness: A retrospective. In: Degano, P., Gorrieri, R., Marchetti-Spaccamela, A. (eds.) ICALP 1997. LNCS, vol. 1256. Springer, Heidelberg (1997)
32. Read, R.C., Tarjan, R.E.: Bounds on backtrack algorithms for listing cycles, paths, and spanning trees. Networks 5, 237–252 (1975)
33. Tamaki, H.: Space-efficient enumeration of minimal transversals of a hypergraph. Technical Report IPSJ-AL 75 (2000)

On Revenue Maximization in Second-Price Ad Auctions*

Yossi Azar[1], Benjamin Birnbaum[2], Anna R. Karlin[2], and C. Thach Nguyen[2]

[1] Microsoft Research and Tel-Aviv University
azar@tau.ac.il
[2] University of Washington
{birnbaum,karlin,ncthach}@cs.washington.edu

Abstract. Most recent papers addressing the algorithmic problem of allocating advertisement space for keywords in sponsored search auctions assume that pricing is done via a first-price auction, which does not realistically model the Generalized Second Price (GSP) auction used in practice. Towards the goal of more realistically modeling these auctions, we introduce the *Second-Price Ad Auctions* problem, in which bidders' payments are determined by the GSP mechanism. We show that the complexity of the Second-Price Ad Auctions problem is quite different than that of the more studied First-Price Ad Auctions problem. First, unlike the first-price variant, for which small constant-factor approximations are known, it is NP-hard to approximate the Second-Price Ad Auctions problem to any non-trivial factor. Second, this discrepancy extends even to the 0-1 special case that we call the *Second-Price Matching* problem (2PM). In particular, offline 2PM is APX-hard, and for online 2PM there is no deterministic algorithm achieving a non-trivial competitive ratio and no randomized algorithm achieving a competitive ratio better than 2. This stands in contrast to the results for the analogous special case in the first-price model, the standard bipartite matching problem, which is solvable in polynomial time and which has deterministic and randomized online algorithms achieving better competitive ratios. On the positive side, we provide a 2-approximation for offline 2PM and a 5.083-competitive randomized algorithm for online 2PM. The latter result makes use of a new generalization of a classic result on the performance of the "Ranking" algorithm for online bipartite matching.

1 Introduction

The rising economic importance of online sponsored search advertising has led to a great deal of research focused on developing its theoretical underpinnings. (See, e.g., [1] for a survey). Since search engines such as Google, Yahoo! and MSN depend on sponsored search for a significant fraction of their revenue, a key problem is how to optimally allocate ads to keywords (user searches) so as to maximize search engine revenue [2,3,4,5,6,7,8,9,10,11,12]. Most of the research on the dynamic version of this problem assumes that once the participants in each

* This research was supported by NSF Grant CCF-0635147, a grant from Yahoo! Research, and an NSF Graduate Research Fellowship.

A. Fiat and P. Sanders (Eds.): ESA 2009, LNCS 5757, pp. 155–166, 2009.

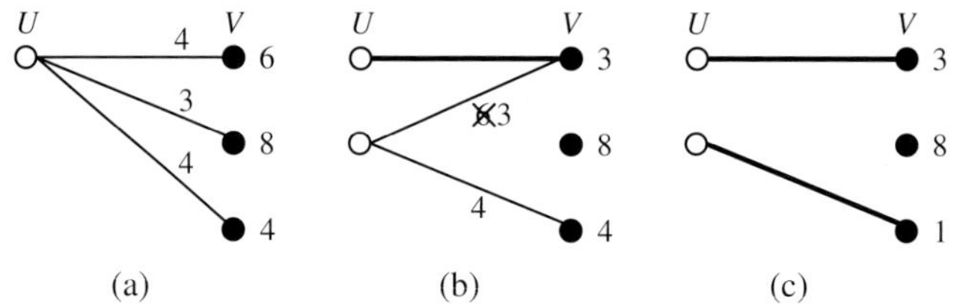

Fig. 1. An example of the Second-Price Ad Auctions problem: the nodes in U are keywords and the nodes in V are bidders. The number immediately to the right of each bidder represents its remaining budget, and the number next to each edge connecting a bidder to a keyword represents the bid of that bidder for that keyword. (a) shows the situation when the first keyword arrives. For this keyword, the search engine selects the first bidder, whose bid is 4, and the second bidder, whose bid is 3. The keyword is allocated to the first bidder at a price of 3, thereby reducing that bidder's budget by 3. (b) shows the situation when the second keyword arrives. The bid of the first bidder for that keyword is adjusted to the minimum of its original bid, 6, and its remaining budget, 3. Then the first and the third bidders are selected, and the keyword is allocated to the third bidder at a price of 3.

keyword auction are determined, the pricing is done via a first-price auction; in other words, bidders pay what they bid. This does not realistically model the standard mechanism used by search engines, called the Generalized Second Price mechanism (GSP) [13,14].

In an attempt to model reality more closely, we study the *Second-Price Ad Auctions* problem, which is the analogue of the above allocation problem when bidders' payments are determined by the GSP mechanism. As in other work [4,5,6,11,12], we make the simplifying assumption that there is only one slot for each keyword. In this case, the GSP mechanism for a given keyword auction reduces to a second-price auction – given the participants in the auction, it allocates the advertisement slot to the highest bidder, charging that bidder the bid of the second-highest bidder.[1]

In the Second-Price Ad Auctions problem, there is a set of keywords U and a set of bidders V, where each bidder $v \in V$ has a known daily budget B_v and a non-negative bid $b_{u,v}$ for every keyword $u \in U$. The keywords are ordered by their arrival time, and as each keyword u arrives, the algorithm (i.e., the search engine) must choose a bidder to allocate it to. The search engine is not required to choose the highest-bidding bidder; in order to optimize the allocation of bidders to keywords, search engines typically use a "throttling" algorithm that chooses which bidders to select to participate in an auction for a given keyword [8].[2]

In the previously-studied first-price version of the problem, allocating a keyword to a bidder meant choosing a single bidder v and allocating u to v at a price of $b_{u,v}$. In the Second-Price Ad Auctions problem, two bidders are selected

[1] This simplication, among others (see [1]), leaves room to improve the accuracy of our model. However, the hardness results clearly hold for the multi-slot case as well.

[2] In this paper, we assume the search engine is optimizing over revenue although it is certainly conceivable that a search engine would consider other objectives.

instead of one. Of these two bidders, the bidder with the higher bid (where bids are always reduced to the minimum of the actual bid and bidders' remaining budgets) is allocated that keyword's advertisement slot at the price of the other bid. (In the GSP mechanism for k slots, $k+1$ bidders are selected, and each of the top k bidders pays the bid of the next-highest bidder.)

This process results in an allocation and pricing of the advertisement slots associated with each of the keywords. The goal is to select the bidders participating in each auction to maximize the total profit extracted by the algorithm. For an example instance of this problem, see Figure 1.

1.1 Our Results

We begin by considering the *offline* version of the Second-Price Ad Auctions problem, in which the algorithm knows all of the original bids of the bidders (Section 3). Our main result here is that it is NP-hard to approximate the optimal solution to this problem to within a factor better than $\Omega(m)$, where m is the number of keywords, even when the bids are small compared to budgets. This strong inapproximability result is matched by the trivial algorithm that selects the single keyword with the highest second-best bidder and allocates only that keyword to its top two bidders. It stands in sharp contrast to the standard First-Price Ad Auctions problem, for which there is a 4/3-approximation to the offline problem [6,12] and an $e/(e-1)$-competitive algorithm to the online problem when bids are small compared to budgets [5,11].

We then turn our attention to a theoretically appealing special case that we call *Second-Price Matching*. In this version of the problem, all bids are either 0 or 1 and all budgets are 1. This can be thought of as a variant on maximum bipartite matching in which the input is a bipartite graph $G = (U \cup V, E)$, and the vertices in U must be matched, in order, to the vertices in V such that the profit of matching $u \in U$ to $v \in V$ is 1 if and only if there is at least one additional vertex $v' \in V$ that is a neighbor of u and is unmatched at that time. One can justify the second-price version of the problem by observing that when we sell an item, we can only charge the full value of the item when there is more than one interested buyer.[3]

Recall that the first-price analogue to the Second-Price Matching problem, the maximum bipartite matching problem, can be solved optimally in polynomial time. The online version has a trivial 2-competitive deterministic greedy algorithm and an $e/(e-1)$-competitive randomized algorithm due to Karp, Vazirani and Vazirani [15], both of which are best possible.

In contrast, we show that the Second-Price Matching problem is APX-hard (Section 4.1). We also give a 2-approximation algorithm for the offline problem (Section 4.2). We then turn to the online version of the problem. Here, we show that no deterministic online algorithm can get a competitive ratio better than m, where m is the number of keywords in the instance, and that no randomized online algorithm can get a competitive ratio better than 2 (Section 5.1). On the

[3] A slightly more amusing motivation is to imagine that the two sets of nodes represent boys and girls respectively and the edges represent mutual interest, but a girl is only interested in a boy if another girl is also actively interested in that boy.

Table 1. A summary of the results in this paper, compared to known results for the first-price case. The upper bound of $e/(e-1)$ for Online 1PAA only holds when when the bids are very small compared to the budgets.

	Offline		Online	
	Upper bound	Lower bound	Upper bound	Lower bound
1PAA	4/3 [6,12]	16/15 [6]	$e(e-1)^*$ [5,11] or 2 [16]	$e/(e-1)$ [11,15]
2PAA	m	$\Omega(m)$	-	-
Matching	poly-time alg.		$e/(e-1)$ [15,9]	$e/(e-1)$ [15]
2PM	2	364/363	$2\sqrt{e}/(\sqrt{e}-1) \approx 5.1$	2

other hand, we present a randomized online algorithm that achieves a competitive ratio of $2\sqrt{e}/(\sqrt{e}-1) \approx 5.08$ (Section 5.2). To obtain this competitive ratio, we prove a generalization of the result due to Karp, Vazirani, and Vazirani [15] and Goel and Mehta [9] that the *Ranking* algorithm for online bipartite matching achieves a competitive ratio of $e/(e-1)$.

Due to space limitations, some proofs are omitted from this version of the paper. Proofs of all results can be found in the full version.

1.2 Related Work

As discussed above, the related First-Price Ad Auctions problem[4] has received a fair amount of attention. Mehta et al. [11] present an algorithm for the online version that achieves an optimal competitive ratio of $e/(e-1)$ for the case when the bids are much smaller than the budgets, a result also proved by Buchbinder et al. [5]. Under similar assumptions, Devanur and Hayes show that when the keywords arrive according to a random permutation, a $(1-\varepsilon)$-approximation is possible [7]. When there is no restriction on the values of the bids relative to the budgets, the best known competitive ratio is 2 [16]. For the offline version of the problem, a sequence of papers [16,3,17,4,12,6] culminating in a paper by Chakrabarty and Goel, and independently, a paper by Srinivasan, show that the offline problem can be approximated to within a factor of 4/3 and that there is no polynomial time approximation algorithm that achieves a ratio better than 16/15 unless $P = NP$ [6].

The most closely related work to ours is the paper of Goel, Mahdian, Nazerzadeh and Saberi [8], which builds on the work of Abrams, Medelvitch, and Tomlin [2]. Goel et al. look at the online allocation problem when the search engine is committed to charging under the GSP scheme, with multiple slots per keyword. They study two models, the "strict" and "non-strict" models, both of which differ from our model even for the one slot case by allowing bidders to keep bidding their orginal bid, even when their budget falls below this amount. Thus, in these models, although bidders are not charged more than their remaining budget when allocated a keyword, a bidder with a negligible amount of remaining budget can keep his bids high indefinitely, and as long as this bidder is

[4] This problem has also been called the *Adwords* problem [7,11] and the *Maximum Budgeted Allocation* problem [4,6,12].

never allocated another slot, this high bid can determine the prices other bidders pay on many keywords. Under the assumption that bids are small compared to budgets, Goel et al. build on the linear programming formulation of Abrams et al. to present an $e/(e-1)$-competitive algorithm for the non-strict model and a 3-competitive algorithm for the strict model.

The significant, qualitative difference between these positive results and the strong hardness we prove for our model suggests that these aspects of the problem formulation are important. We feel that our model, in which bidders are not allowed to bid more than their remaining budget, is more natural because it seems inherently unfair that a bidder with negligible or no budget should be able to indefinitely set high prices for other bidders.

2 Model and Notation

We define the Second-Price Ad Auctions (2PAA) problem formally as follows. The input is a set of ordered keywords U and bidders V. Each bidder $v \in V$ has a budget B_v and a nonnegative bid $b_{u,v}$ for every keyword $u \in U$. We assume that all of bidder v's bids $b_{u,v}$ are less than or equal to B_v.

Let $B_v(t)$ be the remaining budget of bidder v immediately after the t-th keyword is processed (so $B_v(0) = B_v$ for all v), and let $b_{u,v}(t) = \min(b_{u,v}, B_v(t))$. (Both quantities are defined inductively.) A solution (or *second-price matching*) to 2PAA chooses for the t-th keyword u a pair of bidders v_1 and v_2 such that $b_{u,v_1}(t-1) \geq b_{u,v_2}(t-1)$, allocates the slot for keyword u to bidder v_1 and charges bidder v_1 a price of $p(t) = b_{u,v_2}(t-1)$, the bid of v_2. (We say that v_1 acts as the *first-price bidder* for u and v_2 acts as the *second-price bidder* for u.) The budget of v_1 is then reduced by $p(t)$, so $B_{v_1}(t) = B_{v_1}(t-1) - p(t)$. For all other bidders $v \neq v_1$, $B_v(t) = B_v(t-1)$. The final value of the solution is $\sum_t p(t)$, and the goal is to find a solution of maximum value.

In the offline version of the problem, all of the bids are known to the algorithm beforehand, whereas in the online version of the problem, keyword u and the bids $b_{u,v}$ for each $v \in V$ are revealed only when keyword u arrives, at which point the algorithm must irrevocably map u to a pair of bidders without knowing the bids for the keywords that will arrive later.

The special case referred to as Second-Price Matching (2PM) is where $b_{u,v}$ is either 0 or 1 for all (u,v) pairs and $B_v = 1$ for all v. We will think of this as the variant on maximum bipartite matching (with input $G = (U \cup V, E)$) described in Section 1.1. Note that in 2PM, a keyword can only be allocated for profit if its degree is at least two. Therefore, we assume without loss of generality that for all inputs of 2PM, the degree of every keyword is at least two.

For an input to 2PAA, let $R_{min} = \min_{u,v} B_v/b_{u,v}$, and let $m = |U|$ be the number of keywords.

3 Hardness of Approximation of 2PAA

In this section, we present our main hardness result for the Second-Price Ad Auctions problem. For a constant $c \geq 1$, let 2PAA(c) be the version of 2PAA in which we are promised that $R_{min} \geq c$.

Theorem 1. *Let $c \geq 1$ be a constant integer. For any constant $c' > c$, it is NP-hard to approximate 2PAA(c) to a factor of m/c'.*

Hence, even when the bids are guaranteed to be smaller than the budget by a large constant factor, it is NP-hard to approximate 2PAA to a factor better than $\Omega(m)$. As noted in the introduction and shown in the full version of this paper, this hardness is matched by a trivial algorithm.

Proof. Here we provide the reduction. Some of the details of the proof are left for the full version of this paper.

Fix a constant $c' > c$, and let n_0 be the smallest integer such that for all $n \geq n_0$,

$$c' \cdot \frac{c(n^5 + n + 2)}{cn^2 + n + 2} \geq c(n^3 + cn^2 + n + 2) \tag{1}$$

and

$$\frac{n/2 + 1}{2} \geq c \ . \tag{2}$$

Note that since n_0 depends only on c', it is a constant.

We reduce from PARTITION, in which the input is a set of $n \geq n_0$ items, and the weight of the i-th item is given by w_i. If $W = \sum_{i=1}^{n} w_i$, then the question is whether there is a partition of the items into two subsets of size $n/2$ such that the sum of the w_i's in each subset is $W/2$. It is known that this problem (even when the subsets must both have size $n/2$) is NP-hard [18].

Given an instance of PARTITION, we create an instance of 2PAA(c) as follows. (This reduction is illustrated in Figure 2.)

– First, create $n + 2$ keywords $c_1, \ldots, c_n, e_1, e_2$. Second, create an additional set

$$G = \{g_{i,k} \ : \ 1 \leq i \leq n^2 \text{ and } 1 \leq k \leq c\}$$

of cn^2 keywords. The keywords arrive in the order

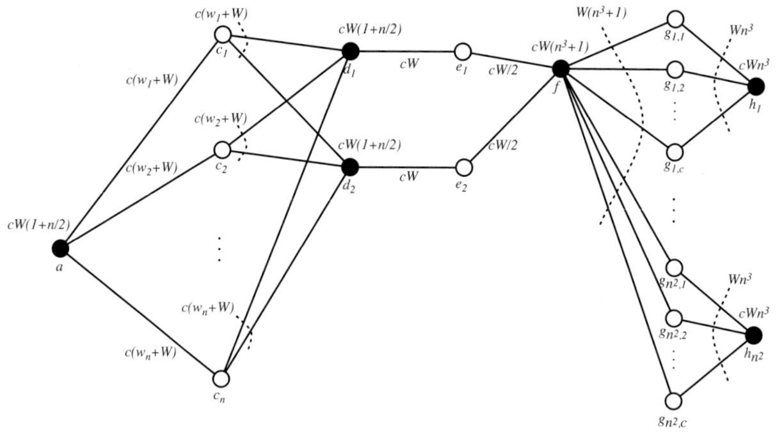

Fig. 2. The 2PAA(c) instance of the reduction. Each bidder's budget is shown above its node, and the bids of bidders for keywords is shown near the corresponding edge.

$$c_1, \ldots, c_n, e_1, e_2, g_{1,1}, \ldots, g_{1,c}, \ldots\ldots, g_{n^2,1}, \ldots, g_{n^2,c} \ .$$

- Create $n^2 + 4$ bidders $a, d_1, d_2, f, h_1, \ldots, h_{n^2}$. Set the budgets of a, d_1, and d_2 to $cW(1 + n/2)$. Set the budget of f to $cW(n^3 + 1)$. For $1 \leq i \leq n^2$, set the budget of h_i to cWn^3.
- For $1 \leq i \leq n$, bidders a, d_1, and d_2 bid $c(w_i + W)$ on keyword c_i.
- For $j \in \{1, 2\}$, bidder d_j bids cW on keyword e_j. Bidder f bids $cW/2$ on both e_1 and e_2.
- For $1 \leq i \leq n^2$ and $1 \leq k \leq c$, keyword $g_{i,k}$ receives a bid of $W(n^3 + 1)$ from bidder f and a bid of Wn^3 from bidder h_i.

This reduction can clearly be performed in polynomial time. Furthermore, it can easily be checked that (2) implies that no bidder bids more than $1/c$ of its budget on any keyword.

We claim (and prove in the full version of the paper) that if the PARTITION instance is a "yes" instance, then there exists a feasible solution to the 2PAA(c) instance of value at least $cW(n^5 + n + 2)$, and if the PARTITION instance is a "no" instance, then every feasible solution to the 2PAA(c) instance must have value less than $cW(n^3 + cn^2 + n + 2)$. Since there are $cn^2 + n + 2$ keywords in the 2PAA(c) instance, equation (1) implies that an m/c'-approximation algorithm can distinguish between the two cases, and hence such an algorithm cannot exist unless $P = NP$. □

4 Offline Second-Price Matching

In this section, we turn our attention to the offline version of the special case of Second-Price Matching (2PM).

4.1 Hardness of Approximation

To prove that 2PM is APX-hard, we reduce from vertex cover, using the following result.

Theorem 2 (Chlebík and Chlebíková [19]). *It is NP-hard to approximate Vertex Cover on 4-regular graphs to within* 53/52.

The precise statement of our hardness result is the following theorem

Theorem 3. *It is NP-hard to approximate 2PM to within a factor of* 364/363.

4.2 A 2-Approximation Algorithm

Consider an instance $G = (U \cup V, E)$ of the 2PM problem. We provide an algorithm that first finds a maximum matching $f : U \to V$ and then uses f to return a second-price matching that contains at least half of the keywords matched by f.[5] Given a matching f, call an edge $(u, v) \in E$ such that $f(u) \neq v$ an *up-edge* if v is matched by f and $f^{-1}(v)$ arrives before u, and a *down-edge*

[5] Note that f is a partial function.

otherwise. Recall that we have assumed without loss of generality that the degree of every keyword in U is at least two. Therefore, every keyword $u \in U$ that is matched by f must have at least one up-edge or down-edge. Theorem 4 shows that the following algorithm, called ReverseMatch, is a 2-approximation for 2PM.

ReverseMatch Algorithm:

Initialization:

Find an arbitrary maximum matching $f : U \to V$ on G.

Constructing a 2nd-price matching:

Consider the matched keywords in reverse order of their arrival.
For each keyword u:
  If keyword u is adjacent to a down-edge (u, v):
    Assign keyword u to bidder $f(u)$ (with v acting as the second-price bidder).
  Else:
    Choose an arbitrary bidder v that is adjacent to keyword u.
    Remove the edge $(f^{-1}(v), v)$ from f.
    Assign keyword u to bidder $f(u)$ (with v acting as the second-price bidder).

Theorem 4. *The ReverseMatch algorithm is a 2-approximation.*

5 Online Second-Price Matching

In this section, we consider the online 2PM problem, in which the keywords arrive one-by-one and must be matched by the algorithm as they arrive. We start, in Section 5.1, by giving a simple lower bound showing that no deterministic algorithm can achieve a competitive ratio better than m, the number of keywords. Then we move to randomized online algorithms and show that no randomized algorithm can achieve a competitive ratio better than 2. In Section 5.2, we provide a randomized online algorithm that achieves a competitive ratio of $2\sqrt{e}/(\sqrt{e} - 1) \approx 5.083$.

5.1 Lower Bounds

The following theorem establishes our lower bound on deterministic algorithms, which matches the trivial algorithm of arbitrarily allocating the first keyword to arrive, and refusing to allocate any of the remaining keywords. The adversary for this lower bound offers the first keyword two bidders; whichever bidder the algorithm chooses for this first keyword will be needed as a second-price bidder for the rest of the keywords.

Theorem 5. *For any m, there is an adversary that creates a graph with m keywords that forces any deterministic algorithm to get a competitive ratio no better than $1/m$.*

We next show that no online (randomized) online algorithm for 2PM can achieve a competitive ratio better than 2. To do this we invoke Yao's Principle [20] and construct a distribution of inputs for which the best deterministic algorithm achieves an expected performance of (asymptotically) 1/2 the value of the optimal solution.

Our distribution is constructed as follows. The first keyword arrives, and it is adjacent to two bidders. Then the second keyword arrives, and it is adjacent to one of the two bidders adjacent to the first keyword, chosen uniformly at random, as well as a new bidder; then the third keyword arrives, and it is adjacent to one of the bidders adjacent to the second keyword, chosen uniformly at random, as well as a new bidder; and so on, until the m-th keyword arrives.

We claim (and prove in the full version) that no deterministic algorithm can achieve an expected performance better than $1/2$ on this distribution. This yields the following theorem.

Theorem 6. *The competitive ratio of any randomized algorithm for 2PM must be at least* 2.

5.2 A Randomized Competitive Algorithm

In this section, we provide an algorithm that achieves a competitive ratio of $2\sqrt{e}/(\sqrt{e}-1) \approx 5.083$. The result builds on a new generalization of the result that the Ranking algorithm for online bipartite matching achieves a competitive ratio of $e/(e-1) \approx 1.582$. This was originally shown by Karp, Vazirani, and Vazirani [15], though a mistake was recently found in their proof by Krohn and Varadarajan and corrected by Goel and Mehta [9].

The online bipartite matching problem is merely the first-price version of 2PM, i.e., the problem in which there is no requirement for there to exist a second-price bidder to get a profit of 1 for a match. The Ranking algorithm chooses a random permutation on the bidders V and uses that to choose matches for the keywords U as they arrive. This is described more precisely below.

Ranking Algorithm:

Initialization:
Choose a random permutation (ranking) σ of the bidders V.

Online Matching:
Upon arrival of keyword $u \in U$:
  Let $N(u)$ be the set of neighbors of u that have not been matched yet.
  If $N(u) \neq \emptyset$, match u to the bidder $v \in N(u)$ that minimizes $\sigma(v)$.

Karp, Vazirani, and Vazirani, and Goel and Mehta prove the following result.

Theorem 7 (Karp, Vazirani, and Vazirani [15] and Goel and Mehta [9]). *The Ranking algorithm for online bipartite matching achieves a competitive ratio of* $e/(e-1) + o(1)$.

In order to state our generalization of this result, we define the notion of a *left* k*-copy* of a bipartite graph $G = (U \cup V, E)$. Intuitively, a left k-copy of G makes k copies of each keyword $u \in U$ such that the neighborhood of a copy of u is the same as the neighborhood of u. More precisely, we have the following definition.

Definition 8. *Given a bipartite graph* $G = (U_G \cup V, E_G)$, *a* **left k-copy** *of* G *is a graph* $H = (U_H \cup V, E_H)$ *for which* $|U_H| = k|U_G|$ *and for which there exists a map* $\zeta : U_H \rightarrow U_G$ *such that*

– *for each* $u_G \in U_G$ *there are exactly* k *vertices* $u_H \in U_H$ *such that* $\zeta(u_H) = u_G$, *and*
– *for all* $u_H \in U_H$ *and* $v \in V$, $(u_H, v) \in E_H$ *if and only if* $(\zeta(u_H), v) \in E_G$.

Our generalization of Theorem 7 describes the competitive ratio of Ranking on a graph H that is a left k-copy of G. Its proof builds on the proof of Theorem 7 presented by Birnbaum and Mathieu [21].

Theorem 9. *Let* $G = (U_G \cup V, E_G)$ *be a bipartite graph that has a maximum matching of size* OPT_{1P}, *and let* $H = (U_H \cup V, E_H)$ *be a left* k*-copy of* G. *Then the expected size of the matching returned by Ranking on* H *is at least*

$$kOPT_{1P}\left(1 - \frac{1}{e^{1/k}} + o(1)\right) .$$

Using this result, we are able to prove that the following algorithm, called RankingSimulate, achieves a competitive ratio of $2\sqrt{e}/(\sqrt{e}-1)$.

RankingSimulate Algorithm:

Initialization:
Set M, the set of *matched* bidders, to $\emptyset$.
Set R, the set of *reserved* bidders, to $\emptyset$.
Choose a random permutation (ranking) σ of the bidders V.

Online Matching:
Upon arrival of keyword $u \in U$:
  Let $N(u)$ be the set of neighbors of u that are not in M or R.
  If $N(u) = \emptyset$, do nothing.
  If $|N(u)| = 1$, let v be the single bidder in $N(u)$.
    With probability 1/2, match u to v and add v to M, and
    With probability 1/2, add v to R.
  If $|N(u)| \geq 2$, let v_1 and v_2 be the two distinct bidders in $N(u)$ that minimize $\sigma(v)$.
    With probability 1/2, match u to v_1, add v_1 to M, and add v_2 to R, and
    With probability 1/2, match u to v_2, add v_1 to R, and add v_2 to M.

Let $G = (U_G \cup V, E_G)$ be the bipartite input graph to 2PM, and let $H = (U_H \cup V, E_H)$ be a left 2-copy of H. In the arrival order for H, the two copies of each keyword $u_G \in U$ arrive in sequential order. The proof of the following lemma is straightforward.

Lemma 10. *Fix a ranking* σ *on* V. *For each bidder* $v \in V$, *let* X_v *be the indicator variable for the event that* v *is matched by Ranking on* H, *when the ranking is* σ.[6] *Let* X'_v *be the indicator variable for the event that* v *is matched by RankingSimulate on* G, *when the ranking is* σ. *Then* $\mathbb{E}(X'_v) = X_v/2$.

With Theorem 9 and Lemma 10, we can now prove the main result of this section.

Theorem 11. *The competitive ratio of RankingSimulate is* $2\sqrt{e}/(\sqrt{e}-1) \approx 5.083$.

[6] Note that once σ is fixed, X_v is deterministic.

Proof. For a permutation σ on V, let RankingSimulate(σ) be the matching of G returned by RankingSimulate, and let Ranking(σ) be the matching of H returned by Ranking. Lemma 10 implies that, conditioned on σ, $\mathbb{E}(|\,\mathrm{RankingSimulate}(\sigma)|) = |\,\mathrm{Ranking}(\sigma)|/2$. By Theorem 9,

$$\mathbb{E}(|\,\mathrm{RankingSimulate}(\sigma)|) = \frac{1}{2}\,\mathbb{E}(|\,\mathrm{Ranking}(\sigma)|) \geq OPT_{1P}\left(1 - 1/e^{1/2} + o(1)\right) .$$

Fix a bidder $v \in V$. Let P_v be the profit from v obtained by RankingSimulate. Suppose that v is matched by RankingSimulate to keyword $u \in U_G$. Recall that we have assumed without loss of generality that the degree of u is at least 2. Let $v' \neq v$ be another bidder adjacent to u. Then, given that v is matched to u, the probability that v' is matched to any keyword is no greater than 1/2. Therefore, $\mathbb{E}(P_v|v \text{ matched}) \geq 1/2$. Hence, the expected value of the second-price matching returned by RankingSimulate is

$$\begin{aligned}\sum_{v\in V}\mathbb{E}(P_v) &= \sum_{v\in V}\mathbb{E}(P_v|v\text{ matched})\,\mathbb{P}\mathrm{r}(v\text{ matched})\\ &\geq \frac{1}{2}\sum_{v\in V}\mathbb{P}\mathrm{r}(v\text{ matched})\\ &= \frac{1}{2}\,\mathbb{E}(|\,\mathrm{RankingSimulate}(\sigma)|)\\ &\geq \frac{1}{2}OPT_{1P}\left(1-1/e^{1/2}+o(1)\right)\\ &\geq \frac{1}{2}OPT_{2P}\left(1-1/e^{1/2}+o(1)\right) ,\end{aligned}$$

where OPT_{2P} is the size of the optimal second-price matching on G. □

References

1. Lahaie, S., Pennock, D.M., Saberi, A., Vohra, R.V.: Sponsored search auctions. In: Nisan, N., Roughgarden, T., Tardos, E., Vazirani, V.V. (eds.) Algorithmic Game Theory, pp. 699–716. Cambridge University Press, Cambridge (2007)
2. Abrams, Z., Mendelevitch, O., Tomlin, J.: Optimal delivery of sponsored search advertisements subject to budget constraints. In: EC 2007 (2007)
3. Andelman, N., Mansour, Y.: Auctions with budget constraints. In: Hagerup, T., Katajainen, J. (eds.) SWAT 2004. LNCS, vol. 3111, pp. 26–38. Springer, Heidelberg (2004)
4. Azar, Y., Birnbaum, B., Karlin, A.R., Mathieu, C., Nguyen, C.T.: Improved approximation algorithms for budgeted allocations. In: Aceto, L., Damgård, I., Goldberg, L.A., Halldórsson, M.M., Ingólfsdóttir, A., Walukiewicz, I. (eds.) ICALP 2008, Part I. LNCS, vol. 5125, pp. 186–197. Springer, Heidelberg (2008)
5. Buchbinder, N., Jain, K., Naor, J.S.: Online primal-dual algorithms for maximizing ad-auctions revenue. In: Arge, L., Hoffmann, M., Welzl, E. (eds.) ESA 2007. LNCS, vol. 4698, pp. 253–264. Springer, Heidelberg (2007)
6. Chakrabarty, D., Goel, G.: On the approximability of budgeted allocations and improved lower bounds for submodular welfare maximization and gap. In: FOCS 2008, pp. 687–696 (2008)

7. Devanur, N.R., Hayes, T.P.: The adwords problem: Online keyword matching with budgeted bidders under random permutations. In: EC 2009 (2009)
8. Goel, A., Mahdian, M., Nazerzadeh, H., Saberi, A.: Advertisement allocation for generalized second pricing schemes. In: Workshop on Sponsored Search Auctions (2008)
9. Goel, G., Mehta, A.: Online budgeted matching in random input models with applications to adwords. In: SODA 2008, pp. 982–991 (2008)
10. Mahdian, M., Nazerzadeh, H., Saberi, A.: Allocating online advertisement space with unreliable estimates. In: EC 2007, pp. 288–294 (2007)
11. Mehta, A., Saberi, A., Vazirani, U., Vazirani, V.: Adwords and generalized online matching. J. ACM 54(5), 22 (2007)
12. Srinivasan, A.: Budgeted allocations in the full-information setting. In: Goel, A., Jansen, K., Rolim, J.D.P., Rubinfeld, R. (eds.) APPROX 2008 and RANDOM 2008. LNCS, vol. 5171, pp. 247–253. Springer, Heidelberg (2008)
13. Edelman, B., Ostrovsky, M., Schwarz, M.: Internet advertising and generalized second-price auction: Selling billions of dollars worth of keywords. American Economic Review 97, 242–259 (2007)
14. Varian, H.R.: Position auctions. International Journal of Industrial Organization 25, 1163–1178 (2007)
15. Karp, R.M., Vazirani, U.V., Vazirani, V.V.: An optimal algorithm for on-line bipartite matching. In: STOC 1990, pp. 352–358 (1990)
16. Lehmann, B., Lehmann, D., Nisan, N.: Combinatorial auctions with decreasing marginal utilities. Games and Economic Behavior 55(2), 270–296 (2006)
17. Feige, U., Vondrak, J.: Approximation algorithms for allocation problems: Improving the factor of 1 - 1/e. In: FOCS 2006, pp. 667–676 (2006)
18. Garey, M.R., Johnson, D.S.: Computers and Intractability. W. H. Freeman and Company, New York (1979)
19. Chlebík, M., Chlebíková, J.: Complexity of approximating bounded variants of optimization problems. Theoretical Computer Science 354(3), 320–338 (2006)
20. Yao, A.: Probabilistic computations: Toward a unified measure of complexity. In: FOCS 1977, pp. 222–227 (1977)
21. Birnbaum, B., Mathieu, C.: On-line bipartite matching made simple. SIGACT News 39(1), 80–87 (2008)

Clustering-Based Bidding Languages for Sponsored Search

Mohammad Mahdian and Grant Wang

Yahoo! Inc., Santa Clara, CA, USA
{mahdian,gjw}yahoo-inc.com

Abstract. Sponsored search auctions provide a marketplace where advertisers can bid for millions of advertising opportunities to promote their products. The main difficulty facing the advertisers in this market is the complexity of picking and evaluating keywords and phrases to bid on. This is due to the sheer number of possible keywords that the advertisers can bid on, and leads to inefficiencies in the market such as lack of coverage for "rare" keywords. Approaches such as *broad matching* have been proposed to alleviate this problem. However, as we will observe in this paper, broad matching has undesirable economic properties (such as the non-existence of equilibria) that can make it hard for an advertiser to determine how much to bid for a broad-matched keyword.

The main contribution of this paper is to introduce a bidding language for sponsored search auctions based on broad-matching keywords to *non-overlapping* clusters that greatly simplifies the bidding problem for the advertisers. We investigate the algorithmic problem of computing the *optimal* clustering given a set of estimated values and give an approximation algorithm for this problem. Furthermore, we present experimental results using real advertisers' data that show that it is possible to extract close to the optimal social welfare with a number of clusters considerably smaller than the number of keywords. This demonstrates the applicability of the clustering scheme and our algorithm in practice.

1 Introduction

Search engine companies such as Google, Yahoo!, and Microsoft run today's largest auction houses, selling millions of advertising opportunities to hundreds of thousands of advertisers every day. Apart from simple budget or capacity constraints, the valuations of bidders in these auctions (known as sponsored search auctions) tend to be additive. Therefore, ignoring such constraints, from a classical auction-theory perspective of maximizing social welfare, the optimal design is to conduct an independent auction for each item, and let the advertisers submit a separate bid for each auction they are interested in. This is, in fact, very close to the mechanisms that are employed in practice.

A major problem with this design is that it ignores the complexity of bidding for the advertisers. The main source of complexity in sponsored search auctions

A. Fiat and P. Sanders (Eds.): ESA 2009, LNCS 5757, pp. 167–178, 2009.

is the sheer size of the market: the number of different types of items that are up for sale in this market is huge (or practically infinite), as each keyword or phrase that a search engine user might potentially search for corresponds to an item that can be sold in the market. This is in contrast to traditional auction houses, where the number of different types of items that are offered for sale is usually not too large, and the main factor that gives rise to complexity is that bidders can have valuations for specific combinations of items.

Ignoring the complexity has lead to inefficiencies such as lack of coverage for rare yet potentially valuable phrases in the sponsored search markets. Approaches such as *broad matching* keywords that an advertiser is interested in to search queries have been applied to alleviate this problem. This means that when an advertiser bids on a keyword a using the broad match option, the search engine applies this bid to a broader set S_a of keywords that are deemed semantically related to a. Broad matching helps the advertisers to expand their bid to other potentially valuable keywords, and the search engine to increase their keyword coverage and thicken the market, thereby increasing the revenue.

However, as we will observe in Section 3, when the sets S_a of broad-matched keywords are allowed to overlap, the mechanism has undesirable economic properties that make it hard for the advertisers to determine their bid, even if we assume only one slot per keyword and a second-price pricing scheme. Intuitively, this is due to the fact that an advertiser bidding on the set S_a using broad-match on keyword a might have different valuation for different keywords in S_a, and hence if a second advertiser overbids them on some (but not all) keywords in S_a, this can change their valuation for the bundle of keywords they receive. Using this intuition, we show examples that demonstrate that in such a broad-match mechanism (with second price auctions for individual keywords), once advertisers fix the set of keywords they are bidding on, the bidding game might not have any Nash equilibrium.[1]

To resolve this issue, we restrict the bidding language to only allow broad-match sets S_a that are non-overlaping. In other words, the broad match is performed through a proper clustering of the set of all keywords into disjoint subsets, and advertisers are only allowed to bid for a cluster (and not for individual keywords). This clearly resolves the problem with the non-existence of equilibria, since an advertiser can compute his expected value for keywords in a cluster independent of other advertisers' bids (since other advertisers either take the whole cluster or none of it), and then bid this value if the underlying payment scheme is truthful (e.g., second price in the case of a single slot or more generally VCG payments), or more generally, according to the equilibria of the mechanism in the case of a single keyword (e.g., as described in [7,16] in the case of the generalized second price auction). Furthermore, to describe a cluster to

[1] Note that the choice of the set of keywords to bid on is not included in this game. If we include this as part of the game, the game always has an equilibrium where no advertiser uses broad match. However, our purpose is to show that assuming the advertisers have chosen to use the broad-match feature, determining the bid values might reduce to a game with no equilibrium.

the advertiser, the search engine only needs to give a random sample of a small number of keywords in the cluster (with sampling probabilities proportional to the historical query volumes) to the advertiser. The advertisers can get an accurate estimate of their value for the (potentially infinite) cluster by evaluating the random sample. Therefore, this scheme greatly simplifies keyword discovery and bidding for the advertiser.

The only drawback of restricting the advertisers to bid for clusters of keywords instead of individual keywords is that this restriction can cause a loss in social welfare. Furthermore, it is not clear how the clustering should be computed. In this paper we study these questions, both from a theoretical and an empirical perspective. First, we give approximation algorithms for the clustering problem (in two models, one deterministic and one stochastic) that given a set of estimated values and a limit k on the number of clusters, compute a clustering with social welfare within a constant factor of the optimal k-clusterings. Next, we give empirical results using real advertisers' bidding data that demonstrate that in practice, with a value of k that is significantly smaller than the total number of keywords, we can achieve a social welfare quite close to the optimal social welfare.[2]

Related work. While there has been extensive research into various aspects of the sponsored search market during the past few years (see, for example, [7,16,10,1,11,3,17,9,14,12]), the topic of broad matching has remained largely unexplored. The only papers we are aware of that study broad match from an economic and optimization perspective are [8,15]. Even-Dar et al. [8] study broad match from an advertiser's point of view, and present algorithms that optimize the advertiser's bids in a broad-match system taking prices as given. Singh and Roychowdhury [15] study the revenue and efficiency impacts of broad match. There are also a number of papers that study the problem of computing relevant queries for broad match or for query expansion from an information retrieval perspective; see, for example, [5]. Our observation on the lack of existence of equilibria in general broad-match bidding languages is related to the concept of *cherry picking* that has been studied in the economic literature [2], in particular in the context of the insurance industry.

Organization. The rest of this paper is organized as follows: in Section 2 we give a formal definition of the model. In Section 3 we give several examples that demonstrate that the bidding game in general broad-matched auctions might not have a full-information Nash equilibrium. In Section 4 we state the clustering problem in a deterministic and a stochastic model, and give approximation algorithms for solving this problem. Section 5 contains a presentation of our experimental results. We conclude in Section 6.

---

[2] Note that here the loss in social welfare due to the restriction to k-clusterings is compounded with the loss due to the fact that our algorithm is only an approximation algorithm. Therefore, our empirical results show that neither of these losses is large.

2 The Model

We study a setting where m advertisers in a set $\mathcal{A}$ bid for a number of impressions.[3] Each impression is indexed by a keyword that comes from a distribution $\mathcal{D}$ over the (potentially infinite) collection $\mathcal{K}$ of all potential keywords. We assume that advertisers have additive valuations, i.e., their value for a set of impressions is the sum of their value for individual impressions in the set.[4] Given this assumption and linearity of expectations, we can focus on the auctioning of a single impression (indexed by a keyword picked from $\mathcal{D}$). The value that an advertiser $i \in \mathcal{A}$ has for an impression indexed by a keyword $j \in \mathcal{K}$ is a fixed number denoted by $v(i, j)$. We assume that given the keyword $j \in \mathcal{K}$, the advertiser i is able to evaluate $v(i, j)$. However, since the number $|\mathcal{K}|$ of potential keywords is huge, or even infinite, we cannot expect the bidders to list all these values. Instead, the bidding language allows for a *broad match* option: for each keyword $a \in \mathcal{K}$, the search engine has a pre-computed set $S_a \subset \mathcal{K}$ of broad-matched keywords, and if an advertiser places a bid on keyword a using the broad-match option, this bid will be applied to all keywords in S_a. This way, the advertiser can bid only on a small set of keywords and the search engine expands his bid to a larger (potentially infinite) set. We assume that the search engine can communicate the sets S_a to the bidder. In practice, this can be done by giving the advertisers a random sample of keywords in S_a.

After the bidders place their bids, the search engine conducts a separate auction for each impression that arrives. That is, as a new query arrives, the search engine runs an auction between the set of bidders that have bid (either directly or through broad match) on this keyword. Throughout this paper, we assume for simplicity that there is only one ad slot on the page, and therefore only a single bidder with the highest bid will win the auction (we will discuss in Section 6 how this assumption can be relaxed). This bidder will be charged an amount equal to the second highest bid on this keyword. The utility of a bidder is the expectation of the value he receives minus the price paid, where the expectation is over the random draw of the keyword from $\mathcal{D}$.

[3] In practice, the commodities sold in sponsored search auctions are often clicks and not impressions. However, one can transform a pay-per-click system to a pay-per-impression one by multiplying each bidder's valuation on a keyword (or set of keywords) by her corresponding *click-through rate*. Therefore, we choose to simplify the model by assuming a pay-per-impression system. Of course, estimating the click-through rate of bidders is a non-trivial problem, and it is worth noting that our proposed clustering-based bidding language can in fact help with this problem by pooling together a number of rare keywords. We leave a formal study of this subject to the future work.

[4] This assumption is not entirely without loss of generality, since some advertisers in sponsored search auctions are budget-constrained. However, given the small fraction of such advertisers, the fact that in practice budget constraints are often flexible, and that incorporating budgets makes even the simplest auctions hard to analyze theoretically [4,6], we feel that it is justified to abstract away such constraints and focus on the complexity problem.

The focus of this paper is mainly on a restriction of the broad-match bidding language, which we call a *clustering-based bidding language*. In this restriction, the advertisers are only allowed to place broad-matched bids (i.e., no exact match), and sets S_a are selected such that for two keywords a and b, the sets S_a and S_b are either disjoint or identical. In other words, the search engine computes a clustering of the set $\mathcal{K}$ of keywords, and each bidder is only allowed to place a bid for all or none of the keywords in a cluster.[5] It is easy to see that the resulting mechanism is dominant-strategy incentive compatible in the sense that for each advertiser, it is a dominant strategy to place a bid for each cluster equal to the expected value of a keyword in that cluster (where the keyword is drawn according to the distribution $\mathcal{D}$ restricted to the cluster).

Clearly, restricting the advertisers to a clustering-based bidding language comes at a loss to the social welfare. To quantify this loss and measure its tradeoff against the complexity of the bidding language, we compare the social welfare of the truthful equilibrium with an idealized situation where each keyword is given to the advertiser with the highest valuation for that keyword.

3 Non-existence of Broad-Match Equilibria

In this section we present examples that show that non-clustering-based broad match languages can have undesirable economic properties; most importantly that the induced bidding game might not have an equilibrium. Before stating the examples, we need to clarify what we mean by the bidding game. We assume that each bidder i has already fixed the set T_i of keywords she is bidding on, and her move in the game consists of picking a single bid for this set. In other words, we do not include the picking of the keywords to bid on as part of the game, and focus on the case that i has fixed a single keyword to bid on, which maps to the set T_i through broad-match. Excluding the keyword selection from the game is essential, as without it the game always has an equilibrium where no bidder uses the broad match option. Our purpose is to show that the bidding problem might not have an equilibrium if the advertisers have already committed to use the broad-match option.

We study full-information Nash equilibria of the bidding game. We start by a very simple example that (even though contains an equilibrium) explains the essence of the problem.

Example 1. (Non-truthfulness) Assume there are two bidders 1 and 2, and two keywords a and b occuring with the same frequency. Bidder 1 bids on $T_1 = \{a, b\}$ and bidder 2 bids on $T_2 = \{a\}$. Bidder 1 has a value of 10 for a and 2 for b, while bidder 2 has a value of 7 for a. If each bidder bids truthfully according to

[5] Although not allowing advertisers to place exact match bids might sound like a harsh constraint, it can help the advertisers at the end by simplifying the bidding process and assuring them that other advertisers can not cherry pick. In fact, this is quite similar to the constraints that states often impose on the insurance industry to disallow cherry picking.

her expected value for the bundle, bidder 1's bid would be $(10+2)/2 = 6$, and hence she loses in the auction for a, receiving only the keyword b which has a value of 2.

The above example still has an equilibrium (in fact even an envy-free one) where bidder 1 overbids and wins both keywords. We now show that there are examples with no equilibrium. We start with a simpler example with no envy-free equilibrium (as defined by Edelman et al. [7]), and then add gadgets to this example to show that sometimes no equilibrium exists.

Example 2. (Non-existence of envy-free equilibria) Assume there are 3 bidders 1, 2, and 3, and three keywords a, b, and c with equal frequencies. The bidders bid on the sets $T_1 = \{a, b\}$, $T_2 = \{b, c\}$, and $T_3 = \{c, a\}$, and have values $v(1, a) = v(2, b) = v(3, c) = 20$ and $v(1, b) = v(2, c) = v(3, a) = 10$. Assume, without loss of generality, that the bids satisfy $b_1 \geq b_2 \geq b_3$ and 1 wins against 2 and 3 and 2 wins against 3.[6] Then bidder 1 wins a and b at prices b_3 and b_2, and bidder 2 wins c at price b_3. For bidder 2 not to envy the set $\{b, c\}$, we must have $(20 + 10) - (b_2 + b_3) \leq 10 - b_3$ and hence $b_2 \geq 20$. Also, for 3 not to envy $\{c\}$, we must have $b_3 \geq 20$. This means that bidder 1 ends up with negative utility.

Note that the above example still has non-envy-free equilibria, for example one where 1 bids infinity and wins $\{a, b\}$, 2 bids 0 and wins nothing, and 3 bids 11 and wins $\{c\}$. However, it is possible to construct a more complicated example where not even a non-envy-free equilibrium exists. This example is omitted due to lack of space.

4 The Clustering Problem

In this section, we define the search engine's problem of computing the clustering of the keywords as an optimization problem. To do this, we need to specify what kind of information the clustering algorirthm gets about the valuations of the bidders. We start with a deterministic formulation of the problem which assumes that the algorithm is given an estimated set of valuations and needs to build a clustering which maximizes social welfare with respect to this set. In practice, such estimates might be obtained by a combination of looking at historical bids on the keywords and using natural language processing or information retrieval techniques (as in [5]) to deduce other valuations. We will use historical bid data in the experiments presented in Section 5.

THE CLUSTERING PROBLEM:
Given a set $\mathcal{A}$ of advertisers, a set $\mathcal{K}$ of keywords, the distribution $\mathcal{D}$ over $\mathcal{K}$, the values $v(i, j)$ for each advertiser $i \in \mathcal{A}$ and each keyword $j \in \mathcal{K}$,

[6] For simplicity we state our argument in the case of a consistent deterministic tie-breaking rule. It is not hard to see that the same reasoning works for a randomized tie breaking rule.

and a number k, compute a clustering of $\mathcal{K}$ into k clusters $S_1, \ldots, S_k$ to maximize the social welfare:

$$\sum_{r=1}^{k} \max_{i \in \mathcal{A}} v(i, S_r).$$

where $v(i, S_r) = \sum_{a \in S_r} v(i, a)$.

Since the set $\mathcal{K}$ can be very large or infinite, we would like an algorithm whose running time does not depend on $|\mathcal{K}|$. The algorithm is given access to an oracle that outputs samples according to $\mathcal{D}$ and another oracle that given i and j, outputs $v(i, j)$. The output of the algorithm also needs to be in the form of an oracle that given a keyword, decides which cluster the keyword belongs to.

It is not hard to show that the above problem is NP-hard, and therefore we cannot hope for an exact efficient solution.

Stochastic formulation. The deterministic formulation of the problem makes the assumption that the algorithm has access to a set of estimated values. This assumption, while makes the clustering problem easy to state and more useful in practice, is somewhat unsatisfactory from a theoretical perspective. Therefore, we also study another formulation of the problem, where the algroithm only knows distributional information about the values, and needs to compute a clustering that optimizes the expected social welfare. We assume the distribution of the values is explicitly given to the algorithm, i.e., there is a fixed number ℓ of different *types* of advertisers, with each type t is associated with a probability p_t and a set of valuations $v(t, j)$ for an advertiser of this type (given using an oracle as described above). There are m advertisers, each having a type picked according to the probabilities p_t. The type of the advertiser determines his valuation. We will discuss this formulation of the problem in Section 4.2, and give an approximation algorithm in cases where probabilities p_t are not too small.

4.1 A $(1 - 1/e)$-Approximation Algorithm for Deterministic Clustering

In this section, we give an approximation algorithm for the deterministic formulation of the clustering problem by reducing it to submodular function maximization. The main component of the solution is the definition of the function.

We define a function f from the domain $\mathcal{A}$ to the set of non-negative reals. The value of the function on a set $R \subset \mathcal{A}$ is defined as follows:

$$f(R) = \mathrm{Exp}_{j \leftarrow \mathcal{D}}[\max_{a \in R} v(a, j)].$$

Corresponding to the set $R \subset \mathcal{A}$, we define a $|R|$-clustering C_R of $\mathcal{K}$ as follows: each cluster S_a in C_R corresponds to an advertiser a in R, and each keyword $j \in \mathcal{K}$ is assigned to the cluster corresponding to an advertiser $a \in \mathrm{argmax}_{a \in R} v(a, j)$.

Note that with this definition and the definition of f, the value of $f(R)$ is precisely the expected value of a keyword in a cluster to the advertiser in R that this cluster is assigned to, i.e.,

$$f(R) = \sum_{a \in R} \Pr_{j \leftarrow \mathcal{D}}[j \in S_a] \cdot \mathrm{Exp}_{j \leftarrow \mathcal{D}|S_a}[v(a, j)].$$

This implies the following Lemma. The proof is omitted here.

Lemma 1. *For every set $R \subseteq \mathcal{A}$, the social welfare of the clustering C_R is at least $f(R)$.*

Note that the inequality in the above lemma can sometimes be strict, i.e., the correspondence between the set R of advertisers and the clustering C_R is not one-to-one. However, we can show that at the social optimum, this correspondence *is* one-to-one.

Lemma 2. *Let OPT denote a partitioning of $\mathcal{K}$ into at most k clusters that maximizes the social welfare among all such clusterings. Let R^* denote the set of advertisers that win (i.e., have the maximum value for) at least one cluster in OPT. Then the social welfare of the clustering OPT is precisely $f(R^*)$.*

We omid the proof of this lemma. By the above two lemmas, it is clear that a c-approximation algorithm for finding the maximum value of $f(R)$ subject to $|R| \leq k$ gives a c-approximation algorithm for the clustering problem: the clustering would simply be C_{R^*}, where R^* is the set that approximately maximizes f. Furthermore, the resulting clustering can be given compactly as a membership oracle: for each given keyword $j \in \mathcal{K}$, the oracle simply evaluates $v(a, j)$ for $a \in R^*$ and assigns j to the cluster with the maximum value.

The only thing that remains is to give an algorithm that maximizes f. Fortunately, this is easy as the function f is submodular.

Lemma 3. *The function f is submodular and non-decreasing.*

The proof of the above lemma is easy and is omitted here. Given the submodularity of f, the maximization problem can be solved using the classical greedy algorithm of Nemhauser et al. [13] with an approximation factor of $1 - 1/e$. This algorithm consists of iteratively picking an element (in this case an advertiser $a \in \mathcal{A}$) with the maximum marginal value. In the case of our problem where the input is given using oracles, instead of computing the marginal value of each $a \in \mathcal{A}$ exactly in each iteration, we can use sampling, with a sample size of $O(\log k)$ to estimate this marginal value to within a small error. It is not hard to bound the total error using the union bound and show that the greedy algorithm with an approximate evaluation of the marginal values can achieve an approximation factor arbitrarily close to $1 - 1/e$ with high probability. The details of the argument are straightforward and are omitted.

Theorem 1. *For every constant $\epsilon > 0$, there is an algorithm that approximates the deterministic formulation of the clustering problem to within a factor of $1 - \frac{1}{e} - \epsilon$. The running time of this algorithm is polynomial in $|\mathcal{A}|$ and k and does not depend on $|\mathcal{K}|$.*

4.2 An Approximation Algorithm for Stochastic Clustering

In this section, we give an algorithm for the stochastic formulation of the clustering problem, in cases where the probability of different types of advertisers is not too small. Our main result is the following.

Theorem 2. *Assume an instance of the stochastic clustering problem satisfies* $\min_t p_t \geq \frac{1}{cm}$ *for a constant* c. *Then there is a polynomial time algorithm that approximately solves this problem on such instance with an approximation ratio of* $\frac{1}{2}(1 - \frac{1}{e})$.

The proof of this theorem is omitted due to lack of space.

5 Experimental Results

In this section, we describe the results of experiments with the greedy algorithm for computing a k-clustering of a set of keywords in the single-slot model. The results of the experiment show that the social welfare of the k-clustering computed by the algorithm is a significant fraction of the total social welfare, even for values of k that are orders of magnitude smaller than the total number of keywords. The resulting k-clusterings are natural and partition the keywords into subsets of roughly the same meaning. Lastly, we show that the k-clusterings obtained by the algorithm are stable over time; the partition of keywords computed from a data set at time t_1 still obtains a large fraction of the social welfare at time $t_2 > t_1$.

5.1 Dataset

The dataset we used in the experiments was a matrix M where the rows are keywords and the advertisers are columns, with M_{ij} being equal to the monthly spend of advertiser j on keyword i. The number of keywords in the dataset was roughly 50,000. We interpreted the spend of advertiser j on keyword i to be the value of the keyword i to advertiser j. Because we are working with the single-slot model, we consider the total social welfare to be the total value obtained when each keyword is assigned to the advertiser with maximum value, i.e.

$$\textsf{total social welfare} = \sum_i \max_j M_{ij}$$

5.2 Experiments

We ran the greedy algorithm for computing a k-clustering on our dataset for $k = 1$ to $5,000$. For each value of k, we computed the percentage of total social welfare that is obtained by the k-clustering. The plot of percentage of total social welfare vs. k is the solid line in Figure 1. When $k = 1,000$, we obtain nearly 80% of the total social welfare, while reducing the complexity of the bidding language (the number of items an advertiser can bid on) by a factor of 50. We also ran

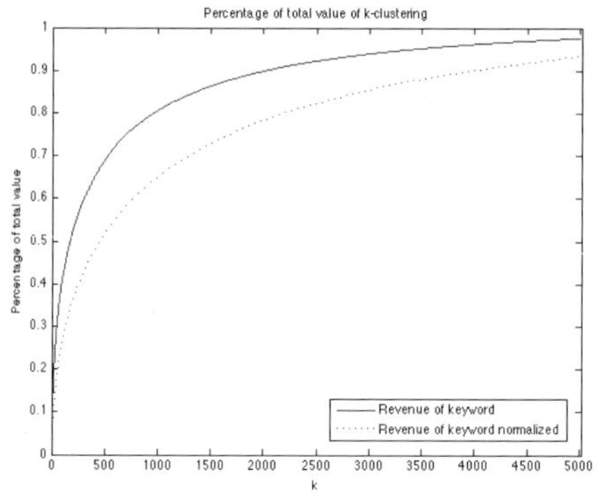

Fig. 1. Percentage of total social welfare vs. k

the greedy algorithm with the row-sums of the matrix M normalized to 1. By normalizing the value of each keyword to 1, we ensure that the values of the keywords are not the primary reason that a k-clustering obtains a large fraction of the total social welfare. The plot of percentage of total social welfare vs. k for the normalized case is the dotted line. While the dotted line is dominated by the solid line, we still see a significant reduction in the complexity of the bidding language while obtaining a significant fraction of social welfare (e.g. at $k = 2,500$, the k-clustering obtains roughly 80% of the total social welfare).

We also compared the social welfare v at time t_2 of the k-clustering C_1 computed on a dataset from time t_1 to the social welfare w at time t_2 of the k-clustering C_2 computed from t_2 for various values of k. The plot of v/w can be found in Figure 2. For small values of k, v/w is quite volatile, but for large enough k, C_1 obtains more than 80% of the social welfare of C_2.

A sample of keywords from the clusters for $k = 1,000$ can be seen in Table 1.

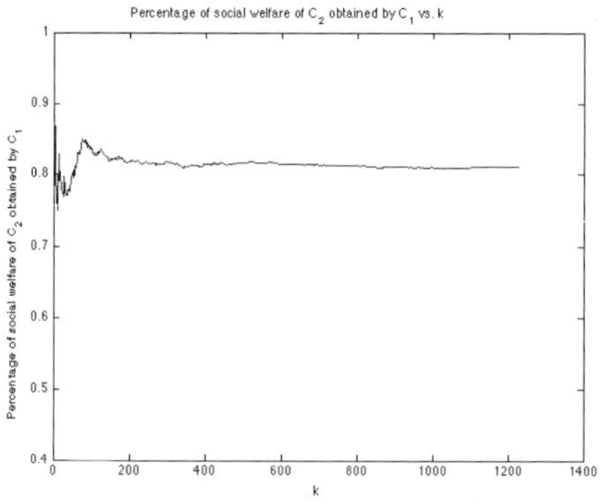

Fig. 2. Time stability of k-clustering

Table 1. Sample keywords and clusters from a 1,000-clustering

Cluster 1	Cluter 2	Cluster 3	Cluster 4
computer fax	septic system	alcohol addiction	brooch pin
download free fax software	septic tank	alcohol addiction treatment	gem stone ring
electronic fax	storm shelter	drug addiction	gem stone
email by fax	timberline shingles	substance abuse	indian jewelry
free computer fax software	tree service	substance abuse treatment	tanzanite jewelry

6 Discussion

We have introduced the idea of a clustering-based bidding language for sponsored search. Clustering-based bidding languages reduce the complexity of bidding for the advertiser. In addition, they avoid the undesirable economic properties of *broad matching*, where advertisers can bid on overlapping sets of keywords. We gave approximation algorithms for computing a k-clustering that maximizes social welfare to within a constant factor in both a deterministic and stochastic formulation of the problem. Experimental results show that a k-clustering obtains a large fraction of the total social welfare when k is several orders of magnitude smaller than the total number of keywords.

It is worth noting that the greedy algorithm we give for the deterministic formulation of the problem with single slots can be extended to the case of multiple slots. The algorithm remains largely the same – instead of greedily choosing the advertiser that obtains the largest incremental gain in social welfare, we choose the *slate* of advertisers that obtain the largest incremental gain. This algorithm gives a $(1 - 1/e)$ approximation, since the underlying function we are maximizing is still submodular. When the number of slots is a constant, this gives us a polynomial-time approximation algorithm. However, if the number of slots is not a constant, we do not know how to implement the greedy step in polynomial time.

While we showed experimentally that the k-clustering is relatively stable over time, eventually the bidding language would need to change. The problem of *evolving* a clustering-based bidding language, both in theory and in practice, is an interesting one.

In theory, increasing the power (and complexity) of a bidding language can only increase social welfare. In practice, there is a tradeoff – as bidding languages become more expressive, social welfare can actually decrease. It would be interesting to both study this experimentally and come up with suitable models.

References

1. Aggarwal, G., Goel, A., Motwani, R.: Truthful auctions for pricing search keywords. In: EC 2006: Proceedings of the 7th ACM conference on Electronic commerce, pp. 1–7. ACM Press, New York (2006)
2. Barzel, Y.: Measurement cost and the organization of markets. Journal of Law and Economics (1982)

3. Christian, B., Jennifer, C., Nicole, I., Kamal, J., Omid, E., Mohammad, M.: Dynamics of bid optimization in online advertisement auctions. In: WWW 2007: Proceedings of the 16th international conference on World Wide Web, pp. 531–540. ACM, New York (2007)
4. Borgs, C., Chayes, J., Immorlica, N., Mahdian, M., Saberi, A.: Multi-unit auctions with budget-constrained bidders. In: EC 2005: Proceedings of the 6th ACM conference on Electronic commerce, pp. 44–51 (2005)
5. Broder, A.Z., Fontoura, M., Gabrilovich, E., Joshi, A., Josifovski, V., Zhang, T.: Robust classification of rare queries using web knowledge. In: SIGIR 2007: Proceedings of the 30th annual international ACM SIGIR conference on Research and development in information retrieval, pp. 231–238. ACM Press, New York (2007)
6. Dobzinski, S., Lavi, R., Nisan, N.: Multi-unit auctions with budget limits. In: Annual IEEE Symposium on Foundations of Computer Science, pp. 260–269 (2008)
7. Edelman, B., Ostrovsky, M., Schwarz, M.: Internet advertising and the generalized second price auction: Selling billions of dollars worth of keywords. American Economic Review (2007)
8. Even-dar, E., Mansour, Y., Mirrokni, V., Muthukrishnan, S., Nadav, U.: Bid optimization for broad match ad auctions. To appear in WWW 2009 (2009), http://arxiv.org/abs/0901.3754
9. Feldman, J., Muthukrishnan, S., Pal, M., Stein, C.: Budget optimization in search-based advertising auctions. In: EC 2007: Proceedings of the 8th ACM conference on Electronic commerce, pp. 40–49. ACM, New York (2007)
10. Martin, D., Gehrke, J., Halpern, J.: Toward expressive and scalable sponsored search auctions. In: ICDE Conference (2008)
11. Mehta, A., Saberi, A., Vazirani, U., Vazirani, V.: Adwords and generalized online matching. J. ACM 54(5), 22 (2007)
12. Muthukrishnan, S.M., Pál, M., Svitkina, Z.: Stochastic models for budget optimization in search-based advertising. In: Deng, X., Graham, F.C. (eds.) WINE 2007. LNCS, vol. 4858, pp. 131–142. Springer, Heidelberg (2007)
13. Nemhauser, G.L., Wolsey, L.A., Fisher, M.L.: An analysis of approximations for maximizing submodular set functions - i. Math. Prog. 14, 265–294 (1978)
14. Rusmevichientong, P., Williamson, D.P.: An adaptive algorithm for selecting profitable keywords for search-based advertising services. In: EC 2006: Proceedings of the 7th ACM conference on Electronic commerce, pp. 260–269. ACM Press, New York (2006)
15. Singh, S.K., Roychowdhury, V.P.: To broad-match or not to broad-match: An auctioneer's dilemma. In: Workshop on Ad Auctions (2008)
16. Varian, H.: Position auctions. To appear in International Journal of Industrial Organization (2006)
17. Zhou, Y., Chakrabarty, D., Lukose, R.: Budget constrained bidding in keyword auctions and online knapsack problems. In: WWW 2008: Proceeding of the 17th international conference on World Wide Web, pp. 1243–1244. ACM Press, New York (2008)

Altruism in Atomic Congestion Games

Martin Hoefer* and Alexander Skopalik**

Dept. of Computer Science, RWTH Aachen University, Germany
{mhoefer,skopalik}@cs.rwth-aachen.de

Abstract. This paper studies the effects of introducing altruistic agents into atomic congestion games. Altruistic behavior is modeled by a trade-off between selfish and social objectives. In particular, we assume agents optimize a linear combination of personal delay of a strategy and the resulting social cost. Stable states are the Nash equilibria of these games, and we examine their existence and the convergence of sequential best-response dynamics. For symmetric singleton games with arbitrary delay functions we provide a polynomial time algorithm to decide existence for symmetric singleton games. Our algorithm can be extended to compute best and worst Nash equilibria if they exist. For more general congestion games existence becomes NP-hard to decide, even for symmetric network games with quadratic delay functions. Perhaps surprisingly, if all delay functions are linear, there exists a Nash equilibrium and any better-response dynamics converges. In addition, we consider a scenario in which a central altruistic institution can motivate agents to act altruistically. We provide constructive and hardness results for finding the minimum number of altruists to stabilize an optimal congestion profile and more general mechanisms to incentivize agents to adopt favorable behavior.

1 Introduction

Algorithmic game theory has been focused on game-theoretic models for a variety of important applications in the Internet. A fundamental assumption in these games, however, is that all agents are *selfish*. Their goals are restricted to optimizing their direct personal benefit, e.g. their personal delay in a routing game. The assumption of selfishness in the preferences of agents is found in the vast majority of present work on economic aspects of the Internet. However, this assumption has been repeatedly questioned by economists and psychologists. In experiments it has been observed that participant behavior can be quite complex and contradictive to selfishness [15,16]. Various explanations have been given for this phenomenon, e.g. senses of fairness [7], reciprocity among agents [10], or spite and altruism [16,5].

* Work was done while visiting the Computer Science Department, Stanford University, USA. Supported by a fellowship within the Postdoc-Program of the German Academic Exchange Service (DAAD) and by UMIC Research Center at RWTH Aachen University.

** Supported in part by the German Israeli Foundation (GIF) under contract 877/05.

A. Fiat and P. Sanders (Eds.): ESA 2009, LNCS 5757, pp. 179–189, 2009.

Prominent developments in the Internet like Wikipedia, open source software development, or Web 2.0 applications involve or explicitly rely on voluntary participation and contributions towards a joint project without direct personal benefit. These examples display forms of *altruism*, in which agents accept certain personal burdens (e.g. by investing time, attention, and money) to improve a common outcome. While malicious behavior has been considered recently for instance in nonatomic routing [14,3,4], virus incoulation [18], or bayesian congestion games [8], a deeper analysis of the effects of altruistic agents on competitive dynamics in algorithmic game theory is still missing.

We consider and analyze a model of altruism inspired by Ledyard [15, p. 154], and recently studied for non-atomic routing games by Chen and Kempe [4]. Each agent i is assumed to be partly selfish and partly altruistic. She optimizes a linear combination of personal cost and social cost, given by the sum of cost values of all agents. The strength of altruism of each agent i is captured by her *altruism level* $\beta_i \in [0, 1]$, where $\beta_i = 0$ results in a purely selfish and $\beta_i = 1$ in a purely altruistic agent.

Chen and Kempe [4] proved that in non-atomic routing games Nash equilibria are always guaranteed to exist and analyzed the price of anarchy for parallel link networks. In our paper, we conduct the first study of altruistic agents in atomic congestion games, a well-studied model for resource sharing. Congestion games received a lot of attention recently, mostly because of the intuitive formulation and their appealing analytical properties. In particular, they always possess a pure Nash equilibrium and every sequential better-response dynamics converges. As one might expect, the presence of altruists can significantly alter the convergence and existence of pure Nash equilibria. After a formal definition of *congestion games with altruists* in Section 2, we concentrate on pure equilibria and leave a study of mixed Nash equilibria for future work. Our results are as follows.

We study singleton games in Section 3, in which every strategy consists of a single resource. In symmetric games with a constant number of different altruism levels, we can decide in polynomial time if a Nash equilibrium exists. Our algorithm can be adapted to compute the Nash equilibrium with best and worst social cost if it exists, for any agent population with a constant number of different altruism levels. For *asymmetric singleton games*, in which strategy spaces of agents differ, deciding the existence of Nash equilibria becomes NP-hard. For the important subclass of convex delay functions, i.e., linear and superlinear functions, previous results imply that for any agent population a Nash equilibrium exists and can be computed in polynomial time. In contrast, we show in Section 4 that convexity of delay functions is not sufficient for more general games. Even for *symmetric network games*, in which strategies represent paths through a network, quadratic delay functions and pure altruists, Nash equilibria can be absent and deciding their existence is NP-hard. Perhaps surprisingly, if all delay functions are *linear*, the game is a potential game.

In Section 5 we consider a slightly more coordinated scenario, in which there is a central institution that can convince agents to act altruistically. In this context a natural question is how many altruists are required to stabilize a social optimum. This has been considered under the name "price of optimum" in [13] for Stackelberg routing in nonatomic congestion games. We consider two measures - an *optimal stability threshold*, which is the minimum number of altruists such that there is *any* optimal Nash equilibrium, and an *optimal anarchy threshold*, which asks for the minimum number of altruists such that *every* Nash equilibrium is optimal. For symmetric singleton games, we adapt our algorithm for computing Nash equilibria to determine both thresholds in polynomial time. The optimal anarchy threshold might not be well-defined even for singleton games, because there are suboptimal Nash equilibria even if all agents are pure altruists. In contrast, we adapt the idea of the optimal stability threshold to a very general scenario, in which each agent has a personalized *stability cost* for accepting a strategy under the given congestions. We provide a truthful mechanism to determine an allocation of agents to strategies with minimum total stability cost. Unfortunately, such a general result is restricted to the case of singleton games. Even for symmetric network games on series-parallel graphs, we show that the problem of determining the optimal stability threshold is NP-hard. Our reduction also yields inapproximability within any finite factor. This resolves an open problem raised in [12] on computing the "price of optimum" in atomic congestion games.

Some proofs have been omitted due to spacial constraints, they will be given in a full version of this paper.

2 Model and Initial Results

We consider congestion games with altruists. A *congestion game with altruists* G is given by a set N of n agents and a set E of m resources. Each agent i has a set $\mathcal{S}_i \subseteq 2^E$ of strategies. In a *singleton* congestion game each agent has only singleton strategies $\mathcal{S}_i \subseteq E$. A vector of strategies $S = (S_1, \ldots, S_n)$ is called a *state*. For a state we denote by n_e the congestion, i.e. the number of agents using a resource e in their strategy. Each resource e has a *latency* or *delay* function $d_e(n_e)$, and the *delay for an agent* i playing S_i in state S is $d_i(S) = \sum_{e \in S_i} d_e(n_e)$. The *social cost* of a state is the total delay of all agents $c(S) = \sum_{i \in N} \sum_{e \in S_i} d_e(n_e) = \sum_{e \in E} n_e d_e(n_e)$. Each agent i has an *altruism level* of $\beta_i \in [0, 1]$, and her *individual cost* is $c_i(S) = \beta_i c(S) + (1 - \beta_i) d_i(S)$. We call an agent i an *egoist* if $\beta_i = 0$ and a β_i*-altruist* otherwise. A *(pure) altruist* has $\beta_i = 1$, a *(pure) egoist* has $\beta_i = 0$. A game G *with only pure altruists and egoists* is a game, in which $\beta_i \in \{0, 1\}$ for all $i \in N$. A game G is said to have β*-uniform* altruists if $\beta_i = \beta \in [0, 1]$ for every agent $i \in N$. A (pure) *Nash equilibrium* is a state S, in which no agent i can unilaterally decrease her individual cost by unilaterally changing her strategy. We exclusively consider pure equilibria in this paper.

If all agents are egoists, the game is a regular congestion game, which has an exact potential function $\Phi(S) = \sum_{e \in E} \sum_{x=1}^{n_e} d_e(x)$ [20]. Thus, existence of Nash equilibria and convergence of iterative better-response dynamics are guaranteed.

Obviously, if all agents are altruists, Nash equilibria correspond to local optima of the social cost function c with respect to a local neighborhood consisting of single player strategy changes. Hence, existence and convergence are also guaranteed. This directly implies the same properties for β-uniform games, in which an exact potential function is $\Phi_\beta(S) = (1-\beta)\Phi(S) + \beta c(S)$. In general, however, Nash equilibria might not exist.

Observation 1. *There are symmetric singleton congestion games with only pure altruists and egoists without a Nash equilibrium.*

Example 2. Consider a game with two resources e and f, three egoists and one (pure) altruist. The delay functions are $d_e(x) = d_f(x)$ with $d_e(1) = 4$, $d_e(2) = 8$, $d_e(3) = 9$, and $d_e(4) = 11$. One can easily check that this game does not posses a Nash equilibrium.

Our interest is thus to characterize the games that have Nash equilibria. Towards this end we observe that an altruistic congestion game can be cast as a congestion game with player-specific latency functions [17]. In such a game the delay of resource e to player i depends on the congestion and on the player, i.e., $c_i(S) = \sum_{e \in S_i} d_e(n_e, i)$. To embed our games within this framework, we consider a game with only pure altruists and egoists for simplicity. An altruist moves from S_i to S_i' if the decrease in total delay $n_e d_e(n_e)$ on the resources $e \in S_i - S_i'$ she is leaving exceeds the increase on resources $e \in S_i' - S_i$ she is migrating to. Hence, altruists can be seen as myopic selfish agents with $c_i(S) = d_i'(S) = \sum_{e \in S_i} d_e'(n_e)$ with $d_e'(n_e) = n_e d_e(n_e) - (n_e - 1) d_e(n_e - 1)$, for $n_e > 0$. We set $d_e'(0) = 0$. Naturally, a β_i-altruist corresponds to a selfish agent with player-specific function $c_i(S) = (1-\beta_i) d_i(S) + \beta_i d_i'(S)$. Thus, our games can be embedded into the class of player-specific congestion games. For some classes of such games it is known that Nash equilibria always exist. In particular, non-existence in Example 2 is due to the fact that the individual delay function for the altruist is not monotone. Monotonicity holds, in particular, if delay functions are convex. In this case, it is known that for matroid games, in which the strategy space of each agent is a matroid, existence of a Nash equilibrium is guaranteed [2].

Corollary 3. *[17,2] For any matroid congestion game with altruists and convex delay functions a Nash equilibrium exists and can be computed in polynomial time.*

3 Singleton Congestion Games

In the previous section we have seen that there are symmetric singleton congestion games with only pure altruists and egoists with and without Nash equilibria. For this class of games we can decide the existence of Nash equilibria in polynomial time. In addition, we can compute a Nash equilibrium with minimum and maximum social cost if they exist.

Theorem 4. *For symmetric singleton games with only pure altruists and egoists there is a polynomial time algorithm to decide if a Nash equilibrium exists and to compute the best and the worst Nash equilibrium.*

The proof relies on the fact that the game is symmetric and the number of resources is polynomial. This allows us to characterize a Nash equilibrium using certain maximum and minimum values for the delays of each resource. Similar to [11] we can implicitly enumerate all states that can be a Nash equilibrium using dynamic programming. The approach can be extended to a constant number k of different altruism levels. In this more general scenario we choose the delay parameters for each level of altruists.

Corollary 5. *For symmetric singleton games with altruists and a constant number of different altruism levels, there is a polynomial time algorithm to decide if a Nash equilibrium exists and to compute the best and the worst Nash equilibrium.*

As a byproduct, our approach also allows us to compute a social optimum state in polynomial time. We simply assume all agents to be pure altruists and compute the best Nash equilibrium.

Corollary 6. *For symmetric singleton congestion games a social optimum state can be obtained in polynomial time.*

In case of asymmetric games, however, deciding the existence of Nash equilibria becomes significantly harder.

Theorem 7. *It is* NP*-hard to decide if a singleton congestion game with only pure altruists and egoists has a Nash equilibrium if G is asymmetric and has concave delay functions.*

Proof. We reduce from 3SAT. Given a formula φ, we construct a congestion game G_φ that has a Nash equilibrium if and only if φ is satisfiable. Let $x_1, \ldots, x_n$ denote the variables and $c_1, \ldots, c_m$ the clauses of a formula φ. Without loss of generality [21], we assume each variable appears at most twice positively and at most twice negatively.

For each variable x_i there is a selfish agent X_i that chooses one of the resources $e^1_{x_i}$, $e^0_{x_i}$, or e_0. The resources $e^1_{x_i}$ and $e^0_{x_i}$ have the delay function $9x$ and resource e_0 has the delay function $7x + 3$. For each clause c_j, there is a selfish agent C_j who can choose one of the following three resources. For every positive literal x_i in c_j he may choose $e^0_{x_i}$. For every negated literal $\bar{x}_i$ in c_j he may choose $e^1_{x_i}$. Note that there is a stable configuration with no variable agent on e_0 if and only if there is a satisfiable assignment for φ. Additionally, there are three selfish agents u_1, u_2, and u_3 who can choose e_1 or e_2. Each of the resources e_1 and e_2 has delay 4 if used by one agent, delay 8 if used by two agents and delay 9 otherwise. The only pure altruist u_0 chooses between e_1, e_2, and e_0. Note that the altruist chooses e_1, e_2 if one of the variable agents is on e_0.

If φ is satisfiable by a bitvector $(x^*_1, \ldots, x^*_n)$, a stable solution for G_φ can be obtained by placing each variable agent x_i on $e^{x^*_i}_{x_i}$. Since $(x^*_1, \ldots, x^*_n)$ satisfies φ

there is one resource for each clause agent that is not used by a variable agent. Thus, we can place each clause agent on this resource, which he then shares with at most one other clause agent. Let the altruist u_0 use e_0 and u_1 and u_2 choose e_1 and u_3 choose e_2. It is easy to check that this is a Nash equilibrium.

If φ is unsatisfiable, there is no stable solution. To prove this it suffices to show that one of the variable agents prefers e_0. In that case the altruist never chooses e_0 and the agent $u_0, \ldots, u_3$ play the sub game of Example 2. For the purpose of contradiction assume that φ is not satisfiable but there is a stable solution in which no variable wants to choose e_0. This implies that there is no other agent, i.e. a clause agent, on a resource that is used by a variable agent. However, if all clause agents are on a resource without a variable agent we can derive a corresponding bit assignment which, by construction, satisfies φ. Therefore, G_φ has a stable solution if and only if φ is satisfiable. □

4 General Games

For any singleton game G with altruists and convex delay functions a Nash equilibrium always exists. For more general network structures, we show that convexity of delay functions is not sufficient. In particular, this holds even for games with only pure altruists and egoists in the case in which almost all delay functions are linear of the form $d_e(x) = a_e x$, except for two edges, which have quadratic delay functions $d_e(x) = a_e x^2$. For simplicity, we use some edges with non-convex constant delay b_e. We can replace these edges by sufficiently many parallel edges with delay $b_e x$. This transformation is of polynomial size and yields an equivalent game with only convex delays.

Theorem 8. *It is* NP*-hard to decide if a symmetric network congestion game with only pure altruists and egoists and quadratic delay functions has a Nash equilibrium.*

Proof. We first reduce from 3SAT to asymmetric congestion games. Again, we assume each variable appears at most twice positively and at most twice negatively. In a second step, we show that the resulting congestion games can be turned into symmetric games while preserving all necessary properties.

Our reduction is similar to the construction that we used in the proof of Theorem 7. The structure of the resulting network congestion game G_Φ is depicted in Figure 1. Table 1 lists the delay functions of the edges. Edges that are not listed there have delay of 0. Due to space limitations we only outline the structure of our construction. The complete proof will appear in the full version.

Each agent X_i chooses one of three paths from his source node s_{x_i} to his target node t'. Each clause agent C_j uses a path from s_{c_j} to t' and uses one of the three edges as described in the proof of Theorem 7. That is, for each positive literal x_i in c_j he may choose a path that includes the edge $e^0_{x_i}$. For every negated literal $\bar{x_i}$ in c_j he may choose a path that contains the edge $e^1_{x_i}$. There is a selfish agent u_1 that chooses a path from s_1 to t' and two selfish agents u_2 and u_3 that

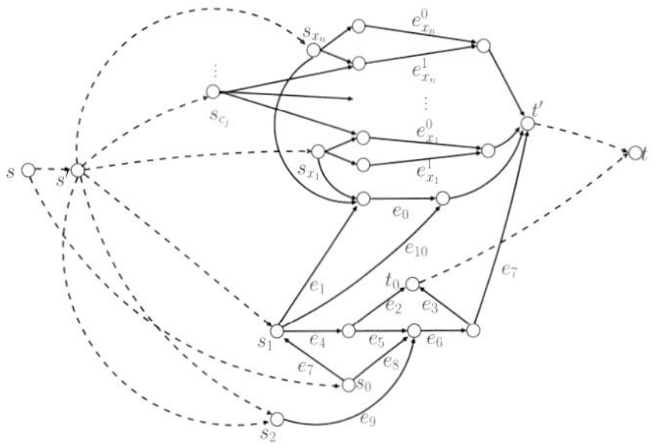

Fig. 1. The structure of the network of G_Φ (solid edges only) and G'_Φ

Table 1. The delay functions on the edges of G_Φ and G'_Φ

Edge	delay function
e_0	$7x+3$
e_1	2
e_2	17
e_4	$2.4x^2$
e_6	x^2
e_{10}	18.5
$e^1_{x_i}, e^1_{\bar{x}_i}$	$9x$
$(s, s_{x_i})\ \forall 1 \le i \le n$	Mx
$(s, s_{c_j})\ \forall 1 \le j \le m$	Mx
(s, s_1), (s, s_2), (s, s')	Mx
(s, s_0)	$(n+m+5)M$
(t_0, t)	$(n+m+5)M$
(t', t)	Mx

allocate the path from s_2 to t'. Finally, one altruistic agent u_0 chooses a path from s_0 to t_0.

The asymmetric network congestion game G_Φ can be turned into a symmetric congestion game G'_Φ. We add a new source node s, a new target node t and a node s' to the network and connect them to G_Φ as depicted by the dashed edges in Figure 1. Note that M is an integer that is larger than the sum of possible delay values in G_Φ. □

Perhaps surprisingly, if *every* delay function is linear $d_e(x) = a_e x + b_e$, then an elegant combination of the Rosenthal potential and the social cost function yields a potential for arbitrary β_i-altruists. Hence, existence of Nash equilibria and convergence of sequential better-response dynamics is always guaranteed. The proof is carefully constructed for altruists, as for congestion games with general player-specific linear latency functions a potential does not exist [9]. We only consider delays $d_e(x) = a_e x$ without offset b_e, but as noted earlier, this is not a restriction.

Theorem 9. *For any congestion game with altruists and linear delay functions there is always a Nash equilibrium and sequential better-response dynamics converges.*

Proof. The theorem follows from the existence of a weighted potential Φ that decreases during every improvement step of any agent i with altruism level β_i.

$$\Phi(S) = \sum_{e\in E}\sum_{j=1}^{n_e} a_e j + \sum_{e\in E} a_e n_e^2 - \sum_{i=1}^{n}\sum_{e\in S_i} \frac{2\beta_i - 1}{\beta_i + 1} a_e$$

Consider a state S and an improving strategy change of an agent i from S_i to S'_i resulting in a strategy profile S'. We show that Φ decreases. For the sake of clarity and brevity we set $\Delta_N = \sum_{e\in S_i\setminus S'_i} a_e n_e - \sum_{e\in S'_i\setminus S_i} a_e(n_e+1)$ and

$\Delta_C = \sum_{e \in S_i \setminus S_i'} (2a_e n_e - a_e) - \sum_{e \in S_i' \setminus S_i} (2a_e n_e + a_e)$. Note that an improving strategy change requires $(1-\beta)\,\Delta_N + \beta \Delta_C > 0$.

$$\Phi(S) - \Phi(S') = \Delta_N + \Delta_C - \sum_{e \in S_i \setminus S_i'} \frac{2\beta_i - 1}{\beta_i + 1} a_e + \sum_{e \in S_i' \setminus S_i} \frac{2\beta_i - 1}{\beta_i + 1} a_e$$

$$= \left(1 - \frac{2(2\beta_i - 1)}{1 + \beta_i}\right) \Delta_N + \Delta_C + \frac{(2\beta_i - 1)}{1 + \beta_i} \Delta_C$$

$$= \frac{3 - 3\beta_i}{1 + \beta_i} \Delta_N + \frac{3\beta_i}{1 + \beta_i} \Delta_C = \frac{3}{1 + \beta_i} \left((1 - \beta_i)\,\Delta_N + \beta_i \Delta_C\right) > 0$$

□

Unfortunately, it follows directly from previous work [6] that the number of iterations to reach a Nash equilibrium can be exponential, and the problem of computing a Nash equilibrium is PLS-hard. For regular congestion games with matriod strategy spaces [1] Nash dynamics converge in polynomial time. It is an interesting open problem if a similar result holds here.

5 Stabilization Methods

This section treats a model in which an institution can convince selfish agents to act as altruists. For simplicity of presentation we first restrict to games with only pure altruists and egoists. A natural question for such an institution to consider is how many altruists are required to guarantee that there is a Nash equilibrium with a certain cost, e.g. a Nash equilibrium as cheap as a social optimum state. We term this number the *optimal stability threshold.* In a more pessimistic direction it is of interest to determine the minimum number of altruists needed to guarantee that the worst-case Nash equilibrium is optimal. We term this number the *optimal anarchy threshold.* Let us denote by n_1^+ and n_1^- the optimal stability and anarchy threshold, respectively. As a consequence from Theorem 4 we can compute both numbers for symmetric singleton congestion games in polynomial time. For each number of altruists we check if the best and/or worst Nash equilibrium is as cheap as the social optimum.

Corollary 10. *For symmetric singleton congestion games with only pure altruists and egoists there is a polynomial time algorithm to compute n_1^+ and n_1^-.*

Note that the optimal anarchy threshold is not well-defined, because the worst Nash equilibrium might always be suboptimal, even for a population of altruists only. In case of symmetric singleton games and convex delay functions, an easy exchange argument serves to show that in this case any local optimum is also a global optimum. However, for concave delay functions or asymmetric singleton games, a local optimum might still be globally suboptimal. Note that for symmetric games, our algorithm is able to detect the cases in which suboptimal local optima exist. In the asymmetric case, however, a similar approach fails, because of the NP-hardness of determining existence of a Nash equilibrium. Thus, in the following we concentrate on the optimal stability threshold.

In asymmetric games, it is also required to determine the identity of agents, so here we strive to find a set (denoted N_e^+) of minimum cardinality. For an optimal set of congestion values $n_E^* = (n_e^*)_{e \in E}$ we can determine $N_1^+(n_E^*)$ such that there is a Nash equilibrium of the game with congestion values n_e^* for all $e \in E$.

Theorem 11. *For singleton games with only pure altruists and egoists and a social optimal congestion vector n_E^* there is a polynomial time algorithm to compute $N_1^+(n_E^*)$.*

The theorem can be shown by constructing a complete bipartite graph. The nodes in one partition correspond to agents, in the other partition there are n_e^* nodes for each resource e. By appropriately assigning costs in $\{0, 1\}$ to the edges we can minimize the number of required altruists with a minimum cost perfect matching. The complete details are deferred to the full version.

This approach can be extended to an even more general natural scenario. Suppose each agent i has a *stability cost* c_{ie} for each strategy $e \in \mathcal{S}_i$. This cost yields the disutility for being forced to play a certain strategy given a congestion vector n_E. Here we redefine $N_1^*(n_E)$ to be the set agents of minimal stability cost. We can compute this set by a minimum weight perfect matching if we set the weights to c_{ie} for all edges connecting i to vertices of e. The stability cost allows for general preferences exceeding categories like altruists and egoists.

Corollary 12. *For singleton games and a congestion vector n_E there is a polynomial time algorithm to compute $N_1^+(n_E)$ with minimal stability cost.*

The underlying problem is a matching problem, which is solved optimally. Hence, it is possible to turn our approach into a truthful mechanism using VCG payments (see e.g. [19, chapter 9]). Our final mechanism (1) learns the stability costs from each agent, (2) determines the allocation, and (3) pays appropriate amounts to agents for truthful revelation of cost values and adaptation of allocated strategies. In addition, it can be verified that all computations needed require only polynomial time.

Corollary 13. *For singleton games and a congestion vector n_E there is a truthful VCG-mechanism to compute $N_1^+(n_E)$ in polynomial time.*

These general results are restricted to the case of singleton games. For more general games we show that it is NP-hard to decide if there is a Nash equilibrium as cheap as the social optimum. Our next theorem establishes this even for symmetric network congestion games with linear delays, in which an arbitrary Nash equilibrium and a social optimum state can be computed in polynomial time [6]. Furthermore, the result requires only a series-parallel network. Thus, even in this restricted case it is NP-hard to decide if the number n_1^+ of pure altruists required is 0 or 1, or equivalently if $N_1^+(n_E^*)$ is empty or not. This directly yields hardness of approximation within any finite factor.

Theorem 14. *For symmetric network congestion games with 3 agents, linear delay functions on series-parallel graphs and optimal congestions n^*_E it is* NP*-hard to decide if there is a Nash equilibrium with congestions n^*_E.*

We remark that the previous theorem contrasts the continuous non-atomic case, in which a minimal fraction of altruistic demand stabilizing an optimum solution can be computed in any symmetric network congestion game [13].

References

1. Ackermann, H., Röglin, H., Vöcking, B.: On the impact of combinatorial structure on congestion games. J. ACM, 55(6) (2008)
2. Ackermann, H., Skopalik, A.: On the Complexity of Pure Nash Equilibria in Player-Specific Network Congestion Games. In: Deng, X., Graham, F.C. (eds.) WINE 2007. LNCS, vol. 4858, pp. 419–430. Springer, Heidelberg (2007)
3. Babaioff, M., Kleinberg, R., Papadimitriou, C.: Congestion games with malicious players. In: Proc 8th Conference on Electronic Commerce (EC 2007), pp. 103–112 (2007)
4. Chen, P.-A., Kempe, D.: Altruism, selfishness, and spite in traffic routing. In: Proc. 9th Conference on Electronic Commerce (EC 2008), pp. 140–149 (2008)
5. Eshel, I., Samuelson, L., Shaked, A.: Altruists, egoists and hooligans in a local interaction model. American Economic Review 88(1), 157–179 (1998)
6. Fabrikant, A., Papadimitriou, C., Talwar, K.: The complexity of pure Nash equilibria. In: Proc 36th Symposium on Theory of Computing (STOC 2004), pp. 604–612 (2004)
7. Fehr, E., Schmidt, K.: A theory of fairness, competition, and cooperation. The Quarterly Journal of Economics 114, 817–868 (1999)
8. Gairing, M.: Malicious bayesian congestion games. In: Bampis, E., Skutella, M. (eds.) WAOA 2008. LNCS, vol. 5426, pp. 119–132. Springer, Heidelberg (2009)
9. Gairing, M., Monien, B., Tiemann, K.: Routing (Un-) splittable flow in games with player-specific linear latency functions. In: Bugliesi, M., Preneel, B., Sassone, V., Wegener, I. (eds.) ICALP 2006. LNCS, vol. 4051, pp. 501–512. Springer, Heidelberg (2006)
10. Gintis, H., Bowles, S., Boyd, R., Fehr, E.: Moral Sentiments and Material Interests: The Foundations of Cooperation in Economic Life. MIT Press, Cambridge (2005)
11. Ieong, S., McGrew, R., Nudelman, E., Shoham, Y., Sun, Q.: Fast and compact: A simple class of congestion games. In: Proc 20th Conf Artificial Intelligence (AAAI 2005), pp. 489–494 (2005)
12. Kaporis, A., Spirakis, P.: Stackelberg games: The price of optimum. In: Encyclopedia of Algorithms. Springer, Heidelberg (2008)
13. Kaporis, A., Spirakis, P.: The price of optimum in Stackelberg games on arbitrary single commodity networks and latency functions. Theoretical Computer Science 410(8–10), 745–755 (2009)
14. Karakostas, G., Viglas, A.: Equilibria for networks with malicious users. Mathematical Programming 110(3), 591–613 (2007)
15. Ledyard, J.: Public goods: A survey of experimental resesarch. In: Kagel, J., Roth, A. (eds.) Handbook of Experimental Economics, pp. 111–194. Princeton University Press, Princeton (1997)

16. Levine, D.: Modeling altruism and spitefulness in experiments. Review of Economic Dynamics 1, 593–622 (1998)
17. Milchtaich, I.: Congestion games with player-specific payoff functions. Games and Economic Behavior 13(1), 111–124 (1996)
18. Moscibroda, T., Schmid, S., Wattenhofer, R.: When selfish meets evil: Byzantine players in a virus inoculation game. In: Proc. 25th Symposium on Principles of Distributed Computing (PODC 2006), pp. 35–44 (2006)
19. Nisan, N., Tardos, É., Roughgarden, T., Vazirani, V. (eds.): Algorithmic Game Theory. Cambridge University Press, Cambridge (2007)
20. Rosenthal, R.W.: A class of games possessing pure-strategy Nash equilibria. Intl. J. Game Theory 2, 65–67 (1973)
21. Tovey, C.: A simplified NP-complete satisfiability problem. Discrete Applied Mathematics 8, 85–89 (1984)

Geometric Spanners for Weighted Point Sets

Mohammad Ali Abam[1,⋆], Mark de Berg[2,⋆⋆], Mohammad Farshi[3,⋆⋆⋆], Joachim Gudmundsson[4], and Michiel Smid[3,⋆⋆⋆]

[1] MADALGO Center, Aarhus University, Denmark
abam@madalgo.au.dk
[2] Department of Computer Science, TU Eindhoven, The Netherlands
m.t.d.berg@tue.nl
[3] School of Computer Science, Carleton University, Ottawa, ON, K1S 5B6, Canada
mfarshi@cg.scs.carleton.ca, michiel@scs.carleton.ca
[4] NICTA, Sydney, Australia
joachim.gudmundsson@nicta.com.au

Abstract. Let $(S, \mathbf{d})$ be a finite metric space, where each element $p \in S$ has a non-negative weight $\mathrm{w}(p)$. We study spanners for the set S with respect to weighted distance function $\mathbf{d}_w$, where $\mathbf{d}_w(p,q)$ is $\mathrm{w}(p) + \mathbf{d}(p,q) + \mathrm{w}(q)$ if $p \neq q$ and 0 otherwise. We present a general method for turning spanners with respect to the $\mathbf{d}$-metric into spanners with respect to the $\mathbf{d}_w$-metric. For any given $\varepsilon > 0$, we can apply our method to obtain $(5+\varepsilon)$-spanners with a linear number of edges for three cases: points in Euclidean space $\mathbb{R}^d$, points in spaces of bounded doubling dimension, and points on the boundary of a convex body in $\mathbb{R}^d$ where $\mathbf{d}$ is the geodesic distance function.

We also describe an alternative method that leads to $(2+\varepsilon)$-spanners for points in $\mathbb{R}^d$ and for points on the boundary of a convex body in $\mathbb{R}^d$. The number of edges in these spanners is $O(n \log n)$. This bound on the stretch factor is nearly optimal: in any finite metric space and for any $\varepsilon > 0$, it is possible to assign weights to the elements such that any non-complete graph has stretch factor larger than $2 - \varepsilon$.

1 Introduction

Motivation. Networks play a central role in numerous applications, and the design of good networks is therefore an important topic of study. In general, a good network has certain desirable properties while not being too expensive. In many applications this means one wants a network providing short paths between its nodes, while not containing too many edges. This leads to the concept of spanners, as defined next in the geometric setting.

⋆ MAA was supported by the MADALGO Center for Massive Data Algorithmics, a Center of the Danish National Research Foundation.
⋆⋆ MdB was supported by the Netherlands' Organisation for Scientific Research (NWO) under project no. 639.023.301.
⋆⋆⋆ MF and MS were supported by NSERC of Canada.

A. Fiat and P. Sanders (Eds.): ESA 2009, LNCS 5757, pp. 190–202, 2009.

Let $\mathcal{G} = (S, E)$ be a geometric graph on a set S of n points in $\mathbb{R}^d$. That is, $\mathcal{G}$ is an edge-weighted graph where the weight of an edge $(p, q) \in E$ is equal to $|pq|$, the Euclidean distance between p and q. The distance in $\mathcal{G}$ between two points p and q, denoted by $\mathbf{d}_{\mathcal{G}}(p, q)$, is defined as the length of a shortest (that is, minimum-weight) path from p to q in $\mathcal{G}$. The graph $\mathcal{G}$ is called a (geometric) *t-spanner*, for some $t \geqslant 1$, if for any two points $p, q \in S$ we have $\mathbf{d}_{\mathcal{G}}(p, q) \leqslant t \cdot |pq|$. The smallest t for which $\mathcal{G}$ is a t-spanner is called the *stretch factor* (or *dilation*, or *spanning ratio*) of $\mathcal{G}$. Geometric spanners have been studied extensively over the past decade. It has been shown that for any set of n points in $\mathbb{R}^d$ and any $\varepsilon > 0$, there is a $(1 + \varepsilon)$-spanner with only $O(n/\varepsilon^{d-1})$ edges—see the recent book by Narasimhan and Smid [1] for this and many other results on spanners. Instead of considering points in Euclidean space, one can also consider points in some other metric space. As it turns out, results similar to the Euclidean setting are possible when the so-called *doubling dimension* of the metric space—see footnote 1 on p. 195 for a definition—is bounded by a constant d: in this case there is a $(1 + \varepsilon)$-spanner with $n/\varepsilon^{O(d)}$ edges [2,3].

Sometimes the cost of traversing a path in a network is not only determined by the lengths of the edges on the path, but also by delays occurring at the nodes on the path: in a (large-scale) road network a node may represent a town and passing through the town will take time, in a computer network a node may need some time to forward a packet to the next node on the path, and so on. The goal of our paper is to study the concept of spanners in this setting.

Problem statement. Let S be a set of n elements—we will refer to the elements as *points* from now on—and let $\mathbf{d}$ be a metric on S. Assume each point $p \in S$ has a non-negative weight, denoted by $\mathrm{w}(p)$. We now define a new distance function on S, denoted by $\mathbf{d}_w$, as follows.

$$\mathbf{d}_w(p, q) = \begin{cases} 0 & \text{if } p = q, \\ \mathrm{w}(p) + \mathbf{d}(p, q) + \mathrm{w}(q) & \text{if } p \neq q. \end{cases}$$

For a graph $\mathcal{G} = (S, E)$ and two points p and q in S, we denote by $\mathbf{d}_{\mathcal{G},w}(p, q)$ the length of a shortest path in $\mathcal{G}$ between p and q, where edge lengths are measured using the distance function $\mathbf{d}_w$; if $p = q$, then we define $\mathbf{d}_{\mathcal{G},w}(p, q) = 0$. For a real number $t > 1$, we say that $\mathcal{G}$ is an *additively weighted t-spanner* of S, if for any two points p and q in S we have $\mathbf{d}_{\mathcal{G},w}(p, q) \leqslant t \cdot \mathbf{d}_w(p, q)$. We want to compute an additively weighted t-spanner of S having few edges and with a small stretch factor. Unfortunately our metric space $(S, \mathbf{d}_w)$ does not necessarily have bounded doubling dimension, even if the underlying metric space $(S, \mathbf{d})$ has bounded doubling dimension. (An easy example is a set S of n points inside a unit disk in the plane, each having unit weight, and when $\mathbf{d}$ is the Euclidean distance function. Then the doubling dimension of the metric space $(S, \mathbf{d}_w)$ will be $\Theta(\log n)$.) This leads us to the main question we want to answer: Is it possible to obtain additively weighted spanners with constant stretch factor—that is, stretch factor independent of n, but also independent of the weights of the points—and a near-linear number of edges?

Recently Bose *et al.* [4] also studied spanners for weighted points. More precisely, they consider points in the plane with positive weights and then define the distance between two points p, q as $|pq| - \mathrm{w}(p) - \mathrm{w}(q)$. The difference between their setting and our setting is thus that they subtract the weights from the Euclidean distance, whereas we add the weights (which in the applications mentioned above is more natural). This is, in fact, a fundamental difference: Bose *et al.* show (under the assumption that the distance between any pair of points is non-negative) that in their setting there exists a $(1+\varepsilon)$-spanner with $O(n/\varepsilon)$ edges, while our lower bounds (see below) imply that such a result is impossible in our setting.

Our results. We present two methods for computing additively weighted spanners. The first method is described in Section 2. It essentially shows that whenever there is a good spanner for the metric space $(S, \mathbf{d})$, there is also a good spanner for the metric space $(S, \mathbf{d}_w)$. This is done by clustering the points in a suitable way, computing a spanner in the $\mathbf{d}$-metric on the cluster centers, and then connecting each point to its cluster center. We apply our method to obtain, for any $0 < \varepsilon < 1$, additively weighted $(5+\varepsilon)$-spanners in $\mathbb{R}^d$ and in spaces of doubling dimension d, with $O(n/\varepsilon^d)$ and $n/\varepsilon^{O(d)}$ edges, respectively.

We also apply our method to points on the boundary of a convex body in $\mathbb{R}^d$, where distances are geodesic distances along the body's boundary. We give a simple and efficient algorithm for computing a well-separated pair decomposition for this metric—we believe this result is interesting in its own right—which proves the existence of a $(1+\varepsilon)$-spanner with $O(n/\varepsilon^d)$ edges. When the points are weighted, we can then use our general method to get an additively weighted $(5+\varepsilon)$-spanner with $O(n/\varepsilon^d)$ edges.

Our second method is described in Section 3. It applies to spaces of bounded doubling dimension for which a semi-separated pair decomposition [5,6] can be constructed. It leads to spanners with a better stretch factor than our first method, but the size of the spanner is larger. In particular, it leads to $(2+\varepsilon)$-spanners with $(n/\varepsilon^{O(d)})\log n$ edges, for points in $\mathbb{R}^d$ and for points on the boundary of a convex body in $\mathbb{R}^d$. We also show that the bound on the stretch factor is nearly optimal: in any finite metric space and for any $\varepsilon > 0$, it is possible to assign weights to the points such that any non-complete graph has stretch factor larger than $2 - \varepsilon$.

2 A Spanner Construction Based on Clustering

Let $(S, \mathbf{d})$ be a finite metric space and let n denote the number of points in S. We assume that each point $p \in S$ has a real weight $\mathrm{w}(p) \geqslant 0$. We will show that if we can find a good spanner for S in the $\mathbf{d}$-metric, we can also find a good additively weighted spanner for S in the $\mathbf{d}_w$-metric.

The main idea is to partition S into clusters, where each cluster has a designated point as its cluster center. The clusters have the following two properties: First, the $\mathbf{d}$-distances and $\mathbf{d}_w$-distances between any two centers are approximately equal. Second, for each point p in the cluster with center c, the distance $\mathbf{d}(p, c)$ is at most proportional to the weight $\mathrm{w}(p)$ of p. We then show that a t-spanner of the cluster centers in the $\mathbf{d}$-metric, while connecting the rest of the points to the center of their clusters, results in an $O(t)$-spanner of S in the $\mathbf{d}_w$-metric.

Clusterings for additively weighted spanners. We start by stating more precisely the properties we require from our clustering. Let k_1 and k_2 be two parameters, with $k_1 > 0$ and $k_2 \geqslant 1$. Define a (k_1, k_2)*-clustering of* S to be a partitioning of S into a collection $\{C_1, ..., C_m\}$ of clusters, each with a center denoted by center(C_i), such that the following three conditions hold:

(I) for every $1 \leqslant i \leqslant m$ and for all $p \in C_i$ we have: $\mathrm{w}(\mathrm{center}(C_i)) \leqslant \mathrm{w}(p)$;

(II) for every $1 \leqslant i \leqslant m$ and for all $p \in C_i$ we have: $\mathbf{d}(\mathrm{center}(C_i), p) \leqslant k_1 \cdot \mathrm{w}(p)$;

(III) for every $1 \leqslant i, j \leqslant m$ we have:

$$\mathbf{d}_w(\mathrm{center}(C_i), \mathrm{center}(C_j)) \leqslant k_2 \cdot \mathbf{d}(\mathrm{center}(C_i), \mathrm{center}(C_j)).$$

Later we will show how to find such clusterings. But first we show how to use such a clustering to obtain a spanner for S in the $\mathbf{d}_w$-metric.

Let $\{C_1, C_2, \ldots, C_m\}$ be a (k_1, k_2)-clustering of S, and let $c_i = \mathrm{center}(C_i)$. Let $\mathcal{C} = \{c_1, c_2, \ldots, c_m\}$ denote the set of cluster centers, and let $\mathcal{G}_1 = (\mathcal{C}, E_1)$ be a t-spanner of the set $\mathcal{C}$ in the $\mathbf{d}$-metric. Finally, let $E_2 = \{(c_i, p) : 1 \leqslant i \leqslant m \text{ and } p \in C_i \text{ and } p \neq c_i\}$. In other words, E_2 contains the edges connecting the points in each cluster to the center of that cluster. The next lemma states that augmenting $\mathcal{G}_1$ with the edges in E_2 gives a spanner in the $\mathbf{d}_w$-metric.

Lemma 1. *The graph $\mathcal{G} = (S, E_1 \cup E_2)$ is a t'-spanner in the $\mathbf{d}_w$-metric, where $t' = \max(2 + k_1 + k_1 k_2 t, k_2 t)$.*

Proof. Let p, q be two distinct points in S. We must show that $\mathbf{d}_{\mathcal{G},w}(p, q) \leqslant t' \cdot \mathbf{d}_w(p, q)$. Let C_i and C_j be the clusters containing p and q, respectively, and consider $c_i = \mathrm{center}(C_i)$ and $c_j = \mathrm{center}(C_j)$. (It can happen that $i = j$, but this will not invalidate the coming argument.) Note that either $p = c_i$ or (p, c_i) is an edge in $\mathcal{G}$; similarly $q = c_j$ or (q, c_j) is an edge in $\mathcal{G}$. Hence,

$$\begin{aligned}
\mathbf{d}_{\mathcal{G},w}(p, q) &= \mathbf{d}_{\mathcal{G},w}(p, c_i) + \mathbf{d}_{\mathcal{G},w}(c_i, c_j) + \mathbf{d}_{\mathcal{G},w}(c_j, q) \\
&= \mathbf{d}_w(p, c_i) + \mathbf{d}_{\mathcal{G},w}(c_i, c_j) + \mathbf{d}_w(c_j, q) \\
&= (\mathrm{w}(p) + \mathbf{d}(p, c_i) + \mathrm{w}(c_i)) + \mathbf{d}_{\mathcal{G},w}(c_i, c_j) + (\mathrm{w}(c_j) + \mathbf{d}(c_j, q) + \mathrm{w}(q)) \\
&\leqslant (2 + k_1) \cdot \mathrm{w}(p) + \mathbf{d}_{\mathcal{G},w}(c_i, c_j) + (2 + k_1) \cdot \mathrm{w}(q),
\end{aligned}$$

where the last inequality follows from properties (I) and (II) of the clustering. Now consider the shortest path in $\mathcal{G}_1$ from c_i to c_j in the $\mathbf{d}$-metric. By property (III) the length of every link on this path—and, hence, its total length—increases by at most a factor k_2 when we measure its length in the $\mathbf{d}_w$-metric. Since $\mathcal{G}_1$ is a t-spanner for $\mathcal{C}$ in the $\mathbf{d}$-metric, we thus have $\mathbf{d}_{\mathcal{G},w}(c_i, c_j) \leqslant k_2 \cdot \mathbf{d}_{\mathcal{G}_1}(c_i, c_j) \leqslant k_2 t \cdot \mathbf{d}(c_i, c_j)$. Finally, we observe that

$$\mathbf{d}(c_i, c_j) \;\leqslant\; \mathbf{d}(c_i, p) + \mathbf{d}(p, q) + \mathbf{d}(q, c_j) \;\leqslant\; k_1 \cdot \mathrm{w}(p) + \mathbf{d}(p, q) + k_1 \cdot \mathrm{w}(q).$$

Combing this with our two earlier derivations, we get

$$\begin{aligned}\mathbf{d}_{\mathcal{G},w}(p,q) &\leqslant (2+k_1)\cdot \mathrm{w}(p) + \mathbf{d}_{\mathcal{G},w}(c_i,c_j) + (2+k_1)\cdot \mathrm{w}(q)\\ &\leqslant (2+k_1)\cdot \mathrm{w}(p) + k_2t\cdot \mathbf{d}(c_i,c_j) + (2+k_1)\cdot \mathrm{w}(q)\\ &\leqslant (2+k_1)\cdot \mathrm{w}(p) + k_2t\cdot (k_1\cdot \mathrm{w}(p)+\mathbf{d}(p,q)+k_1\cdot \mathrm{w}(q)) + (2+k_1)\cdot \mathrm{w}(q)\\ &= (2+k_1+k_1k_2t)\cdot \mathrm{w}(p) + k_2t\cdot \mathbf{d}(p,q) + (2+k_1+k_1k_2t)\cdot \mathrm{w}(q)\\ &\leqslant \max(2+k_1+k_1k_2t, k_2t)\cdot (\mathrm{w}(p)+\mathbf{d}(p,q)+\mathrm{w}(q))\\ &= \max(2+k_1+k_1k_2t, k_2t)\cdot \mathbf{d}_w(p,q).\end{aligned}$$

□

Computing good clusterings and spanners. The following algorithm takes as input the weighted set S and two real numbers k and $\varepsilon > 0$, and computes a clustering $\{C_1, \ldots, C_m\}$ of S.

1. Sort the points of S in nondecreasing order of their weight, and let $p_1, p_2, \ldots, p_n$ be the sorted sequence (ties are broken arbitrarily).
2. Initialize the first cluster C_1: set $C_1 = \{p_1\}$ and $c_1 = \text{center}(C_1) = p_1$. Initialize the set of cluster centers: $\mathcal{C} = \{p_1\}$. Set $m = 1$.
3. For $i = 2$ to n, do the following:
   (a) Compute an index j with $1 \leqslant j \leqslant m$ such that c_j is a $(1+\varepsilon)$-approximate nearest-neighbor of p_i in the set $\mathcal{C}$, in the $\mathbf{d}$-metric.
   (b) If $\mathbf{d}(c_j, p_i) \leqslant k \cdot \mathrm{w}(p_i)$, then set $C_j = C_j \cup \{p_i\}$. Otherwise, start a new cluster: set $m = m + 1$, set $C_m = \{p_i\}$ and $c_m = \text{center}(C_m) = p_i$, and set $\mathcal{C} = \mathcal{C} \cup \{p_i\}$.
4. Return the collection $\{C_1, \ldots, C_m\}$ of clusters.

Lemma 2. *The algorithm above computes a* $(k, 1 + \frac{2(1+\varepsilon)}{k})$*-clustering of* S.

Proof. Since we treat the points in order of increasing weight and the first point put into a cluster is its center, we have $\mathrm{w}(c_j) \leqslant \mathrm{w}(p)$ for every cluster C_j and point $p \in C_j$. Moreover, by step 3 we only put a point p in a cluster C_j if $\mathbf{d}(\text{center}(C_j), p) \leqslant k \cdot \mathrm{w}(p)$. Hence, conditions (I) and (II) are satisfied.

To prove condition (III), consider two distinct cluster centers c and c'. Assume without loss of generality that c was added to $\mathcal{C}$ before c'. Then it follows from the algorithm that $\mathrm{w}(c) \leqslant \mathrm{w}(c')$. Consider the iteration of the for-loop in which $p_i = c'$, and consider the set $\mathcal{C}$ at the beginning of this iteration. Observe that $c \in \mathcal{C}$. Let c_j be the $(1 + \varepsilon)$-approximate nearest-neighbor of c' in $\mathcal{C}$ that is computed by the algorithm. Since c' is added to $\mathcal{C}$, we have $\mathbf{d}(c_j, c') > k \cdot \mathrm{w}(c')$. Let c'' be the exact nearest-neighbor of c' in $\mathcal{C}$. Then, since $c \in \mathcal{C}$, $\mathbf{d}(c_j, c') \leqslant (1+\varepsilon) \cdot \mathbf{d}(c'', c') \leqslant (1+\varepsilon) \cdot \mathbf{d}(c, c')$. It follows that

$$\begin{aligned}\mathbf{d}_w(c,c') &= \mathrm{w}(c) + \mathbf{d}(c,c') + \mathrm{w}(c') \leqslant \mathbf{d}(c,c') + 2\cdot \mathrm{w}(c')\\ &< \mathbf{d}(c,c') + \frac{2}{k}\cdot \mathbf{d}(c_j,c') \leqslant \left(1 + \frac{2(1+\varepsilon)}{k}\right)\cdot \mathbf{d}(c,c').\end{aligned}$$

□

By combining Lemmas 1 and 2, we obtain the following result.

Theorem 1. *Let* $t > 1$ *be a parameter, and let* $(S, \mathbf{d})$ *be a metric space with* n *weighted points such that the following holds:*

- *For any subset $S' \subseteq S$ with m points, we can compute in $T_{\text{sp}}(m)$ time a t-spanner for S' in the* **d***-metric with $E_{\text{sp}}(m)$ edges, where T_{sp} and E_{sp} are non-decreasing functions.*
- *For any $\varepsilon > 0$ there is a semi-dynamic (insertions-only) data structure for $(1+\varepsilon)$-approximate nearest-neighbor queries in the* **d***-metric for S, such that both insertions and queries can be done in $T_{\text{nn}}(\varepsilon, n)$ time, where the function T_{nn} is non-decreasing in n.*

Then we can construct for any $\varepsilon > 0$ a t'-spanner for S in the $\mathbf{d}_w$-metric with $O(E_{\text{sp}}(n))$ edges and $t' = 3t + 2 + 2\varepsilon(t+1)$. The construction can be done in $O(n \log n + T_{\text{sp}}(n) + n \cdot T_{\text{nn}}(\varepsilon, n))$ time.

Due to space limitation, the proof of the theorem is removed in this version.

Applications: Euclidean spaces and spaces of bounded doubling dimension. Theorem 1 can immediately be used to obtain additively weighted spanners in Euclidean spaces and metric spaces of bounded doubling dimension.[1]

Corollary 1. *(i) Given a set S of n points in $\mathbb{R}^d$, each having a non-negative weight, and given a real number $0 < \varepsilon < 1$, we can construct an additively weighted $(5+\varepsilon)$-spanner of S having $O(n/\varepsilon^d)$ edges in $O((n/\varepsilon^d)\log n)$ time.*
(ii) Given a metric space $(S, \mathbf{d})$ of constant doubling dimension d, where S is a set of size n, and in which each point of S has a non-negative real weight, and given a real number $0 < \varepsilon < 1$, we can construct an additively weighted $(5+\varepsilon)$-spanner of S having $n/\varepsilon^{O(d)}$ edges in $O(n \log n) + n/\varepsilon^{O(d)}$ time.

Proof. Callahan and Kosaraju [7] have shown that for any set of n points in $\mathbb{R}^d$ and any $0 < \varepsilon < 1$, one can compute a $(1+\varepsilon)$-spanner with $E_{\text{sp}}(n) = O(n/\varepsilon^d)$ edges in $T_{\text{sp}}(n) = O(n \log n + n/\varepsilon^d)$ time. Moreover, Arya *et al.* [8] presented a data structure for $(1+\varepsilon)$-approximate nearest-neighbor queries in $\mathbb{R}^d$ that has $O((1/\varepsilon^d)\log n)$ query time, and in which insertions can be done in $O(\log n)$ time. Part (i) of the theorem now follows by applying Theorem 1, replacing ε by $\varepsilon/10$ and setting $t = 1 + \varepsilon/10$.

Gottlieb and Roddity [9] have shown that for any metric space $(S, \mathbf{d})$ with n points and doubling dimension d and any $0 < \varepsilon < 1$, one can compute a $(1+\varepsilon)$-spanner with $E_{\text{sp}}(n) = n/\varepsilon^{O(d)}$ edges in $T_{\text{sp}}(n) = O(n \log n) + n/\varepsilon^{O(d)}$ time. Moreover, Cole and Gottlieb [10] presented a data structure for $(1+\varepsilon)$-approximate nearest-neighbor queries in $(S, \mathbf{d})$ that has $2^{O(d)} \log n + 1/\varepsilon^{O(d)}$ query time, and in which insertions can be done in $2^{O(d)} \log n$ time. Part (ii) now follows by applying Theorem 1, replacing ε by $\varepsilon/10$ and setting $t = 1 + \varepsilon/10$. ☐

[1] The doubling dimension of a metric space $(S, \mathbf{d})$ is defined as follows. If p is a point of S and $R > 0$ is a real number, then the **d**-*ball* with center p and radius R is the set $\{q \in S : \mathbf{d}(p,q) \leqslant R\}$. The *doubling dimension* of $(S, \mathbf{d})$ is the smallest real number d such that the following is true: For every real number $R > 0$, every **d**-ball of radius R can be covered by at most 2^d **d**-balls of radius $R/2$.

More applications: the geodesic metric for a convex body. Let S be a set of n points on the boundary $\partial\mathcal{B}$ of a convex body $\mathcal{B}$ in $\mathbb{R}^d$. For any two points $p, q \in S$, let $\mathbf{d}_\mathcal{B}(p, q)$ be the geodesic distance between p and q along ∂B, and let $\mathbf{d}(p, q)$ denote their Euclidean distance. In order to apply Theorem 1 to the metric space $(S, \mathbf{d}_\mathcal{B})$, we need a sparse $(1 + \varepsilon)$-spanner for a set $S' \subseteq S$ based on the distance function $\mathbf{d}_\mathcal{B}$. We will obtain such a spanner using a so-called well-separated pair decomposition (WSPD).

Well-separated pair decompositions were introduced by Callahan and Kosaraju [7] for the Euclidean metric and by Talwar [3] for general metric spaces. They are defined as follows. Let $(S, \mathbf{d})$ be a finite metric space. The *diameter* $\text{diam}_\mathbf{d}(A)$ of any subset A of S is defined as $\text{diam}_\mathbf{d}(A) = \max\{\mathbf{d}(a, b) : a, b \in A\}$, and the *distance* $\mathbf{d}(A, B)$ of any two subsets $A, B \subseteq S$ is defined as $\mathbf{d}(A, B) = \min\{\mathbf{d}(a, b) : a \in A, b \in B\}$. For a real number $s > 0$, we say that the subsets A and B of S are *well-separated with respect to* s, if $\mathbf{d}(A, B) \geqslant s \cdot \max(\text{diam}_\mathbf{d}(A), \text{diam}_\mathbf{d}(B))$.

Definition 1. *Let $(S, \mathbf{d})$ be a finite metric space and let $s > 0$ be a real number. A* well-separated pair decomposition (WSPD) *for $(S, \mathbf{d})$, with respect to s, is a set $\{(A_1, B_1), \ldots, (A_m, B_m)\}$ of pairs of non-empty subsets of S such that*

1. *for each i, A_i and B_i are well-separated with respect to s, and*
2. *for any two distinct points $p, q \in S$, there is exactly one index i with $1 \leqslant i \leqslant m$ such that (i) $p \in A_i$ and $q \in B_i$ or (ii) $p \in B_i$ and $q \in A_i$.*

The following lemma, due to Callahan and Kosaraju [11], shows how a spanner can be obtained from a WSPD. They prove the lemma for Euclidean spaces, but exactly the same proof applies to any metric space.

Lemma 3. **[11]** *Let $(S, \mathbf{d})$ be a finite metric space and let $t > 1$ be a real number. Furthermore, let $\{(A_1, B_1), \ldots, (A_m, B_m)\}$ be a WSPD for $(S, \mathbf{d})$, with respect to $s = \frac{2(t+1)}{t-1}$ and, for each i with $1 \leqslant i \leqslant m$, let a_i be an arbitrary point of A_i and b_i be an arbitrary point of B_i. Then the graph $G = (S, E)$ where $E = \{(a_i, b_i) : 1 \leqslant i \leqslant m\}$ is a t-spanner for S with m edges.*

Lemma 3 tells us that if we have a WSPD for S in the $\mathbf{d}_\mathcal{B}$-metric, we can get a spanner for S in the $\mathbf{d}_\mathcal{B}$-metric. Using Theorem 1 we can then also get a spanner for a weighted point set S. As we show in Lemma 10, the metric space $(S, \mathbf{d}_\mathcal{B})$ has bounded doubling dimension. Using the algorithm of Har-Peled and Mendel [2] we can thus construct a WSPD for this metric space. Unfortunately, their algorithm needs an oracle that returns, for any two points p and q, the geodesic distance $\mathbf{d}_\mathcal{B}(p, q)$ in $O(1)$ time, and computing geodesic distances on a convex body is not so easy. We therefore describe a more direct method for computing a WSPD for points on a convex body. The basic idea behind our method is to compute a WSPD for the Euclidean space $(S, \mathbf{d})$, and then refine this WSPD in a suitable way to obtain a WSPD for $(S, \mathbf{d}_\mathcal{B})$. For the refinement, we only need to know the normal vectors of all points $p \in S$; we do not need any distance computations in the $\mathbf{d}_\mathcal{B}$-metric. An additional advantage of our method over Har-Peled and Mendel's method is that the dependency on ε will be better.

For any point p on ∂B, we denote by $\mathcal{N}_{\mathcal{B}}(p)$ the (outer) normal vector of $\mathcal{B}$ at p. If the tangent plane of p at $\mathcal{B}$ is not unique, then we choose for $\mathcal{N}_{\mathcal{B}}(p)$ the normal vector of an arbitrary tangent plane. We fix a real number σ such that $0 < \sigma < \pi/2$. The following lemma states that $\mathbf{d}_{\mathcal{B}}(p,q)$ and $\mathbf{d}(p,q)$ are approximately equal, provided the angle between the normals of p and q is at most σ. Similar observations have been made in papers on approximate shortest paths on polytopes; see e.g. [12].

Lemma 4. *Let p and q be two points on $\partial\mathcal{B}$ such that $\angle(\mathcal{N}_{\mathcal{B}}(p), \mathcal{N}_{\mathcal{B}}(q)) \leqslant \sigma$. Then $\mathbf{d}(p,q) \leqslant \mathbf{d}_{\mathcal{B}}(p,q) \leqslant \frac{\mathbf{d}(p,q)}{\cos\sigma}$.*

The normal vector of each point of $\partial\mathcal{B}$ at B can be considered to be a point on the sphere of directions, denoted $\mathbb{S}^{d-1}$, in $\mathbb{R}^d$. We partition $\mathbb{S}^{d-1}$ into $O(1/\sigma^{d-1})$ parts such that the angle between any two vectors in the same part is at most σ. Based on this, we partition $\partial\mathcal{B}$ into *patches*: A σ-patch is the set of all points of $\partial\mathcal{B}$ whose normals fall in the same part of the partition of $\mathbb{S}^{d-1}$.

Let $s > 0$ be a real number, and let $\{(A_1, B_1), \ldots, (A_m, B_m)\}$ be a WSPD for the Euclidean metric space $(S, \mathbf{d})$, with respect to s, where $m = O(s^d n)$. We refine the WSPD by partitioning each A_i and B_i into subsets $A_i^1, \ldots, A_i^\ell$ and $B_i^1, \ldots, B_i^\ell$, respectively, where $\ell = O(1/\sigma^{d-1})$. The partitioning is done such that the points in each subset belong to the same σ-patch. Define $\Psi = \{(A_i^j, B_i^k) : 1 \leqslant j \leqslant \ell \text{ and } 1 \leqslant k \leqslant \ell\}$.

Lemma 5. *The set of pairs in Ψ forms a WSPD with respect to $s\cos\sigma$ for the metric space $(S, \mathbf{d}_{\mathcal{B}})$. The number of pairs in this WSPD is $O((s^d/\sigma^{2d-2})n)$.*

Proof. It is clear that Ψ contains $O((s^d/\sigma^{2d-2})n)$ elements. It is also clear that condition 2. in Definition 1 is satisfied. It remains to show that condition 1. is satisfied. Consider a pair $(A_i^j, B_i^k) \in \Psi$. We have to show that

$$\mathbf{d}_{\mathcal{B}}(A_i^j, B_i^k) \geqslant s\cos\sigma \cdot \max(\mathrm{diam}_{\mathbf{d}_{\mathcal{B}}}(A_i^j), \mathrm{diam}_{\mathbf{d}_{\mathcal{B}}}(B_i^k)). \quad (1)$$

We first show that

$$\mathrm{diam}_{\mathbf{d}}(A_i^j) \geqslant \mathrm{diam}_{\mathbf{d}_{\mathcal{B}}}(A_i^j)\cos\sigma. \quad (2)$$

To show this, let a and a' be two arbitrary points in A_i^j. Using Lemma 4, we obtain $\mathbf{d}_{\mathcal{B}}(a,a') \leqslant \frac{\mathbf{d}(a,a')}{\cos\sigma} \leqslant \frac{\mathrm{diam}_{\mathbf{d}}(A_i^j)}{\cos\sigma}$, from which (2) follows. By a symmetric argument, we obtain

$$\mathrm{diam}_{\mathbf{d}}(B_i^k) \geqslant \mathrm{diam}_{\mathbf{d}_{\mathcal{B}}}(B_i^k)\cos\sigma. \quad (3)$$

Let a be an arbitrary point of A_i^j and let b be an arbitrary point of B_i^k. Since $A_i^j \subseteq A_i$ and $B_i^k \subseteq B_i$, and since A_i and B_i are well-separated with respect to s (in the Euclidean metric $\mathbf{d}$), we have $\mathbf{d}_{\mathcal{B}}(a,b) \geqslant \mathbf{d}(a,b) \geqslant s{\cdot}\max(\mathrm{diam}_{\mathbf{d}}(A_i), \mathrm{diam}_{\mathbf{d}}(B_i)) \geqslant s{\cdot}\max(\mathrm{diam}_{\mathbf{d}}(A_i^j), \mathrm{diam}_{\mathbf{d}}(B_i^k))$. Combining this with (2) and (3), it follows that $\mathbf{d}_{\mathcal{B}}(a,b) \geqslant s\cos\sigma \cdot \max(\mathrm{diam}_{\mathbf{d}_{\mathcal{B}}}(A_i^j), \mathrm{diam}_{\mathbf{d}_{\mathcal{B}}}(B_i^k))$. This proves that (1) holds. ◻

Lemmas 3 and 5 now imply the following result (take for instance $\sigma = \pi/3$, so that $\cos\sigma = 1/2$).

Theorem 2. *Let S be a set of n points on the boundary of a convex body $\mathcal{B}$ in $\mathbb{R}^d$, and let $0 < \varepsilon < 1$ be a real number. If we can determine for any $p \in S$ an outward normal of $\mathcal{B}$ at p in $O(1)$ time then we can compute in $O(n \log n + n/\varepsilon^d)$ time a $(1+\varepsilon)$-spanner of S in the $\mathbf{d}_{\mathcal{B}}$-metric, with $O(n/\varepsilon^d)$ edges.*

Corollary 2. *Let S be a set of n points on the boundary of a convex body $\mathcal{B}$ in $\mathbb{R}^d$, each with a non-negative weight. For any $0 < \varepsilon < 1$, there is an additively weighted $(5+\varepsilon)$-spanner of S having $O(n/\varepsilon^d)$ edges.*

3 An Additively Weighted $(2+\varepsilon)$-Spanner

In each of the applications considered in the previous section, our method generated an additively weighted $(5+\varepsilon)$-spanner. The goal of this section is to see if we can obtain additively weighted spanners with a smaller stretch factor. We start with a lower bound.

Lemma 6. *For any finite metric space $(S, \mathbf{d})$ and any real number $\varepsilon > 0$, there exists a set of weights for the points of S, such that every non-complete graph with vertex set S has additively weighted stretch factor larger than $2 - \varepsilon$.*

Proof. Let $D = \text{diam}_{\mathbf{d}}(S)$. Assign each point in S a weight D/ε. Consider a non-complete graph G with vertex set S, and let p and q be two points in S that are not connected by an edge in G. We have $\mathbf{d}_w(p,q) \leqslant (1 + 2/\varepsilon)D$, whereas $\mathbf{d}_{\mathcal{G},w}(p,q) \geqslant 4D/\varepsilon$. Thus $\frac{\mathbf{d}_{\mathcal{G},w}(p,q)}{\mathbf{d}_w(p,q)} \geqslant \frac{4D/\varepsilon}{(1+2/\varepsilon)D} > 2 - \varepsilon$. ∎

In the remainder of this section we will describe a general strategy for computing additively weighted $(2+\varepsilon)$-spanners for spaces of bounded doubling dimension. Given the lower bound, the stretch factor is almost optimal in the worst case. Our method is based on the so-called semi-separated pair decomposition, as introduced by Varadarajan [6]. We use the strategy to obtain additively weighted $(2+\varepsilon)$-spanners for two cases: points in $\mathbb{R}^d$, and points on the boundary of a convex body in $\mathbb{R}^d$.

The semi-separated pair decomposition. Let $(S, \mathbf{d})$ be a metric space, where S is a set of n points, and let d be its doubling dimension. We assume that each point of S has a real weight $\text{w}(p) \geqslant 0$. Our spanner construction will be based on a decomposition $\{(A_1, B_1), \ldots, (A_m, B_m)\}$ having properties similar to those of the WSPD. As we will see, the number of edges in the additively weighted spanner is proportional to $\sum_{i=1}^{m}(|A_i| + |B_i|)$. Thus, we need a decomposition for which this summation is small. Callahan and Kosaraju [7] have shown that, for the WSPD, this summation can be as large as $\Theta(n^2)$; in other words, we cannot use the WSPD to obtain a non-trivial result. By using a decomposition satisfying a weaker condition, it is possible to make sure the summation is only $O(n \log n)$. This decomposition is the semi-separated pair decomposition, as introduced in [6].

For a real number $s > 0$, two subsets $A, B \subseteq S$ are called *semi-separated with respect to s*, if $\mathbf{d}(A, B) \geqslant s \cdot \min(\text{diam}_{\mathbf{d}}(A), \text{diam}_{\mathbf{d}}(B))$. A *semi-separated pair* *decomposition (SSPD)* for the metric space $(S, \mathbf{d})$, with respect to s, is defined to be a set $\Psi = \{(A_1, B_1), \ldots, (A_m, B_m)\}$ of pairs of non-empty subsets of S, having the same properties as in Definition 1, except that in condition 1., the sets A_i and B_i are semi-separated with respect to s. The quantity $\sum_{i=1}^{m}(|A_i| + |B_i|)$ is called the *size* of the SSPD.

The SSPD was introduced by Varadarajan [6]. For the Euclidean distance function in $\mathbb{R}^2$, Abam *et al.* [5] showed that an SSPD with $O(n)$ pairs and size $O(n \log n)$ can be computed in $O(n \log n)$ time. It is known that for any set of n points, any SSPD has size $\Omega(n \log n)$; see [13,14].

From SSPDs to spanners. Let Ψ be an SSPD for S with respect to some $s > 0$. For each pair $(A, B) \in \Psi$ we will add a set $E(A, B)$ of edges to our spanner such that any two points $a \in A$ and $b \in B$ are connected by a path of length at most $(2 + \frac{3}{s}) \cdot \mathbf{d}_w(a, b)$.

The main idea is quite simple. Assume without loss of generality that $\mathrm{diam}_{\mathbf{d}}(A) \leqslant \mathrm{diam}_{\mathbf{d}}(B)$. Thus, we have $\mathbf{d}(A, B) \geqslant s \cdot \mathrm{diam}_{\mathbf{d}}(A)$. Define center$(A)$ to be a point from A of minimum weight (among the points in A), and let $E_1(A, B) = \{(x, \mathrm{center}(A)) : x \in A \cup B \text{ and } x \neq \mathrm{center}(A)\}$. This provides short connections between the points in A and those in B by going via center(A): since $\mathbf{d}(A, B) \geqslant s \cdot \mathrm{diam}_{\mathbf{d}}(A)$, going via center$(A)$ does not create a large detour in the $\mathbf{d}$-metric, and since $\mathrm{w}(\mathrm{center}(A)) \leqslant \mathrm{w}(a)$ the extra path length caused by $\mathrm{w}(\mathrm{center}(A))$ is also limited. In fact, for some pairs of points a, b, the set $E_1(A, B)$ already gives us a path of the required length. The next lemma gives the condition under which this is the case.

Lemma 7. *Let* $c = \mathrm{center}(A)$. *Let* $b \in B$ *be an points such that* $\mathrm{w}(c) \leqslant \mathrm{w}(b) + \mathbf{d}(c, b)$. *Then, for any* $a \in A$, *we have* $\mathbf{d}_w(a, c) + \mathbf{d}_w(c, b) \leqslant \left(2 + \frac{3}{s}\right) \cdot \mathbf{d}_w(a, b)$.

Proof. We have

$$\begin{aligned}
\mathbf{d}_w(a,c) + \mathbf{d}_w(c,b) &= (\mathrm{w}(a) + \mathbf{d}(a,c) + \mathrm{w}(c)) + (\mathrm{w}(c) + \mathbf{d}(c,b) + \mathrm{w}(b)) \\
&\leqslant (2 \cdot \mathrm{w}(a) + \mathrm{diam}_{\mathbf{d}}(A)) + 2 \cdot (\mathbf{d}(c,b) + \mathrm{w}(b)) \\
&\leqslant (2 \cdot \mathrm{w}(a) + \mathrm{diam}_{\mathbf{d}}(A)) + 2 \cdot (\mathbf{d}(c,a) + \mathbf{d}(a,b) + \mathrm{w}(b)) \\
&\leqslant 2 \cdot (\mathrm{w}(a) + \mathbf{d}(a,b) + \mathrm{w}(b)) + 3 \cdot \mathrm{diam}_{\mathbf{d}}(A) \\
&\leqslant 2 \cdot (\mathbf{d}_w(a,b)) + 3 \cdot (\mathbf{d}(a,b)/s) \\
&\leqslant (2 + \tfrac{3}{s}) \cdot \mathbf{d}_w(a,b).
\end{aligned}$$

□

It remains to establish short paths between the points in A and the points $b \in \overline{B}$, where $\overline{B} = \{b \in B : \mathrm{w}(c) > \mathrm{w}(b) + \mathbf{d}(c, b)\}$ with $c = \mathrm{center}(A)$. We cannot use any point from A as an intermediate destination on such paths, because the weights of the points from A are too large compared to those in $\overline{B}$. Hence, we need to go via a point from $\overline{B}$. However, the diameter of $\overline{B}$ can be large. Therefore we first decompose the set $\overline{B}$ into subsets of small diameter.

The points b in $\overline{B}$ have $\mathbf{d}(c, b) < \mathrm{w}(c)$, so they are contained in a $\mathbf{d}$-ball C of radius $\mathrm{w}(c)$. Recall that d is the doubling dimension of $(S, \mathbf{d})$. Thus we can cover C by $s^{O(d)}$ balls of radius $\mathrm{w}(c)/(2s)$. Let $C_1, \ldots, C_\ell$ be such a collection of balls, where $\ell = s^{O(d)}$. We partition $\overline{B}$ into subsets $B_1, \ldots, B_\ell$ in such a way that $B_i \subseteq C_i$ for all $1 \leqslant i \leqslant \ell$. For each B_i, let center(B_i) be a point of minimum weight (among the points in B_i). The next lemma shows that going from any point in A to any point in B_i via center(B_i) gives us a path of the required length.

Lemma 8. *Let* $c_i = \text{center}(B_i)$. *Then, for two points* $a \in A$ *and* $b \in B_i$ *we have* $\mathbf{d}_w(a, c_i) + \mathbf{d}_w(c_i, b) < \left(2 + \frac{2}{s}\right) \cdot \mathbf{d}_w(a, b)$.

The proof of the lemma is similar to the proof of Lemma 7 and is removed because of the space limitation.

We are now ready to define the set of edges for the pair (A, B) in the SSPD Ψ. Namely, we define $E(A, B) = E_1(A, B) \cup \left(\bigcup_{i=1}^{\ell} E_2(A, B_i)\right)$, where $E_2(A, B_i) = \{(x, \text{center}(B_i)) : x \in A \cup B_i \text{ and } x \neq \text{center}(B_i)\}$. For any two points $a \in A$ and $b \in B$, there exists a path in the graph with edge set $E(A, B)$ of length at most $(2 + \varepsilon) \cdot \mathbf{d}_w(a, b)$. This follows by using Lemmas 7 and 8, and setting $s = \varepsilon/3$. Using that $\ell = s^{O(d)} = 1/\varepsilon^{O(d)}$, we get that the total number of edges in $E(A, B)$ is $|E_1(A, B)| + \sum_{i=1}^{\ell} |E_2(A, B_i)| = |A| + |B| + \sum_{i=1}^{\ell}(|A| + |B_i|) = (1/\varepsilon)^{O(d)} \cdot (|A| + |B|)$. By combining the sets $E(A, B)$ for all pairs $(A, B) \in \Psi$ we get our final spanner. Since, by definition of the SSPD, for any two points $a, b \in S$ there is a pair $(A, B) \in \Psi$ such that $a \in A$ and $b \in B$ (or vice versa), we get the following result.

Lemma 9. *The graph* $G = (S, E)$ *with* $E = \bigcup_{(A,B) \in \Psi} E(A, B)$ *is an additively weighted* $(2 + \varepsilon)$*-spanner for* S *with* $(1/\varepsilon)^{O(d)} \cdot \sum_{(A,B) \in \Psi}(|A| + |B|)$ *edges.*

Applications. Let S be a set of n points in $\mathbb{R}^d$ and let $\mathbf{d}(p, q)$ be the Euclidean distance between p and q. Observe that the metric space $(S, \mathbf{d})$ has doubling dimension $\Theta(d)$. Abam *et al.* [5] have shown that in the plane an SSPD of size $O(s^2 n \log n)$ can be computed in $O(n \log n + s^2 n)$ time, for any $s > 1$. Their algorithm in fact also works in higher dimensions; its analysis also goes through, with appropriate changes to the constant factors in certain packing lemmas. This leads to an SSPD of size $O(s^d n \log n)$ that can be computed in $O(n \log n + s^d n)$ time, giving the following result.

Theorem 3. *Given a set* S *of* n *points in* $\mathbb{R}^d$*, each one having a non-negative weight, and given a real number* $0 < \varepsilon < 1$*, we can construct an additively weighted* $(2+\varepsilon)$*-spanner of* S *having* $(n/\varepsilon^{O(d)}) \log n$ *edges in* $(n/\varepsilon^{O(d)}) \log n$ *time.*

We now turn our attention to a set S of points on the boundary of a convex body $\mathcal{B}$. For any two points p and q of S, let $\mathbf{d}_{\mathcal{B}}(p, q)$ be the geodesic distance between p and q along ∂B. The proof of the following lemma is based on the concept of σ-patches introduced earlier, and removed due to space limitation.

Lemma 10. *The metric space* $(S, \mathbf{d}_{\mathcal{B}})$ *has doubling dimension* $\Theta(d)$.

Let $\mathbf{d}$ denote the Euclidean distance function in $\mathbb{R}^d$, let $s > 1$ be a real number, and consider an SSPD $\{(A_1, B_1) \ldots, (A_m, B_m)\}$ for the metric space $(S, \mathbf{d})$, with respect to s, whose size is $O(s^d n \log n)$. We fix a real number σ such that $0 < \sigma < \pi/2$. Let i be an index with $1 \leqslant i \leqslant m$. As before, we partition both A_i and B_i into subsets $A_i^1, \ldots, A_i^\ell$ and $B_i^1, \ldots, B_i^\ell$, respectively, where $\ell = O(1/\sigma^{d-1})$, such that the points in each subset belong to the same σ-patch of $\partial\mathcal{B}$. Now define $\Psi = \{(A_i^j, B_i^k) : 1 \leqslant j \leqslant \ell, 1 \leqslant k \leqslant \ell\}$. The proof of the following lemma is similar to that of Lemma 5.

Lemma 11. *The set Ψ forms an SSPD, with respect to $s\cos\sigma$, for the metric space $(S, \mathbf{d}_\mathcal{B})$. The size of this SSPD is $O((s^d/\sigma^{2d-2})n\log n)$.*

We choose $\sigma = \pi/3$, so that $\cos\sigma = 1/2$. We obtain the following result.

Theorem 4. *Given a convex body $\mathcal{B}$ in $\mathbb{R}^d$ and a set S of n points on the boundary of $\mathcal{B}$. Assume that each point of S has a non-negative real weight. Let $0 < \varepsilon < 1$ be a real number. We can construct an additively weighted $(2+\varepsilon)$-spanner of S having $(n/\varepsilon^{O(d)})\log n$ edges in $(n/\varepsilon^{O(d)})\log n$ time.*

Remark 1. It follows from the proofs of Lemmas 7 and 8 that the graph G has *spanner diameter* 2. That is, for any two points p and q of S, the graph G contains a path between p and q that contains at most two edges and whose $\mathbf{d}_w$-length is at most $(2+\varepsilon)\cdot\mathbf{d}_w(p,q)$. If we want to keep this property, then the number of edges in our spanner is worst-case optimal: For any real number $t > 1$, there exists a metric space $(S, \mathbf{d})$ such that every t-spanner for S having spanner diameter 2 has $\Omega(n\log n)$ edges—see Exercise 12.10 in Narasimhan and Smid [1]. Of course, this then also holds for additively weighted spanners. Note that if all weights are equal and very large compared to the $\mathbf{d}$-diameter of the set, then any additively weighted 2-spanner must have spanner diameter 2. (This does not imply, however, that $\Omega(n\log n)$ is a lower bound on the worst-case size of additively weighted 2-spanners.)

References

1. Narasimhan, G., Smid, M.: Geometric Spanner Networks. Cambridge University Press, Cambridge (2007)
2. Har-Peled, S., Mendel, M.: Fast construction of nets in low-dimensional metrics and their applications. SIAM J. on Computing 35, 1148–1184 (2006)
3. Talwar, K.: Bypassing the embedding: algorithms for low dimensional metrics. In: STOC 2004, pp. 281–290 (2004)
4. Bose, P., Carmi, P., Couture, M.: Spanners of additively weighted point sets. In: Gudmundsson, J. (ed.) SWAT 2008. LNCS, vol. 5124, pp. 367–377. Springer, Heidelberg (2008)
5. Abam, M.A., de Berg, M., Farshi, M., Gudmundsson, J.: Region-fault tolerant geometric spanners. In: SODA 2007, pp. 1–10 (2007)
6. Varadarajan, K.R.: A divide-and-conquer algorithm for min-cost perfect matching in the plane. In: FOCS 1998, pp. 320–331 (1998)
7. Callahan, P.B., Kosaraju, S.R.: A decomposition of multidimensional point sets with applications to k-nearest-neighbors and n-body potential fields. J. of the ACM 42, 67–90 (1995)
8. Arya, S., Mount, D.M., Netanyahu, N.S., Silverman, R., Wu, A.: An optimal algorithm for approximate nearest neighbor searching in fixed dimensions. J. of the ACM 45, 891–923 (1998)

9. Gottlieb, L.A., Roditty, L.: An optimal dynamic spanner for doubling metric spaces. In: Halperin, D., Mehlhorn, K. (eds.) ESA 2008. LNCS, vol. 5193, pp. 478–489. Springer, Heidelberg (2008)
10. Cole, R., Gottlieb, L.A.: Searching dynamic point sets in spaces with bounded doubling dimension. In: STOC 2006, pp. 574–583 (2006)
11. Callahan, P.B., Kosaraju, S.R.: Faster algorithms for some geometric graph problems in higher dimensions. In: SODA 1993, pp. 291–300 (1993)
12. Agarwal, P.K., Har-Peled, S., Sharir, M., Varadarajan, K.R.: Approximate shortest paths on a convex polytope in three dimensions. J. of the ACM 44, 567–584 (1997)
13. Hansel, G.: Nombre minimal de contacts de fermeture nécessaires pour réaliser une fonction booléenne symétrique de n variables. Comptes Rendus de l'Académie des Sciences 258, 6037–6040 (1964)
14. Bollobás, B., Scott, A.D.: On separating systems. European J. of Combinatorics 28, 1068–1071 (2007)

k-Outerplanar Graphs, Planar Duality, and Low Stretch Spanning Trees
(Extended Abstract)

Yuval Emek*

School of Electrical Engineering, Tel Aviv University, Tel Aviv 69978, Israel
yuvale@eng.tau.ac.il

Abstract. Low distortion probabilistic embedding of graphs into approximating trees is an extensively studied topic. Of particular interest is the case where the approximating trees are required to be (subgraph) spanning trees of the given graph (or multigraph), in which case, the focus is usually on the equivalent problem of finding a (single) tree with low average stretch. Among the classes of graphs that received special attention in this context are k-outerplanar graphs (for a fixed k): Chekuri, Gupta, Newman, Rabinovich, and Sinclair show that every k-outerplanar graph can be probabilistically embedded into approximating trees with constant distortion regardless of the size of the graph. The approximating trees in the technique of Chekuri et al. are not necessarily spanning trees, though.

In this paper it is shown that every k-outerplanar multigraph admits a spanning tree with constant average stretch. This immediately translates to a constant bound on the distortion of probabilistically embedding k-outerplanar graphs into their spanning trees. Moreover, a randomized algorithm is presented for constructing such a low average stretch spanning tree in expected linear time. This algorithm relies on some new insights regarding the connection between low average stretch spanning trees and planar duality.

1 Introduction

The Problem. Consider an n-vertex connected graph $G = (V(G), E(G))$ and let $\ell(e)$ be a positive *length* associated with every edge $e \in E(G)$. For any two vertices $u, v \in V(G)$, let $\delta_G(u, v)$ denote the *distance* between u and v in G, namely, the length, taken with respect to ℓ, of a shortest path connecting u and v in G. Given a spanning tree T of G and some edge $e \in E(G)$, the *stretch* of e in T is defined as $\mathrm{str}_T(e) = \delta_T(e)/\ell(e)$. Spanning trees with low stretch for all edges can be very useful in many applications. However, there exist some trivial graphs for which every spanning tree admits an edge with stretch $\Omega(n)$ (e.g., the n-cycle). This motivates the construction of spanning trees with low *average stretch*, denoted by $\text{av-str}_G(T) = \frac{1}{|E(G)|} \sum_{e \in E(G)} \mathrm{str}_T(e)$ [2].

* Supported in part by the Israel Science Foundation, grants 221/07 and 664/05.

A. Fiat and P. Sanders (Eds.): ESA 2009, LNCS 5757, pp. 203–214, 2009.

The following related notion was introduced in [4]. Given a probability distribution $\mathcal{D}$ over a set $\mathcal{T}$ of spanning trees of G, we say that G is *probabilistically embedded* into $\mathcal{T}$ (under $\mathcal{D}$) with *distortion* α if $\mathbb{E}[\delta_T(u,v)] \leq \alpha \cdot \delta_G(u,v)$ for every two vertices $u, v \in V(G)$, where the expectation is with respect to $T \in_{\mathcal{D}} \mathcal{T}$. It is shown in [2] that a graph G can be probabilistically embedded into its spanning trees with distortion α if and only if every multigraph obtained from G by replicating its edges has a spanning tree with average stretch α. Consequently, in the context of constructing low average stretch spanning trees, one usually considers multigraphs rather than simple graphs. (This can be viewed as taking a weighted average of the edge stretch factors.)

A tree T is called a *dominating tree* of the graph G if $V(T) \supseteq V(G)$ and $\delta_T(u,v) \geq \delta_G(u,v)$ for every two vertices $u, v \in V(G)$. Clearly, every spanning tree of G is also a dominating tree of G; the converse is not true as a dominating tree may have vertices and edges that do not exist in the original graph G, and hence it is not necessarily a subgraph of G. The notion of probabilistic embedding can be redefined by allowing the support $\mathcal{T}$ to contain dominating trees that are not subgraphs of G. For many applications and in particular, for those applications mentioned in [4,5], this does not exhibit any obstacle. However, there exist some applications for which it is impossible to use non-subgraph dominating trees in the support of the probabilistic embedding, most notably in the context of networking, where G represents an existing physical graph (e.g., the *minimum communication spanning tree* problem [13]).

k-Outerplanar Graphs. An *outerplanar* graph (or a 1-*outerplanar* graph) is a graph that can be drawn in the plane with all vertices lying on the unbounded face. A planar graph is said to be k-*outerplanar*, $k \geq 2$, if it can be drawn in the plane such that by removing the vertices on the unbounded face we obtain a $(k-1)$-outerplanar graph. A canonical example for a k-outerplanar graph is the $2k \times n$ grid (containing $2k$ rows of vertices with n vertices in each row) which also serves as a canonical example for a graph with tree width proportional to k. When referring to k-outerplanar graphs, we usually assume that k is fixed. However, every planar graph is k-outerplanar for some k (typically, much smaller than n) and this *outerplanarity factor* plays a key role in many polynomial time approximation schemes for NP-hard optimization problems on planar graphs [3].

Related Work. The problem of constructing spanning trees with low average stretch was first studied in [2], where it is proved that every n-vertex multigraph G admits a spanning tree T which satisfies av-str$_G(T) = e^{O(\sqrt{\ln n \ln \ln n})}$. They also show that there exist some graphs, the $\sqrt{n} \times \sqrt{n}$ grid being one of them, for which every spanning tree admits average stretch $\Omega(\log n)$ and conjectured that this lower bound is tight. The upper bound of [2] was improved drastically in [8] by introducing a construction of spanning trees with average stretch $O(\log^2 n \log \log n)$. Very recently, [1] presented a further improvement by establishing an almost tight upper bound of $O(\log n \log \log n \log^3 \log \log n)$.

The notion of probabilistic embedding was explicitly introduced in [4] (although, it was implicitly used in [2]), which initiated a series of papers that developed probabilistic embeddings of arbitrary graphs into non-subgraph dominating trees: in [4] it is shown that every n-vertex graph can be probabilistically embedded into dominating trees with distortion $O(\log^2 n)$, while some graphs must suffer a distortion of $\Omega(\log n)$; the upper bound was improved to $O(\log n \log\log n)$ in [5,6]; and a tight $O(\log n)$ upper bound is proved in [11].

Some papers study probabilistic embeddings of specific graph classes. In [16] it is shown that every planar graph can be probabilistically embedded into its (non-subgraph) dominating trees with distortion $O(\log n)$ using the decomposition technique of [15] for graphs excluding small minors. This is generalized in [14] to graphs of bounded genus by showing that such graphs can be probabilistically embedded with constant distortion into planar graphs. In [12] it is proved that while series-parallel graphs can be embedded into ℓ_1 with constant distortion, there exist some unweighted series-parallel graphs that cannot be probabilistically embedded into dominating trees with distortion $o(\log n)$. This lower bound is matched in [10] by showing that unweighted series-parallel graphs can be probabilistically embedded into their spanning trees with distortion $O(\log n)$.

It is also proved in [12] that every outerplanar graph can be probabilistically embedded into its spanning trees with constant distortion. The construction of [12] for (1-)outerplanar graphs is (partially) generalized to k-outerplanar graphs in [7], where a probabilistic embedding with distortion exponential in k (but independent of n) is presented. Note however, that unlike the construction of [12], the technique of [7] constructs (random) dominating trees which are not necessarily spanning trees of the original graph.

Contribution. In this paper we show that every k-outerplanar multigraph G admits a spanning tree T which satisfies $\text{av-str}_G(T) \leq c^k$, where c is an absolute constant. This immediately implies that every k-outerplanar graph can be probabilistically embedded into its spanning trees with distortion depending solely on k, thus enhancing the result of [7]. Our proof is constructive: we present a randomized algorithm that constructs such spanning trees in expected linear time. Due to lack of space, some of the proofs are omitted from this extended abstract and can be found in the full version [9].

Techniques. The backbone of our algorithm is a rather standard peeling-an-onion decomposition (cf. [7]): on input k-outerplanar graph G, we first peel off the vertices on the unbounded face to obtain a $(k-1)$-outerplanar graph G'; we then recursively construct a good spanning tree T' of G'; next, we insert the missing vertices of G back into T' to obtain the graph H; and finally, we construct a good spanning tree T of H. This framework is formally presented in Section 3. The secret ingredient of the algorithm lies in the last step: constructing a good spanning tree T of H.

As observed in [7], the graph H is essentially a *Halin* graph, which can be viewed as a planar embedding of a tree merged with a cycle. Indeed, our main challenge is to construct low stretch spanning trees for (a generalization of) Halin graphs, as opposed to the non-subgraph dominating trees constructed in [7]. Our construction is completely different than the construction of [7] and it relies on reducing the task of constructing a low stretch spanning tree for a planar graph to that of constructing a low stretch spanning tree for its planar dual (see Theorem 3). This reduction is employed in two distinct occasions within a series of graph manipulations presented in Section 4.

2 Preliminaries

Consider an n-vertex connected graph G. Let $V(G)$ and $E(G)$ denote the vertex and edge sets of G, respectively. Each edge $e \in E(G)$ is associated with some *length* $\ell(e) \in \mathbb{R}_{>0}$. The *length* of a path P in the graph is the sum of lengths of the edges in the path, denoted by $\ell(P) = \sum_{e \in E(P)} \ell(e)$. Given two vertices $u, v \in V(G)$, let $\delta_G(u, v)$ denote the *distance* between them in G, namely, the length of a shortest path from u to v. The *degree* of u, denoted $\deg(u)$, is defined as the number of edges incident on u in G. It will be convenient for us to define the reciprocal of the length of edge e as its *width*, denoted by $w(e) = 1/\ell(e)$.

In what follows we do not distinguish between graphs and multigraphs (namely, a graph may have edge multiplicities). We say that the graph G is *simple*[1] if G contains at most one edge with endpoints u and v for every two vertices $u, v \in V(G)$. Edges that share both endpoints are called *replicas*. Replicas are usually assumed to have the same length (and width). Given two vertices $u, v \in V(G)$, the *multiplicity* of u and v in G, denoted by $\mu_G(u, v)$, is defined to be the number of (u, v)-replicas, i.e., the number of edges connecting u and v. The *skeleton* H of G is the graph obtained from G when all replicas $e_1, \ldots, e_m$ of the edge $(u, v) \in E(G)$, $m = \mu_G(u, v)$, are identified to a single edge $e_{u,v} \in E(H)$ with $w(e_{u,v}) = \sum_{i=1}^{m} w(e_i)$. Clearly, the skeleton H is a simple graph. Given a class $\mathcal{C}$ of graphs, and assuming that $\mathcal{C}$ is not closed under edge replication, the class *replicated-$\mathcal{C}$* consists of every graph whose skeleton is in $\mathcal{C}$.

A path $\pi = (v_1, \ldots, v_k)$ in G is said to be *isolated* if $\deg(v_1), \deg(v_k) \neq 2$ and $\deg(v_i) = 2$ for every $1 < i < k$. The graph H obtained from G by contracting every isolated path π to a single edge e_π with $\ell(e_\pi) = \ell(\pi)$ is referred to as the *core* of G. It is easy to verify that distances between vertices of degree different than 2 in H agree with those in G. Given a class $\mathcal{C}$ of graphs, and assuming that $\mathcal{C}$ is not closed under edge subdivision, the class *subdivided-$\mathcal{C}$* consists of every graph whose core is in $\mathcal{C}$.

A graph is called *biconnected* if the removal of any single vertex does not separate it. A *block* is a maximal biconnected subgraph. Clearly, every spanning tree of G can be edge-partitioned into spanning trees of the blocks of G.

Stretch and Load. Consider some spanning tree T of G and let $e = (u, v)$ be some edge in $E(G)$. The *stretch* of e in T with respect to G is defined to be

$$\mathrm{str}_{T,G}(e) = \delta_T(u, v)/\ell(e) .$$

(Observe that the stretch of e in the spanning tree T does not depend on the graph G, but our notation mentions G to recall that this is the graph that

[1] Self loops are ignored in this paper.

"hosts" T.) The *total stretch* of T with respect to G is denoted by $\text{tot-str}_G(T) = \sum_{e \in E(G)} \text{str}_{T,G}(e)$ and the *average stretch* of T with respect to G is simply $\text{av-str}_G(T) = \text{tot-str}_G(T)/|E(G)|$.

Let $\text{cut}_T(e) \subseteq V(T) \times V(T)$ be the set of all (unordered) vertex pairs which are connected in T via e (if $e \notin E(T)$, then $\text{cut}_T(e) = \emptyset$). The *load* of e in T with respect to G is defined to be

$$\text{load}_{T,G}(e) = \sum_{e' \in E(G) \cap \text{cut}_T(e)} w(e')/w(e) \ .$$

The *total load* of T with respect to G is denoted by $\text{tot-load}_G(T) = \sum_{e \in E(G)} \text{load}_{T,G}(e)$ and the *average load* of T with respect to G is simply $\text{av-load}_G(T) = \text{tot-load}_G(T)/|E(G)|$. Since $\text{load}_{T,G}(e) = 0$ for every edge $e \in E(G) - E(T)$, we can rewrite $\text{tot-load}_G(T) = \sum_{e \in E(T)} \text{load}_{T,G}(e)$. By a simple change of summation, we obtain the following corollary which implies that we may shift our focus from the construction of low average stretch spanning trees to that of low average load spanning trees.

Corollary 1. *For every graph G and spanning tree T of G, we have* $\text{tot-str}_G(T) = \text{tot-load}_G(T)$.

Consider some graph H and let T be a spanning tree of H. The *load-replication* $\widehat{T}$ of T under H is the graph obtained from T if each edge $e \in E(T)$ is replicated $\lceil \text{load}_{T,H}(e) \rceil$ times, namely, $\mu_{\widehat{T}}(e) = \lceil \text{load}_{T,H}(e) \rceil$. Clearly, the cardinality of the edge set of $\widehat{T}$ serves as an upper bound on (and a good estimation of) the total load of T under H. Taking load-replications of spanning trees is a fundamental step in our construction based on the following lemma.

Lemma 2. *Consider some graph G, a vertex induced subgraph H of G, and a spanning tree T of H. Let $\widehat{T}$ be the load-replication of T under H and let $\check{G}$ be the graph resulting from G if H is replaced by $\widehat{T}$, that is, $V(\check{G}) = V(G)$ and $E(\check{G}) = (E(G) - E(H)) \cup E(\widehat{T})$. Consider some spanning tree $\check{T}$ of $\check{G}$ (by definition, $\check{T}$ is also a spanning tree of G). Then $\text{load}_{\check{T},G}(e) \leq \text{load}_{\check{T},\check{G}}(e)$ for every edge $e \in E(\check{T})$.*

Lemma 2 essentially states that if we can construct a spanning tree T of H with low $\text{tot-load}_H(T)$, then for the sake of analysis, we can replace H in G by the load-replication of T under H (a replicated-tree) and continue from there.

Planar Duality. Consider some planar graph G and fix some planar embedding η of G. The *planar dual* $\tilde{G}$ of G under η is the graph which has a vertex v_ϕ corresponding to each face ϕ in η and an edge $\tilde{e} = (v_\phi, v_{\phi'})$ corresponding to each edge $e \in E(G)$ on the boundary of the faces ϕ and ϕ' in η. The planar embedding η uniquely determines a *dual* planar embedding $\tilde{\eta}$ of $\tilde{G}$. It is well known that G is the planar dual of $\tilde{G}$ under $\tilde{\eta}$. We refer to the vertex $v_\phi \in V(\tilde{G})$ as the *dual* of the face ϕ and to the edge $\tilde{e} \in E(\tilde{G})$ as the *dual* of the edge $e \in E(G)$ with respect to the planar duality $\langle \eta, \tilde{\eta} \rangle$. We associate lengths (and widths) with the dual edges by setting $\ell(\tilde{e}) = w(e)$ (and $w(\tilde{e}) = \ell(e)$) for every $\tilde{e} \in E(\tilde{G})$. Clearly, this definition of dual edge lengths does not violate the bi-directionality of the planar duality $\langle \eta, \tilde{\eta} \rangle$, i.e., it is still true that if $\tilde{G}$ is the dual of G under η, then G is the dual of $\tilde{G}$ under $\tilde{\eta}$.

Consider some spanning tree T of G. The *dual* of T with respect to the planar duality $\langle \eta, \tilde{\eta} \rangle$ is the subgraph $\tilde{T}$ of $\tilde{G}$ defined by setting $V(\tilde{T}) = V(\tilde{G})$ and $E(\tilde{T}) = \{\tilde{e} \in E(\tilde{G}) \mid e \in E(G) - E(T)\}$. It is proved in [17] that $\tilde{T}$ is a spanning tree of $\tilde{G}$. Combined with the notion of load, we extend the technique of [17] to establish the following lemma.

Lemma 3. *The dual $\tilde{T}$ of T with respect to the planar duality $\langle \eta, \tilde{\eta} \rangle$ is a spanning tree of $\tilde{G}$. Moreover, $|\mathrm{str}_{T,G}(e) - \mathrm{load}_{\tilde{T},\tilde{G}}(\tilde{e})| \leq 1$ for every edge $e \in E(G)$, and therefore $\text{av-load}_{\tilde{G}}(\tilde{T}) \leq \text{av-load}_{G}(T) + 1$.*

Graph Classes. Given a planar embedding η of G, we say that η is *outerplanar* (or 1-*outerplanar*) if all vertices of G are incident on the unbounded (outer) face in η. Inductively, η is said to be k-*outerplanar*, $k \geq 2$, if by removing the vertices incident on the unbounded face (and the edges incident on these vertices), we obtain a $(k-1)$-outerplanar embedding of the remaining graph. The graph G is called k-*outerplanar* if it admits a k-outerplanar embedding. (An *outerplanar* graph is simply a 1-outerplanar graph.) Observe that outerplanar graphs are closed under edge replication and not closed under edge subdivision.

A *bush* H is a planar graph obtained by taking a planar embedding of a simple cycle C, embedding a forest T in the region enclosed by C (C and T are disjoint), and introducing some new edges, each one of them has at least one endpoint in C. In other words, the (planar) bush H is defined by taking $V(H) = V(T) \cup V(C)$, $V(C) \cap V(T) = \emptyset$, and $E(H) = E(C) \cup E(T) \cup D$, where $D \subseteq V(C) \times (V(C) \cup V(T))$. If each vertex of C has degree at most 3 in H (i.e., it is adjacent to at most one vertex other than its two neighbors in the cycle), then we say that the bush H is a *Halin* graph. Observe that bushes (and Halin graphs) are closed under edge subdivision and not closed under edge replication.

Consider some planar graph G. A vertex $u \in V(G)$ is said to be a *dominating* vertex if it is adjacent to all other vertices of G, that is, if $(u, v) \in E(G)$ for every vertex $v \in V(G) - \{u\}$. The graph is called a *dominated* graph if it has a dominating vertex. A vertex $u \in V(G)$ is said to be a *pivot* vertex if all simple cycles in G go via u. The graph is called a *pivot* graph if it has a pivot vertex. Observe that dominated graphs are closed under edge replication and not closed under edge subdivision. In contrast, pivot graphs are closed under edge subdivision and not closed under edge replication.

Suppose that G is a subdivided-dominated graph and let H be its core. H is a dominated graph, thus it admits a dominating vertex v. Clearly, v is also a vertex of G. Moreover, v is connected by an isolated path (in G) to every vertex of degree different than 2 in $V(G) - \{v\}$. We refer to v as a *weak dominating vertex* of G.

Useful Assumptions. Recall that the graph G may have arbitrary edge multiplicities. In particular, we cannot bound the number of edges $|E(G)|$ as a function of the number of vertices $n = |V(G)|$. Let m be the number of edges in the skeleton of G, that is, the number of (unordered) vertex pairs $(u, v) \in V(G) \times V(G)$ with $\mu_G(u, v) > 0$. The following lemma, which is essentially derived from combining Lemma 5.2 in [2] and Corollary 1, shows that it is sufficient to consider graphs that do not have "too many" edges.

Lemma 4. *For every graph G, there exists some subgraph G' of G on the same vertex set such that (1) $|E(G')| \leq 2m$; and (2) $\text{av-load}_G(T) \leq 2 \cdot \text{av-load}_{G'}(T)$ for every spanning tree T of G'. Moreover, G' can be obtained from G in linear time.*

As G is planar, we know that $m \leq 3n - 6$. Therefore by employing Lemma 4, we can subsequently assume that $|E(G)| = O(n)$ at the price of losing a factor of 2 in the performance guarantee. Another assumption we will have to make is that each vertex in G is adjacent to at most three other vertices (although it may be incident on more than three edges due to edge multiplicities). In that case we say that G is *tri-adjacent*. For the purpose of making such an assumption, we introduce a linear time transformation (based on standard techniques), referred to as the *spreading* transformation. The spreading transformation depends on a real parameter $\tau > 0$ and its properties are stated in the following lemma.

Lemma 5. *Let G' be the outcome of the spreading transformation when applied to G with parameter τ. Then G' satisfies the following properties: (1) G' is tri-adjacent; (2) if G is k-outerplanar, then so is G'; and (3) every spanning tree T' of G' that satisfies $\text{av-load}_{G'}(T') \leq \tau$ can be* translated *in linear time back into a spanning tree T of G such that $\text{av-load}_G(T) \leq 3\tau$.*

Assuming that the input graph G is tri-adjacent, we will construct in the remainder of the paper a spanning tree T of G that satisfies $\text{av-load}_G(T) \leq c^k$. Therefore by employing Lemma 5 with parameter $\tau = c^k$, we may subsequently make this assumption at the price of losing a factor of 3 in the performance guarantee.

3 The Algorithm — Peeling an Onion

Our goal in this section (and in the whole paper) is to prove the following theorem.

Theorem 6. *For every k-outerplanar graph G, there exists a spanning tree T such that $\text{av-load}_G(T) \leq c^k$, where c is a universal constant (independent of k and G).*

The proof of Theorem 6 is constructive: we present a randomized algorithm, referred to as the *onion peeling algorithm*, that given a k-outerplanar graph G with a realizing planar embedding η, constructs the desired spanning tree T of G in expected linear time. Recall our previous assumptions that $|E(G)| = O(n)$, where $n = |V(G)|$ (due to Lemma 4) and that G is tri-adjacent (due to

Lemma 5). The onion peeling algorithm is based on a recursive process similar to that presented in [7] (and essentially, to many other recursive processes on k-outerplanar graphs, cf. [3]). However, the main building block of the onion peeling algorithm, namely, the construction of low stretch spanning trees for (replicated) Halin graphs, is entirely different (see Section 4). This also leads to a different type of analysis.

The onion peeling algorithm works as follows (a formal pseudo-code description is deferred to [9]): (i) remove the vertices on the unbounded face of G (and the edges incident on these vertices) to obtain a $(k-1)$-outerplanar graph G'; (ii) recursively construct a "good" spanning tree T' for G'; (iii) insert the vertices (and edges) that were removed in step (i) back into the planar embedding of T' to compose the graph H; and (iv) construct a "good" spanning tree T of H.

Our algorithm relies on two fundamental constructions. First (implicit in the above description), when the recursion reaches its halting condition on a 1-outerplanar graph G, we have to construct a "good" spanning tree T of G. This is done via the randomized construction of [12] that probabilistically embeds a given outerplanar graph G into its spanning trees with constant distortion. As we will see later on, this randomized construction of [12] is employed by our algorithm in several occasions, and it is subsequently referred to as Procedure `GNRS`. Actually, we shall use a variant of Procedure `GNRS` (the procedure's name is kept, though) whose input may be a subdivided-outerplanar graph[2]. The performance guarantee of Procedure `GNRS` is stated in the following theorem.

Theorem 7. *Procedure* `GNRS`*, when invoked on a subdivided-outerplanar graph G with a realizing planar embedding η, runs in expected linear time and returns a spanning tree T of G that satisfies* $\text{av-load}_G(T) \leq c_1$*, where c_1 is a universal constant (independent of G).*

The existential claim of Theorem 7 is essentially established in [12]. It is trivial to design an expected polynomial time implementation of Procedure `GNRS` and the linear bound on the expected running time is due to a slightly more involved implementation that we omit from this version of the paper.

The second fundamental construction on which the onion peeling algorithm relies is the construction of a "good" spanning tree T of H (step (iv)). A crucial observation in this context is that H is a replicated-Halin graph (actually, if G is simple, then H is strictly a Halin graph). This is due to the assumption that G is tri-adjacent (without which, H would have been a replicated-bush). The technique of [7] probabilistically embeds a Halin graph H into a collection of dominating trees with constant distortion, but these dominating trees are not necessarily spanning trees of H. By contrast, we present a procedure, called Procedure `RH`, which guarantees that T is a spanning tree of H. The input of Procedure `RH` is not assumed to be a (simple) Halin graph, but rather a replicated-Halin graph (hence the name). The performance guarantee of Procedure `RH` is stated in the following theorem, proved in Section 4.

[2] By employing a simple technique presented in [12], one can contract isolated paths at the price of increasing the distortion of the probabilistic embedding by at most 2.

Theorem 8. *Procedure* RH, *when invoked on a replicated-Halin graph* G *with a realizing planar embedding* η, *runs in expected linear time and returns a spanning tree* T *of* G *that satisfies* $\text{av-load}_G(T) \leq c_2$, *where* c_2 *is a universal constant (independent of* G*).*

This leads to the question: what do we mean by a "good" spanning tree? In most of the previous works which considered graph composition based on replacing a subgraph H by a tree T (including [7]), the tree was chosen randomly according to some probability distribution (that may be supported on many trees) and the goal was to guarantee low distortion. In this work we use a different approach: we shall construct a single tree T and our goal is to guarantee low total load. (Corollary 1 stating that the total load is equal to the total stretch, implies that our approach can be viewed as a relaxation of the previous approach.)

Recall that Lemma 2 essentially implies that for the sake of analysis, we may replace the graph G' (the outcome of step (i)) with the load replication of T' (the outcome of step (ii)) under G' before inserting back the vertices and edges that were removed in step (i) and continue with the construction from there. The onion peeling process revolves around this phenomenon.

Analysis (Sketch). Theorem 6 is proved by induction on k. We first employ Theorems 7 (induction's base) and 8 (induction's step) to show that $|E(H)| \leq c^{k-1} \cdot |E(G)|$, where $c = c(c_1, c_2)$ is a universal constant. Next, we use Lemma 2 to argue that $\text{tot-load}_G(T) \leq \text{tot-load}_H(T)$. Finally, Theorem 8 guarantees that $\text{tot-load}_H(T) \leq c \cdot |E(H)|$, which completes the analysis as it implies that $\text{tot-load}_G(T) \leq c^k \cdot |E(G)|$. A full detail of this analysis is deferred to [9].

4 Replicated-Halin Graphs

In this section we present Procedure RH and prove Theorem 8. Recall that the input of Procedure RH is a replicated-Halin graph G with a realizing planar embedding η. The procedure returns a spanning tree T of G which satisfies $\text{av-load}_G(T) \leq c_2$, where c_2 is a universal constant. By Lemma 4, we may assume that $|E(G)| = O(n)$. (This assumption is essentially reflected in the constant c_2, being twice as large as what we obtain in the remainder of this section.)

Taking Planar Duals. Taking planar duals of some special classes of graphs is the main ingredient of our construction. Due to the sensitivity of the definition of load to edge multiplicities, we first want to understand how the operation of identifying two replicas in a planar graph affects its planar dual. To this end, suppose that some two replicas e and e' in the planar primal are identified to form a single edge of width $w(e)+w(e')$. In the planar dual this translates to the contraction of the simple path consisting of $\tilde{e}$ and $\tilde{e}'$ into a single edge of length $\ell(\tilde{e}) + \ell(\tilde{e}') = w(e) + w(e')$. The following observation is a direct consequence of this phenomenon.

Observation 9. *Let* G *and* $\tilde{G}$ *be two planar graphs with planar embeddings* η *and* $\tilde{\eta}$, *respectively. Let* η' *(respectively* $\tilde{\eta}'$*) be the planar embedding of the skeleton* *of G (resp., the core of $\tilde{G}$), naturally derived from η (resp. $\tilde{\eta}$). If η and $\tilde{\eta}$ are duals, then so are η' and $\tilde{\eta}'$.*

We study planar dualities between some specific classes of (planar) graphs. Our insights are cast in the following lemma, whose proof is deferred to [9].

Lemma 10. *Consider a biconnected planar graph G with a planar embedding η and let $\tilde{G}$ be the planar dual of G under η.*

1. *If G is outerplanar with η being a realizing planar embedding, then $\tilde{G}$ is a pivot graph.*
2. *If G is a pivot graph, then $\tilde{G}$ is an outerplanar graph.*
3. *If G is a Halin graph with η being a realizing planar embedding, then $\tilde{G}$ is a dominated graph.*
4. *If G is a dominated graph, then $\tilde{G}$ is a bush.*
5. *Each block of the graph obtained by removing a weak dominating vertex from a subdivided-dominated graph is a subdivided-outerplanar graph.*

Low Load Spanning Trees for Replicated-Halin Graphs. We now turn to describe the operation of Procedure RH on a replicated-Halin graph G with a realizing planar embedding η. As usual, we assume that G is biconnected (otherwise, we can break it and construct a separate spanning tree for each block). The procedure works in 8 steps. The outcome of step i is denoted by T^i if it is (surely) a tree; and by G^i if it is a graph that may contain cycles. (The superscript notation should not be confused with graph powers.) In this spirit, we denote the replicated-Halin graph G by G^0. The 8 steps of Procedure RH are as follows (refer to Figure 1 for a schematic illustration).

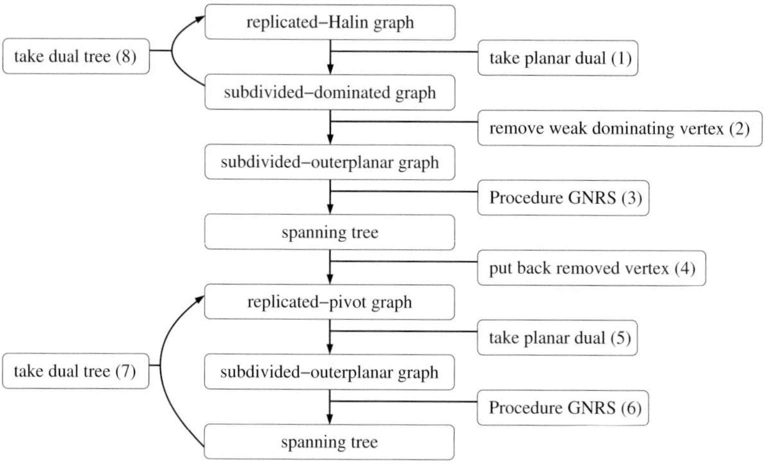

Fig. 1. A schematic illustration of Procedure RH. The step numbers appear in parentheses.

Step 1: Take the planar dual G^1 of G^0 under η. By definition, the skeleton of G^0 is a Halin graph, thus Observation 9 and Lemma 10 imply that G^1 is a subdivided-dominated graph. Let $v \in V(G^1)$ be a weak dominating vertex of G^1.
Step 2: Remove the vertex v and the edges incident on it from G^1 and let G^2 be the remaining graph. Let $G_1^2, \ldots, G_m^2$ be the blocks of G^2. By Lemma 10, G_i^2 is a subdivided-outerplanar graph for every $1 \leq i \leq m$.
Step 3: For $i = 1, \ldots, m$, invoke Procedure `GNRS` on G_i^2 to generate a spanning tree T_i^3. By Theorem 7, we have tot-load$_{G_i^2}(T_i^3) \leq c_1 \cdot |E(G_i^2)|$.
Step 4: For $i = 1, \ldots, m$, construct the load-replication $\widehat{T}_i^3$ of T_i^3 under G_i^2. Insert the vertex and edges that were removed in step 2 back into the planar embedding of the replicated-trees $\widehat{T}_1^3, \ldots, \widehat{T}_m^3$ to compose the graph G^4. Note that $|E(G^4)| \leq (c_1 + 1) \cdot |E(G^1)|$. Since every simple cycle in the skeleton of G^4 must go via v, we conclude that G^4 is a replicated-pivot graph. Let $G_1^4, \ldots, G_{m'}^4$ be the blocks of G^4 (by definition, each of these blocks is also a replicated pivot graph).
Step 5: For $i = 1, \ldots, m'$, fix some arbitrary planar embedding η'_i of G_i^4 and let G_i^5 be the planar dual of G_i^4 under η'_i. By Observation 9 and Lemma 10, G_i^5 is a subdivided-outerplanar graph for every $1 \leq i \leq m'$.
Step 6: For $i = 1, \ldots, m'$, invoke Procedure `GNRS` on G_i^5 to generate a spanning tree T_i^6. By Theorem 7, we have tot-load$_{G_i^5}(T_i^6) \leq c_1 \cdot |E(G_i^5)|$ for every $1 \leq i \leq m'$.
Step 7: For $i = 1, \ldots, m'$, construct the dual T_i^7 of the spanning tree T_i^6 with respect to the planar duality $\langle \tilde{\eta}'_i, \eta'_i \rangle$, where $\tilde{\eta}'_i$ is the dual planar embedding of η'_i. Lemma 3 guarantees that T_i^7 is a spanning tree of G_i^4 and by Lemma 3, we have tot-load$_{G_i^4}(T_i^7) \leq$ tot-load$_{G_i^5}(T_i^6) + |E(G_i^5)| \leq (c_1+1) \cdot |E(G_i^5)| = (c_1+1) \cdot |E(G_i^4)|$ for every $1 \leq i \leq m'$. Let T^7 be the union of the trees $T_1^7, \ldots, T_m^7$. Note that T^7 is a spanning tree of G^4 and tot-load$_{G^4}(T^7) \leq (c_1 + 1) \cdot |E(G^4)|$. Since T^7 is also a spanning tree of G^1, we can apply Lemma 2 to deduce that tot-load$_{G^1}(T^7) \leq$ tot-load$_{G^4}(T^7) \leq (c_1 + 1) \cdot |E(G^4)| \leq (c_1 + 1)^2 \cdot |E(G^1)|$.
Step 8: Construct the dual T^8 of the spanning tree T^7 with respect to the planar duality $\langle \tilde{\eta}, \eta \rangle$, where $\tilde{\eta}$ is the dual planar embedding of η. Lemma 3 guarantees that T^8 is a spanning tree of G^0 and by Lemma 3, we have tot-load$_{G^0}(T^8) \leq$ tot-load$_{G^1}(T^7) + |E(G^1)| \leq ((c_1 + 1)^2 + 1) \cdot |E(G^1)| = ((c_1 + 1)^2 + 1) \cdot |E(G^0)|$.

It follows that upon completion of step 8, we obtain a spanning tree $T = T^8$ which satisfies av-load$_G(T) \leq c_2$, where $c_2 = (c_1+1)^2+1$ is a universal constant. Theorem 8 follows.

5 Conclusions

We prove that every k-outerplanar graph G admits a spanning tree T such that av-load$_G(T) \leq c^k$, where c is an absolute constant. The same bound holds for the average stretch of T with respect to G based on the duality of load and stretch. We find it more convenient to bound the (total) load of the trees we construct, mainly due to the (fairly natural) load-replication representation which enables some sort of an iterative graph decomposition. (In previous works, similar approaches were based on probabilistic embeddings.) Planar duality plays a major role in our construction. We hope that some of the tools we develop here will prove useful in other types of embeddings of planar graphs (e.g., into L_1).

Acknowledgments

I would like to thank Robert Krauthgamer, Manor Mendel, and David Peleg for helpful discussions.

References

1. Abraham, I., Bartal, Y., Neiman, O.: Nearly tight low stretch spanning trees. In: 49th IEEE Symp. on Foundations of Computer Science (FOCS), pp. 781–790 (2008)
2. Alon, N., Karp, R.M., Peleg, D., West, D.: A graph-theoretic game and its application to the k-server problem. SIAM J. Comput. 24, 78–100 (1995)
3. Baker, B.S.: Approximation algorithms for NP-complete problems on planar graphs. J. ACM 41(1), 153–180 (1994)
4. Bartal, Y.: Probabilistic approximations of metric spaces and its algorithmic applications. In: 37th IEEE Symp. on Foundations of Computer Science (FOCS), pp. 184–193 (1996)
5. Bartal, Y.: On approximating arbitrary metrics by tree metrics. In: 30th ACM Symp. on the Theory of Computing (STOC), pp. 161–168 (1998)
6. Charikar, M., Chekuri, C., Goel, A., Guha, S., Plotkin, S.A.: Approximating a finite metric by a small number of tree metrics. In: 39th Symp. on Foundations of Computer Science (FOCS), pp. 379–388 (1998)
7. Chekuri, C., Gupta, A., Newman, I., Rabinovich, Y., Sinclair, A.: Embedding k-outerplanar graphs into ℓ_1. SIAM J. Discrete Math. 20(1), 119–136 (2006)
8. Elkin, M., Emek, Y., Spielman, D.A., Teng, S.-H.: Lower-stretch spanning trees. SIAM J. Comput. 38(2), 608–628 (2008)
9. Emek, Y.: k-Outerplanar graphs, planar duality, and low stretch spanning trees, `http://www.eng.tau.ac.il/~yuvale/Publications/k-outerplanar.pdf`
10. Emek, Y., Peleg, D.: A tight upper bound on the probabilistic embedding of series-parallel graphs. In: 17th ACM-SIAM Symp. on Discrete algorithm (SODA), pp. 1045–1053 (2006)
11. Fakcharoenphol, J., Rao, S., Talwar, K.: A tight bound on approximating arbitrary metrics by tree metrics. J. Comput. Syst. Sci. 69(3), 485–497 (2004)
12. Gupta, A., Newman, I., Rabinovich, Y., Sinclair, A.: Cuts, trees and ℓ_1-embeddings of graphs. Combinatorica 24(2), 233–269 (2004)
13. Hu, T.C.: Optimum communication spanning trees. SIAM J. Comput. 3, 188–195 (1974)
14. Indyk, P., Sidiropoulos, A.: Probabilistic embeddings of bounded genus graphs into planar graphs. In: 23rd ACM Symp. on Computational Geometry (SoCG), pp. 204–209 (2007)
15. Klein, P.N., Plotkin, S.A., Rao, S.: Excluded minors, network decomposition, and multicommodity flow. In: 25th ACM Symp. on Theory of Computing (STOC), pp. 682–690 (1993)
16. Konjevod, G., Ravi, R., Salman, F.S.: On approximating planar metrics by tree metrics. Inf. Process. Lett. 80(4), 213–219 (2001)
17. Whitney, H.: On the abstract properties of linear dependence. Amer. J. Math. 57, 509–533 (1935)

Narrow-Shallow-Low-Light Trees with and without Steiner Points

Michael Elkin and Shay Solomon*

Department of Computer Science, Ben-Gurion University of the Negev,
POB 653, Beer-Sheva 84105, Israel
{elkinm,shayso}@cs.bgu.ac.il

Abstract. We show that for every set $\mathcal{S}$ of n points in the plane and a designated point $rt \in \mathcal{S}$, there exists a tree T that has small maximum degree, depth and weight. Moreover, for every point $v \in \mathcal{S}$, the distance between rt and v in T is within a factor of $(1+\epsilon)$ close to their Euclidean distance $\|rt, v\|$. We call these trees *narrow-shallow-low-light* (NSLLTs). We demonstrate that our construction achieves optimal (up to constant factors) tradeoffs between *all* parameters of NSLLTs. Our construction extends to point sets in $\mathbb{R}^d$, for an arbitrarily large constant d. The running time of our construction is $O(n \cdot \log n)$.

We also study this problem in *general metric spaces*, and show that NSLLTs with small maximum degree, depth and weight can always be constructed if one is willing to compromise the root-distortion. On the other hand, we show that the increased root-distortion is inevitable, even if the point set $\mathcal{S}$ resides in a Euclidean space of dimension $\Theta(\log n)$.

On the bright side, we show that if one is allowed to use Steiner points then it is possible to achieve root-distortion $(1 + \epsilon)$ together with small maximum degree, depth and weight for *general metric spaces*.

Finally, we establish some lower bounds on the power of Steiner points in the context of Euclidean spanning trees and spanners.

1 Introduction

Euclidean Spaces. Given a set $\mathcal{S}$ of n points in the plane and a designated root vertex rt, we want to construct a spanning tree T for $\mathcal{S}$ rooted at rt that enjoys a number of useful properties. First, we want T to be *light*, that is, to be not much heavier than the minimum spanning tree of $\mathcal{S}$ (denoted $MST(\mathcal{S})$). Second, we want it to be *low*, i.e., to have a small depth[1]. Third, we want T to be *shallow*, meaning that for every vertex v in T, the distance $dist_T(rt, v)$ between rt and v in T should not be much greater than the Euclidean distance $\|rt, v\|$. (The maximum ratio $\max\left\{\frac{dist_T(rt,v)}{\|rt,v\|} : v \in \mathcal{S}\right\}$ will be called the *root-stretch* or

* Both authors are supported by the Israeli Academy of Science, grant 483/06, and by the Lynn and William Frankel Center for Computer Science.

[1] The *depth* of a rooted tree (T, rt), denoted $h(T)$, is the maximum number of hops in a path connecting the root rt with a leaf z of T.

A. Fiat and P. Sanders (Eds.): ESA 2009, LNCS 5757, pp. 215–226, 2009.

root-distortion of T.) Fourth, the tree T should be *narrow*, that is, to have a small maximum degree.

Each of these requirements has a natural network-design analogue. The weight of T corresponds to the total cost of building and maintaining the network. The depth and the root-stretch of the tree correspond to communication delays experienced by network end-users. The maximum degree of T corresponds to the load experienced by the relay stations or network routers. Finally, the tree structure of the designed network may be necessary for some applications. In other applications which can be executed in a network that contains cycles, having a cycle-free network may still be very advantageous. Consequently, the problem of designing trees that enjoy all these properties is a basic problem in the area of *geometric network design*. Similar problems arise in the context of the VLSI design [1,6,7], telecommunications and distributed computing [3,4], road network design and medical imaging [9].

Clearly some of these requirements come at the expense of others, and there are inherent tradeoffs between the different parameters. In a seminal STOC'95 paper on Euclidean spanners, Arya et al. [2] have shown that for every set $\mathcal{S}$ of n points in the plane (or even in $\mathbb{R}^d$) there exists a rooted spanning tree (T, rt) with depth $O(\log n)$, constant maximum degree, and an arbitrarily small root-stretch at most $(1+\epsilon)$. However, their trees (called *single-sink spanners*) may have a large weight of $\Omega(n) \cdot w(MST(\mathcal{S}))$. Recently, Dinitz et al. [8] devised a construction that enjoys a small weight (i.e., $O(\log n) \cdot w(MST(\mathcal{S}))$), small depth (i.e., $O(\log n)$) and an arbitrarily small root-stretch at most $(1+\epsilon)$. However, the resulting trees may have vertices of arbitrarily large degree. (The construction of [8] applies to general metric spaces.) In this paper we fill in the gap and devise a single construction that combines all the useful properties of the constructions of [2] and [8]. Specifically, we show that for every n-point set $\mathcal{S}$, a point $rt \in \mathcal{S}$ and parameters ℓ and ϵ, $\ell = O(\log n)$, $\epsilon > 0$, there exists a rooted spanning tree (T, rt) with weight $O(\ell) \cdot w(MST(\mathcal{S}))$ ("light"), depth $O(\ell \cdot n^{1/\ell})$ ("low"), constant maximum degree ("narrow") and root-stretch at most $(1+\epsilon)$ ("shallow"). There also exists a rooted spanning tree (T', rt) with weight $O(\ell \cdot n^{1/\ell}) \cdot w(MST(M))$, depth $O(\ell)$, maximum degree $O(n^{1/\ell})$ and root-stretch at most $(1+\epsilon)$. Moreover, both these trees can be constructed in $O(n \cdot \log n)$ time.

Our results generalize and improve both previous constructions of single-sink spanners [2,8]. Specifically, substituting $\ell = O(\log n)$ in our results we obtain a construction of trees that enjoy all properties of the construction of Arya et al. [2], and, *in addition*, have small weight (specifically, $O(\log n) \cdot w(MST(\mathcal{S}))$. Also, our trees enjoy the same optimal combination between the weight and depth as the trees of Dinitz et al. [8] do, and, *in addition*, enjoy optimal maximum degree[2]. Similarly to the construction of Arya et al. [2], our construction extends to point sets $\mathcal{S} \subseteq \mathbb{R}^d$, for any constant dimension $d \geq 2$. The running time of the extended construction remains $O(n \cdot \log n)$.

[2] The optimality of our tradeoff between weight and depth in the entire range of parameters follows from lower bounds of [8]. The optimality of our tradeoff between depth and maximum degree, again in the entire range of parameters, is obvious.

General Metric Spaces. We also study the problem of constructing trees that satisfy all the aforementioned four properties (henceforth, *narrow-shallow-low-light* trees, or shortly, NSLLTs) in *general* metric spaces. We generalize the results of Dinitz et al. [8], and demonstrate that one can trade maximum degree for root-stretch. Specifically, we show that for every n-point metric space M, a point $rt \in M$ and an integer $\ell = O(\log n)$, there exists a rooted spanning tree (T, rt) with weight $O(\ell) \cdot w(MST(M))$, depth $O(\ell \cdot n^{1/\ell})$, constant maximum degree and root-stretch $O(\log n)$. There also exists a rooted spanning tree (T', rt) with weight $O(\ell \cdot n^{1/\ell}) \cdot w(MST(M))$, depth $O(\ell)$, maximum degree $O(n^{1/\ell})$ and root-stretch $O(\ell)$. In other words, these constructions achieve the optimal tradeoff between the weight and depth, *together with the optimal maximum degree*, at the expense of having root-stretch of $O(\log n)$ and $O(\ell)$, respectively. In addition, we show that this increase in root-stretch is inevitable as long as one considers general (rather than low-dimensional Euclidean) metric spaces. Specifically, we show that our tradeoff between the maximum degree D and root-stretch $O(\frac{\log n}{\log D})$ cannot be improved even if M is a set of n points in Euclidean space of dimension $d = \Omega(\log n)$. We also extend this lower bound and show that in any dimension $d = O(\log n)$, the root-stretch is at least $\Omega(\frac{d}{\log D})$.

On the bright side, we show that this inherent tradeoff between the maximum degree and root-stretch is only valid when considering *spanning trees*. The situation changes drastically if one is allowed to add *Steiner points*, that is, points that do not belong to the original point set of M. In this case the root-stretch can be improved all the way down to $(1+\epsilon)$, without increasing any of the other three parameters! We also show that our lower bounds on the tradeoff between the weight and depth apply to trees that may include Steiner points (henceforth, *Steiner trees*). Consequently, similarly to the case of spanning trees, our tradeoffs between the four involved parameters are optimal with respect to Steiner trees as well.

All our constructions for general metric spaces can be implemented in time $O(n^2)$, which is linear in the size of the input. If an MST, or a constant approximation of an MST, is given as a part of the input, then our constructions can be implemented in time $O(SORT(n)) = O(n \cdot \log n)$, where $SORT(n)$ is the time required to sort n distances. Moreover, if our metric space M is the induced metric of graph G with m edges, and G is given as a part of the input, then our constructions can be implemented in $O(m + n \cdot \log n)$ time.

Lower Bounds for Euclidean Spanners. We have proved two lower bounds on the tradeoffs between different parameters of NSLLTs. These lower bounds were mentioned above. Both these lower bounds have implications for Euclidean spanners. Next, we discuss these implications.

Our lower bound on the tradeoff between the weight and depth parameters of Steiner trees implies directly a lower bound on the tradeoff between these parameters for Euclidean Steiner spanners. Specifically, Dinitz et al. [8] considered the 1-dimensional Euclidean space ϑ_n with n points $1, 2, \ldots, n$ on the x-axis, and showed that any spanning tree of ϑ_n with depth $o(\log n)$ has weight $\omega(n \cdot \log n)$, and vice versa. This result implies that no construction of Euclidean spanners may guarantee hop-diameter[3] $O(\log n)$ and lightness[4] $o(\log n)$, and vice versa. Consequently, the construction of Arya et al. [2] of Euclidean spanners with weight and hop-diameter $O(\log n)$ is optimal. However, the lower bound of [8] does not preclude the existence of *Steiner* spanners with hop-diameter $O(\log n)$ and lightness $o(\log n)$, or vice versa. In the current paper we show that Steiner points do not help in this context, and thus the construction of Arya et al. [2] cannot be improved even if one allows the spanner to use (arbitrarily many) Steiner points.

Our lower bound on the tradeoff between the maximum degree D and root-stretch $\Omega(\frac{d}{\log D})$ of spanning trees for point sets in $\mathbb{R}^d$, $d = O(\log n)$, implies that if $d = \omega(1)$ is super-constant, then either the maximum degree or the root-stretch is super-constant as well. Hence no construction of Euclidean spanners for a super-constant dimension can possibly achieve simultaneously constant maximum degree and stretch. On the other hand, for any constant dimension d, Arya et al. [2] have built spanners with stretch at most $(1 + \epsilon)$ for arbitrarily small $\epsilon > 0$ and constant maximum degree D. Hence our lower bound implies that this result of Arya et al. [2] cannot be extended to super-constant dimension.

Proof Overview. Both our Euclidean and general constructions of NSLLTs are based on the following insight. To construct a tree that enjoys the optimal combination of all four parameters, one can construct two different trees each of which is good with respect to only three out of the four parameters, and combine them into a single tree. Specifically, we start with constructing trees that achieve small maximum degree, root-stretch and depth, henceforth *narrow-shallow-low trees* (NSLoTs). Then we consider the *shallow-low-light trees* (SLLTs) of [8], and observe that in these trees all vertices but the root rt necessarily have small degree. To reduce the degree of the root we manipulate with the star subtree Z rooted at the root of the SLLT T. The vertex set $V(Z)$ of the subtree Z contains the root and all its children $c_1, c_2, \ldots, c_q$, and the edge set of Z is the set $\{(rt, c_1), (rt, c_2), \ldots, (rt, c_q)\}$. Then we construct an NSLoT $\tilde{T}$ for the point set $V(Z)$. Finally, we remove the star Z from the SLLT T, and replace it with the NSLoT $\tilde{T}$. We show that the resulting tree $\widehat{T}$ is an NSLLT, i.e., enjoys *all the four desired properties*. (See Fig. 2 in Sect. 3 for an illustration.)

Our Euclidean construction of NSLoTs is based on the construction of Arya et al. [2], which was, in turn, inspired by the work of Ruppert and Seidel [21]. However, it provides a general tradeoff between the maximum degree D and the depth $O(\log_D n)$, while in the construction of [2] the maximum degree is $O(1)$ and the depth is $O(\log n)$. (Moreover, for $D = \omega(1)$, the running time of our construction is $O(n \cdot \log_D n)$, which is better than the running time $O(n \cdot \log n)$ in [2].) This extension is not difficult, and we provide it for completeness.

[3] *Hop-diameter* or *unweighted diameter* of a possibly weighted graph $G = (V, E, w)$ is the maximum unweighted distance between a pair of vertices in G.

[4] *Lightness* of a spanning subgraph $G' = (V, E, w)$ of the complete Euclidean graph on the point set V is the ratio between $w(G') = \sum_{e \in E} w(e)$ and $w(MST(V))$.

In Sect. 3 we show that the tradeoff of [8] between the hop-diameter and lightness of Euclidean spanners cannot be improved by using Steiner points. To this end we demonstrate that any Steiner tree can be "cleaned" from Steiner points, while increasing the depth and lightness by only a small factor. This result is reminiscent of the work by Gupta [13] that shows that as far as *maximum* stretch and lightness are concerned, one can do without Steiner points. However, our argument is substantially different from that of [13], since, in particular, the hop-diameter parameter exhibits a different behavior than the maximum stretch.

Related Work. Euclidean spanners are being subject of ongoing intensive research since the mid-eighties. See the recent book by Narasimhan and Smid [19] for an excellent survey on this subject. Euclidean single-sink spanners were studied by Arya et al. [2]; see also [19], Chapter 4.2. Lukovszki [16,15] devised fault-tolerant constructions of single-sink spanners. Single-sink spanners were also used in maintenance algorithms for wireless networks [12,17]. Farshi and Gudmundsson [10] conducted an experimental study of single-sink spanners.

Trees that have small weight and guarantee root-stretch at most $(1+\epsilon)$, but do not necessarily have small depth or small maximum degree, are called *shallow-light trees* (henceforth SLTs). SLTs were studied by a number of authors, including Awerbuch et al. [3,4], Khuller et al. [14], Alpert et al. [1] and Cong et al. [6,7]. Salowe et al. [22] studied trees that combine small weight with small "bottleneck" size; see [22] for further details. Papadimitriou and Vazirani [20], Monma and Suri [18], Fekete et al. [11] and Chan [5] devised constructions of light trees with small maximum degree for low-dimensional Euclidean point sets. See also the survey of Eppstein [9] for other references to works that study geometric spanning trees.

The Structure of the Paper. In Sect. 2 we describe our constructions of NSLoTs, and prove lower bounds on the tradeoff between the maximum degree and root-stretch. In Sect. 3 we employ our constructions of NSLoTs from Sect. 2 to devise constructions of NSLLTs, and derive our lower bounds for Euclidean Steiner spanners. Due to space limitations, some proofs are omitted from this extended abstract.

Preliminaries. An *n-point metric space* $M = (V, dist)$ can be viewed as the complete graph $G(M) = (V, \binom{V}{2}, dist)$ in which for every pair of points $x, y \in V$, the weight of the edge $e = (x, y)$ in $G(M)$ is defined by $w(x, y) = dist(x, y)$.

For a rooted tree (T, rt) and a vertex v in T, the *level* of v in T is the hop-distance between the root rt of T and v in T. Denote by $deg(T, v)$ the degree of a vertex v in T and define $\Delta(T) = \max\{deg(T, v) : v \in V\}$. For any two vertices $u, v \in V(T)$, their weighted distance in T is denoted by $dist_T(u, v)$. For a positive integer D, a rooted tree in which every vertex has at most D children is called a *D-ary tree*.

A tree T is called a *Steiner tree* of a metric space $M = (V, dist)$ if it spans a superset of V and if for any pair of points $v, v \in V$, $dist_T(u, v) \geq dist(u, v)$. Let T be either a spanning or a Steiner tree of M rooted at an arbitrary designated vertex rt. We define the *stretch* between two vertices u and v in V to be $\zeta_T(u,v) = \frac{dist_T(u,v)}{dist(u,v)}$, and the *root-stretch* of (T, rt) to be $\varrho(T, rt) = \max\{\zeta_T(rt, v) : v \in V\}$.

For a positive integer n, we denote the set $\{1, 2, \ldots, n\}$ by $[n]$.

2 Narrow-Shallow-Low Trees (NSLoTs)

Upper Bounds. In this section we devise constructions of trees that have small maximum degree, depth and root-stretch, but may be quite heavy. On the other hand, we do require their weight to be bounded by $O(\sum_{v \in M} dist(rt, v))$, where rt is the designated root vertex. We denote the quantity $\sum_{v \in M} dist(rt, v))$ by $W^*(M, rt)$. The following statement summarizes the properties of our construction of NSLoTs for general metric spaces.

Proposition 1. *For any n-point metric space $M = (V, dist)$, an arbitrary designated point rt and a positive integer $2 \le D \le n-1$, there exists a D-ary rooted spanning tree (T, rt) of M with depth at most $\lceil \log_D n \rceil$, root-stretch at most $2 \cdot \lceil \log_D n \rceil$ and weight at most $2 \cdot W^*(M, rt)$.*

Proof. Let $V = (rt = v_0, v_1, \ldots, v_{n-1})$. Without loss of generality assume that the n points $rt = v_0, v_1, \ldots, v_{n-1}$ are ordered by their distance from rt, i.e., $0 = dist(rt, v_0) \le dist(rt, v_1) \le \ldots \le dist(rt, v_{n-1})$. Next, we construct a rooted tree (T, rt) that satisfies the required conditions. The D points $v_1, v_2, \ldots, v_D$ become the children of $rt = v_0$ in T, the next D points $v_{D+1}, v_{D+2}, \ldots, v_{2 \cdot D}$ become the children of v_1, the next D points $v_{2 \cdot D+1}, v_{2 \cdot D+2}, \ldots, v_{3 \cdot D}$ become the children of v_2, and so on. Generally, the point v_i becomes the child of point $v_{\lceil \frac{i}{D} \rceil - 1}$ in T, for each $i \in [n-1]$. (See Fig. 1.a for an illustration.)

Lemma 1. *(1). For any pair of vertices v and w, such that v is an ancestor of w in T, $dist(rt, v) \le dist(rt, w)$. (2) (T, rt) is a D-ary rooted spanning tree of M. (3) The depth $h(T)$ of the rooted tree (T, rt) is no greater than $\lceil \log_D n \rceil$.*

The first assertion of Lemma 1 and triangle inequality imply that $w(T) \le 2 \cdot W^*(M, rt)$. The next lemma provides an upper bound on the root-stretch of the constructed tree.

Lemma 2. *For a vertex v of level i in T, $dist_T(rt, v) \le (2 \cdot i - 1) \cdot dist(rt, v)$.*

The third assertion of Lemma 1 and Lemma 2 imply that the root-stretch of (T, rt) is at most $2 \cdot \lceil \log_D n \rceil$, which concludes the proof of Proposition 1. $\square$

Next, we describe a construction of Steiner NSLoTs for general metric spaces.

Proposition 2. *For any n-point metric space $M = (V, dist)$, an arbitrary designated point rt, a positive integer $2 \le D \le n-1$ and a number $0 < \epsilon' < 1$, there exists a D-ary rooted Steiner tree (T, rt) of M with $O(n/D)$ Steiner points, depth at most $\lceil \log_D n \rceil$, root-stretch at most $(1+\epsilon')$ and weight at most $(1+\epsilon') \cdot W^*(M, rt)$.*

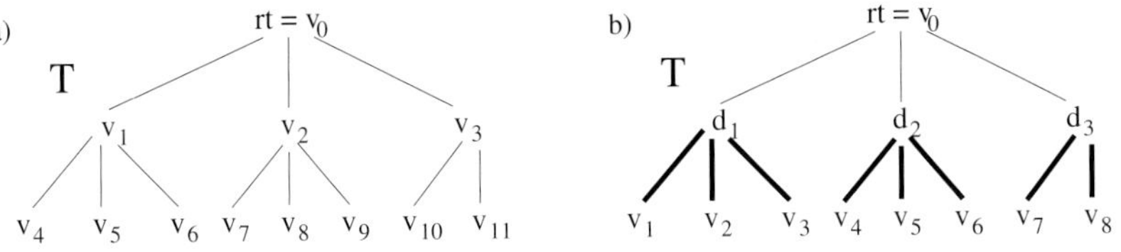

Fig. 1. a) An NSLoT for a 12-point metric space. b) A Steiner NSLoT for a 9-point metric space. The Steiner points are d_1, d_2 and d_3, and the required points are $rt = v_0, v_1, \ldots, v_8$. Edges of weight ϵ are depicted by thin lines. Edges of greater weight (specifically, Edges $(v_i, \pi(v_i))$ of weight $dist_M(rt, v_i)$ are depicted by thick lines.

Proof. Suppose first that $n-1$ is an integer power of D. We form the full D-ary tree T rooted at rt, whose $n-1$ leaves are the $n-1$ points of $V \setminus \{rt\}$. The remaining $\frac{n-2}{D-1}$ vertices of T (excluding rt) are Steiner points. The weight assignment for edges of T is set as follows. For each point $v \in V \setminus \{rt\}$, the weight of the edge $(v, \pi(v))$ that connects it to its parent in T is set as $dist_M(rt, v)$. All other edge weights are set as 0. If one prefers to avoid using weights 0, one can use an arbitrarily small number $\epsilon = \frac{\epsilon'}{2n} \cdot w_{min}$, where w_{min} is the minimum distance between a pair of points in M. It is easy to see that the resulting tree is a D-ary Steiner NSLoT with maximum degree D, depth $\log_D(n-1)$, root-stretch at most $(1+\epsilon')$, and weight at most $(1+\epsilon') \cdot W^*(M, rt)$. This construction generalizes in the obvious way to the case where $n-1$ is not an integer power of D, with the tree depth becoming $\lceil \log_D(n-1) \rceil$. (See Fig. 1.b for an illustration.) □

Next, we show that for point sets in the plane one can construct NSLoTs with significantly smaller root-stretch, without increasing any of the other parameters. The extension of our construction to higher constant dimensions is omitted due to space limitations.

Proposition 3. *Let $k \geq 9$ and $\theta = 2\pi/k$. For any set V of n points in the plane, an arbitrary designated point rt and and a positive integer $2 \leq D \leq n$, there exists a $(2D+k)$-ary rooted Euclidean spanning tree (T_θ, rt) for V with depth at most $\log_D n$, root-stretch at most $\frac{1}{\cos\theta - \sin\theta}$ and weight at most $\frac{1}{\cos\theta - \sin\theta} \cdot W^*(V, rt)$. Moreover, T_θ can be constructed in $O(n \cdot \log_D n)$ time.*

Remark: For large k, $\frac{1}{\cos\theta - \sin\theta} = 1 + O(\theta)$. Hence we get a tree with maximum degree $O(D + \theta^{-1})$, depth at most $\log_D n$, root-stretch $1 + O(\theta)$ and weight $O(W^*(V, rt))$.

Proof. For any $D \geq \frac{n-10}{2}$, the star graph rooted at rt satisfies the conditions of the proposition. We henceforth assume that $D < \frac{n-10}{2}$.

If we rotate the positive x-axis by angles $i \cdot \theta$, $0 \leq i < k$, then we get k rays. Each pair of successive rays defines a *cone* that spans an angle of θ and whose apex is at the origin. Denote by $\mathcal{C} = \{C_1, C_2, \ldots, C_k\}$ the collection of the resulting k cones. For a cone C_i of $\mathcal{C}$ and a point p in the plane, let $C_i(p) = C_i + p = \{x + p : x \in C_i\}$ be the cone obtained from C_i by *translating* it such that its apex is at p, and define $\mathcal{C}(p) = \{C_1(p), C_2(p), \ldots, C_k(p)\}$. We denote by $V_i(p) = V \cap C_i(p)$ the subset of V contained in a cone $C_i(p)$ of $\mathcal{C}(p)$. Note that the collection $\{V_1(p), V_2(p), \ldots, V_k(p)\}$ is a partition of $V \setminus \{p\}$. For each $i \in [k]$, we define $n_i = |V_i(p)|$. Let $\mathcal{P}(p)$ be the collection obtained from $\{V_1(p), V_2(p), \ldots, V_k(p)\}$ by partitioning each set $V_i(p)$ in it (arbitrarily) into $\left\lceil \frac{n_i}{\lfloor n/D \rfloor} \right\rceil$ subsets of size at most $\lfloor n/D \rfloor$ each.

Claim. For any point set V and any point p in the plane, $|\mathcal{P}(p)| \leq 2D + k$.

The tree $T = T_\theta$ is constructed in the following way. First, a partition $\mathcal{P}(rt) = \{P_1(rt), P_2(rt), \ldots, P_m(rt)\}$ of $V \setminus \{rt\}$ is computed, where $m = |\mathcal{P}(rt)| \leq 2D+k$. For each $i \in [m]$, let $rt(i)$ be the point in $P_i(rt)$ whose orthogonal projection onto the bisector of the cone in $\mathcal{C}(rt)$ that contains it is closest to rt. For each $i \in [m]$, $rt(i)$ is set to be a child of rt, and a rooted tree $(T_i, rt(i))$ for the subset $P_i(rt)$ is constructed recursively. The recursion stops if a subset has size one.

Note that T is a $(2D + k)$-ary spanning tree of M rooted at rt, and its depth is at most $\log_D n$. Using arguments from [2] and [19], we show that T can be constructed in $O(n \cdot \log_D n)$ time, and that the root-stretch of T is at most $\frac{1}{\cos\theta - \sin\theta}$. As a corollary, we get that $w(T) \leq \frac{1}{\cos\theta - \sin\theta} \cdot W^*(V, rt)$. □

Lower Bounds. The next statement implies that the upper bound given in Proposition 1 is tight up to constant factors. In particular, it shows that the tradeoff D versus $O(\frac{\log n}{\log D})$ between the maximum degree and root-stretch established there cannot be improved even for Euclidean spaces of dimension $\Theta(\log n)$.

Proposition 4. *There exists a set V of n points in $\mathbb{R}^{O(\log n)}$, such that for any integer $2 \leq D \leq n-1$ and any point $v \in V$, every D-ary spanning tree T of V rooted at $rt = v$ has depth at least $\lfloor \log_D n \rfloor$, weight at least $\Omega(W^*(V, rt))$, and root-stretch at least $\Omega(\log_D n)$.*

Proposition 4 should be compared with Proposition 3. Specifically, as long as the dimension d is constant, one can obtain NSLoTs with root-stretch at most $(1 + \epsilon)$, while for $d = \Omega(\log n)$ it is no longer possible. The next statement extends the lower bound on the tradeoff between the maximum degree D and root-stretch $\Omega(\frac{\log n}{\log D})$ established in Proposition 4 to any dimension $d = O(\log n)$. In particular, it shows that whenever $d = \omega(1)$ is super-constant, it is no longer possible to achieve simultaneously constant maximum degree and root-stretch.

Proposition 5. *For any parameter $d \leq \log n$, there exists a set $\tilde{V}$ of n points in $\mathbb{R}^{O(d)}$, such that for any integer $2 \leq D \leq n-1$ and any point $v \in \tilde{V}$, every D-ary spanning tree $\tilde{T}$ of $\tilde{V}$ rooted at $rt = v$ has root-stretch at least $\Omega(\frac{d}{\log D})$.*

3 Narrow-Shallow-Low-Light Trees

In this section we present a general technique for constructing NSLLTs out of NSLoTs. Then we employ this technique in conjunction with the NSLoTs

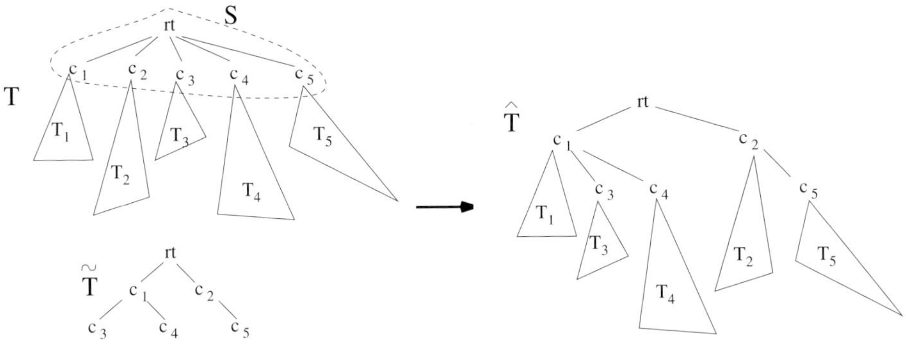

Fig. 2. The root rt of the SLLT T may have a large degree. The star subtree Z is replaced by the NSLoT $\tilde{T}$ to obtain the NSLLT $\hat{T}$.

construction from Sect. 2 to obtain our constructions of NSLLTs, which exhibit optimal tradeoffs between all four parameters.

Consider an n-point metric space M, and let T be a spanning tree for M rooted at some designated point $rt \in M$. Next, we argue that by using NSLoTs one can significantly reduce the degree of rt, while only slightly increasing other parameters of T. Let Z be the star subtree of T rooted at rt. In other words, the vertex set of Z is $V(Z) = \{rt, c_1, c_2, \ldots, c_q\}$, where $c_1, c_2, \ldots, c_q$ are the children of rt in T. Also, the weights of edges (rt, c_i) agree in T and Z, for all indices $i \in [q]$. Let $\tilde{T}$ be some spanning tree rooted at rt for the metric space M_Z induced by the points in $V(Z)$. (Observe that $w(Z) = W^*(M_Z, rt) \leq w(T)$.) Finally, let $\hat{T}$ be the tree obtained from T by replacing the star Z with the tree $\tilde{T}$. (See Fig. 2 for an illustration.) For a tree τ, let $\lambda(\tau) = \max\{deg(\tau, v) : v \in V, v \neq rt\}$ be the degree of a non-root vertex in τ.

The properties of the resulting tree are summarized in the following statement.

Proposition 6. *(1)* $h(\hat{T}) \leq h(T) - 1 + h(\tilde{T})$, *(2)* $w(\hat{T}) = w(T) - w(Z) + w(\tilde{T})$, *(3)* $\lambda(\hat{T}) \leq \lambda(T) + \lambda(\tilde{T})$, *(4)* $deg(\hat{T}, rt) = deg(\tilde{T}, rt)$, *(5)* $\varrho(\hat{T}, rt) \leq \varrho(\tilde{T}, rt) \cdot \varrho(T, rt)$.

Remark: This statement remains valid if $\tilde{T}$ is a Steiner NSLoT of M_Z.
Dinitz et al. [8] devised two constructions of SLLTs for general metric spaces. For a metric space M, a point $rt \in M$ and an integer $\ell = O(\log n)$, the first construction provides a rooted SLLT (T, rt) with depth $h(T) = O(\ell)$, weight $w(T) = O(\ell \cdot n^{1/\ell}) \cdot w(MST(M))$ and root-stretch $\varrho(T, rt) \leq 1 + \epsilon$. Moreover, all vertices of T except its root rt have optimal degree $O(n^{1/\ell})$. The degree of the root may, however, be arbitrarily large. The second construction provides an SLLT T' with depth $h(T') = O(\ell \cdot n^{1/\ell})$, weight $w(T') = O(\ell) \cdot w(MST(M))$, and root-stretch $\varrho(T', rt) \leq 1 + \epsilon$. Similarly to the first construction, all vertices of T' but the root rt have optimal degree $O(1)$, and the root may have arbitrarily large degree.

Next, we reduce the root-degree in the first construction. Reducing the root-degree of the second construction is done similarly. Let Z be the star subtree of

T rooted at rt, and let $\tilde{T}$ be an NSLoT for the $(q+1)$-point metric space M_Z. To construct an NSLLT $\widehat{T}$ out of T and $\tilde{T}$, we replace the star Z by $\tilde{T}$.

Specifically, if M is a set of n points in the plane, then our construction of NSLoTs (Proposition 3) provides a rooted NSLoT $(\tilde{T}, rt)$ for M_Z with depth $h(\tilde{T}) = O(\ell)$, $\Delta(\tilde{T}) = O((q+1)^{1/\ell}) = O(n^{1/\ell})$, weight $w(\tilde{T}) = O(W^*(M_Z, rt))$, and root-stretch $\varrho(\tilde{T}, rt) \leq (1+\epsilon)$. By Proposition 6, replacing the star Z of T with $\tilde{T}$ produces a rooted NSLLT $(\widehat{T}, rt)$ for M with depth $h(\widehat{T}) \leq h(T) - 1 + h(\tilde{T}) = O(\ell)$, weight $w(\widehat{T}) = w(T) - w(Z) + w(\tilde{T}) = w(T) + O(W^*(M_Z, rt)) = O(w(T)) = O(\ell \cdot n^{1/\ell}) \cdot w(MST(M))$, $\Delta(\widehat{T}) = O(n^{1/\ell})$, and root-stretch $\varrho(\widehat{T}, rt) \leq (1+\epsilon)^2 = 1 + O(\epsilon)$. The SLLTs of [8] for Euclidean spaces can be constructed in $O(n \cdot \log n)$ time. By Proposition 3, $\tilde{T}$ can be constructed in $O(n \cdot \log n)$ time. Hence the overall time required to construct $\widehat{T}$ is $O(n \cdot \log n)$. This tradeoff extends to the complementary range of depth $h(\widehat{T}) = \Omega(\log n)$. This argument easily generalizes to point sets in $\mathbb{R}^d$, for any constant $d \geq 2$.

Theorem 1. *Let $d \geq 2$ be an integer constant. For a set M of n points in $\mathbb{R}^d$, an integer $\ell = O(\log n)$, and $\epsilon > 0$, there exists a spanning tree with depth $O(\ell)$, lightness $O(\ell \cdot n^{1/\ell})$, maximum degree $O(n^{1/\ell})$, and root-stretch at most $(1+\epsilon)$. In addition, there exists a spanning tree with depth $O(\ell \cdot n^{1/\ell})$, lightness $O(\ell)$, constant maximum degree, and root-stretch at most $(1+\epsilon)$. Both trees can be constructed in $O(n \cdot \log n)$ time.*

Our construction of NSLoTs for general metric spaces (Proposition 1) provides a rooted NSLoT $(\tilde{T}, rt)$ for M_Z with depth $h(\tilde{T}) = O(\ell)$, $\Delta(\tilde{T}) = O((q+1)^{1/\ell}) = O(n^{1/\ell})$, weight $w(\tilde{T}) = O(W^*(M_Z, rt))$ and root-stretch $\varrho(\tilde{T}, rt) = O(\ell)$. By Proposition 6, replacing the star Z of T with $\tilde{T}$ produces a rooted NSLLT $(\widehat{T}, rt)$ for M with depth $h(\widehat{T}) \leq h(T) - 1 + h(\tilde{T}) = O(\ell)$, weight $w(\widehat{T}) = w(T) - w(Z) + w(\tilde{T}) = w(T) + O(W^*(M_Z, rt)) = O(w(T)) = O(\ell \cdot n^{1/\ell}) \cdot w(MST(M))$, $\Delta(\widehat{T}) = O(n^{1/\ell})$ and root-stretch $\varrho(\widehat{T}, rt) \leq O(\ell) \cdot (1+\epsilon) = O(\ell)$. The SLLTs of [8] for general metric spaces can be constructed in $O(n^2)$ time. Clearly $\tilde{T}$ can be constructed in $O(n^2)$ time, and so the overall time required to construct $\widehat{T}$ is $O(n^2)$. This tradeoff extends to the complementary range of depth $h(\widehat{T}) = \Omega(\log n)$.

Theorem 2. *For a general n-point metric space M, and an integer $\ell = O(\log n)$, there exists a spanning tree with depth $O(\ell)$, lightness $O(\ell \cdot n^{1/\ell})$, maximum degree $O(n^{1/\ell})$, and root-stretch $O(\ell)$. In addition, there exists a spanning tree with depth $O(\ell \cdot n^{1/\ell})$, lightness $O(\ell)$, constant maximum degree, and root-stretch $O(\log n)$. Both trees can be constructed in $O(n^2)$ time.*

Similarly, using our construction of Steiner NSLoTs for general metric spaces (Proposition 2), we construct in $O(n^2)$ time a Steiner rooted NSLLT $(\widehat{T}, rt)$ for M with depth $h(\widehat{T}) = O(\ell)$, weight $w(\widehat{T}) = O(\ell \cdot n^{1/\ell}) \cdot w(MST(M))$, $\Delta(\widehat{T}) = O(n^{1/\ell})$ and root-stretch $\varrho(\widehat{T}, rt) = 1 + O(\epsilon)$. This tradeoff extends to the complementary range of depth $h(\widehat{T}) = \Omega(\log n)$.

Theorem 3. *For a general n-point metric space M, an integer $\ell = O(\log n)$, and $\epsilon > 0$, there exists a Steiner tree with depth $O(\ell)$, lightness $O(\ell \cdot n^{1/\ell})$, maximum degree $O(n^{1/\ell})$, and root-stretch at most $(1 + \epsilon)$. In addition, there exists a Steiner tree with depth $O(\ell \cdot n^{1/\ell})$, lightness $O(\ell)$, constant maximum degree, and root-stretch at most $(1 + \epsilon)$.*

Finally, we extend the lower bounds of Dinitz et al. [8] to Steiner trees.

Theorem 4. *For any metric space M and any Steiner rooted tree (T', rt') of M, there exists a rooted tree (T, rt) spanning only $V(M)$, with depth no greater than that of T' (i.e., $h(T) \leq h(T')$), weight at most twice the weight of T' (i.e., $w(T) \leq 2 \cdot w(T')$), and which also* dominates *T' in the following sense: for any two points u, v in $V(M)$, $dist_T(u, v) \geq dist_{T'}(u, v)$. Moreover, T can be constructed in $O(n)$ time.*

Dinitz et al. [8] analyzed the 1-dimensional metric space ϑ_n with n points $1, 2, \ldots, n$ on the x-axis and have shown that for any parameter $\ell = O(\log n)$, any spanning tree for ϑ_n that has depth $h(T) = O(\ell)$ has weight $w(T) = \Omega(\ell \cdot n^{1+1/\ell})$, and vice versa, i.e., if $w(T) = O(\ell \cdot n)$, then $h(T) = \Omega(\ell \cdot n^{1/\ell})$. Theorem 4 enables us to extend this lower bound to Steiner trees.

Corollary 1. *For a positive integer $\ell = O(\log n)$, any Steiner tree T for ϑ_n that has depth $O(\ell)$ satisfies $w(T) = \Omega(\ell \cdot n^{1+1/\ell}) = \Omega(\ell \cdot n^{1/\ell}) \cdot w(MST(\vartheta_n))$. Also, any Steiner tree T for ϑ_n that has weight $O(\ell \cdot n) = O(\ell) \cdot w(MST(\vartheta_n))$ satisfies $h(T) = \Omega(\ell \cdot n^{1/\ell})$.*

In particular, Corollary 1 implies that any Steiner tree T for ϑ_n has either depth $\Omega(\log n)$ or weight $\Omega(n \cdot \log n) = \Omega(\log n) \cdot w(MST(\vartheta_n))$. On the other hand, Arya et al. [2] devised a construction of Euclidean $(1 + \epsilon)$-spanners with both hop-diameter and lightness (the ratio between the weight and the weight of the MST) at most $O(\log n)$. Corollary 1 implies that the result of [2] cannot be improved even if one allows the spanner to use Steiner points.

Corollary 2. *Any Euclidean (possibly Steiner) spanner for ϑ_n that guarantees hop-diameter $o(\log n)$ has lightness $\omega(\log n)$, and vice versa.*

References

1. Alpert, C.J., Hu, T.C., Huang, J.H., Kahng, A.B., Karger, D.: Prim-Dijkstra tradeoffs for improved performance-driven routing tree design. IEEE Trans. on CAD of Integrated Circuits and Systems 14(7), 890–896 (1995)
2. Arya, S., Das, G., Mount, D.M., Salowe, J.S., Smid, M.H.M.: Euclidean spanners: short, thin, and lanky. In: 27th ACM Symposium on Theory of Computing, pp. 489–498. ACM Press, New York (1995)
3. Awerbuch, B., Baratz, A., Peleg, D.: Cost-sensitive analysis of communication protocols. In: 9th ACM Symposium on Principles of Distributed Computing, pp. 177–187. ACM Press, New York (1990)
4. Awerbuch, B., Baratz, A., Peleg, D.: Efficient Broadcast and Light-Weight Spanners (manuscript) (1991)

5. Chan, T.M.: Euclidean Bounded-Degree Spanning Tree Ratios. Discrete & Computational Geometry 32(2), 177–194 (2004)
6. Cong, J., Kahng, A.B., Robins, G., Sarrafzadeh, M., Wong, C.K.: Performance-Driven Global Routing for Cell Based ICs. In: 9th IEEE International Conference on Computer Design: VLSI in Computer & Processors, pp. 170–173. IEEE press, New York (1991)
7. Cong, J., Kahng, A.B., Robins, G., Sarrafzadeh, M., Wong, C.K.: Provably good performance-driven global routing. IEEE Trans. on CAD of Integrated Circuits and Sys. 11(6), 739–752 (1992)
8. Dinitz, Y., Elkin, M., Solomon, S.: Shallow-Low-Light Trees, and Tight Lower Bounds for Euclidean Spanners. In: 49th IEEE Symposium on Foundations of Computer Science, pp. 519–528. EEE Press, New York (2008)
9. Eppstein, D.: Spanning trees and spanners. Technical report 96–16, Dept. of Information and Computer-Science, University of California, Irvine (1996)
10. Farshi, M., Gudmundsson, J.: Experimental Study of Geometric t-Spanners: A Running Time Comparison. In: Demetrescu, C. (ed.) WEA 2007. LNCS, vol. 4525, pp. 270–284. Springer, Heidelberg (2007)
11. Fekete, S.P., Khuller, S., Klemmstein, M., Raghavachari, B., Young, N.E.: A Network-Flow Technique for Finding Low-Weight Bounded-Degree Spanning Trees. J. Algorithms 24(2), 310–324 (1997)
12. Grünewald, M., Lukovszki, T., Schindelhauer, C., Volbert, K.: Distributed Maintenance of Resource Efficient Wireless Network Topologies. In: Monien, B., Feldmann, R.L. (eds.) Euro-Par 2002. LNCS, vol. 2400, pp. 935–946. Springer, Heidelberg (2002)
13. Gupta, A.: Steiner points in tree metrics don't (really) help. In: 12th ACM-SIAM Symposium on Discrete Algorithms, pp. 220–227. SIAM Press, Philadelphia (2001)
14. Khuller, S., Raghavachari, B., Young, N.E.: Balancing Minimum Spanning and Shortest Path Trees. In: 4th ACM-SIAM Symposium on Discrete Algorithms, pp. 243–250. ACM Press, New York (1993)
15. Lukovszki, T.: New Results on Fault Tolerant Geometric Spanners. In: Dehne, F., Gupta, A., Sack, J.-R., Tamassia, R. (eds.) WADS 1999. LNCS, vol. 1663, pp. 193–204. Springer, Heidelberg (1999)
16. Lukovszki, T.: New Results on Geometric Spanners and Their Applications. Ph.D thesis, Dept. of Computer-Science, University of Paderborn, Paderborn, Germany (1999)
17. Lukovszki, T., Schindelhauer, C., Volbert, K.: Resource Efficient Maintenance of Wireless Network Topologies. J. UCS 12(9), 1292–1311 (2006)
18. Monma, C.L., Suri, S.: Transitions in Geometric Minimum Spanning Trees. Discrete & Computational Geometry 8, 265–293 (1992)
19. Narasimhan, G., Smid, M.: Geometric Spanner Networks. Cambridge University Press, Cambridge (2007)
20. Papadimitriou, C.H., Vazirani, U.V.: On Two Geometric Problems Related to the Traveling Salesman Problem. J. Algorithms 5(2), 231–246 (1984)
21. Ruppert, J., Seidel, R.: Approximating the d-dimensional complete Euclidean graph. In: 3rd Canadian Conference on Computational Geometry, pp. 207–210 (1991)
22. Salowe, J.S., Richards, D.S., Wrege, D.E.: Mixed spanning trees: a technique for performance-driven routing. In: 3rd ACM Great Lakes symposium on VLSI, pp. 62–66. ACM Press, New York (1993)

Bounded Budget Betweenness Centrality Game for Strategic Network Formations

Xiaohui Bei[1], Wei Chen[2], Shang-Hua Teng[3,⋆], Jialin Zhang[1], and Jiajie Zhu[4]

[1] Institute for Theoretical Computer Science, Tsinghua University
{bxh08,zhanggl02}@mails.tsinghua.edu.cn
[2] Microsoft Research Asia
weic@microsoft.com
[3] University of Southern California
shanghua@usc.edu
[4] University of California at Los Angeles
jiajie@cs.ucla.edu

Abstract. In this paper, we introduce the *bounded budget betweenness centrality* game, a strategic network formation game in which nodes build connections subject to a budget constraint in order to maximize their betweenness centrality, a metric introduced in the social network analysis to measure the information flow through a node. To reflect real world scenarios where short paths are more important in information exchange, we generalize the betweenness definition to only consider shortest paths of length at most ℓ. We present both complexity and constructive existence results about Nash equilibria of the game. For the nonuniform version of the game where node budgets, link costs, and pairwise communication weights may vary, we show that Nash equilibria may not exist and it is NP-hard to decide whether Nash equilibria exist in a game instance. For the uniform version of the game where link costs and pairwise communication weights are one and each node can build k links, we construct two families of Nash equilibria based on shift graphs, and study the properties of Nash equilibria. Moreover, we study the complexity of computing best responses and show that the task is polynomial for uniform 2-B^{3}C games and NP-hard for other games.

Keywords: algorithmic game theory, network formation game, Nash equilibrium, betweenness centrality.

1 Introduction

Many network structures in real life are not designed by central authorities. Instead, they are formed by autonomous agents who often have selfish motives [17]. Typical examples of such networks include the Internet where autonomous systems linked together to achieve global connection, peer-to-peer networks where peers connect to one another for online file sharing (e.g. [7,19]), and social networks where individuals connect to one another for information exchange and other social functions [18]. Since these autonomous agents have their selfish motives and are not under any centralized control,

⋆ The work is done when the author is visiting Microsoft Research Asia and Microsoft Research New England.

A. Fiat and P. Sanders (Eds.): ESA 2009, LNCS 5757, pp. 227–238, 2009.

they often act strategically in deciding whom to connect to in order to improve their own benefits. This gives rise to the field of *network formation games*, which studies the game-theoretic properties of the networks formed by these selfish agents as well as the process in which all agents dynamically adjust their strategies [1,9,13,14,15].

A key measure of importance of a node is its *betweenness centrality* (or betweenness for short), which is introduced originally in social network analysis [10,16]. If we view a network as a graph $G = (V, E)$ (directed or undirected), the betweenness of a node (or vertex) i in G is

$$btw_i(G) = \sum_{u \neq v \neq i \in V,\, m(u,v)>0} w(u,v) \frac{m_i(u,v)}{m(u,v)} \tag{1}$$

where $m(u,v)$ is the number of shortest paths from u to v in G, $m_i(u,v)$ is the number of shortest paths from u to v that pass i in G, and $w(u,v)$ is the weight on pair (u,v). Intuitively, if the amount of information from u to v is $w(u,v)$, and the information is passed along all shortest paths from u to v equally, then the betweenness of node i measures the amount of information passing through i among all pair-wise exchanges.

In this paper, we generalize the betweenness definition with a parameter ℓ such that only shortest paths with length at most ℓ are considered in betweenness calculation. Formally, we define

$$btw_i(G,\ell) = \sum_{u \neq v \neq i \in V,\, m(u,v,\ell)>0} w(u,v) \frac{m_i(u,v,\ell)}{m(u,v,\ell)}, \tag{2}$$

where $m(u,v,\ell)$ is the number of shortest paths from u to v in G with length at most ℓ, and $m_i(u,v,\ell)$ is the number of shortest paths from u to v that passes i in G with length at most ℓ. It is easy to see that $btw_i(G) = btw_i(G, n-1)$, where n is the number of vertices in G.

Betweenness with path length constraint is reasonable in real-world scenarios. In peer-to-peer networks such as Gnetella [19], query requests are searched only on nodes with a short graph distance away from the query initiator. In social networks, researches (e.g. [4,5]) show that short connections are much more important than long-range connections.

In a decentralized network with autonomous agents, each agent may have incentive to maximize its betweenness in the network. For example, in computer networks and peer-to-peer networks, a node in the network may be able to charge the traffic that it helps relaying, in which case the revenue of the node is proportional to its betweenness in the network. So the maximization of revenue is consistent with the maximization of the betweenness. In a social network, an individual may want to gain or control the most amount of information travelling in the network by maximizing her betweenness.

In this paper, we introduce a network formation game in which every node in a network is a selfish agent who decides which other nodes to connected to in order to maximize its own betweenness. Building connections with other nodes incur costs. Each node has a budget such that the cost of building its connections cannot exceed its budget. We call this game the *bounded budget betweenness centrality* game or the B^3C game. When distinction is necessary, we use ℓ-B^3C to denote the games using generalized betweenness definition $btw_i(G,\ell)$.

Bounded budget assumption, first incorporated into a network formation game in [14], reflects real world scenarios where there are physical limits to the number of connections one can make. In computer and peer-to-peer networks, each node usually has a connection limit. In social networks, each individual only has a limited time and energy to create and maintain relationships with other individuals. An alternative treatment to connection costs appearing in more studies [1,9,13,15] is to subtract connection costs from the main objectives to be maximized. This treatment, however, restricts the variety of Nash equilibria exhibited by the game, e.g. only allowing dense graphs to be Nash equilibria [13]. Therefore, in this paper we choose to incorporate the bounded budget assumption, even though it makes the game model more complicated.

In this paper, we consider the directed graph variant of the game, in which nodes can only establish outgoing links to other nodes. Since incoming links help increasing nodes' betweenness, nodes should be happy to accept incoming links created by other nodes. This mitigates the concern on many network formation games in which connection creation is one-sided decision. Since the game allows some trivial Nash equilibria (such as a network with no links at all), we study a stronger form called *maximal* Nash equilibria, in which no node can add more outgoing links without exceeding its budget constraint. Adding outgoing links of a node can only help its betweenness, so it is reasonable to study maximal Nash equilibria in the B^3C games.

We present both complexity and existence results about B^3C games. First, we study the existence of maximal Nash equilibria in *nonuniform* ℓ-B^3C game, which is specified by several parameters concerning the node budgets, link costs, and pairwise communication weights (Section 3). We show that a nonuniform ℓ-B^3C game may not have any maximal Nash equilibria for any $\ell \geq 2$. Moreover, given these parameters as input, it is NP-hard to determine whether the game has a maximal Nash equilibrium. The result indicates that finding Nash equilibria in general ℓ-B^3C games is a difficult task.

Second, we address the complexity of computing best responses in ℓ-B^3C games (Section 4). For uniform ℓ-B^3C games where all pair weights are one, all link costs are one, and all node budgets are given as an integer k, we show that with $\ell = 2$, computing a best response takes $O(n^3)$ time. For all other cases (uniform games with $\ell \geq 3$ or nonuniform games with $\ell \geq 2$), the task is NP-hard.

Finally, we turn our attention to the construction and the properties of Nash equilibria in the *uniform* ℓ-B^3C game with n nodes and k outgoing edges from each node (Section 5). We introduce a type of multi-partite graphs that we call *shift graphs*, which are variants of better known De Bruijn graphs and Kautz graphs. Based on these shift graphs, we construct two different families of Nash equilibria for *uniform* ℓ-B^3C games. One family gives a stronger form of Nash equilibria called strict Nash equilibria, while the other family belongs to what we call *ℓ-path-unique graphs* (ℓ-PUGs), which we show are always Nash equilibria for uniform ℓ-B^3C games. We then use ℓ-PUGs to study several properties of Nash equilibria. In particular, we show that (a) for any ℓ, k and large enough n ($n \geq (k+\ell)!/k!$), a maximal Nash equilibrium exists; (b) Nash equilibria may exhibit rich structures, e.g. they may be disconnected or have unbalanced in-degrees and betweenness among nodes; and (c) for 2-B^3C games, all maximal Nash equilibria must be 2-PUGs if the maximum in-degree is $o(n)$ (with k being a constant). The proofs for all results in this paper are available in our full technical report [2].

Related work. There are a number of studies on network formation games with Nash equilibrium as the solution concept [1,3,9,13,14,15]. Most of the above work belong to a class of games in which nodes try to minimize their average shortest distances to other nodes in the network [1,9,14,15], which is called *closeness centrality* in social network analysis [10].

Our research is partly motivated by the work of [13], in which Kleinberg et al. study a different type of network formation games related to the concept of structural holes in organizational social network research. In this game, each node tries to bridge other pairs of nodes that are not directly connected. In a sense, this is a restricted type of betweenness where only length-2 shortest paths are considered. Besides some difference in the game setup, there are two major differences between our work and theirs. First, we consider betweenness with a general path length constraint of ℓ as well as no path length constraints, while they only consider the bridging effect between two immediate neighbors of a node. Second, we incorporate budget constraints to restrict the number of links one node can build, while their work subtracts link costs in the payoff function of each node. As the result of their treatment to link costs, they show that all Nash equilibria are limited to dense graphs with $\Omega(n^2)$ edges where n is the number of vertices. This is what we want to avoid in our study. A couple of other studies [6,11] also address strategic network formations with structural holes, but they do not address the computation issue, and their game formats have their own limitations (e.g. star networks as the only type of equilibria [11] or limited to length-2 paths [6]).

Our game is also inspired by the Bounded Budget Connection game of Laoutaris et al. [14]. This game considers directed links and bounded budgets on nodes, using minimization of average shortest distances to others as the objective for each node. It shows hardness results in determining the existence of Nash equilibria in general games, and provides tree-like structures as Nash equilibria for the uniform version of the game. It also shows that Abelian Cayley graphs cannot be Nash equilibria in large networks.

Solution concepts other than Nash equilibrium are also used in the study of network formation games. Authors in [8,12] consider games in which two end points of a link have to jointly agree on adding the link, and they use pairwise stability as an alternative to Nash equilibrium.

2 Problem Definition

A *(nonuniform) bounded-budget betweenness centrality* ($\mathrm{B^3C}$) game with parameters (n, b, c, w) is a network formation game defined as follows. We consider a set of n players $V = \{1, 2, \ldots, n\}$, which are also nodes in a network. Function $b : V \rightarrow \mathbb{N}$ specifies the budget $b(i)$ for each node $i \in V$ ($\mathbb{N}$ is the set of natural numbers). Function $c : V \times V \rightarrow \mathbb{N}$ specifies the cost $c(i, j)$ for the node i to establish a link to node j, for $i, j \in V$. Function $w : V \times V \rightarrow \mathbb{N}$ specifies the weight $w(i, j)$ from node i to node j for $i, j \in V$, which can be interpreted as the amount of traffic i sends to j, or the importance of the communication from i to j.

The strategy space of player i in $\mathrm{B^3C}$ game is $S_i = \{s_i \subseteq V \setminus \{i\} \mid \sum_{j \in s_i} c(i, j) \leq b(i)\}$, i.e., all possible subsets of outgoing links of node i within i's budget. A strategy profile $s = (s_1, s_2, \ldots, s_n) \in S_1 \times S_2 \times \ldots \times S_n$ is referred to as a *configuration* in this paper. The graph induced by configuration s is denoted as $G_s = (V, E)$, where $E = \{(i, j) \mid i \in V, j \in s_i\}$. For convenience, we will also refer G_s as a configuration.

In the game without path length constraint, the utility of a node i in configuration G is defined by the *betweenness centrality* of i as given in equation (1). When we generalize betweenness centrality and consider only shortest paths of length at most ℓ, the utility of node i is given as in equation (2). We use ℓ-B^{3}C to denote the generalized version of game with parameter ℓ.

In a configuration s, if no node can increase its own utility by changing its own strategy unilaterally, we say that s is a *(pure) Nash equilibrium*, and we also say that s is *stable*. Moreover, if in configuration s any strategy change of any node strictly decreases the utility of the node, we say that s is a *strict* Nash equilibrium.

The following Lemma shows the monotonicity of node betweenness when adding new edges to a node, which motivates our definition of maximal Nash equilibrium.

Lemma 1. *For any graph $G = (V, E)$, let $G' = (V, E \cup \{(i, j)\})$ where $i, j \in V$ and $(i, j) \notin E$. Then $btw_i(G) \leq btw_i(G')$, and $btw_i(G, \ell) \leq btw_i(G', \ell)$ for all $\ell \geq 2$.*

Given a nonuniform B^{3}C game with parameters (n, b, c, w), a *maximal strategy* of a node v is a strategy with which v cannot add any outgoing edges without exceeding its budget. We say that a graph (configuration) is *maximal* if all nodes use maximal strategies in the configuration. A configuration is a *maximal Nash equilibrium* if it is a maximal graph and it is a Nash equilibrium. By Lemma 1, it makes sense to study maximal Nash equilibria where no node can add more edges within its budget limit. Moreover, trivial non-maximal Nash equilibria exist (e.g. graphs with no edges), making it less interesting to study all Nash equilibria. Therefore, for the rest of the paper, we focus on maximal Nash equilibria in B^{3}C games. The following lemma states the relationship between maximal Nash equilibria and strict Nash equilibria, a direct consequence of the monotonicity of betweenness centrality.

Lemma 2. *Given a B^{3}C game with parameters (n, b, c, w), any strict Nash equilibrium in the game is a maximal Nash equilibrium.*

Based on the above lemma, for positive existence of Nash equilibria, we sometimes study the existence of strict Nash equilibria to make our results stronger.

A special case of B^{3}C game is the *uniform* game, which has parameters $n, k \in \mathbb{N}$ such that $b(i) = k$ for all $i \in V$, and $c(i, j) = w(i, j) = 1$ for all $i, j \in V$.

3 Determining Nash Equilibria in Nonuniform Games

In this section we show that a nonuniform B^{3}C game may not have any maximal (or strict) Nash equilibrium, and determining whether a game has a maximal (or strict) Nash equilibrium is NP-hard. For simplicity, we address the B^{3}C game without path length constraint first, and then present the results on the ℓ-B^{3}C game.

3.1 Nonexistence of Maximal Nash Equilibria

We fist show that some B^{3}C game with nonuniform edge cost has no maximal (or strict) Nash equilibrium. We construct a family of graphs, which we refer to as the gadget, and

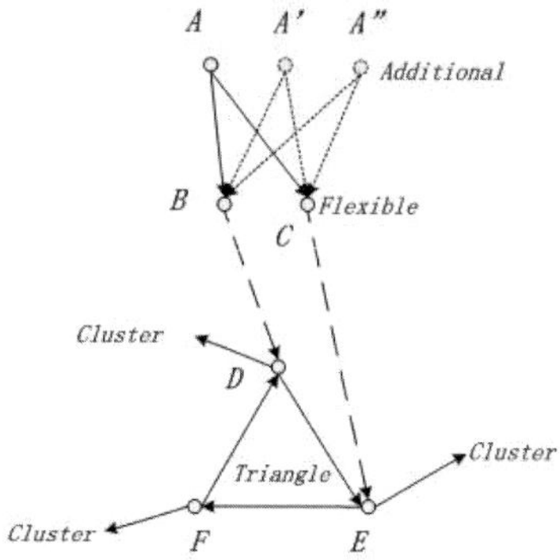

Fig. 1. Main structure of the gadget that has no maximal (or strict) Nash Equilibrium

show that B^{3}C games based on the gadget do not have any maximal Nash equilibrium. The gadget is shown in Figure 1. There are $5 + 3t + r$ nodes in the gadget, where $t \in \mathbb{N}$ and $r = 1, 2, 3$. The values of t and r allow us to construct a graph of any size great than 5. There are r nodes, denoted as A, A', A'' in the figure, which establish edges to B and C. Both B and C can establish at most one edge to a node in $\{D, E, F\}$ respectively. Each node in $\{D, E, F\}$ connects to a cluster of size t each (not shown in the figure). The only requirement for these three clusters is that they are identical to each other and are all strongly connected, so D, E, F can reach all nodes in their corresponding clusters. Nodes in the three clusters do not establish edges to the other clusters or to $A, A', A'', B, C, D, E, F$.

We classify nodes and edges as follows. Nodes B and C are *flexible* nodes since they can choose to connect one node in $\{D, E, F\}$. Nodes D, E, F are *triangle* nodes, nodes in the clusters are *cluster* nodes, and nodes A, A', A'', are *additional* nodes. Edges (i, j) with $i \in \{B, C\}$ and $j \in \{D, E, F\}$ are *flexible* edges. Other edges shown in the figure plus the edges in the clusters are *fixed* edges. The remaining pairs with no edge connected (e.g. (A, D), (A, E), etc.) are referred to as *forbidden* edges.

We use the parameters (n, b, c, w) of a B^{3}C game to realize the gadget. In particular, (a) $n = 5 + 3t + r$; (b) $b(i) = 1$ for all $i \in V$; (c) $c(i, j) = 0$ if (i, j) is a fixed edge, $c(i, j) = 1$ if (i, j) is a flexible edge, $c(i, j) = M > 1$ if (i, j) is a forbidden edge; and (d) $w(i, j) = 1$ for all $i, j \in V$. Note that in the game only the edge costs are nonuniform. With the above construction, we can show the following theorem.

Theorem 1. *The B^{3}C game based on the gadget of Figure 1 does not have any maximal (or strict) Nash equilibrium. This implies that for any $n \geq 6$, there is an instance of B^{3}C game with n players that does not have any maximal (or strict) Nash equilibrium.*

proof (sketch). In any maximal graph, each of the flexible nodes B and C must have exactly one flexible edges pointing to one of the triangle nodes $\{D, E, F\}$. It is mechanical to verify that if one flexible node points to a traingle node $X \in \{D, E, F\}$, the best response of the other flexible node is to point to $Y \in \{D, E, F\}$ that is "downstream" from X, i.e. (X, Y) is a fixed edge. Then the best responses of B and C will cycle through the triangle nodes forever. Therefore, there is no Nash equilibrium for this game. By Lemma 2, there is no strict Nash equilibrium either. □

It is easy to verify that in the proof of Theorem 1 the critical paths that matter for the argument are from A, A' and A'' to nodes D, E, F, and the lengths of these critical paths are at most three. Therefore, with the same argument, we directly know that for all $\ell \geq 3$, the ℓ-B^{3}C game based on Figure 1 does not have maximal Nash equilibrium either. We develop a different gadget in [2] and show that for $\ell = 2, 3$ that ℓ-B^{3}C game based on that gadget has no maximal (or strict) Nash equilibrium. Therefore, we have

Theorem 2. *For any $\ell \geq 2$ and $n \geq 6$, there is an instance of ℓ-B^{3}C game with n players that does not have any maximal (or strict) Nash equilibrium.*

3.2 Hardness of Determining the Existence of Maximal Nash Equilibria

In this section we use the gadget given in Figure 1 as a building block to show that determining the existence of maximal Nash equilibria given a nonuniform B^{3}C game is NP-hard. In fact, we use strict Nash equilibria to obtain a stronger result.

We define a problem TwoExtreme as follows. The input of the problem is (n, b, c, w) as the parameter of a B^{3}C game. The output of the problem is Yes or No, such that (a) if the game has a strict Nash equilibrium, the output is Yes; (b) if the game has no maximal Nash equilibrium, the output is No; and (c) for other cases, the output could be either Yes or No. Both deciding the existence of maximal Nash equilibria and deciding the existence of strict Nash equilibria are stronger problems than TwoExtreme, because their outputs are valid for the TwoExtreme problem by Lemma 2. We show the following result by a reduction from the 3-SAT problem.

Theorem 3. *The problem of* TwoExtreme *is NP-hard.*

Corollary 1. *Both deciding the existence of maximal Nash equilibria and deciding the existence of strict Nash equilibria of a B^{3}C game are NP-hard.*[1]

We obtain the same result for the ℓ-B^{3}C game.

Theorem 4. *For any $\ell \geq 2$, both deciding the existence of maximal Nash equilibria and deciding the existence of strict Nash equilibria in an ℓ-B^{3}C game are NP-hard.*

4 Complexity of Computing Best Responses

The *best response* of a node in a configuration is its strategy that gives the node the best utility (i.e. best betweenness). In this section, we show the complexity of computing best responses for uniform games first, and then extend it for nonuniform games.

[1] In fact, the decision problem for any intermediate concept between maximal Nash equilibrium and strict Nash equilibrium is also NP-hard. For example, deciding the existence of nontransient Nash equilibria [9] is also NP-hard because any strict Nash equilibrium is a nontransient Nash equilibrium while the existence of a nontransient Nash equilibrium implies the existence of a maximal Nash equilibrium in B^{3}C games.

In a uniform game with parameters (n, k), one can exhaustively search all $\binom{n-1}{k}$ strategies and find the one with the largest betweenness. Computing the betweenness of nodes given a fixed graph can be done by all-pair shortest paths algorithms in polynomial time. Therefore, the entire brute-force computation takes polynomial time if k is a constant. However, if k is not a constant, the result depends on ℓ, the parameter bounding the shortest path length in the ℓ-B^{3}C game.

Consider first the case of a uniform 2-B^{3}C game. Let $G = (V, E)$ be a directed graph. For a node v in G, let $G_{v,S}$ be the graph where v has outgoing edges to nodes in $S \subseteq V \setminus \{v\}$ and all other nodes have the same outgoing edges as in G. Then we have

Lemma 3. *For all $S \subseteq V \setminus \{v\}$, $btw_v(G_{v,S}, 2) = \sum_{u \in S} btw_v(G_{v,\{u\}}, 2)$.*

The lemma shows that for a uniform 2-B^{3}C game, the betweenness of a node can be computed by the sum of its betweenness when adding each of its outgoing edges alone.

Theorem 5. *Computing the best response in a uniform 2-B^{3}C game with parameters (n, k) can be done in $O(n^3)$ time.*

proof (sketch). For each $u \in V \setminus \{v\}$, node v computes $btw_v(G_{v,\{u\}}, 2)$ in $O(n^2)$ time. Then v selects the top k nodes u with the largest $btw_v(G_{v,\{u\}}, 2)$ as its strategy, which is guaranteed to be v's best response by Lemma 3. □

For cases other than the uniform 2-B^{3}C game, we show that best response computation is NP-hard, via a reduction from either the knapsack problem (for nonuniform the 2-B^{3}C game) or the set cover problem (the other cases).

Theorem 6. *It is NP-hard to compute the best response in either a nonuniform 2-B^{3}C game, or an ℓ-B^{3}C game with $\ell \geq 3$ (uniform or not), or a B^{3}C game without path length constraint (uniform or not).*

5 Nash Equilibria in Uniform Games

In this section we focus on uniform ℓ-B^{3}C games. we first define a family of graph structures called *shift graphs* and show that they are able to produce Nash equilibria for B^{3}C games. We then study some properties of Nash equilibria in uniform games.

5.1 Construction of Nash Equilibria via Shift Graphs

We first define *shift graphs* and *non-rotational shift graphs*. Then we show that for any ℓ, k and any $\ell' \geq \ell$, the *non-rotational shift graphs* with $n = (\ell' + k)!/k!$ nodes are all Nash equilibria in the uniform ℓ-B^{3}C game with parameter n and k. Moreover, we use shift graphs to construct *strict* Nash equilibria for both ℓ-B^{3}C games and B^{3}C games without path length constraint, for certain combinations of n and k where $k = \Theta(\sqrt{n})$.

Definition 1. *A* shift graph *$G = (V, E)$ with parameters $m, t \in \mathbb{N}_+$ and $t \geq m$, denoted as $SG(m, t)$, is defined as follows. Each vertex of G is labeled by an m-dimensional vector such that each dimension has t symbols and no two dimensions have the same symbol appeared in the label. That is, $V = \{(x_1, x_2, \ldots, x_m) \mid x_i \in [t]$*

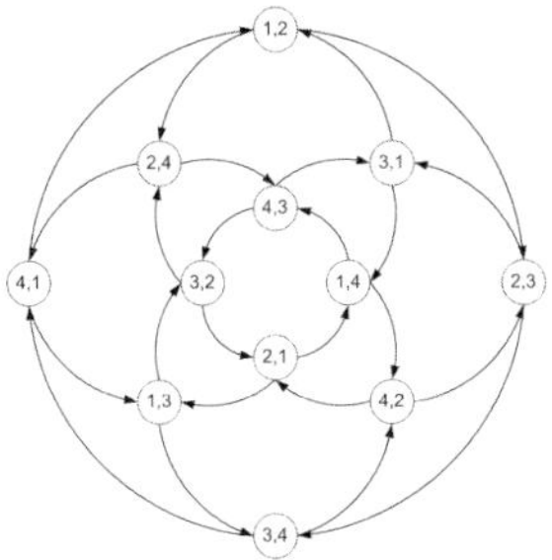

Fig. 2. Non-rotational shift graph $SG_{nr}(2, 4)$

for all $i \in [m]$, *and* $x_i \neq x_j$ *for all* $i, j \in [m], i \neq j\}$. *A vertex* u *has a directed edge pointing to a vertex* v *if we can obtain* v*'s label by shifting* u*'s label to the left by one digit and appending the last digit on the right. That is,* $E = \{(u, v) \mid u, v \in V, u[2 : m] = v[1 : (m-1)]\}$, *where* $u[i : j]$ *denote the sub-vector* $(x_i, x_{i+1}, \ldots, x_j)$ *with* $u = (x_1, x_2, \ldots, x_m)$.

In the shift graph $SG(m, t)$, we know that the number of vertices is $n = t \cdot (t - 1) \cdots (t - m + 1) = t!/(t - m)!$, and each vertex has out-degree $t - m + 1$. Notice that the definition requires that m dimensions have all different symbols. If they are allowed to be the same, then the graphs are the well-known De Bruijn graphs, whereas if we require only that the two adjacent dimensions have different symbols, the graphs are Kautz graphs, which are iterative line graphs of complete graphs.

Definition 2. *A* non-rotational shift graph *with parameter* $m, t \in \mathbb{N}+$ *and* $t \geq m + 1$, *denoted as* $SG_{nr}(m, t)$, *is a shift graph with the further constraint that if* (u, v) *is an edge, then* v*'s label is not a rotation of* u*'s label to the left by one digit. That is,* $E = \{(u, v) \mid u, v \in V, u[2 : m] = v[1 : (m-1)]$ *and* $u[1] \neq v[m]\}$, *where* $u[i]$ *denotes the* i*-th element of* u.

Graph $SG_{nr}(m, t)$ also has $t!/(t-m)!$ vertices but the out-degree of every vertex is $t - m$. A simple non-rotational shift graph $SG_{nr}(2, 4)$ is given in Figure 2 as an example. Non-rotational shift graphs is a class of vertex-transitive graphs that are Eulerian and strongly connected. More importantly, they have one property that makes them Nash equilibria of ℓ-B^{3}C games, as we now explain.

We say that a vertex v in a graph G is *ℓ-path-unique* if any path that passes through v (neither starting nor ending at v) with length no more than ℓ is the unique shortest path from its starting vertex to its ending vertex. A graph is *k-out-regular* if every vertex in the graph has out-degree k. A k-out-regular graph is an *ℓ-path-unique graph* (or *ℓ-PUG* for short) if every vertex in the graph is ℓ-path-unique.

Lemma 4. *Non-rotational shift graph* $SG_{nr}(\ell, k + \ell)$ *is an* ℓ*-PUG.*

Lemma 5. *If a directed graph* G *has* n *nodes and is* k*-out-regular and* ℓ*-path-unique, then* G *is a maximal Nash equilibrium for the uniform* ℓ*-B*^{3}*C game with parameter* n *and* k.

With the above results, we immediately have

Theorem 7. *For any $\ell \geq 2$, $\ell' \geq \ell$, $k \in \mathbb{N}_+$, graph $SG_{nr}(\ell', k+\ell')$ is a maximal Nash equilibrium of the uniform ℓ-B^{3}C game with parameters $n = (k+\ell')!/k!$ and k.*

The above construction of maximal Nash equilibria is based on path-unique graphs. Next we show that shift graphs also lead to another family of Nash equilibria not based on path uniqueness. In fact, we show that they are strict Nash equilibria for uniform ℓ-B^{3}C games for every $\ell \geq 2$ as well as B^{3}C games without path length constraint.

Definition 3. *Given a graph $G = (V, E)$, a* vertex-duplicated graph *$G' = (V', E')$ of G with parameter $d \in \mathbb{N}_+$, denoted as $D(G, d)$, is a new graph such that each vertex of G is duplicated to d copies, and each duplicate inherits all edges incident to the original vertex. That is, $V' = \{(v, i) \mid v \in V, i \in [d]\}$, and $E' = \{((u, i), (v, j)) \mid u, v \in V, (u, v) \in E, i, j \in [d]\}$.*

Theorem 8. *For any $t \geq 2$, $d \geq 2$, graph $D(SG(2, t), d)$ is a strict Nash equilibrium of the uniform ℓ-B^{3}C game with parameters $n = dt(t-1)$ and $k = d(t-1)$. It is also a strict Nash equilibrium of the uniform B^{3}C game without the path length constraint.*

In the simple case of $t = 2$, graph $D(SG(2, 2), d)$ is the complete bipartite graph with d nodes on each side. For larger t, $D(SG(2, t), d)$ is a t-partite graph with more complicated structure. When $d = 2$, we have $n = 2t(t-1)$ and $k = 2(t-1)$. Thus, we have found a family of strict Nash equilibria with $k = \Theta(\sqrt{n})$.

An important remark is that when $d \geq 2$, each node is split into at least two nodes inheriting all incoming and outgoing edges, and thus graphs $D(SG(2, t), d)$ for all $t \geq 2$ and $d \geq 2$ are not ℓ-PUGs for any $\ell \geq 2$. Therefore, the construction by splitting nodes in shift graphs $SG(2, t)$ are a new family of construction not based on path-unique graphs.

5.2 Properties of Nash Equilibria

From Lemma 5, we learn that ℓ-PUGs are good sources for maximal Nash equilibria for uniform ℓ-B^{3}C games. Thus we start by looking into the properties of ℓ-PUGs to obtain more ways of constructing Nash equilibria. The following lemma provides a few ways to construct new ℓ-PUGs given one or more existing ℓ-PUGs.

Lemma 6. *Suppose that G is a k-out-regular ℓ-PUG. The following statements are all true:*

(1) If G' is a k'-out-regular subgraph of G for some $k' \leq k$, then G' is an ℓ-PUG.

(2) Let v be a node of G and $\{v_1, v_2, \ldots, v_k\}$ be v's k outgoing neighbors. We add a new node u to G to obtain a new graph G'. All edges in G remains in G', and u has k edges connecting to $v_1, v_2, \ldots, v_k$. Then G' is also an ℓ-PUG.

(3) If G' is another k-out-regular ℓ-PUG and G' does not shared any node with G, then the new graph G'' simply by putting G together with G' is also an ℓ-PUG.

Lemma 6 has several important implications. First, by repeatedly applying Lemma 6 (2) on an existing ℓ-PUG, we can obtain an ℓ-PUG with an arbitrary size. Combining it with Theorem 7, it immediately implies the following theorem.

Theorem 9. *For any $\ell \geq 2$, $k \in \mathbb{N}_+$, and $n \geq (k+\ell)!/k!$, there is a maximal Nash equilibrium in the uniform ℓ-B^{3}C game with parameters n and k.*

Next, Lemma 6 implies that Nash equilibria of uniform ℓ-B^{3}C games could be disconnected, or weakly connected, or have very unbalanced in-degrees or betweenness among nodes, which implies that there exist rich structures among Nash equilibria.

Finally, we investigate non-PUG maximal Nash equilibria in the uniform 2-B^{3}C game with parameters (n, k), which by Theorem 5 is the most interesting case since its best response computation is polynomial. We want to see that when we fix k, whether we can find non-PUG maximal Nash equilibria for arbitarily large n. Let $maxInd(G)$ denotes the maximum in-degree in graph G. The following result provides the condition under which all maximal Nash equilibria are PUGs.

Theorem 10. *Let G be a k-out-regular graph with n nodes. If $maxInd(G) \leq \frac{n-k}{k^2+k+1}$, then G is a maximal Nash equilibrium for the uniform 2-B^{3}C game with parameter n and k if and only if G is a 2-PUG.*

The above theorem implies that non-PUG equilibria is only possible if $maxInd(G) = \Theta(n)$ when k is a constant, which means that non-PUG equilibria must have very unbalanced in-degrees when n is large. In [2], we show an example of how to construct such a non-PUG equilibria for arbitrarily large n when $k = 2$.

Theorem 10 can also be used to eliminate some families of graphs with balanced in-degrees as maximal Nash equilibria. For example, in [2], we show that when $n \geq k^3 + k^2 + 2k$, a family of symmetrical graphs called Abelian Cayley graphs cannot be maximal Nash equilibria for uniform 2-B^{3}C games.

6 Future Work

There are a number of directions to continue the study of B^{3}C games. First, besides the Nash equilibria we found in the paper, there are other Nash equilibria in the uniform games, some of which have been found by our experiments. We plan to further search for other Nash equilibrium structures and more properties of Nash equilibria. Second, we may also look into other variants of the game and solution concept, such as undirected connections or approximate Nash equilibria. Another direction is to study beyond betweenness definitions based on shortest paths, e.g. betweenness definitions based on network flows or random walks. This can be coupled with enriching the strategy set of the nodes to include fractional weighted edges.

Acknowledgment. We would like to thank Heiko Röglin for his helpful discussions on this research topic.

This work was supported in part by the National Natural Science Foundation of China Grant 60553001, and the National Basic Research Program of China Grants 2007CB807900,2007CB807901.

References

1. Albers, S., Eilts, S., Even-Dar, E., Mansour, Y., Roditty, L.: On nash equilibria for a network creation game. In: Proceedings of the 17th ACM-SIAM Symposium on Discrete Algorithms, pp. 89–98 (2006)
2. Bei, X., Chen, W., Teng, S.-H., Zhang, J., Zhu, J.: Bounded budget betweenness centrality game for strategic network formations, Tech. Report MSR-TR-2009-78, Microsoft Research (June 2009)
3. Berno, B.: Network formation with closeness incentives. In: Naimzada, A.K., Stefani, S., Torriero, A. (eds.) Networks, Topology and Dynamics. Springer, Heidelberg (2008)
4. Burt, R.S.: Structural holes: The social structure of competition, Harvard University Press (1992)
5. Burt, R.S.: Secondhand brokerage: Evidence on the importance of local structure for managers, bankers, and analysts. The Academy of Management Journal 50, 119–148 (2007)
6. Buskens, V., van de Rijt, A.: Dynamics of networks if everyone strives for structural holes. American Journal of Sociology 114(2), 371–407 (2008)
7. Cohen, B.: Incentives build robustness in BitTorrent. In: 1st Workshop on Economics of Peer-to-Peer Systems (2003)
8. Corbo, J., Parkes, D.C.: The price of selfish behavior in bilateral network formation. In: Proceedings of the 24th ACM Symposium on Principles of Distributed Computing, pp. 99–107 (2005)
9. Fabrikant, A., Luthra, A., Maneva, E., Papadimitriou, C.H., Shenker, S.: On a network creation game. In: Proceedings of the 22nd ACM Symposium on Principles of Distributed Computing, pp. 347–351 (2003)
10. Freeman, L.: Centrality in social networks: conceptual clarification. Social Networks 1, 215–239 (1979)
11. Goyal, S., Vega-Redondo, F.: Structural holes in social networks. Journal of Economic Theory 137(1), 460–492 (2007)
12. Jackson, M., Wolinsky, A.: A strategic model of social and economic networks. Journal of Economic Theory 71(1), 44–74 (1996)
13. Kleinberg, J., Suri, S., Tardos, É., Wexler, T.: Strategic newtwork formation with structural holes. In: Proceedings of the 9th ACM Conference on Electronic Commerce (2008)
14. Laoutaris, N., Poplawski, L., Rajaraman, R., Sundaram, R., Teng, S.-H.: Bounded budget connection (bbc) games or how to make friends and influence people, on a budget. In: Proceedings of the 27th ACM Symposium on Principles of Distributed Computing (2008)
15. Moscibroda, T., Schmid, S., Wattenhofer, R.: On the topologies formed by selfish peers. In: Proceedings of the 25th ACM Symposium on Principles of Distributed Computing, pp. 133–142 (2006)
16. Newman, M.E.J., Park, J.: Why social networks are different from other types of networks. Physical Review E 68, 36122 (2003)
17. Papadimitriou, C.H.: Algorithms, games, and the internet. In: Proceedings of the 33rd ACM Symposium on Theory of Computing, 2001, Invited talk, pp. 749–753 (2001)
18. Scott, J.: Social network analysis. Sage, Thousand Oaks (1991)
19. Stutzbach, D., Sen, S.: Characterizing unstructured overlay topologies in modern p2p file-sharing systems. In: ACM/USENIX Internet Measurement Conference (2005)

Exact and Approximate Equilibria for Optimal Group Network Formation

Elliot Anshelevich and Bugra Caskurlu

Computer Science Department, RPI, 110 8th Street, Troy, NY 12180
{eanshel,caskub}@cs.rpi.edu

Abstract. We consider a process called Group Network Formation Game, which represents the scenario when strategic agents are building a network together. In our game, agents can have extremely varied connectivity requirements, and attempt to satisfy those requirements by purchasing links in the network. We show a variety of results about equilibrium properties in such games, including the fact that the price of stability is 1 when all nodes in the network are owned by players, and that doubling the number of players creates an equilibrium as good as the optimum centralized solution. For the most general case, we show the existence of a 2-approximate Nash equilibrium that is as good as the centralized optimum solution, as well as how to compute good approximate equilibria in polynomial time. Our results essentially imply that for a variety of connectivity requirements, giving agents more freedom can paradoxically result in more efficient outcomes.

1 Introduction

Many modern computer networks, including the Internet itself, are constructed and maintained by self-interested agents. This makes network design a fundamental problem for which it is important to understand the effects of strategic behavior. Modeling and understanding of the evolution of nonphysical networks created by many heterogonous agents (like social networks, viral networks, etc.) as well as physical networks (like computer networks, transportation networks, etc.) has been studied extensively in the last several years. In networks constructed by several self-interested agents, the global performance of the system may not be as good as in the case where a central authority can simply dictate a solution; rather, we need to understand the quality of solutions that are consistent with self-interested behavior. Much research in the theoretical computer science community has focused on this performance gap and specifically on the notions of the *price of anarchy* and the *price of stability* — the ratios between the costs of the worst and best Nash equilibrium[1], respectively, and that of the globally optimal solution.

[1] Recall that a (pure-strategy) Nash equilibrium is a solution where no single player can switch her strategy and become better off, given that the other players keep their strategies fixed.

A. Fiat and P. Sanders (Eds.): ESA 2009, LNCS 5757, pp. 239–250, 2009.

In this paper, we study a network design game that we call the *Group Network Formation Game*, which captures the essence of strategic agents building a network together in many scenarios. In this game players correspond to nodes of a graph (although not all nodes need to correspond to players), and the players can have extremely varied connectivity requirements. For example, there might be several different "types" of nodes in the graph, and a player desires to connect to at least one of every type (so that this player's connected component forms a Group Steiner Tree [10]). Or instead, a player might want to connect to at least k other player nodes. The first example above is useful for many applications where a set of players attempt to form groups with "complementary" qualities. The second example corresponds to a network of servers where each server want to be connected to at least k other servers so that it can have a backup of its data; or in the context of IP networks, a set of ISPs that want to increase the reliability of the Internet connection for their customers, and so decide to form multi-homing connections through k other ISPs [21]. Many other types of connectivity requirements fit into our framework, and so the results we give in this paper will be relevant to many different types of network problems.

We now formally define the *Group Network Formation Game* as follows. Let an undirected graph $G = (V, E)$ be given, with each edge e having a nonnegative cost $c(e)$. This graph represents the possible edges that can be built. Each player i corresponds to a single node in this graph (that we call a *player* or *terminal* node), which we will also denote by i. Similarly to [2], a strategy of a player is a payment vector p_i of size $|E|$, where $p_i(e)$ is how much player i is offering to contribute to the cost of edge e. We say that an edge e is *bought*, i.e., it is included in the network, if the sum of payments of all the players for e is at least as much as the cost of e ($\sum_i p_i(e) \geq c(e)$). Let G_p denote the subgraph of bought edges corresponding to the strategy vector $p = (p_1, \ldots, p_N)$. G_p is the outcome of this game, since it is the network which is purchased by the players.

To define the utilities/costs of the players, we must consider their connectivity requirements. Group Network Formation Game considers the class of problems where the players' connectivity requirements can be compactly represented with a function $F : 2^U \rightarrow \{0, 1\}$, where $U \subseteq V$ is the set of player nodes, similar to [11]. This function F has the following meaning. If S is a set of terminals, then $F(S) = 1$ iff the connectivity requirements of all players in S would be satisfied if S formed a connected component in G_p. For the example above, where each player wants to connect to at least one player from each "type", the function $F(S)$ would evaluate to 1 exactly when S contains at least one player of each type. Similarly, for the "data backup" example above, the function $F(S)$ would evaluate to 1 exactly when S contains at least $k + 1$ players. In general, we will assume that the connectivity requirements of the players are represented by a monotone "happiness" function F. The monotonicity of F means that if the connectivity requirements of a player are satisfied in a graph G_p, then they are still satisfied when a player is connected to strictly more nodes. We will call a set of player nodes S a "happy" group if $F(S) = 1$. While not all connectivity requirements can be represented as such a function, it is a reasonably general class that includes the examples given above. Therefore an instance of our game consists of a graph $G = (V, E)$, player nodes $U \subseteq V$, and a function F that states the connectivity requirements of the players. We will say that player i's connectivity requirements are *satisfied* in G_p if and only if $F(S_i(G_p)) = 1$ for $S_i(G_p)$ being the terminals in i's connected component of G_p. While required to connect to a set of terminal nodes satisfying its connectivity requirements, each player also tries to minimize her total payments, $\sum_{e \in E} p_i(e)$ (which we will denote by $|p_i|$). We conclude the definition of our game by defining the cost function for each player i as:

- $cost(i) = \infty$ if $F(S_i(G_p)) = 0$
- $cost(i) = \sum_{e \in E} p_i(e)$ otherwise.

In our game, all players want to be a part of a happy group which can correspond to many connectivity requirements, some of which are mentioned above. The socially optimal solution (which we denote by OPT) for this game is the cheapest possible network where every connected component is a happy group, since this is the solution maximizing social welfare[2]. For our first example above, OPT corresponds to the cheapest forest where every component is a Group Steiner Tree, for the second to the Terminal Backup problem [3], and in general it can correspond to a variety of constrained forest problems [11]. Our goals include understanding the quality of exact and approximate Nash equilibria by comparing them to OPT, and thereby understanding the efficiency gap that results because of the players' self-interest. By studying the price of stability, we also seek to reduce this gap, as the best Nash equilibrium can be thought of as the best outcome possible if we were able to suggest a solution to all the players simultaneously.

In the Group Network Formation Game, we don't assume the existence of a central authority that designs and maintains the network, and decides on appropriate cost-shares for each player. Instead we use a cost-sharing scheme which is sometimes referred to as "arbitrary cost sharing" [2,8] that permits the players to specify the actual amount of payment for each edge. This cost-sharing mechanism is necessary in scenarios where very little control over the players is available, and gives more freedom to players in specifying their strategies, i.e., has a much larger strategy space. The main advantage of such a model is that the players have more freedom in their choices, and less control is required over them. A disadvantage of such a system, however, is that it does not guarantee the existence of Nash equilibria (unlike more constrained systems such as fair sharing [1]). Studying the existence of Nash equilibria under arbitrary cost sharing has been an interesting research problem and researchers have proven existence for many important games [2,8,12,13]. Interestingly, in many of these problems it has been shown that the equilibrium is indeed cheap, i.e., costs as much as the socially optimal network. As we show in this paper, this tells us that in the network design contexts we consider, *arbitrary sharing produces more efficient outcomes while giving the players more freedom.*

[2] The solution that maximizes the social welfare is the one that minimizes the total cost of all the players.

Related Work. Over the last few years, there have been several new papers using arbitrary cost-sharing, e.g., [8,13,14]. Recently, Hoefer [12] proved some interesting results for a generalization of the game in [2], and considered arbitrary sharing in variants of Facility Location.

Unquestionably one of the most important decisions when modeling network design involving strategic agents is to determine how the total cost of the solution is going to be split among the players. Among various alternatives [6], the "fair sharing" mechanism is the most relevant to ours [1,4,5,9]. In this cost sharing mechanism, the cost of each edge of the network is shared equally by the players using that edge. This model has received much attention, mostly because of the following three reasons. Firstly, it nicely quantifies what people mean by "fair" and has an excellent economic motivation since it is strongly related to the concept of Shapley value[1]. Secondly, fair sharing naturally models the congestion effects of network routing games, and so network design games with fair sharing fall into the well-studied class of "congestion games" [4,7,15,20]. Thirdly, this model has many attractive mathematical properties including guarantees on the existence of Nash equilibrium that can be obtained by natural game playing [1].

Despite all of the advantages of congestion games mentioned above, there are extremely important disadvantages as well. Firstly, although congestion games are guaranteed to have Nash equilibria, these equilibria may be very expensive. Anshelevich et al. [1] showed that the cheapest Nash equilibrium solution can be $\Omega(\log n)$ times more expensive than OPT, and that this bound is tight. As we prove in this paper, arbitrary cost-sharing will often guarantee the existence of Nash equilibria that are as cheap as the optimal solution. Secondly, fair sharing inherently assumes the existence of a central authority that regulates the agent interactions or determines the cost shares of the agents, which may not be realistic in many network design scenarios. Arbitrary cost sharing allows the agents to pick their own cost shares, without any requirements by the central authority. Thirdly, although the players are trying to minimize their payments in fair cost sharing, they are not permitted to adjust their payments freely, i.e., a player cannot directly specify her payments on each edge, but is rather asked to specify which edges she wants to use. In the network design contexts that we consider here, we prove that giving players more freedom can often result in better outcomes.

The research on non-cooperative network design and formation games is too much to survey here, see [16,18,20] and the references therein.

Our Results. Our main results are about the existence and computation of cheap approximate equilibria. By an α-approximate Nash equilibrium, we mean that no player in such a solution has a deviation that will improve their cost by a factor of more than α. While our techniques are inspired by [2], our problem and connectivity requirements are much more general, and so require the development of much more general arguments and payment schemes.

- In Section 3, we show that in the case where all nodes are player nodes, there exists a Nash equilibrium as good as OPT, i.e., the price of stability is 1.
- In Section 4, we show that in the general case where some nodes may not be player nodes, there exists a 2-approximate Nash equilibrium as good as OPT.
- We show that if every player is replaced by two players (or if every player node has at least two players associated with it), then the price of stability is 1. This is in the spirit of similar results from selfish routing [1,20], where increasing the total amount of players reduces the price of anarchy.
- Starting with a β-approximation to OPT, we provide poly-time algorithms for computing an $(1+\epsilon)$-approximate equilibrium with cost no more than β times OPT, for the case where all nodes are player nodes. The same holds for the general case with the factor being $(2+\epsilon)$ instead.

Since for monotone happiness functions F, OPT corresponds to a constrained forest problem [11], then the last result gives us a poly-time algorithm with $\beta = 2$. Notice that we assumed that the function F is monotone, i.e., that the addition of more terminals to a component does not hurt. This assumption is necessary, since as we prove in Section 5, if F is not monotone there may not exist *any* approximate Nash equilibria. We also show that the results above are only possible in our model with arbitrary cost-sharing, and not with fair sharing.

Because of its applications to multi-homing [3,21], we are especially interested in the behavior of Terminal Backup connectivity requirements, i.e., when a player node desires to connect to at least k other player nodes. For this special case, we prove a variety of results, such as price of anarchy bounds and the extension of fair sharing results from [1] to this new problem. The lower bounds for Terminal Backup also hold for the general Group Network Formation Game, showing that while the price of stability may be low, the price of anarchy can be as high as the number of players.

2 Properties of the Socially Optimal Network

In this section, we will show some useful properties of the socially optimal network for the Group Network Formation Game, which we refer to as OPT. For notational convenience, we will extend the definition of the happiness function to subgraphs and use $F(S)$ to denote the value of the happiness function for the set of terminal nodes in a subgraph S.

Observation 1. *Since the satisfaction of the players only depends on the terminal nodes they are connected to, OPT is acyclic and therefore, OPT is the minimum cost forest that satisfies all the players.*

Let $e = (i, j)$ be an arbitrary edge of a tree T of OPT. Removal of e will divide T into 2 subtrees, namely T_i and T_j (let T_i be the tree containing node i). After removal of e, connection requirements of some of the players in T will be dissatisfied, i.e., either $F(T_i) = 0$ or $F(T_j) = 0$, since otherwise $OPT - e$ would be a network that is cheaper than OPT and satisfies all the players. Therefore, once e is deleted from OPT, all the players in T_i or T_j or both will be dissatisfied. The players that are dissatisfied upon removal of e are said to *witness* e. If e is witnessed by only the players in T_i or only the players in T_j then e is said to be an edge *witnessed from* 1*-side*. Analogously, we say e is *witnessed from* 2*-sides* if it is witnessed by all the players in T.

In general, some of the edges of a tree T may be witnessed from 1-side whereas some others are witnessed from 2-sides. In the full version of the paper, we show that the edges of T witnessed from 2-sides form a connected component in T. Due to limited space, all our proofs are omitted but the full version of the paper is available online at www.cs.rpi.edu/∼eanshel.

3 When All Nodes Are Terminals

For the *Group Network Formation Game*, we don't know whether there exists an exact Nash equilibrium for all possible instances of the problem. However, for the special case where each node of G is a terminal node, we prove that Nash equilibrium is guaranteed to exist. Specifically, there exists a Nash equilibrium whose cost is as much as OPT, and therefore price of stability is 1. In this section, we will prove this result by explicitly forming the stable payments on the edges of OPT by giving a payment algorithm. The payment algorithm, which will be formally defined below, loops through all the players and decides the payments of them for all their incident edges. The algorithm never asks a player i to pay for the cost of an edge e that is not incident to i.

Since we are trying to form a Nash equilibrium, no player should have an incentive of unilateral deviation when the algorithm terminates. To have an easier analysis we want our algorithm to have a stronger property: we not only want it to ensure stability at termination but also at each intermediate step. To ensure this stronger property, whenever a player i is assigned to make a payment for an edge e during the execution of the algorithm, it should compute $\chi_i(p_i)$, the cheapest deviation of player i from p_i in $G - e$ that satisfies her (assuming the rest of the payments to buy OPT are made by other players), and should ensure that the cost of p_i never exceeds the cost of $\chi_i(p_i)$ at each iteration. The payment for all the edges of OPT will be decided when the algorithm terminates and we will conclude that the resulting strategy profile is a Nash equilibrium since the cost of the strategy p_i of each player i will be at most her cheapest deviation $\chi_i(p_i)$ with respect to p_i.

Let p^* be a strategy vector that buys all the edges of $OPT - e$, i.e., the entry of $p^*(f) = c(f)$ if f is in $OPT - e$ and $p^*(f) = 0$ otherwise. The deviation $\chi_i(p_i)$ is the cheapest strategy of player i that satisfies her connectivity requirements assuming $\sum_{j \neq i} p_j = p^* - p_i$. Observe that all edges of OPT such that i is not contributing any payment to them can be used by i freely in $\chi_i(p_i)$. Therefore, when computing $\chi_i(p_i)$, the algorithm should not use the actual cost of the edges in $G - e$, but instead for each edge f it should use the cost i would face if she is to use f. We call this the *modified cost of f for i*, and denote it by $c'(f)$. Specifically, for f not in OPT, $c'_i(f) = c(f)$, the actual cost of f. For the edges f of $OPT - e$ that i has

```
Input: The socially optimal network OPT
Output: The payment scheme for OPT
Initialize p_i(e) = 0 for all players i and edges e;
Root each tree T of OPT by an arbitrary node incident to an edge
witnessed from 2-sides;
Loop through all trees T of OPT;
    Loop through all nodes i of T in reverse BFS order;
        Loop through all edges of T_i incident to i;
            Let d(e) = c(e) - Σ_{j≠i} p_j(e);
            If χ_i(p_i) - Σ_f p_i(f) ≥ d(e)
                Set p_i(e) = d(e);
            Else break;
        Define g to be the parent edge of node i;
        Set p_i(g) = min{χ_i(p_i) - Σ_f p_i(f), c(g)};
```

Algorithm 1. Algorithm that generates payments on the edges of OPT

not contributed anything to (i.e., $p_i(f) = 0$), we have that $c'_i(f) = 0$, since from i's perspective, she can use these edges for free because other players have paid for them. For all the other edges f that i is paying $p_i(f)$ for, $c'_i(f) = p_i(f)$, since that is how much it costs for i to use f in her deviation from the payment strategy p_i. We use the notation $\chi_i(p_i)$ for both the deviation itself and also the cost of it; in what follows the meaning will be clear from the context.

Recall that the algorithm asks the players to pay for their incident edges only. Therefore, each edge is considered for payment twice. For each edge $e = (u, v)$ where u is the parent of v, first v is asked to pay for e at the maximum amount that will not create an incentive for unilateral deviation for her. At the later iterations of the algorithm, when u is processed, the algorithm asks u to pay for the remaining cost of e. Recall that whenever the algorithm asks a player to contribute to the cost of an edge it also computes her cheapest deviation and ensures that no player makes a payment that will create an incentive of unilateral deviation. Therefore, if the payment algorithm does not break at any of the intermediate stages, then it finds a Nash equilibrium whose cost is as much as OPT. To prove our result all we need to do is prove that the algorithm never breaks at an intermediate stage. We prove this by constructing a network cheaper than OPT which satisfies all the players whenever the algorithm breaks, thus forming a contradiction in the full version of the paper.

4 Good Equilibria in the General Game

In Section 3, we saw that a good equilibrium always exists when all nodes are terminals. In this section, we consider the general Group Network Formation Game, and show that there always exists a 2-approximate Nash equilibrium that is as cheap as the centralized optimum. By a 2-approximate Nash equilibrium, we mean a strategy profile $p = (p_1, p_2, \ldots, p_n)$ such that no player i can reduce her cost by more than a factor of 2 by unilaterally deviating from p_i to p'_i, i.e., $|p'_i| > |p_i/2|$ for any unilateral deviation p'_i of i. To prove this, we first look at an important special case that we call the *Group Network Formation of Couples Game* or *GNFCG*. This game is exactly the same as the Group Network Formation Game, except that every terminal node is guaranteed to have at least two players located at that node (although not all nodes need to be player nodes).

Theorem 1. *If the price of stability for the GNFCG is* 1 *then there exists a* 2*-approximate Nash Equilibrium for the Group Network Formation Game that costs as much as OPT.*

Because of Theorem 1, we will focus on the GNFCG in the rest of the section and prove the existence of a Nash equilibrium as cheap as OPT. This result is interesting in its own right, since it states that to form an equilibrium that is as good as the optimum solution, it is enough to double the number of players. Such results are already known for many variants of congestion games and selfish routing [1,20], but as Theorem 2 shows, we can also prove such results for games with arbitrary sharing.

Given a set of bought edges T; a strategy profile p such that for all players i, p_i is the cheapest strategy satisfying i, assuming rest of the payments to buy all the edges of T are made by other players, is a Nash equilibrium. To prove that price of stability is 1 for GNFCG, we give an algorithm that forms such a strategy profile on the edges of OPT.

Recall that the payment strategies of all the players have to be stable when the algorithm terminates. As in Section 3, to have an easier analysis we not only want our algorithm to ensure stability at termination but also at each intermediate step. To ensure this stronger property, whenever a player i is assigned to make a payment for an edge e during the execution of the algorithm, it should compute $\chi_i(p_i)$, the cheapest deviation of player i from p_i that satisfies her, and should ensure that the cost of p_i never exceeds the cost of $\chi_i(p_i)$ at each iteration by using the modified costs of the edges as in Section 3. In the rest of the section we prove our main theorem for the GNFCG.

Theorem 2. *For GNFCG, there exists a Nash equilibrium as cheap as the socially optimal network, i.e., the price of stability is* 1.

For ease of explanation, we will first consider the case where all the edges of OPT are witnessed from two sides and later illustrate how our algorithm can be modified for the case where some of the edges are witnessed from one side only. We start by rooting each connected component of OPT arbitrarily by a high degree non-player node. Throughout the paper, the term *high degree node* refers to the nodes with degree 3 or more. On each connected component T of OPT, we run a 2-phase algorithm. In the first phase of the algorithm, we assign players to make payments to the edges of T in a bottom-up manner, i.e., we start from a lowest level edge e of T and pick a player i to make some payment for e and continue with the next edge in the reverse BFS order. In the first phase of the algorithm, we ask a player i to contribute only for the cost of edges on the

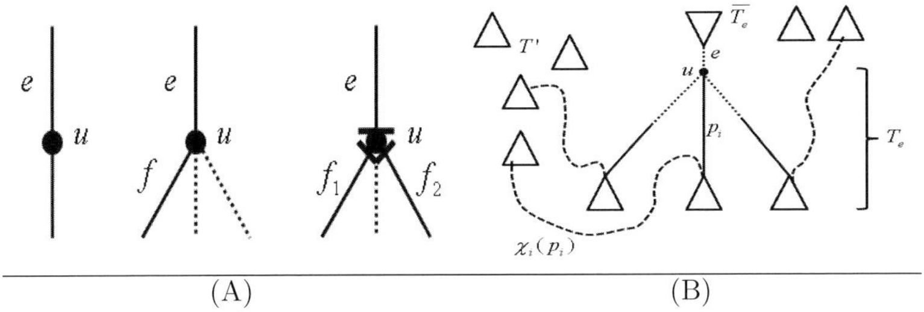

Fig. 1. (A) Illustrates the assignment of the player to pay for the cost of e. (B) Shows how to construct a cheap network that satisfies all the players in T_e by using the deviations of a subset S of them.

unique path between her and the root and furthermore, the payment for each edge is made by only one player.

Algorithm (Phase 1). For an arbitrary edge $e = (u, v)$ where u is the lower level incident node of e, the assignment of the player to pay for e is as follows. If u is a terminal node, we ask a player i located at node u to make maximum amount of payment on e that will not make p_i unstable, i.e., we set $p_i(e) = \min\{\chi_i(p_i) - |p_i|, c(e)\}$. If u is a degree 2 nonterminal node then we ask the player who has completely bought the other incident edge of u, i.e., made a payment equal to $c(e)$, to make maximum amount of payment on e that will not make her strategy unstable as shown on the left of Figure 1(A). Note that it may be the case that no player has bought the other incident edge of u in which case we don't ask any player to pay for e and the payment for e will be postponed to the second phase of the algorithm. If u is a high degree nonterminal then the selection of the player to pay for e is based on the number of lower level incident edges of u that are bought in the previous iterations of the algorithm. If none of the lower level incident edges of u are bought then we postpone the payment on e to the second phase of the algorithm. If exactly one of the lower level incident edges of u, namely f, is bought then we ask the player who bought f to make maximum amount of payment on e that will not make her strategy unstable as shown in the middle of Figure 1(A). If 2 or more of the lower level incident edges of u are already bought, namely $f_1, f_2, \ldots, f_l$, then we fix the strategies of the players $i_1, i_2, \ldots, i_l$ that bought those edges, i.e., the players $i_1, i_2, \ldots, i_l$ are not going to pay any more and therefore the strategies of those players that will be returned at the end of the algorithm are already determined. Since there are two players located at every terminal, pick an arbitrary player located at the same terminal as one of $i_1, i_2, \ldots, i_l$ that has not made any payments yet, and assign her to make maximum amount of payment for e that will not make her strategy unstable as shown on the right of Figure 1(A). We prove in the full version that such a player always exists, i.e., not all of $i_1, i_2, \ldots, i_l$ are the last players to make payment at their respective terminal nodes.

We here present an outline of our analysis of this algorithm. When we are talking about a player i, let T denote the connected component of OPT containing i and let T' denote the set of other connected components of OPT. For an arbitrary edge e of T, we use T_e in order to refer to the subtree of T below e and T_u to refer to the subtree below a node u. To prove the existence of a Nash equilibrium as cheap as OPT, we show that whenever our algorithm cannot form stable payments on the edges of OPT we can find a subgraph of G that is cheaper than OPT and satisfies all the players. Since OPT is the cheapest network satisfying all the players, we will end up with a contradiction.

We give a series of lemmas in the full version that successively proves the following. For every edge e that could not be bought in the first phase of the algorithm by the assigned player to make payment for it, we can connect all the terminal nodes in T_e to the connected components of T' *without using any of the edges of* $T - T_e$ by simply setting $p_i = \chi_i(p_i)$ for a subset S of players in T_e. The deviations of the subset S of the players are depicted in Figure 1(B). The condition that no edges of $T - T_e$ are used by the deviations is crucial, since that is what allows us to have a set of players all deviate at once and still be satisfied afterwards. The fact that such a "re-wiring" exists allows us to argue in our proofs that at least one of the incident edges of the root of T will be bought during the first phase of the algorithm.

Algorithm (Phase 2). In the second phase of the algorithm, we ask the players that have not made any payments yet to make stable payments for the remaining edges and buy them. Let Γ be the set composed of connected components of $G_p - T'$ that include at least one terminal node. In other words, Γ consists of connected components of the edges in T purchased so far by the algorithm (a single terminal node with no adjacent bought edges would also be a connected component in Γ). We call a connected component $C_1 \in \Gamma$ *immediately below* a connected component $C \in \Gamma$ if after contracting the components in Γ, C is above C_1 in the resulting tree and there are no other components of Γ between them. In the second phase of the algorithm, we form payments on the edges in a top-down manner as we explain next. We start from the connected component $C \in \Gamma$ that includes the root of T and assign a player i in C that has not made any payments yet to buy *all* the edges between C and the connected components that are immediately below C. We prove that such a player i always exists in the full version of the paper. Observe that once i buys all the edges between C and the connected components $C_1, C_2, \ldots, C_k$ that are immediately below C, all these $k+1$ connected components form a single connected component C that contains the root. We repeat this procedure, i.e., pick a player i in the top-most connected component C that has not made a payment yet to buy all the edges between C and the connected components that are immediately below C, until all the players in T are in the same connected component and all of T is paid for.

To show that our algorithm forms an equilibrium payment, we need to prove that no player has a deviation from the payment assigned to her. This is true for players making payments during the first phase by construction. To finish the proof, we need to show that a strategy p_i that buys all the edges between a connected component C and the connected components $C_1, C_2, \ldots, C_k$ that are immediately below C is a stable strategy for any player in C, which we show in the full version of the paper.

This concludes the proof of Theorem 2. Recall that for ease of explanation, we only considered the case where all edges of OPT are witnessed from two sides until now. In the full version of the paper, we modify this algorithm to return a Nash equilibrium that purchases OPT even if some of the edges of OPT are witnessed from one side.

The proof of our 2-approximate Nash equilibrium result suggests an algorithm which forms a cheaper network whenever a 2-approximate Nash equilibrium cannot be found. Using techniques similar to [2], this allows us to form efficient algorithms to compute approximate equilibria:

Theorem 3. *Suppose we have an α-approximate socially optimal graph G_α for an instance of the Group Network Formation Game. Then for any $\epsilon > 0$, there is a polynomial time algorithm which returns a $2(1+\epsilon)$-approximate Nash equilibrium on a feasible graph G', where $cost(G') \leq cost(G_\alpha)$. Furthermore, if all the terminal nodes have an associated player or each terminal node is associated with at least 2 players, there is a polynomial time algorithm which returns a $(1+\epsilon)$-approximate Nash equilibrium on a feasible graph G'.*

Since for all monotone functions F, finding OPT is a constrained forest problem [11], then Theorem 3 gives us a poly-time algorithm for $\alpha = 2$.

5 Inapproximability Results and Terminal Backup

Recall that in this paper, we consider games where the happiness functions are monotone. Theorem 4 shows that this property of happiness functions is critical for even approximate stability.

Theorem 4. *For the Group Network Formation Game where the happiness functions may not be monotone, there is no α-approximate Nash equilibrium for any α.*

Recall that congestion games, including our game with fair sharing, are guaranteed to have Nash equilibria, although they may be expensive. The following theorem studies the quality (cost) of approximate Nash equilibrium and shows that there may not be any approximately stable solution that is as cheap as the socially optimal network.

Theorem 5. *For the Group Network Formation Game, there may not be any approximate Nash equilibrium whose cost is as much as OPT if the fair cost-sharing mechanism is used.*

Because of its applications to multi-homing [3,21], we are especially interested in the behavior of Terminal Backup connectivity requirements, i.e., when a player node desires to connect to at least $k-1$ other player nodes.

Theorem 6. *For the Group Network Formation Game and the Terminal Backup problem, the Price of Anarchy is n and $2k-2$ respectively. Furthermore, these bounds are tight. For the Terminal Backup problem, in the Shapley cost-sharing model, the price of stability is at most $H(2k-2)$.*

References

1. Anshelevich, E., Dasgupta, A., Kleinberg, J., Tardos, É., Wexler, T., Roughgarden, T.: The Price of Stability for Network Design with Fair Cost Allocation. SIAM Journal on Computing 38(4), 1602–1623 (2008)
2. Anshelevich, E., Dasgupta, A., Tardos, É., Wexler, T.: Near-Optimal Network Design with Selfish Agents. Theory of Computing 4, 77–109 (2008)
3. Anshelevich, E., Karagiozova, A.: Terminal Backup, 3D Matching, and Covering Cubic Graphs. In: Proc. 39th ACM Symposium on Theory of Computing (2007)
4. Chekuri, C., Chuzhoy, J., Lewin-Eytan, L., Naor, J., Orda, A.: Non-cooperative multicast and facility location games. In: Proceedings of the 7th ACM Conference on Electronic Commerce (EC), Ann Arbor, Michigan, pp. 72–81 (2006)
5. Chen, H., Roughgarden, T.: Network Design with Weighted Players. In: SPAA 2006 (2006)
6. Chen, H., Roughgarden, T., Valiant, G.: Designing Networks with Good Equilibria. In: SODA 2008 (2008)
7. Christodoulou, G., Koutsoupias, E.: On the price of anarchy and stability of correlated equilibria of linear congestion games. In: Brodal, G.S., Leonardi, S. (eds.) ESA 2005. LNCS, vol. 3669, pp. 59–70. Springer, Heidelberg (2005)
8. Epstein, A., Feldman, M., Mansour, Y.: Strong Equilibrium in Cost-Sharing Connection Games. In: EC 2007 (2007)
9. Fiat, A., Kaplan, H., Levy, M., Olonetsky, S., Shabo, R.: On the Price of Stability for Designing Undirected Networks with Fair Cost Allocations. In: Bugliesi, M., Preneel, B., Sassone, V., Wegener, I. (eds.) ICALP 2006. LNCS, vol. 4051, pp. 608–618. Springer, Heidelberg (2006)
10. Garg, N., Konjevod, G., Ravi, R.: A polylogarithmic approximation algorithm for the group Steiner tree problem. In: SODA 2000 (2000)
11. Goemans, M., Williamson, D.: A General Approximation Technique for Constrained Forest Problems. SIAM Journal on Computing 24, 296–317 (1995)
12. Hoefer, M.: Non-cooperative Facility Location and Covering Games. In: Asano, T. (ed.) ISAAC 2006. LNCS, vol. 4288, pp. 369–378. Springer, Heidelberg (2006)
13. Hoefer, M.: Non-cooperative Tree Creation. In: Královič, R., Urzyczyn, P. (eds.) MFCS 2006. LNCS, vol. 4162, pp. 517–527. Springer, Heidelberg (2006)
14. Hoefer, M., Krysta, P.: Geometric Network Design with Selfish Agents. In: Wang, L. (ed.) COCOON 2005. LNCS, vol. 3595, pp. 167–178. Springer, Heidelberg (2005)
15. Holzman, R., Law-Yone, N.: Strong Equilibrium in congestion games. Games and Economic Behavior 21 (1997)
16. Jackson, M.: A survey of models of network formation: stability and efficiency. In: Demange, G., Wooders, M. (eds.) Group Formation in Economics: Networks, Clubs and Coalitions. Cambridge Univ. Press, Cambridge
17. Monderer, D., Shapley, L.: Potential Games. Games and Economic Behavior 14, 124–143 (1996)
18. Nisan, N., Roughgarden, T., Tardos, É., Vazirani, V.V. (eds.): Algorithmic Game Theory. Cambridge University Press, Cambridge
19. Rosenthal, R.W.: The network equilibrium problem in integers. Networks 3, 53–59 (1973)
20. Roughgarden, T.: Selfish Routing and the Price of Anarchy. MIT Press, Cambridge
21. Xu, D., Anshelevich, E., Chiang, M.: On Survivable Access Network Design: Complexity and Algorithms. In: INFOCOM 2008 (2008)

On the Performance of Approximate Equilibria in Congestion Games*

George Christodoulou[1], Elias Koutsoupias[2], and Paul G. Spirakis[3]

[1] Max-Planck-Institut für Informatik, Saarbrücken, Germany
gchristo@mpi-inf.mpg.de
[2] Department of Informatics, University of Athens, Greece
elias@di.uoa.gr
[3] Computer Engineering and Informatics Department, Patras University, Greece
spirakis@cti.gr

Abstract. We study the performance of approximate Nash equilibria for congestion games with polynomial latency functions. We consider how much the price of anarchy worsens and how much the price of stability improves as a function of the approximation factor ϵ. We give almost tight upper and lower bounds for both the price of anarchy and the price of stability for atomic and non-atomic congestion games. Our results not only encompass and generalize the existing results of exact equilibria to ϵ-Nash equilibria, but they also provide a unified approach which reveals the common threads of the atomic and non-atomic price of anarchy results. By expanding the spectrum, we also cast the existing results in a new light.

1 Introduction

Algorithmic Game Theory has studied extensively and with remarkable success the computational issues of Nash equilibria. As a result, we understand almost completely the computational complexity of exact Nash equilibria (they are PPAD-complete for games described explicitly [16,7] and PLS-complete for games described succinctly [19]). The results indicate a long-suspected drawback of Nash equilibria, namely that they cannot be computed effectively. Despite substantial recent progress [21,19,17,33], it is still open whether approximate Nash equilibria share the same drawback. But in any case, they seem to provide a more reasonable equilibrium concept: It usually makes sense to assume that an agent is willing to accept a situation that is almost optimal to him.

In another direction, a large body of research in Algorithmic Game Theory concerns the degree of performance degradation of systems due to the selfish behavior of the users. Central to this area is the notion of price of anarchy (PoA) [20,25] and the price of stability (PoS) [3]. The first notion compares the social cost of the worst-case equilibrium to the social optimum, which could be obtained if every agent followed obediently a central authority. The second notion is very similar but it considers the best Nash equilibrium instead of the worst one.

* Supported in part by IST-15964 (AEOLUS) and IST-2008-215270 (FRONTS).

A. Fiat and P. Sanders (Eds.): ESA 2009, LNCS 5757, pp. 251–262, 2009.

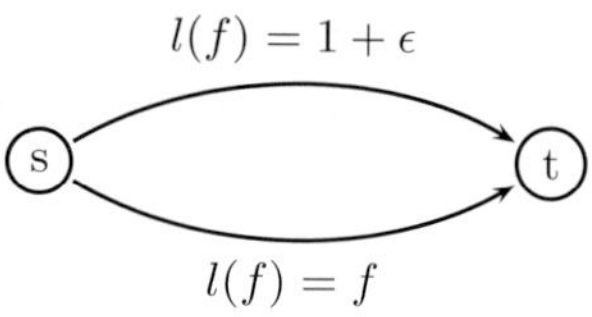

Fig. 1. The Pigou network

A natural question then is how the performance of a system is affected when its users are approximately selfish: What is the *approximate price of anarchy* and the *approximate price of stability*? Clearly, by allowing the players to be almost rational (within an ϵ factor), we expand the equilibrium concept and we expect the PoA to get worse. On the other hand, the PoS should improve. The question is how they change as functions of the parameter ϵ. This is exactly the question that we address in this work.

We study two fundamental classes of games: the class of atomic congestion games [26,23] and the class of non-atomic congestion (or selfish routing) games [5,15,22]. Both classes of games played central role in the development of the area of the PoA [20,27,28]. Although the PoA and PoS of these games for exact equilibria was established long ago [27,11,10,4]—and actually some work [29,32,8] addressed partially the question for the PoA of approximate equilibria—our results add an unexpected understanding of the issues involved.

While the classes of atomic and non-atomic games are conceptually very similar, dissimilar techniques were employed in the past to answer the questions concerning the PoA and PoS. Moreover, the qualitative aspects of the answers were quite different. For instance, the maximum PoA for the non-atomic case is attained by the Pigou network (Figure 1), while for the atomic case it is attained [4,11] by a completely different network (with the structure of Figure 2).

There are two main differences between the atomic and non-atomic games. The first difference is the *"uniqueness"* of equilibria: Non-atomic congestion games have a unique exact Nash (or Wardrop as it is called in these games) equilibrium, and therefore the PoS is not different than the PoA. On the other hand, atomic congestion games may have multiple exact equilibria. Perhaps, because of this, the problem of determining the PoS proved more challenging [10,6] for this case. New techniques needed for upper bounding the PoS for linear latencies which exploited the potential of these games; for polynomial latencies, the problem is still open. Furthermore, the lower bound for linear latencies is quite complicated and, unlike the selfish routing case, it has a dependency on the number of players (it attains the maximum value at the limit).

The second main difference between the two classes of games is *"integrality"*: In the case of atomic games, when a player considers switching to another strategy, he has to take into account the extra cost that he will add to the edges (or facilities) of the new strategy. The number of players on the new edges increases by one and this changes the cost. On the other hand, in the selfish routing games the change of strategies has no additional cost. A simple—although not entirely rigorous—way to think about it, is to consider the effects of a tiny amount of

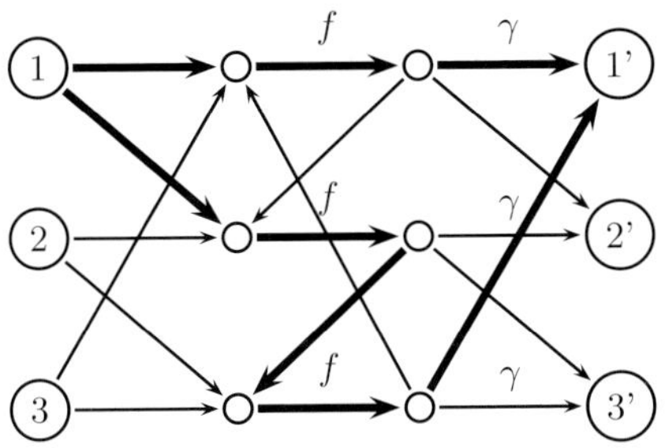

Fig. 2. Lower bound for the PoA for selfish routing. There are n distinct edge latency functions: $l(f) = f$, $l(f) = \gamma$ (a constant which depends on ϵ), $l(f) = 0$ (omitted in the picture). There are n commodities of rate 1 with source i and destination i' and 2 paths that connect them. The two paths for the first commodity are shown in bold lines. A similar construction works for atomic games as well.

flow that ponders whether to change path: it will not really affect the flow on the new edges (at least for continuous cost functions).

When we consider approximate equilibria, the uniqueness difference dissolves and only the integrality difference remains. But still, determining the PoS seems to be a harder problem than determining the PoA.

1.1 Our Contribution and Related Work

Our work encompasses and generalizes some fundamental results in the area of the PoA [27,11,10,4] (see also the recently published book [24] for background information). Our techniques not only provide a unifying approach but they cast the existing results in a new light. For instance, the Pigou network (Figure 1) is still the tight example for the PoS, but not for the PoA. For the latter, the network of Figure 2 is tight for $\epsilon \leq 1$ and the network of Figure 3 is tight for larger ϵ.

We use the *multiplicative definition of approximate equilibria*: In atomic games, a player does not switch to a new strategy as long as his current cost is less than $1 + \epsilon$ times the new cost. In the selfish routing games, we use exactly the same definition [27]: the flow is at an equilibrium when the cost on its paths is less than $1 + \epsilon$ times the cost of every alternate path. There have been other definitions for approximate Nash equilibria in the literature. For example, for algorithmic issues of Nash equilibria, the most-studied one is the additive case [21,16]. However, since the PoA is a ratio, the natural definition is the multiplicative one. A slightly different multiplicative approximate Nash equilibrium definition was used in [9], where they study convergence issues for congestion games.

There is a large body of work on the PoA in various models [24]. More relevant to our work are the following publications: For atomic congestion games, it was shown in [4,11] that the pure PoA of linear latencies is 5/2. Later in [10], it was shown that the ratio 5/2 is tight even for mixed and correlated equilibria. For weighted congestion games, the PoA is $1+\phi \approx 2.618$ [4]. For polynomial latencies of degree p, [11] showed that the PoA of pure equilibria is $p^{\Theta(p)}$, and [4] showed $p^{\Theta(p)}$ bounds for mixed equilibria and for weighted congestion games. Aland et al. [1] gave exact bounds for weighted and unweighted congestion games. For the PoS of the atomic case, it was shown in [10,6] that for linear congestion games it is $1 + \sqrt{3}/3$. Also for the linear selfish routing games, [29,32] give tight bounds for the PoA of ϵ-Nash equilibria when $\epsilon \leq 1$.

For the selfish routing paradigm and exact Nash equilibria, [27] gave the PoA (and consequently the PoS) for polynomial latencies of degree p with nonnegative coefficients and more general functions (see [13] or [24] for a simplified version of the proof); the results are extended to non-atomic games in [28].

In this work, we give almost tight upper and lower bounds for the PoA and PoS of atomic and non-atomic congestion games with polynomial latencies of degree p. We assume throughout that the coefficients of the polynomials are nonnegative. We give tight upper and lower bounds for the PoS and the PoA (as functions of the degree p and the approximation factor ϵ) for both atomic and non-atomic games. The only exception is the PoS for atomic games, where we have weaker bounds; Our results reveal qualitative differences between exact and approximate Nash equilibria for non-atomic congestion games. These games have unique exact Nash equilibria, which makes the PoS to be equal to the PoA. When we consider the larger class of approximate equilibria the uniqueness of equilibria vanishes and the PoS and PoA diverge. Another interesting finding of our work is that for $\epsilon = p$, the PoS drops to 1 in both the atomic and non-atomic case (this was known in the non-atomic case [30,14]).

For all cases we give appropriate *characterizations* of the ϵ-Nash equilibria, which are of independent interest and generalize known characterizations of exact equilibria. These characterizations apply to every nondecreasing and continuous latency functions, not just polynomials. Especially for the PoS, our characterization involves a notion which generalizes the potential function of Rosenthal [26,24].

All upper bounds proofs have the following outline: Using an appropriate characterization for the ϵ-Nash equilibria and suitable algebraic inequalities—which for non-atomic games involve real numbers and for atomic games involve nonnegative integers—we bound the cost of the ϵ-Nash equilibria. The cases of PoS however are more complicated. In these cases, the characterization of ϵ-Nash equilibria involves potential-like quantities. For exact Nash equilibria there is essentially a unique potential; but for approximate equilibria, there are many choices for the potential and we have to figure out which potential gives the best results. This together with the technical challenges of the algebraic inequalities makes the proofs for the PoS harder.

All the games that we consider in this work possess pure equilibria and our upper bound proofs are tuned to these, for simplicity. However, all our proofs can be directly extended to mixed and correlated equilibria—the difference will be an extra outer sum which corresponds to the expectation. The upper bounds of our theorems apply unmodified to the more general classes of *mixed* and *correlated* equilibria.

An unpublished preliminary version of this work [12], studied the same questions for linear latencies. Here we generalize the results to polynomial latencies which are technically much more challenging. Due to space limitations, we refer the reader to the full version of this paper for the missing proofs, as well as for the sections about atomic games.

2 Definitions

A congestion game [26], also called an exact potential game [23], is a tuple $(N, E, (\mathcal{S}_i)_{i\in N}, (f_e)_{e\in E})$, where $N = \{1,\ldots,n\}$ is a set of n players, E is a set of facilities, $\mathcal{S}_i \subseteq 2^E$ is a set of pure strategies for player i: a pure strategy $A_i \in \mathcal{S}_i$ is simply a subset of facilities and l_e is a cost (or latency) function, one for each facility $e \in E$. The cost of player i for the pure strategy profile $A = (A_1,\ldots,A_n)$ is $c_i(A) = \sum_{e\in A_i} l_e(n_e(A))$, where $n_e(A)$ is the number of players who use facility e in the strategy profile A.

Definition 1. *A pure strategy profile A is an ϵ equilibrium iff for every player $i \in N$*

$$c_i(A) \le (1+\epsilon)c_i(A_i, A_{-i}), \qquad \forall A_i \in \mathcal{S}_i \tag{1}$$

We believe that the multiplicative definition of approximate equilibria makes more sense in the framework that we consider: Given that the PoA is a ratio, we need a definition that is insensitive to scaling.

The social cost of a pure strategy profile A is the sum of the players cost

$$\mathrm{C}(A) = \sum_{i\in N} c_i(A).$$

The approximate PoA and PoS, is the social cost of the worst-case and best-case ϵ-equilibrium over the optimal social cost

$$PoA = \max_{A \text{ is an } \epsilon\text{-Nash}} \frac{\mathrm{C}(A)}{\mathrm{OPT}}, \quad PoS = \min_{A \text{ is an } \epsilon\text{-Nash}} \frac{\mathrm{C}(A)}{\mathrm{OPT}}.$$

Instead of defining formally the class of non-atomic congestion games, we prefer to focus on the more illustrative–more restrictive though–class of selfish routing games. The difference in the two models is that in a non-atomic game, there does not exist any network and the strategies of the players are just subsets of facilities (as in the case of atomic congestion games) and they do not necessarily form a path in a network.

Let $G = (V, E)$ be a directed graph, where V is a set of vertices and E is a set of edges. In this network we consider k commodities: source-node pairs (s_i, t_i) with $i = 1\ldots k$, that define the sources and destinations. The set of simple paths in every pair (s_i, t_i) is denoted by $\mathcal{P}_i$, while with $\mathcal{P} = \cup_{i=1}^k \mathcal{P}_i$ we denote their union. A flow f, is a mapping from the set of paths to the set of nonnegative reals $f : \mathcal{P} \to \mathbb{R}^+$. For a given flow f, the flow on an edge is defined as the sum of the flows of all the paths that use this edge $f_e = \sum_{P\in\mathcal{P}, e\in P} f_P$. We relate with every commodity (s_i, t_i) a traffic rate r_i, as the total traffic that needs to move from s_i to t_i. A flow f is feasible, if for every commodity $\{s_i, t_i\}$, the traffic rate equals the flow of every path in $\mathcal{P}_i$, $r_i = \sum_{P\in\mathcal{P}_i} f_P$. Every edge introduces a delay in the network. This delay depends on the load of the edge and is determined by a delay function, $l_e(\cdot)$. An instance of a routing game is denoted by the triple (G, r, l). The latency of a path P, for a given flow f, is defined as the sum of all the latencies of the edges that belong to P, $l_P(f) = \sum_{e \in P} l_e(f_e)$. The social cost that evaluates a given flow f, is the total delay due to f

$$C(f) = \sum_{P \in \mathcal{P}} l_P(f) f_P.$$

The total delay can also be expressed via edge flows $C(f) = \sum_{e \in E} l_e(f_e) f_e$.

From now on, when we refer to flows, we mean feasible flows. We define (as in [27]) the ϵ-Nash or ϵ-Wardrop equilibrium flows:

Definition 2. *A feasible flow f, is an ϵ-Nash (or ϵ-Wardrop) equilibrium, if and only if for every commodity $i \in \{1, \ldots, k\}$ and $P_1, P_2 \in \mathcal{P}_i$ with $f_{P_1} > 0$, $l_{P_1}(f) \leq (1+\epsilon) l_{P_2}(f)$.*

3 Non-atomic PoA

In this section we give tight bounds for the approximate PoA for non-atomic congestion games. We present them for the special case of selfish routing (or network non-atomic congestion games), but they apply unchanged to the more general class of non-atomic congestion games. One of the technical difficulties here is that a different approach is required for small and large values of ϵ.

We start with a condition which relates the cost of ϵ-Nash equilibria to any other flow (and in particular the optimal flow). This is the generalization to approximate equilibria of the inequality established by Beckmann, McGuire, and Winston [5] for exact Wardrop equilibria.

Theorem 1. *If f is an ϵ-Nash flow and f^* is any feasible flow of a non-atomic congestion game:*

$$\sum_{e \in E} l_e(f_e) f_e \leq (1+\epsilon) \sum_{e \in E} l_e(f_e) f_e^*.$$

Proof. Let f be an ϵ-approximate Nash flow, and f^* be the optimum flow (or any other feasible flow). Given a flow f and for a particular commodity i, we denote by f_p^i and f_e^i the corresponding amount flow that i routes through the path p and through edge e respectively. From the definition of approximate Nash equilibria (Inequality (1)), we get that for every commodity i and for every path p with non-zero flow in f and any other path p':

$$\sum_{e \in p} l_e(f_e) \leq (1+\epsilon) \sum_{e \in p'} l_e(f_e).$$

For every commodity i, we sum up these inequalities for all pairs of paths p and p' weighted with the amount of flow of f and f^* on these paths.

$$\sum_{p,p'} f_p^i {f_{p'}^*}^i \sum_{e \in p} l_e(f_e) \leq (1+\epsilon) \sum_{p,p'} f_p^i {f_{p'}^*}^i \sum_{e \in p'} l_e(f_e)$$

$$\sum_{p'} f_{p'}^{*\,i} \sum_{e\in E} l_e(f_e) f_e^i \le (1+\epsilon) \sum_{p} f_p^i \sum_{e\in E} l_e(f_e) f_e^{*i}$$

$$(\sum_{p'} f_{p'}^{*\,i}) \sum_{e\in E} l_e(f_e) f_e^i \le (1+\epsilon) (\sum_{p} f_p^i) \sum_{e\in E} l_e(f_e) f_e^{*i}$$

But $\sum_p f_p^i = \sum_{p'} f_{p'}^{*\,i}$ is equal to the total rate r_i for the feasible flows f and f^*. Simplifying and summing up for all commodities i, we get

$$\sum_{e\in E} l_e(f_e) f_e \le (1+\epsilon) \sum_{e\in E} l_e(f_e) f_e^*.$$

□

To bound the PoA, we will need the following lemma:

Lemma 1. *Let $g(x)$ be a polynomial with nonnegative coefficients of degree p. The inequality*

$$g(x)\,y \le \alpha\, g(x)\, x + \beta\, g(y)\, y$$

holds for the following set of parameters:

$$\alpha = \frac{p}{(p+1)(1+\epsilon)} \qquad \beta = \frac{(1+\epsilon)^p}{p+1} \qquad \textit{for every } \epsilon \ge (1+p)^{1/p} - 1$$

$$\alpha = p/(p+1)^{1+1/p} \qquad \beta = 1 \qquad \textit{for every } \epsilon \ge 0$$

We have now what we need to establish the upper bound of the main theorem of this section. The matching lower bounds are shown in the following two lemmas. Since the lower bounds are based on network non-atomic games, we get that the theorem applies also to the special class of network non-atomic congestion games[1].

Theorem 2. *The PoA of non-atomic congestion games with latency functions polynomials of degree p with nonnegative coefficients is $(1+\epsilon)^{p+1}$, for every $\epsilon \ge (p+1)^{1/p} - 1$, and $(1/(1+\epsilon) - p/(p+1)^{1+1/p})^{-1}$, for every $\epsilon \le (p+1)^{1/p} - 1$.*

Proof. The proof is based entirely on the inequality of Theorem 1:

$$\sum_{e\in E} l_e(f_e) f_e \le (1+\epsilon) \sum_{e\in E} l_e(f_e) f_e^*.$$

We can bound the right hand side using Lemma 1, as follows

$$\sum_{e\in E} l_e(f_e) f_e^* \le \sum_{e\in E} \alpha l_e(f_e) f_e + \beta l_e(f_e^*) f_e^*.$$

It follows that for large ϵ (i.e. $\epsilon \ge (p+1)^{1/p} - 1$), the PoA is at most

$$\frac{(1+\epsilon)\beta}{1-(1+\epsilon)\alpha} = (1+\epsilon)^{p+1},$$

and for small ϵ (i.e. $\epsilon \le (p+1)^{1/p} - 1$), it is at most

[1] This means that one cannot use the special structure of network non-atomic congestion games to show an upper bound of a lower value.

$$\frac{(1+\epsilon)\beta}{1-(1+\epsilon)\alpha} = (1/(1+\epsilon) - p/(p+1)^{1+1/p})^{-1}.$$

The following two lemmas establish that these upper bounds are tight. □

Lemma 2 (Non-Atomic-PoA-Lower-Bound for $\epsilon \leq (p+1)^{1/p} - 1$). *There are instances of non-atomic congestion games with polynomial latencies of degree p, with approximate PoA of at least* $(1/(1+\epsilon) - p/(p+1)^{1+1/p})^{-1}$, *for every* $\epsilon \leq (p+1)^{1/p} - 1$.

Lemma 3 (Non-Atomic-PoA-Lower-Bound for $\epsilon \geq (p+1)^{1/p} - 1$). *There are instances of non-atomic congestion games with polynomial latencies of degree p, with approximate PoA of at least* $(1+\epsilon)^{p+1}$, *for every* $\epsilon \geq (p+1)^{1/p} - 1$.

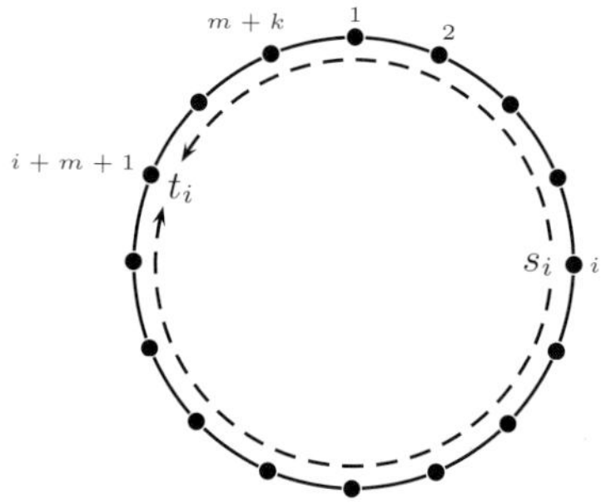

Fig. 3. The lower bound for non-atomic linear latencies and large ϵ

Proof. Consider an undirected cycle of $m+k$ vertices $C = (v_1, \ldots, v_{m+k})$ with edge latency $l_e(f_e) = f_e^p$ on every edge e (Figure 3). We choose m, k so that m/k gives an arbitrarily good approximation of $1+\epsilon$. There are $m+k$ unit-demand commodities each one with source $s_i = v_i$ and destination $t_i = v_{i+m+1}$ (again indices are taken cyclically). Clearly, for each commodity i there are exactly two simple paths that connect s_i to t_i; i.e. the clockwise path using m edges and the counterclockwise path using k edges. If every commodity routes the traffic via the clockwise path, then the load on every edge is m. The latency experienced by players of each commodity is m^{p+1}–each commodity uses m edges each incurring cost m^p. The alternative path has length k and so the latency on that path is km^{p+1}, so this is an m/k-approximate flow. The optimal flow routes all the traffic counterclockwise having latency k^{p+1} per commodity; every edge has load k and every commodity uses k edges. This gives an approximate PoA of $(m/k)^{p+1} \approx (1+\epsilon)^{p+1}$. □

4 Non-atomic PoS

In this section we give tight upper and lower bounds for the non-atomic congestion games with polynomial latencies. We start with a theorem which characterizes the ϵ-Nash (or ϵ-Wardrop) equilibria for every non-atomic congestion game with arbitrary (not necessarily polynomial) latency function. It involves a potential-like function and generalizes a well-known characterization of exact Nash (Wardrop) equilibria [27].

Theorem 3. *In an non-atomic congestion game with latency functions* $l_e(f_e)$*, let* $\phi_e(f_e)$ *be any integrable functions which satisfy*

$$\frac{l_e(f_e)}{(1+\epsilon)} \leq \phi_e(f_e) \leq l_e(f_e), \tag{2}$$

for every $f_e \geq 0$ *and define* $\Phi_e(f_e) = \int_0^{f_e} \phi_e(t)\,dt$*. For a flow* f*, define* $\Phi(f) = \sum_{e \in E} \Phi_e(f_e)$*. If a flow* f *minimizes the potential function* $\Phi(f)$*, it is an* ϵ*-Nash equilibrium.*

Furthermore, when the latency functions are nondecreasing, for any other flow f'*:*

$$\sum_{e \in E} \phi_e(f_e) f_e \leq \sum_{e \in E} \phi_e(f_e) f'_e$$

We now study the PoS of polynomial latency functions of the form $\ell_e(f) = \sum_{k=0}^{p} a_{e,k} f_e^k$. For these latency functions we define the potential function which has derivatives $\phi_e(f_e) = \sum_{k=0}^{p} \zeta_k a_{e,k} f_e^k$, for some ζ_k to be determined later and which satisfy:

$$\frac{1}{1+\epsilon} \leq \zeta_k \leq 1 \tag{3}$$

Theorem 4. *The PoS for non-atomic congestion games with polynomial latencies of degree* p *is exactly*

$$\left((1+\epsilon)\left(1 - \frac{p}{p+1}\left(\frac{1+\epsilon}{p+1}\right)^{1/p}\right)\right)^{-1},$$

for $\epsilon < p$*, and it drops to 1 for* $\epsilon \geq p$*.*

Proof. Our starting point is the second part of Theorem 3. Let f be a flow which minimizes the potential and let $\hat{f}$ be an optimal flow. We have that

$$\sum_{e \in E} \sum_{k=0}^{p} \zeta_k a_{e,k} f_e^{k+1} \leq \sum_{e \in E} \sum_{k=0}^{p} \zeta_k a_{e,k} f_e^k \hat{f}_e.$$

We can bound from above the monomials $f_e^k \hat{f}_e$ as follows:

$$f_e^k \hat{f}_e \leq \alpha_k f_e^{k+1} + \beta_k \hat{f}_e^{k+1},$$

where[2]

$$\alpha_k^k \beta_k = k^k/(k+1)^{k+1} \tag{4}$$

(and in particular $\alpha_0 = 0$ and $\beta_0 = 1$). We rearrange the terms to get:

[2] We use the following fact: for every nonnegative x, y, k, and for every positive α_k, β_k which satisfy $\alpha_k^k \beta_k = k^k/(k+1)^{k+1}$: $x^k y \leq \alpha_k x^{k+1} + \beta_k y^{k+1}$. Proof: Let $z = x/y$. We need to have $\alpha_k z^{k+1} - z^k + \beta_k \geq 0$. Taking the derivative, we see that the expression is minimized when $z = k/((k+1)\alpha_k)$ and the minimum value is $\beta_k - k^k/((k+1)^{k+1}\alpha_k^k)$, which is 0 for our choice of α_k and β_k.

$$\sum_{e \in E} \sum_{k=0}^{p} \zeta_k (1 - \alpha_k) a_{e,k} f_e^{k+1} \leq \sum_{e \in E} \sum_{k=0}^{p} \zeta_k \beta_k a_{e,k} \hat{f}_e^{k+1}.$$

The problem now is to select the parameters α_k, β_k, ζ_k to bound all the coefficients on the left side from below and all the coefficients on the right side from above:

$$\zeta_k \, (1 - \alpha_k) \geq 1 - \alpha_p \qquad\qquad \zeta_k \, \beta_k \leq \beta_p \tag{5}$$

With these, the above inequality becomes

$$(1 - \alpha_p) \sum_{e \in E} \sum_{k=0}^{p} a_{e,k} f_e^{k+1} \leq \beta_p \sum_{e \in E} \sum_{k=0}^{p} a_{e,k} \hat{f}_e^{k+1},$$

which will immediately bound the PoS from above by $\beta_p/(1 - \alpha_p)$.

There is a complication in selecting the parameters α_k, β_k, ζ_k as functions of ϵ, in that we need to have different choices for small and large values of ϵ:

For $k \leq \epsilon$:	$\zeta_k = (k+1)/(1+\epsilon)$	$\beta_k = 1/(k+1)$
For $k > \epsilon$:	$\zeta_k = 1$	$\beta_k = 1/(1+\epsilon)$

The value of α_k is determined by Equation (4)[3]. We need to show that these values satisfy properties (3) and (5). We see immediately that property (3) is satisfied. For property (5), we need to do more work. We first observe that the second part of property (5) is satisfied always with equality. For the first part we get:

Case $\epsilon < k$: We first establish that $\alpha_k \leq k/(k+1)$. Indeed, from $\alpha_k^k \beta_k = k^k/(k+1)^{k+1}$, we get $\alpha_k^k = (1+\epsilon)\, k^k/(k+1)^{k+1} \leq (1+k)\, k^k/(k+1)^{k+1} = k^k/(k+1)^k$. We will also use the fact that the function $x^p/(x+1)^{p+1}$ is increasing in x when $x \leq p$. We then have $\alpha_p^p = (1+\epsilon)\, p^p/(p+1)^{p+1} \geq (1+\epsilon) k^p/(k+1)^{p+1} = (1+\epsilon) k^k/(k+1)^{k+1} k^{p-k}/(k+1)^{p-k} \geq \alpha_k^k \alpha_k^{p-k} = \alpha_k^p$. Therefore, $\alpha_p \geq \alpha_k$, from which the first part of property (5) follows immediately.

Case $\epsilon \geq k$: In this case $\alpha_k = k/(k+1)$ and we need to show that $\zeta_k(1 - \alpha_k) \geq 1 - \alpha_p$ which is equivalent to $\alpha_p \geq \epsilon/(1+\epsilon)$. Since $\alpha_p^p = (1+\epsilon)\, p^p/(p+1)^{p+1}$, we use again the fact that the function $x^p/(x+1)^{p+1}$ is increasing in x when $x \leq p$ to get that $\alpha_p^p \geq (1+\epsilon)\, \epsilon^p/(1+\epsilon)^{1+p} \geq \epsilon^p/(1+\epsilon)^p$. The claim follows. Notice also that this analysis holds also for the special case $k = 0$.

Therefore, the PoS is at most $\beta_p/(1 - \alpha_p)$, where $\beta_p = 1/(1+\epsilon)$ and $\alpha_p^p \beta = p^p/(p+1)^{p+1}$. This establishes the upper bound of the theorem.

---

[3] We can unify the above definitions by letting $q = \min\{k, \epsilon\}$ and

$$\zeta_k = (q+1)/(1+\epsilon) \qquad\qquad \beta_k = 1/(q+1),$$

but this is not very useful in the following analysis.

To establish the lower bound, we use the Pigou network with latencies $1+\epsilon$ and f_e^p and flow rate equal to 1. There is a unique ϵ-Nash equilibrium which uses the second edge with total latency 1. On the other hand, the optimal solution is to send flow x to the second edge and $1-x$ in the first edge and optimize for x. The social cost is $x^{p+1}+(1+\epsilon)(1-x)$. If we minimize this function, we match the upper bound. □

The fact that the PoS drops to 1 when $\epsilon = p$ was first shown in [30] and [14].

5 Conclusions

We considered the PoA and the PoS of approximate Nash equilibria for congestion games (atomic, selfish routing, and non-atomic) with polynomial latencies. We have used a unifying approach and obtained tight upper and lower bounds for all cases except for the PoS for atomic congestion games. This remains a challenging open problem even for the case of exact equilibria ($\epsilon = 0$).

Acknowledgments. The authors would like to thank Ioannis Caragiannis for many helpful discussions and Tim Roughgarden for useful pointers to literature.

References

1. Aland, S., Dumrauf, D., Gairing, M., Monien, B., Schoppmann, F.: Exact price of anarchy for polynomial congestion games. In: Durand, B., Thomas, W. (eds.) STACS 2006. LNCS, vol. 3884, pp. 218–229. Springer, Heidelberg (2006)
2. Albers, S.: On the value of coordination in network design. In: SODA, pp. 294–303 (2008)
3. Anshelevich, E., Dasgupta, A., Kleinberg, J.M., Tardos, É., Wexler, T., Roughgarden, T.: The price of stability for network design with fair cost allocation. In: FOCS, pp. 295–304 (2004)
4. Awerbuch, B., Azar, Y., Epstein, A.: The price of routing unsplittable flow. In: STOC, pp. 57–66 (2005)
5. Beckmann, M., McGuire, C.B., Winston, C.B.: Studies in the Economics of Transportation. Yale University Press (1956)
6. Caragiannis, I., Flammini, M., Kaklamanis, C., Kanellopoulos, P., Moscardelli, L.: Tight bounds for selfish and greedy load balancing. In: Bugliesi, M., Preneel, B., Sassone, V., Wegener, I. (eds.) ICALP 2006. LNCS, vol. 4051, pp. 311–322. Springer, Heidelberg (2006)
7. Chen, X., Deng, X.: Settling the complexity of two-player nash equilibrium. In: FOCS, pp. 261–272 (2006)
8. Chen, H.-L., Roughgarden, T.: Network design with weighted players. In: SPAA, pp. 29–38 (2006)
9. Chien, S., Sinclair, A.: Convergence to approximate nash equilibria in congestion games. In: SODA, pp. 169–178 (2007)
10. Christodoulou, G., Koutsoupias, E.: On the price of anarchy and stability of correlated equilibria of linear congestion games. In: Brodal, G.S., Leonardi, S. (eds.) ESA 2005. LNCS, vol. 3669, pp. 59–70. Springer, Heidelberg (2005)

11. Christodoulou, G., Koutsoupias, E.: The price of anarchy of finite congestion games. In: STOC, pp. 67–73 (2005)
12. Christodoulou, G., Koutsoupias, E., Spirakis, P.G.: On the performance of approximate equilibria in congestion games. CoRR, abs/0804.3160 (2008)
13. Correa, J.R., Schulz, A.S., Stier Moses, N.E.: Selfish routing in capacitated networks. Math. Oper. Res. 29(4), 961–976 (2004)
14. Correa, J.R., Schulz, A.S., Moses, N.E.S.: Fast, fair, and efficient flows in networks. Operations Research 55, 215–225 (2007)
15. Dafermos, S.C., Sparrow, F.T.: The traffic assignment problem for a general network. Journal of Research of the National Bureau of Standards, Series B 73B(2), 91–118 (1969)
16. Daskalakis, C., Goldberg, P.W., Papadimitriou, C.H.: The complexity of computing a nash equilibrium. In: STOC, pp. 71–78 (2006)
17. Daskalakis, C., Mehta, A., Papadimitriou, C.: A note on approximate nash equilibria. In: Spirakis, P.G., Mavronicolas, M., Kontogiannis, S.C. (eds.) WINE 2006. LNCS, vol. 4286, pp. 297–306. Springer, Heidelberg (2006)
18. Fiat, A., Kaplan, H., Levy, M., Olonetsky, S., Shabo, R.: On the price of stability for designing undirected networks with fair cost allocations. In: Bugliesi, M., Preneel, B., Sassone, V., Wegener, I. (eds.) ICALP 2006. LNCS, vol. 4051, pp. 608–618. Springer, Heidelberg (2006)
19. Fabrikant, A., Papadimitriou, C., Talwar, K.: On the complexity of pure equilibria. In: STOC (2004)
20. Koutsoupias, E., Papadimitriou, C.: Worst-case equilibria. In: Meinel, C., Tison, S. (eds.) STACS 1999. LNCS, vol. 1563, pp. 404–413. Springer, Heidelberg (1999)
21. Lipton, R.J., Markakis, E., Mehta, A.: Playing large games using simple strategies. In: EC, pp. 36–41 (2003)
22. Milchtaich, I.: Congestion games with player-specific payoff functions. Games and Economic Behavior 13, 111–124 (1996)
23. Monderer, D., Shapley, L.: Potential games. Games and Economics Behavior 14, 124–143 (1996)
24. Nisan, N., Roughgarden, T., Tardos, E., Vazirani, V.V.: Algorithmic Game Theory. Cambridge University Press, Cambridge (2007)
25. Papadimitriou, C.H.: Algorithms, games, and the internet. In: STOC, pp. 749–753 (2001)
26. Rosenthal, R.W.: A class of games possessing pure-strategy Nash equilibria. International Journal of Game Theory 2, 65–67 (1973)
27. Roughgarden, T., Tardos, E.: How bad is selfish routing? J. ACM 49(2), 236–259 (2002)
28. Roughgarden, T., Tardos, E.: Bounding the inefficiency of equilibria in nonatomic congestion games. Games and Economic Behavior 47(2), 389–403 (2004)
29. Roughgarden, T.: Selfish Routing. PhD. Thesis, Cornell University (May 2002)
30. Roughgarden, T.: How unfair is optimal routing? In: SODA, pp. 203–204 (2002)
31. Roughgarden, T.: The price of anarchy is independent of the network topology. J. Comput. Syst. Sci. 67(2), 341–364 (2003)
32. Roughgarden, T.: Selfish Routing and the Price of Anarchy. MIT Press, Cambridge (2005)
33. Tsaknakis, H., Spirakis, P.G.: An optimization approach for approximate nash equilibria. In: Deng, X., Graham, F.C. (eds.) WINE 2007. LNCS, vol. 4858, pp. 42–56. Springer, Heidelberg (2007)
34. Wardrop, J.G.: Some theoretical aspects of road traffic research. In: Proceedings of the Institute of Civil Engineers, Part II, vol. 1, pp. 325–378 (1952)

Optimality and Competitiveness of Exploring Polygons by Mobile Robots

Jurek Czyzowicz*, Arnaud Labourel**, and Andrzej Pelc***

Département d'informatique, Université du Québec en Outaouais,
Gatineau, Québec J8X 3X7, Canada
jurek@uqo.ca, labourel.arnaud@gmail.com, pelc@uqo.ca

Abstract. A mobile robot, represented by a point moving along a polygonal line in the plane, has to explore an unknown polygon and return to the starting point. The robot has a sensing area which can be a circle or a square centered at the robot. This area shifts while the robot moves inside the polygon, and at each point of its trajectory the robot "sees" (explores) all points for which the segment between the robot and the point is contained in the polygon and in the sensing area. We focus on two tasks: exploring the entire polygon and exploring only its boundary. We consider several scenarios: both shapes of the sensing area and the Manhattan and the Euclidean metrics.

We focus on two quality benchmarks for exploration performance: optimality (the length of the trajectory of the robot is equal to that of the optimal robot *knowing* the polygon) and competitiveness (the length of the trajectory of the robot is at most a constant multiple of that of the optimal robot knowing the polygon). Most of our results concern rectilinear polygons. We show that optimal exploration is possible in only one scenario, that of exploring the boundary by a robot with square sensing area, starting at the boundary and using the Manhattan metric. For this case we give an optimal exploration algorithm, and in all other scenarios we prove impossibility of optimal exploration. For competitiveness the situation is more optimistic: we show a competitive exploration algorithm for rectilinear polygons whenever the sensing area is a square, for both tasks, regardless of the metric and of the starting point. Finally, we show a competitive exploration algorithm for arbitrary convex polygons, for both shapes of the sensing area, regardless of the metric and of the starting point.

1 Introduction

The model and the problem. A mobile robot, represented by a point moving along a polygonal line in the plane, has to explore an unknown polygon and

---

* Partially supported by NSERC discovery grant.

** This work was done during this author's stay at the Université du Québec en Outaouais as a postdoctoral fellow.

*** Partially supported by NSERC discovery grant and by the Research Chair in Distributed Computing at the Université du Québec en Outaouais.

A. Fiat and P. Sanders (Eds.): ESA 2009, LNCS 5757, pp. 263–274, 2009.

return to the starting point. We assume that the boundary is included in the polygon. The robot has a *sensing area* (abbreviated by SA in the sequel) which can be a circle or a square centered at the robot. During the exploration the robot must remain within the polygon, but its SA can partially exceed the boundaries of the polygon. At each point of its trajectory the robot "sees" (explores) all points for which the segment between the robot and the point is contained in the polygon to be explored and in the sensing area. For any explored point the robot is aware of whether this point is on the boundary of the polygon or not. We consider two tasks: exploring the entire polygon and exploring its boundary, for both shapes of the SA and for the Manhattan and the Euclidean metrics. The Manhattan metric will be called L_1 and the Euclidean metric will be called L_2 (Recall that in the L_1-metric the distance between two points is the sum of the differences of their coordinates). We also differentiate the situation when the starting point of the robot is at the boundary and when it is an arbitrary point of the polygon. We assume that the robot remembers what it has explored, i.e., it keeps a partial map of the explored part of the polygon with its trajectory in it, at all times.

The quality measure of an exploration algorithm not knowing the polygon (an on-line algorithm) is the length of the trajectory of the robot, and we seek to minimize this length. We compare it to the smallest length of the trajectory of a robot *knowing* the polygon (an off-line algorithm), executing the same task (exploring the boundary or exploring the entire polygon) and starting at the same point. The ratio between these two lengths, maximized over all pairs (polygon, starting point), is the *competitive ratio* of the on-line exploration algorithm. We focus on two quality benchmarks for exploration performance: optimality (competitive ratio equal 1) and competitiveness (constant competitive ratio).

Our results. Our first set of results concerns the possibility of optimal on-line exploration. Here we consider only rectilinear polygons (those whose angles are either $\pi/2$ or $3\pi/2$). It turns out that optimal exploration is possible only in one scenario, that of exploring the boundary by a robot with square sensing area aligned with the sides of the polygon, starting at the boundary and using the L_1-metric. For this case we give an optimal exploration algorithm. In all other scenarios (when *either* the entire polygon has to be explored, *or* the sensing area is a circle, *or* the metric is L_2, *or* the starting point may be strictly inside the polygon) we prove impossibility of optimal on-line exploration.

For competitiveness, the situation is more optimistic: our optimal boundary exploration algorithm yields a competitive exploration algorithm for rectilinear polygons whenever the sensing area is a square aligned with the sides of the polygon, for both tasks (exploring the boundary or the entire polygon) regardless of the metric and of the starting point. Finally, we show a competitive exploration algorithm for arbitrary convex polygons, for both shapes of the sensing area, regardless of the metric and of the starting point.

To the best of our knowledge we propose the first competitive on-line algorithm to explore arbitrary rectilinear polygons with some limited sensing area.

Related work. Exploration of unknown environments by mobile robots was extensively studied in the literature under many different models. One of the most important works in this domain is [5] where the sensing area is unlimited. The authors gave a 2-competitive algorithm for rectilinear polygon exploration. The competitive ratio was later improved to 5/3 in [8]. It was shown in [13] that there is no deterministic algorithm for this problem better than 5/4-competitive and that there exists a 5/4-competitive randomized algorithm solving it. All these results hold for the L_1-metric. Upper bounds for the L_2-metric can be obtained from the fact that any α-competitive algorithm for the L_1-metric is $\alpha\sqrt{2}$-competitive for the L_2-metric [5]. The case of non-rectilinear polygons was also studied in [4,10] and a competitive algorithm was given in this case.

For polygonal environments with an arbitrary number of polygonal obstacles, it was shown in [5] that no competitive strategy exists, even if all obstacles are parallelograms. Later, this result was improved in [1] by giving a lower bound in $\Omega(\sqrt{k})$ for the competitive ratio of any on-line algorithm exploring a polygon with k obstacles. This bound remains true even for rectangular obstacles. Nevertheless, if the number of obstacles is bounded by a constant m, then there exists a competitive algorithm with competitive ratio in $O(m)$ [4].

Exploration by a robot with a limited sensing area has been studied, e.g., in [6,7,11,12,15]. This model is interesting to study, since it is justified by real world constraints. Indeed, computer vision algorithms based on information obtained by sensors, such as stereo or structured-light finder, can reliably compute visibility scenes only up to a limited range [7]. To the best of our knowledge, there were no previous results concerning competitive on-line exploration for *arbitrary* rectilinear polygons with limited visibility.

The off-line exploration problem with limited SA is related to older problems such as *lawn mowing*, *pocket milling* and *ice rink* problems. All these three problems are concerned with finding an optimal path of a tool moving on a surface (grass area to mow, pocket to mill or ice rink to sweep), such that all points of the surface are covered by the tool (a mower, cutter or ice rink machine) at least once during its travel. The only difference between exploration and the lawn mowing problem is that the robot is not allowed to leave the environment, while the mower can exit the surface. The ice rink problem is the same as the lawn mowing problem, except for the notion of the optimal path. In lawn mowing, only the length of the path is considered, while in the ice rink problem we also need to take into account the number of turns done by the robot, since those turns are costly [14]. In the pocket milling problem, not only the robot cannot leave the surface but also the cutter must not leave it. Here, the goal is to find a shortest path that covers the maximum area possible. The first two problems are NP-hard and the complexity of the third one is unknown [9]. All three problems admit polynomial time approximation algorithms [2,14].

On-line exploration with limited SA has been studied, e.g., in [6,11,12]. Unlike in our model, the robot in [6] can see slightly farther than its tool (six times the tool range). The authors describe an on-line algorithm with competitive ratio $1+3(\Pi D/A)$, where Π is a quantity depending on the perimeter of P, D the size of the tool and A the area of P. Since the ratio $\Pi D/A$ can be arbitrarily large, their algorithm is *not* competitive in the general case. Moreover, the exploration in [6] fails on a certain type of polygons, such as those with narrow corridors.

In [11,12], the authors consider the exploration of a particular class of polygons: those composed of complete identical squares, called cells of size a priori known to the robot. In this model, the robot explores all points in a cell when it enters the cell for the first time, and can move in one step to any adjacent cell. The cost of the exploration is measured by the number of steps. There exists a 2-competitive algorithm for exploration of such polygons with obstacles [11]. For polygons without obstacles, there exists a 4/3-competitive algorithm for exploration and no algorithm can achieve a competitive ratio better than 7/6 [12].

There are only a few papers on how to explore the boundary of a terrain with limited sensing area. This problem was first considered in [15] (in its off-line version) using a reduction to the safari route problem. The *safari route problem* consists in finding a shortest trajectory, starting at the point s of the boundary of a polygon P and going back to s, that visits a specified set of polygons $\mathcal{P}$ contained in P. It is assumed in [15] that the polygons in $\mathcal{P}$ are attached to the boundary of P (share at least one point with the boundary of P), since otherwise the problem is NP-hard [15]. The author gives a $O(mn^2)$ algorithm solving this problem, where m is the cardinality of $\mathcal{P}$ and n is the total number of vertices of P and polygons in $\mathcal{P}$. It is shown that an optimal safari route visiting all the circular sectors of vertices corresponding to the angles of P, (i.e., the region inside P from which the vertex is visible), is an optimal boundary exploration trajectory [15]. To solve the safari route problem, circular sectors are approximated with polygons and the obtained solution is within 0.3% of optimal. It is computed in cubic time.

2 Definitions and Preliminary Results

In this section and in the part of the paper concerning optimality of exploration, we only consider rectilinear polygons. Let P be such a polygon. For convenience, without loss of generality, we assume that all sides of the polygon P are either parallel to the x-axis (east-west sides) or to the y-axis (north-south sides).

A *rectilinear* trajectory path has all its segments parallel to either the x-axis or the y-axis. Since in the L_1-metric there is always a rectilinear path among the shortest paths between two points, we consider only rectilinear paths and we drop the word "rectilinear" in all considerations regarding the L_1-metric. In particular, we use this convention in this section and in Section 3.1.

A segment T contained in a polygon P is *separating*, if it divides P into two simple polygons called the *subpolygons defined by T*. The *foreign polygon* defined by T according to a point u, denoted by $FP_u(T)$, is the subpolygon not containing u. Note that the foreign polygon is undefined if $u \in T$. A separating segment T *dominates* a separating segment T' according to the point u, if $FP_u(T)$ is strictly contained in $FP_u(T')$.

The robot at position r *explores* a point x, if the segment $\overline{rx}$ is included both in the polygon and in the SA centered in r. We consider two types of SA: a *round* SA which is a disc of diameter 2 and a *square* SA which is a 2×2 square. For exploration of rectilinear polygons, we assume that the sides of a square SA are aligned with the sides of the polygons. An exploration trajectory of polygon P is a path contained in P such that each point of P is explored by the robot at some point of this path. A *boundary exploration trajectory* is a trajectory of a robot inside the polygon P, exploring the boundary of P. In both cases, the start and the end of the trajectory are equal and are denoted by r_0.

For each side S of a polygon P, we extend S inside P, possibly from both ends, until it first hits the boundary of P. Each contiguous section of the resulting segment, if any, excluding S itself, is called an *extension segment* (cf. [5]) associated with S. For each side S of a polygon P, we draw the line L parallel to S at distance one from it, on the side of the interior of P. If this line intersects P, we define the *vicinity segment* associated with S, as the part of L between the closest point of $P \cap L$ from S in clockwise order along the boundary and the closest one in anti-clockwise order.

Lemma 1. *Any boundary exploration trajectory is not shorter than GE.*

Each extension or vicinity segment M of side S is a separating segment. In the rest of the paper, any domination relation or foreign polygon $FP(M)$ is defined according to point r_0, if no other point of reference is specified. If $r_0 \in M$, we set $FP(M)$ to be the subpolygon defined by M that contains S. Starting at r_0, if side $S \in FP(M)$, where M is an extension or vicinity segment of S, then S can become explored only if M is visited (i.e., either crossed or touched). If this is the case, we call M a *necessary segment* of S. For two necessary (extension or vicinity) segments M_1 and M_2, if M_1 dominates M_2 then there is no way to visit M_1 without crossing M_2 from r_0. So, we can ignore M_2, since it is automatically visited, if we visit M_1. A non-dominated necessary segment is called *essential.* To see all sides of a polygon, starting at r_0, the robot has to visit every essential segment.

If the starting point r_0 is on the boundary of P, then it induces a *natural order* of essential segments, clockwise along the boundary of the polygon P: $E_1, E_2, \ldots, E_m$, where E_1 is the first essential segment encountered when moving clockwise along the boundary from r_0, and so on. For $i \in 1, \ldots, m$, we denote by x_i the point on E_i at the minimum distance from point x_{i-1}, with the starting point $r_0 = x_0$. As shown in [5], these points are uniquely defined by r_0. This trajectory from x_0 to x_m, and back directly from x_m to x_0, is called *GE* for 'Greedy Essential'.

3 Optimality

3.1 The Optimal Boundary Exploration Algorithm

In this section, we assume that the SA is a 2×2 square aligned with sides of a rectilinear polygon. Our aim is to construct a boundary exploration algorithm starting at a boundary point r_0 and following *GE* as closely as possible. Unfortunately, in the case of a robot with bounded SA (unlike the robot from [5] which had unbounded visibility) it is impossible for an on-line algorithm to visit essential extensions greedily, using shortest paths. The following proposition shows this significant difference between our scenario and that from [5].

Proposition 1. *There is no on-line algorithm that greedily visits the essential segments of every polygon, i.e., that visits the essential segments by following shortest paths between them, even starting at the boundary.*

Since, as shown above, our bounded visibility scenario is more difficult than that from [5], our optimal boundary exploration algorithm must also be more subtle. Its idea is as follows.

The robot tries to increase the contiguous part of the boundary seen to date. The rest of the boundary is not yet explored by the robot for three possible reasons: an obstructing angle limiting the view of the currently explored side, a $3\pi/2$-angle terminating the currently explored side and obstructing the view of the next side, or finally the end of the SA limiting the view of the currently explored side. The strategy of the robot is to move towards the extension corresponding to the obstructing angle (in the first two cases) and to move parallel to the currently explored side (in the third case). Due to limited visibility, no necessary segment is seen by the robot in the third case, which is a crucial difference between our scenario and that from [5]. While it is impossible to move between consecutive essential segments using shortest paths, we prove that for every essential segment there is some essential segment following it (not necessarily the next one) which the robot reaches by a shortest of all paths visiting the intermediate essential segments. Proving this property is the crucial and technically most difficult part of the algorithm analysis.

Algorithm. `BOUNDARY-ON-LINE-EXPLORATION` (*BOE*, for short)
INPUT: A starting point r_0 on the boundary of the polygon to be explored.
OUTPUT: A shortest boundary exploration trajectory, starting and ending at r_0.

We denote by C the contiguous part of the boundary, starting clockwise from r_0, that has been explored so far by the robot, and we call *frontier*, denoted by f, the end of C. The current position of the robot is denoted by r.

Repeat the following strategy until C becomes the boundary of a simple polygon, updating r, f and C whenever any change occurs.

Case 1: *There is an obstructing angle b, i.e., r, b and f are aligned and b is a $3\pi/2$ angle not in C (see Fig. 1(a))*
Move towards the extension $E(b)$ of the side $U(b)$ incident to b and not explored from r. The strategy used to reach $E(b)$ is to move parallel to the other side $S(b)$ incident to b whenever possible, and move towards $S(b)$, parallel to $E(b)$, until it becomes possible again to move parallel to $S(b)$, otherwise.
Case 2: *f is a $3\pi/2$ angle and r is not on the extension $E(f)$ of the side $U(f)$ incident to f and not explored from r. (see Fig. 1(b))*
Same as Case 1 with f instead of b.
Case 3: *There is no obstructing angle, and either f is a $3\pi/2$ angle and r is on the extension $E(f)$ of the side $U(f)$ incident to f and not explored from r, or f is not a $3\pi/2$ angle. (see Fig. 1(c))*

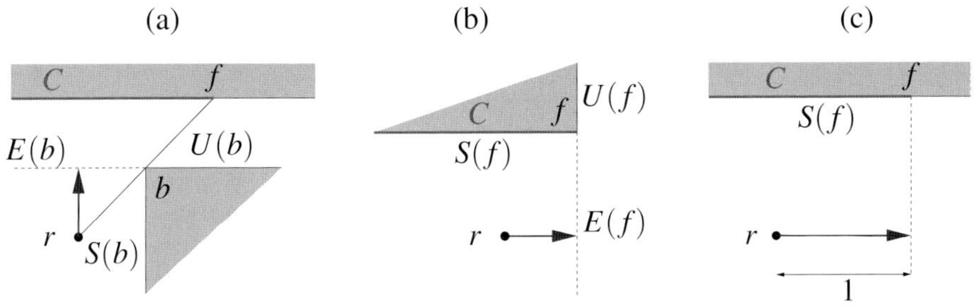

Fig. 1. The three possible configurations during the execution of Algorithm *BOE*

If f is a $3\pi/2$ angle then $S(f) = U(f)$, otherwise $S(f)$ is the side containing f. Move parallel to the side $S(f)$ towards f until:

Case 3.1: *Condition of Case 1 occurs*
Follow Case 1.
Case 3.2: *Condition of Case 2 occurs*
Follow Case 2.
Case 3.3: *A new $\pi/2$ angle a is explored and belongs to C (the robot reaches the vicinity segment $V(a)$ of the new side $U(a)$ incident to a)*
Do nothing (the algorithm proceeds to the next iteration of the repeat loop).

When the above **Repeat** loop is completed (C is a simple polygon), follow a shortest path to r_0 and stop.

The main result of this section is that Algorithm *BOE* is optimal.

Theorem 1. *Algorithm* `BOUNDARY-ON-LINE-EXPLORATION` *is an optimal on-line algorithm for the boundary exploration of rectilinear polygons with square SA in the L_1-metric, starting and ending at a point of the boundary.*

First, we show that Algorithm *BOE* eventually terminates

Lemma 2. *Algorithm BOE eventually terminates with C set to the boundary of the input polygon P.*

Let l be the number of iterations of the main loop of Algorithm *BOE* before terminating. For $i = 1, 2, \ldots, l$, the robot is at point r_i at the end of the i-th iteration of the main loop. The point r_i is either on a vicinity or on an extension segment denoted by M_i. Indeed, at the end of an iteration corresponding to Cases 1 or 3.1, the robot is on the extension segment $E(b)$ of side $U(b)$. For Cases 2 or 3.2, the robot is on the extension segment $E(f)$ of side $U(f)$. Finally, for Case 3.3, the robot is on the vicinity segment $V(a)$ of side $U(a)$.

We define a new trajectory BOE' that reaches segments M_i in a greedy way. For $i \in 1, \ldots, l$, we denote by z_i the point on M_i at the minimum distance (in the L_1-metric) from point z_{i-1}, with $r_0 = z_0 = z_{l+1}$. More formally, the trajectory BOE' is the one following a shortest path from z_{i-1} to z_i, for all $1 \leq i \leq l+1$.

Although BOE might not follow a shortest path between the segment M_i and M_{i+1} for some i, its total length turns out to be equal to that of BOE'.

We denote by $BOE[r_i, r_k]$ (resp. $BOE'[z_i, z_k]$) the part of the trajectory BOE (resp. BOE') between the points r_i and r_k (resp. z_i and z_k).

Lemma 3. *The BOE' trajectory has the same length as the BOE trajectory.*

Proof. We show that for all i, there exists a j, such that $BOE[r_i, r_{i+j}]$ is a shortest path from point r_i to M_{i+j} that visits segments M_{i+k} for $1 \leq k < j$. The proof depends on the type of the $(i+1)$-th iteration of Algorithm BOE.
Case 1: The robot follows a shortest path from r_i to the extension segment $E(b) = M_{i+1}$ as shown in [5]. Hence, the property holds for $j = 1$.
Case 2: Same as Case 1 with f instead of b.
Case 3.1: The robot moves parallel to $S(f)$ and then moves towards the extension segment $E(b)$, where b obstructs the vision to $S(f) = \overline{dv}$ (with d the first vertex of $S(f)$ in clockwise order) from the robot. Assume, without loss of generality, that the robot moves east when moving parallel to $S(f)$ ($S(f)$ is an east-west side) and is south of $S(f)$.

In order to explore the vertex v, the robot has to execute an iteration corresponding to Cases 2, 3.2 or 3.3. Let j denote the number of iterations executed by the robot to fully explore $S(f)$, the last one corresponding to Case 2, 3.2 or 3.3, needed to explore v.

Let S' be the side following the side $S(f)$ in the clockwise order. The segment M_{i+j} is either the extension segment of S', if the angle between $S(f)$ and S' is a $3\pi/2$ angle, or the vicinity segment of S', if the angle between $S(f)$ and S' is a $\pi/2$ angle. In both cases, M_{i+j} is perpendicular to all M_{i+k} for $0 \leq k < j$ and is east of point r_{i+1}.

During the iterations corresponding to Cases 1 or 3.1, the robot moves either north or east, since for all $1 \leq k < j$, M_{i+k} is an east-west segment and the obstructing angle b_k is in the north-east quadrant of the SA of the robot. During the last iteration corresponding to Cases 2, 3.2, or 3.3, the robot moves either north or east, since M_{i+j} is a north-south segment and the angle f (Cases 2 or 3.2) or the angle a (Case 3.3) is in the north-east quadrant of the SA of the robot. Hence, the path from r_i to r_{i+j} is a shortest path.

We show that the point r_{i+j} is the point of M_{i+j} at minimal distance from r_i. Indeed, it is reached by minimal x-axis and y-axis shifting, since the path is monotone and the robot moves north only until reaching the y-coordinate of the angle b_{j-1}. Hence, the path is a shortest path to M_{i+j} visiting all M_{i+k} for $1 \leq k < j$, and the property is verified.
Case 3.2: The robot moves parallel to $S(f)$ and then applies the strategy of Case 2. Since this strategy consists in moving parallel to $S(f)$ whenever it is possible, the property is verified, as in Case 2.
Case 3.3: The robot moves parallel to $S(f)$ from r_i to the vicinity segment M_{i+1} of the side immediately after $S(f)$ in clockwise order. The path followed by the robot to reach M_{i+1} is a shortest path, since $S(f)$ is perpendicular to M_{i+1}. Hence, the property holds for $j = 1$.

Recall that $r_0 = z_0$. We showed that $r_i = z_i$ implies $|BOE[r_i, r_{i+j}]| = |BOE'[z_i, z_{i+j}]|$, and $r_{i+j} = z_{i+j}$, for the index j (depending on i) determined above, since $BOE[r_i, r_{i+j}]$ is a shortest path from r_i to M_{i+j}. It follows by induction that $|\,BOE\,| = |\,BOE'\,|$.

We define a *compatible order* of essential segments as follows. In the natural order of essential segments we choose an arbitrary set of disjoint pairs of consecutive intersecting essential segments, and we swap segments in each pair.

Lemma 4. *The essential segments are visited in a compatible order $D_1, \ldots, D_m$ by the BOE' trajectory.*

In order to compare the BOE trajectory to the GE trajectory, we define a trajectory GC that greedily visits essential segments in the same compatible order as BOE. For $i \in 1, \ldots, l$, we denote by y_i the point on D_i at the minimum distance from point y_{i-1}, with $r_0 = y_0 = y_{m+1}$. More formally, the trajectory GC is the one following a shortest path from y_{i-1} to y_i, for all $1 \leq i \leq m+1$.

Lemma 5. *a) The GC trajectory has the same length as GE.*
b) The BOE' trajectory has the same length as the GC trajectory.

Proof of Theorem 1. Any boundary exploration trajectory (including the optimal one) has length not smaller than that of GE, by Lemma 1. By Lemma 5 (a), we have $|\,GE\,| = |\,GC\,|$. By Lemma 5(b), we have $|\,BOE'\,| = |\,GC\,|$. By Lemma 3, we have $|\,BOE'\,| = |\,BOE\,|$. By Lemma 2, BOE is a boundary exploration trajectory. Hence BOE is an optimal boundary exploration trajectory. □

3.2 Negative Results

In this section we show that in all scenarios except the one covered by Theorem 1, optimal on-line exploration is impossible.

Lemma 6. *There is no optimal on-line algorithm for the exploration of rectilinear polygons, with a square SA, in the L_1-metric, even with the starting point at the boundary.*

Proof. We consider two polygons W and T depicted in Fig. 2, and the exploration problem starting from the point x at the boundary of each of these polygons.

Notice that the visible parts of the two polygons are identical when the robot is at any point inside the rectangle $abkl$, the boundary of the rectangle included. So, the adversary can arbitrarily choose one of the two polygons when the robot leaves this rectangle to explore the rest of the polygon. The adversarial strategy to prevent optimality consists in taking the polygon T, if the robot exits the rectangle $abkl$ through point k or b, and in taking the polygon W otherwise.

We first show that any exploration trajectory passing through point b or k in polygon T is not optimal. Note that the order in which the two angles f and g are explored does not matter because of the symmetry of polygon T. The exploration trajectory R of T depicted in Fig. 2 is optimal, since the trajectory follows shortest paths to explore the angle f (at point y) and then the angle g (at point z) starting from point x.

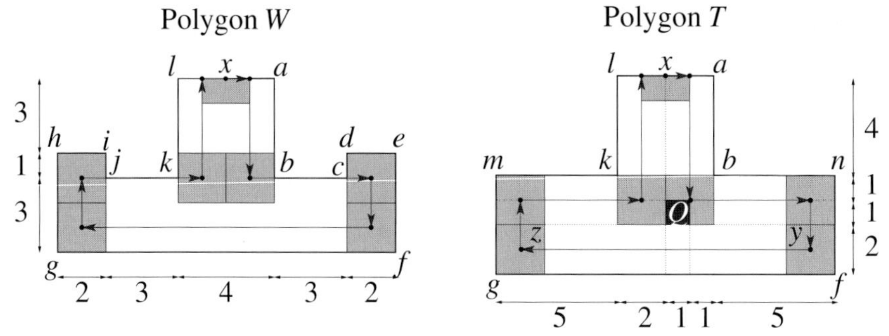

Fig. 2. Optimal solutions in polygon W and T

Let us assume, for contradiction, that there is an optimal exploration trajectory E passing through b. In order to have the same length as R, the trajectory E must follow shortest paths from x to b, from b to y, from y to z and from z to x. Let us consider the square region Q of points at distance at least one from lines ab and bk, and at distance at least two from lines lk and gf. The interior points in Q and the points of side kl cannot be explored by a robot following a shortest path xb, by or xy, since these points are at distance larger than one from any shortest path connecting these pairs of points. Consequently, the points of Q and those in the side kl need to be explored when moving on the trajectory between z and x. To explore the points of Q, the robot has to move past the line kl and continue moving east for a distance strictly greater than one. From the fact that this must be a shortest path to x, the robot cannot move west after this move and so cannot explore points of the side kl. Hence, the trajectory E is not an exploration trajectory and so there is no optimal exploration trajectory passing through b. By symmetry of the polygon, the same is true for point k.

We now show that any optimal exploration trajectory passing through any point t of the segment bk (ends of the segment excluded) in polygon W exits the rectangle $abkl$ through b or k. Note that any optimal trajectory needs to explore the angles e or h before the angles f or g. Indeed, there is a shortest path from the point t to a point from which f is visible (respectively the angle g) that explores the angle e (respectively the angle h). Consequently, any optimal exploration trajectory needs to follow an optimal path from t to explore one of the two angles e or h. These paths exit the rectangle $abkl$ through point b or k, and so no optimal exploration trajectory can exit this rectangle through an inside point of the segment.

Lemma 7. *There is neither an optimal on-line algorithm for the exploration, nor for the boundary exploration of rectilinear polygons with:*

1. *a square SA, in the L_1-metric, starting at an arbitrary point of the polygon.*
2. *a round SA, in the L_1-metric, even with the starting point at the boundary.*
3. *a square or a round SA, in the L_2-metric, even starting at the boundary.*

Theorem 1 and Lemmas 6, 7 imply the following result that completely solves the optimality problem of on-line exploration of rectilinear polygons.

Theorem 2. *The only case where on-line exploration of rectilinear polygons can be optimal is the case of the boundary exploration with square SA in the L_1-metric, starting at the boundary. Algorithm BOE is optimal in this case.*

4 Competitiveness

Algorithm *BOE* can be modified to produce a competitive on-line exploration algorithm under all scenarios with square SA.

Theorem 3. *There exists a competitive on-line algorithm for exploration and for boundary exploration of rectilinear polygons with square SA for both metrics and for any starting point.*

We finally turn attention to exploration of arbitrary convex polygons. We present a competitive exploration algorithm, called Algorithm `Convex`, working for arbitrary convex polygons, for round or square SA and regardless of the starting point. We use the L_2-metric, and the result holds for the L_1-metric as well by changing the competitive constant.

The idea of Algorithm `Convex` is the following. First move along a direction until a boundary point becomes explored. Call this distance δ. This is safe, as the optimal algorithm must travel at least the distance $\delta/\sqrt{2}$. Then move along boundaries of increasing squares centered at the starting point, of sizes 2δ, 4δ, 8δ, and so on, until the entire polygon is explored, or until the size of the square is at least 1. (If the boundary of the polygon to be explored prevents the robot from continuing on the square, then it "slides" on the boundary, returning to the travel on the square when again possible.) Since sizes of squares are doubled at each stage, the total trajectory length is at most the double of traversing the last square. If the whole polygon has been already explored, then the trajectory length is proportional to that of the optimal algorithm. Otherwise, the optimal trajectory length is proportional to the diameter and both these values must be at least 1/4. The trajectory length up to this moment is constant, hence making the tour of the polygon boundary and then applying the optimal off-line algorithm to explore its interior (at this point the polygon is known) is proportional to the diameter and hence competitive.

Theorem 4. *Algorithm* `Convex` *is a competitive algorithm to explore any convex polygon, starting from any point, for round or square SA, and for the L_1 or the L_2-metric.*

5 Conclusion

For the problem of optimality of on-line exploration of rectilinear polygons, our results explain the situation in each of the considered scenarios: we gave an optimal boundary exploration algorithm for a robot with square sensing area starting at the boundary, in the Manhattan metric, while in all other scenarios (exploration of the entire polygon, or arbitrary starting point, or round SA, or the Euclidean metric) we proved that optimal on-line exploration is impossible.

For the problem of competitiveness of on-line exploration of rectilinear polygons, our results are less complete: we showed a competitive algorithm for a robot with square sensing area, regardless of the metric and of the starting point. It is natural to ask if the same result is true for a round sensing area. We conjecture that the answer to this question is positive. It should be noted that competitiveness for the round SA does not immediately follow from competitiveness for the square SA, because there is no bound on the ratio between the lengths of optimal exploration trajectories in these scenarios.

An even bigger challenge would be to show a competitive on-line exploration algorithm for arbitrary polygons, for both shapes of the sensing area. Our competitive algorithm for convex polygons is a step in this direction.

References

1. Albers, S., Kursawe, K., Schuierer, S.: Exploring unknown environments with obstacles. Algorithmica 32(1), 123–143 (2002)
2. Arkin, E.M., Fekete, S.P., Mitchell, J.S.B.: Approximation algorithms for lawn mowing and milling. Comput. Geom. Theory Appl. 17(1-2), 25–50 (2000)
3. Chin, W., Ntafos, S.: Optimum watchman routes. In: Proc. of Symposium on Computational Geometry (SCG 1986), pp. 24–33 (1986)
4. Deng, X., Kameda, T., Papadimitriou, C.: How to learn an unknown environment. In: Proc. of Foundations of Computer Science (FOCS), pp. 298–303 (1991)
5. Deng, X., Kameda, T., Papadimitriou, C.: How to learn an unknown environment: the rectilinear case. J. ACM 45(2), 215–245 (1998)
6. Gabriely, Y., Rimon, E.: Spanning-tree based coverage of continuous areas by a mobile robot. In: Int. Conf. of Robotics and Automaton (ICRA 2001), pp. 1927–1933 (2001)
7. Ghosh, S., Burdick, J., Bhattacharya, A., Sarkar, S.: Online algorithms with discrete visibility - exploring unknown polygonal environments. Robotics & Automation Magazine 15(2), 67–76 (2008)
8. Hammar, M., Nilsson, B.J., Schuierer, S.: Improved exploration of rectilinear polygons. Nordic J. of Computing 9(1), 32–53 (2002)
9. Held, M.: On the computational geometry of pocket machining. Springer-Verlag New York, Inc., New York (1991)
10. Hoffmann, F., Icking, C., Klein, R., Kriegel, K.: The polygon exploration problem. SIAM J. Comput. 31(2), 577–600 (2001)
11. Icking, C., Kamphans, T., Klein, R., Langetepe, E.: On the competitive complexity of navigation tasks. In: Revised Papers from the International Workshop on Sensor Based Intelligent Robots, London, UK, 2002, pp. 245–258. Springer, London (2002)
12. Icking, C., Kamphans, T., Klein, R., Langetepe, E.: Exploring simple grid polygons. In: In 11th Internat. Comput. Combin. Conf., pp. 524–533 (2005)
13. Kleinberg, J.M.: On-line search in a simple polygon. In: Proc. ACM-SIAM Symposium on Discrete Algorithms (SODA 1994), pp. 8–15 (1994)
14. Moret, B.M.E., Collins, M., Saia, J., Yu, L.: The ice rink problem. In: Proc. of the 1st Workshop on Algorithm Engineering (1997)
15. Ntafos, S.: Watchman routes under limited visibility. Comput. Geom. Theory Appl. 1(3), 149–170 (1992)

Tractable Cases of Facility Location on a Network with a Linear Reliability Order of Links

Refael Hassin[1,⋆], R. Ravi[2,⋆⋆], and F. Sibel Salman[3,⋆⋆⋆]

[1] Department of Statistics and Operations Research, Tel Aviv University, Tel Aviv, Israel
hassin@post.tau.ac.il
[2] Tepper School of Business, Carnegie Mellon University
ravi@cmu.edu
[3] College of Engineering, Koç University, Istanbul, Turkey
ssalman@ku.edu.tr

Abstract. In this paper we study the problem of locating k facilities to maximize the expected demand serviced in a network with unreliable links. Given a linear ordering of links, which models the dependencies among link failures, we assume that when a strong link fails, all weaker links with lower reliability also fail. This model is due to Gunnec and Salman [1], and for the single ordering case, an exact algorithm for maximizing expected serviced demand was provided in our earlier work [2] via a greedy method and dynamic programming.

Our main result in this paper is to identify the boundary of hardness of the problem as we extend the model to have more than one disaster scenario and there is a different linear order in each scenario (defining the strength of the links in the scenario). We show that in the case with two disaster scenarios, the resulting facility location problem is polynomial time solvable by proving total unimodularity of a linear programming formulation. We also supply an alternate proof of this fact by the iterative relaxation method. In addition, for the two scenario case, we show that a version maximizing the expected demand served minus the sum of facility opening costs reduces to a bipartite matching problem. We then prove NP-hardness for the case with three orderings, even when all reliability values are one or zero (i.e., every scenario is deterministic, and we have three scenarios). Following the idea of the reduction, we show that the problem with an arbitrary number of orderings generalizes the maximum coverage problem, and hence affords a greedy approximation algorithm with performance ratio $(1 - \frac{1}{e})$. We also consider the distance-bounded version of the problem where a demand point can be covered only if a facility exists within a distance limit, and show that the problem is NP-hard even for a single ordering and is equivalent to the maximum k-facility location problem. Our methods represent the first attempt at finding interesting tractable models of link failure for facility location

⋆ This research is partially supported by a NATO Collaborative Linkage Grant.
⋆⋆ Supported in part by NSF grant CCF-0728841.
⋆⋆⋆ Supported by a TUBITAK Career Grant.

A. Fiat and P. Sanders (Eds.): ESA 2009, LNCS 5757, pp. 275–276, 2009.

planning for disasters. Our earlier results for the case of a single ordering [2] already showed the versatility of various techniques such as the Greedy method and a Dynamic Programming algorithm for addressing the many variants of the problem. This paper reveals an even richer structure for the polynomially solvable two-scenario case by showing non-trivial applications of total unimodularity, the iterative relaxation technique and weighted bipartite perfect matchings.

References

1. Gunnec, D., Salman, F.S.: Assessing the reliability and the expected performance of a network under disaster risk. In: Proceedings of the International Network Optimization Conference (INOC), Spa, Belgium, April 22-25 (2007)
2. Hassin, R., Ravi, R., Salman, F.S.: Facility Location on a Network with a Linear Dependency Order of Unreliable Links. In: Proceedings of the International Network Optimization Conference (INOC), Pisa, Italy, April 26-29 (2009)

Dynamic vs. Oblivious Routing in Network Design

Navin Goyal[1], Neil Olver[2], and F. Bruce Shepherd[3]

[1] Microsoft Reseach India
Bangalore, India
navin001@gmail.com
[2] Department of Mathematics and Statistics
McGill University, Montreal, Canada
olver@math.mcgill.ca
[3] Department of Mathematics and Statistics
McGill University, Montreal, Canada
bruce.shepherd@mcgill.ca

Abstract. Consider the robust network design problem of finding a minimum cost network with enough capacity to route all traffic demand matrices in a given polytope. We investigate the impact of different routing models in this robust setting: in particular, we compare *oblivious* routing, where the routing between each terminal pair must be fixed in advance, to *dynamic* routing, where routings may depend arbitrarily on the current demand. Our main result is a construction that shows that the optimal cost of such a network based on oblivious routing (fractional or integral) may be a factor of $\Omega(\log n)$ more than the cost required when using dynamic routing. This is true even in the important special case of the asymmetric hose model. This answers a question in [4], and is tight up to constant factors. Our proof technique builds on a connection between expander graphs and robust design for single-sink traffic patterns [5].

1 Introduction

One of the most widely studied applications of robustness in discrete optimization has been in the context of network design. This is partly motivated by the fact that traffic demands in modern data networks are often hard to determine and/or are rapidly changing. In one general model (cf. [3]), the input consists of a graph (network topology) where each edge comes with a cost to reserve capacity. In addition, a universe of possible demand matrices is specified as a polyhedron $\mathcal{P}$ (or more generally, as a convex body). In this paper our focus is on undirected demands and so for a demand matrix D, the entries D_{ij} and D_{ji} normally represent the same demand, and are hence equal. The problem is to design a minimum cost network such that each demand matrix in the polytope can be routed (according to routing models we describe shortly) in the resulting capacitated network. Typically we seek to install edge capacities so that the sum of costs is minimized, but other cost measures such as minimizing the maximum

A. Fiat and P. Sanders (Eds.): ESA 2009, LNCS 5757, pp. 277–288, 2009.

congestion are also considered in the literature. We refer to the recent survey by Chekuri [4] for a discussion of these models and previous work.

Since demands are potentially changing, there are two prime natural routing models that are considered. The first is *dynamic routing*: for any given demand $D \in \mathcal{P}$, we may use a traffic routing tailored to this demand. We consider only the case where the routing may be an arbitrary multicommodity flow, i.e. traffic flows may be fractional. We also refer to this routing model as FR. Dynamic routing, out of all possible routing models, clearly leads to the cheapest possible solution. However, this model is typically considered impractical: rerouting demands in realtime is not only computationally infeasible, it would be disruptive to existing demands corresponding to realtime network applications.

On the other extreme, *oblivious routing* models, inspired by routing in packet networks, ask for a *routing template* that defines ahead of time how any future demands will be routed. For each node pair i, j, the template f specifies a unit network flow f_{ij} between i and j. The interpretation is that if there is a future demand of D_{ij} between nodes i, j, then along each ij path P, we should route $D_{ij} f_{ij}(P)$ flow on this path. Flow templates may be either fractional, in which case they are called *multipath routings* (MPR), or we may require the $f(P)$'s to be 0, 1-valued, in which case they are called *single-path routings* (SPR). We also discuss a special case of SPR templates called *tree* templates where the support of f induces a tree in the network; we refer to this model as TR. We can now formally define the robust network design problem (cf. [5]):

Definition 1. *Given a graph G on n nodes, edge costs $c : E \to \mathbb{R}^+$, a polytope $\mathcal{P}$ of demand matrices, and a routing model (*FR, SPR, MPR, TR*), the robust network design problem is defined as follows. Find a minimum cost capacity installation of edge capacities $u : E \to \mathbb{R}^+$ so that all demand matrices in $\mathcal{P}$ can be routed in the given routing model. The cost of capacity installation u is given by $\sum_{e \in E} u(e)c(e)$.*

For a given instance of robust network design $(G, c, \mathcal{P})$, we use $\text{OPT}_{FR}(G, c, \mathcal{P})$, $\text{OPT}_{MPR}(G, c, \mathcal{P})$, $\text{OPT}_{SPR}(G, c, \mathcal{P})$ and $\text{OPT}_{TR}(G, c, \mathcal{P})$ to denote the corresponding cost of an optimally designed robust network for the four routing models. If the context is clear, we may simply write, for instance, OPT_{FR}.

Obviously we have $\text{OPT}_{FR} \leq \text{OPT}_{MPR} \leq \text{OPT}_{SPR} \leq \text{OPT}_{TR}$. It has been previously known that the gap between OPT_{FR} and OPT_{SPR} is $O(\log n)$ (credited to A. Gupta, cf. [4]). This follows by an application of the approximation of arbitrary metrics by tree metrics [6]. One can further show, by similar arguments but now using a theorem of [1] instead, that the gap between OPT_{FR} and OPT_{TR} is at most $\tilde{O}(\log n)$, where $\tilde{O}$ hides an $O(\text{poly} \log \log n)$ factor.

Our Results. In this paper, we seek to understand to what extent these gaps are realizable; in other words, for any pair of routing methods, what is the maximum possible gap between the costs of their optimal solution?

In short, the answer is that the upper bounds of $O(\log n)$ and $\tilde{O}(\log n)$ discussed above are essentially tight: for any pair of optimal solutions from $\{\text{OPT}_{FR}, \text{OPT}_{MPR}, \text{OPT}_{SPR}, \text{OPT}_{TR}\}$, there exists a family of instances of the robust network design problem such that the gap between the two quantities is $\Omega(\log n)$. We prove all of these gaps here except for the gap between OPT_{MPR} and OPT_{SPR}, which follows from an inapproximabilty result in [13].

Discussion. In the robustness paradigm, the question of how large these gaps can be is asked for specific classes of demand polyhedra. A class that has received much attention consists of the so-called "hose models" which come in symmetric and asymmetric flavours. In the symmetric hose model, each terminal v has an associated *marginal* b_v, which represents an upper bound on the *total* amount of traffic that can terminate at v. The demand polytope consists of all symmetric demands which do not violate these "hose" constraints; i.e. $\sum_j D_{ij} \leq b_i$ for each terminal i. The asymmetric hose problem is similar, but the terminals are divided into sources and sinks; all demand is between source and sink nodes, and again, total demand to or from a terminal cannot exceed its marginal. Classes such as the hose model arise naturally in switch design, but they were also motivated by applications in data networks [7,8]; one of these is referred to as the *virtual private network* (VPN) problem.

It is implicit in Fingerhut et al. [7] and explicit in Gupta et al. [8] that in the *symmetric* hose model, $\text{OPT}_{MPR} \leq \text{OPT}_{SPR} \leq 2 \cdot \text{OPT}_{FR}$. However, the gap instance between OPT_{MPR} and OPT_{FR} that we demonstrate in this paper is in fact an instance of the asymmetric hose problem, and hence there is a logarithmic gap for this latter model.[1] We describe a class of graphs G, cost function c, and a demand polytope $\mathcal{P}$, such that $\text{OPT}_{FR}(G, c, \mathcal{P}) = O(n)$ but $\text{OPT}_{SPR}(G, c, \mathcal{P}) = \Omega(n \log n)$ and $\text{OPT}_{MPR}(G, c, \mathcal{P}) = \Omega(n \log n)$. The polytope $\mathcal{P}$ has the property that all demands share a common "sink" node.

This result shows that for at least some robust network design problems of practical interest, the routing model used may have a serious impact on the solution cost. While completely dynamic routing is typically infeasible for reasons mentioned previously, it is plausible that some tradeoff between the two extremes of dynamic and oblivious routing could produce significantly better results while remaining practical.

It turns out that the problem of designing an SPR routing template for our gap instance corresponds to the well-studied rent-or-buy network flow problem (see, e. g., [9]) which is a generalization of the Steiner tree problem. In this problem there is only one demand matrix instead of a polytope of demands, but the cost function is concave. We sketch the lower bound argument for OPT_{SPR} separately in Section 2.3 since it is much simpler; it proceeds by showing that the optimal SPR templates may be assumed to be tree templates for our gap instance.

The lower bound for OPT_{MPR} is more involved. We show that the cost of an MPR template for our gap instance can be characterized by a network design problem that we call *buy-and-rent*. Again there is only one demand to be satisfied, but the cost function is more complex. The buy-and-rent cost function seems to be new and natural: briefly, instead of asking that each edge be either rented or bought, it allows that capacity may be partially bought and the rest rented. This

[1] This rectifies an earlier assertion (cf. Theorem 4.6 in [4]).

new cost function is more amenable to analysis, and leads to our lower bound for OPT_{MPR}.

Relation to congestion lower bounds. We remark that our lower bounds for the total cost model also imply lower bounds for minimizing the maximum congestion, essentially because if every edge had congestion at most α, the total cost would also be bound by a factor α. Since the polytope $\mathcal{P}$ we use is a subset of the single-sink demands routable in G, this also implies a result in [10] which gives an $\Omega(\log n)$ bound for congestion via oblivious routing of single sink demands (although their analysis also extends to the case of lower bounding performance of a general online algorithm). Congestion minimization problems can be seen as equivalent to a robust optimization where one uses maximum edge congestion as a cost function; simply take the polytope consisting of *all* single-sink demands which are routable in G (this is a superset of our choice $\mathcal{P}$). The construction in [10] uses meshes (grids), building on work of [2,11]. This construction does not seem to extend to the total cost model however, and we use instead a construction based on expanders, extending and simplifying a connection shown in earlier work [5].

Note that a gap between OPT_{MPR} (and hence OPT_{SPR}) and OPT_{FR} shows that in general, in order to solve for OPT_{SPR} it is not sufficient to find routings for some constant number of demand matrices, and then simply add up the capacities needed for these routings.

Gaps for tree templates. In Section 3 we give a family of instances (using a different demand polyhedron) showing that the gap between OPT_{SPR} and OPT_{TR} can be $\Omega(\log n)$. This immediately implies that the gaps between OPT_{FR} and OPT_{TR} and between OPT_{MPR} and OPT_{TR} is $\Omega(\log n)$ for this family of instances.

2 A Gap Example

2.1 A Robust Network Design Instance

Let $G = (V, E)$ be a graph on n nodes with constant degree $d \geq 3$ and edge-expansion at least 1; i.e. we have that $|\delta_G(S)| \geq |S|$ for all $S \subseteq V$ with $|S| \leq n/2$. Here $\delta_G(S)$ denotes the set of edges in E with one end-point in S and the other outside S. It is well-known that such edge-expanders with the above parameters exist. Now add a special sink node r to V to obtain our instance $\bar{G} = (\bar{V}, \bar{E}) = (V \cup \{r\}, E \cup \{vr : v \in V\})$; see Figure 1.

We look at a single-sink hose model (cf. [5]), where our demands come from a polytope $\mathcal{P}$ defined as follows. We have a specified marginal capacity b_v at each node: $b_r = \beta n$ (where $0 < \beta < 1$), and $b_v = 1$ for all $v \in V$. Each demand matrix $D_{ij} \in \mathcal{P}$ has the property that $\sum_j D_{ij} \leq b_i$ for each node $i \in \bar{V}$, and $D_{ij} > 0$ only if $r \in \{i, j\}$. Although we often think of nodes routing flow towards the sink, the demands and flows are essentially undirected in this paper.

Thus each demand matrix we must support, identifies a single-sink network flow problem. It is a simple exercise to see that:

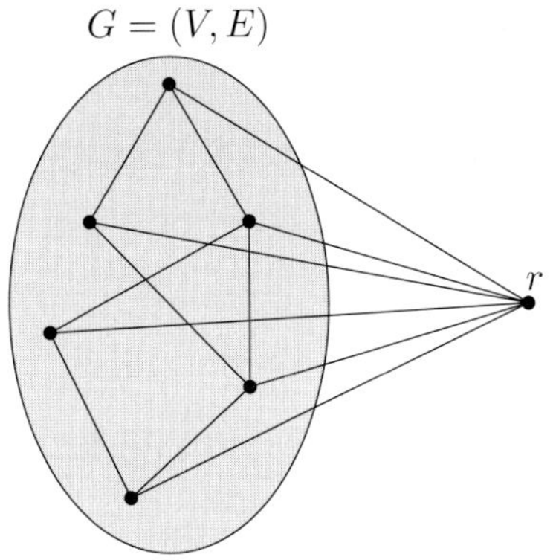

Fig. 1. The gap instance. G is a d-regular expander.

Lemma 1. *If b_r is an integer, then our network is robust for $\mathcal{P}$ and a given routing model if and only if for each subset X of b_r nodes in G, there is enough capacity to route one unit from each node in X to r, using the prescribed routing model.*

We use this fact below. Finally, we also assign costs to the edges: each edge of G has cost 1, and each edge in $\delta_{\bar{G}}(r)$ has cost $1/\beta$.

Our main result is the following theorem:

Theorem 1. *For $\beta = 1/\log n$, there is a dynamic routing for the single-sink hose model instance (defined above) of cost $O(n)$, but every* MPR *solution (and hence every* SPR *solution) has cost $\Omega(n \log n)$.*

The first assertion is proved in the next section. In Section 2.3, we see that determining OPT_{SPR} for single-sink hose models is equivalent to the well-studied single-sink *rent-or-buy* problem (see, e. g., [9]). In Section 2.3, we see that the rent-or-buy problem always has a tree solution. This can be used to show that $\text{OPT}_{SPR} = \Omega(n \log n)$ for our instance with $\beta = 1/\log n$. We give a sketch of a proof of this since it is considerably simpler than (but implied by) the proof of the corresponding bound for MPR. This MPR lower bound is demonstrated in Section 2.4.

We assume throughout the paper that $b_r = \beta n$ is an integer.

2.2 A Solution for the Dynamic Routing Model

Put capacity β on each edge of $\delta_{\bar{G}}(r)$, and capacity 1 on each edge of G. Clearly, the cost of this reservation is $O(n)$ independent of β. We show that this is a valid FR capacity reservation. Using Lemma 1 it suffices to show that for any subset of βn nodes X in G, all nodes in X can simultaneously route a unit flow to r. To this end, we add a new node t to $\bar{G}$ and edges vt for $v \in X$ with unit capacity to form graph G'. We show that G' supports a t-r flow of size $|X| = \beta n$. By the max-flow min-cut theorem it suffices to show that all r-t cuts in G' have size at least βn, i.e. that for each $S \subseteq V$ we have $|\delta_{G'}(S \cup \{t\})| \geq \beta n$.

We have

$$|\delta_{G'}(S \cup \{t\})| = \beta|S| + |X \setminus S| + |\delta_G(S)|.$$

Now, if $|S| \leq n/2$ then using the fact that for G we have $|\delta_G(S)| \geq |S|$ we get

$$\begin{aligned}|\delta_{G'}(S \cup \{t\})| &\geq \beta|S| + |X \setminus S| + |S| \\ &\geq \beta|S| + |X| \\ &\geq |X|.\end{aligned}$$

And if $|S| > n/2$ then using the fact that for G we have $|\delta_G(S)| \geq n - |S|$ we get

$$\begin{aligned}|\delta_{G'}(S \cup \{t\})| &\geq \beta|S| + |X \setminus S| + n - |S| \\ &\geq \beta|S| + |X \setminus S| + \beta(n - |S|) \\ &= \beta n + |X \setminus S| \\ &\geq \beta n.\end{aligned}$$

Hence the above capacity reservation can support the FR routing model and costs $O(n)$.

2.3 Rent-or-Buy: Lower Bounds for SPR Oblivious Routing Solutions

Note that the optimal cost oblivious SPR network can be cast as a minimum cost (unsplittable) flow problem as follows. Each node $v \in V$ must route one unit of flow on a path P_v to r and the overall (truncated) cost of path choices is: $\sum_e c(e) \min\{N(e), b_r\}$, where $N(e)$ is the number of nodes v whose path to r used the edge e. Clearly, if the capacity of each edge is $\min\{N(e), b_r\}$, then we have sufficient capacity to route any demand matrix in $\mathcal{P}$ using as a template the paths P_v. The converse is in fact also true and easy; any template gives rise to a corresponding integer flow whose truncated cost is the same.

This truncated routing cost problem is simply a so-called single-sink *rent-or-buy* (SSROB) problem (cf. [9]). We are given a network G with edge costs $c(e)$, and a special sink node t. A parameter $B \geq 1$ is also given. We also have a list of sources s_i for $i = 1, 2 \ldots, p$; each source needs to route to the sink t. For each edge in the network, we may either purchase it at a cost of $Bc(e)$, in which case it is deemed to have infinite capacity, or we may rent it. In that case, we must pay $c(e)$ per unit of capacity that we use on the edge. The goal is to find which edges to buy and which to rent in order to support a flow from each node to t, at the smallest possible cost. In other words, we seek a fractional flow $\boldsymbol{f}$ of the demands that minimizes $\sum_{e \in E} c(e) \min\{f(e), B\}$. In general, we may also consider such *single-sink flow problems with concave costs* $\sum_e g_e(f(e))$ where each g_e is a concave function.

The following result is immediate from the concavity of the cost function:

Proposition 1. *Any single-sink flow problem with nondecreasing concave costs has an optimal solution whose support is a tree. In particular, such an optimal solution always exists for the* SSROB *problem.*

Proof. (Sketch) Let $\boldsymbol{f}$ be a flow giving an optimal solution to the rent-or-buy instance, chosen so that supp($\boldsymbol{f}$) is setwise minimal. We show that then supp($\boldsymbol{f}$) must form a tree.

Let us consider $\boldsymbol{f}$ as a directed flow, where each terminal sends flow to the sink. If there is any directed cycle in the support of $\boldsymbol{f}$, then we may simply reduce flow on this cycle until some arc becomes zero; this does not increase the cost since our cost function is nondecreasing. So we may assume our support is acyclic in the directed sense. Suppose now that there is some undirected cycle C in the support which by assumption corresponds to some forward (traversing C in order) arcs F and some reverse arcs R. Let $\epsilon = \min\{f(a) : a \in R \cup F\}$. Define two solutions $\boldsymbol{f}^+, \boldsymbol{f}^-$ by $f^{\pm}(a) = f(a) \pm \epsilon$ for $a \in F$, and $f^{\pm}(a) = f(a) \mp \epsilon$ for $a \in R$. By concavity, $C(\boldsymbol{f}) \geq (1/2)[C(\boldsymbol{f}^+) + C(\boldsymbol{f}^-)]$. Then since $\boldsymbol{f}$ was an optimal solution, $C(\boldsymbol{f}^+) = C(\boldsymbol{f}^-) = C(\boldsymbol{f})$. Hence both $\boldsymbol{f}^+$ and $\boldsymbol{f}^-$ are optimal, and one of them must have smaller support than $\boldsymbol{f}$, a contradiction. □

Note that the preceding result shows that in the case of single-sink hose models, $\text{OPT}_{SPR} = \text{OPT}_{TR}$. It is not the case that $\text{OPT}_{MPR} = \text{OPT}_{TR}$ in this setting however. If that were the case, SSROB would be polynomially solvable, but the case where $b_r = 1$ already captures the Steiner Tree problem. Because of this tree structure, arguing why the gap holds in the case of SPR is considerably simpler. The argument contains some intuition as to why the gap also holds for MPR, so we outline this approach now.

Suppose we have an SPR solution where we only use one edge rv from $\delta(r)$. Then in the SPR solution, everyone must route to v in G (think of this as a tree T for now). Since G was bounded degree this means that many nodes (a constant fraction) must use long paths, of length $\log_d(n)$. If these all had to pay one unit along their whole path then this already costs $\Omega(n \log n)$. But it is not as easy as that; if we have a subtree T_w rooted at node w that contains at least $b_r = \beta n$ nodes, then in fact we only need to pay for b_r units on the edge out of w.

Imagine removing the edges of T which are used by more than βn terminals, leaving a number of subtrees, each containing at most βn terminals. If T is fairly balanced, there are around $\Theta(n/(\beta n)) = \Theta(1/\beta)$ such subtrees. (If T is very unbalanced, there could be many more—consider a caterpillar. For the full proof, one must use the larger distances of leaves to the root to get the required bound.) In each such subtree, a good fraction of the leaves are a distance roughly $\log \beta n$ from the root of this subtree. Since there is no cost sharing within this subtree, these nodes really do pay $\beta n \log(\beta n)$. Thus the subtrees combined pay

$$\Omega\left(1/\beta \cdot \beta n \log(\beta n)\right) = \Omega\left(n \log(\beta n)\right).$$

If we set $\beta = \frac{1}{\log n}$, this yields a cost of $\Omega(n \log n)$.

Making the above argument precise requires balancing the use of multiple edges into r, and becomes somewhat technical. It also does not extend to establish the MPR gap, so instead we now turn to the latter problem (which in any case implies the FR versus SPR gap).

2.4 Buy-and-Rent: An $\Omega(\log n)$ Gap between FR and MPR

The main difficulty with analyzing the MPR model is crystalized by the fact that Proposition 1 does not hold for MPR. In particular, we cannot apply the augmentation proof used in this result to the case where nodes are allowed to use fractional routing templates.

Let us first examine more closely the cost on edges induced by an MPR routing template for a single-sink hose design problem. As in Lemma 1, it is sufficient to consider the cases where we wish the network to support the routing of any βn of the nodes in V to the sink r simultaneously. Suppose that $f_i(e)$ represents the flow that node i sends on edge e in a template, then for the single sink hose design problem, the formula for the capacity needed by e is:

$$\max_{D \in \mathcal{P}} \sum_{i \in V} D_{ir} f_i(e) = \max_{W \subseteq V : |W| = \beta n} \sum_{i \in W} f_i(e), \tag{1}$$

where recall that $\mathcal{P}$ is the set of single-sink hose matrices. In other words, the capacity needed on edge e is just the sum of the βn largest values of $f_i(e)$.

We introduce a new routing cost model which we call (single-sink) *buy-and-rent* (BAR). This exactly models the MPR cost model defined above, but is more manageable in terms of analysis. In the buy-and-rent problem, there are costs on the edges, and unit demands from some subset W of nodes called *terminals*. Each terminal wishes to (fractionally) route one unit of demand to the sink r. Apart from the costs $c(e)$ on the edges, we also have a parameter k. The difference from rent-or-buy is that we may now purchase some capacity amount $\gamma(e) \in [0, 1]$ (in rent-or-buy we would buy an infinite capacity link) and the interpretation is that every terminal is allowed to use up to $\gamma(e)$ units of capacity on the edge. If it chooses to route any more on that edge, then it must pay for the additional rental cost. The cost of purchasing capacity on an edge e is $k\gamma(e)c(e)$.

Buy-and-rent can be considered as an LP relaxation of (single-sink) rent-or-buy; this formulation is in fact very similar to the LP relaxation used by Swamy and Kumar [14] to give constant factor approximation algorithms for connected facility location and single-sink rent-or-buy. Their formulation is stronger however (in that the optimum for their LP lies between the BAR and SPR optima), and so does not exactly model the MPR problem. In particular, in buy-and-rent, solutions may conceivably use flow paths that alternate several times between rented capacity and purchased capacity. In contrast, a solution to the LP of Swamy and Kumar [14] always has a connected "core" of purchased edges containing the sink node and terminals use rented capacity to route to that core.

Proposition 2. *The buy-and-rent problem with parameter $k = \beta n$, and the single-sink hose design problem in the* MPR *routing model have the same optimal solution.*

Proof. Suppose that $(\boldsymbol{f}_i)$ is an MPR routing template for the robust hose design problem. Consider the BAR solution for parameter $k = \beta n$ obtained as follows. For each edge e, order the terminals so that $f_{\pi(1)}(e) \geq f_{\pi(2)}(e) \geq \dots f_{\pi(n)}(e)$.

Then we purchase $\gamma(e) = f_{\pi(k)}(e)$ units of capacity on edge e, and we use the same routing $\boldsymbol{f}_i$ as the MPR solution. This guarantees that for any edge, none of the terminals $\pi(j)$ with $j > k$, pays to route on edge e since we purchased enough capacity for them to travel for free. For each terminal $\pi(j)$ with $j \leq k$, it must pay the rental cost to route $f_{\pi(j)}(e) - f_{\pi(k)}(e) \geq 0$. This costs $c(e)$ times the amount $\sum_{j \leq k}(f_{\pi(j)}(e) - f_{\pi(k)}(e)) = \sum_{j \leq k} f_{\pi(j)}(e) - k f_{\pi(k)}(e)$. Since the purchased capacity cost $k f_{\pi(k)}(e)c(e)$, the total buy-and-rent cost is $c(e)\sum_{j \leq k} f_{\pi(j)}(e)$ which is the cost of edge e in the MPR template using (1).

Conversely, suppose that we have a minimum cost solution for BAR and consider the robust design cost for using the same routing as a template. Without loss of generality $\gamma(e) = f_{\pi(k)}(e)$ since if $\gamma(e)$ was larger than this, then by reducing the capacity bought by sufficiently small $\epsilon > 0$, the rental costs are unaffected for terminals $\pi(j)$ for $j \geq k$. And for terminals $\pi(j)$ with $j < k$, their rental cost increases by at most $\epsilon c(e)$. Hence the total rental cost increases by $k\epsilon c(e)$, and the total cost of bought capacity reduces by $k\epsilon c(e)$, thus decreasing the overall cost.

Similarly, if $\gamma(e) < f_{\pi(k)}(e)$, then increasing the bought capacity γ by some small $\epsilon > 0$, has cost of $k\epsilon c(e)$. But the reduction in rental costs is at least the reduction in rental cost of the first k terminals which is $k\epsilon c(e)$, and thus the overall cost does not increase as a result of increasing γ. Hence the cost of edge e is just the purchase cost $c(e) \cdot k f_{\pi(k)}(e)$ plus the rental cost $c(e)\sum_{j \leq k}(f_{\pi(j)}(e) - f_{\pi(k)}(e))$ and this is identical to the robust design cost when using the same template. □

We again take $\beta = 1/\log n$ (so $k = n/\log n$). We now prove that any solution to the BAR problem on our expander instance is expensive; this together with the preceding proposition implies our main result, Theorem 1.

Theorem 2. *Any solution to the* BAR *problem on the expander instance has cost* $\Omega(n \log n)$.

Proof. Consider an arbitrary BAR solution, determined by bought capacity γ_e on each edge, and a flow template ($\boldsymbol{f}_i$: for each terminal i).

Let $\gamma(\delta(r)) := \sum_{v \in V} \gamma_{vr}$ be the total bought capacity on the *port edges* (these are the edges connecting r to the nodes in V), and let $\gamma(E) := \sum_{e \in E} \gamma_e$ be the capacity bought in the expander. The cost of buying capacity in the expander is then $k \cdot \gamma(E)$, so we may assume that $\gamma(E) < \log^2 n$, or else the solution already costs $\Omega(n \log n)$. A similar argument for port edges (but recalling that these edges cost $\log n$) allows us to assume that $\gamma(\delta(r)) < \log n$.

For a terminal v, let $B_i(v)$ be the set of nodes (or sometimes, their induced graph) in the expander that are a distance at most i from v. We are particularly interested in balls of radius $R := \lfloor \log_d \sqrt{n} \rfloor - 1 = \lfloor \log n/(2 \log d) \rfloor - 1$; we use $B(v)$ as shorthand for $B_R(v)$. Note that since G is d-regular,

$$|B(v)| \leq \sum_{i=0}^{R} d^i \leq d^{R+1} \leq n^{1/2}.$$

Let $\gamma^E(v) := \sum_{e \in E: e \subset B(v)} \gamma(e)$ and $\gamma^P(v) := \sum_{w \in B(v)} \gamma(wr)$. A single $\gamma(e)$ for an edge $e = u_1 u_2$ contributes to many $\gamma^E(v)$'s, but not too many:

$$|\{v : e \subset B(v)\}| \le |\{v : u_1 \in B(v)\}| = |B(u_1)| \le n^{1/2}.$$

So we must have that

$$\sum_{v \in V} \gamma^E(v) \le n^{1/2} \gamma(E) \le n^{1/2} \log^2 n. \tag{2}$$

Similarly,

$$\sum_{v \in V} \gamma^P(v) \le n^{1/2} \log n. \tag{3}$$

Consider an arbitrary terminal v. The unit of flow from v can be divided up into three types depending on how the flow enters r:

- A fraction μ_v^r of flow that rents on the port edge it uses.
- A fraction μ_v^b of flow that uses bought port capacity, on a port within a distance R from v.
- A fraction μ_v^t representing all remaining flow; this flow must "travel" and use port edges that are further than R from v.

Clearly $\mu_v^r + \mu_v^b + \mu_v^t = 1$.

We now aim to find a lower bound on the total rental cost paid by the terminals. Flow that rents the port edge must pay $\log n$ just for this edge, giving a cost of $\mu_v^r \log n$. Now consider the μ_v^t fraction of flow that travels outside the ball $B(v)$ in the expander before using a port edge. This flow must cross each of the cuts $C_i := \delta(B_i(v))$, for $0 \le i \le R$.

The maximum amount of flow that can travel across cut C_i for free (using the bought capacity) is $\gamma(C_i)$, and so there is a rental cost of at least $\mu_v^t - \gamma(C_i)$ in crossing cut C_i. Summing over all the cuts, we find that the rental cost associated with this travelling flow is at least

$$\sum_{i=0}^{R-1} (\mu_v^t - \gamma(C_i)) \ge R\mu_v^t - \gamma^E(v).$$

Thus the rental cost associated with terminal v is at least

$$\log n \cdot \mu_v^r + R\mu_v^t - \gamma^E(v).$$

Summing this over all terminals v, we obtain a total rental cost of at least

$$\begin{aligned} C(\text{rent}) &\ge \sum_{v \in V} (\log n \cdot \mu_v^r + R \cdot \mu_v^t) - \sum_{v \in V} \gamma^E(v) \\ &\ge R \sum_{v \in V} (\mu_v^r + \mu_v^t) - \sum_{v \in V} \gamma^E(v) && \text{since } R \le \log n \\ &\ge R \sum_{v \in V} (\mu_v^r + \mu_v^t) - n^{1/2} \log^2 n && \text{by (2).} \end{aligned}$$

Finally, note that

$$\sum_{v \in V} (\mu_v^r + \mu_v^t) = \sum_{v \in V} (1 - \mu_v^b) \geq \sum_{v \in V} (1 - \gamma^P(v))$$
$$\geq n - n^{1/2} \log n \qquad \text{by (3).}$$

Thus

$$C(\text{rent}) \geq R \cdot (n - n^{1/2} \log n) - n^{1/2} \log^2 n$$
$$= \Omega(n \log n),$$

since $R = \Theta(\log n)$. □

3 Single Path Routing vs. Tree Routing

As discussed in the introduction, for any robust network design problem we have $\text{OPT}_{TR} = \tilde{O}(\log n)\text{OPT}_{FR}$. We now show that this is best possible by exhibiting a problem instance such that $\text{OPT}_{TR} = \Omega(\log n)\text{OPT}_{SPR}$, and so also $\text{OPT}_{TR} = \Omega(\log n)\text{OPT}_{FR}$.

Consider a graph on n vertices with girth (length of the shortest cycle in the graph) $\Omega(\log n)$, and with cn edges, where $c > 1$ is a constant. [Note that it's the requirement that $c > 1$ that makes the existence of such graphs nontrivial: for $c = 1$, a cycle on n vertices gives a graph with girth n.] Such graphs exist: e.g., Lemma 15.3.2 in [12] states that there exist graphs with girth ℓ and $\frac{1}{9}n^{1+1/(\ell-1)}$ edges. Taking $\ell := \log n/100$ gives that there exist graphs with girth at least $\log n/100$ and cn edges where $c > 1$, as needed.

This graph defines the network topology for our problem instance: all the nodes are terminals, and all edges have unit cost. The demand polytope is given by a single demand: there is a unit demand between terminals connected by an edge.

Clearly, a good SPR template is the network itself, and its cost is cn, the number of edges. Now, if we take any tree template, then edges of the network that are not included in the tree have to be routed on a path of length $\Omega(\log n)$ because of the girth property of the graph. There are at least $cn - (n - 1) = (c - 1)n + 1$ such edges, and so the total cost of the tree template is $\Omega(n \log n)$.

4 Conclusions

One natural question concerns the gap between MPR and SPR for the general single-sink robust network design problem. We are not aware of any single-sink demand polytopes for which the gap is superconstant. Hence the single-sink robust design problem, i.e. computing OPT_{SPR}, could conceivably have a constant factor approximation algorithm for well-described polytopes. This would be of interest since it generalizes a host of well-known problems such as Steiner tree, single-sink rent-or-buy, and single-sink buy-at-bulk (the last follows from a transformation given in [13]).

Acknowledgements. We would like to thank Gianpaolo Oriolo for some very helpful discussions. We also thank an anonymous reviewer for ESA, for some detailed and useful input.

The second author is supported by a MELS Quebec Merit Scholarship for Foreign Students. The third author is supported by a NSERC Discovery Grant.

References

1. Abraham, I., Bartal, Y., Neiman, O.: Nearly tight low stretch spanning trees. In: Proc. of IEEE FOCS, pp. 781–790 (2008)
2. Bartal, Y., Leonardi, S.: On-line routing in all-optical networks. Theor. Comput. Sci. 221(1-2), 19–39 (1999)
3. Ben-Ameur, W., Kerivin, H.: New economical virtual private networks. Commun. ACM 46(6), 69–73 (2003)
4. Chekuri, C.: Routing and network design with robustness to changing or uncertain traffic demands. SIGACT News 38(3), 106–128 (2007)
5. Chekuri, C., Oriolo, G., Scutella, M.G., Shepherd, F.B.: Hardness of robust network design. Networks 50(1), 50–54 (2007)
6. Fakcharoenphol, J., Rao, S., Talwar, K.: A tight bound on approximating arbitrary metrics by tree metrics. J. Comput. Syst. Sci. 69(3), 485–497 (2004)
7. Fingerhut, A.J., Suri, S., Turner, J.S.: Designing least-cost nonblocking broadband networks. J. Algorithms 24(2), 287–309 (1997)
8. Gupta, A., Kleinberg, J., Kumar, A., Rastogi, R., Yener, B.: Provisioning a virtual private network: a network design problem for multicommodity flow. In: Proc. of ACM STOC, pp. 389–398 (2001)
9. Gupta, A., Kumar, A., Pál, M., Roughgarden, T.: Approximation via cost sharing: Simpler and better approximation algorithms for network design. J. ACM 54(3), 11 (2007)
10. Hajiaghayi, M., Kleinberg, R., Räcke, H., Leighton, T.: Oblivious routing on node-capacitated and directed graphs. ACM Trans. Algorithms 3(4), 51 (2007)
11. Maggs, B.M., Meyer aud der Heide, F., Vöcking, B., Westerman, M.: Exploiting locality for networks of limited bandwidth. In: Proc. of IEEE FOCS, pp. 284–293 (1997).
12. Matousek, J.: Lectures on Dicrete Geometry. Springer, Heidelberg (2002)
13. Olver, N., Shepherd, F.: Approximability of robust network design. (manuscript) (2009)
14. Swamy, C., Kumar, A.: Primal-dual algorithms for connected facility location problems. Algorithmica 40(4), 245–269 (2004)

Algorithms Meet Art, Puzzles, and Magic

Erik D. Demaine

MIT Computer Science and Artificial Intelligence Laboratory
32 Vassar Street, Cambridge, MA 02139, USA
edemaine@mit.edu

Abstract

When I was six years old, my father Martin Demaine and I designed and made puzzles as the *Erik and Dad Puzzle Company*, which distributed to toy stores across Canada. So began our journey into the interactions between algorithms and the arts. More and more, we find that our mathematical research and artistic projects converge, with the artistic side inspiring the mathematical side and vice versa. Mathematics itself is an art form, and through other media such as sculpture, puzzles, and magic, the beauty of mathematics can be brought to a wider audience. These artistic endeavors also provide us with deeper insights into the underlying mathematics, by providing physical realizations of objects under consideration, by pointing to interesting special cases and directions to explore, and by suggesting new problems to solve (such as the metapuzzle of how to solve a puzzle). This talk will give several examples in each category, from how our first font design led to a universality result in hinged dissections, to how studying curved creases in origami led to sculptures at MoMA. The audience will be expected to participate in some live magic demonstrations.

A. Fiat and P. Sanders (Eds.): ESA 2009, LNCS 5757, p. 289, 2009.

Polynomial-Time Algorithm for the Leafage of Chordal Graphs

Michel Habib and Juraj Stacho

LIAFA – CNRS and Université Paris Diderot – Paris VII,
Case 7014, 75205 Paris Cedex 13, France
{habib,jstacho}@liafa.jussieu.fr

Abstract. Every chordal graph G can be represented as the intersection graph of a collection of subtrees of a host tree, the so-called tree model of G. The leafage $l(G)$ of a connected chordal graph G is the minimum number of leaves of the host tree of a tree model of G. This concept was first defined by I.-J. Lin, T.A. McKee, and D.B. West in [9]. In this contribution, we present the first polynomial time algorithm for computing $l(G)$ for a given chordal graph G. In fact, our algorithm runs in time $O(n^3)$ and it also constructs a tree model of G whose host tree has $l(G)$ leaves.

1 Introduction

In this paper, graph is always simple, undirected and loopless.

A graph is *chordal*, if it has no induced cycles of length four or longer. By a result of Gavril [5], a graph G is chordal if and only if G can be represented as the intersection graph of a collection of subtrees of a host tree, the so-called *tree model* of G. The *leafage* $l(G)$ of a connected chordal graph G is defined as the minimum number of leaves of the host tree of a tree model of G. If G is an interval graph (the intersection graph of intervals of the real line), we always have $l(G) = 2$. Hence, the leafage can be seen as a measure of how far a chordal graph is from being an interval graph. This has several algorithmic consequences. For instance, it is shown in [7] that a chordal graph with bounded leafage always has a so-called implicit representation which allows some problems on such graphs to be solved more efficiently. Moreover, efficient solutions to NP-hard problems on interval graphs naturally extend to efficient solutions on chordal graphs whose leafage is bounded; e.g., the k-subcolouring problem [2,12,13].

The leafage of chordal graphs was first introduced by I.-J. Lin, T.A. McKee, and D.B. West in [9] where the authors establish several bounds on this parameter for special cases of chordal graphs such as block graphs, split graphs, and k-trees. Their bounds imply polynomial time algorithms for computing the leafage in some of these special cases; however, the general question of complexity of computing $l(G)$ for a given chordal graph G is not addressed. Since their paper, this question remained unresolved [14] except in special cases such as split graphs [8,9] and the case of deciding $l(G) \leq k$ for $k \in \{2, 3\}$; the case $k = 2$ corresponds to interval graph recognition which is polynomial [1], and $k = 3$ is polynomial by [10].

A. Fiat and P. Sanders (Eds.): ESA 2009, LNCS 5757, pp. 290–300, 2009.

In this paper, we finally resolve this question by providing a polynomial time algorithm computing $l(G)$ for any given chordal graph G. In particular, our algorithm runs in time $O(n^3)$ where n is the number of vertices of G and also outputs a tree model of G with $l(G)$ leaves.

The paper is structured as follows. In Section 2, we define basic notions such as clique tree and the reduced clique graph, and we show some of their useful properties related to the leafage. In Section 3, we introduce the so-called token mappings and explain their relationship to clique trees. In Section 4, we define augmenting paths and sequences and explain how they can be used to decrease the number of leaves in a given clique tree. Finally, in Section 5, we describe our algorithm and analyze its complexity.

2 Basic Concepts

For a graph G and a set $X \subseteq V(G)$, we denote by $G[X]$ the subgraph of G induced on X, and denote by $G - X$ the subgraph $G[V(G) \setminus X]$. A *complete subgraph* or *clique* of G is a (not necessarily maximal) set of pairwise adjacent vertices of G. (For a complete terminology, see [15].)

Let G be a connected chordal graph. A *clique tree* of G is any tree T whose vertices are the maximal cliques of G such that for every two maximal cliques C, C', each clique on the path from C to C' in T contains $C \cap C'$. We shall assume that every edge CC' of T is labeled by $C \cap C'$. (See Figure 1 for an example of a clique tree.)

Any clique tree T can be seen as a tree model of G whose host tree is T. It is shown in [9] that G always has a clique tree with $l(G)$ leaves. Hence, in the rest of the paper, we shall focus on clique trees.

Cliques C, C' of G form a *separating pair*, if every path from a vertex of $C \setminus C'$ to a vertex of $C' \setminus C$ contains a vertex of $C \cap C'$. The *reduced clique graph* $\mathcal{C}_r(G)$ of G is a graph whose vertices are the maximal cliques of G, and whose edges CC' are between cliques C, C' forming separating pairs. In addition, each edge CC' of $\mathcal{C}_r(G)$ is labeled by $C \cap C'$.

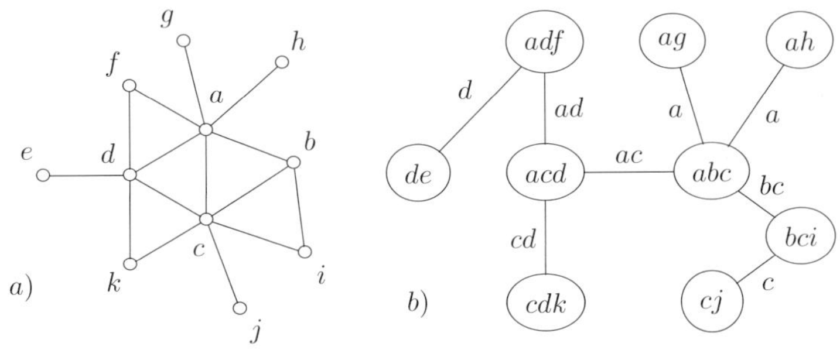

Fig. 1. *a)* Example chordal graph G, *b)* a clique tree T of G

The following is a fundamental result about reduced clique graphs.

Theorem 1. [4] *A tree T is a clique tree of G if and only if T is a maximum weight spanning tree of $\mathcal{C}_r(G)$ where the weight of each edge CC' is defined as $|C \cap C'|$. Moreover, the reduced clique graph $\mathcal{C}_r(G)$ is precisely the union of all clique trees of G.*

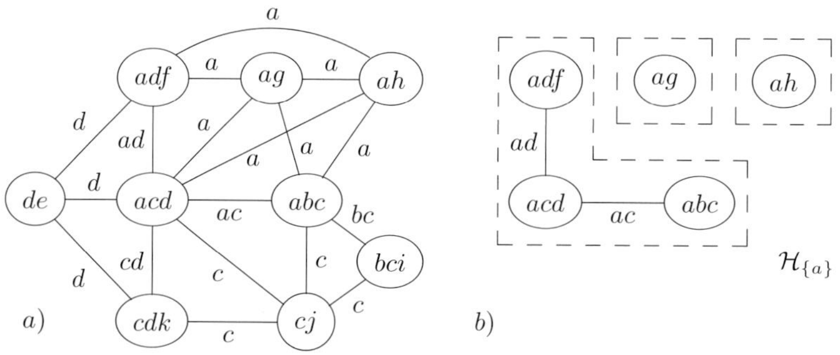

Fig. 2. *a)* The reduced clique graph $\mathcal{C}_r(G)$ of G, *b)* minimal separator graph $\mathcal{H}_{\{a\}}$

2.1 Minimal Separator Graphs

Let a, b be two vertices of G. A subset S of the vertices of G *disconnects* a from b in G, if a and b are in different connected components of $G - S$.

A subset S of the vertices of G is called a *minimal separator*, if there exist vertices a and b such that (i) S disconnects a from b, and (ii) no proper subset of S disconnects a from b in G.

For each minimal separator S, let $\mathcal{R}_S$ denote the set of all maximal cliques C of G with $S \subseteq C$, and let $\mathcal{H}_S$ denote the graph whose vertex set is $\mathcal{R}_S$ and whose edges are between cliques C, C' such that $C \cap C' \supsetneq S$.

Using the graphs $\mathcal{H}_S$ we can characterize the structure of $\mathcal{C}_r(G)$. (The proof is omitted due to the length restriction.)

Theorem 2. *CC' is an edge of $\mathcal{C}_r(G)$ with label $S = C \cap C'$ if and only if C and C' belong to different connected components of $\mathcal{H}_S$.* □

Let $\mathcal{S}(G)$ denote the set of all minimal separators of G. Note that chordality of G implies that for every $S \in \mathcal{S}(G)$, there exist cliques C, C' that form a separating pair such that $C \cap C' = S$. This implies, in particular, that every minimal separator of G appears on some edge of $\mathcal{C}_r(G)$.

2.2 Structure of Clique Trees

Let T be a clique tree of G. By Theorem 1, we have that T is characterized in terms of $\mathcal{C}_r(G)$ as a maximum weight spanning tree of $\mathcal{C}_r(G)$. Conversely, we can characterize the reduced clique graph $\mathcal{C}_r(G)$ in terms of T as follows. (The proof is omitted due to the length restriction.)

Theorem 3. *CC' is an edge of $\mathcal{C}_r(G)$ with label $S = C \cap C'$ if and only if there exists an edge with label S on the path from C to C' in T.* □

Furthermore, we can describe the structure of T in terms of the graphs $\mathcal{H}_S$.

Theorem 4. *Let $S \in \mathcal{S}(G)$ and let k_S denote the number of connected components of $\mathcal{H}_S$. Then T contains exactly $(k_S - 1)$ edges with label S, and each connected component of $\mathcal{H}_S$ induces a connected subgraph in T.*

Proof. Let C, C' be two vertices of $\mathcal{H}_S$, and let P be the path from C to C' in T. Clearly, every vertex on P must belong to $\mathcal{H}_S$ because it contains $C \cap C' \supseteq S$ by the definition of clique tree. Hence, the vertices of $\mathcal{H}_S$ induce a connected subgraph in T. Next, suppose that C, C' belong to some component $\mathcal{K}$ of $\mathcal{H}_S$. Again, it follows that every vertex of P also belongs to $\mathcal{K}$ (details omitted). Hence, both the vertices of $\mathcal{H}_S$ and of each component of $\mathcal{H}_S$ induce a subtree in T. The claim now follows. □

Note that T has at most n vertices by [3], and for each minimal separator S of G, we have $k_S \geq 2$ since there is at least one edge with label S in $\mathcal{C}_r(G)$ (see the remark above). Hence, every minimal separator of G appears on some edge of T, and conversely, for every edge CC' of T, the set $C \cap C'$ is a minimal separator and C, C' form a separating pair (by Theorem 1). In particular, G has at most $n - 1$ minimal separators.

3 Degrees and Tokens

A *degree mapping* assigns to each vertex of a graph its degree.

Theorem 5. [15] *A mapping $f : X \to \mathbb{N}$ is a degree mapping of a tree if and only if*

(i) $1 \leq f(x) \leq |X| - 1$ *for each* $x \in X$, *and*
(ii) $\sum_{x \in X} f(x) = 2|X| - 2$.

We define a similar notion. A *token mapping* is a mapping τ that assigns to each maximal clique C of G a distinct set of *tokens* $\tau(C)$ where each token is labeled by some minimal separator of G. If a token t belongs to $\tau(C)$, we also say that *t is a token of τ* and that *t belongs to C in τ*.

Let T be a clique tree of G. The *extended degree mapping* of T, denoted by ε_T, is a token mapping that assigns to each maximal clique C a set $\varepsilon_T(C)$ of tokens corresponding to the edges incident to C in T where each token is labeled by the label of the corresponding edge. (See Figure 3 for an example of a clique tree T and its extended degree mapping ε_T.)

Using Theorem 4, we now describe necessary and sufficient conditions characterizing extended degree mappings of clique trees.

Theorem 6. *A token mapping τ is an extended degree mapping of a clique tree of G if and only if*

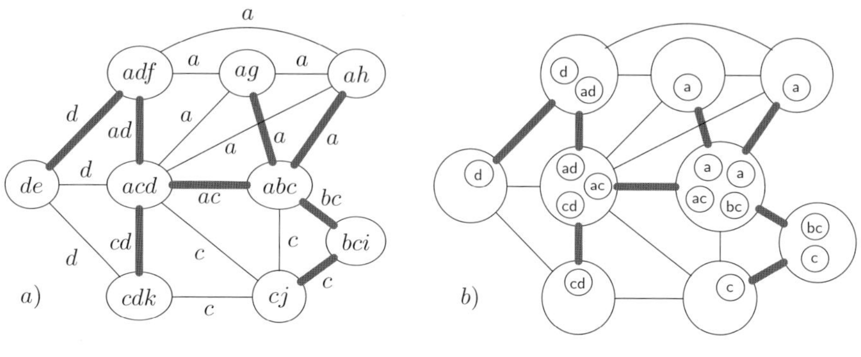

Fig. 3. *a)* The graph $\mathcal{C}_r(G)$ with an example clique tree T, *b)* token mapping $\tau = \varepsilon_T$

(R1) *for each maximal clique C of G, the set $\tau(C)$ is non-empty,*
(R2) *for each minimal separator S of G, if a token of τ with label S belongs to $\tau(C)$ for some maximal clique C of G, then $C \supseteq S$,*
(R3) *for each minimal separator S of G, the number of tokens of τ with label S is exactly $2k_S - 2$ where k_S is the number of components of $\mathcal{H}_S$,*
(R4) *for each minimal separator S of G and every component $\mathcal{K}$ of $\mathcal{H}_S$, there exists a token with label S in $\tau(C)$ for some vertex C of $\mathcal{K}$.*

Proof. The forward direction follows directly from Theorem 4. For the backward direction, let τ be a token mapping satisfying (R1-R4). We construct a subgraph T of $\mathcal{C}_r(G)$ as follows. The vertices of T are the maximal cliques of G, and initially T has no edges.

We consider every minimal separator S of G one by one as follows. We let $\mathcal{K}_1, \ldots, \mathcal{K}_{k_S}$ be the connected components of $\mathcal{H}_S$, and we let $a_1, \ldots, a_{k_S}$ denote the number of tokens of τ with label S in $\mathcal{K}_1, \ldots, \mathcal{K}_{k_S}$, respectively. By (R4), $a_i \geq 1$ for each $1 \leq i \leq k_S$, and by (R2) and (R3), we have $\sum_{i=1}^{k_S} a_i = 2k_S - 2$. Hence, by Theorem 5, there exists a tree $\mathcal{T}$ whose vertices are $\mathcal{K}_1, \ldots \mathcal{K}_S$ such that the degree of $\mathcal{K}_i$ in $\mathcal{T}$ is exactly a_i.

Now, for each $1 \leq i \leq k_S$, we let $t_1^i, \ldots, t_{a_i}^i$ be the tokens of τ on the vertices of $\mathcal{K}_i$ and let $C_1^i, \ldots, C_{a_i}^i$ be the vertices of $\mathcal{K}_i$ that contain tokens $t_1^i, \ldots, t_{a_i}^i$, respectively. (Possibly, $C_j^i = C_{j'}^i$ for some $j \neq j'$.) Also, we let $e_1^i, \ldots, e_{a_i}^i$ be a fixed ordering of edges of $\mathcal{T}$ incident to $\mathcal{K}_i$. Finally, for each edge $e = K_i K_{i'}$ of $\mathcal{T}$, we have j, j' such that $e = e_j^i = e_{j'}^{i'}$ by the above definition, and we add the edge $C_j^i C_{j'}^{i'}$ to T. (See example in Figure 4.)

We now show that T is a clique tree of G. First, we observe that T is a subgraph of $\mathcal{C}_r(G)$ by Theorem 2, and it contains exactly $k_S - 1$ edges with label S for every minimal separator S of G by our construction. By Theorem 4, every clique tree of G also has this property. Therefore, the weight of T is the same as the weight of any clique tree of G. By Theorem 1, it now suffices to show that T is connected. This can be proved by showing that $T[\mathcal{R}_S]$ is connected for each $S \in \mathcal{S}(G)$ which follows by induction on $n - |S|$. (We omit further details.) $\square$

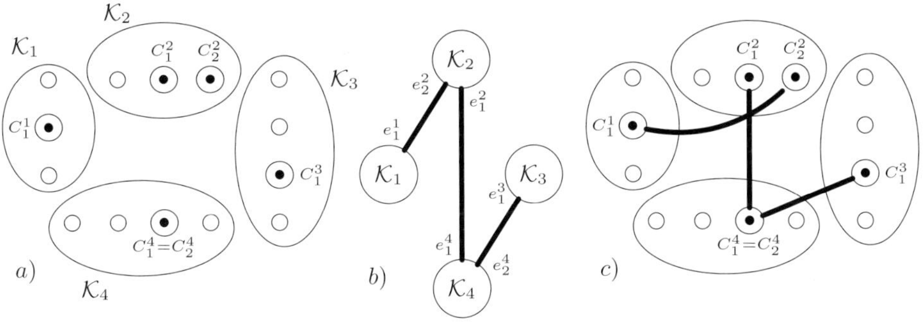

Fig. 4. *a)* Components of $\mathcal{H}_S$ and tokens with label S, *b)* The tree $\mathcal{T}$ with orderings of incident edges, *c)* edges added to T

We say that a token mapping τ is *realizable*, if it satisfies the conditions (R1-R4). In light of the above theorem, we define the *degree* of C in τ, denoted by $\mathsf{deg}_\tau(C)$, to be the value $\mathsf{deg}_\tau(C) = |\tau(C)|$. If $\mathsf{deg}_\tau(C) = 1$, we call C a *leaf* of τ. We denote by $\#\mathsf{leaves}(\tau)$ the number of leaves of τ.

4 Alternation and Augmentation

In this section, we show how to obtain from a realizable token mapping τ another realizable token mapping τ' with less number of leaves (if possible). We do this by moving tokens along certain paths on the maximal cliques; these paths are obtained by exploring an auxiliary graph $\mathcal{D}_\tau$ (described below). In particular, this process will resemble the classical maximum matching algorithm. For this reason, we shall use "alternating" and "augmenting" to describe similar notions in our algorithm.

Let $\mathcal{D}(G)$ denote the multidigraph[1] on the maximal cliques of G with labeled arcs such that $e = CC'$ is an arc of $\mathcal{D}(G)$ labeled with S if and only if C, C' belong to $\mathcal{R}_S$ where S is a minimal separator of G.

Let $e = CC'$ be an arc of $\mathcal{D}(G)$ with label S and τ be a token mapping such that $\tau(C)$ contains a token t labeled with S. We write $\tau \div e$ for the token mapping obtained from τ by removing the token t from $\tau(C)$ and adding t to $\tau(C')$.

4.1 Realizable Arcs

Let τ be a realizable token mapping, and let $\mathcal{D}_\tau$ denote the digraph on the maximal cliques of G with arcs e for which $\tau \div e$ is realizable.

We characterize the digraph $\mathcal{D}_\tau$ as follows.

Proposition 7. *An arc $e = CC'$ with label S belongs to $\mathcal{D}_\tau$ if and only if*

(i) C and C' are both vertices of $\mathcal{H}_S$,

[1] Digraph with multiple arcs between any two points allowed.

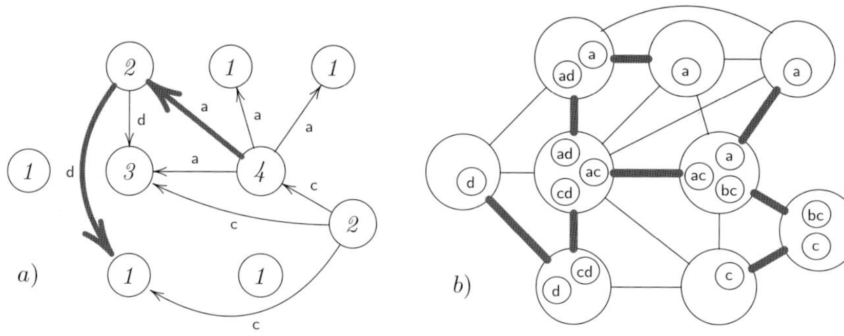

Fig. 5. *a)* the digraph $\mathcal{D}_\tau$ with an augmenting path $P = e_1, e_2$, *b)* the token mapping $\tau' = \tau \div e_1 \div e_2$ and the corresponding clique tree

(ii) $\tau(C)$ *contains a token with label* S,
(iii) $\mathsf{deg}_\tau(C) \geq 2$, *and*
(iv) if $\mathcal{K}$ *is the connected component of* $\mathcal{H}_S$ *that contains* C, *then*
  (a) either C' *belongs to* $\mathcal{K}$,
  (b) or C' *does not belong to* $\mathcal{K}$ *and there exists* $C'' \neq C$ *in* $\mathcal{K}$ *such that* $\tau(C'')$ *contain a token with label* S.

Proof. The claim follows directly from (R1-R4). □

4.2 Sequences

Again, let τ be a realizable token mapping.

We say that a sequence of arcs $e_1, \ldots, e_k$ of $\mathcal{D}(G)$ is a *τ-sequence*, if there exists a sequence of token mappings $\tau_0, \tau_1, \ldots, \tau_k$ where $\tau_0 = \tau$ such that for each $i \in \{1 \ldots k\}$, the arc $e_i = C_i C_i'$ satisfies

(S1) e_i is an arc of $D_{\tau_{i-1}}$,
(S2) $\tau_i = \tau_{i-1} \div e_i$,
(S3) $\mathsf{deg}_{\tau_{i-1}}(C_i) \geq 3$, and
(S4) if $i < k$ then $\mathsf{deg}_{\tau_{i-1}}(C_i') \geq 2$.

In addition, a τ-sequence $e_1, \ldots, e_k$ is an *alternating τ-sequence*, if

(S5) no arc among $e_1, \ldots, e_{k-1}$ is incident to C_k', and
(S6) $\mathsf{deg}_{\tau_{k-1}}(C_k') \leq 2$,

and an alternating τ-sequence $e_1, \ldots, e_k$ is an *augmenting τ-sequence* if

(S7) $\mathsf{deg}_{\tau_{k-1}}(C_k') = 1$.

This definition immediately implies the following statement.

Observation 8. *If* $e_1, \ldots, e_k$ *is an augmenting τ-sequence, then for* $\tau' = \tau \div e_1 \div e_2 \div \ldots \div e_k$, *we have* $\#\mathsf{leaves}(\tau) > \#\mathsf{leaves}(\tau')$. □

Also, we have the following useful observation which we shall need later. (The proof is omitted due to the length restriction.)

Proposition 9. *If a τ-sequence $e_1, \dots, e_k$ satisfies (S7), then $e_1, \dots, e_k$ is an augmenting τ-sequence.* □

We now prove the following theorem which is the first of the two ingredients in our polynomial time algorithm for the leafage.

Theorem 10. *Let T and T^* be two clique trees of G such that T has more leaves than T^*. Then there exists an augmenting ε_T-sequence.*

Proof. We define the *distance* from T to T^* to be the value

$$\mathsf{dist}(T, T^*) = \sum_{CC' \in E(T)} \Big(d_{T^*}(C, C') - 1 \Big)$$

where $d_{T^*}(C, C')$ denotes the distance between C and C' in T^*.

The proof is by induction on $\mathsf{dist}(T, T^*)$. Since T has more leaves than T^* and both have the same number of edges, there must exist C such that the degree of C in T is at least three and is strictly larger than the degree of C in T^*. This implies that if $T_1, \dots, T_k$ are the components of $T - C$, there exists i such that no vertex of T_i is adjacent to C in T^*. Let C' be the vertex of T_i that is adjacent to C in T. Let $\mathcal{A}$ and $\mathcal{B}$ be the vertices of the two connected components we obtain by removing the edge CC' from T; we can assume $C \in \mathcal{A}$ and $C' \in \mathcal{B}$. Let P be the path from C to C' in T^*. Since $C \in \mathcal{A}$ and $C' \in \mathcal{B}$, there exists an edge C^*C^{**} of P such that $C^* \in \mathcal{A}$ and $C^{**} \in \mathcal{B}$. By the definition of T_i, we have $C^* \neq C$. Let $S = C \cap C'$ and $S^* = C^* \cap C^{**}$. Since C^*C^{**} is on the path from C to C' in T^*, we have $S \subseteq S^*$. Also, CC' is on the path from C^* to C^{**} in T, since $C^* \in \mathcal{A}$ and $C^{**} \in \mathcal{B}$. Hence, $S^* \subseteq S$ and consequently $S = S^*$. We observe that $C^* \cap C' \subseteq C^* \cap C^{**} = S$, since the edge C^*C^{**} lies on the path from C^* to C'. Hence, $C^* \cap C' = S$, since $C^* \supseteq S^* = S$ and $C' \supseteq S$. Finally, by Theorem 3, we have that C^*C' is an edge of $\mathcal{C}_\tau(G)$.

Next, let T' denote the tree we obtain by removing the edge CC' from T and adding the edge C^*C'. By Theorem 1, T' is a clique tree of G, since T is. This implies that $e = CC^*$ is an arc of $\mathcal{D}_{\varepsilon_T}$ with label S, since $\varepsilon_{T'} = \varepsilon_T \div e$. Recall that $\mathsf{deg}_{\varepsilon_T}(C) \geq 3$ by the choice of C. If in addition $\mathsf{deg}_{\varepsilon_T}(C^*) = 1$, then e is an augmenting ε_T-sequence, and we are done. Therefore, we can assume that $\mathsf{deg}_{\varepsilon_T}(C^*) \geq 2$. We now observe that $\mathsf{dist}(T', T^*) < \mathsf{dist}(T, T^*)$, since $\mathsf{dist}(T', T^*) - \mathsf{dist}(T, T^*) = d_{T^*}(C^*, C') - d_{T^*}(C, C') < 0$ because $C^* \neq C$ and C^* is on the path from C to C' in T^*. Hence, by induction, there exists an augmenting $\varepsilon_{T'}$-sequence $e_1, \dots, e_k$ which implies that $e, e_1, \dots, e_k$ is an augmenting ε_T-sequence. □

4.3 Paths

A directed path $C_1, C_2, \dots, C_k$ in $\mathcal{D}_\tau$ is an *alternating path* of $\mathcal{D}_\tau$, if

(i) $\mathsf{deg}_\tau(C_1) \geq 3$, and
(ii) $\mathsf{deg}_\tau(C_j) \geq 2$ for each $2 \leq j \leq k - 1$.

An alternating path $C_1, C_2, \dots, C_k$ of $\mathcal{D}_\tau$ is an *augmenting path* of $\mathcal{D}_\tau$, if

(iii) $\mathsf{deg}_\tau(C_k) = 1$.

In what follows, we present the second of the two main ingredients we need for our algorithm for the leafage. In particular, we show that whenever an augmenting τ-sequence exists (for instance, by Theorem 10), we can find a directed path in $\mathcal{D}_\tau$ starting from a vertex of degree at least three to a leaf of τ, i.e., an augmenting path, and conversely, whenever such a path P in $\mathcal{D}_\tau$ exists, we can use it to obtain a τ-sequence starting and ending at the same vertices as P, i.e., an augmenting τ-sequence. In both cases, we construct the sequence (path) by incrementally adding edges one by one; we show that whenever we get stuck, there will be a "shortcut" in the sequence (path) which will allow us to continue this process until a desired path (sequence) is obtained.

Finally, we note that this property (combined with Theorem 10) reduces the problem of leafage to the problem of finding a directed path in a digraph which is what allows us to solve the problem in polynomial time (for the details of the algorithm, see the next section).

Theorem 11. *There exists an augmenting path in $\mathcal{D}_\tau$ if and only if there exists an augmenting τ-sequence.*

Proof. For the forward direction, let $P = C_1, C_2, \ldots, C_k$ be an augmenting path of $\mathcal{D}_\tau$. Let $\mathcal{F}$ denote the subgraph of $\mathcal{D}_\tau$ induced on $C_1, \ldots, C_k$, and let P' be a shortest directed path in $\mathcal{F}$ from C_1 to C_k. It is easy to verify that P' is also an augmenting path of $\mathcal{D}_\tau$. (Possibly $P' = P$.)

Let $e_1, \ldots, e_{k'}$ be the arcs of P' in the order they appear on P'. That is, e_1 is incident to C_1, and for each $1 \leq i < k'$, the arcs e_i and e_{i+1} share an end-point. Now, using Proposition 7 and the minimality of P', it follows that $e_1, \ldots, e_{k'}$ is an augmenting τ-sequence (details omitted).

For the backward direction, we need the following stronger claim $(*)$.

If $e_1, \ldots, e_k$ is a alternating τ-sequence, then there exists an alternating path of $\mathcal{D}_\tau$ that ends in C'_k where $e_k = C_k C'_k$ such that each vertex of this path is incident to some arc among $e_1, \ldots, e_k$. $(*)$

This claim can be proved by induction on k. (We omit further details of this proof due to the length restriction.)

Finally, let $e_1, \ldots, e_k$ be an augmenting τ-sequence where $e_1 = C_1 C'_1, \ldots, e_k = C_k C'_k$, and $\tau_0, \ldots, \tau_k$ are token mappings such that $\tau_0 = \tau$ and $\tau_i = \tau_{i-1} \div e_i$ for each $1 \leq i \leq k$. From (S5-S7), we have $\mathsf{deg}_{\tau_{k-1}}(C'_k) = 1$ and no arc among $e_1, \ldots, e_{k-1}$ is incident to C'_k. This yields $\mathsf{deg}_\tau(C'_k) = \mathsf{deg}_{\tau_0}(C'_k) = 1$. Now, by $(*)$, there exists an alternating path P of $\mathcal{D}_\tau$ that ends in C'_k, and since $\mathsf{deg}_\tau(C'_k) = 1$, the path P is also an augmenting path of $\mathcal{D}_\tau$. That concludes the proof. □

5 Algorithm

Now, we are finally ready to prove the main theorem of this paper.

Theorem 12. *There exists an $O(n^3)$ time algorithm that, given a chordal graph G, computes $l(G)$ and a tree model of G with $l(G)$ leaves.*

Proof. The algorithm goes as follows.

(1) Start by computing some clique tree T of G. Then construct the extended degree mapping ε_T of T, and let $\tau = \varepsilon_T$.
(2) Construct the digraph $\mathcal{D}_\tau$.
(3) Search in $\mathcal{D}_\tau$ for a shortest directed path P from a vertex of degree at least three in τ to a leaf of τ.
(4) If such path P is found, then consider the arcs of P one by one, and for each arc $e = CC'$ of P, if S is the label of e, remove a token with label S from $\tau(C)$ and add it to $\tau(C')$. Then go back to step (2).
(5) If such directed path P is not found, then construct a clique tree T corresponding to τ, that is, a tree T with $\tau = \varepsilon_T$. Then output T and the number of leaves of T.

The correctness of this algorithm follows from Theorems 10, and 11, and the observation that the path P in step (4) is neccessarily an augmenting path of $\mathcal{D}_\tau$, and therefore, Observation 8 implies that the new mapping τ has less leaves. We now discuss the complexity.

Recall that G contains at most n maximal cliques. By [6], we can construct a clique tree T of G and the mapping ε_T in time $O(n^2)$. To simplify processing during the algorithm, we precompute the connected components of $\mathcal{H}_S$ for each minimal separator S. This can be accomplished directly in time $O(n^3)$, since G has $O(n)$ minimal separators.

Next, we observe that there are precisely $2n-2$ tokens of τ and at most $n-1$ ways to move each of them. Hence, $\mathcal{D}_\tau$ contains $O(n^2)$ arcs. In fact, for a token with label S belonging to a clique C, we can find all arcs with label S going out of C in $\mathcal{D}_\tau$ by exploring $\mathcal{H}_S$ and testing conditions of Proposition 7. This can be easily accomplished in time $O(n)$ using the precomputed connected components of $\mathcal{H}_S$. Altogether, constructing $\mathcal{D}_\tau$ takes $O(n^2)$ time. The next step, finding the path P in $\mathcal{D}_\tau$, can be accomplished using a breadth-first search of $\mathcal{D}_\tau$ in time $O(n^2)$. If P is found, we can construct the new mapping τ directly in time $O(n)$.

We repeat the above steps at most n times since T has at most n leaves, and therefore, $O(n^3)$ time altogether.

Finally, if P is not found, we construct a tree T with $\varepsilon_T = \tau$ using the proof of Theorem 6 in time $O(n^2)$. This follows from the fact that G has $O(n)$ minimal separators and from an observation that constructing a tree from a degree mapping (see Theorem 5) takes $O(n)$ time. □

6 Conclusions

We have presented the first polynomial time algorithm for the problem of computing the leafage of a chordal graph. We showed that the algorithm runs in time $O(n^3)$, however, our complexity analysis was not very tight. Therefore, it seems likely that a more efficient, perhaps $O(n^2)$ or linear time implementation can be found. Furthermore, the algorithm provides no certificate for the minimality of the output. We believe that because of the min-max character of the problem it should be possible to characterize the dual of the problem which in turn can be used for certification. Lastly, we remark that our algorithm shows that computing a minimum leaf maximum weight spanning tree is polynomial time solvable for the class of reduced clique graphs whereas this problem is NP-hard in general [11] because the case of at most two leaves is the Hamiltonian path problem.

References

1. Booth, K.S., Lueker, G.S.: Testing for the consecutive ones property, interval graphs, and graph planarity using PQ-tree algorithms. J. Comput. Syst. Sci. 13, 335–379 (1976)
2. Broersma, H., Fomin, F.V., Nešetřil, J., Woeginger, G.J.: More about subcolorings. Computing 69, 187–203 (2002)
3. Fulkerson, D.R., Gross, O.A.: Incidence matrices and interval graphs. Pacific Journal of Mathematics 15, 835–855 (1965)
4. Galinier, P., Habib, M., Paul, C.: Chordal graphs and their clique graphs. In: Nagl, M. (ed.) WG 1995. LNCS, vol. 1017, pp. 358–371. Springer, Heidelberg (1995)
5. Gavril, F.: The intersection graphs of subtrees in trees are exactly the chordal graphs. Journal of Combinatorial Theory B 16, 47–56 (1974)
6. Golumbic, M.C.: Algorithmic Graph Theory and Perfect Graphs. Academic Press, New York (1980)
7. Habib, M., Stacho, J.: Linear algorithms for chordal graphs of bounded directed vertex leafage. In: DIMAP Workshop on Algorithmic Graph Theory, Electronic Notes in Discrete Mathematics, vol. 32, pp. 99–108 (2009)
8. Kloks, T., Kratsch, D., Müller, H.: Asteroidal sets in graphs. In: Möhring, R.H. (ed.) WG 1997. LNCS, vol. 1335, pp. 229–241. Springer, Heidelberg (1997)
9. Lin, I.J., McKee, T.A., West, D.B.: The leafage of a chordal graph. Discussiones Mathematicae Graph Theory 18, 23–48 (1998)
10. Prisner, E.: Representing triangulated graphs in stars. Abhandlungen aus dem Mathematischen Seminar der Universität Hamburg 62, 29–41 (1992)
11. Rahman, M.S., Kaykobad, M.: Complexities of some interesting problems on spanning trees. Information Processing Letters 94, 93–97 (2005)
12. Stacho, J.: On 2-subcolourings of chordal graphs. In: Laber, E.S., Bornstein, C., Nogueira, L.T., Faria, L. (eds.) LATIN 2008. LNCS, vol. 4957, pp. 520–530. Springer, Heidelberg (2008)
13. Stacho, J.: Complexity of subcolourings of chordal graphs (manuscript, 2009)
14. West, D.B.: personal communication (2008)
15. West, D.B.: Introduction to Graph Theory, 2nd edn. Prentice Hall, Englewood Cliffs (2001)

Breaking the $O(m^2 n)$ Barrier for Minimum Cycle Bases

Edoardo Amaldi[1], Claudio Iuliano[1], Tomasz Jurkiewicz[2], Kurt Mehlhorn[2], and Romeo Rizzi[3]

[1] Dipartimento di Elettronica e Informazione, Politecnico di Milano, Milano, Italy
[2] Max-Planck-Institut für Informatik, Saarbrücken, Germany
[3] Dipartimento di Matematica ed Informatica, Universitá degli Studi di Udine, Udine, Italy

Abstract. We give improved algorithms for constructing minimum directed and undirected cycle bases in graphs. For general graphs, the new algorithms are Monte Carlo and have running time $O(m^{\omega})$, where ω is the exponent of matrix multiplication. The previous best algorithm had running time $\tilde{O}(m^2 n)$. For planar graphs, the new algorithm is deterministic and has running time $O(n^2)$. The previous best algorithm had running time $O(n^2 \log n)$. A key ingredient to our improved running times is the insight that the search for minimum bases can be restricted to a set of candidate cycles of total length $O(nm)$.

1 Introduction

Cycles in graphs play an important role in many applications, e.g., analysis of electrical networks, analysis of chemical and biological pathways, periodic scheduling, and graph drawing, see [KLM+09, Section 7]. Cycle bases are a compact description of the set of all cycles of a graph and cycle bases consisting of short cycles or, in weighted graphs, of small weight cycles are to be preferred. We give improved algorithms for computing minimum weight cycle bases. The algorithms run in time $O(m^{\omega})$ for general graphs and $O(n^2)$ for planar graphs, here n and m denote the number of nodes and edges, respectively, and ω is the exponent of matrix multiplication. For planar graphs, this is an improvement by a factor of $O(\log n)$; our result implies a similar improvement for the all-pairs minimum cut problem in planar graphs. For general graphs, our algorithm is the first to run faster than $\tilde{O}(m^2 n)$. We mention that the previous best algorithms already used fast matrix multiplication and our improvement is due to new structural and algorithmic insights. A key ingredient to our improved running times is the insight that the search for minimum bases can be restricted to a set of candidate cycles of total length $O(nm)$.

Let $G = (V,E)$ be a connected undirected graph. We orient the edges of G arbitrarily and obtain a directed graph (V,A) which we denote by either D or G. We use the notation $e = uv$ to denote both directed and undirected edges, i.e., the notation stands for the directed edge (u,v) and the undirected edge $\{u,v\}$. We use $\delta(v)$ to denote the set of edges incident to v and $\delta^+(v)$ and $\delta^-(v)$ for the directed edges leaving and entering v, respectively.

Let κ be a field. A *κ-cycle C* in D is a vector in κ^E such that for any vertex v we have $\sum_{e\in\delta^+(v)} C_e = \sum_{e\in\delta^-(v)} C_e$. In other contexts, cycles are sometimes referred to as

A. Fiat and P. Sanders (Eds.): ESA 2009, LNCS 5757, pp. 301–312, 2009.

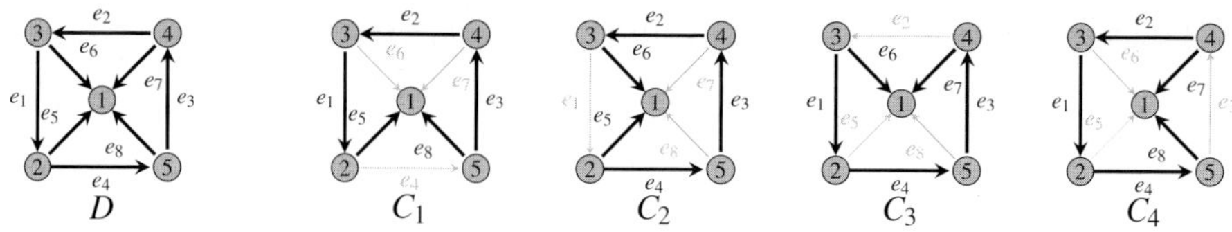

Fig. 1. The figure shows a directed graph D and four circuits C_1 to C_4 in D. The edges of D are e_1 to e_8. The circuit C_1 uses the edges e_1, e_2, e_3, and e_5 in forward direction and the edge e_8 in backward direction. Thus $C_1 = (1,1,1,0,1,0,0,-1)$. The circuits C_1 to C_4 form a directed cycle basis of G. The circuit C consisting of edges 1 to 4 is represented as $C = (1,1,1,1,0,0,0,0) = (C_1+C_2+C_3+C_4)/3$. Let G be the underlying undirected graph, let $\pi(C_i)$ be the undirected circuit corresponding to C_i, and let $\pi(C)$ be the undirected circuit corresponding to C. Then $\pi(C_1) = (1,1,1,0,1,0,0,1)$ and $\pi(C) = \pi(C_1) \oplus \pi(C_2) \oplus \pi(C_3) \oplus \pi(C_4)$, where $\oplus$ is addition modulo 2. The circuits $\pi(C_1)$ to $\pi(C_4)$ form an undirected cycle basis of G. The set $\{C_1, C_2, C_3, 2C_4\}$ is also a directed cycle basis of D. However, $\pi(2C_4) = \mathbf{0}$ and hence $\{\pi(C_1), \pi(C_2), \pi(C_3), \pi(2C_4)\}$ is *not* an undirected cycle basis of G.

circulations and the constraint $\sum_{e\in\delta^+(v)} C_e = \sum_{e\in\delta^-(v)} C_e$ is called flow conservation. Observe that if C is a cycle, then $-C$ is also a cycle, though a different one. The set

$$\{C; C \text{ is a } \kappa\text{-cycle of } G\}$$

forms a vector space over κ, the *κ-cycle space* of G. The support of a cycle is the set of edges e with $C_e \neq 0$. A cycle is *simple* if $C_e \in \{-1,0,+1\}$ for all e, and a simple cycle is a *circuit* if its support is connected and for any v there are most two edges in the support incident to v. A *κ-cycle basis* is a set of circuits forming a basis of the cycle space. Any cycle basis consists of $\nu := m - n + 1$ circuits.

Particularly interesting are the cases $\kappa = GF(2)$, the field of two elements, and $\kappa = \mathbb{Q}$, the field of rationals. In these cases, the cycle space and cycle basis are referred to as *undirected or directed cycle space and basis*, respectively. Let G be an undirected graph and let D be an orientation of it. For any directed circuit $C \in \{-1,0,+1\}^E$ of D, let $\pi(C) := (C_e \bmod 2)_{e\in E}$. Then $\pi(C)$ is an undirected circuit in G, the *projection* of C. Figure 1 illustrates these definitions. In addition, it provides an example showing that directed cycle bases do not necessarily project onto undirected cycle bases. However, if a set of ν directed circuits projects onto an undirected basis, it forms a directed basis.

A *weighted graph* is a graph together with a non-negative weight function $w : E \to \mathbb{R}_{\geq 0}$. The weight of a set of edges is the sum of the weights of its members. The *weight $w(C)$ and length $|C|$ of a simple cycle C* are

$$w(C) := \sum_e |C_e| w(e) \qquad |C| := \sum_e |C_e| \,,$$

and the *weight of a cycle basis B* is the sum of the weights of its circuits, i.e.,

$$w(B) = \sum_{C\in B} w(C) \,.$$

A *minimal κ-cycle basis* of G is a κ-cycle basis with minimum weight.

Horton [Hor87] gave the first polynomial time algorithm for minimum undirected cycle bases. It had running time $O(m^3n)$. In a sequence of papers [DP95, GH02, BGdV04, KMMP08, MM07], the running time was improved to $\tilde{O}(m^2n)$. Kavitha and Mehlhorn [KM07] gave the first polynomial time algorithm for minimum directed cycle bases. It had running time $O(m^4n)$. In a sequence of papers [LR05, Kav05, HKM08, MM07] the running time was improved to $O(m^3n)$ deterministic time and $\tilde{O}(m^2n)$ Monte Carlo time. We improve the running time to $O(m^\omega)$ Monte Carlo time for undirected and directed bases. For planar graphs, we improve the running time from $O(n^2 \log n)$ [HM94] to $O(n^2)$; the algorithm is deterministic.

This paper is structured as follows. In Section 2 we improve upon a result of Horton [Hor87] and show that the search for cycle bases can be restricted to a set of candidate circuits of total length $O(nm)$; Horton had shown that the search can be restricted to a set of $O(nm)$ circuits. In Section 3, we exploit this structural insight to derive the $O(m^\omega)$ Monte Carlo algorithm for minimum undirected and directed bases. In Section 4, we exploit it to derive the $O(n^2)$ algorithm for minimum bases in planar graphs.

2 Structural Results

For any two nodes u and v, let p_{uv} be a minimum weight path from u to v in G with respect to weight function w. We assume that the collection of minimum weight paths is consistent, i.e., if x and z lie on p_{uv} then p_{xz} is a subpath of p_{uv}. This can be guaranteed for instance by lexicographic ordering. Given an arbitrary numbering of the nodes from 1 to n, a path p between two nodes is considered shorter than a path q of the same total weight if the length of p is strictly smaller than the length of q. In case of ties, the shortest path between p and q will be the one that contains the node with minimum index in the non-common part. For a modified minimum weight path algorithm that ensures lex-shortest paths in time $O(mn + n^2 \log n)$, see [HM94].

For any node x, let T_x be the minimum weight path tree rooted at x, i.e., T_x is the union of the paths p_{xv} for all v. In [Hor87] Horton shows that a polynomial subset of all cycles is guaranteed to contain a minimum cycle basis. The set of Horton candidate cycles, denoted by $\mathscr{H}$, contains all cycles of the form $C_{x,e} := p_{xu} e p_{vx}$, for any possible choice of a node x and an edge $e = uv$ not in T_x, i.e., a co-tree edge. Among these $n\nu$ cycles, we have to consider only the circuits, discarding $C_{x,e}$ if p_{xu} and p_{xv} have more than node x in common (see Figure 2(a)). $\mathscr{H}$ is a multi-set because each circuit C can have different representations $C_{x,e}$ for some of its nodes x. Note that there is no representation for a given node x if C contains more than one co-tree edge with respect to T_x. This is equivalent to the existence of a shortcut between x and another node in C, i.e., the shortest path joining them does not belong to the circuit itself.

A circuit C is called *isometric* if for any two nodes u and v on C, p_{uv} is contained in C. See Figure 2 (b) and (c) for examples of non-isometric circuits. The set of isometric circuits will be denoted by $\mathscr{I}$. Clearly each isometric circuit is a Horton candidate cycle, that is $\mathscr{I} \subseteq \mathscr{H}$.

In fact, we just need to consider isometric circuits.

Proposition 2.1 ([Hor87]). *$\mathscr{I}$ contains a minimum undirected (directed) basis.*

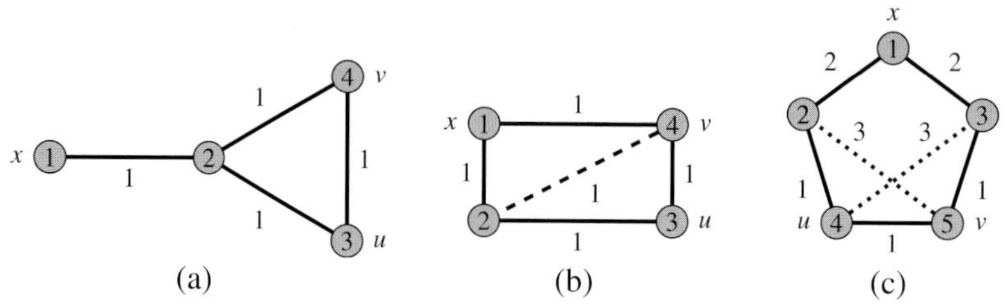

Fig. 2. Examples of non-isometric cycles. (a) $C_{1,\{3,4\}}$ is not a circuit, because p_{13} and p_{14} have also node 2 in common. The contained circuit is obtained as $C_{2,\{3,4\}}$. (b) The minimum weight path connecting 1 and 3 consists of edges $\{1,2\}$ and $\{2,3\}$. $C_{1,\{3,4\}}$ is a non-isometric circuit because of the shortcut, depicted in dashed. The only other representation is $C_{3,\{1,4\}}$. (c) $C_{1,\{4,5\}}$ is a non-isometric circuit with two shortcuts and no representations for any other node.

Moreover, isometric circuits can be characterized in terms of number of representations, namely every isometric circuit C has exactly $|C|$ representations in $\mathscr{H}$.

Property 2.2 (Isometric circuits [Hor87]). *Let C be any isometric circuit and let x be an arbitrary node of C. Then there is an edge $e = uv$ on C such that $C = p_{xu}ep_{vx}$. Conversely, if for every $x \in C$ there is such an edge, then C is isometric.*

Proof: Let $C = (x = v_0, v_1, \ldots, v_k = x)$. Since the empty path is the minimum weight path from x to x and C is not the minimum weight path from x to x, there must be an i such that $p_{xv_i} = (v_0, v_1, \ldots, v_i)$ but $p_{xv_{i+1}} \neq (v_0, v_1, \ldots, v_i, v_{i+1})$. Then $p_{xv_{i+1}} = (v_k, v_{k-1}, \ldots, v_{i+1})$ and hence $e = (v_i, v_{i+1})$ is the desired edge.

For the converse, consider any two nodes x and z on C and let $e = uv$ be such that $C = p_{xu}ep_{vx}$; z lies on one of the paths and hence the minimum weight path from x to z is contained in C. ∎

By considering the set of isometric circuits $\mathscr{I}$ instead of $\mathscr{H}$ we have the following simple but fundamental property.

Property 2.3. *The total length of the isometric circuits is at most $n\nu$.*

Proof: An isometric circuit C occurs $|C|$ times in the Horton multi-set and hence $\sum_{C \in \mathscr{I}} |C|$ can be no larger than the size of the Horton multi-set. ∎

Note that we sum only over the isometric circuits, as we have no control over the number of appearances of non-isometric cycles.

The upper bound of Property 2.3 is tight for instance for the complete graph K_n with n vertices and equal weight on the edges. For any node x, the cycle obtained by adding to T_x any co-tree edge is a triangle. $\mathscr{H}$ consists of $n\nu$ triangles that are clearly isometric. Since there are three representations of each possible triangle, obtained by taking as x each one of its 3 nodes, $\mathscr{I}$ consists of $n\nu/3$ triangles. Therefore, the total length of the isometric circuits is exactly $n\nu$.

The total length of the isometric circuits may be much smaller than $n\nu$. Consider an $s \times s$ grid with equal weights on the edges. Since $n = s^2$ and $m = 2s(s-1)$, we have $\nu = (s-1)^2$, and m and ν are $O(n)$. The isometric circuits are exactly the grid squares and hence their total length is $4(s-1)^2$, that is $O(n)$, whereas the upper bound of Property 2.3 is $n\nu = s^2(s-1)^2$, that is $O(n^2)$.

We will now show that we can extract $\mathcal{I}$ from the Horton multi-set in time $O(nm)$.

For every node $v \neq x$, let $s_x(v)$ be the child of x in T_x containing v in its subtree. In other words, $s_x(v)$ is the first node on the minimum weight path from x to v. The vectors s_x for all $x \in V$ can be the computed in time $O(n^2)$. Note that a candidate cycle $C = C_{x,e}$, for $e = uv$, is a circuit only when p_{xu} and p_{xv} have only node x in common, i.e., $s_x(u) \neq s_x(v)$ (see Figure 2(a)). The next Lemma shows how to identify different representations of the same isometric circuit and how to discover non-isometric circuits. Given a circuit $C_{x,uv}$, the idea is to check for two specific nodes x' and x'' of C whether the minimum weight path $p_{x'x''}$ between them belongs to C. The nodes x' and x'' are chosen so that a negative answer obviously implies that the circuit is non-isometric whereas a positive answer gives a different representation of C for one of x' and x''. This is achieved by taking $x' = s_x(u)$ and $x'' = v$. In fact, if $p_{x'v}$ belongs to C there are only two possibilities: $p_{x'v} = x'xp_{xv}$ and the other representation for C is for the node x' and is given by $C_{x',uv}$; $p_{x'v} = vup_{ux'}$ and the other representation for C is for the node v and is given by $C_{v,xx'}$. When node x' does not exist because node x coincides with node u, the other representation is for node v and is given by $C_{v,uv}$. Lemma 2.4 explains how to check (in constant time) the conditions that allow to identify the different cases, which are illustrated in Figure 3.

Lemma 2.4. *Let $C = C_{x,e}$, let u be an endpoint of e, and let v be the other endpoint.*

1. *If $s_x(u) \neq s_x(v)$ and $x = u$ then $x \neq v$ and $C = ep_{vu} = C_{v,e}$.*
2. *If $s_x(u) \neq s_x(v)$, $x \neq u$, and $x' = s_x(u)$ is the first node on the minimum weight path from x to u then:*
   (a) *if $x = s_{x'}(v)$, then $C = C_{x',e}$,*
   (b) *if $x \neq s_{x'}(v)$ and $u = s_v(x')$ then $C = C_{v,xx'}$, and*
   (c) *if $x \neq s_{x'}(v)$ and $u \neq s_v(x')$ then C is not isometric.*

Proof:

If $x = u$, $C = uvp_{vu} = p_{vu}uv = C_{v,e}$. This proves the first statement.

If $x \neq u$ and x' is the first vertex on the minimum weight path from x to u, we have $p_{xu} = xx'p_{x'u}$.

If x is the first vertex on the minimum weight path from x' to v, then $p_{ux'}p_{x'v} = p_{ux}p_{xv}$. Thus $C = C_{x',e}$. This establishes 2a.

If x is not the first vertex on the minimum weight path from x' to v and u is the first node on the minimum weight path from v to x' then $C = p_{vx}xx'p_{x'v} = C_{v,xx'}$. This establishes 2b.

If x is not the first vertex on the minimum weight path from x' to v and C is isometric, the minimum weight path from x' to v must be $p_{x'u}$ followed by e. Then u is the first vertex on the minimum weight path from v to x'. This establishes 2c. ∎

Lemma 2.4 allows us to identify different representations of the same isometric circuit. It also allows to exclude some circuits as non-isometric.

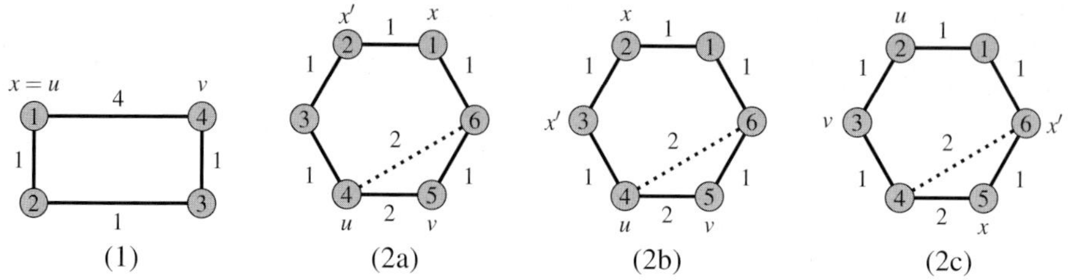

Fig. 3. Examples for the different cases of Lemma 2.4. (1) $C_{1,\{1,4\}}$ where $x = u = 1$. (2a), (2b) and (2c) are three different representations of the same circuit. (2a) $C_{1,\{4,5\}}$ where $s_2(5) = 1$ and we obtain $C_{2,\{4,5\}}$. (2b) $C_{2,\{4,5\}}$ where $s_3(5) \neq 2$ but $s_5(3) = 4$ and we obtain $C_{5,\{2,3\}}$. (2c) $C_{5,\{2,3\}}$ where $s_6(3) \neq 5$ and $s_3(6) \neq 2$, because the minimum weight path connecting 3 and 6 consists of edges $\{3,4\}$ and $\{4,6\}$. This implies that the circuit is not isometric. The shortcut is in dashed line.

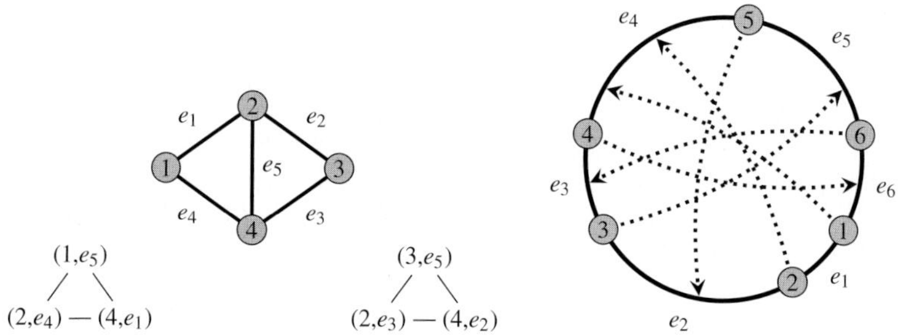

Fig. 4. In the graph on the left all edges have cost one; we select e_1e_2 as the minimum weight path connecting 1 and 3. The circuits C_{1,e_3} and C_{3,e_4} are bad by condition 2c. For the former circuit let $x = 1$, $u = 3$, $v = 4$; then $s_1(3) = 2$ and $s_2(4) \neq 1$ and $s_4(2) \neq 3$. The other circuits are connected as shown below the graph. The figure on the right shows an isometric circuit C embedded on a circle. The edges correspond to the circular arcs between the vertices and the length of an arc is proportional to the weight of the corresponding edge. For any vertex x, we have $C = C_{x,e}$ where e contains the mirror image of x with respect to the center of the circle. We have the following connections: C_{1,e_4} and C_{2,e_4} are connected by condition 2a, C_{2,e_4} and C_{5,e_2} are connected by condition 2b, and so on.

We next show that all representations of an isometric circuit will be identified and all non-isometric circuits will be discovered. We set up a graph whose vertices are the pairs (x,e), $x \in V$, $e \in E$, if (x,e) is a circuit. We label (x,e) as *bad* if condition 2c holds. We connect two pairs if they satisfy condition 1 or 2a or 2b, see Figure 4.

Lemma 2.5. *All representations of an isometric circuit belong to the same connected component.*

Proof: Let $C = (v_0, v_1, \ldots, v_k = v_0)$ be an isometric circuit, let $e_i = v_iv_{i+1}$, and for any i, $0 \leq i < k$, let $j(i)$ be such that $C = C_{v_i,e_{j(i)}}$. Figure 4 shows how the different representations of C are linked together. In this Figure, a representation $C_{v_i,e_{j(i)}}$ is indicated as a dashed arrow from v_i to $e_{j(i)}$. In cases 1 and 2a, v_i and v_{i+1} point to the same edge, i.e., the tail of the arrow advances by one position. In case 2b, we replace the arrow from v_i to $e_{j(i)} = v_{j(i)}v_{j(i)+1}$ by the arrow from $v_{j(i)+1}$ to v_iv_{i+1}, i.e., we reverse the direction of the arrow and it now points from the tail of $e_{j(i)}$ to the edge out of v_i. In this way, the arrow sweeps around the circuit once and links all representations of the same circuit. ∎

Lemma 2.6. *If $C_{x,e}$ is non-isometric then the component of (x,e) contains a bad component.*

Proof: Let $C = (v_0, v_1, \dots, v_k = v_0)$ be a non-isometric circuit and let $e_i = v_i v_{i+1}$. For some, but not all, i, $0 \le i < k$, there will be a $j(i)$ such that $C = C_{v_i, e_{j(i)}}$. Observe, that if $C = C_{v_i, e_{j(i)}}$, the minimum weight paths from v_i to the vertices of C are initial segments of either $p_{v_i v_{j(i)}}$ or $p_{v_i v_{j(i)+1}}$. Also, if the minimum weight path from v_{i+1} to $v_{j(i)+1}$ is contained in C, then either $C = C_{v_{i+1}, e_{j(i)}}$ or $C = C_{v_{j(i)+1}, e_i}$.

Thus if C is non-isometric, there must be i such that the minimum weight path from v_{i+1} to $v_{j(i)+1}$ is not contained in C. For any such i, $C_{v_i, e_{j(i)}}$ will be declared bad. Any non-bad representation of C will be linked to a bad one as described in the preceding Lemma. ∎

Note that checking the conditions of Lemma 2.4 is needed once for each circuit in $\mathscr{H}$.

We summarize the discussion.

Theorem 2.7. *In time $O(nm)$ we can extract for each isometric circuit one pair (x,e) with $C = C_{x,e}$.*

3 Improved Algorithms for General Graphs

We refine de Pina's approach [DP95, KLM$^+$09] for computing minimum cycle bases, see Figure 5. It operates in phases. In each phase, one circuit is added to the basis. The algorithm also maintains a basis of the orthogonal space; more precisely, at the beginning of the i-th iteration is has a set $\{S_i, \dots, S_\nu\}$ of linearly independent vectors $S_j \in \kappa^E$ with $\langle C_j, S_j \rangle = 0$, where $\langle \_, \_ \rangle$ is the inner product of vectors over κ. Throughout this section, $\kappa = GF(p)$ for a prime p with $p = O(m \log m)$. In particular, arithmetic in $GF(p)$ takes constant time. At the start of the computation S_j is initialized to the j-th unit vector for $1 \le j \le \nu$, where the numbering of the edges is such that edges $e_{\nu+1}$ to e_m form a spanning tree of G.

Initialize S_j to the j-th unit vector for $1 \le j \le \nu$.
for $i \leftarrow 1, \dots, \nu$ **do**
  Compute a minimum weight isometric circuit C_i with $\langle C_i, S_i \rangle \neq 0$.
  for $j \leftarrow i+1, \dots, \nu$ **do**
    $S_j = S_j - \frac{\langle C_i, S_j \rangle}{\langle C_i, S_i \rangle} S_i$
  end for
end for
Output $\{C_1, \dots, C_\nu\}$.

Fig. 5. De Pina's algorithm for computing a minimum cycle basis.

Steps (4) and (5) of the algorithm make the S_j, $j > i$, orthogonal to C_i and maintain orthogonality of C_1 to C_{i-1}. Updating the vectors S_j as shown takes time $O(m^2)$ per phase and hence total time $O(m^3)$. In [KMMP08], this was improved to time $O(m^\omega)$. The best known realization of step (3) takes time $\tilde{O}(mn)$ per phase and hence total time $\tilde{O}(m^2 n)$. We describe a Monte Carlo algorithm that improves the total time for step (3) to $o(m^\omega)$. The improved algorithm exploits the new structural result presented in the preceding section.

We start with a simple technical lemma.

Lemma 3.1. *Let $\mathscr{C}$ be a collection of circuits. For each circuit $C \in \mathscr{C}$, let $\lambda_C \in GF(p)$ be chosen randomly and let $D = \sum_{C \in \mathscr{C}} \lambda_C C$. Let S be a nonzero vector in $GF(p)^E$. If all circuits in $\mathscr{C}$ are orthogonal to S, D is orthogonal to S. If $\mathscr{C}$ contains a circuit that is non-orthogonal to S, D is orthogonal to S with probability at most $1/p$.*

Proof: Clearly, if every circuit in $\mathscr{C}$ is orthogonal to S, then D is.

So assume that $C' \in \mathscr{C}$ is non-orthogonal to S and consider a fixed choice of coefficients λ_C for the circuits $C \in \mathscr{C}, C \neq C'$. Also assume that there are two distinct choices α and β for $\lambda_{C'}$ such that $\sum_{C \in \mathscr{C}} \lambda_C C$ are orthogonal to S. Then $\alpha C' + \sum_{C \in \mathscr{C}, C \neq C'} \lambda_C C$ and $\beta C' + \sum_{C \in \mathscr{C}, C \neq C'} \lambda_C C$ are orthogonal to S. Thus $(\beta - \alpha)C'$ is orthogonal to S, a contradiction. Thus the probability that $\langle D, S \rangle = 0$ is at most $1/p$. ∎

Consider the $|\mathscr{I}| \leq nm$ isometric circuits. We sort them by nondecreasing weight and put a binary tree (of depth at most $\log nm$, that is $O(\log n)$) on top of the sorted list. For each node of the tree, we prepare k random linear combinations of the circuits below the node. We find the cheapest circuit that has nonzero inner product with S_i as follows. Assume the search has arrived in some node of the tree. We compute the inner product of S_i with the k linear combinations associated with the left child. If one inner product is nonzero, we proceed to the left child. If all k inner products are zero, we proceed to the right child. The move to the left child is always correct. However, the move to the right child may be incorrect. The probability that any specific decision is incorrect is at most p^{-k}. In any search, we make $\log|\mathscr{I}|$ decisions, and we need to find ν circuits. Thus the total number of decisions is $\nu \log|\mathscr{I}|$ and hence the total probability of error is bounded by $\nu \log|\mathscr{I}| p^{-k}$.

Each step of the binary search is a scalar product and hence selecting one circuit takes time $O(km \log n)$. Selecting all circuits takes time $O(km^2 \log n)$.

How much time does it take to prepare the random linear combinations? We maintain them as sparse vectors, i.e., as the ordered list of their nonzero entries. In order to prepare one linear combination for each node of the search tree, we choose a random multiplier $\lambda_C \in k$ for each isometric circuit C. We then sum the sparse vectors as indicated by the tree. Each nonzero entry of a circuit contributes cost $O(1)$ for each level of the tree and hence the total time to prepare one random linear combination for each node of the search tree is $O(nm \log n)$ by Property 2.3. We want k linear combinations for each node and hence require time $O(knm \log n)$ to prepare all of them.

Theorem 3.2. *There is a Monte Carlo algorithm for finding a minimum $GF(p)$-basis that works in time $O(nm + n^2 \log n + m^\omega + km^2 \log n)$ and errs with probability at most $\nu \log(nm) p^{-k}$. For $k = m^{0.1}$, this is exponentially small, and the running time is $O(m^\omega)$.*

Undirected bases are $GF(2)$-bases and hence we are done. For directed cycle bases we use an observation in [KLM$^+$09, Section 3.5], namely that a minimum $GF(p)$-basis for a random p with $p = \Theta(m \log m)$ is a minimum directed basis with probability at least $1/2$.

Theorem 3.3. *There is a Monte Carlo algorithm for finding a minimum directed cycle basis that works in time $O(m^\omega)$ and errs with probability at most $1/2$.*

4 Planar Graphs

Hartvigsen and Mardon [HM94] have shown that minimum undirected cycle bases in planar graphs can be computed in time $O(n^2 \log n)$. In this section, we summarize their result, improve the running time to $O(n^2)$, and also show that for planar graphs, the notions of minimum directed, undirected, integral, weakly fundamental, and totally unimodular bases coincide, see [KLM$^+$09, Section 3] and the proof of Theorem 4.2 for a definition of the latter terms.

Let G be a plane graph, a planar graph embedded into the plane. A plane graph divides the plane into maximal open connected sets of points that we call *faces*. Any circuit C divides the plane into two maximal open connected sets of points, one bounded and one unbounded. We use $interior(C)$ to denote the bounded set. If $interior(C)$ agrees with one of the faces of G, we call C a *face circuit*. Note that the number of edges and the number of face circuits are both $O(n)$. A collection of circuits is called *nested* if for any two circuits in the collection, the interiors are either disjoint or the interior of one is contained in the interior of the other.

For a collection B of circuits, let F_B be the face circuits that do not belong to B. We define the directed inclusion graph D_B with vertex set $B \cup F_B$ as follows. Let C and C' be circuits in $B \cup F_B$. We have an edge from C to C' if $interior(C) \supset interior(C')$ and there is no circuit $C'' \in B \cup F_B$ such that $interior(C) \supset interior(C'') \supset interior(C')$. The inclusion graph is acyclic; the nodes of D_B with no outgoing edges are precisely the face circuits of G. The inclusion graph is a forest if and only if B is nested.

In [HM94] Hartvigsen and Mardon show that the number of isometric circuits is at most twice the number of face circuits of any planar graph G and there is at least a minimum cycle basis (directed or undirected) that is nested. Moreover, a nested collection of cycles B is a minimum cycle basis iff B is a minimum weight collection of circuits satisfying three properties: (1) every non-leaf in D_B has exactly one child in F_B, (2) the circuits in F_B have parents in D_B, (3) the inclusion graph D_B is a forest.

Our algorithm for finding a minimum weight basis differs from that of [HM94] in two points. First, we use the all-pairs minimum weight paths method for planar graph in $O(n^2)$ proposed in [Fre87]. Then, the main improvement is to exploit the procedure implied by Theorem 2.7 to obtain the set of isometric circuits in $O(n^2)$. This way, the bottleneck of $O(n^2 \log n)$ decreases to $O(n^2)$. The rest of the algorithm proceeds as in [HM94] and we summarize it below for completeness. Recall that the number of isometric circuits is $O(n)$ and that sorting by nondecreasing weight is $O(n \log n)$.

We construct the incidence matrix A between isometric circuits and the faces of G. The entry corresponding to a circuit C and a face R is one if $R \subseteq interior(C)$. This matrix can clearly be computed in time $O(n^2)$.

We initialize the basis B to the empty set and set up the corresponding inclusion graph D_B. The vertices of D_B are the face circuits and there are no edges. As long as B does not have the right number of circuits and hence D_B does not satisfy properties (1) and (2), we do the following.

If there is a non-leaf node C that has more than one child in F_B (case 1), let R_1 and R_2 be two faces of G limited by two face circuits in F_B having C as their common parent. If there is no such non-leaf node, there must be a face circuit in F_B without a parent (case 2). Let R_1 be the face limited by this face circuit and let R_2 be the unbounded face. In either case, we find the least weight circuit D containing exactly one of R_1 or R_2 in its interior. We can find D in time $O(n)$ by scanning the columns of A.

We add D to B and update D_B. If D is a face circuit, we only have to remove D from F_B. The inclusion graph stays the same. So assume that D is not a face circuit. Starting from the face circuits in *interior*(D) (we can find them in matrix A), we determine the maximal subtrees of D_B that are contained in *interior*(D). They become children of D. D either becomes a root (in case 2) or a child of C (in case 1). Updating D_B takes time $O(n)$.

We conclude that we spend time $O(n)$ per base circuit for a total of $O(n^2)$.

Theorem 4.1. *A minimum (directed or undirected) circuit basis of a planar graph can be found in time* $O(n^2)$.

[HM94] observed that the minimum cycle basis problem is dual to the all-pairs minimum cut problem. Hence the all-pairs minimum cut problem in planar graphs can also be solved in time $O(n^2)$.

Theorem 4.2. *Every planar graph has a minimum directed cycle basis that is weakly fundamental, totally unimodular, and integral.*

Proof: Every planar graph has a minimum directed cycle basis that is nested. Let B be such a basis. We first show that B is totally unimodular. We need to show that any circuit is a linear combination of the circuits in B with coefficients in $\{-1,0,+1\}$. Let C be any circuit. Then, C can be obtained as the sum of the face circuits that limit faces in *interior*(C). A face circuit either belongs to B or is equal to the difference of its parent $p(F)$ in D_B and the sum of the other children of $p(F)$ in D_B. Thus

$$C = \sum_{F \in B} F + \sum_{F \in F_B} \left(p(F) - \sum_{D \in B \text{ and } D \text{ is a child of } p(F) \text{ in } D_B} D \right).$$

If a circuit D occurs twice in the representation of C, it occurs once as a parent and once as a child. As a parent, its coefficient is $+1$, and as a child, its coefficient is -1 and hence the two occurrences cancel. Thus every circuit is a linear combination of the circuits in B with coefficients in $\{-1,0,+1\}$.

We next show that B is weakly fundamental. We need to exhibit an ordering $C_1, \ldots, C_\nu$ of the circuits in B such that $C_i \setminus (C_1 \cup \ldots \cup C_{i-1}) \neq \emptyset$ for all i. Let D_B be the inclusion graph corresponding to B. If F_B is empty, every face circuit belongs to B. We determine a reverse ordering of the circuits $C_\nu, \ldots, C_1$ as described in [LR07]. Starting from the circuit C that limits the unbounded face, we add the face circuits with an edge in common with C. After removing the edges of C from G, we proceed in the same way. We now extend the previous result to the general case when F_B is not empty. Since every face circuit in F_B has a parent, we have a non-leaf node D in D_B whose children are all face circuits. One of these face circuits, say F, belongs to F_B and all the others belong to B. The same idea for constructing a reverse ordering is then applied to the circuits in B corresponding to the children of D starting from F. The face circuits among these with an edge in common with F are added and the edges of F that are not in D are removed. Then we proceed in the same way considering the circuit that limits the new face. After that all children of D are added, we delete them from D_B. We repeat this until all nodes in D_B are isolated. By applying the procedure in the remaining graph for the case where there are only face circuits, we find a reverse ordering of the circuits. Thus, the same result holds for general cycle bases.

The proof is completed by the fact that any weakly fundamental basis is integral. ∎

5 Conclusion

We have shown that minimum cycle bases can be computed in time $O(m^{\omega})$ by a Monte Carlo algorithm. A further improvement would have to do away with the maintenance of a basis of the orthogonal subspace.

References

[BGdV04] Berger, F., Gritzmann, P., de Vries, S.: Minimum cycle bases for network graphs. Algorithmica 40(1), 51–62 (2004)

[DP95] De Pina, J.C.: Applications of shortest path methods. PhD thesis, University of Amsterdam, The Netherlands (1995)

[Fre87] Frederickson, G.N.: Fast algorithms for shortest paths in planar graphs, with applications. SIAM J. Computing 16(6), 1004–1022 (1987)

[GH02] Golynski, A., Horton, J.D.: A polynomial time algorithm to find the minimum cycle basis of a regular matroid. In: Penttonen, M., Schmidt, E.M. (eds.) SWAT 2002. LNCS, vol. 2368, pp. 200–209. Springer, Heidelberg (2002)

[HKM08] Hariharan, R., Kavitha, T., Mehlhorn, K.: Faster deterministic and randomized algorithms for minimum cycle basis in directed graphs. SIAM J. Computing 38(4), 1430–1447 (2008)

[HM94] Hartvigsen, D., Mardon, R.: The all-pairs min cut problem and the minimum cycle basis problem on planar graphs. SIAM J. Discrete Math. 7(3), 403–418 (1994)

[Hor87] Horton, J.D.: A polynomial-time algorithm to find the shortest cycle basis of a graph. SIAM J. Computing 16(2), 358–366 (1987)

[Kav05] Kavitha, T.: An $\tilde{O}(m^2n)$ randomized algorithm to compute a minimum cycle basis of a directed graph. In: Caires, L., Italiano, G.F., Monteiro, L., Palamidessi, C., Yung, M. (eds.) ICALP 2005. LNCS, vol. 3580, pp. 273–284. Springer, Heidelberg (2005)

[KLM$^+$09] Kavitha, T., Liebchen, C., Mehlhorn, K., Michail, D., Rizzi, R., Ueckerdt, T., Zweig, K.: Cycle bases in graphs: Characterization, algorithms, complexity, and applications, 78 pages (submitted for publication) (March 2009)

[KM07] Kavitha, T., Mehlhorn, K.: Algorithms to compute minimum cycle basis in directed graphs. Theory of Computing Systems 40(4), 485–505 (2007); A preliminary version of this paper appeared in STACS 2005, vol. 3404, pp. 654–665

[KMMP08] Kavitha, T., Mehlhorn, K., Michail, D., Paluch, K.E.: An $\tilde{O}(m^2n)$ algorithm for minimum cycle basis of graphs. Algorithmica 52(3), 333–349 (2008); A preliminary version of this paper appeared in ICALP 2004, vol. 3142, pp. 846–857

[LR05] Liebchen, C., Rizzi, R.: A greedy approach to compute a minimum cycle basis of a directed graph. Information Processing Letters 94(3), 107–112 (2005)

[LR07] Liebchen, C., Rizzi, R.: Classes of cycle bases. Discrete Applied Mathematics 155(3), 337–355 (2007)

[MM07] Mehlhorn, K., Michail, D.: Minimum cycle bases: Faster and simpler. Accepted for publication in ACM Transactions on Algorithms (2007)

Shape Fitting on Point Sets with Probability Distributions

Maarten Löffler[1] and Jeff M. Phillips[2]

[1] Department of Information and Computing Sciences, Utrecht University
[2] Department of Computer Science, Duke University

Abstract. We consider problems on data sets where each data point has uncertainty described by an individual probability distribution. We develop several frameworks and algorithms for calculating statistics on these uncertain data sets. Our examples focus on geometric shape fitting problems. We prove approximation guarantees for the algorithms with respect to the full probability distributions. We then empirically demonstrate that our algorithms are simple and practical, solving for a constant hidden by asymptotic analysis so that a user can reliably trade speed and size for accuracy.

1 Introduction

In gathering data there is a trade-off between quantity and accuracy. The drop in the price of hard drives and other storage costs has shifted this balance towards gathering enormous quantities of data, yet with noticeable and sometimes intentional imprecision. However, often as a benefit from the large data sets, models are developed to describe the pattern of the data error.

For instance, in the gathering of LIDAR data for GIS applications [17], each data point of a terrain can have error in its x- (longitude), y- (latitude) and z-coordinates (height). Greatly simplifying, we could model the uncertainty as a 3-variate normal distribution centered at its recorded value. Similarly, large data sets are gathered with uncertainty in robotic mapping [12], anonymized medical data [1], spatial databases [24], sensor networks [17], and many other areas.

However, much raw data is not immediately given as a set of probability distributions, rather as a set of points. Approximate algorithms may treat this data as exact, construct an approximate answer, and then postulate that since the raw data is not exact, the approximation errors made by the algorithm may be similar to the errors of the imprecise input data. This is a very dangerous postulation.

An algorithm can only provide answers as good as the raw data *and* the models for error on that data. This paper is not about how to construct error models, but how to take error models into account. While many existing algorithms produce approximations with respect only to the raw input data, algorithms in this paper approximate with respect to the raw input data *and* the error models associated with them.

Geometric error models. An early model for imprecise geometric data, motivated by finite precision of coordinates, is ε-*geometry* [14], where each data point

A. Fiat and P. Sanders (Eds.): ESA 2009, LNCS 5757, pp. 313–324, 2009.

is known to lie within a ball of radius ε. This models has been used to study the robustness of problems such as the Delaunay triangulation [6,19]. This model has been extended to allow different uncertainty regions around each point for object intersection [22] and shape-fitting problems [25]. These approaches give worst case bounds on error, for instance upper and lower bounds on the radius of the minimum enclosing ball. But when uncertainty is given as a probability distribution, then these approaches must use a threshold to truncate the distribution. Furthermore, the answers in this model are quite dependent on the boundary of the uncertainty region, while the true location is likely to be in the interior. This paper thus describes how to use the full probability distribution describing the uncertainty, and to only discretize, as desired, the probability distribution of the final solution.

The database community has focused on similar problems for usually one-dimensional data such as indexing [2], ranking [11], and creating histograms [10].

1.1 Problem Statement

Let $\mu_p : \mathbb{R}^d \to \mathbb{R}^+$ describe the probability distribution of a point p where the integral $\int_{q \in \mathbb{R}^d} \mu_p(q)\, dq = 1$. Let $\mu_P : \mathbb{R}^d \times \mathbb{R}^d \times \ldots \times \mathbb{R}^d \to \mathbb{R}^+$ describe the distribution of a point set P by the joint probability over each $p \in P$. For brevity we write the space $\mathbb{R}^d \times \ldots \times \mathbb{R}^d$ as $\mathbb{R}^{dn}$. For this paper we will assume $\mu_P(q_1, q_2, \ldots, q_n) = \prod_{i=1}^n \mu_{p_i}(q_i)$, so the distribution for each point is independent, although this restriction can be easily circumvented.

Given a distribution μ_P we ask a variety of shape fitting questions about the uncertain point set. For instance, what is the radius of the smallest enclosing ball or what is the smallest axis-aligned bounding box of an uncertain point set. In the presence of imprecision, the answer to such a question is not a single value or structure, but also a *distribution* of answers. The focus of this paper is not just how to answer such shape fitting questions about these distributions, but how to concisely represent them. As a result, we introduce two types of approximate distributions as answers, and a technique to construct coresets for these answers.

ε-Quantizations. Let $f : \mathbb{R}^{dn} \to \mathbb{R}^k$ be a function on a fixed point set. Examples include the radius of the minimum enclosing ball where $k = 1$ and the width of the minimum enclosing axis-aligned rectangle along the x-axis and y-axis where $k = 2$. Define the "dominates" binary operator $\preceq$ so that $(p_1, \ldots, p_k) \preceq (v_1, \ldots, v_k)$ is true if for every coordinate $p_i \leq v_i$. Let $\mathbb{X}_f(v) = \{Q \in \mathbb{R}^{dn} \mid f(Q) \preceq v\}$. For a query value v define, $F_{\mu_P}(v) = \int_{Q \in \mathbb{X}_f(v)} \mu_P(Q)\, dQ$. Then F_{μ_P} is the cumulative density function of the distribution of possible values that f can take[1]. Ideally, we would return the function F_{μ_P} so we could quickly answer any query exactly, however, it is not clear how to calculate $F_{\mu_P}(v)$ exactly for even a single query value v. Rather, we introduce a data structure, which we call an ε-quantization, to answer any such query approximately and efficiently,

[1] For a function f and a distribution of point sets μ_P, we will always represent the cumulative density function of f over μ_P by F_{μ_P}.

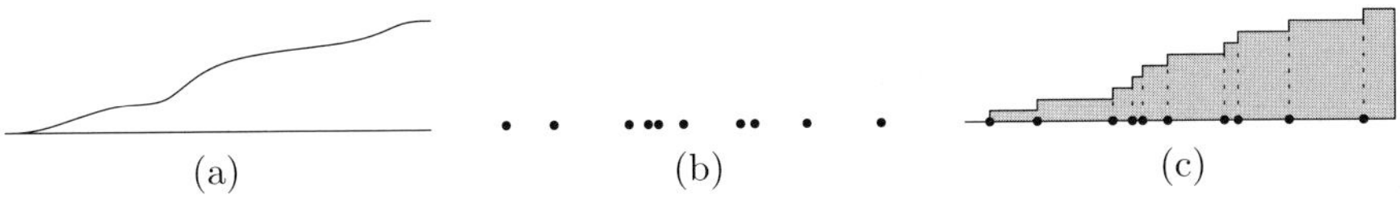

Fig. 1. (a) The true form of a monotonically increasing function from $\mathbb{R} \to \mathbb{R}$. (b) The ε-quantization R as a point set in $\mathbb{R}$. (c) The inferred curve h_R in $\mathbb{R}^2$.

illustrated in Figure 1 for $k = 1$. An *ε-quantization* is a point set $R \subset \mathbb{R}^k$ which induces a function h_R where $h_R(v)$ describes the fraction of points in R that v dominates. Let $R_v = \{r \in R \mid r \preceq v\}$. Then $h_R(v) = |R_v|/|R|$. For an isotonic (monotonically increasing in each coordinate) function F_{μ_P} and any value v, an ε-quantization, R, guarantees that $|h_R(v) - F_{\mu_P}(v)| \leq \varepsilon$. More generally (and, for brevity, usually only when $k > 1$), we say R is a k-variate ε-quantization. An example of a 2-variate ε-quantization is shown in Figure 2. The space required to store the data structure for R is dependent only on ε and k, not on $|P|$ or μ_P.

$(\varepsilon, \delta, \alpha)$-Kernels. Rather than compute a new data structure for each measure we are interested in, we can also compute a single data structure (a coreset) that allows us to answer many types of questions. For an isotonic function $F_{\mu_P} : \mathbb{R}^+ \to [0, 1]$, an *$(\varepsilon, \alpha)$-quantization* data structure M describes a function $h_M : \mathbb{R}^+ \to [0, 1]$ so for any $x \in \mathbb{R}^+$, there is an $x' \in \mathbb{R}^+$ such that (1) $|x - x'| \leq \alpha x$ and (2) $|h_M(x) - F_{\mu_P}(x')| \leq \varepsilon$. An *$(\varepsilon, \delta, \alpha)$-kernel* is a data structure that can produce an (ε, α)-quantization, with probability at least $1 - \delta$, for F_{μ_P} where f measures the width in any direction and whose size depends only on ε, α, and δ. The notion of (ε, α)-quantizations is generalized to a k-variate version, as are $(\varepsilon, \delta, \alpha)$-kernels.

Shape inclusion probabilities. A *summarizing shape* of a point set $P \subset \mathbb{R}^d$ is a Lebesgue-measureable subset of $\mathbb{R}^d$ that is determined by P. I.e. given a class of shapes $\mathcal{S}$, the summarizing shape $S(P) \in \mathcal{S}$ is the shape that optimizes some aspect with respect to P. Examples include the smallest enclosing ball and the minimum-volume axis-aligned bounding box. For a family $\mathcal{S}$ we can study the *shape inclusion probability function* $s_{\mu_P} : \mathbb{R}^d \to [0, 1]$ (or sip function), where $s_{\mu_P}(q)$ describes the probability that a query point $q \in \mathbb{R}^d$ is included in

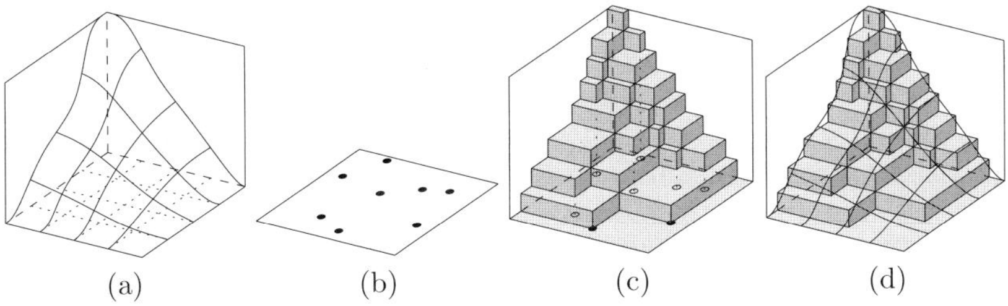

Fig. 2. (a) The true form of a 2-variate function. (b) The ε-quantization R as a point set in $\mathbb{R}^2$. (c) The inferred surface h_R in $\mathbb{R}^3$. (d) Overlay of the two images.

the summarizing shape[2]. There does not seem to be a closed form for many of these functions. Rather we calculate an ε-sip function $\hat{s} : \mathbb{R}^d \to [0,1]$ such that $\forall_{q \in \mathbb{R}^d} |s_{\mu_P}(q) - \hat{s}(q)| \le \varepsilon$. The space required to store an ε-sip function depends only on ε and the complexity of the summarizing shape.

1.2 Contributions

We describe simple and practical randomized algorithms for the computation of ε-quantizations, $(\varepsilon, \delta, \alpha)$-kernels, and ε-sip functions. Let $T_f(n)$ be the time it takes to calculate a summarizing shape of a set of n points $Q \subset \mathbb{R}^d$, which generates a statistic $f(Q)$ (e.g., radius of smallest enclosing ball). We can calculate an ε-quantization of F_{μ_P}, with probability at least $1-\delta$, in $O(T_f(n)(1/\varepsilon^2)\log(1/\delta))$ time. For univariate ε-quantizations the size is $O(1/\varepsilon)$, and for k-variate ε-quantizations the size is $O(k^2(1/\varepsilon)\log^{2k}(1/\varepsilon))$. We can calculate an $(\varepsilon, \delta, \alpha)$-kernel of size $O((1/\alpha^{(d-1)/2}) \cdot (1/\varepsilon^2)\log(1/\delta))$ in time $O((n + (1/\alpha^{d-3/2}))(1/\varepsilon^2) \cdot \log(1/\delta))$. With probability at least $1-\delta$, we can calculate an ε-sip function of size $O((1/\varepsilon^2)\log(1/\delta))$ in time $O(T_f(n)(1/\varepsilon^2)\log(1/\delta))$.

All of these randomized algorithms are simple and practical, as demonstrated by a series of experimental results. In particular, we show that the constant hidden by the big-O notation is in practice at most 0.5 for all algorithms. Due to space restrictions, the results in this abstract had to be compressed. For further explanation and full proofs, please refer to the full version [18].

1.3 Preliminaries: ε-Samples and α-Kernels

ε-Samples. For a set P let $\mathcal{A}$ be a set of subsets of P. In our context usually P will be a point set and the subsets in $\mathcal{A}$ could be induced by containment in a shape from some family of geometric shapes. For example of $\mathcal{A}$, $\mathcal{I}_+$ describes one-sided intervals of the form $(-\infty, t)$. The pair $(P, \mathcal{A})$ is called a *range space*. We say that $Q \subset P$ is an *ε-sample* of $(P, \mathcal{A})$ if

$$\forall_{R \in \mathcal{A}} \left| \frac{\phi(R \cap Q)}{\phi(Q)} - \frac{\phi(R \cap P)}{\phi(P)} \right| \le \varepsilon,$$

where $|\cdot|$ takes the absolute value and $\phi(\cdot)$ returns the measure of a point set. In the discrete case $\phi(Q)$ returns the cardinality of Q. We say $\mathcal{A}$ *shatters* a set S if every subset of S is equal to $R \cap S$ for some $R \in \mathcal{A}$. The cardinality of the largest discrete set $S \subseteq P$ that $\mathcal{A}$ can shatter is the *VC-dimension* of $(P, \mathcal{A})$.

When $(P, \mathcal{A})$ has constant VC-dimension ν, we can create an ε-sample Q of $(P, \mathcal{A})$, with probability $1-\delta$, by uniformly sampling $O((1/\varepsilon^2)(\nu + \log(1/\delta)))$ points from P [26,16]. There exist deterministic techniques to create ε-samples [20,9] of size $O(\nu(1/\varepsilon^2)\log(1/\varepsilon))$ in time $O(\nu^{3\nu}n((1/\varepsilon^2)\log(\nu/\varepsilon))^\nu)$. When P is a point set in $\mathbb{R}^d$ and the family of ranges $\mathcal{R}_d$ is determined by inclusion in axis-aligned boxes, then an ε-sample for $(P, \mathcal{R}_d)$ of size $O((d/\varepsilon)\log^{2d}(1/\varepsilon))$ can be constructed in $O((n/\varepsilon^3)\log^{6d}(1/\varepsilon))$ time [23].

[2] For technical reasons, if there are (degenerately) multiple optimal summarizing shapes, we say each is equally likely to be the summarizing shape of the point set.

For a range space $(P, \mathcal{A})$ the *dual range space* is defined $(\mathcal{A}, P^*)$ where P^* is all subsets $\mathcal{A}_p \subseteq \mathcal{A}$ defined for an element $p \in P$ such that $\mathcal{A}_p = \{A \in \mathcal{A} \mid p \in A\}$. If $(P, \mathcal{A})$ has VC-dimension ν, then $(\mathcal{A}, P^*)$ has VC-dimension $\leq 2^{\nu+1}$. Thus, if the VC-dimension of $(\mathcal{A}, P^*)$ is constant, then the VC-dimension of $(P, \mathcal{A})$ is also constant [21].

When we have a distribution $\mu : \mathbb{R}^d \to \mathbb{R}^+$, such that $\int_{x \in \mathbb{R}} \mu(x)\, dx = 1$, we can think of this as the set P of all points in $\mathbb{R}^d$, where the weight w of a point $p \in \mathbb{R}^d$ is $\mu(p)$. To simplify notation, we write $(\mu, \mathcal{A})$ as a range space where the ground set is this set $P = \mathbb{R}^d$ weighted by the distribution μ.

α-Kernels. Given a point set $P \in \mathbb{R}^d$ of size n and a direction $u \in \mathbb{S}^{d-1}$, let $P[u] = \arg\max_{p \in P} \langle p, u \rangle$, where $\langle \cdot, \cdot \rangle$ is the inner product operator. Let $\omega(P, u) = \langle P[u] - P[-u], u \rangle$ describe the width of P in direction u. We say that $K \subseteq P$ is an *α-kernel* of P if for all $u \in \mathbb{S}^{d-1}$

$$\omega(P, u) - \omega(K, u) \leq \alpha \cdot \omega(P, u).$$

α-kernels of size $O(1/\alpha^{(d-1)/2})$ [4] can be calculated in time $O(n + 1/\alpha^{d-3/2})$ [8,27]. Computing many extent related problems such as diameter and smallest enclosing ball on K approximates the problem on P [4,3,8].

2 Randomized Algorithm for ε-Quantizations

We develop several algorithms with the following basic structure: (1) sample one point from each distribution to get a random point set; (2) construct the summarizing shape of the random point set; (3) repeat the first two steps $O((1/\varepsilon)(\nu + \log(1/\delta)))$ times and calculate a summary data structure. This algorithm only assumes that we can draw a random point from μ_p for each $p \in P$.

2.1 Algorithm for ε-Quantizations

For a function f on a point set P of size n, it takes $T_f(n)$ time to evaluate $f(P)$. We construct an approximation to F_{μ_P} as follows. First draw a sample point q_j from each μ_{p_j} for $p_j \in P$, then evaluate $V_i = f(\{q_1, \ldots, q_n\})$. The fraction of trials of this process that produces a value dominated by v is the estimate of $F_{\mu_P}(v)$. In the univariate case we can reduce the size of $\mathcal{V}$ by returning $2/\varepsilon$ evenly spaced points according to the sorted order.

Theorem 1. *For a distribution μ_P of n points, with success probability at least $1 - \delta$, there exists an ε-quantization of size $O(1/\varepsilon)$ for F_{μ_P}, and it can be constructed in $O(T_f(n)(1/\varepsilon^2)\log(1/\delta))$ time.*

Proof. Because $F_{\mu_P} : \mathbb{R} \to [0, 1]$ is an isotonic function, there exists another function $g : \mathbb{R} \to \mathbb{R}^+$ such that $F_{\mu_P}(t) = \int_{x=-\infty}^{t} g(x)\, dx$ where $\int_{x \in \mathbb{R}} g(x)\, dx = 1$. Thus g is a probability distribution of the values of f given inputs drawn from μ_P. This implies that an ε-sample of $(g, \mathcal{I}_+)$ is an ε-quantization of F_{μ_P}, since both estimate within ε the fraction of points in any range of the form $(-\infty, x)$.

By drawing a random sample q_i from each μ_{p_i} for $p_i \in P$, we are drawing a random point set Q from μ_P. Thus $f(Q)$ is a random sample from g. Hence, using the standard randomized construction for ε-samples, $O((1/\varepsilon^2)\log(1/\delta))$ such samples will generate an $(\varepsilon/2)$-sample for g, and hence an $(\varepsilon/2)$-quantization for F_{μ_P}, with probability at least $1-\delta$.

Since in an $(\varepsilon/2)$-quantization R every value $h_R(v)$ is different from $F_{\mu_P}(v)$ by at most $\varepsilon/2$, then we can take an $(\varepsilon/2)$-quantization of the function described by $h_R(\cdot)$ and still have an ε-quantization of F_{μ_P}. Thus, we can reduce this to an ε-quantization of size $O(1/\varepsilon)$ by taking a subset of $2/\varepsilon$ points spaced evenly according to their sorted order. □

We can construct k-variate ε-quantizations similarly. The output V_i of f is now k-variate and thus results in a k-dimensional point.

Theorem 2. *Given a distribution μ_P of n points, with success probability at least $1-\delta$, we can construct a k-variate ε-quantization for F_{μ_P} of size $O((k/\varepsilon^2)(k+\log(1/\delta)))$ and in time $O(T_f(n)(1/\varepsilon^2)(k+\log(1/\delta)))$.*

Proof. Let $\mathcal{R}_+$ describe the family of ranges where a range $A_p = \{q \in \mathbb{R}^k \mid q \preceq p\}$. In the k-variate case there exists a function $g : \mathbb{R}^k \to \mathbb{R}^+$ such that $F_{\mu_P}(v) = \int_{x \preceq v} g(x)\, dx$ where $\int_{x\in\mathbb{R}^k} g(x)\, dx = 1$. Thus g describes the probability distribution of the values of f, given inputs drawn randomly from μ_P. Hence a random point set Q from μ_P, evaluated as $f(Q)$, is still a random sample from the k-variate distribution described by g. Thus, with probability at least $1-\delta$, a set of $O((1/\varepsilon^2)(k+\log(1/\delta)))$ such samples is an ε-sample of $(g, \mathcal{R}_+)$, which has VC-dimension k, and the samples are also a k-variate ε-quantization of F_{μ_P}. □

We can then reduce the size of the ε-quantization R to $O((k^2/\varepsilon)\log^{2k}(1/\varepsilon))$ in $O(|R|(k/\varepsilon^3)\log^{6k}(1/\varepsilon))$ time [23] or to $O((k^2/\varepsilon^2)\log(1/\varepsilon))$ in $O(|R|(k^{3k}/\varepsilon^{2k})\cdot \log^k(k/\varepsilon))$ time [9], since the VC-dimension is k and each data point requires $O(k)$ storage.

2.2 $(\varepsilon, \delta, \alpha)$-Kernels

The above construction works for a fixed family of summarizing shapes. In this sectin, we show how to build a single data structure, an $(\varepsilon, \delta, \alpha)$-kernel, for a distribution μ_P in $\mathbb{R}^d$ that can be used to construct (ε, α)-quantizations for several families of summarizing shapes. In particular, an $(\varepsilon, \delta, \alpha)$-kernel of μ_P is a data structure such that in any query direction $u \in \mathbb{S}^{d-1}$, with probability at least $1-\delta$, we can create an (ε, α)-quantization for the cumulative density function of $\omega(\cdot, u)$, the width in direction u.

We follow the randomized framework described above as follows. The desired $(\varepsilon, \delta, \alpha)$-kernel $\mathcal{K}$ consists of a set of $m = O((1/\varepsilon^2)\log(1/\delta))$ $(\alpha/2)$-kernels, $\{K_1, K_2, \ldots, K_m\}$, where each K_j is an $(\alpha/2)$-kernel of a point set Q_j drawn randomly from μ_P. Given $\mathcal{K}$, with probability at least $1-\delta$, we can create an (ε, α)-quantization for the cumulative density function of width over μ_P in any direction $u \in \mathbb{S}^{d-1}$. Specifically, let $M = \{\omega(K_j, u)\}_{j=1}^m$.

Lemma 1. *With probability at least* $1 - \delta$, M *is an* (ε, α)*-quantization for the cumulative density function of the width of* μ_P *in direction* u.

Proof. The width $\omega(Q_j, u)$ of a random point set Q_j drawn from μ_P is a random sample from the distribution over widths of μ_P in direction u. Thus, with probability at least $1-\delta$, m such random samples would create an ε-quantization. Using the width of the α-kernels K_j instead of Q_j induces an error on each random sample of at most $2\alpha \cdot \omega(Q_j, u)$. Then for a query width w, say there are γm point sets Q_j that have width at most w and $\gamma' m$ α-kernels K_j with width at most w. Note that $\gamma' > \gamma$. Let $\hat{w} = w - 2\alpha w$. For each point set Q_j that has width greater than w it follows that K_j has width greater than $\hat{w}$. Thus the number of α-kernels K_j that have width at most $\hat{w}$ is at most γm, and thus there is a width w' between w and $\hat{w}$ such that the number of α-kernels at most w' is exactly γm. □

Since each K_j can be computed in $O(n + 1/\alpha^{d-3/2})$ time, we obtain:

Theorem 3. *We can construct an* $(\varepsilon, \delta, \alpha)$*-kernel for* μ_P *on* n *points in* $\mathbb{R}^d$ *of size* $O((1/\alpha^{(d-1)/2})(1/\varepsilon^2) \cdot \log(1/\delta))$ *in* $O((n + 1/\alpha^{d-3/2}) \cdot (1/\varepsilon^2) \log(1/\delta))$ *time.*

The notion of (ε, α)-quantizations and $(\varepsilon, \delta, \alpha)$-kernels can be extended to k-dimensional queries or for a series of up to k queries which all have approximation guarantees with probability $1 - \delta$.

In a similar fashion, coresets of a point set distribution μ_P can be formed using coresets for other problems on discrete point sets. For instance, sample $m = O((1/\varepsilon^2) \log(1/\delta))$ points sets $\{P_1, \ldots, P_m\}$ each from μ_P and then store α-samples $\{Q_1 \subseteq P_1, \ldots, Q_m \subseteq P_m\}$ of each. This results in an $(\varepsilon, \delta, \alpha)$-sample of μ_P, and can, for example, be used to construct (with probability $1-\delta$) an (ε, α)-quantization for the fraction of points expected to fall in a query disk. Similar constructions can be done for other coresets, such as ε-nets [15], k-center [5], or smallest enclosing ball [7].

2.3 Shape Inclusion Probabilities

For a point set $Q \subset \mathbb{R}^d$, let the summarizing shape $S_Q = \mathcal{S}(Q)$ be from some geometric family $\mathcal{S}$ so $(\mathbb{R}^d, \mathcal{S})$ has bounded VC-dimension ν. We randomly sample m point sets $\mathcal{Q} = \{Q_1, \ldots, Q_m\}$ each from μ_P and then find the summarizing shape $S_{Q_j} = \mathcal{S}(Q_j)$ (e.g. minimum enclosing ball) of each Q_j. Let this set of shapes be $S^{\mathcal{Q}}$. If there are multiple shapes from $\mathcal{S}$ which are equally optimal choose one of these shapes at random. For a set of shapes $S' \subseteq \mathcal{S}$, let $S'_p \subseteq S'$ be the subset of shapes that contain $p \in \mathbb{R}^d$. We store $S^{\mathcal{Q}}$ and evaluate a query point $p \in \mathbb{R}^d$ by counting what fraction of the shapes the point is contained in, specifically returning $|S^{\mathcal{Q}}_p|/|S^{\mathcal{Q}}|$ in $O(\nu|S^{\mathcal{Q}}|) = O(\nu m)$ time. In some cases, this evaluation can be sped up with point location data structures.

Theorem 4. *Consider a family of summarizing shapes* $\mathcal{S}$ *where* $(\mathbb{R}^d, \mathcal{S})$ *has VC-dimension* ν *and where it takes* $T_{\mathcal{S}}(n)$ *time to determine the summarizing shape* $\mathcal{S}(Q)$ *for any point set* $Q \subset \mathbb{R}^d$ *of size* n. *For a distribution* μ_P *of a point set of size* n, *with probability at least* $1 - \delta$, *we can construct an* ε-sip *function of size* $O((\nu/\varepsilon^2)(2^{\nu+1} + \log(1/\delta)))$ *and in time* $O(T_{\mathcal{S}}(n)(1/\varepsilon^2) \log(1/\delta))$.

Proof. If $(\mathbb{R}^d, \mathcal{S})$ has VC-dimension ν, then the dual range space $(\mathcal{S}, P^*)$ has VC-dimension $\nu' \leq 2^{\nu+1}$, where P^* is all subsets $\mathcal{S}_p \subseteq \mathcal{S}$, for any $p \in \mathbb{R}^d$, such that $\mathcal{S}_p = \{S \in \mathcal{S} \mid p \in S\}$. Using the above algorithm, sample $m = O((1/\varepsilon^2)(\nu' + \log(1/\delta)))$ point sets Q from μ_P and generate the m summarizing shapes S_Q. Each shape is a random sample from $\mathcal{S}$ according to μ_P, and thus $S^{\mathcal{Q}}$ is an ε-sample of $(\mathcal{S}, P^*)$.

Let $w_{\mu_P}(S)$, for $S \in \mathcal{S}$, be the probability that S is the summarizing shape of a point set Q drawn randomly from μ_P. For any $\mathcal{S}' \subseteq P^*$, let $W_{\mu_P}(\mathcal{S}') = \int_{S \in \mathcal{S}'} w_{\mu_P}(S)\, dS$ be the probability that some shape from the subset $\mathcal{S}'$ is the summarizing shape of Q drawn from μ_P.

We approximate the sip function at $p \in \mathbb{R}^d$ by returning the fraction $|S_p^{\mathcal{Q}}|/m$. The true answer to the sip function at $p \in \mathbb{R}^d$ is $W_{\mu_P}(\mathcal{S}_p)$. Since $S^{\mathcal{Q}}$ is an ε-sample of $(\mathcal{S}, P^*)$, then with probability at least $1 - \delta$

$$\left| \frac{|S_p^{\mathcal{Q}}|}{m} - \frac{W_{\mu_P}(\mathcal{S}_p)}{1} \right| = \left| \frac{|S_p^{\mathcal{Q}}|}{|S^{\mathcal{Q}}|} - \frac{W_{\mu_P}(\mathcal{S}_p)}{W_{\mu_P}(P^*)} \right| \leq \varepsilon.$$

Since for the family of summarizing shapes $\mathcal{S}$ the range space $(\mathbb{R}^d, \mathcal{S})$ has VC-dimension ν, each can be stored using that much space. □

Using deterministic techniques [9] the size can then be reduced to $O(2^{\nu+1}(\nu/\varepsilon^2) \cdot \log(1/\varepsilon))$ in time $O((2^{3(\nu+1)} \cdot (\nu/\varepsilon^2)\log(1/\varepsilon))^{2^{\nu+1}} \cdot 2^{3(\nu+1)}(\nu/\varepsilon^2)\log(1/\delta))$.

Representing ε-sip functions by isolines. Shape inclusion probability functions are density functions. A convenient way of visually representing a density function in $\mathbb{R}^2$ is by drawing the isolines. A γ-*isoline* is a collection of closed curves bounding a region of the plane where the density function is greater than γ.

In each part of Figure 3 a set of 5 circles correspond to points with a probability distribution. In part (a,c) the probability distribution is uniform over the inside of the circles. In part (b,d) it is drawn from a normal distribution with standard deviation given by the radius. We generate ε-sip functions for the smallest enclosing ball in Figure 3(a,b) and for the smallest axis-aligned rectangle in Figure 3(c,d).

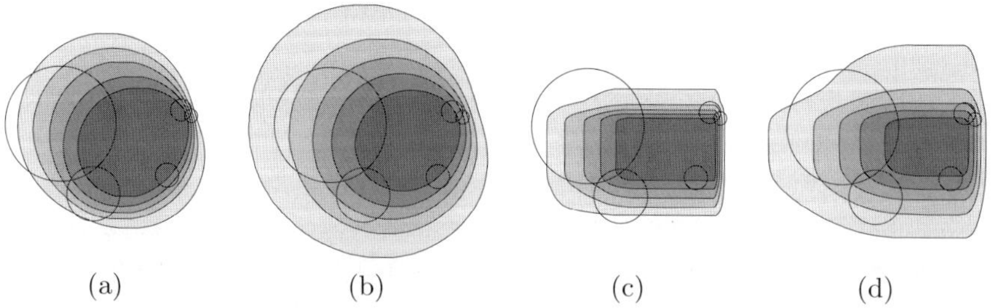

Fig. 3. The sip for the smallest enclosing ball (a,b) or smallest enclosing axis-aligned rectangle (c,d), for uniformly (a,c) or normally (b,d) distributed points

In all figures we draw approximations of $\{.9, .7, .5, .3, .1\}$-isolines. These drawing are generated by randomly selecting $m = 5000$ (Figure 3(a,b)) or $m = 25000$ (Figure 3(c,d)) shapes, counting the number of inclusions at different points in the plane and interpolating to get the isolines. The innermost and darkest region has probability $> 90\%$, the next one probability $> 70\%$, etc., the outermost region has probability $< 10\%$.

3 Measuring the Error

We have established asymptotic bounds of $O((1/\varepsilon^2)(\nu+\log(1/\delta))$ random samples for constructing ε-quantizations and ε-sip functions. In this section we empirically demonstrate that the constant hidden by the big-O notation is approximately 0.5, indicating that these algorithms are indeed quite practical. Additionally, we show that we can reduce the size of ε-quantizations to $2/\varepsilon$ without sacrificing accuracy and with only a factor 4 increase in the runtime. We also briefly compare the (ε, α)-quantizations produced with $(\varepsilon, \delta, \alpha)$-kernels to ε-quantizations. We show that the $(\varepsilon, \delta, \alpha)$-kernels become useful when the number of uncertain points becomes large, i.e. exceeding 1000.

Univariate ε-quantizatons. We consider a set of $n = 50$ sample points in $\mathbb{R}^3$ chosen randomly from the boundary of a cylinder piece of length 10 and radius 1. We let each point represent the center of 3-variate Gaussian distribution with standard deviation 2 to represent the probability distribution of an uncertain point. This set of distributions describes an uncertain point set $\mu_P : \mathbb{R}^{3n} \to \mathbb{R}^+$.

We want to estimate three statistics on μ_P: dwid, the width of the points set in a direction that makes an angle of 75° with the cylinder axis; diam, the diameter of the point set; and seb_2, the radius of the smallest enclosing ball (using code from Bernd Gärtner [13]). We can create ε-quantizations with m samples from μ_P, where the value of m is from the set $\{16, 64, 256, 1024, 4096\}$.

We would like to evaluate the ε-quantizations versus the ground truth function F_{μ_P}; however, it is not clear how to evaluate F_{μ_P}. Instead, we create another ε-quantization Q with $\eta = 100000$ samples from μ_P, and treat this as if it were the ground truth. To evaluate each sample ε-quantization R versus Q we find the maximum deviation (i.e. $d_\infty(R, Q) = \max_{q\in\mathbb{R}} |h_R(q) - h_Q(q)|$) with h defined with respect to diam or dwid. This can be done by for each value $r \in R$ evaluating $|h_R(r) - h_Q(r)|$ and $|(h_R(r) - 1/|R|) - h_Q(r)|$ and returning the maximum of both values over all $r \in R$.

Given a fixed "ground truth" quantization Q we repeat this process for $\tau = 500$ trials of R, each returning a $d_\infty(R, Q)$ value. The set of these τ maximum deviations values results in another quantization S for each of diam and dwid, plotted in Figure 4. Intuitively, the maximum deviation quantization S describes the sample probability that $d_\infty(R, Q)$ will be less than some query value.

Note that the maximum deviation quantizations S are similar for both statistics (and others we tried), and thus we can use these plots to estimate $1-\delta$, the sample probability that $d_\infty(R, Q) \le \varepsilon$, given a value m. We can fit this function

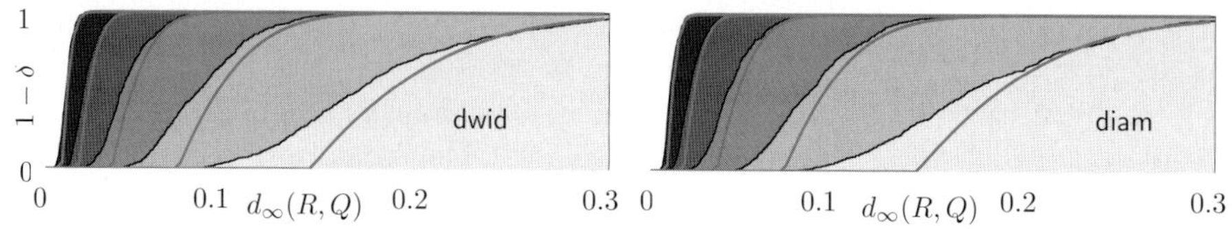

Fig. 4. Shows quantizations of $\tau = 500$ trials for $d_\infty(R, Q)$ where Q and R measure dwid and diam. The size of each R is $m = \{16, 64, 256, 1024, 4096\}$ (from right to left) and the "ground truth" quantization Q has size $\eta = 100000$. Smooth, thick curves are $1 - \delta = 1 - \exp(-2m\varepsilon^2 + 1)$ where $\varepsilon = d_\infty(R, Q)$.

as approximately $1 - \delta = 1 - \exp(-m\varepsilon^2/C + \nu)$ with $C = 0.5$ and $\nu = 1.0$. Thus solving for m in terms of ε, ν, and δ reveals: $m = C(1/\varepsilon^2)(\nu + \log(1/\delta))$. This indicates the big-O notation for the asymptotic bound of $O((1/\varepsilon^2)(\nu + \log(1/\delta))$ [16] for ε-samples only hides a constant of approximately 0.5.

Maximum error in sip functions. We can perform a similar analysis on sip functions. We use the same input data as is used to generate Figure 3(b) and create sip functions R for the smallest enclosing ball using $m = \{16, 36, 81, 182, 410\}$ samples from μ_P. We compare this to a "ground truth" sip function Q formed using $\eta = 5000$ sampled points. The maximum deviation between R and Q in this context is defined $d_\infty(R, Q) = \max_{q \in \mathbb{R}^2} |R(q) - Q(q)|$ and can be found by calculating $|R(q) - Q(q)|$ for all points $q \in \mathbb{R}^2$ at the intersection of boundaries of discs from R or Q.

We repeat this process for $\tau = 100$ trials, for each value of m. This creates a quantization S (for each value of m) measuring the maximum deviation for the sip functions. These maximum deviation quantizations are plotted in Figure 5. We fit these curves with a function $1-\delta = 1-\exp(-m\varepsilon^2/C+\nu)$ with $C = 0.5$ and $\nu = 2.0$, so $m = C(1/\varepsilon^2)(\nu + \log 1/\delta)$. Note that the dual range space $(\mathcal{B}, \mathbb{R}^{2*})$, defined by disks $\mathcal{B}$ has VC-dimension 2, so this is exactly what we would expect.

Maximum error in k-variate quantizations. We extend these experiments to k-variate quantizations by considering the width in k different directions. As

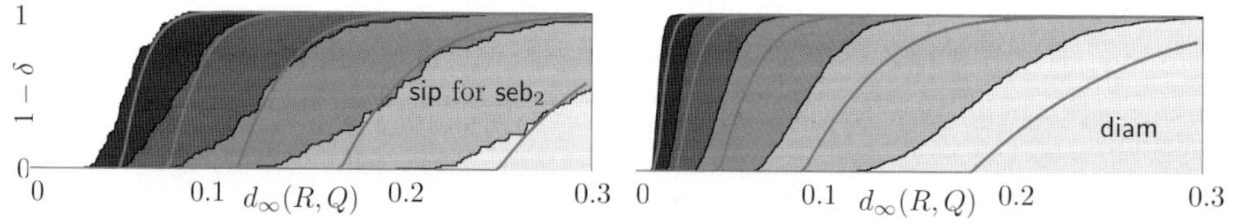

Fig. 5. Left: Quantization of $\tau = 100$ trials of maximum deviation between sip functions for smallest enclosing disc with $m = \{16, 36, 81, 182, 410\}$ (from right to left) sample shapes versus a "ground truth" sip function with $\eta = 5000$ sample shapes. Right: Quantization of $\tau = 500$ trials for $d_\infty(R, Q)$ where Q and R measure diam. Size of each R is $m = \{64, 256, 1024, 4096, 16384\}$, then compressed to size $\{8, 16, 32, 64, 128\}$ (resp., from right to left) and the "ground truth" quantization Q has size $\eta = 100000$.

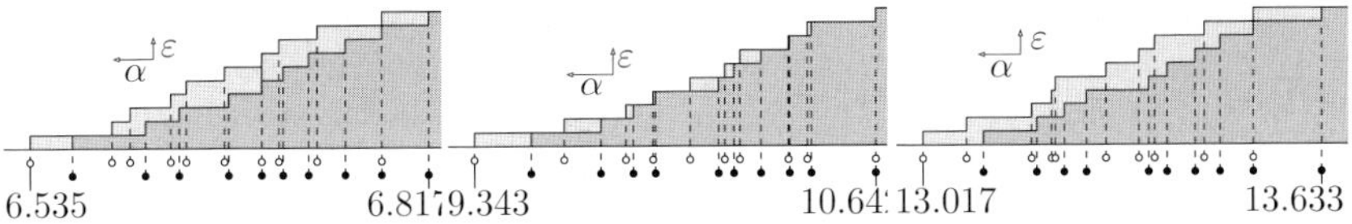

Fig. 6. (ε, α)-quantization (white points) and ε-quantization (black points) for (left) seb_2, (center) dwid, and (right) diam

expected, the quantizations for maximum deviation can be fit with an equation $1 - \delta = 1 - \exp(-m\varepsilon^2/C + k)$ with $C = 0.5$, so $m \leq C(1/\varepsilon^2)(k + \log 1/\delta)$. For $k > 2$, this bound for m becomes too conservative. We omit the resulting figures due to space restrictions.

3.1 Compressing ε-Quantizations

Theorem 1 describes how to compress the size of a univariate ε-quantization to $O(1/\varepsilon)$. We first create an $(\varepsilon/2)$-quantization of size m, then sort the values V_i, and finally take every $(m\varepsilon/2)$th value according to the sorted order. This returns an ε-quantization of size $2/\varepsilon$ and requires creating an initial ε-quantization with 4 times as many samples as we would have without this compression. The results, shown in Figure 5 using the same setup as in Figure 4, confirms that this compression scheme works better than the worst case claims. We only show the plot for diam, but the results for dwid and seb_2 are nearly identical. In particular, the error is smaller than the results in Figure 4, but it takes 4 times as long.

3.2 $(\varepsilon, \delta, \alpha)$-Kernels versus ε-Quantizations

We compare $(\varepsilon, \delta, \alpha)$-kernels to ε-quantizations for diam, dwid, and seb_2, using code from Hai Yu [27] for α-kernels. Using the same setup as in Figure 4 with $n = 5000$ input points, we set $\varepsilon = 0.2$ and $\delta = 0.1$, resulting in $m = 40$ point sets sampled from μ_P. We also generated α-kernels of size at most 40. The $(\varepsilon, \delta, \alpha)$-kernel has a total of 1338 points. We calculated ε-quantizations and (ε, α)-quantizations for diam, dwid, and seb_2, each compressed to a size 10 shown in Figure 6. This method starts becoming useful in compressing μ_P when $n \gg 1000$ (otherwise the total size of the $(\varepsilon, \delta, \alpha)$-kernel may be larger than μ_P) or if computing f_S is very expensive.

Acknowledgements. The authors would like to thanks Pankaj K. Agarwal for many helpful discussions. This research was partially supported by the Netherlands Organisation for Scientific Research (NWO) through the project GOGO.

References

1. Agarwal, C.C., Yu, P.S. (eds.): Privacy Preserving Data Mining: Models and Algorithms. Springer, Heidelberg (2008)
2. Agarwal, P.K., Cheng, S.-W., Tao, Y., Yi, K.: Indexing uncertain data. In: PODS (2009)

3. Agarwal, P.K., Har-Peled, S., Varadarajan, K.: Geometric approximations via coresets. C. Trends Comb. and Comp. Geom. (E. Welzl) (2007)
4. Agarwal, P.K., Har-Peled, S., Varadarajan, K.R.: Approximating extent measure of points. J. ACM 51(4) (2004)
5. Agarwal, P.K., Procopiuc, C.M., Varadarajan, K.R.: Approximation algorithms for k-line center. In: Möhring, R.H., Raman, R. (eds.) ESA 2002. LNCS, vol. 2461, pp. 54–63. Springer, Heidelberg (2002)
6. Bandyopadhyay, D., Snoeyink, J.: Almost-Delaunay simplices: Nearest neighbor relations for imprecise points. In: SODA, pp. 403–412 (2004)
7. Bădoiu, M., Clarkson, K.: Smaller core-sets for balls. In: SODA (2003)
8. Chan, T.: Faster core-set constructions and data-stream algorithms in fixed dimensions. Computational Geometry: Theory and Applications 35, 20–35 (2006)
9. Chazelle, B., Matousek, J.: On linear-time deterministic algorithms for optimization problems in fixed dimensions. J. Algorithms 21, 579–597 (1996)
10. Cormode, G., Garafalakis, M.: Histograms and wavelets of probabilitic data. In: ICDE (2009)
11. Cormode, G., Li, F., Yi, K.: Semantics of ranking queries for probabilistic data and expected ranks. In: ICDE (2009)
12. Eliazar, A., Parr, R.: Dp-slam 2.0. In: ICRA (2004)
13. Gärtner, B.: Fast and robust smallest enclosing balls. In: Nešetřil, J. (ed.) ESA 1999. LNCS, vol. 1643, pp. 325–338. Springer, Heidelberg (1999)
14. Guibas, L.J., Salesin, D., Stolfi, J.: Epsilon geometry: building robust algorithms from imprecise computations. In: SoCG, pp. 208–217 (1989)
15. Haussler, D., Welzl, E.: epsilon-nets and simplex range queries. Disc. & Comp. Geom. 2, 127–151 (1987)
16. Li, Y., Long, P.M., Srinivasan, A.: Improved bounds on the samples complexity of learning. J. Comp. and Sys. Sci. 62, 516–527 (2001)
17. Lillesand, T.M., Kiefer, R.W., Chipman, J.W.: Remote Sensing and Image Interpretaion. John Wiley & Sons, Chichester (2004)
18. Löffler, M., Phillips, J.: Shape fitting on point sets with probability distributions. Technical Report UU-CS-2009-013, Utrecht University, Institute of Information and Computing Sciences (2009)
19. Löffler, M., Snoeyink, J.: Delaunay triangulations of imprecise points in linear time after preprocessing. In: SoCG, pp. 298–304 (2008)
20. Matousek, J.: Approximations and optimal geometric divide-and-conquer. In: SToC, pp. 505–511 (1991)
21. Matousek, J.: Geometric Discrepancy; An Illustrated Guide. Springer, Heidelberg (1999)
22. Nagai, T., Tokura, N.: Tight error bounds of geometric problems on convex objects with imprecise coordinates. In: Akiyama, J., Kano, M., Urabe, M. (eds.) JCDCG 2000. LNCS, vol. 2098, pp. 252–263. Springer, Heidelberg (2001)
23. Phillips, J.M.: Algorithms for ϵ-Approximations of Terrains. In: Aceto, L., Damgård, I., Goldberg, L.A., Halldórsson, M.M., Ingólfsdóttir, A., Walukiewicz, I. (eds.) ICALP 2008, Part I. LNCS, vol. 5125, pp. 447–458. Springer, Heidelberg (2008)
24. Shekhar, S., Chawla, S.: Spatial Databases: A Tour. Pearsons (2001)
25. van Kreveld, M., Löffler, M.: Largest bounding box, smallest diameter, and related problems on imprecise points. Comp. Geom. The. and App. (2009)
26. Vapnik, V., Chervonenkis, A.: On the uniform convergence of relative frequencies of events to their probabilities. The. of Prob. App. 16, 264–280 (1971)
27. Yu, H., Agarwal, P.K., Poreddy, R., Varadarajan, K.R.: Practical methods for shape fitting and kinetic data structures using coresets. In: SoCG (2004)

An Efficient Algorithm for Haplotype Inference on Pedigrees with a Small Number of Recombinants (Extended Abstract)

Jing Xiao[1], Tiancheng Lou[2], and Tao Jiang[3]

[1] IBM China Research Lab, Beijing, China
xiaojing82@gmail.com
[2] Department of Computer Science and Technology, Tsinghua University, Beijing, China
loutiancheng860214@gmail.com
[3] Department of Computer Science and Engineering, University of California, Riverside, CA
jiang@cs.ucr.edu

Abstract. Combinatorial (or rule-based) methods for inferring haplotypes from genotypes on a pedigree have been studied extensively in the recent literature. These methods generally try to reconstruct the haplotypes of each individual so that the total number of recombinants is minimized in the pedigree. The problem is NP-hard, although it is known that the number of recombinants in a practical dataset is usually very small. In this paper, we consider the question of how to efficiently infer haplotypes on a large pedigree when the number of recombinants is bounded by a small constant, *i.e.* the so called k-recombinant haplotype configuration (k-RHC) problem. We introduce a simple probabilistic model for k-RHC where the prior haplotype probability of a founder and the haplotype transmission probability from a parent to a child are all assumed to follow the uniform distribution and k random recombinants are assumed to have taken place uniformly and independently in the pedigree. We present an $O(mn \log^{k+1} n)$ time algorithm for k-RHC on tree pedigrees without mating loops, where m is the number of loci and n is the size of the input pedigree, and prove that when $90 \log n < m < n^3$, the algorithm can correctly find a feasible haplotype configuration that obeys the Mendelian law of inheritance and requires no more than k recombinants with probability $1 - O(k^2 \frac{\log^2 n}{mn} + \frac{1}{n^2})$. The algorithm is efficient when k is of a moderate value and could thus be used to infer haplotypes from genotypes on large tree pedigrees efficiently in practice. We have implemented the algorithm as a C++ program named Tree-k-RHC. The implementation incorporates several ideas for dealing with missing data and data with a large number of recombinants effectively. Our experimental results on both simulated and real datasets show that Tree-k-RHC can reconstruct haplotypes with a high accuracy and is much faster than the best combinatorial method in the literature.

Keywords: computational biology, haplotype inference, pedigree, recombination, combinatorial algorithm, probabilistic model.

1 Introduction

As more progress has been made in science and technology, scientists believe that genetic factors should play a significant role in preventing, diagnosing and treating im-

A. Fiat and P. Sanders (Eds.): ESA 2009, LNCS 5757, pp. 325–336, 2009.

portant human diseases such as diabetes, cancer, stroke, heart disease, depression, and asthma. With the discovery of genetic markers such as microsatellite DNA sequences and single nucleotide polymorphisms (SNPs), it is now possible to provide a unique genetic map to establish connections between diseases and specific genetic variations. One of the main objectives of the International HapMap Project launched in October 2002 [27] is to discover the haplotype structure of human beings and examine the common haplotypes in different populations. This information will greatly facilitate the mapping of many important disease-susceptibility genes. However, the diploid structure of humans makes it very expensive to collect haplotype data directly. Instead, genotype data are collected routinely. Since haplotype data are required (or at least desirable) in many genetic analysis including linkage disequilibrium analysis and disease association mapping, efficient and accurate computational methods for the inference of haplotypes from genotypes, which is also commonly referred to as *phasing*, have been extensively studied in the literature. A recent survey on these methods can be found in [19].

The existing haplotyping algorithms can be classified as either statistical or combinatorial (or rule-based). Both paradigms can be applied to *pedigree* data, *population* data, or *pooled samples*. In this paper, we are interested in pedigree data and the combinatorial paradigm. Although many (statistical or combinatorial) algorithms have been proposed for haplotype inference on pedigrees in the literature [19], they are mostly effective for pedigrees of small to moderate sizes. For example, it took the exact algorithm based on *integer linear programming* (ILP) in PedPhase 5 hours to solve a pedigree with 29 individuals and 51 SNP loci [17,18] on a standard PC. The same data took the popular program SimWalk2 [26] based on a statistical approach 6 days. The well-known Lander-Green algorithm [15] based on the *maximum likelihood* (ML) framework and its subsequent improvements [1,12,14] run in time linear in the number of loci but exponential in the pedigree size [2,19]. These algorithms are thus limited to relatively small pedigrees.

With the advance in sequencing technology, larger and larger pedigrees are being genotyped and scientists are becoming increasingly interested in haplotype inference on large pedigrees. For example, in [2,4], the inference was performed on pedigrees of sizes 368 and 1149, respectively. The existing haplotype inference methods either are very slow (*e.g.* those based on ML or ILP) or have less than desirable accuracies (*e.g.* the block extension heuristic algorithm in PedPhase) when the input pedigree gets large. In fact, the question of how to efficiently and accurately infer haplotypes from genotypes on large pedigrees is one of the challenges raised at the recently held 2008 Haplotype Conference [13].

In general, combinatorial methods for haplotype inference are faster (or intended to be faster) than statistical methods that attempt to maximize the likelihood of the haplotype solution [19]. To our knowledge, all combinatorial algorithms for haplotyping pedigree data aim at solving the *minimum-recombinant haplotype configuration* (MRHC) problem [16,17,18,25] where the goal is to find a haplotype solution requiring the minimum number of recombinants. The problem is sensible since it is known that recombinants are rare in a typical human pedigree [11]. This is especially true when the marker loci considered are from a same haplotype block. For instance, the analysis performed in [17,18] on a HapMap data shows that the average number of recombinants per haplotype block of each chromosome on a (relatively small) pedigree is in fact close to 0 (although not exactly 0). Thus, a minimum-recombinant solution is likely to be the true solution. Unfortunately, MRHC is NP-hard [16]. It remains NP-hard even if the input pedigree is a tree without mating loops [7,21]. The ILP-based exact algorithm for MRHC in PedPhase [19] works well for small pedigrees but its worst case running time is exponential in both the number of loci and pedigree size. The heuristic algorithm in [25] runs for days on a PC even for medium-sized datasets. Hence, recent work on MRHC has been focused on the special case where the number of recombinants is zero, the so called *zero-recombinant haplotype configuration* (ZRHC) problem [5,16,20,22,23,28,30,31]. In particular, Li and Jiang [16] formulated ZRHC as a system of linear equations over the field $F(2)$ and devised an $O(m^3n^3)$ time algorithm using Gaussian elimination, where m is the number of loci and n is the size of the input pedigree. Xiao *et al.* [30,31] improved the running time to $O(mn^2 + n^3 \log^2 n \log\log n)$ by using a compact system of linear equations, taking advantage of some special properties of a pedigree graph, and the low-stretch spanning tree technique. The recent results in [5,20,22] presented linear (*i.e.* $O(mn)$) time algorithms for ZRHC on tree pedigrees using different techniques. Note that tree pedigrees are very common in human pedigrees [2]. They also play important roles in the analysis of general complex pedigrees [3,29].

Since the number of recombinants in a real pedigree is usually very small, a plausible approach to solving MRHC that could potentially be very efficient in practice is to try to infer a haplotype configuration that requires at most k recombinants in the input pedigree, where k is some fixed small constant k. We will refer to this parameterized problem as the *k-recombinant haplotype configuration* (k-RHC) problem. Although ZRHC (or 0-RHC) seems easy to solve [5,20,22,30,31], the general k-RHC problem remains very hard to tackle. Observe that, we could obtain a trivial algorithm for k-RHC with time complexity $O((mn)^k(mn^2 + n^3 \log^2 n \log\log n))$ by using the algorithm in [30,31] for ZRHC and exhaustively enumerating the possible locations of the k recombinants. This is because the k-RHC instance can be easily transformed into a ZRHC instance once the recombinant locations are known. Similarly, one could obtain a trivial algorithm for k-RHC on tree pedigrees with time complexity $O((mn)^{k+1})$ by using the linear time algorithms in [5,20,22] for tree ZRHC. We note in passing that the dynamic programming algorithm in [6] solves MRHC on tree pedigrees in $O(nm^{3k+1}2^m)$ time when each parent-child pair is allowed to have at most k recombinants. This algorithm is inefficient when the number of loci exceeds 30 even if k is very small.

In this paper, we present an algorithm for k-RHC that is efficient in the average sense. More precisely, we consider a simple probabilistic model for k-RHC where the haplotypes of the founders (*i.e.* individuals without parents in the input pedigree) are generated randomly from a uniform distribution, a uniform random haplotype of each parent is passed to a child, and k uniform random recombinants are assumed to have taken place independently in the pedigree. This model is a special case of the general probabilistic model in the genetics literature (see *e.g.* [24]) where the prior founder haplotype probabilities and haplotype transmission probabilities could follow arbitrary distributions. In other words, our model is a primitive Mendelian model. We present an $O(mn \log^{k+1} n)$ time algorithm for k-RHC on tree pedigrees, and prove that when $90 \log n < m < n^3$, the algorithm can correctly find a feasible haplotype configuration that obeys the Mendelian law of inheritance and requires no more than k recombinants with probability $1 - O(k^2 \frac{\log^2 n}{mn} + \frac{1}{n^2})$. (Note that this result does not imply that k-RHC is fixed-parameter tractable as defined in [8].) The algorithm is fast when k is of a moderate value and could thus be used to infer haplotypes from genotypes on large tree pedigrees in practice. We have implemented the algorithm as a C++ program named TREE-k-RHC. The implementation incorporates several effective ideas for dealing with missing data and data with an expectedly large number of recombinants. Our preliminary experimental results on both simulated and real datasets show that TREE-k-RHC can reconstruct haplotypes with a high accuracy and speed. In fact, it runs more than 20 times faster than the ILP-based exact algorithm in PedPahse [17,18] and is more accurate than the heuristic algorithm in PedPhase [16]. We expect that the speed up will grow quickly with m and n as the worst-case time complexity of the ILP-based algorithm is at least $(mn)^k$.

The crux of our algorithm is to formulate k-RHC as an ILP based on the system of linear equations developed in [30,31] (also in [22]). For each instance generated by the probabilistic model, we try to identify small areas of the pedigree where a recombinant might have occurred by comparing the linear (equality) constraints in the ILP. Once the locations of all k recombinants are determined (or enumerated), the instance is transformed to a tree ZRHC instance and solved in linear time by using one of the algorithms in [5,20,22].

The rest of the paper is organized as follows. In Section 2, we present an ILP formulation of k-RHC based on the system of linear equations introduced in [30,31]. Section 3 reviews some graph data structures and constraint generation techniques from [30,31] that can be used to make the ILP more compact. We present the efficient algorithm for k-RHC on tree pedigrees and analyze its success probability in Section 4. Due to the page limit, we will omit all the technical proofs, figures, tables, and pseudocodes as well discussions on the implementation of the algorithm and our experimental results in this extended abstract and present them in the full paper which will soon be submitted to a journal.

2 An Integer Linear Program for k-RHC

In this section, we formulate k-RHC as an ILP based on the system of linear equations in [30,31] for solving ZRHC. All the definitions are the same as in [30,31] except for the definition of the h-variables. Throughout this paper, n denotes the number of the individuals (or members) in the input pedigree and m the number of marker loci. Without loss of generality, suppose that each allele in the given genotypes is numbered numerically as 1 or 2 (*i.e.* the markers are assumed to be *bi-allelic*, which makes the hardest case for MRHC [16]), and the pedigree is free of genotype errors (*i.e.* the two alleles at each locus of a child can always be obtained from its respective parents). Hence, we can represent the genotype of member j as a ternary vector $\boldsymbol{g}_j$ as follows: $g_j[i] = 0$ if locus i of member j is homozygous with both alleles being 1's, $g_j[i] = 1$ if the locus is homozygous with both alleles being 2's, and $g_j[i] = 2$ otherwise (*i.e.* the locus is heterozygous). For any heterozygous locus i of member j, we use a binary variable $p_j[i]$ to denote the *phase* at the locus as follows: $p_j[i] = 0$ if allele 1 is paternal, and

$p_j[i] = 1$ otherwise. When the locus is homozygous, the variable is set to $g_j[i]$ for some technical reasons (so that the equations below involving $p_j[i]$ will hold). Hence, the vector $\boldsymbol{p}_j$ describes the paternal and maternal haplotypes of member j. Observe that the vectors $\boldsymbol{p}_1, \ldots, \boldsymbol{p}_n$ represent a complete haplotype configuration of the pedigree. Also, for technical reasons, define a vector $\boldsymbol{w}_j$ for each member j such that $w_j[i] = 0$ if its i-th locus is homozygous and $w_j[i] = 1$ otherwise.

Suppose that member j_r is a parent of member j. We introduce an auxiliary binary variable $h_{j_r,j}[i]$ to indicate which allele of j_r is passed to j at locus i. If j_r gives its paternal allele to j at locus i, then $h_{j_r,j}[i] = 0$; otherwise $h_{j_r,j}[i] = 1$. Suppose that j is a non-founder member with its father and mother being j_1 and j_2, respectively. We can define two linear (constraint) equations over the field $F(2)$ to describe the inheritance of paternal and maternal haplotypes at j on locus i respectively:

$$\begin{cases} p_{j_1}[i] + h_{j_1,j}[i] \cdot w_{j_1}[i] = p_j[i] \\ p_{j_2}[i] + h_{j_2,j}[i] \cdot w_{j_2}[i] = p_j[i] + w_j[i] \end{cases} \tag{1}$$

Denoting $\boldsymbol{d}_{j_1,j} = \boldsymbol{0}$ and $\boldsymbol{d}_{j_2,j} = \boldsymbol{w}_j$, the above equations can be unified into a single equation as:

$$\boldsymbol{p}_{j_r}[i] + h_{j_r,j}[i] \cdot \boldsymbol{w}_{j_r}[i] = \boldsymbol{p}_j[i] + \boldsymbol{d}_{j_r,j}[i] \qquad (r = 1,\ 2) \tag{2}$$

If there are no recombinants in the pedigree, $h_{j_r,j}[i] = c$ (which is some constant) for all i. Conversely, if $h_{j_r,j}[i] \neq h_{j_r,j}[i+1]$, there must be a recombinant from member j_r to member j between locus i and locus $i+1$. Formally, we can express the k-RHC problem as an ILP:

$$\sum_{\text{for all parent-child pairs } (j_r,\, j)} \sum_{i=1}^{m-1} \left| h_{j_r,j}[i] - h_{j_r,j}[i+1] \right| \leq k$$

$$\begin{cases} p_l[i] + h_{l,j}[i] \cdot w_l[i] = p_j[i] + d_{l,j}[i] & 1 \leq i \leq m,\ 1 \leq j,\ l \leq n,\ l \text{ is a parent of } j \\ p_j[i] = g_j[i] & 1 \leq i \leq m,\ 1 \leq j \leq n,\ g_j[i] \neq 2 \\ w_j[i] = 1 & 1 \leq i \leq m,\ 1 \leq j \leq n,\ g_j[i] = 2 \\ w_j[i] = 0 & 1 \leq i \leq m,\ 1 \leq j \leq n,\ g_j[i] \neq 2 \\ d_{l,j}[i] = w_j[i] & 1 \leq i \leq m,\ 1 \leq j, l \leq n,\ l \text{ is the mother of } j \\ d_{l,j}[i] = 0 & 1 \leq i \leq m,\ 1 \leq j, l \leq n,\ l \text{ is the father of } j \end{cases} \tag{3}$$

where $g_j[i], w_j[i], d_{l,j}[i]$ are all constants depending on the input genotypes, and $p_j[i]$, $h_{l,j}[i]$ are the unknowns. Again, the equality constraints are defined over $F(2)$ whereas the (only) inequality constraint is defined over all integers. Note that, the number of p-variables is exactly mn and the number of h-variables is at most $2mn$. This ILP is different from the ILP for MRHC used in [17,18] which is not based on the system of linear equations. Observe that for any member j, if j or any of its parents are homozygous at locus i, then $p_j[i]$ is fixed based on Equation 3. Such p-variables are called *pre-determined*.

3 Some Graph Structures and a Compact ILP in h-Variables

As in [30,31], the above ILP can be transformed to one concerning only the h-variables. This is not surprising because the h-variables completely describes the inheritance relationship in the pedigree, including the locations of recombinants. In this section, we review some useful graph structures and the generation of a sufficient set of equality constraints on h-variables introduced in [30,31]. Again, all the definitions are the same as in [30,31] except for the definition of the h-variables.

3.1 The Pedigree Graph and Locus Graphs

In [30,31], the input pedigree is transformed into a *pedigree graph* by connecting each parent directly to his children. Although the edges in the pedigree graph representing the inheritance relationship between a parent and a child are directed, we consider them as undirected when dealing with linear constraints. Thus, these edges will sometimes be thought of as directed but other times as undirected according to the context. Clearly, such a pedigree graph $G = (V, E)$ may be cyclic due to mating loops or multiple children shared by a pair of parents. Let $\mathcal{T}(G)$ be any spanning tree on G. $\mathcal{T}(G)$ partitions the edge set E into two subsets: the *tree edges* and the *non-tree edges* (or *cross edges*). Let $E^{\mathbb{X}}$ denote the set of cross edges. Since $|E| \leq 2n$ and the number of edges in $\mathcal{T}(G)$ is $n-1$, we have $|E^{\mathbb{X}}| \leq n+1$.

For any fixed locus i, the value $w_l[i]$ can be viewed as the weight of each edge $(l, j) \in E$, where l is a parent of j. We construct the *i-th locus graph* G_i as the subgraph of G induced by the edges with weight 1. Formally, $G_i = (V, E_i)$, where $E_i = \{(l, j)| \, l \text{ is a parent of } j, w_l[i] = 1\}$. The i-th locus graph G_i induces a subgraph of the spanning tree $\mathcal{T}(G)$. Since the subgraph is a forest, it will be referred to as the *i-th locus forest* and denoted by $\mathcal{T}(G_i)$. The locus graphs can be used to identify some implicit constraints on the h-variables as follows. First, for any edge $(l, j) \in E$, define $h_{l,j}[i] = h_{j,l}[i]$ and $\boldsymbol{d}_{l,j} = \boldsymbol{d}_{j,l}$.

Lemma 1. *[30,31] For any path $P = j_0, \ldots, j_t$ in locus graph G_i connecting vertices j_0 and j_t, we have*

$$p_{j_0}[i] + p_{j_t}[i] + \sum_{r=0}^{t-1} h_{j_r,j_{r+1}}[i] + d_{j_r,j_{r+1}}[i] = 0 \tag{4}$$

Corollary 1. *[30,31] For any cycle $C = j_0, \ldots, j_t, j_0$ in G_i, there exists a binary constant b defined as $b = \sum_{r=0}^{t} d_{j_r,j_{r+1 \bmod t+1}}[i]$ such that $\sum_{r=0}^{t} h_{j_r,j_{r+1 \bmod t+1}}[i] = b$.*

Corollary 2. *[30,31] Suppose that $P = j_0, \ldots, j_t$ is a path in G_i connecting vertices j_0 and j_t, and the variables $p_{j_0}[i]$ and $p_{j_t}[i]$ are pre-determined. There exists a binary constant b defined as $b = p_{j_0}[i] + p_{j_t}[i] + \sum_{r=0}^{t-1} d_{j_r,j_{r+1}}[i]$ such that $\sum_{r=0}^{t-1} h_{j_r,j_{r+1}}[i] = b$.*

3.2 Linear Equality Constraints on the h-Variables

As in [30,31], we generate a sufficient set of linear equality constraints on the h-variables by considering each edge in a locus graph. Such a set of constraints will guarantee a feasible solution to the ILP in Equation 3. Note that since the edges broken in a locus graph involve pre-determined p-variables, we do not have to introduce constraints to cover them. The constraints can be classified into two categories with respect to the spanning tree $\mathcal{T}(G)$: constraints for cross edges and constraints for tree edges.

Cross Edge Constraints. Adding a cross edge e to the spanning tree $\mathcal{T}(G)$ yields a cycle C in the pedigree graph G. Suppose the edge e exists in the i-th locus graph G_i, and consider two cases of the cycle C with respect to G_i.

Case 1: The cycle exists in G_i. We introduce a constraint along the cycle as in Corollary 1. This constraint is called a *cycle constraint*. The set of such cycle constraints for edge e in all locus graphs is denoted by $C^{\mathbb{C}}(e)$, *i.e.*,

$$C^{\mathbb{C}}(e) = \{(b, e) \mid b \text{ is associated with the cycle in } \mathcal{T}(G_i) \cup \{e\},\ 1 \leq i \leq m\}.$$

The set of cycle constraints for all cross edges is denoted by $C^{\mathbb{C}} = \biguplus_{e \in E^{\mathbb{X}}} C^{\mathbb{C}}(e)$.

Case 2: Some of the edges of the cycle do not exist in G_i. This means that the cycle C is broken into several disjoint paths in G_i by the pre-determined vertices. Since e exists in G_i, exactly one of these paths, denoted as P, contains e. Observe that both endpoints of P are pre-determined and thus Corollary 2 could give us a constraint concerning the h-variables along the path. Such a constraint will be called a *path constraint*. The set of such path constraints for e in all locus graphs G_i is denoted by $C^{\mathbb{P}}(e)$, *i.e.*,

$$C^{\mathbb{P}}(e) = \left\{ (l, j, b, e) \,\middle|\, \begin{array}{l} \text{in } \mathcal{T}(G_i) \cup \{e\}, b \text{ is associated with the path containing } e \\ \text{connecting two pre-determined vertices } l \text{ and } j,\ 1 \leq i \leq m \end{array} \right\}.$$

The set of path constraints for all cross edges is denoted by $C^{\mathbb{P}} = \biguplus_{e \in E^{\mathbb{X}}} C^{\mathbb{P}}(e)$.

Tree Edge Constraints. By Corollary 2, there is an implicit constraint concerning the h-variables along each path between two pre-determined vertices in the same connected component of $\mathcal{T}(G_i)$. Therefore, for each connected component $\mathcal{T}$ of $\mathcal{T}(G_i)$, we arbitrarily pick a pre-determined vertex in the component as the *seed* vertex, and generate a constraint for the unique path in $\mathcal{T}(G_i)$ between the seed and each of the other pre-determined vertices in the component, as in Corollary 2. Such a constraint will be called a *tree constraint*.

To conform with the notation of path constraints and for the convenience of presentation, we arbitrarily pick a tree edge denoted as e_0, and write the set of tree constraints at all loci as

$$C^{\mathbb{T}} = \left\{ (l, j, b, e_0) \,\middle|\, \begin{array}{l} \text{in a connected component of } \mathcal{T}(G_i) \text{ with seed } l, b \text{ is associated with} \\ \text{the path connecting vertices } l \text{ and a predetermined vertex } j,\ 1 \leq i \leq m \end{array} \right\}.$$

Define $C = C^{\mathbb{C}} \cup C^{\mathbb{P}} \cup C^{\mathbb{T}}$. The subset of all the constraints in C generated in locus graph G_i will be denoted as C_i. The next two lemmas are easy to prove.

Lemma 2. *[30,31]* $|C| = |C^{\mathbb{C}}| + |C^{\mathbb{P}}| + |C^{\mathbb{T}}| = O(mn)$.

Lemma 3. *None of the constraints in $C^{\mathbb{P}} \cup C^{\mathbb{T}}$ are defined on a path that begins or ends at a founder.*

As in [30,31], we can prove that C forms a sufficient set of constraints, *i.e.* any solution in terms of the h-variables satisfying all these constraints would imply a feasible solution in terms of both the h- and p-variables satisfying Equation 3. The proof is very similar to the corresponding proof in [30,31] and therefore omitted. The following lemma hence follows.

Lemma 4. *The k-RHC problem can be expressed as the following ILP:*

$$\sum_{\textit{for all edges } (j_r, j)} \sum_{i=1}^{m-1} \left|h_{j_r,j}[i] - h_{j_r,j}[i+1]\right| \le k \quad (5)$$

plus

all the linear equality constraints in C

4 An $O(mn \log^{k+1} n)$ Time Algorithm for k-RHC on Tree Pedigrees

As mentioned before, the basic idea of our algorithm is to locate all the k recombinants first. Once we know the locations of all the recombinants, we can define the relationship between h-variables at consecutive loci corresponding to the same edge in the pedigree graph. For example, if there is a recombinant between locus i and locus $i+1$ on edge (u, v), $h_{u,v}[i] = h_{u,v}[i+1] + 1$. If such a recombinant does not exist, $h_{u,v}[i] = h_{u,v}[i+1]$. In this way, all the h-variables at different loci corresponding to the same edge can be represented by a single h-variable in the ILP of Equation 5, and the k-RHC ILP is effectively reduced to a ZRHC instance which can be solved by the linear-time algorithm in [22]. Hence, the challenge here is how to locate the recombinants without exhaustively enumerating all the possibilities in the entire pedigree. The key idea is that we compare the constraints of C (as well as some additional constraints involving one or two h-variables to be constructed in the next two subsections) at different loci to see if they imply the necessity of a recombinants. For example, suppose that we have a constraint along path $P = j_0, \ldots, j_t$ at locus i and another constraint along the same path P at locus l $(l > i)$. By Corollary 2, we have $\sum_{r=0}^{t-1} h_{j_r,j_{r+1}}[i] = b_i$ and $\sum_{r=0}^{t-1} h_{j_r,j_{r+1}}[l] = b_l$. If $b_i \neq b_l$, there is at least one pair of h-variables, say $h_{j_r,j_{r+1}}[i]$ and $h_{j_r,j_{r+1}}[l]$, that do not have the same value. This would suggest a recombinant on the edge (j_r, j_{r+1}) between the loci i and l. Consider the collection of the markers between of loci i and l of each member on the path P as the *region* where this recombinant could occur. The size of the region is $(t+1)(l-i+1)$. If the region is not very large, it contains at most one recombinant with a high probability (since k is a constant). Thus, we could enumerate all the possible locations of this recombinant in the region to locate it exactly.

Before we give the algorithm, we need some notations to describe a random instance of k-RHC. For each founder j, we use the random variable $q_{j,1}[i]$ to represent j's maternal allele at locus i and $q_{j,2}[i]$ to represent j's paternal allele at locus i. These q-variables are independent and they collectively represent the founder haplotypes. Random h-variables are used to represent the random inheritance. Although h-variables concerning different edges in the pedigree are independent, the h-variable concerning the same edge are not. The latter variables are related by the random recombinants. For convenience, we call the edges in the pedigree graph adjacent to the founders the *founder edges*. The other edges are called the *non-founder edges*. In the following subsections, we will show that we can determine many h-variable values (or summations of their values) on these two kinds of edges separately. These determined h-variables and summations will be used as additional constraints besides C to aid the search for the locations of recombinants.

4.1 Determining h-Variables on Non-founder Edges

For each founder j and locus i, the phase $p_j[i]$ is only determined by the random founder allele variables $q_{j,1}[i]$ and $q_{j,2}[i]$. The p-variables of non-founders are determined by both the q-variables and h-variables. When the h-variables are fixed, each phase $p_j[i]$ of a non-founder is only determined by two random q-variables $q_{f,s}[i]$ and $q_{g,t}[i]$. In other words, the paternal allele of member j at locus i is inherited from $q_{f,s}[i]$ and its maternal allele is from $q_{g,t}[i]$. If $(f, s) = (g, t)$, the two alleles of j at locus i are inherited from the same allele of some founder. In this case, the locus i of member j is homozygous no matter what $q_{f,s}[i], q_{g,t}[i]$ are. We say that member j is *pre-homozygous* at locus i. If $(f, s) \neq (g, t)$, the two alleles of member j at locus i are inherited from different alleles of the founders. Then the locus i of member j can be homozygous or heterozygous with equal probability. We say that member j is *pre-heterozygous* at locus i.

Clearly, for a tree pedigree, all its members are pre-heterozygous at every locus. Using this property, the next lemma shows that the phases of many loci are pre-determined around non-founder edges in a random k-RHC instance and thus we can determine the h-variable values on many non-founder edges.

Lemma 5. *Consider a random instance of k-RHC on a tree pedigree. If (u, v) is a non-founder edge in the pedigree graph with u being the parent, then the probability for u to be heterozygous at locus i and both u and v to be pre-determined at locus i (and thus $h_{u,v}[i]$ to be determined) is at least* $1/8$.

4.2 Determining h-Variables on Founder Edges

Without loss of generality, we assume that each founder has at least two children (otherwise recombinants on the edge between the founder and its only child cannot be detected and in fact are unnecessary). For a founder x, if it is homozygous at locus i, all the h-variables concerning locus i and founder edges incident on x will not appear in any constraints. If it is heterozygous at locus i, its phase will not be pre-determined for it has no parents. So, we cannot determine the h variables on founder edges directly like in Lemma 5. However, we can determine the summation of any pair of h-variables concerning the same founder.

Lemma 6. *Consider a random instance of k-RHC on a tree pedigree. If x is a founder with children u and v, then the probability for a locus i to be heterozygous at x but pre-determined at u and v (and thus the summation $h_{x,u}[i] + h_{x,v}[i]$ to be determined) is at least* $1/8$.

Now we are ready to describe how to locate the recombinants. We divide the loci of each member (which could be a haplotype block) into $\frac{m}{a \log n}$ disjoint segments of size $a \log n$ each, where a is a constant to be decided later on, and treat the interior and boundary segments differently. (The boundary segments are the two segments at the end.) It turns out that the boundary segments are much tougher to deal with.

4.3 Locating Recombinants in the Interior Locus Segments

Since we can determine each h-variable with probability $1/8$ for every non-founder edge, we can determine at least one h-variable in each segment with high probability for each non-founder edge. If there is at most one recombinant in any two consecutive segments associated with the same non-founder edge, we can locate such a recombinant in a small region of size $O(\log n)$ by comparing the determined h-variables of both segments. If the values of two neighboring determined h-variables are equal, there is no recombinant between the loci of the h-variables. Otherwise, there exists at least one. Similarly, we can locate recombinants associated with the founder edges. Suppose that u is a founder and $v_1, \ldots, v_l$ are its children. Because we can determine $h_{u,v_s}[i]+h_{u,v_t}[i]$ for each pair of children v_s and v_t at locus i with probability at least $1/8$, we can determine at least one summation $h_{u,v_s} + h_{u,v_t}$ in each segment with high probability. If the values of two neighboring determined summations are equal, there is no recombinant between the associated loci. Otherwise, there exists at least one. The detailed location algorithm, called LOCATE-INTERIOR-RECOMBINANTS, will be given in the full paper.

Lemma 7. *The procedure* LOCATE-INTERIOR-RECOMBINANTS *can locate each recombinant from an interior locus segment in a small region of size at most* $4a \log n$ *correctly with probability at least* $1 - k^2 \frac{6a \log n}{(m-1)n} - \frac{2nm}{a \log n} \left(\frac{7}{8}\right)^{a \log n}$.

4.4 Locating Recombinants in the Boundary Locus Segments

For a non-founder (or founder) edge (u, v), suppose that i_s is the smallest locus such that $h_{u,v}[i_s]$ (or a summation containing $h_{u,v}[i_s]$) can be determined and i_t is the largest such locus. By Lemma 7, each recombinant between loci i_t and i_b on edge (u, v) is located in a small region of size at most $4a \log n$. But the lemma does not show how to decide if there exists a recombinant between loci 1 and i_s or one between loci i_t and m. We call these two regions, which are typically contained in the boundary segments, the *boundary regions* of edge (u, v). In this subsection, we will show how to locate recombinants from the boundary regions in small regions of size $O(\log n)$.

For convenience, define the *length* of a constraint as the number of h-variables in it. First, we give an upper bound on the maximum length of any constraint in the set $C = C^{\mathbb{C}} \cup C^{\mathbb{P}} \cup C^{\mathbb{T}}$.

Lemma 8. *For any constant b, the length of every constraint in the set C is less than* $b \log n$ *with probability* $1 - 2mn^2 \left(\frac{1}{2}\right)^{\frac{1}{4} b \log n}$.

Now we give the basic idea of locating recombinants in the boundary regions. Let us consider two adjacent loci $i-1$ and i. Suppose that all the h-variables at locus i have already been determined. In other words, for the h-variables concerning non-founder edges, their values are known, and for the h-variables corresponding to founder edges, we know the summation of any pair of h-variables concerning edges incident on the same founder vertex. Note that for a founder u with children $v_1, \ldots, v_l$, the summation $h_{u,v_s}[i] + h_{u,v_t}[i]$ for any pair of children v_s, v_t ($s < t$) can be calculated using $\sum_{j=s}^{t-1} h_{u,v_j}[i] + h_{u,v_{j+1}}[i]$. If there is no recombinant between loci $i-1$ and i, all the h-variables at locus $i-1$ will be the same as those at locus i. So, we can set $h_{u,v}[i-1] = h_{u,v}[i]$ for each non-founder edge (u, v) and $h_{u,v_j}[i-1] + h_{x,v_{j+1}}[i-1] = h_{u,v_j}[i] + h_{u,v_{j+1}}[i]$ for each founder u with children $v_1, \ldots, v_l$, and then check if all the constraints in the set C_{i-1} hold. Note that by Lemma 3, each constraint in C_{i-1} contains an even number of founder edges incident on the same founder. So, the validity of each constraint in C_{i-1} can be determined. If any constraint is unsatisfied, there is at least one recombinant on this constraint (path) between loci $i-1$ and i. Since each constraint contains fewer than $b \log n$ h-variables by Lemma 8, it can be regarded as a small region. Thus, each constraint contains no more than one recombinant with high probability. If all the constraints hold, there are no recombinants between these two loci. Otherwise, we can locate each recombinant in a region of size $b \log n$ (*i.e.* some unsatisfied constraint in C_{i-1}). By iterating this towards locus 1 and locus m separately, we can locate all boundary recombinants.

The detailed algorithm for locating boundary recombinants, called LOCATE-BOUNDARY-RECOMBINANTS, will be given in the full paper. It assumes that the procedure LOCATE-INTERIOR-RECOMBINANTS has been run to locate all recombinants in the interior regions. Once all the recombinants have been located, LOCATE-BOUNDARY-RECOMBINANTS in fact returns a feasible (final) solution in terms of the p-variables.

Our main algorithm, TREE k-RHC, first calls a simple preprocessing procedure to set up the constraints and then the procedures LOCATE-BOUNDARY-RECOMBINANTS and LOCATE-INTERIOR-RECOMBINANTS to locate the recombinants and construct a feasible solution. Before we analyze the performance of algorithm TREE k-RHC, we prove two lemmas. An h-variable is called *active* if it appears in some constraints in C. Otherwise, it is *inactive*. Clearly, the values of inactive h-variables have no impact on the constraints.

Lemma 9. *For any edge* (u,v) *and set of* $2a \log n$ *consecutive loci* $i+1, i+2, \ldots, i+2a\log n$, *at least one of* $h_{u,v}[i+1], \ldots, h_{u,v}[i+2a\log n]$ *is active with probability at least* $1 - \frac{2nm}{a \log n}\left(\frac{7}{8}\right)^{a \log n}$.

The next lemma shows that we can focus on active h-variables when trying to locate the recombinants.

Lemma 10. *For each edge* (u,v), *if* $h_{u,v}[i_1] \neq h_{u,v}[i_2]$ *and all the h-variables* $h_{u,v}[i]$ $(i_1 < i < i_2)$ *are inactive, then there is a recombinant between loci* i_1 *and* i_2 *on edge* (u,v). *Moreover, any two consecutive loci in this interval would be a feasible location for this recombinant.*

To prove that the algorithm TREE k-RHC finds a feasible solution in $O(mn \log^{k+1} n)$ time with high probability, we need only show that all the recombinants can be located in the correct regions with high probability.

Theorem 1. *For any* $a > 0, b > 0$ *and* $m > 2a \log n$, *the algorithm* TREE k-RHC *solves the probabilistic k-RHC problem on tree pedigrees in time* $O\left(mn \log n \left(\max\{4a, b\} \log n\right)^k\right)$ *with probability at least* $1 - k^2 \frac{6a \log n}{(m-1)n} - \frac{2nm}{a \log n}\left(\frac{7}{8}\right)^{a \log n} - 2mn^2\left(\frac{1}{2}\right)^{\frac{1}{4} b \log n} - k(k-1)\frac{2ab \log^2 n}{(m-1)n}$.

Corollary 3. *When* $90 \log n < m < n^3$, TREE k-RHC *solves the probabilistic k-RHC problem on tree pedigrees in time* $O(mn \log^{k+1} n)$ *with probability* $1 - O(k^2 \frac{\log^2 n}{mn} + \frac{1}{n^2})$.

Open Problem and Acknowledgement. It remains open to design an efficient algorithm for k-RHC on large general pedigrees with possibly mating loops. The research was supported in part by NIH grant LM008991, NSF grant IIS-0711129, NSFC grant 60553001, and National Key Project for Basic Research (973) grants 2002CB512801 and 2007CB807901.

References

1. Abecasis, G.R., et al.: Nat Genet, 30(1), 97–101 (2002)
2. Albers, C.A., et al.: Genetics 177, 1101–1116 (2007)
3. Axenovich, T.I., et al.: Human Heredity 65(2), 57–65 (2008)
4. Baruch, E., et al.: Genetics 172, 1757–1765 (2006)
5. Chan, M.Y., et al.: SIAM Journal on Computing 38(6), 2179–2197 (2009)
6. Chin, F., et al.: Proc. 5th ICCS, Atlanta, GA, pp. 985–993 (2005)
7. Doi, K., et al.: Minimum recombinant haplotype configuration on tree pedigrees. In: Benson, G., Page, R.D.M., et al. (eds.) WABI 2003. LNCS (LNBI), vol. 2812, pp. 339–353. Springer, Heidelberg (2003)
8. Downey, R., Fellows, M.: Parameterized Complexity. Springer, Heidelberg (1999)
9. Excoffier, L., Slatkin, M.: Mol. Biol. Evol. 12, 921–927 (1995)
10. Gabriel, S.B., et al.: Science 296(5576), 2225–2229
11. Griffiths, A., et al.: Modern Genetic Analysis: Integrating Genes and Genomes. W.H. Freeman and Company, New York (2002)
12. Gudbjartsson, D.F., et al.: Nat. Genet. 25(1), 12–13 (2000)
13. Haplotype Conference (May 2008), `http://www.soph.uab.edu/ssg/nhgri/haplotype2008`
14. Kruglyak, L., et al.: Am. J. Hum. Genet. 58, 1347–1363 (1996)
15. Lander, E.S., Green, P.: Proc. Natl. Acad. Sci. USA. 84, 2363–2367 (1987)
16. Li, J., Jiang, T.: Proc. 7th RECOMB, pp. 197–206 (2003)
17. Li, J., Jiang, T.: Proc. 8th RECOMB, pp. 20–29 (2004)
18. Li, J., Jiang, T.: J. Comput. Biol. 12(6), 719–739 (2005)
19. Li, J., Jiang, T.: J. Bioinformatics and Computational Biology 6(1), 241–259 (2008)
20. Li, X., Li, J.: Proc. 7th CSB, pp. 297–308 (2008)
21. Liu, L., et al.: Complexity and approximation of the minimum recombination haplotype configuration problem. In: Deng, X., Du, D.-Z. (eds.) ISAAC 2005. LNCS, vol. 3827, pp. 370–379. Springer, Heidelberg (2005)
22. Liu, L., Jiang, T.: Proc. 18th GIW, pp. 95–106, Singapore (December 2007)
23. O'Connell, J.R.: Genet. Epidemiol. 19(suppl.1), S64–S70 (2000)
24. Piccolboni, A., Gusfield, D.: Journal of Computational Biololgy 10(5), 763–773 (2003)
25. Qian, D., Beckmann, L.: Am J Hum Genet, 70(6), 1434–1445 (2002)
26. Sobel, E., et al.: In: Speed, T., Waterman, M. (eds.) Genetic Mapping and DNA Sequencing, IMA Vol in Math. and its App., vol. 81, pp. 89–110 (1996)
27. The International HapMap Consortium. Nature 426, 789–796 (2003)
28. Wang, C., et al.: Journal of Chinese Science Bulletin 52(4), 471–476 (2007)
29. Wilson, I.J., Dawson, K.J.: Theor. Popul. Biol. 72(3), 436–458 (2007)
30. Xiao, J., et al.: Proc. 18th SODA, pp. 655–664 (2007)
31. Xiao, J., et al.: SIAM Journal on Computing 38(6), 2198–2219 (2009)

Complete Parsimony Haplotype Inference Problem and Algorithms

Gerold Jäger[1], Sharlee Climer[2], and Weixiong Zhang[3]

[1] Computer Science Institute, University of Halle-Wittenberg
D-06120 Halle (Saale), Germany
jaegerg@informatik.uni-halle.de
[2] School of Medicine, Washington University
St. Louis, Missouri 63110-1093, United States
sharlee@climer.us
[3] Department of Computer Science/Department of Genetics
Washington University
St. Louis, Missouri 63130-4899, United States
weixiong.zhang@wustl.edu

Abstract. Haplotype inference by pure parsimony (HIPP) is a well-known paradigm for haplotype inference. In order to assess the biological significance of this paradigm, we generalize the problem of HIPP to the problem of finding all optimal solutions, which we call complete HIPP. We study intrinsic haplotype features, such as backbone haplotypes and fat genotypes as well as equal columns and decomposability. We explicitly exploit these features in three computational approaches which are based on integer linear programming, depth-first branch-and-bound, and a hybrid algorithm that draws on the diverse strengths of the first two approaches. Our experimental analysis shows that our optimized algorithms are significantly superior to the baseline algorithms, often with orders of magnitude faster running time. Finally, our experiments provide some useful insights to the intrinsic features of this interesting problem.

1 Introduction

In this age of rapid advances in biological and medical fields, a number of compelling computational challenges have arisen and propelled the state-of-the-art. Haplotype inference is one such challenge. A *haplotype* is a set of nucleotides that are in physical proximity on a chromosome strand. Haplotypes do not contain nucleotides that have a common state for all individuals within the given population – only those nucleotides that exhibit variation. Diploid species have pairs of chromosomes and, consequently, pairs of corresponding haplotypes. Current sequencers are capable of producing *genotypes*, which are conflations of haplotype pairs. Thanks to recent advances, genotype data are highly accurate and complete. For example, the International HapMap project has produced genotypes that are 99.7% accurate and 99.3% complete [26]. However, identifying individual haplotypes directly in a laboratory setting is currently infeasible for all but

A. Fiat and P. Sanders (Eds.): ESA 2009, LNCS 5757, pp. 337–348, 2009.

small studies. For this reason, researchers commonly rely on mathematical models and computational algorithms for inferring haplotypes from genotypes. The first widely used algorithm for inferring haplotypes was introduced by Clark in 1990 [4]. This algorithm is based on the assumption that the number of unique haplotypes in a given population is relatively small. Gusfield and Hubbell independently proposed a haplotype inference model using *Pure Parsimony* (HIPP), in which the number of unique haplotypes is absolutely minimized [11]. Given a set of genotypes, HIPP is to find a set of distinct haplotypes with minimum cardinality such that each genotype can be resolved by two haplotypes. HIPP was computationally studied using integer linear programming (IP) by Gusfield [9,10]. Due to its simplicity in formulation and complexity in computation, it has attracted much attention. Many approaches to the problem have been developed, for example those based on IP [2,3,12], depth-first branch-and-bound (DFBnB or BnB for short) [20], and Boolean satisfiability [14,15,16]. It is important to mention that there may exist multiple optimal solutions to a HIPP problem and an algorithm simply returns an arbitrary optimal solution. It is also critical to emphasize that not all optimal solutions in the formulation are *biologically* equal. Furthermore, the true solution, which is the ground truth of the haplotype structure of a given population, may not be a computationally optimal solution, as shown in our previous study [5]. In other words, the mathematical formulation of HIPP captures only some of the biological aspects of haplotype inference (HI). Similar to other computational biology problems, such as RNA folding and multiple sequence alignment, the true solutions to HI are often found within in the set of optimal or even near optimal solutions to the HIPP formulation [5]. Therefore, it is desirable to find all optimal solutions [5], which is the subject of this paper. For convenience, we call the problem of finding all optimal solutions to a HIPP problem *complete HIPP*, or CHIPP. The CHIPP formulation and our new algorithm for CHIPP have helped gaining some initial biological insight into haplotype structures in human population [5]. Note that since it requires to find all optimal solutions, CHIPP is computationally more difficult than HIPP, which is $\mathcal{NP}$-complete [13].

In order to develop efficient algorithms for CHIPP, we consider several intrinsic haplotype features in this paper. These include 1) *backbone haplotypes*, which are haplotypes common to all optimal solutions, 2) *decomposability* of a problem into sub-problems, 3) *fat genotypes*, which are genotypes that reduce the solution size by 2, if they are omitted, and 4) *equal columns* [20], which are sites that appear exactly the same for all haplotypes. We then propose and study three classes of algorithms for CHIPP by exploiting these intrinsic haplotype features. The first two classes of algorithms are based on and extend two existing algorithms for HIPP: Gusfield's IP [10] and Wang and Xu's BnB [20]. We extend beyond these existing algorithms by introducing effective optimization techniques that take advantage of the intrinsic haplotype features. The third class of algorithms strategically hybridizes the greatest strengths of the first two.

Note that exploiting these intrinsic haplotype features can also be viewed in the concept of *fixed parameter tractability* (*FPT*). For an overview of FPT see [7,17]). The motivation of FPT is that even large instances of $\mathcal{NP}$-hard problems can be easy, because they might contain structure that can be exploited. One special idea is *FPT kernelization* which reduces a hard instance by preprocessing to a smaller, equivalent problem kernel [8]. As the intrinsic features *backbone haplotypes, decomposability, fat genotypes* and *equal columns* lead to a reduction technique, which in a preprocessing step reduces the given CHIPP instance to a smaller one, our approach is also a type of problem kernelization technique.

2 Complete Haplotype Inference by Pure Parsimony

Let G represent a set of genotypes, where $G = [g_1, \ldots, g_n]^T = (g_{ij})_{1 \le i \le n, 1 \le j \le m}$ with $g_{ij} \in \{0, 1, 2\}$, for n individuals with m single nucleotide polymorphism (SNP) sites. Haplotype inference is to find a set of haplotypes, $H = [h_1, \ldots, h_p]^T = (h_{ij})_{1 \le i \le p, 1 \le j \le m}$ with $h_{ij} \in \{0, 1\}$, such that the set of genotypes is *explained*, or *covered*, by the haplotypes. If a site g_{ij} has a value of 0 or 1, it can only be explained by two haplotype sites with values of 0 or 1, respectively. These genotypes are referred to as *homozygous*. A *heterozygous* genotype site has a value of 2, and can only be explained by a haplotype pair with values of 0 and 1. For example, $G = \{g_1 = (1, 2), g_2 = (0, 0), g_3 = (2, 1), g_4 = (2, 2)\}$, will be explained by the haplotype set $H = \{h_1 = (0, 0), h_2 = (1, 0), h_3 = (0, 1), h_4 = (1, 1)\}$. We say that haplotypes h_2 and h_4 *explain* or *cover* genotype g_1 and call h_2 and h_4 an *explaining haplotype pair* for g_1. We also say that both h_2 and h_4 *partly* explain genotype g_1. Note that genotype g_4 can be explained by h_1 and h_4 or by h_2 and h_3. Given a set of genotypes, HIPP is to find a minimum set of unique haplotypes that explains the genotypes. For simplicity and without loss of generality, we assume in this research that all individuals (genotypes) are unique. Notice that a HIPP problem may have multiple optimal solutions [5]. This is evident by a trivial example of one genotype $g_1 = (2, 2)$, which has two optimal solutions, $H_1 = \{(0, 0), (1, 1)\}$ and $H_2 = \{(0, 1), (1, 0)\}$. CHIPP is to find all optimal solutions, i.e. all sets of minimal unique haplotypes covering all genotypes.

In principle, any algorithm for HIPP can lend to an algorithm for CHIPP. We now consider two well-established computational paradigms for HIPP, IP and BnB. In this research, we use these as the baseline algorithms for CHIPP. The pseudo-codes of the algorithms are given as Supporting Information [27].

2.1 CHIPP Algorithm Based on Integer Linear Programming

In 2003, Gusfield [10] developed an exponential-size Integer Linear Program (IP) formulation of HIPP. We now briefly describe this model and discuss how to extend it for solving CHIPP.

Consider n genotypes $g_1, \ldots, g_n$, and haplotypes $h_1, \ldots, h_r$ that cover all n genotypes. Let k_i be the number of possible explaining haplotype pairs for genotype g_i for $i = 1, \ldots, n$. It is easy to see that $k_i = \max\{2^{l-1}, 1\}$ if g_i contains l 2's. Let $x = (x_1, \ldots, x_r)$ be a vector defined as

$$x_i = \begin{cases} 1 & \text{if haplotype } h_i \text{ appears in the HIPP solution,} \\ 0 & \text{else} \end{cases}$$

for $i = 1, \ldots, r$. Further, let $\omega = (\omega_{ij})_{1 \le i \le n, 1 \le j \le k_i}$ be a matrix representing haplotype pairs, defined as

$$\omega_{ij} = \begin{cases} 1 & \text{if the } j\text{-th explaining haplotype pair for} \\ & \quad \text{genotype } g_i \text{ is selected in the HIPP solution,} \\ 0 & \text{else} \end{cases}$$

for $i = 1, \ldots, n$ and $j = 1, \ldots, k_i$. In addition, let $f_1(i,j)$ and $f_2(i,j)$ be, respectively, the indices of the first and second haplotype of the j-th explaining haplotype pair for the i-th genotype for $i = 1, \ldots, n$ and $j = 1, \ldots, k_i$. Then HIPP can be represented by the following integer linear program (IP):

$$\min \sum_{s=1}^{r} x_i \text{ subject to} \tag{1}$$

$$\begin{cases} \sum_{j=1}^{k_i} \omega_{ij} = 1 & \text{for } i = 1, \ldots, n \\ x_{f_1(i,j)} \ge \omega_{i,j} & \text{for } i = 1, \ldots, n, \; j = 1, \ldots, k_i \\ x_{f_2(i,j)} \ge \omega_{i,j} & \text{for } i = 1, \ldots, n, \; j = 1, \ldots, k_i \\ x_s, \omega_{i,j} \in \{0,1\} & \text{for } s = 1, \ldots, r, \; i = 1, \ldots, n, \; j = 1, \ldots, k_i \end{cases}$$

In the worst case, this IP needs an exponential number of variables and constraints. In order to solve CHIPP using the IP representation in (1), we implicitly enumerate all optimal solutions to the IP. The key lies in the idea of avoiding the optimal solutions that have been found earlier in the process. In our method, we first solve HIPP using the IP in (1), obtaining one optimal solution. Let p represent the value of the objective function for this solution. In other words, p unique haplotypes are the minimum number of haplotypes that can resolve this set of genotypes. This means that for this solution, there are exactly p entries of the haplotype index vector x having value 1; let $i_1, \ldots, i_p$ be the indices of these haplotypes. In order to avoid the optimal solution that we just found, we introduce to the IP in (1) the following inequality

$$\sum_{s=1}^{p} x_{i_s} \le p - 1. \tag{2}$$

We then solve the newly expanded IP to find the next optimal solution, if any. Notice that (2) can lead to two possibilities. Either the new IP has an objective value larger than p, which means that no new optimal solution exists; or the new IP has an objective value equal to p, which gives rise to another optimal solution. We repeat this process of adding new inequalities in the form of (2) and solving the incrementally expanded and more constrained new IPs to find all optimal solutions.

2.2 CHIPP Algorithm Based on Branch-and-Bound

Wang and Xu [20] introduced an algorithm for HIPP based on BnB. The algorithm starts with a heuristic solution, where for each genotype the explaining haplotype pair is chosen, from which the corresponding haplotypes can partly explain the most genotypes. This provides the initial incumbent (or upper bound) for the BnB search. Now the search implicitly considers all possible explaining haplotype pairs for each genotype, and the best solution found is the optimal solution to be returned. If during the search the node cost is equal to or exceeds the incumbent, the current explaining haplotype pair is discarded and the algorithm continues to the next explaining haplotype pair, if it exists, thus moving on to the next branch of the current search node.

In our implementation of the above BnB algorithm, we first sorted the genotypes in an increasing order of the numbers of 2's that the genotypes have. In other words, we prefer genotypes with less heterozygous sites over ones with more heterozygous sites at nodes near the top of the search tree. For each genotype, we further order the explaining haplotype pairs, in a decreasing order of the number of genotypes that haplotypes can cover. As in Section 2.1, k_i is the number of possible haplotype pairs for genotype g_i for $i = 1, \ldots, n$.

We can directly extend the BnB algorithm for HIPP to an algorithm for CHIPP. In order to find all optimal solutions, pruning is applied only when the node cost strictly exceeds the incumbent. This allows the algorithm to explore a branch that may lead to another optimal solution. Further, if we have found a better solution, all previous "optimal solutions" are discarded.

3 Features of CHIPP and Optimization Techniques

We now consider some important features of CHIPP, and describe how they can be exploited to develop effective methods for solving this challenging problem.

3.1 Backbones

The *backbone* of a combinatorial optimization problem refers to the set of variables that have common values across all optimal solutions for the problem. Backbones are intrinsic features of combinatorial optimization problems, and have been used to characterize many difficult optimization problems [19,21,23], such as the traveling salesman problem and maximum satisfiability, and have been exploited in algorithms for solving these well-studied problems [6,22,24].

For our purpose, the backbone is a set of haplotypes that appear in every optimal solution of HIPP. We call these haplotypes *backbone haplotypes*. One important consequence of using backbone haplotypes is that those genotypes that can be explained by two backbone haplotypes can be omitted in a haplotype inference procedure, as these genotypes can be explained by any solution that the procedure will return. This holds for HIPP and for CHIPP. We call such genotypes *backbone genotypes*. Therefore, solving CHIPP can be accelerated, if we can identify all backbone haplotypes or backbone genotypes.

A special case of backbone haplotypes is when some genotypes have zero or one SNP site, which can be easily identified. A genotype with no 2's gives rise to one backbone haplotype, which is the same as the genotype. A genotype with only one 2 leads to two backbone haplotypes, which are equal to the genotype except for the site with the 2, where one backbone haplotype has entry 0 and the other has entry 1. We call such backbone haplotypes *trivial backbone haplotypes.* In their BnB program, Wang and Xu [20] implicitly considered trivial backbone haplotypes.

A more difficult problem is to identify all backbone haplotypes. At first sight, the problem seems to be as difficult as finding all optimal solutions, but it turns out that all backbone haplotypes can be identified without finding all optimal solutions. Our idea for the problem is based on the fact that when a backbone haplotype is omitted, no optimal solution to a HIPP or CHIPP problem can be found. Therefore, we first find an optimal solution, and then iteratively remove, one and one at a time, the haplotypes in the solution, and repeatedly solve each of the resulting problems to determine if a new optimal solution can be found. If this is the case, the considered haplotype is no backbone haplotype, otherwise it is a backbone haplotype. In the worst case, the complexity of our algorithm is p times the complexity of finding an optimal solution, where p is the cardinality of the set of an optimal solution haplotypes.

3.2 Equal Column Technique

One important technique used in Wang and Xu's BnB program for HIPP is the so called *equal column technique.* The idea of this technique is as follows. Assume we have found an optimal solution for the genotype matrix $G_1 = [g^1, \ldots, g^k]$, where for $l = 1, \ldots, k$, g^l is the l-th column of G_1, and we want to find an optimal solution for the genotype matrix $G_2 = [g^1, \ldots, g^k, g^{k+1}]$, where $g^{k+1} = g^l$ for some $l \in \{1, \ldots, k\}$. We can simply obtain an optimal solution to G_2 by copying the l-th column of all haplotypes in the optimal solution to the $(k+1)$-th column, because the $(k+1)$-th column of all genotypes can be explained in the same way as the l-th column of all genotypes. It is worthwhile to point out that this technique is also applicable in the same spirit to Gusfield's IP algorithm for HIPP.

However, the equal column technique cannot be directly used to find all optimal solutions, since some optimal solutions might be lost by this technique. This can be seen from the simple example of one genotype $g_1 = (2, 2)$. It has two optimal solutions, $H_1 = \{(0, 0), (1, 1)\}$ and $H_2 = \{(0, 1), (1, 0)\}$. However, the equal column technique will only result in the first optimal solution, because only in this case the first and the second column of the optimal solution are equal. This example also shows that the equal column technique cannot be used to find all backbone haplotypes, as $g_0 = (2)$ has two backbone haplotypes $(0), (1)$ and $g_1 = (2, 2)$ has no backbone haplotype. Therefore, special care must be taken when extending the equal column technique to finding all optimal solutions. Again, let $G_1 = [g^1, \ldots, g^k]$ be a genotype matrix with solution value p and $G_2 = [g^1, \ldots, g^k, g^{k+1}]$ another genotype matrix, where $g^{k+1} = g^l$ for some $l \in \{1, \ldots, k\}$. The solution value of G_2 is p as well, and each optimal solution for G_2 can be written as an optimal solution for G_1 plus an additional last column. Therefore, given an optimal solution H_1 to G_1, we have to find all optimal solutions H_2 to G_2 which are equal to H_1 in the first k columns. Using the equal column technique for HIPP, we can obtain one optimal solution of G_2, if we use the l-th column of H_1 as $(k+1)$-th column of H_2. However, there may exist additional optimal solutions of G_2. Each such optimal solution has p haplotypes and trivially, each entry of the last column can be 0 or 1. Thus there are a total of 2^p possible optimal solutions for G_2 from which at least one must be in fact optimal. Let H_0 be such a possible optimal solution. Then H_0 is an optimal solution, if and only if all genotypes can be explained by the haplotypes of H_0. So we can make use of this characteristic. Furthermore, we can reduce the number of all possible optimal solutions to be tested, which is 2^p, by the following idea. Each haplotype of an optimal solution must be used to partly explain at least one genotype. We test this characteristic for all $2p$ haplotypes which can possibly appear in an optimal solution. If this characteristic is not fulfilled for a particular haplotype, we do not have to consider this haplotype and all possible optimal solutions which contain this haplotype. The extended equal column technique for CHIPP may still require up to 2^p steps for each equal column in the worst case. Nevertheless, as we will observe in Section 4, it is rather efficient in practice.

Furthermore, columns of a genotype matrix which contain only 0's (1's) can also be omitted for solving HIPP and CHIPP because of the following reason. If a column g^l of the genotype matrix contains only 0 or 1, then the l-th column of the haplotype matrix of each optimal solution also contains only 0's or 1's, respectively.

3.3 Decomposability

Two genotypes g_1 and g_2 may not share any common explaining haplotypes, in this case we say that g_1 does not overlap with g_2. This happens when g_1 has an entry 0 while g_2 has an entry 1, or vice versa, at one site. The concept of non-overlapping of two genotypes can be generalized to two sets of genotypes $E = \{e_1, \ldots, e_s\}$ and $F = \{f_1, \ldots, f_t\}$. If each genotype $e \in E$ does not overlap with any genotype $f \in F$, E does not overlap with F as well.

A HIPP problem instance can be decomposed, if it contains non-overlapping sets of genotypes, where a sub-problem contains one of these sets. It is evident that it is sufficient to solve each of the sub-problems in order to solve the original problem; a solution of the original problem is a union of the solutions to the sub-problems.

This method can be simply extended to CHIPP. Here a problem instance is decomposable if it contains $l > 1$ non-overlapping sets of genotypes, each of which forms a sub-problem. Assume that each sub-problem has q_i solutions for $i = 1, \ldots, l$. The original problem will have $\prod_{i=1}^{l} q_i$ solutions, corresponding to all combinations of the solutions to the sub-problems. A decomposable HIPP or CHIPP problem can be solved with a significantly reduced complexity due to the potential exponential growth in computation time as a function of the size of the instance.

3.4 Omitting Explaining Haplotype Pairs

RTIP: In order to improve his IP approach, Gusfield [10] used a reduction formulation, called RTIP. RTIP removes from the problem all explaining haplotype pairs in which the two corresponding haplotypes partly explain only one genotype. The genotypes that can only be explained by such haplotype pairs can be explained in the HIPP solution in an arbitrary way. Wang and Xu [20] used a very similar idea in their BnB program. In order to extend this idea to CHIPP, we first introduce the notion of *fat genotypes*.

Fat genotypes: Let G be a set of input genotypes with solution value p, and g a genotype in G such that removing g from G results in a solution to the new instance with solution value $p-2$. We call such a genotype *fat genotype*.

A special case of a fat genotype is a genotype that does not overlap with any other genotype (see Section 3.3) and has at least one entry 2. For example, consider $g_1 = (0,2,0), g_2 = (2,1,0), g_3 = (1,2,1)$, where g_3 does not overlap with g_1 and g_2. Then for $G = \{(g_1, g_2)\}$ the (only) optimal solution is $\{(0,0,0),(0,1,0),(1,1,0)\}$ with solution value 3, whereas for $G' = (g_1, g_2, g_3)$ the (only) optimal solution is $\{(0,0,0), (0,1,0), (1,1,0), (1,0,1), (1,1,1)\}$ with solution value 5. Thus g_3 is a fat genotype. Note that there are cases of fat genotypes which do overlap with other genotypes. For example, let $g_1 = (1,2,2), g_2 = (2,1,1), g_3 = (2,0,1)$, where g_3 overlaps with g_1. Then for $G = (g_1, g_2)$ the (only) optimal solution is $\{(1,0,0),(0,1,1), (1,1,1)\}$ with solution value 3, whereas for $G' = (g_1, g_2, g_3)$ one optimal solution is $\{(1,0,0),(0,1,1), (1,1,1), (0,0,1), (1,0,1)\}$ with solution value 5. Thus g_3 is a fat genotype. It is easy to see that fat genotypes are the only genotypes for which omitting a corresponding explaining haplotype pair by RTIP could cause some optimal solutions to be lost. The fatness of all genotypes can be easily tested.

Our task is to find all optimal solutions that might be lost by RTIP. For this purpose, assume the explaining haplotype pair (h_1, h_2) is omitted by RTIP, but nevertheless is contained in at least one optimal solution. Furthermore, let H_0 be an optimal solution found by RTIP. By the definition of RTIP, h_1 and h_2 partly explain only one genotype g. The only possibility to recover a lost optimal solution which includes h_1 and h_2 is to replace an explaining haplotype pair for g, $(h_3, h_4) \in H_0$, by (h_1, h_2). Let H_0^{new} be H_0 after the replacement. Then H_0^{new} is a new optimal solution, if and only if all genotypes can be explained by haplotype pairs in H_0^{new}. This idea leads to an algorithm which finds all optimal solutions after using RTIP.

If we need to save the (possibly expensive) identification of all fat genotypes, we can try all possible replacements not only for the *fat* genotypes, but for all genotypes. As mentioned, for all non-fat genotypes no further optimal solutions will be found. On the other hand, the time saved by omitting the identification of all fat genotypes can be lost by many more tests for possible new optimal solutions.

Further case. Wang and Xu [20] considered a case of two genotypes g_1 and g_2, where g_1 only has explaining haplotype pairs (h_1, h_2) and (h_4, h_5), and g_2 only has explaining haplotype pairs (h_2, h_3) and (h_5, h_6). In the considered case, all haplotypes h_1, h_2, h_3, h_4, h_5, h_6 do not partly explain any other genotype. For this case when solving HIPP, we can use (h_1, h_2) as explaining haplotype for genotype g_1 and (h_2, h_3) as explaining haplotype for genotype g_2, and omit the pairs (h_4, h_5) and (h_5, h_6).

When extending this case to CHIPP, we only have to replace in each optimal solution the haplotypes h_1, h_2, h_3 by the haplotypes h_4, h_5, h_6 to find a new optimal solution.

4 Experimental Analysis

We have implemented in C++ all the different combinations of HIPP and CHIPP algorithms. All our experiments were carried out on a PC with an Athlon 1900MP dual CPU and 2GB shared memory, while our programs ran on a single processor of the machine. As IP solver we used CPLEX [25]. An individual experiment was terminated after 6 hours of CPU time. The test data come from two types of human biological data: *genotype* data from the International HapMap Project [26], and *known haplotype* pairs [1,18]. Overall, we use 73 test instances: 66 random instances and 7 known instances. Details about these data and the experimental results can be found in the Supporting Information [27] and in [5].

In Section 2 we introduced two baseline CHIPP algorithms, one based on IP (CHIPP-IP) and the other based on BnB (CHIPP-BnB). Furthermore we consider a hybrid algorithm, in which the heuristic for computing the initial upper bound for a BnB algorithm is replaced by a HIPP algorithm (CHIPP-HY). In other words, the initial upper bound for CHIPP-HY is the cost of the optimal solution, which can significantly reduce the search space to be explored and computation time needed. The optimization techniques that we discussed in Section 3 can be combined in different ways with these three algorithms. These possible combinations give rise to many different algorithms. These optimization techniques or algorithmic components include: equal column technique (**E**) or not; trivial backbones (**T**), all backbones (**A**) or no backbones; RTIP with computation of fat genotypes (**F**), RTIP with no computation of fat genotypes (**R**) or no RTIP. In summary, we have a total of $54 = 3 \times 2 \times 3 \times 3$ algorithms for CHIPP to analyze, where each algorithm is written in the following way: **W-XYZ**, where **W** = **IP** or **W** = **BnB** or **W** = **HY** (hybrid); **X** = **E** or **X** = $\emptyset$; **Y** = **T** or **Y** = **A** or **Y** = $\emptyset$; **Z** = **F** or **Z** = **R** or **Z** = $\emptyset$. For example, **IP-EAF** is used to indicate an algorithm based on IP using the equal column technique, all backbones and RTIP with fat genotypes. Note that for all algorithms with the option **A**, i.e. algorithms that exploit all backbones, a HIPP algorithm is needed. For these algorithms that are based on BnB, the heuristic for the initial upper bound can be replaced by a HIPP algorithm. In this regard, the BnB algorithm and the HY algorithm using option **A** are in fact the same. Finally note that with the purpose to simplify the experiments, we decided to use the technique of decomposability (Section 3.3) and the further case of Section 3.4 in all versions except the baseline versions. As the CHIPP-HY algorithm contains a HIPP algorithm and finding a single optimal solution is required by the identification of all backbone haplotypes and by the identification of fat genotypes, an efficient HIPP algorithm is an important component of a CHIPP algorithm. Experiments (not described here due to space limit) show that for both HIPP versions, all of the three features equal column technique, RTIP, and trivial backbones lead to a larger efficiency of the HIPP algorithm, where the most important feature is the equal column technique. This is not surprising, as each of these features can substantially reduce the problem sizes. The equal column technique can help remove some sites that carry redundant information, and the techniques of trivial backbones and RTIP can be used to omit some highly constrained explaining haplotype pairs from the core computation of haplotype inference. Furthermore the mentioned experiments show that the HIPP algorithm based on IP is more efficient than that based on BnB. As a consequence of this comparison, we used this HIPP-IP algorithm as a sub-routine in all CHIPP algorithms, where a HIPP algorithm is needed, except one special case. The exception is the algorithm for finding all backbone haplotypes, where we have to omit the equal column technique. As mentioned in Section 3.2, this technique cannot be used for finding all backbone haplotypes. To reduce the overall computation for testing all 54 algorithms and all 73 instances (the 66 random and the 7 known instances), we first tested 8 small- to medium-difficulty typical instances in the first-stage analysis. We found that except for a few cases, most top performers use the all backbone technique without RTIP, or with RTIP with the computation of fat genotypes, i.e. the algorithms with options **A**, **AF**, **EA**, **EAF**. As mentioned, these four algorithms are the same for BnB and HY. Therefore, we have identified eight top contenders for the champion for solving CHIPP. In order to find the overall champion, we further tested these top contenders on all 73 instances. The results clearly show that the optimized algorithms are superior to the baseline algorithms. On many difficult problem instances, the former run orders of magnitude faster than the latter. The results also show that for the baseline algorithms, IP outperforms BnB, which suggests that replacing the heuristic for computing the upper bound by a HIPP algorithm is important. However, the result comparing the optimized IP-based and HY-based algorithms is mixed. On some instances the optimized HY algorithms were able to solve CHIPP, while the IP algorithms failed within the given 6 hours of running time. On the other hand, there are other instances for which the IP algorithms were faster. One IP-based CHIPP algorithm, algorithm **EA**, which uses the techniques of *all backbones* and *equal columns*, is the champion for 19 of the 73 instances tested. This algorithm is also the overall champion. In addition, among the hybrid versions, the algorithm **EA** is also the best one.

5 Summary

In summary, we made three major contributions in this paper. First, we introduced the problem CHIPP to expand the capability of haplotype inference by pure parsimony for finding all optimal solutions. In [5] extensive experiments showed that CHIPP problem instances can have a large number of optimal solutions and the first optimal solution returned by an algorithm may not necessarily be the true solution. Our results on seven known problem instances also showed that the true solutions to four of these problems are indeed among the optimal solutions. All these results support to find all optimal solutions.

Second, we studied many intrinsic haplotype features, some of which were studied in earlier research on HIPP, particularly by Gusfield [10] and Wang and Xu [20]. However, strategies to exploit these features cannot be directly applied to CHIPP and we formulated methods to recapture solutions lost by these strategies. Furthermore, we introduced the concepts of backbone haplotypes, decomposability and fat genotypes, and formulated and discussed in more detail the concept of equal columns. All these concepts can be viewed as FPT kernelization techniques.

Third, we systematically studied three approaches to CHIPP, one based on integer linear programming [10], another based on depth-first branch-and-bound [20] and one integrating these two. In our algorithms, we explicitly exploited the intrinsic haplotype features that we studied. In our experiments, we analyzed the possible interactions of different problem features and optimization techniques. These studies revealed the best algorithms under these two general problem solving paradigms, as well as the best hybrid algorithms that combines the favorable features of integer linear programming and depth-first branch-and-bound.

Acknowledgement

The research was supported in part by an Olin Fellowship to S.C., an Alzheimers Association grant and two NSF grants (IIS-0535257 and DBI-0743797) to W.Z.

References

1. Andrés, A.M., Clark, A.G., Boerwinkle, E., Sing, C.F., Hixson, J.E.: Assessing the accuracy of statistical haplotype inference with sequence data of known phase. Genet. Epi. 31, 659–671 (2007)
2. Bertolazzi, P., Godi, A., Labbé, M., Tininini, L.: Solving haplotyping inference parsimony problem using a new basic polynomial formulation. Comput. Math. Appl. 55(5), 900–911 (2008)
3. Brown, D.G., Harrower, I.M.: Integer Programming Approaches to Haplotype Inference by Pure Parsimony. IEEE/ACM Transactions on Computational Biology and Bioinformatics 3(2), 141–154 (2006)
4. Clark, A.G.: Inference of Haplotypes from PCR-Amplified Samples of Diploid Populations. Molecular Biology and Evolution 7, 111–122 (1990)
5. Climer, S., Jäger, G., Templeton, A.R., Zhang, W.: How Frugal is Mother Nature with Haplotypes? Bioinformatics 25(1), 68–74 (2009)
6. Climer, S., Zhang, W.: Searching for Backbones and Fat: A Limit-Crossing Approach with Applications. In: Proc. 18th National Conference on Artificial Intelligence (AAAI), pp. 707–712 (2002)

7. Flum, J., Grohe, M.: Parameterized Complexity Theory. Springer, Berlin (2006)
8. Guo, J., Niedermeier, R.: Invitation to data reduction and problem kernelization. SIGACT News 38(1), 31–45 (2007)
9. Gusfield, D.: Inference of Haplotypes from Samples of Diploid Populations: Complexity and Algorithms. J. Computational Biology 8(3), 305–313 (2001)
10. Gusfield, D.: Haplotype Inference by Pure Parsimony. In: Baeza-Yates, R., Chávez, E., Crochemore, M. (eds.) CPM 2003. LNCS, vol. 2676, pp. 144–155. Springer, Heidelberg (2003)
11. Gusfield, D., Orzack, S.H.: Haplotype Inference. In: Handbook on Bioinformatics (2005)
12. Halldórsson, B.V., Bafna, V., Edwards, N., Lippert, R., Yooseph, S., Istrail, S.: A survey of computational methods for determining haplotypes. In: Istrail, S., Waterman, M.S., Clark, A. (eds.) DIMACS/RECOMB Satellite Workshop 2002. LNCS (LNBI), vol. 2983, pp. 26–47. Springer, Heidelberg (2004)
13. Lancia, G., Pinotti, C.M., Rizzi, R.: Haplotype Populations by Pure Parsimony: Complexity of Exact and Approximation Algorithms. INFORMS J. Computing 16(4), 348–359 (2004)
14. Lynce, I., Marques-Silva, J.: Efficient Haplotype Inference with Boolean Satisfiability. In: Proc. 21st National Conference on Artificial Intelligence (AAAI), pp. 104–109 (2006)
15. Lynce, I., Marques-Silva, J.: SAT in Bioinformatics: Making the Case with Haplotype Inference. In: Biere, A., Gomes, C.P. (eds.) SAT 2006. LNCS, vol. 4121, pp. 136–141. Springer, Heidelberg (2006)
16. Lynce, I., Marques-Silva, J., Prestwich, S.: Boosting Haplotype Inference with Local Search. Constraints 13(1-2), 155–179 (2008)
17. Niedermeier, R.: Invitation to Fixed-Parameter Tractability. Oxford University Press, Oxford (2006)
18. Orzack, S.H., Gusfield, D., Olson, J., Nesbitt, S., Subrahmanyan, L., Stanton Jr., V.P.: Analysis and Exploration of the Use of Rule-Based Algorithms and Consensus Methods for the Inferral of Haplotypes. Genetics 165, 915–928 (2003)
19. Slaney, J., Walsh, T.: Backbones in Optimization and Approximation. In: Proc. 17th Intern. Joint Conf. on Artificial Intelligence (IJCAI 2001), pp. 254–259 (2001)
20. Wang, L., Xu, Y.: Haplotype Inference by Maximum Parsimony. Bioinformatics 19(14), 1773–1780 (2003)
21. Zhang, W.: Phase transitions and backbones of 3-SAT and maximum 3-SAT. In: Walsh, T. (ed.) CP 2001. LNCS, vol. 2239, pp. 153–167. Springer, Heidelberg (2001)
22. Zhang, W.: Configuration Landscape Analysis and Backbone Guided Local Search: Part I: Satisfiability and Maximum Satisfiability. Artificial Intelligence 158(1), 1–26 (2004)
23. Zhang, W.: Phase Transitions and Backbones of the Asymmetric Traveling Salesman Problem. J. Artificial Intelligence Research 20, 471–497 (2004)
24. Zhang, W., Looks, M.: A Novel Local Search Algorithm for the Traveling Salesman Problem that Exploits Backbones. In: Proc. 19th International Joint Conference on Artificial Intelligence (IJCAI), pp. 343–350 (2005)
25. Homepage of CPLEX, `http://www.ilog.com/products/optimization/archive.cfm`
26. The International HapMap Consortium: A Haplotype Map of the Human Genome. Nature 437, 1299–1320 (2005)
27. Supporting Information to this paper, `http://www.cse.wustl.edu/~zhang/publications/supplemental/ChippSup.pdf`

Linear-Time Recognition of Probe Interval Graphs

Ross M. McConnell[1] and Yahav Nussbaum[2]

[1] Computer Science Department, Colorado State University, Fort Collins, CO 80528, USA
rmm@cs.colostate.edu
[2] The Blavatnik School of Computer Science, Tel Aviv University, 69978 Tel Aviv, Israel
yahav.nussbaum@cs.tau.ac.il

Abstract. The interval graph for a set of intervals on a line consists of one vertex for each interval, and an edge for each intersecting pair of intervals. A probe interval graph is a variant that is motivated by an application to genomics, where the intervals are partitioned into two sets: probes and non-probes. The graph has an edge between two vertices if they intersect and at least one of them is a probe. We give a linear-time algorithm for determining whether a given graph and partition of vertices into probes and non-probes is a probe interval graph. If it is, we give a layout of intervals that proves that it is. In contrast to previous algorithms for the problem, our algorithm can determine whether the layout is uniquely constrained. As part of the algorithm we solve the consecutive-ones probe matrix problem.

1 Introduction

An *interval graph* is the intersection graph of a set of intervals on a line. The set of intervals constitutes an *interval model* of the graph. Interval graphs play an important role in many problems, see [5,7,9]. The problem of recognizing whether a graph is an interval graph played a key role in the 1950's in proving the linear topology of DNA [1]; the intervals were fragments of genetic material, and it was shown empirically that their intersections form an interval graph.

This gave rise to interest in algorithms for determining whether a graph is an interval graph [6]. Booth and Lueker gave the first linear-time algorithm for recognizing interval graphs and constructing interval models for the graphs in the 1970's [2]. A *consecutive-ones ordering* of columns of a 0-1 matrix is one such that, for every row, the 1's in the row are consecutive. Booth and Lueker's approach was to reduce the problem to that of finding a consecutive-ones ordering of a 0-1 matrix, and to give a linear time bound for finding such an ordering.

A related application for interval graphs is *physical mapping*, which can be used for DNA sequencing. In this process, biologists create *clones*, which are copies of fragments of DNA. The problem is reconstruction of the arrangement of the clones in the genome. For some clones, called *probes*, the intersection data between them and other clones can be collected. If all clones are probes, then

A. Fiat and P. Sanders (Eds.): ESA 2009, LNCS 5757, pp. 349–360, 2009.

we can construct an interval graph from the clones, and an interval model for this graph gives the original sequence.

A *probe interval graphs* [14,17] (also called interval probe graph) is a graph in which the vertex set is partitioned into *probes* and *non-probes*. It is a generalization of intersection graph of an interval model, such that the graph has an edge between two vertices if their intervals intersect *and at least one of them is a probe*. Information about which pairs of non-probe intervals intersect is missing.

There has been recent work on topological and combinatorial properties of these graphs; see [9] for a survey. The problem of recognizing whether a graph is a probe interval graph, and finding a corresponding arrangement of intervals if it is, was first shown to be polynomial by Johnson and Spinrad [10], who gave an $O(V^2)$ algorithm. Using a different approach, McConnell and Spinrad gave an $O(V + E \log V)$ algorithm [13]. The latter algorithm was a critical step in the first linear-time algorithm for recognizing *circular-arc graphs* [12]. Motivated by the biological application, where the partition into probes and non-probes is known in advance, both algorithms get as an input a graph whose vertex set is partitioned into probes and non-probes. Chang et al. [4] consider the problem of recognizing this graph class when this partition is not given.

In this paper, we give the first linear-time algorithm for recognizing whether a graph is a probe interval graph when the partition into probes and non-probes is given, and for finding a corresponding set of intervals. In view of the complexity of the previous work, it is surprising that we are able to reduce the problem to that of finding consecutive-ones orderings of two easily-constructed consecutive-ones matrices, which can then be solved by Booth and Lueker's algorithm.

In the physical mapping problem, the arrangement of clones on the genome is certain to be reconstructed accurately only if there is a unique linear arrangement that is consistent with the probe interval graph. Previous algorithms for finding probe interval arrangements have the defect that they cannot determine whether the graph uniquely constrains the arrangement, and therefore cannot be said to solve the physical mapping problem. Uehara [16] has addressed the issue and gave a polynomial-time algorithm that determines whether a given probe interval graph has a unique model. Our algorithm solves this problem as a by-product of the recognition problem, and it is the first linear-time algorithm that solves it.

A 0-1 matrix is a *consecutive-ones matrix* if it has a consecutive-ones ordering. The consecutive-ones sandwich problem is an extension of this problem where the matrix has 0, 1 or $*$. A $*$ is a "don't care"; it can stand for either a 0 or a 1. This problem is NP-Complete [8]. If we require that the $*$'s form a submatrix then we get the consecutive-ones probe matrix problem (see also [3]). We solve this problem in linear time, for any 0, 1, $*$ probe matrix.

2 Preliminaries

Except for some additional definitions, we use standard terminology and notation from [5]. We will assume the standard adjacency-list representation of a graph. This imposes a *numbering* from 1 to n on the vertices.

A graph $G = (V, E)$ is a *probe graph* if the vertex set is partitioned into P, the set of probes, and N the set of non-probes. In this case, every edge of E is adjacent to at least one probe. We denote this by $G = (P, N, E)$. If X is a nonempty subset of V, let $G[X]$ denote the subgraph of G induced by X, together with the classification of members of X as probes or non-probes.

Let $N(v)$ denote the *open neighborhood* of v, that is, the set of neighbors of v in G, and let $N[v]$ denote its *closed neighborhood*, that is, $\{v\} \cup N(v)$.

An *interval model* of an *interval graph* is a set of intervals, one for each vertex, such that two vertices are adjacent if and only if their intervals intersect. Without loss of generality, we assume that no two endpoints of intervals coincide. Similarly, an *interval model* of a probe interval graph is a set of intervals, one for each vertex, such that two vertices are adjacent if and only if their intervals intersect *and at least one of the vertices is a probe.* If R is an interval model of a (probe) interval graph G and X is a nonempty subset of V, let $R[X]$ denote the set of intervals of members of X. Note that $R[X]$ is an interval model of $G[X]$.

We define the *cliques* of a graph to be its *maximal* complete subgraphs. In an interval model R of a graph G, each clique of G corresponds to the set of vertices whose intervals intersect a unique *clique segment* in R. A clique segment occurs where a right endpoint is immediately to the right of a left endpoint.

The *clique matrix* of a graph is a 0-1 matrix that has one column for each clique, one row for each vertex, and a 1 in row i, column j if and only if vertex i is a member of clique j. Interval graphs are exactly the set of graphs whose clique matrices have consecutive-ones orderings [6].

An interval model consists of alternating *blocks* of consecutive left endpoints and of consecutive right endpoints. The order of endpoints within a block does not change the realized graph. Therefore, we represent an interval model combinatorially by giving an ordered list of blocks, listing for each block the endpoints inside of it. In fact, a consecutive-ones ordering of the clique matrix of an interval graph is such a model, where the set of left endpoints in a column and the set of right endpoints in the column are each interpreted to be a block, where the block of left endpoints is implicitly to the left of the block of right endpoints.

A *chordal graph* is a is a graph with no induced cycle of size greater than three. Every interval graph is a chordal graph. A chordal graph has $O(V)$ cliques. It is possible to find a sparse representation of the clique matrix of a chordal graph in $O(V)$ time [15]. Booth and Lueker's algorithm [2] for recognizing interval graphs uses this to find the clique matrix or else determine that the graph is not chordal, hence not an interval graph. If it is chordal, it reduces the problem to that of finding a consecutive-ones ordering of this clique matrix.

The algorithm of [2] gives a compact representation of *all* consecutive-ones orderings of a matrix, called a *PQ-tree*. The leaves of the PQ-tree are the columns of the matrix. The PQ-tree gives all consecutive-ones orderings by constraining the orderings of children of internal nodes as follows. Some of the internal nodes are labeled *P nodes*. For such a node there is no constraint on the order of its children. Others are labeled *Q nodes*. For such a node an ordering of its children is given; the only permissible orderings of its children are the given ordering and its reverse. For a PQ-tree T, let $\Pi(T)$ denote the set of all possible orderings of its leaves, given these constraints. The algorithm of [2] either finds the PQ-tree for consecutive-ones orderings of columns of a matrix, or determines that the matrix is not a consecutive-ones matrix. Given a sparse representation of a 0-1 matrix, this takes $O(i + j + k)$ time, where i is the number of rows, j is the number of columns, and k is the number of 1's in the matrix.

Let $\Pi = \Pi(T)$. We may consider each $\pi \in \Pi$ to be a bijective function that maps elements of C, the set of columns, to elements of $\{1, 2, \ldots, |C|\}$, where for all $c \in C$, $\pi(c)$ tells the position of c in a consecutive-ones ordering represented by π. If X is a nonempty subset of C, let π_X be the bijective function that maps elements of X to $\{1, 2, \ldots, |X|\}$, giving the relative order of elements of X in π. Let $\Pi[X]$ denote $\{\pi_X | \pi \in \Pi\}$, namely, the relative orderings of elements of X given by orderings in Π. It is not hard to show that $\Pi[X]$ is the set of orderings of a PQ-tree with leaf set X; let us call this tree the *restriction* $T[X]$ *of* T *to* X. If T_1 and T_2 are two PQ-trees whose leaf sets are both C, it is not hard to show that $\Pi(T_1) \cap \Pi(T_2)$ is a set of permutations that can also be represented by a PQ-tree. Let us call this tree the *intersection* $T_1 \cap T_2$ *of* T_1 *and* T_2.

A *probe matrix* is a generalization of 0-1 matrix, which has the values $0, 1, *$, such that the $*$'s form a submatrix. The *consecutive-ones probe matrix* problem is a generalization of the consecutive-ones problem. In this problem we look for an ordering of the columns of the matrix such that there is an interpretation of the values of the $*$'s such that the 1's in every row are consecutive.

We represent a probe matrix in space proportional to the size of the matrix and the number of 1's in it, we do not represent the $*$'s explicitly. To do that, we split a probe matrix M into two submatrices. Let M_R be the submatrix of M whose rows are the rows that do not have $*$'s, and whose columns are all columns of M. Let M_C be the submatrix of M whose columns are the columns that do not have $*$'s, and whose rows are all rows of M. We represent M using sparse representations of M_R and M_C.

The rest of the paper is organized as follows. In Sect. 3 we construct a probe matrix for the input graph that generalizes the clique matrix. In Sect. 4 we construct an interval model from a consecutive-ones ordering of this matrix. In Sect. 5 we present a linear-time algorithm for the consecutive-ones probe matrix problem. Last, in Sect. 6 we determine if the interval model is unique.

3 Extension of the Clique Matrix

In this section we show how to build a probe matrix M that has the consecutive-ones property if G is a probe interval graph. The basis of this matrix is M_P, the clique matrix of $G[P]$. In addition, for every non-probe we define either new columns or a new row. A new column has a value of 0 or 1 for every row of M_P. Similarly, a new row has a value of 0 or 1 for every column of M_P. The submatrix of M induced by the new rows and the new columns consists exclusively of $*$'s, and so M is a probe matrix. We view each row of M as a *constraint*, since it limits the possible consecutive-ones orderings of M.

The graph $G[P]$ has no non-probes. Therefore, if G is a probe interval graph, then $G[P]$ is an interval graph and M_P has the consecutive-ones property. Let $x \in N$, If G is a probe interval graph then $G[P \cup \{x\}]$ is an interval graph, because a pair of non-probes is required to give rise to an interval intersection that is not an edge. This last observation is the basis of our probe matrix construction.

Therefore, we begin by finding a consecutive-ones ordering of M_P. Using [2] we can either find such an ordering or determine that G is not a probe interval graph, in $O(V + E)$ time.

Let $\mathcal{C}$ denote the set of cliques of $G[P]$. For each probe p, let $\mathcal{Q}(p)$ denote the set of cliques of $\mathcal{C}$ that contain p. In an interval model of G, the interval for p must intersect the clique segments of members of $\mathcal{Q}(p)$ and these clique segments must be consecutive. In M we represent these constraints for all P by the rows of M_P. We call these constraints *probe - clique* constraints.

Similarly, for each non-probe x, let $\mathcal{Q}(x)$ denote the set of cliques of $\mathcal{C}$ that are subsets of $N(x)$, and Q_x denote $\bigcup \mathcal{Q}(x)$. A vertex v is *simplicial* if $N(x)$ induces a complete subgraph. We split the set of non-probes into three sets: N_1 is the set of non-probes x such that $|\mathcal{Q}(x)| \geq 1$; N_2 is the set of non-simplicial vertices with $\mathcal{Q}(x) = \emptyset$; and N_3 is the set of simplicial vertices. Note that, according to this definition, a simplicial non-probe x such that $|\mathcal{Q}(x)| = 1$ is contained both in N_1 and in N_3; it does not matter into which of the two sets we put x.

The vertices of N_1 and N_2 add three kinds of rows (constraints) to M, while the vertices of N_3 add columns. We show the details below, but first we show how to partition N into these three sets. We must find for every $x \in N$ the set $\mathcal{Q}(x)$ and determine whether x is simplicial or not.

Let the *left endpoint* of a row of a consecutive-ones ordering of M_P be the column of the leftmost 1 in the row, and the *right endpoint* be the rightmost. Let $x \in N$, and assume that $G[P]$ is an interval graph. In the consecutive-ones ordering of M_P we find for every $p \in N(x)$ the left endpoint and the right endpoint of the row of p. We keep the column numbers of these two endpoints, together with their side (left or right) in a list L_x. We sort L_x for all x in linear time using a single radix sort, with x as the primary sort key, column number as the secondary sort key, and left *versus* right endpoint as the tertiary key so that if a left endpoint and right endpoint have the same primary and secondary key, the left endpoint goes to the left of the right.

We sweep through L_x from left to right, keeping a running count of the number of neighbors of x in the current column. Each time we encounter a left endpoint in L_x we increment the counter, and each time we encounter a right endpoint we decrement it. Each time we encounter a right endpoint e that follows a left endpoint, we compare the counter with the size of the clique C represented by the column of e, and include C in $\mathcal{Q}(x)$ if they are equal.

Every time we change the value of the counter, we compare it to $|N(x)|$, if these values are equal at some column C, then $N(x) \subseteq C$ and so x is simplicial.

The procedure for x takes time proportional to $|N[x]|$ for every non-probe x. Summing over all x, we have an $O(N + E)$ bound for splitting N into N_1, N_2 and N_3. We conclude with the following lemma:

Lemma 1. *In linear time we can either split N into N_1, N_2 and N_3 and find $\mathcal{Q}(x)$ for every $x \in N$, or else determine that G is not a probe interval graph.*

3.1 Non-probe - Clique Constraints

Assume that G is indeed a probe interval graph and consider the interval of $x \in N_1$ in an interval model R. The interval of x must intersect the clique segments that correspond to members of $\mathcal{Q}(x)$, and these clique segments must be consecutive in the ordering of clique segments of $G[P]$ given by any interval model of G.

The number of cliques containing a vertex v in an interval graph is bounded by $|N[v]|$, since a neighbor of v ends at the clique segment for each clique that contains v. Since $G[P \cup \{x\}]$ is an interval graph, for any $x \in N_1$, the number of cliques in $\mathcal{Q}(x)$ is bounded by $|N[x]|$.

We therefore add to M a row for each $x \in N_1$ that has 1's in the columns of $\mathcal{Q}(x)$. This adds $O(|N[x]|)$ to the size of the matrix. We call these new rows *non-probe - clique constraints.*

3.2 Non-probe - Probe Binding Constraints

The non-probe - clique constraints defined for members of N_1 are not enough. These constraints allow the interval of $x \in N_1$ to intersect the clique segments of $\mathcal{Q}(x)$ and thus intersect the intervals of Q_x, but there may be some vertices in $N(x) \setminus Q_x$. For these we add more constraints to M.

Let $x \in N_1$ and let $p \in N(x) \setminus Q_x$. Since x and p are adjacent, we know that their intervals must intersect in any model of G, and therefore $\mathcal{Q}(x) \cup \mathcal{Q}(p)$ must be consecutive. Let us call this additional constraint a *non-probe - probe binding constraint* imposed by x and p. Adding such a constraint for every such x and p will make M too large. We show that a set of new rows with a linear number of 1's is enough to enforce the non-probe - probe binding constraints.

We know that $\mathcal{Q}(x) \cap \mathcal{Q}(p) = \emptyset$, because $p \notin Q_x$. Therefore, in any interval model of G, the interval of p covers exactly one endpoint of the interval of x. Moreover, in the order of the clique segments either the rightmost member of $\mathcal{Q}(p)$ must be consecutive with the leftmost member of $\mathcal{Q}(x)$ or vice versa.

The set of probes that x is bound to is $N(x) \setminus Q_x$. In an interval model of G, we can divide this set into the set Y_1 that covers the left endpoint of x and the set Y_2 that covers the right endpoint of x. Note that although we used a specific model of G, the same Y_1 and Y_2 arise in every model (up to interchange).

Recall that the vertices are numbered arbitrarily from 1 through n. For two vertices v and u, let $v \prec u$ denote that either $\mathcal{Q}(v) \subset \mathcal{Q}(u)$ or that $\mathcal{Q}(v) = \mathcal{Q}(u)$ and v has a smaller vertex number than u does.

Since the members of Y_1 all end at the clique segment to the left of x's left endpoint and they all occupy consecutive cliques, it follows that for any two $y, y' \in Y_1$, either $\mathcal{Q}(y) \subseteq \mathcal{Q}(y')$ or $\mathcal{Q}(y') \subseteq \mathcal{Q}(y)$. It follows that Y_1 induces a linear order in the $\prec$ relation, so it has a unique a minimal member y_1 in this relation. Similarly, Y_2 has a unique minimal member y_2 in the $\prec$ relation.

By similar reasoning, each for each probe p, the $\prec$ relation on non-probes that p is bound to has at most two nonadjacent minimal members x_1 and x_2. Let us say that x and p are a *representative bound pair* if p is a minimal bound neighbor of x and x is a minimal bound neighbor of p in the $\prec$ relation.

Consider the current status of the matrix M. The matrix includes the probe - clique constraints and the non-probe - clique constraints. In $O(V+E)$ time we can either find a consecutive-ones ordering of M or determine that G is not a probe interval graph, since we cannot satisfy all the constraints. Using this ordering of M, we can determine in $O(1)$ time for two vertices $v, u \in P \cup N_1$ whether $\mathcal{Q}(v) \subseteq \mathcal{Q}(u)$ by examining the position of leftmost and rightmost 1's in the rows of u and v. Thus, the relation $\prec$ for two vertices can be determined in $O(1)$ time as well. We get that we can find the minimal bound neighbors of every vertex, and thus all the representative bound pairs, in $O(V+E)$ time.

We add to M a row for any representative pair $\{x, p\}$ that has 1's in the columns of $\mathcal{Q}(x) \cup \mathcal{Q}(p)$. This adds $O(|N[x]| + |N[p]|)$ to the size of the matrix. Since every vertex adds at most two new rows to M, the size of M remains linear in the size of G.

A consecutive-ones ordering of M satisfies the binding constraint not just for representative bound pairs, but for all such bound pairs of vertices. This is true since M already has a row with the characteristic vector of $\mathcal{Q}(v)$ for each probe or non-probe vertex v because of the probe - clique constraints and the non-probe - clique constraints.

3.3 Probe - Probe Binding Constraints

Consider $x \in N_2$. In this case, $\mathcal{Q}(x) = \emptyset$ and $N(x)$ is not a complete subgraph. If G is a probe interval graph, then in an interval model R of G, the interval of x lies between two consecutive clique segments of $R[P]$. Let C_1 and C_2 be the corresponding cliques of $G[P]$, such that C_1's segment lies to the left of C_2's. Let $Y_1 = N(x) \setminus C_2$ and let $Y_2 = N(x) \setminus C_1$. The sets Y_1 and Y_2 satisfy $Y_1 \subseteq C_1$, $Y_2 \subseteq C_2$ and $Y_1 \cap Y_2 = \emptyset$. Also, since x is not simplicial, neither Y_1 nor Y_2 is empty. Note that although we used a specific model to define Y_1 and Y_2 for x, these sets are unique for every $x \in N_2$, up to interchange between the two.

Let $y \in Y_1$ and $y' \in Y_2$. Since x is adjacent to both y and y', and does not intersect any clique segment, we know that $\mathcal{Q}(y) \cup \mathcal{Q}(y')$ must be consecutive in any interval model of G. We call this additional constraint a *probe - probe binding constraint* imposed by y and y'. As with the non-probe - probe binding constraints, we can use the same relation $\prec$ and add to M such a constraint only for a pair of minimal bound neighbors. To find these in $O(V+E)$ time, we find for each $x \in N_2$, the sets Y_1 and Y_2. All elements of Y_1 are bound to elements of Y_2, but it is enough to bind only the minimal members of the two sets to each other. This gives $O(|N_2|)$ candidate pairs for bindings. We proceed on these as in the case of probe - non-probe bindings to find representative pairs.

We add to M a row for each representative pair p, p' that has 1's in the columns of $\mathcal{Q}(p) \cup \mathcal{Q}(p')$. This adds $O(|N[p]| + |N[p']|)$ to the size of the matrix. Again, the size of M remains linear in the size of G.

3.4 Additional Segments

Last, we consider N_3. For this set of probes we do not define further constraints, but refine the probe - clique constraints. This is done by adding columns to M. As mentioned earlier, the new columns have 0 or 1 in rows of M_P and $*$ in rows that we added for N_1 and N_2.

Let $x \in N_3$, and assume that G is a probe interval graph. Let R be an interval model of G. The set $N[x]$ is a clique in G. Therefore there is a clique segment in R that is intersected by the intervals of $N[x]$.

Let $\mathcal{C}' = \{N[x] \mid x \in N_3\}$. The members of $\mathcal{C}'$ are the cliques of G that are not in $\mathcal{C}$. For each vertex v, let $\mathcal{Q}'(v)$ denote the set of members of $\mathcal{C}'$ that contain v. In an interval model of G, the interval for a probe p must intersect the clique segments that correspond to members of $\mathcal{Q}(p) \cup \mathcal{Q}'(p)$, and the clique segments of $\mathcal{Q}(p) \cup \mathcal{Q}'(p)$ must be consecutive in the left-to-right ordering of clique segments. This gives us a refinement of the probe - clique constraints.

To represent the cliques of $\mathcal{C}'$, we add a new column for every $x \in N_3$, that has a 1 in the row of a probe p if $p \in N(x)$, and 0 otherwise. Using a sparse representation of the matrix, this adds $O(|N[x]|)$ to the size of the matrix.

This concludes the construction of M. If G is a probe interval graph, then there must be a consecutive-ones ordering of the columns of M that obeys all constraints. We summarize the section in the following lemma:

Lemma 2. *It takes $O(V + E)$ time to construct the probe matrix M or else decide that G is not a probe interval graph. Moreover, if G is a probe interval graph then M is a consecutive-ones probe matrix.*

We use the algorithm of Sect. 5 to find a consecutive-ones ordering of M. If such an ordering does not exist then G is not a probe interval graph.

4 Constructing an Interval Model

In this section we use the consecutive-ones ordering of M that we found in the previous section to find an interval model of G, if one exists. The construction is similar to the construction of [2] of an interval model from the clique matrix of an interval graph. Each interval must intersect the clique segments it belongs to. In addition, realizing bindings requires some differential stretching of endpoints inside the zone between two consecutive clique segments.

Recall that we represent an interval model combinatorially by a list of alternating blocks of left and right endpoints. We begin by defining two sets of endpoints for every column C of M: C_ℓ and C_r. We show below how we populate these sets. We order the sets according to the consecutive-ones ordering of the columns of M, such that C_ℓ is to the left of C_r.

Let $v \in V \setminus N_2$. In this case, $\mathcal{Q}(v) \cup \mathcal{Q}'(v)$ is not empty. If $v \in P \cup N_1$, then $\mathcal{Q}(v) \cup \mathcal{Q}'(v)$ has a row in M. Let C be the leftmost column with 1 in this row and D be the rightmost column with 1 in this row. If $v \in N_3$, we let C and D both be the column of the clique $N[v]$. We put the left endpoint of v in C_ℓ and the right endpoint of v in D_r. Because M has a consecutive-ones ordering, this takes linear time. Let us denote the resulting interval model by R_1.

In R_1, for every column C, the segment on the line between C_ℓ and C_r is the clique segment of C. If the intervals of v and u intersect in R_1, and at least one of the two vertices is a probe, then v and u are adjacent.

However, there still might be some edges in E that are not realized by R_1. These edges are between $x \in N_1 \cup N_2$ and $p \in N(x) \setminus Q_x$. In order to realize these adjacencies we place the endpoints of vertices of N_2 and stretch the intervals of N_1, N_2 and $N(x) \setminus Q_x$ for $x \in N_1 \cup N_2$ between the clique segments of $\mathcal{C} \cup \mathcal{C}'$.

Let $x \in N_2$ and let Y_1 and Y_2 be as defined in Sect. 3.3, that is, the two sets for which x defines probe - probe constraints, such that the intervals of Y_1 are to the left of the intervals of Y_2. Let C be the rightmost column in which all rows of members of Y_1 have a 1. Let D be the leftmost column in which all rows of members of Y_2 have a 1. The columns C and D exist and are consecutive because of the probe - probe constraints. We place the left endpoint of x in D_ℓ and the right endpoint of x in C_r. Denote the construction so far by R_2. Note that R_2 is not an interval model, since we place the left endpoint of x to the right of its right endpoint. We will resolve this problem when we stretch the intervals.

The last step of the construction is to stretch intervals of vertices of N_1, N_2 and $N(x) \setminus Q_x$ for $x \in N_1 \cup N_2$. Consider two vertices v and u that are adjacent in G, but whose adjacency is not realized in R_2. Assume that v is to the left of u. (If one of them is in N_2, then it does not have a real interval, but it is still clear which one is to the left.) Because of the non-probe - probe constraints and the probe - probe constraint, we know that the set C_r, which contains the right endpoint of v, is immediately to the left of D_ℓ, which contains the left endpoint of u. We must stretch the endpoints of intervals that have unrealized intersections, between the clique segments.

For every C_r and the set to its right, D_ℓ, we split the two sets and order them as follows. We split C_r into subsets $A_0, A_1, \ldots, A_{|D_\ell|}$ and A' such that an endpoint $f \in C_r$ is in A_i if it is an endpoint of an interval of a probe p with $|N(p) \cap D_\ell| = i$, and in A' if it is an endpoint of an interval of a non-probe. Similarly we split D_ℓ into subsets $B_0, B_1, \ldots, B_{|C_r|}$ and B'. Note that some of the subsets might be empty. We replace C_r with the A_i's, where A_0 is the leftmost. We replace D_ℓ with B_i's where B_0 is the rightmost. For every endpoint $f \in B'$, we place f in a set to the right of A_j where j is the largest index such that the vertex of f is non-adjacent to all vertices of $A_0, A_1, \ldots, A_j$ and adjacent to all vertices of $A_{j+1}, A_{j+2}, \ldots, A_{|D_\ell|}$. Similarly we place every endpoint $f \in A'$ in a set on the left of the appropriate B_j. Note that the set between $A_{|D_\ell|}$ and $B_{|C_r|}$ contains both right and left endpoints. We split this set F into a set F_ℓ of left endpoints and a set F_r of right endpoints. Let us denote the resulting construction by R. See Fig. 1.

If G is a probe interval graph, then we can place every member of A' and B', and therefore we can construct R. This is because if we cannot place $f \in B'$ on the right side of any A_i or $f \in A'$ on the left side of any B_i, then any interval model of G must contain an induced chordless cycle, which is impossible.

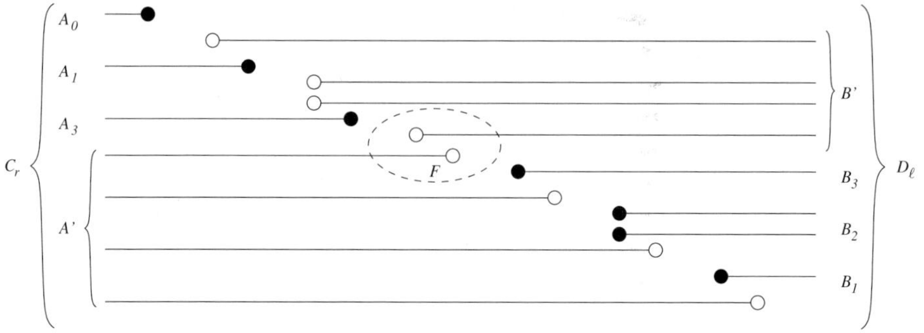

Fig. 1. Reordering the endpoints of C_r and D_ℓ. The endpoints of C_r are right endpoints and the endpoints of D_ℓ are left endpoints. Endpoints of non-probes are empty, endpoints of probes are full.

Since every interval has two endpoints, and the splitting of C_r and D_ℓ takes time proportional to the number of edges between vertices that have endpoints in these sets, in $O(V + E)$ time we can either construct R or decide that R cannot be constructed, and thus that G is not a probe interval graph.

If we manage to construct R, then it is an interval model of G. First, note that the endpoints of every vertex of N_2 are now ordered properly, that is, the left endpoint is to the left of the right endpoint. To show that R realizes G, we consider the following cases for a probe p and a vertex v, and show that their intervals in R intersect if and only if they are adjacent. If $v \in P$ then the claim is true because $R[P]$ is a model of $G[P]$. If $v \in N_1$ and $p \in Q_v$ or if $v \in N_3$ then the claim is true because the intervals intersect if and only if $(\mathcal{Q}(v) \cup \mathcal{Q}'(v)) \cap (\mathcal{Q}(p) \cup \mathcal{Q}'(p)) \neq \emptyset$. Otherwise, the claim follows by the way we stretch intervals into the region between the clique segments.

We conclude with the main theorem:

Theorem 3. *Let G be a probe graph. In $O(V + E)$ time we can construct an interval model for G, or decide that G is not a probe interval graph.*

5 Consecutive-Ones Probe Matrices

In this section we present a linear-time algorithm for the consecutive-ones probe matrix problem. Let M be a probe matrix with i rows, j columns and k 1's. We determine if M is a consecutive-ones probe matrix in $O(i + j + k)$ time. We do so by finding the PQ-tree of two 0-1 submatrices of M using [2], and combining the trees using tools of [11]. If M is a consecutive-ones probe matrix then we find a consecutive-ones ordering of it. With a modification of [11] we can find a PQ-tree that represents all consecutive-ones orderings of M. We do not present it here, because a single consecutive-ones ordering is enough, and we want to keep the description simple.

Let M_R be the submatrix of M whose rows are the rows that do not have $*$'s, and whose columns are all columns of M. Let M_C be the submatrix of M whose columns are the columns that do not have $*$'s, and whose rows are all rows of M. Let X be the set of columns of M_C.

Let T_R be the PQ-tree of M_R and let T_C be the PQ-tree of M_C. Let $T' = T_R[X] \cap T_C$. Each permutation $\sigma \in \Pi(T')$ is a permutation in both $\Pi(T_R[X])$ and in $\Pi(T_C)$. This means that $\sigma = \pi_X$ where $\pi \in \Pi(T_R)$. Assume that such a permutation $\sigma \in \Pi(T')$ exists. We can order T_R so that the relative order of leaves that are members of X is σ, because it is a restriction of some $\pi \in \Pi(T_R)$. At a P node, we order the set of children that contain leaf descendants in X according to the order of those members in σ. At a Q node, if two children contain leaf descendants in X, from the two allowed linear orders of children, we choose the one that is consistent with σ. The result is π, a consecutive-ones ordering of columns of M_R, such that $\sigma = \pi_X$ is a consecutive-ones ordering of M_C. Therefore, π is a consecutive-ones ordering of M.

On the other hand, if $\Pi(T') = \emptyset$, then there is no ordering of X that imposes a consecutive-ones ordering both for $T_R[X]$ and for T_C, and therefore M is not a probe interval matrix.

Using [2] we can find T_R and T_C in $O(i + j + k)$ time, and using [11] we can find $T_R[X]$ and from it T' in the same time bound. Choosing σ and π takes $O(j)$ time.

Theorem 4. *Let M be a probe matrix. In time linear in the size of the matrix and the number of 1's in it we can either find a consecutive-ones ordering of M, or else decide that M is not a consecutive-ones probe matrix.*

6 Determining Whether a Model Is Uniquely Constrained

Recall that we represent an interval model combinatorially by a list of alternating blocks of left and right endpoints, as the order of endpoints within a block is inconsequential. Let R and R' be two interval models. We say that R and R' are *equivalent* if they are identical, or if we can get R' from R by reversing the order of its blocks and exchanging the blocks between the two endpoints of each interval. If every model of G is equivalent to R, then R is a *unique* model of G.

Let T' be as in Sect. 5, for the matrix M, found in Sect. 3, and let R be the model found in Sect. 4.

The model R is unique only if M has a unique consecutive-ones ordering up to reversal. A consecutive-ones ordering of a matrix is unique up to reversal if and only if the PQ-tree has a single internal node that is a Q node, if this is not the case for T', then R is not a unique model. The same happens also if there is more one way to produce π from σ. Specifically, this happens if there is a P node with a child that does not contain an element of X or if there is a Q node for which less than two children contain elements of X. This can be checked in time linear in the size of T'. Otherwise, the order for the columns of M is unique.

Even if M does have a unique consecutive-ones ordering, different models are possible. The matrix M defines a unique order for the cliques of G, and therefore a unique order of the clique segments in any interval model of G. Thus, for every two vertices of V such that at least one is a probe, there is a unique order defined among their endpoints. However, if there are two non-probes x and x' such that the set that contains the left endpoint of x in the model R is next to the set that contains the right endpoint of x', then we can change the order between the two endpoints. In this case as well, R is not a unique model of G. This last case can also be detected in time linear in the size of G.

References

1. Benzer, S.: On the topology of the genetic fine structure. Proc. Nat. Acad. Sci. U.S.A. 45, 1607–1620 (1959)
2. Booth, K.S., Lueker, G.S.: Testing for the consecutive ones property, interval graphs, and graph planarity using PQ-tree algorithms. J. Comput. Syst. Sci. 13, 335–379 (1976)
3. Chandler, D.B., Guo, J., Kloks, T., Niedermeier, R.: Probe matrix problems: Totally balanced matrices. In: Kao, M.-Y., Li, X.-Y. (eds.) AAIM 2007. LNCS, vol. 4508, pp. 368–377. Springer, Heidelberg (2007)
4. Chang, G.J., Kloks, T., Liu, J., Peng, S.-L.: The PIGSs full monty - a floor show of minimal separators. In: Diekert, V., Durand, B. (eds.) STACS 2005. LNCS, vol. 3404, pp. 521–532. Springer, Heidelberg (2005)
5. Cormen, T.H., Leiserson, C.E., Rivest, R.L., Stein, C.: Introduction to Algorithms. McGraw Hill, Boston (2001)
6. Fulkerson, D.R., Gross, O.: Incidence matrices and interval graphs. Pacific J. Math. 15, 835–855 (1965)
7. Golumbic, M.C.: Algorithmic Graph Theory and Perfect Graphs. Academic Press, New York (1980)
8. Golumbic, M.C.: Matrix sandwich problems. Linear Algebra and Applications 277, 239–251 (1998)
9. Golumbic, M.C., Trenk, A.N.: Tolerance Graphs. Cambridge studies in advanced mathematics 89, New York (2004)
10. Johnson, J.L., Spinrad, J.P.: A polynomial time recognition algorithm for probe interval graphs. In: SODA 2001, pp. 477–486. Association for Computing Machinery, New York (2001)
11. McConnell, R.M., de Montgolfier, F.: Algebraic operations on PQ trees and modular decomposition trees. In: Kratsch, D. (ed.) WG 2005. LNCS, vol. 3787, pp. 421–432. Springer, Heidelberg (2005)
12. McConnell, R.M.: Linear-time recognition of circular-arc graphs. Algorithmica 37, 93–147 (2003)
13. McConnell, R.M., Spinrad, J.P.: Construction of probe interval models. In: SODA 2002, pp. 866–875. Association for Computing Machinery, New York (2002)
14. McMorris, F.R., Wang, C., Zhang, P.: On probe interval graphs. Discrete Applied Mathematics 88, 315–324 (1998)
15. Rose, D., Tarjan, R.E., Lueker, G.S.: Algorithmic aspects of vertex elimination on graphs. SIAM J. Comput. 5, 266–283 (1976)
16. Uehara, R.: Canonical data structure for interval probe graphs. In: Fleischer, R., Trippen, G. (eds.) ISAAC 2004. LNCS, vol. 3341, pp. 859–870. Springer, Heidelberg (2004)
17. Zhang, P.: United states patent 5667970: Method of mapping DNA fragments (July 3, 2000)

Wireless Scheduling with Power Control*

Magnús M. Halldórsson

School of Computer Science, Reykjavik University & Icelandic Center of Excellence in Theoretical Computer Science (ICE-TCS)
www.ru.is/faculty/mmh

Abstract. We consider the scheduling of arbitrary wireless links in the physical model of interference to minimize the time for satisfying all requests. We study here the combined problem of scheduling and power control, where we seek both an assignment of power settings and a partition of the links so that each set satisfies the signal-to-interference-plus-noise (SINR) constraints.

We give an algorithm that attains an approximation ratio of $O(\log n \cdot \log\log \Lambda)$, where Λ is the ratio between the longest and the shortest linklength. Under the natural assumption that lengths are represented in binary, this gives the first *polylog*(n)-approximation. The algorithm has the desirable property of using an oblivious power assignment, where the power assigned to a sender depends only on the length of the link. We show this dependence on Λ to be unavoidable, giving a construction for which any oblivious power assignment results in a $\Omega(\log\log \Lambda)$-approximation.

We also give a simple *online* algorithm that yields a $O(\log \Lambda)$-approximation, by a reduction to the coloring of unit-disc graphs. In addition, we obtain improved approximation for a bidirectional variant of the scheduling problem, give partial answers to questions about the utility of graphs for modeling physical interference, and generalize the setting from the standard 2-dimensional Euclidean plane to doubling metrics.

1 Introduction

We are interested in fundamental limits on communication in wireless networks. How much communication throughput is possible? This is an issue of efficient spatial separation, keeping the interference from simultaneously communicating links sufficiently low. The interference scheduling problem is then to schedule an arbitrary set of communication links in the least amount of time while satisfying interference constraints. In this paper, we focus on the power control version, where we also choose the power settings for the links.

The scheduling problem depends strongly on the model of interference. Until recently, previous algorithmic work has revolved around various graph-based

* Work done in part while visiting the Research Institute for Mathematical Sciences (RIMS) at Kyoto University. Supported by Icelandic Research Fund grant 90032021.

A. Fiat and P. Sanders (Eds.): ESA 2009, LNCS 5757, pp. 361–372, 2009.

models, where interference is modeled as a pairwise constraint. This, however, fails to capture the accumulative property of actual radio signals. In contrast, researchers in information, communication, or network theory ("EE") are working with wireless models that sum up interference and respect attenuation. The standard model is the signal-to-interference-plus-noise (SINR) model, to be formally introduced in Section 1.3. The SINR model reflects physical reality more accurately and is therefore often simply called the physical model. On the other hand, "EE researchers" tend to propose heuristics that are evaluated by simulation, which neither give insights into the complexity of the problem nor give algorithmic results that may ultimately lead to new protocols.

In a seminal work, Moscibroda and Wattenhofer [15] initated the study of scheduling in the SINR model. Formally, given is an arbitrary set of links, each a sender-receiver pair of points in the plane. We seek an assignment of power settings to the senders and a partition of the linkset into minimum number of slots, so that the links in each slot satisfy the SINR-constraints. We refer to this as the PC-Scheduling problem. In the related bidirectional scheduling problem, both nodes in a link may be transmitting, which implies a stronger, symmetric form of interference.

We seek algorithms that result in good schedules. Beyond this rather obvious objective, we seek to map the landscape of the issues surrounding SINR-scheduling. In particular, we shall study in this paper two relevant questions: the utility of "simple" power allocation strategies, and the extent to which graphs can capture interference in the SINR-model.

For reasons of simplicity of use, it is desirable to use power assignments that are precomputable independent of other links. Such *oblivious* assignments depend only on the length of the given link. In fact, oblivious assignments appear unavoidable in the distributed setting. The two most frequently used power assignment strategies are indeed of this type, using either *uniform* (or fixed) power for all the links, or *linear* assignment that ensures that the signals received at the intended receivers are identical.

The other issue of particular interest is the utility of graphs for modeling interference. It is *a priori* given that graphs are imperfect models, given both the non-locality and the additive nature of interference in the SINR model. The perceived difficulty in reasoning analytically about these additional complications has been cited as a factor against SINR model. Still, graphs have proved to be highly versatile tools for analysis and algorithm design, and pairwise constraints are in general much easier to handle than many-to-many constraints. We would therefore like to quantify the cost of doing business using graphs, or the overhead that amenable graph models have over non-graphic models, as well as pinpointing particular situations where graphs work especially well.

1.1 Our Contributions

We present a simple scheduling algorithm that works for any oblivious power assignment strategy, resulting in a $O(\log \Lambda)$ approximation ratio, where Λ is the ratio between the maximum and minimum link length. In particular, when all links are of nearly equal length, we obtain the first constant approximation. This matches the constructions in [14] that show that both fixed and linear assignments can be as much as $\Omega(\log \Lambda)$ factor from optimal.

We then examine a new oblivious assignment, the *mean* assignment studied recently in [7], which is the geometric mean of the uniform and linear power assignments. We give a simple scheduling algorithm that uses mean assignment and obtains a $O(\log\log \Lambda \cdot \log n)$-approximation. Under the natural assumption that lengths can be represented in binary, this implies also $O(\log^2 N)$-approximation, where N is the size of the input. In the bidirectional version, the algorithm results in a $O(\log n)$-approximation, improving on the previous $O(\log^c n)$-factor with $c > 6$ [6] using considerably simpler arguments.

We show that the dependence on Λ is unavoidable for oblivious power assignment strategies. Namely, any oblivious assignment forces $\Omega(\log\log \Lambda)$-approximate schedules. Thus, within the framework of power assignments that are oblivious of the link instance, our results are best possible up to logarithmic factor. Our results also pinpoint the issues of essential difficulty in the scheduling of wireless links. When the input instance is "well-behaving" in that no sender is much closer to the receiver of another link than its sender, or when the criteria is changed from unidirectional to bidirectional scheduling, the bound improves to a single logarithmic factor. Thus, we can characterize more precisely the structural property of link arrangements that make scheduling hard.

Our results apply to the standard setting of the two-dimensional Euclidean plane with the path-loss constant $\alpha > 2$ (see Section 1.3). More generally, they hold for a general class of distance metrics that are *doubling metrics*, for the case when α is greater than the doubling constant of the metric. The requirement on α is to ensure that the cumulative power of a transmission fades away. This is a natural assumption, since preservation or amplification would contradict the second law of thermodynamics. We can also extend all the results to general metrics, with a roughly logarithmic increase in the approximation factor.

The simplicity of our algorithms leaves them suitable for distributed implementation. For links of nearly equal length, we can use uniform power, and we show that the problem reduces, within a constant factor, to the coloring of unit-disc graphs, a very well-studied problem. This also implies an $O(\log \Lambda)$-competitive online scheduling algorithm.

Our work also gives partial answer to a nagging question regarding the utility of graphs in representing physical models of interference. Our results indicate that graphs can still play a useful role. For nearly equal length links, the basic class of unit-disc graphs is in fact sufficient. The $O(\log n \cdot \log\log \Lambda)$-approximation result is also relative to the underlying graph.

Finally, it can be verified that our algorithms give equivalent results for throughput maximization, or the Single-Slot scheduling problem.

1.2 Related Work

Most work in wireless scheduling in the physical (SINR) model has been of heuristic nature, e.g. [5]. Only after the work of Gupta and Kumar [11] did analytical results became *en vogue*, but were largely non-algorithmic and restricted to networks with a well-behaving topology and traffic pattern such as uniform geometric distribution.

In contrast, the body of algorithmic work is mostly on graph-based models that ultimately abstract away the nature of wireless communication. The inefficiency of graph-based protocols in the SINR model is well documented and has been shown theoretically as well as experimentally [10,13,16].

Approximation algorithms for the problem of scheduling wireless links in the SINR model were given in [17], [14] and [3]. In all cases the performance ratios obtained consist of the product of structural properties and a function of the number of nodes. The structural properties are different but can all grow linearly with the size of the network.

A number of recent related results have featured a $O(\log \Lambda)$-approximation. Andrews and Dinitz [1] gave a $O(\log \Lambda)$ approximation for the Single-Slot scheduling problem. Fanghänel, Kesselheim and Vöcking [7] gave a randomized algorithm for the scheduling problem using linear power assignment that uses $O(OPT \log \Lambda + \log^2 n)$ slots, matching our results for dense instances. And finally, Avin, Lotker and Pignolet [2] show that assumption of $\alpha > 2$ used by all previous work may not be necessary, in that the ratio between non-oblivious and oblivious Single-Slot schedules is $O(\log \Lambda)$, at least in the 1-dimensional metric.

In [6], Fanghänel et al. give a construction that shows that any schedule based on any oblivious power assignment can be a factor of n from optimal. We show that in terms of Λ, the gap is actually $\Omega(\log \log \Lambda)$, using similar constructions. They also introduce the bidirectional version of the scheduling problem and give a $O(\log^{4.5+\alpha} n)$-approximation factor using the mean power assignment in general metrics. Their proof involves non-trivial embeddings into tree metric spaces.

In contrast, the scheduling complexity of arbitrary links in the case of fixed, uniform power is now fairly well understood. Constant factor approximations were recently obtained for the Single-Slot scheduling problem [8] and the scheduling problem [12]. Both of these problems are known to be NP-complete [9]. The results obtained here for power control build on and extend the techniques and properties derived in the case of uniform power in [8,12].

1.3 Notation and Preliminaries

Given is a set $L = \{\ell_1, \ell_2, \ldots, \ell_n\}$ of links, where each link ℓ_v represents a communication request from a sender s_v to a receiver r_v. The distance between two points x and y is denoted $d(x, y)$. The asymmetric distance from link v to link w is the distance from v's sender to w's receiver, denoted $d_{vw} = d(s_v, r_w)$. The length of link ℓ_v is denoted simply ℓ_v. We shall assume for simplicity of exposition that all links are of different length; this does not affect the results materially. We assume that each link has a unit-traffic demand, and model the case of non-unit traffic demands by replicating the links.

The nodes can transmit with different power. Let P_v denote the power assigned to node v. We assume the *path loss radio propagation* model for the reception of signals, where the signal received from w at receiver v is P_w/d_{wv}^{α} and $\alpha > 2$ denotes the path-loss exponent. We adopt the *physical interference model*, in which a node r_v successfully receives a message from a sender s_v if and only if the following condition holds:

$$\frac{P_v/\ell_v^{\alpha}}{\sum_{\ell_w \in S \setminus \{\ell_v\}} P_w/d_{wv}^{\alpha} + N} \geq \beta, \tag{1}$$

where N is the ambient noise, $\beta \geq 1$ denotes the minimum SINR (signal-to-noise-ratio) required for a message to be successfully received, and S is the set of concurrently scheduled links in the same *slot*. Note that by scaling the power of all the senders, the effect of the noise can be made arbitrarily small, thus we ignore this term. Of course, in real situations, there are upper bounds on maximum power, etc, which we ignore here. We shall also assume that $\beta = 1$; by Prop. 1 this does not affect the results materially. We say that S is *SINR-feasible* if (1) is satisfied for each link in S.

The *affectance* of link ℓ_v caused by a set S of links, is the sum of the interferences of the links in S on ℓ_v relative to the power received, or

$$a_S(\ell_v) = \sum_{\ell_w \in S \setminus \{v\}} \frac{P_w/d_{wv}^{\alpha}}{P_v/\ell_v^{\alpha}} = \sum_{\ell_w \in S \setminus \{v\}} \frac{P_w}{P_v} \cdot \left(\frac{\ell_v}{d_{wv}}\right)^{\alpha}$$

For a single link ℓ_w, we use the shorthand $a_w(v) = a_{\{\ell_w\}}(\ell_v)$. Note that affectance is additive in that for disjoint sets of links S_1, S_2, $a_{S_1 \cup S_2}(\ell_v) = a_{S_1}(\ell_v) + a_{S_2}(\ell_v)$. For convenience, let $a_v(v) = 0$.

Let $OPT = OPT(L)$ denote a SINR-feasible schedule with minimum number of slots. Let $\Gamma = \Gamma(L)$ denote the number of slots in OPT. A *p-signal* set or a schedule is one where the affectance of any link is at most $1/p$. A set is SINR-feasible iff it is a 1-signal set. Let OPT_p be a p-signal schedule with minimum number of slots, and let Γ_p denote its number of slots. Let Λ denote the ratio between the maximum and minimum length of a link.

For a graph G, let $\Delta(G)$ denote the maximum degree of a vertex, and $\chi(G)$ denote the chromatic number.

We extend the setting from the Euclidean plane to *doubling metrics* (see Clarkson [4]). We require that the path loss exponent α be strictly greater than the doubling dimension of the metric. We shall refer to such a combination of distance metric and path loss function as a *fading metric*.

Preliminaries We shall refer to two links ℓ_v and ℓ_w as *q-independent* if they satisfy the constraint

$$d_{vw} \cdot d_{wv} \geq q^2 \cdot \ell_w \ell_v \ .$$

A set S of links is a *q-independent set* if the links in S are mutually q-independent.

Define the *link graph* $G_q(L)$ on a link set L, parameterized by a constant q such that a pair of links are adjacent in G_q iff they are not q-independent. The following observation shows that a schedule of a linkset forms a coloring of the corresponding link graph. The converse, however, does not necessarily hold, as we shall see. Thus, the graph representation is more relaxed than required.

Lemma 1. *Links that belong to the same q^α-signal slot are q-independent.*

Proof. Since the links belong to the same p-signal slot, they satisfy

$$\frac{P_v/\ell_v^\alpha}{P_w/d_{wv}^\alpha} \geq p, \quad \text{and} \quad \frac{P_w/\ell_w^\alpha}{P_v/d_{vw}^\alpha} \geq p \ .$$

By multiplying these inequalities together and rearranging, we get that

$$d_{vw} \cdot d_{wv} \geq p^{2/\alpha} \cdot \ell_w \ell_v = q^2 \cdot \ell_w \ell_v \ .$$

One of the main difficulty in scheduling is the asymmetry of the links. In more stringent schedules, the links are kept further apart, which diminishes the problems of asymmetry. For this purpose, the following result from [12] is crucial, which shows that increasing stringency affects only the constant in the approximation ratio.

Proposition 1 ([12]). *For any $p \geq 1$ and any linkset L, $\Gamma_p(L) \leq \lceil 2p \rceil^2 \Gamma(L)$.*

2 Uniform Power Assignment

One of the most widely used power assignment is the uniform one, where senders use the same power setting. This might be viewed as ultra-oblivious, as transmissions are now independent of link length.

We show in Sec. 2.1 that uniform power assignment performs very well when links are of nearly equal lengths. This results in a $O(\log \Lambda)$-approximation of PC-Scheduling, using any oblivious power assignment. Additionally, the global nature of the problem disappears, and local strategies become sufficient. In fact, as we show in Sec. 2.2, it suffices to color a unit-disc graph, with discs of radius proportional to the link lengths, to obtain a constant approximation.

2.1 Nearly Equal Linklengths

We say that a set of links is *nearly equilength* if lengths of any pair of links in the set differ by a factor of at most 2. We first observe that the scheduling complexity of a linkset is at least proportional to the degree of its link graph.

Lemma 2. *Let L be a set of nearly equilength links and q be a constant. Then, $\Gamma(L) = \Omega(\Delta(G_q))$.*

Proof. Let ℓ_v be a link with a set N_v of $\Delta(G_q)$ neighbors in G_q. We shall argue that the links in N_v must belong to distinct slots in any p'-signal schedule, for some $p' = O(q^\alpha)$. The theorem then follows from the signal-strengthening Proposition 1.

Let ℓ_u and ℓ_w be links in N_v and let D be the longest length of a link in N_v. By the link relationship in G_q, we have that $d_{uv} \cdot d_{vu} \leq q^2 \ell_v \ell_u \leq q^2 D^2$ and $d_{wv} \cdot d_{vw} \leq q^2 D^2$. For any pair of links ℓ_x, ℓ_y in N_v, we have by the triangular inequality that $d_{xy} \leq d_{yx} + 2D$. Thus, we have that $\max(d_{uv}, d_{vu}, d_{wv}, d_{vw}) \leq (q+2)D$. Again by the triangular inequality, $d_{uw} \leq d(s_u, r_v) + d(r_v, s_w) + d(s_w, r_w) = d_{uv} + d_{wv} + \ell_w \leq (2q+5)D$. Similarly, $d_{wu} \leq (2q+5)D$. Thus,

$$d_{wu} \cdot d_{uw} \leq (2q+5)^2 D^2 \leq (4q+10)^2 \ell_w \ell_u \ .$$

Hence, N_v forms a clique in G_{4q+10}. Hence, by Lemma 1 $\Gamma_{p'} \geq \Delta(G_q)$, for $p' = (4q+10)^\alpha$. By Proposition 1, the theorem now follows.

The following results extends similar lemmas in previous works (see [8,12]) from the setting of the Euclidean plane to the more general class of doubling metrics. It yields a converse of Lemma 1 for the case of nearly equilength links. This is the only place where we use the fading property of the metric, i.e., that α is strictly greater than the doubling dimension.

Lemma 3. *Let S be a z-independent set of nearly equilength links in a fading metric, with uniform power assignment. Then, S is a $\Omega(z^\alpha)$-signal set.*

The two preceding lemmas imply the following result. Lemma 3 shows that when q is sufficiently large, any coloring of $G_q(L)$ gives a SINR-feasible schedule, while Lemma 2 gives a matching lower bound on the optimal solution.

Theorem 1. *Let L be a set of nearly equilength links, and let q be appropriately chosen constant. A coloring of $G_q(L)$ with $O(\Delta(G_q(L)))$ yields a uniform-power schedule that is a constant approximation of PC-Scheduling of L.*

We can handle links of arbitrary lengths by partitioning them into groups, where lengths of links in each group differ by a factor of at most 2. A simple approach is to schedule each group separately using Theorem 1. We can choose an arbitrary fixed power to apply to each length class, or modify the powers within each class up to a constant factor. Thus, we can apply any length-consistent power assignment.

Let $g(L)$ denote the *length diversity* of the link set L, or the number of length groups. Note that $g(L) \leq \log \Lambda$.

Theorem 2. *The PC-Scheduling problem is $O(g(L))$-approximable, using any oblivious power assignment.*

Moscibroda and Wattenhofer [15] showed that uniform and linear power scheduling can be highly suboptimal, and Moscibroda, Oswald and Wattenhofer [14] showed that they can can be as far as n or $\Omega(g(L))$ from optimal.

2.2 Unit Disc Graphs and SINR Scheduling

We can represent the link graph $G_q = G_q(L)$ approximately with a unit-disc graph (UDG). Suppose the links have lengths in the range $[d, 2d)$. Let G'_q be the UDG formed by the points r_v, for $\ell_v \in L$, with radius $\frac{q}{2} \cdot d$. We find that the link graphs and UDGs are closely related, in that pairs of graphs of one type sandwich graphs of the other type.

Lemma 4. *For any $q \geq 1$ and any linkset L, $G'_q \subseteq G_{q+1}$ and $G_q \subseteq G'_{2(q+1)}$.*

Proof. Let v and w be neighbors in G'_q. Then, $d(r_v, r_w) \leq q \cdot d$. Thus, $d_{vw} \leq \ell_v + d(r_v, r_w) \leq (q+1)\ell_v$, and similarly $d_{wv} \leq (q+1)\ell_w$. Hence, $d_{vw} \cdot d_{wv} \leq (q+1)^2 \ell_v \ell_w$, so ℓ_v and ℓ_w are neighbors in G_{q+1}.

On the other hand, suppose we have neighbors ℓ_u and ℓ_w in G_q. Notice that $d(r_u, r_w) \leq d_{uw} + \ell_v \leq d_{uw} + 2d$, and similarly $d(r_u, r_w) \leq d_{wu} + 2d$. Then,

$$(d(r_u, r_w) - 2d)^2 \leq d_{uw} \cdot d_{wu} \leq q^2 \ell_v \ell_w \leq (2qd)^2 .$$

Thus, $d(r_u, r_w) \leq 2(q+1)d$. Hence, ℓ_u and ℓ_w are neighbors in $G'_{2(q+1)}$.

The scheduling problem reduces then, within a constant factor, to the coloring of UDGs. Using Lemma 4, we can now simply find an ordinary graph coloring of the UDG graph $G'_{2(q+1)}$, where q is as in Theorem 1. Any minimal coloring suffices, leading to an easy online algorithm.

Theorem 3. *Applying an algorithm that colors unit disc graphs with $O(\Delta(G))$ colors yields a constant approximation for* PC-Scheduling *on nearly equilength links. In particular, there is an online algorithm that is constant competitive on nearly equilength links and $O(\log \Lambda)$-competitive in general.*

3 Oblivious Power Assignments

We present in this section an scheduling algorithm using a certain oblivious power assignment that achieves a ratio of $O(\log\log \Lambda \cdot \log n)$. In the bidirectional setting, the algorithm obtains an improved $O(\log n)$-ratio, as shown in Sec. 3.1. We also give a construction that shows a $\Omega(\log\log \Lambda)$-separation between the lengths of optimal schedules with or without oblivious power assignments.

We consider the *mean* power assignment (or, square-root assignment [7]) given by $P_v = \ell_v^{\alpha/2}$. The affectance of link ℓ_w on link ℓ_v under mean power assignment is

$$a_w(v) = \frac{P_w/d_{wv}^\alpha}{P_v/\ell_v^\alpha} = \left(\frac{\ell_w}{\ell_v}\right)^{\alpha/2} \left(\frac{\ell_v}{d_{wv}}\right)^\alpha = \left(\frac{\sqrt{\ell_v \ell_w}}{d_{wv}}\right)^\alpha .$$

We say that a set S of links is *well-separated* if any pair of links differ by a factor that is either less than 2 or greater than $8n^{2/\alpha}$. We say that a link ℓ_v and ℓ_w are *τ-close* under mean power assignment if, $\max(a_v(w), a_w(v)) \geq \tau$.

The key observation that we make is that each link affects (or is affected by) very few links that are of widely different length. We can then treat those affectance relationships in a graph-theoretic manner.

Lemma 5. *Let Q be a well-separated SINR-feasible set of links, and let ℓ_v be a link that is shorter than the links in Q (by a factor of at least $n^{2/\alpha}$). Suppose all the links in Q are $\frac{1}{2n}$-close to ℓ_v under mean power assignment. Then, $|Q| = O(\log\log \Lambda)$.*

Proof. Let Q' be a maximum 3^α-signal subset of Q. By Prop. 1, $|Q'| \geq |Q|/9^\alpha$. Q' consists of two types of links: those that affect ℓ_v by at least $\frac{1}{2n}$ under mean power, and those that are affected by ℓ_v by that amount. We shall consider the former type; the argument is nearly identical for the latter type, and will be omitted.

Consider a pair $\ell_w, \ell_{w'}$ in Q' that affect ℓ_v by at least $\frac{1}{2n}$, and suppose without loss of generality that $\ell_w \geq \ell_{w'}$. Thus, $\sqrt{\ell_v \ell_w}^{\alpha} \geq d_{wv}^\alpha \cdot \frac{1}{2n}$, which implies that $d_{wv} \leq \sqrt{\ell_v \ell_w}(2n)^{1/\alpha}$. Similarly, $d_{w'v} \leq \sqrt{\ell_v \ell_{w'}}(2n)^{1/\alpha}$. By the triangular inequality we have that

$$d_{w'w} \leq d(s_{w'}, r_v) + d(r_v, s_w) + d(s_w, r_w) = d_{w'v} + d_{wv} + \ell_w \leq \ell_w + 2^{1+1/\alpha} n^{1/\alpha} \sqrt{\ell_v \ell_w} \leq 3\ell_w,$$

where the last inequality holds because $\ell_w \geq 8n^{2/\alpha}\ell_v$ and $\alpha \geq 1$. Similarly,

$$d_{ww'} \leq d_{vw} + d_{vw'} + \ell_{w'} \leq \ell_{w'} + 2^{1+1/\alpha} n^{1/\alpha} \sqrt{\ell_v \ell_w} \,.$$

Multiplying together, we obtain that

$$d_{w'w} \cdot d_{ww'} \leq 3\ell_{w'}\ell_w + 12 n^{1/\alpha} \sqrt{\ell_v \ell_w} \ell_w \,.$$

However, since Q' is a 3^α-signal set, $d_{w'w} \cdot d_{ww'} \geq 9\ell_w \ell_{w'}$. By combining the last two inequalities and cancelling a $6\ell_w$ factor, we have that

$$\ell_{w'} \leq 2n^{1/\alpha} \sqrt{\ell_v \ell_w} \,. \tag{2}$$

Note that (2) implies that $\ell_w \geq 2\ell_{w'}$, and thus, by well-separation, that $\ell_w \geq 8n^{2/\alpha}\ell_{w'}$.

Label the links in Q' by $\ell_1, \ell_2, \ldots, \ell_t$ in increasing order of length. Equation (2) implies that

$$\ell_{i+1} \geq \frac{\ell_i^2}{\ell_v n^{2/\alpha}} \geq 2\frac{\ell_i^2}{\ell_1},$$

for any $i = 2, 3, \ldots, t$. Thus, if we let $\lambda_i = \ell_i/\ell_1$, we get that $\lambda_{i+1} \geq 2\lambda_i^2$, and by induction that $\lambda_t \geq 2^{2^{t-1}-1}$. Hence, $|Q'| = t \leq \lg\lg \lambda_t + 2 = \lg\lg \Lambda + 2$, and the lemma follows.

We find that links of widely different lengths can be scheduled together rather easily.

Lemma 6. *Let L be a set of links partitioned into length groups $L_1, L_2, \ldots, L_t$ such that links in the same group differ by a factor of at most 2 but links in different groups differ by a factor of at least n^2. Suppose each group L_i has been scheduled with uniform power using Γ_i slots. Then, there is an algorithm that produces a combined schedule of L with mean power assignment using $O(\log\log \Lambda \cdot \max_i \Gamma_i)$ slots.*

Proof. Let $p = O(\log\log \Lambda)$ denote the bound of Lemma 5 on the size of the set Q. First, we transform the schedules of the length groups into f-signal schedules, where $f = 2^{\alpha/2+1}$. By Proposition 1, this stretches each schedule by a factor of at most $(f+1)^2$. Let S_i be some slot in the resulting schedule for L_i, for $i = 1, 2, \ldots, t$. We show that $S = \cup_i S_i$ can be scheduled in $p+1$ slots, resulting in a total of $(p+1) \cdot (f+1)^2 \cdot \max_i \Gamma_i$ slots for L. This yields the claimed result.

Process the links in S in decreasing order of length, and consider a link ℓ_v. By Lemma 7, there are at most p longer links ℓ_w in S that are $\frac{1}{2n}$-close to ℓ_v. Assign ℓ_v to a slot T_j, $j \in \{1, 2, \ldots, p+1\}$, that does not contain a $\frac{1}{2n}$-close link. This completes the specification of the algorithm.

It remains to argue that this assignment yields a SINR-feasible schedule. Consider a link ℓ_v in slot T_j, that originally came from slot S_k. The affectance $a_{S_k \cap T_j}(\ell_v)$ by links of nearly equal length in T_j is at most $1/f$ by the f-signal property. Changing the power assignment in the length group S_k from uniform to mean power assignment increases affectance by at most a factor of $2^{\alpha/2}$, for a total of $2^{\alpha/2}/f = 1/2$. The affectance of all other links (from different length classes) in T_j on ℓ_v is at most $1/(2n)$ each, by construction, or at most $1/2$ in total. Hence, the total affectance is at most one, resulting in a SINR-feasible schedule.

We obtain an algorithm that processes the length groups in decreasing order, greedily assigning the links to the first class in which it affects no link by a non-trivial amount.

Proposition 2. *Suppose there exists a ρ-approximate algorithm for PC-Scheduling on equi-length links that uses uniform power. Then, there exists a $O(\rho \cdot \log\log \Lambda \cdot \log n)$-approximate algorithm for PC-Scheduling that uses mean power assignment.*

Proof. Given a set L of links, divide L into length groups $S_1, S_2, \ldots$, such that $S_i = \{\ell_v \in S | \ell_v \in [2^{i-1}\ell_{min}, 2^i \ell_{min})\}$, where ℓ_{min} denotes the length of the shortest link in S. Then, partition S into classes $B_i = \cup_j S_{i+j \cdot \frac{2}{\alpha} \log n}$, for $i = 1, 2, \ldots, \frac{2}{\alpha} \log n$. The theorem follows from applying Lemma 6 on each class B_i separately.

Using our result of Thm. 3 on equi-length links, we obtain the following result.

Theorem 4. *PC-Scheduling is $O(\log\log \Lambda \cdot \log n)$-approximable in fading metrics.*

For general metrics, we obtain approximations with higher logarithmic factors. Fanghänel et al [7] gave an algorithm using linear power assignment that yields a schedule of length $O(\log \Lambda \log n(\Gamma(L) + \log^2 n))$. On nearly equilength links, this implies a schedule length of $O(\log n(\Gamma(L) + \log^2 n))$. They also gave a simpler algorithm with a ratio of $O(\log^2 n)$.

Corollary 1. *PC-Scheduling is $O(\log\log \Lambda \cdot \log^3 n)$-approximable in general metrics, and within a factor of $O(\log\log \Lambda \cdot \log^2 n)$ when $\Gamma(L) = \Omega(\log^2 n)$.*

We obtain as corollary, a relationship between schedule length and the chromatic number of the link graph.

Corollary 2. *For any link set L, there is a schedule using $O(\log\log \Lambda \cdot \log n)$ $\chi(G_p)$ slots in fading metrics.*

3.1 Bidirectional Scheduling

In the bidirectional variant introduced by Fanghänel et al [6], a stronger separation criteria applies, since communication along each link can occur in either direction. The distance between two links is now the shortest distance between any endpoints of the links. Thus, $d_{uv} = d_{vu} = \min(d(r_v, r_u), d(r_v, s_u), d(s_v, s_u), d(s_v, r_u))$. Other definitions are unchanged.

We can obtain a better approximation ratio for this problem, with essentially the same algorithm, via the following stronger version of Lemma 5.

Lemma 7. *Let S be a set of links inducing an independent set in G_p and let ℓ_v be a link. Then, there is at most one link ℓ_w in S with $\ell_w \geq n^{2/\alpha} \cdot \ell_v$ such that $a_v(w) \geq 1/(2n)$ under mean power assignment and bidirectional measure.*

Proof. Suppose the lemma is false and let $\ell_w, \ell_{w'}$ be two links in S that are longer than $n^{2/\alpha}$ times ℓ_v and affect it by at least $1/(2n)$ each. Suppose without loss of generality that $\ell_w \geq \ell_{w'}$. The assumption of affectance under mean power assignment implies that $\left(\frac{\sqrt{\ell_v \ell_u}}{d_{vu}}\right)^\alpha \geq 1/(2n)$, for $u \in \{w, w'\}$. Thus, $d_{vu} \leq (2n)^{1/\alpha}\sqrt{\ell_v \ell_u}$. In the bidirectional case, $d_{vu} = d_{uv}$. Thus, by the triangular inequality, we have that

$$d_{ww'} \leq d_{wv} + d_{vw'} \leq 2(2n)^{1/\alpha}\sqrt{\ell_v \ell_w} \leq 2(2n)^{1/\alpha}\sqrt{(\ell_{w'}/n^{2/\alpha})\ell_w} \leq 2^{1+1/\alpha}\sqrt{\ell_{w'}\ell_w} \ .$$

Then, ℓ_w and ℓ'_w are not independent in G_p, for $p \geq 2$, which contradicts our assumption.

The rest of the argument is identical to the unidirectional case.

Theorem 5. *Suppose there exists a ρ-approximate algorithm for the bidirectional scheduling problem on equi-length links. Then, there exists a $O(\rho \log n)$-approximate algorithm for the general bidirectional problem. In particular, there is a $O(\log n)$-approximation algorithm for bidirectional scheduling in fading metrics.*

We obtain as corollary a connection between length of bidirectional schedules and the chromatic number of the link graph representation.

Corollary 3. *For any link set L with link graph G_p, there is a bidirectional schedule using $O(\log n)\chi(G_p)$ slots in fading metrics.*

We can show that the bound obtained is best possible for oblivious power functions, up to the logarithmic factor. The following result follows also from the constructions in [7] by analyzing the dependence on Λ.

Theorem 6. *For any length-consistent power function ϕ, there is a SINR-feasible instance for which any schedule under ϕ requires $\Omega(\log\log \Lambda)$ slots.*

Acknowledgement

The author is grateful to Roger Wattenhofer for posing the questions of scheduling complexity in the physical model, as well as for helpful comments and advice.

References

1. Andrews, M., Dinitz, M.: Maximizing capacity in arbitrary wireless networks in the sinr model: Complexity and game theory. In: INFOCOM (April 2009)
2. Avin, C., Lotker, Z., Pignolet, Y.A.: On the power of uniform power: Capacity of wireless networks with bounded resources. In: These proceedings (2009)
3. Chafekar, D., Kumar, V., Marathe, M., Parthasarathy, S., Srinivasan, A.: Cross-layer Latency Minimization for Wireless Networks using SINR Constraints. In: Mobihoc (2007)
4. Clarkson, K.L.: Nearest neighbor queries in metric spaces. Discrete and Computational Geometry 22, 63–93 (1999)
5. ElBatt, T.A., Ephremides, A.: Joint scheduling and power control for wireless ad hoc networks. IEEE Transactions on Wireless Communications 3(1), 74–85 (2004)
6. Fanghänel, A., Keßelheim, T., Räcke, H., Vöcking, B.: Oblivious interference scheduling. In: PODC (August 2009)
7. Fanghänel, A., Keßelheim, T., Vöcking, B.: Improved algorithms for latency minimization in wireless networks. In: Albers, S., Marchetti-Spaccamela, A., Matias, Y., Nikoletseas, S., Thomas, W. (eds.) ICALP 2009. Part II. LNCS, vol. 5556, pp. 447–458. Springer, Heidelberg (2009)
8. Goussevskaia, O., Halldórsson, M.M., Wattenhofer, R., Welzl, E.: Capacity of Arbitrary Wireless Networks. In: INFOCOM (April 2009)
9. Goussevskaia, O., Oswald, Y.A., Wattenhofer, R.: Complexity in Geometric SINR. In: Mobihoc, pp. 100–109 (2007)
10. Gronkvist, J., Hansson, A.: Comparison between graph-based and interference-based STDMA scheduling. In: Mobihoc, pp. 255–258 (2001)
11. Gupta, P., Kumar, P.R.: The Capacity of Wireless Networks. IEEE Trans. Information Theory 46(2), 388–404 (2000)
12. Halldórsson, M.M., Wattenhofer, R.: Wireless Communication is in APX. In: Albers, S., Marchetti-Spaccamela, A., Matias, Y., Nikoletseas, S., Thomas, W. (eds.) ICALP 2009. Part II. LNCS, vol. 5556, pp. 447–458. Springer, Heidelberg (2009)
13. Maheshwari, R., Jain, S., Das, S.R.: A measurement study of interference modeling and scheduling in low-power wireless networks. In: SenSys, pp. 141–154 (2008)
14. Moscibroda, T., Oswald, Y.A., Wattenhofer, R.: How optimal are wireless scheduling protocols? In: INFOCOM, pp. 1433–1441 (2007)
15. Moscibroda, T., Wattenhofer, R.: The Complexity of Connectivity in Wireless Networks. In: INFOCOM (2006)
16. Moscibroda, T., Wattenhofer, R., Weber, Y.: Protocol Design Beyond Graph-Based Models. In: Hotnets (November 2006)
17. Moscibroda, T., Wattenhofer, R., Zollinger, A.: Topology Control meets SINR: The Scheduling Complexity of Arbitrary Topologies. In: Mobihoc (2006)

On the Power of Uniform Power: Capacity of Wireless Networks with Bounded Resources

Chen Avin[1], Zvi Lotker[1], and Yvonne-Anne Pignolet[2]

[1] Ben Gurion University of the Negev, Israel
{avin,zvilo}@cse.bgu.ac.il
[2] ETH Zurich, Switzerland
pignolet@tik.ee.ethz.ch

Abstract. The *throughput capacity* of arbitrary wireless networks under the physical *Signal to Interference Plus Noise Ratio* (SINR) model has received much attention in recent years. In this paper, we investigate the question of how much the worst-case performance of uniform and non-uniform power assignments differ under constraints such as a bound on the area where nodes are distributed or restrictions on the maximum power available. We determine the maximum factor by which a non-uniform power assignment can outperform the uniform case in the SINR model. More precisely, we prove that in one-dimensional settings the capacity of a non-uniform assignment exceeds a uniform assignment by at most a factor of $O(\log L_{\max})$ when the length of the network is $L_{\max}$. In two-dimensional settings, the uniform assignment is at most a factor of $O(\log P_{\max})$ worse than the non-uniform assignment if the maximum power is $P_{\max}$. We provide algorithms that reach this capacity in both cases. Due to lower bound examples in previous work, these results are tight in the sense that there are networks where the lack of power control causes a performance loss in the order of these factors. As a consequence, engineers and researchers may prefer the uniform model due to its simplicity if this degree of performance deterioration is acceptable.

1 Introduction

The great success of wireless networks is mainly due to the fact that any device can exchange information with any other device in its reception range. However, this advantage is also the most problematic characteristic of wireless networks. Since the communication medium is shared by all participants, it is necessary to address the problem of interference. Simultaneous communication attempts cause interference and might even prevent the correct reception of a signal. To tap the full potential of a network, algorithms that coordinate the transmission of messages are necessary. To reach the *throughput capacity* of a network, we have to solve the problem of assigning time slots, frequencies and (depending on hardware capabilities) transmitting power levels to a set of n pairs of wireless transmitters (senders) and receivers distributed in a given area.

A. Fiat and P. Sanders (Eds.): ESA 2009, LNCS 5757, pp. 373–384, 2009.

When attempting to solve this and related problems, we must first choose the appropriate interference model. A standard interference model that captures some of the key characteristics of wireless communication and is sufficiently concise for rigorous reasoning is the physical SINR model [7]. It describes interference as continuous property, decreasing polynomially with the distance from the sender. In this model, a message is received successfully if the ratio between the strength of the sender signal at the receiving location and the sum of interferences created by all other simultaneous senders plus ambient noise is larger than some hardware-defined threshold. The fading speed depends on the value of the so-called path-loss exponent α.

The analysis of problems in the SINR model is intricate, due to the non-binary and accumulative features of interference. Only recently have some theoretical guarantees for SINR-based algorithms been provided. One of problems under scrutiny is the *scheduling problem*. Given a set of n pairs of senders and receivers along with the power level of the transmitters, the goal is to devise a scheduling scheme that minimizes the total number of rounds that will satisfy all the communication requests of every pair. In addition to the timing, the signal strengths of the transmitting nodes greatly influence the performance of wireless networks, since the number of simultaneous transmissions can be increased if the nodes are able to emit signals of different power levels. Thus, power control constitutes an additional aspect of interest. Orthogonally to the scheduling problem, it is necessary to address the *power assignment problem*, i.e., determining a power assignment for each sender of a given set of communication pairs in such a way that the total number of communication requests in one round is maximized. The two problems are often combined, and many algorithms addressing the problem of joint power control and scheduling of a set of links have been devised (see related work section). Since *power control*, i.e., the possibility to assign a different power level to each sender, may play a major role in the complexity of the problems or the performance of the algorithms, we distinguish between two settings: *non-uniform power* (i.e., power control), where each transmitter can transmit with a different power, and *uniform power*, where there is only one power level. It has been demonstrated [12,13] that uniform power has significant performance disadvantages compared to the non-uniform case. However, examples of situations in which power control algorithms outperform uniform energy assignment schemes usually position the nodes in an area of exponential size in the number of nodes and require transmission power levels that differ by a factor exponential in the number of nodes.

On the other hand, a uniform power assignment has several important advantages due to its simplicity. Most importantly, the production cost of wireless devices that always transmit at the same power is lower. Therefore, the uniform power assignment has been widely adopted in practical systems. The lack of power control implies that a device only has to decide whether or not it should send a message at the certain point of time, and not at which power level. As a consequence, there are fewer possibilities to consider which makes reaching a decision much

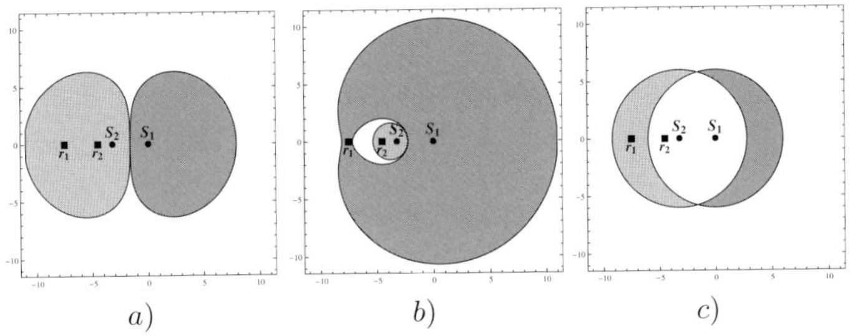

Fig. 1. Impact of the choice of the interference model and power assignment. Given two communication pairs, $l_1 = (s_1, r_1)$ and $l_2 = (s_2, r_2)$, the shaded areas indicate where the signal of a sender can be received (the area in the lighter gray belongs to sender s_2). White areas imply that the received signal power is too weak for reception. *a)* Uniform power: only node r_2 receives a message from its sender, the interference is to high at r_1. *b)* Non-uniform power: both transmissions are successful. *c)* Unit Disk Graph model: neither r_1 nor r_2 receive a message from their corresponding senders.

simpler. Moreover, recently a study of *SINR diagrams*[1] [1] showed that the reception zones of all senders are *convex* for a uniform scheme but not necessarily for non-uniform power assignments. This finding suggests that designing algorithms may be much simpler for uniform networks than for non-uniform networks.

In this paper, we compare the uniform and non-uniform cases and study the trade-off involved between the two, i.e., simplicity vs. performance. As mentioned above, in the absence of restrictions, the performance of the non-uniform model clearly exceeds the uniform model. However, by taking a closer look, we notice that this conclusion is based on examples that involve unbounded resources. Of course, resources are restricted in reality, e.g., the maximum available power for a transmitter or the space where nodes are distributed may be limited. This observation has motivated us to ask the following question:

> *In a resource-constrained setting (i.e., bounded area, bounded power), what is the worst-case performance difference between the uniform and non-uniform case?*

In a nutshell, we show that with bounded resources the two cases are not significantly different; therefore, engineers and researchers may actually prefer the uniform model due to its simplicity.

1.1 Problem Statement and Overview of Our Results

In order to quantify the gap induced by the ability to adjust the transmission power when the available resources are bounded, we consider the following game

[1] The SINR diagram of a set of transmitters divides the plane into $n+1$ *regions* or reception zones, one region for each transmitter that indicates the set of locations in which it can be heard successfully, and one more region that indicates the set of locations in which no sender can be heard. This concept is perhaps analogous to the role played by Voronoi diagrams in computational geometry.

between two players, a *non-uniform (power control) player* and a *uniform player*. The non-uniform player begins by setting up a *configuration* by selecting n communication pairs where each pair consists of a sender and a receiver and their locations. For these pairs the following two conditions must be met:

1. The distance between a sender and its intended receiver is at least one.
2. There exists a power assignment such that the receivers are able to decode the messages of their senders when all senders transmit simultaneously with the same frequency, i.e., the non-uniform configuration is *feasible*.

After the first player has reached its decision, the uniform player can, choose a subset of these pairs that can transmit simultaneously with uniform power, i.e., a feasible uniform configuration. The non-uniform player tries to position the sender/receivers in such a way that the uniform player can pick only a small subset of the available pairs, without causing too much interference. On the other hand, the uniform player tries to select as many pairs as possible.[2]

Let the size of a configuration be the number of pairs in the configuration. Our first result concerns the one-dimensional case. If the non-uniform player can place its configuration on a interval of length at most L_{max}, then we can state the following:

Theorem 1. *For any feasible non-uniform configuration of size n, there is a feasible uniform configuration of a size of at least $\Omega(n/\log L_{max})$ if $\alpha = 2$.*

This result is tight in the sense that from previously known examples [11] there are feasible non-uniform configurations of size n for which the size of any feasible uniform configuration is at most $O(n/\log L_{max})$. Our proof is constructive and we present an algorithm that achieves our bound. This algorithm, combined with previous algorithms, can be used as an approximation scheme for the *uniform scheduling problem* and the *uniform power assignment problem*. In both cases, one can take the output a non-uniform scheduling and/or power assignment algorithm produces, which is basically a non-uniform configuration, and obtain a uniform configuration by "paying" an additional approximation price of $\log L_{\max}$.

In the two-dimensional case, we consider power assignments rendering the non-uniform player's configuration feasible with transmission power levels in the range of $[1, P_{max}]$. In this case we prove:

Theorem 2. *For any feasible non-uniform configuration of size n, there is a feasible uniform configuration of a size of at least $\Omega(n/\log P_{max})$ if $0 < \alpha$.*

This result is tight since there is a non-uniform configuration of size n for which the uniform player can at most select a configuration with a size of at most $O(n/\log P_{max})$. As for Theorem 1, we devise an algorithm that achieves this bound. Note that even if the ratio of the lowest and the highest power level is

[2] Observe that the players are assumed to have unlimited computational power, since the problem of selecting the largest subset of nodes transmitting with fixed power levels has been shown to be NP-hard [6].

constant, it is not immediately obvious that in this case the capacity for the uniform power assignment is in the same order as in setting with power control; namely because there are infinitely many possible power assignments and nested pairs of links are feasible.

It follows from Theorem 1 and Theorem 2 that if we bound either $P_{\max}$ or $L_{\max}$ to n in the one-dimensional case, then a uniform power assignment is at most $\log n$ worse than the best non-uniform power assignment. In the two-dimensional case, this is true only if we bound $P_{\max}$.

The number of links able to transmit simultaneously crucially depends on the path-loss exponent α. The faster the signal strength falls, the smaller an amount of interference is caused. In [15], measurements of indoor and outdoor path-loss exponents at various frequencies are reported, ranging from 1.6 to 6. Most existing work relies on the assumption that $\alpha > 2$, exploiting the fact that in this case the interference of far away nodes can be bounded easily. For $\alpha \leq 2$ the situation changes dramatically and different arguments are necessary. In this paper, the results for two dimension hold for all $\alpha > 0$, the results for one dimension for $\alpha = 2$.

2 Related Work

The study of the capacity of wireless metworks has been initiated by the seminal work of Gupta and Kumar [7]. The authors bounded the throughput capacity in the best-case (i.e., optimal configurations) for the physical models for $\alpha > 2$. More recently, a worst-case view point was adopted [10] by proving lower bounds. We also use this approach in the current paper. The fact that interference is continuous and accumulative as well as the geometric constraints render the scheduling task difficult in the physical model, even if the transmission power of the nodes is fixed. See [5,6,9] for the analysis of such scheduling algorithms. The complexity of connectivity of a uniform power network is examined in [2].

Depending on the hardware, nodes are able to adjust their transmission power. This capability can increase the number of links that are able to transmit successfully at the same time. To exploit this fact, efficient power control algorithms are necessary. For a given set of links, the highest achievable signal to noise ratio can be computed in polynomial time [16], yet the complexity of the problem of joint scheduling and power control in the physical model taking into account the geometry of the problem is unknown. Nevertheless many algorithms and heuristics have been suggested, see [11] for a classification and more detailed discussion of these approaches. Very recent work, [3,4,8] gives upper and lower bounds for power-controlled oblivious scheduling.

Non-uniform power assignment can clearly outperform a uniform assignment [13,12] and increase the capacity of the network, therefore the majority of the work on capacity and scheduling addressed non-uniform power. As we discussed earlier, the study of the uniform case is still worthwhile, due to its simplicity. To the best of our knowledge, the gap between these two models has not been investigated under restricted resources.

The proof of the scheduling algorithm for fixed power levels in [5] can be adapted to conclude that the number of links that are able to communicate concurrently with uniform power is bounded logarithmically in the ratio of the highest and the lowest power, yet the analysis of their algorithm depends on the fact that $\alpha > 2$. In addition, the authors adopt a different viewpoint: Given a set of links, they try to find an approximation of the shortest schedule without power control. In contrast, we are concerned with the lower bound of the size of the largest subset of links that are able to communicate simultaneously with uniform power under the assumption that the original set was feasible with a non-uniform assignment.

Very few papers have been devoted to the one-dimensional case, as the capacity is more restricted than in two dimensions, especially for randomly distributed nodes. Nevertheless, Moscibroda et al. [12,11] showed that the capacity of one-dimensional networks can be linear in the number of the links, at the expense of exponentially long links and and exponentially high power. Hence, we are among the first to study the capacity for uniform and non-uniform power assignments in one dimension.

Another paper published at ESA addresses power control and scheduling in the SINR model. It proposes an oblivious $O(\log n \log\log \Lambda)$-approximation algorithm for the scheduling problem, where Λ is the ratio between the longest and the shortest link length. Moreover it considers the approximation ratio uniform power algorithms can achieve. Using a different tool set from ours, [8] shows that if the Assouad dimension A of the underlying metric is strictly less than α, uniform power assignments are at most a $O(\log(\Lambda))$-factor worse than unconstrained power control. In other words, this more general result works well for $\alpha > A$, while our results for the two-dimensional case hold for any $\alpha > 0$. We believe that the techniques of [8] and the approaches of this paper are complementary and their combination might help to understand the remaining open problems in the SINR interference model.

3 Model and Preliminaries

Let (M, d) be a metric space and $V \subseteq M$ a finite set of $|V|$ *nodes*. A node v_j successfully receives a message from node v_i depending on the set of concurrently transmitting nodes and the applied interference model. In this paper, we adopt the physical SINR model [7], where the successful reception of a transmission depends on the strength of the received signal, the interference caused by nodes transmitting simultaneously, and the ambient noise level. The received power $P_{r_i}(s_i)$ of a signal transmitted by a sender s_i at an intended receiver r_i is $P_{r_i}(s_i) = P(s_i) \cdot g(s_i, r_i)$, where $P(s_i)$ is the transmission power of s_i and $g(s_i, r_i)$ is the propagation attenuation (link gain) modeled as $g(s_i, r_i) = d(s_i, r_i)^{-\alpha}$. The *path-loss exponent* $\alpha \geq 1$ is a constant typically between 1.6 and 6. The exact value of α depends on external conditions of the medium (humidity, obstacles, etc.) and on the exact sender-receiver distance. Measurements for indoor and outdoor path-loss exponents can be found in [15].

Given a sender and a receiver pair $l_i = (s_i, r_i)$, we use the notation $I_{r_i}(s_j) = P_{r_i}(s_j)$ for any other sender s_j concurrent to s_i in order to emphasize that the signal power transmitted by s_j is perceived at r_i as interference. The *total interference* $I_{r_i}(L)$ experienced by a receiver r_i is the sum of the interference power values created by the set L of nodes transmitting simultaneously (except the intending sender s_i), i.e., , $I_{r_i}(L) := \sum_{l_j \in L \setminus \{l_i\}} I_{r_i}(s_j)$. Finally, let N denote the ambient noise power level. Then, r_i receives s_i's transmission if and only if

$$\mathrm{SINR}(l_i) = \frac{P_{r_i}(s_i)}{N + I_{r_i}(L)} = \frac{P(s_i)g(s_i, r_i)}{N + \sum_{j \neq i} P(s_j)g(s_j, r_i)} = \frac{\frac{P(s_i)}{d(s_i, r_i)^\alpha}}{N + \sum_{j \neq i} \frac{P(s_j)}{d(s_j, r_i)^\alpha}} \geq \beta,$$

where $\beta \geq 1$ is the minimum SINR required for a successful message reception. In the sequel we assume $\beta = 1$ we set $N = 0$ and ignore the influence of noise in the calculation of the SINR, for the sake of simplicity. However, this has no significant effect on the results: by scaling the power of all senders, the influence of ambience noise can be made arbitrarily small. Observe that for real scenarios with upper bounds on the maximum transmission power this is not possible, however, for our asymptotic calculations we can neglect this term.

For a uniform power assignment, we say a set of links $L = \{l_1, \ldots, l_n\}$ is a *uniformly feasible configuration of size* n if $P(s_i) = 1$ and $\mathrm{SINR}(l_i) \geq \beta$ for all links $l_i \in L$. If the power level of a device is adjustable, we denote a set L to be a *PC feasible configuration of size* n if there exists a power assignment such that $\mathrm{SINR}(l_i) \geq \beta$ for all links $l_i \in L$.

Zander [16] showed that the *maximum achievable SINR* (denoted SINR^*) for wireless networks can be computed efficiently. Solving the eigenvalue problem for the matrix $Z = \left[\frac{g(s_i, r_j)}{g(s_i, r_i)}\right]$ yields an eigenvalue λ^* for which all elements of the corresponding eigenvector have the same sign. Then, the maximum achievable SINR, is given by $\mathrm{SINR}^* = 1/(\lambda^* - 1)$. Furthermore, the corresponding eigenvector $\mathbf{P}^*$ is a power vector reaching this maximum for all links, i.e., they all have the same SINR level. Since we defined $\beta = 1$, this means that the largest eigenvalue of Z has to be less than 2, otherwise the successful concurrent transmission of all links is impossible.[3]

Theorem 3 (Zander [16]). *A set of senders can transmit simultaneously if the largest eigenvalue of the normalized link gain matrix is less than 2.*

We use the following equations from linear algebra repeatedly.

Theorem 4 (Eigenvalue relationships). *Given an n-by-n matrix with real or complex entries where* $\lambda_1, \ldots, \lambda_n$ *are the (complex and distinct) eigenvalues of* A*, then it holds for* $k \in \mathbb{N}$ *that for the trace of* A^k, $\mathrm{tr}(A^k) = \sum \lambda_i^k$. *In contrast, the determinant of* A *is the product of its eigenvalues; i.e.,* $\det(A) = \prod \lambda_i$.

4 One Dimension: Length Constraint

In this section, we aim at determining the advantage of power control in one-dimensional settings. We prove that at most a factor of $\log L_{max}$ more links can

[3] [16] ignores the influence of noise. See [14] for an approach that handles noise as well.

be scheduled with power control than without, if the senders and receivers are located on a line of length L_{max}. Moreover, we present an algorithm that given a configuration feasible with power control, selects a subset of these links that can be scheduled with uniform power and contains at least a $1/\log L_{max}$-fraction of the links in the original configuration. As a first step, we show that, even when power is adjustable, two links transmitting concurrently must not cross, otherwise at least one of the receiver cannot decode the message.

Lemma 1 (Crossings). *Two senders s_1 and s_2 cannot transmit successfully at the same time if their respective receiver is closer to the other sender, i.e., if $d(s_1, r_1) > d(s_1, r_2)$ and $d(s_2, r_2) > d(s_2, r_1)$. [Proof in full version, applies Theorem 3]*

Nested pairs however are possible. But, as soon as there are more than two nested pairs, they cannot be too close to each other, since the interference is too high otherwise. More precisely, we can show that no matter how we position three nested sender/receiver pairs in an interval of length three and regardless of the power levels we assign to them, they cannot transmit simultaneously.

Lemma 2 (Nestings). *Three nested communication pairs need at least an interval of two times the shortest link distance, otherwise they cannot transmit simultaneously if $\alpha = 2$.*

Proof. (Sketch) We compute the normalized gain matrix Z for three nested links sending in the same direction. We show that for $\alpha = 2$ the following holds

- $\text{tr}(Z)) = 3$
- $\text{tr}(Z^2) > 5$
- $\det(Z) > 0$

Applying Theorem 4 we can derive three conditions the eigenvalues of Z have to satify and we can show that they conflict with the requirement that the largest eigenvalue has be less than two (otherwise no feasible solution, Theorem 3). Thus there is no configuration with three links transmitting in the same direction simultaneously if the longest link is at most twice as long as the shortes. For the other scenarios, where at least one link transmits in the other direction, $\text{tr}(Z)) = 3$ and $\text{tr}(Z^2))$ exceeds 9 and the arguments carry over.

We know from Lemma 2 that three nested links require an interval of more than three times the shortest link distance. Let the shortest distance be one. If we want to add three additional nested links that include the first three links we need at least an interval of length 3^2, if we neglect the interference from the three inner most links. We can repeat this procedure at most $O(\log L_{max})$ times, before we cover the entire interval length L_{max}. Thus Corollary 1 follows.

Corollary 1 (Nestings). *At most $O(\log L_{max})$ links can be nested on a interval of length L_{max}, otherwise there exists no power assignment allowing them to transmit simultaneously if $\alpha = 2$.*

Apart from crossing and nested links, we need to consider parallel links as well. We can show the following lemma.

Lemma 3. *Let $\zeta(x) = \sum_i^{\infty} i^{-x}$ the Riemann zeta function. Out of m parallel links (no crossings, no nestings) where the length of the longest link is at most twice the length of the shortest link, there are at least $k = \lceil 2^{1/\alpha-1}\zeta(\alpha)^{(1/\alpha)} \rceil$ senders that can transmit successfully at the same time with a uniform power assignment if $\alpha > 1$.*

Since $\zeta(x)$ konverges for all real $x > 1$, we have now all the ingredients necessary to conclude how much the capacity in a uniform power setting suffers from the lack of power control. We rephrase Theorem 1:

Theorem 5. *Consider an interval of length L_{max}. Given a PC-feasible configuration of size n, there exists a uniformly feasible configuration of size at least $\Omega(n/\log L_{max})$ if $\alpha = 2$.*

Proof. Given a set of links, we divide the links into $\log L_{max}$ length classes such that the class k contains the links of length in the interval $[2^{k-1}, 2^k]$. We pick the class containing the largest number of links. At least half of them transmit in the same direction. If there are any nested links, we know thanks to Lemma 2 that there are at most two nested links in the same length class. As a next step we can apply Lemma 3 by picking the first and every $\lceil 2^{1/\alpha-1}\zeta(\alpha)^{(1/\alpha)} \rceil^{th}$ of these links and let them transmit concurrently with a uniform power assignment. Due to this procedure a power control algorithm can schedule at most $O(\log n)$ more links simultaneously and the claim follows. □

5 Power Restriction

Even if the transmission power of a device is adjustable, there is typically a bounded range for the power or, even more restricted, a set of available power levels. In these situations, a uniform assignment can achieve a substantial fraction of the capacity a power control scheme can reach. In the following we examine the difference between the two strategies if the first player can place the nodes in arbitrary positions and use transmission power levels in the range $[1, P_{max}]$. We rephrase Theorem 2:

Theorem 6. *Given a PC-feasible configuration of size n in the two-dimensional Euclidean space, there exists a uniformly feasible configuration of size at least $\Omega(n/\log P_{max})$*

We construct an algorithm that given a PC-feasible set of n links placed in the two-dimensional Euclidean plane returns a uniformly feasible set that contains at least $\Omega(n/\log P_{max})$ links. Before we start with the description of the algorithm we elaborate on the problems that our algorithm faces. Since the senders in the resulting set of links have to be able to transmit with the same power and since there is typically background noise we are forced to increase the power levels (of a selected subset of transmitters) to the same power level. Consequently, the interference raises and we need to push the interference back to the level it was. We can do this by further reducing the number of senders. The main challenge is to keep the number of senders as large as possible. So far we have only considered the senders. In order to ensure that the signal to interference ratio is high enough at the receivers as well, we use the fact that from far away the position of the sender and the receiver almost coincide.

Proof (Sketch). Our algorithm (cf. Algorithm 2 for a description in pseudo code) starts by computing an optimal power assignment for the given set of links. Then, we divide the links into classes of similar power levels, i.e. we build subsets of links where the highest transmission power is at most twice the lowest power. Among these sets, we choose the one of greatest cardinality, T_1. The intuition for this step is that we now have a fairly homogeneous set of links and the difference between a uniform power assignment and arbitrary power levels is negligible. We remove some links of this set T_1 to guarantee that all remaining senders can transmit concurrently. To this end, we start with the longest link l_1 and we first clean an area around its receiver, i.e., we delete links with senders too close to r_1 from the set T_1. We assign l_1 to the candidate set T_2 and repeat this procedure with the remaining links until no links are left. These steps ensure that the interference at s_j and at r_j is roughly the same. Next, we remove more links to guarantee that the remaining set is uniformly feasible. We pick one link l_j of the candidate set T_2 and we partition the plane into six sectors of $60°$ around s_j. In each sector we remove the links whose senders are closest to s_j. Thereafter we add l_i to the set T_3 and recursively repeat the partitioning and removal with the remaining links. Before finally declaring the set of the surviving links to be simultaneously schedulable, we repeat the procedure with T_3, this time considering the links in the opposite sequence. We show now that the configuration S produced by this algorithm is uniformly feasible[4] and at most a factor of $O(1/\log P_{max})$ smaller than the original set L. Line 4 determines the largest set of similar power levels. A set of links where the feasible power levels differ by at most a factor of two is very similar to a uniformly feasible set. In particular, the number of senders in close proximity to a receiver is limited. This can be shown by adapting Lemma 4.2 from [5] to this case.

Lemma 4 (Extension of Lemma 4.2 from [5]). *For any link l_i of a PC-feasible set L the number of senders within distance $c \cdot d(s_i, r_i)$ from the receiver r_i is at most $2c^\alpha/\beta$ if $P_{max}/P_{min} < 2$. [Proof in full version]*

Thus we can conclude that this algorithm runs in polynomial time and the original set size is reduced by a factor of $\mathrm{O}(1/\log P_{max})$ in line 3, by $O(\frac{\beta}{2\mu^\alpha})$ in lines

[4] This algorithm can be generalized to higher dimensions at the expense of a higher constant in its approximation guarantee. The only adjustments affect the lines 11-12 and 16-17, where cones instead of sectors are considered. We omit the explicit treatment of higher dimensional cases to increase the clarity of the arguments and more than three dimension are unlikely to be of practical importance.

4-7 (due to Lemma 4), by $O(\frac{1}{6\nu})$ in lines 8-12 and by $O(\frac{1}{6\nu})$ in lines 13-17. Consequently, the ratio between the input and the output set is $\frac{L}{S} \leq O(\frac{n\beta}{72\mu^\alpha\nu^2\log P_{max}}) \in O\left(\frac{n}{\log P_{max}}\right)$.

It remains to demonstrate that the resulting set is indeed uniformly feasible. By setting the transmission power to two for every sender, the strength of the interference at the receivers is at most doubled. As a consequence we have to reduce the number of simultaneous transmissions such that the interference is halved in order to obtain a uniformly feasible set. Clearing a disk around each receiver and removing close senders diminishes the interference by half. We prove this in two steps. First we show how much the interference experienced at the senders is decreased and then we derive the resulting amount of interference at their respective receivers.

Algorithm 2 2D $\log(p_{max})$-approximation

Require: PC-feasible set $L = \{l_1, \ldots, l_n\}$
Ensure: uniformly feasible set $S \subset L$

1: set $\mu := 1 + 2\alpha$ and $\nu := \left\lceil (2 \cdot \frac{\mu}{\mu-1})^\alpha \right\rceil - 1$
2: determine power assignment for L (Zander)
3: partition L into subsets $L_i := \{l_j | 2^i \leq P(s_j) < 2^{i+1}\}$, $i = 1, \ldots, \log P_{max} - 1$
4: set $T_1 := \arg\max_{0 \leq i < \log(p_{max})} |L_i|, T_2 := \emptyset, T_3 := \emptyset$
5: **repeat**
6: move longest link l_j from T_1 to T_2
7: remove l_i from T_1 if $d(s_i, r_j) < \mu d(s_j, r_j)$
8: **until** $T_1 = \emptyset$
9: **repeat**
10: move longest link l_j from T_2 to T_3
11: partion plane into 6 sectors of $60°$ around s_j
12: in each sector remove the ν links with the closest senders to s_j from T_2
13: **until** $T_2 = \emptyset$
14: **repeat**
15: move shortest link l_j from T_3 to S
16: partion plane into 6 sectors of $60°$ around s_j
17: in each sector remove the ν links with the closest senders to s_j from T_3
18: **until** $T_3 = \emptyset$
19: **return** S;

Lemma 5. *$I_{s_i}(S) < \frac{1}{\nu+1} I_{s_i}(L)$ and $I_{r_i}(S) < (\frac{\mu}{\mu-1})^\alpha I_{s_i}(S)$ for all $l_i \in S$. [Proof in full version]*

As a consequence the algorithm has reduced the interference at the receivers by at least $\Omega(\frac{1}{\nu+1} \cdot \left(\frac{\mu}{\mu-1}\right)^\alpha) = \frac{1}{\lceil 2(\frac{\mu}{\mu-1})^\alpha \rceil} \cdot \left(\frac{\mu}{\mu-1}\right)^\alpha \geq \frac{1}{2} \cdot \left(\frac{\mu - 1}{\mu}\right)^\alpha \cdot \left(\frac{\mu}{\mu-1}\right)^\alpha = \frac{1}{2}$. Therefore all transmitters that survive can transmit at power 2, while their receivers are guaranteed to be able to decode the message successfully. This concludes the proof that there always exists a uniform power $O(\log P_{max})$-approximation of a power control problem. □

6 Conclusion

In this paper we show that for limited resources, e.g., an upper bound on the maximum transmission power or the maximum distance between a sender and a receiver, a uniform power assignment provides a log-approximation for the achievable capacity by a non-uniform power assignment. These results can be understood in two ways. We can design and solve algorithmic problems in the uniform power model instead of the non-uniform power model and lose a log factor in the solution. The result presented in this paper suggest the following methodology for solving algorithmic problems in the non uniform power models.

First solve the same problem in the uniform power model (this task is usually simpler and less general). Use this solution as a guide line for the general case involving power control and try to eliminate the logarithmic factor.

References

1. Avin, C., Emek, Y., Kantor, E., Lotker, Z., Peleg, D., Roditty, L.: SINR Diagrams: Towards Algorithmically Usable SINR Models of Wireless Networks. In: Proc. 28th Symposium on Principles of Distributed Computing, PODC (2009)
2. Avin, C., Lotker, Z., Pasquale, F., Pignolet, Y.A.: A Note on Uniform Power Connectivity in the SINR Model. In: Proc.5th Intl Workshop on Algorithmic Aspects of Wireless Sensor Networks, ALGOSENSOR (2009)
3. Fanghänel, A., Kesselheim, T., Räcke, H., Vöcking, B.: Oblivious Interference Scheduling. In: Proc. 28th Symposium on Principles of Distributed Computing (PODC) (2009)
4. Fanghänel, A., Keßelheim, T., Vöcking, B.: Improved Algorithms for Latency Minimization in Wireless Networks. In: Albers, S., Marchetti-Spaccamela, A., Matias, Y., Nikoletseas, S., Thomas, W. (eds.) ICALP 2009. Part I. LNCS, vol. 5555, pp. 525–536. Springer, Heidelberg (2009)
5. Goussevskaia, O., Halldorsson, M., Wattenhofer, R., Welzl, E.: Capacity of Arbitrary Wireless Networks. In: 28th Annual IEEE Conference on Computer Communications (INFOCOM) (2009)
6. Goussevskaia, O., Oswald, Y.A., Wattenhofer, R.: Complexity in Geometric SINR. In: ACM International Symposium on Mobile Ad Hoc Networking and Computing (MOBIHOC), Montreal, Canada (September 2007)
7. Gupta, P., Kumar, P.R.: The Capacity of Wireless Networks. IEEE Trans. Inf. Theory 46(2), 388–404 (2000)
8. Halldórsson, M.: Wireless Scheduling with Power Control. In: Proc. 17th annual European Symposium on Algorithms (ESA), pp. 368–380 (2009)
9. Halldorsson, M., Wattenhofer, R.: Wireless Communication is in APX. In: Albers, S., Marchetti-Spaccamela, A., Matias, Y., Nikoletseas, S., Thomas, W. (eds.) ICALP 2009. Part I. LNCS, vol. 5555, pp. 525–536. Springer, Heidelberg (2009)
10. Moscibroda, T.: The worst-case capacity of wireless sensor networks. In: Proc. 6th Conference on Information Processing in Sensor Networks, IPSN (2007)
11. Moscibroda, T., Oswald, Y.A., Wattenhofer, R.: How Optimal are Wireless Scheduling Protocols? In: Proc. 26th IEEE Conference on Computer Communications (INFOCOM) (2007)
12. Moscibroda, T., Wattenhofer, R.: The Complexity of Connectivity in Wireless Networks. In: Proc. 25th Conference of the IEEE Computer and Communications Societies, INFOCOM (2006)
13. Moscibroda, T., Wattenhofer, R., Weber, Y.: Protocol Design Beyond Graph-based Models. In: Proc. 5th ACM SIGCOMM Workshop on Hot Topics in Networks, HOTNETS (2006)
14. Pillai, S., Suel, T., Cha, S.: The Perron-Frobenius theorem. IEEE Signal Processing Magazine 22(2), 62–75 (2005)
15. Rappaport, T.: Wireless communications. Prentice-Hall, Englewood Cliffs (2002)
16. Zander, J.: Performance of optimum transmitter power control in cellular radio systems. IEEE Trans. Veh. Technol. 41 (1992)

Approximability of OFDMA Scheduling

Marcel Ochel and Berthold Vöcking

Department of Computer Science, RWTH Aachen University, Germany

Abstract. In this paper, we study the complexity and approximability of the orthogonal frequency-division multiple-access (OFDMA) scheduling problem with uniformly related communication channels.

One is given $n \geq 1$ terminals each coming with a demand $d_i > 0$ and $m \geq n$ communication channels each coming with a cost parameter $p_j > 0$. The channels shall be assigned to the terminals in a way that each channel is mapped to at most one terminal and each terminal receives at least one channel. Additionally, each channel j needs to be assigned a communication rate $r_j > 0$ such that the sum of the rates of the channels mapped to terminal i satisfies at least the demand d_i. Using the Shannon rate-power function, the energy requirement for channel j is assumed to be $p_j(2^{r_j} - 1)$. The objective is to minimize the sum of the energy requirements over all channels.

We prove that the problem is NP-hard and cannot be approximated with approximation factor α, unless P = NP, where $\alpha > 1$ is any polynomial time computable function. We then consider a complementary problem setting in which one is given a threshold on the energy requirement and the objective is to maximize λ such that each terminal receives a rate of at least λd_i. We show that this maximin version of the problem admits a PTAS if all demands are identical and a $\frac{1}{2}$-approximation for general demands.

1 Introduction

Orthogonal frequency division multiplexing (OFDM) has become an increasingly popular technology in the area of broadband communication and its inherent optimization problems were intensively studied by the engineering sciences [8, 11, 5].

We study the problem of downlink OFDMA (orthogonal frequency-division multiple-access), which is a central problem faced by implementing wireless or other modes of broadband communication in practice [6, 2, 7]. In this problem data has to be send from a base station to $n \geq 1$ many *terminals* using a set of $m \geq n$ orthogonal subcarriers also called *channels*. We consider the model of *uniformely related channels*, where each channel comes with an associated *cost* parameter p_j which is the reciprocal value to the signal-to-noise ratio of the channel. By applying the formulas for the Shannon capacity, the energy required to sent data with a rate of r_j over channel j is assumed to be $p_j(2^{r_j} - 1)$. For given data rates d_i for each terminal i the total transmit power has to be minimized.

A. Fiat and P. Sanders (Eds.): ESA 2009, LNCS 5757, pp. 385–396, 2009.

In the general OFDM model a channel can be used for transmissions to more than one terminal at a time, but then interference could be caused, resulting in a possible loss of data. In order to avoid this, the sharing of channels is explicitly forbidden in OFDMA (a composition of OFDM and FDMA, frequency-division multiple-access), so each channel can be assigned to at most one terminal. Some theoretical justification for this can be found in [5].

We formalize the problem of OFDMA scheduling by the following nonlinear program:

Min-Power OFDMA

$$\min \sum_{j=1}^{m} p_j \cdot (2^{r_j} - 1) \tag{1}$$

subject to

$$\forall i \in [n] \quad \sum_{j=1}^{m} x_{ij} r_j \geq d_i \tag{2}$$

$$\forall j \in [m] \quad \sum_{i=1}^{n} x_{ij} \leq 1 \tag{3}$$

$$\forall i \in [n], \forall j \in [m] \quad x_{ij} \in \{0,1\}, \quad r_j \geq 0 \tag{4}$$

A feasible solution (X, r) with the *assignment matrix* $X = (x_{ij})_{i \in [n], j \in [m]}$ and the *rate vector* $r = (r_1, \ldots, r_m)$ will be called a *schedule*.

As in [11] we will also consider a natural complementary variant of the problem. Instead of minimizing the total transmit power satisfying given demands, one strives to maximize the minimal fraction of demand that can be achieved without violating a specified energy bound.

Max-Rate OFDMA

$$\max \lambda \tag{5}$$

subject to

$$\sum_{j=1}^{m} p_j \cdot (2^{r_j} - 1) \leq E \tag{6}$$

$$\forall i \in [n] \quad \sum_{j=1}^{m} x_{ij} r_j \geq \lambda d_i \tag{7}$$

$$\forall j \in [m] \quad \sum_{i=1}^{n} x_{ij} \leq 1 \tag{8}$$

$$\forall i \in [n], \forall j \in [m] \quad x_{ij} \in \{0,1\}, \quad r_j \geq 0 \tag{9}$$

In literature many different variants of OFDMA are covered. A common generalization is to allow more general rate-power functions to replace the one implied by the Shannon capacity. Also independent cost parameters p_{ij} instead of the uniformely related p_j can be introduced to model user dependend signal-to-noise ratios.

There have been numerous attempts to solve OFDMA scheduling, but because of the widely assumed intractability of the problem these were mostly heuristics not guaranteeing any bound on the quality of the found solutions or the running time. Also several optimal superpolynomial time algorithms were developed.

In 2006 a proof for NP-hardness of Min-Power OFDMA was published [9] using a reduction from the Subset-Sum problem. Unfortunately the authors overlooked an exponential blowup in their construction. Nevertheless, the problem is NP-complete, as we will show in Section 2. Moreover, even an approximation of Min-Power OFDMA within any polynomial time computable factor α is not possible.

In contrast to this, we will examine approximability for the case of the Max-Rate variant, which does not suffer from the handicap of an exponential objective function. In Section 3 we develop a polynomial time approximation scheme (PTAS) for the Max-Rate OFDMA problem with uniform demands, that for any given constant ϵ computes a schedule satisfying a minimum rate of at least $(1-\epsilon)\lambda^{\text{opt}}d$. Up to our knowledge no polynomial time approximation algorithm for Max-Rate OFDMA with a theoretical worst-case bound has been known to date. The scheme will be constructed by adapting techniques from the area of makespan scheduling [3]. As a byproduct a PTAS different from the one in [10] will be given for the Santa Claus Scheduling problem on identical machines.

In Section 4 we will then look at the case of Max-Rate OFDMA with general demands which contains the case of slightly more general cost parameters of the form $p_{ij} = p_i \cdot p_j$. As our PTAS does not apply there, we use a different approach and obtain a simple 1/2-approximation algorithm.

2 Hardness of Approximation

We will show approximation-hardness of Min-Power OFDMA for any polynomial computable factor α. This will be done by a reduction from 3-Partition. As an intermediate step we reduce this problem to Min-3π-Partition which is defined below.

Definition 1 (Min-3π-Partition). *Given a set $A = \{a_1, \ldots, a_{3n}\}$ of nonnegative integers. Find a partition P of A into triples such that $\sum_{(x,y,z)\in P} xyz$ is minimal.*

Theorem 1. *Let $\alpha(k)$ be any polynomial time computable function. Then Min-3π-Partition can not be approximated in polynomial time within a factor of $\alpha(|A|)$, unless $P = NP$.*

Proof. Assume there is a polynomial time algorithm which approximates Min-3π-Partition within a factor of α. We will reduce the classic 3-Partition problem to Min-3π-Partition in a gap-introducing way.

Let $B = \{b_1, \ldots, b_{3n}\}$ be an instance of 3-Partition. Since the problem is known to be strongly NP-hard, we can assume the b_i to be unary coded. For every integer b_i we construct an element $a_i = (\alpha n)^{b_i}$. Note that for every a_i the size of the binary representation is bounded by $b_i \cdot \log(\alpha n)$ and thereby polynomial in the size of the 3-Partition instance. Then the Min-3π-Partition instance is given by $A = \{a_1, \ldots, a_{3n}\}$.

If B is a yes-instance of 3-Partition, the cost of the optimal Min-3π-Partition is bounded by $c_{\text{yes}} \leq n \cdot (\alpha n)^s$ where $s := \sum_{i=1}^{3n} b_i/n$. A no-instance B however, would incur a cost of $c_{\text{no}} > (\alpha n)^{s+1} \geq \alpha \cdot c_{\text{yes}}$. Thus the α-approximation algorithm for Min-3π-Partition would solve 3-Partition, which leads to a contradiction (unless P = NP). □

The following lemma will help us to identify a substructure of the OFDMA scheduling problem that basically behaves like Min-3π-Partition.

Definition 2. *For a given set C of channels, a demand d, and a channel $c \in C$ we define the function*

$$\rho_c^{(d)}(C) := \frac{d + \sum_{j \in C} \log p_j}{|C|} - \log p_c \quad .$$

We call a set C of channels ρ-feasible *with respect to d if for all $c \in C$ we have $\rho_c^{(d)}(C) > 0$.*

We say a schedule *uses* a channel c if in this schedule the rate on this particular channel is strictly positive.

Lemma 1. *Consider the subproblem of OFDMA scheduling in which there is only one terminal with demand d and a set $C = \{1, \ldots, m\}$ of channels with $p_{min} > 0$. W.l.o.g. assume $p_1 \leq p_2 \leq \cdots \leq p_m$. Then the subset of channels used by the optimal solution is exactly the set $C' := \{1, \ldots, k\}$, where k is the largest index such that C' is ρ-feasible, and the optimal assignment is given by*

$$r_c^{opt} = \begin{cases} \rho_c^{(d)}(C') & c \in C' \\ 0 & c \in C \setminus C' \end{cases}$$

with a total cost of

$$\text{optcost}^{(d)}(C) := k \cdot \sqrt[k]{2^d \prod_{j \in C'} p_j} - \sum_{j \in C'} p_j \quad .$$

Proof. Clearly the subset of channels used by an optimal solution has the form $C^* = \{1, \ldots, h\}$ for an index $h \geq 1$, otherwise there are channels $c^* \in C^*$ and $\bar{c} \in C \setminus C^*$ with $p_{c^*} > p_{\bar{c}}$ which can be swapped in order to decrease the energy.

For the set $\{1, \ldots, k\}$ we can minimize the function

$$f(r_1, \ldots, r_{k-1}) = \sum_{i=1}^{k-1} p_i(2^{r_i} - 1) + p_k(2^{d - \sum_{i=1}^{k-1} r_i})$$

which is just the relaxed version of the original problem without nonnegativity constraint restricted to the channels $1, \ldots, k$. Note that f is convex and any local minimum is a global minimum. Setting the partial derivatives to zero leads to the solution

$$r_i = (d + \sum_{j=1}^{k} \log p_j)/k - \log p_i = \rho_i^{(d)}(\{1, \ldots, k\})$$

for every $i \in [k]$ including $i = k$ with $r_k = 1 - \sum_{i=1}^{k-1} r_i$.

Since the set $\{1, \ldots, k\}$ is ρ-feasible by definition of k, the above computed optimal solution on this set is also a feasible optimal solution to the original problem restricted to the channels $1, \ldots, k$ implying that $h \leq k$. From the maximality of k follows that any rate vector $\bar{r}$ with $\bar{r}_h > 0$ for some $h > k$ would have a positive partial derivative at $\bar{r}_h$ and thereby could not be optimal.

Knowing that the optimal rates are defined by r_c^{opt} the optimal cost for the terminal can be easily computed. □

Now we show approximation-hardness of Min-Power OFDMA by a gap-preserving reduction from Min-3π-Partition.

Theorem 2. *Min-Power OFDMA with uniform demands can not be approximated within a factor of α, where α is any polynomial time computable funtion in the size of the input.*

Proof. Let $A = \{a_1, \ldots, a_{3n}\}$ be the instance of Min-3π-Partition. For every element of A we will add a channel to our constructed OFDM instance with cost-coefficient $p_i := a_i^3$. The number of terminals is given by n. We choose the demand d to be bigger than $6 \cdot (\log m + \log(p_{\max}/p_{\min}) + \log \alpha)$.

Distributing all channels equally among the terminals incurs a cost smaller than $n \cdot p_{\max} \cdot 2^{d/3}$, so this can be used as an upper bound on the optimal solution. Any solution where there is a terminal which gets at most two channels has cost larger than $2 \cdot p_{\min} \cdot 2^{d/2} - \sum_{j=1}^{m} p_j$ which is more than $\alpha \cdot n \cdot p_{\max} \cdot 2^{d/3}$ for the specified d. Thus any solution which is within a factor of at most α times the optimal solution has the property that every terminal gets exactly three channels.

Together with Lemma 1 it follows, that the optimal cost of a terminal is given by $3 \cdot \sqrt[3]{2^d \cdot p_i \cdot p_j \cdot p_k} - (p_i + p_j + p_k)$ if the terminal gets assigned channels i, j and k. The optimal solution yields a partition P of the channel set into triples such that

$$3 \cdot 2^{d/3} \cdot \sum_{(i,j,k) \in P} \sqrt[3]{p_i \cdot p_j \cdot p_k} - \sum_{i=1}^{m} p_i$$

is minimal. Now consider an α-approximate solution of the instance. The cost of this solution is less than or equal to

$$3 \cdot 2^{d/3} \cdot \left(\alpha \cdot \sum_{(i,j,k) \in P} a_i \cdot a_j \cdot a_k \right) - \alpha \cdot \sum_{i=1}^{n} a_i^3$$

which gives at least an α-approximation for Min-3π-Partition. This proves the theorem. □

3 A PTAS for Max-Rate OFDMA

In this section we will develop a PTAS for the Max-Rate OFDMA problem with uniform demands. Our approach roughly follows [3] for makespan scheduling. We will only consider uniform demands throughout this section unless stated otherwise.

There are some similarities between Max-Rate OFDMA, Makespan scheduling [3,4] and the Santa Claus problem [1,10] also known as the max-min allocation problem with indivisible goods. Consider an optimal solution of Max-Rate OFDMA given by the data rates $r^* = (r_1^*, \ldots, r_m^*)$ and the assignment matrix $X^* = (x_{ij}^*)_{i \in [n], j \in [m]}$. Then X^* corresponds to an optimal machine schedule, which assigns jobs of size $r_1^*, \ldots, r_m^*$ to n machines in order to maximize the minimum processing time. Thus the problem of computing an assignment that maximizes λd under knowledge of the optimal data rates is indeed Santa Claus Scheduling on identical machines.

Unfortunately the 'object sizes' in OFDMA are not known a priori — they are variables that also have to be optimized. This results in some difficulties for the application of known techniques that were developed for machine scheduling. However there is a structure in optimal data rates that can be exploited in order to construct a PTAS.

As a consequence we will first give a polynomial time approximation scheme for the Santa Claus Scheduling problem on identical machines (which differs from the approach in [10]) and then modify it in order to solve the Max-Rate OFDMA problem.

3.1 PTAS for Santa Claus Scheduling

Definition 3 (Santa Claus Scheduling Problem). *Given n identical machines and m jobs with processing times $r_1, \ldots, r_m$. The Santa Claus Scheduling problem is to find an assignment of jobs to machines that maximizes the minimum processing time.*

We can assume that the optimal minimum processing time d is known. Otherwise we can loop the dynamic program below in a binary search for d. For every given constant ϵ the following PTAS finds a schedule with minimum processing time at least $(1-\epsilon)d$.

In analogy to [3] we distinguish between the set of jobs with small size (smaller than $\bar{\epsilon}d$, where $\bar{\epsilon} = \epsilon/2$) and the large ones. The large jobs are rounded down to the nearest multiple of $\bar{\epsilon}^2 d$. Then a dynamic program will search the optimal of all possible schedules of rounded large jobs, taking into account that afterwards the small jobs are filled in by the least loaded (LL) heuristic. Using LL, each small job gets assigned to the machine with smallest processing time at the time of its assignment.

W.l.o.g. let the index set of large jobs be $\{1, \ldots, \bar{m}\}$ (with rounded sizes $\tilde{r}_1, \ldots \tilde{r}_{\bar{m}}$) and the index set of small jobs be $\{\bar{m}, \ldots, m\}$. Note that there are at most $z := (1-\bar{\epsilon})/\bar{\epsilon}^2 + 1$ many possible loads for a machine considering only rounded large jobs. The important information that is kept track of in our dynamic program is, whether a schedule of k_ℓ many machines with load $(\ell-1)\cdot\bar{\epsilon}^2 d$ for $1 \le \ell \le z$ can be extended by scheduling the remaining large jobs $\{i,\dots,\bar{m}\}$ to have minimum processing time at least $(1-2\bar{\epsilon})d$ after filling it up with small jobs using LL. More precisely, the dynamic program computes the entries of a boolean matrix B with index set $[n]^z \times (\bar{m}+1)$ sequentially by the recursive rule

$$b(k_1,\dots,k_z,i) := \max_{\ell\in[z]}\{b(k_1,\dots,k_\ell-1,\dots,k_{\ell+\bar{r}_i}+1,\dots,k_z,i+1) \mid k_\ell > 0\}$$

where $\bar{r}_i := \min\{\tilde{r}_i, z-\ell\}$. The actual schedule can be reconstructed by recording the corresponding assignments in a second matrix.

We will see that if $b(n,0,\dots,0,1) = 1$, then a feasible schedule satisfying a minimum rate of $(1-\epsilon)d$ is found.

Let $V_s := \sum_{j=\bar{m}}^{m} r_j$ denote the total processing time of all small jobs, $V_k := \sum_{\ell=1}^{z} k_\ell\cdot(\ell-1)\bar{\epsilon}^2 d$ denote the total load of all machines according to $k_1,\dots,k_z$, and let $V_g := n(1-\bar{\epsilon})d - V_k$ be the corresponding total gap that needs to be filled to reach $(1-\bar{\epsilon})d$. Then the starting entries $b(k_1,\dots,k_z,\bar{m}+1)$ for the dynamic program are defined by

$$b(k_1,\dots,k_z,\bar{m}+1) := \begin{cases} 1 & \text{if } V_s \ge V_g, \\ 0 & \text{otherwise,} \end{cases}$$

for all $k_i \in [z]$, $1 \le i \le n$. This is based on an observation formalized by the following lemma.

Lemma 2. *Consider a partial schedule for the Santa Claus problem where each machine $i \in [n]$ has load $d - g_i$ ($g_i \ge 0$). Let the volume of unassigned jobs be at least as big as $\sum_{i=1}^{n} g_i$ and let $r_{\max}$ be the largest yet unassigned job. Then LL computes a solution where each machine $i \in [n]$ has a processing time larger than $d - r_{\max}$.*

Proof. We will investigate the dynamic sets X_i of assigned jobs for each machine $i \in [n]$ during the execution of the LL heuristic. At every step LL chooses the machine where the gap $g_i := d - \sum_{j\in X_i} r_j$ is largest. We define the volume of the remaining gaps to be $V_g := \sum_{i=1}^{n} \max\{0, g_i\}$. We will show that in every step of the algorithm the invariant $V_r \ge V_g$ holds unless $\max_i g_i < r_{\max}$.

Obviously at the beginning of the algorithm this is true by assumption. Now consider a step where $\max_i g_i \ge r_{\max}$. Then by assigning a job with size r to a machine, the volume V_r is decreased by r but also V_g since $\max_i g_i \ge r_{\max}$.

If $\max_i g_i < r_{\max}$ or $V_g = 0$, we already have an assignment with processing time greater than $d - r_{\max}$ on each machine $i \in [n]$. Otherwise $V_r \ge V_g > 0$ holds and LL can continue assigning jobs. □

Theorem 3. *If all large jobs are rounded down to the nearest smaller multiple of $\bar{\epsilon}^2 d$, then there is a schedule of large jobs that can be filled by small jobs applying the LL-heuristic to yield a minimum processing time of at least $(1-\epsilon)d$.*

Proof. Consider an optimal schedule that by assumption has minimum processing time d. There can be at most $1/\bar{\epsilon}$ many large jobs on a machine. Each looses at most $\bar{\epsilon}^2 d$ size by rounding. Thus the rounded solution still guarantees a minimum processing time of at least $(1-\bar{\epsilon})d$. By removing all small jobs from this schedule we introduce gaps that can further decrease the minimum processing time. Lemma 2 ensures that these gaps can be covered by LL loosing at most an additive term of size $\bar{\epsilon}d$. □

Since the stated dynamic program tries all possible assignments of rounded large jobs eventually the right schedule is found.

3.2 PTAS for Max-Rate OFDMA with Uniform Demands

It is not straightforward to apply the above PTAS to the Max-Rate OFDMA problem as the optimal data rates corresponding to the fixed job sizes are unknown. Nevertheless, the construction was done in a way that allows us to compensate for this lack of knowledge with a few modifications.

The first obstacle we have to overcome is the division into 'large' and 'small' data rates without knowing their actual size in the optimal solution. We exploit a property described in the following lemma.

Lemma 3. *Consider an optimal rate vector $r = (r_1, \ldots, r_m)$ for the Max-Rate OFDMA problem. Pick any two channels $i, j \in [m]$. If $r_i > r_j$, then $p_i \le p_j$.*

Proof. Assume $r_i > r_j$ and $p_i > p_j$. Swapping r_i with r_j and the respective terminal assignments does not change the data rates on these two terminals but the energy cost of the modified allocation decreases by $p_i(2^{r_i}-1)+p_j(2^{r_j}-1))-p_i(2^{r_j}-1)+p_j(2^{r_i}-1)$. Now the terminal with the minimal data rate can increase its rate by an ϵ without violating E contradicting optimality. □

As a result, the large jobs correspond to data rates that can be achieved on cheap channels while the small jobs are related to the costly channels. All we need to do in order to find the right partition is to try all possibilities. W.l.o.g we assume $p_1 \le \cdots \le p_m$, so each partition is given by a split index $\bar{m}$ that separates costly from cheap channels, thus large from small data rates. The exhaustive search for the optimal partition can be done in linear time.

Like before, the minimal rate d^* that can be achieved for every machine in the optimal solution can be found by a binary search containing the exhaustive search for $\bar{m}$ as an outer loop for the following dynamic program.

Our dynamic program searches for the minimum amount $e(k_1, \ldots, k_z, i)$ of energy that is needed to schedule the as yet unassigned cheap channels $\{i, \ldots, \bar{m}\}$ on the partial schedule implied by $k_1, \ldots, k_z$ while ensuring that the total rate on the costly channels is big enough to cover the total gap $V_g = n(1-\bar{\epsilon})d^* - V_k$ left to reach a minimum rate of $(1-\bar{\epsilon})d^*$ on every terminal.

Therefore $e(k_1, \ldots, k_z, i)$ is computed by

$$\min_{\ell \in [z], 1/\bar{\epsilon} \le t \le z-\ell} \{p_i(2^{t \cdot \bar{\epsilon}^2 d^*}-1)+e(k_1, \ldots, k_\ell-1, \ldots, k_{\ell+t}+1, \ldots, k_z, i+1) \mid k_\ell > 0\}$$

where the starting elements $e(k_1, \ldots, k_z, \bar{m}+1)$ indicate the energy needed to fill the total remaining gap with small data rates on the costly channels and therefore are defined as the optimal value of the convex program Δ minimizing $\sum_{j=\bar{m}+1}^{m} p_j \cdot (2^{r_j} - 1)$ subject to $\sum_{j=\bar{m}+1}^{m} r_j \geq V_g$ and $0 \leq r_j \leq \bar{\epsilon} d^*$ for all $j \in [m]$.

In Δ a slightly modified Min-Power OFDMA problem has to be solved. Note that the given formulation does not use any assignment variables x_{ij} as the channel assignment will be done later by the LL heuristic once the optimal data rates have been computed. Therefore the upper bound on the data rates ensures that the condition of Lemma 2 is satisfied. It is easy to see that the simple analytic approach from Lemma 1 can be used to efficiently compute the *exact* optimal solution of Δ. Just omit the upper bound on the data rates. If the data rate of the cheapest costly channel exceeds the upper bound it is set to $\bar{\epsilon} d^*$ and the problem is solved recursively for the decreased demand with the remaining channels. Since by this approach the directional derivatives of the objective funtion in every feasible direction is nonnegative the computed solution is optimal. Although this optimality only refers to the formulation given by Δ and not to the original OFDMA problem with rounded cheap channels we can distribute these data rates using LL while the proof of the approximation factor still works.

After execution of the dynamic program it is checked if the final entry $e(n, 0, \ldots, 0, 1)$ satisfies the energy bound E. Then a feasible schedule guaranteeing a minimum data rate of $(1-\epsilon)d^*$ can be efficiently constructed. The required backtracking information has to be stored independently during the execution of the dynamic program.

Theorem 4. *The above approximation scheme computes a feasible* $(1-\epsilon)$*-approximation of the optimal Max-Rate OFDMA solution.*

Proof. As argued before, we can efficiently find d^* and $\bar{m}$ and thus assume them to be given. Note that if the dynamic program finds a schedule, it always has a minimum data rate of $(1-\epsilon)d^*$. Also, after the execution of the dynamic program only feasible solutions are accepted. So we only have to show that the dynamic program always finds a solution if it is possible to reach a minimum data rate of d^*.

Consider any feasible schedule S with minimum data rate d^* and split index $\bar{m}$. As in Theorem 3 we know that there is a schedule with large data rates beeing multiples of $\bar{\epsilon}^2 d^*$ and small data rates with total size at least as big as the total remaining gap in order to reach $(1-\bar{\epsilon})d^*$. Moreover using Lemma 2 we know that independent of the actual small data rates on the costly channels we can always schedule them by LL to reach a minimum data rate of $(1-2\bar{\epsilon})d^*$, so we can use the possibly different rates computed by the convex program Δ. Note that the optimal rates computed by Δ do not need more energy than is used by the small rates on costly channels in schedule S — since S satisfies all the constraints of the convex program. Only the data rates on cheap channels could violate the energy constraint, but because we try all possible rounded rates, eventually we find a solution in which the energy bound holds. □

If we consider d^* and $\bar{m}$ to be given, the running time of the above algorithm can be bounded by $O(\epsilon^{-4} \cdot n^{(\epsilon^{-2})} \cdot m)$. The exhaustive search for the split index causes an extra factor of $O(m)$ and the number of iterations of the binary search for d^* is linear in the input size and logarithmic in ϵ^{-1} because d^* is bounded by $\log(E/(m \cdot p_{\min}) + 1) \cdot m/n$.

4 A 1/2-Approximation-Algorithm for OFDMA with Non-uniform Demands

The PTAS developed in the last section fails when it comes to non-uniform demands because in this case the notion of small and large jobs differs for each terminal. However we can give a fast and simple structured approximation algorithm which achieves an approximation factor of 1/2. For this result we need the optimal fraction λ^* to be known. We treat the case of unknown λ^* afterwards.

Definition 4 (ordering constraint). *Assume $p_1 \leq \cdots \leq p_m$ and $d_1 \geq \cdots \geq d_n$. Let X_i denote the set of channels assigned to terminal i. Then we say that an assignment $X = (X_1, \ldots, X_n)$ satisfies the* ordering constraint *if for all terminals i, all channels $c \in X_i$ and $c' \in \bigcup_{j=i+1}^{n} X_j$ it holds that $c < c'$.*

A schedule with an assignment satisfying the ordering constraint is said to be *ordered.*

Theorem 5. *If there exists a feasible schedule S with a minimum rate of $\lambda^* d_i$ on each terminal i, then there also is a feasible ordered schedule S' that guarantees at least a rate of $\lambda^* d_i/2$ on each terminal i.*

Proof. We use the data rates $r_1 \geq \cdots \geq r_m$ determined by the schedule S to build a modified schedule S' that satisfies the ordering constraint. W.l.o.g. we assume $\lambda^* = 1$. Depending on the dynamic sets $X_1, \ldots, X_n$ that will be updated at every step of our construction, we define the sets $Y_1 \supseteq \cdots \supseteq Y_n$ by

$$Y_i := \{c \in [m] \setminus \bigcup_{j=1}^{i-1} X_j \mid r_c \leq d_i\} \quad .$$

We consider all sets to be ordered sets according to the natural ordering of the contained numbers. The collection of the first q elements of a set Y will be called a *prefix* of Y (of cardinality q).

Our construction starts with $X_1 := \cdots := X_n := 0$. We will show that the invariants $\sum_{c \in Y_i} r_c \geq \sum_{j=i}^{n} d_j$ and $c < c'$ for all $c \in X_i$ and $c' \in \bigcup_{j=i+1}^{n} Y_j$ hold for every $i \in [n]$ at every step. This is trivially true at the beginning.

At step t we assign to terminal t the maximal prefix M of Y_t with a total data rate not bigger than d_t. Note that by induction assumption $\sum_{c \in Y_t} r_c \geq d_t$ and because of its maximality, M has a rate of at least $d_t/2$. The prefix property guarantees the invariant $c < c'$ for all $c \in X_t$ and $c' \in \bigcup_{j=t+1}^{n} Y_j$. Moreover since no elements are added to any set except from X_t this invariant remains satisfied for all other sets which proves that the ordering constraint holds at every step.

Let Y_i^b denote the dynamic set Y_i before execution of the step k and Y_i^a after. To show $\sum_{c \in Y_i^a} r_c \geq \sum_{j=i}^{n} d_j$ we distinguish between two cases. The first case is trivial: if $M \subseteq Y_t^b \setminus Y_{t+1}^b$ then the Y-sets are not changed and the invariant follows by the induction assumption. Otherwise $Y_{t+1}^b \supseteq Y_t^b \setminus M$ is true resulting in $Y_{t+1}^a = Y_{t+1}^b \setminus M = Y_t^b \setminus M$. We pick the smallest $k > t$ with $Y_k^a \setminus Y_{k+1}^a \neq \emptyset$. By the prefix property only the sets $Y_{t+1}, \ldots, Y_k$ could have been affected by the assignment step t. Using $Y_k^a = Y_{t+1}^a = Y_t^b \setminus M$ we get $\sum_{c \in Y_k^a} r_c \geq \sum_{j=t}^{n} d_j - d_t = \sum_{j=t+1}^{n} d_j$ thus proving the invariant and thereby the theorem. □

The following algorithm always computes an optimal ordered min-power OFDMA schedule satisfying a data rate of exactly $\lambda^* d_i/2$ for known λ^* and runs in time $O(m^2 n)$.

Algorithm Opt-Ordered: Let $p_1 \leq \cdots \leq p_m$. Set $\bar{\lambda} = \lambda^*/2$. Compute the matrix $F = (f(i,j))_{i \in [n], j \in [m+1]}$ according to

$$\begin{aligned} f(i,j) &:= \min_{1 \leq k \leq m-j} \{ f(i+1, j+k) + \mathrm{optcost}^{(\bar{\lambda} d_i)}(\{j, \ldots, j+k-1\}) \} \\ f(i, m+1) &:= \infty \\ f(n,j) &:= \mathrm{optcost}^{(\bar{\lambda} d_n)}(\{j, \ldots, m\}) \end{aligned}$$

where $\mathrm{optcost}^{(d)}\{C\}$ is computed according to Lemma 1 and indicates the energy value of an optimal Min-Power schedule of the single terminal variant with demand d and set of allowed channels C. If $f(1,1) \leq E$, then a feasible schedule which guarantees a data rate of at least $\bar{\lambda} d_i$ for every terminal i can be found by backtracking the corresponding assignments.

If the optimal λ^* is not known, Opt-Ordered can be looped in a binary search for the maximal $\bar{\lambda}$ that allows a feasible ordered schedule. By Theorem 5 this $\bar{\lambda}$ is at least $\lambda^*/2$. When the maximal value $\bar{\lambda}$ is determined up to an accuracy of 2ϵ, an approximate solution with guaranteed rate λd_i for each terminal i with λ at least $(1-2\epsilon) \cdot \bar{\lambda} = (1/2 - \epsilon)\lambda^*$ is found. This binary search can be done in time $O(\log(m/(n \cdot \epsilon)) + \log\log(E/(m \cdot p_{\min})))$ which leads to an algorithm that runs in polynomial time even when ϵ is not considered to be constant but is given as a part of the input.

5 Discussion

We examined the problem of OFDMA scheduling and gave a proof for its hardness showing inapproximability for the variant where the objective is to minimize the overall energy. However, this inapproximability result does not carry over to the Max-Rate formulation, where the minimal fraction of demand that can be satisfied while not violating a given energy bound has to be maximized. We gave a polynomial time $(1/2-\epsilon)$-approximation algorithm for this problem in the general case and derived a PTAS for the case of uniform demands.

It is an open question if for the given problem even a *fully* polynomial time approximation scheme could be obtained or if instead a hardness proof showing *strong* NP-hardness could be constructed. Because of the similarities to Makespan Scheduling and the Santa Claus problem, we conjecture the latter. Also, it is left open for further research whether the given PTAS could be extended to the case of general demands using known techniques from makespan scheduling for machines with different speeds [4]. Moreover, considering the generalization of unrelated channel costs should provide further interesting problems.

References

1. Bansal, N., Sviridenko, M.: The santa claus problem. In: STOC 2006: Proceedings of the thirty-eighth annual ACM symposium on Theory of computing, pp. 31–40. ACM, New York (2006)
2. Feiten, A., Mathar, R., Reyer, M.: Rate and power allocation for multiuser ofdm: An effective heuristic verified by branch-and-bound. IEEE Transactions on Wireless Communications 7(1), 60–64 (2008)
3. Hochbaum, D.S., Shmoys, D.B.: Using dual approximation algorithms for scheduling problems: Theoretical and practical results. J. ACM 34(1), 144–162 (1987)
4. Hochbaum, D.S., Shmoys, D.B.: A polynomial approximation scheme for scheduling on uniform processors: Using the dual approximation approach. SIAM Journal on Computing 17(3), 539–551 (1988)
5. Jang, J., Lee, K.B.: Transmit power adaptation for multiuser ofdm systems. IEEE Journal on Selected Areas in Communications 21(2), 171–178 (2003)
6. Kim, I., Park, I.-S., Lee, Y.H.: Use of linear programming for dynamic subcarrier and bit allocation in multiuser ofdm. IEEE Transactions on Vehicular Technology 55(4), 1195–1207 (2006)
7. Kivanc, D., Li, G., Liu, H.: Computationally efficient bandwidth allocation and power control for ofdma. IEEE Transactions on Wireless Communications 2(6), 1150–1158 (2003)
8. Seong, K., Yu, D.D., Kim, Y., Cioffi, J.M.: Optimal resource allocation via geometric programming for ofdm broadcast and multiple access channels. In: GLOBECOM (2006)
9. Vemulapalli, M., Dasgupta, S.: Np-hardness of bit allocation in multiuser multicarrier communications. In: Proceedings of EUSIPCO 2006 (2006)
10. Woeginger, G.J.: A polynomial-time approximation scheme for maximizing the minimum machine completion time. Operations Research Letters 20(4), 149–154 (1997)
11. Wong, C.Y., Cheng, R.S., Lataief, K.B., Murch, R.D.: Multiuser ofdm with adaptive subcarrier, bit, and power allocation. IEEE Journal on Selected Areas in Communications 17(10), 1747–1758 (1999)

Maximum Flow in Directed Planar Graphs with Vertex Capacities*

Haim Kaplan and Yahav Nussbaum

The Blavatnik School of Computer Science,
Tel Aviv University, 69978 Tel Aviv, Israel
{haimk,yahav.nussbaum}@cs.tau.ac.il

Abstract. In this paper we present an $O(n \log n)$ algorithm for finding a maximum flow in a directed planar graph, where the vertices are subject to capacity constraints, in addition to the arcs. If the source and the sink are on the same face, then our algorithm can be implemented in $O(n)$ time.

For general (not planar) graphs, vertex capacities do not make the maximum flow problem more difficult, as there is a simple reduction that eliminates vertex capacities. However, this reduction does not preserve the planarity of the graph. The essence of our algorithm is a different reduction that does preserve the planarity, and can be implemented in linear time. For the special case of undirected planar graph, an algorithm with the same time complexity was recently claimed, but we show that it has a flaw.

1 Introduction

The problem of finding a maximum flow in a graph, or in a network, is a well-studied problem with applications in many fields, see the book of Ahuja, Magnanti and Orlin [1] for a survey. The maximum flow problem is also interesting if we restrict it to *planar* graphs, which are graphs that have an embedding in the plane without crossing edges. The case of planar graphs appears in many applications of the problem, for example road traffic or VLSI design. The special structure of planar graphs allows us to get simpler and more efficient algorithms for the maximum flow and related problems.

In the maximum flow problem, usually the arcs of the graph have capacities which limit the amount of flow that may go through each arc. We study a version of the problem in which the vertices of the graph also have capacities, which limit the amount of flow that may enter each vertex. This version appears for example when computing vertex disjoint paths in graphs, and in other problems where the vertices model objects which have a capacity.

Ford and Fulkerson [3, Chapter I.11] studied this version of the problem. They suggested the following simple reduction to eliminate vertex capacities.

* This research was partially supported by the United States - Israel Binational Science Foundation, project number 2006204 and by the Laura Schwarz-Kipp Institute of Computer Networks.

A. Fiat and P. Sanders (Eds.): ESA 2009, LNCS 5757, pp. 397–407, 2009.

We replace every vertex v with a finite capacity c by two vertices v' and v''. The arcs that were directed into v now enter v', and the arcs that were directed out of v now leave v''. We also add a new arc with capacity c from v' to v''. Unfortunately, this reduction does not preserve the planarity of the graph [7]. Consider for example the complete graph of four vertices, where one of the vertices has finite capacity. This graph is planar. If we apply the construction of Ford and Fulkerson we get a graph whose underlying undirected graph is the compete graph with 5 vertices. This graph is not planar by Kuratowski's Theorem.

The most efficient algorithm for maximum flow in directed planar graphs without vertex capacities, to date, was given by Borradaile and Klein [2] (Weihe [11] gave an algorithm with the same time bound but assuming a certain connectivity condition on the graph). Their paper also contains a survey of the history of the maximum flow problem on planar graphs. The time bound of the algorithm of [2] is $O(n \log n)$ where n is the number of vertices in the input graph. Borradaile and Klein ask whether their algorithm can be generalized to the case where the flow is subject to vertex capacities.

A planar graph is a *st-planar* graph if the source and the sink are on the same face. Hassin [4] gave an algorithm for the maximum flow problem in directed *st*-planar graphs without vertex capacities. The bottleneck of the algorithm is the computation of single-source shortest-path distances, which takes $O(n)$ time in a planar graph, using the algorithm of Henzinger et al. [5].

Khuller and Naor [7] were the first to study the problem of maximum flow with vertex capacities in planar graphs. They gave various results, including an $O(n\sqrt{\log n})$ time algorithm for finding the value of the maximum flow in *st*-planar graphs (which can be improved to $O(n)$ time using the algorithm of [5]), an $O(n \log n)$ time algorithm for finding the maximum flow in *st*-planar graphs, an $O(n \log n)$ time algorithm for finding the value of the maximum flow in undirected planar graphs, and an $O(n^{1.5} \log n)$ time algorithm for the same problem on directed planar graphs. If all vertices have unit capacities, then we get the *vertex-disjoint paths problem.* Ripphausen-Lipa et al. [10] solved this problem in $O(n)$ time for undirected planar graphs.

Recently, Zhang, Liang and Chen [13], used a construction similar to the one of [7] to obtain a maximum flow for undirected planar graphs with vertex capacities that runs in $O(n \log n)$ time. Their algorithm first constructs a planar graph without vertex capacities, and then uses the algorithm of [2] to find a maximum flow in it, which is modified in $O(n \log n)$ time to a flow in the original graph with vertex capacities. They also gave a different $O(n)$ time algorithm for finding a maximum flow in undirected *st*-planar graphs. Zhang et al. also ask in their paper if there is an algorithm that solves the problem for directed planar graphs.

In this paper we answer [2] and [13], and show a linear time reduction of the problem of finding maximum flow in directed planar graphs with arc and vertex capacities, to the problem of finding maximum flow in directed graphs with only arc capacities. This problem is more general than the one for undirected planar graphs, since an undirected planar graph can be viewed as a special case of a directed planar graph, in which there are two opposite arcs between any pair of adjacent vertices.

We show how to apply the constructions of [7] and [13] to directed planar graphs. Given directed planar graph, we construct another directed planar graph without vertex capacities, such that we can transform a maximum flow in the new graph back to a maximum flow in the original graph. Since the new graph does not have vertex capacities we can find a maximum flow in it using the algorithm of [2] (or of [4] if it is an *st*-planar graph). The time bound of our reduction is linear in the size of the graph, therefore we show that vertex capacities do not increase the time complexity of the maximum flow problem also for planar graphs.

In addition, we show that the algorithm of [13] unfortunately has a flaw. We give an undirected graph in which this algorithm does not find a correct maximum flow. Therefore, in fact our algorithm is also the first to solve the problem for undirected graphs.

The outline of the paper is as follows: In the next section we give some background and terminology. In Sect. 3 we describe the construction of [7] that we use, and in Sect. 4 we describe the one of [13]. In Sect. 5 we characterize when a maximum flow in the constructed graph induces a maximum flow in the original graph and show how to efficiently find such a flow in the constructed graph. Finally, in the last section we combine all the pieces together to get our algorithm.

2 Preliminaries

We consider a simple directed planar graph $G = (V, E)$, where V is the set of vertices and E is the set of arcs, with a given planar embedding. The planar embedding of the graph G is represented combinatorially, see [9] for survey on planar graphs. An arc $e = (u, v) \in E$ is directed from $u \in V$ to $v \in V$. We denote the number of vertices by n, since the graph is planar we have $|E| = O(n)$.

A *path* $P = (e_0, e_1, \ldots, e_{k-1})$ is a sequence of arcs $e_i = (u_i, v_i)$ such that for $0 \leq i < k - 1$ we have $v_i = u_{i+1}$. If in addition $v_{k-1} = u_0$ then P is a *cycle*. We say that a path P *contains* a vertex v, if either (u, v) or (v, u) is in P, for some vertex u. The path $P = (e_0, e_1, \ldots, e_{k-1})$ *starts* at u_0 and *ends* at v_{k-1}. For $v \in V$, $in(v) = \{(u, v) \mid (u, v) \in E\}$ is the set of incoming arcs and $out(v) = \{(v, u) \mid (v, u) \in E\}$ is the set of outgoing arcs.

The graph G has two distinguished vertices, $s \in V$ is the *source* and $t \in V$ is the *sink*. The source s has no incoming arcs, and the sink t has no outgoing arcs. Every arc $e \in E$, has a capacity $c(e) \geq 0$, and in addition every vertex $v \in V \setminus \{s, t\}$ has a capacity $c(v) \geq 0$. A capacity might be ∞. We assume that the source and the sink have no capacities, if we wish to allow them to have capacities, we can add a vertex s' that will be the source instead of s, and an arc (s', s) with the desired capacity, and similarly add a new sink t', and an arc (t, t') with the desired capacity. Note that this transformation keeps the graph planar, and even *st*-planar if it was so. It is easy to extend the given embedding to accommodate s', t', and the arcs (s', s) and (t, t'). A graph without vertex capacities can be viewed as a special case in which $c(v) = \infty$ for every vertex.

A function $f : E \rightarrow R$ is a *flow function* if and only if it satisfies the following three constraints:

$$0 \leq f(e) \leq c(e) \quad \forall e \in E \ , \tag{1}$$

$$\sum_{e \in in(v)} f(e) \leq c(v) \quad \forall v \in V \setminus \{s, t\} \ , \tag{2}$$

$$\sum_{e \in in(v)} f(e) = \sum_{e \in out(v)} f(e) \quad \forall v \in V \setminus \{s, t\} \ . \tag{3}$$

Constraints (1) are the *arc capacity constraints*, Constraints (2) are the *vertex capacity constraints* and Constraints (3) are the *flow conservation constraints.*

We say that $e \in in(v)$ *carries flow into* v if $f(e) > 0$, and that $e' \in out(v)$ *carries flow out of* v if $f(e') > 0$.

The *value* of a flow f is $\sum_{e \in in(t)} f(e)$, the amount of flow which enters the sink. If the value of f is 0 then f is a *circulation.* Our goal, in the *maximum flow problem*, is to find a flow function of maximum value.

For a flow function f we define a cycle C to be a *flow-cycle* if $f(e) > 0$ for every arc in C. We extend this definition to every function $f : E \rightarrow R$, even if it is not a flow. If a function f has no flow-cycles we say that f is *acyclic.* An *acyclic flow* is a flow function which is acyclic.

Let $e = (u, v)$ we denote $rev(e) = (v, u)$. For a flow function f, we may assume that f does not contain an arc e such that both $f(e) > 0$ and $f(rev(e)) > 0$, because otherwise the flows in both directions can cancel each other. For a path $P = (e_0, e_1, \ldots, e_{k-1})$ we let $rev(P) = (rev(e_{k-1}), rev(e_{k-2}), \ldots, rev(e_0))$.

The planar embedding of G partitions the plane into connected regions called *faces*. For a simpler description of our algorithm, we fix an embedding of G such that t is on the boundary of the infinite face. It is easy to convert any given embedding to such an embedding [9].

The *dual graph* G^* of G has a vertex $D(h)$ for every face h of G, and an arc $D(e)$ for every arc e of G. The arc $D(e)$ connects the two vertices corresponding to the faces incident to e. The arc $D(e)$ is directed from the vertex that corresponds to the face on the left side of e to the one of the face on the right side of e. Intuitively, G^* is obtained from G by turning the arcs clockwise. The dual graph G^* is planar, but it is may have loops or parallel arcs. Every face h of G^* corresponds to a vertex v in G, such that the arcs that bound h are dual to the arcs that are incident to v. See Fig. 1. The capacity of $e \in E$, $c(e)$, is interpreted in G^* as the *length* of $D(e)$.

In the construction we present below we add undirected edges to directed graphs. Each such undirected edge uv can be represented by two antiparallel directed arcs (u, v) and (v, u), with the same capacity. If e is an undirected edge, then $D(e)$ is also undirected.

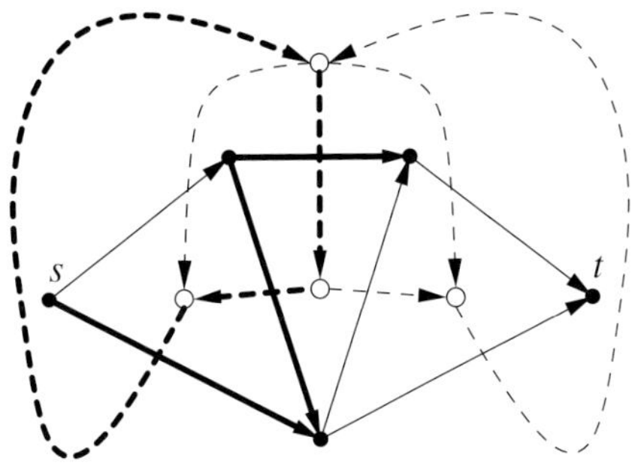

Fig. 1. A planar graph and its dual graph. The vertices of G are dots, and its arcs are solid. The vertices of G^* are circles, and its arcs are dashed. The bold arcs are an arc-cut in G and a cut-cycle in G^*. Capacities are not shown in this figure.

2.1 Residual Cycles

Let f be a flow in G. The *residual capacity* of an arc e with respect to f is defined as $c_r(e) = c(e) - f(e) + f(rev(e))$. In other words, the residual capacity of e is the amount of flow that we can add to e, or reduce from $rev(e)$. The *residual graph* of G with respect to f has the same vertex set and arc set as G, and the capacity for each edge e is $c_r(e)$. A *residual arc* with respect to f is an arc e with a positive residual capacity. A *residual path* is a path made of residual arcs. A *residual cycle* is a cycle made of residual arcs.

Khuller, Naor and Klein [8] presented an algorithm that finds a circulation such that there are no clockwise residual cycles with respect to this circulation, in a directed planar graph. The bottleneck of the algorithm of [8] is the computation of single-source shortest-path distances. Henzinger et al. [5] showed how to find these distances in a planar graph in $O(n)$ time, so the algorithm of [8] can be implemented in the same time bound. The complete details of the algorithm can be found also in [2]. We present an extension of this algorithm that changes a given flow into another flow, with the same value, without clockwise residual cycles with respect to it. We use this algorithm later to get the linear time bound for our reduction. The algorithm of [8] is for directed planar graphs without vertex capacities, so for the rest of this section assume that the graph G does not have vertex capacities.

In this section we also assume that if $e \in E$ then also $rev(e) \in E$. This assumption can be satisfied, without changing the problem, by adding the arc $rev(e)$ with capacity 0 for every arc e such that $rev(e)$ is not in E.

Given a flow f in G, we wish to find a flow f' with the same value, such that there are no clockwise residual cycles with respect to f'.

Let G' be the residual graph of G with respect to f. We find a circulation f_r in G', such that G' does not have clockwise residual cycles with respect to f_r, using the algorithm of [8]. Define f' to be the sum of f and f_r, that is $f'(e) = \max\{0, f(e) + f_r(e) - [f(rev(e)) + f_r(rev(e))]\}$. In other words, we add $f(e)$ and $f_r(e)$, and let the flows on e and $rev(e)$ cancel each other.

The function f' satisfies the two constraints of a flow without vertex capacities. The capacity of an arc e in G' is $c(e) - f(e) + f(rev(e))$ and therefore $f_r(e)$ is smaller than this capacity, therefore $f(e) + f_r(e) - [f(rev(e)) + f_r(rev(e))] \leq c(e)$, so $f'(e) \leq c(e)$ and the arc capacity constraints are satisfied in f'. The conservation constraints are satisfied, because these constraints are satisfied for f and for f_r, and f' is the sum of these two flows.

The value of the flow f' is the sum of the values of f and f_r. Since f_r is a circulation, its value is 0, and so the value of f' is the same as the value of f.

The flow f' has the desired property that G has no clockwise residual cycles with respect to f'. To show that, we show that if C is a clockwise residual cycle in G with respect to f', then C is also a residual cycle in G' with respect to f_r, contrary to the way we find f_r. Let e be an arc of C, and assume for contradiction that e is not residual in G' with respect to f_r. From our assumption $f_r(e) = c_r(e) = c(e) - f(e) + f(rev(e))$ and $f_r(rev(e)) = 0$. Therefore, $f'(e) = c(e)$ and e is not residual in G with respect to f', contradicting the fact that e is a member of C.

Lemma 1. *Let G be a directed planar graph without vertex capacities, and let f be a flow in G. We can find a flow f' in G, with the same value as f, such that there are no clockwise residual cycles with respect to f', in $O(n)$ time.*

3 Minimum Cut

In a graph without vertex capacities, a *cut* S is a minimal subset of E such that every path from s to t contains an arc in S. To avoid ambiguity later, when we introduce cuts that may contain vertices, we call such a cut an *arc-cut*. See Fig. 1. The *value* of an arc-cut S is $\sum_{e \in S} c(e)$. The *minimum cut problem* asks to find an arc-cut of minimum value. The fundamental connection between maximum flow and the minimum cut problems was given by Ford and Fulkerson [3] in the Max-Flow Min-Cut Theorem:

Theorem 2. *[3] The value of the maximum flow (in a graph without vertex capacities) is equal to the value of the minimum arc-cut in the same graph.*

Let C be a cycle in G^*. We say that C is a *cut-cycle* if it separates the faces corresponding to s and t, and goes counterclockwise around s (or equivalently, clockwise around t). See Fig. 1. The length of C is the sum of the lengths of its arcs. Johnson [6] showed the following relation between the value of minimum arc-cut and the value of shortest cut-cycle:

Lemma 3. *[6] Let G be a directed planar graph without vertex capacities. Then the value of the minimum arc-cut of G, is the same as the length of the shortest cut-cycle in G^*.*

Ford and Fulkerson [3, Chapter I.11] extended the definition of cuts to graphs with vertex capacities. In such a graph, a cut S is a minimal subset of $E \cup V$ such that every path from s to t contains an arc or a vertex in S. The value of a cut S is similarly defined as $\sum_{x \in S} c(x)$. Ford and Fulkerson also presented a

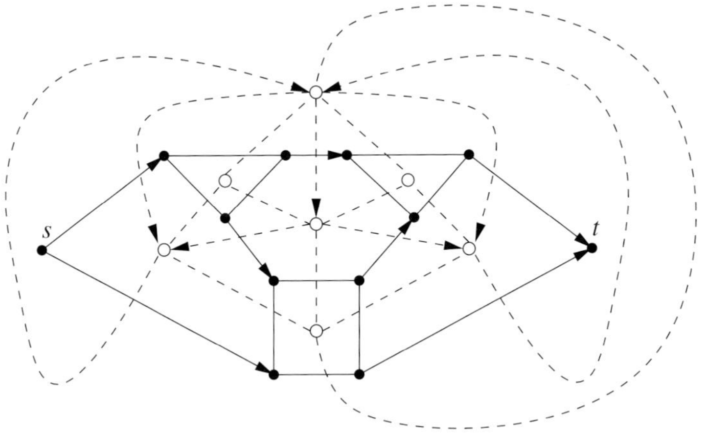

Fig. 2. Construction of G_C and G_E for the graph in Fig. 1. The graph G_C is presented as the dual graph of G_E. The newly added (undirected) edges are without arrowheads. Capacities are not shown in this figure.

version of the Max-Flow Min-Cut Theorem for graphs with vertex capacities, in this case the value of maximal flow (subject to both arc and vertex capacities) is equal to the value of the minimum cut (which contains both arcs and vertices).

Khuller and Naor [7] extended Lemma 3 using a supergraph G_C of G^* which they construct as follows. Let h be a face of G^* that corresponds to a vertex v of G with finite capacity. We add a new vertex v_h inside h and connect it by an (undirected) edge of length $c(v)/2$ to every vertex on the boundary of h. See Fig. 2.

Lemma 4. *[7] The values of the maximum flow and minimum cut in G are equal to the length of the shortest cut-cycle in G_C.*

4 The Extended Graph

Zhang, Liang and Jiang [12] and Zhang, Liang and Chen [13] construct the *extended graph* for an undirected graph with vertex capacities, based on the construction of Khuller and Naor [7]. We use the same construction for directed planar graphs with vertex capacities. The extended graph is defined as follows. We replace every vertex $v \in V$ which has a finite capacity with d vertices $v_0, \cdots, v_{d-1}$, where $d = |in(v)| + |out(v)|$ is the degree of v. We connect every v_i to $v_{(i+1) \bmod d}$ with an (undirected) edge of capacity of $c(v)/2$. We make every arc that was adjacent to v, adjacent to some vertex v_i instead, such that each arc is connected to a different vertex v_i, and the clockwise order of the arcs is preserved. We identify the new arc (u, v_i) or (v_i, u) with the original arc (u, v) or (v, u). We denote the resulting graph by G_E, and the cycle that replaces $v \in V$ in G_E by C_v. The graph G_E is a simple directed planar graph without vertex capacities. The arc set of G_E contains the arc set of G. See Fig. 2.

From the construction of G_E and G_C follows that G_C is the dual of G_E. Let v be a vertex with finite capacity and let h be the corresponding face in G^*. Then, in G_E we replaced v with C_v, and in G_C we placed v_h inside h. The edges which connects v_h to the boundary of h are dual to the edges of C_v.

From Theorem 2 we get that the value of the maximum flow in G_E is the same as the value of minimum arc-cut in G_E. By Lemma 3 this value is the same as the value of the shortest cut-cycle in G_C. Lemma 4 implies that this value equals to the value of the maximum flow in G, so the next lemma follows.

Lemma 5. *The value of the maximum flow of G is equal to the value of the maximum flow of G_E.*

5 Reduction from the Extended Graph to the Original Graph

We denote by f_E a flow function in G_E, and by f the *restriction* of f_E to the arcs of G, that is for every arc e of G, $f(e) = f_E(e)$. The value of f is the same as the value of f_E. The next lemma generalizes the result of Zhang et al. [13, Theorem 3], and its proof is similar.

Lemma 6. *Let f_E be a flow function in G_E. If f is acyclic then f is a flow function in G.*

In order to use Lemma 6 we must find a maximum flow f_E in G_E such that f is acyclic. In this section we show how to do that.

The algorithm of Zhang et al. [13, Section 3] for undirected planar graphs finds a flow f_E in G_E and than cancels flow-cycles in f in an arbitrary order. They call the resulting flow f_a and claim that this flow satisfies vertex capacities constraints. This approach is flawed. Fig. 3 shows an example on which the algorithm of [13] fails. After we cancel flow-cycles in f in an arbitrary order it is possible that there is no flow f'_E in G_E such that the restriction f' of f'_E to G is f_a, and therefore Lemma 6 does not apply.

As the example in Fig. 3 shows, it is not enough to cancel arbitrary flow-cycles in f. We can cancel a flow-cycle in f only if there is a cycle C in G_E that contains it, such that we can reduce flow along the cycle C. In this case the cycle $rev(C)$ in G_E is a residual cycle with respect to f_E. Therefore, in order to cancel a flow-cycle in G with respect to f we must cancel a residual cycle in G_E with respect to f_E. Canceling a arbitrary residual cycle is not enough, since we always want to reduce the flow that f assigns to arcs, and never increase it.

Let f_E be a flow in G_E. We define a new capacity function c' on the arcs of G_E which guarantees that the flow in an arc e of G never increases beyond the value of $f(e)$. For $e \in E$ we let $c'(e) = f(e)$. The arcs of G_E which are not in G are arcs of C_v for some vertex v, for these arcs we do not have to limit the flow to the amount in f_E, so we set $c'(e) = c(e) = c(v)/2$. The flow function f_E is also a flow function in G_E with the new capacity function c', by the way we defined c'. Since $c'(e) \leq c(e)$ for every arc e, every flow in G_E with the capacity function c' is also a flow in G_E with the original capacity function c.

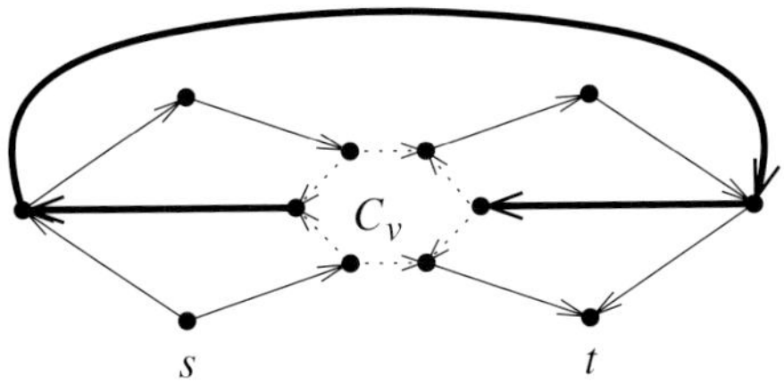

Fig. 3. A counterexample to the algorithm of [13]. The edges of the original undirected graph G are solid. The vertex v has capacity 1, the edges of C_v are dotted. The flow in every solid edge is 1, the flow in every dotted edges is 1/2, in the specified direction (the edges are undirected). The bold edges form a flow-cycle in G, after we cancel it we remain with an acyclic flow in G, but the amount of flow that enters v is 2. The correct solution is to cancel the flow in the two internal flow-cycles.

Instead of canceling the residual cycles one by one, we apply to G_E and c' the algorithm in Sect. 2.1 and find a new flow f'_E with the same value as f_E, such that there are no clockwise residual cycles in G_E with respect to f'_E and c'. The following lemma shows the crucial property of f'_E.

Lemma 7. *The restriction f' of f'_E to G does not contain counterclockwise flow-cycles.*

Proof. Assume, for a contradiction, that there is a counterclockwise flow-cycle C with respect to f' in G. We choose C such that C does not contain any other counterclockwise flow-cycle inside its embedding in the plane. We show that we can extend $rev(C)$ to a clockwise residual cycle with respect to f'_E in the graph G_E with capacity function c', in contradiction to the way we constructed f'_E.

For every arc e of C, $f'_E(e) > 0$, and therefore $rev(e)$ is a residual arc with respect to f'_E and c'. If C does not contain a vertex $v \in V$ with $c(v) \neq \infty$ then $rev(C)$ is a clockwise residual cycle with respect to f'_E and c', and we obtain a contradiction.

Let v be a vertex in C with $c(v) \neq \infty$. Let (u, v) and (v, u') be the arcs of C which are incident to v. These arcs correspond to arcs (u, v_i) and (v_j, u') in G_E, where v_i and v_j are in C_v. Let P be the path from v_j to v_i which goes counterclockwise around C_v (recall that C_v is undirected in G_E). To show a complete residual cycle in G_E, we argue that P is a residual path in G_E with respect to f'_E and c', so we can use it to fill the gap between v_j and v_i in $rev(C)$. An arc e of P is not residual if and only if $f'_E(e) = c'(e) = c(v)/2$. Without loss of generality we assume that the vertex v_k in the path P is followed by v_{k+1}.

Let $e_i = (v_{i-1}, v_i)$ be the last arc in the path P from v_j to v_i. Assume for contradiction that the arc e_i is not residual with respect to f'_E and c'. Then $f'_E(e_i) = c(v)/2$ and so the total flow into v_i in f'_E is $\sum_{e \in in(v_i)} f'_E(v_i) \geq f'_E((u, v_i)) + f'_E(e_i) > c(v)/2$. The only remaining arc that can carry flow out of v_i is $e_{i+1} = (v_i, v_{i+1})$. Because f'_E satisfies flow conservation constraints $f'_E(e_{i+1}) > c(v)/2$. But this is impossible since the capacity of the arc e_{i+1} is $c(v)/2$. Therefore, $f'_E(e_i) < c(v)/2$ and e_i is residual with respect to f'_E and c'.

We now proceed by induction. Assume by induction that we already know that the arc $e_{k+1} = (v_k, v_{k+1})$ on the path P from v_j to v_i is residual with respect to f'_E and c'. If $k = j$ then we are done. Otherwise, we prove that $e_k = (v_{k-1}, v_k) \in P$ is also residual with respect to f'_E and c'. Since e_{k+1} is residual it follows that $f'_E(e_{k+1}) < c(v)/2$. Let e' be the single arc of E incident to v_k. If e' is directed out of v_k and $f'_E(e') > 0$ then since C is a cycle and e' is inside C in the embedding of G, there must be a path carrying flow that starts with e' and continues to another vertex on C. (Recall that t is on the boundary of the outer face.) This implies that there is a counterclockwise flow-cycle with respect to f' and G inside the embedding of C, in contradiction to the choice of C. Therefore e' does not carry flow of f'_E out of v_k in G_E. This implies, by the conservation constraint on v_k, that $f'_E(e_k) \leq f'_E(e_{k+1}) < c(v)/2$, so e_k is indeed residual with respect to f'_E and c'.

We showed that there is a residual path from v_j to v_i. Since v was an arbitrary vertex with $c(v) > 0$ on C it follows that we can extend $rev(C)$ to a residual cycle in G_E with respect to f'_E and c'. Since C is a counterclockwise cycle, the residual cycle we get from $rev(C)$ is a clockwise cycle. This contradicts the definition of f'_E, and therefore a counterclockwise flow-cycle C with respect to f' and G does not exist. □

We repeat the previous procedure symmetrically, by defining a new capacity c'' which restricts the flow in G to the flow in f', and applying a symmetric version of the algorithm of Sect. 2.1. This way we get from f'_E a flow f''_E of the same value, such that f'' does not contain clockwise flow-cycles in G. For every $e \in E$ we changed the flow such that $f''(e) \leq f'(e) \leq f(e)$, so we did not create any new flow-cycles. Therefore we have the following lemma.

Lemma 8. *The flow function f''_E has the same value as the flow function f_E. The restriction f'' of f''_E to G is acyclic.*

6 The Algorithm

Combining together the results of the previous sections we get an algorithm for finding maximum flow in a directed planar graph with vertex capacities.

First, we construct G_E from G by replacing each vertex that has a finite capacity with C_v as defined in Sect. 4. Next, we find a maximum flow f_E in G_E, which is a directed planar graph without vertex capacities. Last, we change f_E to another flow f''_E as in Sect. 5.

According to Lemma 8, the flow f''_E is a maximum flow in G_E, and its restriction f'' is acyclic. By Lemma 6, the function f'' is a flow in G. Since the value of f'' is the same as the value of f_E, Lemma 5 implies that f'' is a maximum flow.

The construction of G_E from G takes $O(n)$ time. The computation of f''_E from f_E also takes $O(n)$ time by Lemma 1. Therefore, the only bottleneck of our algorithm is finding f_E, a maximum flow in a directed planar graph without vertex capacities.

Theorem 9. *The maximum flow in a directed planar graph with both arc capacities and vertex capacities can be computed within the same time bound as the maximum flow in a directed planar graph with arc capacities only.*

The algorithm of Borradaile and Klein [2] finds a maximum flow in directed planar graph with arcs capacities in $O(n \log n)$ time. If G is a st-planar graph, then G_E preserves this property. In this case the algorithm of Hassin [4], using the algorithm of [5] for single-source shortest-path distances, finds a maximum flow in $O(n)$ time.

References

1. Ahuja, R.K., Magnanti, T.L., Orlin, J.B.: Network Flows: Theory, Algorithms and Applications. Prentice-Hall, New Jersey (1993)
2. Borradaile, G., Klein, P.: An $O(n \log n)$ algorithm for maximum st-flow in a directed planar graph. J. ACM 56 (2009)
3. Ford, L.R., Fulkerson, D.R.: Flows in Networks. Princeton University Press, New Jersey (1962)
4. Hassin, R.: Maximum flow in (s, t) planar networks. Information Processing Letters 13, 107 (1981)
5. Henzinger, M.R., Klein, P., Rao, S., Subramania, S.: Faster shortest-path algorithms for planar graphs. J. Comput. Syst. Sci. 55, 3–23 (1997)
6. Johnson, D.B.: Parallel algorithms for minimum cuts and maximum flows in planar networks. J. ACM 34, 950–967 (1987)
7. Khuller, S., Naor, J.: Flow in planar graphs with vertex capacities. Algorithmica 11, 200–225 (1994)
8. Khuller, S., Naor, J., Klein, P.: The lattice structure of flow in planar graphs. SIAM J. Disc. Math. 63, 477–490 (1993)
9. Nishizwki, T., Chiba, N.: Planar Graphs: Theory and Algorithms. Ann. Discrete Math, vol. 32. North-Holland, Amsterdam (1988)
10. Ripphausen-Lipa, H., Wagner, D., Weihe, K.: The vertex-disjoint Menger problem in planar graphs. SIAM J. Comput. 26, 331–349 (1997)
11. Weihe, K.: Maximum (s, t)-flows in planar networks in $O(|V| \log |V|)$-time. J. Comput. Syst. Sci. 55, 454–476 (1997)
12. Zhang, X., Liang, W., Jiang, H.: Flow equivalent trees in node-edge-capacitated undirected planar graphs. Information Processing Letters 100, 100–115 (2006)
13. Zhang, X., Liang, W., Chen, G.: Computing maximum flows in undirected planar networks with both edge and vertex capacities. In: Hu, X., Wang, J. (eds.) COCOON 2008. LNCS, vol. 5092, pp. 577–586. Springer, Heidelberg (2008)

A Fast Output-Sensitive Algorithm for Boolean Matrix Multiplication*

Andrzej Lingas

Department of Computer Science, Lund University, 22100 Lund, Sweden
Andrzej.Lingas@cs.lth.se

Abstract. We use randomness to exploit the potential sparsity of the Boolean matrix product in order to speed up the computation of the product. Our new fast output-sensitive algorithm for Boolean matrix product and its witnesses is randomized and provides the Boolean product and its witnesses almost certainly. Its worst-case time performance is expressed in terms of the input size and the number of non-zero entries of the product matrix. It runs in time $\widetilde{O}(n^2 s^{\omega/2-1})$, where the input matrices have size $n \times n$, the number of non-zero entries in the product matrix is at most s, ω is the exponent of the fast matrix multiplication and $\widetilde{O}(f(n))$ denotes $O(f(n) \log^d n)$ for some constant d. By the currently best bound on ω, its running time can be also expressed as $\widetilde{O}(n^2 s^{0.188})$. Our algorithm is substantially faster than the output-sensitive column-row method for Boolean matrix product for s larger than $n^{1.232}$ and it is never slower than the fast $\widetilde{O}(n^{\omega})$-time algorithm for this problem.

We also present a partial derandomization of our algorithm as well as its generalization to include the Boolean product of rectangular Boolean matrices. Finally, we show several applications of our output-sensitive algorithms.

1 Introduction

Boolean matrix multiplication is a basic tool in the design of efficient algorithms, especially graph algorithms (see, e.g., [5,17,18]). Just squaring a Boolean matrix is equivalent to the following fundamental problem: for n subsets of $\{1, 2, ..., n\}$, determine all pairs of the subsets that have a non-empty intersection. This problem has numerous applications, among other things, in databases and data mining [2].

By the standard definition of Boolean matrix multiplication, it is sufficient to use $O(n^3)$ conjunctions and disjunctions to compute the Boolean product of two Boolean $n \times n$ matrices. Slightly less known is another simple combinatorial way of computing the Boolean product termed as a *column-row* method [21].

Consider two Boolean $n \times n$ matrices A and B, and their Boolean product C. If $C[i,j] = 1$ then any index k such that $A[i,k] = 1$ and $B[k,j] = 1$ is called a *witness* for $C[i,j]$. In the column-row approach, for each k, one considers the set of indices of rows in A that have 1 in the k-th column of A as well as the set of indices of columns in B that have 1 in the k-th row of B. Note that for any i in the former set and any j in the latter one, k is a witness for $C[i,j]$. It follows that the column-row method can

* Research supported in part by the VR grant 621-2005-4085.

A. Fiat and P. Sanders (Eds.): ESA 2009, LNCS 5757, pp. 408–419, 2009.

be implemented in time $O(W + n^2)$ where W is the total number of witnesses for all non-zero entries of C [21].

The column-row approach is output sensitive in terms of the number s of non-zero entries of the product matrix. Simply, each non-zero entry can have at most n witnesses and therefore the total number of witnesses cannot exceed ns. In this way, we obtain the output sensitive upper time-bound of $O(ns + n^2)$ for Boolean square matrix multiplication.

This upper bound is still attractive in comparison with the running time of the fastest known combinatorial algorithms for Boolean matrix multiplication, which is $O(n^3/\log^2 n)$ (see [3,5,18]). However, it seems less attractive when compared with the running time of the fastest known algorithm for Boolean matrix multiplication. The latter is simply obtained by treating the Boolean 0 and 1 as arithmetic ones and using the fast arithmetic matrix multiplication. Thus, its running time is $O(n^\omega)$, where ω is the smallest constant for which the product of two $n \times n$ arithmetic matrices can be computed in time $O(n^\omega)$. The best currently known asymptotic upper bound on ω is 2.376, due to Coppersmith and Winograd [9]. Thus, the fast algorithm for Boolean matrix multiplication subsumes the output-sensitive column-row method whenever s is substantially larger than $n^{1.376}$.

In the mathematical programming literature, the assumption of *no cancellation*, i.e., that the inner product of two vectors with overlapping non-zero entries never cancels to zero is very common [7]. The justification of the assumption comes from the unlikeness of cancellations and the aggregation of rounding errors. For this reason, the important problem of the structure prediction of the product of two arithmetic matrices reduces to that of the Boolean product of the two corresponding Boolean matrices, where the Boolean 1 entries correspond to the non-zero ones.

The prediction of the structure of the product of two matrices is important both in efficient memory block allocation as well as in determining an optimal order of chains of matrix products [7,12]. Cohen provided an efficient procedure for close estimation of the size of rows and columns in the Boolean product [7]. The problem of more precise efficient prediction of the structure of arithmetic matrix product has remained open.

Of course one can simply compute the Boolean product of the corresponding Boolean matrices. Unfortunately, the fast method for Boolean matrix multiplication takes no fewer operations than the multiplication of the original arithmetic matrices itself, i.e., $O(n^\omega)$. What about the situation when the Boolean matrix product is moderately dense or sparse, i.e., it has a substantially sub-quadratic and substantially superlinear in n number of non-zero entries?

Very recently, Amossen and Pagh [2], extending the input sensitive algorithm of Yuster and Zwick [22], proposed an input- and output-sensitive algorithm for Boolean matrix product running in time $O(t^{0.86}s^{0.41} + (ts)^{2/3})$, where t stands for the number of non-zero entries in the input matrices and s for the number of non-zero entries in the output matrix. While their upper time-bound is very interesting, it can easily exceed n^ω, when the input matrices are quite dense and the ouput one is moderately dense.

Summarizing, a natural question arises if there is a fast output-sensitive algorithm for Boolean matrix multiplication, whose running time is substantially better than that of the output-sensitive column-row method, and than $O(n^\omega)$ when the product matrix is moderately dense/sparse?

We answer this question in the affirmative by providing a randomized algorithm for Boolean matrix product and its witnesses running in time $\widetilde{O}(n^2 s^{\omega/2-1})$.

Thus, using the upper bound of Coppersmith-Winograd on ω, our algorithm runs in time $\widetilde{O}(n^2 s^{0.188})$. Since $n^2 s^{\omega/2-1} \leq n^\omega$ for all $s \leq n^2$, our algorithm is never slower than that of Coppersmith-Winograd and it is faster than the column-row method for $s > n^{1/(2-w/2)} \approx n^{1.232}$.

It is not an easy task to compare our algorithm with the algorithm of Amossen and Pagh which is also input sensitive. For instance, for moderately dense/sparse random input Boolean matrices, the Boolean product might be expected to be already dense. Then the input sensitive algorithms of Yuster and Zwick, and of Amossen and Pagh will be superior. *However, the sparsity of the product matrix does not imply the sparsity of any of the input matrices generally.* For instance, if only the first halves of rows in the first input matrix are filled with ones and similarly only the second halves of columns in the other input matrix are filled with ones, then we have $t = n^2$ and $s = 0$. Assuming a worst-case scenario for input sensitivity, i.e., $t = \Omega(n^2)$, our algorithm is faster than that of Amossen and Pagh for $s > n^{1.244}$.

The use of randomness to exploit the potential sparsity of the Boolean product is crucial in the design of our algorithm. However, if upper bounds on the number of non-zero entries in the respective rows and columns of the product matrix summing to at most $2s$ are given *a priori* then we can partially derandomize our algorithm so it uses only $O(\log^2 n)$ random bits and runs in the same asymptotic time. In both cases, our algorithms provide the Boolean matrix products and their witnesses almost certainly. We also present generalizations of our algorithms to include the Boolean product of rectangular Boolean matrices.

Finally, we show applications of our algorithms and its generalizations to the derivation of output-sensitive upper time-bounds for the following problems: composition of binary relations, maximum witnesses of Boolean matrix product, lowest common ancestors in directed acyclic graphs and prediction of the structure of arithmetic matrix product.

1.1 Other Related Results

The column-row approach to (non-necessarily Boolean) matrix multiplication has been known for a long time. In the arithmetic case, it can be expressed as the fact that the matrix product of two $n \times n$ matrices A and B is equal to the sum of the outer products of the i-th column of A with the i-th row of B, over $i = 1, ..., n$ [7,13]. Thus, the column-row approach takes $O(\sum_{k=1}^{n} a_k b_k)$ multiplications and additions, where a_k is the number of non-zero elements in the k-th column of A and b_k is the number of elements in the k-th row of B. Additionally, an $O(n^2)$ time is sufficient to compute the list of non-zero elements for each column of A and each row of B. Hence, under the no cancellation assumption the $O(ns+n^2)$ time-bound, where s is the number of non-zero entries in the product, easily follows in the arithmetic case.

Sparse arithmetic matrices, and hence also, sparse Boolean matrices, are well known to admit more efficient multiplication algorithms than those aforementioned for the general case, see [7,22]. Note that if A or B has at most m non-zero entries then by $\sum_{k=1}^{n} a_k b_k \leq \max\{(\sum_{k=1}^{n} a_k)n, (\sum_{k=1}^{n} b_k)n\} \leq mn$, one obtains an $O(mn + n^2)$ upper bound for sparse matrix multiplication by the column-row approach [22]. For the case where both A and B have at most m non-zero entries, Yuster and Zwick provide an algorithm using $O(m^{0.7}n^{1.2} + n^{2+o(1)})$ multiplications, additions and subtractions over R [22]. Their algorithm consists in a nice and simple reduction to the sum of two rectangular matrix multiplications, the first fast one corresponding to the indices that are witnesses for relatively many entries and the other column-row one corresponding to the indices that are witnesses for relatively few entries.

One of the drawbacks of the fast Boolean matrix multiplication is that it does not yield explicitly witnesses for the non-zero entries of the product matrix in contrast with the aforementioned combinatorial methods. However, at the beginning of the 90s, efficient randomized and deterministic methods for Boolean product witnesses have been developed [4,11,20]. They typically provide a single witness per non-zero entry and run in time $\widetilde{O}(n^{\omega})$.

1.2 Organization

In the next section, we present our fast output-sensitive randomized algorithm for the Boolean product of two square Boolean matrices and its analysis. In Section 3, we discuss its partial derandomization. In section 4, we outline a generalization of our algorithm and its analysis to include rectangular input Boolean matrices. In section 5, we show a number of applications of our output-sensitive algorithms for Boolean matrix product.

2 The Output-Sensitive Algorithm for Boolean Product

The following definition and two lemmata lie behind the idea of our algorithm.

Definition 1. *Let C be an $n \times n$ Boolean matrix. A row of C is r-*sparse *if the number of non-zero entries in it is at most r, and otherwise, the row is r-*dense*. Similarly, a column of C is r-sparse if the number of non-zero entries in it is at most r, and otherwise, the column is r-*dense*. An entry of C is r-sparse if both its row and column are r-sparse, otherwise the entry is r-*dense.

Lemma 1. *If an $n \times n$ Boolean matrix C has at most r^2 non-zero entries then the number of r-dense rows and r-dense columns in C is $O(r)$.*

Proof. Otherwise, there would be more than r^2 non-zero entries in C. □

Lemma 2. *Let C be an $n \times n$ Boolean matrix with at most r^2 non-zero entries. Let C' be the matrix resulting from permuting uniformly at random the rows and the columns of C. For a given constant $c > 1$, divide C' into $cr \times cr$ sub-arrays of size $\frac{n}{cr} \times \frac{n}{cr}$. For any given r-sparse non-zero entry $C[i, j]$ of C the probability that another non-zero entry of C is in the same sub-array of C' is $O(\frac{1}{c})$.*

Proof. The probability that another non-zero entry $C[i',j']$ of C, where $i \neq i'$ and $j \neq j'$, is in the same sub-array of C' is at most

$$(r^2-1)\frac{(\frac{n}{cr}-1)}{n-1}\frac{(\frac{n}{cr}-1)}{n-1} = O(\frac{1}{c^2})$$

The probability that another non-zero entry $C[i',j']$ of C, where either $i=i'$ or $j=j'$, is in the same sub-array of C' is at most

$$2(r-1)\frac{(\frac{n}{cr}-1)}{n-1} = O(\frac{1}{c})$$

□

Before presenting our algorithm, we recall the concept of witnesses in Boolean matrix multiplication.

Definition 2. *If an entry $C[i,j]$ of the Boolean product of two Boolean matrices A and B is equal to 1 then any index l such that $A[i,l]$ and $B[l,j]$ are equal to 1 is a* ***witness*** *for $C[i,j]$. The problem of* **computing witnesses** *for the product matrix C consists in determining for each non-zero entry of C a single witness.*

Our algorithm consists of two main stages. The first stage takes care of the r-sparse non-zero entries. It consists of iterations of Algorithm 1 which is an efficient algorithmic counterpart of Lemma 2, where the matrix C is the Boolean product of two Boolean matrices A and B. In Algorithm 1, the sub-arrays of C' in Lemma 2 correspond to single entries of a smaller product matrix resulting from gluing batches of $\Omega(\frac{n}{r})$ rows in the row-permuted matrix A and batches of $\Omega(\frac{n}{r})$ columns in the column-permuted matrix B.

Algorithm 1

Input: Two $n \times n$ Boolean matrices A and B whose Boolean product has at most r^2 non-zero entries.
Output: An $n \times n$ Boolean matrix C whose non-zero entries are a subset of the non-zero entries of the Boolean product of A and B, and witnesses for the non-zero entries of C.

1. Permute uniformly at random the rows of A and the columns of B.
2. Contract each block of $\frac{n}{r}$ consecutive rows of row-permuted A into a single super-row by taking the "or" along the $\frac{n}{r}$-long column fragments. Similarly, contract each block of $\frac{n}{r}$ consecutive columns of column permuted B into a single super-column by taking the "or" along the $\frac{n}{r}$-long row fragments.
3. Form the $r \times n$ matrix A^* composed of the super-rows, and similarly form the $n \times r$ matrix B^* composed of the super-columns.
4. Compute the witnesses of the Boolean product C^* of A^* with B^* (one for each non-zero entry of C^*).
5. For each witness k of an entry $C^*[i^*,j^*]$ computed in the previous step, find the first row i' in the block of the row-permuted A corresponding to the i^* (super)row of A^* satisfying $A[i',k]=1$, as well as the first column j' in the block of column-permuted B corresponding to the j^* (super)column of B^* satisfying $B[k,j']=1$.

Let i be the row number of i' in the original matrix A, and similarly, let j be the column number of j' in the original matrix B. Set $C[i,j]$ to 1 and the witness of $C[i,j]$ to k.

Lemma 3. *Algorithm 1 is partially correct, i.e., it sets to* 1 *only non-zero entries of the product matrix and it provides true witnesses (one for each non-zero entry) for them.*

Proof. It is sufficient to prove the correctness of the settings in Step 5. The i^*-th row of A^* is the result of column-wise "or" of the rows $(i^*-1)\frac{n}{r}+1$ through $i^*\frac{n}{r}$. Similarly, the j^*-th column of B^* is the result of row-wise "or" of the columns $(j^*-1)\frac{n}{r}+1$ through $j^*\frac{n}{r}$. Consequently, a witness k for the entry $C^*[i^*,j^*]$ is an index k which satisfies $\bigvee_{i''=(i^*-1)\frac{n}{r}+1}^{i^*\frac{n}{r}} A[i'',k] = 1$ and $\bigvee_{j''=(j^*-1)\frac{n}{r}+1}^{j^*\frac{n}{r}} B[k,j''] = 1$. Hence, the indices i' and j' are well defined in Step 5 and k is also a witness of $C[i,j]$. □

Throughout this section, we shall assume that there is an available method for square $(m \times m)$ Boolean matrix product also producing a witness for each non-zero entry of the product, running in time $f(m)$. Clearly, we have $f(m) \geq m^2$, and we may assume w.l.o.g that $f(qm) \leq q^3 f(m)$ for any positive integer q.

For instance, $f(m)$ could be $O(m^3)$ in case of the straightforward Boolean product method following from the definition, or $\widetilde{O}(n^\omega)$ in case of the method based on the fast arithmetic matrix multiplication and the following fact due to Alon and Naor [4] (see also [20] for an earlier randomized version of this fact).

Fact 1. For all non-zero entries of the Boolean product of two $n \times n$ Boolean matrices representatives of their witnesses can be computed deterministically in total time $\widetilde{O}(n^\omega)$.

Lemma 4. *Algorithm 1 can be implemented in time* $O(\frac{n}{r}f(r)+n^2)$.

Proof. Steps 1,2,3 can be easily implemented in time linear in the input size, i.e., $O(n^2)$. To implement Step 4, we divide vertically A^* into square $r \times r$ sub-arrays A^*_p, $p=1,\ldots,\frac{n}{r}$, as well as B^* horizontally into $r \times r$ sub-arrays B^*_p, $p=1,\ldots,\frac{n}{r}$. In this way, we reduce the computation of the witnesses for the product C^* of A^* and B^* to the computation of witnesses for the products $A^*_p \times B^*_p$, $p=1,\ldots,\frac{n}{r}$, and taking their union. This all takes time $\frac{n}{r} \times O(f(r)+r^2) = O(\frac{n}{r}f(r)+nr)$. Finally, Step 5 takes time $O(\frac{n}{r}r^2)=O(nr)$. □

Lemma 5. *Suppose that the Boolean product C of A and B has at most r^2 non-zero entries. For sufficiently large constant c, after $O(\log n)$ iterations of Algorithm 1, the non-zero r-sparse entries of the Boolean product C will be set to* 1 *and their witnesses will be provided, almost certainly, i.e., with probability not less than* $1-\frac{1}{n^\alpha}$ *for some* $\alpha > 1$.

Proof. Set c to the smallest positive integer such that the probability bound $O(\frac{1}{c})$ in Lemma 2 is at most $\frac{1}{2}$. It follows then from the specification of Algorithm 1 and Lemma 2 that for a given non-zero r-sparse entry of C the probability that it is not set to 1 (and its witness is not provided) after $d\log n$ iterations is at most $2^{-d\log n}$. Hence, for a given $\alpha > 1$, there is a sufficiently large constant d so the probability that at least one non-zero r-sparse entry is not set to 1 after $d\log n$ iterations is at most $n^{-\alpha}$. □

The second stage of our algorithm takes care of, i.e., provides witnesses for, the non-zero r-dense entries of the Boolean product of A and B. It relies on Lemma 1 and the following fact due to Cohen (see pp. 312-315 in [7]).

Fact 2. *Let A and B be two $n \times n$ Boolean matrices. For any fixed $\epsilon > 0$, there is an $O(n^2 \log n)$-time randomized algorithm that estimates the number of non-zero entries in the columns and rows of the Boolean product of A and B with relative error $< \epsilon$ almost certainly.*

Lemma 6. *The witnesses for a superset of the non-zero r-dense entries of the Boolean product of A and B can be computed in time $O((\frac{n}{r})^2 f(r) + n^2 \log n)$, almost certainly.*

Proof. We can detect a superset of the rows of the matrix A corresponding to the r-dense rows of the product of A and B, as well as the columns of the matrix B corresponding to the r-dense columns of the product matrix as follows. We run the algorithm of Cohen [7] with ϵ set, say to $\frac{1}{2}$, to estimate the number of non-zero entries in each row and each column of the product matrix in time $O(n^2 \log n)$. If the estimation for a row or a column of the product matrix exceeds $\frac{r}{2}$ then we mark the corresponding row of A or the corresponding column of B.

Note that in particular all r-dense rows and columns of the product matrix are marked in this way. On the other hand, each of the marked rows or columns is $\frac{r}{4}$ dense.

Let A^* be the matrix composed of the marked rows of the matrix A. Similarly, let B^* be the matrix composed of the marked rows of B. Now, it is sufficient to determine witnesses for the Boolean products of A^* with B and A with B^*. The union of the two resulting sets of witnesses has to include witnesses for all r-dense non-zero entries of the product of A with B. Since the rows of A^* correspond to some $\frac{r}{4}$ dense rows of the product matrix, A^* has at most $O(r)$ rows by Lemma 1. Analogously, B^* has at most $O(r)$ columns. Hence, by dividing A^* vertically into $O(\frac{n}{r})$ sub-arrays of size $O(r) \times O(r)$ and B both horizontally and vertically into $O((\frac{n}{r})^2)$ sub-arrays of size $O(r) \times O(r)$, we can compute the witnesses of the product of A^* with B by computing the witnesses of $O((\frac{n}{r})^2)$ products of $O(r) \times O(r)$ sub-arrays. This takes time $O((\frac{n}{r})^2 f(r))$. Symmetrically, we can compute the witnesses for the product of A with B^* in time $O((\frac{n}{r})^2 f(r))$. □

We can also use the application of Fact 2 in Lemma 6 to estimate the number r^2 of non-zero entries in the Boolean product matrix for the sake of the iterations of Algorithm 1. Note that if we set $s = r^2$ then the upper bound $O((\frac{n}{r})^2 f(r) + n^2 \log n)$ given in Lemma 6 can be rewritten as $O(\frac{n^2}{s} f(\sqrt{s}) + n^2 \log n) = \widetilde{O}(\frac{n^2}{s} f(\sqrt{s}))$. Similarly the time taken by $O(\log n)$ iterations of Algorithm 1 can be expressed as $\widetilde{O}(\frac{n}{\sqrt{s}} f(\sqrt{s}) + n^2)$ by Lemma 4. Hence, by combining Lemmata 3, 4, 5, 6, we obtain our main theorem.

Theorem 1. *Let A and B be two $n \times n$ Boolean matrices whose Boolean product has at most s non-zero entries. The Boolean product of A and B, and the witnesses for the product can be computed by a randomized algorithm in time $\widetilde{O}(\frac{n^2}{s} f(\sqrt{s}))$, almost certainly. In particular, if we apply the fastest algorithm for the Boolean product and its witnesses, i.e., $f(m) = \widetilde{O}(m^\omega)$, the running time is $\widetilde{O}(n^2 s^{\omega/2-1})$.*

Note that if we apply the straightforward cubic-time algorithm for the Boolean product and its witnesses, i.e., $f(m) = O(m^3)$, we obtain another output-sensitive algorithm for this problem running in time $\widetilde{O}(n^2\sqrt{s})$.

3 Partial Derandomization

Let S_n be the set of all permutations of $0, ..., n-1$. A family of permutations $F \subseteq S_n$ is *pairwise independent* (cf. [6]) if for any π chosen at random in F and any $\{i_1, i_2, k_1, k_2\} \subseteq \{0, ..., n-1\}$ where $i_1 \neq i_2$ and $k_1 \neq k_2$,

$$Pr(\pi(i_1) = k_1 \; \& \; \pi(i_2) = k_2) = \frac{1}{n(n-1)}$$

Assuming that n is a prime number, the family of linear transformations of the form $\pi(i) = ai + b \mod n$, where $a \neq 0$, is known to be pairwise independent (e.g., see Exercise in [16]).

Under the assumption that n is a prime number, we can replace uniform at random permutations in step 1 in Algorithm 1 with such randomly chosen linear transformations. Due to the property of pairwise independence, Lemma 5 can be proved analogously after the replacement. Then, if additionally we use the deterministic version of the algorithm due to Alon and Naor for witnesses of Boolean matrix product [4], one iteration of the so modified Algorithm 1 will require only $O(\log n)$ random bits. Thus, the total number of random bits used by our method, but for the application of Cohen's algorithm to estimate the number of non-zero entries in the rows and columns of the product, can be decreased to $O(\log^2 n)$ at the cost of increasing its time complexity by a poly-logarithmic factor. The increase is caused by patching the input matrices to the size $n' \times n'$, where n' is the smallest prime number not larger than n, and the use of the deterministic algorithm for witnesses [4]. (To find such a prime number n', one can use the AKS primality testing or its improved versions running in time polynomial in the length of bit representation [1].)

Hence, we obtain the following variant of Theorem 1.

Theorem 2. *Let A and B be two $n \times n$ Boolean matrices. Suppose that there are given* a priori *upper bounds on the number of non-zero entries in the respective rows and columns of the Boolean product of A and B summing to at most $2s$. The Boolean product of A and B, and the witnesses for the product can be computed almost certainly by a randomized algorithm, using $O(\log^2 n)$ random bits and $\widetilde{O}(n^2 s^{\omega/2-1})$ time.*

4 Rectangular Boolean Matrix Multiplication

Analogous output sensitive algorithms can be derived for rectangular Boolean matrix multiplication.

Denote by $\omega(p, q, t)$ the exponent of the fast multiplication of an $n^p \times n^q$ matrix by an $n^q \times n^t$ matrix.

We obtain the following partial generalization of Theorems 1, 2.

Theorem 3. *Let A and B be two Boolean matrices of size $n^p \times n^q$ and $n^q \times n^t$, respectively, such that their product has at most $n^{p(1-\delta_1)+t(1-\delta_2)}$ non-zero entries. The Boolean product of A and B and its witnesses can be computed by a randomized algorithm running in time $\widetilde{O}(n^{\omega(|p-\delta_1|,q,t)} + n^{\omega(p,q,|t-\delta_2|)} + n^{p+q} + n^{q+t})$, almost certainly. Furthermore, if upper bounds on the number of non-zero entries in the respective rows and columns of the Boolean product of A and B summing to at most $n^{q(1-\delta_1)+t(1-\delta_2)}$ are known a priori then the algorithm requires only $O(\log^2 n)$ random bits.*

Proof. sketch. Let C denote the $n^p \times n^t$ Boolean product of A and B. Consider $n^{t(1-\delta_2)}$-sparse rows of C, i.e., the rows of C with at most $n^{t(1-\delta_2)}$ non-zero entries as well as $n^{p(1-\delta_1)}$-sparse columns of C, i.e., the columns of C with at most $n^{p(1-\delta_1)}$ non-zero entries. Next, define $(n^{p(1-\delta_1)}, n^{t(1-\delta_2)})$-sparse non-zero entries as those non-zero entries which lie within a $n^{t(1-\delta_2)}$-sparse row and a $n^{p(1-\delta_1)}$-sparse column of C. The remaining entries are $(n^{p(1-\delta_1)}, n^{t(1-\delta_2)})$-dense.

The consecutive steps of the proof are straightforward generalizations of those in the proof of Theorems 1, 2, resulting from the use of the generalized concepts of sparse (and, dense) rows, columns and entries of C. For instance, the generalization of Lemma 1 states that if C has at most $n^{p(1-\delta_1)+t(1-\delta_2)}$ non-zero entries then the number of $n^{t(1-\delta_2)}$-dense rows is $O(n^{p(1-\delta_1)})$ while the number of $n^{p(1-\delta_1)}$-dense columns is $O(n^{t(1-\delta_2)})$. Next, in the generalization of Lemma 2, C' is divided into $cn^{p(1-\delta_1)} \times cn^{t(1-\delta_2)}$ sub-arrays of size $\frac{n^{p\delta_1}}{c} \times \frac{n^{t\delta_2}}{c}$, and so on. The further analogous details in the generalizations of consecutive lemmata as well as in Algorithm 1 are left to the full version.

The terms $n^{\omega(|p-\delta_1|,q,t)}$ and $n^{\omega(p,q,|t-\delta_2|)}$ in the claimed upper time-bound come from the final step (the generalization of Lemma 6), taking care of the non-zero $(n^{p(1-\delta_1)}, n^{t(1-\delta_2)})$-dense entries. They overshadow the term $n^{\omega(|p-\delta_1|,q,|t-\delta_2|)}$ coming from the iterations of a generalization of Algorithm 1. □

Note that since the sizes of A and B are not necessarily symmetric, forcing $\delta_1 = \delta_2$ might weaken the upper time-bound in Theorem 3.

5 Applications

Theorems 1, 2 yield corresponding output-sensitive upper time-bounds for several applications of Boolean matrix product and its witnesses, e.g., composition of binary relations, maximum witnesses of Boolean product, lowest common ancestors in directed acyclic graphs, prediction of the structure of the arithmetic matrix product, etc.

A *binary relation* is a set of ordered pairs over some universe. The composition of two binary relations R_1 and R_2 over a common universe U is a binary relation R_3 over U defined by $R_3 = \{(v, w) | \exists_{u \in U} (v, u) \in R_1 \wedge (u, w) \in R_2\}$.

Corollary 1. *If the composition of two binary relations over an n-element universe has at most s elements then it can be computed from these two relations by a randomized algorithm in time $\widetilde{O}(n^2 s^{\omega/2-1})$, almost certainly.*

A *maximum witness* for a non-zero entry $C[i, j]$ of the Boolean product of two $n \times n$ Boolean matrices A, B is the largest witness for $C[i, j]$, i.e., the largest index k such that $A[i,k] = 1$ and $B[k,j] = 1$. The next corollary is concerned with computing maximum witnesses (one per each non-zero entry) in a sparse Boolean matrix product.

Corollary 2. *Let A and B be two $n \times n$ Boolean matrices whose product has at most $n^{2-2\delta}$ non-zero entries, and let $\lambda(\delta)$ be such that $\omega(1, \lambda(\delta), |1-\delta|) = 1 + \lambda(\delta)$. The maximum witnesses of the product of A and B can be computed by a randomized algorithm running in time $\widetilde{O}(n^{2+\lambda(\delta)})$, almost certainly. By augmenting this upper time bound with the additive term $O(n^\omega)$, we can decrease the number of required random bits to $O(\log^2 n)$.*

Proof. The proof requires solely a slight modification of the proof of Theorem 16 in [10]. The key observation is that in the aforementioned proof the matrices C_p, $p = 1, ..., n/l$, have also at most $n^{2-2\delta}$ non-zero entries like the Boolean product C of A and B. Hence, since such a matrix C_p is the Boolean product of an $n \times n^r$ Boolean matrix with an $n^r \times n$ Boolean matrix (see [10]), where $r = \log_n l$, it can be computed in time $\widetilde{O}(n^{\omega(1,r,|1-\delta|)} + n^{\omega(|1-\delta|,r,1)} + n^{1+r})$ by Theorem 3. Therefore, by the proof of Theorem 16 in [10] and the equality $\omega(1,r,|1-\delta|) = \omega(|1-\delta|,r,1)$, the total cost of computing the maximum witnesses is $\widetilde{O}(n^{1-r+\omega(1,r,|1-\delta|)} + n^{3-r} + n^{2+r})$. Analogously as in [10], we get rid of the additive term n^{3-r} assuming $r \geq \frac{1}{2}$, and solve the equation $1 - \lambda(\delta) + \omega(1, \lambda(\delta), |1-\delta|) = 2 + \lambda(\delta)$, implying $\lambda(\delta) \geq \frac{1}{2}$ by $\omega(1, \lambda(\delta), |1-\delta|) \geq 2$, in order to balance the two remaining exponent terms. This yields the first claimed upper time-bound. To estimate the number of non-zero entries in the rows and columns of the product matrix we can simply compute the product in time $O(n^\omega)$ instead of using the algorithm of Cohen [7]. □

By [8,14], the best known upper bound on $\lambda(0)$ is 0.575 (see, e.g., [10]). Hence, the partially derandomized version of Corollary 2 should be interesting at least for small δ.

In turn, the problem of maximum witnesses has many applications [19]. Here, we just consider its original application to computing lowest common ancestors in directed acyclic graphs [10,15]. By combining Corollary 2 with Theorem 11 in [10], we obtain the corollary

Corollary 3. *Let G be a directed acyclic graph on n vertices such that at most $n^{2-2\delta}$ pairs of vertices in G have a common ancestor. Next, let $\lambda(\delta)$ be such that $\omega(1, \lambda, |1-\delta|) = 1 + \lambda(\delta)$. There is a randomized algorithm almost certainly reporting for all pairs of vertices in G with a common ancestor their lowest common ancestor, running in total time $\widetilde{O}(n^{2+\lambda(\delta)})$. By augmenting this upper time bound with the additive term $O(n^\omega)$, we can decrease the number of required random bits to $O(\log^2 n)$.*

The methods of Theorems 1 and 2 applied to Boolean counterparts of two $n \times n$ arithmetic matrices yield the complete information on the location of non-zero entries in the product of these two arithmetic matrices under the assumption of no cancellation.

Corollary 4. *Let A and B be two $n \times n$ arithmetic matrices whose product has at most s non-zero entries. Under the assumption of no cancellation, the exact non-zero structure of the product of A and B can be determined by a randomized algorithm running in time $\widetilde{O}(n^2 s^{\omega/2-1})$, almost certainly. Furthermore, if the upper bounds on* *the number of non-zero entries in the respective rows and columns of the product matrix summing to at most* $2s$ *are known* a priori *then the algorithm requires only* $O(\log^2 n)$ *random bits.*

6 Conclusions and Extensions

Our main new idea is the use of randomness to compute sparse Boolean product. We have applied the idea to the fast algorithms for Boolean product and its witnesses, treating them as black boxes.

The next natural step would be to derive general upper time bounds sensitive both to the sparsity of the input matrices as well as to that of the product matrix by combining the ideas from this paper with those from [2,22].

Acknowledgments

The author is grateful to all those who provided useful comments, in particular Mirosław Kowaluk, Christos Levcopoulos, Rasmus Pagh and several anonymous referees.

References

1. Agrawal, M., Kayal, N., Saxena, N.: PRIMES is in P. Annals of Mathematics 160(2), 781–793 (2004)
2. Amossen, R.R., Pagh, R.: Faster Join-Projects and Sparse Matrix Multiplication. To appear in proc. ACM International Conference on Database Theory (ICDT 2009), St. Petersburg (March 2009)
3. Arlazow, V.L., Dinic, E.A., Kronrod, M.A., Faradzev, I.A.: On economical construction of the transitive closure of an oriented graph. Soviet Math. Dokl. 11, 1209–1210 (1970)
4. Alon, N., Naor, M.: Derandomization, Witnesses for Boolean Matrix Multiplication and Construction of Perfect hash functions. Algorithmica 16, 434–449 (1996)
5. Bash, J., Khanna, S., Motwani, R.: On Diameter Verification and Boolean Matrix Multiplication. Technical Report, Stanford University CS department (1995)
6. Broder, A.Z., Charikar, M., Frieze, A.M., Mitzenmacher, M.: Min-Wise Independent Permutations. Journal of Computer and System Sciences 60, 630–659 (2000)
7. Cohen, E.: Structure Prediction and Computation of Sparse Matrix Products. Journal of Combinatorial Optimization 2, 307–332 (1999)
8. Coppersmith, D.: Rectangular matrix multiplication revisited. Journal of Symbolic Computation 13, 42–49 (1997)
9. Coppersmith, D., Winograd, S.: Matrix Multiplication via Arithmetic Progressions. J. of Symbolic Computation 9, 251–280 (1990)
10. Czumaj, A., Kowaluk, M., Lingas, A.: Faster algorithms for finding lowest common ancestors in directed acyclic graphs. The special ICALP 2005 issue of Theoretical Computer Science 380(1-2), 37–46 (2005)
11. Galil, Z., Margalit, O.: Witnesses for Boolean Matrix Multiplication and Shortest Paths. Journal of Complexity, 417–426 (1993)
12. George, A., Gilbert, J., Liu, J.W.H. (eds.): Graph Theory and Sparse Matrix Computation. The IMA Volumes in Mathematics and its Applications, vol. 56. Springer, Heidelberg (1993)

13. Golub, G., Van Loan, C. (eds.): Matrix Computations. The Johns Hopkins U. Press, Baltimore (1989)
14. Huang, X., Pan, V.Y.: Fast rectangular matrix multiplications and applications. Journal of Complexity 14, 257–299 (1998)
15. Kowaluk, M., Lingas, A.: LCA queries in directed acyclic graphs. In: Caires, L., Italiano, G.F., Monteiro, L., Palamidessi, C., Yung, M. (eds.) ICALP 2005. LNCS, vol. 3580, pp. 241–248. Springer, Heidelberg (2005)
16. Motwani, R., Raghavan, P.: Randomized Algorithms. Cambridge University Press, Cambridge (1995)
17. Munro, J.I.: Efficient determination of the transitive closure of a directed graph. Information Processing Letters 1(2), 56–58 (1971)
18. Rytter, W.: Fast recognition of pushdown automaton and context-free languages. Information and Control 67(1-3), 12–22 (1985)
19. Shapira, A., Yuster, R., Zwick, U.: All-pairs bottleneck paths in vertex weighted graphs. In: Proc. SODA 2007, pp. 978–985 (2007)
20. Seidel, R.: On the All-Pairs-Shortest-Path Problem. In: Proc. 24th annual ACM Symposium on Theory of Computing, pp. 745–749 (1992)
21. Schnorr, C.P., Subramanian, C.R.: Almost Optimal (on the average) Combinatorial Algorithms for Boolean Matrix Product Witnesses, Computing the Diameter. In: Rolim, J.D.P., Serna, M., Luby, M. (eds.) RANDOM 1998. LNCS, vol. 1518, pp. 218–231. Springer, Heidelberg (1998)
22. Yuster, R., Zwick, U.: Fast sparse matrix multiplication. ACM Transactions on Algorithms (TALG) 1(1), 2–13 (2004) (Preliminary version in proc. ESA 2004)

On Optimally Partitioning a Text to Improve Its Compression*

Paolo Ferragina, Igor Nitto, and Rossano Venturini

Department of Computer Science, University of Pisa
{ferragina,nitto,rossano}@di.unipi.it

Abstract. In this paper we investigate the problem of partitioning an input string T in such a way that compressing individually its parts via a base-compressor $\mathcal{C}$ gets a compressed output that is shorter than applying $\mathcal{C}$ over the entire T at once. This problem was introduced in [2,3] in the context of table compression, and further elaborated and extended to strings and trees by [10,11,20], but it is still open how to efficiently compute the optimal partition [4]. In this paper we provide the first algorithm which is guaranteed to compute in $O(n\ \mathtt{polylog}(n))$ time a partition of T whose compressed output is guaranteed to be no more than $(1+\epsilon)$-worse the optimal one, where ϵ is any positive constant.

1 Introduction

Reorganizing data in order to improve the performance of a given compressor $\mathcal{C}$ is an important paradigm of data compression (see e.g. [3,10]). The basic idea consist of *permuting* the input data T to form a new string T' which is then *partitioned* into substrings that are finally compressed *individually* by the base compressor $\mathcal{C}$. The goal is to find the best instantiation of the two steps Permuting+Partitioning which minimizes the total length of the compressed output. This approach (hereafter abbreviated as PPC) is clearly *at least* as powerful as the classic data compression approach that applies $\mathcal{C}$ to the entire T: just take the identity permutation and set $k = 1$. The question is whether it can be *more powerful* than that!

Intuition leads to think favorably about it: by grouping together objects that are "related", one can hope to obtain better compression even using a very weak compressor $\mathcal{C}$. Surprisingly enough, this intuition has been sustained by convincing theoretical and experimental results only recently. These results have investigated the PPC-paradigm under various angles by considering: different data formats (strings [10], trees [11], tables [3], etc.), different granularities for the items of T to be permuted (chars, node labels, columns, blocks [1,19], files [6,23], etc.), different permutations (see e.g. [14,25]), different base compressors to be boosted (0-th order compressors, `gzip`, `bzip2`, etc.). Among these plethora

* It has been partially supported by Yahoo! Research, and two Italian MIUR FIRB Projects: Italy-Israel and 2006-Linguistica. The authors' address is Dipartimento di Informatica, L.go B. Pontecorvo 3, 56127 Pisa, Italy.

A. Fiat and P. Sanders (Eds.): ESA 2009, LNCS 5757, pp. 420–431, 2009.

of proposals, we survey below the most notable examples which are useful to introduce the problem we attack in this paper, and refer the reader to the cited bibliography for other interesting results.

The PPC-paradigm was introduced in [2], and further elaborated upon in [3]. In these papers, T is a *table* formed by fixed-size columns, and the goal is to permute them in such a way that individually compressing contiguous groups of columns gives the shortest compressed output. The authors of [3] showed that the PPC-problem in its full generality is MAX-SNP hard, devised a link between PPC and the classical asymmetric TSP problem, and then resorted known *heuristics* to find approximate solutions for the A-TSP based on several measures of correlations between the table's columns. For the grouping they proposed either an optimal but very slow approach, based on Dynamic Programming (see below), or some fast and very simply heuristics.

When T is a text string, the most famous instantiation of the PPC-paradigm has been obtained by combining the Burrows and Wheeler Transform [5] (shortly `BWT`) with a context-based grouping of the input characters, which are finally compressed via proper 0-th order-entropy compressors (like `MTF`, `RLE`, Huffman, Arithmetic, or their combinations, see e.g. [26]). In this scenario the permutation acts on single characters, and the partitioning/permuting steps deploy the context (substring) following each symbol in the original string. Several papers have given an analytic account of this phenomenon [21,9,17,20] and have shown, also experimentally [8], that the partitioning of the BW-transformed text is a key step for achieving effective compression ratios. Starting from these premises, [15] attacked the computation of the optimal partition of T via a DP-approach, which turned out to be very costly; then [10] (and subsequently many other authors, see e.g. [9,20,11]) proposed a *boosting* solution which is *not* optimal but, nonetheless, achieves interesting k-th order-entropy bounds. This is indeed a subtle point, frequently neglected. In fact, we are able to show [12] that there exists an infinite class of strings for which the compression achieved by the *booster* is far from the optimal-partitioning by a multiplicative factor $\Omega(\sqrt{\log n})$

There is another interesting scenario in which the PPC-paradigm occurs and this is when T is a single (long) file, eventually obtained by concatenating a collection of (smaller) files via any permutation of them: think to the serialization induced by the Unix `tar` command, or other more sophisticated heuristics like the ones discussed in e.g. [23,6]. In these cases, the partitioning step looks for *homogeneous* groups of contiguous files (or even within-file blocks of chars) which can be effectively compressed together by a base-compressor $\mathcal{C}$ — typically `gzip`, `bzip2`, `ppm`, etc. [26]. It is clear that how much redundancy can be detected and exploited by these compressors depends on their ability to "look back" at the previously seen data, and this has a cost in terms of memory usage and running time. Thus most compression systems provide a facility that controls the amount of data that may be processed at once — usually called the *block size* — ranging from few hundreds KBs [26] to few hundreds MBs [8]. Subtly, using larger blocks to be compressed at once does not necessarily induce a better compression ratio! As an example, let us take $\mathcal{C}$ as the simple Huffman or Arithmetic coders and use them to compress the text $T = 0^{n/2}1^{n/2}$: there is a clear difference whether we compress individually the two halves of T (achieving an output size of about $O(\log n)$ bits) or we compress T as a whole (achieving $n + O(\log n)$ bits). A similar example can be shown [12] for more powerful compressors, such as the k-th order entropy encoder `ppm` which compresses each symbol according to its preceding k-long context. Therefore the impact of the choice of the block size cannot be underestimated and may be problematic, since it may change along the whole file we are compressing.

In summary, fast and effective algorithms or heuristics for the permuting step are known (see e.g. [3,10,11,24]), but not yet known is how to compute *efficiently* the *optimal* partition of the permuted data for compression boosting (see [4]). The goal of this paper is to provide the first efficient approximation algorithm for this problem, formally stated as follows.

Let $\mathcal{C}$ be the base compressor we wish to boost, and let $T[1, n]$ be the input string we wish to partition and then compress by $\mathcal{C}$. So, we are assuming that T has been (possibly) permuted in advance, and we are concentrating on the last two steps of the PPC-paradigm. Now, given a partition $\mathcal{P}$ of the input string into contiguous substrings, say $T = T_1 T_2 \cdots T_k$, we denote by $\mathtt{Cost}(\mathcal{P})$ the cost of this partition and measure it as $\sum_{i=1}^{l} |\mathcal{C}(T_i)|$, where $|\mathcal{C}(\alpha)|$ is the length in bit of the string α compressed by $\mathcal{C}$. The problem of *optimally partitioning* T according to the base-compressor $\mathcal{C}$ consists then of computing the partition $\mathcal{P}_{\mathtt{opt}}$ which achieves the shortest compressed output, namely $\mathcal{P}_{\mathtt{opt}} = \min_{\mathcal{P}} \mathtt{Cost}(\mathcal{P})$. As we mentioned above, $\mathcal{P}_{\mathtt{opt}}$ might be computed via a Dynamic-Programming approach [3,15] in $\Theta(n^3)$ time, which is clearly unfeasible even on small input sizes n. In this paper we provide the first algorithm which is guaranteed to compute in $O(n(\log_{1+\epsilon} n)\,\mathtt{polylog}(n)))$ time a partition of T whose compressed output is guaranteed to be no more than $(1+\epsilon)$-worse than the optimal one, where ϵ may be any positive constant. Due to the lack of space, proofs and many details are omitted from this paper but can be found in [12].

2 Notation

In this paper we will use entropy-based upper bounds for the estimation of $|\mathcal{C}(T[i, j])|$, so we need to recall some basic notation and terminology about entropies. Let $T[1, n]$ be a string drawn from the alphabet Σ of size σ. For each $c \in \Sigma$, we let n_c be the number of occurrences of c in T. The zero-th order *empirical* entropy of T is defined as $H_0(T) = \sum_{c\in\Sigma}^{h} \frac{n_c}{n} \log \frac{n}{n_c}$. Recall that $|T|H_0(T)$ provides an information-theoretic lower bound to the output size of any compressor that encodes each symbol of T with a fixed code [26]. The so-called zero-th order statistical compressors (such as Huffman or Arithmetic [26]) achieve an output size which is very close to this bound. However, they require to know information about frequencies of input symbols (called the *model* of the source). Those frequencies can be either known in advance (*static* model) or computed by scanning the input text (*semi-static* model). In both cases the model must be stored in the compressed file to be used by the decompressor.

In the following we will bound the compressed size achieved by zero-th order compressors by $|\mathcal{C}_0(T)| \leq \lambda n H_0(T) + f_0(n,\sigma)$ bits, where λ is a positive constant and $f_0(n,\sigma)$ is a function including the extra costs of encoding the source model and/or other inefficiencies of $\mathcal{C}$. We will also assume that $f_0(n,\sigma)$ can be computed in constant time. As an example, Huffman has $f_0(n,\sigma) = \sigma \log \sigma + n$ and $\lambda = 1$, whereas Arithmetic has $f_0(n,\sigma) = \sigma \log n + \log n/n$ and $\lambda = 1$.

Let us now come to more powerful compressors. For any string u of length k, we denote by u_T the string of single symbols following the occurrences of u in T, taken from left to right. For example, if $T =$ `mississippi` and $u =$ `si`, we have $u_T =$ `sp` since the two occurrences of `si` in T are followed by the symbols `s` and `p`, respectively. The k-th order *empirical* entropy of T is defined as $H_k(T) = \frac{1}{|T|}\sum_{u \in \Sigma^k} |u_T|\, H_0(u_T)$. We have $H_k(T) \geq H_{k+1}(T)$ for any $k \geq 0$. As usual in data compression [21], the value $nH_k(T)$ is an information-theoretic lower bound to the output size of any compressor that encodes each symbol of T with a fixed code that depends on the symbol itself and on the k immediately preceding symbols. Recently (see e.g. [18,21,10,9,20,11] and refs therein) authors have provided *upper bounds* in terms of $H_k(T)$ for sophisticated data-compression algorithms, such as `gzip` [18], `bzip2` [21,10,17], and `ppm`. These bounds have the form $|\mathcal{C}(T)| \leq \lambda |T|\, H_k(T) + f_k(|T|,\sigma)$, where λ is a positive constant and $f_k(|T|,\sigma)$ is a function including the extra-cost of encoding the source model and/or other inefficiencies of $\mathcal{C}$. The smaller are λ and $f_k()$, the better is the compressor $\mathcal{C}$. As an example, the bound of the compressor in [20] has $\lambda = 1$ and $f(|T|,\sigma) = O(\sigma^{k+1} \log |T| + |T| \log \sigma \log\log |T| / \log |T|)$.[1]

In our paper we will use these entropy-based bounds for the estimation of $|\mathcal{C}(T[i,j])|$, but of course this will not be enough to achieve a fast DP-based algorithm for our optimal-partitioning problem. We cannot re-compute from scratch those estimates for every substring $T[i,j]$ of T, being them $\Theta(n^2)$ in number. So we will show some structural properties of our problem (Sect 3) and introduce few novel technicalities (Sect 4–5) that will allow us to compute $H_k(T[i,j])$ only on a *reduced* subset of T's substrings, having size $O(n \log_{1+\epsilon} n)$, by taking $O(\texttt{polylog}(n))$ time per substring and $O(n)$ space overall.

3 The Problem and Our Solution

The optimal partitioning problem, stated in Sect 1, can be reduced to a single source shortest path computation (SSSP) over a directed acyclic graph $\mathcal{G}(T)$ defined as follows. The graph $\mathcal{G}(T)$ has a vertex v_i for each text position i of T, plus an additional vertex v_{n+1} marking the end of the text, and an edge connecting vertex v_i to vertex v_j for any pair of indices i and j such that $i < j$. Each edge (v_i, v_j) has associated the cost $c(v_i, v_j) = |\mathcal{C}(T[i, j-1])|$ that corresponds to the size in bits of the substring $T[i, j-1]$ compressed by $\mathcal{C}$. We remark the following crucial, but easy to prove, property of the cost function:

[1] Many results (see [21,10,9] and refs therein) are expressed in term of k-th order modified empirical entropy of T and have the form $\lambda |T| H_k^*(T) + g_k(\sigma)$. In [12] we show how to extend our results to this entropy too.

Fact 1 *For any vertex* v_i*, it is* $0 < c(v_i, v_{i+1}) \leq c(v_i, v_{i+2}) \leq \ldots \leq c(v_i, v_{n+1})$

There is a one-to-one correspondence between paths from v_1 to v_{n+1} in $\mathcal{G}(T)$ and partitions of T: every edge (v_i, v_j) in the path identifies a contiguous substring $T[i, j-1]$ of the corresponding partition. Therefore the cost of a path equals the (compression-)cost of the corresponding partition. Thus we can find the optimal partition of T by computing the shortest path in $\mathcal{G}(T)$ from v_1 to v_{n+1}. Unfortunately this simple approach has two main drawbacks:

1. the number of edges in $\mathcal{G}(T)$ is $\Theta(n^2)$, thus making the SSSP computation inefficient if executed directly over $\mathcal{G}(T)$;
2. the computation of the each edge cost might take $\Theta(n)$ time over most T's substrings, if $\mathcal{C}$ is run on each of them from scratch.

In the following sections we will successfully address both these two drawbacks. First, we sensibly reduce the number of edges in the graph $\mathcal{G}(T)$ to be examined during the SSSP computation and show that we can obtain a $(1+\epsilon)$ approximation using only $O(n \log_{1+\epsilon} n)$ edges, where $\epsilon > 0$ is a user-defined parameter (Sect 3.1). Second, we show some sufficient properties that $\mathcal{C}$ needs to satisfy in order to compute every edge's cost efficiently. These properties hold for some well-known compressors— e.g. 0-order compressors, `PPM`-like and `bzip`-like compressors— so, for them, we will show how to compute each edge cost in constant or polylogarithmic time (Sect 4–6).

3.1 A Pruning Strategy

The aim of this section is to design a *pruning* strategy that produces a subgraph $\mathcal{G}_\epsilon(T)$ of the original DAG $\mathcal{G}(T)$ in which the shortest path distance between its leftmost and rightmost nodes, v_1 and v_{n+1}, increases by no more than a factor $(1+\epsilon)$. We define $\mathcal{G}_\epsilon(T)$ to contain all edges (v_i, v_j) of $\mathcal{G}(T)$, recall $i < j$, such that at least one of the following two conditions holds:

1. there exists a positive integer k such that $c(v_i, v_j) \leq (1+\epsilon)^k < c(v_i, v_{j+1})$;
2. $j = n+1$.

In other words, by property 1, we are keeping for each integer k the edge of $\mathcal{G}(T)$ that approximates at the best the value $(1+\epsilon)^k$ from below. Given this, we will call ϵ-*maximal* the edges of $\mathcal{G}_\epsilon(T)$. Clearly, each vertex of $\mathcal{G}_\epsilon(T)$ has at most $\log_{1+\epsilon} n = O(\frac{1}{\epsilon} \log n)$ outgoing edges, which are ϵ-maximal by definition. Therefore the total size of $\mathcal{G}_\epsilon(T)$ is $O(n \log_{1+\epsilon} n)$. Hereafter, we denote with $d_G(-,-)$ the shortest path distance between any two nodes in a graph G.

The following lemma states a basic property of shortest path distances over our special DAG $\mathcal{G}(T)$:

Lemma 1. *For any triple of indices* $1 \leq i \leq j \leq q \leq n+1$ *we have* $d_{\mathcal{G}(T)}(v_j, v_q) \leq d_{\mathcal{G}(T)}(v_i, v_q)$ *and* $d_{\mathcal{G}(T)}(v_i, v_j) \leq d_{\mathcal{G}(T)}(v_i, v_q)$.

The correctness of our pruning strategy relies on the following theorem:

Theorem 2. *For any text T, the shortest path in $\mathcal{G}_\epsilon(T)$ from v_1 to v_{n+1} has a total cost of at most $(1+\epsilon)\ d_{\mathcal{G}(T)}(v_1, v_{n+1})$.*

Proof: We prove a stronger assertion: $d_{\mathcal{G}_\epsilon(T)}(v_i, v_{n+1}) \leq (1+\epsilon)\ d_{\mathcal{G}(T)}(v_i, v_{n+1})$ for any index $1 \leq i \leq n+1$. This is clearly true for $i = n+1$, because in that case the distance is 0. Now let us inductively consider the shortest path π in $\mathcal{G}(T)$ from v_i to v_{n+1} and let $(v_k, v_{t_1})(v_{t_1}, v_{t_2}) \ldots (v_{t_h} v_{n+1})$ be its edges. By the definition of ϵ-maximal edge, it is possible to find an ϵ-maximal edge (v_k, v_r) with $t_1 \leq r$, such that $c(v_k, v_r) \leq (1+\epsilon)\ c(v_k, v_{t_1})$. By Lemma 1, $d_{\mathcal{G}(T)}(v_r, v_{n+1}) \leq d_{\mathcal{G}(T)}(v_{t_1}, v_{n+1})$. By induction, $d_{\mathcal{G}_\epsilon(T)}(v_r, v_{n+1}) \leq (1+\epsilon)\ d_{\mathcal{G}(T)}(v_r, v_{n+1})$. Combining this with the triangle inequality we get the thesis. □

3.2 Space and Time Efficient Algorithms for Generating $\mathcal{G}_\epsilon(T)$

Theorem 2 ensures that, in order to compute a $(1+\epsilon)$ approximation of the optimal partition of T, it suffices to compute the SSSP in $\mathcal{G}_\epsilon(T)$ from v_1 to v_{n+1}. This can be easily computed in $O(|\mathcal{G}_\epsilon(T)|) = O(n \log_\epsilon n)$ time since $\mathcal{G}_\epsilon(T)$ is a DAG [7], by making a single pass over its vertices and relaxing all edges going out from the current one.

However, generating $\mathcal{G}_\epsilon(T)$ in efficient time is a non-trivial task for three main reasons. First, the original graph $\mathcal{G}(T)$ contains $\Omega(n^2)$ edges, so that we cannot check each of them to determine whether it is ϵ-maximal or not, because this would take $\Omega(n^2)$ time. Second, we cannot compute the cost of an edge (v_i, v_j) by executing $\mathcal{C}(T[i, j-1])$ *from scratch*, since this would require time linear in the substring length, and thus $\Omega(n^3)$ time over all T's substrings. Third, we cannot materialize $\mathcal{G}_\epsilon(T)$ (e.g. its adjacency lists) because it consists of $\Theta(n\ \texttt{polylog}(n))$ edges, and thus its space occupancy would be super-linear in the input size.

The rest of this section is devoted to design an algorithm which overcomes the three limitations above. The specialty of our algorithm consists of materializing $\mathcal{G}_\epsilon(T)$ on-the-fly, as its vertices are examined during the SSSP-computation, by spending only polylogarithmic time per edge. The actual time complexity per edge will depend on the entropy-based cost function we will use to estimate $|\mathcal{C}(T[i, j-1])|$ (see Sect 2) and on the dynamic data structure we will deploy to compute that estimation efficiently.

The key tool we use to make a fast estimation of the edge costs is a dynamic data structure built over the input text T and requiring $O(|T|)$ space. We state the main properties of this data structure in an abstract form, in order to design a general framework for solving our problem; in the next sections we will then provide implementations of this data structure and thus obtain real time/space bounds for our problem. So, let us assume to have a dynamic data structure that maintains a set of *sliding windows* over T denoted by $w_1, w_2, \ldots, w_{\log_{1+\epsilon} n}$. The sliding windows are substrings of T which start at the same text position l but have different lengths: namely, $w_i = T[l, r_i]$ and $r_1 \leq r_2 \leq \ldots \leq r_{\log_{1+\epsilon} n}$. The data structure must support the following three operations:

1. `Remove`() moves the starting position l of all windows one position to the right (i.e. $l+1$);
2. `Append`(w_i) moves the ending position of the window w_i one position to the right (i.e. r_i+1);
3. `Size`(w_i) computes and returns the value $|\mathcal{C}(T[l,r_i])|$.

This data structure is enough to generate ϵ-maximal edges via a single pass over T, using $O(|T|)$ space. More precisely, let v_l be the vertex of $\mathcal{G}(T)$ currently examined by our SSSP computation, and thus l is the current position reached by our scan of T. We maintain the following invariant: the sliding windows correspond to all ϵ-maximal edges going out from v_l, that is, the edge (v_l, v_{1+r_t}) is the ϵ-maximal edge satisfying $c(v_l, v_{1+r_t}) \leq (1+\epsilon)^t < c(v_l, v_{1+(r_t+1)})$. Initially all indices are set to 0. To maintain the invariant, when the text scan advances to the next position $l+1$, we call operation `Remove`() once to increment index l and, for each $t = 1, \ldots, \log_{1+\epsilon}(n)$, we call operation `Append`(w_t) until we find the largest r_t such that `Size`(w_t) $= c(v_l, v_{1+r_t}) \leq (1+\epsilon)^t$. The key issue here is that `Append` and `Size` are paired so that our data structure should take advantage of the rightward sliding of r_t for computing $c(v_l, v_{1+r_t})$ efficiently. Just one character is entering w_t to its right, so we need to deploy this fact for making the computation of `Size`(w_t) fast (given its previous value). Here comes into play the second contribution of our paper that consists of adopting the entropy-bounded estimates for the compressibility of a string, mentioned in Sect 2, to estimate indeed the edge costs `Size`(w_t) $= |\mathcal{C}(w_t)|$. This idea is crucial because we will be able to show that these functions do satisfy some structural properties that admit a *fast incremental computation*, as the one required by `Append` + `Size`. These issues will be discussed in the following sections, here we just state that, overall, the SSSP computation over $\mathcal{G}_\epsilon(T)$ takes $O(n)$ calls to operation `Remove`, and $O(n \log_{1+\epsilon} n)$ calls to operations `Append` and `Size`.

Theorem 3. *If we have a dynamic data structure occupying $O(n)$ space and supporting operation* `Remove` *in time $L(n)$, and operations* `Append` *and* `Size` *in time $R(n)$, then we can compute the shortest path in $\mathcal{G}_\epsilon(T)$ from v_1 to v_{n+1} taking $O(n\,L(n) + (n \log_{1+\epsilon} n)\,R(n))$ time and $O(n)$ space.*

4 On Zero-th Order Compressors

In this section we explain how to implement the data structure above whenever $\mathcal{C}$ is a 0-th order compressor, and thus H_0 is used to provide a bound to the compression cost of $\mathcal{G}(T)$'s edges (see Sect 2). The key point is actually to show how to efficiently compute `Size`(w_i) as the sum of $|T[l,r_i]|H_0(T[l,r_i]) = \sum_{c\in\Sigma} n_c \log((r_i-l+1)/n_c)$ (see its definition in Sect 2) plus $f_0(r_i-l+1, |\Sigma_{T[l,r_i]}|)$, where n_c is the number of occurrences of symbol c in $T[l,r_i]$ and $|\Sigma_{T[l,r_i]}|$ denotes the number of different symbols in $T[l,r_i]$.

The first solution we are going to present is very simple and uses $O(\sigma)$ space per window. The idea is the following: for each window w_i we keep in memory an array of counters $A_i[c]$ indexed by symbol c in Σ. At any step of our algorithm, the counter $A_i[c]$ stores the number of occurrences of symbol c in $T[l, r_i]$. For any window w_i, we also use a variable E_i that stores the value of $\sum_{c\in\Sigma} A_i[c]\log A_i[c]$. It is easy to notice that:

$$|T[l, r_i]|\, H_0(T[l, r_i]) = (r_i - l + 1)\log(r_i - l + 1) - E_i. \qquad (1)$$

Therefore, if we know the value of E_i, we can answer to a query Size(w_i) in constant time. So, we are left with showing how to implement efficiently the two operations that modify l or any rs value and, thus, modify appropriately the E's value. This can be done as follows:

1. Remove(): For each window w_i, we subtract from the appropriate counter and from variable E_i the contribution of the symbol $T[l]$ which has been evicted from the window. That is, we decrease $A_i[T[l]]$ by one, and update E_i by subtracting $(A_i[T[l]]+1)\log(A_i[T[l]]+1)$ and then summing $A_i[T[l]]\log A_i[T[l]]$. Finally we set $l = l + 1$.
2. Append(w_i): We add to the appropriate counter and variable E_i the contribution of the symbol $T[r_i + 1]$ which has been appended to window w_i. That is, we increase $A_i[T[r + 1]]$ by one, then we update E_i by subtracting $(A[T[r_i+1]]-1)\log(A[T[r_i+1]]-1)$ and summing $A[T[r_i+1]]\log A[T[r_i+1]]$. Finally we set $r_i = r_i + 1$.

In this way, operation Remove requires constant time per window, hence $O(\log_{1+\epsilon} n)$ time overall. Append(w_i) takes constant time. The space required by the counters A_i is $O(\sigma \log_{1+\epsilon} n)$ words. Unfortunately, the space complexity of this solution can be too much when it is used as the basic-block for computing the k-th order entropy of T (see Sect 2) as we will do in Sect 5. In fact, we would achieve $\min(\sigma^{k+1}\log_{1+\epsilon} n, n\log_{1+\epsilon} n)$ space, which may be superlinear in n depending on σ and k. We can extend the previous scheme to obtain an efficient solution up to $\sigma = O(\texttt{poly}(n))$ but, because of space limitations, its technicalities are reported in the full version of this paper [12]. The following lemma states the final result.

Lemma 2. *Let $T[1, n]$ be a text drawn from an alphabet of size $\sigma = \texttt{poly}(n)$. If we estimate* Size() *via 0-th order entropy (as detailed in Sect 2), then we can design a dynamic data structure that takes $O(n)$ space and supports the operations* Remove *in $R(n) = O(\log_{1+\epsilon} n)$ time, and* Append *and* Size *in $L(n) = O(1)$ time.*

In order to evict the cost of the model from the compressed output (see Sect 2), authors typically resort to zero-th order *adaptive* compressors which do not store the symbols' frequencies, since they are computed *incrementally* during the compression [16]. An approach similar to the previous one (but with little more technicalities, given in [12]) can be used to solve the case in which we use a 0-th order adaptive compressor $\mathcal{C}$ for providing the edge-costs of $\mathcal{G}(T)$. For this case we can prove the same time and space bounds of Lemma 2. Combining this with Theorem 3 we obtain:

Theorem 4. *Given a text $T[1,n]$ drawn from an alphabet of size $\sigma = \texttt{poly}(n)$, we can find an $(1+\epsilon)$-optimal partition of T with respect to a 0-th order (adaptive) compressor in $O(n \log_{1+\epsilon} n)$ time and $O(n)$ space, where ϵ is any positive constant.*

We point out that this result can be applied to the compression booster of [10] to fast obtain an approximation of the optimal partition of $\texttt{BWT}(T)$. This may be better than the algorithm of [10] both in time complexity, since that algorithm takes $O(n\sigma)$ time, and in compression ratio (details in [12]). The case of a large alphabet (namely, $\sigma = \Omega(\texttt{polylog}(n))$) is particularly interesting whenever we consider either a word-based BWT [22] or the XBW-transform over labeled trees [10] for which Σ is either the set of words or tags in a text. We notice that our result is interesting also for the *Huffword* compressor which is the standard choice for the storage of Web pages [26]; here Σ consists of the distinct words constituting the Web-page collection.

5 On k-th Order Compressors

In this section we make one step further and consider the more powerful k-th order compressors, for which do exist H_k bounds for estimating the size of their compressed output. Here $\texttt{Size}(w_i)$ must compute $|\mathcal{C}(T[l, r_i])|$ which is estimated, as detailed in Sect 2, by $(r_i - l + 1)H_k(T[l, r_i]) + f_k(r_i - l + 1, |\Sigma_{T[l,r_i]}|)$, where $\Sigma_{T[l,r_i]}$ denotes the number of different symbols in $T[l, r_i]$..

Let us denote with $T_q[1, n-q]$ the text whose i-th symbol $T_q[i]$ is equal to the q-gram $T[i, i+q-1]$. Actually, we can remap the symbols of T_q to integers in $[1, n]$. In fact the number of distinct q-grams occurring in T_q is less than $n = |T|$. Thus T_q's symbols take $O(\log n)$ bits and T_q can be stored in $O(n)$ space. This remapping takes linear time and space, whenever σ is polynomial in n.

It is well-known that the k-th order entropy of a string (see definition Sect 2) can be expressed as the difference between the zero-th order entropy of its $(k+1)$-grams and the one of its k-grams. This suggests that we can use the solution of the previous section in order to compute the zero-th order entropy of the appropriate substrings of T_{k+1} and T_k. More precisely, we use two instances of the data structure of Theorem 4 (one for T_{k+1} and one for T_k), which are kept *synchronized* in the sense that, when operations are performed on one data structure, then they are also executed on the other.

Lemma 3. *Let $T[1,n]$ be a text drawn from an alphabet of size $\sigma = \texttt{poly}(n)$. If we estimate* Size() *via k-th order entropy (as detailed in Sect 2), then we can design a dynamic data structure that takes $O(n)$ space and supports the operations* Remove *in $R(n) = O(\log_{1+\epsilon} n)$ time, and* Append *and* Size *in $L(n) = O(1)$ time.*

Essentially the same technique is applicable to the case of k-th order *adaptive* compressor $\mathcal{C}$, in this case we keep up-to-date the 0-th order *adaptive* entropies of the strings T_{k+1} and T_k (details in [12]).

Theorem 5. *Given a text $T[1,n]$ drawn from an alphabet of size $\sigma = \texttt{poly}(n)$, we can find an $(1+\epsilon)$-optimal partition of T with respect to a k-th order (adaptive) compressor in $O(n \log_{1+\epsilon} n)$ time and $O(n)$ space, where ϵ is any positive constant.*

We point out that this result applies also to the practical case in which the base compressor $\mathcal{C}$ has a maximum (block) size B of data it can process at once (this is the typical scenario for gzip, bzip2, etc.). In this situation the time performance of our solution reduces to $O(n \log_{1+\epsilon}(B \log \sigma))$.

6 On BWT-based Compressors

As we mentioned in Sect 2 we know entropy-bounded estimates for the output size of BWT-based compressors. So we could apply Theorem 5 to compute the optimal partitioning of T for such a type of compressors. Nevertheless, it is also known [8] that such compression-estimates are rough in practice because of the features of the compressors that are applied to the BWT(T)-string. Typically, BWT is encoded via a sequence of simple compressors such as MTF, RLE (which is optional), and finally a 0-order encoder like Huffman or Arithmetic [26]. For each of these compression steps, a 0-th entropy bound is known [10,9], but the combination of these bounds may result far from the final compressed size produced by the overall sequence of compressors in practice [8].

We propose a solution to the optimal partitioning problem for BWT-based compressors that introduces a $\Theta(\sigma \log n)$ slowdown in the time complexity of Theorem 5, but with the advantage of computing the $(1+\epsilon)$-optimal solution wrt the real compressed size, thus without any estimation of it by entropy-cost functions. Since in practice it is $\sigma = \texttt{polylog}(n)$, this slowdown is negligible. In order to achieve this result, we need to address a slightly different problem which is defined as follows. The input string T has the form $S[1]\#_1 S[2]\#_2 \ldots S[m]\#_n$ where each $S[i]$ is a text (called *page*) drawn from an alphabet Σ, and $\#_1, \#_2, \ldots, \#_n$ are special characters greater than any symbol of Σ. A partition of T must be page-aligned, in that it must form *groups of contiguous pages* $S[i]\#_i \ldots S[j]\#_j$, denoted also $S[i,j]$. Our aim is to find a page-aligned partition whose compressed output-size is a factor at most $(1+\epsilon)$ the minimum possible one, for any fixed $\epsilon > 0$. We notice that this problem generalizes the table partitioning problem [3], since we can assume that $S[i]$ is a column of the table.

To simplify things we will drop the RLE encoding step of a BWT-based algorithm, and defer the complete solution to the full version of this paper. We start by noticing that an analog of Theorem 3 holds for this variant of the optimal partitioning problem, which implies that a $(1+\epsilon)$-approximation of the optimum cost (and the corresponding partition) can be computed using a data structure supporting operations Append, Remove, and Size; with the only difference that the windows $w_1, w_2, \ldots, w_m$ subject to the operations are groups of contiguous pages of the form $w_i = S[l, r_i]$.

It goes without saying that there exist data structures designed to maintain a dynamic text compressed with a BWT-based compressor under insertions and deletions of symbols (see [13] and references therein). But they do not fit our context for two reasons: (1) their underlying compressor is significantly different from the scheme above; (2) in the worst case, they would spend linear space per window yielding a super-linear overall space complexity.

Instead of keeping a given window w in compressed form, our approach will only store the frequency distribution of the integers in the string $w' =$ MTF(BWT(w)) since this is enough to compute the compressed output size produced by the final step of the BWT-based algorithm, which is usually implemented via Huffman or Arithmetic [26]. Indeed, since MTF produces a sequence of integers from 0 to σ, we can store their number of occurrences for each window w_i into an array F_{w_i} of size σ. The update of F_{w_i} due to the insertion or the removal of a page in w_i incurs two main difficulties: (1) how to update w_i' as pages are added/removed from the extremes of the window w_i, (2) perform this update implicitly over F_{w_i}, because of the space reasons mentioned above. Our solution relies on two key facts about BWT and MTF:

1. Since the pages are separated in T by distinct separators, inserting or removing one page into a window w does not alter the relative lexicographic order of the original suffixes of w (see [13]).
2. If a string s' is obtained from string s by inserting or removing a char c into an arbitrary position, then MTF(s') differs from MTF(s) in at most σ symbols. More precisely, if c' is the next occurrence in s of the newly inserted (or removed) symbol c, then the MTF has to be updated only in the first occurrence of each symbol of Σ among c and c'.

Due to space limitations we defer to [12] for details, and state here the result we are able to achieve.

Theorem 6. *Given a sequence of texts of total length n and alphabet size $\sigma =$ poly(n), we can compute an $(1+\epsilon)$-approximate solution to the optimal partitioning problem for a* BWT*-based compressor, in $O(n(\log_{1+\epsilon} n)\ \sigma\ \log n)$ time and $O(n + \sigma \log_{1+\epsilon} n)$ space.*

We conclude this paper by devising two possible directions of research, which consist either in investigating the design of $o(n^2)$-time algorithms for computing the *exact* optimal partition, and/or experimenting our solution over large textual datasets.

References

1. Bentley, J.L., McIlroy, M.D.: Data compression with long repeated strings. Information Sciences 135(1-2), 1–11 (2001)
2. Buchsbaum, A.L., Caldwell, D.F., Church, K.W., Fowler, G.S., Muthukrishnan, S.: Engineering the compression of massive tables: an experimental approach. In: Proc. ACM-SIAM SODA, pp. 175–184 (2000)
3. Buchsbaum, A.L., Fowler, G.S., Giancarlo, R.: Improving table compression with combinatorial optimization. J. ACM 50(6), 825–851 (2003)

4. Buchsbaum, A.L., Giancarlo, R.: Table compression. In: Kao, M.Y. (ed.) Encyclopedia of Algorithms, pp. 939–942. Springer, Heidelberg (2008)
5. Burrows, M., Wheeler, D.: A block-sorting lossless data compression algorithm. Technical Report 124, Digital Equipment Corporation (1994)
6. Chang, F., Dean, J., Ghemawat, S., et al.: Bigtable: A distributed storage system for structured data. ACM Trans. Comput. Syst. 26(2) (2008)
7. Cormen, T.H., Leiserson, C.E., Rivest, R.L., Stein, C.: Introduction to Algorithms, 2nd edn. The MIT Press, McGraw-Hill Book Company (2001)
8. Ferragina, P., Giancarlo, R., Manzini, G.: The engineering of a compression boosting library: Theory vs practice in BWT compression. In: Azar, Y., Erlebach, T. (eds.) ESA 2006. LNCS, vol. 4168, pp. 756–767. Springer, Heidelberg (2006)
9. Ferragina, P., Giancarlo, R., Manzini, G.: The myriad virtues of wavelet trees. Information and Computation 207, 849–866 (2009)
10. Ferragina, P., Giancarlo, R., Manzini, G., Sciortino, M.: Boosting textual compression in optimal linear time. J. ACM 52, 688–713 (2005)
11. Ferragina, P., Luccio, F., Manzini, G., Muthukrishnan, S.: Structuring labeled trees for optimal succinctness, and beyond. In: Proc. FOCS, pp. 184–193 (2005)
12. Ferragina, P., Nitto, I., Venturini, R.: On optimally partitioning a text to improve its compression. CoRR, abs/0906.4692 (2009)
13. Ferragina, P., Venturini, R.: The compressed permuterm index. ACM Transactions on Algorithms (to appear, 2009)
14. Giancarlo, R., Restivo, A., Sciortino, M.: From first principles to the burrows and wheeler transform and beyond, via combinatorial optimization. Theoretical Computer Science 387(3), 236–248 (2007)
15. Giancarlo, R., Sciortino, M.: Optimal partitions of strings: A new class of Burrows-Wheeler compression algorithms. In: Baeza-Yates, R., Chávez, E., Crochemore, M. (eds.) CPM 2003. LNCS, vol. 2676, pp. 129–143. Springer, Heidelberg (2003)
16. Howard, P.G., Vitter, J.S.: Analysis of arithmetic coding for data compression. Information Processing Management 28(6), 749–764 (1992)
17. Kaplan, H., Landau, S., Verbin, E.: A simpler analysis of burrows-wheeler-based compression. Theoretical Computer Science 387(3), 220–235 (2007)
18. Kosaraju, R., Manzini, G.: Compression of low entropy strings with Lempel–Ziv algorithms. SIAM Journal on Computing 29(3), 893–911 (1999)
19. Kulkarni, P., Douglis, F., LaVoie, J.D., Tracey, J.M.: Redundancy elimination within large collections of files. In: USENIX, pp. 59–72 (2004)
20. Mäkinen, V., Navarro, G.: Implicit compression boosting with applications to self-indexing. In: Ziviani, N., Baeza-Yates, R. (eds.) SPIRE 2007. LNCS, vol. 4726, pp. 229–241. Springer, Heidelberg (2007)
21. Manzini, G.: An analysis of the Burrows-Wheeler transform. J. ACM 48(3), 407–430 (2001)
22. Moffat, A., Isal, R.Y.: Word-based text compression using the Burrows-Wheeler transform. Information Processing Management 41(5), 1175–1192 (2005)
23. Suel, T., Memon, N.: Algorithms for delta compression and remote file synchronization. In: Lossless Compression Handbook. Academic Press, London (2002)
24. Trendafilov, D., Memon, N., Suel, T.: Compressing file collections with a TSP-based approach. Technical report, TR-CIS-2004-02, Polytechnic University (2004)
25. Vo, B.D., Vo, K.-P.: Compressing table data with column dependency. Theoretical Computer Science 387(3), 273–283 (2007)
26. Witten, I.H., Moffat, A., Bell, T.C.: Managing Gigabytes: Compressing and Indexing Documents and Images. Morgan Kaufmann, San Francisco (1999)

An Average-Case Analysis for Rate-Monotonic Multiprocessor Real-Time Scheduling

Andreas Karrenbauer* and Thomas Rothvoß

Institute of Mathematics
EPFL, Lausanne, Switzerland
{andreas.karrenbauer,thomas.rothvoss}@epfl.ch

Abstract. We introduce the *First Fit Matching Periods* algorithm for rate-monotonic multiprocessor scheduling of periodic tasks with implicit deadlines and show that it yields asymptotically optimal processor assignments if utilization values are chosen uniformly at random. More precisely we prove that the *expected waste* is upper bounded by $\mathcal{O}(n^{3/4}(\log n)^{3/8})$. Here the waste denotes the ratio of idle times, cumulated over all processors and n gives the number of tasks.

The algorithm can be implemented to run in time $\mathcal{O}(n \log n)$ and even in the worst case, an asymptotic approximation ratio of 2 is guaranteed. Experiments yield an average waste proportional to $n^{0.70}$, indicating that the above upper bound on the expected waste is almost tight.

While such average-case analyses are a classical topic of Bin Packing, to the best of our knowledge, this is the first result dealing with a theoretical average-case analysis for this scheduling problem, which was described by Liu and Layland more than 35 years ago and has received a lot of attention, especially in the real-time and embedded-systems community.

1 Introduction

In this paper, we are concerned with a scheduling problem introduced by Liu and Layland [1], which is of fundamental importance in the real-time and embedded-systems community. Here one is given a set of *tasks* $\mathcal{S} = \{\tau_1, \dots, \tau_n\}$, where each task τ is characterized by two positive values, its *period* $p(\tau)$ and its *running time* $c(\tau)$. The task τ releases a *job* requiring running time $c(\tau)$ at each integer multiple of its period. Each job has a relative deadline of $p(\tau)$, thus we have *implicit deadlines*. The *utilization* of a task τ is defined as $u(\tau) = c(\tau)/p(\tau)$, thus it gives the average fraction of processor cycles, which are consumed by τ. More general for a set $\mathcal{S}$, we denote $u(\mathcal{S}) = \sum_{\tau \in \mathcal{S}} u(\tau)$.

We consider *fixed-priority, preemptive* scheduling, i.e. priorities are assigned to the tasks and the arrival of a job of a higher priority task, preempts the execution of lower priority tasks. Liu and Layland [1] have proven that the *rate-monotonic* (RM) scheduling policy is optimal, meaning that if there is a feasible priority assignment, then the one in which the priority of a task τ equals $1/p(\tau)$ is also feasible (i.e. larger periods imply lower priorities). Therefore we only consider rate-monotonic priorities.

---

* Supported by the Deutsche Forschungsgemeinschaft (DFG) within Priority Programme 1307 "Algorithm Engineering".

A. Fiat and P. Sanders (Eds.): ESA 2009, LNCS 5757, pp. 432–443, 2009.

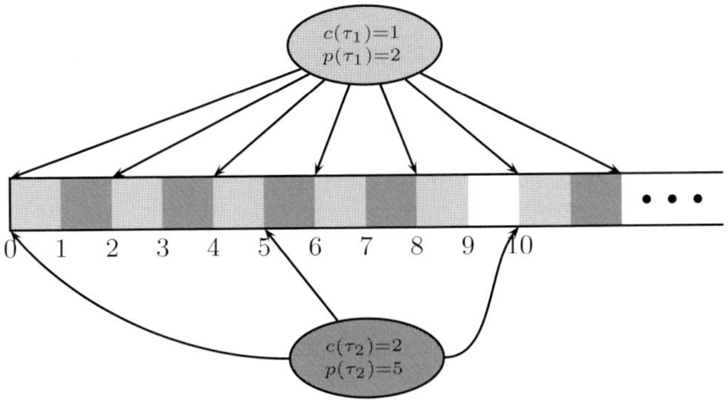

Fig. 1. The picture shows a set $\mathcal{S} = \{\tau_1, \tau_2\}$ of tasks. The arrows indicate the points in time, where the two tasks τ_1 and τ_2 release jobs. At time 0, the first job of τ_1 as well as the first job of τ_2 are released. Since the period of τ_1 is smaller than the period of τ_2, the first job of τ_1 is executed, until it is finished at time 1. Now the first job of τ_2 is executed, but interrupted by the second job of τ_1 at time 2. The execution of the first job of τ_2 is resumed at time 3 and finished at time 4. Notice that the processor is idle for one time unit at time 9 and that the schedule repeats at the least common multiple of the periods which is 10. All jobs finish in time. The set $\mathcal{S}$ is feasible.

If several tasks $\mathcal{S}' \subseteq \mathcal{S}$ are assigned to one processor, then we call this assignment *feasible* (or RM-schedulable) if in the rate-monotonic schedule all jobs of all tasks always meet their deadlines. See Figure 1 for an example.

In a multiprocessor environment, the algorithmic challenge is to determine a partition of a task-set $\mathcal{S}$ into $\mathcal{S}_1, \ldots, \mathcal{S}_k$, such that each $\mathcal{S}_i$ is a feasible set of tasks for one processor and the number k of processors is minimized. The minimum possible value for k is denoted by OPT. The *rate-monotonic multiprocessor scheduling problem* has received considerable attention in the real-time and embedded-systems community [2,3,4,5,6,7,8,9,10,11,12,13]. This popularity is due to the fact that more and more safety-critical control applications are carried out by microprocessors and in particular by multiprocessor environments. Such scheduling problems are for example relevant in the automotive and aviation industry.

A measure for the quality of a solution is the so-called *waste*, which is frequently used concerning the related *Bin Packing* problem. That is, the waste of a solution with k processors is the ratio of idles times, cumulated over all processors, i.e. $k - u(\mathcal{S})$. Clearly, minimizing the waste is equivalent to minimization of the number of partitions.

For the rate-monotonic single-processor scheduling Lehoczky et al. [2] gave a probabilistic analysis, indicating that the reachable processor utilization on average is much better, than the worst-case value of $\ln(2) \approx 69\%$. For example, if periods are drawn from $[1, 100]$ and the running times are scaled by the largest value, such that the system is barely schedulable, then the utilization tends to 88% for $n \to \infty$.

This motivates us to study also the average-case behavior in the multiprocessor case. Our analysis will work for an arbitrary distribution of the periods, as long as the utilization values are drawn independently and uniformly from $[0, 1]$.

Related work. For the famous *Bin Packing* problem a list of items $a_1, \ldots, a_n \in [0, 1]$ is given. The goal is to assign these items to a minimal number of bins such that the total sizes of items, assigned to each bin does not exceed 1.

We will see that if for the considered scheduling problem all periods $p(\tau)$ were multiples of each other, then the problem would be exactly Bin Packing, where the utilization values correspond to the item sizes. This is because a set of tasks $\mathcal{S}' \subseteq \mathcal{S}$ would be feasible on one processor in this case if and only if the sum of their utilization is bounded by one.

Successful heuristics for Bin Packing are *First Fit*, *Next Fit* and *Best Fit*. In all variants the items are assigned in a consecutive manner to a bin, which has enough space (or a new one is opened). For First Fit the current item is put in the bin with the smallest index, in Best Fit it is assigned to the bin, whose item sum is maximal. For Next Fit an active bin is maintained. If the current item does not fit into it, a new bin is opened, now being the active one; old bins are never considered again. In *First Fit Decreasing* the items are first sorted by decreasing sizes and then distributed via First Fit. In the worst case Next Fit produces a 2-approximation, while First Fit needs $\lceil \frac{17}{10} OPT_{\text{BinPacking}} \rceil + 1$ [14] many bins. Asymptotically both, Best and First Fit Decreasing have an approximation ratio of $11/9$ [15].

If the items are generated randomly, the heuristics perform much better, than in the worst-case scenarios. For item sizes drawn uniformly at random from $[0, 1]$ the Best Fit algorithm yields an expected waste of $\Theta(\sqrt{n}\log^{3/4} n)$ [16], while for First Fit this value is lower bounded by $\Omega(n^{2/3})$ and upper bounded by $\mathcal{O}(n^{2/3}\sqrt{\log n})$ [16]. The upper bound even holds if First Fit is restricted to never assign more than 2 items per bin. Later we will refer to this algorithm as *Matching First Fit* (`MFF`). First Fit Decreasing yields an even smaller waste of $\Theta(\sqrt{n})$ [17,18,19]. If item sizes are drawn uniformly from $[0, \alpha]$, for any constant $\alpha \leq 1/2$, the waste of First Fit Decreasing is even constant with high probability.

Note that here the waste is defined similar to multiprocessor scheduling, namely as the number of bins minus the sum of all item sizes. But for Bin Packing also in the worst-case nearly optimal solutions can be computed, for example there is an asymptotic PTAS [20] and even an asymptotic FPTAS exists [21]. More on Bin Packing can be found in the excellent survey of Coffman, Garey and Johnson [22].

One major difference between rate-monotonic scheduling and Bin Packing is that for the latter it can be checked easily whether given items fit into one bin, whereas it is conjectured that this does not hold for a set of tasks and one processor. If a set $\mathcal{S}$ of implicit-deadline tasks is feasible (i.e. RM-schedulable), then the utilization $u(\mathcal{S})$ is at most 1. However, $\mathcal{S}$ can be infeasible, even if $u(\mathcal{S}) < 1$. Consider, for example, again the task system $\mathcal{S}$ in Figure 1. If we increase the running time of τ_2 by any $\varepsilon > 0$, then the set $\mathcal{S}$ is no longer feasible and its utilization is $u(\mathcal{S}) = (9 + 2\varepsilon)/10$. Liu and Layland [1] have shown that $\mathcal{S}$ is feasible, if $u(\mathcal{S})$ is bounded by $n(2^{1/n} - 1)$, where $n = |\mathcal{S}|$. This bound tends to $\ln(2) \approx 0.69$ and the condition is not necessary for feasibility, as the example in Figure 1 shows. Stronger, but still not necessary conditions for feasibility are given in [8,7,6].

Note that the first job of each task is the *critical instance* [1], thus if $p(\tau_1) \leq \ldots \leq p(\tau_n)$ then response times for τ_i in a rate-monotonic, single-processor schedule are given by the smallest value $r(\tau_i) \geq 0$ with

$$r(\tau_i) = c(\tau_i) + \sum_{j<i} \left\lceil \frac{r(\tau_i)}{p(\tau_j)} \right\rceil c(\tau_j).$$

Of course $\tau_1, \ldots, \tau_n$ are feasible if and only if $r(\tau_i) \leq p(\tau_i)$ for $i = 1, \ldots, n$ [2]. But it was proven in [23] that such response times cannot even be approximated in polynomial time within a factor of $n^{c/\log\log n}$ for a fixed constant $c > 0$, unless **NP** = **P**. Nevertheless in practice response times can be efficiently computed using a fix-point iteration approach [5]. Furthermore Baruah and Fisher [12] showed that there is an FPTAS for computing the minimum processor speed, which is needed to make a task system RM-schedulable.

Baker and Oh [10] showed that if m processors are needed to schedule $\mathcal{S}$, one must have $u(\mathcal{S}) \geq m \cdot (\sqrt{2} - 1) \approx 0.41m$. This quantity was later improved by Liebeherr et al. [8] to $m \cdot \frac{1}{1+\sqrt[m]{2}} \approx 0.5m$ (for large m).

Most popular algorithms for rate-monotonic periodic multiprocessor scheduling first sort the tasks in a suitable way and then distribute them in a First Fit or Next Fit manner using a sufficient feasibility criterion. See the following table for an overview (with our algorithm in the last row, for the sake of comparability).

Name	**sorting**	**distribution**	**ratio**	**run time**
RMNF	inc. $p(\tau)$	Next Fit	2.67	$\mathcal{O}(n \log n)$
RMFF	inc. $p(\tau)$	First Fit	2.00	$\mathcal{O}(n \log n)$
FFDU	dec. $u(\tau)$	First Fit	2.00	$\mathcal{O}(n \log n)$
RMST	inc. $\alpha(\tau)$	Next Fit	$\frac{1}{1-u_{\max}}$	$\mathcal{O}(n \log n)$
RMGT	-	First Fit + RMST	1.75	$\mathcal{O}(n^2)$
FFMP	inc. $\alpha(\tau)$	First Fit	2.00	$\mathcal{O}(n \log n)$

Here $\alpha(\tau) = \log_2 p(\tau) - \lfloor \log_2 p(\tau) \rfloor$ and $u_{\max} = \max_{\tau \in \mathcal{S}} u(\tau)$. In the table, the column "ratio" gives the best known upper bounds on the asymptotic approximation ratio. The *rate-monotonic general task* algorithm (RMGT) [8] distributes tasks with utilization at most $1/3$ using RMST and the rest separately with First Fit. A more detailed description can be found in [13].

Furthermore there is an asymptotic PTAS under resource augmentation, computing for any fixed $\varepsilon > 0$ a solution with $(1+\varepsilon)OPT + 1$ processors, where the tasks on each processor can be feasibly scheduled after increasing the processor speed by a factor of $1 + \varepsilon$ [24]. In the same paper it was proven that unless **P** $\neq$ **NP** no asymptotic FPTAS can exist for this multiprocessor scheduling problem. But it is still an open question whether there might be an asymptotic PTAS and thus an algorithm that is asymptotically optimal and does not depend on any assumption about the input. Here we call an algorithm asymptotically optimal, if the approximation ratio tends to 1 for $OPT \to \infty$. We refer to the article of Baruah and Goossens [11] for an overview on complexity issues of real-time scheduling.

Our contribution We introduce an efficient and easy to implement algorithm for the multiprocessor rate-monotonic scheduling problem called *First Fit Matching Periods* (FFMP) . We proof that it is asymptotically optimal for arbitrary periods provided that the utilizations follow a uniform distribution[1]. To this end, we show that our algorithm produces a solution with expected waste of $\mathcal{O}(n^{3/4}(\log n)^{3/8})$. Since the expected approximation ratio of $1 + \mathcal{O}(n^{-1/4}(\log n)^{3/8})$ tends to 1 for $n \to \infty$, the solution is asymptotically optimal on average. To the best of our knowledge this is the first proof that any algorithm for this problem admits this property w.r.t. a reasonable probability distribution.

To achieve our results, we use the following technique: We introduce an auxiliary algorithm FFMP* and prove that for any task set it needs at least as many processors as FFMP. Thus it suffices to derive an upper bound on the waste of this easier algorithm. We then point out that for suitable subsets of the input tasks, FFMP* behaves like a well studied Bin Packing algorithm MFF. Eventually this allows to bound the waste for FFMP* in terms of the waste of MFF.

In addition to the proof of the asymptotic optimality of our algorithm, we present experimental results showing that FFMP outperforms the algorithms known from literature already on random instances with a small number of tasks. We thereby provide an example of an algorithm that has been designed for asymptotic optimality and which is, in addition, competitive on reasonably small instances. Moreover, we present a family of instances where the average waste scales with $n^{0.70}$, which is almost tight to our theoretical upper bound and thus showing that our technique is suitable for sharp analyses.

2 Preliminaries

For our algorithm we need the following sufficient (but still not necessary) schedulability condition of Burchard et al.

Lemma 1. *[8] For tasks $\mathcal{S} = \{\tau_1, \dots, \tau_n\}$ define*

$$\alpha(\tau_i) := \log_2 p(\tau_i) - \lfloor \log_2 p(\tau_i) \rfloor \quad \textit{and} \quad \beta(\mathcal{S}) := \max_{i=1,\dots,n} \alpha(\tau_i) - \min_{i=1,\dots,n} \alpha(\tau_i).$$

Then the tasks can be RM-scheduled on a single processor if $u(\mathcal{S}) \leq 1 - \beta(\mathcal{S}) \ln(2)$.

The intuition behind this is that a small value of $\beta(\mathcal{S})$ indicates that the periods of tasks in $\mathcal{S}$ are nearly multiples of each other and consequently the tasks are guaranteed to "harmonize".

The idea for our heuristic is now as follows: Sort the tasks w.r.t. their α-values. Then assign them in a First Fit manner using the sufficient feasibility test from Lemma 1. See Algorithm 1 for a formal description.

Note that the *Rate-monotonic small tasks* algorithm (RMST) of Burchard et al. [8] is similar, just that a Next Fit assignment is used instead of First Fit. But already from average case analysis of Bin Packing, it is well know that Next Fit approaches generate linear waste for uniformly distributed item sizes [25].

[1] To be exact, we assume that first arbitrary periods may be given and then the utilizations are chosen randomly.

Algorithm 1. FFMP

Input: Set $\tau_1, \ldots, \tau_n$ of implicit-deadline tasks

(1) Sort tasks such that $0 \leq \alpha(\tau_1) \leq \alpha(\tau_2) \leq \ldots \leq \alpha(\tau_n) < 1$
(2) FOR $i = 1, \ldots n$ DO
  (3) Assign τ_i to the processor P_j with least index j such that $u(P_j \cup \{\tau_i\}) \leq 1 - \beta(P_j \cup \{\tau_i\}) \cdot \ln(2)$

3 The Result of Shor

It is our aim to convey known bounds on the waste of Bin Packing algorithms to the waste of our algorithm. To this end we consider the following auxiliary algorithm *Matching First Fit* (MFF) of Shor [16], which distributes a list $L = (a_1, \ldots, a_n)$ of items to bins B_j. Denote $\text{size}(B_j) = \sum_{i \in B_j} a_i$.

Algorithm 2. Matching First Fit (MFF)

Input: Set $a_1, \ldots, a_n$ of items

(1) FOR $i = 1, \ldots, n$ DO
  (2) Assign item a_i to the bin B_j with the least index j such that either B_j is empty or both of the following conditions hold
    - B_j contains one item and this item has size at least $1/2$
    - $\text{size}(B_j) + a_i \leq 1$

Shor [16] proved that MFF is monotonic, i.e. for all Bin Packing instances I and all items $a_i \in I$ one has

$$\texttt{MFF}(I) \geq \texttt{MFF}(I \backslash \{a_i\}) \geq \texttt{MFF}(I) - 1$$

where $\texttt{MFF}(I)$ denotes the number of bins used by MFF if applied to instance I. Furthermore MFF is never better than the pure First Fit algorithm and it has an expected waste of $\mathcal{O}(n^{2/3}\sqrt{\log n})$ for Bin Packing instances, whose item sizes are taken uniformly from $[0, 1]$.

Like MFF is a restriction to First Fit, we now state a restricted version of FFMP.

4 An Auxiliary Algorithm

Let $\gamma := \gamma(n)$ be an integer value, which we are going to choose later. We now define a simplified version FFMP* of FFMP which can be analyzed more easily. First the tasks are partitioned into *groups* $\mathcal{S}_1, \ldots, \mathcal{S}_\gamma$ with $\mathcal{S}_j = \{\tau_i \in \mathcal{S} \mid \frac{j-1}{\gamma} \leq \alpha(\tau_i) < \frac{j}{\gamma}\}$, thus the α-values of tasks from the same group differ only slightly. Next, FFMP* never assigns more than 2 tasks to each processor and tasks from different periods are never mixed. Here we say that an algorithm *mixes* two tasks τ_1, τ_2, if they are assigned to the same processor. The algorithm even considers a processor to be full if the first assigned task has a utilization of at most $\approx 1/2$. Note that this algorithm is precisely tailored for the used probability distribution. A formal definition of $\mathtt{FFMP}^*$ now follows

Algorithm 3. $\mathtt{FFMP}^*$

Input: Set $\tau_1, \ldots, \tau_n$ of implicit-deadline tasks

(1) Sort tasks such that $0 \leq \alpha(\tau_1) \leq \ldots \leq \alpha(\tau_n) < 1$
(2) Partition tasks into groups $\mathcal{S}_1, \ldots, \mathcal{S}_\gamma$, with $\mathcal{S}_j = \{\tau_i \in \mathcal{S} \mid \frac{j-1}{\gamma} \leq \alpha(\tau_i) < \frac{j}{\gamma}\}$.
(3) FOR $i = 1, \ldots n$ DO
   (4) Assign τ_i to the processor P_j with the least index j such that either P_j is empty or all following conditions are satisfied
      (a) P_j contains only one item and this item is from the same group as τ_i
      (b) the item on P_j has utilization $\geq (1 - \frac{\ln(2)}{\gamma})/2$
      (c) $u(P_j \cup \{\tau_i\}) \leq 1 - \frac{\ln(2)}{\gamma}$

Note that $1 - \frac{\ln(2)}{\gamma}$ is just slightly below 1. Observe that $\mathtt{FFMP}^*$ assigns either 1 or 2 tasks to each processor. Let $\mathtt{FFMP}^*(\mathcal{S})$ be the number of processors, needed when scheduling tasks $\mathcal{S}$ with algorithm $\mathtt{FFMP}^*$. As a slight abuse of notation $\mathtt{FFMP}^*(\mathcal{S})$ means as well the schedule, obtained when applying $\mathtt{FFMP}^*$ to $\mathcal{S}$, however the meaning will be clear from the context. From Lemma 1 we see that the produced solution is always feasible since either a single task is assigned to a processor or in case that two tasks are assigned, their α-values differ by at most $1/\gamma$ and their cumulated utilization is upper bounded by $1 - \ln(2)/\gamma$.

The following observation is crucial for our analysis and allows to link the expected waste of $\mathtt{FFMP}^*$ to $\mathtt{MFF}$.

Observation 1. *Consider tasks $\tau_1, \ldots, \tau_m$ such that one has $\frac{j-1}{\gamma} \leq \alpha(\tau_i) < \frac{j}{\gamma}$ (i.e. all tasks fall into the same group) and $0 \leq u(\tau_i) \leq 1 - \frac{\ln(2)}{\gamma}$ for all $i = 1, \ldots, m$. Create m Bin Packing items $a_1, \ldots, a_m$ with item sizes $a_i := u(\tau_i) \cdot /(1 - \ln(2)/\gamma)$, i.e. $a_i \in [0, 1]$. Then $\mathtt{FFMP}^*$ schedules $\tau_1, \ldots, \tau_m$ in exactly the same way, that $\mathtt{MFF}$ distributes $a_1, \ldots, a_m$, i.e. task τ_i is assigned to the ℓth processor if and only if item a_i is assigned to the ℓth bin. Especially $\mathtt{FFMP}^*(\{\tau_1, \ldots, \tau_m\}) = \mathtt{MFF}(\{a_1, \ldots, a_m\})$.*

The main result of this section will be to show that $\mathtt{FFMP}^*(\mathcal{S}) \geq \mathtt{FFMP}(\mathcal{S})$ for any set of tasks $\mathcal{S}$. The simplicity of $\mathtt{FFMP}^*$ will enable us to prove *monotonicity* for it, meaning that removing tasks from $\mathcal{S}$ can only lower the value of $\mathtt{FFMP}^*(\mathcal{S})$. Although this is trivially true for algorithms yielding optimal solutions, for approximation algorithms with a complex behavior this does not necessarily hold.

Lemma 2. *For any set of tasks $\mathcal{S}$ and $\tau^* \in \mathcal{S}$ one has*

$$\mathtt{FFMP}^*(\mathcal{S}) \geq \mathtt{FFMP}^*(\mathcal{S} \backslash \{\tau^*\}) \geq \mathtt{FFMP}^*(\mathcal{S}) - 1$$

Proof. Denote $\mathcal{S}' = \mathcal{S} \backslash \{\tau^*\}$ and let $\mathcal{S}_1, \ldots, \mathcal{S}_\gamma$ $[\mathcal{S}'_1, \ldots, \mathcal{S}'_\gamma]$ be the groups of $\mathcal{S}$ $[\mathcal{S}'$, resp.]. Let i^* be the index such that $\tau^* \in \mathcal{S}_{i^*}$. Since the algorithm never mixes tasks from different groups one has $\mathtt{FFMP}^*(\mathcal{S}_i') = \mathtt{FFMP}^*(\mathcal{S}_i)$ for all $i \neq i^*$ and $\mathtt{FFMP}^*(\mathcal{S}) = \sum_{i=1}^{\gamma} \mathtt{FFMP}^*(\mathcal{S}_i)$. Thus we may assume that all groups but $\mathcal{S}_{i^*}$ are empty. Furthermore tasks with utilization larger than $1 - \frac{\ln(2)}{\gamma}$ are never mixed with other tasks, thus their removal does not change the claim. Due to this we may assume that such tasks are not contained in $\mathcal{S} = \mathcal{S}_{i^*}$, hence $\mathcal{S}$ contains just tasks from the same group, all with utilization at most $1 - \frac{\ln(2)}{\gamma}$. Sticking together Observation 1 and the monotonicity of MFF [16] yields the claim.

By iteratively applying Lemma 2 we obtain

Corollary 1. *For all task sets $\mathcal{S}$ and $\mathcal{S}' \subseteq \mathcal{S}$ one has*

$$\mathtt{FFMP}^*(\mathcal{S}) \geq \mathtt{FFMP}^*(\mathcal{S}').$$

We may now conclude that the restricted variant of FFMP never produces better solutions than FFMP itself.

Theorem 2. *For all task sets $\mathcal{S}$ one has*

$$\mathtt{FFMP}^*(\mathcal{S}) \geq \mathtt{FFMP}(\mathcal{S}).$$

Proof. Let $P_1 \dot\cup \ldots \dot\cup P_m = \mathcal{S}$ be the solution computed by FFMP and denote the groups of $\mathcal{S}$ by $\mathcal{S}_1, \ldots \mathcal{S}_\gamma$. Consider an arbitrary processor P_j and after renaming let $\tau_1, \ldots, \tau_p$ be the tasks on P_j in incoming order ($p \geq 1$). Remove $\tau_3, \ldots, \tau_p$. Given that $p \geq 2$, remove τ_2 if at least one of the following conditions is true

- τ_1 and τ_2 stem from different groups
- $u(\tau_1) < \frac{1}{2}(1 - \frac{\ln(2)}{\gamma})$
- $u(\{\tau_1, \tau_2\}) > 1 - \frac{\ln(2)}{\gamma}$

Let $\mathcal{S}' \subseteq \mathcal{S}$ the remaining tasks. Clearly $\mathtt{FFMP}^*$ schedules $\mathcal{S}'$ in exactly the same way that FFMP schedules them in the solution leading to $\mathtt{FFMP}(\mathcal{S})$. Thus $\mathtt{FFMP}^*(\mathcal{S}') = \mathtt{FFMP}(\mathcal{S})$. From Corollary 1 we gain $\mathtt{FFMP}^*(\mathcal{S}) \geq \mathtt{FFMP}^*(\mathcal{S}')$. Plugging both equations/inequalities together, yields the claim.

5 An Upper Bound for FFMP*

In this section we will give an upper bound on the expected waste of FFMP*, by exploiting the bound on the waste of MFF. Again Observation 1 will be crucial.

Theorem 3. *Let $f : \mathbb{R}_{\geq 1} \to \mathbb{R}$ be a concave and monotonic increasing function, such that $f(n)$ yields an upper bound on the expected waste of* MFF *applied to n items drawn uniformly at random from $[0, 1]$. Then the expected waste of* FFMP* *is bounded by $\frac{n}{\gamma} + \gamma \cdot f(n/\gamma)$ for n tasks with arbitrary periods, but utilization values drawn uniformly at random from $[0, 1]$.*

Proof. Let $\mathcal{S}_1, \ldots, \mathcal{S}_\gamma$ be the partition of the tasks $\mathcal{S}$ into groups. Denote $n = |\mathcal{S}|$ and $n_i = |\mathcal{S}_i|$. $\texttt{FFMP}^*$ never mixes tasks from different groups, thus

$$\texttt{FFMP}^*(\mathcal{S}) = \sum_{i=1}^{\gamma} \texttt{FFMP}^*(\mathcal{S}_i).$$

Consider an arbitrary group $\mathcal{S}_i$. Call tasks τ with a utilization of $u(\tau) > 1 - \frac{\ln(2)}{\gamma}$ *full tasks* and *ordinary tasks* otherwise. Let $\mathcal{S}_i^{\text{full}}$ be the set of full tasks from $\mathcal{S}_i$ and let $\mathcal{S}_i' = \mathcal{S}_i \backslash \mathcal{S}_i^{\text{full}}$ be the ordinary tasks. Condition that $|\mathcal{S}_i'| = n_i^o$. Clearly the algorithm $\texttt{FFMP}^*$ does not mix ordinary and full tasks, thus

$$\texttt{FFMP}^*(\mathcal{S}_i) = \texttt{FFMP}^*(\mathcal{S}_i^{\text{full}}) + \texttt{FFMP}^*(\mathcal{S}_i').$$

A full task has a utilization of at least $1 - \frac{\ln(2)}{\gamma}$, thus for each full task it suffices to account a waste of $\frac{\ln(2)}{\gamma} \leq \frac{1}{\gamma}$. The expected waste stemming from the processors, owning the full tasks of group i is then

$$E[\texttt{FFMP}^*(\mathcal{S}_i^{\text{full}}) - u(\mathcal{S}_i^{\text{full}})] \leq \frac{n_i - n_i^o}{\gamma}.$$

It remains to bound the waste from the ordinary tasks. The utilization values of tasks in $\mathcal{S}_i'$ are conditioned to be in $[0, 1 - \frac{\ln(2)}{\gamma}]$. It is not difficult to see that the distribution of $u(\tau)$ for $\tau \in \mathcal{S}_i'$ is uniformly w.r.t. $[0, 1 - \frac{\ln(2)}{\gamma}]$. If we define a Bin Packing instance I_i' with an item of size $u(\tau)/(1 - \frac{\ln(2)}{\gamma})$ for each $\tau \in \mathcal{S}_i'$, then the item sizes in I_i' are distributed uniformly w.r.t. $[0, 1]$. By Observation 1

$$E[\texttt{FFMP}^*(\mathcal{S}_i')] = E[\texttt{MFF}(I_i')] \leq \frac{n_i^o}{2} + f(n_i^o).$$

The rest of the proof simply consists of summing up the achieved bounds on the waste. We can express the expected waste, stemming from the processors owning ordinary tasks from the ith group as

$$\begin{aligned} E[\texttt{FFMP}^*(\mathcal{S}_i') - u(\mathcal{S}_i')] &\leq (\frac{n_i^o}{2} + f(n_i^o)) - E[u(\mathcal{S}_i')] \\ &= (\frac{n_i^o}{2} + f(n_i^o)) - n_i^o \frac{1 - \ln(2)/\gamma}{2} \leq f(n_i^o) + \frac{n_i^o}{\gamma} \end{aligned}$$

Combining ordinary and full tasks yields

$$E[\texttt{FFMP}^*(\mathcal{S}_i) - u(\mathcal{S}_i)] \leq \frac{n_i - n_i^o}{\gamma} + (f(n_i^o) + \frac{n_i^o}{\gamma}) \leq f(n_i) + \frac{n_i}{\gamma}$$

using monotonicity of f. Hence the total expected waste for solution $\texttt{FFMP}^*(\mathcal{S})$ can be written as

$$E[\texttt{FFMP}^*(\mathcal{S}) - u(\mathcal{S})] \stackrel{(*)}{=} \sum_{i=1}^{\gamma} E[\texttt{FFMP}^*(\mathcal{S}_i) - u(\mathcal{S}_i)] \leq \sum_{i=1}^{\gamma} (f(n_i) + \frac{n_i}{\gamma}) \stackrel{(**)}{\leq} \frac{n}{\gamma} + \gamma \cdot f(\frac{n}{\gamma})$$

For $(*)$ we used linearity of expectation and $(**)$ follows by Jensen's inequality and concaveness of f.

Applying the best known bound on $f(n)$ we obtain

Theorem 4. *For the expected waste of* FFMP *one has*

$$E[\texttt{FFMP}(\mathcal{S}) - u(\mathcal{S})] = \mathcal{O}(n^{3/4}(\log n)^{3/8})$$

if $\mathcal{S}$ *consists of* n *tasks, whose utilization values are drawn uniformly at random from* $[0, 1]$.

Proof. Theorem 2 provides that bounding the waste of FFMP* is sufficient. Choosing $\gamma(n) = \lceil n^{1/4}/(\log n)^{3/8} \rceil$ and using the bound of $f(n) = \mathcal{O}(n^{2/3}(\log n)^{1/2})$ [16] together with Theorem 3 yields the claim (observe that $c \cdot n^{2/3} \cdot (\log n)^{1/2}$ is concave and monotonic).

Observing that $OPT(\mathcal{S}) = \Omega(n)$ with very high probability, we conclude that

Corollary 2. *Let* $\mathcal{S}$ *consist of* n *tasks, whose utilization values are drawn uniformly at random from* $[0, 1]$. *Then the expected approximation ratio of* FFMP *is*

$$E\left[\frac{\texttt{FFMP}(\mathcal{S})}{OPT(\mathcal{S})}\right] \leq 1 + \mathcal{O}(n^{-1/4}(\log n)^{3/8})$$

Using essentially the same proof as [8] (see also [13]) one can easily show that even in the worst-case one has $\texttt{FFMP}(\mathcal{S}) \leq 2u(\mathcal{S}) + 4$, i.e. the asymptotic worst-case approximation ratio of FFMP is 2. A proof can be found in the full version of this paper.

6 Experimental Results

We have performed simulations of our FFMP algorithm and compared it with RMFF, FFDU, and RMGT. The experimental setting is as follows. We choose the periods $p(\tau_i) \in [0, 500]$ and the utilizations $u(\tau_i) \in [0, 1]$ uniformly at random. We create random instances in the range of 10 to 100000 tasks. For each given n, we generate 100 random samples to get a good estimate of the expected value of the waste. We use the same instances to test each algorithm to allow also a direct comparison of their performance.

The log-log-plot in Fig. 2 shows the power law behavior of the average waste of FFMP as predicted by Theorem 4. The regression yields an exponent of 0.70 which is close to $\frac{3}{4}$ from the Theorem showing that the theoretical analysis is almost tight, i.e. that we do not loose much by analyzing the dominated algorithm FFMP*. In contrast to that, the average waste produced by the other algorithms shows an almost linear dependence on the number of tasks. In fact, we believe that the dependence is linear since the measurements of their average waste show a slight curvature to the left, indicating that the averages are actually growing faster than the fitted straight lines.

The simulated average processor load shown in Fig. 2 supports this claim. By *average processor load*, we mean the expected value of the mean utilization of the processors. The closer this value is to 1 the less processor cycles are wasted. Hence, it comes to no surprise that the average load for FFMP tends to 1 with increasing n. For the other algorithms, there is strong evidence that they converge to respective constants strictly smaller than 1 and likely even not more than 0.9.

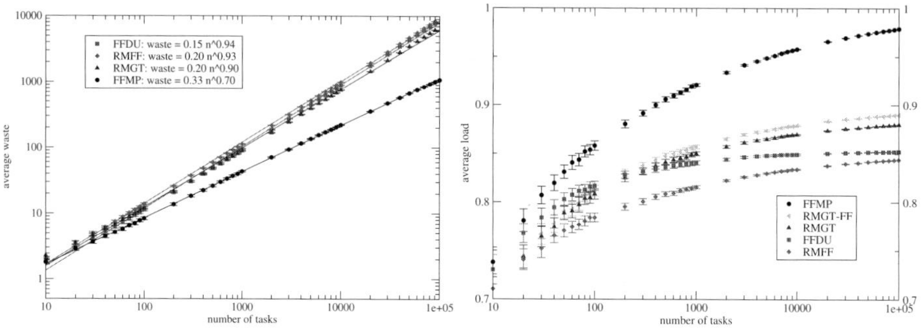

Fig. 2. The average waste depending on the number of tasks are shown with 3σ error bars for FFMP and three algorithms from literature. For each algorithm, we fit a power function (straight lines in the log-log plot) and display the results in the legend box of (a). The average load of the processors is shown in (b) with 3σ error bars. In the algorithm RMGT-FF, the Next Fit distribution for the small tasks is replaced by First Fit.

Interestingly, the quadratic running time of RMGT, which is due to the exact feasibility test for the large tasks, does not pay off in comparison with FFMP, which runs in $\mathcal{O}(n \log n)$ time. This does not only hold for the average waste, but also on a per instance basis: FFMP performs better than RMGT for 94 out of 100 random instances with 10 tasks and always better on our random test instances with a larger number of tasks. This is due to the splitting of the tasks into small tasks (i.e. utilization at most $1/3$) and large tasks. Thus all tasks with utilization at least $2/3$ are deterministically scheduled alone on a processor. For example in expectation 10% of all tasks have a utilization between 0.7 and 0.8. Each of those tasks contributes at least 0.2 to the total waste. Therefore the expected waste of RMGT must be at least $0.1 \cdot 0.2 \cdot n = \Omega(n)$, even if after splitting, the algorithm would find an optimum solution for both parts. FFMP can be implemented in $\mathcal{O}(n \log n)$ using a heap data structure (see the full version of this paper).

Acknowledgement

The authors want to thank Sanjoy K. Baruah, for reading a preliminary version of this paper and giving very helpful advices.

References

1. Liu, C.L., Layland, J.W.: Scheduling algorithms for multiprogramming in a hard-real-time environment. J. ACM 20(1), 46–61 (1973)
2. Lehoczky, J.P., Sha, L., Ding, Y.: The rate monotonic scheduling algorithm: Exact characterization and average case behavior. In: IEEE Real-Time Systems Symposium (1989)
3. Lehoczky, J.P.: Fixed priority scheduling of periodic task sets with arbitrary deadlines. In: IEEE Real-Time Systems Symposium, pp. 201–213 (1990)
4. Korst, J., Aarts, E.H.L., Lenstra, J.K., Wessels, J.: Periodic multiprocessor scheduling. In: Aarts, E.H.L., van Leeuwen, J., Rem, M. (eds.) PARLE 1991. LNCS, vol. 505, pp. 166–178. Springer, Heidelberg (1991)

5. Audsley, A.N., Burns, A., Richardson, M., Tindell, K.: Applying new scheduling theory to static priority pre-emptive scheduling. Software Engineering Journal, 284–292 (1993)
6. Oh, Y., Son, S.H.: Allocating fixed-priority periodic tasks on multiprocessor systems. Real-Time Syst. 9(3), 207–239 (1995)
7. Davari, S., Dhall, S.K.: On-line algorithms for allocating periodic-time-critical tasks on multiprocessor systems. Informatica (Slovenia) 19(1) (1995)
8. Liebeherr, J., Burchard, A., Oh, Y., Son, S.H.: New strategies for assigning real-time tasks to multiprocessor systems. IEEE Trans. Comput. 44(12), 1429–1442 (1995)
9. Korst, J., Aarts, E., Lenstra, J.K.: Scheduling periodic tasks with slack. INFORMS J. Comput. 9(4), 351–362 (1997)
10. Oh, D.I., Baker, T.P.: Utilization bounds for N-processor rate monotone scheduling with static processor assignment. Real-Time Systems (1998)
11. Baruah, S., Goossens, J.: Scheduling real-time tasks: Algorithms and complexity. In: Leung, J.Y.T. (ed.) Handbook of Scheduling — Algorithms, Models, and Performance Analysis. Computer and Information Science Series, vol. 28. Chapman & Hall/CRC, Boca Raton (2004)
12. Fisher, N., Baruah, S.: A fully polynomial-time approximation scheme for feasibility analysis in static-priority systems with arbitrary relative deadlines. In: ECRTS 2005. IEEE Computer Society, Los Alamitos (2005)
13. Leung, J., Kelly, L., Anderson, J.H.: Handbook of Scheduling: Algorithms, Models, and Performance Analysis. CRC Press, Inc., Boca Raton (2004)
14. Garey, M.R., Graham, R.L., Johnson, D.S., Yao, A.C.C.: Resource constrained scheduling as generalized bin packing. J. Combin. Theory Ser. A 21, 257–298 (1976)
15. Johnson, D.S.: Near-optimal bin packing algorithms. PhD thesis, MIT, Cambridge, MA (1973)
16. Shor, P.W.: The average-case analysis of some on-line algorithms for bin packing. In: FOCS 1984, Singer Island, FL. IEEE, Los Alamitos (1984)
17. Frederickson, G.N.: Probabilistic analysis for simple one- and two-dimensional bin packing algorithms. Information Processing Letters 11(4/5), 156–161 (1980)
18. Knödel, W.: A bin packing algorithm with complexity $O(n \log n)$ and performance 1 in the stochastic limit. In: Gruska, J., Chytil, M.P. (eds.) MFCS 1981. LNCS, vol. 118, pp. 369–378. Springer, Heidelberg (1981)
19. Lueker, G.S.: An average-case analysis of bin packing with uniformly distributed item sizes. Technical Report 181, Dept. Inf. and CS, University of California at Irvine (1982)
20. Fernandez de la Vega, W., Lueker, G.S.: Bin packing can be solved within $1 + \varepsilon$ in linear time. Combinatorica 1(4), 349–355 (1981)
21. Karmarkar, N., Karp, R.M.: An efficient approximation scheme for the one-dimensional bin-packing problem. In: FOCS 1982, pp. 312–320. IEEE, Los Alamitos (1982)
22. Coffman Jr., E.G., Garey, M.R., Johnson, D.S.: Approximation algorithms for bin-packing—an updated survey. In: Algorithm design for computer system design. CISM Courses and Lectures, vol. 284, pp. 49–106. Springer, Vienna (1984)
23. Eisenbrand, F., Rothvoß, T.: Static-priority Real-time Scheduling: Response Time Computation is NP-hard. In: IEEE Real-Time Systems Symposium, RTSS (2008)
24. Eisenbrand, F., Rothvoß, T.: A PTAS for static priority real-time scheduling with resource augmentation. In: Aceto, L., Damgård, I., Goldberg, L.A., Halldórsson, M.M., Ingólfsdóttir, A., Walukiewicz, I. (eds.) ICALP 2008, Part I. LNCS, vol. 5125, pp. 246–257. Springer, Heidelberg (2008)
25. Coffman, E.G.J., So, K., Hofri, M., Yao, A.C.: A stochastic model of bin-packing. Inf. Control 44, 105–115 (1980)

Minimizing Maximum Response Time and Delay Factor in Broadcast Scheduling

Chandra Chekuri*, Sungjin Im**, and Benjamin Moseley***

Dept. of Computer Science, University of Illinois, Urbana, IL 61801
{chekuri,im3,bmosele2}@cs.illinois.edu

Abstract. We consider *online* algorithms for pull-based broadcast scheduling. In this setting there are n pages of information at a server and requests for pages arrive online. When the server serves (broadcasts) a page p, all outstanding requests for that page are satisfied. We study two related metrics, namely maximum response time (waiting time) and maximum delay-factor and their *weighted* versions. We obtain the following results in the worst-case online competitive model.

- We show that FIFO (first-in first-out) is 2-competitive even when the page sizes are different. Previously this was known only for unit-sized pages [10] via a delicate argument. Our proof differs from [10] and is perhaps more intuitive.
- We give an online algorithm for maximum delay-factor that is $O(1/\epsilon^2)$-competitive with $(1+\epsilon)$-speed for unit-sized pages and with $(2+\epsilon)$-speed for different sized pages. This improves on the algorithm in [13] which required $(2+\epsilon)$-speed and $(4+\epsilon)$-speed respectively. In addition we show that the algorithm and analysis can be extended to obtain the same results for maximum *weighted* response time and delay factor.
- We show that a natural greedy algorithm modeled after LWF (Longest-Wait-First) is not $O(1)$-competitive for maximum delay factor with any constant speed even in the setting of standard scheduling with unit-sized jobs. This complements our upper bound and demonstrates the importance of the tradeoff made in our algorithm.

1 Introduction

We consider *online* algorithms in pull-based broadcasting. In this model there are n pages (representing some form of useful information) available at a server and clients request a page that they are interested in. When the server transmits a page p, all outstanding requests for that page p are satisfied since it is assumed that all clients can simultaneously receive the information. It is in this respect that broadcast scheduling differs crucially from standard scheduling where each job needs its own service from the server. We distinguish two cases: all the pages

* Partially supported by NSF grants CCF-0728782 and CNS-0721899.
** Partially supported by NSF grant CNS-0721899 and Samsung Fellowship.
*** Partially supported by NSF grant CNS-0721899.

A. Fiat and P. Sanders (Eds.): ESA 2009, LNCS 5757, pp. 444–455, 2009.

are of same size (unit-size without loss of generality) and when the pages can be of different size. Broadcast scheduling is motivated by several applications in wireless and LAN based systems [1,2,26]. It has seen a substantial interest in the algorithmic scheduling literature starting with the work of Bartal and Muthukrishanan [5]; see [21]. In addition to the applications, broadcast scheduling has sustained interest due to the significant technical challenges that basic problems in this setting have posed for algorithm design and analysis. To distinguish broadcast scheduling from "standard" job scheduling, we refer to the latter as unicast scheduling — we use requests in the context of broadcast and jobs in the context of unicast scheduling.

In this paper, we focus on scheduling to minimize two related objectives: the maximum response time and the maximum delay factor. We also consider their *weighted* versions. Interestingly, the maximum response time metric was studied in the (short) paper of Bartal and Muthukrishnan [5] where they claimed that the online algorithm FIFO (for First In First Out) is 2-competitive for broadcast scheduling, and moreover that no deterministic online algorithm is $(2-\epsilon)$-competitive. (It is easy to see that FIFO is optimal in unicast scheduling). Despite the claim, no proof was published. It is only recently, almost a decade later, that Chang et al. [10] gave formal proofs for these claims for unit-sizes pages. This simple problem illustrates the difficulty of broadcast scheduling: the ability to satisfy multiple requests for a page p with a single transmission makes it difficult to relate the total "work" that the online algorithm and the offline adversary do. The upper bound proof for FIFO in [10] is short but delicate. In fact, [5] claimed 2-competitiveness for FIFO even when pages have different sizes. As noted in previous work [5,15,25], when pages have different sizes, one needs to carefully define how a request for a page p gets satisfied if it arrives midway during the transmission of the page. In this paper we consider the sequential model [15], the most restrictive one, in which the server broadcasts each page sequentially and a client receives the page sequentially without buffering; see [25] on the relationship between different models. The claim in [5] regarding FIFO for different pages is in a less restrictive model in which clients can buffer and take advantage of partial transmissions and the server is allowed to preempt. The FIFO analysis in [10] for unit-sized pages does not appear to generalize for different page sizes. Our first contribution in this paper is the following.

Theorem 1. *In the sequential model, FIFO is 2-competitive for minimizing maximum response time in broadcast scheduling even with different page sizes.*

Remark 1. In the model where buffering is allowed, a variant of FIFO can be shown to be 2-competitive. We defer the proof to the full version of this paper.

Note that FIFO, whenever the server is free, picks the page p with the earliest request and *non-preemptively* broadcasts it. Our bound matches the lower bound shown even for unit-sized pages, thus closing one aspect of the problem. Our proof differs from that of Chang et al.; it does not explicitly use the unit-size assumption and this is what enables the generalization to different page sizes. The analysis is inspired by our previous work on maximum delay factor [13] which we discuss next.

Maximum (Weighted) Delay Factor and Weighted Response Time: The delay factor of a schedule is a metric recently introduced in [10] (and implicitly in [7]) when requests have deadlines. Delay factor captures how much a request is delayed compared to its deadline. More formally, let $J_{p,i}$ denote the i'th request of page p. Each request $J_{p,i}$ arrives at $a_{p,i}$ and has a deadline $d_{p,i}$. The finish time $f_{p,i}$ of a request $J_{p,i}$ is defined to be the earliest time after $a_{p,i}$ when the page p is sequentially transmitted by the scheduler starting from the beginning of the page. Note that multiple requests for the same page can have the same finish time. Formally, the delay factor of the job $J_{p,i}$ is defined as $\max\{1, \frac{f_{p,i}-a_{p,i}}{d_{p,i}-a_{p,i}}\}$; we refer to the quantity $S_{p,i} = d_{p,i} - a_{p,i}$ as the *slack* of $J_{p,i}$. For a more detailed motivation of delay factor, see [13]. Note that for unit-sized pages, delay factor generalizes response time since one could set $d_{p,i} = a_{p,i} + 1$ for each request $J_{p,i}$ in which case its delay factor equals its response time. In this paper we are interested in online algorithms that minimize the maximum delay factor, in other words the objective function is $\min \max_{p,i}\{1, \frac{f_{p,i}-a_{p,i}}{d_{p,i}-a_{p,i}}\}$. We also consider a related metric, namely *weighted* response time. Let $w_{p,i}$ be a non-negative weight associated with $J_{p,i}$; the weighted response time is then $w_{p,i}(f_{p,i} - a_{p,i})$ and the goal is to minimize the maximum weighted response time. Delay factor and weighted response time have syntactic similarity if we ignore the 1 term in the definition of delay factor — one can think of the weight as the inverse of the slack. Although the metrics are some what similar we note that there is no direct way to reduce one to the other. On the other hand, we observe that upper bounds for one appear to translate to the other. We also consider the problem of minimizing the maximum weighted delay factor $\min \max_{p,i} w_{p,i}\{1, \frac{f_{p,i}-a_{p,i}}{d_{p,i}-a_{p,i}}\}$.

Surprisingly, the maximum weighted response time metric appears to not have been studied formally even in classical unicast scheduling; however a special case, namely maximum *stretch* has received attention. The stretch of a job is its response time divided by its processing time; essentially the weight of a job is the inverse of its processing time. Bender et al. [6,8], motivated by applications to web-server scheduling, studied maximum stretch and showed very strong lower bounds in the online setting. Using similar ideas, in some previous work [13], we showed strong lower bounds for minimizing maximum delay factor even for unit-time jobs. In [13], constant competitive algorithms were given for minimizing maximum delay factor in both unicast and broadcast scheduling; the algorithms are based on resource augmentation [20] wherein the algorithm is given a speed $s > 1$ server while the offline adversary is given a speed 1 server. They showed that **SSF** (shortest slack first) is $O(1/\epsilon)$-competitive with $(1+\epsilon)$-speed in unicast scheduling. **SSF** does not work well in the broadcast scheduling. A different algorithm that involves waiting, **SSF-W** (shortest slack first with waiting) was developed and analyzed in [13]; the algorithm is $O(1/\epsilon^2)$-competitive for unit-size pages with $(2 + \epsilon)$-speed and with $(4 + \epsilon)$-speed for different sized pages. In this paper we obtain improved results by altering the analysis of **SSF-W** in a subtle and important way. In addition we show that the algorithm and analysis can be altered in an easy fashion to obtain the same bounds for weighted response time and delay factor.

Theorem 2. *There is an algorithm that is $(1+\epsilon)$-speed $O(1/\epsilon^2)$-competitive for minimizing maximum delay factor in broadcast scheduling with unit-sized pages. For different sized pages there is a $(2+\epsilon)$-speed $O(1/\epsilon^2)$-competitive algorithm. The same bounds apply for minimizing maximum weighted response time and maximum weighted delay factor.*

Remark 2. Minimizing maximum delay factor is NP-hard and there is no $(2-\epsilon)$-approximation unless $P = NP$ for any $\epsilon > 0$ in the *offline* setting for unit-sized pages. There is a polynomial time computable 2-speed schedule with the optimal delay factor (with 1-speed) [10]. Theorem 2 gives a polynomial time computable $(1+\epsilon)$-speed schedule that is $O(1/\epsilon^2)$-optimal (with 1-speed).

We remark that the algorithm **SSF-W** makes an interesting tradeoff between two competing metrics and we explain this tradeoff in the context of weighted response time and a lower bound we prove in this paper for a simple greedy algorithm. Recall that FIFO is 2-competitive for maximum response time in broadcast scheduling and is optimal for job scheduling. What are natural ways to generalize FIFO to delay factor and weighted response time? As shown in [13], **SSF** (which is equivalent to maximum weight first for weighted response time) is $O(1/\epsilon)$-competitive with $(1+\epsilon)$-speed for job scheduling but is not competitive for broadcast scheduling — it may end up doing much more work than necessary by transmitting a page repeatedly instead of waiting and accumulating requests for a page. One natural algorithm that extends FIFO for delay factor or weighted response time is to schedule the request in the queue that has the largest current delay factor (or weighted wait time). This greedy algorithm was labeled **LF** (longest first) since it can be seen as an extension of the well-studied **LWF** (longest-wait-first) for average flow time. Since **LWF** is known to be $O(1)$-competitive with $O(1)$-speed for average flow time, it was suggested in [11] that **LF** may be $O(1)$-speed $O(1)$-competitive for maximum delay factor. We show that this is not the case even for unicast scheduling.

Theorem 3. *For any constants $s, c > 1$, **LF** is not c-competitive with s-speed for minimizing maximum delay factor (or weighted response time) in unicast scheduling of unit-time jobs.*

Our algorithm **SSF-W** can be viewed as an interesting tradeoff between **SSF** and **LF**. **SSF** gives preference to small slack requests while the **LF** strategy helps avoid doing too much extra work in broadcast scheduling by giving preference to pages that have waited sufficiently long even if they have large slack. The algorithm **SSF-W** considers all requests whose delay factor at time t (or weighted wait time) is within a constant factor of the largest delay factor at t and amongst those requests schedules the one with the smallest slack. This algorithmic principle may be of interest in other settings and is worth exploring in the future.

Other Related Work: We have focussed on maximum response time and its variants and have already discussed closely related work. Other metrics that have received substantial attention in broadcast scheduling are minimizing average flow time and maximizing throughput of satisfied requests when requests have deadlines. We refer the reader to a comprehensive survey on online scheduling algorithms by Pruhs, Sgall and Torng [24] (see also [23]). The recent paper of Chang et al. [10] addresses, among other things, the offline complexity of several basic problems in broadcast scheduling. Average flow-time received substantial attention in both the offline and online settings [21,17,18,19,3,4]. For average flow time, there are three $O(1)$-speed $O(1)$-competitive online algorithms. **LWF** is one of them [15,11] and the others are BEQUI [15] and its extension [16]. Our recent work [11] has investigated L_k norms of flow-time and showed that **LWF** is $O(k)$-speed $O(k)$-competitive algorithm. Constant competitive online algorithms for maximizing throughput for unit-sized pages can be found in [22,9,27,14]. A more thorough description of related work is deferred to a full version of the paper.

Organization: We prove each of the theorems mentioned above in a different section. The algorithm and analysis for weighted response time and weighted delay factor are very similar to that for delay factor and hence, in this version, we omit the analysis and only describe the algorithm for weighted response time. Due to space limitations, we prove Theorem 2 only with unit-sized pages and briefly discuss Theorem 3. See [12] for a more detailed version.

Notation: We denote the length of page p by ℓ_p. That is, ℓ_p is the amount of time a 1-speed server takes to broadcast page p non-preemptively. We assume without loss of generality that for any request $J_{p,i}$, $S_{p,i} \geq \ell_p$. For an algorithm A we let α^A denote the maximum delay factor witnessed by A for a given sequence of requests. We let α^* denote the optimal delay factor of an offline schedule. Likewise, we let ρ^A denote the maximum response time witnessed by A and ρ^* the optimal response time of an offline schedule. For a time interval $I = [a, b]$ we define $|I| = b - a$ to be the length of interval I.

2 Minimizing the Maximum Response Time

In this section we analyze **FIFO** for minimizing maximum response time when page sizes are different. We first describe the algorithm **FIFO**. **FIFO** broadcasts pages *non-preemptively*. Consider a time t when **FIFO** finished broadcasting a page. Let $J_{p,i}$ be the request in **FIFO**'s queue with earliest arrival time breaking ties arbitrarily. **FIFO** begins broadcasting page p at time t. At any time during this broadcast, we will say that $J_{p,i}$ *forced* **FIFO** to broadcast page p at this time. When broadcasting a page p all requests for page p that arrived before the start of the broadcast are simultaneously satisfied when the broadcast completes. Any request for page p that arrive during the broadcast are not satisfied until the next full transmission of p.

We consider **FIFO** when given a 1-speed machine. Let σ be an arbitrary sequence of requests. Let OPT denote some fixed offline optimum schedule and let ρ^* denote the optimum maximum response time. We will show that $\rho^{\mathbf{FIFO}} \leq 2\rho^*$. For the sake of contradiction, assume that **FIFO** witnesses a response time $c\rho^*$ by some job $J_{q,k}$ for some $c > 2$. Let t^* be the time $J_{q,k}$ is satisfied, that is $t^* = f_{q,k}$. Let t_1 be the smallest time less than t^* such that at any time t during the interval $[t_1, t^*]$ the request which forces **FIFO** to broadcast a page at time t has response time at least ρ^* when satisfied. Throughout the rest of this section we let $I = [t_1, t^*]$. Let $\mathcal{J}_I$ denote the requests which forced **FIFO** to broadcast during I. Notice that during the interval I, all requests in $\mathcal{J}_I$ are completely satisfied during this interval. In other words, any request in $\mathcal{J}_I$ starts being satisfied during I and is finished during I.

We say that OPT *merges* two distinct requests for a page p if they are satisfied by the same broadcast.

Lemma 1. OPT *cannot merge any two requests in* $\mathcal{J}_I$ *into a single broadcast.*

Proof. Let $J_{p,i}, J_{p,j} \in \mathcal{J}_I$ s.t. $i < j$. Note that that $J_{p,i}$ is satisfied before $J_{p,j}$. Let t' be the time that **FIFO** *starts* satisfying request $J_{p,i}$. By the definition of I, request $J_{p,i}$ has response time at least ρ^*. The request $J_{p,j}$ must arrive after time t', that is $a_{p,j} > t'$, otherwise $J_{p,j}$ is satisfied by the same broadcast of page p that satisfied $J_{p,i}$. Hence, it follows that if OPT merges $J_{p,i}$ and $J_{p,j}$ then the finish time of $J_{p,i}$ in OPT is strictly greater than its finish time in **FIFO** which is already at least ρ^*; this is a contradiction to the definition of ρ^*. □

Lemma 2. *All requests in* $\mathcal{J}_I$ *arrived no earlier than time* $t_1 - \rho^*$.

Proof. For the sake of contradiction, suppose some request $J_{p,i} \in \mathcal{J}_I$ arrived at time $a_{p,i} < t_1 - \rho^*$. During the interval $[a_{p,i} + \rho^*, t_1]$ the request $J_{p,i}$ must have wait time at least ρ^*. However, then any request which forces **FIFO** to broadcast during $[a_{p,i} + \rho^*, t_1]$ must have response time at least ρ^*, contradicting the definition of t_1. □

We are now ready to prove Theorem 1, stating that **FIFO** is 2-competitive.

Proof. Recall that all requests in $\mathcal{J}_I$ are completely satisfied during I. Thus we have that the total size of requests in $\mathcal{J}_I$ is $|I|$. By definition $J_{q,k}$ witnesses a response time greater than $2\rho^*$ and therefore $t^* - a_{q,k} > 2\rho^*$. Since $J_{q,k} \in \mathcal{J}_I$ is the last request done by **FIFO** during I, all requests in $\mathcal{J}_I$ must arrive no later than $a_{q,k}$. Therefore, these requests must be finished by time $a_{q,k} + \rho^*$ by the optimal solution. From Lemma 2, all the requests $\mathcal{J}_I$ arrived no earlier than $t_1 - \rho^*$. Thus OPT must finish all requests in $\mathcal{J}_I$, whose total volume is $|I|$, during $I_{opt} = [t_1 - \rho^*, a_{q,k} + \rho^*]$. Thus it follows that $|I| \leq |[t_1 - \rho^*, a_{q,k} + \rho^*]|$, which simplifies to $t^* \leq a_{q,k} + 2\rho^*$. This is a contradiction to the fact that $t^* - a_{q,k} > 2\rho^*$. □

We now discuss the differences between our proof of **FIFO** for varying sized pages and the proof given by Chang et al. in [10] showing that **FIFO** is 2-competitive for unit sized pages. In [10] it is shown that at anytime t, $F(t)$, the set of *unique* pages in **FIFO**'s queue satisfies the following property: $|F(t) \setminus O(t)| \leq |O(t)|$ where $O(t)$ is the set of unique pages in OPT's queue. This easily implies the desired bound. To establish this, they use a slot model in which unit-sized pages arrive only during integer times which allows one to define unique pages. This may appear to be a

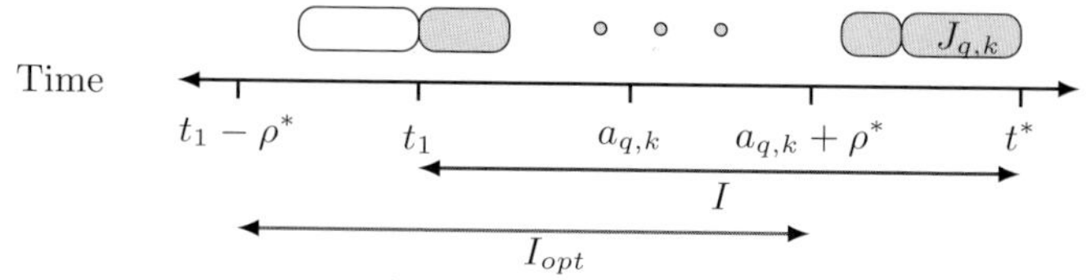

Fig. 1. Broadcasts by **FIFO** satisfying requests in $\mathcal{J}_I$ are shown in blue. Note that $a_{q,k}$ and $a_{q,k} + \rho^*$ are not necessarily contained in I.

technicality, however when considering different sized pages, it is not so clear how one even defines unique pages since this number varies during the transmission of p as requests accumulate. Our approach avoids this issue in a clean manner by not assuming a slot model or unit-sized pages.

3 Minimizing Maximum Delay Factor and Weighted Response Time

In this section we consider the problem of minimizing maximum delay factor and prove Theorem 2.

3.1 Unit Sized Pages

In this section we consider the problem of minimizing the maximum delay factor when all pages are of unit size. In this setting we assume preemption is not allowed. In the standard unicast scheduling setting where each broadcast satisfies exactly one request, it is known that the algorithm which always schedules the request with smallest slack at any time is $(1+\epsilon)$-speed $O(\frac{1}{\epsilon})$-competitive [13]. However, in the broadcast setting this algorithm, along with other simple greedy algorithms, do not provide constant competitive ratios even with extra speed. The reason for this is that the adversary can force these algorithm to repeatedly broadcast the same page even though the adversary can satisfy each of these requests in a singe broadcast.

Due to this, we consider a more sophisticated algorithm called **SSF-W** (Shortest-Slack-First with Waiting). This algorithm was developed and analyzed in [13]. In this paper we alter the algorithm in a slight but practically important way. The main contribution is, however, a new analysis that is at a high-level similar in outline to the one in [13] but is subtly different and leads to much improved bound on its performance. **SSF-W** *adaptively* forces requests to wait after their arrival before they are considered for scheduling. The algorithm is parameterized by a real value $c > 1$ which is used to determine how long a request should wait. Before scheduling a page at time t, the algorithm determines the largest current delay factor of any request that is unsatisfied at time t, α_t. Amongst the unsatisfied requests that have a current delay factor at least $\frac{1}{c}\alpha_t$, the page corresponding to the request with smallest slack is broadcasted. Note that in the algorithm, each request is forced to wait to be scheduled until it has delay factor at least $\frac{1}{c}\alpha_t$. Thus **SSF-W** can be seen an adaptation of the algorithm which schedules the request with smallest slack in broadcasting setting with explicit waiting. Waiting is used to potentially satisfy multiple requests with similar arrival times in a single broadcast. Another interpretation, that we mentioned earlier, is that **SSF-W** is a balance between **LF** and **SSF**.

Algorithm. SSF-W
- Let α_t be the maximum delay factor of any request in **SSF-W**'s queue at time t.
- At time t, let $Q(t) = \{J_{p,i} \mid J_{p,i}$ has not been satisfied and $\frac{t-a_{p,i}}{S_{p,i}} \geq \frac{1}{c}\alpha_t\}$.
- If the machine is free at t, schedule the request in $Q(t)$ with the smallest slack *non-preemptively*.

First we note the difference between **SSF-W** above and the one described in [13]. Let α'_t denote the maximum delay factor witnessed so far by **SSF-W** at time t over all requests seen by t including satisfied and unsatisfied requests. In [13], a request $J_{p,i}$ is in $Q(t)$ if $\frac{t-a_{p,i}}{S_{p,i}} \geq \frac{1}{c}\alpha'_t$. Note that α'_t is monotonically increases with t while α_t can increase and decrease with t and is never more than α'_t. In the old algorithm it is possible that $Q(t)$ is empty and no request is scheduled at t even though there are outstanding requests! Our new version of **SSF-W** can be seen as more practical since there will always be requests in $Q(t)$ if there are outstanding requests and moreover it adapts and may reduce α_t as the request sequence changes with time. It is important to note that our analysis and the analysis given in [13] hold for both definitions of **SSF-W** with some adjustments.

We analyze **SSF-W** when it is given a $(1+\epsilon)$-speed machine. Let $c > 1 + \frac{2}{\epsilon}$ be the constant which parameterizes **SSF-W**. Let σ be an arbitrary sequence of requests. We let OPT denote some fixed offline optimum schedule and let α^* and $\alpha^{\mathbf{SSF\text{-}W}}$ denote the maximum delay factor achieved by OPT and **SSF-W**, respectively. We will show that $\alpha^{\mathbf{SSF\text{-}W}} \leq c^2\alpha^*$. For the sake of contradiction, suppose that **SSF-W** witnesses a delay factor greater than $c^2\alpha^*$. We consider the *first* time t^* when **SSF-W** has some request in its queue with delay factor $c^2\alpha^*$. Let the request $J_{q,k}$ be a request which achieves the delay factor $c^2\alpha^*$ at time t^*. Let t_1 be the smallest time less than t^* such that at each time t during the interval $[t_1, t^*]$ if **SSF-W** is forced to broadcast by request $J_{p,i}$ at time t it is the case that $\frac{t-a_{p,i}}{S_{p,i}} \geq \alpha^*$ and $S_{p,i} \leq S_{q,k}$. Throughout this section we let $I = [t_1, t^*]$. The main difference between the analysis in [13] and the one here is in the definition of t_1. In [13], t_1 was implicitly defined to be $a_{q,k} + c(f_{q,k} - a_{q,k})$.

We let $\mathcal{J}_I$ denote the requests which forced **SSF-W** to schedule broadcasts during the interval $[t_1, t^*]$. We now show that any two request in $\mathcal{J}_I$ cannot be satisfied with a single broadcast by the optimal solution. Intuitively, the most effective way the adversary to performs better than **SSF-W** is to merge requests of the same page into a single broadcast. Here we will show this is not possible for the requests in $\mathcal{J}_I$. We omit the proof of Lemma 3, since the proof is similar to that of Lemma 1.

Lemma 3. OPT *cannot merge any two requests in $\mathcal{J}_I$ into a single broadcast.*

To fully exploit the advantage of speed augmentation, we need to ensure that the length of the interval I is sufficiently long.

Lemma 4. $|I| = |[t_1, t^*]| \geq (c^2 - c)S_{q,k}\alpha^*$.

Proof. The request $J_{q,k}$ has delay factor at least $c\alpha^*$ at any time during $I' = [t', t^*]$, where $t' = t^* - (c^2 - c)S_{q,k}\alpha^*$. Let $\tau \in I'$. The largest delay factor any request can have at time τ is less than $c^2\alpha^*$ by definition of t^* being the first time **SSF-W** witnesses delay factor $c^2\alpha^*$. Hence, $\alpha_\tau \leq c^2\alpha^*$. Thus, the request $J_{q,k}$ is in the queue $Q(\tau)$ because $c\alpha^* \geq \frac{1}{c}\alpha_\tau$. Moreover, this means that any request that forced **SSF-W** to broadcast during I', must have delay factor at least α^* and since $J_{q,k} \in Q(\tau)$ for any $\tau \in I'$, the requests scheduled during I' must have slack at most $S_{q,k}$. □

We now explain a high level view of how we lead to a contradiction. From Lemma 3, we know any two requests in $\mathcal{J}_I$ cannot be merged by OPT. Thus if we show that OPT must finish all these requests during an interval which is not long enough to include all of them, we can draw a contradiction. More precisely, we will show that all requests in $\mathcal{J}_I$ must be finished during I_{opt} by OPT, where $I_{opt} = [t_1 - 2S_{q,k}\alpha^* c, t^*]$. It is easy to see that all these requests already have delay factor α^* by time t^*, thus the optimal solution must finish them by time t^*. For the starting point, we will bound the arrival times of the requests in $\mathcal{J}_I$ in the following lemma. After that, we will draw a contradiction in Lemma 6.

Lemma 5. *Any request in $\mathcal{J}_I$ must have arrived after time $t_1 - 2S_{q,k}\alpha^* c$.*

Proof. For the sake of contradiction, suppose that some request $J_{p,i} \in \mathcal{J}_I$ arrived at time $t' < t_1 - 2S_{q,k}\alpha^* c$. Recall that $J_{p,i}$ has a slack no bigger than $S_{q,k}$ by the definition of I. Therefore at time $t_1 - S_{q,k}\alpha^* c$, $J_{p,i}$ has a delay factor of at least $c\alpha^*$. Thus any request scheduled during the interval $I' = [t_1 - S_{q,k}\alpha^* c, t_1]$ has a delay factor no less than α^*. We observe that $J_{p,i}$ is in $Q(\tau)$ for $\tau \in I'$; otherwise there must be a request with a delay factor bigger than $c^2\alpha^*$ at time τ and this is a contradiction to the assumption that t^* is the first time that **SSF-W** witnessed a delay factor of $c^2\alpha^*$. Therefore any request scheduled during I' has a slack no bigger than $S_{p,i}$. Also we know that $S_{p,i} \leq S_{q,k}$. In sum, we showed that any request done during I' had slack no bigger than $S_{q,k}$ and a delay factor no smaller than α^*, which is a contradiction to the definition of t_1. □

Now we are ready to prove the competitiveness of **SSF-W**.

Lemma 6. *Suppose c is a constant s.t. $c > 1 + 2/\epsilon$. If* **SSF-W** *has $(1+\epsilon)$-speed then $\alpha^{\textbf{SSF-W}} \leq c^2\alpha^*$.*

Proof. For the sake of contradiction, suppose that $\alpha^{\textbf{SSF-W}} > c^2\alpha^*$. During the interval I, the number of broadcasts which **SSF-W** transmits is $(1+\epsilon)|I|$. From Lemma 5, all the requests processed during I have arrived no earlier than $t_1 - 2c\alpha^* S_{q,k}$. We know that the optimal solution must process these requests before time t^* because these requests have delay factor at least α^* by t^*. By Lemma 3 the optimal solution must make a unique broadcast for each of these requests. Thus, the optimal solution must finish all of these requests in $2c\alpha^* S_{q,k} + |I|$ time steps. Thus, the it must hold that $(1+\epsilon)|I| \leq 2c\alpha^* S_{q,k} + |I|$. Using Lemma 4, this simplifies to $c \leq 1 + 2/\epsilon$, which is a contradiction to $c > 1 + 2/\epsilon$. □

The previous lemmas prove the first part of Theorem 2 when $c = 1+3/\epsilon$. Namely that **SSF-W** is a $(1+\epsilon)$-speed $O(\frac{1}{\epsilon^2})$-competitive algorithm for minimizing the maximum delay factor in broadcast scheduling with unit sized pages.

We now compare proof of Theorem 2 and the proof of Theorem 1 with the analysis given in [13]. The central technique used in [13] and in our analysis is to draw a contradiction by showing that the optimal solution must complete more requests than possible on some time interval I. This technique is well known in unicast scheduling. At the heart of this technique is to find the which requests to consider and bounding the length of the interval I. This is where our proof and the one given in [13] differ. Here we are more careful on how I is defined and how we find requests the optimal solution must broadcast during I. This allows us to show tighter bounds on the speed and competitive ratios while simplifying the analysis. In fact, our analysis of **FIFO** and **SSF-W** shows the importance of these definitions. Our analysis of **FIFO** shows that a tight bound on the length of I can force a contradiction without allowing extra speed-up given to the algorithm. Our analysis of **SSF-W** shows that when the length of I varies how resource augmentation can be used to force the contradiction.

3.2 Weighted Response Time and Weighed Delay Factor

In this section, we discuss the connection of our analysis of **SSF-W** to the problem of minimizing *weighted* response time. In this setting a request $J_{p,i}$ has a weight $w_{p,i}$ instead of a slack. The goal is to minimize the maximum weighted response time $\max_{p,i} w_{p,i}(f_{p,i} - a_{p,i})$. We develop an algorithm which we call **BWF-W** for Biggest-Wait-First with Waiting. This algorithm is defined analogously to the definition of **SSF-W**. The algorithm is parameterized by a constant $c > 1$. At any time t before broadcasting a page, **BWF-W** determines the largest weighted wait time of any request which has yet to be satisfied. Let this value be ρ_t. The algorithm then chooses to broadcast a page corresponding to the request with largest weight amongst the requests whose current weighted wait time at time t is larger than $\frac{1}{c}\rho_t$.

Algorithm. BWF-W

- Let ρ_t be the maximum weighted wait time of any request in **BWF-W**'s queue at time t.
- At time t,
  let $Q(t) = \{J_{p,i} \mid J_{p,i}$ has not been satisfied and $w_{p,i}(t - a_{p,i}) \geq \frac{1}{c}\rho_t\}$.
- If the machine is free at t, schedule the request in $Q(t)$ with largest weight *non-preemptively*.

Although minimizing the maximum delay factor and minimizing the maximum weighted flow time are very similar metrics, the problems are not equivalent. For the problems of minimizing the maximum weighted response time and weighted delay factor, the upper bounds shown for **SSF-W** in this paper also hold for **BWF-W**. The analysis of **BWF-W** is very similar to that of **SSF-W** and the proof is omitted.

4 Lower Bound for a Natural Greedy Algorithm LF

We briefly discuss the algorithm **LF** which always schedules the page with the largest delay factor and show that it is not $O(1)$-competitive with any constant speed. This is interesting, since **LF** can be seen as a natural generalization of **FIFO** to the problem of minimizing maximum delay factor. This demonstrates the importance of the tradeoff between scheduling a request with smallest slack and forcing requests to wait; notice that our algorithm **LF** is the same as **SSF-W** when $c = 1$. Interestingly, our lowerbound instance σ holds even in the standard unicast scheduling with unit sized jobs. In the instance σ, a series of job groups $\mathcal{J}_i$ for $0 \le i \le k$ arrive, with all jobs in each group having the same arrival time and the same slack. The slack and the number of jobs in $\mathcal{J}_i$ exponentially decreases with i. Here the optimal solution is to process $\mathcal{J}_1$ through $\mathcal{J}_k$, and finally $\mathcal{J}_0$ where jobs have a very big slack. However, **LF**, which makes a scheduling decision only based on the current largest delay factor ignoring slack sizes, processes jobs from $\mathcal{J}_0$ through $\mathcal{J}_k$, thereby resulting in a large delay factor for some jobs in $\mathcal{J}_k$. The full proof can be found in [12].

5 Conclusion

In this paper, we showed an almost fully scalable algorithm[1] for minimizing the maximum delay factor in broadcasting for unit sized jobs. The slight modification we make to **SSF-W** from [13] makes the algorithm more practical. Using the intuition developed for the maximum delay factor, we proved that **FIFO** is in fact 2-competitive for varying sized jobs closing the problem for minimizing the maximum response time online in broadcast scheduling.

We close this paper with the following open problems. Although the new algorithm for the maximum delay factor with unit sized jobs is almost fully scalable, it explicitly depends on speed given to the algorithm. Can one get another algorithm independent of this dependency? For different sized pages, it is still open on whether there exists a $(1+\epsilon)$-speed algorithm that is $O(1)$-competitive. For minimizing the maximum response time offline it is of theoretical interest to show a lower bound on the approximation ratio that can be achieved or to show an algorithm that is a c-approximation for some $c < 2$, improving upon **FIFO**.

References

1. Acharya, S., Franklin, M., Zdonik, S.: Dissemination-based data delivery using broadcast disks. IEEE Pers. Commun. 2(6), 50–60 (1995)
2. Aksoy, D., Franklin, M.J.: rxw: A scheduling approach for large-scale on-demand data broadcast. IEEE/ACM Trans. Netw. 7(6), 846–860 (1999)
3. Bansal, N., Charikar, M., Khanna, S., Naor, J.S.: Approximating the average response time in broadcast scheduling. In: SODA, pp. 215–221 (2005)

[1] An algorithm is said to be almost fully scalable if for any fixed $\epsilon > 0$, it is $O(1+\epsilon)$-speed $O(1)$-competitive.

4. Bansal, N., Coppersmith, D., Sviridenko, M.: Improved approximation algorithms for broadcast scheduling. In: SODA, pp. 344–353 (2006)
5. Bartal, Y., Muthukrishnan, S.: Minimizing maximum response time in scheduling broadcasts. In: SODA, pp. 558–559 (2000)
6. Bender, M.A., Chakrabarti, S., Muthukrishnan, S.: Flow and stretch metrics for scheduling continuous job streams. In: SODA, pp. 270–279 (1998)
7. Bender, M.A., Clifford, R., Tsichlas, K.: Scheduling algorithms for procrastinators. J. Scheduling 11(2), 95–104 (2008)
8. Bender, M.A., Muthukrishnan, S., Rajaraman, R.: Improved algorithms for stretch scheduling. In: SODA, pp. 762–771 (2002)
9. Chan, W.-T., Lam, T.W., Ting, H.-F., Wong, P.W.H.: New results on on-demand broadcasting with deadline via job scheduling with cancellation. In: Chwa, K.-Y., Munro, J.I.J. (eds.) COCOON 2004. LNCS, vol. 3106, pp. 210–218. Springer, Heidelberg (2004)
10. Chang, J., Erlebach, T., Gailis, R., Khuller, S.: Broadcast scheduling: algorithms and complexity. In: SODA, pp. 473–482 (2008)
11. Chekuri, C., Im, S., Moseley, B.: Longest wait first for broadcast scheduling (manuscript, 2009)
12. Chekuri, C., Im, S., Moseley, B.: Minimizing maximum response time and delay factor in broadcast scheduling. CoRR, abs/0906.2048 (2009)
13. Chekuri, C., Moseley, B.: Online scheduling to minimize the maximum delay factor. In: SODA, pp. 1116–1125 (2009)
14. Chrobak, M., Dürr, C., Jawor, W., Kowalik, L., Kurowski, M.: A note on scheduling equal-length jobs to maximize throughput. J. Scheduling 9(1), 71–73 (2006)
15. Edmonds, J., Pruhs, K.: Multicast pull scheduling: When fairness is fine. Algorithmica 36(3), 315–330 (2003)
16. Edmonds, J., Pruhs, K.: Scalably scheduling processes with arbitrary speedup curves. In: SODA, pp. 685–692 (2009)
17. Erlebach, T., Hall, A.: Np-hardness of broadcast scheduling and inapproximability of single-source unsplittable min-cost flow. In: SODA, pp. 194–202 (2002)
18. Gandhi, R., Khuller, S., Kim, Y.-A., Wan, Y.-C.J.: Algorithms for minimizing response time in broadcast scheduling. Algorithmica 38(4), 597–608 (2004)
19. Gandhi, R., Khuller, S., Parthasarathy, S., Srinivasan, A.: Dependent rounding and its applications to approximation algorithms. J. ACM 53(3), 324–360 (2006)
20. Kalyanasundaram, B., Pruhs, K.: Speed is as powerful as clairvoyance. J. ACM 47(4), 617–643 (2000)
21. Kalyanasundaram, B., Pruhs, K., Velauthapillai, M.: Scheduling broadcasts in wireless networks. J. Scheduling 4(6), 339–354 (2000)
22. Kim, J.-H., Chwa, K.-Y.: Scheduling broadcasts with deadlines. Theor. Comput. Sci. 325(3), 479–488 (2004)
23. Pruhs, K.: Competitive online scheduling for server systems. SIGMETRICS Perform. Eval. Rev. 34(4), 52–58 (2007)
24. Pruhs, K., Sgall, J., Torng, E.: Online Scheduling. In: Handbook of Scheduling: Algorithms, Models, and Performance Analysis (2004)
25. Pruhs, K., Uthaisombut, P.: A comparison of multicast pull models. Algorithmica 42(3-4), 289–307 (2005)
26. Wong, J.: Broadcast delivery. Proc. IEEE 76(12), 1566–1577 (1988)
27. Zheng, F., Fung, S.P.Y., Chan, W.-T., Chin, F.Y.L., Poon, C.K., Wong, P.W.H.: Improved on-line broadcast scheduling with deadlines. In: Chen, D.Z., Lee, D.T. (eds.) COCOON 2006. LNCS, vol. 4112, pp. 320–329. Springer, Heidelberg (2006)

Preemptive Online Scheduling with Reordering

György Dósa[1] and Leah Epstein[2]

[1] Department of Mathematics, University of Pannonia, Veszprém, Hungary
dosagy@almos.vein.hu

[2] Department of Mathematics, University of Haifa, 31905 Haifa, Israel
lea@math.haifa.ac.il

Abstract. We consider online preemptive scheduling of jobs, arriving one by one, on m identical parallel machines. A buffer of a positive fixed size, K, which assists in partial reordering of the input, is available for the storage of at most K unscheduled jobs. We study the effect of using a fixed sized buffer (of an arbitrary size) on the supremum competitive ratio over all numbers of machines (the overall competitive ratio), as well as the effect on the competitive ratio as a function of m.

We find a tight bound on the competitive ratio for any m. This bound is $\frac{4}{3}$ for even values of m and slightly lower for odd values of m. We show that a buffer of size $\Theta(m)$ is sufficient to achieve this bound, but using $K = o(m)$ does not reduce the best overall competitive ratio which is known for the case without reordering, $\frac{e}{e-1}$. We further consider the semi-online variant where jobs arrive sorted by non-increasing processing time requirements. In this case we show that it is possible to achieve a competitive ratio of 1. In addition, we find tight bounds as a function of both K and m.

1 Introduction

Scheduling of jobs arriving one by one (or *over list*) is a basic model in online scheduling [17]. The system consists of a set of m identical machines that can process a sequence of arriving jobs. Each job j, which has a processing time p_j associated with it (also called *size*), needs to be assigned upon arrival. The completion time, or load, of a machine is the total time needed to process the jobs assigned to it, including idle time in which the machine is waiting for a job to be executed (if idle time exists). The goal is to minimize the maximum completion time of any machine, also known as the *makespan*.

We consider online and semi-online preemptive scheduling of jobs. An arriving job can be split into parts, which need to be assigned to non-overlapping time slots, possibly on different machines. Idle time is allowed, and each machine can process at most one job at each time. In the online scenario, a job must be treated before the next job is revealed. For an algorithm $\mathcal{A}$, we denote its cost by $\mathcal{A}$ as well. An optimal offline algorithm that knows the complete sequence of jobs in advance is denoted by OPT. In this paper we measure the performance quality of algorithms using the (absolute) competitive ratio, which is the most common

A. Fiat and P. Sanders (Eds.): ESA 2009, LNCS 5757, pp. 456–467, 2009.

measure for the performance evaluation of online algorithms. The competitive ratio of $\mathcal{A}$ is the infimum $\mathcal{R}$ such that for any input, $\mathcal{A} \leq \mathcal{R} \cdot \text{OPT}$.

We consider a model where a reordering buffer, of a fixed size $K > 0$, is available. This buffer can store up to K unassigned jobs and thus assists in partial reordering of the input. Upon the arrival of a job, it is possible to either assign it completely to machines and time slots, or otherwise it is possible to store it in the buffer rather than assigning it. If the buffer already contains K jobs, at least one of these jobs must be assigned to the machines in order to make room for the new job, or else the new job must be assigned.

Non-preemptive scheduling (i.e., the case where a job cannot be split into parts and must be processed continuously on one machine), with a reordering buffer, was previously studied in several papers [12,19,13,6,3]. The main research question in these papers was to find the effect of using a reordering buffer on the competitive ratio, that is, finding the lowest competitive ratio which can be achieved if the online algorithm is supplied with a buffer, and whether this competitive ratio is achievable only in the limit, or whether there exists a size of a buffer which allows to achieve this bound. This competitive ratio can then be compared to the best possible competitive ratio which can be achieved without a buffer. Clearly, an offline algorithm can be seen as an algorithm which uses an unbounded buffer. Limiting the online algorithm to a fixed sized buffer still means in most cases that the algorithm cannot perform as well as an optimal offline algorithm. Consequently, the competitive ratio for every value of m is of interest, as well as the *overall* competitive ratio, which is the supremum competitive ratio over all values of m.

In all (non-preemptive) variants studied in the past, a finite length buffer already allows to achieve the best competitive ratio. In particular, for two identical machines, a buffer of size 1 is sufficient, as was proved by Kellerer et al. [12] and independently by Zhang [19]. For m identical machines, Englert, Özmen and Westermann [6] showed that a buffer of size $O(m)$ is sufficient. For the more general case of uniformly related machines, where machines may have different speeds, it was shown [3] that for two machines, a buffer of size 2 allows to achieve the best competitive ratio. In fact, for some speed ratios between the two machines, a buffer of size 1 is sufficient, while for some other speed ratios, a buffer of size 1 provably does not allow to achieve the best bound. Note that it was shown by [6] that a buffer of size $m-1$ reduces the competitive ratio for uniformly related machines below the lower bound of the case without reordering. In this paper we answer analogous questions for preemptive scheduling.

Preemptive online scheduling without reordering was studied by Chen, van Vliet, and Woeginger [2] (see also [14,16,1]). They designed an algorithm of the best possible competitive ratio for any number of machines m. This competitive ratio is a monotonically increasing function of m, $\frac{m^m}{m^m-(m-1)^m}$, which implies an overall competitive ratio of $\frac{e}{e-1} \approx 1.58$. A number of papers generalized this result for uniformly related machines [18,9,7,4,10].

We study an additional variant where it is known in advance that jobs arrive sorted by size, in a non-increasing order. This common semi-online variant of preemptive semi-online scheduling was analyzed by Seiden, Sgall and Woeginger for identical machines [15]. The overall tight bound on the competitive ratio shown by [15] is $\frac{1+\sqrt{3}}{2} \approx 1.366$. Semi-online preemptive scheduling on uniformly related machines was considered in [8,5].

Our results. We study the case of m identical machines. We find a tight bound on the competitive ratio for general inputs for any number of machines, m. We show that a buffer of size $\lceil \frac{m-2}{2} \rceil$ is sufficient to achieve this bound. In fact, reordering via the usage of a buffer allows to reduce the tight competitive ratio to $\frac{4}{3}$ for even m, and to $\frac{4m^2}{3m^2+1} < \frac{4}{3}$ for odd m, whereas $K = o(m)$ does not reduce the overall competitive ratio, which remains $\frac{e}{e-1}$, as in [2]. Surprisingly, we find that the best competitive ratio, as a function of m, is not monotone, and the overall competitive ratio is $\frac{4}{3}$. Note that this value is the tight competitive ratio for $m = 2$, the only case where a buffer is not necessary. This is different from the non-preemptive problem, where for $m = 2$ the usage of a buffer reduces the best competitive ratio from $\frac{3}{2}$ to $\frac{4}{3}$ [12,19,6]. As a motivation for using this specific size of buffer, $\lceil \frac{m-2}{2} \rceil$, we show that for $m = 6$ machines, where our general result uses $K = 2$, a buffer of size 1 leads to a larger competitive ratio. We show tight bounds of $\frac{19}{14} \approx 1.35714$ on the competitive ratio for this case.

We further consider the semi-online variant where jobs arrive sorted by non-increasing processing time requirements. In this case we show that a buffer of size $m-1$ is sufficient to achieve a competitive ratio of 1, whereas a buffer of size $m-2$ is not sufficient. That is, the combination of a reordering buffer with jobs arriving in a sorted order is as good as receiving the entire set of jobs in advance. We show that the tight bound for this last case, where the buffer has a size of $m-2$, is $1 + \frac{1}{m^2+m-1}$. Finally, we find tight bounds for all buffer sizes $K < m-2$, in these cases, the competitive ratio is the maximum of $m-K-1$ values $\max_{1 \le \mu \le m-K-1} \frac{2m(m+\mu)}{2m^2+2K\mu+\mu^2+\mu}$.

Our algorithms are based on a unified approach where largest jobs are kept in the buffer, and the created schedule is as imbalanced as possible keeping less loaded machines free to receive new jobs. Using the master algorithm with different parameters results in distinct algorithms for the multiple cases. Lower bounds are based on a unified approach where one sequence is used, and the considered inputs are its subsequences. This approach is general and allows the usage of several types of inputs for the different cases. These general approaches were used in the past for preemptive problems, but some important adaptations were required to be able to deal with the existence of a buffer.

Due to space constraints, some of the proofs are omitted.

2 Algorithms

2.1 The Master Algorithm

All our algorithms have a common structure which is explained in this section. These algorithms avoid the usage of idle time, and try to assign as much work as possible to the more loaded machines, while keeping K jobs is the buffer. Let n denote the number of jobs (which is unknown to the online algorithm).

The algorithm can be used for any $K \leq m$ (we will see later that larger values of K are not useful). In addition to the number of machines m and the size of the buffer K, the master algorithm uses a parameter $\mathcal{R}$, which is the required competitive ratio. In the case $K = 0$, the algorithm reduces to that used by [2]. We note that in the case of identical machines, idle time gives no advantage.

After initialization, and as long as at most K jobs have arrived, all jobs are stored in the buffer. If the input consists of at most K jobs, then each job is assigned to a separate machine, to run on this machine non-preemptively, starting from time 0, which results in an optimal solution. Otherwise, after $K + 1$ jobs have arrived, and as long as jobs keep arriving, the algorithm keeps the K largest jobs seen so far in the buffer. After jobs stop arriving, the algorithm keeps removing a smallest job from the buffer and assigning it using the same algorithm, until all jobs have been assigned. The loads after the assignment of i jobs are denoted by $L_1^i \leq L_2^i \leq \ldots \leq L_m^i$. Note that these are the loads after the arrival of $\min\{K+i, n\}$ jobs. Specifically, if $K + i \leq n$, then these are the loads after the arrival of $K + i$ jobs, out of which K are stored in the buffer. Otherwise, if $K + i > n$, these are the loads after all jobs have arrived and $n - i$ of them are in the buffer. We also use $Q_i = \sum_{k=1}^{m} L_k^i$, i.e., Q_i is the total size assigned jobs after i jobs have been assigned. This amount includes all scheduled jobs in both of the described cases. Let OPT_i denote the cost of OPT at this time, after i jobs have been assigned.

We assume that each machine has an index in $\{1, 2, \ldots, m\}$. We first explain how to assign jobs to machines in a way that the sorted order of machines does not change, that is, the load L_g^i is always the load of machine g. Later we show how to modify the algorithm so that it uses at most one preemption per job, the sequence of loads remains as in the first variant of the algorithm, but the sorted order of machines changes frequently (i.e., L_g^i is the g-th load in the sorted order of loads, but it does not necessarily belong to machine g).

We say that the buffer is *full*, if it contains exactly K jobs. Since the algorithm for the case $n \leq K$ is completely defined above, we assume in what follows that $n > K$, i.e., there is at least one case where a job needs to be assigned while the buffer is full, and we describe the assignment of jobs for the cases where the buffer is full, or was full at some previous time. This last option means that jobs no longer arrive, and the jobs which remained in the buffer need to be assigned.

To assign a job, non-overlapping slots are reserved on the machines, and the job is assigned into these slots, one by one, by a decreasing order of indices of machines, until the job is completely assigned, or until all slots are full. In each case for which we use this master algorithm with specific parameters, we will show that the second option never occurs. If one of the used slots is not filled completely, then the earliest part of this slot is used, so that no idle time is created. The slots for the $(i + 1)$-th job ever assigned are $[L_m^i, \mathcal{R} \cdot \text{OPT}_{i+1}]$ on machine m, and $[L_j^i, L_{j+1}^i]$ for on each other machine, $1 \leq j < m$. The slots are clearly non-overlapping. Note that some of the slots may be empty, if there are at least two consecutive identical loads.

In order to prove upper bounds, we use two lower bounds on the cost of an optimal solution, which are the average load, implied by the sum of all jobs (including those in the buffer), and the maximum size of any job. The algorithm, however, needs to compute the exact value of OPT_i. We exploit the property that OPT_i is in fact equal to the maximum of these two bounds [11]. Therefore, calculating the slot on machine m for the assignment of a job can be done in constant time.

Clearly, as long as every job is assigned successfully, the competitive ratio of the master algorithm is at most $\mathcal{R}$. Therefore, in each case we consider, it is necessary to show that the algorithm never fails. In most cases we derive a set of invariants which are proved by induction and allow to prove this property. In one case we use a small number of invariants and a direct proof for the most important invariant rather than induction. In the latter case, the usage of a similar structure of proof to the former case, i.e., additional similar invariants together with induction, does not seem to be helpful. Since the algorithm is a generalization of the algorithm of [2], the main technical contribution here is the design of the correct set of invariants, or the design of a more direct proof. This is done for each case separately, since the exact value of $\mathcal{R}$ affects the execution of the algorithm and leads to very different schedules in the different cases. Note that for both variants and all values of m, the largest jobs are always the jobs which are kept in the buffer. In the case of general inputs, it is either the case that the new job is stored in the buffer, or that it is the job which is assigned. In the case of non-increasing sequences, the first K jobs are kept in the buffer until the sequence ends, and then they are assigned in an order which is opposite to the order of their arrival.

In previous work on scheduling with a buffer, in many cases, the largest jobs (or largest job, in the case $K = 1$) were those which are stored in the buffer. Intuitively, this seems to be the correct approach; the algorithm is aware of the exact sizes of the largest jobs and takes them into account in the other scheduling decisions, but it postpones their assignment until the later. Nevertheless, in [3] one of the algorithms of optimal competitive ratio, which uses $K = 1$, has two cases, where in one of the cases the larger available job is assigned while the smaller job is stored in the buffer. We note an interesting difference with the algorithms of [6]. Our algorithm uses the same method of assignment for all jobs, even after no additional jobs arrive. It is possible in fact to use the same algorithm also in the case $n \leq K$ and avoid cases in the definition of the algorithm. However, since this case is very simple, so we prefer the current presentation, to avoid cases in the proof.

Finally, we show how to modify the algorithm to use at most one preemption per job instead of at most $m - 1$ preemptions. Assume that machine j has the j-th load before the assignment of a job. If the job is assigned to slots on machines $j, j+1, \ldots, m$, where $j < m - 1$, i.e., it was assigned using at least two preemptions, then instead of using the slots on machines $j + 2, \ldots, m$, it is possible to assign all these parts continuously on machine $j + 1$. Due to the definition of the algorithm, the only idle time in the processing of the job (but not in the schedule) may occur between the end of the time slot in which it is assigned to run on machine j and the beginning of the time slot on machine $j+1$, since the slot on machine j is not necessarily completely occupied. In this case the assignment of the part to machine j remains unchanged. The load of machine $j+1$ becomes the largest load (i.e., the m-th load), while the load of a machine y (for $j+2 \leq y \leq m$) becomes the $(y-1)$-th load. If the slot on machine j it is occupied completely, and $j < m$, then it is possible to schedule the job on machine j non-preemptively and without idle time (if $j = m$, then the job is already scheduled non-preemptively). In this case, the resulting load of machine j becomes the m-th load, while the load of machine y (for $j+1 \leq y \leq m$) becomes the $(y-1)$-th load. In both cases, the resulting sorted list of loads is the same as before, but some machines may switch places in this list.

2.2 General Inputs

The algorithm uses a buffer of size $K = \lfloor \frac{m-1}{2} \rfloor = \lceil \frac{m-2}{2} \rceil$, that is, $K = \frac{m}{2} - 1$ for even values of m, and $K = \frac{m-1}{2}$ for odd values of m. Let $\mathcal{R} = \frac{4}{3}$ if m is even, and $\mathcal{R} = \frac{4m^2}{3m^2+1}$ if m is odd. Thus the bound for odd m is strictly below $\frac{4}{3}$, and achieves it lowest value for $m = 3$, for which the bound equals to $\frac{9}{7} \approx 1.28571$. For $m = 2$ the competitive ratio is $\frac{4}{3}$, but the size of the buffer which we use is zero, thus the result of [2] for $m = 2$ already covers this case. By the results of Section 3, this result is best possible, i.e., the usage of a buffer does not improve the performance in this case.

We define a set of invariants which must hold after every assignment. Invariant j (for $1 \leq j \leq \lfloor \frac{m}{2} \rfloor$) at time i is defined by $\sum\limits_{k=1}^{j} L_k^i \leq (\mathcal{R}-1)\frac{j \cdot Q_i}{m-j}$.

The motivation of these invariants comes from a situation where very small jobs of a total size of x arrive first, after which some number $j \leq \lfloor \frac{m}{2} \rfloor$ of identical jobs of size $\frac{x}{m-j}$ arrive. An optimal solution spreads the small jobs over $m-j$ machines, while the best response of the algorithm would be to use the j least loaded machines. The invariants make sure that in such a case the competitive ratio is not exceeded. Note that even though such a construction could be a good candidate for a lower bound proof, the lower bound proof of Section 3 considers an input with a much smaller number of cases.

We use some further definitions and notations. After $i \geq 0$ jobs were assigned, no matter whether a new job arrives, or if no additional jobs arrive, but the buffer is non-empty, let the sorted list of sizes of jobs, which includes the jobs in the buffer and the new job (if exists) be $Y_0^{i+1} \leq Y_1^{i+1} \leq \ldots \leq Y_{\ell-1}^{i+1}$, where $\ell \leq K+1$. If a new job has just arrived (and thus the buffer is full) then $\ell = K+1$. We are going to assign the smallest job, thus $Q_{i+1} = Q_i + Y_0^{i+1}$. In addition, as explained in Section 2.1, $\text{OPT}_{i+1} = \max\{Y_{\ell-1}^{i+1}, \frac{1}{m}(Q_i + \sum\limits_{k=0}^{\ell-1} Y_k^{i+1})\}$, i.e., OPT_{i+1} is the optimal cost for a schedule for all received jobs so far, just before Y_0^{i+1} is scheduled. Using $Y_0^{i+1} \leq Y_k^{i+1}$ for any $1 \leq k \leq \ell-1$, we have $\text{OPT}_{i+1} \geq Y_0^{i+1}$ and $\text{OPT}_{i+1} \geq \frac{1}{m}(Q_i + jY_0^{i+1})$, which holds for any $1 \leq j \leq \ell$.

Lemma 1. *For a given value of i, if after i jobs were assigned the buffer is full, then the j-th invariant holds for time i and every $1 \le j \le \lfloor \frac{m}{2} \rfloor$. If the buffer contains $\ell < K$ jobs at this time, then the j-th invariant holds for time i and every $1 \le j \le \ell$.*

As mentioned above, the main technical difficulty lies in finding the correct invariants.

Lemma 2. *For every $i+1$ ($i \ge 0$), if the invariants hold for time i ($i < n$), then the algorithm assigns the $(i+1)$-th job successfully.*

Proof. We first show that there is enough space for the job, that is, the total size of the slots is no smaller than the size of the job to be assigned. This gives the following condition: $\mathcal{R} \cdot \text{OPT}_{i+1} - L_1^i \ge Y_0^{i+1}$.

Using invariant 1 at time i, $L_1^i \le (\mathcal{R}-1)\frac{Q_i}{m-1}$, and the following bounds on the optimal cost: $\text{OPT}_{i+1} \ge Y_0^{i+1}$ and $m\text{OPT}_{i+1} \ge Q_i + Y_0^{i+1}$, we get

$$L_1^i + Y_0^{i+1} \le (\mathcal{R}-1)\frac{Q_i}{m-1} + Y_0^{i+1} = \frac{\mathcal{R}-1}{m-1}(Q_i + Y_0^{i+1}) + Y_0^{i+1}(\frac{m-\mathcal{R}}{m-1})$$
$$\le \frac{\mathcal{R}-1}{m-1} m\text{OPT}_{i+1} + \text{OPT}_{i+1}(\frac{m-\mathcal{R}}{m-1}) \le \mathcal{R}\text{OPT}_{i+1} ,$$

for any $\mathcal{R} \le 2$. □

Theorem 5 will show that the obtained bounds are tight within algorithms for a fixed size buffer. We summarize these bounds in the following.

Theorem 1. *An application of the master algorithm, using $K = \frac{m}{2} - 1$ and $\mathcal{R} = \frac{4}{3}$ for even values of m, and $K = \frac{m-1}{2}$ and $\mathcal{R} = \frac{4m^2}{3m^2+1}$, for odd values of m, is successful, i.e., results in an algorithm of a competitive ratio of at most $\mathcal{R}$.*

One special case. We investigate the special case $m = 6$ separately. In this case, the application of the master algorithm with $K = \frac{m-2}{2} = 2$ leads to a competitive ratio of $\frac{4}{3}$. We show that a buffer of size 1 does not allow to achieve the best possible competitive ratio, $\frac{4}{3}$. In this last case, $K = 1$, we show that the best competitive ratio which can be achieved is $\mathcal{R} = \frac{19}{14} \approx 1.3571 > \frac{4}{3}$. To prove the upper bound, the algorithm of Section 2.2 is applied with $\mathcal{R} = \frac{19}{14}$. Using a different method of analysis we prove the following theorem. The proved bound is tight, by Theorem 6.

Theorem 2. *An application of the master algorithm for $m = 6$ using $K = 1$ and $\mathcal{R} = \frac{19}{14}$ is successful.*

In this proof we use two invariants. The first one is $L_5^i + L_6^i \ge \mathcal{R} \cdot \frac{Q_i}{3} = \frac{19}{42}Q_i$ and the second one is $L_1^i \le \frac{\mathcal{R}-1}{5}Q_i = \frac{Q_i}{14}$. The first invariant is proved by induction while the second one is proved directly.

Lemma 3. *The first invariant holds for any $i \ge 0$, for which after i jobs have been assigned, the buffer contains a job.*

Lemma 4. *The second invariant holds for any $i \geq 0$, for which after i jobs have been assigned, the buffer contains a job, as long as the first i jobs ever assigned are assigned successfully.*

Lemma 5. *For every time $i + 1$ $(i \geq 0)$, if the invariants hold at time i, then the algorithm assigns the next job successfully.*

Note that it is not assumed in Lemma 5 that there is an additional job in the buffer, thus it holds for the very last job which is assigned as well. The next corollary completes the proof.

Corollary 1. *All jobs are scheduled successfully.*

2.3 Non-increasing Job Sizes

We assume that jobs arrive sorted by non-increasing sizes. We use the master algorithm with the parameters $\mathcal{R} = \max\limits_{\mu \in \{0,1,2,\ldots,m-K-1\}} \frac{2m(m+\mu)}{2m^2+2K\mu+\mu^2+\mu}$, for $1 \leq K \leq m-1$. In the case $K = m-1$, $\mathcal{R} = 1$, so the algorithm finds an optimal schedule.

Recall that in this semi-online variant the algorithm keeps the first K jobs (which are the largest jobs) in the buffer, and that we assume $n > K$. Let $Z_1 \geq Z_2 \geq \ldots \geq Z_K$ be the sizes of jobs which are stored in the buffer. After $n - s$ jobs are assigned, for $0 \leq s < K$, the buffer contains only the jobs of sizes $Z_1, Z_2, \ldots, Z_s$. Let X_i denote the size of the i-th job which is assigned. Since Z_1 is the size of the largest job in the sequence, if the buffer contains ℓ jobs after the i-th job is assigned, then we have $\text{OPT}_i = \max\{Z_1, \frac{1}{m}(Q_i + \sum\limits_{k=1}^{\ell} Z_k)\}$. Thus there exists an integer $1 \leq f \leq n+1$ so that $\text{OPT}_i = Z_1$ for all $i < f$ and $\text{OPT}_i > Z_1$ for all $i \geq f$.

We next define a set of invariants which must hold after every assignment, which are addressed in Lemma 6 below. Invariant j at time i is defined by $\sum\limits_{k=1}^{j} L_k^i + \sum\limits_{k=1}^{j} Z_k \leq j \cdot \mathcal{R}\text{OPT}_i$.

Lemma 6. *For a given i, let ℓ denote the number of jobs in the buffer after i jobs were assigned. The j-th invariant holds for i and every $1 \leq j \leq \ell$.*

The motivation of these invariants comes from the possibilities of assigning the jobs which are stored in the buffer, if no additional jobs arrive. Every set of largest j jobs cannot run in parallel on more than j machines. The invariants consider the load resulting from assigning the largest j jobs to the least loaded j machines. To prove the K-th invariant, if the invariant does not hold immediately using induction, we use a direct proof, similar to the direct proof of Lemma 4.

Lemma 7. *For every time $i+1$ $(i \geq 0$ and $i + 1 \leq n)$, if the invariants hold at time i, then the algorithm assigns the next job successfully.*

Theorem 3. *The applications of the master algorithm using the parameter* $1 \leq K \leq m-1$ *and* $\mathcal{R} = \max_{0 \leq \mu \leq m-K-1} \frac{2m(m+\mu)}{2m^2+2K\mu+\mu^2+\mu}$ *is successful. These bounds are best possible.*

3 Lower Bounds

In order to prove lower bounds, we provide a "recipe" for designing lower bounds for the problem. This method is an adaptation of the method used in [10,16], which takes into account the existence of a buffer.

We restrict ourselves to inputs which have a specific form. All the considered inputs are prefixes of one sequence. The sequence consists of a non-negative number of blocks $b \geq 0$, where each block contains identical sized jobs, and finally, the sequence of blocks may be followed by t additional jobs. Let n_i be the number of jobs in the i-th block, and s_i be the size of each such job. We require $n_i \geq K+1$. The additional jobs arriving after all blocks are called *further* jobs. Let q_j denote the size of the j-th further job, for $1 \leq j \leq t$. We also require that $t + \sum_{i=1}^{b} n_i \geq m$. We denote such a sequence, for a specific value of K, by σ.

We use OPT_j to denote the optimal makespan for the sequence of jobs up to (and including) the j-th **further** job. In addition, we let $\text{OPT}_{i,j}$ denote the optimal makespan for the sequence of jobs up to (and including) the j-th job of the i-th block.

For each block i, we define the last $n_i - K$ optimal costs ($\text{OPT}_{i,j}$ for $K+1 \leq j \leq n_i$) as *crucial*.

We define a sequence of costs $\mathcal{C}_j$ for $1 \leq j \leq m$ as follows. Consider first the sequence of all crucial optimal costs of all blocks (i.e., neglecting the first K optimal costs of each block), followed by the optimal costs OPT_j for $K+1 \leq j \leq t$, and finally, K times the cost OPT_t. The last m costs in this sequence are denoted by $\mathcal{C}_1, \mathcal{C}_2, \ldots, \mathcal{C}_m$.

For the analysis, we define a modified sequence, which is based on a specific output of the online algorithm, as follows. For each block i, remove from the schedule the last K jobs of this block (i.e., of size s_i) which were ever assigned. If the execution of the algorithm on the original sequence is restricted to the modified sequence (and the treatment of the removed jobs is simply neglected), then at the time of arrival of the first job of a block, and also at the time of arrival of the first further job, the buffer is empty. This holds since the buffer can contain at most K such jobs, which are the last jobs of this block which are assigned, and thus all these jobs were removed from the schedule. Therefore, the order of assignment is according to blocks, and the further jobs are assigned last, in some order. We consider the relation between the $\mathcal{C}_j$ values and the optimal costs at the time of assignment of jobs. Let $a_1, a_2, \ldots, a_m$ be the last m jobs of the modified sequence ever assigned. Let O_j denote the optimal cost at the time of assignment of a_j.

Lemma 8. *For every* $1 \leq j \leq m$, $O_j \leq \mathcal{C}_j$.

Theorem 4. *Given a sequence σ as defined above. The competitive ratio of any algorithm with a buffer of size K is at least*

$$(\sum_{j=1}^{t} q_j + \sum_{i=1}^{b} (n_i - K) \cdot s_i)/(\sum_{k=1}^{m} C_k) .$$

We use Theorem 4 to prove all the lower bounds in the section. We start with a lower bound for general inputs which is not specific, but it uses a single block before the further jobs. This lower bound can be used for different values of K. This lower bound sequence σ_1 contains a block of very small jobs, and $t < m$ further jobs. Let N be a large integer, and let $\delta = \frac{1}{N}$. The first block contains KN jobs of size $\frac{1}{KN} = \frac{\delta}{K}$.

Lemma 9. *For any fixed value of K, the competitive ratio of any algorithm with a buffer of size K is at least $\frac{4}{3}$ for even m and at least $\frac{4m^2}{3m^2+1}$ for odd m.*

Lemma 10. *Let $t < m$. The competitive ratio of any algorithm which uses a buffer of size $K < t$ is at least*

$$(1 + \sum_{i=1}^{t} q_i)/(\frac{m-t}{m} + \sum_{i=1}^{t-K} \text{OPT}_{i+K} + K\text{OPT}_t) ,$$

where q_j is the size of the j-th further job in the sequence σ_1, and OPT_j is the optimal cost computed for σ_1 excluding the last $t - j$ jobs.

We next consider a special case of σ_1, where the list of further jobs consists of $K+1$ identical jobs, followed by a sequence of jobs with increasing sizes. The sequence is constructed so that an optimal schedule for the sequence up to the $(K+1)$-th further job is flat, that is, each further job is assigned to a dedicated machine, and spreading the small jobs on the other machines equally gives the same load to each machine. The additional further jobs form an increasing sequence of sizes, where the size of each job is exactly of the size that would cause an optimal schedule to assign it to a separate machine, where all previous jobs are spread over the other machines, giving a flat schedule.

Corollary 2. *The competitive ratio of any algorithm which uses a buffer of size $K \leq \lceil \frac{m-2}{2} \rceil$ is at least*

$$(m^t)/((t + tK - tm - Km)m^{K-1}(m-1)^{t-K-1} + (K+m)m^{t-1}) .$$

for any $K + 1 \leq t \leq m - 1$.

We next show as a corollary of the lower bound above that using a buffer of size $o(m)$ gives a competitive ratio which tends to $\frac{e}{e-1}$ for $m \to \infty$. Thus, the size of the buffer must be a linear function of m in order to improve over the upper bound of the case where no buffer is used.

Corollary 3. *Any algorithm using a buffer of size* $o(m)$ *has an overall competitive ratio of at least* $\frac{e}{e-1}$.

We summarize the results which show the optimality of the results of Section 2.2 (excluding the special case, which is considered later).

Theorem 5. *No algorithm for general inputs which uses a fixed size buffer can have a smaller competitive ratio than* $\frac{4}{3}$ *for even* m*, and no algorithm which uses a fixed size buffer can have a smaller competitive ratio than* $\frac{4m^2}{3m^2+1}$ *for odd* m*. An algorithm which uses a buffer of size* $o(m)$ *has an overall competitive ratio of* $\frac{e}{e-1}$*, that is, the usage of a buffer of this size is not helpful.*

In the construction of the sequence of the following proof, two blocks are used before the further jobs.

Theorem 6. *Any algorithm with* $K = 1$ *has a competitive ratio of at least* $\frac{4m^3-12m^2+4m}{3m^3-11m^2+18m-24}$*, which gives a lower bound of* $\frac{19}{14}$ *for* $m = 6$.

Note that the lower bound for the case $m = 6$, $K = 1$ resulting from Corollary 2 (with $t = 4$) is only $\frac{648}{481} \approx 1.3472$. The lower bound for $m = 7$ given by Theorem 6 is $\frac{203}{148} \approx 1.3716216$, which is an improvement over the lower bound which is implied of Corollary 2 as well. Next, we consider the case of non-increasing job sizes and prove the following.

Corollary 4. *For the case of non-increasing job sizes, let* $0 \le K \le m-2$*. The competitive ratio of any algorithm is at least* $\max_{1\le\mu\le m-K-1} \frac{2m(m+\mu)}{2m^2+2K\mu+\mu^2+\mu}$ *(where* μ *takes integer values). Specifically, if* $K \le m-2$*, then no algorithm can compute an optimal solution. The competitive ratio of an algorithm with* $K = m-2$ *is at least* $\frac{m^2+m}{m^2+m-1}$*. For* $K = m-1$ *(or for any other value of* K*), no algorithm can have a competitive ratio below 1. Thus, the algorithms of Section 2.3 are best possible both in terms of competitive ratio and the size of the used buffer.*

4 Conclusion

We studied preemptive scheduling with reordering and showed that a buffer of size $\Theta(m)$ is necessary and sufficient to achieve the best competitive ratios for both general sequences and for non-increasing sequences. All the algorithms do not use idle time, which is not helpful in the case of identical machines.

One direction for future research is to find the tight competitive ratio for every pair K, m of a buffer size and number of machines. This goal is already achieved here for the case of non-increasing job sequences.

The algorithms considered here are deterministic. Allowing randomization would not be helpful since even though the lower bounds are stated deterministically, all the lower bounds of Section 3 can be extended for randomized algorithms by considering expected loads of machines rather than the loads.

References

1. Chen, B., van Vliet, A., Woeginger, G.J.: Lower bounds for randomized online scheduling. Information Processing Letters 51, 219–222 (1994)
2. Chen, B., van Vliet, A., Woeginger, G.J.: An optimal algorithm for preemptive on-line scheduling. Operations Research Letters 18, 127–131 (1995); Also in ESA 1994
3. Dósa, G., Epstein, L.: Online scheduling with a buffer on related machines. Journal of Combinatorial Optimization (2008) (to appear)
4. Ebenlendr, T., Jawor, W., Sgall, J.: Preemptive online scheduling: Optimal algorithms for all speeds. In: Azar, Y., Erlebach, T. (eds.) ESA 2006. LNCS, vol. 4168, pp. 327–339. Springer, Heidelberg (2006)
5. Ebenlendr, T., Sgall, J.: Semi-online preemptive scheduling: One algorithm for all variants. In: Proc. of the 26th Annual Symposium on Theoretical Aspects of Computer Science, STACS 2009 (to appear, 2009)
6. Englert, M., Özmen, D., Westermann, M.: The power of reordering for online minimum makespan scheduling. In: Proc. 48th Symp. Foundations of Computer Science (FOCS), pp. 603–612 (2008)
7. Epstein, L.: Optimal preemptive on-line scheduling on uniform processors with non-decreasing speed ratios. Operations Research Letters 29(2), 93–98 (2001); Also in STACS 2001
8. Epstein, L., Favrholdt, L.M.: Optimal preemptive semi-online scheduling to minimize makespan on two related machines. Operations Research Letters 30(4), 269–275 (2002)
9. Epstein, L., Noga, J., Seiden, S.S., Sgall, J., Woeginger, G.J.: Randomized online scheduling on two uniform machines. Journal of Scheduling 4(2), 71–92 (2001)
10. Epstein, L., Sgall, J.: A lower bound for on-line scheduling on uniformly related machines. Operations Research Letters 26(1), 17–22 (2000)
11. Horwath, E., Lam, E.C., Sethi, R.: A level algorithm for preemptive scheduling. Journal of the ACM 24, 32–43 (1977)
12. Kellerer, H., Kotov, V., Speranza, M.G., Tuza, Z.: Semi online algorithms for the partition problem. Operations Research Letters 21, 235–242 (1997)
13. Li, S., Zhou, Y., Sun, G., Chen, G.: Study on parallel machine scheduling problem with buffer. In: Proc. of the 2nd International Multisymposium on Computer and Computational Sciences (IMSCCS 2007), pp. 278–281 (2007)
14. Seiden, S.: Preemptive multiprocessor scheduling with rejection. Theoretical Computer Science 262(1-2), 437–458 (2001)
15. Seiden, S., Sgall, J., Woeginger, G.: Semi-online scheduling with decreasing job sizes. Operations Research Letters 27(5), 215–221 (2000)
16. Sgall, J.: A lower bound for randomized on-line multiprocessor scheduling. Information Processing Letters 63(1), 51–55 (1997)
17. Sgall, J.: On-line scheduling. In: Fiat, A., Woeginger, G. (eds.) Online Algorithms - The State of the Art, ch. 9, pp. 196–231. Springer, Heidelberg (1998)
18. Wen, J., Du, D.: Preemptive on-line scheduling for two uniform processors. Operations Research Letters 23, 113–116 (1998)
19. Zhang, G.: A simple semi on-line algorithm for $P2//C_{\max}$ with a buffer. Information Processing Letters 61, 145–148 (1997)

d-Dimensional Knapsack in the Streaming Model

Sumit Ganguly[1] and Christian Sohler[2,⋆]

[1] Indian Institute of Technology, Kanpur, India
[2] Technical University Dortmund, Dortmund, Germany

Abstract. We study the d-dimensional knapsack problem in the data streaming model. The knapsack is modelled as a d-dimensional integer vector of capacities. For simplicity, we assume that the input is scaled such that all capacities are 1. There is an input stream of n items, each item is modelled as a d-dimensional integer column of non-negative integer weights and a scalar profit. The input instance has to be processed in an online fashion using sub-linear space. After the items have arrived, an approximation for the cost of an optimal solution as well as a template for an approximate solution is output.

Our algorithm achieves an approximation ratio $(2(\frac{1}{2}+\sqrt{2d+\frac{1}{4}}))^{-1}$ using space $O(2^{O(d)}\cdot \log^{d+1} d \cdot \log^{d+1} \Delta \cdot \log n)$ bits, where $\{\frac{1}{\Delta}, \frac{2}{\Delta}, \ldots, 1\}$, $\Delta \geq 2$ is the set of possible profits and weights in any dimension. We also show that any data streaming algorithm for the $t(t-1)$-dimensional knapsack problem that uses space $o(\sqrt{\Delta}/t^2)$ cannot achieve an approximation ratio that is better than $1/t$. Thus, even using space Δ^{γ}, for $\gamma < 1/2$, i.e. space polynomial in Δ, will not help to break the $1/t \approx 1/\sqrt{d}$ barrier in the approximation ratio.

1 Introduction

The 0/1 knapsack problem is a popular and well-studied combinatorial problem with applications in many different areas. Its basic form is as follows. Given n items numbered $i = 1, 2, \ldots, n$ and their weights w_i and profits p_i, find a subset of the items with maximum profit whose sum of weights does not exceed a given sack size R. The problem is well-known to be an NP-hard problem [7] with a classical FPTAS approximation algorithm by Ibarra and Kim [4].

A well-studied generalization is the d-dimensional knapsack problem (see, for example, [2], whose input is the set of items indexed by $\{1, 2, \ldots, n\}$, where the ith item is associated with (a) a non-negative d-dimensional vector $A_i = [A_{1,i}, \ldots, A_{d,i}]^T$ denoting the weight of the item along each of the d dimensions, and, (b) a profit p_i. The knapsack dimensions is given by the column vector R with R_s being the sack dimension along dimension s. The problem may be specified as follows. $\max_{S \subset \{1,2,\ldots,n\}} \sum_{j\in S} p_j$ subject to $\sum_{j\in S} A_{s,j} \leq R_s, \quad s = 1, 2, \ldots, d$. We consider the above problem in the data stream setting. For simplicity, we assume that the knapsack capacities are scaled to be 1. This is equivalent to assuming that the knapsack capacities are given as input to the algorithm

⋆ Supported by DFG project So 514/1-2.

A. Fiat and P. Sanders (Eds.): ESA 2009, LNCS 5757, pp. 468–479, 2009.

instead of being read from the stream. In this setting, the items arrive in a sequence whose entries are of the form (item, profit, d-dimensional column of weights)= (i, p_i, A_i). The profits and weights are chosen from $\{\frac{1}{\Delta}, \frac{2}{\Delta}, \ldots, 1\}$, $\Delta \geq 2$. The algorithm may only use a small amount of space, say $s(n, d, \Delta)$ bits, for storing the input. In the streaming scenario we aim at $s(n, d, \Delta)$ being sublinear or even polylogarithmic in the input length. Each item is processed in an online fashion. After all the items have been processed, the streaming algorithm may use its $s(n, d, \Delta)$-bit state to report a template for a packing of the items into the knapsack. The following constraints must be satisfied: (1) Any template packing of items reported by the algorithm must be feasible for the original instance, and, (2) the corresponding profit reported must not be higher than the sum of the profits of the items in the sack.

The knapsack problem in the data stream setting is substantially different than the online resource knapsack problem with resource augmentations, studied by Iwama and Taketomi [5] and by Iwama and Zhang [6]. In the online knapsack problem, each item may be seen only once, and a decision regarding whether to include it in the knapsack must be made. Additionally, items arriving later may cause a currently included item in the sack to be evicted; evicted items are not recoverable. Iwama and Zhang show that the problem is not approximable (i.e., ratio is ∞). However, if one allows resource augmentation, that is, the algorithm is allowed to store a set of items whose weight is at most $\alpha \geq 1$ knapsacks, then, a greedy algorithm attains a $\alpha - 1$ approximation ratio and this is optimal [6].

The main difference between the online resource augmented version and the streaming version is that the online version allows a set of items to be stored as long as its weight is at most αR. Since items may have small weight, it is possible that $\Theta(n)$ items are stored, which would not be allowed in the data stream setting.

Contributions and Overview. In this work, we study deterministic solutions to the d-dimensional knapsack problem in the streaming model. We assume that the input is scaled such that the sack capacities are 1. We show that for $\epsilon = O(1/\log(d))$, there is an algorithm that gives a $(2(\frac{1}{2} + \sqrt{2d + \frac{1}{4}}))^{-1}$-approximate packing using space $O(2^{O(d)}\epsilon^{-(d+1)}(\log^{d+1} \Delta)(\log n))$, where weights and profits are taken from $\{\frac{1}{\Delta}, \frac{2}{\Delta}, \ldots, 1\}$, $\Delta \geq 2$. Further, we show that for any $d = t(t-1)$, any algorithm that uses $o(\Delta/d)$ space has approximation ratio no better than $1/t$. Thus, our technique yields an approximation ratio that is within a small constant factor of the best possible. In fact, even using space *polynomial* in Δ will not help to break the $1/t \approx 1/\sqrt{d}$ barrier in the approximation ratio.

Our technique is as follows. We compress the input instance by rounding up the weights[1] and rounding down the profits, respectively, to the nearest power of $1+\epsilon$. The input instance is now approximated as a $d+1$-dimensional array H, where there are d coordinates for each of the weight dimensions and one profit coordinate. An entry $H[w_1, \ldots, w_d, p]$ in this array is the count of the number of

[1] The actual scheme for rounding up weights is slightly more involved.

items that have profit p and weights $w_1, \ldots, w_d$ respectively in the d-dimensions, after the rounding operation.

Our approach for the analysis is to take any given feasible solution F for the original instance and show that it can be packed into at most $2(\frac{1}{2} + \sqrt{2d + \frac{1}{4}})$ sacks using the modified weights. This is done using a 2-level hierarchical packing method. The first level is obtained as the color classes of an optimal coloring of a certain hypergraph. In the second stage, we use the probabilistic method to prove that the items of each of the color classes can be packed into 2 sacks. We show that the first level coloring has at most $\frac{1}{2} + \sqrt{2d + \frac{1}{4}}$ color classes. Since each such class is packed using at most 2 sacks, we obtain a $2(\frac{1}{2} + \sqrt{2d + \frac{1}{4}})$-sack packing of the given feasible solution F. Therefore, one among these packings has cost at least $1/(2(\frac{1}{2} + \sqrt{2d + \frac{1}{4}})$ of the original feasible solution F. By choosing the feasible solution to be the optimal solution, we obtain a $1/(2(\frac{1}{2} + \sqrt{2d + \frac{1}{4}})$-approximate solution for the modified weights instance.

Our lower bound is obtained as a reduction from the t-party set disjointness problem in one-way communication complexity.

2 Compressed Representation

An instance $\Pi = (A, p)$ of the d-dimensional knapsack problem is a $d \times n$-dimensional weights matrix A and an n-dimensional row vector p. To simplify notation, we will assume that the item weights along each dimension, namely the $A_{i,j}$'s are from $\{\frac{1}{\Delta}, \frac{1}{\Delta}, \ldots, 1\}$, $\Delta \geq 2$. The knapsack is assumed to have unit capacity in each of its dimensions. Thus our problem is the following.

$$\max_{F \subset \{1,2,\ldots,n\}} \sum_{j \in F} p_j \quad \text{subject to} \sum_{j \in F} A_{ij} \leq 1, \quad i = 1, 2, \ldots d \ .$$

Let $a = 1/\Delta$. In order to represent the input instance using compact space, we discretize the weights and profits as follows. The interval $[a, 1/2]$ is divided into bins using a logarithmic scale: $(a, a(1+\epsilon)], (a(1+\epsilon), a(1+\epsilon)^2], \ldots, (a(1+\epsilon)^{j_0}, 1/2]$. The last interval ends at $1/2$. For $x \in [a, 1/2]$, we map x to the right end of the interval in powers of $(1+\epsilon)$ that contains x; however, we never cross $1/2$. Call this mapping $x \mapsto up_{a,\epsilon}(x)$, that is, $up_{a,\epsilon}(x) = \min(a(1+\epsilon)^{\lceil \log_{1+\epsilon}(x/a) \rceil}, 1/2), a \leq x \leq 1/2$. This gives $\lceil \log_{1+\epsilon} 1/(2a) \rceil$ distinct discrete weights. Define the mapping $x \mapsto dn_{a,\epsilon}(x)$, where, $dn_{a,\epsilon}(x)$ is the left end of the interval in the above list of intervals that contains x, that is, $dn_{a,\epsilon}(x) = a(1+\epsilon)^{\lfloor \log_{1+\epsilon}(x/a) \rfloor}, a \leq x < 1/2$. The mapping for the interval $[1/2, 1 \quad a]$ is obtained by dividing the interval backwards, that is, $x \mapsto up_{a,\epsilon}(x)$, where, $up_{a,\epsilon}(x) = 1 - dn_{a,\epsilon}(1-x), 1/2 < x < 1-a$. In the interval $[a, 1/2]$, $up_{a,\epsilon}(x)$ rounds x upwards to the nearest value of the form $a(1+\epsilon)^j$. Hence, $a \leq up_{a,\epsilon}(x) \leq 1/2, x \in [a, 1/2]$ and $0 \leq up_{a,\epsilon}(x) - x \leq \epsilon x$, for $x \in [a, 1/2]$. In the interval $[1/2, 1]$, $up_{a,\epsilon}(x)$ rounds x upwards to $1 - dn_{a,\epsilon}(1 - x)$. Therefore, $(1 + \epsilon)/2 \leq up_{a,\epsilon}(x) \leq 1 - a, x \in (1/2, 1 - a]$, and $0 \leq up_{a,\epsilon}(x) - x \leq \epsilon(1 - x)$, for $x \in (1/2, 1 - a]$. A round-down discretization is performed for profits by mapping p_i to $dn_{p_{\min},\epsilon}(p_i)$, for the entire range of profit values, where, $p_{\min}$ is a lower bound on the smallest profit and ϵ is a parameter. This ensures that $0 \leq p_j - dn_{p_{\min},\epsilon}(p_j) \leq \epsilon p_j$, which is sufficient for our purposes.

Compressed instance. Let A' be the $d \times n$ matrix obtained by mapping each entry A_{ij} to $up_{a,\epsilon}(A_{ij})$, for $1 \leq i \leq d, 1 \leq j \leq n$. We represent A' as a histogram H in $d+1$-dimensions as follows, where, the first d dimensions are the weights and the last dimension is the profit. H contains $2\lceil \log_{1+\epsilon}(1/2a) \rceil + 2$ entries in each of first d dimensions and $\lceil \log_{1+\epsilon}(\Delta) \rceil + 1$ entries for each the profit dimension. Each cell of H is initialized to 0. When an item with profit p_j and weight column A_j appears, we map A_j to the vector $A_j \mapsto A'_j = up_{a,\epsilon}(A_j) = [up_{a,\epsilon}(A_{1,j}), up_{a,\epsilon}(A_{2,j}), \ldots, up_{a,\epsilon}(A_{d,j})]^T$ and the profit p_j to $dn_{p_{\min},\epsilon}(p_j)$. The histogram entry corresponding to $H[A'_j, p'_j]$ is incremented by 1.

*Space requirement.*Each entry of the histogram stores a count of the number of items with the same rounded weights and profit. It suffices to use $\log n$ bits for each entry. Thus, the histogram requires space $O((\lceil \log_{1+\epsilon}(2/a) \rceil + 2)^d (\log_{1+\epsilon}(\Delta) + 1)(\log n)) = O(2^{O(d)} \epsilon^{-d+1} \log^{d+1}(\Delta)(\log n))$ bits where, n is the number of items and $\Delta \geq 2$.

A simple but important property of the rounding procedure is that, (a) a feasible solution for the modified (i.e., the rounded) instance is a feasible solution for the original instance–this is a consequence of the rounding up of the weights, and, (b) the profit of a feasible solution of the modified instance is at most the profit of the same solution for the original instance–this is a consequence of rounding down of the profits. The profit of a feasible solution of the modified instance is at least $1/(1+\epsilon)$ of the profit of the same solution for the original instance.

Definition 1. *Let $\Pi = (A, p)$ be an instance of the d-dimensional knapsack problem. A pair-wise non-intersecting family of subsets $S_1, \ldots, S_k$ is called a feasible packing using k knapsacks or, in short, a feasible k-sack solution, if $\sum_{l \in S_j} A_{i,l} \leq 1$, for each $j = 1, 2, \ldots, k$ and $1 \leq i \leq d$.*

We use feasible k-sack solutions as follows. Given an instance Π of a knapsack problem, let $\mathrm{OPT}(\Pi)$ denote the optimal profit feasible.

Lemma 1. *Let g be a function that maps an instance Π of the knapsack to another instance $g(\Pi)$ that does not change the profit vector p but may change the weight matrix A to A'. Suppose that for every instance Π and every feasible solution F of Π, there is a feasible k-sack solution for $g(\Pi)$ whose union of sets is F. Then, $\mathrm{OPT}(g(\Pi)) \geq \mathrm{OPT}(\Pi)/k$, for all instances Π.*

Proof. There is a k-sack feasible solution for $g(\Pi)$ whose union is $\mathrm{OPT}(\Pi)$. One of these k-sacks has profit at least $\mathrm{OPT}(\Pi)/k$. Thus $\mathrm{OPT}(g(\Pi)) \geq \mathrm{OPT}(\Pi)/k$.

Lemma 1 guides our approach towards obtaining a bound of k on the approximation ratio of the original versus the compressed instance. The bound is obtained by showing that for each feasible solution of the original instance there is a k-sack feasible solution of the compressed instance. It then follows that the optimal solution for the compressed instance is $1/k$-approximation. The problem now reduces to making k as small as possible.

3 Conflict Hypergraph and Its Coloring

Given an instance Π of the knapsack problem, we denote by Π' the instance that replaces all input weights $A_{t,j}$ by $A'_{t,j} := up_{a,\epsilon}(A_{t,j})$ and profits p_j by $p'_j := dn_{p_{\min},\epsilon}(p_j)$. Let F be an arbitrary feasbile solution for Π. We define a conflict hypergraph that captures all sets of items of F that violate the sack size in any dimension for instance Π'.

Definition 2. *The hypergraph $H_F = (V, E_F)$ with $V := F$ and $E_F := \{h \subseteq V : \exists s, 1 \leq s \leq d, \sum_{j \in h} A'_{s,j} > 1\}$ is called* conflict hypergraph *of a feasible solution F. An edge h has label s, if $\sum_{j \in h} A'_{s,j} > 1$. Note that an edge can have different labels.*

A hypergraph k-coloring is an assignment $\chi : V \to \{1, \ldots, k\}$ of the vertices to one of the k colors. We call the sets $\chi^{-1}(i)$ the color classes of the coloring. A hyperedge h is called monochromatic (under a k-coloring χ), if there exists a color $c, 1 \leq c \leq k$ such that $\chi(v) = c$ for all $v \in h$. A coloring is called *proper*, if there are no monochromatic edges. A hypergraph is called k-colorable, if it has a proper k-coloring. A hypergraph has chromatic number k, if it is k-colorable but not $k-1$-colorable. We can now formulate a relationship between the error introduced by our rounding procedure and the problem of coloring the conflict hypergraph.

Lemma 2. *Let F be a feasible solution for $\Pi = (A, p)$. A feasible k-sack solution $S'_1, \ldots, S'_k$ for $\Pi' = (A', p')$ with $F = \bigcup_{i \in \{1,\ldots,k\}} S'_i$ exists, iff H_F is k-colorable.*

Proof. Let $S'_1, \ldots, S'_k$ be a feasible k-sack solution for $\Pi' = (A', p')$. We define a k-coloring χ for H_F by setting $\chi^{-1}(j) = S'_j$ for $1 \leq j \leq k$. The coloring is well-defined since by definition of a k-sack solution the sets S'_j do not intersect and since $F = \bigcup_{j \in \{1,\ldots,k\}} S'_j$. By the definition of a feasible k-sack solution we have that $\sum_{l \in S'_j} A'_{i,l} \leq 1, \quad$ for each $j = 1, 2, \ldots, k$ and $1 \leq i \leq d$. Now assume that there is a hyperedge h and a color c with $\chi(v) = c$ for all $v \in h$. Then we have that $v \in S'_c$ for all $v \in h$ and so $\sum_{l \in h} A'_{i,l} \leq \sum_{l \in S'_c} A'_{i,l} \leq 1$ for $1 \leq i \leq d$. But this is a contradiction to the fact that $h \in E_F$, which states that for some dimension s we have $\sum_{l \in h} A'_{s,l} > 1$.

Now let us assume that there is a proper coloring of H_F. Then we define $S'_j := \chi^{-1}(j)$. By definition of a proper coloring, there is no edge $h \subseteq S'_j$. Hence, $\sum_{l \in S'_j} A'_{i,l} \leq 1$ for each $1 \leq j \leq k$ and $1 \leq i \leq d$. Thus $S_1, \ldots, S_k$ is a feasible k-sack solution. □

Our approach will be to show that, for every feasible solution F, the conflict hypergraph H_F has chromatic number at most $k := 2(\frac{1}{2} + \sqrt{2d + \frac{1}{4}})$. By the above lemma this implies that a k-sack feasible solution exists. By Lemma 1 this will give a $1/k$-approximate solution.

We first give a folklore result that any hypergraph with E edges can be colored using $\frac{1}{2} + \sqrt{2|E| + \frac{1}{4}}$ colors. The result follows from the observation that a coloring with minimum number of colors must have a hyperedge for every pair of colors. The proof for the graph version (see, for example, [3], page 124) extends immediately to hypergraphs.

Lemma 3 (folklore). *Let* $H = (V, E)$ *be a hypergraph. Then,* H *is* $\frac{1}{2} + \sqrt{2|E| + \frac{1}{4}}$*-colorable.* □

Lemma 3 will prove useful in obtaining desirable packings. However, an immediate application of Lemma 3 is not useful, since, the number of conflict edges could be exponential in the number of vertices. Thus, the number k of sacks required to obtain a k-feasible packing by Lemma 3 may be exponential. It is therefore necessary to apply a two step approach to construct a proper $(2(\frac{1}{2} + \sqrt{2d + \frac{1}{4}}))$-coloring of H_F. Our first step will be to construct a restricted conflict graph H'_F over vertex set F and obtain a $\frac{1}{2} + \sqrt{2d + \frac{1}{4}}$-coloring of H'_F. This coloring is later refined to a $(2(\frac{1}{2} + \sqrt{2d + \frac{1}{4}}))$-coloring of H_F.

Definition 3. *Let* $Z_{F,s} = \bigcap_{g \in E_F \,:\, g \text{ has label } s} g$ *and* $E'_F := \{Z_{F,s} : |Z_{F,s}| > 1\}$. *The hypergraph* $H'_F = (F, E'_F)$ *is called* restricted conflict hypergraph.

An immediate motivation for Definition 3 is to reduce the number of hyperedges. Corresponding to any feasible solution F of the original instance, by construction, H'_F has at most one hyperedge per dimension. Hence, $|E'_F| \leq d$, and by Lemma 3, there exists a $\frac{1}{2} + \sqrt{2d + \frac{1}{4}}$-coloring χ of H'_F. However, not all conflicts in H_F are reflected in H'_F and thus, there may be monochromatic edges in H_F under the coloring χ. However, an edge with label s can only be monochromatic, if $|Z_{F,s}| \leq 1$.

Therefore we revise our strategy to a 2-level *hierarchical packing*. Given a feasible packing of the original instance, we first obtain a decomposition of the vertices into color classes that are a proper coloring of the restricted conflict hypergraph H'_F. In the second stage, we take the items in each of the color classes obtained and show that these items can be packed into 2 sacks. In order to prove this, we only have to show that for every dimension s with $|Z_{F,s}| \leq 1$ the items can be packed into two sacks. We use the probabilistic method to show this, namely, we show that a random packing is a feasible packing with constant probability. This implies that there exists a refined coloring that has no monochromatic edges.

From now on, let us assume that we are given a subset S of items that forms a color class of the restricted conflict hypergraph H'_F corresponding to some given feasible packing F of the original instance of the knapsack problem. By construction, we only have to consider dimensions s with $|Z_{F,s}| = 0$ or 1. Clearly, in the worst case we have $S = F$ and so we will show that even in this case, our random packing will succeed with large probability. We will first analyze consequences of $|Z_{F,s}|$ being 0 or 1. To simplify notation, let $A_{i,J} := \sum_{j \in J} A_{i,j}$.

Lemma 4. *Let $\epsilon < 1/3$. Suppose that we are given a feasible solution F for the original instance of the knapsack problem and a dimension $s \in \{1, 2, \ldots, d\}$ such that $|Z_{F,s}| = 1$ and $Z_{F,s} = \{u_s\}$. Let g, h be hyperedges in the conflict hypergraph H_F that have label s and contain u_s. Let $I = g - \{u_s\}$ and $J = h - \{u_s\}$.*

1. *If $A_{s,u_s} > 1/2$ then, $A_{s,I \cap J} > \frac{1-3\epsilon}{1+\epsilon}(1 - A_{s,u_s})$.*
2. *If $A_{s,u_s} \le 1/2$, then, $A_{s,I \cap J} > \frac{1-\epsilon}{1+\epsilon} - A_{s,u_s}$.*

Proof. *Case 1*: $A_{s,u_s} > 1/2$. Since g is not feasible for the modified weights, $A'_{s,g} > 1$. So, $1 < A'_{s,g} = A'_{s,u_s} + A'_{s,I} \le \epsilon(1 - A_{s,u_s}) + A_{s,u_s} + (1+\epsilon)A_{s,I}$. Therefore, we get, $(1+\epsilon)A_{s,I} > (1-\epsilon)\beta_s$, where, $\beta_s = (1 - A_{s,u_s})$. Applying the above inequality to the hyperedge h, we obtain similarly that $(1+\epsilon)A_{s,J} > (1-\epsilon)\beta_s$. Adding, we have, $(1+\epsilon)(A_{s,I} + A_{s,J}) = (1+\epsilon)(A_{s,I\cap J} + A_{s,I\cup J}) > 2(1-\epsilon)\beta_s$. However, since F is a feasible set, the set of elements $\{u_s\} \cup I \cup J$ fit in the sack along dimension s. Thus, $A_{s,u_s} + A_{s,I\cup J} \le 1$ or, $A_{s,I\cup J} \le 1 - A_{s,u_s} = \beta_s$ and therefore, $(1+\epsilon)(A_{s,I\cap J} + \beta_s) > 2(1-\epsilon)\beta_s$ or, $A_{s,I\cap J} > \frac{2(1-\epsilon)\beta_s}{1+\epsilon} - \beta_s = \frac{(1-3\epsilon)\beta_s}{1+\epsilon}$.

Case 2: $A_{s,u_s} \le 1/2$. We have $1 < A'_{s,g} = A'_{s,u_s} + A'_{s,I} \le (1+\epsilon)A_{s,u_s} + (1+\epsilon)A_{s,I}$, and $1 < A'_{s,h} = A'_{s,u_s} + A'_{s,J} \le (1+\epsilon)A_{s,u_s} + (1+\epsilon)A_{s,J}$ Adding, we obtain $2 < 2(1+\epsilon)A_{s,u_s} + (1+\epsilon)(A_{s,I\cup J} + A_{s,I\cap J}) = (1+\epsilon)(A_{s,u_s} + A_{s,I\cap J}) + (1+\epsilon)(A_{s,u_s} + A_{s,I\cup J})$. Since F is a feasible packing, the elements $\{s\} \cup I \cup J$ all fit in the sack in terms of their original weights, that is, $A_{s,u_s} + A_{s,I\cup J} \le 1$. Thus, $2 < (1+\epsilon)(A_{s,u_s} + A_{s,I\cap J}) + (1+\epsilon)$, or, $A_{s,I\cap J} > \frac{1-\epsilon}{1+\epsilon} - A_{s,u_s}$. □

Lemma 5. *Let $\epsilon < 1/3$ and F be a feasible solution for the original instance of the knapsack problem. Let s be a dimension $s \in \{1, 2, \ldots, d\}$ such that $|Z_{F,s}| = 1$ and $Z_{F,s} = \{u_s\}$. Suppose j is a member of some hyperedge in H_F with label s.Then, the following holds.*

1. *For $j \ne u_s$ and $A_{s,u_s} > 1/2$, $A_{s,j} \le (1 - A_{s,u_s})(4\epsilon)/(1+\epsilon)$.*
2. *For $j \ne u_s$ and $A_{s,u_s} \le 1/2$, $A_{s,j} \le 2\epsilon/(1+\epsilon)$.*

If j is not a member of any hyperedge in H_F with label s, then, $A_{s,j} \le \epsilon/(1+\epsilon)$.

Proof. Let $\beta_s = 1 - A_{s,u_s}$. We will consider two cases with regard to an item $j \ne u_s$; Case 1 when an item j is a member of some hyperedge in H_F with label s, and, Case 2 when an item j is not a member of any hyperedge in H_F labeled s.

Case 1: Suppose $j \ne u_s$ and j is a member of some hyperedge $\{u_s\} \cup I$ in H_F. Since, $j \notin Z_s$, there exists another hyperedge $h = \{u_s\} \cup J$ in H_U such that $j \notin h$. We get $A_{s,j} \le \beta_s - A_{s,I\cap J}$. *Case 1.1*: $A_{s,u_s} > 1/2$. By Lemma 4 part (1), $A_{s,I\cap J} \ge \beta_s(1-3\epsilon)/(1+\epsilon)$. Thus, $A_{s,j} \le \beta_s - \beta_s\frac{1-3\epsilon}{1+\epsilon} = \frac{4\epsilon\beta_s}{1+\epsilon}$. *Case 1.2*: $A_{s,u_s} \le 1/2$. By Lemma 4 part (2), $A_{s,I\cap J} \ge \frac{1-\epsilon}{1+\epsilon} - A_{s,u_s}$. Therefore,

$A_{s,j} \le 1 - A_{s,u_s} - \frac{1-\epsilon}{1+\epsilon} + A_{s,u_s} = \frac{2\epsilon}{1+\epsilon}$. *Case 2*: Suppose j is not a member of any hyperedge in H_F with label s. Then, for any hyperedge g with label s (and there is at least one since $|Z_s| = 1$), $1 < A'_{s,g} \le (1+\epsilon)A_{s,g} \le (1+\epsilon)(1 - A_{s,j})$ or, $A_{s,j} < \frac{\epsilon}{1+\epsilon}$. □

Lemma 5 implies that for the case when $|Z_{F,s}| = 1$, all elements except the largest element are of size $O(\epsilon)$. This follows from Lemma 5 by noting that $A_{s,j} \le (1 - A_{s,u_s})(4\epsilon)/(1+\epsilon) \le 4\epsilon/(1+\epsilon)$, since, $0 \le 1 - A_{s,u_s} \le 1$.

Lemma 6. *Given a feasible solution F for the original instance of the knapsack problem and a dimension $s \in \{1, 2, \ldots, d\}$ such that $|Z_{F,s}| = 0$. Then, either there are no conflicting edges along dimension s, or, all items have weight at most $\epsilon/(1+\epsilon)$ along dimension s.*

Proof. If there are no conflicting edges in H_F with label s then there is nothing to prove. So suppose there exists at least one conflicting edge g in H_F with label s.

Case 1: Suppose $j \in g$. Since $Z_{F,s} = \phi$, it follows that there is at least one other conflicting edge $h \in H_F$ such that $j \notin h$. Therefore, $A'_{s,h} > 1$, or, $1 < A'_{s,h} \le (1+\epsilon)A_{s,h} \le (1+\epsilon)(1 - A_{s,j})$. Simplifying, we obtain $A_{s,j} \le \frac{\epsilon}{1+\epsilon}$.

Case 2: Suppose j does not appear in any hyperedge of H_F with label s, then, the argument of Lemma 5 can be used to show that $A_{s,j} \le \epsilon/(1+\epsilon)$. □

Packing Items in a Color Class

In this section, we consider the remaining portion of the strategy of the 2-level hierarchical packing outlined above. In the hierarchical packing strategy, given a feasible solution F for the original knapsack instance, we first form the conflict hypergraph H_F and then derive the restricted conflict hypergraph H'_F from it as explained earlier. The vertex set F of the restricted conflict hypergraph H'_F is partitioned into the color classes of a proper coloring. The final step is to pack the items comprising each color class using as few sacks as possible. In this section, we present this step.

We will pack the items in an independent set of H'_F using a random strategy. The random packing strategy picks each item at random with probability p and attempts to place it in the sack. We will show that for $p = 1/2$ with some probability the sack does not overflow.

Lemma 7. *Let $p = 1/2$ and $\epsilon \le \min(1/24, 1/(64\log(4d)))$ and F be a feasible solution to the original knapsack instance. Let S be a set of items that forms a color class of the restricted conflict hypergraph H'_F. If we put each item with probability p into the sack, the probability that the sack overflows along any given dimension is at most $1/(4d)$.*

Proof. Consider a random packing that chooses each item with probability p. Define an indicator variable x_j to be 1 if item j is selected and 0 otherwise. Let u_t be an item with the maximum weight along dimension t, that is $A_{t,u_t} = \max_i A_{t,i}$. Let $\beta'_t = 1 - A'_{t,u_t}$ and for $j \ne u_t$, let $w'_{t,j} = \frac{A'_{t,j}}{\max_{k \ne u_t} A'_{t,k}} x_j, j \in \{1, 2, \ldots, n\} - \{u_t\}$ and let $X_t = \sum_{j \ne u_t} w'_{t,j}$. Thus, $\mathsf{E}[X_t] = \sum_{j \ne u_t} \frac{A'_{s,j}}{\max_{k \ne u_t} A'_{t,k}} p$

Case 1.1: $|Z_{F,t}| = 1$ and $A_{t,u_t} > 1/2$. By Lemma 5, if $|Z_{F,t}| = 1$ then, $\max_{k \neq u_t} A'_{t,k} \leq 4\epsilon(1 - A_{t,u_t})/(1+\epsilon)$. We wish to obtain an upper bound for the probability that the items selected do not fit into the sack along dimension t, while leaving room for u_t. A sufficient condition for this is $\sum_{j \neq u_t} A'_{t,j} x_j \leq \beta'_t$, or, equivalently, $X_t \leq \frac{\beta'_t}{\max_{k \neq u_t} A'_{t,k}}$.

$$\Pr\{\text{Overflow along dimension } t\} = \Pr\left\{X_t > \frac{\beta'_t}{\max_{k \neq u_t} A'_{t,k}}\right\}.$$

By Hoeffding's bound applied to the sum X_t of random variables $w'_{t,j}$, with $p \geq 1/2$, by Lemma 5 this is

$$\Pr\left\{X_t > \frac{\beta'_t}{\max_{k \neq u_t} A'_{t,k}}\right\} \leq \exp\left\{-(1/6)\frac{((1-\epsilon)-(1+\epsilon)p)^2}{\epsilon(1+(1+\epsilon)p)}\right\}. \tag{1}$$

Case 1.2: $|Z_{F,t}| = 1$ and $A_{t,u_t} \leq 1/2$. By Lemma 5, $\max_{k \neq u_t} A'_{t,k} \leq (1+\epsilon)A_{t,k} \leq 2\epsilon$. It follows

$$\Pr\left\{X_t > \frac{\beta'_t}{\max_{k \neq u_t} A'_{t,k}}\right\} < \exp\left\{-(1/3)\frac{((1-\epsilon)-(1+\epsilon)p)^2}{\epsilon(1+(1+\epsilon)p)}\right\}. \tag{2}$$

Case 2: $|Z_{F,t}| = 0$. By Lemma 6, if $|Z_{F,t}| = 0$, then, $\max_k A'_{t,k} \leq (1+\epsilon)\max_k A_{t,k} \leq \epsilon$. Applying Hoeffding's bound yields,

$$\Pr\left\{X_t > \frac{\beta'_t}{\max_{k \neq u_t} A'_{t,k}}\right\} < \exp\left\{-(2/3)\frac{((1-\epsilon)-(1+\epsilon)p)^2}{\epsilon(1+(1+\epsilon)p)}\right\}. \tag{3}$$

Combining (1), (2) and (3), we have under all cases,

$$\Pr\left\{X_t > \frac{\beta'_t}{\max_{k \neq u_t} A'_{t,k}}\right\} \leq \exp\left\{-(1/6)\frac{((1-\epsilon)-(1+\epsilon)p)^2}{\epsilon(1+(1+\epsilon)p)}\right\}. \tag{4}$$

Therefore, $\Pr\left\{X_t > \frac{\beta'_t}{\max_{k \neq u_t} A'_{t,k}}\right\} \leq 1/(4d)$ provided $\frac{((1-\epsilon)-(1+\epsilon)p)^2}{\epsilon(1+(1+\epsilon)p)} > 6\log(4d)$. Equation (3) is satisfied if $p = 1/2$ and $\epsilon = \min(1/24, 1/(64\log(4d)))$. □

We can now use Lemma 7, to prove that a 2-sack packing of the items in each color class can be obtained.

Lemma 8. *Let $p = 1/2$ and $\epsilon = \min(1/24, 1/(64\log(4d)))$ and F be a feasible solution to the original knapsack instance. Let S be a set of items that forms a color class of the restricted conflict hypergraph H'_F. Then, there exists a 2-sack feasible packing of the items in S.*

Proof. Suppose we sample items with probability 1/2. In this case, the probability that an item is not sampled is also 1/2. Thus, we can also apply the previous analysis (Lemma 7) to the set of items that were not selected in the sample. The packing of this set of items succeeds with probability $1 - d/(4d)$. Thus, by the union bound, the probability that for every dimension and both the sets of selected and not selected items are feasible is at least $1 - (2d)/(4d) = 1/2 > 0$ by the union bound. Thus a 2-sack feasible packing of the items in S exists. □

We can now state the resulting property of our hierarchical packing analysis.

Theorem 1. *For $\epsilon \leq \min(1/24, 1/(64\log(4d)))$, our algorithm computes a rounded instance such that this instance has a $(2(\frac{1}{2} + \sqrt{2d + \frac{1}{4}}))^{-1}$-approximate solution. Our rounded instance is stored in a histogram that requires $O(2^{O(d)} \log^{d+1}(d) \log^{d+1}(\Delta)(\log n))$ bits of space, for some constant c.*

Proof. The algorithm for filling the knapsack is as follows. We round up the weights as explained in Section 2 and round down the profits. In this manner we create a histogram of multi-dimensional weights. The size of this histogram is $O(2^{O(d)}(\epsilon^{-(d+1)}) \log^d(\Delta) \log n)$ bits. From the histogram we can obtain a solution using exhaustive search.

The approximation factor is derived as follows. Let F be any feasible solution for the original instance. Let H_F be the conflict hypergraph and let H'_F denote the restricted conflict hypergraph (Definition 3. We then color H'_F using the smallest number of colors. Lemma 3 shows that this can be done using at most $\frac{1}{2} + \sqrt{2d + \frac{1}{4}}$ colors. Since, there are at most d edges in H'_F, we obtain at most $\frac{1}{2} + \sqrt{2d + \frac{1}{4}}$ color classes with no hyperedges from H'_F.

By Lemma 8 each color class S can be packed using $t = 2$ sacks, where, $\epsilon = O(1/(\log(d))$. This means that the set of items in F can be packed into at most $2(\frac{1}{2} + \sqrt{2d + \frac{1}{4}})$ sacks. By Lemma 1, this implies that there is a $1/(t(\frac{1}{2} + \sqrt{2d + \frac{1}{4}}))$-approximation solution. The final two statements of the theorem are special cases obtained from the corresponding special cases of Lemma 8. □

4 Lower Bounds: Space versus Approximation Ratio

In this section, we present a lower bound on space versus approximation ratio of streaming algorithms for the d-dimensional knapsack problem.

Theorem 2. *For $t \geq 1$, any streaming algorithm for the $t(t-1)$-dimensional knapsack problem with items from the universe $\{\frac{1}{\Delta}, \frac{2}{\Delta}, \ldots, 1\}$ that uses $o(\sqrt{\Delta}/t^2)$ bits has an approximation ratio of at most $1/t$.*

Let t be a positive integer such that $d = t(t-1)/2$. We reduce the t-party set disjointness problem to $2d$-dimensional knapsack problem. An instance of the t-party set disjointness problem consists of subsets $S_1, \ldots, S_t$ of $[n]$ provided to each of t players. It is given that the sets are either pair-wise disjoint, or, there is exactly one element in the common intersection. The players follow a pre-specified communication protocol at the end of which a designated player can distinguish between the two kinds of input. If the protocol is randomized, then, the final answer must be correct with probability say 7/8. The communication complexity of a protocol is the maximum over all legal inputs, of the total number of bits communicated during the execution of the protocol. The communication complexity of the problem is the complexity of the best possible protocol for this problem. It was shown by Chakrabarti, Khot and Sun [1] that the randomized communication complexity of this problem is $\Omega(n/(t \log t))$. Assuming the one-way communication model, where, the players communicate in a certain order, that is, player 1 sends to player 2, player 2 to player 3 and so on, the communication complexity is $\Omega(n/t)$ [1].

Proof. Consider an input instance of the t-player set disjointness problem, namely, t subsets $S_1, \ldots, S_t$ of the universe $\{1, 2, \ldots, n\}$ that is provided to each of the t-parties, together with the promise that either, (a) the sets are pair-wise disjoint, or, (b) they have exactly one common intersection element and are otherwise pair-wise disjoint. We view the set $\{1, \ldots, t(t-1)\}$ as being isomorphic to the set of triples $\{(a, b, 0/1)\}$, where $1 \le a < b \le t$. Let $d = t(t-1)$. For each $c = 1, 2, \ldots, t$, player c maps each element x of S_c to a d-dimensional vector denoted by $x^{(c)}$ and whose coordinates are referred to as $[x^{(c)}_{a,b,e}]$, $1 \le a < b \le t$ and $e \in \{0, 1\}$. The vector $x^{(c)}$ is constructed as follows.

1. $x^{(c)}_{c,b,0} = 2x, x^{(c)}_{c,b,1} = 2(4n - x)$
2. $x^{(c)}_{a,c,0} = 2(4n - x), x^{(c)}_{a,c,1} = 2x$
3. $x^{(c)}_{a,b,0} = x^{(c)}_{a,b,1} = \epsilon$, if $a \neq c$ and $b \neq c$.

ϵ is chosen to be a sufficiently small number, say $1/(2n)$, so that $n\epsilon < 1$. Suppose that the knapsack size is $8n + 1$ along each dimension.

Case: non-empty intersection. Suppose there is a common element x in each of the S_c's. Then, $x^{(a)}_{a,b,0} = 2x, x^{(b)}_{a,b,0} = 2(4n - x)$ and $x^{(t)}_{a,b,0} = \epsilon$, for all $t \neq a, t \neq b$. So the sum of coordinate $(a, b, 0)$ of the d-dimensional vectors corresponding to x across the t parties is $2x + 2(4n - x) + (n - 2)\epsilon < 8n + 1$. Similarly, the coordinate $(a, b, 1)$ of the sum of the d-dimensional vectors corresponding to x across the t parties is less than $8n + 1$. That is, the vectors $\{x^{(1)}, \ldots, x^{(t)}\}$ form a feasible packing into one knapsack.

Case: empty intersection. Choose a pair of distinct elements $x \in S_c$ and $y \in S_{c'}$, and suppose that $c < c'$. Then, $x^{(c)}_{c,c',0} + y^{(c')}_{c,c',0} = 2x + 2(4n - y)$ and $x^{(c)}_{c,c',1} + y^{(c)}_{c,c',1} = 2(4n - x) + 2y$ Both $2x + 2(4n - y)$ and $2(4n - x) + 2y$ cannot be at most $8n$ unless $x = y$, hence, at least one of the two is at least $8n + 2$ (being even), and therefore does not fit in the knapsack of size $8n + 1$.

We also note that no two items from the same set S_c can fit in one sack. Let $x, y \in S_c$. Then, for any c' with $c < c'$, $x^{(c)}_{c,c',0} + y^{(c)}_{c,c',0} = 2x + 2y$ and $x^{(c)}_{c,c',1} + y^{(c)}_{c,c',1} = 2(4n - x) + 2(4n - y)$ Both $2x + 2y$ and $2(4n - x) + 2(4n - y)$ cannot be at most $8n$ unless $x + y = 4n$. [If $2x + 2y \le 8n$, then, $x + y \le 4n$, and

$2(4n - x) + 2(4n - y) \leq 8n$ implies that $x + y \geq 4n$, or, $x + y = 4n$.] However, x, y are each at most n, and hence $x + y < 2n - 1$. Thus, no two items from the same set S_c can fit in one knapsack.

So, in the case of a common intersection $\{x\}$ of the sets, the d elements $x^{(c)}$, $c = 1, 2, \ldots, t$ fit together in a sack. Assuming all profits of all items to be 1, the total profit is t. In case when the sets are disjoint, then, at most one element from any of the sets may be placed in the sack, with resulting profit 1.

Suppose there is a streaming algorithm for the $t(t-1)$-dimensional knapsack problem with approximation ratio less than $1/t$ and that uses s bits. Player 1 presents the input $\{x^{(1)} : x \in S_1\}$ to this algorithm with all profits set to 1. The state of the algorithm is then sent to player 2, which in turn inserts its input $\{x^{(2)} : x \in S_2\}$ to the state of the algorithm, relays it to player 3 and so on. Finally, player t calls the profit function of the streaming knapsack algorithm. If this profit is 1 or less, it concludes that the sets are disjoint, otherwise, it concludes that the sets have unique common intersection. This protocol correctly determines set disjointness, since, as argued above, for the disjoint case the optimal profit is 1 and is reported as no more than 1 by the streaming knapsack solution. For the unique intersection case, the optimal profit is t and is reported to be greater than 1, since, the approximation ratio is $1/t$. The total communication is $(t-1)$ times the space requirement of the streaming algorithm, namely, s bits. Note that the entries of the vector $x^{(c)}$ are at most $8n$ and at least $epsilon = 1/(2n)$ and the sack size of $8n+1$. Therefore, we can scale the input in such a way that the entries come from $\{\frac{1}{\Delta}, \frac{2}{\Delta}, \ldots, 1\}$ by setting $\Delta = 2n \cdot (8n+1)$.

By the lower bound of the t-party set disjointness problem, $(t-1)s = \Omega(n/t)$ or, $s = \Omega(n/t^2) = \Omega(\sqrt{\Delta}/t^2)$. Thus any streaming algorithm for the $t(t-1)$-dimensional knapsack problem that uses $o(\sqrt{\Delta}/t^2)$ bits must have an approximation ratio at best $1/t$. □

References

1. Chakrabarti, A., Khot, S., Sun, X.: Near-Optimal Lower Bounds on the Multi-Party Communication Complexity of Set Disjointness. In: Proceedings of International Conference on Computational Complexity, pp. 107–117 (2003)
2. Chekuri, C., Khanna, S.: On multi-dimensional packing problems. In: Proceedings of the 10th ACM-SIAM Symposium on Discrete Algorithms, pp. 185–194 (1999)
3. Diestel, R.: Graphentheorie. Springer, Heidelberg (2006)
4. Ibarra, O.H., Kim, C.E.: Fast Approximation Algorithms for the Knapsack and the Sum of Subset Problems. J. ACM 22(4) (October 1975)
5. Iwama, K., Taketomi, S.: Removable online knapsack problems. In: Proc. of the 29th Intl. Conf. on Automata, Languages and Programming, pp. 293–305 (2002)
6. Iwama, K., Zhang, G.: Optimal Resource Augmentations for Online Knapsack. In: Proceedings of the 10th Intl. Workshop on Approximation Algorithms for Combinatorial Optimization Problems, pp. 180–188 (2007)
7. Karp, R.: Reducibility among Combinatorial Problems. In: Complexity of Computer Computations, pp. 85–103 (1972)

Sparse Cut Projections in Graph Streams

Atish Das Sarma[1], Sreenivas Gollapudi[2], and Rina Panigrahy[2]

[1] Georgia Institute of Technology
atish@cc.gatech.edu
[2] Microsoft Research, Silicon Valley
{sreenig,rina}@microsoft.com

Abstract. Finding sparse cuts is an important tool for analyzing large graphs that arise in practice, such as the web graph, online social communities, and VLSI circuits. When dealing with such graphs having billions of nodes, it is often hard to visualize global partitions. While studies on sparse cuts have traditionally looked at cuts with respect to all the nodes in the graph, some recent works analyze graph properties projected onto a small subset of vertices that may be of interest in a given context, e.g., relevant documents to a query in a search engine. In this paper, we study how sparse cuts in a graph partition a certain subset of nodes. We call this partition a *cut projection*. We study the problem of finding cut projections in the streaming model that is appropriate in this context as the input graph is too large to store in main memory. Specifically, for a d-regular graph G on n nodes with a cut of conductance Φ and constant balance, we show how to partition a randomly chosen set of k nodes in $\tilde{O}(\frac{1}{\sqrt{\alpha\Phi}})$ passes over the graph stream and space $\tilde{O}(n\alpha + \frac{n^{3/4}k^{1/4}}{\sqrt{\alpha}\Phi^{19/4}})$, for any choice of $\alpha \leq 1$. The resulting partition is the projection of a cut of conductance of at most $\tilde{O}(\sqrt{\Phi})$. We note that for $k < n\alpha^6\Phi^{O(1)}$, this can be done in $\tilde{O}(1/\sqrt{\alpha\Phi})$ passes and space $\tilde{O}(n\alpha)$ that is sublinear in the number of nodes.

1 Introduction

The problem of finding sparse cuts on graphs has been studied extensively [6,5,13,4,22,20]. Sparse cuts form an important tool for analyzing/partitioning real world graphs, such as the web graph, click graphs from search engine query logs and online social communities [8]. While traditionally studies on sparse cuts have looked at cuts with respect to all the nodes in the graph, more recent works [17,16] study graph properties projected onto a small subset of nodes that may be of interest. For example, while the web graph may consist of several billions of nodes, in a given context, one may only be interested in the most important nodes such as those with high PageRank or those nodes (representing web pages) relevant to a specific search query. Specifically, we may be interested in finding how connected components of the graph partition these nodes, or we may wish to compute the diameter of the graph with respect to these nodes, or perhaps compute the distance between these nodes with respect to the original graph. Such operations not only enable us to understand the structure of a facet of the graph, but also make it feasible to visualize the graph using a much smaller set of nodes. This approach has been taken in several studies including HITS [14], SALSA [15], and web

A. Fiat and P. Sanders (Eds.): ESA 2009, LNCS 5757, pp. 480–491, 2009.

projections [16], where they study the properties of the web graph restricted to a set of documents that match a given query.

A well developed framework for studying large graphs under memory constraints is the streaming model wherein the input graph is assumed to be on disk, and the algorithm is allowed to make few sequential passes over the input while using a small amount of space in main memory. Our approach works on the streaming model with sub-linear space. The space and pass requirements on streaming algorithms can vary significantly depending on the problem. Demetrescu et. al. [10] give an excellent exposition on the space-passes trade off in graph streaming problems. Henzinger et. al. [12] showed linear lower bounds on the "space $\times$ passes" product for several graph problems including connectivity and shortest path problems. In this work, we present a streaming algorithm for finding how a sparse cut partitions a small random set of nodes when the graph is presented as a stream of edges in no particular order. Our streaming algorithm uses space sublinear in the number of nodes. We also provide an algorithm for finding a sparse cut on the entire graph. We now introduce some definitions below.

Definition 1 (Conductance and Sparsity). *The conductance of a graph $G = (V, E)$ of n nodes $1, 2, \ldots, n$ is defined as $\Phi(G) = \min_{S:E(S)\leq E(V)/2} \frac{E(S,V\setminus S)}{E(S)}$ where $E(S, V \setminus S)$ is the number of edges crossing the cut $(S, V \setminus S)$ and $E(S)$ is the number of edges with at least one end point incident on S. For d-regular graphs, $\Phi(G) = \min_{S:|S|\leq|V|/2} \frac{E(S,V\setminus S)}{d|S|}$. Further, this is within a factor two of $\min_S \frac{nE(S,V\setminus S)}{d|S||V\setminus S|}$. We also note that the sparsity of a d-regular graph is related to the conductance by a factor d.*

Definition 2 (Balance). *The balance of a cut $(S, V \setminus S)$ is defined as $\min\{\frac{|V\setminus S|}{|V|}, \frac{|S|}{|V|}\}$.*

Definition 3 (Cut Projections). *Given a cut $(S, V \setminus S)$, we will say that $(S \cap U, V \setminus S \cap U)$ is a projection of the cut $(S, V \setminus S)$ on U. Further, we will say that a cut $(C, U \setminus C)$, where $C \subseteq U$, is a projected cut of conductance Φ if it is a projection of a cut $(S, V \setminus S)$ with conductance Φ.*

1.1 Contributions of This Study

Our approach builds on the streaming algorithms presented in [9] for performing a large number of random walks efficiently on a graph stream. These random walks are used to estimate the probability distribution of the random walk that is in turn be used to find a sparse cut by adapting the method of Lovasz and Simonovits [18,22].

One of the main contributions of this paper is an algorithm to estimate the probability distributions on an arbitrarily chosen subset of k nodes in a d-regular graph. To obtain the probability of reaching destination t from source s after a walk of length l, the algorithm runs multiple walks (starting with length $l/2$) from source-destination pairs, and recursively estimates probability distributions of mid-points, by looking at the "collisions" of these walks. A similar idea has been used in property testing for expander graphs in [11]. However, in their case, they just need to run walks of length $l/2$ and investigate the collisions. Since we need a good estimate of the probability distribution at t, the algorithm needs to run walks recursively of shorter lengths. All our techniques depend on the reversibility of the random walk, and hence only work for

d-regular, unweighted graphs. We now describe our results beginning with a definition of some notation.

Definition 4. *Let $P_l[st]$ denote the probability of landing at node t after a random walk of length l starting from s. Further, let $p_l(i) = P_l[si]$. We drop the subscript l when it is clear from context.*

The following theorem, proved in Section 3, shows how to compute the approximate distribution on a arbitrarily chosen subset K of k nodes.

Theorem 1. *Given an arbitrarily chosen subset K of k nodes, one can compute an estimate $\tilde{p}(i)$ for $p(i)$ (the probability distribution after a walk of length l) for all $i \in K$ in $\tilde{O}(\sqrt{\frac{l}{\alpha}})$ passes and $\tilde{O}(n\alpha + \frac{1}{\epsilon}\sqrt{\frac{nkl}{\alpha}})$ space for any choice of $\alpha \leq 1$, such that the error in the estimate $|\tilde{p}(i) - p(i)|$ is at most $\tilde{O}(l\sqrt{\frac{p(i)\epsilon}{n}} + \frac{l\epsilon}{n} + (l\sqrt{\epsilon})p(i))$.*

Our main results for computing projected cuts (described in Section 4) with sparsity at most $\tilde{O}(\sqrt{\Phi})$ are stated below.

Theorem 2. *For any d-regular graph G that has a cut of balance b and conductance at most Φ, given a set K of randomly chosen k nodes, we show that there is an algorithm that achieves the following on a graph stream (for any choice of $\alpha \leq 1$).*
(a) Partitions K into two sets such that the partitioning is a projected cut of conductance at most $\tilde{O}(\sqrt{\Phi})$, in $\tilde{O}(\frac{1}{\sqrt{\alpha\Phi}})$ passes and $\tilde{O}(n\alpha + \frac{n^{3/4}k^{1/4}}{b\sqrt{\alpha}\Phi^{19/4}})$ space.
(b) Outputs k candidate partitions such that at least one of them is a projected cut of conductance at most $\tilde{O}(\sqrt{\Phi})$, in $\tilde{O}(\frac{1}{\sqrt{\alpha\Phi}})$ passes and $\tilde{O}(n\alpha + \frac{\sqrt{nk}}{b\sqrt{\alpha}\Phi^{\frac{9}{2}}})$ space.

Corollary 1. *Given a set of randomly chosen $k \leq n\alpha^6 b^4 \Phi^{O(1)}$ nodes, there is an algorithm that partitions them, in $\tilde{O}(\frac{1}{\sqrt{\Phi\alpha}})$ passes and $\tilde{O}(n\alpha)$ space, into a projected cut of conductance at most $\tilde{O}(\sqrt{\Phi})$ w.h.p.*

Observe that the space required is sublinear in the number of nodes if k satisfies the bound in the above corollary. Our algorithms can also be extended to partition all the n nodes in the graph; that is, find the entire (approximate, sparse) cut. The following theorem shows how find an approximate sparse cut in (possibly) sublinear space. The proof is detailed in the full version of the paper.

Theorem 3. *For any d-regular graph G that has a cut of conductance at most Φ and balance b, there is an algorithm that performs $\tilde{O}(\sqrt{\frac{1}{\Phi\alpha}})$ passes over the graph stream and using space $\tilde{O}(\min\{n\alpha + \frac{1}{b}\left(\frac{n\alpha}{d\Phi^3} + \frac{n}{d\sqrt{\alpha}\Phi^{5/2}}\right), (n\alpha + \frac{1}{b}\frac{n}{d\alpha\Phi^2})\sqrt{\frac{1}{\Phi\alpha}} + \frac{1}{\Phi}\})$, for any choice of $\alpha \leq 1$. and outputs, with high probability, a cut of conductance at most $\tilde{O}(\sqrt{\Phi})$.*

1.2 Related Work

A well-known approach for graph partitioning is to compute the second eigenvector that can be used to compute a sparse cut by ordering the nodes in increasing order of coordinate value in the eigenvector. The second eigenvector technique has been analyzed in a series of results [2,7,21] relating the gap between the first and second eigenvalue.

The best known approximation algorithm to compute the sparsest cut in a graph is due to Arora, Rao, and Vazirani [4]. They provide $O(\sqrt{\log n})$-approximation algorithm using semi-definite programming techniques. While their algorithm guarantees good approximation ratios, it is slower than algorithms based on spectral methods and random walks.

Lovasz and Simonovits [18,19] proposed another approach to finding cuts of small conductance. They showed how random walks can be used to find sparse cuts. Specifically, they show that if you start a random walk from a certain node and order the nodes by the probability of reaching them, then this ordering contains a sparse cut. They prove that if the sparsest cut has conductance ϕ, then their method can be used to find a cut with conductance at most $O(\sqrt{\phi})$.

Spielman and Teng [22] build upon the work of Lovasz and Simonovits and show how it can be implemented more efficiently by sparsifying the graph. They show that for a dense graph, it is possible to look at a near linear number of edges and only compute the sparse cuts on the sampled set of vertices. Given a graph $G = (V, E)$ with a cut $(S, V \setminus S)$ with sparsity ϕ and balance $b(S) = |e(S)|/2|E| \geq 1/2$ where $e(.)$ denotes the set of edges incident on nodes in S, their algorithm finds a cut $(D, V \setminus D)$ with sparsity $O(\phi^{1/3} \log^{O(1)} n)$ and balance of the cut $(D, V \setminus D)$, $b(D) \geq b(S)/2$.

Andersen, Chung, and Lang [3] proposed a local partitioning algorithm to find cuts near a specified vertex and global cuts. The running time of their algorithm was proportional to the size of small side of the cut. Their results improve upon those in [22];

In a more recent work [9], the authors proposed algorithms to perform several random walks efficiently on graphs presented as edge streams using a small number of passes. A recent study [1] shows how to find $1 + \epsilon$-approximate sparse cuts in $\tilde{O}(n)$ space by making use of graph sparsifiers. In contrast, our algorithm requires sublinear space for a certain range of parameters, but provides much a weaker approximation to the sparsest cut.

2 Cuts from Approximate Probability Distributions of Random Walks

In this section we will show how one can compute candidate sparse cuts from approximate probability distributions of random walks. We start from a random source s from the smaller side of the best cut with conductance Φ and perform a random walk of length about $1/\Phi$. We extend the algorithm of Lovasz and Simonovits [18] to find sparse cuts using approximate distributions. This is similar to the work by Spielman and Teng [22] that also works with estimates of $p(i)$. But the magnitude of error allowed in our work is larger than in theirs. We adapt a set of lemmas from their work to prove Theorem 4 below. The proof is detailed in the full version of this paper.

Definition 5. *For a probability distribution $p(i)$ on nodes, let $\rho_p(i) = p(i)/d(i)$. Let π_p denote the ordering of nodes in decreasing order of ρ_p; that is, $\rho_p(\pi_p(i)) \geq \rho_p(\pi_p(i+1))$.*

Recall that $p(i)$ denotes the probability of ending at node i after a random walk of length l. The following theorem shows how one can find candidate sparse cuts using approximate values $\tilde{p}(i)$ of $p(i)$. It looks at the n candidate cuts obtained by ordering the nodes in the order $\pi_{\tilde{p}}$.

Theorem 4. *Let $(U, V \setminus U)$ (with $|U| \leq |V|/2$) be a cut of conductance at most Φ. Let $\tilde{p}(i)$ denote an estimate for the probability $p(i)$ of a random walk of length l from a source s from U. Assume that $|\tilde{p}(i) - p(i)| \leq \epsilon(p(i) + 1/n)$, where $\epsilon \leq o(\Phi)$. Consider the n candidate cuts obtained by ordering the vertices in decreasing order of $\rho_{\tilde{p}}(i)$; each candidate cut $(S, V \setminus S)$ is obtained by setting S equal to a prefix $S_j = \pi_{\tilde{p}}\{1, 2, \ldots, j\}$. If the source node s is chosen randomly from U and the length l is chosen randomly in the range $\{1, \ldots, O(1/\Phi)\}$, then with constant probability, one of these n candidate cuts has conductance $\Phi(S_j) \leq \tilde{O}(\sqrt{\Phi})$*

Note that the source node s needs to be sampled from U, the smaller side of the cut. To obtain such a source, we have to sample several sources from V, since U is not known, and execute the algorithm in parallel (so as not to increases the number of passes required). Given a cut of balance b, this increases the number of walks required by a factor of $\tilde{O}(\frac{1}{b})$, and therefore the space required accordingly; in all our space bounds, the first term of $n\alpha$, however, does not depend on the number of walks performed.

In section 4 we will show how Theorem 4 in conjunction with Theorem 1 is used to prove Theorem 2. The essential idea is to look at the k candidate cuts obtained by arranging the nodes in decreasing order of $\pi_{\tilde{p}}$ and then estimate the conductance across these candidate cuts to pick the best one.

In proving these theorems, we use the techniques presented in [9] for performing a large number of random walks efficiently on a graph stream. They show how to perform $O(n/l)$ independent random walks using $\tilde{O}(n\alpha)$ space and $\tilde{O}(\sqrt{\frac{l}{\alpha}})$ passes over the graph stream. Their main result is stated below.

Theorem 5 ([9]). *One can perform k independent random walks from a given source distribution, on a graph stream, in $\tilde{O}(\sqrt{l/\alpha})$ passes and $\tilde{O}(\min\{n\alpha + kl\alpha + k\sqrt{l/\alpha}, n\alpha\sqrt{l/\alpha} + k\sqrt{l/\alpha} + l\})$ space, for any choice of $\alpha \leq 1$. For $k = \frac{n}{l}$ walk, this requires $\tilde{O}(\sqrt{l/\alpha})$ passes and $\tilde{O}(n\alpha)$ space for $1/l \leq \alpha \leq 1$.*

Next we will show how performing random walks can be used to compute the probability distribution approximately so that we may apply theorem 4. To get good approximations, we need to perform random walks recursively as shown in the next section.

3 Estimating Probability Distribution p_i on a Small Set of Nodes

In this section we show how to estimate the probability distribution on a small set of k nodes resulting in Theorem 1. The distribution is required for the endpoint of a random walk of length l from a specific source node s (or more generally a source distribution). The naïve approach would be to perform several random walks of length l from s and look at the end points of these walks to see how many times each of the k nodes occurs. This can be inefficient as k may be much smaller than n and most of the random walks may end up at nodes other than the k nodes we are interested. So we seek a more efficient approach tailored towards estimating the distribution of a specific small set of nodes.

We will begin by stating the following technical lemma. The lemma is later used to approximate distributions. It bounds the error in estimating a_{ij} for a matrix A, where i and j are drawn from two different probability distributions. The guarantee is stated as a trade-off with the number of samples drawn for i and for j.

Lemma 1. *Let $A = \{a_{ij}\}_{m \times n}$ denote a matrix of non-negative entries a_{ij}. Let $\mu_{XY} = E_{i \in X, j \in Y}[a_{ij}]$ denote the expected value of a_{ij} where i and j are drawn independently from probability distributions $X = \{x_i, x_2, \ldots, x_m\}$ and $Y = \{y_1, y_2, \ldots, y_n\}$ on the rows and columns respectively. Assuming $||A^t x||_\infty \leq \tilde{O}(\frac{1}{\epsilon n_x})$ and $||Ay||_\infty \leq \tilde{O}(\frac{1}{\epsilon n_y})$, one can obtain an estimate $\mu_{\tilde{X}\tilde{Y}}$ for μ_{XY} by drawing $\tilde{O}(n_x)$ samples from X and $\tilde{O}(n_y)$ samples from Y. Here $\tilde{X}$ and $\tilde{Y}$ are the distributions induced by the $\tilde{O}(n_x)$ and $\tilde{O}(n_y)$ samples respectively. The error $|\mu_{\tilde{X}\tilde{Y}} - \mu_{XY}|$ is at most $\tilde{O}(\sqrt{\frac{\mu_{XY}}{\epsilon n_x n_y}} + \frac{1}{\epsilon n_x n_y})$ w.h.p.*

Proof. Let $\mu_D = E_{i \in D}[c_i]$. For a distribution D and a vector c with non-negative entries between $[0, 1]$, $\tilde{O}(n)$ samples are drawn from D to estimate μ_D w.h.p such that $|\mu_{\tilde{D}} - \mu_D| \leq \sqrt{\frac{\mu}{n}} + \frac{1}{n}$ by Chernoff bounds. More generally, $|\mu_{\tilde{D}} - \mu_D| \leq \sqrt{\frac{\mu ||c||_\infty}{n}} + \frac{||c||_\infty}{n}$.

We need to bound $|\mu_{XY} - \mu_{\tilde{X}\tilde{Y}}| \leq |\mu_{XY} - \mu_{X\tilde{Y}}| + |\mu_{\tilde{X}Y} - \mu_{\tilde{X}\tilde{Y}}|$. Set $c_i = E_{j \in Y}[a_{ij}]$, i.e., $c = Ay$. This gives $|\mu_{XY} - \mu_{\tilde{X}Y}| \leq \sqrt{\frac{\mu_{XY}||Ay||_\infty}{n_x}} + \frac{||Ay||_\infty}{n_x} \leq \sqrt{\frac{\mu_{XY}}{\epsilon n_x n_y}} + \frac{1}{\epsilon n_x n_y}$.

Further, setting $c_j = E_{i \in \tilde{X}}[a_{ij}]$ or $c = A^t \tilde{x}$ gives $|\mu_{\tilde{X}Y} - \mu_{\tilde{X}\tilde{Y}}| \leq \sqrt{\frac{\mu_{\tilde{X}Y}||A^t\tilde{x}||_\infty}{n_y}} + \frac{||A^t\tilde{x}||_\infty}{n_y}$ w.h.p. But, note that $||A^t\tilde{x}||_\infty \leq ||A^t x||_\infty + \tilde{O}(\sqrt{\frac{A^t||x||_\infty}{n_x}} + \frac{A^t||x||_\infty}{n_x}) \leq \tilde{O}(\frac{1}{\epsilon n_x})$ as $||A^t x||_\infty \leq \tilde{O}(\frac{1}{\epsilon n_x})$. And since $\mu_{\tilde{X}Y} \leq \mu_{XY} + \tilde{O}(\sqrt{\frac{\mu_{XY}}{\epsilon n_x n_y}} + \frac{1}{\epsilon n_x n_y}) \leq \tilde{O}(\mu_{XY} + \frac{1}{\epsilon n_x n_y})$, the difference is at most $\tilde{O}(\sqrt{\frac{(\mu_{XY} + \frac{1}{\epsilon n_x n_y})1/\epsilon n_x}{n_y}} + \frac{1}{\epsilon n_x n_y}) \leq \tilde{O}(\sqrt{\frac{\mu_{XY}}{\epsilon n_x n_y}} + \frac{1}{\epsilon n_x n_y})$. Combining the two, the lemma follows. ∎

We first describe the main idea in estimating the probabilities after a random walk for a subset of nodes as Algorithm RECURSIVERANDOMWALK. The algorithm uses a parameter m that controls the accuracy of error in estimation. Given a set of nodes K, with $|K| = k$, and a source node s, we wish to estimate $P_l[st]$ for all nodes in $t \in K$ up to an additive accuracy of about $O(\sqrt{P_l[st]/m})$. Rather than performing m walks of length l from s, $\tilde{\Theta}(\sqrt{mk})$ walks are performed from s and $\tilde{\Theta}(\sqrt{\frac{m}{k}})$ walks from each node in K, all of length $l/2$. Note that the product of the number of walks performed is m. We then use collisions in the end points of these walks of length $l/2$ to estimate the probabilities (since it is a reversible random walk). The key insight is that $P_l[st] = \sum_i P_{l/2}[su_i].P_{l/2}[tu_i] = \sum_i x_i y_i$ where $x_i = P_{l/2}[su_i]$ and $y_i = P_{l/2}[tu_i]$. That is, one can break all paths from s to t at half way, and sum over all $l/2$ length paths from s to u_i and u_i to t. Any node u_i may either have a *small* or a *large* probability of being reached from s after a random walk of length $l/2$; the same observation holds for for t as well. Roughly, a node u has a small probability for source s if $P_{l/2}[su]$ is $o(1/\sqrt{mk})$, and large probability otherwise. In the formal description of the algorithm we denote these sets of nodes by S_a and S_b respectively. Notice that $o(1/\sqrt{mk})$ is less than the reciprocal of the number of walks run from s. We denote the number of walks of length $l/2$ performed from s by n_s. Similarly, a node u has small probability estimate from t if $P_{l/2}[tu]$ is $o(\sqrt{m/k})$, and a large probability otherwise. In the algorithm, these sets of nodes are denoted by T_a and t_b respectively. The total number of walks of length $l/2$ performed from each node t is denoted by n_t. Now, four cases arise for every u_i.

- $x_i y_i$ is $\Omega(1/m)$ and x_i is $\Omega(\frac{1}{\sqrt{mk}})$ and y_i is $\Omega(\sqrt{\frac{k}{m}})$.
- $x_i y_i$ is $o(1/m)$ and x_i is $o(\frac{1}{\sqrt{mk}})$ and y_i is $o(\sqrt{\frac{k}{m}})$.
- $x_i y_i$ is $\Omega(1/m)$ but x_i is $o(\frac{1}{\sqrt{mk}})$ and y_i is $\Omega(\sqrt{\frac{k}{m}})$.
- $x_i y_i$ is $\Omega(1/m)$ but x_i is $\Omega(\frac{1}{\sqrt{mk}})$ and y_i is $o(\sqrt{\frac{k}{m}})$.

The first case is when u_i has large $P_{l/2}[su_i]$ and $P_{l/2}[tu_i]$, and therefore, it will be seen in walks from both ends and gives us a good estimate of u_i's contribution to $\sum_i P_{l/2}[su_i].P_{l/2}[tu_i]$. The second case is when u_i has small probability for both s and t. In this case, w.h.p., u_i will not be seen in either set of walks. Therefore, u_i's contribution to $\sum_i P_{l/2}[su_i].P_{l/2}[tu_i]$ cannot be estimated. However, since $P_{l/2}[su_i].P_{l/2}[tu_i]$ itself is $o(1/m)$, u_i's contribution to the estimate of $P_l[st]$ is *negligible*.

The difficulty arises in estimating the product for u_i in the third and fourth cases. In both these scenarios, just the walks described above aren't sufficient to estimate the product and yet the contribution may be significant. This is because u_i has a large probability from one end, but a small probability from the other end. The small probability cannot be estimated with just these walks, but the product could be significant, in particular the product could be $\Omega(1/m)$. Hence one needs to resort to a recursive estimation algorithm where the small estimate is captured by performing further walks.

For any node u_i with a large value of $P_{l/2}[su_i]$ and a small value of $P_{l/2}[tu_i]$, we adopt a recursive approach between u_i and t by performing random walks of length $l/4$ from all such u_i and from t. Similarly, for all nodes u_i that have a large value of $P_{l/2}[tu_i]$ and a small value of $P_{l/2}[su_i]$, we perform random walks of length $l/4$ from s and all these u_i to get better estimates of $P_{l/2}[su_i]$ and consequently the product $P_l[st]$. These probabilities may themselves be estimated by random walks of length $l/8$, in another depth of the recursion, and so on. Eventually, combining all of these carefully gives us the probability distribution of t from s after a random walk of length l (notice that we use the reversibility of the random walk). The exact details are stated in Algorithm 1.

We now state and prove the guarantee of RECURSIVERANDOMWALK in estimating probabilities by making use of Lemma 1.

Lemma 2. *Algorithm* RECURSIVERANDOMWALK *gives an estimate of* $\mu = P_l[st]$ *within an additive error of* $\tilde{O}(l\sqrt{\epsilon}\mu + l\sqrt{\frac{\mu}{\epsilon m}} + \frac{l}{\epsilon m})$.

Algorithm 1. RECURSIVERANDOMWALK(s, K, l)

1: **Input:** Starting node/distribution s, length l, and chosen set of k nodes K. Set K need not necessarily be chosen at random.
2: **Output:** $p(t) = \tilde{P}_l[st]$ for each $t \in K$, an estimate of $P_l[st]$ with explicit bound on additive error.
3: Perform $n_s = \tilde{\Theta}(\sqrt{mk})$ walks from s and $n_t = \tilde{\Theta}(\sqrt{m/k})$ walks from each $t \in K$, all of length $l/2$.
4: Denote by S_a the set of nodes seen at most $\tilde{O}(\frac{1}{\epsilon})$ times (small number of times) as endpoints of the n_s walks from s and denote the remaining nodes (seen large number of times) by S_b. Similarly, for each t, partition the nodes into t_a(seen small number of times from t) and t_b(seen large number of times from t). Denote by $\tilde{x}$ and $\tilde{y}$ the distributions of nodes in the end points of the walks from s and t respectively. Thus, $\tilde{x}_i$ is the fraction of walks from s that end up at node i.
5: Let $w(S_b) = \sum_{i \in S_b} \tilde{x}_i$ and $w(t_b) = \sum_{i \in t_b} \tilde{y}_i$. Denote by D_{S_b} the distribution of end points of walks over the nodes in S_b, i.e., the probability of i in D_{S_b} is $\tilde{x}_i / w(S_b)$ if $i \in S_b$ and 0 otherwise. Similarly, denote by D_{t_b} the distribution of end points of walks over the nodes in t_b.
6: For each $t \in K$, set $P_l[st] = \sum_{i \in S_a \cap t_a} \tilde{x}_i \tilde{y}_i + w(S_b) P_{l/2}[D_{S_b} t] + w(t_b) P_{l/2}[s D_{t_b}] - \sum_{i \in S_b \cap t_b} \tilde{x}_i \tilde{y}_i$.
7: In the above expression, $P_{l/2}[D_{S_b} t]$ and $P_{l/2}[s D_{t_b}]$ are estimated recursively for all the $t \in K$. Note that if length is 1, $P_1[st]$ can be computed exactly in one pass.

Proof. $P_l[st] = \sum_{i=1}^{n} x_i y_i$. If all x_i are smaller than $\tilde{O}(\frac{1}{\epsilon\sqrt{km}})$ and y_i is smaller than $\tilde{O}(\frac{1}{\epsilon\sqrt{m/k}})$, then by Lemma 1, the algorithm estimates $\mu = P_l[st]$ within error of $\tilde{O}(\sqrt{\frac{\mu}{\epsilon m}} + \frac{1}{\epsilon m})$. More generally, $P_l[st]$ can be computed as a sum over four sets as $P_l[st] = \sum_{i \in S_a \cap t_a} x_i y_i + \sum_{i \in t_b} x_i y_i + \sum_{i \in S_b} x_i y_i - \sum_{i \in S_b \cap t_b} x_i y_i$; here $i \in S_a$ if x_i is $\tilde{O}(\frac{1}{\epsilon\sqrt{km}})$ and $i \in S_b$ otherwise. $-\sum_{i \in S_b \cap t_b} x_i y_i$ is required as it is counted once in each of $\sum_{i \in t_b} x_i y_i$ and $\sum_{i \in S_b} x_i y_i$. Similarly $j \in t_a$ if y_j is $\tilde{O}(\frac{1}{\epsilon\sqrt{m/k}})$, and $j \in t_b$ otherwise. We will argue that step 6 of the algorithm RECURSIVERANDOMWALK is summing the estimates of each term.

Let $\mu_{aa}, \mu_{*b}, \mu_{b*}, \mu_{bb}$ respectively denote $\sum_{i \in S_a \cap t_a} x_i y_i$, $\sum_{i \in t_b} x_i y_i$, $\sum_{i \in S_b} x_i y_i$ and $\sum_{i \in S_b \cap t_b} x_i y_i$. Note that it is not known which of x_i's or y_i's are small or large. The number of walks performed, however, are sufficient to obtain the right classification to small or large, by standard Chernoff bounds.

Let $\tilde{x}$ and $\tilde{y}$ denote the distributions induced by the end points of the n_s walks and n_t walks respectively. Note that by Lemma 1, $\sum_{i \in S_a \cap t_a} x_i y_i$ can be estimated as $\sum_{i \in S_a \cap t_a} \tilde{x}_i \tilde{y}_i$ with error at most $\tilde{O}(\sqrt{\frac{\mu_{aa}}{\epsilon m}} + \frac{1}{\epsilon m})$. Also note that if $i \in S_b$, then $\tilde{x}_i$ is within $(1 \pm \sqrt{\epsilon})\tilde{x}_i$ w.h.p. from Chernoff bounds. Thus, μ_{bb} can be estimated with error at most $\sqrt{\epsilon}\mu_{bb}$. Again, μ_{*b} can be estimated as $\mu_{*\tilde{b}} = \sum_{i \in t_b} x_i \tilde{y}_i$ where the error $|\mu_{*b} - \mu_{*\tilde{b}}| \leq \sqrt{\epsilon}\mu_{*b}$. Similarly, μ_{b*} can be estimated as $\mu_{\tilde{b}*} = \sum_{i \in S_b} \tilde{x}_i y_i$ with additive error at most $\sqrt{\epsilon}\mu_{b*}$.

Observe that the $\mu_{*\tilde{b}} = \sum_{i \in t_b} x_i \tilde{y}_i = w(S_b) P_{l/2}[D_{S_b} t]$ and $\mu_{b\tilde{*}} = \sum_{i \in S_b} \tilde{x}_i y_i = w(t_b) P_{l/2}[s D_{t_b}]$ are estimated recursively by computing $P_{l/2}[D_{S_b} t]$ and $P_{l/2}[s D_{t_b}]$.

Let $\delta_l(P_l[st])$ denote the error in estimating the probability $P_l[st]$. Then $\delta_l(\mu) = \sqrt{\frac{\mu_{aa}}{\epsilon m}} + \frac{1}{\epsilon m} + \sqrt{\epsilon \mu_{bb}} + \sqrt{\epsilon \mu_{b*}} + \sqrt{\epsilon \mu_{*b}} + w(S_b)\delta_{l/2}(P_{l/2}[D_{S_b}t]) + w(t_b)\delta_{l/2}$ $(P_{l/2}[sD_{t_b}]) \leq \sqrt{\frac{\mu}{\epsilon m}} + \frac{1}{\epsilon m} + 3\sqrt{\epsilon\mu} + w(S_b)\delta_{l/2}(\frac{\mu}{w(S_b)}) + w(t_b)\delta_{l/2}(\frac{\mu}{w(t_b)})$.

The branching factor of the recursion is 2 and has depth $\log l$ with l leaves. Also note that at the leaf, $\delta_1(P_1[st]) = 0$. It follows from standard methods for solving recursion that $\delta_l(\mu) = \tilde{O}(l\sqrt{\epsilon\mu} + l\sqrt{\frac{\mu}{\epsilon m}} + \frac{l}{\epsilon m})$. ∎

We now use the approach in [9] to bound the number of passes and space required in performing RECURSIVERANDOMWALK. The analysis is simple and only requires a careful calculation of the number of walks performed for each length $l/2, l/4, l/8, \ldots$.

Lemma 3. *Algorithm* RECURSIVERANDOMWALK *can be implemented on a graph stream in* $\tilde{O}(\sqrt{\frac{l}{\alpha}})$ *passes and* $\tilde{O}(n\alpha + \sqrt{\frac{mkl}{\alpha}})$ *space for any choice of* $\alpha \leq 1$.

Proof. Notice that in the first phase of RECURSIVERANDOMWALK, $O(\sqrt{mk})$ walks are performed from s of length $l/2$ and $O(\sqrt{m/k})$ walks from each t of length $l/2$. Whenever recursively calculating the distribution, $O(\sqrt{mk})$ walks are required for any source distribution to destination distribution pair. Since the length of the walks halve with every recursive depth, the number of levels is $O(\log l)$ before the length of the walks required becomes a constant. However, the pairs for which the distributions need to be estimated keeps doubling. So after, say, t phases, we perform $O(2^t\sqrt{mk})$ walks of length $l/2^t$.

From [9], the space required for k walks is $\tilde{O}(n\alpha + kl\alpha + k\sqrt{l/\alpha})$. Although [9] assumes that the source distribution was the same for all the k walks, it is easy to extend their result to perform a specific number of walks from different source distributions (to a total of k walks), with only a logarithmic factor increase in the space (by Chernoff bounds).

Summing this over t phases, the first term remains $\tilde{O}(n\alpha)$. The second term of $\tilde{O}(kl\alpha)$ is always $\tilde{O}(\sqrt{mkl\alpha})$, as only a $\log l$ factor increases. The third term of $k\sqrt{l/\alpha}$ is dominated by the last phase (since this term depends linearly in k but only as square-root in the length of the walks), where $l\sqrt{mk}$ walks of length $O(1)$ are performed. The dominating term therefore is $\tilde{O}(\sqrt{mkl\alpha})$, completing the proof. ∎

Remark 1. If algorithm RECURSIVERANDOMWALK is to be performed from r different sources, this would increase the space requirement to $\tilde{O}(n\alpha + r\sqrt{\frac{mkl}{\alpha}})$. This follows from the fact that the first term of $\tilde{O}(n\alpha)$ in the space requirement does not depend on the number of walks performed.

Notice that combining Lemmas 2, 3, and choosing $m = \frac{n}{\epsilon^2}$ immediately gives Theorem 1. Observe that by setting $\epsilon = o(\Phi^2/l^2)$ we satisfy the conditions of Theorem 4 and can thus compute candidate cuts.

Algorithm 2. CUTPROJECTIONCANDIDATES(G, K, s)

1: **Input:** Graph G, set K of k randomly sampled nodes, and a source node s.
2: **Output:** k partitions on these k nodes.
3: Estimate probabilities on the k nodes using RECURSIVERANDOMWALK for source s and walk of length l, where l is chosen at random between 1 and $O(1/\phi^2)$.
4: Order the k nodes in decreasing order of the probability estimates. Return the k candidate cuts implied by taking prefixes of this ordering.

Algorithm 3. CUTPROJECTION

1: **Input:** Graph G with cut of conductance at most Φ and source s from the smaller side of this cut; set K of k randomly sampled nodes.
2: **Output:** Partition of these k nodes such that this is a projected cut of conductance is at most $\phi = \tilde{O}(\sqrt{\Phi})$ with constant probability.
3: Sample additional nodes randomly and add it to the set K so that the resulting set K' is of size $k' = \sqrt{\frac{kn}{\phi}}$.
4: Call CUTPROJECTIONCANDIDATES with G', K', s.
5: Consider all of the k' cuts returned by CUTPROJECTIONCANDIDATES that have at least $\frac{k'}{k}$ nodes on either side of the cut. Notice that each of these cuts has at least one of the k nodes in K on either side, with constant probability.
6: For each of these cuts, compute the conductance on the induced subgraph over these k' nodes.
7: Output the cut induced on the k nodes by the cut on the k' nodes that has the minimum conductance in the induced subgraph.

4 Finding Sparse Cut Projections on a Small Set of Nodes

In this section, we prove Theorem 2. Approximate values of $p(i)$ are known from Theorem 1. By setting $\epsilon = \tilde{O}(\frac{\Phi^2}{l^2})$ in theorem 4, we find probability estimates $\tilde{p}(i)$ that satisfy the condition required in theorem 1. Notice that part (b) of Theorem 2 follows directly by using walks of length $l = O(\frac{1}{\Phi})$. If we order all the n vertices by the probability estimates at least one cut has conductance at most $\tilde{O}(\sqrt{\Phi})$. Naturally this ordering induces an ordering on the k vertices that results in k candidate cuts. Note that in our algorithm we need to sample $\tilde{\Omega}(1/b)$ sources so that at least one falls on the smaller side of the optimal cut with high probability. The factor $1/b$ is not applied to the $n\alpha$ term as we can perform all the random walks concurrently as stated in remark 1.

We now prove Theorem 2 part (a). We need to estimate the projected cut conductance for each of the k candidate cuts from the ordering of approximate probabilities. This is done by boosting the number of random nodes from k to k'. It turns out that one can estimate the projected cut conductance value of a specific cut on the k nodes by looking at the conductance on the induced subgraph and cuts on the k' nodes (for an appropriate choice of k'), as stated in Lemma 4. The formal description is in algorithm CUTPROJECTIONCANDIDATES and algorithm CUTPROJECTION. Let $\Phi_{K'}(U, K' \setminus U)$ denote the conductance of the cut $(U, K' \setminus U)$ on the induced subgraph on nodes in K'.

Lemma 4. *For a set K' of randomly chosen nodes with $|K'| = k' \geq \sqrt{\frac{kn}{\phi}}$ and $|U| \geq \frac{k'}{k}$ and $|K' \setminus U| \geq \frac{k'}{k}$, and let $(U, K' \setminus U)$ be a projected cut of conductance of be Ψ. Then $\Phi_{K'}(U, K' \setminus U)$ gives a constant factor approximation to Ψ with high probability.*

Proof (sketch). The essential idea is that considering the conductance of the induced cut on $\sqrt{\frac{kn}{\phi}}$ nodes, is identical to estimating $\sum_{(i,j)\in E} na_{ij}x_iy_j/(|X||Y|)$ by sampling each x_i with probability $k'|X|/n$ and each y_i with probability $k'|Y|/n$ and a_{ij} is $\frac{1}{d}$ for an edge (i, j). The proof is then completed using Lemma 1. A more detailed exposition is presented in the full version of the paper. ∎

This lemma automatically gives a method for estimating the projected cut conductance for a partition of a subset of k nodes. We are now ready to prove the main theorem 2 part (b).

Proof (Proof of Theorem 2(b)). By setting $\epsilon = \tilde{O}(\frac{\Phi^2}{l^2})$ in theorem 4, we find probability estimates $\tilde{p}(i)$ that satisfy the condition required in theorem 1. So if we order all the n vertices by the probability estimates at least one cut has conductance at most $\tilde{O}(\sqrt{\Phi})$. Naturally this ordering induces an ordering on the k' vertices that contain the set K. We are simply estimating the conductance of all the cuts in this ordering that has at least one vertex from K in the smaller side. If none of these cuts give the required conductance of $\tilde{O}(\sqrt{\Phi})$, then all k nodes are put on the same side of the cut and output. Note that in our algorithm we need to sample $\tilde{\Omega}(1/b)$ sources so that at least one falls on the smaller side of the optimal cut with high probability. This gives $\sqrt{\frac{l}{\Phi\alpha}}$ passes and $\tilde{O}(n\alpha + \frac{1}{b}\frac{1}{\epsilon}\sqrt{\frac{nk'}{\phi\alpha}}) = \tilde{O}(n\alpha + \frac{n^{3/4}k^{1/4}}{b\sqrt{\alpha}\Phi^{19/4}})$ space. The factor $1/b$ is not applied to the $n\alpha$ term as the we can perform all the random walks concurrently as stated in remark 1. ∎

5 Conclusions

In this work, we present an approach for finding cuts that approximate the conductance of a graph presented as a stream of edges. In particular, we show that this problem can be solved more efficiently if we are only required to partition a small set of k random nodes with respect to a sparse cut. The streaming algorithms we present require space that is sublinear in the number of nodes for a certain range of parameters.

Acknowledgments. We thank D. Sivakumar, S. Vempala, and K. Munagala for comments on the paper.

References

1. Ahn, K.J., Guha, S.: Graph sparsification in the semi-streaming model. In: Albers, S., Marchetti-Spaccamela, A., Matias, Y., Nikoletseas, S., Thomas, W. (eds.) ICALP 2009. Part II. LNCS, vol. 5556, pp. 328–338. Springer, Heidelberg (2009)
2. Alon, N.: Eigenvalues and expanders. Combinatorica 6(2), 83–96 (1986)

3. Andersen, R., Chung, F.R.K., Lang, K.J.: Local Graph Partitioning using PageRank Vectors. In: FOCS, pp. 475–486 (2006)
4. Arora, S., Rao, S., Vazirani, U.V.: Expander flows, geometric embeddings and graph partitioning. In: STOC, pp. 222–231 (2004)
5. Benczúr, A.A., Karger, D.R.: Approximating s-t Minimum Cuts in $\tilde{O}(n^2)$ Time. In: STOC, pp. 47–55 (1996)
6. Bhatt, S.N., Leighton, F.T.: A Framework for Solving VLSI Graph Layout Problems. J. Comput. Syst. Sci. 28(2), 300–343 (1984)
7. Boppana, R.: Eigenvalues and graph bisection: an average case analysis. In: 28th IEEE Symposium on Foundations of Computer Science, FOCS (1987)
8. Borgs, C., Chayes, J.T., Mahdian, M., Saberi, A.: Exploring the community structure of newsgroups. In: KDD, pp. 783–787 (2004)
9. Sarma, A.D., Gollapudi, S., Panigrahy, R.: Estimating PageRank on graph streams. In: PODS, pp. 69–78 (2008)
10. Demetrescu, C., Finocchi, I., Ribichini, A.: Trading of space for passes in graph streaming problems. In: ACM-SIAM Symposium on Discrete Algorithms, SODA, pp. 714–723 (2006)
11. Goldreich, O., Ron, D.: Property Testing in Bounded Degree Graphs. In: STOC, pp. 406–415 (1997)
12. Henzinger, M., Raghavan, P., Rajagopalan, S.: Computing on data streams. In: External Memory Algorithms. DIMACS Series in Discrete Mathematics and Theoretical Computer Science, vol. 50, pp. 107–118 (1999)
13. Karger, D.R.: Minimum cuts in near-linear time. J. ACM 47(1), 46–76 (2000)
14. Kleinberg, J.M.: Authoritative sources in a hyperlinked environment. In: SODA, pp. 668–677 (1998)
15. Lempel, R., Moran, S.: Salsa: the stochastic approach for link-structure analysis. ACM Trans. Inf. Syst. 19(2), 131–160 (2001)
16. Leskovec, J., Dumais, S., Horvitz, E.: Web projections: learning from contextual subgraphs of the web. In: WWW 2007: Proceedings of the 16th international conference on World Wide Web, pp. 471–480. ACM, New York (2007)
17. Leskovec, J., Faloutsos, C.: Sampling from large graphs. In: KDD, pp. 631–636 (2006)
18. Lovász, L., Simonovits, M.: The Mixing Rate of Markov Chains, an Isoperimetric Inequality, and Computing the Volume. In: FOCS, pp. 346–354 (1990)
19. Lovász, L., Simonovits, M.: Random Walks in a Convex Body and an Improved Volume Algorithm. Random Struct. Algorithms 4(4), 359–412 (1993)
20. Manokaran, R., Naor, J., Raghavendra, P., Schwartz, R.: SDP gaps and UGC hardness for multiway cut, 0-extension, and metric labeling. In: STOC, pp. 11–20 (2008)
21. Sinclair, A., Jerrum, M.: Conductance and the mixing property of markov chains; the approximation of the permanent resolved. In: Proc. of the 20th annual ACM Symposium on Theory of computing, pp. 235–244 (1988)
22. Spielman, D.A., Teng, S.-H.: Nearly-linear time algorithms for graph partitioning, graph sparsification, and solving linear systems. In: STOC, pp. 81–90 (2004)

Bipartite Graph Matchings in the Semi-streaming Model (Extended Abstract)

Sebastian Eggert, Lasse Kliemann*, and Anand Srivastav

Institut für Informatik
Christian-Albrechts-Universität Kiel
Christian-Albrechts-Platz 4
24118 Kiel
{see,lki,asr}@informatik.uni-kiel.de

Abstract. We present an algorithm for finding a large matching in a bipartite graph in the semi-streaming model. In this model, the input graph $G = (V, E)$ is represented as a stream of its edges in some arbitrary order, and storage of the algorithm is bounded by $O(n \operatorname{polylog} n)$ bits, where $n = |V|$. For $\epsilon > 0$, our algorithm finds a $\frac{1}{1+\epsilon}$-approximation of a maximum-cardinality matching and uses $O\left((\frac{1}{\epsilon})^8\right)$ passes over the input stream. The only previously known algorithm with such arbitrarily good approximation – though for general graphs – required exponentially many $\Omega\left((\frac{1}{\epsilon})^{\frac{1}{\epsilon}}\right)$ passes (McGregor 2005).

Keywords: bipartite graph matching, streaming algorithms, approx. algorithms.

1 Introduction

Given a bipartite graph $G = (A, B, E)$, $n := |A \dot{\cup} B|$, the maximum-cardinality matching problem (or *maximum matching problem*, as we will refer to it) is to find a cardinality-maximal set of edges such that no two intersecting edges are selected. An exact solution for this problem can be found in $O(n^{2.5})$ time with the algorithm of Hopcroft and Karp [3]. There is also a faster randomized algorithm by Mucha and Sankowski [5], which runs in $O(n^{\omega})$, where ω depends on the running time of the best known matrix multiplication algorithm; it is $\omega < 2.38$. For massive graphs, not only the time complexity, but also the space complexity plays an important role. In the streaming model, we assume that the entire graph cannot be stored in random access memory (RAM). Therefore the algorithm has no random access to the whole input, which is required for most of the algorithms. The set of edges has to be stored on an external device, like a disk or a tape, and is only presented as a stream. In this stream, each edge

* Supported by Deutsche Forschungsgemeinschaft, Priority Program 1307 *Algorithm Engineering*, Grant Sr7/12-1.

A. Fiat and P. Sanders (Eds.): ESA 2009, LNCS 5757, pp. 492–503, 2009.

is presented exactly once, and each time the stream is read, edges may appear in an arbitrary order. The stream can only be read as a whole and reading the whole stream once is called a pass (over the input). It sometimes is assumed that the order in which edges appear is arbitrary, but the same for each pass. However, our result works without such an assumption.

The number of passes used by the algorithm should be independent of the size n of the input graph. One is interested in a good approximation of the optimum, and there the number of passes can and does depend on the approximation parameter, e.g., [2] or [4].

Standard streaming models, for example a (poly)log-space streaming model, are too restrictive for graph problems. For example, it is impossible to decide if there is a path of length 2 between two given vertices x and y [2]. This shows that even easy graph problems are not solvable in this model. For solving graph problems, Muthukrishnan [6] proposed the semi-streaming model: memory of the algorithm is restricted to $O(n \operatorname{polylog} n)$, which is not enough to store the entire graph if the graph is sufficiently dense, but enough to store $O(n)$ edges.

Feigenbaum et al. [2] showed that finding an exact solution of the maximum matching problem in one pass requires $\Omega(m)$ bits of space. Therefore, with the given memory restriction, we aim for an *approximation* of a maximum matching. We speak of an α-approximation, or α-factor approximation, for $0 < \alpha \leq 1$, if the algorithm delivers a matching of cardinality at least $\alpha\,\mathsf{OPT}$, where OPT is the cardinality of a maximum matching. It is desirable to achieve an arbitrarily good, say a $\frac{1}{1+\epsilon}$-approximation, where $\epsilon > 0$, called approximation parameter, can be arbitrarily small. An algorithm for finding a $(\frac{2}{3} - \epsilon)$-approximation of a maximum matching in bipartite graphs was given by Feigenbaum et al. [2]. This algorithm requires $O\left(\frac{1}{\epsilon}\log\frac{1}{\epsilon}\right)$ passes over the input stream. For general graphs, McGregor in his pioneering paper [4] gave the first randomized algorithm in the semi-streaming model for finding a $\frac{1}{1+\epsilon}$-approximation of maximum matching with a constant number of passes over the input stream, where ϵ is considered a constant. The dependence on $\frac{1}{\epsilon}$ is rather strong, namely $\Omega\left((\frac{1}{\epsilon})^{\frac{1}{\epsilon}}\right)$ passes are needed. In Sec. 3 we show that this requirement essentially also holds for the special case of bipartite graphs. A more efficient approximation would insist on a *polynomial* dependence on relevant input parameters. In the last years this has been a key issue in the qualification of efficiency, for example in research on parameterized complexity. For the maximum (bipartite) matching problem it was not known whether there is an approximation algorithm in the semi-streaming model requiring only a polynomial number of passes in $\frac{1}{\epsilon}$.

Our Contribution. We present in the semi-streaming model a deterministic $\frac{1}{1+\epsilon}$-approximation algorithm for the maximum matching problem in bipartite graphs with only $O\left((\frac{1}{\epsilon})^8\right)$ passes over the input stream, and thus break the exponential barrier on the number of passes.

Both, McGregor's and our algorithm, build augmenting paths in a layer-wise fashion. McGregor repeatedly uses randomization to decide which matching edge may occur in which layer. As we show in Thm. 1, unfavorable random decisions can be responsible for the requirement of a large number of passes. The novelty of our approach is to introduce a new *dynamic and deterministic* assignment of matching edges to layers which is responsible for the complexity reduction. We will give a detailed description of our algorithm that can be easily implemented.

2 Preliminaries

We denote by $G = (A, B, E)$ a bipartite graph with vertex set $V = A \cup B$, where $A \cap B = \emptyset$, and edge set E. A set $M \subseteq E$ is called a *matching* if no two edges in M intersect. A matching M is a *maximal matching* if $M \cup \{e\}$ is not a matching for any $e \in M^{\mathsf{C}}$ $(= E \setminus M)$. A matching is called a *maximum matching* if it has maximum cardinality among all matchings. Obviously, a maximum matching is also a maximal matching. A vertex is called *free* if it does not appear in any edge from M, and *matched* otherwise. The free vertices of a subset $X \subseteq V$ are denoted $\mathrm{free}(X)$. We often denote free vertices with small Greek letters, e.g., $\alpha \in \mathrm{free}(A)$ for a free vertex of partition A. For a matched vertex v denote $\Gamma_M(v) := u$ where $\{v, u\} \in M$. We denote paths as sequences of vertices and edges; this notation has redundancy, but will be helpful in some places. Denote $E(P)$ the edges in a path P and $V(P)$ the vertices. The *length* of a path P, denoted $|P|$, is its number of edges. An *M-augmenting* path of length $2k + 1$ is a path with $2k + 1$ edges and $2k+2$ vertices which is (M^{C}, M)-alternating and starts and ends with a free vertex, formally: $(v_0, e_1, v_1, m_1, v_2, \ldots, v_{2k-1}, m_k, v_{2k}, e_{k+1}, v_{2k+1})$, where v_0 and v_{2k+1} are free vertices, $e_j = \{v_{2j-2}, v_{2j-1}\} \in M^{\mathsf{C}}$ for $j \in [k+1]$ and $m_j = \{v_{2j-1}, v_{2j}\} \in M$ for $j \in [k]$. For a matching M and an M-augmenting path P, the symmetric difference $M \bigtriangleup E(P) = (M \cup E(P)) \setminus (M \cap E(P))$ is a matching of size $|M| + 1$.

Motivated by Berge's famous theorem [1], stating that a matching is a maximum matching if and only if there is no M-augmenting path, a majority of matching algorithms uses augmenting paths for finding a maximum matching. As well, our algorithm uses augmenting paths for increasing the size of the matching constructed so far. It starts with a maximal matching, which can be found easily in one pass over the input by selecting an edge iff this edge is not incident with any already selected edge. It is a well-known fact that a maximal matching is already a $\frac{1}{2}$-approximation of a maximum matching. But for our result, this approximation factor is not good enough. On the other hand, we will not aim for eliminating *all* augmenting paths, but only up to a certain limit. We will prove two lemmas, Lem. 3 and 4, which guarantee a good approximation if there are only few augmenting paths left. We will elaborated on this fact in Sec. 5.

3 McGregor's Algorithm

We briefly describe McGregor's Algorithm [4]. It is a randomized algorithm and finds a $\frac{1}{1+\epsilon}$-approximation of a maximum matching in general graphs in the semi-streaming model. We then show that even on a *bipartite* input, it requires

$\Omega\left((\frac{1}{\epsilon}-1)^{\frac{1}{\epsilon}-1}\right)$ passes over the input stream in order to achieve the claimed success probability of $1-e^{-1}$, cf. [4, Th. 2]. It is hence – on bipartite graphs – outperformed by our algorithm, which only needs a number of passes *polynomial* in $\frac{1}{\epsilon}$, and moreover is deterministic.

McGregor's algorithm takes a general graph G and a parameter ϵ as input. It first computes a maximal matching M, which is done in one pass over the input stream. Define $k := \lceil \frac{1}{\epsilon}+1 \rceil$ and $r := \Omega(k^k)$. The algorithm performs r iterations, each of them called a *phase*. A phase consists of k iterations, $i := 1, \ldots, k$, of a sub-routine that tries to find a large set of pairwise vertex-disjoint M-augmenting paths, each of length $2i+1$. Of those k sets of M-augmenting paths, one set $\mathcal{P}$ is chosen which yields the largest augmented matching $M \triangle E(\mathcal{P})$, and then M is replaced by that augmented matching, and the next phase starts. Each phase needs at least one pass over the input, and so we have $\Omega(k^k)$ passes, which means $\Omega\left((\frac{1}{\epsilon})^{\frac{1}{\epsilon}}\right)$ passes. For each i in each phase, a bipartite graph G' is (implicitly) constructed, that captures certain aspects of the relation of M to G and helps finding M-augmenting paths. We describe the construction of the bipartite graph. The set of free vertices is partitioned randomly into F_{left} and F_{right}, with each vertex being independently assigned to one of the sets with probability $\frac{1}{2}$. Then matching edges M are randomly partitioned into $M_1, \ldots, M_i$, each one independently assigned to one of these sets with probability $\frac{1}{i}$. Each matching edge $\{u, v\}$ is given an orientation randomly: either we have (u, v) or we have (v, u), each chosen with probability $\frac{1}{2}$ and independently for each edge.[1] Denote the sets of oriented edges by $\overrightarrow{M}_1, \ldots, \overrightarrow{M}_i$. The random choices reflected by sets $F_{\text{left}}, F_{\text{right}}$ and $\overrightarrow{M}_1, \ldots, \overrightarrow{M}_i$ fully specify the bipartite graph $G' = (V', E')$ in the following way. We interpret the free vertices and the directed matching edges as vertices of G', i.e., $V' = F_{\text{left}} \cup F_{\text{right}} \cup \bigcup_{j=1}^{i} \overrightarrow{M}_j$. Edges run as follows:

$$\begin{aligned} E' = &\{\{x, (u,v)\};\ x \in F_{\text{left}},\ (u,v) \in \overrightarrow{M}_1,\ \{x,u\} \in E\} \\ &\cup \bigcup_{j=1}^{i-1} \{\{(s,t),(u,v)\};\ (s,t) \in \overrightarrow{M}_j,\ (u,v) \in \overrightarrow{M}_{j+1},\ \{t,u\} \in E\} \\ &\cup \{\{(u,v), y\};\ (u,v) \in \overrightarrow{M}_i,\ y \in F_{\text{right}},\ \{v,y\} \in E\}\ . \end{aligned}$$

We can think of G' being in layers: on the left is F_{left}, then we have[2] $\overrightarrow{M}_1, \ldots, \overrightarrow{M}_i$, and on the right we have F_{right}. All directed edges (which are vertices of G') point to the right. Edges in G' only run between neighboring layers.

Storing E' in memory is impossible in general, and so we only store V' and determine parts of E' as needed by passes over the input stream. A path in G' between the 'end' layers F_{left} and F_{right} yields an M-augmenting path of length $2i+1$ in G. McGregor's algorithm tries to find a large collection of such paths once G' is constructed; we skip those details here.

[1] In [4] symbols "a" and "b" are used to mark the orientation.

[2] In [4] layers are enumerated with *decreasing* indices from left to right.

Theorem 1. *There are bipartite instances on which McGregor's algorithm requires* $\Omega\left((\frac{1}{\epsilon}-1)^{\frac{1}{\epsilon}-1}\right)$ *passes over the input stream to succeed with probability at least* $1-e^{-1}$.

Proof. Let $l \in \mathbb{N}$ and G be the path on $2l-1$ edges, and $\epsilon := \frac{1}{l}$. Suppose the algorithm chooses an initial maximal matching M such that there exists exactly one M-augmenting path of length $2l-1$ in G, namely G itself. Then $|M| = l-1$, and the size of a maximum matching is l. This gives $(1+\epsilon)|M| = (1+\frac{1}{l})(l-1)$ $= l-1+\frac{l-1}{l} < l = \mathsf{OPT}$. Hence M is not a $\frac{1}{1+\epsilon}$-approximation. The only way to find the desired approximation is to find that one M-augmenting path, which is G itself. To this end, the two free vertices must be assigned to different layers, and all matching edges must be assigned to the layers and oriented in a way that G occurs as a path in G'. This can only happen in an iteration where $i = l-1$, and then the probability that it happens is $p := \frac{1}{2}\frac{1}{(l-1)^{(l-1)}}\frac{1}{2^{(l-1)}}$. The probability that it does *not* happen within d phases is $(1-p)^d$. To push this probability below e^{-1}, we need $d\ln(1-p) \le -1$, hence, using that for all $-1 < x$ we have $\ln(1+x) \le x$,

$$d \ge \frac{-1}{\ln(1-p)} = \frac{1}{\ln\left(\frac{1}{1-p}\right)} = \frac{1}{\ln\left(1+\left(\frac{1}{1-p}-1\right)\right)}$$

$$\ge \frac{1}{\frac{1}{1-p}-1} = \frac{1}{p} - 1 \ge (l-1)^{l-1} - 1 = \left(\frac{1}{\epsilon}-1\right)^{\frac{1}{\epsilon}-1} - 1\,. \qquad \square$$

This lower bound still stands if we do a straightforward modification of the algorithm, allowing it to use the knowledge that it is working on a bipartite graph $G = (A, B, E)$. Consider the following modification. We put the free vertices of A into F_{left} and the free vertices of B into F_{right}. Each matching edge $\{a, b\}$ with $a \in A$ and $b \in B$ is oriented (b, a). Probability p in the above example is now $p := \frac{1}{(l-1)^{(l-1)}}$, yielding the same lower bound for d.

4 Our Matching Algorithm

The main algorithm is approx-maximum-matching, shown on page 499. It implements the usual scheme of starting with an arbitrary maximal matching M, then repeatedly computing a set of vertex-disjoint M-augmenting paths and replacing M by an augmented version. We stop when the number of augmenting paths found falls below the threshold of $2\delta|M|$. We only consider M-augmenting paths of length at most $2k+1$. These are computed by sub-routine disjoint-paths. That sub-routine is the most complex part; it is displayed on page 499. We now give an explanatory description of it; an illustration is given in Fig. 1 on page 500. Input is the bipartite graph G, given as a stream, a threshold parameter δ, a length parameter k, and a maximal matching M. The task is to find many pairwise vertex-disjoint M-augmenting paths of length at most $2k+1$. The state of its main loop, starting in line 4, is (essentially) given by:

- a position number $i \in \{1, \dots, k+1\}$;
- for each matching edge $m \in M$ a position limit $\ell(m) \in \{1, \dots, k+1\}$;
- the remaining vertices V';
- for each $\alpha \in \text{free}(A)$ a path $P(\alpha) = (\alpha, \dots)$ of length at most $2k+1$, being called a *constructed* path.

The set of constructed paths is partitioned into *incomplete paths* $\mathcal{I}$ and *augmenting paths* $\mathcal{A}$. Set $\mathcal{I}$ consists of (M^{C}, M)-alternating paths which could not (yet) be completed to M-augmenting paths. Set $\mathcal{A}$ consist of M-augmenting paths; it is the result of disjoint-paths when it terminates. Once an incomplete path is completed to an augmenting path, it is moved from $\mathcal{I}$ to $\mathcal{A}$ and never touched again until the end of this run of disjoint-paths; its vertices are removed from V'. Denote $\mathcal{I}_{>0}$ the set of all incomplete paths that consist of at least one edge (they in fact consist of at least two edges then). Several invariants hold for the constructed paths during execution: Any two constructed paths are vertex-disjoint. All constructed paths are (M^{C}, M)-alternating. Their vertices are alternately from A and B. Incomplete paths end with a matching edge and with a vertex from A.

We think of constructed paths starting at the left and proceeding to the right. So, in each incomplete path, each vertex from B has a matching edge to its right. If a matching edge m is contained in an incomplete path P, its position limit $\ell(m)$ is exactly so that m is matching edge number $\ell(m)$ counted from the left of P, that is edge number $2\,\ell(m)$.

Initialization is done by setting position limits $\ell(m) := k+1$ for each $m \in M$ (which is an impossible value for an edge inside a constructed path) and constructed paths $P(\alpha) := (\alpha)$ for each $\alpha \in \text{free}(A)$. Then the algorithm does several passes over the input. It cycles the position i after each pass, realized by the **for** loop in line 5. We call one execution of that **for** loop a *sweep*. Hence, a sweep consists of $k+1$ passes, with positions $i = 1, \dots, k+1$.

We explain what happens for each edge during a pass. Let the current edge be $e = \{a, b\}$, $a \in A$, $b \in B$, between two remaining vertices. It is tested whether we can – and wish to – use e to extend an incomplete path. Two conditions have to be met for a $P(\alpha)$ to be eligible for extension:

(i) $P(\alpha)$ must have length $2(i-1)$;

(ii) $P(\alpha)$ must have a as its endpoint, i.e., $P(\alpha) = (\alpha, \dots, a)$.

Since all paths are pairwise vertex-disjoint, there can be at most one path that fulfills condition (ii). If there is none, we discard e and continue the pass with the next edge. Otherwise, let $P(\alpha^*) = (\alpha^*, \dots, a)$ be that path.

If b is a free vertex, we have found an augmenting path, namely $A := (\alpha^*, \dots, a, e, b)$. We set $\mathcal{A} := \mathcal{A} \cup \{A\}$, $\mathcal{I} := \mathcal{I} \setminus \{P(\alpha^*)\}$, and update the set of remaining vertices V'.

The other case is that b is a matched vertex, let $m := \{b, \Gamma_M(b)\} \in M$. Then we check whether we are below m's position limit, i.e., $i < \ell(m)$. If so, there are two more cases to consider. The first is that m is in no incomplete path (it then is in no constructed path at all). Then we set[3] $P(\alpha^*) := (\alpha^*, \ldots, a, e, b, m, \Gamma_M(b))$ and $\ell(m) := i$. That is, we append e and its matching edge m to the incomplete path and update m's position limit. The second case is that m is included in another incomplete path $P(\tilde{\alpha}) = (\tilde{\alpha}, \ldots, \tilde{a}, \tilde{e}, b, m, \Gamma_M(b), \ldots)$. The order is in fact always $(\ldots, b, m, \Gamma_M(b), \ldots)$, since $b \in B$ and $\Gamma_M(b) \in A$; moreover it is always $e \neq \tilde{e}$. We now move b and its right wing in $P(\tilde{\alpha})$ to $P(\alpha^*)$, i.e., we set $P(\tilde{\alpha}) := (\tilde{\alpha}, \ldots, \tilde{a})$ and $P(\alpha^*) := (\alpha^*, \ldots, a, e, b, m, \Gamma_M(b), \ldots)$. We set $\ell(m) := i$ and also update the position limits of any matching edges to the right of m in the sub-path which we just moved, i.e., the next matching edge m' will get $\ell(m') := i + 1$, and so on.

When the pass is over and $i = 1$, a test is done whether we shall carry on or finish and return $\mathcal{A}$ to the main algorithm.[4] We finish when $|\mathcal{I}_{>0}| \leq \delta |M|$. We will later show that disjoint-paths cannot find more then $2|\mathcal{I}_{>0}|$ additional augmenting paths. When the pass is over and $i = k + 1$, i.e., we just completed a pass at maximal position, we do backtracking. This means to remove the last two edges from each incomplete path in $\mathcal{I}_{>0}$, i.e., if $P(\alpha) = (\alpha, \ldots, a, e, b, m, a')$ we set $P(\alpha) := (\alpha, \ldots, a)$ for each $\alpha \in \text{free}(A)$ such that $P(\alpha) \in \mathcal{I}_{>0}$. The justification for this is that an incomplete path $P(\alpha) \in \mathcal{I}_{>0}$ that could not be completed in the sweep which just finished will also not be completed in any following sweep – any admissible way of extending $P(\alpha)$ has already been tried. The removed edges will never be put at that position in $P(\alpha)$ again, due to their position limits. This gives a chance to other edges to be included there instead.

Then the next pass begins, with the next position, which is either $i + 1$ if $i < k + 1$, or 1 if $i = k + 1$.

We can think of all matching edges being to the right, at position $k + 1$, in the beginning. This is an impossible position to obtain inside of a constructed path. Then, matching edges move to the left. Each time a matching edge m is inserted into a constructed path, it moves at least one position to the left, accompanied by a decrement of its position limit $\ell(m)$. When it is removed by backtracking, its position limit is not changed, hence the edge is still eligible for being inserted in any position left of its last one. If a matching edge has reached the left end, that is, its position limit has been decreased to 1, and if it then is removed by backtracking, it will not be inserted into any constructed path again.

If an edge is in some constructed path, the only way by which it is made a non-member of any constructed path is that it is captured by backtracking in line 23. During a pass with position i, incomplete paths of length at most $2(i-2)$ or length exactly $2i$ are not changed in any way. Incomplete paths with length at least $2(i + 1)$ may be reduced, namely when they contain a matching edge (at some position right of i) that can be used to extend some incomplete path of length $2(i - 1)$.

---

[3] Strictly, we should write $\mathcal{I} := (\mathcal{I} \setminus (\alpha^*, \ldots, a)) \cup \{(\alpha^*, \ldots, a, e, b, m, \Gamma_M(b))\}$. However, in order to simplify notation, we treat $\mathcal{I}$ as a set of 'mutable' objects, to which we refer by $P(\cdot)$.

[4] We could as well do this test for all i, but it suffices for $i = 1$.

Algorithm 1. approx-maximum-matching

Input: bipartite graph $G = (A, B, E)$, parameter $\epsilon > 0$
Output: matching M

```
set k := ⌈1/ε⌉ and set δ := 1/(8k(k+1)(k+2))
set M := maximal matching
repeat
    set c := |M|
    set 𝒜 := disjoint-paths(G, δ, k, M)
    set M := M △ E(𝒜)
until |𝒜| ≤ 2δc
return M
```

Algorithm 2. disjoint-paths

Input: bipartite $G = (A, B, E), \delta > 0, k \in \mathbb{N}$, maximal matching M
Output: M-augmenting paths $\mathcal{A}$ of length at most $2k + 1$

```
forall m ∈ M do set ℓ(m) := k + 1
forall α ∈ free(A) do set P(α) := (α); set ℐ := ℐ ∪ {P(α)}
set 𝒜 := ∅ and set V' := A ∪ B
repeat
    for i := 1 to k + 1 do
        forall e = {a, b} ∈ E do /* assume a ∈ A, b ∈ B */
            if a, b ∈ V' and ∃P(α*) = (α*, ..., a) ∈ ℐ : |P(α*)| = 2(i − 1) then
                if b is a free vertex then
                    set P(α*) := (α*, ..., a, e, b)
                    set 𝒜 := 𝒜 ∪ {P(α*)} /* store */
                    set ℐ := ℐ \ {P(α*)}
                    set V' := V' \ V(P(α*))
                else if i < ℓ(m) with m = {b, Γ_M(b)} then
                    if b is in no incomplete path then
                        set P(α*) := (α*, ..., a, e, b, m, Γ_M(b))
                        set ℓ(m) := i
                    else
                        let P(α̃) = (α̃, ..., ã, ẽ, b, m, Γ_M(b), ...) ∈ ℐ
                        /* move right wing of b from P(α̃) to P(α) */
                        set P(α̃) := (α̃, ..., ã)
                        set P(α*) := (α*, ..., a, e, b, m, Γ_M(b), ...)
                        adjust ℓ-values on new edges of P(α*)
        if i = 1 and |ℐ_{>0}| ≤ δ|M| then break and return 𝒜
    forall P ∈ ℐ_{>0} do remove last two edges from P
until forever
```

5 Analysis – Approximation

The key idea is to consider a relaxation of the semi-streaming model restrictions in the following way: we allow more than a constant number of passes. This gives us the possibility to find a *maximal* set of vertex-disjoint M-augmenting paths of length at most $2k + 1$.

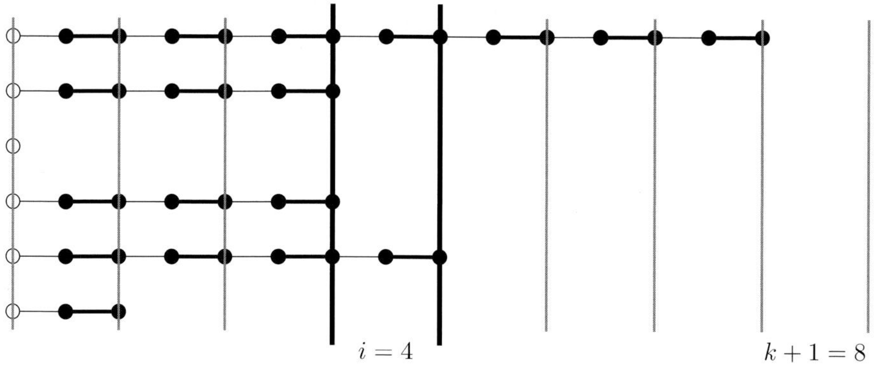

Fig. 1. A state of disjoint-paths with position $i = 4$. We call the region between the two thick vertical lines the *focus* of the sweep; the other vertical lines mark possible other positions for the sweep. There are 6 incomplete paths, 5 of them in $\mathcal{I}_{>0}$. The non-filled vertices on the left are free vertices of A. Thick edges are matching edges. Length parameter is $k = 7$, so we are looking for augmenting paths of length at most $2k + 1 = 15$, and the maximal position for the sweep is $k + 1 = 8$. Edges may only be inserted in the focus of the sweep, so for the current position $i = 4$, only paths number 2 and 4 can be extended. Edges may only be removed from paths provided that they are positioned outside of the focus and to its right. So, we can only remove edges from path number 1 here. In particular, any edges inserted in the focus will not be removed during the rest of this sweep (unless their path is completed).

Lemma 1. *If condition "$|\mathcal{I}_{>0}| \leq \delta|M|$" in line 22 is substituted by "$|\mathcal{I}_{>0}| = 0$," then algorithm* disjoint-paths *finds a maximal set of pairwise vertex-disjoint M-augmenting paths of length at most $2k + 1$.*

Proof. Let the stop condition in line 22 be "$i = 1$ and $|\mathcal{I}_{>0}| = 0$." Consider the point of time when this condition is reached and the algorithm terminates. Let us assume that there exists a path $(\alpha, e_1, v_1, m_1, \ldots, m_r, v_{2r}, e_{r+1}, \beta)$ in G with $1 \leq r \leq k$, $e_j \in M^{\complement}$ for all $j \in [r + 1]$, $m_j \in M$ for all $j \in [r]$, $\alpha \in \text{free}(A)$ and $\beta \in \text{free}(B)$, which is vertex-disjoint to all paths in $\mathcal{A}$. In the following we show that this assumption leads to a contradiction.

We show by induction over j, from $j = r$ downwards to 1, that for all $1 \leq j \leq r$ it holds that $\ell(m_j) > j$. In particular, $\ell(m_1)$ is thus greater than 1. We can then conclude as follows. In the beginning of the algorithm, free vertex α is made the starting vertex of an incomplete path. Since α is in no augmenting path in $\mathcal{A}$, its path remains incomplete until the end. In the last sweep, the stop condition is fulfilled, meaning that there are no incomplete paths of positive length left at the end of the pass with $i = 1$. Due to the test in line 13, once a matching edge is added to some incomplete path at position 1 during a sweep S, or if it is there at the start of a sweep S, it remains there for the rest of S, unless its path is completed and stored in $\mathcal{A}$. In the given situation, this means that at the start of the last pass (with $i = 1$), the incomplete path containing α has length 0, or otherwise it would still be in $\mathcal{I}_{>0}$ when the test in line 22 is made. Moreover, e_1, which has α as an endpoint, and m_1 are still in $G[V']$. So, α occurs as vertex a in some pass with b such that $\{b, \Gamma_M(b)\} = m_1$. Since, by induction, $\ell(m_1) > 1$, the test in line 13 is true in this pass, resulting in e_1 and m_1 being added behind α. This is a contradiction and the proof is finished.

We start the induction. Consider case $j = r$, i.e., the induction base. If $\ell(m_j) = k+1$, then we are done. Assume hence for contradiction that $\ell(m_j) \leq k$, which means that m_j was inserted into some constructed path – and removed again by backtracking, since in the end it is neither in any path of $\mathcal{A}$ nor in any path of $\mathcal{I}$ (since $\mathcal{I}_{>0}$ is empty). Let S be a sweep directly after which m_j is removed from its path the last time, i.e., made a non-member of any path. The removal can only happen by backtracking, line 23. This means that m_j has to be at the end of its path after the last pass of S. We consider all possible developments that can lead to this, and show that each induces a contradiction. The reader is encouraged to frequently look at Fig. 1 while reading the proof. We need three general claims. The first follows directly from the test in line 13:

> Let $m \in M$ be in some incomplete path. Position limit $\ell(m)$ does not change in a pass with position i for which $\ell(m) \leq i$. This holds in particular for such a pass in which m is inserted into an incomplete path. (+)

The second claim is a direct consequence of where the algorithm inserts edges into incomplete paths:

> For any matching edge m: if during a pass with position i, position limit $\ell(m)$ is reduced at all, it is reduced to no less than i. (++)

The third claim in particular implies that no edges may be added behind m_j in sweep S:

> Once some edges are added behind m_j (which would happen in line 15 or line 20) in some sweep, edge m_j will never be at the end of some incomplete path again during the same sweep. (∗)

Proof of (∗). Let e and $m \in M$ be edges added behind m_j in a pass with position i_0. Then $\ell(m) = i_0$ for the rest of this sweep by (+). These edges could, *during* this sweep, only be removed from this incomplete path in line 19. For this to happen in a pass with position $i \geq i_0$, it is required that $\ell(m) > i$. This is impossible since $i \geq i_0 = \ell(m)$. This proves (∗).

Recall that we consider a sweep S directly after which m_j is removed. By (∗), the following three cases are left to consider:

(1) m_j is at the end of an incomplete path when S starts and it remains at the end of an incomplete path until the end of S.

(2) $m_j = \{b_j, a_j\}$ is somewhere in the middle of an incomplete path $P(\tilde{\alpha})$ when S starts, say $(\tilde{\alpha}, \ldots, b_j, m_j, a_j, e', b', m', a', \ldots)$, and then $(e', b', m', a', \ldots)$ is removed from $P(\tilde{\alpha})$ in line 19, and henceforth m_j remains at the end of $P(\tilde{\alpha})$ until the end of S.

(3) m_j is inserted into an incomplete path *during* S and remains at its end until the end of S.

We can bring all three cases to a contradiction by the following claim:

m_j cannot be at the end of an incomplete path at the start of a pass with position $\ell(m_j) + 1$ in sweep S. (#)

Proof of (#). Assume m_j is at the end of an incomplete path $P(\alpha^*)$ at the start of a pass with position $i = \ell(m_j) + 1$. Then $|P(\alpha^*)| = 2\ell(m_j) = 2(i-1)$, and so it is eligible for being extended. However, m_j remains at the end of this path during this pass because of $(*)$ (so the path is in fact not extended), and also m_j's position limit $\ell(m_j)$ does not change because of $(+)$. This implies a contradiction immediately: when we reach e_{j+1} in the input stream, the algorithm *will* use it to complete m_j's incomplete path, with β as endpoint. This is a contradiction to that m_j does not occur in any augmenting path in $\mathcal{A}$. We have proved (#).

This immediately rules out case (1). Case (2) is also ruled out: the removal of $(e', b', m', a', \ldots)$ could only happen in a pass with position $i_0 < \ell(m') = \ell(m_j) + 1$, and then m_j would be at the end of an incomplete path at the start of all later passes, i.e., passes with positions $i_0 + 1, i_0 + 2, \ldots$. By $(++)$ this is true in particular for that pass with position $\ell(m_j) + 1$, with $\ell(m_j)$ taken at the time when the pass occurs; a contradiction by (#).

The final possibility to consider is case (3). If m_j is inserted into its path during S in a pass with position i_0, henceforward we have $\ell(m_j) = i_0$ for the rest of S by $(+)$, and so in particular the next pass has position $i_0 + 1 = \ell(m_j) + 1$. This is a contradiction by (#).

We have completed the induction base. The induction step works similar: claim (#) holds with a different proof, which uses the induction hypothesis, and the treatment of the three cases works the same. □

We can show that with the relaxation of the semi-streaming model, the algorithm would find only a small number of additional M-augmenting paths.

Lemma 2. *Let $r \geq 0$. If algorithm disjoint-paths terminates when $|\mathcal{I}_{>0}| = 0$ instead of $|\mathcal{I}_{>0}| \leq r$, then it finds at most $2r$ additional pairwise vertex-disjoint M-augmenting paths of length at most $2k + 1$.*

We apply this lemma with $r := \delta|M|$ and consider the set $\mathcal{Y}$ of M-augmenting paths which would be constructed by disjoint-paths with stop condition "$|\mathcal{I}_{>0}| = 0$".

Corollary 1. *Let $\mathcal{A}$ be the output of disjoint-paths(G, δ, k, M). Then there exists a maximal set $\mathcal{Y}$ of pairwise vertex-disjoint M-augmenting paths of length at most $2k + 1$ such that $|\mathcal{Y}| \leq |\mathcal{A}| + 2\delta|M|$.*

The following lemma, similar to [4, Lem. 1], is used in the proof of Lem. 4, which allows us to deduce the main approximation result in Thm. 2 from Cor. 1.

Lemma 3. *Let M be a maximal and M^* a maximum matching. Let $k \in \mathbb{N}$. Choose $\alpha_i \in [0, 1]$ for each $i \geq 1$ such that $\alpha_i|M|$ is the number of connected components in $M \triangle M^*$ with i edges from M and $i + 1$ edges from M^*. If $\sum_{i=1}^{k} \alpha_i \leq \frac{1}{2k(k+1)}$ then $(1 + \frac{1}{k})|M| \geq |M^*|$.*

Lemma 4. *Let M be a maximal and M^* a maximum matching. Let $\mathcal{Y}$ be a maximal set of pairwise vertex-disjoint M-augmenting paths of length at most $2k+1$ such that $|\mathcal{Y}| \leq \frac{1}{2k(k+1)(k+2)}|M|$. Then $(1+\frac{1}{k})|M| \geq |M^*|$.*

Theorem 2. *Let M^* be a maximum matching and $\epsilon > 0$. Algorithm approx-maximum-matching finds a matching M with $(1+\epsilon)|M| \geq |M^*|$.*

6 Analysis – Running Time and Space Requirements

Lemma 5. *Algorithm disjoint-paths needs at most $(k+1)^2\frac{1}{\delta}$ passes.*

Proof (Idea). It is sufficient to bound the number of sweeps. For each sweep (except the last one), we can guarantee a minimum number of edges being removed from incomplete paths. This implies a certain demand for edges being *inserted* into those paths. Each insertion, however, decreases a position limit, and this can only happen a limited number of times. □

Using this lemma, we can show the claimed bound on the number of passes. Since we only have to store $O(n)$ edges at a time, we also get the necessary bound on the space requirement.

Theorem 3. *Algorithm approx-maximum-matching needs $O\left((\frac{1}{\epsilon})^8\right)$ passes and $O(n \log n)$ bits of space.*

7 Open Questions

The most intriguing question is whether our approach can be adapted to the maximum matching problem in general graphs, while maintaining a substantially lower number of passes than McGregor's algorithm needs. It is also worth asking whether the number of passes can be reduced further for the bipartite case.

References

1. Berge, C.: Two theorems in graph theory. Proceedings of the National Academy of Sciences of the United States of America 43(9), 842–844 (1957)
2. Feigenbaum, J., Kannan, S., McGregor, A., Suri, S., Zhang, J.: On graph problems in a semi-streaming model. In: Díaz, J., Karhumäki, J., Lepistö, A., Sannella, D. (eds.) ICALP 2004. LNCS, vol. 3142, pp. 531–543. Springer, Heidelberg (2004)
3. Hopcroft, J.E., Karp, R.M.: An $n^{5/2}$ algorithm for maximum matchings in bipartite graphs. SIAM J. Comput. 2(4), 225–231 (1973)
4. McGregor, A.: Finding graph matchings in data streams. In: Chekuri, C., Jansen, K., Rolim, J.D.P., Trevisan, L. (eds.) APPROX 2005 and RANDOM 2005. LNCS, vol. 3624, pp. 170–181. Springer, Heidelberg (2005)
5. Mucha, M., Sankowski, P.: Maximum matchings via Gaussian elimination. In: 45th Symposium on Foundations of Computer Science (FOCS 2004), Rome, Italy, October 2004, pp. 248–255 (2004)
6. Muthukrishnan, S.M.: Data streams: Algorithms and applications. Foundations and Trends in Theoretical Computer Science 1(2), 67 (2005)

The Oil Searching Problem

Andrew McGregor[1], Krzysztof Onak[2,⋆], and Rina Panigrahy[3]

[1] University of Massachusetts, Amherst
mcgregor@cs.umass.edu
[2] Massachusetts Institute of Technology
konak@mit.edu
[3] Microsoft Research Silicon Valley
rina@microsoft.com

Abstract. Given n potential oil locations, where each has oil at a certain depth, we seek good trade-offs between the number of oil sources found and the total amount of drilling performed. The cost of exploring a location is proportional to the depth to which it is drilled. The algorithm has no clue about the depths of the oil sources at the different locations or even if there are any. Abstraction of the oil searching problem applies naturally to several life contexts. Consider a researcher who wants to decide which research problems to invest time into. A natural dilemma whether to invest all the time into a few problems, or share time across many problems. When you have spent a lot of time on one problem with no success, should you continue or move to another problem?

One could study this problem using a competitive analysis that compares the cost of an algorithm to that of an adversary that knows the depths of the oil sources, but the competitive ratio of the best algorithm for this problem is $\Omega(n)$. Instead we measure the performance of a strategy by comparing it to a weaker adversary that knows the set of depths of the oil sources but does not know which location has what depth. Surprisingly, we find that to find k oil sources there is a strategy whose cost is close to that of any adversary that has this limited knowledge of only the set of depths. In particular, we show that if any adversary can find k oil sources with drilling cost B while knowing the set of depths, our strategy finds $k - \tilde{O}(k^{5/6})$ sources with drilling cost $B(1 + o(1))$. This proves that our strategy is very close to the best possible strategy in the total absence of information.

1 Introduction

Consider n potential oil locations where each has oil at a certain depth. The cost of exploring a location is proportional to the depth to which it is drilled. If a location has been drilled to depth d_1 at some point in the past, then drilling it do depth $d_2 > d_1$ costs $d_2 - d_1$. The algorithm has no clue about the depths of the oil sources at the different locations or even if there are any. The natural question then is what is a good strategy to go about drilling for oil. Our abstraction of the oil searching problem

⋆ The research was initiated during a summer internship at Microsoft Research Silicon Valley. The author is supported in part by a Symantec Research Fellowship, NSF grant 0732334, and NSF grant 0728645.

A. Fiat and P. Sanders (Eds.): ESA 2009, LNCS 5757, pp. 504–515, 2009.

applies naturally to several life contexts. Consider a researcher who wants to decide which research problems to invest time into. A natural dilemma is that of whether one should invest all the time into few problems, or share time across many problems. When you have spent a lot of time at one problem with no success, should you continue or move to another problem? Is it better to continue since you already invested so much time or should you cut your losses and move to another problem? These dilemmas cut across several common decision-making processes including management strategies, investment decisions, and career changes.

While it is typical to study such an 'online' problem under competitive analysis that compares the cost incurred by an algorithm to that of an adversary that knows the depths of the oil sources, we note that in this problem the competitive ratio of the best algorithm is $\Omega(n)$; e.g., consider one location with oil at depth 1 and the other $n-1$ locations having no oil. While giving some insight, competitive analysis doesn't fully capture what constitutes a good strategy in practice. When should we give up on a research problem (drilling location) and move on to other? The approach we take here to measure the performance of a strategy is to compare its performance to a weaker adversary that knows the set of depths of the oil sources but does not know which location has what depth. Surprisingly, we find that to find k oil sources there is a strategy whose cost is close to that of any adversary that has this limited knowledge of only the set of depths. In particular, we show that if any adversary can find k oil sources in drilling cost B while knowing the set of depths, our strategy finds $k - \tilde{O}(k^{5/6})$ sources of oil in budget $B(1+o(1))$. This essentially proves our strategy is 'very close' in performance to the best possible strategy in the total absence of information. When k is constant, our strategy incurs at most $O(\log n)$ times the cost of any strategy that knows the set of depths. We show that this is the best possible ratio.

Search Games: Our problem is closely related to the class of problems known as *search games*. In general, there are two players: the Hider and the Searcher. The Hider is hiding in a space (which can be a weighted graph, or some continuous metric space), and the goal of the Searcher is to locate the Hider by traversing the space and wants to minimize the traversed distance. An extensive description of search games can be found in the book of Gal [1] or the book of Alpern and Gal [2]. A well-known example of a search game is looking for lost key on a road when we do not know in which direction the key is and at what distance. It is easy to show that a near-optimal strategy is to alternate search direction doubling the search distance iteratively. Search games find applications in robotics when a robot searches for an object in unknown terrain.

A special case of the oil searching problem, where we are only looking for one oil source has been studied as the *cow-path problem*, where a cow searches for a grass field starting at a junction of several roads. A sequence of papers resulted in an optimal solution to this problem [3,4,5] in terms of the competitive-ratio of distance travelled to the distance from the field. As we stated before, our analysis style is very different as we do not compare ourselves to an all knowing adversary. Lastly we note that, in the same proceedings, Kirkpatrick [6] also studies variations of the cow-path and oil searching problems. His focus is on finding a single oil source.

Multi-armed bandits: Our problem is also related to work on multi-armed bandits (see, e.g., [7,8,9]) and there has been work in the multi-armed bandit setting that attempts to model some of the issues that arise when drilling for oil (see, e.g., [10,11,12]). In this problem we consider several arms of a bandit each of which is has a known "state." At each time step, one of the arms can be played resulting in a random payoff according to a distribution determined by the current state of the arm. Given a current state and the payoff, the arm may deterministically move into a new state. Various objective functions are considered including maximizing the discounted reward over an infinite horizon or minimizing the regret over a finite number of steps. Our objective functions do not map to either of these objective function although, in spirit, the goal is similar.

The Problems: We are given n locations and each contains oil at some depth from $\{1, 2, 3, \ldots\}$. The set of depths of the oil is $\{d_1, \ldots, d_n\}$. The algorithm has no information about the set of depths. The adversary on the other hand knows the set of depths but does not know which location has which depth – the n locations are assigned a random permutation of these depths which is unknown to the adversary. Let $n_{\leq d}$ be the number of oil wells whose depth is at most d. The *cost of obliviousness* is the ratio between the performance (appropriately defined) of an optimal algorithm that knows the set of depths to an optimal algorithm that has no information of the depths whatsoever.

Our Contributions: We present an algorithm that expects to find $k - \tilde{O}(k^{5/6})$ sources of oil in budget $B(1 + o(1))$, where the adversary, who knows the set of depths, would expect to find at most k sources in expected budget B. We develop the algorithm in stages, first studying the problem for one oil source (in Section 2) getting a cost of obliviousness of $O(\log n)$; we also show that this is tight. Next in Section 3 we study a variant of the problem where the set of depths are chosen from a distribution; again the adversary knows the distribution and the algorithm doesn't. Section 3.1 investigates the strategy the adversary who knows the distribution should use; Section 3.2 shows how the algorithm can emulate this strategy by first trying to learn the distribution during the initial few drills. In Section 4, we generalize our algorithm to the case when the set of depths is simply a set and is not necessarily chosen from a distribution (note the subtle difference between the two cases as picking from a distribution corresponds to selecting the depth of each location independently with replacement whereas when there is a set of n depths, these depths are matched to the locations in some permutation.) Again the algorithm essentially treats the set as a distribution and makes use of the algorithm for the distribution case. In the full version we extend the problem to trees.

2 Warm-Up: Finding a Single Oil Source

In this section we consider the expected amount of drilling required to find one oil source or to find one oil source with constant probability. We show that there exists an oblivious algorithm that in expectation uses at most a factor $O(\log n)$ more drilling than an algorithm that knows the set of the oil depths. This is best possible.

We start by showing that if we wish to maximize the probability of finding a single oil source given some budget, we may restrict to algorithms that choose $k \leq n$ locations at random and drill to some set of depths at these locations.

Lemma 1. *Consider a fixed set of depths. Let $\mathcal{A}$ be an algorithm that for any assignment of depths to locations, finds oil with probability p using budget B. There is a set*

Algorithm 1. For finding a single source of oil when the depths are not known.

```
b := 1
while oil not found do
    for i = 0, ..., ⌈log n⌉ do
        Drill in min{2^i, n} random locations to depth b/2^i
    b := 2b
```

of k depths $b_1, \ldots, b_k$ satisfying $\sum b_i \leq B$ such that the probability that one finds oil by drilling to depths b_i in k different random locations is at least p.

Proof. Assume that the locations are randomly permuted before $\mathcal{A}$ starts solving the problem. This does not impact the probability with which $\mathcal{A}$ solves the problem. Consider coin tosses of $\mathcal{A}$. For some setting of coin tosses the probability of finding oil is at least p. We simulate $\mathcal{A}$ for this setting of coin tosses. We tell the algorithm that it hasn't found oil, as long as it hasn't drilled enough to find oil for all permutations of locations. We stop when $\mathcal{A}$ stops, or we know it must have found oil. This exhibits $\mathcal{A}$'s exploration pattern. We set k to the number of locations $\mathcal{A}$ drilled in. We also set b_i, $1 \leq i \leq k$, to the depths $\mathcal{A}$ drilled to in consecutive locations. Clearly $\sum b_i \leq B$, because $\mathcal{A}$ does at most B drilling. By using the same exploration pattern as $\mathcal{A}$ with locations permuted at random, and therefore, with drilling applied to random locations, one can also find oil with probability at least p. □

Lemma 2. *Drilling to depths $b_1 \geq \ldots \geq b_k$ at $k \leq n$ random, distinct locations finds oil with probability $1 - \prod_{i=1}^{k} (1 - n_{\leq b_i}/(n - i + 1))$.*[1]

Proof is deferred to the full version. Hence finding an oil source with constant probability requires $\Omega(\min_d (dn/n_{\leq d}))$ drilling.

Theorem 3. *Algorithm 1 finds a single oil source with constant probability using at most a factor $O(\log n)$ more drilling than that required by an algorithm that knows the set of the oil depths.*

Proof. Suppose all depths to powers of 2 and note that this assumption at most doubles the amount of drilling. The cost of the k-th round in Algorithm 1 is $O(2^k \log n)$, and therefore, the total cost of the first k rounds is also $O(2^k \log n)$. Let $d^* = \operatorname{argmin}_d (dn/n_{\leq d})$ and note that $\mathrm{OPT} \geq d^* n/n_{\leq d^*}$. The algorithm has found a source of oil with constant probability by the time it reaches level $\log(d^* n/n_{\leq d^*})$. But at this point at most $O(\mathrm{OPT} \log n)$ drilling has been performed. □

Theorem 4. *Any oblivious algorithm that finds a single oil source with constant probability may need to use a factor $\Omega(\log n)$ more drilling than that required by an algorithm that knows the set of the oil depths.*

Proof. Consider the following sets $\mathcal{I}_i$ of depths, for $i \in [\log n]$. $\mathcal{I}_i$ consists of 2^i oil sources at depth 2^i, and the other oil sources are at depth ∞. For each i, there is an

[1] Dynamic programming can maximize these probabilities for $\sum b_i \leq B$ in $\mathrm{poly}(B, n)$ time.

algorithm that finds oil with probability $1/2$ by drilling to depth 2^i in $n/2^i$ locations, which gives n drilling in total. Assume that each $\mathcal{I}_i$ appears with the same probability. Suppose that the oblivious algorithm finds oil with constant probability by drilling only d units in total. By Lemma 1, we can assume that the algorithm chooses a few depths d_1 to d_k that it drills to in different locations. We know that $\sum d_i = d$. The expected number of oil sources found by this strategy bounds the probability of finding at least one oil source. Let X_i be the event that drilling to depth d_i exhibits oil. We have

$$\Pr[\exists X_i = 1] \le \sum_i E[X_i] = \sum_i \sum_{1\le j\le \log d_i} \frac{1}{\lfloor \log n \rfloor} \cdot \frac{2^j}{n} \le \frac{\sum_i 2d_i}{n \cdot \lfloor \log n \rfloor} = \frac{2d}{n \cdot \lfloor \log n \rfloor}.$$

Hence, the oblivious algorithm must drill $\Omega(n \log n)$ in total, which is a factor $\Omega(\log n)$ more than the optimal strategy drills for any of the inputs. □

Remark: Similarly, one can show that for a given set of depths, Algorithm 1 uses in expectation at most $O(\log n)$ factor more than an algorithm that knows the input and minimizes the expected amount of drilling to find one oil source. Any oblivious algorithm must use $\Omega(\log n)$ more drilling in some cases.

3 Multiple Sources When Depths Are Chosen from a Distribution

If the adversary can find k oil sources in B drilling cost, the objective of the algorithm is to match this cost as well as possible. In this section, we will consider a special case where the depths of the oil sources are independently chosen from the same distribution and there are infinitely many locations to dig; again the adversary knows the distribution and the algorithm does not. In the next section we will generalize such an algorithm to the case when there are finitely many locations and the depths are not chosen from a distribution but simply assigned using a set of n depths that is known to the adversary.

3.1 Adversary's Strategy When Distribution Is Known

In this section we will show a simple procedure for minimizing the expected amount of drilling to find k oil sources for the adversary who knows the distribution. It turns out that when there are infinitely many locations, the best algorithm for the adversary is to simply to dig upto a fixed depth repeatedly; that fixed depth depends on the distribution. Consider an algorithm for the adversary that in a given location drills until a fixed depth d, unless it finds oil earlier. Let X be a random variable equal to the depth at which oil occurs. The expected payoff, i.e., the number of oil sources found by the algorithm, divided by the expected amount of drilling equals

$$g(d) = \frac{E[\text{payoff}]}{E[\text{amount of drilling}]} = \frac{\Pr[X \le d]}{d \cdot \Pr[X > d] + \sum_{1\le j\le d} j \cdot \Pr[X = j]}.$$

Define $x^\star$ to be the smallest $d \in \mathbb{Z}_+ \cup \{\infty\}$ that maximizes the above value. The next lemma shows that no algorithm exploring one location can achieve a better ratio of the expected payoff to the expected amount of drilling than $g(x^\star)$. The proof (deffered

Algorithm 2. For finding k oil sources when the depths are known.

```
while less than k oil sources found do
  pick a new location, and drill until depth x* or until oil found
```

to full version) starts by showing that any algorithm that explores one oil source is equivalent to an algorithm that selects a depth d from some distribution, and drills to d. Then, the ratio of the expected payoff to the expected amount of drilling cannot be greater than the maximum such ratio for depths in the distributions, which is bounded by $g(x^\star)$.

Lemma 5. $E[\text{payoff}]/E[\text{amount of drilling}] \leq g(x^\star)$ *for any algorithm* $\mathcal{A}$ *that explores only one location.*

The proof of the next lemma (deferred to full version) shows that even if drilling in multiple locations, the ratio of the expected payoff to the expected amount of drilling is still at most $g(x^\star)$. This gives a lower bound on the expected amount of drilling to find k oil sources.

Lemma 6. *Let* $\mathcal{A}$ *be an algorithm that finds exactly* k *oil sources.* $\mathcal{A}$*'s expected amount of drilling is at least* $k/g(x^\star)$.

Theorem 7. *Algorithm 2 minimizes the expected amount of drilling to find* k *oil sources, when the distribution is known.*

Proof. By Lemma 6 we know that the expected amount of drilling to find k oil sources is at least $k/g(x^\star)$. It suffices to show that the expected time to find one oil source is exactly $1/g(x^\star)$. Let $p = \Pr[X \leq x^\star]$, and let D be the expected amount of drilling in a new location. By definition, $p/D = g(x^\star)$. Since the depth at each location is independent, by Wald's theorem, the expected amount of drilling until one oil source is found is exactly $D \cdot \sum_{i=0}^{\infty}(1-p)^i = p/g(x^\star) \cdot 1/p = 1/g(x^\star)$. □

3.2 Extending to an Algorithm When Distribution Is Unknown

We will now show how the adversary's strategy can be extended to an algorithm that does not know the distribution. The essential idea is to estimate the distribution in the initial few drillings, and then emulate the adversary's strategy. Our oblivious algorithm finds k oil sources with probability $1-\delta$, and performs in expectation at most a factor $(1+\tilde{O}(\sqrt[5]{k^{-1}\log\delta^{-1}}))$ more drilling than the expected amount of drilling for the adversary's algorithm. Let B_1 be the min. expected amount of drilling to find one oil source.

Lemma 8. *There is an algorithm that approximates* B_1 *up to a factor of* $O(\log B_1)$ *with probability* $1-\delta$. *The expected amount of drilling is* $O(B_1 \cdot \log B_1 \cdot \log(1/\delta))$.

Proof. Note that the minimum expected budget to get one oil source and the minimum budget to get a single oil source with probability $1/2$ are within a constant factor. Thus,

Algorithm 3. For finding a single source of oil when the distribution is unknown.

```
Let b := 1
while no oil found do
    For each i ∈ {1, ..., log b}, drill 2^i locations up to depth b/2^i
    b := √2 · b
Return b
```

it suffices to approximate the latter. To do this we run Algorithm 3 a total of $O(\log \delta^{-1})$ times. There is a constant C_1 such that the probability that we find oil using a budget $b \leq C_1 \cdot B_1 / \log B_1$ is less than $1/4$. Otherwise, we could get a better minimum expected budget to find a single oil source. On the other hand, there is a constant C_2 such that the probability that $b \geq C_2 \cdot B_1$ is less than $1/4$. Hence, by the Chernoff bounds, if we have $O(\log(1/\delta))$ samples and take the median of them, we get a value that is between $C_1 \cdot B_1 / \log B_1$ and $b \geq C_2 \cdot B_1$ with probability $1 - \delta$. Once b gets greater than some constant times B_1, it finds oil in each iteration with probability greater than $1/2$. Since the total budget spent grows by a factor smaller than and bounded away from 2, the expected amount of drilling in each run of the auxiliary algorithm is $O(B_1 \log B_1)$. □

Fact 9. *Let X be a Bernoulli variable and let $\delta, \epsilon \in (0,1)$. $O(\delta^{-1}\epsilon^{-2})$ samples of X suffice to distinguish* $\Pr[X = 0] \leq \delta$ *from* $\Pr[X = 0] \geq \delta(1+\epsilon)$ *with probability* $2/3$.

Lemma 10. *Let t satisfy* $\Pr[X > t] \leq \epsilon$ *and $x^\star > t$. Then, $g(t) > g(x^\star) \cdot (1 - \epsilon)$.*

Proof. Let T and D be the expected amount of drilling if we stop at depth t and $x^\star$ respectively. Then $g(t) = (1 - \Pr[X > t])/T \geq (1-\epsilon)/D \geq (1-\epsilon)g(x^\star)$. □

Lemma 11. *There is an algorithm that finds t such that $g(t) \geq g(x^\star)(1-\epsilon)$ with probability $1-\delta$. The expected amount of drilling used is $\tilde{O}(B_1\epsilon^{-4} \cdot \log(1/\delta))$.*

Proof. We first run the algorithm of Lemma 8 to approximate the minimum expected budget B_1 required to discover a single oil source. We get a budget upper-bound $b = O(B_1 \log B_1)$. We now argue that our desired t is $O(B_1 \log B_1 \log(1/\epsilon))$. By Lemma 10, we need not care about big depths. Since we can find oil with probability at least $1/2$ up to depth $O(B_1 \log B_1)$ using a budget of at most b, it makes no sense to drill deeper than $d = O(B_1 \log B_1 \log(1/\epsilon))$ to find oil with probability higher than $1-\epsilon$ in any location, since there is a more effective method that explores only small depths. Next we round up each drilling depth to a power of $(1+\epsilon)$. This decreases $g(x)$ by at most a factor of $1+\epsilon$ for any x, We now have $s = \log_{1+\epsilon} d$ possible values for t. We have

$$\frac{g((1+\epsilon)^i)}{1+\epsilon} \leq \frac{\Pr[X \leq (1+\epsilon)^i]}{1 + \sum_{0 \leq j \leq i-1} \epsilon(1+\epsilon)^j \cdot \Pr[X \geq (1+\epsilon)^j]} \leq g((1+\epsilon)^i) .$$

Now using $O(d\epsilon^{-4} \log(s/\delta))$ drilling, we learn for each of the possible choices $(1+\epsilon)^i$ for t the probability $\Pr[X \geq (1+\epsilon)^i]$ up to an additive term of ϵ^2. If the probability is smaller than ϵ for some $(1+\epsilon)^i$, we can assume by Lemma 10 that t is bounded by $(1+\epsilon)^i$. Otherwise, our approximation gives a good multiplicative approximation for the denominator of $g((1+\epsilon)^i)$.

It remains to estimate the numerator. We are only interested in i such that the probability of finding oil up to depth $(1+\epsilon)^i$ using a total budget of b is $\Omega(1)$. This means that we are interested in $(1+\epsilon)^i$ such that $\Pr[X \leq (1+\epsilon)^i] = \Omega((1+\epsilon)^i/b)$. To approximate each $\Pr[X \geq (1+\epsilon)^i]$ up to a factor of $1+\epsilon$, given it is larger than $\Omega((1+\epsilon)^i/b)$, it suffices to use $O\left(s(\log s)b\epsilon^{-2}\log(s\delta^{-1})\right)$ drilling by Fact 9. Hence with $\tilde{O}(B_1\epsilon^{-4}\log(1/\delta))$ drilling we find t with $g(t) \geq g(x^\star)/(1+\epsilon)^{O(1)}$ with probability $1-\delta$. Rescaling ϵ we get the required approximation. □

Corollary 12. *There is an algorithm that with probability* $1-\delta$ *uses* $B\left(1+\tilde{O}\left(\sqrt[5]{\log(1/\delta)k^{-1}}\right)\right)$ *drilling in expectation to find* k *oil sources, while the minimum expected amount of drilling to find* k *oil sources is* $B = kB_1$.

Proof. We use the algorithm described in the proof of Lemma 11 to find a t such that $g(t)$ approximates $g(x^\star)$. By running an algorithm that drills up to depth t, with probability $1-\delta$, the expected amount of drilling we use to find k oil sources is $kB_1(1+\epsilon)+\tilde{O}\left(B_1\epsilon^{-4}\cdot\log(1/\delta)\right)$. Setting $\epsilon = \sqrt[5]{\log(1/\delta)/k}$ gives the result. □

4 Generalizing to an Arbitrary Set of Depths

We will now genaralize the algorithms to the case when the depths of the locations are a (random) permutation of a set of n depths that is known only to the adversary.

4.1 Adversary's Strategy When Set of Depths Is Known

In this section we will study algorithms for the case when the set of depths are known but not the depth of each oil source separately. We will provide an efficient (polynomial time) algorithm for the adversary whose perfomance is close to that of the best possible (perhaps exponential time) algorithm. If any algorithm finds k' oil sources in expectation using B amount of drilling in expectation then the proposed algorithm can find k sources in expectation using $B(1+\epsilon)$ expected drilling where $k' \leq k + \tilde{O}(\sqrt{k/\epsilon})$.

Assume that all the depths in the input instance are rounded up to powers of $(1+\epsilon)$; this only increases the budget by a factor of $(1+\epsilon)$. Denote different depths by $H_i = (1+\epsilon)^i$ for $i \leq r = O(\epsilon^{-1}\log B)$. Define $h_i = H_i - H_{i-1}$, where $H_{-1} = 0$.

We will say that (B,k) is an *achievable* solution if there exists an algorithm whose expected drilling is B and the expected number of oil sources found is k. We say that the pair (B,k) is *feasible* in the following program if the program has a solution. We write n_d and $n_{\geq d}$ to denote the number of oil sources at depth exactly d and at least d, respectively. Here $m_i = n_{H_i}/n_{\geq H_i}$. x_i corresponds to the number of locations that we drill from depth H_{i-1} to H_i, which results in expectation in $x_i m_i$ oil source discoveries.

$$k \leq \sum \min(m_i x_i, x_i - x_{i+1}); \quad B \geq \sum h_i x_i; \quad x_r \leq \ldots \leq x_0 \leq n \qquad (1)$$

The first inequality corresponds to stating that at least k oil sources are found in expectation. Note that $m_i x_i$ is the expected number of oil sources found while drilling from depth H_{i-1} to H_i; however it cannot exceed $x_i - x_{i+1}$. The second inequality

Algorithm 4. For finding k oil with a known set of depths in budget B.

1 Compute a (B, k) solution to Program 2.
2 Start drilling x_0 random locations to depth h_0.
3 Let A_i be the number of locations drilled to depth H_i where no oil was found. Among these, choose $\min(x_{i+1}, A_i)$ at random and drill these to depth H_{i+1}.

ensures that the total budget is at most B. The third simply ensures that the solution is meaningful as the number of wells drilled to increasing depths must be decreasing. We show (proof deferred to the full version) that Program 1 is equivalent to:

$$k \leq \sum m_i x_i; \quad B \geq \sum h_i x_i; \quad \forall\, i \in [r-1] : x_{i+1} \leq (1 - m_i)x_i; \quad x_0 \leq n \quad (2)$$

Lemma 13. *If (B, k) is feasible in Program 2 then Algorithm 4 finds k oil sources in expectation while spending budget B.*

Proof. Consider Algorithm 4 and let Y_i be the number of locations that we drill from depth H_{i-1} to H_i. Since $Y_i \leq x_i$ the total amount of drilling is at most B. We will argue that the expected number of oil sources it finds is at least k. To prove this, we alter the algorithm so that even if less than x_i locations are available to drill to depth H_i as we have found oil in some of them, we will pretend to also continue drilling $x_i - Y_i$ locations where we found oil earlier. This can only increase the cost but will not change the number of oil sources found. The number of locations drilled to depth H_i is exactly x_i. The number of oil sources G_i found at depth H_i is $x_i - A_i$. Now, $E[G_i] = x_i - E[A_i]$. But $E[A_i] = (1 - m_i)E[Y_i] \leq (1 - m_i)x_i$. So $E[G_i] \geq m_i x_i$, and hence, $\sum_i E[G_i] \geq \sum_i m_i x_i \geq k$. □

Next we will lower bound the performance of the best possible algorithm. We will show that if (B, k) is achievable, then it must be feasible in the following program.

$$k \leq \sum_i \min((m_i + \epsilon_i)x_i + \epsilon'_i, x_i - x_{i+1}); \; B = \sum_i h_i x_i; \; x_{i+1} \leq \ldots \leq x_0 \leq n \quad (3)$$

where $\epsilon_i = \tilde{O}(\sqrt{m_i/n_{\geq H_i}} + 1/n_{\geq H_i})$ and $\epsilon'_i = m_i/(n_{\geq H_i})^{\Omega(1)}$ as guaranteed by the following lemma (proof deferred to the full version) for $n_{\geq H_i}$ boxes with $m_i n_{\geq H_i}$ of them containing gold.

Lemma 14 (Boxes of Gold). *Consider n boxes of which pn contain a gold ingot. For any randomized algorithm that opens B boxes and finds G ingots, $E[G] \leq (p + \epsilon)E[B] + \epsilon'$ where $\epsilon = \tilde{O}(\sqrt{p/n} + 1/n)$, $\epsilon' = p/n^{\Omega(1)}$.*

Lemma 15. *If (B, k') is an achievable solution for an instance of the oil searching problem with known depths then it is a feasible solution for Program 3*

Proof. Assume that there is an algorithm $\mathcal{A}^*$ that spends budget B in expectation and finds k' oil in expectation. Let $x_i = E$[number of wells drilled to depth at least H_i]. Let $g_i = E$[number of wells found at depth H_i]. Then, by Lemma 14, $g_i \leq (m_i + \epsilon_i)x_i + \epsilon'_i$

as we can think of each location with the depth at least H_i as a box and the ones with oil at depth H_i as boxes containing gold. Also clearly $g_i \leq x_i - x_{i+1}$ as we can continue drilling to depth H_{i+1} only if we have not found gold already. So $k' = \sum g_i \leq \sum_i \min((m_i + \epsilon_i)x_i + \epsilon'_i, x_i - x_{i+1})$ and $B = \sum_i h_i x_i$. □

Lemma 16. *If (B, k') is feasible for Program 3 then (B, k) is feasible for Program 1 where $k' \leq k + \tilde{O}(\sqrt{k/\epsilon})$.*

Proof. Consider the feasible solution in Program 3 that satisfies $k' \leq \sum_i \min((m_i + \epsilon_i)x_i + \epsilon'_i, x_i - x_{i+1})$. Substituting this solution in Program 1 is feasible if $k = \sum_i \min(m_i x_i, x_i - x_{i+1})$. Let Q denote the set $\{i : (m_i+\epsilon_i)x_i+\epsilon'_i \leq x_i-x_{i+1}\}$. Now $k' - k \leq \sum_{i\in Q} x_i \epsilon_i + \epsilon'_i \leq \tilde{O}(\sum_{i\in Q}\sqrt{m_i x_i})) + o(1)$. Since $\sum_{i\in Q} m_i x_i \leq k'$, this is at most $\tilde{O}(\sqrt{k'}\, r) = \tilde{O}(\sqrt{k/\epsilon})$. □

The above lemmas, with the depth rounding, gives the main theorem of this section:

Theorem 17. *If (B, k') is an achievable solution for an instance of the oil searching problem with known depths then Algorithm 4 finds in expectation k oil sources in expected cost $B(1 + \epsilon)$ where $k' \leq k + \tilde{O}(\sqrt{k/\epsilon})$.*

4.2 Extending to an Algorithm That Does Not Know the Set of Depths

We now study algorithms that are oblivious to the set of depths. We compare our solution to that which can be achieved by the adversary with knowledge of the set of depths but not the depth of each source separately. We show that if an adversary that knows the set of depths, expects to find k' oil sources, and expected to perform B drilling in expectation, then our algorithm expects to find k oil sources where $k' \leq k + \tilde{O}(k^{5/6})$ sources and peforms $B(1 + o(1))$ drilling in expectation.

Let us view the set of depths S as a distribution $D(S)$ obtained by picking a random location from the set S. Any algorithm that drills one location at random from the set S is an algorithm that drills one location with depth from the distribution $D(S)$. Now we know from the results in Section 3.1 that the optimal k/B is obtained by drilling up to depth $d = x^*$, which maximizes $g(d)$. Let us say that a solution to Program 2 is tight if the first two inequalities involving B and k are equalities. Any tight solution to Program 2 can be converted to an algorithm for drilling locations from distribution $D(S)$ achieving and vice-versa. For instance, an algorithm that drills all locations up to depth $d = H_s$ can be realized by setting $x_0 = n$ and $x_{i+1} = (1-m_i)x_i$, for all $i+1 \leq s$, and $x_{i+1} = 0$, for all $i + 1 \geq s + 1$, in Program 2. Let (B_s, k_s) denote the values of B and k in this tight solution. Similarly we can view a tight solution to Program 2 as a strategy to drill one location drawn from the distribution $D(S)$ by scaling all x_i by x_0 resulting in $(B/x_0, k/x_0)$ expected budget and payoff. The strategy continues drilling from H_i to H_{i+1} with probability $x_{i+1}/((1-m_i)x_i)$. Clearly $k_s/B_s = g(H_s)$. The algorithm of Lemma 11 can be used to compute a near optimal depth H_t so that $g(H_t) \geq g(x^*)(1 - \epsilon)$. This is done by sampling locations from S with replacement.

Lemma 18. *(B, k) is a tight solution for Program 2 if and only if it can be expressed as a linear combination $\sum_i \alpha_i(B_i, k_i)$ such that $\sum_i \alpha_i \leq 1$.*

Proof. Without loss of generality we may assume that $x_0 = n$ as any solution to Program 2 can be scaled to satisfy this. We know that any such solution (B, k) to Program 2 can be viewed as an algorithm to drill one location chosen from distribution $D(S)$ achieving expected budget and payoff: $(B/n, k/n)$. We know from the proof of Lemma 5 that any such algorithm can be viewed as a distribution or linear combination of the strategies that always drill a location up to H_i. The latter result corresponds to $(B_s/n, k_s/n)$ expected budget and payoff. The only if part is trivial. □

Theorem 19. *If (B, k) is a tight solution for Program 2, then there is a $(B, k(1-\epsilon))$ solution for Program 2 such that $x_{i+1} = (1 - m_i)x_i$, for all $i \leq t-1$ and either $x_0 = n$, or $x_t > 0$. This essentially corresponds to a solution that first explores all locations to depth up to H_t, and only then drills to greater depths.*

Proof. Express (B, k) as $\sum_i \alpha_i(B_i, k_i)$. The essential observation is that shifting budget to α_t from other α_i can only increase k. It suffices to show that we can convert the (B, k) solution to a $(B, k(1-\epsilon))$ solution, where either $\alpha_i = 0$, for all $i < t$, and $\sum_i \alpha_i = 1$, or only α_t is non-zero and others 0. Assume first for simplicity that $g(H_t) = g(x^*)$. Hence transferring budget from any other (B_i, k_i) to (B_t, k_t) only increases k. However we need to respect the constraint $\sum_i \alpha_i \leq 1$. Also note that transferring budget from components $i < t$ only decreases $\sum_i \alpha_i$ as B_i is non-decreasing in i. So we can always move to an improved (or same quality) solution where $\alpha_i = 0$ for all $i < t$. Further, as long as $\sum_i \alpha_i < 1$ we can transfer budget from higher components $i > t$ till $\sum_i \alpha_i$ hits 1. We stop only if either $\alpha_i = 0$ for all $i \neq t$ or $\sum_i \alpha_i = 1$ and $\alpha_i = 0$ for all $i < t$. Now if $g(H_t) = g(x^*)(1-\epsilon)$ then each transfer of budget will have a rate of return that is smaller by a factor of ϵ. So total payoff loss is at most ϵk. □

Remark: Observe that the above theorem also implies that there is an optimal solution to 2 that first spends all its budget drilling to x^* and then recursively solves the remaining instance. The solution is thus to drill according to a sequence of depths x^*'s for the different instances until either all locations are drilled or budget is exhausted. Also (assuming $g(H_t) = g(x^*)$) the $g(x^*)$'s found in the different recursions is non-increasing as otherwise there is a better value of x^* at the recursion after which it increased. Further if we use H_t at each recursive step instead of x^* the total loss in payoff is at most ϵk as the budget moved in the different recursions is disjoint (the budget moved in a recursion is never moved again in another recursion). If $\tilde{B}$ is the expected budget and $\tilde{k}$ is the expected payoff used in a recursion then $\tilde{B}/\tilde{k} = g(H_t)$.

See Algorithm 5. The depth of the recursion is at most $r = \log G/\epsilon$ as there are only r distinct depths. This gives:

Lemma 20. *If (B, k) is a feasible solution for Program 2 then Algorithm 5 finds $k(1-\epsilon)$ oil sources in expectation with budget B (ignoring the cost of computing $g(H_t)$).*

We also need to take into account the cost $O(\epsilon^{-4} \log(1/\delta)/g(x^*))$ of computing $g(H_t)$ at the beginning each recursive step. Besides this the only non-determinism in the budget used and payoff is in the last step of the recursion; at previous steps both are fixed as we are dealing with a set of depths.

Algorithm 5. Finding multiple sources with budget B and unknown set of depths.

1 Treating the set of depths as a distribution, compute t such that $g(H_t) \geq g(x^*)(1-\epsilon)$. This is done with the algorithm (of Lemma 11) for estimating $g(x^*)$ on the distribution $D(S)$ by sampling locations with replacement.
2 Drill all locations to depth H_t (except those that hit oil earlier) unless budget is exhausted. If budget is left over we have found all oil sources at depth at most H_t. The remaining locations are all drilled to depth H_t.
3 In such a case recursively explore the remaining locations with the remaining budget.

Theorem 21. *If* (B, k') *is an achievable solution for the oil searching problem with known depths then with probability* $1 - r\delta$, *algorithm 5 finds in expectation* $k(1-\epsilon) - \tilde{O}(\log(1/\delta)/\epsilon^5)$ *sources in expectation, where* $k' \leq k + \tilde{(}\sqrt{k/\epsilon})$, *and spends* $B(1+\epsilon)$ *budget. For* $\epsilon = o(k^{-1/6})$, *this amounts to* $k - \tilde{O}(k^{5/6})$ *sources in budget* $B(1+o(1))$.

Proof. If (B, k') is a feasible solution for the oil searching problem with known depths then there is a (B, k) solution to Program 2 where $k' \leq k + O(\min\{k, \sqrt{kr}\})$. To bound the cost of computing $g(H_t)$, we increase B by a factor of ϵ and in each recursion only allocate at most ϵ fraction of the remaining budget to computing $g(H_t)$. If cost of computing $g(H_t)$ is more than ϵ fraction of remaining budget in a certain recursion, then we can stop the algorithm as this means $O(\log(1/\delta)/(\epsilon^5 g(x^*)) \geq \epsilon B$, which means we can find at most $g(x^\star)B \leq O(\log(1/\delta)/\epsilon^5)$ more sources in expectation. Thus, if we stop, we are only giving up $O(\log(1/\delta)/\epsilon^5)$ sources. □

References

1. Gal, S.: Search Games. Academic Press, London (1980)
2. Alpern, S., Gal, S.: The Theory of Search Games and Rendezvous. Kluwer Academic Publishers, Dordrecht (2003)
3. Baeza-Yates, R.A., Culberson, J.C., Rawlins, G.J.E.: Searching in the plane. Inf. Comput. 106(2), 234–252 (1993)
4. Kao, M.Y., Reif, J.H., Tate, S.R.: Searching in an unknown environment: An optimal randomized algorithm for the cow-path problem. Inf. Comput. 131(1), 63–79 (1996)
5. Kao, M.Y., Ma, Y., Sipser, M., Yin, Y.L.: Optimal constructions of hybrid algorithms. J. Algorithms 29(1), 142–164 (1998)
6. Kirkpatrick, D.: Hyperbolic dovetailing. In: Fiat, A., Sanders, P. (eds.) ESA 2009. LNCS, vol. 5757, pp. 516–527. Springer, Heidelberg (2009)
7. Robbins, H.: Some aspects of the sequential design of experiments. Bull. Amer. Math. Soc. 58, 527–535 (1952)
8. Gittins, J.C., Jones, D.M.: A dynamic allocation index for the sequential design of experiments. In: Progress in Statistics: European Meeting of Statisticians, vol. 1, pp. 241–266 (1974)
9. Whittle, P.: Arm-acquiring bandits. The Annals of Probability 9, 284–292 (1981)
10. Benkherouf, L., Pitts, S.: On a multidimensional oil exploration problem. Journal of Applied Mathematics and Stochastic Analysis 2005(2), 97–118 (2005)
11. Benkherouf, L., Glazebrook, K., Owen, R.: Gittins indices and oil exploration. J. Roy. Statist. Soc. Ser. B 54, 229–241 (1992)
12. Grayson, C.: Decisions Under Uncertainty: Drilling Decisions by Oil and Gas Operators. Harvard, Division of Research, Graduate School of Business Administration (1960)

Hyperbolic Dovetailing

David Kirkpatrick

University of British Columbia, Vancouver, Canada
kirk@cs.ubc.ca

Abstract. A familiar quandary arises when there are several *possible* alternatives for the solution of a problem, but no way of knowing which, if any, are viable for a particular problem instance. Faced with this uncertainty, one is forced to simulate the parallel exploration of alternatives through some kind of co-ordinated interleaving (*dovetailing*) process. As usual, the goal is to find a solution with low total cost. Much of the existing work on such problems has assumed, implicitly or explicitly, that at most one of the alternatives is viable, providing support for a competitive analysis of algorithms (using the cost of the unique viable alternative as a benchmark). In this paper, we relax this worst-case assumption in revisiting several familiar dovetailing problems.

Our main contribution is the introduction of a novel process interleaving technique, called *hyperbolic dovetailing* that achieves a competitive ratio that is within a logarithmic factor of optimal on *all* inputs in the worst, average and expected cases, over all possible deterministic (and randomized) dovetailing schemes. We also show that no other dovetailing strategy can guarantee an asymptotically smaller competitive ratio for all inputs.

An interesting application of hyperbolic dovetailing arises in the design of what we call *input-thrifty* algorithms, algorithms that are designed to minimize the total precision of the input requested in order to evaluate some given predicate. We show that for some very basic predicates involving real numbers we can use hyperbolic dovetailing to provide input-thrifty algorithms that are competitive, in this novel cost measure, with the best algorithms that solve these problems.

1 Introduction

You are trapped underground in a mine following a massive earthquake. While there are many potential escape routes, you have no way of knowing how much effort will be required to clear any one of them. You quickly realize the first exploration strategies that come to mind might be poor choices for your particular situation. If it happens that all of the escape routes have about the same amount of debris, then you may as well simply choose one and start digging. On the other hand, if only a few of routes are viable, it makes sense to keep trying all of the routes, more or less equitably, until one of these few is discovered.

Upon further reflection, you recognize that being trapped in this way is not such an uncommon occurrence. Being more adept at thinking than digging, you wonder if there are strategies that are arguably good, or even best, to adopt in situations like this...

A. Fiat and P. Sanders (Eds.): ESA 2009, LNCS 5757, pp. 516–527, 2009.

1.1 Dovetailed Execution of Multiply-Viable Process Sets

Finding an escape route in a mine has much in common with problems that arise in many computational settings where several possible avenues are available for the solution of a problem but there is no way of knowing which, if any, are viable (or adequately efficient) for a particular problem instance. Faced with this uncertainty, we elect to simulate their parallel exploration through some kind of co-ordinated interleaving (*dovetailing*) of their associated processes. The goal, of course, is to find a solution with low total cost. (Examples include geometric or graph search, the synthesis of hybrid algorithms based on a suite of heuristics, and adaptive raising strategies.) Existing work on such problems invariably makes the assumption that at most one of the processes is viable. This provides support for a competitive analysis of algorithms (using the cost of running the unique viable process alone as a benchmark). Algorithms with optimal competitive ratios, based on variants of round-robin doubling search, have been formulated for a large variety of such problems [2,6,8,9,10,13,16].

Competitive analysis was introduced to provide a more realistic alternative to worst-case analysis, which in our setting would make all strategies equivalent since each could, with sufficiently bad input, be forced to run an arbitrarily long time. However, it is similarly unrealistic in many scenarios, including those mentioned above, to adopt the worst-case assumption that at most one of the underlying process is viable.

In this paper, we relax this assumption in revisiting some of these dovetailing problems. Our contributions are of four types: (i) we formulate a natural notion of intrinsic cost that provides a basis for a modified form of competitive analysis that can be applied in this more general setting; (ii) we introduce a non-uniform process interleaving technique, called *hyperbolic dovetailing*, that can be viewed as a hybrid (or mixture) of a family of bounded depth-first strategies covering the full spectrum between a pure breadth-first (i.e. round robin) and a pure depth-first strategy, (iii) we prove that hyperbolic dovetailing achieves a competitive ratio that is within a logarithmic factor of optimal on *all* inputs in the worst, average and expected cases, over all possible deterministic (and randomized) dovetailing schemes, and (iv) we prove that no other dovetailing strategy can guarantee an asymptotically smaller competitive ratio for all inputs.

Our work was motivated, in part, by the need to develop efficient algorithms in computational settings in which knowledge about the input is incomplete but extendable. For instance, input numbers may initially be known only up to some precision, and additional precision may be obtainable, but only at a high cost. Here are two examples for such a scenario: 1) the input numbers are initially the result of cheap measurements, and with further, more elaborate measurements additional precision can be obtained; 2) the input numbers are produced bit by bit by a computational process such as root-finding via bisection. In such scenarios it is of interest to design algorithms that solve the problem using as little total input precision as possible.

We show that the hyperbolic dovetailing technique can be applied to the design of algorithms for certifying certain simple properties of a set of input numbers, with the goal of minimizing the number of input bits that need to be examined by the algorithm, for each set of possible inputs. We refer to this as the *leading-input-bits-cost*, or LIB-cost, of the algorithm. Algorithms whose LIB-cost is "small", in some quantifiable sense, relative to the intrinsic LIB-cost for every input, are said to be *input-thrifty* algorithms.

1.2 Multi-list and Cow-Path Traversal Problems

The following *multi-list traversal* problem captures the essence of the mine-escape-route search problem; it is a simplified version of other multi-process dovetailing problems that we wish to study and, as such, it allows us to introduce our strategies and analyses in their most basic setting. Let $\mathcal{L}$ be a given multi-list, that is a sequence of m not-necessarily-finite-length lists. Let λ_i denote the length of the i-th longest list in $\mathcal{L}$, so $\infty \geq \lambda_1 \geq \lambda_2 \geq \ldots \geq \lambda_m$. The goal is to traverse at least one list to its end, while minimizing the total *cost* (the number of list positions examined). In general, we assume that the (multi)set of list lengths $\Lambda = \{\lambda_1, \lambda_2, \ldots, \lambda_m\}$ is *not* known to the strategy, and we do worst-case or average-case analysis over the set of all multi-lists with associated lengths set Λ (what we refer to as *presentations* of Λ). However, for the sake of comparison, we also consider the the behaviour of strategies that know Λ (so, they are not constrained to run efficiently, or even terminate correctly, for multi-lists outside of this restricted class).

Our multi-list traversal problem bears a strong resemblance to what has come to be known as the *m-lane cow-path problem*. An instance of the cow-path problem specifies a sequence of m rays (*lanes*) of unbounded length incident on a common origin (*crossroad*). A goal (*pasture*) lies at some unknown distance d from the origin along some (unknown) ray. The objective is to to formulate a provably good strategy for an agent (*cow*) to reach the goal, starting from the origin.

The cow-path problem was introduced by Baeza-Yates *et al.* [2] as a simple graph-theoretic abstraction of search problems in the plane. It has been studied in several variations including directionally dependent traversal costs, turnaround penalties, shortcuts and dead-ends [6,8,12,13,15]. It has also been analysed in terms of worst-case and average-case competitive ratio (using the distance d as a benchmark), as well as in a game-theoretic framework [2,9,10,16,17,18].

Essentially the same ideas as those used in solving the cow-path problem have been used in the synthesis of deterministic and randomized hybrid algorithms with optimal (or near optimal) competitive ratios [1,9]. The setting is one in which there are a number of *basic* algorithms which might (or might not) be useful in solving some problem. The goal is to synthesize a hybrid algorithm from these basic components by some kind of dovetailing process. In this context, memory limitations may impose restrictions on the number of processes that can be suspended at any given time (the alternative being a complete restart with successively larger computation bounds).

As we have already suggested, the multi-list traversal problem provides a natural generalization of the cow-path problem in which *every ray* leads to a goal (perhaps arbitrarily far from the origin) Of course, the list traversal cost is not the same as that employed in the cow-path problem; in particular, it does not take into account the cost of re-traversal of paths. However, as we shall see later, our multi-list traversal algorithm can be implemented in such a way that the two traversal cost functions agree to within a small constant factor, without increasing the total list traversal cost by more than a small constant factor.

1.3 Competitive Analysis

In order to formulate a compelling notion of competitiveness (in the spirit of [19]) for our multi-list traversal problems, we need to specify some measure of *inherent complexity* that permits us to distinguish easy from more difficult problem instances. The objective, of course, is to formulate *adaptive* strategies that run relatively more efficiently on instances of relatively low inherent complexity.

One natural candidate is simply the length of the shortest list. Since this corresponds to the length of the shortest proof that some specified list has some specified length, we refer to this as the *minimum certification cost* of the instance. Clearly this provides a lower bound on the cost of any multi-list traversal strategy, and is realizable by a strategy that knows (or correctly guesses) the identity of the shortest list. It also coincides with the notion of inherent complexity that underlies the competitive analysis used for the conventional m-lane cow-path problem.

In general, however, traversal strategies must deal with uncertainty in both the *set* Λ of numbers that correspond to the path lengths (and completely determine the minimum certification cost) and the *presentation* of these numbers as a sequence (that is their assignment to individual paths). For a given set Λ, we specify the maximum-traversal-cost of an algorithm $\mathcal{A}$ *for* Λ, and then define the intrinsic maximum-traversal-cost of Λ to be the minimum, over all algorithms $\mathcal{A}$ that are only guaranteed to solve the problem for multi-lists that are presentations of Λ, of the maximum-traversal-cost of $\mathcal{A}$ for Λ. These definitions resemble refinements of competitive analysis (in particular, the so-called relative worst order ratio) introduced to provide a more meaningful/sensitive analysis for certain on-line algorithms [3,4,5,11]. (See [7] for a comprehensive overview of these alternative measures.)

It turns out to be reasonably straightforward to specify the intrinsic maximum-traversal-cost of an arbitrary input set Λ in this way. With this in hand, we do competitive analysis of two types: (i) with respect to the family of algorithms that are constrained to answer correctly only when the input multi-list is a presentation of Λ (i.e. relative to the intrinsic maximum traversal-cost); and (ii) with respect to the family of algorithms that are constrained to answer correctly only when its input multi-list is a presentation of some input set Λ' whose intrinsic maximum-traversal-cost is the same as that of Λ.

In the former case, we show that our algorithms are competitive to within a logarithmic factor. More precisely, for every set Λ, if $\xi(\Lambda)$ denotes the intrinsic maximum-traversal-cost of input set Λ, then our algorithm achieves maximum-traversal-cost $O(\xi(\Lambda) \log \min\{\xi(\Lambda), |\Lambda|\})$. In the latter case, we show that our algorithms are competitive in maximum-traversal-cost to within a constant factor. We also develop similar results for average-traversal-cost.

Results that have striking similarities to those in this paper were presented by Luby *et al.* [14] for the problem of minimizing the expected time to complete the execution of Las Vegas algorithms (which can be viewed as an infinite sequence of deterministic algorithms with unknown completion times).

2 Multi-list Traversal Strategies

Let Λ be a (multi)set of m (positive) list lengths, and let λ_i denote the ith largest element of Λ, for $1 \leq i \leq m$. We denote by $\mathcal{L}$ a generic presentation of Λ (that is, a multi-list whose associated set of list lengths coincides with Λ).

2.1 Intrinsic Maximum-Traversal-Cost and Average-Traversal-Cost

We define the maximum-traversal-cost of some multi-list traversal strategy $\mathcal{A}$ on Λ to be the maximum cost of $\mathcal{A}$ over all multi-list presentations of Λ. The intrinsic maximum-traversal-cost of Λ, denoted $\xi(\Lambda)$, is the minimum, over all multi-list traversal strategies $\mathcal{A}$ of the maximum-traversal-cost of $\mathcal{A}$ on Λ.

Theorem 1. $\xi(\Lambda) = \min_{1 \leq i \leq m} i\lambda_i$.

Proof. Consider any multi-list traversal strategy $\mathcal{A}$ that works correctly on all presentations of Λ. Such a strategy specifies a sequence of traversal steps that culminates in the discovery of the end of some list. At any point in time, up to termination, the strategy has traversed the i-th list to some depth d_i. Provided that $\hat{d}_i$, the i-th largest element of $\{d_1, \ldots, d_m\}$, satisfies $\hat{d}_i < \lambda_i$, for $1 \leq i \leq m$, the strategy will have not terminated for at least one multi-list presentation of Λ. Thus, at termination, we must have $\hat{d}_i = \lambda_i$, for some i, and $\hat{d}_j \geq \lambda_i$, for all j, $1 \leq j < i$. It follows that $\mathcal{A}$ must explore at least $\min_{1 \leq i \leq m} i\lambda_i$ list positions.

On the other hand, if $i_\Lambda = \text{argmin}_{1 \leq i \leq m} i\lambda_i$, then the strategy that explores lists to fixed depth λ_{i_Λ}, in any sequence, will never explore more than $\min_{1 \leq i \leq m} i\lambda_i$ list positions. □

In the average case, where the average is taken over all presentations of the given length set Λ, we can hope to get away with something considerably less than the intrinsic maximum-traversal cost. We note that the average case behaviour of any deterministic multi-list traversal strategy is realized as the expected behaviour of a two-phased randomized algorithm that first randomly permutes the indices of the lists in the input multi-list and then continues in an entirely deterministic fashion. Thus, a lower bound on the expected cost of any randomized list-traversal strategy is also a lower bound on the average-case cost of (deterministic) list traversal.

Consider first the fixed-depth traversal strategy (denoted FDT(d)), already encountered, that explores the lists, one after another, to some fixed depth d, stopping if and when some list is completely traversed.

Lemma 1. *Strategy* FDT(λ_k) *solves the multi-list traversal problem, with average cost (over all presentations of* Λ*) at most* $\lambda_k(m+1)/(m-k+2)$.

Proof. By definition, at least $m-k+1$ of the lists have length at most λ_k. Since lists are explored to depth λ_k, it suffices to argue that, among all possible permutations of the input lists (i.e. all possible presentations of Λ) the average position of the first list with length at most λ_k is at most $(m+1)/(m-k+2)$. (Equivalently, this is the expected position of the first 1 in a random permutation of a binary string of length m containing $n-k+1$ 1's.) □

Let $\bar{i}_\Lambda = \arg\min_{1\leq i\leq n}\{\lambda_i/(m-i+2)\}$. Then, assuming that Λ is known, a strategy, namely FDT$(\lambda_{\bar{i}_\Lambda})$, exists that solves the list traversal problem with average cost at most $m\lambda_{\bar{i}_\Lambda}/(m-\bar{i}_\Lambda+2)$. Thus the *intrinsic average-traversal-cost* of the list length set Λ, which we denote by $\bar{\xi}(\Lambda)$, is at most $m\lambda_{\bar{i}_\Lambda}/(m-\bar{i}_\Lambda+2)$.

As it turns out, this bound is essentially tight:

Lemma 2. *Any randomized multi-list traversal strategy* $\mathcal{B}$ *that terminates after at most* $m\lambda_{\bar{i}_\Lambda}/3(m-\bar{i}_\Lambda+2)$ *steps on all presentations of* Λ*, fails with probability at least one half on a random presentation of* Λ.

Proof. Since an adversary is free to choose the least favourable list indexing, the expected cost of strategy $\mathcal{B}$, forced by an adversary, is at least that which is required for a random permutation of the input lists. Hence, we can assume that $\mathcal{B}$ behaves the same on all list orderings (without loss of generality, it begins by randomly permuting the lists) and thus the i-th list has length $\lambda_{\pi(i)}$, for some random permutation π.

Denote the expression $m\lambda_{\bar{i}_\Lambda}/(m-\bar{i}_\Lambda+2)$ by c_Λ. To prove the desired result, it suffices to argue that any randomized list-traversal strategy that has been modified to terminate after exploring at most $c_\Lambda/3$ list positions, must fail (i.e. not complete the traversal of any list) with probability at least 1/2. We note that any such truncated list-traversal strategy can be interpreted as a probability distribution over the set of all (deterministic) traversals where list i is traversed to depth d_i and $\sum_{1\leq i\leq n} d_i \leq c_\Lambda/3$.

Because the list indices are assigned randomly, we can assume, without loss of generality, that $d_1 \geq d_2 \geq \ldots \geq d_m$. We will argue that every such $(d_1, d_2, \ldots, d_m)$-traversal fails with probability at least 1/2. Let $\hat{\Lambda} = \{\hat{\lambda}_1, \ldots, \hat{\lambda}_m\}$, where $\hat{\lambda}_i = (m-i+2)c_\Lambda/m$, for $1 \leq i \leq m$. Since $\hat{\lambda}_i \leq \lambda_i$, for $1 \leq i \leq m$, and $\bar{\xi}(\hat{\Lambda}) = \bar{\xi}(\Lambda)$, it suffices to prove the result under the assumption that $\Lambda = \hat{\Lambda}$. Furthermore, since the strategy need only work for presentations of Λ, there is no loss of generality in assuming that the strategy exploits the knowledge that all of the λ_i values are integral multiples of c_Λ/m and thus the exploration depth values $d_1, d_2, \ldots, d_m$ satisfy $d_i = k_i c_\Lambda/m$, for some integers $k_1 \geq k_2 \ldots \geq k_m \geq 0$.

Since $\sum_i k_i = \sum_i d_i m/c_\Lambda \leq m/3$, it follows that $k_i = 0$, for $m/3 < i \leq m$. Furthermore, since $\lambda_j > d_i$ just when $j < m - k_i + 2$, at least $m - k_i$ of the lists have length greater than d_i. Thus, Pr(failure) $= \prod_{j=1}^{m} \Pr(\lambda_{\pi(i)} > d_j) \geq \prod_{j=1}^{m}\left(\frac{m-j+1-k_j}{m-j+1}\right) = \prod_{j=1}^{m/3}\left(\frac{m-j+1-k_j}{m-j+1}\right)$. If we relax the constraint that $k_i \geq k_j$, for $i \leq j$, the expression $\prod_{j=1}^{m/3}\left(\frac{m-j+1-k_j}{m-j+1}\right)$ is minimized when $k_{m/3} = m/3$ and $k_j = 0$, for $j < m/3$. Hence, Pr(failure) $\geq 1/2$. □

Corollary 1. *The average cost of any deterministic multi-list traversal strategy, over all presentations of Λ, is at least $m\lambda_{\bar{i}_\Lambda}/6(m - \bar{i}_\Lambda + 2)$, even if Λ is known.*

Theorem 2. $\bar{\xi}(\Lambda) = \Theta(m\lambda_{\bar{i}_\Lambda}/(m - \bar{i}_\Lambda + 2))$.

2.2 Competitive Ratio of Conventional Dovetailing Strategies

It is instructive to analyse the competitive ratios achieved by two familiar dovetailing strategies applied to multi-list traversal. These correspond to the extremes of uniformity in traversal of lists; breadth-first (usually called round-robin) traversal explores all lists in a completely equitable fashion and depth-first traversal chooses one list and explores it to its end.

Theorem 3. *For both the maximum and average-traversal-costs, the competitive ratio of breadth-first traversal is $\Omega(m)$ and the competitive ratio of depth-first traversal is unbounded.*

Proof. Suppose first that we have a multi-list whose associated length set Λ consists of one finite value d and $m-1$ infinite (or arbitrarily large) values. In this case, it is easy to confirm that the intrinsic maximum-traversal-cost of Λ, $\xi(\Lambda)$, satisfies $\xi(\Lambda) = md$ and the intrinsic average-traversal-cost of Λ, $\bar{\xi}(\Lambda)$, satisfies $\xi(\Lambda) = \Theta(d)$. This coincides with the familiar observation that an adversary can "hide" the identity of the sole finite list until all other lists have been explored to depth at least d. Breadth-first traversal achieves maximum-traversal-cost and average-traversal-cost $\Theta(md)$, but depth-first traversal has an unbounded cost in this case, for both maximum-traversal-cost and average-traversal-cost.

On the other hand, if we have a multi-list whose associated length set Λ consists of m lists all of which have the same length d, then $\xi(\Lambda)$ and $\bar{\xi}(\Lambda)$ are both $\Theta(d)$ (despite the fact that the minimum certification cost d is the same as in the previous example). In this case the intrinsic maximum-traversal-cost is achieved by any depth-first traversal but breadth-first traversal has maximum-traversal-cost and average-traversal-cost $\Theta(md)$. □

2.3 Hyperbolic Dovetailing

We now introduce a novel multi-list traversal strategy that makes use of a technique that we call *hyperbolic dovetailing*. The strategy maintains an indexing of all the list positions in the input multi-list and, in terms of this indexing, defines a *rank* function on the next unexplored position of each list. The traversal explores positions in order of increasing rank until some list is exhausted. In general, the rank of the t-th position of the i-th list (in the input order) is just the product ti. Thus if we view the list positions as points in the plane (where the t-th position of the i-th list has coordinates (i, t)), the positions are examined in the order encountered by a hyperbola $t = c/i$, for increasing values of c (see Fig. 1(b)). This interpretation explains the term "hyperbolic dovetailing".

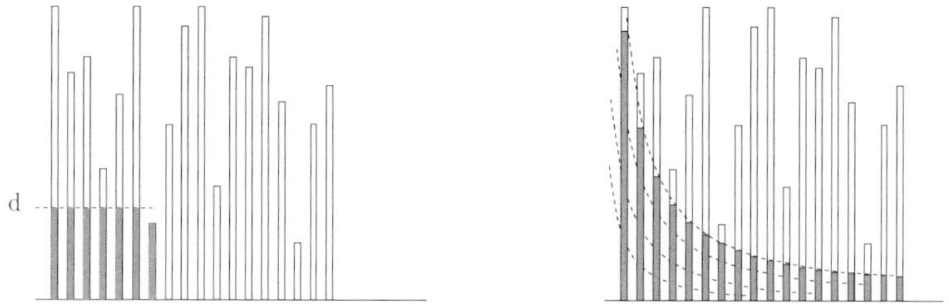

(a) Bounded depth-first traversal (b) Hyperbolic traversal

Fig. 1.

HYPERBOLIC-TRAVERSAL

$c \leftarrow 1$;
repeat until some list is fully explored
  for $i = 1$ to m
    do continue exploration of list i up to depth $\lfloor c/i \rfloor$;
  increment c;

2.4 Worst-Case Competitive Case Analysis of Hyperbolic Traversal

Here we argue that hyperbolic traversal is competitive, up to a logarithmic factor, even against strategies that know the set of list lengths.

Theorem 4. *The hyperbolic-traversal strategy solves the multi-list traversal problem for a multi-list with associated length set Λ, with maximum-traversal-cost $O(\xi(\Lambda) \lg \min\{|\Lambda|, \xi(\Lambda)\})$.*

Proof. Suppose that the last list position explored in the hyperbolic traversal has rank c. Then (i) $c \leq \xi(\Lambda)$, and (ii) the total number of list positions traversed is $O(c \lg \min\{m, c\})$. Point (i) follows from Theorem 1. Point (ii) follows from the fact that, according to our hyperbolic traversal, all of the explored positions in each list have rank at most c and, since there are at most $\lfloor c/t \rfloor$ positions of rank at most c in the t-th list, the number of list positions explored is bounded above by $\sum_{t=1}^{m} \lfloor c/t \rfloor \leq \sum_{t=1}^{\min\{m,c\}} c/t \leq c + \int_{1}^{\min\{m,c\}} \frac{c}{t} dt = c(1 + \ln \min\{m, c\})$. □

Furthermore, the logarithmic competitiveness bound is the best that one could hope for in a general strategy.

Theorem 5. *Any deterministic list-traversal strategy that behaves correctly on all multi-list presentations of length sets Λ satisfying $\xi(\Lambda) = \xi_0$, must have maximum-traversal-cost at least $\xi_0 \ln \min\{|\Lambda|, \xi_0\}$, on at least one such presentation.*

Proof (Sketch). We define a family of canonical input length sets Λ, with $\xi(\Lambda) = \xi_0$ and argue that any deterministic traversal strategy that explores fewer than $\xi_0 \ln \xi_0$ list positions must, in the worst case, fail to complete the traversal of at least one presentation of some Λ in this family. □

2.5 Average and Expected Case Competitive Analysis of Hyperbolic Traversal

We now turn to the average case behaviour of our hyperbolic multi-list traversal strategy. As in the worst case, hyperbolic traversal is competitive, up to a logarithmic factor, even against strategies that know the set of list lengths.

Theorem 6. *The hyperbolic-traversal strategy solves the multi-list traversal problem for a multi-list with associated length set* Λ *with average-traversal-cost* $O(\bar{\xi}(\Lambda) \lg \min\{|\Lambda|, \bar{\xi}(\Lambda)\})$.

Proof. Let $T^{(j)} = \{\delta_1^{(j)}, \ldots, \delta_m^{(j)}\}$, where $\delta_i^{(j)} = \lambda_j$, for $j \le i \le m$ and $\delta_i^{(j)} = \infty$, for $i < j$. By the monotonicity of average traversal cost, avg-trav-cost$(\Lambda) \le \min_{1 \le j \le m}$ avg-trav-cost$(T^{(j)})$. Given this is suffices to prove that for the hyperbolic-traversal strategy avg-trav-cost$(T^{(j)})$ is $O(\hat{c}_j \ln \hat{c}_j)$, where $\hat{c}_j = \lambda_j m/(m-j+1)$.

By the nature of $T^{(j)}$, any strategy discovers the end of a list at depth exactly λ_j. But, since exactly $j-1$ lists have length greater than λ_j, the probability that the number of lists that need to be explored exceeds $m/(m-j+1)$ (which happens just when the first $m/(m-j+1)$ lists all have length exceeding λ_j) is just $\binom{j-1}{m/(m-j+1)}/\binom{m}{m/(m-j+1)} < \left(\frac{j-1}{m}\right)^{m/(m-j+1)} < \left(\frac{1}{e}\right)$.

Since the hyperbolic-traversal strategy explores $\lfloor c/\lambda_j \rfloor$ lists to depth λ_j when its traversal parameter has the value c, it follows that the event, denoted τ_t, that it fails to terminate by the time its traversal parameter reaches $t\hat{c}_j$ has probability at most $\left(\frac{1}{e}\right)^t$. As we have already seen, if τ_t holds, the total exploration cost is $O(t\hat{c}_j \lg(t\hat{c}_j))$. Thus, the total expected cost is $\sum_{r=-\infty}^{\infty} \sum_{t=2^r}^{2^{r+1}} \Pr(\tau_t) t\hat{c}_j \lg(t\hat{c}_j)) = O(\hat{c}_j \lg(t\hat{c}_j))$. □

If we precede the hyperbolic traversal strategy with a step that randomly permutes the indices of the input lists, we produce a randomized algorithm whose expected cost coincides with the average cost of deterministic hyperbolic traversal. This randomized hyperbolic traversal strategy, which works correctly on *all* input list sets, is essentially optimal even among randomized list traversal strategies whose only constraint is that they succeed with probability at least 1/2 on multi-lists whose associated length sets Λ satisfy $\bar{\xi}(\Lambda) = \bar{\xi}_0$, for some fixed $\bar{\xi}_0$.

Theorem 7. *Any randomized list-traversal strategy that succeeds with probability at least* 1/2 *on all multi-list presentations of length sets* Λ *satisfying* $\bar{\xi}(\Lambda) = \bar{\xi}_0$, *must have expected cost at least* $c\bar{\xi}_0 \lg \bar{\xi}_0$, *for some fixed constant* $c > 0$, *on at least one such presentation.*

Proof (Sketch). We view a randomized algorithm as being a distribution over deterministic strategies. Suppose that $\sum_{1 \le i \le n} d_i = \bar{\xi}_0 \lg \bar{\xi}_0/3$. We argue that any $(d_1, d_2, \ldots, d_m)$-traversal, with $d_1 \le d_2 \le \ldots \le d_m$, will fail on at least half of the presentations of at least half of the members of a family of canonical input length sets Λ, with $\bar{\xi}(\Lambda) = \bar{\xi}_0$. It follows that at least one of these sets will, for at least half of its presentations, be incompletely traversed by deterministic strategies whose total assigned probability is at least 1/2.

Corollary 2. *Any deterministic list-traversal strategy that behaves correctly on all multi-list presentations of length sets* Λ *satisfying* $\bar{\xi}(\Lambda) = \bar{\xi}_0$*, must have average cost at least* $c\bar{\xi}_0 \lg \bar{\xi}_0$*, for some fixed constant* $c > 0$*, on at least one such presentation.*

3 Applications of Hyperbolic Dovetailing

3.1 Generalized Cow-Path Search and Hybrid Algorithm Synthesis

As discussed in the introduction, the multi-list traversal problem provides a well-motivated generalization of cow-path search and hybrid algorithm synthesis. The variations in search cost functions, while significant for the exact competitive analysis of interest in the single goal versions of these problems, are essentially negligible when we consider asymptotic competitiveness bounds. For example, our lower bounds for multi-list traversal obviously still hold in the setting of cow-path search (where search cost is counted for revisiting path locations) but our hyperbolic traversal algorithm can be modified (by re-exploring a list only when the hyperbolic rank function doubles its value from the preceding exploration) so that the total search cost (counting revisits) can be assigned in such a way that every explored location has $O(1)$ charges. Thus all of our multi-list traversal results carry over (with modified asymptotic constants) to these other problems.

Our results also extend to a slightly more general form of multi-list search in which individual lists may contain zero or more *goal* locations and the objective is to traverse at least one list to the location of its first goal. By truncating lists at their first goal, this reduces to what we call the *signed* multi-list traversal problem in which some lists are *positive* (i.e. they have a goal location at their end) and some are *negative* (i.e. they terminate, if at all, in a dead-end). Signed multi-list traversal provides a natural model of dovetailing processes that may terminate without success.

3.2 Input-Thrifty Algorithms

In many natural problem settings the knowledge about the input is incomplete. For instance, input numbers may initially be known only up to some precision, and additional precision may be obtainable, but only at a high cost. In such situations it is of interest to design algorithms that solve the problem using as little total input precision as possible.

Our results on list searching provide a modest but non-trivial contribution to the study of such algorithms, taking the extreme point of view that the computation within such an algorithm is free, and the only cost incurred is the number of input bits that need to be examined by the algorithm. Specifically, for a given algorithm and input sequence, we define the *leading-input-bits-cost*, or LIB-cost, for short, to be the number of input bits that the algorithm examines. We are interested in *input-thrifty algorithms*, i.e. algorithms whose LIB-cost is "small" in some quantifiable sense.

We consider problems whose input is a sequence $p_1, \ldots, p_n$ of real numbers in the half-open interval $[0, 1)$. An algorithm can access each such number $p = \sum_{j>0} p^{(j)} 2^{-j}$ via its binary representation $p^{(1)}, p^{(2)}, \cdots$, and this happens by examining the bits $p^{(j)}$ individually *in order of decreasing significance*, i.e. an algorithm can examine bit $p^{(j)}$ only after it has examined bits $p^{(1)}$ through $p^{(j-1)}$ already. In practice, we will assume that each p_i has a finite, although arbitrarily long, representation (thereby guaranteeing finite cost for all inputs).

As a concrete application, suppose we are given a set of s numbers $\{p_1, \ldots, p_s\}$. Our goal is to determine if there exists a pair (p_i, p_j), such that $p_i \neq p_j$. We can map this to an instance of the multi-list traversal problem as follows: each number p_i is interpreted as a list of length λ_i, where $\lambda_i = \arg\min\{j \mid p_i^{(j)} \neq p_1^{(j)}\}$, (that is the position of the most significant bit on which p_i and p_1 differ). It should be clear that (i) if the input set contains at least two distinct numbers then the associated multi-list contains at least one finite length list, and (ii) any multi-list traversal scheme corresponds to an not-all-equal certification algorithm in the LIB-cost model.

Note that any not-all-equal certification algorithm can be reformulated in such a way that at least as many bits of input p_1 are explored as any other input, with at most a constant factor increase in the LIB-cost. (In effect this says that in the LIB-cost model the cost of certifying not-all-equal is essentially the same as certifying that some input number differs from one specific input, p_1.) It follows that competitive algorithms for multi-list traversal translate directly to input-thrifty not-all-equal certification algorithms.

Acknowledgements

The author gratefully acknowledges helpful discussions concerning the material of this paper with Rolf Klein, Robert Tseng, and especially Raimund Seidel.

References

1. Azar, Y., Broder, A.Z., Manasse, M.S.: On-line choice of on-line algorithms. In: Proc. 4th Annual ACM-SIAM Symposium on Discrete Algorithms, pp. 432–440 (1993)
2. Baeza-Yates, R.A., Culberson, J.C., Rawlins, G.J.E.: Searching in the plane. Information and Computation 106(2), 234–252 (1993)
3. Ben-David, S., Borodin, A.: A new measure for the study of on-line algorithms. Algorithmica 11, 73–91 (1994)
4. Boyar, J., Favrholdt, L.M.: The relative worst order ratio for on-line algorithms. ACM Trans. on Algorithms 3(2) (2007)
5. Boyar, J., Favrholdt, L.M., Larsen, K.S.: The relative worst order ratio applied to paging. In: Proc. 16th Annual ACM-SIAM Symposium on Discrete Algorithms, pp. 718–727 (2005)
6. Demaine, E., Fekete, S., Gal, S.: Online searching with turn cost. Theoretical Computer Science 361, 342–355 (2006)

7. Dorrigiv, R., Lopez-Ortiz, A.: A survey of performance measures for on-line algorithms. ACM SIGACT News 36(3), 67–81 (2005)
8. Kao, M.-Y., Littman, M.L.: Algorithms for informed cows. In: AAAI 1997 Workshop on On-Line Search (1997)
9. Kao, M.-Y., Ma, Y., Sipser, M., Yin, Y.: Optimal constructions of hybrid algorithms. J. Algorithms 29(1), 142–164 (1998)
10. Kao, M.-Y., Reif, J.H., Tate, S.R.: Searching in an unknown environment: An optimal randomized algorithm for the cow-path problem. Information and Computation 131(1), 63–79 (1996)
11. Kenyon, C.: Best-fit bin-packing with random order. In: Proc. 7th Annual ACM-SIAM Symposium on Discrete Algorithms, pp. 359–364 (1996)
12. Koutsoupias, E., Papadimitriou, C., Yannakakis, M.: Searching a fixed graph. In: Proc. 23rd International Colloquium on Automata, Languages and Programming, pp. 280–289 (1996)
13. Lopez-Ortiz, A., Schuierer, S.: The ultimate strategy to search on m rays. Theoretical Computer Science 2(28), 267–295 (2001)
14. Luby, M., Sinclair, A., Zuckerman, D.: Optimal speedup of Las Vegas algorithms. In: Proc. Second Israel Symposium on Theory of Computing and Systems, June 1993, pp. 128–133 (1993)
15. Papadimitriou, C.H., Yannakakis, M.: Shortest path without a map. In: Proc. 16th International Colloquium on Automata, Languages and Programming, pp. 610–620 (1989)
16. Schonhage, A.: Adaptive raising strategies optimizing relative efficiency. In: Proc. 30th International Colloquium on Automata, Languages and Programming, pp. 611–623 (2003)
17. Schuierer, S.: Lower bounds in on-line geometric searching. Computational Geometry: Theory and Applications 18(1), 37–53 (2001)
18. Schuierer, S.: A lower bound for randomized searching on m rays. In: Klein, R., Six, H.-W., Wegner, L. (eds.) Computer Science in Perspective. LNCS, vol. 2598, pp. 264–277. Springer, Heidelberg (2003)
19. Sleator, D.D., Tarjan, R.E.: Amortized efficiency of list update and paging rules. Comm. ACM, 202–208 (February 1985)

On the Expansion and Diameter of Bluetooth-Like Topologies*

Alberto Pettarin, Andrea Pietracaprina, and Geppino Pucci

Department of Information Engineering,
University of Padova, Padova, Italy
{pettarin,capri,geppo}@dei.unipd.it

Abstract. The routing capabilities of an interconnection network are strictly related to its bandwidth and latency characteristics, which are in turn quantifiable through the graph-theoretic concepts of expansion and diameter. This paper studies expansion and diameter of a family of subgraphs of the random geometric graph, which closely model the topology induced by the device discovery phase of Bluetooth-based ad hoc networks. The main feature modeled by any such graph, denoted as $BT(r(n), c(n))$, is the small number $c(n)$ of links that each of the n devices (vertices) may establish with those located within its communication range $r(n)$. First, tight bounds are proved on the expansion of $BT(r(n), c(n))$ for the whole set of functions $r(n)$ and $c(n)$ for which connectivity has been established in previous works. Then, by leveraging on the expansion result, tight (up to a logarithmic additive term) upper and lower bounds on the diameter of $BT(r(n), c(n))$ are derived.

1 Introduction

Random graph models have been employed in the literature for the analytical characterization of topological properties of ad hoc wireless networks governed by a variety of network-formation protocols. One such case concerns networks based on the *Bluetooth* technology [1,2]. A Bluetooth network connects n devices, each endowed with a wireless transmitter/receiver able to communicate within a certain *visibility range*. The network is obtained by means of the following process: each device attempts at discovering other devices contained within its visibility range and at establishing reliable communication channels with them, in order to form a connected topology, called the *Bluetooth topology*. Subsequently, a hierarchical organization is superimposed on this initial topology. Since requiring each device to discover *all* of its neighbors is too time-consuming [3], the device discovery phase is terminated by a suitable time-out, hence only a limited number of neighbors are actually discovered.

The following random graph model for the Bluetooth topology has been proposed in [4] and subsequently generalized in [5]. The devices are represented by

* Support for the authors was provided in part by the European Union under the FP6-IST/IP Project AEOLUS.

A. Fiat and P. Sanders (Eds.): ESA 2009, LNCS 5757, pp. 528–539, 2009.

n nodes, whose coordinates are randomly chosen within the unit square $[0,1]^2$; each node selects $c(n)$ *neighbors* among all *visible nodes*, that is, among all nodes within Euclidean distance $r(n)$, where $r(n)$ models the visibility range, which is assumed to be the same for all devices. The resulting graph, called $BT\,(r(n), c(n))$, is the one where there is an undirected edge for each pair of neighbors. Note that such a graph is a subgraph of the well-known random geometric graph in $[0,1]^2$, where all pairs of visible nodes are connected by an edge [6]. Experimental evidence shows that $BT\,(r(n), c(n))$ is a good model for the Bluetooth topology [4]. Moreover, $BT\,(r(n), c(n))$ may be employed as a model for network scenarios where nodes are constrained to maintain a small number of simultaneous connections, because of limited resources, both energetic and computational, or where establishing links to every visible node is, by far, too costly either in time or energy.

Properties of $BT\,(r(n), c(n))$ have been investigated in a number of recent works. In [7] the authors show that for any fixed constant $r > 0$ there exists a (large) constant c such that $BT(r, c)$ is an expander with high probability. In [8] it is proved that with high probability $BT(r, c)$ is connected for any fixed constant $r > 0$ and $c \geq 2$ whenever n becomes sufficiently large. These results require that the visibility range be a constant, which implies that every node can choose its neighbors among a constant fraction of all of the nodes in the system. Such an assumption becomes unfeasible as the number of devices grows large.

To overcome the latter problem, a more general setting has been analyzed in [5], where it has been proved that $BT\,(r(n), c(n))$ stays connected, with high probability, also for vanishing values of $r(n)$ (as $n \to \infty$), as long as each node selects a suitable number of neighbors. Precisely, if $r(n) = \Omega\left(\sqrt{\log n/n}\right)$, just allowing $c(n) = O\left(\log\left(1/r(n)\right)\right)$ neighbor selections per node ensures that the resulting graph be connected with high probability. The lower bound on $r(n)$ cannot be improved: in fact, when $r(n) \leq \delta\sqrt{\log n/n}$, for some constant $0 < \delta < 1$, the visibility graph obtained connecting every node to *all* visible ones (i.e., the random geometric graph $RGG\,(r(n))$ of [6] with radius $r(n)$) is disconnected with high probability [9]. The tightness of the lower bound on $c(n)$ is instead an open problem.

Most of the previous research focuses on the connectivity of the Bluetooth topology, with the exception of the expansion result of [7] which only considers the extreme case of constant visibility range. In this paper, we contribute to a deeper understanding of the Bluetooth topology by providing upper and lower bounds for two crucial structural properties, namely, expansion and diameter, for the values of $r(n)$ and $c(n)$ for which connectivity has been established by previous works. All of our bounds are tight, except for an additive logarithmic term in the upper bound on the diameter. To emphasize the relevance of our results, observe that the bandwidth and latency characteristics of a network, which determine its ability to perform efficient routing, are closely related to the expansion and diameter properties of its underlying topology [10].

The rest of the paper is organized as follows. Section 2 introduces key definitions and properties which will be used throughout the paper. The lower and upper bounds on the expansion of $BT(r(n), c(n))$ are presented in Section 3, while those on the diameter are obtained in Section 4. Section 5 concludes the paper with some final remarks.

2 Preliminaries

In this section we formally define the Bluetooth topology, illustrate the notation and recall some facts for later use.

Definition 1 (Bluetooth topology). *Given an integer n, a real-valued function $r(n) \to (0, \sqrt{2}]$ and a positive integer function $c(n)$, the Bluetooth topology, denoted by $BT(r(n), c(n))$, is the undirected random graph $G = (V_n, E_n)$, defined as follows.*

- *The vertex set V_n is a set of n points chosen uniformly and independently at random in $[0,1]^2$.*
- *The edge set E_n is obtained through the following process: independently, each node selects a random subset of $c(n)$ neighbors among all nodes within distance $r(n)$ (all of them, if they are less than $c(n)$). An edge $\{u, v\} \in E_n$ exists if and only if u has selected v, or viceversa.*

In the next sections, we assume the following setting. Consider the standard tessellation of $[0,1]^2$ into k^2 square *cells* of side $1/k$ where $k = \lceil \sqrt{5}/r(n) \rceil$. We say that two cells are *adjacent* if they share a side. Thus, any two nodes residing in the same or in two adjacent cells are at distance at most $r(n)$ (i.e., each node is within the range of the other) and we say that they can *see* each other. When the context is clear, with a slight abuse of notation, we identify a cell with the set of nodes residing therein.

Recall that an event occurs *with high probability* (in brief, *w.h.p.*) if its probability is at least $1 - 1/\operatorname{poly}(n)$. Let $m = n/k^2$ be the expected number of nodes residing in a cell. The following proposition will be exploited several times.

Proposition 1 ([5]). *Let $\alpha = 9/10$, $\beta = 11/10$. There exists a constant $\gamma_1 > 0$ such that for every $r(n) \geq \gamma_1 \sqrt{\log n/n}$ the following two events occur w.h.p.:*

1. *every cell contains at least αm and at most βm nodes;*
2. *every node has at least $(\alpha/4)\pi n r^2(n)$ and at most $\beta \pi n r^2(n)$ nodes in its visibility range.*

Let $G = (V, E)$ be an undirected graph. Below, we define the quantities at the center of our analysis.

Definition 2 (Neighborhood). *Given a set of vertices $X \subseteq V$, its neighborhood is the set $\Gamma(X) = \{ u \in V(G) : \exists e = \{u, v\} \in E(G), v \in X \}$.*

Definition 3 (Expansion). *The expansion of G is a function $\lambda(s)$, for $1 \leq s \leq |V|/2$, such that*

$$\lambda(s) = \min_{S \subseteq V : |S| = s} \frac{|\Gamma(S) - S|}{|S|}.$$

We remark that, in some works, the term "expansion" is used to refer to a global property of the graph, that is, the minimum value of the function $\lambda(s)$ [10]. In contrast, in this paper we offer a finer characterization of the expansion properties of $BT\,(r(n), c(n))$ by proving explicit bounds on all values of $\lambda(s)$.

Definition 4 (Diameter). *The* diameter *of G, denoted as* $\mathrm{diam}(G)$, *is the maximum distance between any two nodes $u, v \in V$, where the distance between two nodes is the number of edges of a shortest path connecting them.*

Observe that, under any reasonable cost model for communication, the maximum latency to be expected of a point-to-point communication in a network is proportional to the diameter of its underlying topology.

In the rest of the paper we focus on $BT\,(r(n), c(n))$ and we study its expansion and diameter for those ranges of the parameters for which the connectivity is guaranteed by the results of [5], that is, $r(n) \geq \gamma_1 \sqrt{\log n/n}$ and $c(n) = \gamma_2 \log(1/r(n))$ for two suitable positive real constants γ_1, γ_2.

3 Expansion of $BT\,(r(n), c(n))$

In this section we study the node expansion of $BT\,(r(n), c(n))$. Specifically, in Section 3.1 and Section 3.2 we establish, respectively, a lower bound and an upper bound to the node expansion of this family of random graphs. Recall that $m = n/k^2 = \Theta\left(nr^2(n)\right)$ denotes the expected number of nodes residing in a cell.

3.1 Lower Bound

The main result of this section is the following theorem.

Theorem 1. *Consider an instance of $BT\,(r(n), c(n))$ with $r(n) \geq \gamma_1 \sqrt{\log n/n}$ and $c(n) = \gamma_2 \log(1/r(n))$, for two suitable positive constants γ_1 and γ_2. With high probability, for every integer s, $1 \leq s \leq n/2$, we have*

$$\lambda(s) = \begin{cases} \Omega\left(\min\{c(n), m/s\}\right) & \text{if } 1 \leq s \leq \alpha m \\ \Omega\left(\sqrt{m/s}\right) & \text{if } \alpha m < s \leq n/2. \end{cases}$$

The proof of Theorem 1 relies on three technical lemmas, which characterize the expansion of certain types of node subsets confined within a single cell. Consider a given subset of vertices S of size s. For any cell Q, we call the set $P = S \cap Q$ the *pocket* of S in Q.

Lemma 1. *Let α' and ε' be two suitable positive constants, with $\alpha' \leq \min\{\varepsilon', \alpha\}$. When $|P| \geq \log n$ or $r(n) = O\left(n^{-1/8}\right)$, then with high probability, for any cell Q and for every pocket $P \subseteq Q$, with $1 \leq |P| \leq \alpha' m$, we have $|\Gamma(P) - P| \geq \varepsilon' \min\{c(n)\,|P|, m\}$.*

Proof. Fix a cell Q and a size p for P, with $1 \le p \le \alpha' m$. We bound the probability that, for every $P \subseteq Q$ its neighborhood is contained in $P \cup T$, where T is a set of nodes not belonging to P with a certain (small) size t. For notational convenience, we abbreviate $c = c(n) = \gamma_2 \log(1/r(n))$ and introduce the following quantities:

- q is the number of nodes in Q;
- v is the total number of nodes visible by at least one node in Q;
- w is the minimum number of nodes visible by any node;
- w' is the maximum number of nodes visible by any node;
- z is the minimum number of nodes visible by all nodes in P.

Conditioning on the events of Proposition 1, we have that $q, v, w, w', z = \Theta(m)$.

Let $\mathcal{E}$ be the union, over all the cells Q and all the choices of the pocket $P \subseteq Q$, of the events $|\Gamma(P) - P| \le t$. We can bound the probability of $\mathcal{E}$:

$$
\begin{aligned}
\Pr[\mathcal{E}] &\le \binom{q}{p}\binom{v}{t}\left(\frac{\binom{t+p}{c}}{\binom{w}{c}}\right)^{p}\left(\frac{\binom{w'-p}{c}}{\binom{w'}{c}}\right)^{z-(t+p)} \\
&\le \left(\frac{eq}{p}\right)^{p}\left(\frac{ev}{t}\right)^{t}\left(\frac{t+p}{w}\right)^{cp}\left(\frac{w'-p}{w'}\right)^{c(z-(t+p))} \\
&\le \left(\frac{eq}{p}\right)^{p}\left(\frac{ev}{t}\right)^{t}\left(\frac{t+p}{w}\right)^{cp} e^{-\frac{cp}{w'}(z-(t+p))}.
\end{aligned}
$$

We distinguish between two cases, depending on the value of p.

Case 1: $1 \le p \le m/c$. Let $t = \varepsilon' cp$. We rewrite the bound on $\Pr[\mathcal{E}]$ as

$$
\Pr[\mathcal{E}] \le \left(\left(\frac{eqc}{cp}\right)^{1/c}\left(\frac{ev}{\varepsilon' cp}\right)^{\varepsilon'}\left(\frac{\varepsilon' cp}{aw}\right)\right)^{cp},
$$

where a is a positive constant, since $p = O(t)$ and $(z-(t+p))/w' = \Theta(1)$. By regrouping the factors, we obtain:

$$
\Pr[\mathcal{E}] \le \left(\frac{c^{1/c}}{a\varepsilon'^{\varepsilon'}}\frac{(eq)^{1/c}(ev)^{\varepsilon'}}{w}(cp)^{1-\varepsilon'-1/c}\varepsilon'\right)^{cp} < \frac{1}{n^3},
$$

where the last inequality holds for a sufficiently large γ_2 in $c = \gamma_2 \log(1/r(n))$, and for a sufficiently small ε', since $cp = \Omega(\log n)$. The claim follows by invoking the union bound over the $O(n)$ cells and the $O(n)$ choices of $p = |P|$.

Case 2: $m/c < p \le \alpha' m$. Note that in this case $cp > m$, whence we set $t = \varepsilon' m$. We rewrite the upper bound on $\Pr[\mathcal{E}]$ as

$$
\begin{aligned}
\Pr[\mathcal{E}] &\le \left(\frac{eq}{p}\right)^{p}\left(\frac{ev}{\varepsilon' m}\right)^{\varepsilon' m}\left(\frac{\varepsilon' m + p}{aw}\right)^{cp} \\
&\le \left(\left(\frac{eq}{p}\right)^{1/c}\left(\frac{ev}{\varepsilon' m}\right)^{\varepsilon' m/(cp)}\left(\frac{\varepsilon' m + p}{aw}\right)\right)^{cp}.
\end{aligned}
$$

The first and the second factor of the latter bound are bounded by a constant, for a suitable choice of c and ε'. By our choice of α', letting ε' be a sufficiently small value, we can make the product of the three factors at most a constant less than 1, so that $\Pr[\mathcal{E}] < \frac{1}{n^3}$ since $cp = \Omega(\log n)$. The claim then follows by applying the union bound as done for Case 1. □

Lemmas 2 and 3 are proved via counting arguments which are similar in spirit to the one employed in the proof of Lemma 1, and are omitted due to space limitations. Detailed proofs are reported in [11].

Lemma 2. *Let $r(n) = \Omega\left(n^{-1/8}\right)$, and $c(n) \geq 3$. With high probability, for any cell Q and for every pocket $P \subseteq Q$, with $|P| < \log n$, we have $|\Gamma(P)| > \frac{1}{3}c(n)\,|P|$.*

Lemma 3. *Let α'' and ε'' be two suitable positive constants, with $\alpha'' \leq \alpha/(2(1+\varepsilon''))$. With high probability, for any pair of distinct adjacent cells Q and Q' and for every pocket $P \subseteq Q$, with $m/c(n) \leq |P| \leq \alpha'' m$, we have $|\Gamma(P) \cap Q'| \geq (1+\varepsilon'')\,|P|$.*

We are now ready to prove the main result of this section.

Proof (Theorem 1). Throughout the proof, we condition on the events stated in Proposition 1 and in the three previous lemmas. We also define $\bar{\alpha} = \min\{\alpha', \alpha''\}$ and $\bar{\varepsilon} = \min\{\varepsilon', \varepsilon'', 1/3\}$ where $\alpha', \alpha'', \varepsilon', \varepsilon''$ are the constants appearing in the statements of Lemma 1 and Lemma 3, respectively. Consider an arbitrary set S of s vertices of $BT(r(n), c(n))$, with $1 \leq s \leq n/2$. We classify the cells according to the size of the pockets of S that they contain: namely, a cell Q such that $Q \cap S \neq \emptyset$ is said to be *black* if it contains at least $\bar{\alpha}m$ nodes of S, and *gray* otherwise. Two cases are possible: either a majority of nodes of S resides in black cells or a majority of nodes of S resides in gray cells.

In the first case, there are at least $\lceil s/(2\beta m)\rceil$ black cells. By well-known topological properties of two-dimensional meshes [12], we have that at least $\Omega\left(\sqrt{s/m}\right)$ black cells are adjacent to distinct non-black cells. From Lemma 3, every subset $P' \subseteq S$ of size $\bar{\alpha}m$ contained in one of these black cells expands into the corresponding adjacent non-black cell Q' of at least $(1+\bar{\varepsilon})$ times its cardinality, hence $|\Gamma(P) - S| \geq |(\Gamma(P') \cap Q') - S| \geq \bar{\varepsilon}\bar{\alpha}m$, and thus $\lambda(s) = \Omega\left(\sqrt{m/s}\right)$, which is the correct bound since $s = \Omega(m)$ in this case.

In the second case, we resort to a proof strategy inspired by the one employed in [7]. Referring to the tessellation of $[0,1]^2$ into k^2 cells, let us index the cells as Q_{ij}, with $1 \leq i, j \leq k$. Define the *sector* $\mathcal{S}_{ij}$ of a cell Q_{ij} as

$$\mathcal{S}_{ij} = \bigcup_{\substack{\max\{i-6,1\} \leq x \leq \min\{i+6,k\} \\ \max\{j-6,1\} \leq y \leq \min\{j+6,k\}}} Q_{xy}.$$

The *active area* $\mathcal{A}_{ij}$ of sector $\mathcal{S}_{ij}$ is defined as

$$\mathcal{A}_{ij} = \bigcup_{\substack{\max\{i-3,1\} \leq x \leq \min\{i+3,k\} \\ \max\{j-3,1\} \leq y \leq \min\{j+3,k\}}} Q_{xy}.$$

Cell Q_{ij} is called the *center* of both sector $\mathcal{S}_{ij}$ and its active area $\mathcal{A}_{ij}$. Note that the neighborhood of the pocket $P_{ij} = Q_{ij} \cap S$ is entirely contained in $\mathcal{A}_{ij}$ and that the definition of a sector ensures that given two sectors $\mathcal{S}_{ij}$ and $\mathcal{S}_{i'j'}$, with $Q_{i'j'} \cap \mathcal{S}_{ij} = \emptyset$, their active areas are non-overlapping.

Let G be the set of at least $s/2$ nodes of S belonging to gray cells. To estimate the expansion of S, we first execute a greedy procedure, which selects a number of gray cells which are centers of nonoverlapping active areas, and then obtain a lower bound on the expansion by adding up the contributions related to these selected cells. The selection of the centers is obtained via the following marking strategy. Initially all of the gray cells are unmarked. Then, iteratively, the center of the next active area is selected as the unmarked gray cell Q containing the largest pocket of S, and all of the unmarked cells of the sector centered at Q are marked. The procedure terminates as soon as every gray cell becomes marked. The procedure is described by the following pseudocode, where sets I and U maintain, respectively, the indices of the selected centers and the indices of unmarked cells, and subroutine LargestPocket(U) returns the pair (i, j) corresponding to the unmarked cell with the largest pocket (ties broken arbitrarily).

```
CENTER SELECTION
I ← ∅; U ← {(i, j) : Q_ij is a gray cell}
while U ≠ ∅ do
   (i, j) ← LARGESTPOCKET(U)
   I ← I ∪ (i, j)
   for each Q_xy ∈ S_ij do U ← U − (x, y)
```

Let $\langle c_1, c_2, \ldots, c_w \rangle$ be the list of w centers picked by Center Selection, where $c_t = (i_t, j_t)$ was chosen at the t-th iteration of the **while** loop. Let $p_t = |P_{c_t}|$, and let g_t be the number of nodes residing in unmarked gray cells of $\mathcal{S}_{c_t}$ at the beginning of iteration t. Clearly, we have that $\sum_{t=1}^{w} g_t = |G|$ and, by the greedy choice of the centers, $g_t \leq 169p_t$.

In order to lower bound the expansion of S, we proceed as follows. For each t, with $1 \leq t \leq w$, we determine a suitable set of nodes $N_t \subseteq \Gamma(S)$, which belong to gray cells of $\mathcal{A}_{c_t}$. We distinguish between two different cases. If $\mathcal{A}_{c_t}$ contains a black cell, since Q_{c_t} is gray, there must exists a pair of adjacent black-gray cells in $\mathcal{A}_{c_t}$, and we pick N_t as a set of $(1 + \bar{\varepsilon})\bar{\alpha}m$ nodes in the gray cell reached by nodes of S in the black cell, which exists by virtue of Lemma 3. Otherwise, we let $N_t = \Gamma(P_{c_t}) - P_{c_t}$ and observe that by Lemmas 1 and 2, $|N_t| \geq \bar{\varepsilon} \min\{c(n)p_t, m\}$. Note that the N_t's are all disjoint, but the sum of their sizes does not immediately yield a lower bound to $|\Gamma(S) - S|$, since each N_t may itself contain nodes of S, which have to be subtracted from the overall count.

Let us first consider the special case when no active area $\mathcal{A}_{c_t}$ contains black cells. In this case, the number of *external neighbors of* S (i.e., nodes of $\Gamma(S) - S$) accounted for by the N_t's is

$$\left(\sum_{t=1}^{w} |N_t|\right) - |G| = \sum_{t=1}^{w} (|N_t| - g_t) \geq \sum_{t=1}^{w} (|N_t| - 169p_t).$$

Since $p_t \leq \bar{\alpha}m$ and $|N_t| \geq \bar{\varepsilon}\min\{c(n)p_t, m\}$, then for a sufficiently large choice of γ_2 in $c(n) = \gamma_2 \log(1/r(n))$ and a sufficiently small value of $\bar{\alpha}$, we have that $|N_t| - 169p_t \geq \mu|N_t|$ for a certain constant $\mu > 0$. Hence,

$$\sum_{t=1}^{w}(|N_t| - 169p_t) = \Omega\left(\sum_{t=1}^{w}\bar{\varepsilon}\min\{c(n)p_t, m\}\right) = \Omega(\min\{c(n)s, m\}),$$

and the theorem follows.

Consider now the general case where some active areas contain black cells, which implies that $s = \Omega(m)$. Observe that $\sum_{t=1}^{w}|N_t| = \Omega(|G|) = \Omega(s)$, and note that it is sufficient to show that the number of external neighbors of S is $\Omega(\sum_{t=1}^{w}|N_t|)$. Partition the index set $I = \{1, 2, \ldots, t\}$ into two disjoint subsets B_1 and B_2, such that $t \in B_1$ if $\mathcal{A}_{c_t}$ contains no black cells, and $t \in B_2$ otherwise. Suppose that $\sum_{t\in B_2}|N_t| \geq \tau\sum_{t\in B_1}|N_t|$, for a suitable positive constant τ which will be specified later. For each $t \in B_2$ the set N_t contains $(1+\bar{\varepsilon})\bar{\alpha}m$ nodes, and at least $\bar{\varepsilon}\bar{\alpha}m$ of these nodes are external neighbors of S. Hence, the total number of external neighbors of S is at least

$$\sum_{t\in B_2}\bar{\varepsilon}\bar{\alpha}m = \frac{\bar{\varepsilon}}{1+\bar{\varepsilon}}\sum_{t\in B_2}|N_t| \geq \frac{\bar{\varepsilon}}{1+\bar{\varepsilon}}\frac{\tau}{1+\tau}\sum_{t=1}^{w}|N_t|,$$

and the theorem follows. Finally, if $\sum_{t\in B_2}|N_t| < \tau\sum_{t\in B_1}|N_t|$, the number of external neighbors of S accounted for by the nodes in the N_t's is

$$\begin{aligned}\left(\sum_{t=1}^{w}|N_t|\right) - |G| &= \sum_{t\in B_1}(|N_t| - 169p_t) + \sum_{t\in B_2}(|N_t| - 169p_t)\\ &\geq \sum_{t\in B_1}\mu|N_t| + \sum_{t\in B_2}((1+\bar{\varepsilon})\bar{\alpha}m - 169\bar{\alpha}m)\\ &> \sum_{t\in B_1}\mu|N_t| - \sum_{t\in B_1}\left(\frac{169}{1+\bar{\varepsilon}} - 1\right)\tau|N_t|.\end{aligned}$$

By fixing τ such that $((169/(1+\bar{\varepsilon})) - 1)\tau = \mu/2$, we get

$$\sum_{t\in B_1}\mu|N_t| - \sum_{t\in B_1}\left(\frac{169}{1+\bar{\varepsilon}} - 1\right)\tau|N_t| = \frac{\mu}{2}\sum_{t\in B_1}|N_t| = \Omega\left(\sum_{t=1}^{w}|N_t|\right),$$

and the theorem follows. □

3.2 Upper Bound

We now prove that the lower bound of Theorem 1 is asymptotically tight.

Theorem 2. *Consider an instance of BT $(r(n), c(n))$ with $r(n) \geq \gamma_1\sqrt{\log n/n}$ and $c(n) = \gamma_2\log(1/r(n))$, for two suitable positive constants γ_1 and γ_2. With* *high probability, for every integer s, $1 \leq s \leq n/2$, there exists a set of vertices S of size s whose expansion is*

$$\lambda(s) = \begin{cases} O\left(\min\{c(n), m/s\}\right) & \text{if } 1 \leq s \leq \alpha m \\ O\left(\sqrt{m/s}\right) & \text{if } \alpha m < s \leq n/2. \end{cases}$$

Proof. Suppose that the events stated in Proposition 1 occur. If $s \leq \alpha m$, we can choose any subset S of the nodes in a single corner cell Q, so that a total of at most $13\beta m$ nodes are visible from S. Hence, $\lambda(s) = O(m/s)$. Consider a list $\langle v_1, v_2, \ldots, v_n \rangle$ of the vertices of V, sorted by nondecreasing node degree. If we take $S = \{v_1, v_2, \ldots, v_s\}$, we are guaranteed that the sum of the degrees of all nodes in S is not greater than $2c(n)s$, or otherwise the sum of the degrees of the n nodes would exceed $2c(n)n$, which is impossible. Combining the two cases above yields the thesis for the case $s \leq \alpha m$. Let $s > \alpha m$ and consider a set S which occupies an approximately square area of $\Theta(s/m)$ cells in a corner of $[0,1]^2$. Since only the nodes in $O\left(\sqrt{s/m}\right)$ cells are visible from S, we have that $\lambda(s) = O(\sqrt{ms}/s) = O\left(\sqrt{m/s}\right)$, and the theorem follows. □

We remark that the tight bounds on the expansion of $BT(r(n), c(n))$ provided by Theorems 1 and 2 extend the results in [7] from the specific case of $r(n) = \Theta(1)$ to any value of $r(n)$ which guarantees the connectivity of the graph. Note also that if we consider the minimum expansion $\lambda = \min_{1 \leq s \leq n/2} \lambda(s)$, we obtain that for the Bluetooth topology $\lambda = \Theta(r(n))$.

Similar techniques may be applied to prove that $RGG(r(n))$ features an expansion of $\lambda(s) = \Theta(m/s)$ for $1 \leq s \leq \alpha m$, and $\lambda(s) = \Theta\left(\sqrt{m/s}\right)$ for $\alpha m < s \leq n/2$ (details are given in [11]). Hence, quite surprisingly, the expansion of $BT(r(n), c(n))$ is, within a constant factor, equal to the expansion of $RGG(r(n))$ whenever $s = \Omega(m/c(n))$.

4 Diameter of $BT(r(n), c(n))$

In this section, we provide upper and lower bounds on the diameter of $BT(r(n), c(n))$ by leveraging on the expansion result of Section 3. Specifically, the upper bound relies on the following lemma, which relates diameter and expansion.

Lemma 4. *Given a connected undirected graph $G = (V, E)$ with n nodes and expansion $\lambda(s)$, for $1 \leq s \leq n/2$, consider the following recurrence:*

$$\begin{aligned} N_0 &= 1 \\ N_i &= (1 + \lambda(N_{i-1}))\, N_{i-1}. \end{aligned} \tag{1}$$

Define $i^\star$ as the smallest index such that $N_{i^\star} > n/2$. Then, $\operatorname{diam}(G) \leq 2i^\star$.

Proof. Let $d = \mathrm{diam}(G)$ and let u and v be two nodes at distance d in G. Consider a breadth-first tree rooted at u. For $0 \le i \le d$, let W_i denote the set of nodes at level i in the tree, and $Y_i = \bigcup_{\ell=0}^{i} W_\ell$. Note that the expansion properties of G imply that $|Y_i| \ge N_i$. Define now $j^\star$ as the smallest index such that $|Y_{j^\star}| > n/2$, which implies that $j^\star \le i^\star$. Also, w.l.o.g., we can assume that $j^\star \ge \lceil d/2 \rceil$, or otherwise we repeat the argument considering the breadth-first tree rooted at v. Indeed, since u and v are at distance d, one of the two breadth-first trees must reach at most $n/2$ nodes within the first $\lceil d/2 \rceil - 1$ levels, or there would be a path shorter than d connecting u and v. The lemma follows. □

Theorem 3. *Consider an instance of* $BT(r(n), c(n))$ *with* $r(n) \ge \gamma_1 \sqrt{\log n / n}$ *and* $c(n) = \gamma_2 \log(1/r(n))$, *for two suitable positive constants* γ_1 *and* γ_2. *With high probability,*

$$\mathrm{diam}(BT(r(n), c(n))) = O\left(\frac{1}{r(n)} + \log n\right).$$

Proof. We apply Lemma 4 by estimating the value $i^\star$ for the graph $BT(r(n), c(n))$, conditioning on the fact that the expansion of $BT(r(n), c(n))$ is $\lambda(s) = \Omega(\min\{c(n), m/s\})$ for $s \le \alpha m$, and $\lambda(s) = \Omega\left(\sqrt{m/s}\right)$ for $s > \alpha m$, which happens w.h.p. (see Theorem 1).

In order to account for these two different expansion regimes, we proceed as follows. Let $K(j) = \min\{i : N_i \ge 2^j\}$, so that $i^\star = K(\log n - 1)$ and let j_1 be such that $2^{j_1} = \Theta(m)$. Since $\lambda(N_i) = \Omega(1)$ for $0 \le i < K(j_1)$, it follows that $K(j_1) = O(\log n)$. Observe that for $i > K(j_1)$, there exists a constant σ such that $\lambda(N_i) \ge \sigma\sqrt{m/N_i}$. As a consequence, for $j > j_1$, we have:

$$\begin{aligned} N_{K(j)} &\ge N_{K(j-1)} \prod_{s=K(j-1)}^{K(j)-1} \left(1 + \frac{\sigma\sqrt{m}}{\sqrt{N_s}}\right) \\ &\ge N_{K(j-1)} \left(1 + \frac{\sigma\sqrt{m}}{\sqrt{N_{K(j)-1}}}\right)^{K(j)-K(j-1)} \\ &\ge 2^{j-1} \left(1 + \frac{\sigma\sqrt{m}}{2^{j/2}}\right)^{K(j)-K(j-1)}. \end{aligned}$$

Since $K(j)$ is defined as the smallest index for which $N_{K(j)} \ge 2^j$, from the above inequalities it follows that $K(j) - K(j-1) = O\left(\frac{2^{j/2}}{r(n)\sqrt{n}}\right)$. Therefore,

$$\begin{aligned} i^\star = K(\log n - 1) &= \sum_{j=1}^{\log n - 1} (K(j) - K(j-1)) \\ &= \sum_{j=1}^{j_1} (K(j) - K(j-1)) + \sum_{j=j_1+1}^{\log n - 1} (K(j) - K(j-1)) \\ &= O(\log n) + O\left(\frac{1}{r(n)}\right), \end{aligned}$$

and the theorem follows from Lemma 4. □

We now show that Theorem 3 gives a tight estimate for the diameter of $BT\,(r(n), c(n))$ when $r(n) = O\,(1/\log n)$.

Theorem 4. *Consider an instance of* $BT\,(r(n), c(n))$ *with* $r(n) \geq \gamma_1 \sqrt{\log n/n}$ *and* $c(n) \geq \gamma_2 \log\,(1/r(n))$, *for two suitable positive constants* γ_1 *and* γ_2. *With high probability,*

$$\mathrm{diam}(BT\,(r(n), c(n))) = \Omega\left(\frac{1}{r(n)}\right).$$

Proof. Consider the natural tessellation introduced in Section 2. By Proposition 1, with high probability the top leftmost cell and the bottom rightmost cell contain at least one node each, hence the Euclidean distance between these two nodes is $\Theta\,(1)$. Therefore, any path in $BT\,(r(n), c(n))$ connecting them must contain at least $\Omega\,(1/r(n))$ nodes. □

We point out that the lower bound for the case $r(n) = \Theta\,(1)$ can be improved to $\Omega\,(\log n/\log\log n)$. (Full details will be given in the full version of this extended abstract.)

5 Conclusions

The main result of this paper is a tight characterization of the node expansion properties of the Bluetooth topology. Since expansion is essentially a measure of bandwidth, being able to provide a quantitative estimate of this property is useful for the design and analysis of routing strategies [10]. Our result is valid for the entire set of visibility ranges $r(n)$ and number of neighbor choices $c(n)$ which are known to produce a connected graph, as opposed to the result of [7] which holds only for the extreme case $r(n) = \Theta\,(1)$.

By leveraging on the expansion properties, we also derive nearly tight bounds on the diameter of the same topology, which is again an important measure for routing, related to the latency of the network. Our bounds are tight for a large spectrum of visibility ranges (i.e., $r(n) = O\,(1/\log n)$), which includes "small ranges", that is, those which are most interesting for the large scale deployment of the technology. In fact, for the larger ranges $r(n) = \omega\,(1/\log n)$ the upper and lower bounds differ by a mere additive logarithmic term. Closing this gap is still an open problem.

A consequence of our results is that for subsets of $s = \Omega\,(m/c(n))$ vertices, $BT\,(r(n), c(n))$ exhibits roughly the same expansion as the much denser random geometric graph $RGG\,(r(n))$ of [6]. Also, the diameters of the two graphs differ by at most a logarithmic additive term. These are important considerations for real ad hoc networks, especially for what concerns routing capabilities, since they imply that $BT\,(r(n), c(n))$ features similar bandwidth and latency characteristics of $RGG\,(r(n))$ at only a fraction of the costs.

Finally, we recall that it is still an open problem to establish, for every given visibility range $r(n) = \Omega\left(\sqrt{\log n/n}\right)$, the *minimum* number $c(n)$ of neighbor choices which yield connectivity and to assess the corresponding diameter and expansion properties.

References

1. Whitaker, R., Hodge, L., Chlamtac, I.: Bluetooth scatternet formation: a survey. Ad Hoc Networks 3, 403–450 (2005)
2. Stojmenovic, I., Zaguia, N.: Bluetooth scatternet formation in ad hoc wireless networks. In: Misic, J., Misic, V. (eds.) Performance Modeling and Analysis of Bluetooth Networks, pp. 147–171. Auerbach Publications (2006)
3. Basagni, S., Bruno, R., Mambrini, G., Petrioli, C.: Comparative performance evaluation of scatternet formation protocols for networks of Bluetooth devices. Wireless Networks 10(2), 197–213 (2004)
4. Ferraguto, F., Mambrini, G., Panconesi, A., Petrioli, C.: A new approach to device discovery and scatternet formation in Bluetooth networks. In: Proc. of the 18th Int. Parallel and Distributed Processing Symposium (2004)
5. Crescenzi, P., Nocentini, C., Pietracaprina, A., Pucci, G.: On the connectivity of Bluetooth-based ad hoc networks. Concurrency and Computation: Practice and Experience 21(7), 875–887 (2009)
6. Penrose, M.: Random Geometric Graphs. Oxford University Press, Oxford (2003)
7. Panconesi, A., Radhakrishnan, J.: Expansion properties of (secure) wireless networks. In: Proc. of the 16th ACM Symp. on Parallelism in Algorithms and Architectures, pp. 281–285 (2004)
8. Dubhashi, D., Johansson, C., Häggström, O., Panconesi, A., Sozio, M.: Irrigating ad hoc networks in constant time. In: Proc. of the 17th ACM Symp. on Parallelism in Algorithms and Architectures, pp. 106–115 (2005)
9. Ellis, R., Jia, X., Yan, C.: On random points in the unit disk. Random Structures and Algorithms 29(1), 14–25 (2005)
10. Leighton, T., Rao, S.: Multicommodity max-flow min-cut theorems and their use in designing approximation algorithms. J. ACM 46(6), 787–832 (1999)
11. Pettarin, A., Pietracaprina, A., Pucci, G.: On the expansion and diameter of Bluetooth-like topologies. Technical report, `http://www.dei.unipd.it/~pettarin/pubb/manu/PettarinPP09.pdf`
12. Bilardi, G., Preparata, F.P.: Area-time lower-bound techniques with applications to sorting. Algorithmica 1(1), 65–91 (1986)

Minimum Makespan Multi-vehicle Dial-a-Ride

Inge Li Gørtz[1,⋆], Viswanath Nagarajan[2], and R. Ravi[3,⋆⋆]

[1] Technical University of Denmark
ilg@imm.dtu.dk
[2] IBM T.J. Watson Research Center
viswanath@us.ibm.com
[3] Tepper School of Business, Carnegie Mellon University
ravi@cmu.edu

Abstract. Dial-a-Ride problems consist of a set V of n vertices in a metric space (denoting travel time between vertices) and a set of m objects represented as source-destination pairs $\{(s_i, t_i)\}_{i=1}^m$, where each object requires to be moved from its source to destination vertex. In the *multi-vehicle Dial-a-Ride* problem, there are q vehicles each having capacity k and where each vehicle $j \in [q]$ has its own depot-vertex $r_j \in V$. A feasible schedule consists of a capacitated route for each vehicle (where vehicle j originates and ends at its depot r_j) that together move all objects from their sources to destinations. The objective is to find a feasible schedule that minimizes the maximum completion time (i.e. *makespan*) of vehicles, where the completion time of vehicle j is the time when it returns to its depot r_j at the end of its route. We consider the *preemptive* version of multi-vehicle Dial-a-Ride, where an object may be left at intermediate vertices and transported by more than one vehicle, while being moved from source to destination. Approximation algorithms for the single vehicle Dial-a-Ride problem ($q = 1$) have been considered in [3,10].

Our main results are an $O(\log^3 n)$-approximation algorithm for *preemptive multi-vehicle Dial-a-Ride*, and an improved $O(\log t)$-approximation for its special case when there is no capacity constraint (here $t \leq n$ is the number of distinct depot-vertices). There is an $\Omega(\log^{1/4} n)$ hardness of approximation known [9] even for single vehicle capacitated preemptive Dial-a-Ride. We also obtain an improved constant factor approximation algorithm for the uncapacitated multi-vehicle problem on metrics induced by graphs excluding any fixed minor.

1 Introduction

The *multi-vehicle Dial-a-Ride* problem involves routing a set of m objects from their sources to respective destinations using a set of q vehicles starting at t distinct depot nodes in an n-node metric. Each vehicle has a *capacity* k which is the maximum number of objects it can carry at any time. Two versions arise based on whether or not the vehicle can use any node in the metric as a preemption (a.k.a. transshipment) point - we study the less-examined *preemptive version* in this paper. The objective in these problems is either the total completion time or the makespan (maximum completion time) over the q vehicles, and again we study the more challenging *makespan* version

⋆ Sponsored in part by a grant from the Carlsberg Foundation.
⋆⋆ Supported in part by NSF grant CCF-0728841.

A. Fiat and P. Sanders (Eds.): ESA 2009, LNCS 5757, pp. 540–552, 2009.

of the problem. Thus this paper studies the preemptive, capacitated minimum makespan multi-vehicle Dial-a-Ride problem.

While the multiple qualifications may make the problem appear contrived, this is exactly the problem that models courier or mail delivery over a day from several city depots: preemption is cheap and useful for packages, trucks are capacitated and the makespan reflects the daily working time limit for each truck. Despite its ubiquity, this problem has not been as well studied as other Dial-a-Ride versions. One reason from the empirical side is the difficulty in handling the possibility of preemptions in a clean mathematical programming model. On the theoretical side which is the focus of this paper, the difficulty of using preemption in a meaningful way in an approximation algorithm persists. It is further compounded by the hardness of the makespan objective.

The requirement in preemptive Dial-a-Ride, that preemptions are allowed at all vertices, may seem unrealistic. In practice, a subset P of the vertex-set V represents the vertices where preemption is permitted: the two extremes of this general problem are non-preemptive Dial-a-Ride ($P = \emptyset$) and preemptive Dial-a-Ride ($P = V$). However preemptive Dial-a-Ride is more generally applicable: specifically in situations where the preemption-points P form a *net* of the underlying metric (i.e. every vertex in V has a nearby preemption-point). I.e., approximation algorithms for preemptive Dial-a-Ride imply good approximations even in this general setting, wherein the precise guarantee depends on how well P covers V.

We note that although our model allows any number of preemptions and preemptions at all vertices our algorithms do not use this possibility to its full extent. Our algorithm for the capacitated case preempts each object at most once and our algorithm for the uncapacitated case only preempts objects at depot vertices.

The preemptive Dial-a-Ride problem has been considered earlier with a single vehicle, for which an $O(\log n)$ approximation [3] and an $\Omega(\log^{1/4-\epsilon} n)$ hardness of approximation (for any constant $\epsilon > 0$) [9] are known. Note that the completion time and makespan objectives coincide for this case.

Moving to multiple vehicles, the total completion time objective admits a straightforward $O(\log n)$ approximation along the lines of the single vehicle problem [3]: Using the FRT tree embedding [7], one can reduce to tree-metrics at the loss of an expected $O(\log n)$ factor, and there is a simple constant approximation for this problem on trees. The maximum completion time or makespan objective, which we consider in this paper turns out to be considerably harder. Due to non-linearity of the makespan objective, the above reduction to tree-metrics does not hold. Furthermore, the makespan objective does not appear easy to solve even on trees.

Unlike in the single-vehicle case, note that an object in multi-vehicle Dial-a-Ride may be transported by several vehicles one after the other. Hence it is important for the vehicle routes to be coordinated so that the objects trace valid paths from respective sources to destinations. For example, a vehicle may have to wait at a vertex for other vehicles carrying common objects to arrive. Interestingly, the multi-vehicle Dial-a-Ride problem captures aspects of both machine scheduling and network design problems.

Results and Paper Outline. We first consider the special case of multi-vehicle Dial-a-Ride (*uncapacitated* mDaR) where the vehicles have no capacity constraints (i.e. $k \geq m$). This problem is interesting in itself, and serves as a good starting point before we present the algorithm for the general case. The uncapacitated mDaR problem itself highlights differences from the single vehicle case: For example, in single vehicle Dial-a-Ride, preemption plays no role in the absence of capacity constraints; however in uncapacitated mDaR, an optimal non-preemptive schedule may take $\Omega(\sqrt{q})$ longer than the optimal preemptive schedule (see the full version of the paper). We prove the following theorem in Section 2.

Theorem 1. *There is an $O(\log t)$-approximation algorithm for uncapacitated preemptive* mDaR *obtaining a tour where objects are only preempted at depot vertices.*

The above algorithm has two main steps: the first one reduces the instance (at a constant factor loss in the performance guarantee) to one in which all demands are between depots (a "depot-demand" instance). In the second step, we use a *sparse spanner* on the demand graph to construct routes for moving objects across depots.

We also obtain an improved guarantee for the following special class of metrics using the notion of *sparse covers* in such metrics [14].

Theorem 2. *There is an $O(1)$-approximation algorithm for uncapacitated* mDaR *on metrics induced by graphs that exclude any fixed minor.*

In Section 3, we study the capacitated preemptive mDaR problem, and obtain our main result. Recall that there is an $\Omega(\log^{1/4-\epsilon} n)$ hardness of approximation for even single vehicle Dial-a-Ride [9]. A feasible solution to preemptive mDaR is said to be *1-preemptive* if every object is preempted at most once while being moved from its source to destination.

Theorem 3. *There is an $O(\log^3 n)$ approximation algorithm for preemptive* mDaR *obtaining a 1-preemptive tour.*

This algorithm has four key steps: (1) We *preprocess* the input so that demand points that are sufficiently far away from each other can be essentially decomposed into separate instances for the algorithm to handle independently. (2) We then solve a single-vehicle instance of the problem that obeys some additional bounded-delay property (Theorem 6) that we prove; This property combines ideas from algorithms for *light approximate shortest path trees* [13] and *capacitated vehicle routing* [11]. The bounded-delay property is useful in *randomly partitioning* the single vehicle solution among the q vehicles available to share this load. This random partitioning scheme is reminiscent of the work of Hochbaum-Maass [12], Baker [1] and Klein-Plotkin-Rao [14], in trying to average out the effect of the cutting in the objective function. (3) The partitioned segments of the single vehicle tour are assigned to the available vehicles; However, to check if this assignment is feasible we solve a matching problem that identifies cases when this load assignment must be *rebalanced*. This is perhaps the most interesting step in the algorithm since it identifies stronger lower bounds for subproblems where the current load assignment is not balanced. (4) We finish up by *recursing* on the load rebalanced subproblem; An interesting feature of the recursion is that the fraction of demands that are processed recursively is not a fixed value (as is more common in such recursive algorithms) but is a carefully chosen function of the number of vehicles on which these demands have to be served.

Due to lack of space some proofs are omitted from this paper. The proofs can be found in the full version.

Related Work. Dial-a-Ride problems form an interesting subclass of Vehicle Routing Problems that are well studied in the operations research literature. Paepe et al. [5] provide a classification of Dial-a-Ride problems using a notation similar to that for scheduling and queuing problems: preemption is one aspect in this classification. Savelsberg and Sol [18] and Cordeau and Laporte [4] survey several variants of non-preemptive Dial-a-Ride problems that have been studied in the literature. Most Dial-a-Ride problems arising in practice involve making routing decisions for multiple vehicles.

Dial-a-Ride problems with transshipment (the preemptive version) have been studied in [15,16,17]. These papers consider a more general model where preemption is allowed only at a specified subset of vertices. Our model (and that of [3]) is the special case when every vertex can serve as a preemption point. It is clear that preemption only reduces the cost of serving demands: [17] studied the maximum decrease in the optimal cost upon introducing one preemption point. [15,16] also model time-windows on the demands, and study heuristics and a column-generation based approach; they also describe applications (eg. courier service) that allow for preemptions. The *truck and trailer routing problem* has been studied in [2,19]. Here a number of capacitated trucks and trailers are used to deliver all objects. Some customers are only accessible without the trailer. The trailers can be parked at any point accessible with a trailer and it is possible to shift demand loads between the truck and the trailer at the parking places.

For single vehicle Dial-a-Ride, the best known approximation guarantee for the preemptive version is $O(\log n)$ (Charikar and Raghavachari [3]), and an $\Omega(\log^{1/4-\epsilon} n)$ hardness of approximation (for any constant $\epsilon > 0$) is shown in Gørtz [9]. The non-preemptive version appears much harder and the best known approximation ratio is $\min\{\sqrt{k}\log n, \sqrt{n}\log^2 n\}$ (Charikar and Raghavachari [3], Gupta et al. [10]); however to the best of our knowledge, APX-hardness is the best lower bound. There are known instances of single vehicle Dial-a-Ride where the ratio between optimal non-preemptive and preemptive tours is $\Omega(\sqrt{n})$ in general metrics [3], and $\tilde{\Omega}(n^{1/8})$ in the Euclidean plane [10]. A 1.8-approximation is known for the $k = 1$ special case of single vehicle Dial-a-Ride (a.k.a. *stacker-crane* problem) [8].

The uncapacitated case of preemptive mDaR is also a generalization of a problem called *nurse-station-location* that was studied in Even et al. [6] (where a 4-approximation algorithm was given). Nurse-station-location is a special case of uncapacitated mDaR when each source-destination pair coincides on a single vertex. In this paper, we handle not only the case with arbitrary pairs (uncapacitated mDaR), but also the more general problem with finite capacity restriction.

Problem Definition and Preliminaries We represent a finite metric as (V, d) where V is the set of vertices and d is a symmetric distance function satisfying the triangle inequality. For subsets $A, B \subseteq V$ we denote by $d(A, B)$ the minimum distance between a vertex in A and another in B, so $d(A, B) = \min\{d(u, v) \mid u \in A, v \in B\}$. For a subset $E \subseteq \binom{V}{2}$ of edges, $d(E) := \sum_{e\in E} d_e$ denotes the total length of edges in E.

The *multi-vehicle Dial-a-Ride problem* (mDaR) consists of an n-vertex metric (V, d), m objects specified as source-destination pairs $\{s_i, t_i\}_{i=1}^m$, q vehicles having respective depot-vertices $\{r_j\}_{j=1}^q$, and a common vehicle capacity k. A feasible schedule is a set of q routes, one for each vehicle (where the route for vehicle $j \in [q]$ starts and ends at r_j), such that no vehicle carries more than k objects at any time and each object is moved from its source to destination. The completion time C_j of any vehicle $j \in [q]$

is the time when vehicle j returns to its depot r_j at the end of its route (the schedule is assumed to start at time 0). The objective in mDaR is to minimize the makespan, i.e., $\min \max_{j\in[q]} C_j$. We denote by $S := \{s_i \mid i \in [m]\}$ the set of sources, $T := \{t_i \mid i \in [m]\}$ the set of destinations, $R := \{r_j \mid j \in [q]\}$ the set of distinct depot-vertices, and $t := |R|$ the number of distinct depots. Unless mentioned otherwise, we only consider the *preemptive* version, where objects may be left at intermediate vertices while being moved from source to destination.

Single vehicle Dial-a-Ride. The following are lower bounds for the single vehicle problem: the minimum length TSP tour on the depot and all source/destination vertices (*Steiner* lower bound), and $\frac{\sum_{i=1}^{m} d(s_i,t_i)}{k}$ (*flow* lower bound). Charikar and Raghavachari [3] gave an $O(\log n)$ approximation algorithm for this problem based on the above lower bounds. Gupta et al. [10] showed that the single vehicle preemptive Dial-a-Ride problem always has a 1-preemptive tour of length $O(\log^2 n)$ times the Steiner and flow lower-bounds.

Lower bounds for mDaR. The quantity $\frac{\sum_{i=1}^{m} d(s_i,t_i)}{qk}$ is a lower bound similar to the flow bound for single vehicle Dial-a-Ride. Analogous to the Steiner lower bound above, is the optimal value of an induced *nurse-station-location* instance. In the nurse-station-location problem [6], we are given a metric (V, d), a set $\mathcal{T}$ of terminals and a multi-set $\{r_j\}_{j=1}^{q}$ of depot-vertices; the goal is to find a collection $\{F_j\}_{j=1}^{q}$ of trees that collectively contain all terminals $\mathcal{T}$ such that each tree F_j is rooted at vertex r_j and $\max_{j=1}^{q} d(F_j)$ is minimized. Even et al. [6] gave a 4-approximation algorithm for this problem. The optimal value of the nurse-station-location instance with depots $\{r_j\}_{j=1}^{q}$ (depots of vehicles in mDaR) and terminals $\mathcal{T} = S \cup T$ is a lower bound for mDaR. The following are some lower bounds implied by nurse-station-location: (a) $1/q$ times the minimum length forest that connects every vertex in $S \cup T$ to some depot vertex, (b) $\max_{i\in[m]} d(R, s_i)$, and (c) $\max_{i\in[m]} d(R, t_i)$. Finally, it is easy to see that $\max_{i\in[m]} d(s_i, t_i)$ is also a lower bound for mDaR.

2 Uncapacitated Preemptive mDaR

In this section we study the uncapacitated special case of preemptive mDaR, where vehicles have no capacity constraints (i.e., capacity $k \geq m$). We give an algorithm that achieves an $O(\log t)$ approximation ratio for this problem (recall $t \leq n$ is the number of distinct depots). Unlike in the single vehicle case, preemptive and non-preemptive versions of mDaR are very different even without capacity constraints (there exists an $\Omega(\sqrt{q})$ factor gap, where q is number of vehicles). The algorithm for uncapacitated preemptive mDaR proceeds in two stages. Given any instance, it is first reduced (at the loss of a constant factor) to a depot-demand instance, where all demands are between depot vertices. Then the depot-demand instance is solved using an $O(\log t)$ approximation algorithm.

Reduction to depot-demand instances. We define *depot-demand instances* as those instances of uncapacitated mDaR where all demands are between depot vertices. Given any instance $\mathcal{I}$ of uncapacitated mDaR, the algorithm UncapMulti (given below) reduces $\mathcal{I}$ to a depot-demand instance.

Input: instance $\mathcal{I}$ of uncapacitated preemptive mDaR.

1. Solve the nurse-station-location instance with depots $\{r_j\}_{j=1}^q$ and all sources/ destinations $S \cup T$ as terminals, using the 4-approximation algorithm [6]. Let $\{F_j\}_{j=1}^q$ be the resulting trees covering $S \cup T$ such that each tree F_j is rooted at depot r_j.
2. Define a depot-demand instance $\mathcal{J}$ of uncapacitated mDaR on the same metric and set of vehicles, where the demands are $\{(r_j, r_l) \mid s_i \in F_j \;\&\; t_i \in F_l,\ 1 \le i \le m\}$. For any object $i \in [m]$ let the *source depot* be the depot r_j for which $s_i \in F_j$ and the *destination depot* be the depot r_l for which $t_i \in F_l$.
3. **Output** the following schedule for $\mathcal{I}$:
   (a) Each vehicle $j \in [q]$ traverses tree F_j by an Euler tour, picks up all objects from sources in F_j and brings them to their source-depot r_j.
   (b) Vehicles implement a schedule for *depot-demand instance* $\mathcal{J}$, and all objects are moved from their source-depot to destination-depot (see Section 2).
   (c) Each vehicle $j \in [q]$ traverses tree F_j by an Euler tour, picks up all objects having destination-depot r_j and brings them to their destinations in F_j.

Note that objects only are preempted at depot vertices. We now argue that the reduction in UncapMulti only loses a constant approximation factor. Let B denote the optimal makespan of instance $\mathcal{I}$. Since the optimal value of the nurse-station-location instance solved in the first step of UncapMulti is a lower bound for $\mathcal{I}$, we have $\max_{j=1}^q d(F_j) \le 4B$.

Claim. The optimal makespan for the depot-demand instance $\mathcal{J}$ is at most $17B$.

Assuming a feasible schedule for $\mathcal{J}$, it is clear that the schedule returned by UncapMulti is feasible for the original instance $\mathcal{I}$. The first and third rounds in $\mathcal{I}$'s schedule require at most $8B$ time each. Thus an approximation ratio α for depot-demand instances implies an approximation ratio of $17\alpha + 16$ for general instances. Next we show an $O(\log t)$-approximation algorithm for depot-demand instances (here t is the number of depots), which implies Theorem 1.

Algorithm for depot-demand instances. Let $\mathcal{J}$ be any depot-demand instance: note that the instance defined in the second step of UncapMulti is of this form. It suffices to restrict the algorithm to the induced metric (R, d) on only depot vertices, and use only one vehicle at each depot in R. Consider an undirected graph H consisting of vertex set R and edges corresponding to demands: there is an edge between vertices r and s iff there is an object going from *either* r to s *or* s to r. Note that the metric length of any edge in H is at most the optimal makespan $\tilde{B}$ of instance $\mathcal{J}$. In the schedule produced by our algorithm, vehicles will only use edges of H. Thus in order to obtain an $O(\log t)$ approximation, it suffices to show that each vehicle only traverses $O(\log t)$ edges. Based on this, we further reduce $\mathcal{J}$ to the following instance $\mathcal{H}$ of uncapacitated mDaR: the underlying metric is shortest paths in the graph H (on vertices R), with one vehicle at each R-vertex, and for every edge $(u, v) \in H$ there is a demand from u to v and one from v to u. Clearly any schedule for $\mathcal{H}$ having makespan β implies one for $\mathcal{J}$ of makespan $\beta \cdot \tilde{B}$. The next lemma implies an $O(\log |R|)$ approximation for depot-demand instances.

Lemma 4. *There exists a poly-time computable schedule for $\mathcal{H}$ with makespan $O(\log t)$, where $t = |R|$.*

Proof: Let $\alpha = \lceil \lg t \rceil + 1$. We first construct a *sparse spanner* A of H as follows: consider edges of H in an arbitrary order, and add an edge $(u, v) \in H$ to A iff the shortest path between u and v using current edges of A is more than 2α. It is clear from this construction that the girth of A (length of its shortest cycle) is at least 2α, and that for every edge $(u, v) \in H$, the shortest path between u and v in A is at most 2α.

We now assign each edge of A to one of its end-points such that each vertex is assigned at most two edges. Repeatedly pick any vertex v of degree at most two in A, assign its adjacent edges to v, and remove these edges and v from A. We claim that at the end of this procedure (when no vertex has degree at most 2), all edges of A would have been removed (i.e. assigned to some vertex). Suppose for a contradiction that this is not the case. Let $\tilde{A} \neq \phi$ be the remaining graph; note that $\tilde{A} \subseteq A$, so the girth of $\tilde{A}$ is at least 2α. Every vertex in $\tilde{A}$ has degree at least 3, and there is at least one such vertex w. Consider performing a breadth-first search in $\tilde{A}$ from w. Since the girth of $\tilde{A}$ is at least 2α, the first α levels of the breadth-first search is a tree. Furthermore every vertex has degree at least 3, so each vertex in the first $\alpha - 1$ levels has at least 2 children. This implies that $\tilde{A}$ has at least $1 + 2^{\alpha-1} > t$ vertices, which is a contradiction! For each vertex $v \in R$, let A_v denote the edges of A assigned to v by the above procedure; we argued that $\cup_{v \in R} A_v = A$, and $|A_v| \leq 2$ for all $v \in R$.

The schedule for $\mathcal{H}$ involves 2α rounds as follows. In each round, every vehicle $v \in R$ traverses the edges in A_v (in both directions) and returns to v. Since $|A_v| \leq 2$ for all vertices v, each round takes 4 units of time; so the makespan of this schedule is $8\alpha = O(\log t)$. The route followed by each object in this schedule is the shortest path from its source to destination in spanner A; note that the length of any such path is at most 2α. To see that this is indeed feasible, observe that every edge of A is traversed by some vehicle in each round. Hence in each round, every object traverses one edge along its shortest path (unless it is already at its destination). Thus after 2α rounds, all objects are at their destinations. ∎

Tight example for uncapacitated mDaR lower bounds. We note that known lower bounds for uncapacitated preemptive mDaR are insufficient to obtain a sub-logarithmic approximation guarantee. The lower bounds we used in our algorithm are the following: $\max_{i \in [m]} d(s_i, t_i)$, and the optimal value of a nurse-station-location instance with depots $\{r_j\}_{j=1}^q$ and terminals $S \cup T$. We are not aware of any lower bounds stronger than these two bounds. There exist instances of uncapacitated mDaR where the optimal makespan is a factor $\Omega(\frac{\log t}{\log\log t})$ larger than both the above lower bounds.

3 Preemptive Multi-vehicle Dial-a-Ride

In this section we prove our main result: an $O(\log^2 m \cdot \log n)$ approximation algorithm for the preemptive mDaR problem. In the full version we remove this dependence on m, to obtain Theorem 3. We first prove a new structure theorem on single-vehicle Dial-a-Ride tours (Subsection 3.1) that preempts each object at most once, and where the total time spent by objects in the vehicle is small. Obtaining such a single vehicle tour is crucial in our algorithm for preemptive mDaR, which appears in Section 3.2. The algorithm for mDaR relies on a partial coverage algorithm Partial that, given subsets Q of vehicles and D of demands, outputs a schedule for Q of near-optimal makespan that covers some *fraction* of demands in D. Algorithm Partial follows an interesting recursive framework where the fraction of satisfied demands is not a fixed value but some function of the number $|Q|$ of vehicles (Lemma 7). The main steps in Partial are as follows. **(1)** Obtain a *single-vehicle* tour satisfying 1-preemptive and bounded-delay properties (Theorem 6), **(2)** Randomly partition the single vehicle tour into $|Q|$ equally spaced pieces, **(3)** Solve a matching problem to assign *some* of these pieces to vehicles of Q that satisfy a subset of demands D, **(4)** A suitable fraction of the residual demands in D are covered recursively by unused vehicles of Q.

3.1 Capacitated Vehicle Routing with Bounded Delay

Before we present the structural result on Dial-a-Ride tours, we consider the classic *capacitated vehicle routing problem* [11] with an additional constraint on object 'delays'. In the capacitated vehicle routing problem (CVRP) we are given a metric (V, d), specified depot-vertex $r \in V$, and m objects each having source r and respective destinations $\{t_i\}_{i\in[m]}$. The goal is to compute a minimum length *non-preemptive* tour of a capacity k vehicle originating at r that moves all objects from r to their destinations. In *CVRP with bounded delay*, we are additionally given a *delay parameter* $\beta > 1$, and the goal is to find a minimum length capacitated non-preemptive tour serving all objects such that the time spent by each object $i \in [m]$ in the vehicle is at most $\beta \cdot d(r, t_i)$. The following are natural lower bounds [11], even without the bounded delay constraint: (i) the minimum length TSP tour on $\{r\} \cup \{t_i \mid i \in [m]\}$ (cf. Steiner lower bound), and (ii) the quantity $\frac{2}{k}\sum_{i=1}^{m} d(r, t_i)$ (cf. flow lower bound).

Theorem 5. *There is a* $(2.5 + \frac{3}{\beta-1})$ *approximation algorithm for CVRP with bounded delay, where* $\beta > 1$ *is the delay parameter. This guarantee is relative to the Steiner and flow lower bounds.*

We now consider the *single vehicle* preemptive Dial-a-Ride problem given by metric (V, d), set D of demands, and a vehicle of capacity k. We prove the following structural result which extends a result from [10].

Theorem 6. *There is a randomized poly-time computable 1-preemptive tour* τ *servicing* D *that satisfies the following conditions (where* LB_{pmt} *is the maximum of the Steiner and flow lower bounds):*

1. **Total length:** $d(\tau) \le O(\log^2 n) \cdot \mathsf{LB}_{pmt}$.
2. **Bounded delay:** $\sum_{i\in D} T_i \le O(\log n) \sum_{i\in D} d(s_i, t_i)$ *where* T_i *is the total time spent by object* $i \in D$ *in the vehicle under the schedule given by* τ.

3.2 Algorithm for Preemptive mDaR

The algorithm first guesses the optimal makespan B of the given instance of preemptive mDaR (it suffices to know B within a constant factor for a polynomial-time algorithm). Let $\alpha = 1 - \frac{1}{1+\lg m}$. For any subset $Q \subseteq [q]$, we abuse the notation and use Q to denote both the set of vehicles Q and the multi-set of depots corresponding to vehicles Q.

We give an algorithm Partial that takes as input a tuple $\langle Q, D, B\rangle$ where $Q \subseteq [q]$ is a subset of vehicles, $D \subseteq [m]$ a subset of demands and $B \in \mathbb{R}_+$, with the *promise* that vehicles Q (originating at their respective depots) suffice to completely serve the demands D at a makespan of B. Given such a promise, Partial $\langle Q, D, B\rangle$ returns a schedule of makespan $O(\log n \log m) \cdot B$ that serves a good fraction of D. Algorithm Partial$\langle Q, D, B\rangle$ is given in below. We set parameter $\rho = \Theta(\log n \log m)$, the precise constant in the Θ-notation comes from the analysis.

Input: Vehicles $Q \subseteq [q]$, demands $D \subseteq [m]$, bound $B \geq 0$ *such that* Q can serve D at makespan B.

Preprocessing

1. If the minimum spanning tree (MST) on vertices Q contains an edge of length greater than $3B$, there is a non-trivial partition $\{Q_1, Q_2\}$ of Q with $d(Q_1, Q_2) > 3B$. For $j \in \{1, 2\}$, let $V_j = \{v \in V \mid d(Q_j, v) \leq B\}$ and D_j be all demands of D induced on V_j. Run in parallel the schedules from Partial$\langle Q_1, D_1, B\rangle$ and Partial$\langle Q_2, D_2, B\rangle$. Assume there is no such long edge in the following.

Random partitioning

2. Obtain single-vehicle 1-preemptive tour τ using capacity k and serving demands D (Theorem 6).
3. Choose a uniformly random offset $\eta \in [0, 2\rho B]$ and cut edges of tour τ at distances $\{2p\rho B + \eta \mid p = 1, 2, \cdots\}$ along the tour to obtain a set $\mathcal{P}$ of pieces of τ.
4. C'' is the set of objects $i \in D$ such that i is carried by the vehicle in τ over some edge that is cut in Step (3); and $C' := D \setminus C''$. Ignore the cut objects C'' in the rest of the algorithm.

Load rebalancing

5. Construct a bipartite graph H with vertex sets $\mathcal{P}$ and Q and an edge between piece $P \in \mathcal{P}$ and depot $f \in Q$ iff $d(f, P) \leq 2B$. For any subset $A \subseteq \mathcal{P}$, $\Gamma(A) \subseteq Q$ denotes the neighborhood of A in graph H. Let $\mathcal{S} \subseteq \mathcal{P}$ be any *maximal* set that satisfies $|\Gamma(\mathcal{S})| \leq \frac{|\mathcal{S}|}{2}$.
6. Compute a *2-matching* $\pi : \mathcal{P} \setminus \mathcal{S} \to Q \setminus \Gamma(\mathcal{S})$, i.e., a function such that the number of pieces mapping to any $f \in Q \setminus \Gamma(\mathcal{S})$ is $|\pi^{-1}(f)| \leq 2$.

Recursion

7. Define $C_1 := \{i \in C' \mid$ either $s_i \in \mathcal{S}$ or $t_i \in \mathcal{S}\}$; and $C_2 := C' \setminus C_1$.
8. Run in parallel the *recursive* schedule Partial$\langle \Gamma(\mathcal{S}), C_1, B\rangle$ for C_1 and the following for C_2:
   (a) Each vehicle $f \in Q \setminus \Gamma(\mathcal{S})$ traverses the pieces $\pi^{-1}(f)$, moving all C_2-objects in them from their source to preemption-vertex, and returns to its depot.
   (b) Each vehicle $f \in Q \setminus \Gamma(\mathcal{S})$ again traverses the pieces $\pi^{-1}(f)$, this time moving all C_2-objects in them from their preemption-vertex to destination, and returns to its depot.

Output: A schedule for vehicles Q of makespan $(16 + 16\rho) \cdot B$ that serves an $\alpha^{\lg \min\{|Q|, 2m\}}$ fraction of D.

Lemma 7. *If there exists a schedule for vehicles Q covering all demands D, having makespan at most B, then* Partial *invoked on $\langle Q, D, B\rangle$ returns a schedule of vehicles Q of makespan at most $(16 + 16\rho) \cdot B$ that covers at least an $\alpha^{\lg z}$ fraction of D (here $z := \min\{|Q|, 2m\} \leq 2m$).*

The final algorithm invokes Partial iteratively until all demands are covered: each time with the entire set $[q]$ of vehicles, all uncovered demands, and bound B. If $D \subseteq [m]$ is the set of uncovered demands at any iteration, Lemma 7 implies that Partial$\langle[q], D, B\rangle$ returns a schedule of makespan $O(\log m \log n) \cdot B$ that serves at least $\frac{1}{4}|D|$ demands. Hence a standard set-cover analysis implies that all demands will be covered in $O(\log m)$ rounds, resulting in a makespan of $O(\log^2 m \log n) \cdot B$.

It remains to prove Lemma 7. We proceed by induction on the number $|Q|$ of vehicles. The base case $|Q| = 1$ reduces to the single vehicle preemptive Dial-a-Ride, where we can serve *all* the demands D in a 1-preemptive fashion at makespan $O(\log^2 n) \cdot B$ using the algorithm from [10]. In the rest of this section, we prove the inductive step.

Preprocessing. Suppose Step (1) applies. Note that $d(V_1, V_2) > B$ and hence there is no demand with source in one of $\{V_1, V_2\}$ and destination in the other. So demands D_1 and D_2 partition D. Furthermore in the optimal schedule, vehicles Q_j (any $j = 1, 2$) only visit vertices in V_j (otherwise the makespan would be greater than B). Thus the two recursive calls to Partial satisfy the assumption: there is some schedule of vehicles Q_j serving D_j having makespan B. Inductively, the schedule returned by Partial for each $j = 1, 2$ has makespan at most $(16 + 16\rho) \cdot B$ and covers at least $\alpha^{\lg c} \cdot |D_j|$ demands from D_j, where $c \leq \min\{|Q| - 1, 2m\} \leq z$. The schedules returned by the two recursive calls to Partial can clearly be run in parallel and this covers at least $\alpha^{\lg z}(|D_1| + |D_2|)$ demands, i.e. an $\alpha^{\lg z}$ fraction of D. So we have the desired performance in this case.

Random partitioning. The harder part of the analysis is when Step (1) does not apply: so the MST length on Q is at most $3|Q| \cdot B$. Note that when the depots Q are contracted to a single vertex, the MST on the end-points of D plus the contracted depot-vertex has length at most $|Q| \cdot B$ (the optimal makespan schedule induces such a tree). Thus the MST on the depots Q along with end-points of D has length at most $4|Q| \cdot B$. Based on the assumption in Lemma 7 and the flow lower bound, we have $\sum_{i\in D} d(s_i, t_i) \leq k|Q| \cdot B$. It follows that for the single vehicle Dial-a-Ride instance solved in Step (2), the Steiner and flow lower-bounds (denoted LB_{pmt} in Theorem 6) are $O(1) \cdot |Q|B$. Theorem 6 now implies that τ is a 1-preemptive tour τ servicing D, of length at most $O(\log^2 n)|Q| \cdot B$ such that $\sum_{i\in D} T_i \leq O(\log n) \cdot |D|B$, where T_i denotes the total time spent in the vehicle by demand $i \in D$. The bound on the delay uses the fact that $\max_{i=1}^m d(s_i, t_i) \leq B$.

Choosing a large enough constant corresponding to $\rho = \Theta(\log n \log m)$, the length of τ is upper bounded by $\rho|Q| \cdot B$ (since $n \leq 2m$). So the cutting procedure in Step (3) results in at most $|Q|$ pieces of τ, each of length at most $2\rho B$. The objects $i \in C''$ (as defined in Step (4)) are called *cut objects*. We restrict our attention to the other objects $C' = D \setminus C''$ that are not 'cut'. For each object $i \in C'$, the path traced by it (under single vehicle tour τ) from its source s_i to preemption-point and the path (under τ) from its preemption-point to t_i are each completely contained in pieces of $\mathcal{P}$. Figure 1 gives an example of objects in C' and C'', and the cutting procedure.

Claim. The expected number of objects in C'' is at most $\sum_{i\in D} \frac{T_i}{2\rho B} \leq O(\frac{1}{\log m}) \cdot |D|$.

We can derandomize Step (3) and pick the best offset η (there are at most polynomially many combinatorially distinct offsets). Claim 3.2 implies (again choosing large enough constant in $\rho = \Theta(\log n \log m)$) that $|C'| \geq (1 - \frac{1}{2\lg m})|D| \geq \alpha \cdot |D|$ demands are *not cut*. From now on we only consider the set C' of uncut demands. Let $\mathcal{P}$ denote the pieces obtained by cutting τ as above, recall $|\mathcal{P}| \leq |Q|$. A piece $P \in \mathcal{P}$ is said to be non-trivial if the vehicle in the 1-preemptive tour τ carries some C'-object while traversing P. Note that the number of non-trivial pieces in $\mathcal{P}$ is at most $2|C'| \le 2m$: each C'-object appears in at most 2 pieces, one where it is moved from source to preemption-vertex and other from preemption-vertex to destination. Retain only the non-trivial pieces in $\mathcal{P}$; so $|\mathcal{P}| \le \min\{|Q|, 2m\} = z$. The pieces in $\mathcal{P}$ may not be one-one assignable to the depots since the algorithm has not taken the depot locations into account. We determine which pieces may be assigned to depots by considering a matching problem between $\mathcal{P}$ and the depots in Step (5) and (6).

Load rebalancing. The bipartite graph H (defined in Step (5)) represents which pieces and depots may be assigned to each other. Piece $P \in \mathcal{P}$ and depot $f \in Q$ are assignable iff $d(f, P) \le 2B$, and in this case graph H contains an edge (P, f). We claim that corresponding to the 'maximal contracting' set $\mathcal{S}$ (defined in Step (5)), the 2-matching π (in Step (6)) is guaranteed to exist. Note that $|\Gamma(\mathcal{S})| \le \frac{|\mathcal{S}|}{2}$, but $|\Gamma(\mathcal{T})| > \frac{|\mathcal{T}|}{2}$ for all $\mathcal{T} \supset \mathcal{S}$. For any $T' \subseteq \mathcal{P} \setminus \mathcal{S}$, let $\tilde{\Gamma}(T')$ denote the neighborhood of T' in $Q \setminus \Gamma(\mathcal{S})$. The maximality of $\mathcal{S}$ implies: for any non-empty $T' \subseteq \mathcal{P} \setminus \mathcal{S}$, $\frac{|\mathcal{S}|}{2} + \frac{|T'|}{2} = \frac{|\mathcal{S} \cup T'|}{2} < |\Gamma(\mathcal{S} \cup T')| = |\Gamma(\mathcal{S})| + |\tilde{\Gamma}(T')|$, i.e. $|\tilde{\Gamma}(T')| \ge \frac{|T'|}{2}$. Hence by *Hall's condition*, there is a 2-matching $\pi : \mathcal{P} \setminus \mathcal{S} \to Q \setminus \Gamma(\mathcal{S})$. The set $\mathcal{S}$ and 2-matching π can be easily computed in polynomial time.

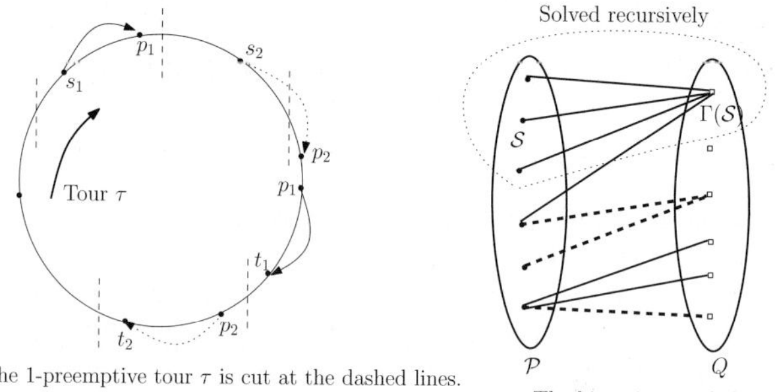

The 1-preemptive tour τ is cut at the dashed lines.
Object 1 is in C', it is not cut.
Object 2 is not in C', it is a cut object.

The bipartite graph H
The 2-matching π is shown by dashed edges.

Fig. 1. Cutting and patching steps in algorithm Partial

Recursion. In Step (7), demands C' are further partitioned into two sets: C_1 consists of objects that are *either* picked-up *or* dropped-off in some piece of $\mathcal{S}$; and C_2-objects are picked-up *and* dropped-off in pieces of $\mathcal{P} \setminus \mathcal{S}$. The vehicles $\Gamma(\mathcal{S})$ suffice to serve all C_1 objects, as shown below.

Claim. There exists a schedule of vehicles $\Gamma(\mathcal{S})$ serving C_1, with makespan B.

In the final schedule, a *large fraction* of C_1 demands are served by vehicles $\Gamma(\mathcal{S})$, and *all* the C_2 demands are served by vehicles $Q \setminus \Gamma(\mathcal{S})$. Figure 1 shows an example of this partition.

Serving C_1 demands. Based on Claim 3.2, the recursive call Partial $\langle \Gamma(\mathcal{S}), C_1, B \rangle$ (made in Step (8)) satisfies the assumption required in Lemma 7. Since $|\Gamma(\mathcal{S})| \le$

$\frac{|\mathcal{P}|}{2} \leq \frac{|Q|}{2} < |Q|$, we obtain inductively that Partial $\langle \Gamma(\mathcal{S}), C_1, B \rangle$ returns a schedule of makespan $(16 + 16\rho) \cdot B$ covering at least $\alpha^{\lg y} \cdot |C_1|$ demands of C_1, where $y = \min\{|\Gamma(\mathcal{S})|, 2m\}$. Note that $y \leq |\Gamma(\mathcal{S})| \leq |\mathcal{P}|/2 \leq z/2$ (as $|\mathcal{P}| \leq z$), which implies that at least $\alpha^{\lg z - 1}|C_1|$ demands are covered.

Serving C_2 demands. These are served by vehicles $Q \setminus \Gamma(\mathcal{S})$ using the 2-matching π, in two rounds as specified in Step (8). This suffices to serve all objects in C_2 since for any $i \in C_2$, the paths traversed by object i under τ, namely $s_i \rightsquigarrow p_i$ (its preemption-point) and $p_i \rightsquigarrow t_i$ are contained in pieces of $\mathcal{P} \setminus \mathcal{S}$. Furthermore, since $|\pi^{-1}(f)| \leq 2$ for all $f \in Q \setminus \Gamma(\mathcal{S})$, the distance traveled by vehicle f in one round is at most $2 \cdot 2(2B + 2\rho B)$. So the time taken by this schedule is at most $2 \cdot 4(2B + 2\rho B) = (16 + 16\rho) \cdot B$.

The schedule of vehicles $\Gamma(\mathcal{S})$ (serving C_1) and vehicles $Q \setminus \Gamma(\mathcal{S})$ (serving C_2) can clearly be run in parallel. This takes time $(16 + 16\rho) \cdot B$ and covers in total at least $|C_2| + \alpha^{\lg z - 1}|C_1| \geq \alpha^{\lg z - 1}|C'| \geq \alpha^{\lg z}|D|$ demands of D. This proves the inductive step of Lemma 7.

Using Lemma 7 repeatedly as mentioned earlier, we obtain an $O(\log^2 m \cdot \log n)$ approximation algorithm for capacitated preemptive mDaR.

References

1. Baker, B.S.: Approximation Algorithms for NP-Complete Problems on Planar Graphs. J. ACM 41, 153–180 (1994)
2. Chao, I.-M.: A tabu search method for the truck and trailer routing problem. Computer & Operations Research 29, 469–488 (2002)
3. Charikar, M., Raghavachari, B.: The Finite Capacity Dial-A-Ride Problem. In: FOCS, pp. 458–467 (1998)
4. Cordeau, J.-F., Laporte, G.: The Dial-a-Ride Problem (DARP): Variants, modeling issues and algorithms. 4OR: A Quarterly Journal of Operations Research 1(2) (2003)
5. de Paepe, W.E., Lenstra, J.K., Sgall, J., Sitters, R.A., Stougie, L.: Computer-Aided Complexity Classification of Dial-a-Ride Problems. Informs J. Comp. 16(2), 120–132 (2004)
6. Even, G., Garg, N., Könemann, J., Ravi, R., Sinha, A.: Covering Graphs Using Trees and Stars. In: Arora, S., Jansen, K., Rolim, J.D.P., Sahai, A. (eds.) RANDOM 2003 and APPROX 2003. LNCS, vol. 2764, pp. 24–35. Springer, Heidelberg (2003)
7. Fakcharoenphol, J., Rao, S., Talwar, K.: A tight bound on approximating arbitrary metrics by tree metrics. In: STOC, pp. 448–455 (2003)
8. Frederickson, G.N., Hecht, M.S., Kim, C.E.: Approximation algorithms for some routing problems. SIAM J. Comput. 7(2), 178–193 (1978)
9. Gørtz, I.L.: Hardness of Preemptive Finite Capacity Dial-a-Ride. In: Díaz, J., Jansen, K., Rolim, J.D.P., Zwick, U. (eds.) APPROX 2006 and RANDOM 2006. LNCS, vol. 4110, pp. 200–211. Springer, Heidelberg (2006)
10. Gupta, A., Hajiaghayi, M., Nagarajan, V., Ravi, R.: Dial a Ride from k-forest. In: Arge, L., Hoffmann, M., Welzl, E. (eds.) ESA 2007. LNCS, vol. 4698, pp. 241–252. Springer, Heidelberg (2007)
11. Haimovich, M., Kan, A.H.G.R.: Bounds and heuristics for capacitated routing problems. Math. Oper. Res. 10, 527–542 (1985)
12. Hochbaum, D., Maass, W.: Approximation Schemes for Covering and Packing Problems in Image Processing and VLSI. J. ACM 32, 130–136 (1985)
13. Khuller, S., Raghavachari, B., Young, N.: Balancing minimum spanning and shortest path trees. In: SODA, pp. 243–250 (1993)

14. Klein, P., Plotkin, S.A., Rao, S.: Excluded minors, network decomposition, and multicommodity flow. In: STOC, pp. 682–690 (1993)
15. Mitrović-Minić, S., Laporte, G.: The Pickup and Delivery Problem with Time Windows and Transshipment. INFOR Inf. Syst. Oper. Res. 44, 217–227 (2006)
16. Mues, C., Pickl, S.: Transshipment and time windows in vehicle routing. In: 8th Int. Symp. on Parallel Architectures, Algorithms and Networks, pp. 113–119 (2005)
17. Nakao, Y., Nagamochi, H.: Worst case analysis for pickup and delivery problems with transfer. IEICE Trans. on Fund. of Electronics, Comm. and Computer Sci. E91-A(9) (2008)
18. Savelsbergh, M., Sol, M.: The general pickup and delivery problem. Transportation Science 29, 17–29 (1995)
19. Scheuerer, S.: A tabu search heuristic for the truck and trailer routing problem. Computer & Operations Research 33, 894–909 (2006)

Google's Auction for TV Ads

Noam Nisan

The Selim and Rachel Benin School of Computer Science and Engineering
The Hebrew University of Jerusalem
and Google, Tel-Aviv
noam@cs.huji.ac.il

Abstract. This talk describes the auction system used by Google for allocation and pricing of TV ads. It is based on a simultaneous ascending auction, and has been in use since September 2008.

A. Fiat and P. Sanders (Eds.): ESA 2009, LNCS 5757, p. 553, 2009.

Inclusion/Exclusion Meets Measure and Conquer

Exact Algorithms for Counting Dominating Sets

Johan M.M. van Rooij[1], Jesper Nederlof[2,⋆], and Thomas C. van Dijk[1]

[1] Department of Information and Computing Sciences, Utrecht University, P.O.Box 80.089, 3508 TB Utrecht, The Netherlands
{jmmrooij,thomasd}@cs.uu.nl
[2] Department of Informatics, University of Bergen, N-5020 Bergen, Norway
Jesper.Nederlof@ii.uib.no

Abstract. In this paper, two central techniques from the field of exponential time algorithms are combined for the first time: inclusion/exclusion and branching with measure and conquer analysis.

In this way, we have obtained an algorithm that, for each κ, counts the number of dominating sets of size κ in $\mathcal{O}(1.5048^n)$ time. This algorithm improves the previously fastest algorithm that counts the number of minimum dominating sets. The algorithm is even slightly faster than the previous fastest algorithm for minimum dominating set, thus improving this result while computing much more information.

When restricted to c-dense graphs, circle graphs, 4-chordal graphs or weakly chordal graphs, our combination of branching with inclusion/exclusion leads to significantly faster counting and decision algorithms than the previously fastest algorithms for dominating set.

All results can be extended to counting (minimum) weight dominating sets when the size of the set of possible weight sums is polynomially bounded.

1 Introduction

Recently, the field of exact exponential time algorithms has been an area of growing interest. Maybe the most notable recent developments are *measure and conquer* [10,11] and *inclusion/exclusion* [1,5,17]. Both techniques have been demonstrated on Set Cover problems in early stages: measure and conquer was introduced on a set cover formulation of Dominating Set, and in [5] inclusion/exclusion was used for counting sets coverings and set partitionings.

In this paper, we show that it is possible to use both techniques in one combined approach. This allows for fast measure and conquer running times on inclusion/exclusion based algorithms.

⋆ This research was done while all authors were associated with Utrecht University.

A. Fiat and P. Sanders (Eds.): ESA 2009, LNCS 5757, pp. 554–565, 2009.

The best known shape of inclusion/exclusion is the formula summing over some powerset; see [3,5,17]. However, the fundamental branching perspective from [1] is more direct and powerful. In this paper, we will apply this branching perspective to set cover instances obtained from the set cover formulation of dominating set that has been used to introduce measure and conquer [10,16].

In this setting, we use a traditional branching rule to branch on a set, or an application of inclusion/exclusion to branch on an element. The sole application of either one of these strategies gives a typical exhaustive search or the aforementioned shape of inclusion/exclusion sum, respectively. We use both branching strategies in unity obtaining a *mixed inclusion/exclusion branching* algorithm that can be analysed using measure and conquer.

Until 2004, no exact algorithm for Dominating Set beating the trivial $\mathcal{O}(2^n n^{\mathcal{O}(1)})$ was known. In that year, three algorithms were published [13,16,20], the fastest of which is Grandoni's running in time $\mathcal{O}(1.8019^n)$ [16]. One year later, the algorithm of Grandoni was analysed using measure and conquer giving a bound of $\mathcal{O}(1.5137^n)$ on the running time [10]. This was later improved by Van Rooij and Bodlaender [21] to $\mathcal{O}(1.5063^n)$.

When we want to *count minimum dominating sets*, there is an algorithm by Fomin et al. running in time $\mathcal{O}(1.5535^n)$ [9]. This algorithm combines branching with path decomposition techniques: something we will use for our own algorithm as well. Also related is a result by Björklund and Husfeldt solving this problem on cubic graphs in $\mathcal{O}(1.3161^n)$ using path decompositions in combination with inclusion/exclusion [3]. To our knowledge, there are no existing algorithms combining measure and conquer with inclusion/exclusion.

Our algorithm is more general. It counts the number of dominating sets in an n-vertex graph of each size $0 \leq \kappa \leq n$, with an upper bound on the running time of $\mathcal{O}(1.5048^n)$. This is slightly faster than even the current fastest algorithm that computes a minimum dominating set.

Gaspers et al. [14] show that algorithms for the set cover formulation of dominating set can be combined with dynamic programming over tree decompositions to obtain faster running times for the dominating set problem restricted to some graph classes. These classes are c-dense graphs, chordal graphs, circle graphs, 4-chordal graphs and weakly chordal graphs. We show that our mixed branching approach with inclusion/exclusion branches works even better on four of these graph classes; we do not only improve these results because we have a faster algorithm for the underlying set cover problem, but do so more significantly by exploiting vertices of high degree twice by using both techniques. Moreover, we can count the number of dominating sets of each size, in contrast to the previous results that only compute a single minimum dominating set.

2 Preliminaries

We consider the #κ-Dominating Set problem: how many dominating sets of size κ exist for G, i.e., how many subsets $V' \subseteq V$ with $|V'| = \kappa$ such that for all $u \in V \backslash V'$ there is a $v \in V'$ for which $(u, v) \in E$?

We formulate the problem as the set cover variant #κ-SET COVER [16]: given a collection of subsets $\mathcal{S}$ of a finite universe $\mathcal{U}$ and a positive integer κ, how many set covers for $\mathcal{U}$ of size κ does $\mathcal{S}$ contain? The transformation between this problem and our original problem is straightforward: for every vertex in $v \in V$ we introduce both an element in $\mathcal{U}$ and a set in $\mathcal{S}$ corresponding to $N[v]$. We now use the *cardinality* $|S|$ of the set S and the *frequency* $f(e)$ of the element e instead of the degree of a vertex. The *dimension* of a set cover instance is defined as $\dim(\mathcal{S},\mathcal{U}) = |\mathcal{S}|+|\mathcal{U}|$. Hence, an n-vertex dominating set instance is transformed into a set cover instance of dimension $d = 2n$.

We also look at the problem as a #κ-RED/BLUE DOMINATING SET problem in the incidence graph of the set cover instance [9]. The *incidence graph* is the bipartite graph with *red vertices* $V_{Red} = \mathcal{S}$ and *blue vertices* $V_{Blue} = \mathcal{U}$. Vertices $S \in V_{Red}$ and $u \in V_{Blue}$ are adjacent if and only if $u \in S$. In this problem, we count the number of ways to take κ red vertices to dominate all the blue vertices. It is easy to see that this perspective is equivalent to the set cover variant.

Finally, we assume the reader to be familiar with the concepts of a (nice) tree decomposition and a (nice) path decompositions of a graph, and how to perform dynamic programming over these structures. For a good overview see [6,7].

3 Inclusion/Exclusion Based Branching

We will begin by showing that one can look at Inclusion/Exclusion as a branching rule [2]. In this way, we can Inclusion/Exclusion-branch on an element in a SET COVER instance in the same way as one would normally branch on a set.

A set S is *optional* in an instance, if either S is in the solution, or S is not. Branching on this choice is straightforward: the total number of set covers of size κ equals the number of set covers of size $\kappa - 1$ after we take S (*require*), plus the number of set covers of size κ after we discard S (*forbid*). I.e.:

$$\text{OPTIONAL} = \text{REQUIRED} + \text{FORBIDDEN}$$

We now consider branching on an element [2]. This may appear strange at first as elements are not optional. Inspired by Inclusion/Exclusion techniques, we can, however, rearrange the above formula to give:

$$\text{REQUIRED} = \text{OPTIONAL} - \text{FORBIDDEN}$$

That is, the number of solutions that cover an element e is equal to the number of solutions in which covering e is *optional* (maybe cover e), minus the number of solutions in which covering e is *forbidden* (that do not cover e). We call this type of branching *inclusion/exclusion based branching* or simply *IE-branching*.

Notice that both branching rules are symmetric when applied to the incidence graph representation of our problem: in one branch a (red or blue) vertex is removed, and in the other, this vertex and its neighbours are removed.

An algorithm that branches on every set is called *exhaustive search*, while an algorithm that solely use IE-branching is an *inclusion/exclusion* algorithm.

To see the relation to the inclusion/exclusion formula [5], let c_κ be the number of set covers of cardinality κ, and let $a(X)$ be the number of sets in $\mathcal{S}$ that do not contain any element in X. Consider the branching tree after exhaustively applying IE-branching and look at the contribution of a leaf to the total number computed. In each leaf of the tree, each element is either optional or forbidden; the $2^{|\mathcal{U}|}$ leaves represent the possible subsets $X \subseteq \mathcal{U}$ of forbidden elements. The contribution of this leaf equals the number of set covers of cardinality κ where it is optional to cover each element not in X and forbidden to cover an element in X, i.e. $\binom{a(X)}{\kappa}$. A minus sign is added for each time we have entered a forbidden branch, so the total contribution equals $(-1)^{|X|}\binom{a(X)}{\kappa}$. This leads to the formula given below on the left. Compare this to the inclusion/exclusion formula [5] on the right: the difference comes from the fact that Björklund et al. allow a single set to be picked multiple times while we do not.

$$c_\kappa = \sum_{X \subseteq \mathcal{U}} (-1)^{|X|} \binom{a(X)}{\kappa} \qquad\qquad c'_\kappa = \sum_{X \subseteq \mathcal{U}} (-1)^{|X|} a(X)^\kappa$$

4 An Algorithm for Counting Dominating Sets

We start by applying our combined technique to the problem of counting dominating sets. The previously fastest algorithm that counts the number of *minimum* dominating sets is by Fomin et al. [9]. Their algorithm combines pathwidth techniques with branching and measure and conquer analysis. We present a modification of this algorithm (Algorithm 1) that solves the #κ-Dominating Set problem. This algorithm computes the number of dominating sets of each size κ ($0 \leq \kappa \leq n$) in $O(1.5048^n)$ time improving both the results of [9] and the previously fastest algorithm for minimum dominating set [21].

Algorithm 1 works on the #κ-Set Cover transformation of the problem and returns a list containing the number of set covers $\boldsymbol{n}_\kappa$ of size κ for each $0 \leq \kappa \leq n$. It is a branch and reduce algorithm, branching both on sets and elements as discussed in Section 3. When an instance generated by the branching is sparse enough, the algorithm will compute a path decomposition of the incidence graph of the instance. The algorithm then solves this instance by dynamic programming on this path decomposition.

Algorithm 1 takes as input a collection of sets $\mathcal{S}$ forming a #κ-Set Cover instance, and a multiplicity function m. This function m exists because we want to avoid identical sets to be created by the algorithm; the algorithm deals with multiple identical sets by using the multiplicity counters in m. We will begin by describing a series of polynomial time reduction rules that Algorithm 1 applies before branching or applying pathwidth techniques.

Base Case

Some inputs can be completely reduced to a collection of m' empty sets by the reduction rules below. There are no elements left, and we only have empty sets to choose from, therefore the algorithm returns $\boldsymbol{n}_\kappa = \binom{m'}{k}$ for each $0 \leq \kappa \leq n$.

Algorithm 1. Count-SC($\mathcal{S}$,m)

Input: A collection of sets $\mathcal{S}$ over the universe $\mathcal{U} = \cup\mathcal{S}$ and a multiplicity function m.
Output: A list $\boldsymbol{n}$ of length $\#(\mathcal{S})+1$ containing the number of set covers of $(\mathcal{S},\mathcal{U})$ of each size $0 \leq \kappa \leq \#(\mathcal{S})$.

//reduction rules
if $\mathcal{S}$ equals a collection of $m' = m(\emptyset)$ empty sets **then** //base case
    return $(\binom{m'}{0}, \binom{m'}{1}, \ldots, \binom{m'}{m'})$
else if there exist identical sets in $\mathcal{S}$ **then** //identical sets
    Remove the identical sets from $\mathcal{S}$ and update the multiplicity function m.
    return Count-SC($\mathcal{S}$,m)
else if there exists an element $e \in \mathcal{U}$ of frequency one **then** //unique elements
    Let $\mathcal{S}'$ be $\mathcal{S}$ after removing the set S with $e \in S$ and let $\boldsymbol{n}_{\text{take}}$ = Count-SC($\mathcal{S}$,m)

    return $\boldsymbol{n}_{\text{take}}$ after updating it using the multiplicity of S in formula 1
else if there exist two elements $e, e' \in \mathcal{U}$ such that for all $S \in \mathcal{S}$: if $e \in S$, then $e' \in S$ **then** //subsumption
    Let $\mathcal{S}' = \{S \backslash \{e'\} \mid S \in \mathcal{S}\}$, and update m such that it now works on $\mathcal{S}'$
    return Count-SC($\mathcal{S}'$,m)
else if if $\mathcal{S}$ can be partitioned in two subcollections $\mathcal{C}, \bar{\mathcal{C}}$ such that every element of $e \in \mathcal{U}$ occurs either in $\mathcal{C}$ or in $\bar{\mathcal{C}}$ and not in both **then** //connected components
    Let $\boldsymbol{n}_{\mathcal{C}}$ = Count-SC($\mathcal{C}$,m), and $\boldsymbol{n}_{\bar{\mathcal{C}}}$ = Count-SC($\bar{\mathcal{C}}$,m)
    return The solution to $\mathcal{S}$ by merging $\boldsymbol{n}_{\mathcal{C}}$ and $\boldsymbol{n}_{\bar{\mathcal{C}}}$ using formula 2
end if

//branching or path decomposition
Let $S \in \mathcal{S}$ be of maximum cardinality and not an exceptional case[1]
Let $e \in \mathcal{U}$ be of maximum frequency, also not an exceptional case[1]
Preference order P: $\mathsf{S}_4 < \mathsf{S}_5 < \mathsf{S}_6 < \mathsf{E}_5 < \mathsf{E}_6 < \mathsf{S}_7 < \mathsf{E}_7 < \mathsf{E}_{\geq 8} = S_{\geq 8}$
if $\mathsf{S}_{|S|}$ and $\mathsf{E}_{\text{freq}(e)}$ are too small to be in P **then** //path decomposition
    Compute a path decomposition P_I of the incidence graph of $(\mathcal{S},\mathcal{U})$
    return The solution to $\mathcal{S}$ obtained by dynamic programming over P_I.
else if $\mathsf{E}_{\text{freq}(e)}$ is in the order P and $\mathsf{E}_{\text{freq}(e)} \not< \mathsf{S}_{|S|}$ **then** //element branch
    Let $\mathcal{S}' = \{S \backslash \{e'\} \mid S \in \mathcal{S}\}$, and update m such that it now works on $\mathcal{S}'$
    Let $\boldsymbol{n}_{\text{optional}}$ = Count-SC($\mathcal{S}'$,m)
    Let $\boldsymbol{n}_{\text{forbidden}}$ = Count-SC($\mathcal{S} \setminus \{S \in \mathcal{S} \mid e \in S\}$,$m$)
    return $\boldsymbol{n}_{\text{optional}} - \boldsymbol{n}_{\text{forbidden}}$
else //$\mathsf{S}_{|S|}$ is in the order P and $\mathsf{S}_{|S|} \not< \mathsf{E}_{\text{freq}(e)}$ //set branch
    Let $\mathcal{S}' = \{S' \backslash S \mid S' \in (\mathcal{S} \setminus \{S\})\}$, and update m such that it now works on $\mathcal{S}'$
    Let $\boldsymbol{n}_{\text{take}}$ = Count-SC($\mathcal{S}'$,m)
    Update $\boldsymbol{n}_{\text{take}}$ using the multiplicity of S in formula 1
    Let $\boldsymbol{n}_{\text{discard}}$ = Count-SC($\mathcal{S} \setminus \{S\}$,$m$)
    return $\boldsymbol{n}_{\text{take}} + \boldsymbol{n}_{\text{discard}}$
end if

[1] There are some exceptional combinations of cardinalities of sets and frequencies of elements on which the algorithm will not branch. These will be handled by the path decomposition phase. For a complete list of these cases see Overview 1.

Identical Sets
When $\mathcal{S}$ contains identical sets, we remove all but one copies of this set and keep track of this using multiplicity counters in m. We can do this because taking at least one copy in a solution will result in the same subproblem regardless of the number of copies chosen. Whenever the set is explicitly taken in a solution later on, we compute the required result from the values from the recursive call n'_κ using the formula below.

$$\boldsymbol{n}_\kappa = \sum_{i=1}^{m} \binom{m}{i} \boldsymbol{n}'_{\kappa-i} \tag{1}$$

To avoid confusion, we consider copies of sets to be removed when considering the frequency of its elements or the number of sets in $\mathcal{S}$.

Unique Elements
Whenever there exists an element e of frequency one in $\mathcal{U}$, the set S containing e must belong to every set cover. Therefore, the algorithm acts as if it takes this set and goes in recursion on the instance with S and all its elements removed, counting the number of set covers of size κ-1. Notice that it is not a problem if the set taken has multiplicity greater than one: simply use the above formula.

Subsumption
If there exists an element e which occurs in every set (and possibly more) in which another element e' occurs, then every set cover that covers e also covers e'. Thus, we can remove e' from the instance and recursively apply our algorithm.

Connected Components
If the incidence graph of the instance contains multiple connected components, then we can solve the problem on each component separately and merge the results. In this case, there exist two disjoint sets $\mathcal{C}, \bar{\mathcal{C}}$ with $\mathcal{C} \cup \bar{\mathcal{C}} = \mathcal{S}$ and with the property that every element of $e \in \mathcal{U}$ occurs either in $\mathcal{C}$ or in $\bar{\mathcal{C}}$ and not in both. Let $\boldsymbol{n}(\mathcal{C})_\kappa$, $\boldsymbol{n}(\bar{\mathcal{C}})_\kappa$ be the number of solutions of size κ to these two subproblems. In order to compute the total number of size covers n_κ of size κ in $\mathcal{C} \cup \bar{\mathcal{C}}$ we evaluate the following sum:

$$\boldsymbol{n}_\kappa = \sum_{i=0}^{\kappa} \boldsymbol{n}(\mathcal{C})_i \times \boldsymbol{n}(\bar{\mathcal{C}})_{\kappa-i} \quad \text{where: } \boldsymbol{n}(\mathcal{C})_i = 0 \text{ if } i > |\mathcal{C}| \tag{2}$$

Branching
When no reduction rules are applicable, the algorithm chooses a set or an element to branch on. From the instance, it chooses a set of maximum cardinality and an element of maximum frequency that are both not exceptional cases. We postpone the discussion of these exceptional cases for a moment. In order to choose between branching on the chosen set and branching on the chosen element, the algorithm uses the following preference order P:

$$\mathsf{P}: \quad \mathsf{S}_4 < \mathsf{S}_5 < \mathsf{S}_6 < \mathsf{E}_5 < \mathsf{E}_6 < \mathsf{S}_7 < \mathsf{E}_7 < \mathsf{E}_{\geq 8} = S_{\geq 8}$$

There are exceptional cases of elements on which, despite the preference order, our algorithm does not branch. These cases represent local neighbourhoods of sets or elements which would increase the running time of the algorithm when branched on, but can be handled by dynamic programming on a path decomposition quite effectively. The exceptional cases are:

1. Elements of frequency five that occur in many sets of small cardinality.
   Let the 5-tuple $(s_1, s_2, s_3, s_4, s_5, s_6)$ represent a frequency five element occurring s_i times in a cardinality i set. In this way, our special cases can be denoted as:

(1, 4, 0, 0, 0, 0) - (0, 5, 0, 0, 0, 0) - (1, 3, 1, 0, 0, 0) - (0, 4, 1, 0, 0, 0) - (1, 2, 2, 0, 0, 0)
(0, 3, 2, 0, 0, 0) - (1, 1, 3, 0, 0, 0) - (0, 2, 3, 0, 0, 0) - (0, 1, 4, 0, 0, 0) - (1, 0, 4, 0, 0, 0)
(1, 3, 0, 1, 0, 0) - (0, 4, 0, 1, 0, 0) - (1, 2, 1, 1, 0, 0) - (0, 3, 1, 1, 0, 0) - (1, 1, 2, 1, 0, 0)
(1, 0, 3, 1, 0, 0) - (1, 2, 0, 2, 0, 0) - (1, 3, 0, 0, 1, 0) - (1, 2, 1, 0, 1, 0) - (1, 3, 0, 0, 0, 1)

2. Sets of cardinality four, five or six, containing one of the elements described above.

Overview 1. Exceptional cases for our algorithm

In this ordering, $\mathsf{S}_i < \mathsf{E}_j$ means that the algorithm prefers to branch on an element of frequency j over branching on a set of cardinality i.

Sets of cardinality at most three and elements of frequency at most four do not occur in the preference order P. These are considered too small for efficient branching since branching on them would remove or reduce too few elements and sets. The remaining instances are handled by dynamic programming over a path decomposition of the incidence graph, similar to [9].

The exceptional cases are described in Overview 1. These exceptional cases represent local neighbourhoods around a set or an element which, despite the general rule imposed by the preference order, can be handled more efficiently by the path decomposition phase of our algorithm than by branching. They exist to properly balance the two parts of the algorithm.

Theorem 1. *There is an algorithm that solves the #κ-*DOMINATING SET *for all* $0 \leq \kappa \leq n$ *in an n-vertex graph G in* $\mathcal{O}(1.5048^n)$.

Proof (Sketch). The proof consists of a measure and conquer analysis of Algorithm 1 and is an extension of the proof in [9]. Due to space restrictions, we will only sketch it here. For the full proof, see [23].

We analyse our algorithm using measure and conquer [10,11]; see also [15,21]. Let $v, w : \mathbb{N} \to [0,1]$ be weight functions giving weight $v(i)$ to an element of frequency i and weight $w(i)$ to a set of cardinality i, respectively. With these functions we define the following complexity measure (identical to [10,21]):

$$k(\mathcal{S},\mathcal{U}) = \sum_{S\in\mathcal{S}} w(|S|) + \sum_{e\in\mathcal{U}} v(f(e)) \qquad \text{notice: } k(\mathcal{S},\mathcal{U}) \leq \dim(\mathcal{S},\mathcal{U})$$

We derive recurrence relations for the number of subproblems generated by the branching of the algorithm expressed in this complexity measure k. Given fuctions v, w, we can solve these recurrences and obtain an upper bound on the number of subproblems generated. The proof the comes down to computing the functions v, w that minimise the running time. This is a quasiconvex program [8] that we solve by computer. In this way, we prove:

Let $N_h(k)$ be the number of subproblem of complexity h generated by our algorithm on an input of complexity k. Then, $N_h(k) < 1.22670^{k-h}$.

Next, we use that by standard dynamic programming techniques we can solve the problem on a path decomposition of width p in $\mathcal{O}^*(2^p)$. We compute an upper bound on the maximum width p any path decomposition that is computed by our algorithm can have. Using upper bound on the pathwidth of sparse graphs [9,12], we formulate a linear program that computes the maximum pathwidth that any instance of complexity k. As a result we find that $p < 0.28991k$, and thus this part of the algorithm runs in $\mathcal{O}(2^{0.28991k}) \subset \mathcal{O}(1.2226^k)$.

By combining the time bound on both parts of the algorithm and using that initially $k \leq 2n$, we conclude that Algorithm 1 runs in $\mathcal{O}(1.22670^{2n}) \subset \mathcal{O}(1.5048^n)$. □

5 Dominating Set Restricted to Some Graph Classes

The algorithm from the previous section, not only gives the currently fastest algorithm to compute the number of dominating sets of given sizes, but also is the currently fastest algorithm for the minimum dominating set problem. However, the improvement over the previous fastest minimum dominating set algorithm [21] is small. When we consider the dominating set problem on specific graph classes, we get a larger improvement with our approach. This also extends the results on these graph classes to the counting variant of Dominating Set.

Gaspers et al. [14] consider exact algorithms for the dominating set problem on c-dense graphs, circle graphs, chordal graphs, 4-chordal graphs, and weakly chordal graphs. On these graph classes the problem is still NP-complete. They show that if we restrict ourselves to such a graph class, then there are either many vertices of high degree allowing more efficient branching, or the graph has low treewidth allowing the problem to be efficiently solved by dynamic programming over a tree decomposition. Using our approach, we will show that less vertices of high degree are required to obtain the same effect by branching on them with both branching rules. This leads to faster algorithms.

If we combine the results of the previous section with a result from Gaspers el al. [14], then we have the following proposition:

Proposition 1 ([14], Theorem 1). *Let $t > 0$ be a fixed integer, and let $\mathcal{G}_t$ be a class of graphs with for all $G \in \mathcal{G}_t$: $|\{v \in V(G) : d(v) \geq t-2\} \geq t$. Then, there is a $\mathcal{O}(1.22670^{2n-t})$ time algorithm to solve the minimum dominating set problem on graphs in $\mathcal{G}_t$.*

Using our two branching rules, we prove a stronger variant of this proposition.

Lemma 1. *Let $t > 0$ be a fixed integer, and let $\mathcal{G}_t$ be a class of graphs with for all $G \in \mathcal{G}_t$: $|\{v \in V(G) : d(v) \geq t-2\}| \geq \frac{1}{2}t$. Then, there is a $\mathcal{O}(1.22670^{2n-t})$ time algorithm to solve the minimum dominating set problem on graphs in $\mathcal{G}_t$.*

Table 1. Effect of two branching rules on the running times on some graph classes

Graph class	[14] (+[21]+[22])	[14] (+[22]) + Lemma 1
c-dense graphs	$\mathcal{O}\left(1.2273^{(1+\sqrt{1-2c})n}\right)$	$\mathcal{O}\left(1.2267^{(\frac{1}{2}+\frac{1}{2}\sqrt{9-16c})n}\right)$
circle graphs	$\mathcal{O}(1.4843^n)$	$\mathcal{O}(1.4806^n)$
chordal graphs	$\mathcal{O}(1.3720^n)$	$\mathcal{O}(1.3712^n)$ *
4-chordal graphs	$\mathcal{O}(1.4791^n)$	$\mathcal{O}(1.4741^n)$
weakly chordal graphs	$\mathcal{O}(1.4706^n)$	$\mathcal{O}(1.4629^n)$

* This result does not use Lemma 1; the improvement comes only from Theorem 1.

Proof. Let H be the set of vertices of degree at least $t-2$ from the statement of the lemma, and consider the set cover formulation of the dominating set problem.

Let S be a set corresponding to a vertex in H. We branch on this set and consider the branch in which we take this set in the set cover: the set is removed and all its elements are covered and hence removed also. These are at least $t-1$ elements, and therefore this branch results in a problem of dimension at most $2n-t$. Only a single set is removed in the other branch, in which case we repeat this process and branch on the next set represented by another vertex in H. This gives us $\frac{1}{2}t$ problem instances of dimension at most $2n-t$ and one problem instance of dimension $2n-\frac{1}{2}t$ because here $\frac{1}{2}t$ sets are removed.

In this latter instance, we use our new inclusion/exclusion based branching rule on the elements corresponding to the vertices in H. These elements still have frequency at least $\frac{1}{2}t-1$, since only $\frac{1}{2}t$ sets have been discarded until now. When branching on an element and forbidding it, a subproblem of dimension at most $2n-t$ is created because at least an additional element and $\frac{1}{2}t-1$ sets are removed in this branch. What remains is one subproblem generated in the branch after discarding $\frac{1}{2}t$ sets and making $\frac{1}{2}t$ elements optional. Since all these sets and elements are removed in these branches, this also gives us a problem of dimension $2n-t$.

The above procedure generates $t+1$ problems of dimension $2n-t$, which can all be solved by Algorithm 1 in $\mathcal{O}(1.22670^{2n-t})$ time. These are only a linear number of instances giving us a total running time of $\mathcal{O}(1.22670^{2n-t})$. □

Using Lemma 1 and following the computations in [14] we have obtained the following result. Due to space restrictions we refer to [23] for the full proof.

Theorem 2. *There exist algorithms that count the number of dominating sets of each size in a c-dense graph in* $\mathcal{O}\left(1.2267^{(\frac{1}{2}+\frac{1}{2}\sqrt{9-16c})n}\right)$ *time, in a circle graph in* $\mathcal{O}(1.4806^n)$ *time, in a 4-chordal graph in* $\mathcal{O}(1.4741^n)$ *time, and in a weakly chordal graph in* $\mathcal{O}(1.4629^n)$ *time.*

See Figure 1 and Table 1 for a comparison of our results with [14].

6 Further Applications

In principle, we could take any inclusion/exclusion algorithm and reformulate it into a branching algorithm. Then, we can look for reduction rules transforming

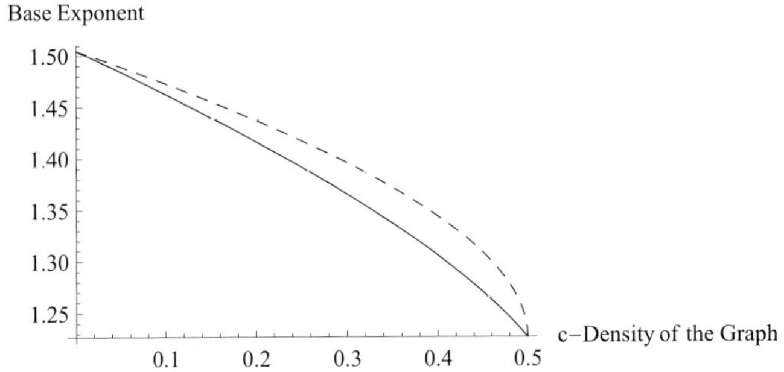

The solid line represents the upper bound on the running time of our algorithm, and the dashed line represents the upper bound obtained from [14] after plugging in our faster algorithm for dominating set.

Fig. 1. Comparison of bounds on the running time on c-dense graphs

it into a branch and reduce algorithm. Finding these reduction rules is often not very hard. However, finding any reduction rules for which you can prove that it improves the worst case behaviour of the algorithm is often very hard.

For example, consider the problem of counting the number of perfect matchings in a graph. This is easily modified into a $\#(n/2)$-Set Cover instance with n elements and possibly $O(n^2)$ sets to which we can apply a branching algorithm using the reduction rules of Algorithm 1. However, for such an algorithm, it is of no use to branch on a set since their cardinalities are too small, and we obtain an $\mathcal{O}^*(2^n)$ algorithm using polynomial space as in [3].

What this approach does accomplish, is that a branch and reduce inclusion/exclusion algorithm no longer has the property that its worst and best case behaviour conincide. When using the inclusion/exclusion formula one always evaluates every term of the sum, while if we are branching, then the number of leaves of the search tree can very well be a lot smaller due to the reduction rules.

To apply our combined approach, branching both on sets and elements, we need to consider problems that can be transformed into variations of set cover instances having a linear number of sets and elements. In search of such problems, we, very recently, obtained several results when considering the problem of whether there exists a locally subjective homomorphism form a fixed graph H into the input graph G (also known as role-assignment problems) [19].

7 Conclusion

While the improvements of the running times for the studied problems are interesting, we believe that the most important contribution of our paper is the novel combination of inclusion/exclusion and branching with a measure and conquer analysis. This gives a nice way to create inclusion/exclusion algorithms without the usual $\mathcal{O}(2^n n^{\mathcal{O}(1)})$ running time.

Many counting and decision variants of dominating set can be translated to set cover problems and solved by our algorithms in the same time. For example: directed dominating set, total dominating set[2], k-distance dominating set[2], weak/strong dominating, and combinations of these. We can also solve the weighted versions of all these problems as long as the size of the set of possible weight sums Σ is polynomially bounded: modify the algorithm such that it computes the number of set covers of each possible weight $w \in \Sigma$ at each step.

Our running times are highly dependent on the current best known upper bounds on the pathwidth of bounded degree graphs [9,12]. Any result that would improve these bounds would also improve our algorithm.

We use path decompositions while tree decompositions are more general and allow the same running times when using fast subset convolutions [4] to perform join operations [22]. We consider it to be an important open problem to give stronger (or even tight) bounds on the treewidth [18] or pathwidth of bounded degree graphs for which decompositions can be computed efficiently.

Acknowledgements

We would like to thank our advisor Hans L. Bodlaender for his enthusiasm for this research and for useful comments on an earlier draft of this paper. We would also like to thank Alexey A. Stepanov for useful discussions on [9].

References

1. Bax, E.T.: Inclusion and exclusion algorithm for the hamiltonian path problem. Information Processing Letters 47(4), 203–207 (1993)
2. Bax, E.T.: Recurrence-based reductions for inclusion and exclusion algorithms applied to #P problems (1996)
3. Björklund, A., Husfeldt, T.: Exact algorithms for exact satisfiability and number of perfect matchings. Algorithmica 52(2), 226–249 (2008)
4. Björklund, A., Husfeldt, T., Kaski, P., Koivisto, M.: Fourier meets Möbius: fast subset convolution. In: STOC 2007: Proceedings of the thirty-ninth annual ACM symposium on Theory of computing, pp. 67–74. ACM, New York (2007)
5. Björklund, A., Husfeldt, T., Koivisto, M.: Set partitioning via inclusion-exclusion. SIAM Journal of Computing, Special Issue for FOCS 2006 (to appear)
6. Bodlaender, H.L.: A tourist guide through treewidth. Acta Cybernetica 11, 1–23 (1993)
7. Bodlaender, H.L., Koster, A.M.C.A.: Combinatorial optimization on graphs of bounded treewidth. The Computer Journal 51(3), 255–269 (2008)
8. Eppstein, D.: Quasiconvex analysis of backtracking algorithms. In: Proceedings of the 15th Annual ACM-SIAM Symposium on Discrete Algorithms, SODA 2004, pp. 781–790 (2004)

[2] Total dominating set requires a $\mathcal{O}(4^t n^{\mathcal{O}(1)})$ algorithm on tree decompositions [22]. k-distance dominating set can also not be solved on tree decompositions in $\mathcal{O}(3^t n^{\mathcal{O}(1)})$ if $k > 1$. Hence the results from Section 5 do not extend to these problems.

9. Fomin, F.V., Gaspers, S., Saurabh, S., Stepanov, A.A.: On two techniques of combining branching and treewidth. In: Algorithmica Special issue of ISAAC 2006 (2006) (to appear)
10. Fomin, F.V., Grandoni, F., Kratsch, D.: Measure and conquer: Domination — a case study. In: Caires, L., Italiano, G.F., Monteiro, L., Palamidessi, C., Yung, M. (eds.) ICALP 2005. LNCS, vol. 3580, pp. 191–203. Springer, Heidelberg (2005)
11. Fomin, F.V., Grandoni, F., Kratsch, D.: Measure and conquer: a simple $O(2^{0.288n})$ independent set algorithm. In: Proceedings of the 16th Annual ACM-SIAM Symposium on Discrete Algorithms, SODA 2006, pp. 18–25 (2006)
12. Fomin, F.V., Høie, K.: Pathwidth of cubic graphs and exact algorithms. Information Processing Letters 97, 191–196 (2006)
13. Fomin, F.V., Kratsch, D., Woeginger, G.J.: Exact (exponential) algorithms for the dominating set problem. In: Hromkovič, J., Nagl, M., Westfechtel, B. (eds.) WG 2004. LNCS, vol. 3353, pp. 245–256. Springer, Heidelberg (2004)
14. Gaspers, S., Kratsch, D., Liedloff, M.: Exponential time algorithms for the minimum dominating set problem on some graph classes. In: Arge, L., Freivalds, R. (eds.) SWAT 2006. LNCS, vol. 4059, pp. 148–159. Springer, Heidelberg (2006)
15. Gaspers, S., Sorkin, G.B.: A universally fastest algorithm for max 2-sat, max 2-csp, and everything in between. In: Proceedings of the 20th annual ACM-SIAM Symposium on Discrete Algorithms (SODA 2009). SIAM, Philadelphia (to appear, 2009)
16. Grandoni, F.: A note on the complexity of minimum dominating set. J. Disc. Alg. 4, 209–214 (2006)
17. Karp, R.M.: Dynamic programming meets the principle of inclusion-exclusion. Operations Research Letters 1, 49–51 (1982)
18. Kneis, J., Mölle, D., Richter, S., Rossmanith, P.: Algorithms based on the treewidth of sparse graphs. In: Kratsch, D. (ed.) WG 2005. LNCS, vol. 3787, pp. 385–396. Springer, Heidelberg (2005)
19. Paulusma, D., van Rooij, J.M.M.: A fast exact algorithm for the 2-role assignment problem (submitted)
20. Randerath, B., Schiermeyer, I.: Exact algorithms for minimum dominating set. Technical Report zaik2004-469, Universität zu Köln, Cologne, Germany (2005)
21. van Rooij, J.M.M., Bodlaender, H.L.: Design by measure and conquer – a faster exact algorithm for dominating set. In: Albers, S., Weil, P. (eds.) Proceedings of the 25th Annual Symposium on Theoretical Aspects of Computer Science, STACS 2008, pp. 657–668. IBFI Schloss Dagstuhl (2008)
22. van Rooij, J.M.M., Bodlaender, H.L., Rossmanith, P.: Dynamic programming on tree decompositions using generalised fast subset convolution. In: Fiat, A., Sanders, P. (eds.) ESA 2009. LNCS, vol. 5757, pp. 566–577. Springer, Heidelberg (2009)
23. van Rooij, J.M.M., Nederlof, J., van Dijk, T.C.: Inclusion/exclusion meets measure and conquer: Exact algorithms for counting dominating set. Technical Report UU-CS-2008-043, Department of Information and Computing Sciences, Utrecht University, Utrecht, The Netherlands (2008)

Dynamic Programming on Tree Decompositions Using Generalised Fast Subset Convolution

Johan M.M. van Rooij[1], Hans L. Bodlaender[1], and Peter Rossmanith[2]

[1] Department of Information and Computing Sciences, Utrecht University,
P.O.Box 80.089, 3508 TB Utrecht, The Netherlands
{jmmrooij,hansb}@cs.uu.nl
[2] Computer Science Department, RWTH Aachen University,
52056 Aachen, Germany
rossmani@cs.rwth-aachen.de

Abstract. In this paper, we show that algorithms on tree decompositions can be made faster with the use of generalisations of fast subset convolution. Amongst others, this gives algorithms that, for a graph, given with a tree decomposition of width k, solve the DOMINATED SET problem in $O(nk^2 3^k)$ time and the problem to count the number of perfect matchings in $O^*(2^k)$ time. Using a generalisation of fast subset convolution, we obtain faster algorithms for all $[\rho, \sigma]$-domination problems with finite or cofinite ρ and σ on tree decompositions. These include many well known graph problems. We give additional results on many more graph covering and partitioning problems.

1 Introduction

Many recent investigations show that problems that are intractable on general graphs (e.g., NP-hard) become tractable (e.g., linear time solvable) when restricted to graphs of bounded treewidth. However, the constant factors involved in such algorithms are often large, and thus, for practical considerations, it is useful to find algorithms on graphs of small treewidth where the factor in the time, as function of treewidth, grows as slow as possible.

Most algorithms on graphs of bounded treewidth consist of two steps:

1. Find a tree decomposition of small width of the input graph.
2. Solve the problem on this tree decomposition using dynamic programming.

Concerning the first of these steps, one can find in linear time a tree decomposition of width at most k, for fixed k, if existing [4], but the constant factor of this algorithm is very high. However, in several cases, tree decompositions of small width can be obtained for special graph classes (see [5]), and there are also several good heuristics that often work well in practice (see [7]).

In this paper, we improve on the running time of the second of these steps, as a function of the treewidth. In particular, we show that *covering products* and *fast subset convolutions* [3] can be used to speed up these algorithms. In some cases,

A. Fiat and P. Sanders (Eds.): ESA 2009, LNCS 5757, pp. 566–577, 2009.

our algorithms use a generalisation of these algorithms. While covering products and subset convolutions are defined on sets with subsets, we generalise these to multiple states. Our algorithms are optimal in a certain sense: the running times are a polynomial in the size of the graph times the size of tables in any known tree decomposition based dynamic programming algorithm for the problems.

There are several recent papers that analyse the running time of algorithms on tree decompositions. For several vertex partitioning problems, Telle and Proskurowski [16] show that there are algorithms that solve these problems in $O(c^k n)$ time, with c a constant only depending on the problem at hand. For the Dominating Set problem, Alber and Niedermeier [2] give an improved algorithm that runs in $O(4^k n)$ time; similar results are given in [2] for related problems: Independent Dominating Set, Total Dominating Set, Perfect Dominating Set, Perfect Code, Total Perfect Dominating Set, weighted versions, and Red-Blue Dominating Set. See also [1]. The improvement in [2] is used by exploiting a notion of *monotonicity*. In this paper, we introduce two new techniques, the first using a variant of convolutions, and the second a simple way of partitioned table handling, that speed up these results and can be used for several other problems as well.

Another technique to speed up algorithms on tree decompositions or on the related branch decompositions was introduced by Dorn [9], who employed fast matrix multiplication. If the input graph is planar, then other improvements are possible, see [10,12]. In [10], Dorn showed amongst others, that Dominating Set for planar graphs with a given tree decomposition of width k can be solved in $O^*(3^k)$ time. Some of these results can be generalised to graphs that avoid a minor, see [11]. We obtain the same result without requiring planarity.

Many dynamic programming algorithms on tree decompositions compute a table for each node in the tree decomposition. These tables have an entry for each assignment of a state from some fixed set to each vertex in the bag. It is possible to use different sets of states for the same problem that lead to different running times. In Section 3, we discuss how we can transfer a table for one set of states to the equivalent table for a different set of states. In Sections 4 and 5, we obtain faster tree decomposition algorithms for Dominating Set and #Perfect Matching; the switch technique of Section 3 is used as a subroutine. These results serve as examples of a general method, as the techniques can be applied to many problems. In Section 6, we obtain faster tree decomposition algorithms for a large set of problems, namely all $[\rho, \sigma]$-domination problems with finite or cofinite ρ and σ, as introduced by Telle and Proskurowski [16].

2 Preliminaries

Let $G = (V, E)$ be an n-vertex graph. A *tree decomposition* of G is a tree T in which each node $i \in T$ has an assigned set of vertices $X_i \subseteq V$ (called a *bag*) such that $\bigcup_{i \in T} X_i = V$ with the following properties:

1. for all $\{u, v\} \in E$, there exists an $i \in T$ such that $\{u, v\} \subseteq X_i$.
2. if $v \in X_i$ and $v \in X_j$, then $v \in X_k$ for all X_k on the path from X_i to X_j in T.

Table 1. $[\rho, \sigma]$-domination problems (taken from [15,16])

ρ	σ	Standard problem description
$\{0, 1, \ldots\}$	$\{0\}$	Independent Set
$\{1, 2, \ldots\}$	$\{0, 1, \ldots\}$	Dominating Set
$\{0, 1\}$	$\{0\}$	Strong Stable Set/2-Packing/Distance-2 Independent Set
$\{1\}$	$\{0\}$	Perfect Code/Efficient Dominating Set
$\{1, 2, \ldots\}$	$\{0\}$	Independent Dominating Set
$\{1\}$	$\{0, 1, \ldots\}$	Perfect Dominating Set
$\{1, 2, \ldots\}$	$\{1, 2, \ldots\}$	Total Dominating Set
$\{1\}$	$\{1\}$	Total Perfect Dominating Set
$\{0, 1\}$	$\{0, 1, \ldots\}$	Nearly Perfect Set
$\{0, 1\}$	$\{0, 1\}$	Total Nearly Perfect Set
$\{1\}$	$\{0, 1\}$	Weakly Perfect Dominating Set
$\{0, 1, \ldots\}$	$\{0, 1, \ldots, p\}$	Induced Bounded Degree Subgraph
$\{p, p+1, \ldots\}$	$\{0, 1, \ldots\}$	p-Dominating Set
$\{0, 1, \ldots\}$	$\{p\}$	Induced p-Regular Subgraph

The *treewidth* of a tree decomposition is the size of the largest bag of T minus one: $\max_{i \in T} |X_i| - 1$. The treewidth of a graph G is the minimum treewidth over all possible tree decompositions of G. In this paper, we will always assume that tree decompositions of the appropriate width are given.

Dynamic programming algorithms solving $\mathcal{NP}$-hard problems on graphs of bounded treewidth are often presented on *nice* tree decompositions. These are rooted tree decompositions in which each node is of one of the following types:

1. *Leaf* node: a leaf of T.
2. *Introduce* node: an internal node i with one child node c for which $X_i = X_c \cup \{v\}$. This node is said to introduce the vertex v.
3. *Forget* node: an internal node i with one child node c for which $X_i = X_c \setminus \{v\}$. This node is said to forget the vertex v.
4. *Join* node: an internal node i with two child nodes l, r with $X_i = X_l = X_r$.

Given a tree decomposition, a *nice* tree decomposition of equal width can be found in polynomial time [14]. For more information on tree decompositions and dynamic programming over tree decompositions, see [6]. We associate to each node i in the tree decomposition a subgraph $G_i = (V_i, E_i)$ of G. A vertex $v \in V$ belongs to G_i, if and only if there is a bag j with $j = i$ or j a descendant of i with $v \in X_j$, and $\{v, w\} \in E$ belongs to G_i iff $v, w \in V_i$.

A *dominating set* in a graph G is a set of vertices $D \subseteq V$ such that for every $v \in V \setminus D$ there exists a $(v, d) \in E$ with $d \in D$. Finding a dominating set of minimum size in G is a classical $\mathcal{NP}$-complete problem (Dominating Set).

A *perfect matching* in G is a set of edges $M \subseteq E$ such that every $v \in V$ is contained in exactly one edge $e \in M$. Counting the number of perfect matchings in a graph is a classical $\#\mathcal{P}$-complete problem (#Perfect Matching).

Let $\rho, \sigma \subseteq \mathbb{N}$, a $[\rho, \sigma]$-dominating set is a subset $D \subseteq V$ such that: for every $v \in V \setminus D$: $|N(v) \cap D| \in \rho$, and for every $v \in D$: $|N(v) \cap D| \in \sigma$, where $N(v)$ denotes the open neighbourhood of a vertex $v \in V$ in G.

The $[\rho, \sigma]$-domination problems were introduced by Telle in [15,16] and form a large class of graph covering problems: see Table 1. Throughout this paper, we assume that ρ and σ are either finite or cofinite.

Throughout the paper, tables are denoted with their parameters, e.g., $A(c)$ is a table that maps each c (from a domain that is clear from the context) to a value (with the range also clear from the context).

3 Switching between State Representations

A dynamic programming algorithm on a tree decomposition T traverses T in a bottom up fashion. In each visited node $i \in T$, partial solutions to the problem on G_i are stored. These partial solutions must satisfy all problem specific constraints everywhere except on the vertices in X_i. To store these partial solutions and identify them by their behaviour on the current bag, for each node of T, a table $A(c)$ is computed containing the solutions corresponding to an assignment of states c (also called a *colouring*) to the vertices in X_i. During the traversal, the table for each node is computed from the tables from its child nodes. This results in an algorithm that is polynomial in the size of T, but exponential in the treewidth of T (the size of the largest bag).

For example, when solving #PERFECT MATCHING, the obvious tree decomposition based dynamic programming algorithm uses states $\{0, 1\}$, where 1 means this vertex is matched and 0 means that it is not. The table entry $A(c)$ now contains the number of matchings in G_i such that the only vertices that are not matched are exactly the vertices in the current bag X_i with state 0 in c.

We make the following observation: the described table $A(c)$ contains exactly the same information as a table $A'(c)$ using states $\{0, ?\}$, where ? represents a vertex where we do not specify if it is matched. I.e., for a specific assignment of states c, we count the number of matchings in G_i, where all vertices in $V_i \setminus X_i$ are matched, all vertices with state 0 in c are unmatched, and all vertices with state ? can either be matched or not. In this second representation, there no longer is a direct relation between the partial solutions and the states: a matching where a vertex $v \in X_i$ is not matched is counted both in the cases where v has state ? and where v has state 0. We will use such alternative representations throughout this paper to obtain exponentially faster algorithms.

Lemma 1. *A table containing the number of perfect matchings corresponding to states $\{0, 1\}$ contains the same information as a table using states $\{0, ?\}$. Transformations using $\mathcal{O}(k2^k)$ operations between both tables exists.*

If one defines a vertex with state 1 or ? to be in S, and a vertex with state 0 not to be in S, then the state changes are Möbius transforms and inversions, see [3]. The proof then directly follows from their fast evaluation algorithms.

Proof. The fast transformations work in k steps. In step $1 \leq i \leq k$, we assume that the first $i-1$ coordinates of the colouring c in our table use one set of states, and the other $k - i + 1$ coordinates use the other set of states. Using this as an invariant, we change the set of states used for the i-th coordinate at step i.

Table 2. Vertex states for the Dominating Set Problem

state	meaning
1	this vertex is in the dominating set.
0_1	this vertex is not in the dominating set and has already been dominated.
0_0	this vertex is not in the dominating set and has not yet been dominated.
$0_?$	this vertex is not in the dominating set and may or may not be dominated.

Transforming from $\{0,1\}$ to $\{0,?\}$ can be done using the following formula in which $A(c)$ represents our table for colouring c, c_1 is a subcolouring of size i using states $\{0,1\}$, and c_2 is a subcolouring of size $k-i-1$ using states $\{0,?\}$.

$$A(c_1 \times \{?\} \times c_2) = A(c_1 \times \{0\} \times c_2) + A(c_1 \times \{1\} \times c_2)$$

In words, the number of matchings that may contain some vertex equals the sum of those that do and those that do not contain it.

For the reverse transformation from $\{0,?\}$ to $\{0,1\}$ there is a similar formula:

$$A(c_1 \times \{1\} \times c_2) = A(c_1 \times \{?\} \times c_2) - A(c_1 \times \{0\} \times c_2)$$

In words, the number of matchings that contain some vertex equals all matchings minus those that do not contain the vertex.

Each step computes 2^k values resulting in $\mathcal{O}(k2^k)$ time transformations. □

For Dominating Set the obvious set of states to use would be $\{1, 0_1, 0_0\}$: see Table 2. The tables of the dynamic programming algorithm would then, for each colouring c, store the minimum size of a vertex set dominating all vertices seen this far except those with state 0_0 in the current bag and that contain exactly those vertices in the current bag that have state 1.

We take a slightly different approach and use a table $A(c, \kappa)$ containing the number of dominating sets of each size $0 \leq \kappa \leq n$ corresponding to the colouring c on the current bag. With this modification, we can also use different sets of states capturing exactly the same information for Dominating Set.

Lemma 2. *A table $A(c,\kappa)$ containing the number of dominating sets of size $0 \leq \kappa \leq n$ corresponding to a colouring c contains the same information independent of the following three choices of states (see Table 2): $\{1, 0_1, 0_0\}$, $\{1, 0_1, 0_?\}$, $\{1, 0_?, 0_0\}$. Transformations using $\mathcal{O}(nk3^k)$ operations between the tables exist.*

Proof. Repeat the following for each $0 \leq \kappa \leq n$. Fix the 1 states and then apply the same transformations as described in Lemma 1 on the 0 states. We can use either one of the three sets of states since the $0_?$ state can be created while preserving either the 0_0 state or the 0_1 state. □

4 Minimum Dominating Set

The previously fastest algorithm for Dominating Set on graphs of treewidth k runs in $\mathcal{O}(n4^k)$ and due is to Alber et al. [1,2]. Their dynamic programming

Table 3. Join tables for the DOMINATING SET and #PERFECT MATCHING problems

$\times$	1	0_1	0_0
1	1		
0_1		0_1	0_1
0_0		0_1	0_0

$\times$	1	0_1	$0_?$
1	1		
0_1			0_1
$0_?$		0_1	$0_?$

$\times$	1	$0_?$	0_0
1	1		
$0_?$		$0_?$	
0_0			0_0

$\times$	0	1
0	0	1
1	1	

$\times$	0	$?$
0	0	
$?$		$\not{?}$

algorithm uses states $\{1, 0_1, 0_?\}$ (see Table 2), and stores the sizes of the minimum dominating sets corresponding to each state colouring on the bag. This algorithm uses $\mathcal{O}^*(3^k)$ time on leaf, introduce and forget bags, but $\mathcal{O}^*(4^k)$ time on a join bag. Using the results from Section 3, we will improve this to $\mathcal{O}^*(3^k)$.

First of all, we observe that the Alber et al. algorithm [1,2] can be extended quite easily to *count* the *number* of dominating sets of each size $0 \leq \kappa \leq n$; we will use these procedures to compute tables for leaf, introduce, and forget bags.

We begin with our main theorem on DOMINATING SET.

Theorem 1. *There is an algorithm counting the number of dominating set of each size* $0 \leq \kappa \leq n$ *on a graph of treewidth* k *in* $\mathcal{O}(n^3 3^k i_\times(n))$ *time.*

Here, $i_\times(n)$ is the time required to work with (multiply) n-bit integers: currently $i_\times(n) = n \log(n) 2^{\log^*(n)}$ [13].

Proof. We prove the result by showing that we can perform a join operation in $\mathcal{O}(n^2 3^k i_\times(n))$ time. For this operation, two child bags with given tables $A_L(c, \kappa)$ and $A_R(c, \kappa)$ are provided. Using Lemma 2, we make sure that they are given using states $\{1, 0_?, 0_0\}$. Now, the table for the join bag $A(c, \kappa)$ can be computed using the following formula, where $\#_1(c)$ stand for the number of 1 states in c:

$$A(c, \kappa) = \sum_{\kappa_L + \kappa_R - \#_1(c) = \kappa} A_L(c, \kappa_L) \cdot A_R(c, \kappa_R)$$

We sum over all ways to obtain a set of size κ by combining both child tables. The sum is evaluated in $\mathcal{O}(n i_\times(n))$ time, giving an $\mathcal{O}(n^2 3^k i_\times(n))$ time join.

This is correct because each vertex with one of three states in $\{1, 0_?, 0_0\}$ in the join bag requires that the corresponding vertices in both child bags have the same state: a vertex is in the dominating set if it is so in both child bags (1), a vertex may or may not be dominated if it is so in both child bags ($0_?$), and a vertex is not dominated if it is so in both child bags (0_0), see Table 3. □

A similar join with states $\{1, 0_1, 0_0\}$ in which a 0_1 state can be created from a 0_1 from both children, or a 0_1 from one child and a 0_0 from the other, gives an $\mathcal{O}^*(5^k)$ algorithm. The decision procedure of Alber et al. with states $\{1, 0_1, 0_?\}$ leads to an $\mathcal{O}^*(4^k)$ algorithm. See Table 3.

The proof of Theorem 1 uses the principle of the *covering product* [3]. We presented it using mostly terminology from tree decomposition algorithms. We choose this manner of presentation because it easily generalises to more than two states. In this way, one, for example, easily proves the following proposition.

Proposition 1. *There is an algorithm counting the number of partitions of a graph of treewidth k into two total dominating sets in $\mathcal{O}^*(6^k)$.*

We proceed by improving the polynomial factors for DOMINATING SET.

Corollary 1. *There is an algorithm counting the number of minimum dominating sets on a graph of treewidth k in $\mathcal{O}(nk^2 3^k i_\times(n))$ time.*

Proof. Dominating set has the *finite integer index* property [8]: the size difference between any two *minimum* dominating sets stored in a table is at most k.

For the leaf, introduce and forget bags, we store for each colouring c the size of the corresponding minimum dominating set $B(c)$ and the number of such sets $A(c)$ using states $\{1, 0_1, 0_0\}$. At a join node, we expand the tables to:

$$A(c,\kappa) = \begin{cases} A(c) \text{ if } B(c) = \kappa \\ 0 \quad\quad \text{otherwise} \end{cases}$$

Then, change states as in Theorem 1 and perform the join. To extract $A(c)$ and $B(c)$ for the join table, let $A(c) = A(c,\kappa')$ and let $B(c) = \kappa'$, where κ' is the smallest value of κ for which $A(c,\kappa)$ is non-zero. The sum now has at most k terms on child tables of size only $k3^k$ giving the running time. □

Corollary 2. *There is an algorithm solving the minimum dominating set problem on graphs of treewidth k in $\mathcal{O}(nk^2 3^k)$ time.*

Proof. We improve on Corollary 1 by only keeping track of the size of the dominating set $B(c)$ using states $\{1, 0_1, 0_0\}$. At a join node, expand the tables to:

$$A(c,\kappa) = \begin{cases} 1 \text{ if } B(c) = \kappa \\ 0 \text{ otherwise} \end{cases}$$

Then, proceed as in Corollary 1 and perform the join. Now, we do not count dominating sets, but simply the number of 1-entries in the table $A(c,\kappa)$. □

5 Counting the Number of Perfect Matchings

We need a more involved algorithm to count the number of perfect matchings on a graph of treewidth k. This algorithm uses fast subset convolutions [3].

Given a set U and functions $f, g : 2^U \to \mathbb{Z}$ their subset convolution is defined:

$$(f * g)(S) = \sum_{X \subseteq S} f(X) g(S \setminus X) \tag{1}$$

The fast subset convolution algorithm computes this in $\mathcal{O}(k^2 2^k)$ operations [3].

Our algorithm uses states $\{0,1\}$ with the invariant that vertices with state 1 are matched only with vertices outside of the bag, i.e. vertices that have already been forgotten by the algorithm. This prevents vertices to be matched within the bag and greatly simplifies the join. To obtain the final solution from the root of the tree decomposition, we add a series of forget nodes to the root.

Our proof uses the fast subset convolution algorithm without mentioning it. Instead, we use state changes: this approach will be generalised in Section 6.

Theorem 2. *There is an algorithm counting the number of perfect matchings in a graph of treewidth k in $\mathcal{O}(nk^2 2^k i_\times(n))$ time.*

Proof. Due to space restrictions, we will only explain the join operation.

We cannot simply change states to $\{0, ?\}$ and perform the join similar to dominating set as suggested by Table 3. This is because two ? states do not combine to a new ? state since this would allow a vertex to be matched twice.

To compensate for this, one can expand the tables and index them by the number of matched vertices. Let $A_C(c)$ be one of the child tables $A_L(c)$ or $A_R(c)$.

$$A_C(c, i) = \begin{cases} A_C(c) & \text{if } \#_1(c) = i \\ 0 & \text{otherwise} \end{cases}$$

Now we do the following. First change the state representation in the tables $A_C(c, i)$ to $\{0, ?\}$ obtaining new tables $A'_C(c, i)$; these tables does not use state 1, but are still indexed by the number of 1 states used in the previous representation. Then, join the tables by combining all possibilities that arise from i 1 states in the old representation (stored in the index i) using the following formula:

$$A'(c, i) = \sum_{i = i_L + i_R} A'_L(c, i_L) \cdot A'_R(c, i_R)$$

Next, change states back to $\{0, 1\}$ obtaining a table $A(c, i)$; notice that a 1 state can now represent a vertex that is matched twice. Finally, we can find those entries that do not match a vertex twice by the following observation: the total number of 1 states in c should equal the sum of those in its child tables, now stored in the index i. The join table follows $A(c)$ from extracting these entries:

$$A(c) = A(c, \#_1(c))$$

The running time is dominated by the time required to compute $A'(c, i)$: $k2^k$ entries, each requiring a k-term sum, giving the $\mathcal{O}(nk^2 2^k i_\times(n))$ time bound. □

6 $[\rho, \sigma]$-Domination Problems

We have shown how to solve two elementary problems in $\mathcal{O}^*(s^k)$ on graphs of treewidth k, where s is the number of states per vertex used in representations of solutions. In this section, we generalise our results and show that we can solve all $[\rho, \sigma]$-domination problems with finite or cofinite ρ and σ in $\mathcal{O}^*(s^k)$; this includes existence, minimisation, maximisation, and counting variants of the problems.

For the $[\rho, \sigma]$-dominating problems one could use states ρ_j and σ_j, where ρ_j and σ_j represent vertices in or not in the $[\sigma, \rho]$-dominating set D, respectively, that have j neighbours in D. For finite ρ, σ we can suffice with states $\{\rho_0, \ldots, \rho_p, \sigma_0, \ldots, \sigma_q\}$. If ρ or σ is cofinite, we replace the last states by $\rho_{\geq p}$ or $\rho_{\geq q}$, respectively. For readability reasons, we restrict ourselves to finite ρ and σ.

In our algorithm, the meaning of these states is slightly different: the subscript j of states ρ_j and σ_j will count only the number of neighbours in the

$[\rho, \sigma]$-dominating set D that have already been forgotten by the algorithm. This prevents us from having to keep track of any adjacencies within a bag during a join operation; we can update these subscripts j easily in each forget bag.

Dynamic programming tables for the $[\rho, \sigma]$-domination problems can also be represented using different sets of states that contain the same information.

Lemma 3. *A table $A(c, \kappa)$ containing the number of $[\rho, \sigma]$-dominating sets of size κ corresponding to the colouring c on a bag of size k contains the same information independent of the following choices of states:*

- *Set I: $\{\rho_0, \rho_1, \rho_2, \ldots, \rho_p, \sigma_0, \sigma_1, \sigma_2, \ldots, \sigma_q\}$.*
- *Set II: $\{\rho_0, \rho_{\leq 1}, \rho_{\leq 2}, \ldots, \rho_{\leq p}, \sigma_0, \sigma_{\leq 1}, \sigma_{\leq 2}, \ldots, \sigma_{\leq q}\}$.*

The identifiers of the states are self explaining: $\rho_{condition}$, $\sigma_{condition}$ considers the number of $[\rho, \sigma]$-dominating sets D that do not contain (ρ state) or do contain (σ state) this vertex with a number of neighbours in D satisfying the *condition*.

Proof (Sketch). Repeatedly apply transformations as in Lemma 2. □

In the proof of Theorem 2, we expanded the child tables to $A(c, i)$, where i was an index indicating the number of 1 states used to create the ? states in c. We need to generalise this to more states to prove our result. To this end, we introduce an *index vector* $\boldsymbol{i} = (i_{\rho 1}, i_{\rho 2}, \ldots, i_{\rho p}, i_{\sigma 1}, i_{\sigma 2}, \ldots, i_{\sigma q})$, where $i_{\rho j}$ and $i_{\sigma j}$ index the sum of the number of neighbours in D of the vertices with state $\rho_{\leq j}$ and $\sigma_{\leq j}$, respectively. We say that a solution corresponding to a colouring c using state set I (Lemma 3) *satisfies* a combination of a colouring c' using state set II and an index vector $\boldsymbol{i}$ if: c is counted in c', and for each $i_{\rho j}$, $i_{\sigma j}$, the sum of the number of neighbours in D in the colouring c for vertices which have colour $\rho_{\leq j}$ and $\sigma_{\leq j}$ in c' equal $i_{\rho j}$ and $i_{\sigma j}$, respectively. We explain this by example.

Suppose that we have a bag of size 3 and use states $\{\rho_0, \rho_1, \rho_2, \sigma_0\}$ which we want to transform to states $\{\rho_0, \rho_{\leq 1}, \rho_{\leq 2}, \sigma_0\}$; thus $\boldsymbol{i} = (i_{\rho 1}, i_{\rho 2})$. A partial solution with states $c = (\rho_0, \rho_1, \rho_2)$ will be counted in both $c'_1 = (\rho_0, \rho_{\leq 2}, \rho_{\leq 2})$ and $c'_2 = (\rho_{\leq 1}, \rho_{\leq 1}, \rho_{\leq 2})$. In this case, c satisfies the combination $(c'_1, \boldsymbol{i} = (0, 3))$ but no combination of c'_1 with an other index vector. Also, c satisfies the combination $(c'_2, \boldsymbol{i} = (1, 2))$ and no other combination involving c'_2.

What we need is a table containing, for each possible combination of a colouring using state set II with an index vector, the number of partial solutions that satisfy these. We can do this by using the following lemma.

Lemma 4. *Given a table $A(c, \kappa)$ containing the number of $[\rho, \sigma]$-dominating sets of size κ corresponding to the colouring c on a bag of size k using states set I from Lemma 3, we can compute a table $A(c, \kappa, \boldsymbol{i})$ containing the number of solutions of size κ satisfying the combination of a colouring using state set II from Lemma 3 and the index vector $\boldsymbol{i}$ in $\mathcal{O}(nk^2(sk)^{s-2}s^k)$ time.*

Proof. We start with the following table $A(c, \kappa, \boldsymbol{i})$ using state set I:

$$A(c, \kappa, \boldsymbol{i}) = \begin{cases} A(c, \kappa) & \text{if } \boldsymbol{i} \text{ is the all-0 vector} \\ 0 & \text{otherwise} \end{cases}$$

Then, we change the states of the j-th coordinate at step j, similar to previous state changes, but now changing all states at this coordinate in one pass using the following formulas that also updates the index vector $\boldsymbol{i}$:

$$A(c_1 \times \{\rho_{\leq j}\} \times c_2, \kappa, \boldsymbol{i}) - \sum_{k=0}^{j} A(c_1 \times \{\rho_k\} \times c_2, \kappa, \boldsymbol{i}_{i_{\rho j} \to (i_{\rho j}-k)})$$

$$A(c_1 \times \{\sigma_{\leq j}\} \times c_2, \kappa, \boldsymbol{i}) = \sum_{k=0}^{j} A(c_1 \times \{\sigma_k\} \times c_2, \kappa, \boldsymbol{i}_{i_{\sigma j} \to (i_{\sigma j}-k)})$$

Here, $\boldsymbol{i}_{i_{\rho j} \to (i_{\rho j}-k)}$ is the vector $\boldsymbol{i}$ with the value of $i_{\rho j}$ set to $i_{\rho j} - k$.

The algorithm runs in k steps, computing a value for each of the s^k colourings, n sizes, and $s-2$ indices that range over sk values in a k-term sum. This gives a running time bounded by $\mathcal{O}(nk^2(sk)^{s-2}s^k)$. □

Since we fix the specific $[\rho, \sigma]$-domination problem, this is $\mathcal{O}^*(s^k)$ time.

We are now ready to prove our main result of this section.

Theorem 3. *Let $\rho, \sigma \subseteq \mathbb{N}$ finite or cofinite. There exist an algorithm counting the number of $[\rho, \sigma]$-dominating sets of each size $0 \leq \kappa \leq n$ of a fixed $[\rho, \sigma]$-domination problem involving s states in $\mathcal{O}^*(s^k)$ time on graphs of treewidth k.*

Proof. We will only describe the join operation of the algorithm for finite σ, ρ.

First, we transform the child tables to state set II (Lemma 3) using Lemma 4 obtaining tables $A'_L(c, \kappa, \boldsymbol{l})$, $A'_R(c, \kappa, \boldsymbol{r})$, where we have the extra index vectors indicating, for each state, the sum of the number of neighbours in D of the set of vertices with this state. Then, compute the table $A(c, \kappa, \boldsymbol{i})$ by joining both tables combining identical states from both tables using the formula below. In this formula, we sum over all ways of getting a set of size κ by combining the sizes in the child tables and all ways of getting index vector $\boldsymbol{i}$ from $\boldsymbol{i} = \boldsymbol{r} + \boldsymbol{l}$.

$$A'(c, \kappa, \boldsymbol{i}) = \sum_{\kappa_L + \kappa_R - \#_1(c) = \kappa} \left(\sum_{i_{\rho 1} = r_{\rho 1} + l_{\rho 1}} \cdots \sum_{i_{\sigma q} = r_{\sigma q} + l_{\sigma q}} A'_R(c, \kappa_R, \boldsymbol{r}) \cdot A'_L(c, \kappa_L, \boldsymbol{l}) \right)$$

A joined partial solution is counted in an entry in $A'(c, \kappa, \boldsymbol{i})$ if and only if it satisfies the following three conditions.

- The sum of the number of neighbours in D of this joined solution on the $\rho_{\leq j}$ and $\sigma_{\leq j}$ states equals $i_{\rho j}$ and $i_{\sigma j}$, respectively.
- The number of neighbours in D in both partial solutions used to create this joined solution on each of the $\rho_{\leq j}$ and $\sigma_{\leq j}$ states is at most j.
- The total number of vertices in D in this joined solution is κ.

Let $\Sigma^i_\rho(c)$, $\Sigma^i_\sigma(c)$ be the weighted sums of the number of ρ_j and σ_j states with $0 \leq j \leq i$ in c, respectively, defined by:

$$\Sigma^i_\rho(c) = \sum_{j=1}^{i} j \cdot \#_{\rho_i}(c) \qquad \Sigma^i_\sigma(c) = \sum_{j=1}^{i} j \cdot \#_{\sigma_i}(c)$$

Now, using Lemma 3 we change states back to state set I to obtain a table $A(c, \kappa, \boldsymbol{i})$ and extract the join table in the following way:

$$A(c, \kappa) = A\left(c, \kappa, (\Sigma_\rho^1(c), \Sigma_\rho^2(c), \ldots, \Sigma_\rho^p(c), \Sigma_\sigma^1(c), \Sigma_\sigma^2(c), \ldots, \Sigma_\sigma^q(c))\right)$$

We will now prove correctness of the entries in $A(c, \kappa)$.

An entry in $A(c, \kappa)$ with $c \in \{\rho_0, \sigma_0\}^k$ is correct: these states are unaffected by the state changes and the index vector is not used.

Now consider an entry in $A(c, \kappa)$ with $c \in \{\rho_0, \rho_1, \sigma_0\}^k$. Each ρ_1 state comes from a $\rho_{\leq 1}$ state in $A'(c, \kappa, \boldsymbol{i})$ and is a join of partial solutions with the following combinations of states on this vertex: (ρ_0, ρ_0), (ρ_0, ρ_1), (ρ_1, ρ_0), (ρ_1, ρ_1). Because we have changed states back to set I, each (ρ_0, ρ_0) combination is counted in the ρ_0 state on this vertex, and subtracted from the combinations used to from state ρ_1: the other tree combinations remain counted in the ρ_1 state. Since we consider only those solutions with index vector $i_{\rho_1} = \Sigma_\rho^1(c)$, the total number of ρ_1 states used to form this joined solutions equals $\Sigma_\rho^1(c) = \#_{\rho_1}(c)$. Therefore no (ρ_1, ρ_1) combination could have been used, and each partial solution counted in $A(c, \kappa)$ has exactly one neighbour in D on each of the ρ_1 states.

We repeat this argument for the other states, but omit the details. □

This proof generalises ideas from the fast subset convolution algorithm; while convolutions use ranked Möbius transforms, we use transformations with multiple ranks and multiple states. The polynomial factors in the proof of Theorem 3 can be improved in several ways, some of which resemble Corollaries 1 and 2.

7 Conclusion

In this paper, we have introduced the use of fast subset convolutions and generalisations of these to speed up the running time of algorithms on tree decompositions. Our techniques can be applied to a large number of problems, and in our paper, we have given a few of such examples, but it is not hard to find more applications. We give one set of such results without proof. Because each of the problems is finite integer index, the running time is linear in n.

Proposition 2. *The* MAXIMUM TRIANGLE PACKING, PARTITION INTO TRIANGLES, PARTITION INTO ℓ-CLIQUES *for fixed ℓ and* MINIMUM CLIQUE PARTITION *can be solved in $O(poly(k)2^k n)$ time on graphs of treewidth k.*

Our algorithms are in a certain sense optimal concerning the exponential dependency on the width of the tree decomposition, as the running times equal a polynomial in n and k times the 'usual' size of the dynamic programming tables. However, we have to pay: this exponential improvement goes with extra polynomial factors. In some cases, the pay is little, e.g., for DOMINATING SET, we improve the $O(4^k n)$ time [2] to $O(k^2 3^k n)$, but this is not always the case.

We end with a few directions for further research. An experimental study on these algorithms would be very interesting. Other questions are: is it possible to use similar techniques to speed up algorithms on branch decompositions? Is it possible to use generalisations of fast subset convolution to speed up algorithms on (tree or branch decompositions of) planar graphs, as in [12]?

Acknowledgements. The first author thanks Jan Arne Telle and Jesper Nederlof for several very useful discussions.

References

1. Alber, J., Bodlaender, H.L., Fernau, H., Kloks, T., Niedermeier, R.: Fixed parameter algorithms for dominating set and related problems on planar graphs. Algorithmica 33, 461–493 (2002)
2. Alber, J., Niedermeier, R.: Improved tree decomposition based algorithms for domination-like problems. In: Rajsbaum, S. (ed.) LATIN 2002. LNCS, vol. 2286, pp. 613–627. Springer, Heidelberg (2002)
3. Björklund, A., Husfeldt, T., Kaski, P., Koivisto, M.: Fourier meets Möbius: Fast subset convolution. In: Proceedings of the 39th Annual Symposium on Theory of Computing, STOC 2007, pp. 67–74 (2007)
4. Bodlaender, H.L.: A linear time algorithm for finding tree-decompositions of small treewidth. SIAM J. Comput. 25, 1305–1317 (1996)
5. Bodlaender, H.L.: A partial k-arboretum of graphs with bounded treewidth. Theor. Comp. Sc. 209, 1–45 (1998)
6. Bodlaender, H.L., Koster, A.M.C.A.: Combinatorial optimization on graphs of bounded treewidth. The Computer Journal 51(3), 255–269 (2008)
7. Bodlaender, H.L., Koster, A.M.C.A.: Treewidth computations I. upper bounds. Technical Report UU-CS-2008-032, Department of Information and Computing Sciences, Utrecht University, Utrecht, the Netherlands, Submitted (2008)
8. Bodlaender, H.L., van Antwerpen-de Fluiter, B.: Reduction algorithms for graphs of small treewidth. Information and Computation 167, 86–119 (2001)
9. Dorn, F.: Dynamic programming and fast matrix multiplication. In: Azar, Y., Erlebach, T. (eds.) ESA 2006. LNCS, vol. 4168, pp. 280–291. Springer, Heidelberg (2006)
10. Dorn, F.: How to use planarity efficiently: New tree-decomposition based algorithms. In: Brandstädt, A., Kratsch, D., Müller, H. (eds.) WG 2007. LNCS, vol. 4769, pp. 280–291. Springer, Heidelberg (2007)
11. Dorn, F., Fomin, F.V., Thilikos, D.M.: Catalan structures and dynamic programming in h-minor-free graphs. In: Proceedings of the 18th Annual ACM-SIAM Symposium on Discrete Algorithms, SODA 2008, pp. 631–640 (2008)
12. Dorn, F., Penninkx, E., Bodlaender, H.L., Fomin, F.V.: Efficient exact algorithms on planar graphs: Exploiting sphere cut branch decompositions. In: Brodal, G.S., Leonardi, S. (eds.) ESA 2005. LNCS, vol. 3669, pp. 95–106. Springer, Heidelberg (2005)
13. Fürer, M.: Faster integer multiplication. In: Proceedings of the 39th Annual Symposium on Theory of Computing, STOC 2007 (2007)
14. Kloks, T.: Treewidth. Computations and Approximations. LNCS, vol. 842. Springer, Heidelberg (1994)
15. Telle, J.A.: Complexity of domination-type problems in graphs. Nordic J. Comput. 1, 157–171 (1994)
16. Telle, J.A., Proskurowski, A.: Algorithms for vertex partitioning problems on partial k-trees. SIAM J. Disc. Math. 10, 529–550 (1997)

Counting Paths and Packings in Halves

Andreas Björklund[1], Thore Husfeldt[1,2], Petteri Kaski[3,⋆], and Mikko Koivisto[3,⋆]

[1] Lund University, Department of Computer Science,
P.O. Box 118, SE-22100 Lund, Sweden
andreas.bjorklund@yahoo.se, thore.husfeldt@cs.lu.se
[2] IT University of Copenhagen,
Rued Langgaards Vej 7, 2300, København S, Denmark
[3] Helsinki Institute for Information Technology HIIT,
Department of Computer Science, University of Helsinki,
P.O. Box 68, FI-00014 University of Helsinki, Finland
petteri.kaski@cs.helsinki.fi, mikko.koivisto@cs.helsinki.fi

Abstract. We show that one can count k-edge paths in an n-vertex graph and m-set k-packings on an n-element universe, respectively, in time $\binom{n}{k/2}$ and $\binom{n}{mk/2}$, up to a factor polynomial in n, k, and m; in polynomial space, the bounds hold if multiplied by $3^{k/2}$ or $5^{mk/2}$, respectively. These are implications of a more general result: given two set families on an n-element universe, one can count the disjoint pairs of sets in the Cartesian product of the two families with $O(n\ell)$ basic operations, where ℓ is the number of members in the two families and their subsets.

1 Introduction

Some combinatorial structures can be viewed as two halves that meet in the middle. For example, a k-edge path is a combination of two $k/2$-edge paths. *Bidirectional search* [10,25] is a general approach to find such structures by searching the two halves simultaneously until the two search frontiers meet. In instantiations of this idea, it is crucial to efficiently *join* the two frontiers to obtain a valid or optimal solution. For instance, the meet-in-the-middle algorithm for the Subset Sum problem, by Horowitz and Sahni [15], implements the join operation via a clever pass through two sorted lists of subset sums.

In the present paper, we take the meet-in-the-middle approach to counting problems, in particular, to counting paths and packings. Here, the join operation amounts to consideration of pairs of *disjoint* subsets of a finite universe, each subset weighted by the number of structures that span the subset. We begin in Sect. 2 by formalizing this as the *Disjoint Sum* problem and providing an algorithm for it based on inclusion–exclusion techniques [5,6,7,17,18,20]. In Sect. 3 we apply the method to count paths of k edges in a given n-vertex graph in time $O^*\left(\binom{n}{k/2}\right)$; throughout the paper, O^* suppresses a factor polynomial in the

⋆ This research was supported in part by the Academy of Finland, Grants 117499 (P.K.) and 125637 (M.K.).

A. Fiat and P. Sanders (Eds.): ESA 2009, LNCS 5757, pp. 578–586, 2009.

mentioned parameters (here, n and k). In Sect. 4 we give another application, to count k-packings in a given family of m-element subsets of an n-element universe in time $O^*\left(\binom{n}{mk/2}\right)$. For both problems we also present slightly slower algorithms that require only polynomial space.

We note that an earlier report on this work [8] already introduces a somewhat more general technique and an application to counting paths. The results on the application in particular have attracted some recent interest [1,26]; these results are subsumed by the present work. We discuss related work in more detail below.

1.1 Related Work and Discussion

Deciding whether a given n-vertex graph contains a Hamiltonian path, that is, a simple path of $n-1$ edges, is well known to be NP-hard. The fastest known algorithms, due to Bellman [3,4] and, independently, Held and Karp [14], are based on dynamic programming across the vertex subsets and run in time $O^*(2^n)$. Equally fast polynomial-space variants that actually count all Hamiltonian paths via inclusion–exclusion were discovered later by Kohn, Gottlieb, and Kohn [20], and independently, Karp [17]. Our algorithm (cf. Theorem 2, for $k = n - 1$), too, runs in time $O^*(2^n)$, if allowing exponential space.

In this light, it is intriguing that the parameterized problem of counting paths of k edges seems harder than the corresponding decision problem; this is the present understanding that has emerged from a series of works, starting perhaps in Papadimitriou and Yannakakis's [24] conjecture that for $k = O(\log n)$ the decision problem can be solved in polynomial time. The conjecture was proved by Alon, Yuster, and Zwick's [2] color-coding technique that gave a randomized algorithm with expected running time $O^*(5.44^k)$ and a derandomized variant with running time $O^*(c^k)$ for a large constant c. With a more efficient color-coding scheme, Chen, Lu, Sze, and Zhang [9] improved the latter bound to $O^*(12.8^k)$; see also Kneis, Mölle, Richter, and Rossmanith [19]. Using completely different techniques, Koutis [22], followed by Williams [27], developed a randomized algorithm that runs in expected time $O^*(2^k)$. Unfortunately, it is unlikely that the randomization based techniques extend to counting. For instance, very recently Alon and Gutner [1] showed that color-coding is doomed to fail as every "balanced" family of hash functions from a k-set to an n-set is of size at least $c(k)n^{\lfloor k/2 \rfloor}$ for some function c. Flum and Grohe [12] proved another negative result, namely that the counting problem is #W[1]-hard with respect to the parameter k. Thus it is not expected that the problem, unlike its decision counterpart, is fixed–parameter tractable. From the positive side, a very recent result of Vassilevska and Williams [26] implies that k-edge paths can be counted in time $O^*\left(2^k(k/2)!\binom{n}{k/2}\right)$ in polynomial space; our polynomial-space algorithm (cf. Theorem 3) is faster still, by a factor of $(4/3)^{k/2}(k/2)!$.

Concerning set packings the situation is analogous, albeit the research has been somewhat less extensive. Deciding whether a given family of f subsets of an n-element universe contains a k-packing is known to be W[1]-hard [11], and thus it is unlikely that the problem is fixed–parameter tractable, that is, solvable in time $c(k)f^d$ for some function c and constant d. If f is fairly large, say exponential in n, the fastest known algorithms actually count the packings by employing the inclusion–exclusion machinery [5,6] and run in time $O^*(2^n)$. This bound holds also for the presented algorithm (cf. Theorem 4).

Again, it is interesting that there is a natural parameterization under which counting k-packing seems harder than the corresponding decision problem. Indeed, Jia, Zhang, and Chen [16] showed that the decision problem is fixed–parameter tractable with respect to the total size mk of the packing, assuming each member is of size m. Koutis [21], followed by Chen, Lu, Sze, and Zhang [9], gave faster algorithms with running time $O^*(c^{mk})$ for some constant c; we note that here the running time also grows about linearly in the input size f, which can be as large as $\binom{n}{m}$. For *counting* m-set k-packings, previous techniques [5,6] alone only give a running time bound of $O^*\left(\binom{n}{mk}\right)$ if $mk \le n/2$ and $O^*(2^n)$ otherwise. Besides the present work, we are aware of two recent improvements: For the special case of counting t-matchings, that is 2-set $t/2$-packings,[1] Vassilevska and Williams [26] give a time bound of $O^*\left(2^{t+c(t)}\binom{n}{t/2}\right)$, where $c(t)$ is of order $o(t)$; our polynomial-space algorithm (cf. Corollary 1) turns out to be slightly faster, by a factor of about $(4/3)^{t/2}$. For the general case, Koutis and Williams [23] give a time bound of $O^*\left(n^{mk/2}\right)$; our bounds (Theorem 4) appear to be superior, e.g., when mk grows linearly in n.

The presented meet-in-the-middle approach resembles the randomized divide-and-conquer technique by Chen, Lu, Sze, and Zhang [9] and the similar divide-and-color method by Kneis, Mölle, Richter, and Rossmanith [19], designed for parameterized decision problems. These can, in turn, be viewed as extensions of the recursive partitioning technique of Gurevich and Shelah [13] for the Hamiltonian Path problem. That said, the contribution of the present paper is mainly in the observation that, in the counting context, the join operation can be implemented efficiently using the inclusion–exclusion machinery. While our formalization of the problem as the Disjoint Sum problem is new, the solution itself can, in essence, already be found in Kennes [18], even though in terms of possibility calculus and without the idea of "trimming," that is, restricting the computations to small subsets. Kennes's results were rediscovered in a dual form and extended to accommodate trimming in the authors' recent works [5,6,7].

2 The Disjoint Sum Problem

Given two set families $\mathcal{A}$ and $\mathcal{B}$, and functions α and β that associate each member of $\mathcal{A}$ and $\mathcal{B}$, respectively, an element from a ring R, the *Disjoint Sum* problem is to find the sum of the products $\alpha(A)\beta(B)$ over all *disjoint* pairs of subsets (A, B) in the Cartesian product $\mathcal{A} \times \mathcal{B}$; denote the sum by $\alpha \boxtimes \beta$. In applications, the ring R is typically the set of integers equipped with the usual addition and multiplication operation. Note that, had the condition of disjointness removed, the problem could be easily solved using about $|\mathcal{A}| + |\mathcal{B}|$

[1] Whether counting t-matchings is fixed–parameter tractable remains a major open question in parameterized complexity.

additions and one multiplication. However, to respect the disjointness condition, the straightforward algorithm appears to require about $|\mathcal{A}||\mathcal{B}|$ ring operations and tests of disjointness.

In many cases, we fortunately can do better by applying the principle of inclusion and exclusion. The basic idea is to compute the sum over pairs (A, B) with $A \cap B = \emptyset$ by subtracting the sum over pairs with $A \cap B = X \neq \emptyset$ from the sum over pairs with no constraints. For a precise treatment, it is handy to denote by N the union of all the members in the families $\mathcal{A}$ and $\mathcal{B}$, and extend the functions α and β to all subsets of N by letting them evaluate to 0 outside $\mathcal{A}$ and $\mathcal{B}$, respectively. We also use the Iverson bracket notation: $[P] = 1$ if P is true, and $[P] = 0$ otherwise. Now, by elementary manipulation,

$$\begin{aligned} \alpha \boxtimes \beta &= \sum_A \sum_B [A \cap B = \emptyset]\, \alpha(A)\, \beta(B) \\ &= \sum_A \sum_B \sum_X (-1)^{|X|}\, [X \subseteq A \cap B]\, \alpha(A)\, \beta(B) \\ &= \sum_X (-1)^{|X|} \sum_A \sum_B [X \subseteq A]\, [X \subseteq B]\, \alpha(A)\, \beta(B) \\ &= \sum_X (-1)^{|X|} \Big(\sum_{A \supseteq X} \alpha(A) \Big) \Big(\sum_{B \supseteq X} \beta(B) \Big) . \end{aligned} \tag{1}$$

Here we understand that A, B, and X run through all subsets of N unless otherwise specified. Note also that the second equality holds because every *nonempty* set has exactly as many subsets of even size as subsets of odd size.

To analyze the complexity of evaluating the inclusion–exclusion expression (1), we define the *lower set* of a set family $\mathcal{F}$, denoted by $\downarrow\mathcal{F}$, as the family consisting of all the sets in $\mathcal{F}$ and their subsets. We first observe that in (1) it suffices to let X run over the intersection of $\downarrow\mathcal{A}$ and $\downarrow\mathcal{B}$, for any other X has no supersets in $\mathcal{A}$ or in $\mathcal{B}$. Second, we observe that the values

$$\hat{\alpha}(X) \doteq \sum_{A \supseteq X} \alpha(A) ,$$

for all $X \in \downarrow\mathcal{A}$, can be computed in a total of $|\downarrow\mathcal{A}|\, n$ ring and set operations, as follows. Let $a_1, a_2, \ldots, a_n$ be the n elements of N. For any $i = 0, 1, \ldots, n$ and $X \in \downarrow\mathcal{A}$ define $\hat{\alpha}_i(X)$ as the sum of the $\alpha(A)$ over all sets $A \in \downarrow\mathcal{A}$ with $A \supseteq X$ and $A \cap \{a_1, a_2, \ldots, a_i\} = X \cap \{a_1, a_2, \ldots, a_i\}$. In particular, $\hat{\alpha}_n(X) = \alpha(X)$ and $\hat{\alpha}_0(X) = \hat{\alpha}(X)$. Furthermore, by induction on i one can prove the recurrence

$$\hat{\alpha}_{i-1}(X) = [a_i \notin X]\, \hat{\alpha}_i(X) + [X \cup \{a_i\} \in \downarrow\mathcal{A}]\, \hat{\alpha}_i(X \cup \{a_i\}) ;$$

for details, see closely related recent work on trimmed zeta transform and Moebius inversion [6,7]. Thus, for each i, the values $\hat{\alpha}_i(X)$ for all $X \in \downarrow\mathcal{A}$ can be computed with $|\downarrow\mathcal{A}|$ ring and set operations.

We have shown the following.

Theorem 1. *The Disjoint Sum problem can be solved with* $O\big(n\,(\,|{\downarrow}\mathcal{A}| + |{\downarrow}\mathcal{B}|\,)\big)$ *ring and set operations, and with a storage for* $O\big(\,|{\downarrow}\mathcal{A}| + |{\downarrow}\mathcal{B}|\,\big)$ *ring elements, where* n *is the number of distinct elements covered by the members of* $\mathcal{A}$ *and* $\mathcal{B}$.

3 Paths

Consider paths in an undirected graph with vertex set V and edge set E. Define a *k-edge path* as a sequence of $k+1$ distinct vertices $v_0 v_1 \cdots v_k$ such that the adjacent vertices v_{i-1} and v_i are connected by an edge $v_{i-1}v_i$ in E, for $i = 1, 2, \ldots, k$. We call the set $\{v_0, v_1, \ldots, v_k\}$ the *support* of the path and v_0 and v_k the *ends* of the path. For any vertex v and a subset of j vertices $S \subseteq V$, let $p_j(S, v)$ denote the number of j-edge paths with an end v and support $S \cup \{v\}$. Clearly, the values can be computed by dynamic programming using the recurrence

$$p_0(S, v) = [S = \emptyset]\,, \quad p_j(S, v) = \sum_{u \in S} p_{j-1}(S \setminus \{u\}, u)\,[uv \in E] \quad \text{for } j > 0\,.$$

Alternatively, one may use the inclusion–exclusion formula [17,20]

$$p_j(S, v) = \sum_{Y \subseteq S} (-1)^{|S \setminus Y|}\, w_j(Y, v)\,,$$

where $w_j(Y, v)$ is the number of j-edge walks starting from v and visiting some vertices of Y, that is, sequences $u_0 u_1 \cdots u_j$ with $u_0 = v$, each $u_{i-1}u_i \in E$, and $u_1, u_2, \ldots, u_j \in Y$. Note that for any given Y, v, and j, the term $w_j(Y, v)$ can be computed in time polynomial in n. Using either of the above two formulas, the values $p_j(S, v)$ for all $v \in V$ and sets $S \subseteq V \setminus \{v\}$ of size j, can be computed in time $O^*\big(\binom{n}{\downarrow j}\big)$; here and henceforth, $\binom{q}{\downarrow r}$ denotes the sum of the binomial coefficients $\binom{q}{0} + \binom{q}{1} + \cdots + \binom{q}{r}$. In particular, the number of k-edge paths in the graph is obtained as the sum of $p_k(S, v)$ over all $v \in V$ and $S \subseteq V \setminus \{v\}$ of size k, in time $O^*\big(\binom{n}{\downarrow k}\big)$.

However, meet-in-the-middle yields a much faster algorithm. Assuming for simplicity that k is even, the path has a mid-vertex, $v_{k/2}$, at which the path uniquely decomposes into two $k/2$-edge paths, namely $v_0 v_1 \cdots v_{k/2}$ and $v_{k/2} v_{k/2+1} \cdots v_k$, with almost disjoint supports. Thus, the number of k-edge paths is obtained as the sum of the products

$$p_{k/2}(S, v)\, p_{k/2}(T, v) \big/ 2$$

over all vertices $v \in V$ and disjoint pairs of subsets $S, T \subseteq V \setminus \{v\}$ of size $k/2$. Applying Theorem 1, once for each $v \in V$, with $\mathcal{A} \doteq \mathcal{B} \doteq \{S \subseteq V \setminus \{v\} : |S| = k/2\}$ and $\alpha \doteq \beta \doteq p_{k/2}$ gives the following.

Theorem 2. *The* k*-edge paths in a given graph on* n *vertices can be counted in time* $O^*\big(\binom{n}{k/2}\big)$.

In the remainder of this section we present a polynomial-space variant of the above described algorithm. Let the mid-vertex v be fixed. Then the task is to compute, for each $X \subseteq V \setminus \{v\}$ of size at most $k/2$, the sum

$$\sum_{S \supseteq X} p_{k/2}(S, v) = \sum_{S \supseteq X} \sum_{Y \subseteq S} (-1)^{|S \setminus Y|} w_{k/2}(Y, v)$$

in space polynomial in n and k. If done in a straightforward manner, the running time, ignoring polynomial factors, becomes proportional to the number of triplets (X, S, Y) with $X, Y \subseteq S \subseteq V \setminus \{v\}$ and $|S| = k/2$. This number is $\binom{n-1}{k/2}2^k$ because there are $\binom{n-1}{k/2}$ choices for S and for any fixed S, there are $2^{k/2}$ choices for X and $2^{k/2}$ choices for Y.

A faster algorithm is obtained by reversing the order of summation:

$$\begin{aligned}\sum_{S \supseteq X} p_{k/2}(S, v) &= \sum_{Y} w_{k/2}(Y, v) \sum_{S} (-1)^{|S \setminus Y|} [X, Y \subseteq S] \\ &= \sum_{Y} w_{k/2}(Y, v) \, (-1)^{k/2 - |Y|} \binom{n - |X \cup Y|}{k/2 - |X \cup Y|} ;\end{aligned}$$

here Y and S run through all subsets of $V \setminus \{v\}$ of size at most $k/2$ and exactly $k/2$, respectively. The latter equality holds because S is of size $k/2$ and contains $X \cup Y$. It remains to find in how many ways one can choose the sets X and Y such that the union $U \doteq X \cup Y$ is of size at most $k/2$. This number is

$$\sum_{s=0}^{k/2} \binom{n-1}{s} 3^s \le \frac{3}{2} \binom{n-1}{k/2} 3^{k/2} ,$$

because there are $\binom{n-1}{s}$ ways to choose U of size s, and one can put each element in U either to X or Y or both.

Theorem 3. *The k-edge paths in a given graph on n vertices can be counted in time $O^*\left(3^{k/2}\binom{n}{k/2}\right)$ in space polynomial in n and k.*

4 Set Packing

Next, consider packings in a set family $\mathcal{F}$ consisting of subsets of a universe N. We will assume that each member of $\mathcal{F}$ is of size m. A *k-packing* in $\mathcal{F}$ is a set of k mutually disjoint members of $\mathcal{F}$. The members $F_1, F_2, \ldots, F_k$ of a k-packing can be ordered in $k!$ different ways to an *ordered k-packing* $F_1 F_2 \cdots F_k$. Define the *support* of the ordered k-packing as the union of its members. For any $S \subseteq N$, let $\pi_j(S)$ denote the number of ordered j-packings in $\mathcal{F}$ with support S. The values can be computing by dynamic programming using the recurrence

$$\pi_0(S) = [S = \emptyset] , \quad \pi_j(S) = \sum_{F \subseteq S} \pi_{j-1}(S \setminus F) \, [F \in \mathcal{F}] \quad \text{for } j > 0 .$$

Alternatively, one may use the inclusion–exclusion formula

$$\pi_j(S) = \sum_{Y\subseteq S} (-1)^{|S\setminus Y|} \Big(\sum_{F\subseteq Y} [F \in \mathcal{F}] \Big)^j$$

(here we use the assumption that every member of $\mathcal{F}$ is of size m) [5,6]. Using the inclusion–exclusion formula, the values $\pi_j(S)$ for all $S \subseteq N$ of size mj can be computed in time $O^*\big(\binom{n}{\downarrow mj}\big)$, where n is the cardinality of N; a straightforward implementation of the dynamic programming algorithm yields the same bound, provided that m is a constant. In particular, the number of k-packings in $\mathcal{F}$ is obtained as the sum of $\pi_k(S)/k!$ over all $S \subseteq N$ of size mk, in time $O^*\big(\binom{n}{\downarrow mk}\big)$.

Again, meet-in-the-middle gives a much faster algorithm. Assuming for simplicity that k is even, we observe that the ordered k-packing decomposes uniquely into two ordered $k/2$-packings $F_1F_2\cdots F_{k/2}$ and $F_{k/2+1}F_{k/2+2}\cdots F_k$ with disjoint supports. Thus the number of ordered k-packings in $\mathcal{F}$ is obtained as the sum of the products

$$\pi_{k/2}(S)\,\pi_{k/2}(T)\Big/2$$

over all disjoint pairs of subsets $S, T \subseteq N$ of size $mk/2$. Applying Theorem 1 with $\mathcal{A} \doteq \mathcal{B} \doteq \{S \subseteq N : |S| = mk/2\}$ and $\alpha \doteq \beta \doteq \pi_{k/2}$ gives the following.

Theorem 4. *The k-packings in a given family of m-element subsets of an n-element set can be counted in time $O^*\big(\binom{n}{mk/2}\big)$.*

We next present a polynomial-space variant. The task is, in essence, to compute for each $X \subseteq N$ of size at most $mk/2$ the sum

$$\sum_{S\supseteq X} \pi_{k/2}(S) = \sum_{S\supseteq X}\sum_{Y\subseteq S} (-1)^{|S\setminus Y|} \Big(\sum_{F\subseteq Y} [F \in \mathcal{F}] \Big)^{k/2}$$

in space polynomial in n, k, and m.

As with counting paths in the previous section, a faster than the straightforward algorithm is obtained by reversing the order of summation:

$$\begin{aligned}\sum_{S\supseteq X} \pi_{k/2}(S) &= \sum_{Y} \Big(\sum_{F\subseteq Y} [F \in \mathcal{F}] \Big)^{k/2} \sum_{S} (-1)^{|S\setminus Y|} [X, Y \subseteq S] \\ &= \sum_{Y} \Big(\sum_{F\subseteq Y} [F \in \mathcal{F}] \Big)^{k/2} (-1)^{k/2-|Y|} \binom{n - |X\cup Y|}{mk/2 - |X \cup Y|}\,;\end{aligned}$$

here Y and S run through all subsets of N of size at most $mk/2$ and exactly $mk/2$, respectively. It remains to find the number of triplets (X, Y, F) satisfying $|X \cup Y| \le mk/2$, $|F| = m$, and $F \subseteq Y$. This number is

$$\begin{aligned}\sum_{s=m}^{mk/2} \binom{n}{s}\binom{s}{m} 2^m 3^{s-m} &< \frac{3}{2}\binom{n}{mk/2}\binom{mk/2}{m} 2^m 3^{mk/2-m} \\ &\le \frac{3}{2}\binom{n}{mk/2} 5^{mk/2}\,, \end{aligned} \qquad (2)$$

because there are $\binom{n}{mk/2}$ choices for the union $U \doteq X \cup Y$ of size s, within which there are $\binom{s}{m}$ choices for F; the elements in F can be put to only Y or to both X and Y, whereas each of the remaining $s - m$ elements in U is put to either X or Y or both.

Theorem 5. *The k-packings in a given family of m-element subsets of an n-element set can be counted in time $O^*\left(5^{mk/2}\binom{n}{mk/2}\right)$ in space polynomial in n, k, and m.*

We remark that the upper bound (2) is rather crude for small values of m. In particular, provided that m is a constant, we can replace the constant 5 by 3.

Corollary 1. *The k-packings in a given family of 2-element subsets of an n-element set can be counted in time $O^*\left(3^k\binom{n}{k}\right)$ in space polynomial in n and k.*

References

1. Alon, N., Gutner, S.: Balanced hashing, color coding and approximate counting. Electronic Colloquium on Computational Complexity, Report TR09-12 (2009)
2. Alon, N., Yuster, R., Zwick, U.: Color-coding. J. Assoc. Comput. Mach. 42, 844–856 (1995)
3. Bellman, R.: Combinatorial processes and dynamic programming. In: Bellman, R., Hall Jr., M. (eds.) Combinatorial Analysis. Proceedings of Symposia in Applied Mathematics 10, pp. 217–249. American Mathematical Society (1960)
4. Bellman, R.: Dynamic programming treatment of the travelling Salesman Problem. J. Assoc. Comput. Mach. 9, 61–63 (1962)
5. Björklund, A., Husfeldt, T., Koivisto, M.: Set partitioning via inclusion–exclusion. SIAM J. Comput. (to appear) Special Issue for FOCS 2006
6. Björklund, A., Husfeldt, T., Kaski, P., Koivisto, M.: Fourier meets Möbius: fast subset convolution. In: 39th ACM Symposium on Theory of Computing (STOC 2007), pp. 67–74. ACM Press, New York (2007)
7. Björklund, A., Husfeldt, T., Kaski, P., Koivisto, M.: Trimmed Moebius inversion and graphs of bounded degree. In: 25th International Symposium on Theoretical Aspects of Computer Science (STACS 2008). Dagstuhl Seminar Proceedings 08001, pp. 85–96. IBFI Schloss Dagstuhl (2008)
8. Björklund, A., Husfeldt, T., Kaski, P., Koivisto, M.: The fast intersection transform with applications to counting paths. CoRR, abs/0809.2489 (2008)
9. Chen, J., Lu, S., Sze, S.-H., Zhang, F.: Improved algorithms for path, matching, and packing problems. In: 18th Annual ACM-SIAM Symposium on Discrete Algorithms (SODA 2007), Philadelphia, PA, USA, pp. 298–307. Society for Industrial and Applied Mathematics (2007)
10. Danzig, G.: Linear Programming and Extensions. Princeton University Press, Princeton (1963)
11. Downey, R.G., Fellows, M.R.: Parameterized Complexity. Springer, Berlin (1999)
12. Flum, J., Grohe, M.: The parameterized complexity of counting problems. SIAM J. Comput. 33, 892–922 (2004)
13. Gurevich, Y., Shelah, S.: Expected computation time for Hamiltonian path problem. SIAM J. Comput.16, 486–502 (1987)

14. Held, M., Karp, R.M.: A dynamic programming approach to sequencing problems. J. Soc. Indust. Appl. Math.$\tilde{1}$0, 196–210 (1962)
15. Horowitz, E., Sahni, S.: Computing partitions with applications to the knapsack Problem. J. Assoc. Comput. Mach.$\tilde{2}$1, 277–292 (1974)
16. Jia, W., Zhang, C., Chen, J.: An efficient parameterized algorithm for m-set packing. J. Algorithms 50, 106–117 (2004)
17. Karp, R.M.: Dynamic programming meets the principle of inclusion and exclusion. Oper. Res. Lett. 1, 49–51 (1982)
18. Kennes, R.: Computational aspects of the Moebius transform on a graph. IEEE Transactions on System, Man, and Cybernetics 22, 201–223 (1991)
19. Kneis, J., Mölle, D., Richter, S., Rossmanith, P.: Divide-and-color. In: Fomin, F.V. (ed.) WG 2006. LNCS, vol. 4271, pp. 58–67. Springer, Heidelberg (2006)
20. Kohn, S., Gottlieb, A., Kohn, M.: A generating function approach to the traveling salesman problem. In: ACM Annual Conference (ACM 1977), pp. 294–300. ACM Press, New York (1977)
21. Koutis, I.: A faster parameterized algorithm for set packing. Information Processing Letters 94, 4–7 (2005)
22. Koutis, I.: Faster algebraic algorithms for path and packing problems. In: Aceto, L., Damgård, I., Goldberg, L.A., Halldórsson, M.M., Ingólfsdóttir, A., Walukiewicz, I. (eds.) ICALP 2008, Part I. LNCS, vol. 5125, pp. 575–586. Springer, Heidelberg (2008)
23. Koutis, I., Williams, R.: Limitations and applications of group algebras for parameterized problems. In: Albers, S., Marchetti-Spaccamela, A., Matias, Y., Nikoletseas, S., Thomas, W. (eds.) ICALP 2009. Part I. LNCS, vol. 5555, pp. 653–664. Springer, Heidelberg (2009)
24. Papadimitriou, C.H., Yannakakis, M.: On limited nondeterminism and the complexity of the V-C dimension. J. Comput. Syst. Sci. 53, 161–170 (1996)
25. Pohl, I.: Bi-directional and heuristic search in path problems. PhD thesis, Report SLAC-104, Stanford University (1969)
26. Vassilevska, V., Williams, R.: Finding, minimizing, and counting weighted subgraphs. In: 41st ACM Symposium on Theory of Computing (STOC 2009), pp. 455–464 (2009)
27. Williams, R.: Finding paths of length k in O*$O^*(2^k)$ time. Information Processing Letters 109, 315–318 (2009)

Accelerating Multi-modal Route Planning by Access-Nodes*

Daniel Delling, Thomas Pajor, and Dorothea Wagner

Department of Computer Science, Universität Karlsruhe (TH), P.O. Box 6980, 76128 Karlsruhe, Germany
{delling,pajor,wagner}@informatik.uni-karlsruhe.de

Abstract. Recent research on fast route planning algorithms focused *either* on road networks *or* on public transportation. However, on the long run, we are interested in planning routes in a *multi-modal* scenario: we start by car to reach the nearest train station, ride the train to the airport, fly to an airport near our destination and finally take a taxi. In other words, we need to *incorporate* public transportation into road networks. However, we do not want to switch the type of transportation too often. We end up in a *label constrained* variant of the shortest path problem. In this work, we present a first efficient solution to a restricted variant of this problem including experimental results for transportation networks with up to 125 Mio. edges.

1 Introduction

Computing the quickest path in graphs modeling transportation networks is one of the showpieces of applied algorithms. In general, DIJKSTRA's algorithm [1] finds a shortest (or quickest) path between a given source s and target t. Unfortunately, the algorithm is far too slow to be used on huge datasets which appear frequently in route planning. Thus, several speed-up techniques have been developed that compute additional data during a *preprocessing* step in order to speed-up queries during the *online* phase.

However, all developed techniques so far suffer from one drawback: they only work either in road or railway networks. On the long run, we are interested in multi-modal queries where we change the type of transportation along our journey. Unfortunately, it is not sufficient to merge all networks and compute

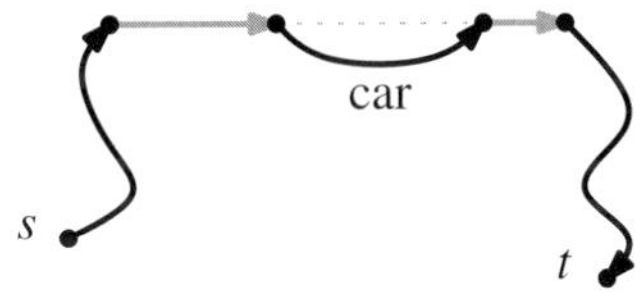

(a) Suboptimal path.

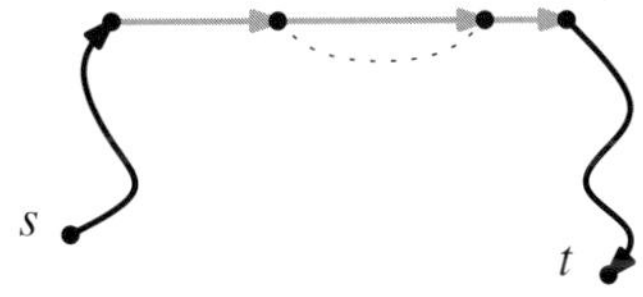

(b) Valid path.

Fig. 1. Motivation of a label-constrained shortest path: while the path at the top is the quickest connection, it requires us to use a car (black) between two trains (gray). The path to the bottom, however, is preferable since we do not need to leave the train.

* Partially supported by the Future and Emerging Technologies Unit of EC (IST priority – 6th FP), under contract no. FP6-021235-2 (project ARRIVAL) and the DFG (project WA 654/16-1).

A. Fiat and P. Sanders (Eds.): ESA 2009, LNCS 5757, pp. 587–598, 2009.

quickest paths in the resulting bigger network: the quickest path may force us to change the type of transportation too frequently. See Figure 1 for an example. A possible approach to this problem is the LABEL CONSTRAINED SHORTEST PATH PROBLEM. The idea is as follows. Each edge gets a label assigned depicting the type of transportation network it represents. Then, only a path between s and t is valid if certain constraints are fulfilled by the labels along the path. In this work, we present an approach how to accelerate multi-modal queries by skipping main parts of the network without losing correctness.

Related Work. Theoretical results on the hardness of the LABEL CONSTRAINED SHORTEST PATH PROBLEM can be found in [3,4]. Experimental evaluations of basic algorithms are given in [5], while basic speed-up techniques like A^* [6] and bidirectional search [7] have systematically been examined in [8]. Route planning in uni-modal scenarios has been undergoing a rapid development in recent years with the fastest techniques yielding query times of several microseconds in road networks [9]. For a recent overview over uni-modal speed-up techniques, we direct the interested reader to [2]. However, to the best of our knowledge, there exists no route planning algorithm that can answer a multi-modal query in a huge combined transportation network within milliseconds. Since our approach is closely related to (uni-modal static) Transit-Node Routing [9], we briefly recap this technique. TNR selects a subset $\mathcal{T} \subset V$ (normally 10 000 nodes) of so called transit nodes and stores distances between them in a table. Moreover, each node $v \in V$ stores the distances to all *relevant* transit nodes, called *access-nodes*. Then, with good choice of $\mathcal{T}$, a long-range query can be reduced to three table lookups. In order to decide whether a query is a long-range query, a *locality* filter is introduced. In case s and t are too close to each other, an arbitrary speed-up technique is applied. The percentage of global queries can be increased by introducing several *layers* of transit nodes.

Our Contribution. In this work, we present an efficient approach to a special case of multi-modal route planning. We assume that we want to use the road network only at the beginning and the end of a journey and that the public transportation network is much smaller than the road network. Then, by adapting some ideas from Transit-Node Routing, we may "skip" the road network with a table lookup and restrict the search to the much smaller public transportation network (PTN). We present how to compute so-called *access-nodes* to the PTN for each node efficiently. With this information at hand, we are able to present Access-Node Routing, accelerating multi-modal queries (in our scenario) by more than 4 orders of magnitude. However, the main achievement is that we are able to separate the public transportation network from the road network in a multi-modal context. This allows us to run different query algorithms on the public transportation and the road networks. Although, in this work we use an augmentation of DIJKSTRA's algorithm on the public transportation network, it would also be possible to use a speed-up technique or even multi-criteria search.

This work is organized as follows. Section 2 settles basic definitions we need for this work. In Section 3 we briefly recapt existing and new approaches for modeling transportation networks. Moreover, we show how to obtain a multi-modal network and present that a *label-constrained* variant of the shortest path problem is a possible approach for better routes in multi-modal networks. It turns out that we need an automaton that restricts the number of network changes. Starting from analyzing our networks and automata, we develop our main contribution (Section 4) of this work: Access-Node Routing. By running several experiments on transportation networks with up to 125 Mio. edges (Section 5), we show that Access-Node Routing is up to 31 000 times faster than a label-constrained variant of DIJKSTRA. This value is achieved without using a speed-up technique within the public transportation network. Section 6 concludes our work with a brief summary and interesting open questions.

2 Preliminaries

A *graph* is a tuple $G = (V, E)$ consisting of a finite set V of *nodes* and a set $E \subseteq V \times V$ of *edges* which are ordered pairs (u, v) if the graph is *directed*. The node u is called the *tail* of the edge, v the *head*. The reverse graph $\overleftarrow{G} = (V, \overleftarrow{E})$ is the graph obtained from G by substituting each $(u, v) \in E$ by (v, u). The main difference between uni- and multi-modal route planning is that the nodes and edges of a graph are *labeled* by a finite set Σ of symbols which is often called an *alphabet*. The node-label is denoted by $\mathrm{lab} : V \rightarrow \Sigma$, the edge-label by $\mathrm{lab} : E \rightarrow \Sigma$. Throughout the whole work we restrict ourselves to directed *labeled* graphs which are weighted by a length function len depicting the travel time from u to v. Depending on the edge-label, edge weights may be time-independent or time-dependent. For time-independent weights, we use a positive length function $\mathrm{len} : E \rightarrow \mathbb{R}^+$, while for time-dependent edges, we use a time-dependent length function $\mathrm{len} : E \rightarrow \mathbb{F}$ with the function space $\mathbb{F}$ consisting of positive *periodic* functions $f : \Pi \rightarrow \mathbb{R}^+, \Pi = [0, p], p \in \mathbb{N}$ such that $f(0) = f(p)$ and $f(x) + x \leq f(y) + y$ for any $x, y \in \Pi, x \leq y$. Note that these functions respect the FIFO property [10], and hence our networks fulfill the FIFO property as well. In the following, we call Π the *period* of the input. The upper bound of f is denoted by $\overline{f} = \max_{x \in \Pi} f(x)$, the lower by $\underline{f} = \min_{x \in \Pi} f(x)$. Note that we can obtain a time-independent labeled lower/upper bound graph $\underline{G}/\overline{G}$ from G by replacing each edge function with their lower/upper bounds.

A sequence $w := \sigma_1, \sigma_2, \ldots, \sigma_k$ of symbols from Σ is called a *word*. The *length* of a word is the number of symbols it is composed of. A not necessarily finite set L of words over Σ is called a *language* over Σ. A *non-deterministic finite automaton* $\mathscr{A} := (Q, \Sigma, \delta, S, F)$ consists of a finite set Q of states, an alphabet Σ, the transition function $\delta : Q \times \Sigma \rightarrow \mathscr{P}(Q)$, a set S of initial states and a set F of final states. A language L is called *regular* if it can be accepted by a non-deterministic finite automaton [11,12]. Throughout the whole work we restrict ourselves to regular languages.

3 Models and Basic Algorithms

We now briefly present how to model transportation networks as graphs and how to construct a multi-modal graph from these ingredients. Finally, we present why the LABEL CONSTRAINED SHORTEST PATH PROBLEM is useful for our application of reasonable route planning in multi-modal scenarios.

3.1 Models

Modeling a *road network* or *foot networks* as a graph $G_{\texttt{road}}$ is straightforward. Junctions are modeled as nodes and an edge $e = (u,v)$ between two junctions $u,v \in V$ is inserted if and only if a road segment from u to v exists in the road network. The edge weights $\text{len}(e)$ in the road network represent the average travel time on the specific road segment. For foot edges, we assign a weight based on the assumption that we walk the segment with 4 km/h on average. Note that our foot networks compared to our road networks have the same node set whereas the edge sets may differ due to motorways (not open to pedestrians), one-ways and pedestrian zones (not open to traffic). For *railway networks* $G_{\texttt{rail}}$, we use the *realistic* time-dependent approach as presented in [13] where edges are time-dependent and depict several trains running on the same route from one station to another. We model *flight networks* $G_{\texttt{flight}}$ as follows. Our realistic raw data incorporates two major flight alliances: StarAlliance [14] and Oneworld [15]. Thus, for each airport we introduce a supernode and two departure and arrival nodes—one for each flight alliance. Edges between supernodes and departure nodes model check-in, arrival-departure edges the changing of planes (with different weights for transfers between flight alliances) and arrival-supernode edges the check-out. Time-dependent edges between two airports A and B depict direct flights between A and B. Depending on the fact which flight alliance operates the respective flight, head and tail are chosen as the corresponding departure and arrival nodes. See [16] for details.

Let $\Sigma = \{\texttt{road}, \texttt{foot}, \texttt{rail}, \texttt{flight}\}$ be an alphabet. We construct a *multi-modal* network from our four types of transportation networks $G_\sigma = (V_\sigma, E_\sigma), \sigma \in \Sigma$ by first labeling each node u and each edge e with a label $\text{lab}(u), \text{lab}(e) \in \Sigma$. Then, we unify the node and edge-sets to a graph $G_{\texttt{multi}} = (\bigcup_{\sigma\in\Sigma} V_\sigma, \bigcup_{\sigma\in\Sigma} E_\sigma) =: (V_{\texttt{multi}}, E_{\texttt{multi}})$. In order to connect the networks among each other, we introduce so called *link-edges* (u,v) with $\text{lab}((u,v)) = \texttt{link}$ and $\text{lab}(u) \neq \text{lab}(v)$. These edges depict possible switches from one transportation type to another. So, we have $\Sigma = \{\texttt{road}, \texttt{foot}, \texttt{rail}, \texttt{flight}, \texttt{link}\}$. We connect each station node (railways) and supernode (flights) with its closest node from the road network. Moreover, we connect each airport with its closest train stations by a link edge. However, head and tail must not be more than 5 km away from each other.

3.2 The Label-Constrained Shortest Path Problem

The LABEL CONSTRAINED SHORTEST PATH PROBLEM is an augmentation of the classic SHORTEST PATH PROBLEM. The idea is that only such paths between s and t are valid that form a word of a language L. More precisely, given an alphabet Σ, a language $L \subset \Sigma^*$, a weighted, directed graph $G = (V,E)$ with Σ-labeled edges and source and target nodes $s,t \in V$, we ask for a shortest path $P = (u_1, \ldots, u_k)$ from $s = u_1$ to $t = u_k$, where the sequence of labels along the edges of the path forms a word of L. In other words, $\text{lab}((u_1,u_2)) \ldots \text{lab}((u_{k-1},u_k)) \in L$ must hold.

It turns out that the complexity of this problem depends on the restrictions to L [4]. In our case, where L is regular, the problem remains polynomially solvable. Then, we can use a straightforward adaption of DIJKSTRA's algorithm for computing a label-constrained shortest s–t path. Besides the inputs for a normal DIJKSTRA, we also require an automaton $\mathscr{A} := (Q, \Sigma, \delta, S, F)$ that accepts our language L. We initialize pairs (s,q), $q \in S$ with $\text{dist}_s((s,q)) = 0$ and insert them with key 0 into a priority queue. Any other node-state tuple (u,q') is initialized with $\text{dist}_s((u,q')) = \infty$. Then, we remove a node-state tuple (u,q') with minimum key from the queue, relax all outgoing edges $e = (u,v)$, determine all states $q'' \in \delta(q', \text{lab}(e))$ and check whether $\text{dist}_s((u,q')) + \text{len}(e) < \text{dist}_s((v,q''))$ holds. If so, we update the distance label and enqueue (v,q''). We may stop the search as soon as we settle a tuple (t,q') with a final state $q' \in F$.

Note that this procedure can be adapted to a time-dependent scenario easily: if we want to compute the shortest path for a given departure time τ, we simply have to evaluate edge weights for their correct departure time when relaxing them. If we want to compute the shortest path for all departure times, we have to use a label-correcting algorithm that propagates functions instead of constants through the network (see [17,18] for details).

A Special Case. When planning a typical multi-modal voyage, we observe that in most cases we start in the road or foot network. Then, we enter the public transportation network without using the road network again (except for transfers) until the end of the journey. There, we either use a taxi or rental car or go by foot to reach our final destination. This observation is characterized by the following definition on languages.

Definition 1 (Enclose Property). *Let L be a regular language over the alphabet Σ of edge-labels. If L is of the form $L = \sigma_{r_1}^* l \sigma_t^* l \sigma_{r_2}^*$, where $\sigma_{r_1}, \sigma_{r_2} \in \{$`road`, `foot`$\}$ and $\sigma_t \in \{$`rail`, `flight`$\}$ and $l =$ `link`, we say that L fulfills the* enclose *property.*

In other words, the public transportation is enclosed by the road network part. An example for an automaton for such languages only allows railway connections enclosed by foot-edges. In the following, we denote this automaton by `foot-and-rail`. By substituting `foot` by `road` and `rail` by `flight`, we obtain a second automaton which we call `road-and-flight`.

Trees and Profile-Graphs. In the following, we build label-constrained shortest path trees and profile graphs. As discussed in Section 3.1, our railway- and flight-networks are time-dependent. So, running a label-constrained DIJKSTRA with a given departure time τ from a node u until the priority queue is empty yields a tree rooted at u. Similarly, running a label-constrained label-correcting algorithm (cf. [17,18]) for all possible departure times from u yields a so-called profile-graph. For each node v, we obtain a function depicting the travel time from v to u (the profile) with changing parent node during the period.

4 Access-Node Routing

Analyzing the networks obtained from our multi-modal approach, we observe that most part of the graph is made up of road segments and, hence, a multi-modal DIJKSTRA spends most its time settling road nodes. Access-Node Routing (ANR) accelerates queries with an automaton fulfilling the enclose property as described in Section 3.2.

The main idea is to precompute distances to all *relevant* access points to the public transportation network. With this information at hand, we may "skip" the road network and restrict the search to a much smaller network. In the following, we define access-nodes, how they can be computed efficiently and the resulting query algorithm. It turns out that space consumption of this approach is rather high, hence, we present how to reduce space consumption by using the concept of contraction [19]. Note that in the following, we explain how to skip the road network, however, this approach can also be used to skip a foot network. Experiments for both variants can be found in Section 5.

4.1 Access-Nodes

Not every node in the public transportation network is suited as "access-node". For example, in the flight network the departure and arrival nodes are used to model internal procedures at airports and should not be accessed directly from the road network. More precisely, a node v is called *access-node candidate* if $\text{lab}(v) \neq \texttt{road}$ and at least one incident edge is a link-edge. The set of all access-node candidates is denoted by A. In our case the set A includes exactly all station nodes regarding the railway network and all supernodes of the flight network (cf. Section 3). Nodes $v \in A$ can be interpreted as entry (or exit) points to/from the public transportation network. Computing distances from every road network node to every access-node would require too much space. Moreover, not every entry point to the public transportation network is mandatory for correct shortest paths. More precisely, a node $v \in A$ with $\text{lab}(v) \neq \texttt{road}$ is an access-node for all nodes u with $\text{lab}(u) = \texttt{road}$ if there exists another node $w \in A$ with $\text{lab}(w) \neq \texttt{road}$ for which the shortest u–w path (for at least one departure time) uses v to enter the public transportation network (i.e., all ancestors of v are road-labeled). The set of (forward) access-nodes for a node u is denoted by $\overrightarrow{\text{A}}(u)$. Note that we can define *backward* access-nodes $\overleftarrow{\text{A}}(u)$ analogously.

4.2 Computing Access-Nodes

In the following, we explain how to compute forward access-nodes for each $v \in V_{\texttt{multi}}$ with $\text{lab}(v) = \texttt{road}$ efficiently. Computing backward access-nodes is done analogously. In general, we present two approaches for computing access-nodes, a forward and an inverse approach, which we both explain in detail.

For the *forward* approach, we construct a profile-graph (with an automaton fulfilling the enclose property) T_v from each v with $\text{lab}(v) = \texttt{road}$. Whenever we settle a candidate node $a \in A$, we add a to $\overrightarrow{\text{A}}(v)$ if a parent—with respect to T_v—of a is a road node. We may stop the search as soon as all nodes u in the priority queue are *covered*. A node u is called covered if at least one of its ancestors—with respect to T_v—is a non-road node. Note that this approach requires two profile-graph constructions (forward and backward) per road node. Hence, this approach is only useful if the number of access-node candidates is high and thus, the construction terminates early.

For the *backward* approach, we compute for each access-node candidate $a \in A$ all nodes $u \in A^{-1}(a)$. Therefore, we construct a backward profile-graph *not* relaxing edges whose tail is a road node. By this, we obtain travel time functions from each $a' \in A \setminus \{a\}$ to $a \in A$. Next, we construct a backward shortest path forest: we initialize a with distance 0 and any a' with an upper bound on the travel time functions to a. This time, we do not relax public transportation edges. We may stop the search as soon as all nodes u in the priority queue have a node $a' \in A \setminus \{a\}$ as ancestor. Finally, we add all nodes with ancestor a to $A^{-1}(a)$. After performing this task for all candidates, we obtain $\overrightarrow{A}(v)$ for all road nodes by inverting the relation $A^{-1}(\cdot)$. Note that by bounding the functions, we may compute a *superset* of the actual access-nodes.

4.3 Query

With the preprocessed data at hand, we can skip the road network part for multi-modal queries with automata fulfilling the enclose property. Assume $\text{lab}(s) = \text{lab}(t) = \texttt{road}$. In a first step, we add node-state tuples (a_s, q_s) (with corresponding distances) for all $a_s \in \overrightarrow{A}(s)$ and all $q_s \in Q$ obtained by a `link`-labeled transition originating from an arbitrary initial state of the automaton to the priority queue. Node-state tuples (a_f, q_f) for all $a_f \in \overleftarrow{A}(t)$ and all $q_f \in Q$ such that there is a `link`-labeled transition from q_f to a final state in the automaton are accumulated in a target node set T. In a second step, we run a multi-modal query in the public transportation network to T not relaxing edges whose head is a road node. We may stop the search if all $(a_f, q_f) \in T$ have a final label assigned or the priority queue runs empty. We end up having distances $\text{dist}_s((a_f, q_f))$ for all $(a_f, q_f) \in T$. Then, the length of the shortest path from s to t is $\min_{(a_f, q_f) \in T}\{\text{dist}_s((a_f, q_f)) + \text{dist}(a_f, t)\}$. Note that we can run time-, profile-, or even multi-criteria-queries within the public transportation network. The main gain we obtain is skipping the road network.

Note that the paths found by our access-node query algorithm must contain a public transportation node. While for long-range queries this may be a meaningful restriction (nobody wants to drive 30 hours by car), low-range queries may contain *only* road nodes. Fortunately, computations of quickest paths in road networks can be done in microseconds. So, we run a *check-query* with the CHASE-algorithm [20] that outputs the length of the quickest path if only the road network is used.

4.4 Core Access-Node Routing

One of the main drawbacks of pure Access-Node Routing is its high space consumption. For each node v of the road network we store two sets of access-nodes together with corresponding distances. Although the number of access-nodes per road node is relatively small, it is not necessary to store these sets for all nodes but only for important ones, i.e., the *core* of the graph. We use a contraction routine that removes unimportant nodes and adds shortcuts in order to preserve distances between core nodes.

Preprocessing for Core-Based Access-Node Routing is done in two steps. First, we contract the graph by a node-reduction followed by an edge-reduction which we obtain by a straightforward adaption of the concepts presented in [20]. However, in order to preserve correctness, we may only remove road nodes having no incident link-edges. As a result, the embedded time-dependent public transportation network is fully contained in the core. See [19] for more details on uni-modal contraction. In a second step, we compute forward and backward access-nodes for all core nodes u with $\text{lab}(u) = \texttt{road}$.

The *query* is a three-phase algorithm. During phase 1, we run a bidirectional DIJKSTRA, not relaxing edges outgoing from core-nodes. Whenever the forward search settles a core node, we add it to a set S. A set T is maintained for backward search analogously. Phase 1 ends if both priority queues are empty. Then, a (two-phase) access-node query is started with S as source nodes and T as target nodes.

4.5 Comparison to Transit-Node Routing

As already mentioned, Access-Node Routing (ANR) adapts ideas from Transit-Node Routing (TNR). Recall that TNR has three ingredients: a table for storing distances between transit nodes, stored distances for each node to its relevant transit (access) nodes, and a locality filter for deciding whether s and t are far away enough from each other (cf. Section 1). In fact, ANR can be interpreted as a multi-modal variant of TNR. Our access-node candidates become transit nodes and we store distances to the relevant access-nodes. Unfortunately, due to poor upper bounds on the travel time functions of the time-dependent edges of the public transportation network, we cannot make up an efficient locality filter as for TNR: we have to run a local path query in almost all cases. In fact, the only time we do not need to run a check query is when s and t are not in the same component of the road network.

Note that TNR uses a (sophisticated) forward approach for determining the relevant access-nodes. Unfortunately, our public transportation networks tend to be sparse and they are time-dependent as well. As a result, the inverse approach turns out to be better for our scenario. Summarizing, ANR can be interpreted as a variant of TNR with higher flexibility but worse query performance. However, TNR has neither been adapted to time-dependent nor label-constrained route planning so far. Both aspects are crucial preconditions for route planning in our scenario.

5 Experiments

We conducted our experiments on one core of an AMD Opteron 2218 running SUSE Linux 10.3. The machine is clocked at 2.6 GHz, has 32 GB of RAM and 2 x 1 MB of L2 cache. The program was compiled with GCC 4.2, using optimization level 3. Our implementation is written in C++ using solely the STL at some points. As priority queue we use a binary heap.

Inputs. We use two different networks and two automata. The first network depicts the foot network of Germany ($|V| \approx 4.5$ Mio, $|E| \approx 11.2$ Mio) merged with all long distance trains from the timetable of the winter period 2000/2001 (498 stations and 18069 connections). This includes InterRegio (IR), InterCity (IC) and InterCityExpress (ICE) trains. We call the resulting graph `de-road-rail(long)` and apply the `foot-and-rail` automaton to this input. Our second input consists of the road network of Western Europe ($|V| \approx 30$ Mio, $|E| \approx 73$ Mio) and North America (including Canada, $|V| \approx 20$ Mio, $|E| \approx 51$ Mio) and the flight network of both flight alliances (359 airports with 32 621 flights). We use this graph together with the `road-and-flight` automaton. In the following, this input is referenced to by `na-eur-road-flight`. Note that second input has about 50 Mio. nodes and 125 Mio. edges.

Methodology. In the following, we report preprocessing times and the overhead of the preprocessed data in terms of *additional* bytes per node. We evaluate query performance by *random queries*, i.e., the nodes s and t are picked uniformly at random. Since public transportation networks are time-dependent, we additionally need a departure time τ, which we pick uniformly at random as well. We provide the average number of settled nodes, i.e., the number of nodes extracted from the priority queue and the average query time. Unless otherwise stated, all figures in this paper are based on 1 000 000 random s-t queries and refer to the scenario that only the lengths of the shortest paths have to be determined, without outputting a complete description of the paths. Efficient methods for unpacking table lookups have been published in [9]. Local paths are computed by CHASE [20]which is a combination of Contraction Hierarchies [19] and Arc-Flags [21].

Preprocessing. In Section 4, we presented two approaches to compute access-nodes: a forward variant and an inverse approach. It turns out that the former is too slow for our multi-modal inputs, hence, we only use the inverse approach. Table 1 reports key figures for computing access-nodes. Note that `na-eur-road-flight`, we only use the core-based approach (cf. Section 4.4). Also note that we report the preprocessing effort for CHASE. Since we need to keep the road network as a *uni-modal* graph in memory, the space consumption here *includes* the graph.

Regarding `de-road-rail(long)`, the average number of forward and backward access-nodes per road node is 32.4 resp. 20.9. Thus, 32.4 railway stations are important (on average) to enter the railway network. While this seems to be a very high number (even more if we consider that these railway stations have to be reached by foot), there are two good reasons for this. First, the railway network is sparsely embedded into the road network, thus, for a single road network node a lot of stations are important at least once a day. Second, long distance trains do not operate very frequently on some parts of the network. As a consequence, the upper bounds are of poor quality yielding many unnecessary access-nodes. The same effect can be observed in the `na-eur-road-flight` network, as flights are even more infrequent. This makes it attractive to cover far distances by car, thus, including many (also far away) airports into the set of relevant access-nodes.

Preprocessing times are in the range of several hours (between 26 minutes for the core of the German network and almost three hours for the core of the continental

Table 1. Preprocessing Figures for (Core-Based) Access-Node Routing. We report the number of access-node candidates, the average number of forward and backward access-nodes per road node and the preprocessing time for computing access-nodes as well as the additionally required space per node. We also report preprocessing effort for CHASE. Note that for CHASE we report the space consumption *including* the graph.

	Access-Node Routing						CHASE	
network	core-based	AN-cand.	forward access-nodes	backward access-nodes	time [min]	space [B/n]	time [min]	space [B/n]
`de-road-rail(long)`		473	32.4 (6.8%)	20.9 (4.4%)	143	435	17	56
`de-road-rail(long)`	✓	473	31.0 (6.5%)	19.7 (4.1%)	26	56	17	56
`na-eur-road-flight`	✓	359	118.7 (33.0%)	119.1 (33.1%)	161	224	233	57

network). As expected, switching to Core-Based Access-Node Routing drastically reduces both preprocessing time and the required space for the access-nodes. The additional effort for preprocessing CHASE is comparable to ANR. We need 56 bytes per node (including a uni-modal graph) and preprocessing times are within a reasonable range.

Query Performance. Table 2 reports query performance of a multi-modal DIJKSTRA, ANR, and of our CHASE check-query for all our inputs. Note that the figures for plain DIJKSTRA are based on 1 000 random queries. We observe a drastic drop in both the number of settled nodes and the query time when using Access-Node Routing. In `de-road-rail(long)` we observe that the query time increases from 3.9 to 5.8 milliseconds when switching to the core-based variant. This is due to the computational overhead of the initialization phase. The performance of Access-Node Routing on `na-eur-road-flight` is better than on `de-road-rail(long)`. This is due to the fact that the flight network is significantly smaller than the railway network embedded in `de-road-rail(long)`. Moreover, the number of access-nodes per road node is only 14.2 whereas in Germany it is twice that much. The highest speed-up of 31 551 is achieved when applying our biggest input. We are able to perform intercontinental queries with an average time of 2.3 ms compared to over 72 sec when the standard algorithm is used. We also observe that the running time for CHASE is negligible compared to the execution time of ANR: query times are between 51 and 111 μs for settling only very few nodes.

Three Phases. The query algorithm for Core-Based Access-Node Routing is made up of three distinct phases (cf. Section 4.4). Table 3 reports the distribution of the running time among the particular phases of the query algorithm. We observe that the public transportation query makes up the major part of the running time (between 73.9% and 96.2% depending on the network). This is expected since we do not use a speed-up technique within the public transportation network. The time for looking up the access-nodes is negligible as it is less than 2% on `de-road-rail(long)` and 7.9% on `na-eur-road-flight` of the running time. This is due to the fact that the number of average access-nodes per road node is higher than on the other two networks (cf. Table 1).

Furthermore, we report the relative number of local paths, i.e., how many quickest paths do not use the public transportation network. We observe a great variation depending on the network. While in `de-road-rail(long)` only 2.3% of the queries do

Table 2. Query performance of (Core-Based) Access-Node Routing without local queries (i.e., all shortest paths use the transportation network) compared to plain multi-modal DIJKSTRA. Note that the figures for the latter are based on only 1 000 random queries.

	DIJKSTRA		Access-Node Routing				Local (CHASE)	
network	settled nodes	time [ms]	core-based	#settled nodes	time [ms]	speed-up	#settled nodes	time [ms]
de-road-rail(long)	2 483 030	3 492		13 295	3.9	895	168	0.111
de-road-rail(long)	2 483 030	3 492	✓	13 524	5.8	602	168	0.111
na-eur-road-flight	46 244 703	72 566	✓	4 200	2.3	31 551	75	0.051

Table 3. In-depth analysis of *Core-Based* Access-Node Routing. This table reports the distribution of query time among the particular phases of the query algorithm: the bidirectional initialization, the table-lookups of access-nodes, and DIJKSTRA on the public transportation network. Furthermore, we report the amount of local queries (paths that do not use the transportation network) when generating 1000 (`de-road-rail(long)`, `ny-de-road-flight`) and 100 (`na-eur-road-flight`) random queries.

	QUERY				
Network	initialization phase	access-node lookup	public transport	total [ms]	local queries
`de-road-rail(long)`	0.15 (2.4%)	0.08 (1.4%)	5.87 (96.2%)	5.8	2.3%
`na-eur-road-flight`	0.42 (18.2%)	0.18 (7.9%)	1.70 (73.9%)	2.3	24%

not use the railway, while for `na-eur-road-flight` the amount of local queries is 24%. Still, as observable in Table 2, running a CHASE query comes almost for free compared to the running time of the multi-modal query. In general, the high number of local queries is also due to the fact that we pick τ randomly. By computing a departure time with minimal travel time via profile-searches [18], the amount of local queries most probably will decrease. Note that in our scenario, it is sufficient to use profile-searches within the public-transportation network in order to answer such requests.

6 Conclusion

In this work we presented a first efficient approach to a special variant of multi-modal routing in large transportation networks. Using the reasonable assumption that we want to use a car only at the beginning and the end of the journey, we can split the search by adapting some ideas from Transit-Node Routing. The key idea is to skip the road network during the query. Experiments on real-world multi-modal networks with up to 125 Mio. edges confirm the feasibility of our approach: random queries are up to 31 000 times faster than with a multi-modal variant of DIJKSTRA. We want to stress out that we could achieve further speed-ups by using a better routing algorithm than DIJKSTRA within the public transportation network.

Regarding future work, it would be interesting to develop a multi-modal speed-up technique that does not restrict the choice of the automaton combined with reasonable speed-ups. Preliminary results from [16] confirm that the adaption of some speed-up techniques, e.g., Contraction Hierarchies, Arc-Flags or Landmarks, is much more challenging than one might expect. So, such a technique is non-trivial. Furthermore, we want to develop better foot networks and accelerate public transportation queries.

Acknowledgments. We would like to thank Martin Holzer and Christos Zaroliagis for interesting discussions on multi-modal route planning. Moreover, we thank PTV AG and HaCon for providing us with real-world data for scientific use.

References

1. Dijkstra, E.W.: A Note on Two Problems in Connexion with Graphs. Numerische Mathematik 1, 269–271 (1959)
2. Delling, D., Sanders, P., Schultes, D., Wagner, D.: Engineering Route Planning Algorithms. In: Lerner, J., Wagner, D., Zweig, K.A. (eds.) Algorithmics of Large and Complex Networks. LNCS, vol. 5515, pp. 117–139. Springer, Heidelberg (2009)
3. Mendelzon, A.O., Wood, P.T.: Finding Regular Simple Paths in Graph Databases. SIAM Journal on Computing 24(6), 1235–1258 (1995)
4. Barrett, C., Jacob, R., Marathe, M.V.: Formal-Language-Constrained Path Problems. SIAM Journal on Computing 30(3), 809–837 (2000)
5. Barrett, C., Bisset, K., Jacob, R., Konjevod, G., Marathe, M.V.: Classical and Contemporary Shortest Path Problems in Road Networks: Implementation and Experimental Analysis of the TRANSIMS Router. In: Möhring, R.H., Raman, R. (eds.) ESA 2002. LNCS, vol. 2461, p. 126. Springer, Heidelberg (2002)
6. Hart, P.E., Nilsson, N., Raphael, B.: A Formal Basis for the Heuristic Determination of Minimum Cost Paths. IEEE Transactions on Systems Science and Cybernetics 4, 100–107 (1968)
7. Dantzig, G.B.: Linear Programming and Extensions. Princeton University Press, Princeton (1962)
8. Barrett, C., Bisset, K., Holzer, M., Konjevod, G., Marathe, M.V., Wagner, D.: Engineering Label-Constrained Shortest-Path Algorithms. In: Shortest Paths: Ninth DIMACS Implementation Challenge. DIMACS Book. American Mathematical Society (to appear, 2009)
9. Bast, H., Funke, S., Sanders, P., Schultes, D.: Fast Routing in Road Networks with Transit Nodes. Science 316(5824), 566 (2007)
10. Orda, A., Rom, R.: Shortest-Path and Minimum Delay Algorithms in Networks with Time-Dependent Edge-Length. Journal of the ACM 37(3), 607–625 (1990)
11. Kleene, S.C.: Representation of Events in Nerve Nets and Finite Automata. In: Shannon, C.E., McCarthy, J. (eds.) Automata Studies. Annals of Mathematics Studies, pp. 3–42. Princeton University Press, Princeton (1956)
12. Rabin, M.O., Scott, D.: Finite Automata and their Decision Problems. IBM Journal of Research and Development 3(1559), 114–125
13. Pyrga, E., Schulz, F., Wagner, D., Zaroliagis, C.: Efficient Models for Timetable Information in Public Transportation Systems. ACM J. of Exp. Algorithmics 12, Article 2.4 (2007)
14. Star Alliance (1997), `http://www.staralliance.com`
15. Oneworld Management Ltd. (1999), `http://www.oneworld.com`
16. Pajor, T.: Multi-Modal Route Planning. Master's thesis, Universität Karlsruhe (TH), Fakultät für Informatik (2009)
17. Dean, B.C.: Continuous-Time Dynamic Shortest Path Algorithms. Master's thesis, Massachusetts Institute of Technology (1999)
18. Delling, D.: Time-Dependent SHARC-Routing. Algorithmica (to appear, 2009)
19. Geisberger, R., Sanders, P., Schultes, D., Delling, D.: Contraction Hierarchies: Faster and Simpler Hierarchical Routing in Road Networks. In: McGeoch, C.C. (ed.) WEA 2008. LNCS, vol. 5038, pp. 319–333. Springer, Heidelberg (2008)
20. Bauer, R., Delling, D., Sanders, P., Schieferdecker, D., Schultes, D., Wagner, D.: Combining Hierarchical and Goal-Directed Speed-Up Techniques for Dijkstra's Algorithm. In: McGeoch, C.C. (ed.) WEA 2008. LNCS, vol. 5038, pp. 303–318. Springer, Heidelberg (2008)
21. Hilger, M., Köhler, E., Möhring, R.H., Schilling, H.: Fast Point-to-Point Shortest Path Computations with Arc-Flags. In: Shortest Paths: Ninth DIMACS Implementation Challenge. DIMACS Book. American Mathematical Society (to appear, 2009)

Parallel Algorithms for Mean-Payoff Games: An Experimental Evaluation*

Jakub Chaloupka

Faculty of Informatics, Masaryk University,
Botanická 68a, 60200 Brno, Czech Republic
xchalou1@fi.muni.cz

Abstract. Mean-payoff games (MPGs) have many applications, especially in the synthesis, analysis and verification of computer systems. Because of the size of these systems, there is a need to solve very large MPGs. Existing algorithms for solving MPGs are sequential, hence limited by the power of a single computer. In this paper, we propose several parallel algorithms based on the sequential ones. We also evaluate and compare the parallel algorithms experimentally.

Keywords: mean-payoff games, parallel algorithms, experimental evaluation.

1 Introduction

A *Mean-Payoff Game (MPG)* [7,9,15] is a two-player infinite game on a finite weighted directed graph. The two players, named Max and Min, move a token along the edges of the graph ad infinitum. Roughly speaking, Max wants to maximize the average weight of the traversed edges whereas Min wants to minimize it. The exact statement of the problem is given in Section 2.

Mean-payoff games have many applications, especially in the synthesis, analysis and verification of reactive (non-terminating) systems. Many natural models of such systems include quantitative information, and the corresponding question requires the solution of quantitative games, like MPGs. Quantities may represent, for example, the power usage of an embedded component, or the buffer size of a networking element [3].

Examples of applications include various kinds of scheduling, finite-window online string matching, or more generally, analysis of online problems and algorithms, and selection with limited storage [15]. Moreover, μ-calculus model-checking is polynomial-time reducible to MPGs via parity games [10]. MPGs can even be used for solving the max-plus algebra $Ax = Bx$ problem, which in turn has further applications [6].

Because of their importance, MPGs have attracted many researchers, especially in the last decade, and several algorithms for solving MPGs have been proposed. They can be roughly divided into two categories. In the first category are algorithms based on linear programming [1,14]. This category also includes algorithms based on reduction to discounted payoff games [12] and simple stochastic games [5]. In the second category are pure combinatorial graph algorithms [15,2,11,6,9,13].

* This work has been partially supported by the Grant Agency of the Czech Republic grant No. 201/09/1389.

A. Fiat and P. Sanders (Eds.): ESA 2009, LNCS 5757, pp. 599–610, 2009.

The systems that need to be analyzed are often very complex, which leads to large MPGs. Solving such games is very time and memory consuming. Natural way to expand our possibilities is to employ parallel computers. However, the algorithms mentioned above are sequential, they were not designed to run in parallel environment. To the best of our knowledge, to date, there is no work on parallel algorithms for solving MPGs.

We considered all the sequential combinatorial algorithms from [15,2,11,6,9,13] for parallelization and we proposed parallel versions of all of them except for the algorithms of Lifshits and Pavlov [11] (LP) and Schewe [13] (SCH). LP was excluded because of its exponential space complexity and SCH was excluded because we were unable to design an efficient parallelization of it. SCH uses a technique which Schewe calls a "generalization" of Dijkstra's single source shortest path algorithm. Although there are methods to make Dijkstra's algorithm parallel, they are not usable for SCH. The parallelization of the remaining algorithms also was not always completely straightforward. We had to adjust some of the original algorithms, which sometimes led to more variants of one algorithm. This is described in detail in Section 3. To evaluate the parallel algorithms, we implemented them and carried out an experimental study.

According to the experimental study, the best algorithm is the algorithm of Dhingra and Gaubert [6] (DG) with a modification that we proposed. Interestingly enough, the algorithm of Björklund and Vorobyov [2] (BV), which is not much worse than DG on completely random MPGs, dramatically fails on more structured instances. The algorithms of Zwick and Paterson [15] and Gurvich, Karzanov, and Khachivan [9] turned out to be practically unusable.

We did not include the algorithms based on linear programming in the study. The reasons are the following. First, the algorithms based on linear programming are numerically unstable due to round-off errors. Second, the use of well-established LP solver would make the comparison unfair, since the combinatorial algorithms were implemented from scratch. Both problems could be alleviated by using arbitrary precision arithmetic, and developing our own LP solver or improving our implementations of the combinatorial algorithms, but for space reasons we decided not to do so.

2 Algorithms

This section presents the algorithms included in our study. The next section shows how to parallelize them. Before the actual presentation, we give basic terms and definitions.

A *Mean-Payoff Game* (MPG) [7,9,15] is given by a triple $(\mathcal{G}, V_{\mathrm{Max}}, V_{\mathrm{Min}})$, where $\mathcal{G} = (V, E, w)$ is a finite weighted directed graph such that V is a disjoint union of the sets V_{Max} and V_{Min}, $w : E \to \mathbb{Z}$ is the weight function, and each $v \in V$ has out-degree at least one. The game is played by two opposing players, named Max and Min. A play starts by placing a token on some given vertex and the players then move the token along the edges of $\mathcal{G}$ ad infinitum. If the token is on vertex $v \in V_{\mathrm{Max}}$, Max moves it. If the token is on vertex $v \in V_{\mathrm{Min}}$, Min moves it. This way an infinite path $p = (v_0, v_1, v_2, \ldots)$ is formed. Max's aim is to maximize his gain: $\liminf_{n\to\infty} \frac{1}{n} \sum_{i=0}^{n-1} w(v_i, v_{i+1})$, and Min's aim is to minimize her loss: $\limsup_{n\to\infty} \frac{1}{n} \sum_{i=0}^{n-1} w(v_i, v_{i+1})$. For each vertex $v \in V$, we define its *value*, denoted by $\nu(v)$, as the maximum gain that Max can ensure if the play starts at vertex v. It was proved that it is equal to the minimum loss that Min can ensure.

Moreover, both players can ensure $\nu(v)$ by using positional strategies defined below [7]. A strategy that ensures $\nu(v)$, for all $v \in V$ is called an *optimal* strategy. To solve an MPG is to find the values of all vertices and, optionally, also optimal positional strategies for both players.

A *positional strategy* for Max is a function $\sigma : V_{\mathrm{Max}} \to V$ such that $(v, \sigma(v)) \in E$, for each $v \in V_{\mathrm{Max}}$. A positional strategy for Min is defined analogously. We define $\mathcal{G}_\sigma$, the restriction of $\mathcal{G}$ to σ, as the graph (V, E_σ, w_σ), where $E_\sigma = \{(u,v) \in E \mid u \in V_{\mathrm{Min}} \vee \sigma(u) = v\}$, and $w_\sigma = w|_{E_\sigma}$. That is, we get $\mathcal{G}_\sigma$ from $\mathcal{G}$ by deleting all the edges emanating from Max's vertices that do not follow σ. Let now σ be a strategy of Max and let π be a strategy of Min. $\mathcal{G}_\sigma$ has just been defined, $\mathcal{G}_\pi$ is defined analogously, and $\mathcal{G}_{\sigma\cup\pi}$ is the intersection of $\mathcal{G}_\sigma$ and $\mathcal{G}_\pi$, i.e., $\mathcal{G}_{\sigma\cup\pi} = (V, E_{\sigma\cup\pi}, w_{\sigma\cup\pi})$, where $E_{\sigma\cup\pi} = \{(u,v) \in E \mid (u \in V_{\mathrm{Min}} \wedge \pi(u) = v) \vee (u \in V_{\mathrm{Max}} \wedge \sigma(u) = v)\}$, and $w_{\sigma\cup\pi} = w|_{E_{\sigma\cup\pi}}$.

In this paper, we often talk about paths and cycles. The weight of a path is the sum of the weights of its edges, and the length of a path is the number of its edges. Similarly for cycles. For the whole paper, let $(\mathcal{G} = (V,E,w), V_{\mathrm{Max}}, V_{\mathrm{Min}})$ be an MPG, and $W = \max_{e \in E} |w(e)|$. The presentation of the algorithms follows.

2.1 Zwick-Paterson (ZP)

The algorithm of Zwick and Paterson [15] (ZP) uses the following facts. We can define, for each $v \in V$, the value $\nu_k(v)$ as the maximal sum of weights of traversed edges that Max can ensure in a k-step game on the graph $\mathcal{G}$. This value is, of course, equal to the the minimal sum of weights of traversed edges that Min can ensure, and it can be defined inductively in the following way:

$$\begin{array}{lll} \nu_0(v) & = 0 & \forall v \in V \\ \nu_{k+1}(v) & = \max_{(v,u)\in E}(\nu_k(u) + w(v,u)) & \forall v \in V_{\mathrm{Max}} \\ \nu_{k+1}(v) & = \min_{(v,u)\in E}(\nu_k(u) + w(v,u)) & \forall v \in V_{\mathrm{Min}} \end{array}$$

From the existence of positional strategies in the original infinite game, it follows that $\nu_k(v)/k$ converges to $\nu(v)$. Moreover, if k is large enough, the exact value of $\nu(v)$ can be computed from $\nu_k(v)$. This follows from the fact that $\nu(v)$ is always a fraction with denominator at most $|V|$. Zwick and Paterson showed that $k = 4 \cdot |V|^3 \cdot W$ is sufficient for the exact computation of $\nu(v)$, for each $v \in V$. The ν_k values can be easily computed using the inductive definition, which yields an algorithm with complexity $\Theta(|V|^3 \cdot |E| \cdot W)$.

2.2 Gurvich-Karzanov-Khachivan (GKK)

The basic building block of the algorithm of Gurvich, Karzanov and Khachivan [9] (GKK) is a method to divide V into vertices with $\nu \geq 0$ and vertices with $\nu < 0$. This method can also be used to divide V into vertices with $\nu \geq c$ and vertices with $\nu < c$, for arbitrary $c \in \mathbb{Q}$, because by subtracting c from all edge-weights, the ν value of all vertices also decreases by c. The exact values are then computed by binary search. Each branch of the search is finished when there is only one fraction with denominator at most $|V|$ in the search interval, which has to be the ν value of all vertices in that branch. It remains to show how the vertices are divided. For this task, GKK employs potential transformation.

If we assign a potential $d(v)$ to each vertex $v \in V$ and use the potentials to create a new weight function w', $w'(x,y) = w(x,y) + d(x) - d(y)$, then for each vertex $v \in V$, $\nu(v)$ in the modified game $(\mathcal{G} = (V, E, w'), V_{\mathrm{Max}}, V_{\mathrm{Min}})$ is the same as $\nu(v)$ in the original game. It follows from the fact that the weights of all cycles remain the same. It is possible to select the potentials in such a way that Max has a strategy to force a play starting in a vertex with $\nu \geq 0$ to traverse only edges with $w' \geq 0$, and Min has a strategy to force a play starting in a vertex with $\nu < 0$ to traverse only edges with $w' \leq 0$ and to traverse at least one edge with $w' < 0$ infinitely many times. The potentials with the above properties can then be considered as a proof of the fact that a certain vertex has $\nu \geq 0$ or $\nu < 0$. GKK computes these potentials and it computes them iteratively.

GKK starts with $d(v) = 0$, for each vertex $v \in V$, and computes a set of vertices from which Min can force a play to traverse at least one (strictly) negative edge without traversing a (strictly) positive edge first. Let's denote the set by X. It follows that Max has a strategy to force a play starting in any vertex from $\neg X$ to traverse at least one positive edge without traversing a negative edge first. The next step is to increase the potentials of all vertices in X by the same amount $\varepsilon > 0$. As a result, the weights of the edges from $\neg X$ to X decrease and the weights of the edges from X to $\neg X$ increase. The value of ε is defined as the maximal value that keeps Min a strategy to force a non-positive first step in a play starting from X and that keeps Max a strategy to force a non-negative first step in a play starting from $\neg X$. Then the whole process is repeated with the new weight function w'. The algorithm terminates when ε can be arbitrarily large, in which case the set X is the set of all vertices with $\nu < 0$.

The overall complexity of GKK is $O(|V|^3 \cdot |E| \cdot (|V| + \log(W)) \cdot W)$.

2.3 Björklund-Vorobyov (BV)

Like the algorithm GKK, the algorithm of Björklund and Vorobyov [2] (BV) builds on a method that divides V into vertices with $\nu > 0$ and vertices with $\nu \leq 0$, and the exact values are computed by binary search. The inequalities defining the division are actually slightly different than in GKK, where the first is not strict and the second is. However, this is not significant, because each of these two algorithms can be adjusted so that it divides the vertices as the other algorithm. To divide the vertices, BV uses a strategy improvement technique. It starts with a certain strategy σ of Max and then improves it until it ensures that Max's gain in a play starting from any vertex with $\nu > 0$ will be (strictly) positive, which is equivalent to the statement that all cycles reachable from vertices from $\{v \in V \mid \nu(v) > 0\}$ in $\mathcal{G}_\sigma$ are (strictly) positive.

Before the actual strategy improvement, BV introduces an auxiliary vertex r called "retreat" and adds a zero-weight edge from each Max's vertex to r. Then, it removes all negative and zero-weight cycles consisting of Min's vertices only. It also removes the vertices from which Min has a strategy that forces a play to the removed cycles. It is obvious that these vertices have $\nu \leq 0$ and they have to be eliminated because otherwise BV could give us incorrect results.

The iterative process of strategy improvement starts with a strategy σ that for each vertex $v \in V_{\mathrm{Max}}$, selects the auxiliary edge, i.e., $\sigma(v) = r$. Each iteration consists of strategy evaluation and strategy improvement. To evaluate σ, BV computes the weight $d(v)$ of the minimum-weight path from v to r in $\mathcal{G}_\sigma$, for each vertex $v \in V$. The vector d is then used to improve σ: Each edge $(x,y) \in E$, where $x \in V_{\text{Max}}$, is checked to see whether $d(x) < d(y) + w(x,y)$. If there is such an edge, then $\sigma(x)$ is set to y. If there are more such edges, any combination of improvements is acceptable. After the improvement, another iteration is started. If no improvement is possible, the algorithm terminates and the vertices from which r is not reachable in $\mathcal{G}_\sigma$ are exactly the vertices with $\nu > 0$.

The original algorithm is a sub-exponential randomized algorithm. To prove that the algorithm is sub-exponential, some restrictions had to be imposed. If these restrictions are not obeyed, BV runs faster. Therefore, we decided not to obey the restrictions. We used only the modified BV algorithm in our experimental study, its complexity is $O(|V|^3 \cdot |E| \cdot \log(|V| \cdot W) \cdot W)$ and it is no longer a randomized algorithm.

2.4 Dhingra-Gaubert (DG)

The algorithm of Dhingra and Gaubert [6] (DG) could be called a double strategy improvement algorithm. It is because it alternately improves Min's and Max's strategy until they both become optimal.

DG starts with an arbitrary strategy π of Min and an arbitrary strategy σ of Max. Then it improves π, so that for each vertex $v \in V$, the unique cycle reachable from v in $\mathcal{G}_{\sigma\cup\pi}$ has the minimum mean-weight among all cycles reachable from v in $\mathcal{G}_\sigma$. It means that π is an optimal counter-strategy to σ. The improvement of π proceeds iteratively. In each iteration, the strategy is first evaluated and then improved. The evaluation consists of two vectors of size $|V|$, d and mw. For each $v \in V$, $mw(v)$ is the mean-weight of the unique cycle reachable from v in $\mathcal{G}_{\sigma\cup\pi}$, and $d(v)$ is the "distance" to the cycle. More precisely, one vertex in each cycle in $\mathcal{G}_{\sigma\cup\pi}$ is selected and $d(v)$ is a reduced weight of the unique path p from v to the appropriate selected vertex. The reduced weight is the weight of p with respect to the reduced weight function w', $w'(x,y) = w(x,y) - mw(y)$. In other words, the weight of each edge $(x,y) \in E_{\sigma\cup\pi}$ is reduced by the mean-weight of the unique cycle reachable from x in $\mathcal{G}_{\sigma\cup\pi}$. After d and mw are computed, the strategy π is improved. To that end, each edge $(v,u) \in E \cap V_{\text{Min}} \times V$ is checked to see whether it satisfies the *strategy improvement condition of Min*: $mw(v) > mw(u) \vee (mw(v) = mw(u) \wedge d(v) > d(u) + w(v,u) - mw(u))$. If yes, then $\pi(v)$ is set to u. If there is a vertex $v \in V_{\text{Min}}$ such that more than one edge emanating from v satisfies the condition, anyone of them can be used for the improvement. Then, another iteration is started. If there are no edges satisfying the condition, DG proceeds to the improvement of σ for which it will use the final values of d and mw.

Actually, it may be necessary to recompute d and mw before they are used to improve σ. The recomputation is needed if the vector mw didn't increase due to the last improvement of σ, i.e., $mw = mw'$, where mw' is the previous vector of mean-weights. Let d' be the vector of distances corresponding to mw'. To recompute mw and d, DG first finds all vertices with the property that their mw value is equal to the mean-weight of some cycle in $\mathcal{G}_\sigma$ they are part of. These vertices are assigned their d' values as their new d values. The new d values of the other vertices are then computed using a Bellman-Ford-like algorithm which works with the reduced weights, which guarantees absence of negative cycles. With or without the recomputation, we now have vectors d and mw that are used to improve σ. The vectors are also stored and will be used for testing if recomputation is needed before the next improvement of σ.

To improve σ, each edge $(v,u) \in E \cap V_{\text{Max}} \times V$ is checked to see whether it satisfies the *strategy improvement condition of Max*: $mw(v) < mw(u) \vee (mw(v) = mw(u) \wedge d(v) < d(u) + w(v,u) - mw(u))$. Max's improvement technique is analogous to the improvement technique of Min. If σ is improved, DG goes again to the improvement of π and repeats the whole process. If there are no edges satisfying the condition, σ is an optimal strategy, and for each $v \in V$, $mw(v) = \nu(v)$.

The authors of DG do not state its overall complexity in [6]. We derived a rough upper bound on the complexity of the algorithm from the lower and upper bounds on the vectors d and mw. The complexity is pseudo-polynomial, namely $O(|V|^{10} \cdot |E| \cdot |W|^3)$.

3 Parallelization

We adapted the algorithms to the multi-processor/multi-core environment with shared memory. Let's suppose we have r processing nodes numbered $0, \ldots, r-1$. The input graph is partitioned into r disjoint, roughly equal-sized parts, $V_0, V_1, \ldots, V_{r-1}$. Each one of the r processing nodes owns one of the parts. There is a function $owner : V \rightarrow \{0, \ldots, r-1\}$ that for a vertex $v \in V$ gives the number of the node that owns v, i.e. $owner(v) = i \Leftrightarrow v \in V_i$. This function is available to all nodes. The partitioning divides the edges into local edges, those with both ends on the same node, and cross edges, those with its ends on different nodes. Each node executes the same code. The rest of this section discusses the parallelization of the individual algorithms.

ZP. The easiest algorithm to parallelize is the algorithm ZP. Each node computes the k-step game values of its vertices. For that purpose, values of vertices belonging to other nodes may be needed, but since we have a shared memory, each node has access to the values computed by the other nodes. Therefore, the only thing that we have to add is a barrier synchronization that separates distinct iterations of the algorithm to ensure that one processing node does not start a new iteration too early, which could result in usage of variables that have not been assigned the correct values yet. The synchronization can be implemented using a shared variable with mutually exclusive access.

GKK. The parallel version of GKK also has to separate distinct iterations by barrier synchronization, but the rest is not as easy as in ZP. In each iteration, GKK has to compute the set X from which Min can force plays with certain properties. This is done by a procedure similar to backward reachability analysis. It starts from vertices which are either Max's vertices and all of their emanating edges are negative or Min's vertices with at least one negative emanating edge. The procedure then traverses only zero-weight edges in their reversed direction and proceeds through Max's vertex only if it was visited from all it's successors except for those connected with negative edges. Fortunately, this procedure can also be parallelized quite easily. Each node executes the sequential algorithm on its subset of vertices. When a node encounters a cross edge (u,v), it stops this branch of the search and tells to the owner of v to continue computation from v. To this end, for each pair of nodes (A,B), there is a queue q_{AB} designated for passing messages from A to B. A enqueues messages to q_{AB}, and B then takes the messages from q_{AB}. We also need a method to detect a point when all r processing nodes finished their computation and the algorithm can proceed to adjustment of potentials. For space reasons, we do not discuss this in detail. It is accomplished by counting sent/received messages and storing this information in shared variables.

BV. To parallelize the algorithm BV, we had to make a modification to it (other than the one mentioned in Section 2.3). BV needs to eliminate negative and zero-weight cycles consisting of Min's vertices only. This complicates the parallelization. Therefore, we modified the algorithm, so that the cycle elimination is not necessary. Instead, new vertices are introduced that ensure that each cycle in the game graph contains at least one Max's vertex. This modification does not increase the complexity of the algorithm.

We already described everything that is needed to parallelize the modified version of BV. Distinct iterations are, again, separated by barrier synchronization and minimum-weight paths are computed using Bellman-Ford, the parallelization of which uses the same techniques as reachability analysis in the algorithm GKK.

DG. The crucial part of the algorithm DG with respect to parallelization is the computation of the vectors d and mw in the graph $\mathcal{G}_{\sigma\cup\pi}$. The original algorithm finds all cycles in $\mathcal{G}_{\sigma\cup\pi}$ by repeatedly picking an arbitrary unvisited vertex v and following the unique path that leads from v until a cycle is closed. When all vertices are visited all cycles are found. This gives us the vector mw. The vector d is computed using a backward reachability from the selected vertices in the cycles. This method is "too sequential", especially the part that searches for cycles, because it explores only one path at a moment. Moreover, if there are not many cycles, which is typically the case, the backward reachability analysis also does not provide a lot of space for parallelism, because the graph has only $|V|$ edges. Therefore, we modified the algorithm.

Not one, but all processing nodes start a cycle search and each node proceeds not only through its own vertices, but through all vertices it encounters until a vertex it visited earlier is reached. To this end, each node has a visited flag for all $v \in V$. Therefore, each vertex has r visited flags assigned to it. This increases memory consumption, but not in a serious way. Also, since each vertex can be visited up to r times, the complexity of cycle searching increases from $O(|V|)$ to $O(r \cdot |V|)$. However, in practice, each vertex is almost always visited much less than r times, because, for each node i, significantly less that $|V|$ vertices are reachable from V_i.

The computation of the vector d is done similarly as the cycle searching. Each node selects one of its unvisited vertices and follows the unique path that starts at that vertex. This time, all visited vertices are pushed into a stack. The traversal stops either at a previously visited vertex or at the unique reachable selected vertex. The d and mw values of the vertex at which the search stops are then used to compute the d and mw values of all vertices in the stack that belong to the node that started the search.

We included two versions of the algorithm DG in our study. Both of them use our method for the computation of the vectors d and mw described above. The difference is that one version does not have to recompute the vectors as described in Section 2.4. We will first describe the version that has to recompute the vectors. The first problem is that the computation of vertices that keep the d value from the previous iteration involves SCC decomposition, which is difficult to parallelize. However, according to our experiments, the SCC decomposition is not a bottleneck of the algorithm. Therefore, we implemented it sequentially (All but one processing node are stopped and the whole graph is decomposed by the remaining node). The Bellman-Ford is then parallelized in the same way as in BV.

The second version of DG avoids the recomputation by assigning the previous d values to the selected vertices in the cycles in $\mathcal{G}_{\sigma \cup \pi}$ already during the improvement of π, but only in those cycles, the mean-weighs of which are equal to the previous *mw* values of their vertices. It is not trivial to prove the correctness of this modification, but it can indeed be done. We note that the two versions are not equivalent, they may produce different vectors d.

4 Experimental Evaluation

The experiments were carried out on a machine equipped with eight dual-core AMD Opteron™ 880 processors and 32GB of RAM, running GNU/Linux kernel version 2.6.28. All algorithms were implemented in C++ on the top of the POSIX Threads standard and compiled with GCC version 4.3.2 with the "-O2"option.

4.1 Input MPGs

This paper presents the results of the algorithms on synthetic MPGs generated by two generators, namely SPRAND and TOR [4], downloadable from [8]. The outputs of these generators are only directed weighted graphs, and so we had to divide vertices between Max and Min ourselves. We divided them uniformly at random.

SPRAND was used to generate the "rand$x$" MPG family. Each of these MPGs contains $|E| = x \cdot |V|$ edges and consist of a random Hamiltonian cycle and $|E| - |V|$ additional random edges, with weights chosen uniformly at random from $[1, 1000]$.

TOR was used for generation of the families "sqnc", "lnc", and "pnc", which are more structured. The families sqnc and lnc are *2-dimensional grids* with wrap-around, where each vertex is connected to its neighbor above by a short edge (with weight in $[1, 100]$) and to its neighbor to the right by a long edge (with weight in $[1000, 10000]$). The two families differ in dimensions of the grids, sqnc are *square grids* with $\sqrt{|V|}$ columns and rows, and lnc are *long grids*, with $|V|/16$ columns and 16 rows. The family denoted by pnc contains *layered networks* embedded on a torus. Graphs are partitioned into layers, each containing a cycle of 32 edges of weight 1, plus 64 random edges (all with weight in the range $[1, 100]$). Each vertex also has five edges to forward layers (with wrap-around): an edge going x layers forward has length picked uniformly at random from $[1, 10000]$ and multiplied by x^2.

We also created subfamilies of the families generated by TOR by adding negative cycles of weight -1 to the graphs. The subfamily (01) contains no added cycles, (02) contains one added cycle of length 3, (03) contains $\sqrt{|V|}$ added cycles of length 3, (04) contains eight added cycles of length $\sqrt{|V|}$, and (05) contains one added cycle of length $|V|$. After the cycles are added to each subfamily, potentials are used to transform edge-weights. Each vertex is assigned a random value in $[0, 16384)$ and the weight of each edge is changed by the difference of the potentials of its endpoints.

Our implementations were also tested on MPGs modeling simple reactive systems. However, for space reasons, we did not include the results on these MPGs in the paper. They are "easy" for the algorithms and the runtimes relative to number of vertices were very similar to the runtimes on the SPRAND generated MPGs.

4.2 Results

We will first comment on the results of the algorithms ZP and GKK. ZP has a great potential for speedup, because the computation of ν_k can be almost perfectly divided among the processing nodes and the only thing that slows it down is the barrier synchronization that separates distinct iterations. However, the enormous number of iterations necessary to finish the computation makes it practically unusable. Moreover, when ZP is ran on a very small MPG which can be solved in reasonable time, the speedup gained from the perfect division of work is lost due to too frequent synchronizations.

The algorithm GKK also has a great potential for speedup, because most of the time is spent on the computation of ε, which is done by examining all edges, which in turn can be almost perfectly divided among the processing nodes. GKK also achieves much better results than ZP, but its runtimes are still too high for it to be practically usable.

We illustrate the behavior of ZP and GKK on two small MPGs from the sqnc01 subfamily. On one core, it took ZP more than 22 minutes to solve an MPG with only 64 vertices and the time increased as more cores were added, to more than 15 hours on eight cores. The reason for this was given above. GKK is much better than ZP. It solves the game with 64 vertices instantly and on a game with 4096 vertices, it achieves a very good speedup: from 71 minutes on one core to 9 minutes on 16 cores. However, 71 minutes on an MPG with 4096 vertices is still too much in comparison with the best algorithms. Therefore, the rest of this section focuses only on BV and the two versions of DG. In the following, the version of DG that recomputes the vectors d and mw is denoted by "DG". The version that avoids the recomputation is denoted by "MDG".

Tables 1 and 2 give the results of our experiments. The column heading "nc" stands for number of processor cores. Each column headed by a name of an algorithm contains runtimes of that algorithm. Each MPG used in the experiments has five rows in each table. For the runtimes of all algorithms on one, two, four, eight, and sixteen cores. Each runtime is an average of 5 runs, rounded to whole seconds. All MPGs in Table 1 have 65536 vertices. In Table 2, the MPGs with no suffix have 65536 vertices, the MPGs with suffix "b" have 131072 vertices, and the MPGs with suffix "h" have 262144 vertices.

The results are very clear. On each MPG and number of cores, the runtimes satisfy MDG $<$ DG $<$ BV, with the exception of lnc01 and sixteen cores, where the first inequality is not strict. This clearly suggests that MDG is the best algorithm. In addition to that, there are several other interesting points.

While on the SPRAND families, BV is worse than MDG only by a factor of 5, on average, the difference on the TOR families is an order of magnitude, and for lnc01 and lnc02, even three orders of magnitude. The reason for this is that these MPGs contain many paths of the same length and similar weight between a lot of pairs of vertices, which causes a lot of updates in Bellman-Ford used by BV. The additional cycles with small edge-weights added to the subfamilies (02)-(05) create "shortcuts" in the graphs, which leads to better runtimes.

The runtimes of MDG and DG are mostly very similar. DG is always worse, but not more than by a factor of 2 on most MPGs. This is because the recomputations of the vectors d and mw are usually not frequent. However, on lnc02, pnc01, and pnc02, the difference is an order of magnitude. Recall that, like BV, DG uses Bellman-Ford and this is again the reason for the bad runtimes.

Table 1. Runtimes for the TOR input families (in seconds)

MPG	nc	MDG	DG	BV
sqnc01	1	133	141	3702
	2	87	98	2209
	4	61	64	985
	8	48	50	1003
	16	49	55	1094
sqnc02	1	206	230	4363
	2	137	164	2256
	4	90	116	1123
	8	75	88	1064
	16	79	96	1180
sqnc03	1	90	96	852
	2	63	85	449
	4	48	67	237
	8	38	53	232
	16	39	57	229
sqnc04	1	84	105	683
	2	55	76	366
	4	40	55	218
	8	30	42	177
	16	31	45	191
sqnc05	1	49	70	207
	2	35	43	125
	4	22	31	73
	8	16	23	55
	16	15	24	52

MPG	nc	MDG	DG	BV
lnc01	1	34	35	52633
	2	22	23	22511
	4	14	17	7683
	8	10	12	6952
	16	13	13	8507
lnc02	1	29	284	21333
	2	20	296	9659
	4	15	144	3539
	8	10	105	4049
	16	13	112	3103
lnc03	1	70	156	879
	2	48	136	452
	4	35	87	243
	8	26	72	215
	16	31	81	248
lnc04	1	96	109	1078
	2	65	77	534
	4	47	57	322
	8	36	44	291
	16	40	47	309
lnc05	1	43	64	227
	2	29	43	133
	4	19	30	74
	8	15	25	57
	16	14	24	61

MPG	nc	MDG	DG	BV
pnc01	1	77	611	1743
	2	47	352	999
	4	32	212	384
	8	21	217	333
	16	20	242	293
pnc02	1	83	903	10346
	2	51	532	5590
	4	30	341	1963
	8	25	364	1871
	16	20	396	1681
pnc03	1	59	90	1388
	2	39	55	740
	4	23	37	349
	8	16	27	271
	16	14	27	270
pnc04	1	67	144	1355
	2	40	95	748
	4	25	61	356
	8	19	56	266
	16	16	55	277
pnc05	1	73	128	960
	2	45	86	458
	4	27	52	253
	8	21	47	186
	16	16	42	190

Table 2. Runtimes for the SPRAND input families (in seconds)

MPG	nc	MDG	DG	BV
rand5	1	65	94	260
	2	39	61	152
	4	23	43	85
	8	15	31	64
	16	13	26	59
rand5b	1	203	306	808
	2	122	195	481
	4	75	130	262
	8	47	91	175
	16	41	86	173
rand5h	1	456	630	2016
	2	271	430	1145
	4	177	284	655
	8	123	187	462
	16	96	173	426

MPG	nc	MDG	DG	BV
rand10	1	98	145	529
	2	55	88	314
	4	31	56	174
	8	19	41	114
	16	15	36	117
rand10b	1	172	299	955
	2	101	169	539
	4	61	110	308
	8	37	91	211
	16	30	91	200
rand10h	1	476	857	2932
	2	283	489	1709
	4	176	364	873
	8	97	260	661
	16	81	245	597

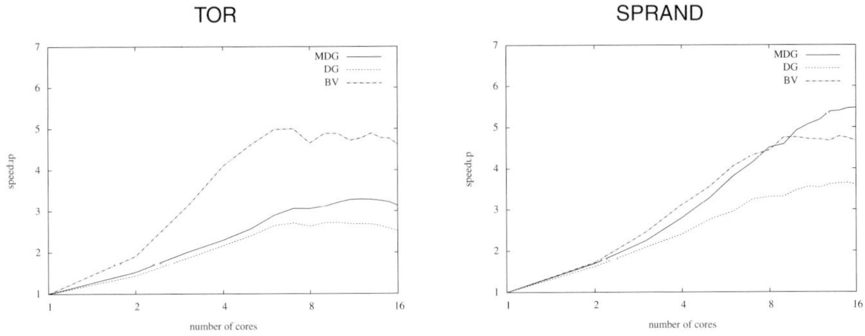

Fig. 1. Average speedup of the algorithms (relative to one core)

It remains to comment on the scalability of the algorithms. This includes asymptotic behavior and speedup as more cores are added. As for the asymptotic behavior, Table 2 shows that doubling the number of vertices (and edges) increased the runtime by a factor of 4, at most. This was also the case for our experiments on the TOR families, the results of which are not included for space reasons. For the SPRAND families, we also tried doubling the number of edges only, which resulted in an increase of runtime by a factor of 2, at most. These experiments suggest that the runtime is roughly proportional to $|V| \cdot |E|$, and so the algorithms are able to scale up to very large MPGs.

As for the speedup, all algorithms achieve quite a good speedup up to 4 cores. BV on some of the TOR generated inputs achieved even a super-linear speedup. The speedup on 8 cores relative to 4 cores is not so good, and the speedup on 16 cores relative to 8 cores is very poor, sometimes even negative. However, the results can still be considered satisfactory. We believe that the bad results on 8 and 16 cores are caused by the nature of the inputs, imperfectness of our implementations and limitations of the hardware, not by inability of the algorithms to scale up to a larger number of processors.

Figure 1 shows the average speedup relative to one core, separately for the TOR and the SPRAND families. The behavior of the parallel algorithms on one core is practically the same as the behavior of the original sequential algorithms, which justifies our decision not to measure the speedups relative to runtimes of the sequential algorithms.

The speedup of BV on the TOR families is better than that of the other two algorithms. This is because these inputs are "too small" for DG and MDG, and so the initializations and synchronizations consume a considerable portion of runtime. However, the speedups achieved by BV are still too small to make it competitive. The best speedups on the SPRAND families were achieved by MDG, which is the only algorithm that achieved a decent speedup from 8 to 16 cores.

5 Conclusion

We parallelized all the sequential combinatorial algorithms from [15,2,6,9] to find algorithms for practical solving of large mean-payoff games in parallel. We implemented the algorithms and carried out an experimental study to evaluate them. Only two of the algorithms turned out to be suitable. Namely, the algorithm of Björklund and Vorobyov [2]

(BV), and the algorithm of Dhingra and Gaubert [6] (DG). We designed two parallel versions of DG, one is more or less a straightforward parallelization of the original algorithm and the other one changes the operation of the algorithm slightly. The latter version is the clear winner of our experimental study. Its runtime was the best on all input instances, and for some instances, it was better than all the other algorithms by at least an order of magnitude.

References

1. Björklund, H., Svensson, O., Vorobyov, S.: Linear complementarity algorithms for mean payoff games. Technical Report DIMACS-2005-13, DIMACS, New Jersey, USA (2005)
2. Björklund, H., Vorobyov, S.: A combinatorial strongly subexponential strategy improvement algorithm for mean payoff games. Discrete Applied Math. 155(2), 210–229 (2007)
3. Chakrabarti, A., de Alfaro, L., Henzinger, T., Stoelinga, M.: Resource interfaces. In: Alur, R., Lee, I. (eds.) EMSOFT 2003. LNCS, vol. 2855, pp. 117–133. Springer, Heidelberg (2003)
4. Cherkassky, B.V., Goldberg, A.V.: Negative-cycle detection algorithms. Mathematical Programming 85, 277–311 (1999)
5. Condon, A.: On algorithms for simple stochastic games. In: Advances in Computational Complexity Theory. DIMACS Series in Discrete Mathematics and Theoretical Computer Science, vol. 13, pp. 51–73. American Mathematical Society (1993)
6. Dhingra, V., Gaubert, S.: How to solve large scale deterministic games with mean payoff by policy iteration. In: Proc. Performance evaluation methodolgies and tools, article no. 12. ACM, New York (2006)
7. Ehrenfeucht, A., Mycielski, J.: Positional strategies for mean payoff games. International Journal of Game Theory 8(2), 109–113 (1979)
8. Andrew Goldberg's network optimization library (June 2009), `http://www.avglab.com/andrew/soft.html`
9. Gurvich, V.A., Karzanov, A.V., Khachivan, L.G.: Cyclic games and an algorithm to find minimax cycle means in directed graphs. USSR Comput. Math. and Math. Phys. 28(5), 85–91 (1988)
10. Jurdziński, M.: Deciding the winner in parity games is in UP $\cap$ co-UP. Inf. Process. Lett. 68(3), 119–124 (1998)
11. Lifshits, Y., Pavlov, D.: Fast exponential deterministic algorithm for mean payoff games. Zapiski Nauchnyh Seminarov POMI 340, 61–75 (2006)
12. Puri, A.: Theory of hybrid systems and discrete event systems. Phd thesis, EECS University of Berkeley, Berkeley, CA, USA (1995)
13. Schewe, S.: An optimal strategy improvement algorithm for solving parity and payoff games. In: Kaminski, M., Martini, S. (eds.) CSL 2008. LNCS, vol. 5213, pp. 369–384. Springer, Heidelberg (2008)
14. Svensson, O., Vorobyov, S.: Linear programming polytope and algorithm for mean payoff games. In: Cheng, S.-W., Poon, C.K. (eds.) AAIM 2006. LNCS, vol. 4041, pp. 64–78. Springer, Heidelberg (2006)
15. Zwick, U., Paterson, M.S.: The complexity of mean payoff games on graphs. Theoretical Computer Science 158(1-2), 343–359 (1996)

Experimental Study of FPT Algorithms for the Directed Feedback Vertex Set Problem

Rudolf Fleischer, Xi Wu, and Liwei Yuan*

Fudan University; Software School, CSE Department and IIPL; Shanghai, China
{rudolf,wuxi,yuanliwei}@fudan.edu.cn

Abstract. We evaluate the performance of FPT algorithms for the directed feedback vertex set problem (DFVS). We propose several new data reduction rules for DFVS. which can significantly reduce the search space. We also propose various heuristics to accelerate the FPT search. Finally, we demonstrate that DFVS is not more helpful for deadlock recovery (with mutex locks) than simple cycle detection.

1 Introduction

In the *minimum feedback vertex set problem (*FVS*)* we are given a graph $G = (V, E)$ and we want to find a minimum number of nodes in V whose removal would make the graph acyclic. The problem is NP-complete on undirected (UFVS) and directed (DFVS) graphs. Since DFVS has some applications in compiler optimization [14,16] and database deadlock recovery (see [4]), it is important to solve this problem quickly. Therefore, the parameterized versions of FVS have recently been studied. In k-DFVS the input is a pair (G, k), where G is the graph and $k > 0$ is a parameter, and the task is to either find a set of at most k nodes blocking all cycles or to report that such a set does not exist. A parameterized problem is *fixed parameter tractable* (see [6]) if it can be solved in time $O(f(k)poly(n))$, where n is the number of nodes of G and f is an arbitrary computable function. Intuitively, a problem is in the class FPT if it can be solved efficiently for fixed parameter k. We denote the runtime by $O^\star(f(k))$, omitting the less interesting polynomial dependency on n. Obviously, we do not expect $f(k)$ to be polynomial for NP-hard problems.

If we can reduce (G, k) in time $g(k)$ to an equivalent parameterized problem (H, k'), where g is some computable function and k' only depends on k, such that the size of H is bounded by some function $h(k)$, then (H, k') is a *kernel* of (G, k). It is known that being in FPT is equivalent to having a kernel [6]. While a polynomial-size kernel immediately implies the existence of an FPT algorithm

* According to international standard, the authors are listed in alphabetic order; first-author order would be X. Wu, L. Yuan, and R. Fleischer. This research was supported by the Shanghai Leading Academic Discipline Project (project number B114), the Shanghai Committee of Science and Technology of China (nos. 08DZ2271800 and 09DZ2272800), and the Robert Bosch Foundation (Science Bridge China 32.5.8003.0040.0).

A. Fiat and P. Sanders (Eds.): ESA 2009, LNCS 5757, pp. 611–622, 2009.

(reduce the problem to its kernel, then solve the problem on the kernel by brute-force search), it is not known how to derive a polynomial-size kernel from a given FPT algorithm.

Wang et al. [16] first observed that DFVS is more difficult than UFVS. Bodlaender showed that UFVS is in FPT [1], and Thomasse recently gave a quadratic size kernel [15]. For planar graphs, there is even a linear size kernel [2]. On the other hand, the FPT status of DFVS had been an open problem for fifteen years [5] until Chen et al. recently proposed an FPT algorithm [4] based on branching. It is not known whether the problem has a polynomial size kernel.

Our work. We implemented the DFVS algorithm by Chen et al. [4] and tested it on random graphs with various values for edge density ed and optimum DFVS size k. Fig. 1(a) shows the runtime of the algorithm for graphs with $ed = 2$ and $k = 2, 4, 6, 8$. As expected, the performance degrades dramatically when the parameter k increases; for $k = 8$ we often observed a time-out (set to three hours in our experiments), in particular on low-density graphs. Data reduction rules are very attractive to speed up FPT algorithms because no branching is involved and they may sometimes lead to a good kernelization. Inspired by the recent kernelization techniques for UFVS by Thomasse [15], we propose four new data reduction (preprocessing) rules for DFVS: *Dummy Nodes*, *Chaining Nodes*, *Flower*, and *Shortcut*. We achieved another speed-up by starting Chen's FPT search not in an arbitrary initial configuration. We propose three heuristics to choose a good initial configuration: a simple greedy heuristic based on picking large-degree nodes, and two more sophisticated heuristics based on computing good approximations to the optimal DFVS solution [7].

We evaluated the performance of our data reduction rules and heuristics on randomly generated graphs with varying number of nodes, edge density, and optimal solution size. In particular, we analyzed the impact of each parameter and heuristic on the total runtime. We measured runtime, kernel size (since we do not have a kernelization yet, we actually measure the maximum problem size sent to standard min cut when solving the skew separator problem, an important step in Chen's algorithm [4]), memory usage, and recursion depth. Our data suggest that *Chaining Nodes*, *Flower*, and *Shortcut* reductions can significantly reduce the FPT search space enabling us to solve DFVS on graphs that are several orders of magnitude larger than what Chen's algorithm can handle (see Fig. 1). Overall, we think best approach for solving DFVS is to first apply our reduction rules, then use the *Big-Degree* heuristic to compute an initial configuration for Chen's FPT algorithm, which is then used to solve the problem.

When we started this study, our main goal was actually to evaluate the algorithms on real data from the purported main application of DFVS [4,7], namely deadlock recovery in multi-thread computing environments. However, due to restrictions of the parallel programming model, cycle detection, rather than DFVS search, is already sufficient for efficient deadlock recovery. For example, it was reported in a recent paper on deadlock immunity [10] that cycle detection only incurs a loss of efficiency of 6% in state-of-the-art multi-core systems, so there is no need to employ complicated FPT algorithms.

Structure of the paper. In Section 2, we give some basic definitions. In Section 3, we briefly review Chen's FPT algorithm for DFVS and propose our new reduction rules and heuristics to speed-up the algorithm. In Section 4 we discuss implementation details of the algorithms, the random graph generator, and we present and discuss our experimental data. In Section 5 we discuss the suitability of DFVS for the deadlock recovery problem for concurrent threads. We conclude the paper in Section 6 with some thoughts about future work.

2 Preliminaries

Let $G = (V, E)$ be a directed graph with nodes V and edges E. A *simple path* is a path with no repeated nodes. Two paths are *internally disjoint* if no node appears on both paths except maybe the end nodes. A *directed cycle* is a path ending at its start node. Similarly, a *simple cycle* is cycle in which no node appears twice. A directed graph is *acyclic* (DAG) if and only if it contains no directed cycle.

For a node u, a *u-flower of order k* is a set of k directed cycles that are pairwise disjoint except for their common node u. We call these cycles *petals*. To compute u-flowers in G, we split u into two nodes s and t, with s incident to all outgoing edges and t to all incoming edges. Then, the maximum flower size at u equals the maximum number of node disjoint paths from s to t in G, which we can easily compute using standard min-cut techniques. and *Menger's Theorem* for directed graphs [3] which states that the maximum number of internally disjoint paths from s to t equals the size of a minimum (s, t) node cut if s is not adjacent to t. In the following, we use $petal(u)$ to denote the maximum flower size at u. Note that $petal(u) = 1$ implies that there exists another node v such that *all* cycles containing u also contain v (i.e., v dominates u); v is the single cut node between s and t in Menger's Theorem.

3 Chen's Algorithm and Speed-Ups

Chen's algorithm is based on *iterative compression* [9]. Given an input (G, k), we arbitrarily choose an induced $(k + 1)$-node subgraph H of G and an arbitrary k-node subset I of the nodes in H. Note that I is a k-node feedback vertex set of H. Then we repeatedly add another node of G to H and I, until $H = G$. Note that whenever we just added a new node, I is a $(k + 1)$-node feedback vertex set of H. We then use the subroutine *FVS-Reduction* to *compress* I into a k-node feedback vertex set of H. Eventually, $H = G$ and I is a k-node feedback vertex set of G. If in some iteration we cannot find a k-node feedback vertex set for H, we conclude that G does not have a k-node feedback vertex set. We can improve the runtime in two ways: use data reduction rules, and use heuristics to find a good initial configuration for the algorithm.

Data reduction rules. We preprocess the given graph by applying different rules (in arbitrary order and as long as possible) to reduce the size of the graph and the parameter k. Essentially, this reduces the search space for the FPT algorithm later, whose runtime is exponential in k. Initially, we set the feedback vertex set $I = \emptyset$.

Rule 1 (*Self-Loop*). If u has a self-loop, we add u to I, decrease k by 1, and delete u and adjacent edges from the graph.

Rule 2 (*Edge Canonicalization*). If there are two nodes u and v with multiple edges from u to v we remove all but one edge.

Rule 3 (*Dummy Nodes*). If there is a node u without incoming edges or without outgoing edges, then we remove u and its incident edges from the graph.

Rule 4 (*Chaining-Nodes*). If there is a node u such that (v, u) $((u, v))$ is the only incoming (outgoing) edge, then we merge u with v.

It is easy to see that these four rules are *safe*, i.e., the reduced instance has a solution if and only if the original problem has a solution. Rule 1 and Rule 2 are already implicit in Chen's algorithm. Rule 3 and Rule 4 are the directed versions of similar rules for the undirected case [15]. The next rules are new.

Rule 5 (*Shortcut*). If there is a node u with $petal(u) = 1$, then we delete u and all incident edges from the graph, but we add all shortcuts bypassing u as new edges, i.e., for any path $v \to u \to w$ we add the edge (v, w).

Rule 6 (*Flower*). If there exists a node u with $|petal(u)| > k$, we add u to I, decrease k by 1, and delete u and adjacent edges from the graph.

Rule 5 is safe because any node u with $petal(u) = 1$ is dominated by another node (see above). Rule 6 is safe because not choosing u for the feedback vertex set would imply we must take one node in each of the at least $k + 1$ disjoint cycles through u. Note that Rule 4 is actually a special case of Rule 5. However, Rule 4 can be tested in constant time, while Rule 5 requires a min-cut computation. Therefore, we list them as two rules. Rule 6 is the directed version of the the undirected flower reduction proposed by Thomasse [15]. In contrast to the undirected case, this rule can be tested efficiently using standard min-cut techniques.

Note that we have reduction rules for nodes with small petal size and large petal size, but no rules for petal size between 2 and k. Such rules might be necessary to find a true kernelization for `DFVS`.

Initial Heuristics. It may be helpful to choose the initial subgraph H and its k-node feedback vertex set I more carefully. This was already suggested by Chen [4]. Assume a heuristic gives us a set F of nodes whose removal makes G acyclic. If $|F| \leq k$, we can choose $H = G$ and $I = F$ and we are done. Otherwise, we pick a random k-node subset F_0 of F and start Chen's algorithm with $H = G - (F - F_0)$ and $I = F_0$. Clearly, F_0 is a feedback vertex set of H, and if F is a good approximation to the optimal solution, H may be close to G. We propose three heuristics: Chen et al. had suggested to use [4], which unfortunately exhibited the worst performance in our experiments.

Heuristic 1 (*Big Degree*). We greedily add nodes of maximum undirected degree (i.e., *indegree* + *outdegree*) to F until the graph becomes acyclic.

Heuristic 2 (*Fractional Approximation*). We first compute a $(1 + \epsilon)$-approximation for fractional DFVS [7], then greedily add nodes with heaviest fractional weight to F until the graph becomes acyclic.

Heuristic 3 (*Full Approximation*). We compute a DFVS approximation F with factor $O(\min\{\log \tau^* \log\log \tau^*, \log n \log\log n\})$ [7], where τ^* is the cost of a minimum fractional feedback vertex set.

4 Experiments

We used a PC with Intel(R) Xeon(TM) CPU (3.20GHz), 4 processors, 1 core per processor. The machine had 2 GB main memory (DDR2 SDRAM), 2 MB L2 cache memory, and an 80 GB serial ATA hard drive. We tested the algorithms on random graphs (see Section 4.1). For each set of parameters we used ten random graphs and recorded minimum, maximum, and average performance. For all runs of the algorithms we set a time-out threshold of three hours. We implemented the algorithms in `C++` using `LEDA-6.2` [12] on the `Fedora Core 8` operating system. We compiled the programs with `g++-4.1.2 -O3`. We have approximately `4,000 LOC` in total.

For Chen's algorithm, we found it non-trivial to map some of the basic graph operations to program code. For example, there is no way for two different graphs in `LEDA` to share common nodes, which is required when we want to compute induced graphs. We also found that some key operations common in FPT algorithms are not well supported by `LEDA`. For example, FPT algorithms often apply kernelization rules and then use brute-force search on the kernel to solve the problem. The brute-force search is usually done by iterating over all subsets of size k or each topological order of the graph. Therefore, we extended `LEDA` with two interfaces:

- `bool foreach_subset(list<node> &sub, graph &G, subset_prop_func pf, void *pars)`

- `bool foreach_topord(list<node> &sub, graph &G, topord_prop_func pf, void *pars)`

In the subset enumerator, `sub` specifies a subset of nodes in G. It comes together with a property function that tests whether the subset has a certain property. Each time we generate a new subset, we call the property function with the new subset and the user-supplied parameters `pars`. This function will stop and return `true` once the property function returns `true`. If all subsets have been tried, we simply return `false` to indicate that for any subset of `sub` the property cannot be satisfied.

In subsection 4.1, we first briefly discuss how we generated the random graph instances. Then we present our experimental data in four subsections. Complete experimental data and the source code of our programs are available online [8]. In

Section 4.2 we present data on the runtime performance and kernel size of Chen's original algorithm. In Section 4.3 we show the effectiveness of our new data reduction rules in improving the FPT search and reducing large input instances. In Section 4.4 we evaluate and compare three heuristics to obtain an initial configuration for Chen's algorithm. Finally, in Section 4.5 we evaluate the runtime performance of Chen's algorithm with respect to different parameters k for a graph with fixed optimum solution.

4.1 The Random Graph Generator

Our goal was to generate sufficiently difficult "random" instances while still precisely controlling three parameters: n, the number of nodes in the graph; k, the size of the optimum DFVS solution; and ed, the edge density ($m = ed \cdot n$). It seems difficult to generate truly random graphs with these parameters fixed, so we developed our own methods to generate graphs that seem to be quite random. We generate the graphs in two stages. First, we generate a random spanning tree [17] from which we then obtain a random connected DAG [13]. Finally, we add cycles to the graph, which poses two challenges: how to precisely control the minimum feedback vertex set size, and how to ensure that the newly generated graph is difficult to solve. We first randomly choose k nodes as the optimum solution, and then generate k node-independent cycles passing through each of them. We do not generate self-loops since they can easily be removed. Note that this method cannot generate k node-independent cycles for $k > \frac{n}{2}$. To create more overlapping cycles, we randomly fix a topological order for the nodes not in the solution. Each time we generate a cycle, we first randomly choose a subset A of the solution set and a subset B of the remaining nodes. We keep the nodes in B in their topological order when building cycles through these nodes. This ensures that the optimum solution size cannot exceed k. To ensure that our graphs are difficult to solve, we randomly add more cycles until we reach the required *edge density* ed, where the number of edges in a cycle is within $\frac{1}{4}$ of the total *edge bound.* Also, we try not to generate cycles that are too big by limiting the cycle size to at most $\frac{n}{4}$. Although we know one optimal solution of the generated graphs, they usually did not have a unique solution in our experiments.

4.2 Chen's Algorithm

Since FPT search scales poorly with n and k, we only considered graphs with $n = 40, 60, 80, \ldots, 200$, $k = 2, 4, 6, 8$, and edge density $ed = 2, 3, 3.5, 4$. We generated ten graphs for each triple (n, k, ed). Fig. 1(a) shows the average search time as a function of n and k for $ed = 2.0$. For all experiments, the memory usage was constant with approximately 10 MB.

We see from the data that the runtime of the algorithm scales poorly with k, which is typical for FPT algorithms. For $k = 8$ and $ed = 2.0$, the algorithm never finished within 3 hours. For $ed = 3.0$ and $k = 8$, the algorithm showed time-outs when n was larger than 140.

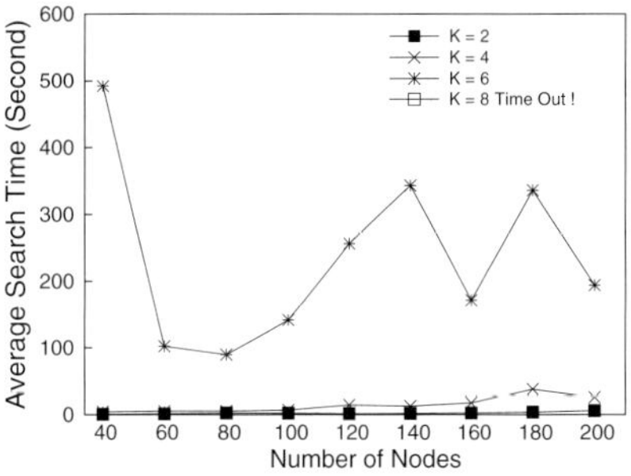

(a) No reductions.

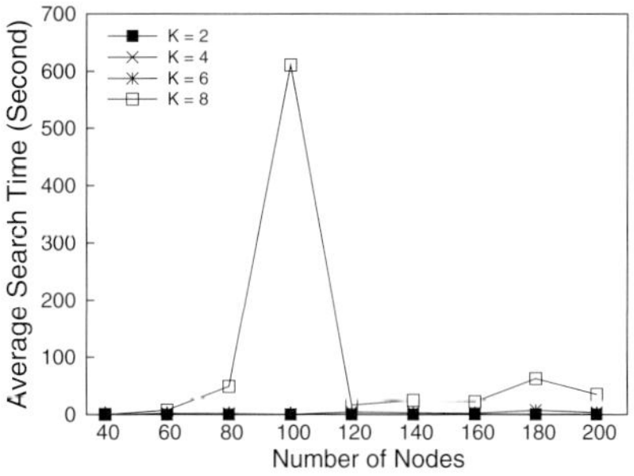

(b) With our new reductions.

Fig. 1. Runtime of Chen's algorithm, $ed = 2.0$. Note that the top graph in (a) corresponds to the case $k = 6$ (because $k = 8$ never finished within three hours), while the top graph in (b) is for the case $k = 8$.

It seems that the runtime is not monotonely increasing in n, probably because it is dominated by the exponential dependency on k. The high runtime for low values of n (for example, $(40, 8, 2.0)$) is due to the fact that these graphs have many independent cycles, and Chen's algorithm seems to perform poorly on such graphs. A possible explanation may be found in the iterative compression technique. If we have many nodes on a few cycles and must add an additional node which is on a not yet blocked independent cycle, we must keep the new node and kick out one of the previously chosen nodes. Identifying such a node might be time-consuming.

Also, the runtime does not increase monotonely in the edge density. The algorithm can still quickly solve all instances with small parameter k when the number of edges increases significantly. Also, for bigger k, the runtime does not increase monotonely with the number of edges. For example, for $k = 8$, the problem becomes most difficult for the FPT algorithm when $ed = 2$. Most graphs time-out within three hours for this configuration.

As we do not have kernelization rules for `DFVS`, we evaluated the kernel size by recording the maximum size of an instance of the standard min cut problem when solving the skew separator problem (in Chen's algorithm). However, we found this size is roughly the same as the size of the original graph (the deviation is usually two or three nodes).

4.3 Data Reduction Rules

Fig. 1(b) shows the average runtime of Chen's algorithm after applying the new reduction rules for the same graphs as in Section 4.2. Now we can solve all the graphs quickly (the average longest time was 10 minutes, most often even less than 1 minute). However, the runtime is still not monotone with regard to n or ed; it is actually closely related to the reduced size after preprocessing. For example, there is a peak in the runtime when $ed = 2.0$ and $n = 100$. For this configuration, some instances got solved directly by preprocessing, while others needed hundreds

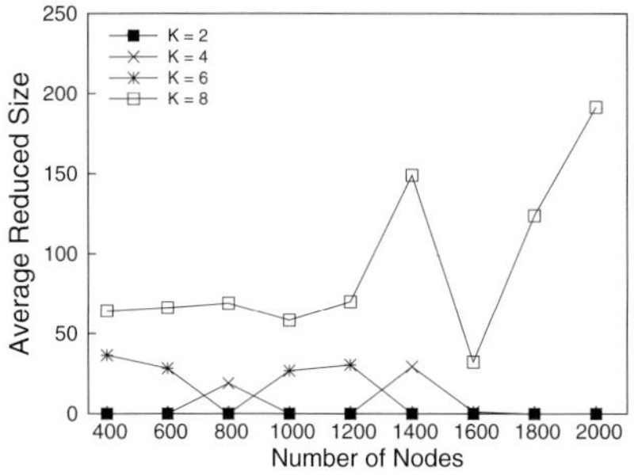

(a) Reduced size, small k.

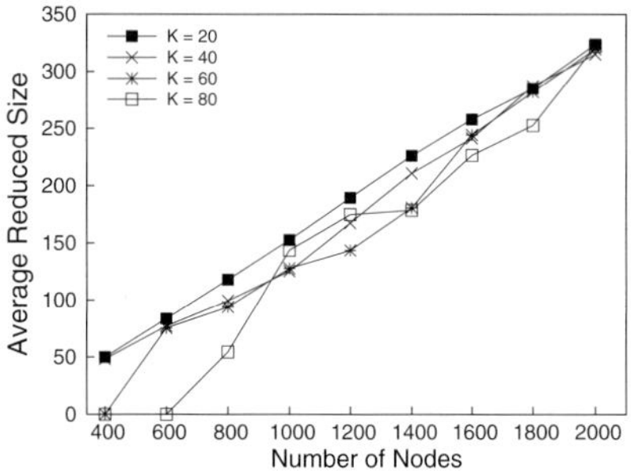

(b) Reduced size, large k.

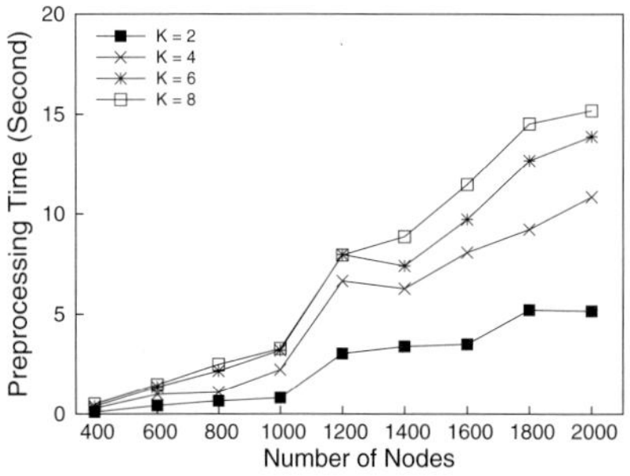

(c) Preprocessing time.

Rules	$k = 4$	$k = 40$
Chain	607	377
Dummy	785	679
Cut	996	1000
Chain+Dummy	522	377
Chain+Cut	366	377
Dummy+Cut	151	377
Dummy+Dummy+Cut	0	377

(d) Reduced size by combining different rules, $n = 1000$, $ed = 3.0$.

Fig. 2. Chen's algorithm with data reductions, $ed = 2.0$

of seconds to be solved. In particular, there is one instance with a big kernel (33 nodes) which was only solved after more than 2,000 seconds.

We designed two more experiments to further test the power of data reductions. In the first experiment, we kept the small values for k and varied n between 400 and 2,000, with step length 200. Fig. 2(a) and 2(c) show the average size reductions and preprocessing times, respectively. Surprisingly, we found that the reduction rules could directly handle most of the input instances when $ed > 2$. There are two possible explanations: The parameter k may be too small, or our "random" graphs actually have many overlapping cycles. In both cases, the *Flower* reduction could reduce the input size considerably. The preprocessing time grows linearly in n and ed. On the average, it was just 3 minutes for all graphs together.

In the second experiment, we varied k in $\{20, 40, 60, 80\}$ for the same n as in the first experiment. Fig. 2(b) shows the average reduced size for these configurations. We observe that for fixed edge density the reduced size scales linearly with the number of nodes, which indicates that our reduction rules work consistently for different graph sizes. Also, the reduced size increases with increasing edge density. This is quite different from the first experiment where the preprocessing could directly solve almost all graphs when ed is large. The reason is that the Flower reduction rule does not work well in these cases if k gets larger. Thus, it would be

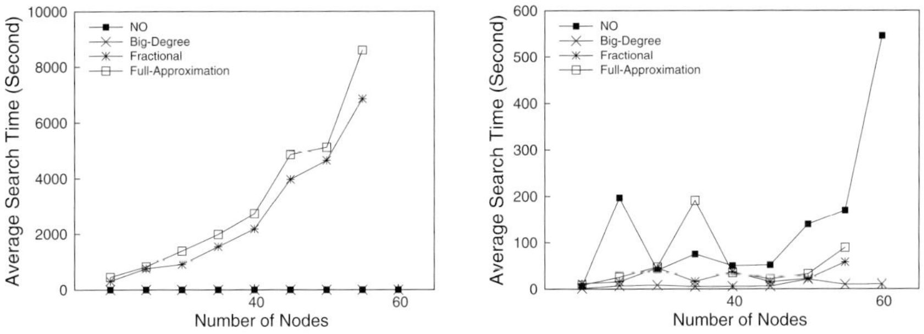

(a) Heuristics. (b) Chen's algorithm after heuristic.

Fig. 3. Runtimes of heuristics for Chen's algorithm, $k = 8$ and $ed = 3.0$

important to devise reduction rules that work well for relatively large parameter k as well as large edge density.

Individual reduction rules. We tried to find out which rules are more important than others. We studied two groups of graphs, the first group had parameters $n = 1000$, $k = 4$, $ed = 3.0$, and the second group had $n = 1000$, $k = 40$, $ed = 3.0$. We evaluated three types of rules: *Chain*, *Dummy*, and *Cut*. Type *Chain* rules consisted of the *Chaining Nodes* and *Shortcut* rules; Type *Dummy* rules contained the *Dummy Nodes* rule and the reduction that removes all nodes with *petal* size 0; Type *Cut* rules were just the *Flower* rule. Table 2(d) summarizes the average reduced sizes for these two configurations. We see that the *Cut* rules do not reduce the graph significantly, but they are useful in determining the nodes in the minimum feedback vertex cover. For example, the *Cut* rules may reduce a graph from 1000 to 996 nodes when $k = 4$, but these 4 nodes are in the feedback vertex set, so we have already solved the problem even though there are still 996 unprocessed nodes. The *Flower* rules might help other rules to further reduce the graph Our data show that the graphs are reduced significantly when we combine *Chain* or *Dummy* rules with *Flower* rules. This is because a *Flower* rule actually selects a node for the feedback vertex set and then deletes it from the graph, which may trigger other rules. The *Flower* rule becomes useless when k grows, because graphs usually do not have large flowers.

4.4 Heuristics for the Initial Configuration

We evaluated the three heuristics mentioned in Section 4 to find a good initial configuration for Chen's algorithm. Since the FPT algorithm does not scale well with n and k and the approximation algorithms are very slow even on small graphs, we only considered graphs with $n = 20, 25, 30, \ldots, 60$ nodes, with $k \in \{2, 4, 6, 8\}$ and $ed = 3.0$.

Fig. 3 shows the runtimes for $k = 8$ (including the times for no heuristic). The graphs would be similar for other values of k. Note that we distinguish between runtime of the heuristic and runtime of Chen's algorithm afterwards. The heuristics

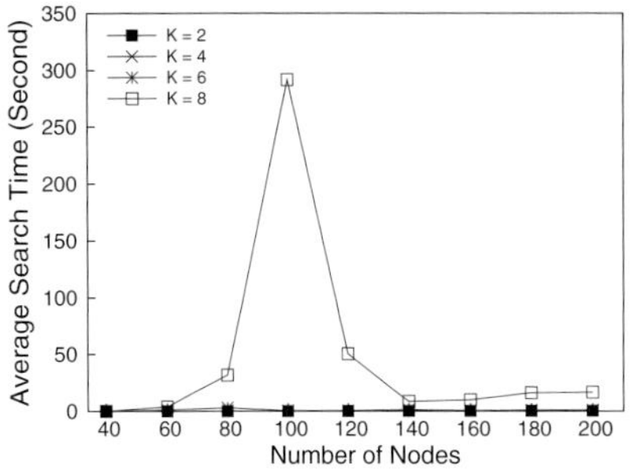

Fig. 4. FPT search time with preprocessing and *Big-Degree* heuristic

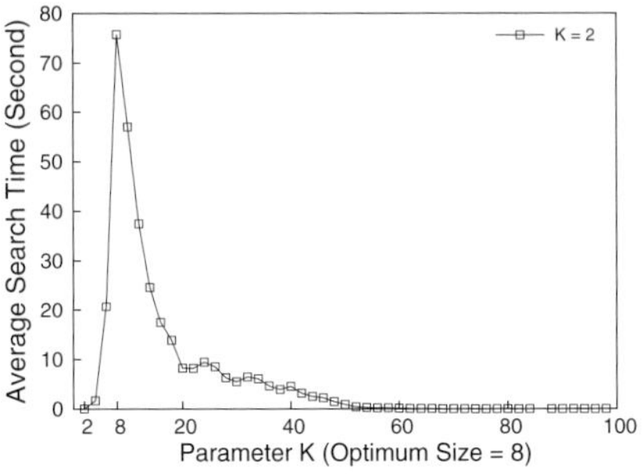

Fig. 5. The FPT search time with respect to different parameter k

seem to become very useful for larger parameters k. On all our graphs, the *Big-Degree* heuristic achieved the most significant acceleration (4x to 50x speedup). Moreover, the *Big-Degree* heuristic (which usually takes only a few seconds) runs thousands of times faster than *Fractional Approximation* and *Full-Approximation* (which usually need thousands of seconds). Overall, we conclude that the best approach for solving DFVS is to first apply our reduction rules, then use *Big-Degree* to compute the initial configuration, and then use Chen's algorithm [4] to solve the problem. For example, Fig. 4 shows that, with the *Big-Degree* heuristic, we can achieve another speed-up of $2 - -3$x compared to reduction rules without heuristic (Fig. 1(b)).

4.5 The Impact of the Parameter k

Intuitively, we may quickly find a feedback vertex set of size k if k is much larger than the optimal one. We may also quickly reject the input if k is much smaller than the optimal parameter. To test this hypothesis, we generated graphs with 100 nodes for $k = 8$ and $ed = 3.0$. Then we used Chen's algorithm to search a feedback vertex set of size $k = 2, 4, 6, 8, 10, \ldots, 98$. Fig 5 summarizes the average search times for this experiment. As expected, the runtime reaches its peak when the parameter k is near the size of the minimum feedback vertex set. The runtime is quite fast when k is much larger or much smaller than the optimum.

5 Deadlock Detection

Deadlock recovery in concurrent programs has always been considered an important application of DFVS [4,7]. Indeed, deadlock recovery is a very important topic in our modern *multi-core/many-core* computing era. Solving the concurrency problem has recently seen tremendous research interest in operating systems, programming languages and computer architecture communities.

For simplicity of analysis we mainly focus on `mutex` locks in the POSIX Thread library (`Pthread`). We will argue that, with the restrictions of the programming model, `DFVS` is not more helpful than cycle detection for deadlock recovery. This proposition is further confirmed by a report on a deadlock immunity system [10] in a recent systems conference.

In a concurrent program, the nodes of a Resource Allocation Graphs (`RAG`) are resources and threads. Resources are simply mutex locks. There are three types of edges in a `RAG`: *request*, *grant* and *own*. The *request* and *grant* edges are edges from thread to lock, while the *own* edges go from lock to thread. *request* edge means a thread is requesting a lock, and *grant* edge means the thread library allows the thread to wait on the lock (note that the thread may yield its execution to another one before being allowed to wait on the resource). The *own* edge from lock to thread means the thread currently owns this resource exclusively. A deadlock appears as a cycle in the `RAG`. In such a cycle, all edges are exclusively the *grant* and *own* edges. One lock can be owned by *only one* thread, even for *recursive* locks, though certain threads could acquire the lock multiple times. One thread, at one time, can be granted to wait on *only one* lock because threads are executed sequentially, and we cannot wait on two resources simultaneously. Thus, we cannot have *overlapping* cycles in a `RAG`, and therefore simple cycle detection suffices to resolve the deadlocks.

6 Conclusions

We presented a comprehensive experimental study on the feedback vertex set problem in digraphs. We proposed new data reduction rules to efficiently reduce the FPT search space. Finally, we demonstrated that `DFVS` search is not more helpful than cycle detection in efficient deadlock detection in modern concurrent systems.

We would like to find better data reduction rules for `DFVS`, in particular a polynomial-size kernel. We remark that though `UFVS` has a small problem kernel, the parameter k is potentially very large for dense graphs, as in undirected graphs it is quite easy to have a cycle. We may also study other parameters than "solution size", for example, "edge density". Our reduction rules seem to perform well when the edge density is low. Better approximation algorithms for `DFVS` might also help to speed up the FPT search. We demonstrated that `DFVS` is not more helpful than cycle detection in detecting deadlocks with mutex locks, but there are also other types of locks, such as read/write locks, that may generate complex overlapping deadlocks. However, we found that other lock types, for example spin locks, behave essentially the same as mutex locks. Read/write locks may behave differently. but most thread libraries (e.g. `NPTL` (Native POSIX Thread Library) restrict the number of threads in read mode, and once a thread is waiting in write mode, later read requests are pending until completion of the write. We are now investigating the practicability of FPT `DFVS` algorithms in circuit testing to reduce the hardware overhead required for "scan registers" [11].

References

1. Bodlaender, H.L.: On linear time minor tests with depth first search. Journal of Algorithms 14, 1–23 (1993)
2. Bodlaender, H.L., Penninkx, E.: A linear kernel for planar feedback vertex set. In: Grohe, M., Niedermeier, R. (eds.) IWPEC 2008. LNCS, vol. 5018, pp. 160–171. Springer, Heidelberg (2008)
3. Chatrand, G., Lesniak, L.: Graphs & Digraphs, 2nd edn. The Wadsworth and Brooks/Cole Mathematics Series (1986)
4. Chen, J., Liu, Y., Lu, S., Sullivan, B., Razgon, I.: A fixed-parameter algorithm for the directed feedback vertex set problem. Journal of the ACM 55(5), 1–19 (2008)
5. Downey, R.G., Fellows, M.R.: Fixed-parameter tractability and completeness I: basic results. SIAM Journal on Computing 24, 873–921 (1995)
6. Downey, R.G., Fellows, M.R.: Parameterized Complexity. Springer, Heidelberg (1999)
7. Even, G., Naor, J., Schieber, B.: Approximating minimum feedback sets and multicuts in directed graphs. Algorithmica 20, 151–174 (1998)
8. Fleischer, R., Xi, W., Yuan, L.: DFVS Project (2009), `http://www.tcs.fudan.edu.cn/rudolf/Projects/DFVS/dfvs.html`
9. Huffner, F., Niedermeier, R., Wernicke, S.: Techniques for practical fixed-parameter algorithms. The Computer Journal 1(51), 7–25 (2008)
10. Jula, H., Tralamazza, D.M., Zamfir, C., Candea, G.: Deadlock immunity: Enabling systems to defend against deadlocks. In: Proceedings of the 8th USENIX Symposium on Operating System Design and Implementation (OSDI 2008), pp. 295–308 (2008)
11. Kunzamann, A., Wunderlich, H.J.: An analytical approach to the partial scan problem. Journal of Electronic Tesing: Theory and Applications 1(5), 163–1741 (1990)
12. LEDA: A library of the data types and algorithms of combinatorial computing, `http://www.mpi-inf.mpg.de/LEDA/`
13. Melancon, G., Dutour, I., Bousquet-Melou, M.: Random generation of directed acyclic graphs. Technical report, CWI Amsterdam (2006), `http://www.cwi.nl/InfoVisu`
14. Seidl, H.: Personal communication (2000)
15. Thomasse, S.: A quadratic kernel for feedback vertex set. In: Proceedings of the 20th ACM-SIAM Symposium on Discrete Algorithms (SODA 2009), pp. 115–119 (2009)
16. Wang, C.-C., Lloyd, E.L., Soffa, M.L.: Feedback vertex sets and cyclically reducible graphs. Journal of the ACM 32(2), 2960–2913 (1985)
17. Wilson, D.B.: Generating random spanning trees more quickly than the cover time. In: Proceedings of the 28th ACM Symposium on the Theory of Computation (STOC 1996), pp. 296–303 (1996)

Fast Evaluation of Interlace Polynomials on Graphs of Bounded Treewidth

Markus Bläser and Christian Hoffmann

Saarland University, Germany

Abstract. We consider the multivariate interlace polynomial introduced by Courcelle (2008), which generalizes several interlace polynomials defined by Arratia, Bollobás, and Sorkin (2004) and by Aigner and van der Holst (2004). We present an algorithm to evaluate the multivariate interlace polynomial of a graph with n vertices given a tree decomposition of the graph of width k. The best previously known result (Courcelle 2008) employs a general logical framework and leads to an algorithm with running time $f(k) \cdot n$, where $f(k)$ is doubly exponential in k. Analyzing the $GF(2)$-rank of adjacency matrices in the context of tree decompositions, we give a faster and more direct algorithm. Our algorithm uses $2^{3k^2+O(k)} \cdot n$ arithmetic operations and can be efficiently implemented in parallel.

1 Introduction

Inspired by some counting problem arising from DNA sequencing [1], Arratia, Bollobás, and Sorkin defined a graph polynomial which they called interlace polynomial [2]. It turned out that the interlace polynomial is related [2, Theorem 24] to the Martin polynomial, which counts the number of edge partitions of a graph into circuits. This polynomial has been defined in Martin's thesis from 1977 [3] and generalized by Las Vergnas [4]. Further work on the Martin polynomial has been pursued [5, 6, 7, 8, 9, 10], including a generalization to isotropic systems [11, 12, 13, 14]. In particular, the Tutte polynomial of a planar graph and the Martin polynomial of its medial graph are related. This implies a connection between the Tutte polynomial and the interlace polynomial (see [15] for an explanation).

One way to define the interlace polynomial is by a recursion that uses a graph operation. Arratia et. al. used a pivot operation for edges [2]. This operation is a composition of local complementations to neighbour vertices (see [16], where the operations are called switch operations). The orbits of graphs under local complementation are related to error-correcting codes and quantum states, and so is the interlace polynomial as well [17].

The interlace polynomial can also be defined by a closed expression using the $GF(2)$-rank of adjacency matrices [16, 18, 19]. This linear algebra approach has been used in several generalizations of the interlace polynomial. In this paper, we consider the multivariate interlace polynomial $C(G)$ defined by Courcelle [20] (see Definition 1 below) as it subsumes the two-variable interlace polynomial of Arratia, Bollobás, and Sorkin [21] and the weighted versions of Traldi [22], as well as the interlace polynomials defined by Aigner and van der Holst [16]. Furthermore, the interlace polynomials

A. Fiat and P. Sanders (Eds.): ESA 2009, LNCS 5757, pp. 623–634, 2009.

$Q(x, y)$ and Q_n^{HN}, which have emerged from a spectral view on the interlace polynomials [23], are also special cases of Courcelle's multivariate interlace polynomial.

Results and Related Work. Our aim is to present an algorithm that, given a graph $G = (V, E)$ and an evaluation point, i.e. a tupel of numbers $((x_a)_{a\in V}, (y_a)_{a\in V}, u, v)$, evaluates the multivariate interlace polynomial $C(G)$ at $((x_a)_{a\in V}, (y_a)_{a\in V}, u, v)$. Whereas this is a #P-hard problem in general [24], it is fixed parameter tractable with cliquewidth as parameter [20, Theorem 23, Corollary 33]. This is a consequence of the fact that the interlace polynomial is a monadic second order logic definable polynomial. Such graph polynomials can be evaluated in time $f(k) \cdot n$, where n is the number of vertices of the graph and k is the cliquewidth. The function $f(k)$ can be very large and is not explicitly stated in most cases. In general, it grows as fast as a tower of exponentials the height of which is proportional to the number of quantifier alternations in the formula describing the graph polynomial [20, Page 34]. In the case of the interlace polynomial, this formula involves two quantifier alternations [20, Lemma 24], [25]. If a graph has tree width k, its cliquewidth is bounded by 2^{k+1} [26]. Thus, the machinery of monadic second order logic implies the existence of an algorithm that evaluates the interlace polynomial of an n-vertex graph in time $f(k) \cdot n$, where k is the tree width of the graph and $f(k)$ is at least doubly exponential in k. (In particular, the interlace polynomial of graphs of treewidth 1, that is, trees, can be evaluated in polynomial time, which also has been observed by Traldi [22].)

The monadic second order logic approach is very general and can be applied not only to the interlace polynomial but to a much wider class of graph polynomials [27]. However, it does not consider characteristic properties of the actual graph polynomial. In this paper, we restrict ourselves to the interlace polynomial so as to exploit its specific properties and to gain a more efficient algorithm (Algorithm 1). Our algorithm performs $2^{3k^2+O(k)}n$ arithmetic operations to evaluate Courcelle's multivariate interlace polynomial (and thus any other version of interlace polynomial mentioned above) on an n-vertex graph given a tree decomposition of width k (Theorem 13). The algorithm can be implemented in parallel using depth polylogarithmic in n, see the full version of this work [28, Section 7]. Apart from evaluating the interlace polynomial, our approach can also be used to compute coefficients of the interlace polynomial [28, Section 7], for example so called d-truncations [20, Section 5]. Our approach is not via logic but via the $GF(2)$-rank of adjacency matrices, which is specific to the interlace polynomial.

Obstacles. It has been noticed that the Tutte polynomial and the interlace polynomial are similar in some respect. These similarities suggest that evaluating the interlace polynomial using tree decompositions might work completely analogously to the respective approaches for the Tutte polynomial [29, 30]. This is not the case because of the following facts that distinguish the interlace polynomial from the Tutte polynomial: First, the graph operation that is used in the recursive definition of the interlace polynomial is not compliant with tree decomposition (i.e. can increase the treewidth). Second, the recurrence for the multivariate interlace polynomial is very complicated. Third, the "behaviour" of the rank involved in the definition of the interlace polynomial is not as benign as the rank used for the Tutte polynomial (cf. also the beginning of Sect. 3). We discuss these issues in more detail in the full version of this work [28].

Outline. Section 2 contains the definition of Courcelle's multivariate interlace polynomial, which we will consider in this work. We will also fix our notation for tree decompositions there. In Sect. 3 we give the general idea of our approach. Section 4 and 5 provide technical details, and in Sect. 6 we describe our algorithm. Due to space limitations we have to omit some details and most of the proofs. These can be found in the full version of this work [28].

2 Preliminaries

We consider undirected graphs without multiple edges but with self loops allowed. Let $G = (V, E)$ be such a graph and $A \subseteq V$. By $G[A]$ we denote the subgraph of G induced by A, i.e. $(A, \{e \mid e \in E, e \subseteq A\})$. $G\nabla A$ denotes the graph G with self loops in A toggled, i.e. the graph obtained from G by performing the following operation for each vertex $a \in A$: if a has a self loop, remove it; if a does not have a self loop, add one.

The adjacency matrix of G is a symmetric square matrix with entries from $\{0, 1\}$. As the matrices that we will consider are adjacency matrices of graphs, we use vertices as column/row indices. Thus, the adjacency matrix of G is a $V \times V$ matrix $M = (m_{uv})$ over $\{0, 1\}$ with $m_{uv} = 1$ iff $uv \in E$. Furthermore, we will refer to entries and submatrices by specifying first the rows and then the columns: the (u, v)-entry of $M = (m_{uv})$ is m_{uv}, the $A \times B$ submatrix of M is the submatrix of the entries of M with row index in A and column index in B. All matrix ranks will be ranks over the field with two elements, $\{0, 1\} = GF(2)$, i.e. $+$ is XOR and $\cdot$ is AND. Slightly abusing notation we write $\mathrm{rk}(G)$ for the rank of the adjacency matrix of the graph G. The nullity (or co-rank) of an $n \times n$ matrix M is $\mathrm{n}(M) = n - \mathrm{rk}(M)$. If G is a graph, we write $\mathrm{n}(G)$ for the nullity of the adjacency matrix of G.

Graph polynomials are, from a formal perspective, mappings of graphs to polynomials that respect graph isomorphism. We will consider a *multivariate* graph polynomial, the multivariate interlace polynomial. To define such a polynomial, one has to distinguish "ordinary" indeterminates from *G-indexed indeterminates*. For instance, x being a G-indexed indeterminate means that for each vertex a of G there is a different indeterminate x_a. If $A \subseteq V$, we write x_A for $\prod_{a \in A} x_a$. Also, if S is a set, we write $\sum S$ for the sum of all the elements in the set.

Definition 1 (Courcelle [20]). *Let $G = (V, E)$ be an undirected graph. The multivariate interlace polynomial is defined as*

$$C(G) = \sum \{x_A y_B u^{\mathrm{rk}((G\nabla B)[A\cup B])} v^{\mathrm{n}((G\nabla B)[A\cup B])} \mid A, B \subseteq V, A \cap B = \emptyset\},$$

where u, v are called ordinary indeterminates and x, y G-indexed indeterminates.

Tree Decompositions. We borrow most of our notation from Bodlaender and Koster [31]. A *tree decomposition* of a graph $G = (V, E)$ is a pair $(\{X_i \mid i \in I\}, T = (I, F))$ where T is a tree and each node $i \in I$ has a subset of vertices $X_i \subseteq V$ associated to it, called the bag of i, such that the following holds:

1. Each vertex belongs to at least one bag, that is $\bigcup_{i \in I} X_i = V$.

2. Each edge is represented by at least one bag, i.e. for all $e = vw \in E$ there is an $i \in I$ with $v, w \in X_i$.
3. For all vertices $v \in V$, the set of nodes $\{i \in I \mid v \in X_i\}$ induces a subtree of T.

The width of a tree decomposition $(\{X_i\}, T)$ is $\max\{|X_i| \mid i \in I\} - 1$. The treewidth of a graph G, $\operatorname{tw}(G)$, is the minimum width over all tree decompositions of G.

Computing the treewidth of a graph is NP-complete. But given a graph with n vertices, we can compute a tree decomposition of width k (or detect that none exists) using Bodlaender's algorithm in time $2^{O(k^3)}n$ [32].

To evaluate the interlace polynomial we will use *nice* tree decompositions. Note that our definition slightly deviates from the usual one[1]. This has no substantial influence on the running time of the algorithms discussed in this work but it simplifies the presentation of our algorithm. In a nice tree decomposition $(\{X_i\}, T)$, one node r with $|X_r| = 0$ is considered to be the root of T, and each node i of T is of one of the following types:

- Leaf: node i is a leaf of T and $|X_i| = 0$.
- Join: node i has exactly two children j_1 and j_2, and $X_i = X_{j_1} = X_{j_2}$.
- Introduce: node i has exactly one child j, and there is a vertex $a \in V$ with $X_i = X_j \cup \{a\}$.
- Forget: node i has exactly one child j, and there is a vertex $a \in V$ with $X_j = X_i \cup \{v\}$.

A tree decomposition of width k with n nodes can be converted into a nice tree decomposition of width k with $O(n)$ nodes in time $O(n) \cdot \mathsf{poly}(k)$ [33, Lemma 13.1.2, 13.1.3].

For a graph G with a nice tree decomposition $(\{X_i\}, T)$, we define

$$V_i = \left(\bigcup\{X_j \mid j \text{ is in the subtree of } T \text{ with root } i\}\right) \setminus X_i \quad \text{and} \quad G_i = G[V_i].$$

We can think of G_i as the subgraph of G induced by all vertices that have already been forgotten below node i.

3 Idea

We will now sketch our idea how to evaluate the interlace polynomial. Our approach is dynamic programming. Let G be a graph for which we want to evaluate the interlace polynomial and $(\{X_i\}, T)$ a nice tree decomposition of G. For each node i of the tree decomposition, we have defined the graph G_i that consists of all vertices in the bags below i that are not in X_i. We will compute "parts" of the interlace polynomial of G_i. These parts are essentially defined by the answer to the following question: How does the rank of the adjacency matrix of some subgraph of G_i increase when we add (some or all) vertices of X_i? For the leaves these parts are trivial. Our algorithm traverses the tree decomposition bottom-up. We will show how to compute the parts of an introduce,

[1] Usually, there is no special restriction on the bag size of the root node, and the leaf nodes contain exactly *one* vertex.

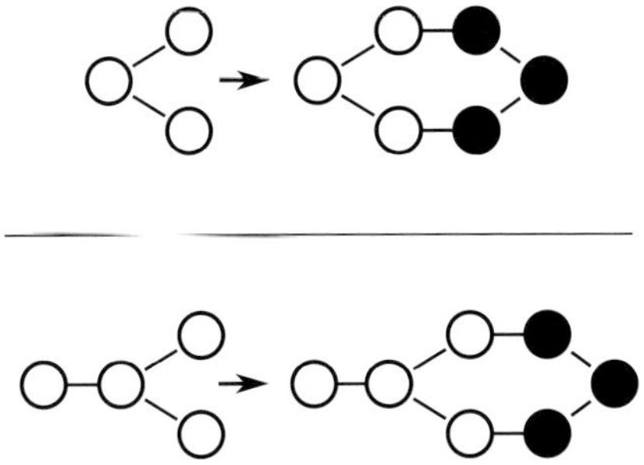

Fig. 1. Rank over $GF(2)$ of the adjacency matrix may increase by 2 (from 2 to 4, upper half) or by 4 (from 2 to 6, lower half), even if the same extension is used

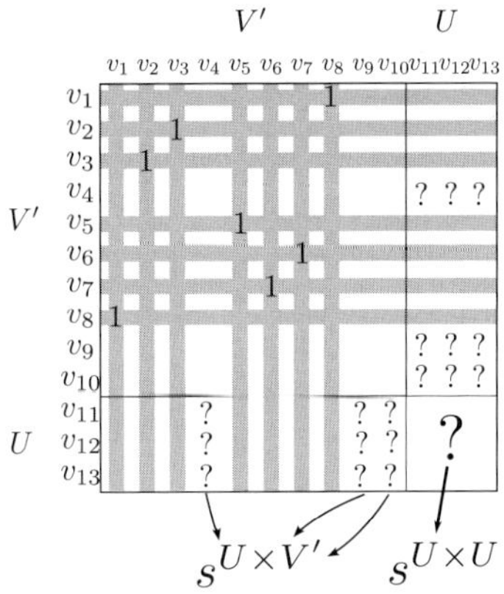

Fig. 2. Adjacency matrix of $G[V' \cup U]$ after symmetric Gaussian elimination using V'. Empty entries are zero.

forget, or join node from the parts of its child node (children nodes, resp.). At the root node, there is only one part left. This part is the interlace polynomial of G.

Before we go into details, let us remark that the answer to the above question ("How does the rank of the adjacency matrix increase when adding some vertices?") depends on the internal structure of the graph being extended. Consider the situation in the upper half of Fig. 1. If we extend the graph by the black vertices, the rank increases by 2. But in the situation depicted in the lower half of Fig. 1, the *same extension* causes a rank increase by 4.[2]

Let us see how we handle this issue. We start with the following definition.

Definition 2 (Extended graph). *Let $G = (V, E)$ be some graph, $V', U \subseteq V$, $V' \cap U = \emptyset$. Then we define $G[V', U]$ to denote $G[V' \cup U]$ and call $G[V', U]$ an* extended graph, *the graph obtained by* extending $G[V']$ by U according to G. *We call U the* extension *of $G[V', U]$.*

Let us fix an extension U. We consider all $V' \subseteq V(G)$ such that $G[V']$ may be extended by U according to the input graph G. For any such extended graph we ask: "How does the rank of $G[V']$ increase when adding some vertices of U?". Our key observation is that the answer to this question can be given without inspecting the actual G if we are provided with a compact description (of size independent of $n = |V(G)|$), which we call the scenario of $G[V', U]$.

The scenario of $G[V', U]$ (Definition 5) will be constructed in the following way. Consider M, the adjacency matrix of $G[V' \cup U]$. Perform symmetric Gaussian elimination on M *using only the vertices in* V' (for the details see Sect. 4). The resulting matrix M' is symmetric again and has the same rank as M. Furthermore, M' is of a form as in Fig. 2: The $V' \times V'$ submatrix is a symmetric permutation matrix with some

[2] This phenomenon distinguishes the interlace polynomial from the Tutte polynomial. In the case of the Tutte polynomial, the rank increase would be the same in both situations as it only depends on the extension and how it is connected to the graph being extended.

additional zero columns/rows. The nonzero entries correspond to edges or self loops (not of the original graph G but of some modified graph that is obtained from G in a well-defined way) "ruling" over their respective columns/rows: The edge between v_1 and v_8 rules over columns and rows v_1 and v_8. Here, "to rule" means that the only 1s in these columns and rows are the 1s at (v_1, v_8) and (v_8, v_1). Similarly, the self loop at vertex v_5 rules over column and row v_5. The columns and rows that are ruled by some edge or self loop in V' are also empty (i.e. entirely zero) in the $U \times V'$ submatrix of M'. Some columns/rows are not ruled by any edge or self loop in V', for instance column/row v_4. This is because there is neither a self loop at vertex v_4 nor does it have a neighbor in V'. However, v_4 may have neighbors in U. Thus, column v_4 of the $U \times V'$ submatrix may be any value from $\{0,1\}^U$, which is indicated by the question marks. Also, the contents of the $U \times U$ submatrix is not known to us.

Let us choose a basis of the subspace spanned by the nonzero columns of the $U \times V'$ submatrix and call it $s^{U \times V'}$. Let $s^{U \times U}$ be contents of the $U \times U$ submatrix. By this construction, we are able to describe the rank of M' as the rank of its $V' \times V'$ submatrix plus a value that can be computed solely from $s^{U \times V'}$ and $s^{U \times U}$.

This will solve our problem that the rank increase depends on the internal structure of the graph $G[V']$ being extended: all we need to know is the scenario $s = (s^{U \times V'}, s^{U \times U})$ of $G[V', U]$. From s, without considering $G[V']$, we can compute in time $\mathsf{poly}(|U|)$ how the rank of the adjacency matrix of $G[V']$ increases when we add some vertices from U. This motivates the following definition.

Definition 3 (Scenario). *Let U be an extension, i.e. a finite set of vertices. A* scenario of U *is a tuple $s = (s^{U \times V'}, s^{U \times U})$ where $s^{U \times V'}$ is an ordered set of linear independent vectors spanning a subspace of $\{0,1\}^U$ and $s^{U \times U}$ is a symmetric $(U \times U)$-matrix with entries from $\{0,1\}$. A* scenario for k vertices *is a scenario of some vertex set U with $|U| = k$.*

Let us come back to the evaluation of the interlace polynomial of G using a tree decomposition. Recall that at a node i of the tree decomposition we want to compute "parts" of the interlace polynomial of $G[V_i]$. Essentially every scenario s of X_i will define such a part: The interlace polynomial itself is a sum over *all* induced subgraphs with self loops toggled for some vertices. The part of the interlace polynomial corresponding to scenario s will be the respective sum not over all these graphs but only over the ones such that s is the scenario of $G[V_i, X_i]$. This will lead us to (1) in Sect. 6.

The time bound of our algorithm stems from the following observation: The number of parts managed at a node i of the tree decomposition is essentially bounded by the number of scenarios of its bag X_i. This number is independent of the size of G and single exponential in the bag size (and thus single exponential in the treewidth of G):

Lemma 4. *Let U be an extension, i.e. a finite set of vertices, $|U| = k$. There are less than $2^{(3k+1)k/2}$ scenarios of U.*

4 Symmetric Gaussian Elimination

We want to convert adjacency matrices into matrices of a form as in Fig. 2 without touching the rank. In order to achieve this, we introduce a special way of performing

Gaussian elimination that differs from standard Gaussian elimination in the following way. First, it is symmetric, as in general every column operation is followed by a corresponding row operation. In this way, we maintain the correspondence between rows/columns of the matrix we are manipulating and vertices of a graph. Second, we adhere to a particular order when deciding which entry to use for the next pivot operation. This order is (partially) fixed by the tree decomposition. It is crucial for our proofs of the statements in Sect. 5 that the elimination process proceeds according to this order. Third, we perform symmetric Gaussian elimination using only vertices in a *subset* V' of the vertices: When seeking a pivot entry in a particular row/column, we do not consider all entries of the row/column but only the ones that correspond to edges between vertices in V'.

Eliminating with a Single Vertex. Assume we are given a graph G, its adjacency matrix M, and a vertex v. A symmetric Gaussian elimination step on M using v is defined in the following way:

- If v is an isolated vertex without a self loop, the result of the elimination step is just M (cf. (1) in Fig. 3).
- If v has a self loop, there is a 1 in the (v, v)-entry of M. We add column v and row v to the column and row of every neighbor of v. This transfers the matrix in a form as depicted as (2) in Fig. 3.
- If v is neither isolated nor has a self loop, there is a neighbor u of v. The elimination step is performed in a similar manner as elimination steps using self loops. The result is of a form as (3) in Fig. 3. (We do not swap columns/rows, as we must keep the vertices in a particular order that is determined by the tree decomposition.)

Eliminating with a Set of Vertices. Eliminating with a *set* of vertices V' means that we perform an elimination step as described above for each vertex $v \in V'$. But any edge vu is used for an elimination step only if both, v and u, are in V'. In the setting of Fig. 2, this protects us from using the entries of the $U \times V'$ submatrix for elimination steps.

The order the vertices are processed in is given by a fixed vertex order of the graph computed in the beginning. Also, if an unlooped vertex v has more than one neighbor u in V', the minimum u with respect to the vertex order is chosen for the elimination step.

The vertex order we use is obtained in the following way. Let a graph G and a nice tree decomposition $(\{X_i\}, T)$ of G be given. We traverse the tree decomposition

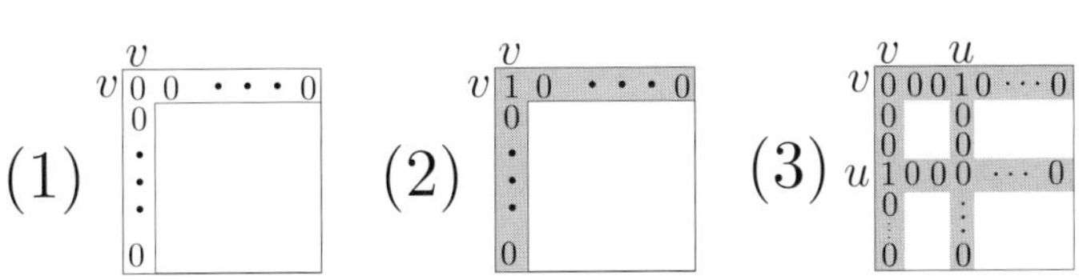

Fig. 3. Effect of a symmetric Gaussian elimination step. Adjacency matrix with isolated unlooped vertex v (1), adjacency matrix after eliminating with a self loop at v (2), adjacency matrix after eliminating with edge vu (3).

bottom up. Whenever we reach a forget node that forgets vertex v, we assign the next free position in the vertex order to v. This ensures that for all extended graphs $G[V_i, X_i]$ the vertices in X_i are greater then the vertices in V_i.

We also order vertex vectors (i.e. elements from $\{0,1\}^U$, U some vertex set) and sets of vertex vectors according to the vertex order (lexicographically). This induced order is used for choosing a "minimal" basis in the following definition.

Definition 5 (Scenario of an extended graph). *Let $G[V', U]$ be an extended graph obtained by extending $G[V']$ by U according to graph $G = (V, E)$. Let the vertex order be such that $v' < u$ for all $v' \in V'$ and $u \in U$. Then the scenario* $\mathrm{scen}(G[V', U])$ *of $G[V', U]$ is defined as follows: Let M be the adjacency matrix of $G[V' \cup U]$. Perform symmetric Gaussian elimination on M using V' to obtain M'. Let $M'_{UV'}$ be the $U \times V'$ submatrix of M'. Consider the column space W of $M'_{UV'}$. We can choose a basis of W from the column vectors of $M'_{UV'}$. Let $s^{U\times V'}$ be the minimal such basis. Let $s^{U\times U}$ be the contents of the $U \times U$ submatrix of M'. We define* $\mathrm{scen}(G[V', U])$ *to be $(s^{U\times V'}, s^{U\times U})$.*

5 Scenarios and Nice Tree Decompositions

In this section we will collect lemmas that allow us to compute the parts of a join, forget, and introduce node from the parts of its children nodes (child node, resp.).

Lemma 6 (Join). *Let $G = (V, E)$ be a graph, $U \subseteq V$, and s_1, s_2 two scenarios of U. Then there is a unique scenario s_3 of U such that the following holds: If $G[V_1]$ and $G[V_2]$ are disjoint subgraphs of G that may be extended by U according to G,* $\mathrm{scen}(G[V_1, U]) = s_1$, *and* $\mathrm{scen}(G[V_2, U]) = s_2$, *then* $\mathrm{scen}(G[V_1 \cup V_2, U]) = s_3$. *Moreover, s_3 can be computed from s_1, s_2 and $G[U]$ within* $\mathsf{poly}(|U|)$ *steps.*

Definition 7. *In the situation of Lemma 6 we write $s_{\mathrm{join}}(s_1, s_2, G[U])$ for s_3.*

Lemma 8 (Introduce vertex). *Let $G = (V, E)$ be a graph, $U \subseteq V$, s a scenario of U, $u \in V \setminus U$. Then there is a unique scenario $\tilde{s}$ of $\tilde{U} = U \cup \{u\}$ such that the following holds: If $G[V']$ may be extended by $\tilde{U}$ according to G, u is not connected to V' in G, and* $\mathrm{scen}(G[V', U]) = s$, *then* $\mathrm{scen}(G[V', \tilde{U}]) = \tilde{s}$. *Moreover, $\tilde{s}$ can be computed from s and $G[\tilde{U}]$ in* $\mathsf{poly}(|U|)$ *steps.*

Definition 9. *In the situation of Lemma 8 we write $s_{\mathrm{introduce}}(s, u, G[\tilde{U}])$ for $\tilde{s}$.*

Lemma 10 (Forget vertex). *Let $G = (V, E)$ be a graph, $u \in U \subseteq V$, $\tilde{U} = U \setminus \{u\}$, $\tilde{V} = V' \cup \{u\}$, and s a scenario of U. Then there is a unique scenario $\tilde{s}$ of $\tilde{U}$ and $r, n \in \{0, 1, 2\}$ such that the following holds: If $G[V']$ is a subgraph of G that may be extended by U according to G, $u > v'$ for all $v' \in V'$, and* $\mathrm{scen}(G[V', U]) = s$, *then* $\mathrm{scen}(G[\tilde{V}, \tilde{U}]) = \tilde{s}$ *and the rank (nullity) of the adjacency matrix of $G[\tilde{V}]$ equals the rank (nullity, resp.) of the adjacency matrix of $G[V']$ plus r (n, resp.). Moreover, $\tilde{s}$ and r can be computed from s and $G[U]$ in* $\mathsf{poly}(|U|)$ *steps, and we have $n = 2 - r$.*

Definition 11. *In the situation of Lemma 10 we write $s_{\mathrm{forget}}(s, u, G[U])$ for $\tilde{s}$, $\Delta r_{\mathrm{forget}}(s, u, G[U])$ for r, and $\Delta n_{\mathrm{forget}}(s, u, G[U])$ for n.*

The operation defined in Definition 11 deletes a vertex u from a scenario in the sense that u is deleted from the extension but added to the graph being extended. We also need a notation for deleting a vertex completely from a scenario, i. e. ignoring some vertex of the extension.

Definition 12. *Let $s = (s^{U\times V'}, s^{U\times U})$ be a scenario of an extension U and $u \in U$. Then $s_{\text{ignore}}(s,u)$ is the scenario obtained from s in the following way: Delete the u-components from the elements of $s^{U\times V'}$ to obtain s_1. Choose the minimum (according to the vertex order) basis s_1' for the span of s_1 from the elements of s_1 using standard Gaussian elimination. Delete the u-column and u-row from $s^{U\times U}$ to obtain s_2. We define $s_{\text{ignore}}(s,u) = (s_1', s_2)$.*

6 The Algorithm

Algorithm 1 evaluates the interlace polynomial using a tree decomposition. The input for the algorithm is $G = (V,E)$, the graph of which we want to evaluate the interlace polynomial, and a nice tree decomposition $(\{X_i\}_I, (I,F))$ of G with $O(n)$ nodes, $n = |V|$. In Sect. 2 we discussed how to obtain a nice tree decomposition. Let $k-1$ be the width of the tree decomposition, i.e. k is the maximum bag size.

Algorithm 1 essentially traverses the tree decomposition bottom-up and computes parts $S(i,D,s)$ of the interlace polynomial for each node i. One such part is defined in the following way:

$$\begin{aligned} S(i,D,s) = \sum \{ & x_A y_B u^{\text{rk}((G_i\nabla B)[A\cup B])} v^{\text{n}((G_i\nabla B)[A\cup B])} \mid \\ & A, B \subseteq V_i,\ A\cap B = \emptyset,\ \text{scen}(G'[A\cup B, X_i]) = s, \\ & \text{where } G' = G\nabla(B\cup D)\}, \end{aligned} \tag{1}$$

where $D \subseteq X_i$ and s is a scenario of X_i. Recall that we write $\sum S$ for the sum of all the elements in S and that V_i is the set of vertices that have been forgotten below node i. Thus, $S(i,D,s)$ is the part of the interlace polynomial of $G[V_i]$ corresponding to D and s. Equation (1) shows that $S(i,\emptyset,((),())) = 1$ for leaves i. The value $S(r,\emptyset,((),()))$, where r is the root node, is just the interlace polynomial of G.

The parts of join and forget nodes are computed by Algor. 2 and 3. The parts of an introduce node can be computed similarly.

Running Time. There are $O(n)$ nodes i in the tree decomposition and for each of it at most 2^k subsets D of X_i. The number of scenarios (Line 2 of Algor. 2, Lines 2 and 6 of Algor. 3) is dominated by the number of *pairs* of scenarios (Line 8 of Algor. 2). By Lemma 4, there are at most $(2^{(3k+1)k/2})^2$ such pairs. Converting the scenarios (Line 9 of Algorithm 2 and Lines 7, 10 and 14 of Algorithm 3) takes time polynomial in k (Sect. 5). Thus, the running time is at most

$$O(n)\cdot 2^k \cdot 2 \cdot (2^{(3k+1)k/2})^2 \cdot \mathsf{poly}(k).$$

Algorithm 1. Evaluating the interlace polynomial using a tree decomposition.

Input: Graph G, nice tree decomposition $(\{X_i\}_i, (I, F))$ of G, k such that any bag X_i of the tree decomposition contains at most k vertices

Compute a vertex order as described in Sect. 4
for all nodes i of the tree decomposition, in the order they appear in bottom-up traversal **do**
  for all $D \subseteq X_i$ **do**
    if i is a leaf **then**
      $S(i, D, ((), ())) \leftarrow 1$
    else if i is a join node **then**
      JOIN(i, D)
    else if i is an introduce node **then**
      INTRODUCE(i, D)
    else if i is a forget node **then**
      FORGET(i, D)
return $S(\text{root}, \emptyset, ((), ()))$ ▷ $X_{\text{root}} = \emptyset$

Algorithm 2. Computing the parts at a join node.

procedure JOIN(i, D)
  for all scenarios s for $|X_i|$ vertices **do**
    ▷ i.e., enumerate all pairs $s = (s^{X_i \times V'}, s^{X_i \times X_i})$ with $s^{X_i \times V'}$ being a list of linear independent vectors from $\{0,1\}^{X_i}$ and $s^{X_i \times X_i}$ a symmetric $X_i \times X_i$ matrix with entries from $\{0,1\}$ – cf. Def. 3
    $S(i, D, s) \leftarrow 0$
  $(j_1, j_2) \leftarrow$ (left child of i, right child of i)
  for all scenarios s_1, s_2 for $|X_i|$ vertices **do**
    $s \leftarrow s_{\text{join}}(s_1, s_2, G \nabla D[X_i])$ ▷ Definition 7
    $S(i, D, s) \leftarrow S(i, D, s) + S(j_1, D, s_1) \cdot S(j_2, D, s_2)$

Algorithm 3. Computing the parts at a forget node.

procedure FORGET(i, D)
  for all scenarios s for $|X_i|$ vertices **do**
    $S(i, D, s) \leftarrow 0$
  $j \leftarrow$ child of i
  $a \leftarrow$ vertex being forgotten in X_i
  for all scenarios s' for $|X_j|$ vertices **do**
    $s \leftarrow s_{\text{ignore}}(s', a)$ ▷ Definition 12
    $S(i, D, s) \leftarrow S(i, D, s) + S(j, D, s')$
    $G' \leftarrow G \nabla D[X_j]$
    $s \leftarrow s_{\text{forget}}(s', a, G')$ ▷ Definition 11
    $S(i, D, s) \leftarrow S(i, D, s) + x_a u^{\Delta r_{\text{forget}}(s', a, G')} v^{\Delta n_{\text{forget}}(s', a, G')} S(j, D, s')$
    $D' \leftarrow D \cup \{a\}$
    $G' \leftarrow G \nabla D'[X_j]$
    $s \leftarrow s_{\text{forget}}(s', a, G')$
    $S(i, D, s) \leftarrow S(i, D, s) + y_a u^{\Delta r_{\text{forget}}(s', a, G')} v^{\Delta n_{\text{forget}}(s', a, G')} S(j, D', s')$

Theorem 13. *Let $G = (V, E)$ be a graph with n vertices. Let a nice tree decomposition of G with $O(n)$ nodes and width k be given, as well as numbers u, v and, for each $a \in V$, x_a, y_a. Then Algorithm 1 evaluates the multivariate interlace polynomial $C(G)$ at $((x_a)_{a\in V}, (y_a)_{a\in V}, u, v)$ using $2^{3k^2+O(k)} \cdot n$ arithmetic operations. If the bit length of u, v, and $x_a, y_a, a \in V$, is at most ℓ, the operands occurring during the computation are of bit length $O(\ell n)$.*

Proof. The time bound stems from the discussion above. The statement about operand length follows as the degree of the interlace polynomial is at most n in each variable.

Acknowledgement. We would like to thank Bruno Courcelle and the anonymous referees for their helpful comments on previous versions of the paper, which led to a reduction of the running time bound.

References

[1] Arratia, R., Bollobás, B., Coppersmith, D., Sorkin, G.B.: Euler circuits and DNA sequencing by hybridization. Discrete Appl. Math. 104(1-3), 63–96 (2000)
[2] Arratia, R., Bollobás, B., Sorkin, G.B.: The interlace polynomial of a graph. J. Comb. Theory Ser. B 92(2), 199–233 (2004)
[3] Martin, P.: Enumérations Eulériennes dans le multigraphes et invariants de Tutte–Grothendieck. PhD thesis, Grenoble, France (1977)
[4] Las Vergnas, M.: Le polynôme de martin d'un graphe eulerian. Ann. Discrete Math. 17, 397–411 (1983)
[5] Las Vergnas, M.: Eulerian circuits of 4-valent graphs imbedded in surfaces. In: Algebraic Methods in Graph Theory, Szeged, Hungary, 1978. Colloq. Math. Soc. János Bolyai, vol. 25, pp. 451–477. North-Holland, Amsterdam (1981)
[6] Las Vergnas, M.: On the evaluation at (3,3) of the Tutte polynomial of a graph. J. Comb. Theory Ser. B 45(3), 367–372 (1988)
[7] Jaeger, F.: On Tutte polynomials and cycles of plane graphs. J. Comb. Theory Ser. B 44(2), 127–146 (1988)
[8] Ellis-Monaghan, J.A.: New results for the Martin polynomial. J. Comb. Theory Ser. B 74(2), 326–352 (1998)
[9] Ellis-Monaghan, J.A.: Martin polynomial miscellanea. In: Proceedings of the 30th Southeastern International Conference on Combinatorics, Graph Theory, and Computing, Boca Raton, FL, pp. 19–31 (1999)
[10] Bollobás, B.: Evaluations of the circuit partition polynomial. J. Comb. Theory Ser. B 85(2), 261–268 (2002)
[11] Bouchet, A.: Isotropic systems. Eur. J. Comb. 8(3), 231–244 (1987)
[12] Bouchet, A.: Graphic presentations of isotropic systems. J. Comb. Theory Ser. B 45(1), 58–76 (1988)
[13] Bouchet, A.: Tutte Martin polynomials and orienting vectors of isotropic systems. Graphs Combin. 7, 235–252 (1991)
[14] Bénard, D., Bouchet, A., Duchamp, A.: On the Martin and Tutte polynomials. Technical report, Département d'Infornmatique, Université du Maine, Le Mans, France (1997)
[15] Ellis-Monaghan, J.A., Sarmiento, I.: Distance hereditary graphs and the interlace polynomial. Comb. Probab. Comput. 16(6), 947–973 (2007)
[16] Aigner, M., van der Holst, H.: Interlace polynomials. Linear Algebra and its Applications 377, 11–30 (2004)

[17] Danielsen, L.E., Parker, M.G.: Interlace polynomials: Enumeration, unimodality, and connections to codes (2008) preprint, arXiv:0804.2576v1

[18] Bouchet, A.: Graph polynomials derived from Tutte–Martin polynomials. Discrete Mathematics 302(1-3), 32–38 (2005)

[19] Ellis-Monaghan, J.A., Sarmiento, I.: Isotropic systems and the interlace polynomial (2006) preprint, arXiv:math/0606641v2

[20] Courcelle, B.: A multivariate interlace polynomial and its computation for graphs of bounded clique-width. The Electronic Journal of Combinatorics 15(1) (2008)

[21] Arratia, R., Bollobás, B., Sorkin, G.B.: A two-variable interlace polynomial. Combinatorica 24(4), 567–584 (2004)

[22] Traldi, L.: Weighted interlace polynomials (2008) preprint, arXiv:0808.1888v1

[23] Riera, C., Parker, M.G.: One and two-variable interlace polynomials: A spectral interpretation. In: Ytrehus, Ø. (ed.) WCC 2005. LNCS, vol. 3969, pp. 397–411. Springer, Heidelberg (2006)

[24] Bläser, M., Hoffmann, C.: On the complexity of the interlace polynomial. In: Albers, S., Weil, P. (eds.) 25th International Symposium on Theoretical Aspects of Computer Science (STACS 2008), Dagstuhl, Germany, Internationales Begegnungs- und Forschungszentrum für Informatik (IBFI), Schloss Dagstuhl, Germany, pp. 97–108 (2008)

[25] Courcelle, B., il Oum, S.: Vertex-minors, monadic second-order logic, and a conjecture by seese. J. Comb. Theory, Ser. B 97(1), 91–126 (2007)

[26] Courcelle, B., Olariu, S.: Upper bounds to the clique width of graphs. Discrete Applied Mathematics 101(1-3), 77–114 (2000)

[27] Courcelle, B., Makowsky, J.A., Rotics, U.: On the fixed parameter complexity of graph enumeration problems definable in monadic second-order logic. Discrete Applied Mathematics 108(1-2), 23–52 (2001)

[28] Bläser, M., Hoffmann, C.: Fast computation of interlace polynomials on graphs of bounded treewidth (2009) preprint, arXiv:0902.1693

[29] Andrzejak, A.: An algorithm for the Tutte polynomials of graphs of bounded treewidth. Discrete Mathematics 190(1-3), 39–54 (1998)

[30] Noble, S.D.: Evaluating the Tutte polynomial for graphs of bounded tree-width. Combinatorics, Probability & Computing 7(3), 307–321 (1998)

[31] Bodlaender, H.L., Koster, A.M.C.A.: Combinatorial optimization on graphs of bounded treewidth. Comput. J. 51(3), 255–269 (2008)

[32] Bodlaender, H.L.: A linear-time algorithm for finding tree-decompositions of small treewidth. SIAM Journal on Computing 25(6), 1305–1317 (1996)

[33] Kloks, T.: Treewidth. Computations and Approximations. LNCS, vol. 842. Springer, Heidelberg (1994)

Kernel Bounds for Disjoint Cycles and Disjoint Paths

Hans L. Bodlaender[1], Stéphan Thomassé[2], and Anders Yeo[3]

[1] Institute of Information and Computing Sciences, Utrecht University, Utrecht, the Netherlands
hansb@cs.uu.nl

[2] LIRMM-Université Montpellier II, 161 Rue Ada, 34392 Montpellier Cedex, France
thomasse@lirmm.fr

[3] Departmente of Computer Science, Royal Holloway University of London, Egham, Surrey TW20 OEX, United Kingdom
anders@cs.rhul.ac.uk

Abstract. In this paper, we give evidence for the problems DISJOINT CYCLES and DISJOINT PATHS that they cannot be preprocessed in polynomial time such that resulting instances always have a size bounded by a polynomial in a specified parameter (or, in short: do not have a polynomial kernel); these results are assuming the validity of certain complexity theoretic assumptions. We build upon recent results by Bodlaender et al. [3] and Fortnow and Santhanam [13], that show that NP-complete problems that are or-compositional do not have polynomial kernels, unless $NP \subseteq coNP/poly$. To this machinery, we add a notion of transformation, and thus obtain that DISJOINT CYCLES and DISJOINT PATHS do not have polynomial kernels, unless $NP \subseteq coNP/poly$. We also show that the related DISJOINT CYCLES PACKING problem has a kernel of size $O(k \log k)$.

1 Introduction

In many practical settings, exact solutions to NP-hard problems are needed. A common approach in such cases is to start with a preprocessing or data reduction algorithm: before employing a slow exact algorithm (e.g., ILP, branch and bound), we try to transform the input to an equivalent, smaller input.

Currently, the theory of fixed parameter complexity gives us tools to make a theoretical analysis of such data reduction or preprocessing algorithms. A *kernelisation* algorithm is an algorithm, that uses polynomial time, and transforms an input for a specific problem to an equivalent input whose size is bounded by some function of a parameter. The resulting instance is also called a *kernel.* Questions of both theoretical and practical interests are for a specific problem: does it have a kernel, and if so, how large can this kernel be? An excellent overview of much recent work on kernelisation was made by Guo and Niedermeier [15].

For the question, whether a specific (parameterised) problem has a kernel, the fixed parameter tractability theory introduced by Downey and Fellows gives

A. Fiat and P. Sanders (Eds.): ESA 2009, LNCS 5757, pp. 635–646, 2009.

Table 1. Size of kernels and evidence. ETH = Exponential Time Hypothesis. ADC = And-distillation Conjecture. NP-c = NP-completeness.

size	positive evidence	negative evidence	conjecture
$O(1)$	P-time algorithm	NP-hardness	$P \neq NP$
polynomial	poly-kernel algorithm	compositionality & NP-c **transformations**	$NP \not\subseteq coNP/poly$, ADC
any kernel	$\in$ FPT	$W[1]$-hardness	$FPT \neq W[1]$ ETH

good tools to answer these. We say a problem is fixed parameter tractable (in FPT), if it has an algorithm that runs in time $O(n^c f(k))$, with n the input size, k the parameter, c a constant, and f any function. It is well known that a decidable problem is in FPT, if and only if it has a kernel. (See [7].)

Recently, Bodlaender et al. [3] gave a framework to give evidence that problems (in FPT) do not have a kernel of polynomial size. The framework is based upon the notion of *compositionality*. There are actually two forms: and-compositionality, and or-compositionality. We have a parameterised problem, whose variant as a decision problem is NP-complete, and it is and-compositional, then it does not have a kernel whose size is bounded by a polynomial, unless the and-distillation conjecture does not hold. Similarly for or-compositionality, and the or-distillation conjecture, but in this case, one can use a result by Fortnow and Santhanam, and strengthen the conjecture to $NP \not\subseteq coNP/poly$ [13]. In this paper, we extend the framework by introducing a notion of *transformation*. While the main idea parallels classic notions of transformation, we think that our contribution is a new important tool for the theory of data reduction/kernelisation.We use our framework to show for the following problems that they do not have a kernel of polynomial size unless $NP \subseteq coNP/poly$: Disjoint Cycles, Disjoint Paths. The result for Disjoint Paths is not so surprising, given the very complicated FPT algorithms for this problem [19]. The result for Disjoint Cycles came unexpected to us, also because polynomial kernels are known for the closely related problems: Feedback Vertex Set (see [21]) and Disjoint Cycle Packing (see Section 4).

Concerning the size of a kernel of a parameterised problem, we can summarise the situation in Table 1. Assuming that the problem is decidable, the second and third column give the main available positive or negative, respectively, evidence that the problem has a kernel of the size given in the first column. E.g., $W[1]$-hardness indicates that a problem is not in FPT; a problem is in FPT, if and only if it has a kernel of any size (i.e., bounded by a function of k.)

2 Notions

In this section, we give several results, mostly from [3], and introduce some new notation. We also give some basic notions from fixed parameter tractability, as introduced by Downey and Fellows, see e.g., [11].

A parameterised problem is a subset of $L^* \times \mathbf{N}$ for some finite alphabet L: the second part of the input is called the *parameter*. (We assume here that the second parameter is an integer. It is not hard to modify the techniques such that it works with other types of parameters, e.g., pairs of integers.)

A parameterised problem $Q \subseteq L^* \times \mathbf{N}$ is said to belong to the class FPT (to be *fixed parameter tractable*), if there is an algorithm A, a polynomial p, and a function $f : \mathbf{N} \to \mathbf{N}$, such that A determines for a given pair $(x, k) \in L^* \times \mathbf{N}$ whether $(x, k) \in Q$ in time at most $p(|x|) \cdot f(k)$. ($|x|$ is the length of input x).

A *kernelisation algorithm* for a parameterised problem $Q \subseteq L^* \times \mathbf{N}$ computes a function $K : L^* \times \mathbf{N} \to L^* \times \mathbf{N}$, such that

- For all $(x, k) \in L^* \times \mathbf{N}$, the algorithm takes time polynomial in $|x| + k$.
- For all $(x, k) \in L^* \times \mathbf{N}$: $(x, k) \in Q \Leftrightarrow K(x, k) \in Q$.
- There is a function $g : \mathbf{N} \to \mathbf{N}$, such that for all $(x, k) \in L^* \times \mathbf{N}$: $|K(x, k)| + k \leq g(k)$.

We say that Q has a *kernel of size* g. If g is bounded by a polynomial in k, we say that Q has a *polynomial kernel.*

Sometimes, in the literature one requires that the kernelisation algorithm does not increase the parameter, i.e., when we write $K(x, k) = (x', k')$, then $k' \leq k$. This assumption is not necessary for our results and deleting it slightly strengthens them. Bodlaender et al. [3] give the following two conjectures.

Conjecture 1 (And-distillation conjecture [3]). Let R be an NP-complete problem. There is no algorithm D, that gets as input a series of m instances of R, and outputs one instance of R, such that

- If D has as input m instances, each of size at most n, then D uses time polynomial in m and n, and its output is bounded by a function that is polynomial in n.
- If D has as input instances $x_1, \ldots, x_m$, then $D(x_1, \ldots, x_m) \in R$, if and only if $\forall_{1 \leq i \leq m} x_i \in R$.

Conjecture 2 (Or-distillation conjecture [3]). Let R be an NP-complete problem. There is no algorithm D, that gets as input a series of m instances of R, and outputs one instance of R, such that

- If D has as input m instances, each of size at most n, then D uses time polynomial in m and n, and its output is bounded by a function that is polynomial in n.
- If D has as input instances $x_1, \ldots, x_m$, then $D(x_1, \ldots, x_m) \in R$, if and only if $\exists_{1 \leq i \leq m} x_i \in R$.

Note that in these definitions, the output of D cannot be polynomial in m, and thus the algorithm D should map an input of size $O(mn)$ to an input of size polynomial in n only.

Theorem 1 (Fortnow and Santhanam [13]). *If the or-distillation conjecture does not hold, then $NP \subseteq coNP/poly$.*

There is no equivalent to Theorem 1 known for and-distillation. This is an important open problem in this area.

The main tool to give evidence for the non-existence of polynomial kernels for specific parameterised problems from [3] is the notion of compositionality. Compositionality allows us to build one instance from a collection of instances. There are two different notions: and-compositionality and or-compositionality. In the first case, the new instance is a yes-instance, if and only if each instance in the collection is a yes-instance; in the second case, this happens, if and only if at least one instance in the collection is a yes-instance.

An *and-composition* algorithm for a parameterised problem $Q \subseteq L^* \times \mathbf{N}$ is an algorithm, that gets as input a sequence $((x_1, k), \ldots, (x_r, k))$, with each $(x_i, k) \in L^* \times \mathbf{N}$, and outputs a pair (x', k'), such that

- the algorithm uses time polynomial in $\sum_{1 \leq i \leq r} |x_i| + k$;
- k' is bounded by a polynomial in k;
- $(x', k') \in Q$, if and only if for all i, $1 \leq i \leq r$, $(x_i, k) \in Q$.

The definition for *or-composition* is identical, except that the last condition becomes:

- $(x', k') \in Q$, if and only if there exists an i, $1 \leq i \leq r$, $(x_i, k) \in Q$.

Note that we allow the first argument x' of the output of a composition algorithm to be polynomial both in r and $\max_{1 \leq i \leq r} |x_i|$. A difference with the notion of distillation is that there, we assume no dependency on the number of inputs. Indeed, many problems have an and- or or-composition algorithm, but we expect that there is no distillation algorithm for the derived classic variant.

Many problems have natural composition algorithms. For many graph problems, the only operation needed is the disjoint union of connected components. E.g., consider the LONGEST CYCLE problem: does G have a cycle of length at least k? As a graph has a cycle of length at least k, if and only if at least one of its connected components has such a cycle, the problem is trivially or-compositional.

We need one further notion: for a parameterised problem, we have the *derived classic problem*. Formally, if $R \subseteq L^* \times \mathbf{N}$ is a parameterised problem, we take a symbol $\mathbf{1} \notin L$, and take as derived classic problem the set $\{x\mathbf{1}^k \mid (x, k) \in R\}$. Here, we associate in a natural way a classic one-argument input problem with a parameterised problem; note that we assume that the parameter is given in unary. For instance, the DISJOINT CYCLES problem as parameterised problem belongs to FPT, and its derived classic problem is NP-complete. Often, we use the same name for the derived classic problem as for the parameterised version. We now give here some results from [3] and other papers, and introduce some notation (the classes NPK^0_{or} and NPK^0_{and}) that will be helpful for further presentation of the results.

Definition 1. *The class NPK^0_{and} is the class of parameterised problems, that are and-compositional and whose derived classical problem is NP-complete.*

Definition 2. *The class NPK^0_{or} is the class of parameterised problems, that are or-compositional and whose derived classical problem is NP-complete.*

Theorem 2 (Bodlaender et al. [3]).

1. *If a problem in the class NPK^0_{and} has a polynomial kernel, then the and-distillation conjecture does not hold.*
2. *If a problem in the class NPK^0_{or} has a polynomial kernel, then the or-distillation conjecture does not hold.*

As a corollary of Theorems 2 and Theorem 1, we have

Corollary 1 (Bodlaender et al. [3], Fortnow and Santhanam [13]). *If a problem in the class NPK^0_{or} has a polynomial kernel, then $NP \subseteq coNP/poly$.*

In turn, $NP \subseteq coNP/poly$ would imply a collapse of the polynomial time hierarchy to the third level. Currently, there is no equivalent result to Theorem 1 known for the and-distillation conjecture.

3 Polynomial Time and Parameter Transformations

We now introduce a notion of transformation, that allows us to prove results for problems that do not obviously have compositionality.

Definition 3. *Let P and Q be parameterised problems. We say that P is polynomial time and parameter reducible to Q, written $P \leq_{Ptp} Q$, if there exists a polynomial time computable function $f : \{0,1\}^* \times \mathbf{N} \to \{0,1\}^* \times \mathbf{N}$, and a polynomial $p : \mathbf{N} \to \mathbf{N}$, and for all $x \in \{0,1\}^*$ and $k \in \mathbf{N}$, if $f((x,k)) = (x',k')$, then $(x,k) \in P$, if and only if $(x',k') \in Q$, and $k' \leq p(k)$. We call f a polynomial time and parameter transformation from P to Q.*

If P and Q are parameterised problems, and P^c and Q^c are the derived classical problems, then f can also be used as a polynomial time transformation (in the usual sense of the theory of NP-completeness) from P^c to Q^c. As an additional condition to polynomial time transformations, we have that the size of the parameter can grow at most polynomially. Note that the fixed parameter reductions by Downey and Fellows (see e.g., [9,10,11]) are similar, but allow non-polynomial growth of the parameter, and are used for a different purpose: to show hardness for $W[1]$ or a related class.

Theorem 3. *Let P and Q be parameterised problems, and suppose that P^c and Q^c are the derived classical problems. Suppose that P^c is NP-complete, and $Q^c \in NP$. Suppose that f is a polynomial time and parameter transformation from P to Q. Then, if Q has a polynomial kernel, then P has a polynomial kernel.*

Note that the conditions of Theorem 3 imply that Q^c is NP-complete, as f is also a "classic" polynomial time transformation. The proof follows standard lines of arguments and is omitted here.

Corollary 2. 1. *Let P and Q be parameterised problems, and suppose that P^c and Q^c are the derived classical problems. Suppose that P^c and Q^c are NP-complete, that P is and-compositional, and that $P \leq_{Ptp} Q$. If Q has a polynomial kernel, then the and-distillation conjecture does not hold.*

2. *Let P and Q be parameterised problems, and suppose that P^c and Q^c are the derived classical problems. Suppose that P^c and Q^c are NP-complete, that P is or-compositional, and that $P \leq_{Ptp} Q$. If Q has a polynomial kernel, then the or-distillation conjecture does not hold, and thus $coNP \subseteq NP/poly$, and thus the polynomial time hierarchy collapses to the third level.*

We now define the classes NPK_{or} and NPK_{and} as the closures of NPK^0_{or} and NPK^0_{and} under polynomial time and parameter transformations. Membership in these classes gives a strong indication that there is no polynomial kernel for the problem. We can reformulate the discussion above as (using results from [3] and [13] and our own observations):

Corollary 3. 1. *If parameterised problem $P \in NPK_{or}$, then P has no polynomial kernel, unless $coNP \subseteq NP/poly$.*
2. *If parameterised problem $P \in NPK_{and}$, then P has no polynomial kernel, unless the and-distillation conjecture does not hold.*

The polynomial time and parameter transformations thus give us a nice method to show unlikeliness of the existence of polynomial kernels.

4 A Small Kernel for Disjoint Cycle Packing

We first give a short proof that the Disjoint Cycle Packing problem has a polynomial kernel. The current proof is due to Saket Saurabh [20]. In the Disjoint Cycle Packing we are given a graph $G = (V, E)$ with an integer parameter k, and ask if G contains at least k edge disjoint cycles.

Theorem 4. Disjoint Cycle Packing *has a kernel with $O(k \log k)$ vertices.*

We first reduce the graph, by deleting all vertices of degree 0 and 1, and by contracting each vertex of degree 2 to a neighbour. Clearly, these operations do not change the number of edge disjoint cycles in the graph. By Lemma 1, there is a constant c, such that if the resulting graph (which has minimum degree at least three) has at least $ck \log k$ vertices, we can decide positively.

Lemma 1. *Let G be a graph with n vertices with minimum degree at least three. Then G has $\Omega(n/ \log n)$ edge disjoint cycles.*

Proof. Alon et al. [1] showed that a graph G with average degree d and n vertices has a cycle of length at most $2(\log_{d-1} n)+2$. Now, consider the greedy algorithm, where we repeatedly choose a minimum length cycle that is edge disjoint from any other chosen cycle. Run this algorithm while there are at least $4n/3$ edges not on one of these cycles. At each point, the average degree is at least $8/3$, and hence each chosen cycle contains at most $O(\log n)$ edges. So we find $\Omega(n/ \log n)$ edge disjoint cycles. □

It is not hard to strengthen the proof slightly, and obtain a kernel with $O(k \log k)$ vertices and $O(k \log k)$ edges. A related result is a kernel for the version where each cycle must be of length exactly three (Edge Disjoint Triangle Packing) by Mathieson et al. [18].

5 No Polynomial Kernels for Disjoint Cycles and Disjoint Paths

The Disjoint Cycles problem has as input an undirected graph $G = (V, E)$ and integer parameter k, and asks if G contains at least k vertex disjoint cycles. This problem is strongly related to the Feedback Vertex Set problem, which has a kernel of size $O(k^2)$ [21] by Thomassé, who improved upon a kernel of size $O(k^3)$ [2]. Feedback Vertex Set, Disjoint Cycles and Disjoint Cycle Packing have linear kernels, when restricted to planar graphs, see [17,4,5]. We also consider the following well known problem, also known as k-Linkage.

Disjoint Paths
Input: Undirected graph $G = (V, E)$, vertices $s_1, \ldots, s_k, t_1, \ldots, t_k \in V$
Parameter: k
Question: Is there a collection of k paths $P_1, \ldots, P_k$ that are vertex disjoint, such that P_i is a path from s_i to t_i?

The result that the Disjoint Paths problem is fixed parameter tractable is a famous result by Robertson and Seymour as part of their fundamental work on graph minors: in [19], they show that for each fixed k, the problem can be solved in $O(n^3)$ time. It is well known that the derived classic variants of Disjoint Cycles and Disjoint Paths are NP-complete, see [14,16].

We introduce a new problem, which is used as an intermediate problem: we show that it is or-compositional and (its derived classic variant) NP-complete, and then give a polynomial time and parameter transformation to Disjoint Cycles, respectively Disjoint Paths.

Let L_k be the alphabet consisting of the letters $\{1, 2, \ldots, k\}$. We denote by L_k^* the set of words on L_k. A *factor* of a word $w_1 \cdots w_r \in L_k^*$ is a substring $w_i \cdots w_j \in L_k^*$, with $1 \leq i < j \leq r$, which starts and ends with the same letter, i.e., the factor has length at least two and $w_i = w_j$.

A word $W \in L_k^*$ has the *disjoint factor property* if one can find disjoint factors $F_1, \ldots, F_k$ in W such that the factor F_i starts and ends by the letter i. Observe that the difficulty lies in the fact that the factors F_i do not necessarily appear in increasing order, otherwise detecting them would be obviously computable in $O(n)$, where n is the length of W. We now introduce the parameterised problem Disjoint Factors.

The input of the Disjoint Factors problem is an integer $k \geq 1$ and a word W of L_k^*. The output is true if W has the disjoint factor property, otherwise false. This problem is clearly FPT since one can try all the $k!$ possible orders of the F_i's, and compute each of them linearly. A slightly more involved dynamic programming algorithm gives an $O(nk \cdot 2^k)$ algorithm.

Proposition 1. *The* Disjoint Factors *problem can be solved in* $O(nk \cdot 2^k)$ *time.*

Theorem 5. *The* Disjoint Factors *is NP-complete.*

Proof. Clearly, the problem belongs to NP. We show NP-hardness by a transformation from 3-satisfiability. Let F be a 3-SAT formula with $c + 1$ clauses

$C_0, \ldots, C_c$. We start our construction of our word W with the prefix 1231231234 56456456... $(3c+1)(3c+2)(3c+3)(3c+1)(3c+2)(3c+3)(3c+1)(3c+2)(3c+3)$. Here the factor $(3i+1)(3i+2)(3i+3)(3i+1)(3i+2)(3i+3)(3i+1)(3i+2)(3i+3)$ corresponds to the clause C_i, for all $i = 0, \ldots, c$.

Note that the string 123123123 does not have the disjoint factor property, but fails it only by one. Indeed, one can find two disjoint factors $\{F_1, F_2\}$, $\{F_1, F_3\}$, or $\{F_2, F_3\}$, but not three disjoint factors F_1, F_2, F_3. Hence, in this prefix of W, one can find two disjoint factors for each clause, but not three.

Each variable appearing in F is a letter of our alphabet. In addition to W, we concatenate for each variable x a factor to W of the form $xp_1p_1p_2p_2 \ldots p_lp_lxq_1 q_1 \ldots q_mq_mx$ where the p_i's are the position in which x appears as a positive literal, and the q_i's are the position in which x appears as a negative literal.

We feel that an example will clarify our discourse. To the formula $F = (\overline{x} \vee \overline{y} \vee z) \wedge (y \vee \overline{x} \vee \overline{z}) \wedge (x \vee y \vee z)$, we associate the word W_F

$123123123456456456789789789x77x1155xy4488y22yz3399z66z$

Observe that the solution $x = 1, y = 0, z = 0$ which satisfies F corresponds to the disjoint factors appearing in this order in W_F: 1231, 3123, 4564, 5645, 8978, 9789, 77, $x1155x$, $y4488y$, 22, $z3399z$, 66.

Finally, F is satisfiable iff W_F has the disjoint factor property. This proves our result. □

Lemma 2. Disjoint Factors *is or-compositional.*

Proof. Suppose a collection of inputs $(W_1, k), \ldots, (W_t, k)$ for Disjoint Factors is given.

First we look at the case that $t > 2^k$. In this case, we solve each instance by the dynamic programming algorithm of Proposition 1. Note that the time to do this is polynomial in $\sum_1^t |W_i| + k$, as $2^k < \sum_1^t |W_i|$ here. So, we completely solve the problem, and can then transform to a trivial $O(1)$-size yes- or no-instance.

Now, suppose $t \leq 2^k$. We can assume that t is a power of two, say $t = 2^\ell$; if necessary, we add trivial no-instances ($k > 0$, W the empty string). For $1 \leq i \leq t$, $0 \leq j < \ell$, $i + 2^{j+1} \leq t + 1$, we define the word $W_{i,j}$ recursively as follows. If $j = 0$, $W_{i,0}$ is the word $(k+1)W_i(k+1)W_{i+1}(k+1)$. If $j > 0$, $W_{i,j}$ is the word $(k+1+j)W_{i,j-1}(k+1+j)W_{i+2^j,j-1}(k+1+j)$.

Note that $W_{i,j}$ contains each of the instances $W_i, \ldots, W_{i+2^{j+1}-1}$ as substrings. As result of the composition, we take the word $W' = W_{1,\ell-1}$.

In other words, W' is the limit word of

- $(k+1)W_1(k+1)W_2(k+1)$
- $(k+2)(k+1)W_1(k+1)W_2(k+1)(k+2)(k+1)W_3(k+1)W_4(k+1)(k+2)$
- $(k+3)(k+2)(k+1)W_1(k+1)W_2(k+1)(k+2)(k+1)W_3(k+1)W_4(k+1)(k+2)(k+3)(k+2)(k+1)W_5(k+1)W_6(k+1)(k+2)(k+1)W_7(k+1)W_8(k+1)(k+2)(k+3)$ ...

The resulting instance is $(W', k+\ell)$.

By construction, (W', k') has a solution, if and only if at least one of the (W_i, k) has a solution. Suppose (W', k') has a solution. Then, there are two possibilities for the factor $F_{k+\ell}$: either it is $(k+\ell)W_{1,\ell-2}(k+\ell)$, or $(k+\ell)$ $W_{1+2^{\ell-1},\ell-2}(k+\ell)$. In the first case, it 'shields' the instances $W_1, \ldots, W_{2^{\ell-1}}$; in the other case, it 'shields' the instances $W_{1+2^{\ell}-1}, \ldots, W_{2^\ell}$. No other factors can be taken in the shielded part. This repeats with the other symbols above k: $F_{k+\ell-1}$ shields half of what was left by $F_{k+\ell}$, and one can see that there remains exactly one substring W_i that does not belong to any of the F_i with $i > k$. In this substring, we must find the factors $F_1, \ldots, F_k$, and thus there is at least one (W_i, k) which has a solution.

Suppose (W_i, k) has a solution. We take from W' the factors $F_1, \ldots, F_k$ from W_i. The other factors can be easily chosen: take factors $F_{k+\ell}$, $F_{k+\ell-1}$, etc., in this order, each time taking the unique possibility which does not overlap already chosen factors.

Finally, note that $k + \ell \leq 2k$, so we have a polynomial time and parameter transformation. □

Corollary 4. Disjoint Factors *belongs to the class* NPK^0_{or}*, and thus has no polynomial kernel, unless* $NP \subseteq coNP/poly$.

We now show that Disjoint Cycles belongs to NPK_{or} (and hence has no polynomial kernel, unless $NP \subseteq coNP/poly$), by giving a polynomial time and parameter transformation from Disjoint Factors to Disjoint Cycles.

Theorem 6. Disjoint Cycles $\in NPK_{or}$*, and hence has no polynomial kernel unless* $NP \subseteq coNP/poly$.

Proof. We use the following polynomial time and parameter transformation from Disjoint Factors. Given an input (W, k) of Disjoint Factors, with $W = w_1 \cdots w_n$ a word in L_k^*, we build a graph $G = (V, E)$ as follows. First, we take n vertices $v_1, \ldots, v_n$, and edges $\{v_i, v_{i+1}\}$ for $1 \leq i < n$, i.e., these vertices form a path of length n. Call this subgraph of G P. Then, for each $i \in L_k$, we add a vertex x_i, and make x_i incident to each vertex v_j with $w_j = i$, i.e., to each vertex representing the letter i.

G has k disjoint cycles, if and only if (W, k) has the requested k disjoint factors.

Suppose G has k disjoint cycles $c_1, \ldots, c_k$. As P is a path, each of these cycles must contain at least one vertex not on P, i.e., of the form x_j, and hence each of these cycles contains exactly one vertex x_j. For $1 \leq j \leq k$, the cycle c_j thus consists of x_j and a subpath of P. This subpath must start and end with a vertex incident to x_j. These both represent letters in W equal to j. Let F_j be the factor of W corresponding to the vertices on P in c_j. Now, $F_1, \ldots, F_k$ are disjoint factors, each of length at least two (as the cycles have length at least three), and F_j starts and ends with j, for all j, $1 \leq j \leq k$.

Conversely, if we have disjoint factors $F_1, \ldots, F_k$ with the properties as in the Disjoint Factors problem, we build k vertex disjoint cycles as follows: for each j, $1 \leq j \leq k$, take the cycle consisting of x_j and the vertices corresponding to factor F_j.

As Disjoint Cycles in an NP-complete problem, the transformation just given and the membership of Disjoint Factors in NPK^0_{or} show that Disjoint Cycles $\in NPK_{or}$.

By Corollary 1, it follows that Disjoint Cycles has no polynomial kernel unless $NP \subseteq coNP/poly$. □

A simple modification of the proof above gives the result for Disjoint Paths.

Theorem 7. Disjoint Paths $\in NPK_{or}$, *and hence has no polynomial kernel unless* $NP \subseteq coNP/poly$.

Proof. Suppose we have input (W, k) for the Disjoint Factors problem, W a word in L_k^*. Build a graph G as follows. Start with a path P with vertices $v_1, \ldots, v_n$ (as in the previous proof). For each $i \in L_k$, take a new vertex x_i and a new vertex y_i. Make x_i incident to each vertex representing the first, third, fifth, etc., occurrence of the letter i in W, and make y_i incident to each vertex representing the second, fourth, sixth, etc., occurrence of the letter i in W. As instance of Disjoint Paths we take G with the pairs (x_i, y_i), $i \in L_k$.

Now, the Disjoint Factors problem has a solution, if and only if it has a solution where each factor F_i starts and ends with an i but has no other i (otherwise we can replace the solution by an equivalent one where F_i is shorter), and hence is between an even and an odd occurrence of the letter i. With a proof, similar to the proof of Theorem 6, correctness of the construction follows. □

6 Conclusions

In this paper, we showed that the Disjoint Cycles and Disjoint Paths problems do not have polynomial kernels, assuming that $NP \not\subseteq coNP/poly$. The result for Disjoint Cycles came unexpected to us, given the similarity of the problem to Feedback Vertex Set and Disjoint Cycle Packing, which do have polynomial kernels. Our initial expectation that techniques for these problems would carry over to Disjoint Cycles proved to be false. Thus, the result in our paper for e.g. Disjoint Cycles plays the role that it can direct further research efforts, i.e., it appears not to be useful to aim at finding a polynomial kernel for the problem; this is somewhat comparable to stating that an NP-completeness proof directs our research efforts away from finding a polynomial time algorithm for a problem.

Transformations are a powerful mechanism to derive no-polynomial-kernel results. In a longer and earlier version of this paper [6], we showed that Hamiltonian Path and Hamiltonian Circuit parameterised by treewidth has no polynomial kernel, unless the AND-distillation conjecture does not hold. Fernau et al. [12] show that k-Leaf-Out-Branching has no polynomial kernel, unless $NP \subseteq coNP/poly$. A large collection of no-polynomial-kernel results were obtained by Dom et al. [8], using intricate and clever reductions.

The further development of the theory of kernel sizes is an interesting topic for further research. An important topic with many recent results (see e.g., the overview paper by Guo and Niedermeier [15]) is to find kernels of sizes as small as possible for concrete combinatorial problems. Some questions we want to add to this are: The theory so far allows to distinguish between constant size, polynomial size, and any size kernels: can we refine this? Are the classes NPK_{and} and NPK_{or} the same? (There are some problems that belong to both classes.) Are there complete or "hardest" problems for these classes? Another still open problem is whether there exists a result for and-distillation that is similar to the result of Fortnow and Santhanam for the or-distillation conjecture, i.e., can we relate the and-distillation conjecture to more widely known and believed complexity theoretic conjectures?

Finally, we mention a few concrete open problems: does Disjoint Paths have a polynomial kernel when restricted to planar graphs? Is there a polynomial kernel for Feedback Arc Set or Directed Feedback Vertex Set? Or, if not, do these problems have a polynomial kernel when restricted to directed planar graphs?

Acknowledgement

We thank Saket Saurabh for his proof of the kernel for the Disjoint Cycle Packing problem, which greatly improved our earlier, larger kernel.

References

1. Alon, N., Hoory, S., Linial, N.: The Moore bound for irregular graphs. Graphs and Combinatorics 18, 53–57 (2002)
2. Bodlaender, H.L.: A cubic kernel for feedback vertex set. In: Thomas, W., Weil, P. (eds.) STACS 2007. LNCS, vol. 4393, pp. 320–331. Springer, Heidelberg (2007)
3. Bodlaender, H.L., Downey, R.G., Fellows, M.R., Hermelin, D.: On problems without polynomial kernels (extended abstract). In: Aceto, L., Damgård, I., Goldberg, L.A., Halldórsson, M.M., Ingólfsdóttir, A., Walukiewicz, I. (eds.) ICALP 2008, Part I. LNCS, vol. 5125, pp. 563–574. Springer, Heidelberg (2008)
4. Bodlaender, H.L., Penninkx, E.: A linear kernel for planar feedback vertex set. In: Grohe, M., Niedermeier, R. (eds.) IWPEC 2008. LNCS, vol. 5018, pp. 160–171. Springer, Heidelberg (2008)
5. Bodlaender, H.L., Penninkx, E., Tan, R.B.: A linear kernel for the k-disjoint cycle problem on planar graphs. In: Hong, S.-H., Nagamochi, H., Fukunaga, T. (eds.) ISAAC 2008. LNCS, vol. 5369, pp. 306–317. Springer, Heidelberg (2008)
6. Bodlaender, H.L., Thomassé, S., Yeo, A.: Analysis of data reduction: Transformations give evidence for non-existence of polynomial kernels. Technical Report CS-UU-2008-030, Department of Information and Computer Sciences, Utrecht University, Utrecht, The Netherlands (2008)
7. Cai, L., Chen, J., Downey, R.G., Fellows, M.R.: Advice classes of parameterized tractability. Annals of Pure and Applied Logic 84, 119–138 (1997)
8. Dom, M., Lokshtanov, D., Saurabh, S.: Incompressibility through colors and IDs. In: Albers, S., Marchetti-Spaccamela, A., Matias, Y., Nikoletseas, S., Thomas, W. (eds.) ICALP 2009. Part I. LNCS, vol. 5555, pp. 378–389. Springer, Heidelberg (2009)
9. Downey, R.G., Fellows, M.R.: Fixed-parameter tractability and completeness I: Basic results. SIAM J. Comput. 24, 873–921 (1995)

10. Downey, R.G., Fellows, M.R.: Fixed-parameter tractability and completeness II: On completeness for $W[1]$. Theor. Comp. Sc. 141, 109–131 (1995)
11. Downey, R.G., Fellows, M.R.: Parameterized Complexity. Springer, Heidelberg (1999)
12. Fernau, H., Fomin, F.V., Lokshtanov, D., Raible, D., Saurabh, S., Villanger, Y.: Kernel(s) for problems with no kernel: On out-trees with many leaves. In: Albers, S., Marion, J.-Y. (eds.) Proceedings 26th International Symposium on Theoretical Aspects of Computer Science, STACS 2009, pp. 421–432. Schloss Dagstuhl - Leibniz-Zentrum fuer Informatik, Germany (2009)
13. Fortnow, L., Santhanam, R.: Infeasibility of instance compression and succinct PCPs for NP. In: Proceedings of the 40th Annual Symposium on Theory of Computing, STOC 2008, pp. 133–142. ACM Press, New York (2008)
14. Garey, M.R., Johnson, D.S.: Computers and Intractability, A Guide to the Theory of NP-Completeness. W.H. Freeman and Company, New York (1979)
15. Guo, J., Niedermeier, R.: Invitation to data reduction and problem kernelization. ACM SIGACT News 38, 31–45 (2007)
16. Karp, R.M.: On the complexity of combinatorial problems. Networks 5, 45–68 (1975)
17. Kloks, T., Lee, C.M., Liu, J.: New algorithms for k-face cover, k-feedback vertex set, and k-disjoint cycles on plane and planar graphs. In: Kučera, L. (ed.) WG 2002. LNCS, vol. 2573, pp. 282–295. Springer, Heidelberg (2002)
18. Mathieson, L., Prieto, E., Shaw, P.: Packing edge disjoint triangles: A parameterized view. In: Downey, R.G., Fellows, M.R., Dehne, F. (eds.) IWPEC 2004. LNCS, vol. 3162, pp. 127–137. Springer, Heidelberg (2004)
19. Robertson, N., Seymour, P.D.: Graph minors. XIII. The disjoint paths problem. J. Comb. Theory Series B 63, 65–110 (1995)
20. Saurabh, S.: Personal communication (2009)
21. Thomassé, S.: A quadratic kernel for feedback vertex set. In: Proceedings of the 19th Annual ACM-SIAM Symposium on Discrete Algorithms, SODA 2009, pp. 115–119 (2009)

Constant Ratio Fixed-Parameter Approximation of the Edge Multicut Problem

Dániel Marx[1,⋆] and Igor Razgon[2,⋆⋆]

[1] Department of Computer Science and Information Theory, Budapest University of Technology and Economics
dmarx@cs.bme.hu

[2] Cork Constraint Computation Centre, University College Cork
i.razgon@cs.ucc.ie

Abstract. The input of the Edge Multicut problem consists of an undirected graph G and pairs of terminals $\{s_1, t_1\}, \ldots, \{s_m, t_m\}$; the task is to remove a minimum set of edges such that s_i and t_i are disconnected for every $1 \leq i \leq m$. The parameterized complexity of the problem, parameterized by the maximum number k of edges that are allowed to be removed, is currently open. The main result of the paper is a parameterized 2-approximation algorithm: in time $f(k) \cdot n^{O(1)}$, we can either find a solution of size $2k$ or correctly conclude that there is no solution of size k.

The proposed algorithm is based on a transformation of the Edge Multicut problem into a variant of parameterized Max-2-SAT problem, where the parameter is related to the number of clauses that are *not* satisfied. It follows from previous results that the latter problem can be 2-approximated in a fixed-parameter time; on the other hand, we show here that it is W[1]-hard. Thus the additional contribution of the present paper is introducing the first natural W[1]-hard problem that is constant-ratio fixed-parameter approximable.

1 Introduction

The minimum cut problem and its variants are among the most well-studied combinatorial optimization problems. The focus of the paper is Edge Multicut: given a graph G and pairs of vertices $\{s_1, t_1\}, \ldots, \{s_m, t_m\}$, remove a minimum set of edges such that s_i and t_i are disconnected for every $1 \leq i \leq m$. Edge Multicut generalizes the classical $s - t$ cut problem (disconnect s and t) and the Multiway Cut problem (disconnect all the terminals from each other). Edge Multicut can be approximated within a factor of $O(\log m)$ in polynomial time [13] (even in the weighted case where the goal is to minimize the total weight of the removed edges). However, under the Unique Games Conjecture of Khot [17], no constant factor approximation is possible for Edge Multicut [5].

⋆ Supported by the Magyary Zoltán Felsőoktatási Közalapítvány and the Hungarian National Research Fund (OTKA grant 67651).

⋆⋆ Supported by Science Foundation Ireland (Grant Number 05/IN/I886).

A. Fiat and P. Sanders (Eds.): ESA 2009, LNCS 5757, pp. 647–658, 2009.

Parameterized complexity approaches hard computational problems through a multivariate analysis of the running time. Instead of expressing the running time as a function of the input size n only, the running time is expressed as a function of n and k, where k is a well-defined parameter of the input instance. We say that a problem (with a particular parameter k) is *fixed-parameter tractable (FPT)* if it can be solved in time $f(k) \cdot n^{O(1)}$, where f is an arbitrary function depending only on k. Thus we relax polynomial time by allowing exponential (or worse!) dependence on the parameter k. For more background on parameterized complexity, the reader is referred to the monographs [10,12,22].

Edge Multicut on trees is FPT, parameterized by the maximum number k of edges that can be deleted [3,15]. The problem and its vertex-cut version were studied in [14] for other classes of graphs. For general graphs, Edge Multicut is FPT if both k and m are chosen as parameters (i.e, the problem can be solved in time $f(k,m) \cdot n^{O(1)}$) [19,26]. However, it is an open question whether Edge Multicut is FPT in general graphs parameterized by k only. Besides the fundamental nature of the problem, there are other reasons why this question is important. It has been observed that Edge Multicut is equivalent to Fuzzy Cluster Editing, a correlation clustering problem [2,8,1]. Furthermore, it seems that cut problems are important ingredients in the solution of certain parameterized problems. For example, the fixed-parameter tractability of Directed Feedback Vertex Set [6] was a longstanding open question and solving a variant of directed multicut was an important step in its solution.

Recently, it has been proposed that the notion of approximability can be investigated in the framework of fixed-parameter tractability as well [4,7,9,21]. Here we follow this approach and present a parameterized 2-approximation for Edge Multicut: the main result of the paper is an algorithm with running time $f(k) \cdot n^{O(1)}$ that, given an instance of the Edge Multicut problem and an integer k, either finds a solution of size $2k$ or correctly concludes that no solution of size k exists. As surveyed in [21], so far there are very few natural problems where a parameterized approximation is possible, but the problem is not known to be fixed-parameter tractable.

The main idea of our approximation algorithm is to reduce Edge Multicut to a variant of Almost 2SAT (delete k clauses to make a 2-CNF formula satisfiable). The reduction is nontrivial: it consists of several steps and requires the use of iterative compression. Almost 2SAT is known to be fixed-parameter tractable [24] and this immediately implies a parameterized 2-approximation for the variant we use here. Proving that this variant is FPT would seem an obvious approach for proving that Edge Multicut is FPT. However, we rule out this possibility by showing the W[1]-hardness of the Almost 2SAT variant. This might be of independent interest, as it is the first natural W[1]-hard problem having a constant-ratio parameterized approximation.

Besides giving an algorithm for a particular problem, the paper has a conceptual contribution as well by introducing a new technique: we demonstrate that reduction to Almost 2SAT can be a useful approach in the design of fixed-parameter algorithms. We believe that this technique will find uses for other problems in the future. However, it is not obvious what type of problems can be handled this way: for example, it was not apparent that Multicut has any connections with 2SAT.

2 Preliminaries

The objects considered in the present paper are (simple undirected) graphs and 2-CNF formulas. We define the related notation that will be used further in the paper. For a graph G, we denote by $V(G)$ and $E(G)$ its set of vertices and edges, respectively. For C such that either $C \subseteq V(G)$ or $C \subseteq E(G)$, $G \setminus C$ is the graph obtained from G by removal of the elements of C (if $C \subseteq V(G)$ then the edges incident to C are removed from G as well). For $E^* \subseteq E(G)$, $G[E^*]$ is a graph whose set of edges is E^* and the set of vertices is the set of end points of E^*.

Let us specify two sets $\{s_1, \dots, s_\ell\}$ and $\{t_1, \dots, t_\ell\}$ of vertices of G and call their union the set of *terminal* vertices. Let $\mathbf{T} = \{\{s_1, t_1\}, \dots, \{s_l, t_\ell\}\}$ and let C be either a set of *non-terminal* vertices or a set of edges of G such that $G \setminus C$ has no path between s_i and t_i for each i from 1 to l. In this case, we say that C *separates* $\mathbf{T}$ in G. If C is a set of edges, we also say that C is an edge multicut (EMC) of $(G, \mathbf{T})$. Let $Y \subseteq V(G)$. We say that C separates $\mathbf{T}$ and Y in G if C separates $\mathbf{T}$ in G and $G \setminus C$ has no path between any two distinct vertices of Y. If C is a set of edges, we also say that C is an EMC of $(G, \mathbf{T}, Y)$. Note that in the Edge Multicut problem we have to find a set of edges that separates a set $\mathbf{T}$ of terminal pairs.

Now we define the central problem considered in the present paper.

The emc problem
Input: A graph G, an integer k, and a set $\mathbf{T}$ of pairs of terminal vertices of G
Parameter: k
Output: An EMC of $(G, \mathbf{T})$ of size at most k or 'NO' if no such EMC exists.

We will also need the auxiliary problems defined below and referred as AEMC1 and AEMC2.

The aemc1 problem
Instance: A graph G, an integer k, a set $\mathbf{T}$ of pairs of terminal vertices of G, a set Y of at most $2k+1$ non-terminal vertices separating $\mathbf{T}$ in G
Parameter: k
Output: An EMC of $(G, \mathbf{T}, Y)$ of size at most k or 'NO' if no such EMC exists.

The aemc2 problem
Instance: The same as in the AEMC1 problem
Parameter: k
Output: An EMC of $(G, \mathbf{T})$ of size at most k or 'NO' if no such EMC exists.

Finally, we use the following modification of the Almost 2-SAT problem [24]. Let F be a 2-CNF formula and let $C = (\ell_1 \vee \ell_2)$ be a clause of F. A literal l *satisfies* C if $\ell = \ell_1$ or $\ell = \ell_2$. A set L of literals satisfies F or, in other words, L is a *satisfying assignment* of F if L does not contain a literal together with its negation and each clause of F is satisfied by at least one literal of L. Let $L = \{\ell_1, \ldots, \ell_r\}$. We denote the 2-CNF formula $\bigwedge_{i=1}^{r}(\ell_i \vee \ell_i)$ by $\bigwedge L$. The clause $(\ell_1 \rightarrow \ell_2)$ is shorthand for $(\neg\ell_1 \vee \ell_2)$.

Almost 2-sat problem with blocks and fixed literals (2-asat-bfl)
Instance: (F, P, L, k) where

- F is a *satisfiable* 2-CNF formula with possible repeated occurrences of clauses;
- P is a family of (not necessarily disjoint) subsets (*blocks*) of at most 2 clauses covering all the clauses of F;
- L is a set of literals;
- k is a non-negative integer.

Parameter: k.
Output: A set B of at most k blocks of P such that $F' \wedge \bigwedge L$ is satisfiable (F' is the formula obtained from F as a result of removal of the clauses of B) or 'NO' if no such set of blocks exists.

Let P be a parameterized problem where the parameter k is an integer appearing in the input and the task is to find some object of size at most k or report 'NO' if no such object exists. Following [7,21], we say that problem P is *fixed-parameter approximable* (FPA) with ratio c if there is an $f(k) \cdot n^{O(1)}$ time algorithm that either returns an object satisfying all output specifications except that its size is at most ck, or 'NO' and in the latter case it is guaranteed that there is no object of size at most k satisfying the output specifications.

Proposition 1. *The* 2-ASAT-BFL *problem is FPA with ratio 2 and the approximation can be achieved in time $O(25^k km^2)$ where m is the number of clauses of F.*

Proof. Claim 8 in [23] (this is the full version of [24]) states that in time $O(5^k km^2)$ it is possible either to compute a set S of at most k clauses so that $F' \wedge \bigwedge L$ is satisfiable, where F' is the formula obtained from F as a result of the removal of S, or to conclude that no such set of clauses exists. Run this algorithm with parameter $2k$ (thus raising the exponential part to 25^k). Assume that the algorithm returns a set S of clauses of size at most $2k$. Then return an arbitrary minimal set B of blocks covering these clauses. Otherwise return 'NO'. Clearly, if a set of blocks B is returned and F' is the formula resulting from removal of the clauses of these blocks from F then $F' \wedge \bigwedge L$ is satisfiable: in particular, all clauses of S are removed. If the resulting algorithm returns 'NO', it follows that removal of $2k$ clauses cannot make F satisfiable. Since each block consists of at most 2 clauses, it follows that removal of k blocks cannot make F satisfiable, implying that the 'NO' answer is correct. □

3 Reduction to Almost 2-Sat

Let $(G, \mathbf{T}, Y, k)$ be an instance of the AEMC1 problem. We define the instance (F, P, L, k) of the 2-ASAT-BFL problem corresponding to $(G, \mathbf{T}, Y, k)$. Then we show that $(G, \mathbf{T}, Y, k)$ is a 'YES' instance of the AEMC1 problem[1] if and only if (F, P, L, k) is a 'YES' instance of the 2-ASAT-BFL problem. The fixed-parameter approximability of the AEMC1 will then follow from Proposition 1.

The set of variables of F is $\{z_{u,v} | u \in V(G), v \in Y\}$. The variable $z_{u,v}$ represents the truth of the ground statement "u and v belong to the same connected component". The clauses of F can be partitioned into the following 3 groups.

Group 1. For each $\{s_i, t_i\} \in \mathbf{T}$ and for each $v \in Y$, the group contains $2k+1$ copies of clause $(\neg z_{s_i,v} \vee \neg z_{t_i,v})$. The purpose of these clauses is to forbid two terminals to be separated to belong to the same connected component.

Group 2. For each pair $\{v_1, v_2\}$ of vertices of Y such that $v_1 \neq v_2$, and for each $u \in V(G)$ the group contains $2k+1$ copies of clause $(\neg z_{u,v_1} \vee \neg z_{u,v_2})$. The purpose of these clauses is to forbid two different vertices of Y to belong to the same connected component.

Group 3. For each $\{u, v\} \in E(G)$ and for each $w \in Y$, the group contains clause $(z_{u,w} \rightarrow z_{v,w})$ and clause $(z_{v,w} \rightarrow z_{u,w})$. These clauses show that vertices u and v belong to the same connected component.

Observe that F is satisfiable, for instance, by an assignment including the negative literals of all the variables. The set P of blocks is defined as follows. For each clause of F there is a block containing this clause only. Also for all possible pairs $\{(u, w_1), (v, w_2)\}$ where $\{u, v\} \in E(G)$, $w_1 \in Y$, $w_2 \in Y$, there is a block $\{(z_{u,w_1} \rightarrow z_{v,w_1}), (z_{v,w_2} \rightarrow z_{u,w_2})\}$. Finally, $L = \{z_{v,v} | v \in Y\}$.

Lemma 2. *If $(G, \mathbf{T}, Y, k)$ is a 'YES' instance of the* AEMC1 *problem then (F, P, L, k) is a 'YES' instance of the* 2-ASAT-BFL *problem.*

Proof. Let C be an EMC of $(G, \mathbf{T}, Y, k)$ of size at most k. We associate with each $\{u, v\} \in C$ the block $B(\{u, v\})$ corresponding to the 'location' of u and v. In particular, if in $G \setminus C$ there are two different vertices w_1 and w_2 of Y such that u belongs to the component of w_1 while v belongs to the component of w_2 then $B(\{u, v\}) = \{(z_{u,w_1} \rightarrow z_{v,w_1}), (z_{v,w_2} \rightarrow z_{u,w_2})\}$. Otherwise if exactly one of $\{u, v\}$, say, u belongs to the connected component of some vertex $w \in Y$ while the connected component of v contains no vertex of Y then $B(\{u, v\}) = \{(z_{u,w} \rightarrow z_{v,w})\}$. Finally, if neither u nor v belong to the same component with a vertex of Y then $B(\{u, v\})$ can be an arbitrarily chosen block.

Let F' be a 2-CNF formula obtained from F by removal of the union of all $B(\{u, v\})$. We claim that F' is satisfiable by an assignment including L as a subset. In particular, let L^* be the set of literals of the variables of V created as follows: $z_{u,w} \in L^*$ whenever u belongs to the same component with w in $G \setminus C$, otherwise $\neg z_{u,w} \in L^*$. Clearly, $L \subseteq L^*$. We claim that L^* satisfies F'. Clauses of

[1] A 'YES' instance is one whose output is not 'NO'.

Group 1 are satisfied because otherwise there is a pair $\{s_i, t_i\}$ of terminals that belong to the same component (with some $w \in Y$) in $G \setminus C$ in contradiction to being C an EMC of $(G, \mathbf{T}, Y)$. Clauses of Group 2 are satisfied because otherwise there is a vertex of $G \setminus C$ that belongs to two distinct connected components, which is absurd. Finally, assume that a clause $c = (z_{u,w} \rightarrow z_{v,w})$ is not satisfied by L^*. It can happen only if u belongs to the same component with w, while v does not. By description of Group 3, $\{u, v\} \in E(G)$ and, since u and v belong to different connected components of $G \setminus C$, $\{u, v\} \in C$. Observe that the first or the second condition of creation of $B(\{u, v\})$ is satisfied and hence $c \in B(\{u, v\})$, i.e. c is not a clause of F'. The proof is now complete. □

Lemma 3. *If* (F, P, L, k) *is a 'YES' instance of the* 2-ASAT-BFL *problem then* $(G, \mathbf{T}, Y, k)$ *is a 'YES' instance of the* AEMC1 *problem.*

Proof. Let B ($|B| \leq k$) be a set of blocks whose removal from F makes the resulting 2-CNF formula F' satisfiable by an assignment L^* such that $L \subseteq L^*$. Observe that it makes no sense to include in B any of the blocks containing only a single clause from Group 1 or 2: those clauses are present in $2k + 1$ copies in F, hence they have to be satisfied in L^* as well. Thus we can safely assume that every block in B contains one or two clauses from Group 3.

By construction, each block of B corresponds to exactly one edge of G. Let C be the set of all such edges. We claim that C is an EMC of $(G, \mathbf{T}, Y)$. Assume first that C does not separate Y, i.e., there are vertices w_1 and w_2 of Y such that $G \setminus C$ has a path p from w_1 to w_2. If the length of p is 1 then let $S = \{(z_{w_1,w_1} \rightarrow z_{w_2,w_1})\}$. Otherwise, let $u_1, \ldots, u_q$ be the intermediate vertices of p listed in the order of their occurrence when p is traversed from w_1 to w_2 and let $S = \{(z_{w_1,w_1} \rightarrow z_{u_1,w_1}), (z_{u_1,w_1} \rightarrow z_{u_2,w_1}), \ldots, (z_{u_{q-1},w_1} \rightarrow z_{u_q,w_1}), (z_{u_q,w_1} \rightarrow z_{w_2,w_1})\}$. Since $L \subseteq L^*$, $z_{w_1,w_1} \in L^*$ as well as $z_{w_2,w_2} \in L^*$. If all the clauses of S are contained in F', then $z_{w_2,w_1} \in L^*$ would follow from this chain of implications, contradicting $(\neg z_{w_2,w_1} \vee \neg z_{w_2,w_2})$ (that necessarily belongs to F'). Hence at least one clause of S belongs to a block of B, implying that at least one edge of p belongs to C.

Thus, if C is not an EMC of $(G, \mathbf{T}, Y)$, it remains to assume that C does not separate $\mathbf{T}$, i.e., there is $\{s, t\} \in \mathbf{T}$ such that $G \setminus C$ has a path p between s and t. By definition of Y, p contains at least one vertex $w \in Y$.[2] If s and w are adjacent in p then let $S = \{(z_{w,w} \rightarrow z_{s,w})\}$. Otherwise, let $u_1, \ldots u_q$ be the intermediate vertices of p occurring in p between w and s listed in the order they occur if p is traversed from w to s. Then $S = \{(z_{w,w} \rightarrow z_{u_1,w}), (z_{u_1,w} \rightarrow z_{u_2,w}), \ldots, (z_{u_{q-1},w} \rightarrow z_{u_q,w}), (z_{u_q,w} \rightarrow z_{s,w})\}$. Arguing as in the previous case, we derive that either $z_{s,w} \in L^*$ or one of the edges corresponding to S belongs to C. Arguing analogously regarding the subpath of p between w and t we derive that either $z_{t,w} \in L^*$ or at least one edge of this subpath belongs to C. It follows that if no edge of p belongs to C then both $z_{s,w} \in L^*$ and $z_{t,w} \in L^*$ hold. But

[2] Notice that this is the only place where it is essential that Y separates all the pairs of terminals of $\mathbf{T}$.

this is a contradiction since in this case the clause $(\neg z_{s,w} \vee \neg z_{t,w})$ is not satisfied. We conclude that C is an EMC of $(G, \mathbf{T}, Y)$. □

The following theorem is an immediate consequence of Proposition 1, Lemma 2, and Lemma 3.

Theorem 4. *The* AEMC1 *problem is FPA with ratio 2.*

4 Fixed-Parameter Approximability of the Emc Problem

We prove the main result of the paper in this section: EMC is fixed-parameter approximable with ratio 2. First, we reduce the AEMC2 problem to the AEMC1 problem. The only difference between the two problems is that in the instance $(G, \mathbf{T}, Z, k)$ of the AEMC2 problem, the solution does not have to separate Z. However, the algorithm can be extended by trying all possible ways in which the solution partitions the set Z.

Lemma 5. *The* AEMC2 *problem is FPA with ratio 2.*

Proof. Apply the following algorithm. Explore all possible partitions of vertices of Y into subsets. For the given partition $Z = Z_1 \cup \cdots \cup Z_q$, let G^* be the graph obtained from G by *contracting* each Z_i into a vertex y_i (loops produced by the contraction are removed, multiple edges are subdivided). Let $Y = \{y_1, \ldots, y_q\}$. Using Theorem 4, we can obtain a 2-approximation for the instance $(G^*, \mathbf{T}, Y, k)$ of the AEMC1. If for at least one such instance an EMC S of $(G^*, \mathbf{T}, Y)$ is returned then return S. Otherwise, return 'NO'.

Since the number of partitions of Z depends on $|Z| \leq 2k + 1$, the above algorithm is an FPT algorithm with parameter k. It is easy to see that if the algorithm returns an EMC S of $(G^*, \mathbf{T}, Y, k)$, then S is an EMC of $(G, \mathbf{T})$ as well. Conversely, assume that $(G, \mathbf{T})$ has an EMC C of size at most k. Let $Z_1, \ldots, Z_q$ be the partition of Z so that two vertices get into the same partition class if and only if they belong to the same connected component of $G \setminus C$. According to Theorem 4, being applied to the tuple $(G^*, \mathbf{T}, Y, k)$ resulting from this partition, the above algorithm necessarily produces an EMC S of $(G^*, \mathbf{T})$ having size at most $2k$. Consequently, if the above algorithm returns 'NO' an EMC of $(G, \mathbf{T})$ of size at most k cannot exist and the answer 'NO' is valid. □

The problem AEMC2 is easier than EMC, since the input contains more information, namely the set Z separating $\mathbf{T}$. We apply a methodology known under the name 'iterative compression' which essentially gives us such a set Z 'for free.' Iterative compression was first used by Reed et al. [25] and has become a very useful technique in the design of parameterized algorithms [6,18,16,20,24].

Theorem 6. *The* EMC *problem is FPA with ratio 2.*

Proof. Let $(G, \mathbf{T}, k)$ be an instance of the EMC problem. Let $e_1, \ldots, e_m$ be the edges of G. Let $G_0, \ldots, G_m$ be the graphs defined as follows. For each G_i,

$V(G_i) = V(G)$. $E(G_0) = \emptyset$ and for each $i > 0$, $E(G_i) = \{e_1, \ldots, e_i\}$. One by one, we consider the $(G_i, \mathbf{T}, k)$ instances of the EMC problem in ascending order of i, and for each instance we find a 2-approximation. The approximation for each $(G_i, \mathbf{T}, k)$ results in output S_i, where S_i is either a set of edges or 'NO'. In particular, $S_0 = \emptyset$. Consider computing of S_i, $i > 0$ provided that S_{i-1} is already known. If $S_{i-1} =$ 'NO' then $S_i =$ 'NO', as S_i is a supergraph of S_{i-1}. Otherwise, S_{i-1} is an EMC of size at most $2k$ for $\mathbf{T}$ in G_{i-1}, hence $S_{i-1} \cup \{e_i\}$ is an EMC of size at most $2k+1$ for $\mathbf{T}$ in G_i. Subdivide each edge of $S_{i-1} \cup \{e_i\}$ with a new vertex; clearly, subdivisions does not change the existence of an EMC. Let G^* be the graph obtained this way and let Z be the set of new vertices. It follows that Z has size at most $2k+1$ and separates $\mathbf{T}$ in G^*. Thus we can use the algorithm for AEMC2 on the instance $(G^*, \mathbf{T}, Z, k)$. It either returns an EMC of $\mathbf{T}$ in G^* of size at most $2k$ (which can be modified to obtain a EMC S_i of $\mathbf{T}$ in G by replacing each subdivided edge by the corresponding edge of $S_{i-1} \cup \{e_i\}$) or returns 'NO', in which case we can set $S_i =$ 'NO'. The validity of the algorithm is easy to verify by induction on i combined with Lemma 5. □

We conclude the section with computing the runtime of the algorithm achieving the ratio 2 approximation of the EMC problem. Denote $|V(G)|$ by n, $|E(G)|$ by m and $|\mathbf{T}|$ by ℓ. The iterative compression process described in the proof of Theorem 6 takes $O(m)$ iterations of solving the AEMC2 problem. The algorithm for the AEMC2 problem takes $P(2k+1, k)$ iterations of solving the AEMC1 problem, where $P(2k+1, k)$ is the number of partitions of a $2k+1$-element set into at most k classes. Finally, in order to solve the AEMC1 problem the graph is transformed into a 2-CNF formula. The number of clauses of this formula is $m_1 = O(\ell k^2 + nk^3 + mk) = O(nk^3 + mk)$ (the term ℓk^2 corresponding to the number of clauses of Group 1 is absorbed by nk^3). Then the 2-ASAT-BFL problem is solved for the obtained formula,which takes $O(25^k k m_1^2) = O(25^k k^3(n^2k^4 + m^2))$. Thus the overall complexity is $O(25^k P(2k+1, k) \cdot k^3(n^2k^4 + m^2))$.

5 Hardness of the 2-ASAT-BFL Problem

It is easy to see from the above discussion that the fixed-parameter tractability of the 2-ASAT-BFL problem would imply the fixed-parameter tractability of the EMC problem. In this section we show that the latter is very unlikely to be derived in this way because the 2-ASAT-BFL problem turns out to be W[1]-hard. To the best of our knowledge, this is the first problem known to be both W[1]-hard and FPA with a constant ratio.

Theorem 7. 2-ASAT-BFL *problem is* W[1]*-hard even if the blocks are disjoint.*

Proof. The proof is by reduction from Multicolored Clique, where given a graph G, an integer k, and a proper k-coloring of the vertices of G, the task is to decide whether there is a k-clique in G. (Proper k-coloring is a mapping from $V(G)$ to $\{1, \ldots, k\}$ such that adjacent vertices have different colors.) Multicolored Clique is known to be W[1]-hard [11]. We can assume that every

c-colored vertex has at least one neighbor from every color class except c: otherwise the vertex cannot be part of a k-clique and can be safely deleted. Let n_c be the number of vertices of color c. Let $v_{c,i}$ $(1 \le i \le n_c)$ be the vertices with color c. Let $d(c,i,c') \ge 1$ be the number of neighbors of $v_{c,i}$ having color c'. Let us fix an ordering of these neighbors and let $n(c,i,c',j)$ be the j-th neighbor of $v_{c,i}$ having color c' in this ordering.

Set $k' := \binom{k}{2}$. We construct a satisfiable 2-CNF formula F and a set of literals L such that deletion of k' blocks makes F satisfiable by an assignment including L if and only if G has a multicolored clique of size k. For every $1 \le c \le k$ and $0 \le i \le n_c$, we introduce a variable $x_{c,i}$. For every $1 \le c, c' \le k$, $c \ne c'$, $1 \le i \le n_c$, $0 < j < d(c,i,c')$, we introduce a variable $y_{c,i,c',j}$. For ease of notation, we define $y_{c,i,c',0} := x_{c,i-1}$ and $y_{c,i,c',d(c,i,c')} := x_{c,i}$ (note that the second index of $x_{c,i}$ can be 0, while it is at least 1 for $y_{c,i,c',j}$).

The clauses of F are the union of the disjoint blocks B_e for each edge e of G. Suppose that edge e connects $v_{c,i}$ and $v_{c',i'}$ and $n(c,i,c',j) = v_{c',i'}$ as well as $n(c',i',c,j') = v_{c,i}$ hold for some j, j'. Block B_e consists of the clauses $(y_{c,i,c',j-1} \to y_{c,i,c',j})$ and $(y_{c',i',c,j'-1} \to y_{c',i',c,j'})$. It is easy to see that F is satisfiable by setting all the variables to 0. The set L of literals is defined as follows: $L = \{x_{c,0} | 1 \le c \le k\} \cup \{\neg x_{c,n_c} | 1 \le c \le k\}$.

Before introducing the formal proof, we give an intuitive explanation. Formula F can be considered as containing k *components*, one for each color. The component corresponding to a color c consists of n_c *fragments*, one for each vertex colored in c. The fragment corresponding to vertex $v_{c,i}$ consists of $k-1$ sets of implications one for each $c' \ne c$, and it is convenient to imagine that each such set is a *sequence of implications* of the form $y_{c,i,c',0} \to y_{c,i,c',1} \to \dots \to y_{c,i,c',d_{c,i,c'}}$. Due to the settings $y_{c,i,c',0} := x_{c,i-1}$ and $y_{c,i,c',d(c,i,c')} := x_{c,i}$ and the literals of L, F can be made satisfiable if and only if for each component of color c we identify a fragment corresponding to vertex $v_{c,i}$ and remove a clause from each sequence of implications of this fragment. That is, to make the formula satisfiable, it is necessary and sufficient to remove $k(k-1)$ clauses. Since we want to remove only $k' = k(k-1)/2$ blocks, we have to find out such fragments whose sequences of implications can be partitioned into pairs so that for each pair there is a block 'covering' both sequences of this pair. The blocks are designed in such a way that two fragments can be 'connected' by at most one block and even this can happen only in the case when the vertices corresponding to these fragments are adjacent. It follows that removal of k' blocks can make F' satisfiable if and only if the considered fragments correspond to a set of mutually adjacent vertices, one for each color, i.e., a multicolored clique.

Now we introduce the formal proof. Suppose that G has a multicolored clique K of size k; let v_{c,i_c} be the vertex of K having color c. For every $1 \le c, c' \le k$, $c \ne c'$, there is an integer $j_{c,c'}$ such that $n(c, i_c, c', j_{c,c'}) = v_{c',i_{c'}}$. Let F' be the formula obtained from F by deleting the blocks corresponding to the edges of K.

Consider a set L^* of literals of variables $y_{c,i,c',j}$ (for every $1 \le c, c' \le k$, $c \ne c'$, $1 \le i \le n_c$, $0 \le j \le d(c,i,c')$) created as follows: $y_{c,i,c',j} \in L^*$ if $i < i_c$

(independently of the values of c' and j), or $i = i_c$, provided that $j < j_{c,c'}$. Otherwise $\neg y_{c,i,c',j} \in L^*$.

Since L^* contains literals of all variables $y_{c,i,c',0}$ and $y_{c,i,c',d(c,i,c')}$, it in fact contains the literals of all variables $x_{c,i}$. Let us verify that all variables $x_{c,i}$ are consistently assigned. In addition, to ensure that $L \subseteq L^*$, we check that L^* contains $x_{c,0}$ and $\neg x_{c,n_c}$ for $1 \leq c \leq k$. Consider first a variable $x_{c,0}$. By definition its value equals the value of $y_{c,1,c',0}$ for all possible values of c'. If $i_c > 1$ then $y_{c,1,c',0} \in L^*$. Otherwise, $i_c = 1$ and in this case, as $j_{c,c'} \geq 1$, it follows again that $y_{c,1,c',0} \in L^*$. Thus we have verified the validity of assigning $x_{c,0}$. Now, consider x_{c,n_c}. By definition, $x_{c,n_c} = y_{c,n_c,c',d(c,n_c,c')}$ for all possible values of c'. Clearly $n_c \geq i_c$. If $n_c > i_c$ then $\neg y_{c,n_c,c',d(c,n_c,c')} \in L^*$. Otherwise, $\neg y_{c,n_c,c',d(c,n_c,c')} \in L^*$ because $n_c \geq j_{c,c'}$, implying the validity of assigning x_{c,n_c}. Finally, consider $x_{c,i}$ when $0 < i < n_c$. The value of $x_{c,i}$ is equal to the value of $y_{c,i,c',d(c,i,c')}$ and the value of $y_{c,i+1,c',0}$ for all the values of c'. Using the description of L^*, it is not hard to verify the consistency of instantiation of $x_{c,i}$ by considering first $i < i_c$ then $i = i_c$ and finally $i > i_c$. It remains to verify that each clause of F' is satisfied by L^*. Assume that a clause $(y_{c,i,c',j-1} \rightarrow y_{c,i,c',j})$ is not satisfied. This is only possible if $i = i_c$ and $j - 1 < j_{c,c'}$ and $j \geq j_{c,c'}$, i.e., $j = j_{c,c'}$, but in that case the clause was deleted from F', a contradiction.

For the other direction of the proof, suppose that it is possible to obtain, by the deletion of at most k' blocks, a formula F' that has a satisfying assignment L^* such that $L \subseteq L^*$. In particular this means that $x_{c,0} \in L^*$ and $\neg x_{c,n_c} \in L^*$ for every $1 \leq c \leq k$. Thus for every $1 \leq c \leq k$, there is a smallest $1 \leq i_c \leq n_c$ such that $x_{c,i_c-1} \in L^*$ and $\neg x_{c,i_c} \in L^*$. We claim that $K := \{v_{c,i_c} : 1 \leq c \leq k\}$ is a clique of size k. Let E^* be the set of edges corresponding to the deleted blocks. We show that for every $1 \leq c, c' \leq k$ and $c \neq c'$, v_{c,i_c} is adjacent to a c'-colored vertex in $G[E^*]$. It follows that $G[E^*]$ has k vertices of degree $k-1$. On the other hand $|E^*| = \binom{k}{2}$. It only possible if $G[E^*]$ is a complete graph and $V(G[E^*]) = K$. In other words, K is a clique of size k in G.

Suppose that v_{c,i_c} is not adjacent to a c'-colored vertex in $G[E^*]$. That is, E^* does not contain any of the edges $\{v_{c,i_c}, n(c, i_c, c', j)\}$ for $1 \leq j \leq d(c, i_c, c')$. This means that none of the clauses $(y_{c,i_c,c',j-1} \rightarrow y_{c,i_c,c',j})$ $(1 \leq j \leq d(c, i_c, c'))$ are deleted. Since $d(c, i_c, c') \geq 1$, these clauses ensure that if $y_{c,i_c,c',0} = 1$, then $y_{c,i_c,c',d(c,i_c,c')} = 1$ as well. However, by the definition of i_c, we have $y_{c,i_c,c',0} = x_{c,i_c-1} = 1$ and $y_{c,i_c,c',d(c,i_c,c')}) = x_{c,i_c} = 0$, which gives a contradiction. □

References

1. Bansal, N., Blum, A., Chawla, S.: Correlation clustering. Machine Learning 56(1-3), 89–113 (2004)
2. Bodlaender, H.L., Fellows, M.R., Heggernes, P., Mancini, F., Papadopoulos, C., Rosamond, F.: Clustering with partial information. In: MFCS 2008: Proceedings of the 33rd international symposium on Mathematical Foundations of Computer Science, pp. 144–155. Springer, Heidelberg (2008)

3. Bousquet, N., Daligault, J., Thomasse, S., Yeo, A.: A polynomial kernel for multicut in trees. In: Albers, S., Marion, J.-Y. (eds.) 26th International Symposium on Theoretical Aspects of Computer Science (STACS 2009). Leibniz International Proceedings in Informatics, vol. 3, pp. 183–194. Schloss Dagstuhl - Leibniz-Zentrum fuer Informatik, Germany (2009)
4. Cai, L., Huang, X.: Fixed-parameter approximation: conceptual framework and approximability results. In: Bodlaender, H.L., Langston, M.A. (eds.) IWPEC 2006. LNCS, vol. 4169, pp. 96–108. Springer, Heidelberg (2006)
5. Chawla, S., Krauthgamer, R., Kumar, R., Rabani, Y., Sivakumar, D.: On the hardness of approximating multicut and sparsest-cut. Comput. Complexity 15(2), 94–114 (2006)
6. Chen, J., Liu, Y., Lu, S., O'Sullivan, B., Razgon, I.: A fixed-parameter algorithm for the directed feedback vertex set problem. J. ACM 55(5) (2008)
7. Chen, Y., Grohe, M., Grüber, M.: On parameterized approximability. In: Bodlaender, H.L., Langston, M.A. (eds.) IWPEC 2006. LNCS, vol. 4169, pp. 109–120. Springer, Heidelberg (2006)
8. Demaine, E.D., Emanuel, D., Fiat, A., Immorlica, N.: Correlation clustering in general weighted graphs. Theor. Comput. Sci. 361(2-3), 172–187 (2006)
9. Downey, R., Fellows, M., McCartin, C.: Parameterized approximation algorithms. In: Bodlaender, H.L., Langston, M.A. (eds.) IWPEC 2006. LNCS, vol. 4169, pp. 121–129. Springer, Heidelberg (2006)
10. Downey, R.G., Fellows, M.R.: Parameterized Complexity. Monographs in Computer Science. Springer, New York (1999)
11. Fellows, M.R., Hermelin, D., Rosamond, F.A., Vialette, S.: On the parameterized complexity of multiple-interval graph problems. Theor. Comput. Sci. 410(1), 53–61 (2009)
12. Flum, J., Grohe, M.: Parameterized Complexity Theory. Springer, Berlin (2006)
13. Garg, N., Vazirani, V.V., Yannakakis, M.: Approximate max-flow min-(multi)cut theorems and their applications. SIAM J. Comput. 25(2), 235–251 (1996)
14. Guo, J., Hüffner, F., Kenar, E., Niedermeier, R., Uhlmann, J.: Complexity and exact algorithms for vertex multicut in interval and bounded treewidth graphs. European J. Oper. Res. 186(2), 542–553 (2008)
15. Guo, J., Niedermeier, R.: Fixed-parameter tractability and data reduction for multicut in trees. Networks 46(3), 124–135 (2005)
16. Hüffner, F., Niedermeier, R., Wernicke, S.: Techniques for practical fixed-parameter algorithms. The Computer Journal 51(1), 7–25 (2008)
17. Khot, S.: On the power of unique 2-prover 1-round games. In: Proceedings of the Thirty-Fourth Annual ACM Symposium on Theory of Computing, pp. 767–775. ACM, New York (2002) (electronic)
18. Liu, Y., Lu, S., Chen, J., Sze, S.-H.: Greedy localization and color-coding: improved matching and packing algorithms. In: Bodlaender, H.L., Langston, M.A. (eds.) IWPEC 2006. LNCS, vol. 4169, pp. 84–95. Springer, Heidelberg (2006)
19. Marx, D.: Parameterized graph separation problems. Theoretical Computer Science 351(3), 394–406 (2006)
20. Marx, D.: Chordal deletion is fixed-parameter tractable. To appear in Algorithmica (2008)
21. Marx, D.: Parameterized complexity and approximation algorithms. The Computer Journal 51(1), 60–78 (2008)

22. Niedermeier, R.: Invitation to Fixed-Parameter Algorithms. Oxford Lecture Series in Mathematics and Its Applications, vol. 31 (2006)
23. Razgon, I., O'Sullivan, B.: Almost 2-sat is fixed-parameter tractable. CoRR, abs/0801.1300 (2008)
24. Razgon, I., O'Sullivan, B.: Almost 2-sat is fixed-parameter tractable (extended abstract). In: Aceto, L., Damgård, I., Goldberg, L.A., Halldórsson, M.M., Ingólfsdóttir, A., Walukiewicz, I. (eds.) ICALP 2008, Part I. LNCS, vol. 5125, pp. 551–562. Springer, Heidelberg (2008)
25. Reed, B., Smith, K., Vetta, A.: Finding odd cycle transversals. Operations Research Letters 32(4), 299–301 (2004)
26. Xiao, M.: Algorithms for multiterminal cuts. In: Hirsch, E.A., Razborov, A.A., Semenov, A., Slissenko, A. (eds.) Computer Science – Theory and Applications. LNCS, vol. 5010, pp. 314–325. Springer, Heidelberg (2008)

Rank-Pairing Heaps

Bernhard Haeupler[1], Siddhartha Sen[2,⋆], and Robert E. Tarjan[2,3,⋆]

[1] Massachusetts Institute of Technology
haeupler@mit.edu
[2] Princeton University
{sssix,ret}@cs.princeton.edu
[3] HP Laboratories, Palo Alto CA 94304

Abstract. We introduce the *rank-pairing heap*, a heap (priority queue) implementation that combines the asymptotic efficiency of Fibonacci heaps with much of the simplicity of pairing heaps. Unlike all other heap implementations that match the bounds of Fibonacci heaps, our structure needs only one cut and no other structural changes per key decrease; the trees representing the heap can evolve to have arbitrary structure. Our initial experiments indicate that rank-pairing heaps perform almost as well as pairing heaps on typical input sequences and better on worst-case sequences.

1 Introduction

A *meldable heap* (henceforth just a *heap*) is a data structure consisting of a set of items, each with a distinct real-valued key, that supports the following operations:

- $make\text{-}heap$: return a new, empty heap.
- $insert(x, H)$: insert item x, with predefined key and currently in no heap, into heap H.
- $find\text{-}min(H)$: return the item in heap H of minimum key.
- $delete\text{-}min(h)$: if heap H is not empty, delete from H the item of minimum key.
- $meld(H_1, H_2)$: return a heap containing all the items in disjoint heaps H_1 and H_2, destroying H_1 and H_2.

Some applications of heaps need either or both of the following additional operations.

- $decrease\text{-}key(x, \Delta, H)$: decrease the key of item x in heap H by amount $\Delta > 0$.
- $delete(x, H)$: delete item x from heap H.

We break ties between equal keys using any total order of the items. We allow only binary comparisons of keys, and we study the amortized efficiency [26] of heap operations. We assign to each configuration of the data structure a non-negative *potential*, initially zero. We define the *amortized time* of an operation to be its actual time plus the

⋆ Research at Princeton University partially supported by NSF grants CCF-0830676 and CCF-0832797 and US-Israel Binational Science Foundation grant 2006204. The information contained herein does not necessarily reflect the opinion or policy of the federal government and no official endorsement should be inferred.

A. Fiat and P. Sanders (Eds.): ESA 2009, LNCS 5757, pp. 659–670, 2009.

change in potential it causes. Then for any sequence of operations the sum of the actual times is at most the sum of the amortized times.

Since a heap can be used to sort n numbers, the classical $\Omega(n \log n)$ lower bound [19, p. 183] on comparisons implies that either insertion or minimum deletion must take $\Omega(\log n)$ amortized time, where n is the number of items currently in the heap. For simplicity in stating bounds we assume $n \geq 2$. We investigate simple data structures such that minimum deletion (or deletion of an arbitrary item if this operation is supported) takes $O(\log n)$ amortized time, and each of the other heap operations takes $O(1)$ amortized time. These bounds match the lower bound.

Many heap implementations have been proposed over the years. See e.g. [17]. We mention only those directly related to our work. The *binomial queue* of Vuillemin [28] supports all the heap operations in $O(\log n)$ worst-case time per operation and performs well in practice [2]. Fredman and Tarjan [11,12] invented the *Fibonacci heap* specifically to support key decrease operations in $O(1)$ time, which allows efficient implementation of Dijkstra's shortest path algorithm [4,12] and several other algorithms [6,12,13]. Fibonacci heaps support deletion of the minimum or of an arbitrary item in $O(\log n)$ amortized time and the other heap operations in $O(1)$ amortized time.

Several years later, Fredman et al. [10] introduced a self-adjusting heap implementation called the *pairing heap* that supports all the heap operations in $O(\log n)$ amortized time. Fibonacci heaps do not perform well in practice [20,21], but pairing heaps do [20,21]. Fredman et al. conjectured that pairing heaps have the same amortized efficiency as Fibonacci heaps, but despite empirical evidence supporting the conjecture [20,25], Fredman [9] showed that it is not true: pairing heaps and related data structures that do not store subtree size information require $\Omega(\log \log n)$ amortized time per key decrease. In contrast, the best known upper bound is $O(2^{2\sqrt{\lg \lg n}})$ [24] amortized time [1].

These results motivated work to improve Fibonacci heaps and pairing heaps. Some of this work obtained better bounds but only by making the data structure more complicated. In particular, the bounds of Fibonacci heaps can be made worst-case: run-relaxed heaps [5] and fat heaps [16] achieve the bounds except for melding, which takes $O(\log n)$ time worst-case, and a very complicated data structure of Brodal [1] achieves all the bounds worst-case. Also, Fredman's lower bound can be matched: very recently Elmasry [8] has proposed an alternative to pairing heaps that does not store subtree size information but takes $O(\log \log n)$ amortized time for a key decrease.

Working in a different direction, several authors proposed data structures with the same amortized efficiency as Fibonacci heaps but intended to be simpler. Peterson [23] gave a structure based on AVL trees. Høyer [14] gave several structures, including ones based on red-black trees, AVL trees, and a, b-trees. Høyer's simplest structure is one he calls a *one-step heap*. Kaplan and Tarjan [17] filled a lacuna in Høyer's presentation of this heap and gave a related structure, the *thin heap*. Independently of our own work but concurrently, Elmasry [7] developed *violation heaps* and Chan [3] *quake heaps*.

In all these structures, the trees representing the heap satisfy a balance property. As a result, a key decrease must in general do restructuring to restore balance. Our insight is that such restructuring is unnecessary: all that is needed is a way to control the size of

[1] We denote by lg the base-two logarithm.

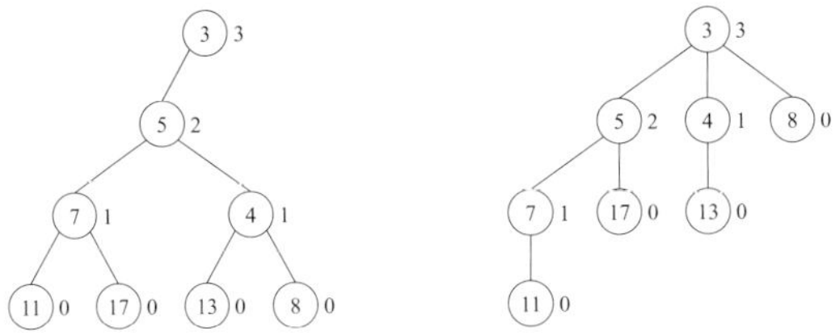

Fig. 1. A half-ordered half tree and the equivalent heap-ordered tree

trees that are combined. Our new data structure, the *rank-pairing heap*, does (at most) one cut and no other restructuring per key decrease, allowing trees to evolve to have arbitrary structure. We store a rank for each node and only combine trees whose roots have equal rank. After a key decrease, rank changes (decreases) can propagate up the tree. Rank-pairing heaps have the same amortized efficiency as Fibonacci heaps and are, at least in our view, the simplest such structure so far proposed. Our initial experiments suggest that rank-pairing heaps compare well with pairing heaps in practice.

The remainder of our paper has six sections. Section 2 describes a one-pass version of binomial queues on which we base our data structure. Section 3 extends this version to obtain two types of rank-pairing heaps, which we analyze in Section 4. Section 5 shows that some even simpler ways of implementing key decrease have worse amortized efficiency. Section 6 describes our initial experiments comparing rank-pairing heaps with pairing heaps. Section 7 concludes and mentions some open problems.

2 One-Pass Binomial Queues

We represent a heap by a set of half-ordered half trees [10,18] whose nodes are the items in the heap. A *half tree* is a binary tree whose right subtree is missing. A *half-ordered* binary tree is a binary tree in which each node has a key, such that the key of a node is less than that of all nodes in its left subtree. Thus in a half-ordered half tree the root has minimum key. A half-ordered half tree is just the binary tree representation [18] of a heap-ordered tree, corresponding to the latter's implementation. (See Figure 1.)

We represent a half tree by storing with each node x pointers to $left(x)$ and $right(x)$, its left and right child, respectively. We represent a set of half trees by a singly-linked circular list of the tree roots, with the root of minimum key first. Access to the list is via the first root. This representation allows us to find the minimum, insert a new half tree, or catenate two such lists in $O(1)$ time.

The basic operation on half trees is *linking*, which combines two half trees into one, in $O(1)$ time. To link two half trees with roots x and y, compare the keys of x and y. Assume x has smaller key; the other case is symmetric. Detach the left subtree of x and make it the right subtree of y. Then make the tree rooted at y the left subtree of x.

To make our data structure efficient, we restrict linking by using *ranks*. A *ranked half tree* is a half tree in which each node has an integer *rank*. The *rank* of a half tree is the rank of its root. We only link half trees of equal rank. After a link, we increase the rank of the new root by one. (See Figure 2.)

We implement the various heap operations as follows. To find the minimum in a heap, return the first root. To make a heap, create an empty list of roots. To insert an

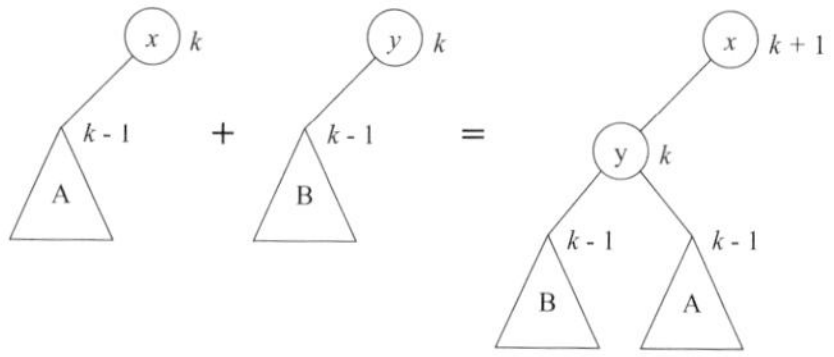

Fig. 2. A link of two half trees of rank k. Ranks are to the right of nodes.

item, make it a one-node half-tree of rank zero and insert it into the list of roots, in first or second position depending on whether it has minimum key or not. To meld two heaps, catenate their lists of roots, making the root of minimum key first on the new list. Each of these operations takes $O(1)$ time worst-case.

To delete the minimum, take apart the half tree rooted at the node of minimum key by deleting this node and walking down the path from its left (only) child through right children, making each node on the path together with its left subtree into a half tree. The set of half trees now consists of the original half trees, minus the one disassembled, plus the new ones formed by the disassembly. Group these half trees into a maximum number of pairs of equal rank and link the half trees in each pair. To do this, maintain a set of buckets, one per rank. Process each half tree in turn by putting it into the bucket for its rank if this bucket is empty; if not, link the half tree with the half tree in the bucket, and add the root of the new half tree to the list representing the updated heap, leaving the bucket empty. Once all the half trees have been processed, add the half trees remaining in buckets to the list of roots, leaving all the buckets empty.

This data structure is a one-pass version of binomial queues. Linking only half trees of equal rank guarantees that every child has rank exactly one less than that of its parent, and every half tree is *perfect*: its left subtree is a perfect binary tree. Thus a half tree of rank k contains exactly 2^k nodes, and the maximum rank is at most $\lg n$.

In the original version of binomial queues, there is never more than one tree per rank. Maintaining this invariant requires doing links during insertions and melds and doing additional links during minimum deletions, until no two half trees have equal rank. The original version has a worst-case time bound of $O(1)$ for make-heap and $O(\log n)$ for the other operations. Since we are interested in amortized efficiency, we prefer to avoid linking during insertions and melds and to link lazily during minimum deletions; extra links of equal-rank half trees done during minimum deletions do not affect our bounds.

To analyze one-pass binomial queues, we define the potential of a heap to be the number of half trees it contains. A make-heap, find-min, or meld operation takes $O(1)$ time and does not change the potential. An insertion takes $O(1)$ time and increases the potential by one. Thus each of these operations takes $O(1)$ amortized time. Consider a minimum deletion. Disassembling the half tree rooted at the node of minimum key increases the number of trees and thus the potential by at most $\lg n$. Let h be the number of half trees after the disassembly but before the links. The total time for the minimum deletion is $O(h + 1)$, including the time to pair the trees by rank. There are at least $(h - \lg n - 1)/2$ links, reducing the potential by at least this amount. If we scale the running time so that it is at most $h/2 + O(1)$, then the amortized time of the minimum deletion is $O(\log n)$.

3 Rank-Pairing Heaps

Our main goal is to implement key decrease so that it takes $O(1)$ amortized time. Once key decrease is supported, one can delete an arbitrary item by decreasing its key to $-\infty$ and doing a minimum deletion. A parameter of both key decrease and arbitrary deletion is the heap containing the given item. If the application does not provide this information and melds occur, one needs a separate disjoint set data structure to maintain the partition of items into heaps. To find an item's heap takes $\omega(1)$ time [15].

We shall modify one-pass binomial queues to support key decrease, obtaining *rank-pairing heaps*, or *rp-heaps*. We make two changes to the data structure. First, we need to provide access from each node to its parent as well as to its children. We add parent pointers, resulting in three pointers per node; this can be reduced to two pointers if we are willing to trade time for space, as observed by Fredman et al. [10].

Second, we relax the constraint on ranks. Let $p(x)$ and $r(x)$ be the parent and rank of node x, respectively. If x is a child, the *rank difference* of x is $r(p(x)) - r(x)$. A node of rank difference i is an i*-child*; a node whose children have rank differences i and j is an i, j*-node*. The latter definition does not distinguish between left and right children. We adopt the convention that a missing child has rank -1.

We shall define two types of rank-pairing heaps. In both, every child of a root is a 1-child, and every leaf has rank zero and is thus a 1,1-node. In a *type-1 rank-pairing heap*, every child is a 1,1-node or a $0, i$-node for some $i > 0$. We call this the *rank rule*. Ranks give a lower bound but not an upper bound on subtree sizes.

Lemma 1. *In a type-1 rp-heap, every node of rank k has at least 2^k descendants including itself, at least $2^{k+1} - 1$ if it is a child.*

Proof. The second part of the lemma implies the first part. We prove the second part by induction on the height of a node. A leaf has rank zero and satisfies the second part. Consider a non-root x of rank k whose children satisfy the second part. If x is a $0, i$-node, by the induction hypothesis it has at least $2^{k+1} - 1$ descendants. If x is a 1,1-node, by the induction hypothesis it has at least $2(2^k - 1) + 1 = 2^{k+1} - 1$ descendants. □

We implement make-heap, find-min, insert, meld, and delete-min exactly as on one-pass binomial queues. Links preserve the rank rule: if a link makes a root into a child, the new child is a 1-child and a 1,1-node. (See Figure 2.)

We implement key decrease as follows. (See Figure 3.) To decrease the key of item x in rp-heap H by Δ, subtract Δ from the key of x. If x is a root, make it first on the root list if it now has minimum key, and stop. Otherwise, let $y = right(x)$; detach the subtrees rooted at x and y; reattach the subtree rooted at y in place of the original subtree rooted at x; add x to the list of roots, making it first if it has minimum key; set $r(x) = r(left(x)) + 1$. There may now be a violation of the rank rule at $p(y)$, whose new child, y, may have lower rank than x, the node it replaces. To restore the rank rule, let $u = p(y)$ and repeat the following step until it stops:

Decrease rank (type 1): If u is the root, set $r(u) = r(left(u)) + 1$ and stop. Otherwise, let v and w be the children of u. Let k equal $r(v)$ if $r(v) > r(w)$, $r(w)$ if $r(w) > r(v)$, or $r(w) + 1$ if $r(v) = r(w)$. If $k = r(u)$, stop. Otherwise, let $r(u) = k$ and $u = p(u)$.

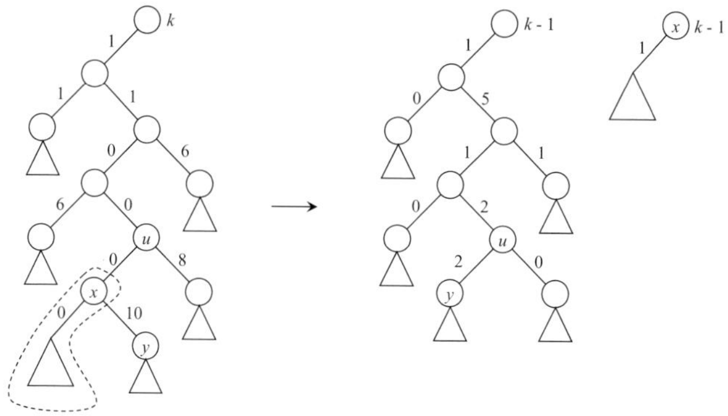

Fig. 3. Key decrease in a type-1 rp-heap

If u breaks the rank rule, it obeys the rule after its rank decreases, but $p(u)$ may not. Rank decreases propagate up along a path through the tree until all nodes obey the rule. Each successive rank decrease is by the same or a smaller amount.

Before analyzing type-1 rp-heaps, we introduce a relaxed version. In a *type-2 rank-pairing heap*, every child is a 1,1-node, a 1-2-node, or a $0, i$-node for some $i > 1$. The implementations of the heap operations are identical to those of type-1 rp-heaps, except for key decrease, which is the same except that the rank decrease step becomes the following:

Decrease rank (*type 2*): If u is the root, set $r(u) = r(left(u)) + 1$ and stop. Otherwise, let v and w be the children of u. Let k equal $r(v)$ if $r(v) > r(w) + 1$, $r(w)$ if $r(w) > r(v) + 1$, or $\max\{r(v) + 1, r(w) + 1\}$ otherwise. If $k = r(u)$, stop. Otherwise, let $r(u) = k$ and $u = p(u)$.

We denote by F_k the kth Fibonacci number, defined by the recurrence $F_0 = 0$, $F_1 = 1$, $F_k = F_{k-1} + F_{k-2}$ for $k > 1$, and by $\phi = (1 + \sqrt{5})/2$ the golden ratio.

Lemma 2. *In a type-2 rp-heap, every node of rank* k *has at least* $F_{k+2} \geq \phi^k$ *descendants including itself, at least* $F_{k+3} - 1$ *if it is a child.*

Proof. The inequality $F_{k+2} \geq \phi^k$ is well-known [18, p. 18]. The second part of the lemma implies the first part. We prove the second part by induction on the height of a node. A leaf has rank zero and satisfies the second part of the lemma. A missing node also satisfies the second part. Consider a non-root x of rank k whose children satisfy the second part. If x is a $0, i$-node, by the induction hypothesis it has at least $F_{k+3} - 1$ descendants. If x is a 1,1-node, by the induction hypothesis it has at least $2(F_{k+2} - 1) + 1 \geq F_{k+3} - 1$ descendants. If x is a 1,2-node, by the induction hypothesis it has at least $F_{k+2} - 1 + F_{k+1} - 1 + 1 = F_{k+3} - 1$ descendants. □

The worst-case time for a key decrease in an rp-heap of either type is $\Theta(n)$, as in Fibonacci heaps. We can reduce this to $O(1)$ by delaying key decreases until the next minimum deletion and maintaining the set of nodes that might have minimum key. This complicates the data structure, however, and may worsen its performance in practice.

4 Amortized Efficiency of Rank-Pairing Heaps

We begin by analyzing type-2 heaps, which is easier than analyzing type-1 heaps. We define the potential of a node to be the sum of the rank differences of its children, minus one if it is a 1,2- or $0, i$-node, minus two if it is a 1,1-node. That is, its potential is zero if it is a 1,1-node, one if it is a root, two if it is a 1,2-node, or $i - 1$ if it is a $0, i$-node. We define the potential of a heap to be the sum of the potentials of its nodes.

Theorem 1. *The amortized time for an operation on a type-2 rp-heap is* $O(1)$ *for a make-heap, find-min, insert, meld, or decrease-key, and* $O(\log n)$ *for a delete-min.*

Proof. A make-heap, find-min, or meld operation takes $O(1)$ actual time and does not change the potential; an insertion takes $O(1)$ time and increases the potential by one. Hence each of these operations takes $O(1)$ amortized time. Consider a minimum deletion. The disassembly increases the potential by at most the number of 1,1-nodes that become roots. By Lemma 2 there are at most $\log_\phi n$ such conversions, so the disassembly increases the potential by at most $\log_\phi n$. Let h be the number of half trees after the disassembly. The entire minimum deletion takes $O(h + 1)$ time. Scale this time to be at most $h/2 + O(1)$. Each link after the disassembly converts a root into a 1,1-node, which reduces the potential by one. At most $\log_\phi n + 1$ half trees do not participate in a link, so there are at least $(h - \log_\phi n - 1)/2$ links. The minimum deletion thus decreases the potential by at least $h/2 - O(\log n)$, which implies that its amortized time is $O(\log n)$.

The novel part of the analysis is that of key decrease. Consider decreasing the key of a node x. If x is a root, the key decrease takes $O(1)$ actual time and does not change the potential. Otherwise, let $u_0 = left(x)$, $u_1 = x$, and $u_2, \ldots, u_k$ be the successive nodes at which rank decrease steps take place, omitting the root if its rank decreases ($u_j = p(u_{j-1})$ for $2 \le j \le k$). For $1 \le j \le k$ let v_j be the child of u_j other than u_{j-1}. Let r, r' denote ranks before and after the key decrease, respectively. The only nodes whose potential changes as a result of the key decrease are $u_1, \ldots, u_k$. At most two of these nodes are 1,1-nodes before the key decrease; if there is a second, it must be u_k, which must become a 1,2-node. The sum of the rank differences of $v_1, u_1, \ldots, u_{k-1}$ before the key decrease telescopes to $r(u_k) - r(v_1)$; the sum of the rank differences of $v_1, u_2, \ldots, u_{k-1}$ after the key decrease telescopes to $r'(u_k) - r'(v_1) \le r(u_k) - r(v_1)$ since $r'(u_k) \le r(u_k)$ and $r'(v_1) = r(v_1)$. The key decrease reduces the rank difference of each v_j, $2 \le j < k$ by at least one. It follows that the key decrease reduces the potential by at least $k - 6$; the "−6" accounts for one additional root and up to two fewer 1,1-nodes. If we scale the time of a rank decrease step so that it is at most one, the amortized time of the key decrease is $O(1)$. □

To analyze type-1 rp-heaps, we assign potentials based on the children of a node. We call a non-leaf node *good* if it is a root whose left child is a 1,1-node, or it and both of its children are 1,1-nodes, or it is a 0,1-node whose 0-child is a 1,1-node; otherwise, the node is *bad*. We define the potential of a leaf to be zero if it is a non-root or 3/2 if it is a root. We define the potential of a non-leaf node to be the sum of the rank differences of its children, plus two if it is a root, minus one if it is good, plus three if it is bad. Thus the potential of a 0,1-node is zero if it is good or four if it is bad, of a 1,1-node is one if it is good or five if it is bad, of a root is two if it is good or six if it is bad, and of a $0, i$-node with $i > 1$ is $i + 3$. We define the potential of a heap to be the sum of the potentials of its nodes. If we restrict the links during a minimum deletion to preferentially pair the half trees produced by the disassembly, the following theorem holds:

Theorem 2. *The amortized time for an operation on a type-1 rp-heap is* $O(1)$ *for a make-heap, find-min, insert, meld, or decrease-key, and* $O(\log n)$ *for a delete-min.*

Proof. A make-heap, find-min, or meld operation takes $O(1)$ actual time and does not change the potential; an insertion takes $O(1)$ time and increases the potential by 3/2. Hence each of these operations takes $O(1)$ amortized time.

Our analysis of minimum deletion relies on the fact that for each rank at most one of the half trees formed by the disassembly is not linked to another such half tree. During the disassembly, we give each new root of rank one or more a potential of four whether it is good or bad. We give the correct potential (two if good, six if bad) to each root of a half tree formed by a link; and, once all half trees formed by the disassembly have been inserted into buckets, we give the correct potential to the root of each such half tree remaining in a bucket. This correction increases the potential by two for each such root that is bad. We charge these two units against the rank of the bad root.

Consider the effect of the disassembly on the potential. Deletion of the root reduces the potential by at least 3/2. The only nodes whose potential can increase are the new roots. At most one leaf can become a root, increasing the potential by 3/2. Each bad node that becomes a root already has the four units of potential it needs as a root. Consider a good node x that becomes a root. There are three cases. If x is a 1,1-node, it needs three additional units of potential as a root. We charge these to $r(x)$. If x is a 0,1-node whose right child is a 0-child, we walk down the path of right children from x until reaching a node y that is either bad or a leaf. Each node along the path must be a 1,1-node. If y is a bad 1,1-node, it has five units of potential, four of which it needs as a root, leaving one for x. We charge the remaining three needed by x to $r(y)$. If y is a leaf, it has rank zero; we charge the units needed by x (3/2 or 4) to rank zero. Node y is reached only from one such node x. If x is a 0,1-node whose left child is a 0-child, we charge the four units x needs as a root to $r(x)$. In this case x is the last new root of its rank; since x is good, its rank is not charged two units to correct the rank of a bad root. Each rank can only be charged once for a new root. The total increase in potential caused by the disassembly and correction of the potential is thus at most $5 \lg n$: each rank up to the maximum rank minus one can be charged two units for a correction and three for a new root, or zero for a correction and four for a new root.

Each link of two rank-0 roots reduces the potential by one. Each link of two roots of positive rank converts a root into a 1,1-node and makes the remaining root good. After the link, these two nodes have total potential at most seven. Before the link, they have potential at least eight (four plus four or at least two plus six), unless they were both good before the link, in which case the new 1,1-node is good after the link and the link reduces the total potential of the two nodes from four to three. Thus each link reduces the potential by one. Let h be the number of half trees after the disassembly. The entire minimum deletion takes $O(h+1)$ time. Scale this time to be at most $h/2+O(1)$. At most $\lg n+1$ half trees do not participate in a link, so there are at least $(h-\lg n-1)/2$ links. The minimum deletion thus decreases the potential by at least $h/2-(11/2)\lg n-1/2$, which implies that its amortized time is $O(\log n)$.

The analysis of a key decrease at a node x is just like that for type-2 heaps, except we must show that the key decrease can make only $O(1)$ nodes bad. A good 1,1-node cannot become bad; it can only become a good 0,1-node. A good 0,1-node cannot decrease in rank, so if it becomes bad it is the last node at which a rank decrease step occurs. If x becomes a root, it can only become bad if it was previously a good 0,1-node with a right 0-child, in which case no ranks decrease and x is the only node that becomes bad. For the root of the old half tree containing x to become bad, its left child must be a 1,1-node, and the old root is the only node that becomes bad. We conclude that the key decrease can make only one node bad, increasing the potential by at most four, and it can create a new root, increasing the potential by two. An argument like that in the proof of Theorem 1 shows that if there are k rank decrease steps, the potential drops by $k - O(1)$. Thus the key decrease takes $O(1)$ amortized time. □

5 Can Key Decrease Be Made Simpler?

It is natural to ask whether there is an even simpler way to decrease keys while retaining the amortized efficiency of Fibonacci heaps. We give two answers: "no" and "maybe". We answer "no" by describing two possible methods that fail. The first method allows arbitrarily negative but bounded positive rank differences. With such a rank rule, the rank decrease process need only examine the ancestors of the node whose key decreases, not their siblings. This method can take $\Omega(\log n)$ amortized time per key decrease, however. The second, even simpler method spends only $O(1)$ time worst-case on each key decrease. By doing enough operations, however, one can build a half tree of each possible rank up to a rank that is $\omega(\log n)$. Then, repeatedly doing an insertion of minimum key followed by a minimum deletion will result in each minimum deletion taking $\omega(\log n)$ time. We omit the details of these counterexamples.

One limitation of the second construction is that building the initial trees takes a number of operations exponential in the size of the heap. Thus it is not a counterexample to the following question: is there a fixed d such that if each key decrease performs at most d rank decrease steps (say of type 1), then the amortized time is $O(1)$ per insert, meld, and key decrease, and $O(\log m)$ per deletion, where m is the total number of insertions? A related question is whether Fibonacci heaps without cascading cuts have these bounds. We conjecture that the answer is yes for some positive d, perhaps even $d = 1$. The following counterexample shows the answer is no for $d = 0$; that is, if key decrease changes the rank only of the node whose key decreases. For arbitrary k, build a half tree of each rank from 0 through k, each consisting of a root and a path of left children, inductively as follows. Given such half trees of ranks 0 through $k - 1$, insert an item less than all those in the heap and then do k repetitions of an insertion of minimum key followed by a minimum deletion. The result will be one half tree of rank k consisting of the root, a path of left children descending from the root, a path P of right children descending from the left child of the root, and a path of left children descending from each node of P; every child has rank difference 1. (See Figure 4.) Do a key decrease on each node of P. Now there is a collection of half trees of rank 0 through k except for $k - 1$, each a path. Repeat this process on the set of half trees up to rank $k-2$, resulting in a set of half trees of ranks 0 through k with $k-2$ missing. Continue in

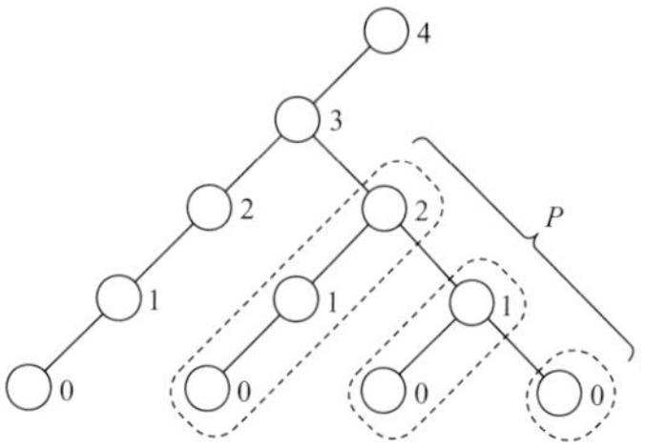

Fig. 4. A half tree of rank $k = 4$ buildable in $O(k^3)$ operations if key decreases do not change ranks. Key decreases on the right children detach the circled subtrees.

this way until only rank 0 is missing, and then do a single insertion. Now there is a half tree of each rank, 0 through k. The total number of heap operations required to increase the maximum rank from $k - 1$ to k is $O(k^2)$, so in m heap operations one can build a set of half trees of each possible rank up to a rank that is $\Omega(m^{1/3})$. Each successive cycle of an insertion followed by a minimum deletion takes $\Omega(m^{1/3})$ time.

6 Experiments

In our preliminary experiments, rank-pairing heaps performed almost as well as pairing heaps on typical input sequences and faster on worst-case sequences. We compared rp-heaps to the standard two-pass version of pairing heaps on typical sequences and to the *auxiliary two-pass* version [25] on worst-case sequences; Stasko and Vitter [25] claimed the latter outperforms other versions on the class of worst-case sequences we used. We compared them to four versions of rp-heaps: type-1 and type-2, performing as many links as possible during minimum deletion ("multipass") and performing only a single pass of links ("one-pass"). All heap implementations were written in C and compiled with the `gcc -O2` option; all tests were performed on a 2.13 Ghz Intel CPU running Windows Vista.

For typical input sequences, we performed three sets of experiments (Table 1). Our measure of performance is the total number of field updates performed by each heap implementation. The first set of experiments consists of publicly available heap operation sequences [27] obtained by running Dijkstra's shortest paths algorithm [4] on graphs that maximize heap size and the number of key decreases, respectively; the Nagamochi-Ibaraki minimum-cut algorithm [22] on a random graph; and heapsort. The second set consists of heap operation sequences obtained by running two-way Dijkstra between random pairs of nodes on real road networks. The third set tests key decreases by running $2n$ rounds of the following operations on a binomial heap of n nodes: an insertion, followed by $\lg n - 1$ key decreases, followed by a minimum deletion.

The results in Table 1 show that rp-heaps performed fewer field updates than pairing heaps on sequences with many key decreases (2,3,7,8), and more updates on sequences with few key decreases. The multipass versions outperformed the one-pass versions overall, with an average speedup of over 1.8% versus an average slowdown of under 5% relative to pairing heaps, respectively. The one-pass versions, while benefiting from smaller trees (and hence fewer rank updates), experienced greater overhead in maintaining longer lists of half trees during minimum deletions.

Table 1. Performance of pairing heaps versus rp-heaps on typical input sequences. n is the number of vertices or items in the heap; m is the number of arcs. All results are in millions of updates.

Test	Parameters		P-heap	Type-2 rp-heap		Type-1 rp-heap	
	n	m		m-pass	1-pass	m-pass	1-pass
1. Dijkstra (max heap size)	8388609	25165824	339.40	346.38	323.05	352.59	325.07
2. Dijkstra (max key decreases)	65536	655360	5.06	5.49	4.27	5.66	4.27
3. Nagamochi-Ibaraki	32768	2684272	136.96	96.15	119.66	98.78	121.77
4. Sorting	100000	-	23.65	26.89	30.95	26.89	30.95
5. Two-way Dijkstra, E. USA	3598623	8778114	2371.04	2603.39	2878.15	2611.77	2896.85
6. Two-way Dijkstra, NYC	264346	733846	1070.66	1316.20	1430.82	1314.26	1425.03
7. Key decrease	262144	-	421.30	322.62	390.00	322.04	389.31
8. Key decrease	4096	-	4.32	3.32	3.91	3.40	3.95

To investigate the worst-case behavior of key decreases, we ran Fredman's [9] version of an experiment of Stasko and Vitter [25]. The experiment is identical to the third set of experiments in Table 1, but uses information-theoretic heuristics in the heap operations. In particular, the winner of a link is always the root of the larger tree, and the candidates for key decreases are chosen to detach links of high efficiency, where efficiency is defined as the ratio of the child and parent tree sizes prior to linking. (See [9].) The cost of a round is the number of links performed—plus, for rp-heaps, the number of rank updates in a key decrease and the number of unpaired half trees in a minimum deletion—divided by $\lg n$. The round cost for pairing heaps converges to 2.76 and 2.97 for $n = 2^{12}$ and $n = 2^{18}$, respectively, showing positive growth. The round cost for type-2 multipass rp-heaps converges to 2.36 and 2.32, showing slightly negative growth.

7 Remarks

We have presented a new data structure, the rank-pairing heap, that combines the performance guarantees of Fibonacci heaps with simplicity approaching that of pairing heaps. Our results build on previous work by Peterson, Høyer, and Kaplan and Tarjan, and may be the natural conclusion of this work: we have shown that simpler methods of doing key decreases do not have the desired efficiency. In our preliminary experiments, rp-heaps are competitive with pairing heaps on typical input sequences and better on worst-case sequences. Type-1 rp-heaps, although simple, are not simple to analyze.

Several interesting theoretical questions remain. Is there a simpler analysis of type-1 rp-heaps? Do type-1 rp-heaps still have the efficiency of Fibonacci heaps if the restrictions on linking in Section 4 are removed? More interestingly, can one obtain an $O(1)$ amortized time bound for insert, meld, and key decrease and $O(\log m)$ for minimum deletion (where m is the total number of insertions) if only $O(1)$ rank changes are made after a key decrease? (See Section 5.)

Acknowledgement

We thank Haim Kaplan and Uri Zwick for discussions that helped to clarify the ideas in this paper, and for pointing out an error in our original analysis of type-1 rp-heaps.

References

1. Brodal, G.: Worst-case efficient priority queues. In: SODA, pp. 52–58 (1996)
2. Brown, M.R.: Implementation and analysis of binomial queue algorithms. SIAM J. Comput., 298–319 (1978)
3. Chan, T.M.: Quake heaps: a simple alternative to Fibonacci heaps (2009)
4. Dijkstra, E.W.: A note on two problems in connexion with graphs. Numer. Math. 1, 269–271 (1959)
5. Driscoll, J.R., Gabow, H.N., Shrairman, R., Tarjan, R.E.: Relaxed heaps: an alternative to Fibonacci heaps with applications to parallel computation. Comm. ACM 31(11), 1343–1354 (1988)
6. Edmonds, J.: Optimum branchings. J. Res. Nat. Bur. Standards B71, 233–240 (1967)
7. Elmasry, A.: Violation heaps: A better substitute for Fibonacci heaps. CoRR (2008)
8. Elmasry, A.: Pairing heaps with $O(\log \log n)$ decrease cost. In: SODA, pp. 471–476 (2009)
9. Fredman, M.L.: On the efficiency of pairing heaps and related data structures. J. ACM 46(4), 473–501 (1999)
10. Fredman, M.L., Sedgewick, R., Sleator, D.D., Tarjan, R.E.: The pairing heap: a new form of self-adjusting heap. Algorithmica 1(1), 111–129 (1986)
11. Fredman, M.L., Tarjan, R.E.: Fibonacci heaps and their uses in improved network optimization algorithms. In: FOCS, pp. 338–346, 24–26 (1984)
12. Fredman, M.L., Tarjan, R.E.: Fibonacci heaps and their uses in improved network optimization algorithms. J. ACM 34(3), 596–615 (1987)
13. Gabow, H.N., Galil, Z., Spencer, T.H., Tarjan, R.E.: Efficient algorithms for finding minimum spanning trees in undirected and directed graphs. Combinatorica 6(2), 109–122 (1986)
14. Høyer, P.: A general technique for implementation of efficient priority queues. In: ISTCS, pp. 57–66 (1995)
15. Kaplan, H., Shafrir, N., Tarjan, R.E.: Meldable heaps and boolean union-find. In: STOC, pp. 573–582 (2002)
16. Kaplan, H., Tarjan, R.E.: New heap data structures. Technical Report TR-597-99, Princeton Univ. (1999)
17. Kaplan, H., Tarjan, R.E.: Thin heaps, thick heaps. ACM Trans. Alg. 4(1), 1–14 (2008)
18. Knuth, D.E.: The Art of Computer Programming. Fundamental Algorithms, vol. 1. Addison-Wesley, Reading (1973)
19. Knuth, D.E.: The Art of Computer Programming. Sorting and Searching, vol. 3. Addison-Wesley, Reading (1973)
20. Liao, A.M.: Three priority queue applications revisited. Algorithmica 7, 415–427 (1992)
21. Moret, B., Shapiro, H.: An empirical analysis of algorithms for constructing a minimum spanning tree. In: Dehne, F., Sack, J.-R., Santoro, N. (eds.) WADS 1991. LNCS, vol. 519, pp. 400–411. Springer, Heidelberg (1991)
22. Nagamochi, H., Ibaraki, T.: Computing edge-connectivity in multigraphs and capacitated graphs. J. Disc. Math. 5(1), 54–66 (1992)
23. Peterson, G.L.: A balanced tree scheme for meldable heaps with updates. Technical Report GIT-ICS-87-23, Georgia Inst. of Tech (1987)
24. Pettie, S.: Towards a final analysis of pairing heaps. In: FOCS, pp. 174–183 (2005)
25. Stasko, J.T., Vitter, J.S.: Pairing heaps: experiments and analysis. Comm. ACM 30(3), 234–249 (1987)
26. Tarjan, R.E.: Amortized computational complexity. J. Alg. Disc. Methods 6, 306–318 (1985)
27. Various. The Fifth DIMACS Challenge—Priority Queue Tests (1996)
28. Vuillemin, J.: A data structure for manipulating priority queues. Comm. ACM 21(4), 309–315 (1978)

3.5-Way Cuckoo Hashing for the Price of 2-and-a-Bit

Eric Lehman[1] and Rina Panigrahy[2]

[1] Google, Mountain View, CA
elehman@google.com
[2] Microsoft Research, Mountain View, CA
rina@microsoft.com

Abstract. The study of hashing is closely related to the analysis of balls and bins; items are hashed to memory locations much as balls are thrown into bins. In particular, Azar et. al. [2] considered putting each ball in the less-full of two random bins. This lowers the probability that a bin exceeds a certain load from exponentially small to doubly exponential, giving maximum load $\log\log n + O(1)$ with high probability. Cuckoo hashing [20] draws on this idea. Each item is hashed to two buckets of capacity k. If both are full, then the insertion procedure moves previously-inserted items to their alternate buckets to make space for the new item. In a natural implementation, the buckets are represented by partitioning a fixed array of memory into non-overlapping blocks of size k. An item is hashed to two such blocks and may be stored at any location within either one. We analyze a simple twist in which each item is hashed to two *arbitrary* size-k memory blocks. (So consecutive blocks are no longer disjoint, but rather overlap by $k-1$ locations.) This twist increases the space utilization from $1-(2/e+o(1))^k$ to $1-(1/e+o(1))^{1.59k}$ in general. For $k=2$, the new method improves utilization from 89.7% to 96.5%, yet lookups access only two items at each of two random locations. This result is surprising because the opposite happens in the non-cuckoo setting; if items are not moved during later insertions, then shifting from non-overlapping to overlapping blocks makes the distribution less uniform.

1 Introduction

The study of hashing is closely related to the analysis of balls and bins; items are hashed to memory locations much as balls are thrown into bins. Simple twists on balls and bins processes have produced surprising observations and led to breakthroughs in hashing methods.

In particular, it is well-known that if n balls are thrown into n bins independently and randomly, then the largest bin gets $(1+o(1))\ln n/\ln\ln n$ balls with high probability. Azar et. al. [2] showed that assigning each ball to the less-full of two random bins makes the final distribution far more uniform. In fact, the probability that a bin exceeds a certain load drops from exponentially small to doubly exponential. This leads to the concept of two-way hashing, where the most-loaded bucket gets $\log\log n+O(1)$ items with high probability. So dramatic is this improvement that it can be used in practice to efficiently implement hash lookups in packet routing hardware [5]. The lookup operation must search for an item in two buckets, but these operations can be parallelized in hardware by placing two different hash tables in separate memory components. More

A. Fiat and P. Sanders (Eds.): ESA 2009, LNCS 5757, pp. 671–681, 2009.

generally, if each item is hashed to $d \geq 2$ buckets, then the maximum load improves to $\log\log n/\log d + O(1)$.

Cuckoo hashing [20,12] extends two-way hashing by moving previously-inserted items to their alternate buckets to make space for a new item. Pagh and Rodler [20] showed that even with buckets of capacity one, moving items during inserts gives a space utilization of 50% with high probability. Several generalizations of cuckoo hashing perform even better. Fotakis et al [12] suggested hashing each item to $d > 2$ buckets, and Dietzfelbinger and Weidling [9] suggested using buckets with capacity $k > 1$. One appealing choice is to hash items to $d = 2$ buckets of capacity $k = 2$, which gives 89.7% [21,14,6] space utilization. (The latter two references improve upon the earlier, weaker estimate.) More generally, the analysis of cuckoo hashing is related to the appearance of dense subgraphs in random graphs. For example, the space utilization achieved by cuckoo hashing where items are hashed to $d = 2$ buckets of capacity k is directly related to the threshold at which a dense subgraph appears in the random graph $G(n,p)$. The space utilization is precisely p/k where p is the threshold at which a dense subgraph with ratio of edges to vertices exceeding k appears. Analysis of this threshold in [14,6] implies a space utilization of about $1-(2/e+o(1))^k$.

In a natural implementation of two-way cuckoo hashing, the buckets are represented by partitioning a fixed array of memory into non-overlapping blocks of size k. Each item is hashed to two such blocks and may be stored at any of the $2k$ memory locations within those blocks. This implementation avoids expensive dynamic memory allocation. Furthermore, a lookup searches just two contiguous memory segments, which is highly desirable in practice. For example, after an initial read from a random location in main memory or on a disk, subsequent bytes can often be read orders of magnitude faster. (This is a heuristic, not a certainty; for example, the extra bytes might lie beyond a cached portion of memory.) And all $2k$ memory locations can be probed in parallel in a hardware implementation.

We suggest another simple twist that significantly improves space utilization while preserving the desirable property of only two random memory accesses per lookup. Previously, the hash table memory was partitioned into disjoint blocks of size k. Now, we regard every set of k consecutive memory locations as a block. So consecutive blocks are no longer disjoint, but rather overlap by $k-1$ memory locations. As before, each item is hashed to two blocks and may be stored at any memory location within those blocks. We show that this simple change improves the space utilization from $1-(2/e+o(1))^k$ to $1-(1/e+o(1))^{1.59k}$. Experimentally, we demonstrate that space utilization improves from 89.7% to 96.5% in the practically-important case where each item is hashed to two blocks of capacity $k = 2$. This result is surprising because the opposite happens in the non-cuckoo setting; if items are not moved during later insertions, then shifting from non-overlapping to overlapping blocks actually makes the distribution of items less uniform.

2 Related Work

Balls and bins analysis still continues to produce surprising results. Vöcking [24] observed that asymmetry helps in load balancing. If each ball is mapped to d bins with equal load, then the ball should be inserted in the leftmost bin. With this simple change, the maximum load drops to $O(\log\log n/d)$. He also showed that breaking ties in this way is the best possible policy to minimize the maximum load.

Berenbrink *et al.* [4] extended the balls and bins analysis to the case where the number of balls m is greater than the number of bins n, showing that the difference in the height of the minimum and maximum bin is independent of m. Precisely, it is $\frac{\log\log n}{\ln d} + O(1)$. Corresponding results hold when ties are broken asymmetrically. Balls and bins on graphs has been analyzed in [17]. Weighted analysis of balls and bins was studied in [23]. Multi-choice hashing with a limited number of moves was studied in [21]. Extensive work has been done in the area of parallel balls and bins [1] and the related study of algorithms to emulate shared memory machines (as for example, PRAMs) on distributed memory machines (DMMs) [8,7,13,22].

More recent works have studied the idea of using a small CAM (content-addressable memory) in conjunction with cuckoo hashing to lower the insert/delete time [18,19]. Arbitman *et al.* [25] proved that it is possible to get constant time for all operations with about 50% space utilization by using a small auxiliary hash table. The appearance of dense subgraphs in random graphs was studied in [14,6]. These build upon earlier works that investigate the appearance of a k-core – a subgraph with minimum degree at least k – in random graphs [3,16].

Other related work includes the first static dictionary data structure with constant look up time by Fredman, Komlos and Szemeredi [15] that was generalized to a dynamic data structure by Dietzfelbinger et al. in [11] and [10]. In practice, however, these algorithms are more complex to implement than cuckoo hashing.

3 Our Contribution

We propose a new twist on multi-choice hashing that significantly improves memory utilization, yet accesses only two small regions of memory. Our main theorem compares this new variation to the algorithm analyzed in [14,6]. In both algorithms, each item is hashed to two memory blocks of size k. The item may be stored at any location in either block, and previously-inserted items may be moved to their alternate locations to make space for the new item. The lookup operation searches the k locations in each of the two blocks associated with the item sought. The distinction is that the earlier ALG-DISJOINT-CUCKOO algorithm hashes items to only a restricted set of memory blocks; specifically, the hash table memory is initially partitioned into disjoint blocks of size k, and items are hashed only to those blocks. In our new twist, ALG-OVERLAP-CUCKOO, an item may be hashed to any two size-k memory blocks. Our main result states that this new algorithm has better space utilization for large k. Let α_k denote the utilization for ALG-CUCKOO-DISJOINT and β_k for ALG-CUCKOO-OVERLAP.

Theorem 1. *For large k, $\alpha_k < \beta_k$. Specifically,*

- $\alpha_k \leq 1 - (2/e - o(1))^k$
- $\beta_k \geq 1 - (1/e + o(1))^{(2-\gamma)k}$, *where γ is the maximum value of the function $-x + x\log(2(1+1/x))$, which is about* 0.41.

Experimentally, we show that memory utilization improves significantly for small, practical values of k as well. For example, $\alpha_2 = 89.7\%$ while $\beta_2 = 96.5\%$.

This result is surprising, because the opposite effect is observed in the "non-cuckoo" setting; that is, when previously-inserted items are not allowed to be moved to make space for a new item. Again, we compare two algorithms. In both cases, there are n balls and n bins in a line. For each ball, we randomly pick two blocks of k consecutive bins and throw the ball into the least-loaded bin in the less-loaded block. As before, ALG-NOMOVE-DISJOINT uses only blocks from an initial, disjoint partition, while ALG-NOMOVE-OVERLAP uses all blocks. In this case, using overlapping blocks actually leads to a less uniform distribution:

Theorem 2. *[17] For large k,*

- *with high probability, ALG-NOMOVE-DISJOINT gives a maximum load on a bin of $O(\log\log n/k)$ and*
- *with high probability, ALG-NOMOVE-OVERLAP gives a maximum load on a bin of $\Omega(\log\log n/\log k)$.*

Intuition: Here is a simple intuition as to why overlapping blocks give higher space utilization for cuckoo-hashing than disjoint blocks. Consider the case $k = 2$ with disjoint blocks. Note that in cuckoo-hashing, we perform a breadth-first-search for an empty bin by first looking at the two blocks where a new ball hashes. If these are full, then we look at the 4 alternate blocks where the 4 balls in those blocks could go. Continuing recursively, we visit 2^k blocks at a depth of k. Thus the search tree is binary for ALG-CUCKOO-DISJOINT. We will argue that this search tree has a slightly higher degree for ALG-CUCKOO-OVERLAP, which uses overlapping blocks. The faster this search tree branches, the more likely we are to find an empty bin before getting stuck; that is, before reaching leaves whose potential children are all already in the tree. In the ALG-CUCKOO-OVERLAP variant, the two balls in a full block B can potentially be moved to other bins besides those in their alternate blocks. This happens if one of the blocks of these balls overlaps partially with B – in this case, such a ball can also be displaced to the other bin in this partially-overlapping block. Thus the branching factor of the search tree is slightly more than 2. We will demonstrate this phenomena in the experiment section.

Our analysis assumes that each item is hashed to two blocks in a single hash table. But essentially the same analysis applies to the case where an item is hashed to one block in each of two, separate tables.

4 Theoretical Analysis

We will now prove the main theorem 1. We are comparing two cuckoo-based algorithms that access two random blocks of size k each. Algorithm ALG-CUCKOO-DISJOINT accesses from a collection of disjoint blocks at offsets that are multiples of k; whereas ALG-CUCKOO-OVERLAP picks blocks at random offsets. Let us say we have nk bins and we are adding balls one by one till we overflow.

For any balls and bins process, a configuration of balls and bins can be viewed as a hypergraph G where each bin is a node and each ball is a hyperedge connecting its bin choices. The following lemma is well known.

Lemma 3. *For any cuckoo algorithm (ALG-CUCKOO-DISJOINT or ALG-CUCKOO-OVERLAP) a set of ball insertions succeeds iff there is no subgraph with more hyper-edges then vertices.*

Remark 4. For ALG-CUCKOO-DISJOINT, we can think of a block as a single vertex. Thus α_k corresponds to the number of edges in a random graph $G(n, p)$ when a subgraph with density (ratio of edges to vertices) more than k appears.

Proof. The "only if" part is straightforward. For the "if" part, consider the case when a ball insertion fails. We will look at another bipartite graph with balls on one side and bins on another and an edge between them if the ball is allowed to be placed in the bin. So each ball has degree k. Further, we mark an edge between a ball and a bin red if the ball actually chooses that bin and blue otherwise. Now, for a new ball insert, if there is an alternating path of red and blue edges that leads to an empty bin, then we can successfully insert the ball using a sequence of cuckoo moves. So a ball insert fails iff all bins reachable through such alternating paths are full. In such a case, look at the set of bins reachable using such alternating paths from a new ball whose insertion failed. This set of bins is a subgraph with more balls than bins because all bins have a ball plus there is the new ball that could not be inserted. This subgraph of the bipartite graph corresponds to a subgraph in the hypergraph G with more hyperedges than nodes. □

Our main theorem follows from the following two claims. We will sometimes drop the subscript k for convenience. Let $\overline{\alpha}, \overline{\beta}$ denote $1-\alpha$ and $1-\beta$ respectively. Similarly for other variables.

Claim. [6,16] $\alpha_k \leq 1 - (2/e - o(1))^k$

Proof. Although the exact threshold for α_k has been computed before in [6,16], we present a simpler analysis of the asymptotic formula. For a random graph on n nodes with $k(1-\overline{\alpha})n$ edges, let us find the fraction f of nodes that have degree at most $k-1$. This degree distribution of the nodes is given by the Poisson distribution with mean $2k(1-\overline{\alpha})$ and f is at least the fraction of nodes with degree exactly $k-1$.

$$\begin{aligned} f &\geq e^{-2k(1-\overline{\alpha})} \frac{(2k(1-\overline{\alpha}))^{k-1}}{(k-1)!} \\ &= e^{2k\overline{\alpha}} \frac{1}{2(1-\overline{\alpha})} (1-\overline{\alpha})^k e^{-2k} \frac{(2k)^k}{k!} \\ &= e^{k\overline{\alpha}} \frac{1}{k^{O(1)}} e^{-2k} \frac{(2k)^k}{(k/e)^k} \\ &= e^{k\overline{\alpha}} k^{-O(1)} (2/e)^k \end{aligned}$$

So if we ignore this f fraction of the nodes, the remaining $(1-f)n$ nodes have at least $k(1-\overline{\alpha})n + fn$ edges; let us check when the density (ratio of edges to vertices) in the remaining subgraph is more than k. This happens if $k(1-\overline{\alpha})n + fn > k(1-f)n$ or $(k+1)f > k\overline{\alpha}$ or $f > \frac{k}{k+1}\overline{\alpha}$

Plugging in the previous expression for f, we need $e^{k\overline{\alpha}} k^{-O(1)} (2/e)^k > \frac{k}{k+1}\overline{\alpha}$ or $\overline{\alpha} < k^{-O(1)} e^{k\overline{\alpha}} (2/e)^k = e^{k\overline{\alpha}} (2/e + o(1))^k$

Clearly $\overline{\alpha} = (2/e - o(1))^k$ satisfies this. Note that although we used the expected value of f, the actual fraction is concentrated close to this with high probability by Chernoff bounds. □

Next we will show a lower bound for β_k. We will make use of the following simple claims.

Claim. If $0 \le p \le q \le 1$ and q is allowed to vary, the function $q\log(\frac{p}{q}) + (1-q)\log(\frac{1-p}{1-q})$ is decreasing in q.

Proof. Taking derivative with respect to q, we get $\log(\frac{p}{q}) - \log(\frac{1-p}{1-q})$ which is negative. □

Claim. Let γ denote the maximum value of the function $g(x) = -x + x\log(2(1+1/x))$ when $x > 0$. Then $\gamma \le 0.41$.

The claim can be verified by a simple plot of the function.

Claim. $\beta_k \ge 1 - (1/e + o(1))^{(2-\gamma)k}$

Proof. We need to demonstrate a value of $\overline{\beta} = (1/e + o(1))^{(2-\gamma)k}$ for which there is almost surely no subgraph in the hypergraph with more edges than vertices. We will upper bound the probability of finding a subset of bins with more edges than bins and argue that this is unlikely with high probability. Consider nk nodes (bins) and βnk hyperedges (balls) where each hyperedge chooses two sets of k contiguous bins. Although in our algorithm the bins are arranged in a line, for simplicity of analysis we will think of them as arranged in a circle. Any subset of bins can be viewed as a union of contiguous regions in this circle. Look at a subset S of nodes, say it consists of $r = \epsilon n$ contiguous regions and suppose $t = \delta nk$ bins fall outside S. A sequence of k bins can be viewed as a segment of length k. If this is to lie in S, then both its endpoints must be in one of the r contiguous regions in S. Of the nk possible segments of length k at most $nk - r(k-1) - t = nk(1 - \delta - \epsilon(1 - 1/k))$ lie in S. The probability that a hyperedge falls in the nodes in S is $p = (1 - \delta - \epsilon(1 - 1/k))^2$. For S to have more edges than nodes, the required fraction of edges to fall in S is at least $q = (1-\delta)/\beta$ (note that $\delta \ge \overline{\beta}$). Let $x = \delta/\epsilon$.

First, a simple calculation will show that unless $\delta = O(1/\sqrt{k})$ and $\epsilon = \tilde{\Omega}(k\beta^2)$, the probability of finding such a high-density subgraph on S is exponentially small. To see that, note that the subset of $(1-\delta)nk$ bins in S is expected to get no more than $(1-\delta)^2\beta nk$ edges. To get $(1-\delta)nk$ edges, it has to get at least factor $\frac{1}{\beta(1-\delta)} \ge 1 + \delta + \overline{\beta}$ of the expectation. By Chernoff bounds, the probability this happens it at most $e^{-\Omega(\delta+\overline{\beta})^2 nk}$. The number of ways of choosing S is at most the number of ways of choosing the $2\epsilon n$ endpoints of the regions, which is $\binom{n}{2\epsilon n} \le (\frac{e}{2\epsilon})^{2\epsilon n} \le e^{2\epsilon \log \frac{e}{2\epsilon} n}$. So the total probability is at most $e^{n(2\epsilon \log \frac{e}{2\epsilon} - k\Omega(\delta+\overline{\beta})^2)}$. This is exponentially small unless the exponent is nonnegative, which happens when $\epsilon\log(1/\epsilon) \ge k\Omega(\delta + \overline{\beta})^2$. Since $\delta > 0$, we get $\epsilon > \tilde{\Omega}(k\beta^2)$. Also since $\epsilon\log(1/\epsilon) < 1$, we get $k\delta^2 \le O(1)$ or $\delta \le O(1/\sqrt{k})$.

Next, we will do a more detailed calculation of the probability. The number of ways of choosing the subset S is at most the number of ways of first choosing the r starting points of the r regions which is at most $\binom{nk}{\epsilon n}$, and then choosing the r closing points of the regions so that the total size of the r segments is δnk. This can be done in at most $\binom{\delta nk}{\epsilon n}$ ways.

For S to have more edges than nodes, it must get at least qN edges, where $N = n\beta k$ is the total number of edges and $q \geq (1-\delta)/\beta \geq 1-\delta$. The probability L, that a set S gets qN edges is $\binom{N}{qN} p^{qN}(1-p)^{N(1-q)}$. Taking natural log, we get.:

$$\begin{aligned}
\log L &= \log \binom{N}{qN} + qN \log p + \overline{q}N \log \overline{p} \\
&\leq N(-q \log q - \overline{q} \log \overline{q}) + qN \log p + \overline{q}N \log \overline{p} \\
&= N(q \log(p/q) + \overline{q} \log(\overline{p}/\overline{q})) \\
&= \beta nk(q \log(p/q) + \overline{q} \log(\overline{p}/\overline{q})) \\
&= \beta nka
\end{aligned}$$

where $a = q \log(p/q) + \overline{q} \log(\overline{p}/\overline{q})$. Since, $1 - \delta \leq q \leq 1$, a as a function of q is maximized when $q = 1 - \delta$, so

$$\begin{aligned}
a &\leq (1-\delta) \log \frac{(1-\delta-\epsilon(1-1/k))^2}{1-\delta} + \delta \log \frac{2(\delta+\epsilon)}{\delta} \\
&\leq (1-\delta)(-2(\delta+\epsilon)) - (1-\delta)\log(1-\delta) + \delta \log \frac{2(\delta+\epsilon)}{\delta} \\
&\leq -2(1-\delta)(\delta+\epsilon) + \delta + \delta \log \frac{2(\delta+\epsilon)}{\delta} \\
&\leq (1-\delta)(-2\epsilon - 2\delta) + (1-\delta)(\delta + \delta \log \frac{2(\delta+\epsilon)}{\delta}) \\
&\leq (1-o(1))(-2\epsilon - \delta + \delta \log \frac{2(\delta+\epsilon)}{\delta}) \\
&\leq (1-o(1))\epsilon(-2 - x + x \log 2(1+1/x))
\end{aligned}$$

The total log probability of finding some component S that has more edges than vertices is at most

$$\begin{aligned}
&\log[\binom{nk}{n\epsilon}\binom{\delta nk}{\epsilon n} L] \\
&\leq \log \binom{nk}{n\epsilon} + \log \binom{\delta nk}{\epsilon n} + nk\beta a \\
&= n\epsilon \log \frac{ek}{\epsilon} + n\epsilon \log \frac{e\delta k}{\epsilon} + nk(1-o(1))\epsilon(-2 - x + x \log 2(1+1/x)) \\
&= n\epsilon \log \frac{ek}{\epsilon} + n\epsilon \log ekx + nk(1-o(1))\epsilon(-2 - x + x \log 2(1+1/x)) \\
&= nf
\end{aligned}$$

where

$$f = \epsilon(\log \frac{ek}{\epsilon} + \log(ekx) + k(1-o(1))(-2 - x + x \log 2(1+1/x)))$$

$$= \epsilon(\log \frac{e^2k^2}{\epsilon} + k(1 - o(1))(o(x) - 2 - x + x \log 2(1 + 1/x)))$$
$$\leq \epsilon(\log \frac{e^2k^2}{\epsilon} + k(1 - o(1))(-2 + \gamma + o(1)))$$
$$= \epsilon(\log \frac{e^2k^2}{\epsilon} + k(1 - o(1))(-2 + \gamma)).$$

Observe that the expression bounding f is independent of n. So if for a given β for all ϵ and δ, f is negative and less than some fixed value independent of n, then it means that the probability that there is a high density set S is exponentially small. Also ϵ cannot be arbitrarily small as we know that it is $\tilde{\Omega}(k\beta^2)$. Now for f to be non negative we need $\log \frac{e^2k^2}{\epsilon} > (1 - o(1))k(2 - \gamma)$ or $\epsilon < e^{-(1-o(1))(2-\gamma)k}$. We also need to add the probability over all possible values of r and t for S but there are only n^2 possible values which cannot compensate for an exponentially decreasing function; so w.h.p. no such set S exists.

Also from the expression $f = \epsilon(\log \frac{e^2k^2}{\epsilon} + k(1 - o(1))(o(x) - 2 - x + x \log 2(1 + 1/x))) \leq \epsilon(\log \frac{e^2k^2}{\epsilon} + k(1 - o(1))(-x + x \log 2 + x \log(1 + 1/x)) \leq \epsilon(\log \frac{e^2k^2}{\epsilon} + k(1 - o(1))(-x \log e/2 + 1)$ it is clear that for f to be non negative $x \leq O(\frac{1}{k} \log \frac{e^2k^2}{2\epsilon})$ giving $\delta \leq O(\frac{1}{k} \log \frac{e^2k^2}{2\epsilon})\epsilon \leq e^{-(1-o(1))(2-\gamma)k}$. This gives, $\overline{\beta} \leq \delta < e^{-(1-o(1))(2-\gamma)k} \leq (1/e + o(1))^{(2-\gamma)k}$. □

5 Experiments

Experiments suggest that using overlapping blocks improves memory utilization substantially even for small k. This implies that our twist gives a significant practical improvement. The situation is summarized in the figure below:

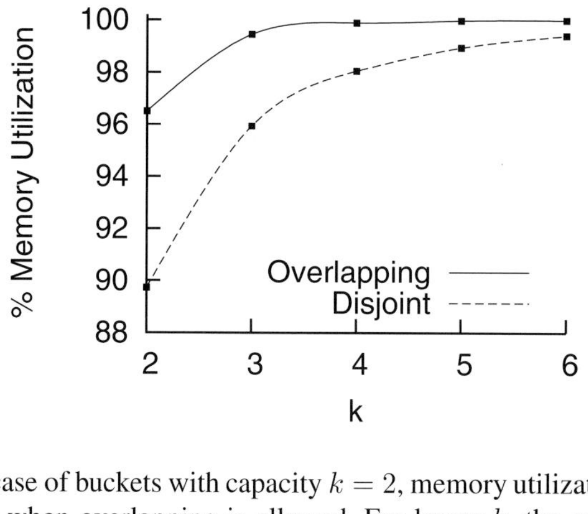

For the basic case of buckets with capacity $k = 2$, memory utilization increases from 89.7% to 96.5% when overlapping is allowed. For larger k, the overlapping scheme rapidly approaches full memory utilization: 99.44% for $k = 3$ and 99.90% for $k = 4$. Each percentage is from twenty trials using hash tables with an absolute capacity of 2^{20}

items and "random" hash functions based on a cryptographic-quality pseudorandom number generator. Items were inserted into the hash table one-by-one until some item could not be added. The results were notably stable. In each case, the standard deviation was a few hundredths of a percent, so error bars would be invisible in the diagram. Such strongly-predictable behavior is appealing from a practical standpoint.

Other experimental data gives additional insight into this performance gap. Recall that the cuckoo insertion algorithm performs a breadth-first search for an empty location in the hash table. The table below shows the number of search tree nodes at each depth for a typical insertion using various hashing schemes. The structure of these trees is random in all cases, but much more noticeably for the new scheme, so three examples are given for that case:

	# of Nodes at Depth					
Each Item is Hashed to:	1	2	3	4	5	6
3 buckets, capacity 1	3	6	12	24	48	96
4 buckets, capacity 1	4	12	36	108	324	972
2 disjoint buckets, capacity 2	4	8	16	32	64	128
2 overlapping blocks, capacity 2	4	11	24	55	136	330
same as above	4	9	20	49	122	305
same as above	4	10	24	62	150	364

For example, if each item is hashed to 4 buckets of capacity 1, as on the second line, then the root of the search tree has 4 children. Items occupying the corresponding locations can each move to 3 other slots, so nodes lower in the search tree have 3 children each. This gives the sequence 4, 12, 36, 108, 324, 972. In contrast, suppose each item is hashed to two disjoint buckets of capacity $k = 2$ as on the third line. Again, the root of the search tree has 4 children. But items in those slots can be moved to at most *two* slots not already explored. Thus, lower nodes in the search tree have only 2 children instead of 3.

The last three lines show data for the new scheme introduced here. Once again, the root has 4 children. But now items occupying those location can be moved to either *two or three* other slots with equal probability. Thus, the width of the search tree is greater than for the disjoint-buckets approach and lies somewhere between the 3- and 4-bucket schemes shown on the first two lines. Since the new scheme accesses only two small, contiguous regions of memory, one might regard it as 3.5-way cuckoo hashing for the price of 2-and-a-bit.

References

1. Adler, M., Chakrabarti, S., Mitzenmacher, M., Rasmussen, L.: Parallel randomized load balancing. In: Proceedings of the Twenty-Seventh Annual ACM Symposium on the Theory of Computing, May 1995, pp. 238–247 (1995)
2. Azar, Y., Broder, A.Z., Karlin, A.R., Upfal, E.: Balanced allocations. SIAM Journal on Computing 29, 180–200 (1999); A preliminary version of this paper appeared in Proceedings of the Twenty-Sixth Annual ACM Symposium on the Theory of Computing (1994)

3. Spencer, S., Pittel, B., Wormald, N.: Sudden emergence of a giant k-core in a random graph. J. Combin. Theory Ser. B 67(1), 111–151 (1996)
4. Berenbrink, P., Czumaj, A., Steger, A., Vöcking, B.: Balanced allocations: The heavily loaded case. SIAM J. Comput. 35(6), 1350–1385 (2006)
5. Broder, A., Mitzenmacher, M.: Using multiple hash functions to improve ip lookups. In: Proceedings of IEEE INFOCOM 2001, pp. 1454–1463 (2001)
6. Cain, J.A., Sanders, P., Wormald, N.: The random graph threshold for k-orientiability and a fast algorithm for optimal multiple-choice allocation. In: SODA 2007: Proceedings of the eighteenth annual ACM-SIAM symposium on Discrete algorithms, Philadelphia, PA, USA, pp. 469–476 (2007)
7. Czumaj, A., Meyer auf der Heide, F., Stemann, V.: Shared memory simulations with triple-logarithmic delay. LNCS, vol. 979, pp. 46–59. Springer, Heidelberg (1995)
8. Dietzfelbinger, M., Meyer auf der Heide, F.: Simple efficient shared memory simulations. In: Proc. of the 5th SPAA, pp. 110–119 (1993)
9. Dietzfelbinger, M., Woelfel, P.: Almost random graphs with simple hash functions. In: 35th STOC, pp. 629–638 (2003)
10. Dietzfelbinger, M., Meyer auf der Heide, F.: A new universal class of hash functions and dynamic hashing in real time. In: Paterson, M. (ed.) ICALP 1990. LNCS, vol. 443, pp. 6–19. Springer, Heidelberg (1990)
11. Dietzfelbinger, M., Karlin, A., Mehlhorn, K., Meyer auf der Heide, F., Rohnert, H., Tarjan, R.E.: Dynamic perfect hashing: Upper and lower bounds. SIAM J. Comput. 23(4), 738–761 (1994)
12. Fotakis, P.S.D., Pagh, R., Spirakis, P.: Space efficient hash tables with worst case constant access time. In: 20th Annual Symposium on Theoretical Aspects of Computer Science (2003)
13. Scheideler, C., Meyer auf der Heide, F., Stemann, V.: Exploiting storage redundancy to speed up randomized shared memory simulations. Theoretical Computer Science 162(2), 245–281 (1996); Preliminary version in Proc. of the 12th STACS, pp. 267–278 (1995)
14. Fernholz, D., Ramachandran, V.: The k-orientability thresholds for gn, p. In: SODA 2007: Proceedings of the eighteenth annual ACM-SIAM symposium on Discrete algorithms, Philadelphia, PA, USA, pp. 459–468. Society for Industrial and Applied Mathematics (2007)
15. Fredman, M.L., Komlos, J., Szemeredi, E.: Storing a sparse table with o(1) worst case access time. J. Assoc. Comput. Mach. 31(3), 538–544 (1984)
16. Janson, S., Luczak, M.J.: A simple solution to the k-core problem. Random Struct. Algorithms 30(1-2), 50–62 (2007)
17. Kenthapadi, K., Panigrahy, R.: Balanced allocation on graphs. In: SODA 2006: Proceedings of the seventeenth annual ACM-SIAM symposium on Discrete algorithm, pp. 434–443. ACM, New York (2006)
18. Kirsch, A., Mitzenmacher, M.: Using a queue to de-amortize cuckoo hashing in hardware. In: 45th Allerton Conference on Communication, Control, and Computing, pp. 751–758 (2007)
19. Kirsch, A., Mitzenmacher, M., Wieder, U.: More robust hashing: Cuckoo hashing with a stash. In: Halperin, D., Mehlhorn, K. (eds.) Esa 2008. LNCS, vol. 5193, pp. 611–622. Springer, Heidelberg (2008)
20. Pagh, R., Rodler, F.: Cuckoo hashing. Journal of Algorithms 51, 122–144 (2004); A preliminary version appeared in proceedings of the 9th Annual European Symposium on Algorithms, pp. 121–133 (2001)

21. Panigrahy, R.: Efficient hashing with lookups in two memory accesses. In: SODA 2005: Proceedings of the sixteenth annual ACM-SIAM symposium on Discrete algorithms, Philadelphia, PA, USA, pp. 830–839. Society for Industrial and Applied Mathematics (2005)
22. Jan, H., Korst Peter Sanders, M., Egner, S.: Fast concurrent access to parallel disks. Algorithmica 35(1), 21–55 (2003); A Preliminary version appeared in SODA 2000
23. Talwar, K., Wieder, U.: Balanced allocations: the weighted case. In: STOC 2007: Proceedings of the thirty-ninth annual ACM symposium on Theory of computing, pp. 256–265. ACM, New York (2007)
24. Vöcking, B.: How asymmetry helps load balancing. J. ACM 50(4), 568–589 (2003)
25. Arbitman, M.N.Y., Segev, G.: De-amortized cuckoo hashing: Provable worst-case performance and experimental results (2009)

Hash, Displace, and Compress

Djamal Belazzougui[1], Fabiano C. Botelho[2,⋆], and Martin Dietzfelbinger[3,⋆⋆]

[1] Ecole Nationale Supérieure d'Informatique, ESI Oued Smar, Algiers, Algeria
d_belazzougui@esi.dz

[2] Department of Computer Engineering, Federal Center for Technological Education of Minas Gerais, Belo Horizonte, Brazil
fabiano@decom.cefetmg.br

[3] Faculty of Computer Science and Automation, Technische Universität Ilmenau, P.O. Box 100565, 98684 Ilmenau, Germany
martin.dietzfelbinger@tu-ilmenau.de

Abstract. A hash function h, i.e., a function from the set U of all keys to the range range $[m] = \{0, \dots, m-1\}$ is called a perfect hash function (PHF) for a subset $S \subseteq U$ of size $n \leq m$ if h is 1–1 on S. The important performance parameters of a PHF are representation size, evaluation time and construction time. In this paper, we present an algorithm that permits to obtain PHFs with expected representation size very close to optimal while retaining $O(n)$ expected construction time and $O(1)$ evaluation time in the worst case. For example in the case $m = 1.23n$ we obtain a PHF that uses space 1.4 bits per key, and for $m = 1.01n$ we obtain space 1.98 bits per key, which was not achievable with previously known methods. Our algorithm is inspired by several known algorithms; the main new feature is that we combine a modification of Pagh's "hash-and-displace" approach with data compression on a sequence of hash function indices. Our algorithm can also be used for k-perfect hashing, where at most k keys may be mapped to the same value.

1 Introduction

In this paper, we study the problem of providing perfect hash functions, minimal perfect hash functions, and k-perfect hash functions. In all situations, a "universe" U, $|U| = u$, of possible keys is given, and a set $S \subseteq U$ of size $n = |S|$ of relevant keys is given as input. The range is $[m] = \{0, 1, \dots, m-1\}$.

Definition 1. (a) *A function $h\colon U \to [m]$ is called a* ***perfect hash function (PHF)*** *for $S \subseteq U$ if h is one-to-one on S.* (b) *A function $h\colon U \to [m]$ is called a* ***minimal perfect hash function (MPHF)*** *for $S \subseteq U$ if h is a PHF and $m = n = |S|$.* (c) *For integer $k \geq 1$, a function $h\colon U \to [m]$ is called a*

⋆ Research partially supported by the Brazilian National Institute of Science and Technology for the Web (grant MCT/CNPq 573871/2008-6), and for the InfoWeb Project (grant MCT/CNPq/CT-INFO 550874/2007-0).

⋆⋆ Research partially supported by DFG grant DI 412/10-1.

A. Fiat and P. Sanders (Eds.): ESA 2009, LNCS 5757, pp. 682–693, 2009.

***k*-perfect hash function (*k*-PHF)** *for $S \subseteq U$ if for every $j \in [m]$ we have $|\{x \in S \mid h(x) = j\}| \leq k$.*

A "hash function construction" is an algorithm that for a given set S construct a (static) data structure D_S such that using D_S on input $x \in U$ one can calculate a value $h(x) \in [m]$, with the property that h is a PHF (MPHF, k-PHF) for S. The evaluation time of h should be constant.

Space Lower Bounds. One of the most important metrics regarding PHFs is the space required to describe such a function. The information theoretic lower bound to describe a PHF was studied in [10,14]. A simpler proof of such a bound was later given in [17]. There is an easy way to obtain quite good lower bounds starting from the simple formulas in [14, Theorem III.2.3.6 (a)]. There it was noted that the bit length of the description of a perfect hash function for S must be at least[1] $\log\left(\binom{u}{n} \,/\, \left(\left(\frac{u}{m}\right)^n \cdot \binom{m}{n}\right)\right)$. Using a simple argument that involves estimating sums of the form $\sum_{0 \leq i < n} \log(1 - \frac{i}{m})$ by an integral one obtains the lower bounds $(m-n)\log(1-\frac{n}{m}) - \log n - (u-n)\log(1-\frac{n}{u})$ (for PHFs) and $-(u-n)\log(1-\frac{n}{u}) - \log n$ for MPHFs.

Considering $u \gg n$, for PHFs where $m = 1.23n$ this gives a value of approximately $0.89n$ bits per key, and the lower bound for MPHFs ($n = m$) is approximately $n/\ln 2 \approx 1.44n$ bits.

It does not seem to be easy to derive similar bounds for k-perfect hashing where $km = (1+\varepsilon)n$ and $\varepsilon \geq 0$. For $n = km$ (corresponding to an MPHF with $k > 1$), one can argue similarly as in [14, Theorem III.2.3.6 (a)] to obtain the space lower bound $\log\left(\binom{u}{n} \,/\, \binom{u/(n/k)}{k}^{n/k}\right)$. This can be used to derive the lower bound $-(u-n)\log(1-\frac{n}{u}) - \log n + (n/k)\cdot\log(k!/k^k)$, or, for $u \gg n$ and n large, the lower bound $n \cdot (\log e + \log(k!/k^k)/k - o(1))$. For example for $k = 4$ we have the lower bound $n \cdot (\log e - 0.854) \approx 0.59n$ bits, for $k = 8$ the bound $\approx 0.36n$ bits, and for $k = 16$ the bound $\approx 0.21n$ bits.

Known Constructions. Many constructions for PHFs are known. The first functions that need $O(n)$ bits of space were proposed by Schmidt and Siegel [19]. Hagerup and Tholey [12] gave an asymptotically optimal construction for MPHF, without assumptions about random hash functions, but the scheme does not seem to be useful for set sizes n that occur in practice. In 1999, Pagh [16] suggested an extremely clear suboptimal scheme (using only very simple hash functions) that needed $(2+\varepsilon)\log n$ bits per key (analyzed). In experiments it was shown that values near $0.35 \log n$ bits per key could be achieved. By splitting tricks (see e.g. [5]) the space may be reduced to $\delta \log n$ bits per key, for arbitrary constant $\delta > 0$. The past five years have seen some surprising new developments. In [3] it was implicit how simple perfect hash functions can be constructed, and finally in [2] an explicit construction was given that can construct a PHF with fewer than 2 bits per key in a very simple and lucid way when $m = cn$, $c > 1.22$ and n is large. The last two constructions assume that random hash functions are given for free.

[1] Log stands for $\log_2$ throughout the paper.

The Full Randomness Assumption. We assume throughout that for all ranges $[r]$ that occur in the construction we have access to a family of fully random hash functions with range $[r]$. This means that naming r and an index $q \geq 1$, say, there is a hash function $h_{r,q} \colon U \to [r]$ that is fully random on U, and so that all the hash functions with distinct pairs (r, q) are independent. Moreover, given $x \in U$ and r and q, the value $h_{r,q}(x)$ can be found in $O(1)$ time. This looks like quite a strong assumption, but there are simple ways (the "split-and-share trick") to justify such an assumption without wasting more than $n^{1-\Omega(1)}$ (bits of) space. For details, see [5,6].

In our experimental implementations, we did not use split-and-share, but just used simple indexed families of hash functions that are reported to behave very well in practice. "Indexed" here means that one can choose new hash functions whenever they are needed by just switching an index, while not using too much space. (The hash functions used are described in the full paper [1].) In all our constructions, success can be checked at runtime. Whenever this is the case one may try to get by with cheap hash functions for several attempts and switch to the more sophisticated functions that guarantee the analysis to go through only if this did not succeed. (No such case occurred in the experiments.)

Our Results. We present an algorithm that, for realistic set sizes n, permits to obtain PHFs with expected representation size that exceed the optimal value only by a factor of approximately 1.43 while retaining $O(n)$ expected construction time and $O(1)$ evaluation time in the worst case. For example, in the case $m = 1.23n$ we obtain a PHF that uses space 1.4 bits per key, and for $m = 1.01n$ we obtain space 1.98 bits per key, which was not achievable with previously known methods. Our algorithm is inspired by several known algorithms; the main new feature is that we combine a modification of Pagh's "hash-and-displace" approach with data compression on a sequence of hash function indices. In experiments we demonstrate that our algorithm also generates fairly compact k-perfect hash functions.

2 The Data Structure and Its Construction

We start with the description of the PHF construction. (The MPHF construction is derived from a PHF with range $[(1 + \varepsilon)n]$ by applying some standard compression tricks on the range, just as in [2].) The data structure consists of two levels. We choose some size r of an "intermediate" table. A "first level hash function" g maps U into $[r]$, and thus splits S into r "buckets"

$$B_i = \{x \in S \mid g(x) = i\},\ 0 \leq i < r.$$

For each bucket there is a second hash function $f_i \colon U \to [m]$, picked by the construction algorithm from a sequence $(\phi_1, \phi_2, \phi_3, \ldots)$ of independent fully random hash functions. To name f_i, one only has to know the index $\sigma(i)$ such that $f_i = \phi_{\sigma(i)}$. We want to obtain a mapping $h \colon U \to [m]$ defined by

$$h(x) = f_{g(x)}(x) = \phi_{\sigma(g(x))}(x)$$

that is perfect for S. Since g and the family $(\phi_1, \phi_2, \phi_3, \ldots)$ are assumed to be given for free, the data structure only has to store the sequence $\Sigma = (\sigma(i), 0 \leq i < r)$, and make sure that $\sigma(i)$ can be retrieved in $O(1)$ time.

Following Pagh [16], whose construction was based on an idea of Tarjan and Yao [20], we now sort the buckets in falling order according to their size, and then find a "displacement value" for the buckets one after the other, in this order. But while in Pagh's construction (with $n = m$) the displacements were values in $[n]$, and $r > n$ was required for the analysis, we deviate here from his approach, utilizing the power of fully random functions. For each bucket B_i we find a suitable index $\sigma(i)$, in the following manner:

Algorithm 1. ***Hash, displace, and compress***

(1) Split S into buckets $B_i = g^{-1}(\{i\}) \cap S$, $0 \leq i < r = n/\lambda$ for $\lambda \geq 1$;
(2) Sort buckets B_i in falling order according to size $|B_i|$ ($O(n)$ time);
(3) Initialize array $\texttt{T[}0 \ldots m-1\texttt{]}$ with 0's;
(4) for all $i \in [r]$, in the order from (2), do
(5) for $\ell = 1, 2, \ldots$ repeat forming $K_i = \{\phi_\ell(x) \mid x \in B_i\}$
(6) until $|K_i| = |B_i|$ and $K_i \cap \{j \mid \texttt{T[}j\texttt{]} = 1\} = \emptyset$;
(7) let $\sigma(i) =$ the successful ℓ;
(8) for all $j \in K_i$ let $\texttt{T[}j\texttt{]} = 1$;
(9) Transform $(\sigma(i))_{0 \leq i < r}$ into compressed form, retaining $O(1)$ access.

(Clearly, if $B_i = \emptyset$, then $\sigma(i) = 1$.) The output of this procedure is the sequence $\Sigma = (\sigma(i), 0 \leq i < r)$. By replacing the 0-1-valued array $\texttt{T[}0 \ldots m-1\texttt{]}$ with an array of counters for counting from 0 to k and the obvious modifications we obtain an algorithm for constructing a k-perfect hash function.

It can be shown (Theorem 2 below, proof in full paper [1]) that, if $m = (1+\varepsilon)n$ for some constant $\varepsilon > 0$, then computing the sequence Σ will succeed in expected linear time. Also, from that analysis it follows directly that with high probability the values $\sigma(i)$ can be bounded by $C \log n$ for some constant C, so that each number $\sigma(i)$ can be represented using $\log\log n + O(1)$ bits. Packing the numbers $\sigma(i)$ in fixed size frames of size $\log\log n + O(1)$ will lead to space requirements of $n(\log\log n + O(1))$ bits, while constant evaluation time for h is retained. (This idea for obtaining an MPHF, communicated to the third author by Peter Sanders [18], was explored in the master's thesis of Lars Dietzel [4].)

However, we can go one step further. Clearly, for each i the random variable $\sigma(i)$ is geometrically distributed. It can be shown (Theorem 2 below) that for $\varepsilon > 0$ the expectations $\mathbf{E}(\sigma(i))$ can be bounded by a constant, and hence the expected construction time is $O(n)$. Moreover we show that the expected sum of the bit lengths $\mathbf{E}(\sum_i \log(\sigma(i)))$ is linear, so that by using a compression scheme like those in [11] or [9] the whole sequence $(\sigma(i), 0 \leq i < r)$ can be coded in $O(n)$ bits, in a way that random access is possible, and so that the property that h can be evaluated in $O(1)$ time is retained.

3 Analysis

We show that our scheme has the theoretical potential to approximate the optimum space of $n \cdot \log e$ bits for an MPHF up to a small constant factor—if we use sufficiently strong coding schemes. The calculation is carried out under the assumption that $\varepsilon = 0$. Introducing $\varepsilon > 0$ in the calculations is easy and only has the effect that the space requirements *decrease*. (But note that the algorithm cannot simply be used for $\varepsilon = 0$, because then the construction time is $\Theta(n \log n)$ (coupon's collector's problem!) and we need $\Theta(n)$ hash functions. In order to obtain an MPHF, one has to use $\varepsilon > 0$ and add some compression method as mentioned in [2] or use $\varepsilon = 0$ and treat buckets of size 1 separately.)— The following is a special case of Jensen's inequality (since $\log x$ is concave):

Fact 1. $\mathbf{E}(\log(\sigma(i))) \leq \log(\mathbf{E}(\sigma(i)))$.

3.1 Heavy Buckets

We first note that keys from heavy buckets are easy to place. Let $\lambda = n/r$ be the load factor of the intermediate table. Only for notational convenience we will assume that λ is an integer. Then the average size of B_i is λ, and $|B_i|$ is $\mathrm{Bi}(n, 1/r)$-distributed, with expectation λ. Let $\mathsf{Poisson}(\lambda)$ denote the distribution of a random variable X with $\mathbf{Pr}(X = t) = e^{-\lambda} \cdot \lambda^t/t!$. It is well known that for $n, r \to \infty$ with $n/r = \lambda$ the binomial distribution $\mathrm{Bi}(n, 1/r)$ converges to $\mathsf{Poisson}(\lambda)$. We will assume n and r to be large. In this case, for each fixed t and each $i \in [r]$:

$$\mathbf{Pr}(|B_i| = t) = \frac{e^{-\lambda} \cdot \lambda^t}{t!} \cdot (1 + o(1)). \tag{1}$$

Lemma 1. $\mathbf{Pr}(X \geq t) \leq e^{-\lambda}(e\lambda/t)^t$, *for* $t \geq \lambda$. (For a proof see [15, p. 97].)

Using (1) and Azuma's inequality, it is easy to establish the following estimates.

Lemma 2. *For each fixed* $t \leq 2\lambda + 2$, *with high probability* $(1 - n^{-c}$ *for an arbitrary* $c > 0)$ *we have:*

$$\begin{aligned} a_{\geq}(\lambda, t) &= |\{i \mid |B_i| \geq t\}| &&= r \cdot \mathsf{Poisson}(\lambda, \geq t) \cdot (1 + o(1)); \\ b_{\geq}(\lambda, t) &= |\{x \in S \mid |B_{g(x)}| \geq t\}| &&= n \cdot \mathsf{Poisson}(\lambda, \geq t - 1) \cdot (1 + o(1)). \end{aligned}$$

For simplicity of notation, we omit the factor $1 + o(1)$ in the following. Assume $t = |B_i| \geq 2\lambda + 1$. At the moment when the algorithm tries to find a function $f_i = \phi_{\sigma(i)}$ for this bucket, at most $b_{\geq}(\lambda, t)$ keys are stored in $\mathtt{T[0}\ldots m-1\mathtt{]}$. By the remarks above and Lemmas 1 and 2 we have that $b_{\geq}(\lambda, t) = n \cdot \mathsf{Poisson}(\lambda, \geq t - 1) \leq n \cdot e^{-\lambda}(e\lambda/(t-1))^{t-1}$.

This entails that the success probability when we try some ϕ_ℓ is bounded below by $\left(1 - e^{-\lambda}(e\lambda/(t-1))^{t-1}\right)^t$. The latter bound is increasing in t, for $t \geq 2\lambda + 1$. So the success probability is bounded below by $(1 - (e/4)^\lambda)^{2\lambda+1}$. Thus,

$\mathbf{E}(\sigma(i)) < (1-(e/4)^\lambda)^{-(2\lambda+1)}$, which is bounded by the constant $(1-(e/4))^{-3} < 31$. By Fact 1 we conclude $\mathbf{E}(\log(\sigma(i))) < \log(31) < 5$. Both the expected time for finding $\sigma(i)$ and the space for storing $\sigma(i)$ for these heavy buckets is bounded by $a_{\geq}(\lambda, t)\cdot O(1) < r\cdot \mathsf{Poisson}(\lambda, \geq 2\lambda)\cdot O(1) \leq e^{-d\lambda}\cdot n$, for some constant $d > 0$, which shows that for larger λ the influence of these buckets is negligible.

3.2 Overall Analysis

In the following, we estimate $\mathbf{E}(\sum_{1\leq i<r}\log(\sigma(i)))$ under the assumption $n = m$. Assuming that g has been applied and the buckets B_i have been formed we define:

$$\begin{aligned}
T_t &= \text{number of buckets of size } t;\\
L_{=t} &= t\cdot T_t = \text{number of keys in buckets of size } t;\\
L_{\geq t} &= \text{number of keys in buckets of size } t \text{ or larger};\\
\beta_{=t} &= L_{=t}/n = \text{fraction of keys in buckets of size } t;\\
\beta_{\geq t} &= L_{\geq t}/n = \text{fraction of keys in buckets of size } t \text{ or larger}.
\end{aligned}$$

Assume in Algorithm 1 $s-1$ buckets of size t have been treated already, and bucket B_i is next. In the table `T` exactly $L_{\geq t+1} + (s-1)t$ keys have been placed. The probability that all keys in B_i hit distinct empty places in `T` is at least $\left(\frac{n-(L_{\geq t+1}+(s-1)t+(t-1))}{n}\right)^t = \left(1-\beta_{\geq t+1}-\frac{st-1}{n}\right)^t$. Thus, $\mathbf{E}(\sigma(i)) \leq \left(1-\beta_{\geq t+1}-\frac{st-1}{n}\right)^{-t}$. By Fact 1 we conclude $\mathbf{E}(\log(\sigma(i))) \leq t\cdot \log\left(\left(1-\beta_{\geq t+1}-\frac{st-1}{n}\right)^{-1}\right)$. By summing over all buckets of size t we get

$$\sum_{i:|B_i|=t}\mathbf{E}(\log(\sigma(i))) \leq t\cdot \sum_{1\leq s\leq T_t}\log\left(\frac{1}{1-\beta_{\geq t+1}-\frac{st-1}{n}}\right). \tag{2}$$

The sum in (2) we may estimate by an integral plus a correction term:

$$t\cdot \int_0^{T_t}\log\left(\frac{1}{1-\beta_{\geq t+1}-\frac{xt-1}{n}}\right)dx + t\cdot(C_t - C_{t+1}), \tag{3}$$

where $C_t = \log(1/(1-\beta_{\geq t}+\frac{1}{n}))$.

To evaluate the integral, we substitute $y = 1-\beta_{\geq t+1}-\frac{xt-1}{n}$ and obtain

$$t\cdot \int_{1-\beta_{\geq t+1}+1/n}^{1-\beta_{\geq t}+1/n}\log\left(\frac{1}{y}\right)\cdot\left(-\frac{n}{t}\right)dy = n\cdot \int_{1-\beta_{\geq t}+1/n}^{1-\beta_{\geq t+1}+1/n}\log\left(\frac{1}{y}\right)dy. \tag{4}$$

Since $\int\log(1/y)\,dy = -y(\log y - \log e)$, we get from (2), (3), and (4) that

$$\sum_{i:|B_i|=t}\mathbf{E}(\log(\sigma(i))) \leq n\cdot\left[-y(\log y-\log e)\right]_{1-\beta_{\geq t}+1/n}^{1-\beta_{\geq t+1}+1/n} + t\cdot(C_t - C_{t+1}). \tag{5}$$

We sum (5) up to the first t_0 with $\beta_{\geq t_0+1} = 0$, hence $C_{t_0+1} = -\log(1+\frac{1}{n})$, to obtain

$$\sum_{0 \leq i < r} \mathbf{E}(\log(\sigma(i))) \leq n \cdot [-y(\log y - \log e)]_{1/n}^{1+1/n} + \sum_{1 \leq t \leq t_0+1} C_t. \qquad (6)$$

A little calculation, using $(1+\frac{1}{n})\ln(1+\frac{1}{n}) > \frac{1}{n}$, shows that

$$n \cdot [-y(\log y - \log e)]_{1/n}^{1+1/n} < n \log e - \log n - 2\log e. \qquad (7)$$

This yields

$$\sum_{0 \leq i < r} \mathbf{E}(\log(\sigma(i))) < n \log e - \log n - 2 \log e + \sum_{1 \leq t \leq t_0+1} C_t. \qquad (8)$$

The C_t-sum depends on the distribution of the sizes of the buckets (and on ε). For $\varepsilon = 0$, the first summand C_1 equals $\log n$, which cancels against the term $-\log n$ in (8). (If $\varepsilon > 0$, it is not hard to verify that all C_t are $O(\log(1/\varepsilon))$.) For estimating $\sum_{2 \leq t \leq t_0+1} C_t$ we must make an assumption about the sequence $\beta_2, \beta_3, \ldots$. In case of the Poisson distribution we have $C_t < \log(1/(1-\beta_{\geq t})) \leq \log(1/(1-\beta_{\geq 2})) = \log(1/e^{-\lambda}) = \lambda \log e$. As seen in Section 3.1, the total contribution of buckets of size $2\lambda+1$ and larger to $\sum_{0 \leq i < r} \mathbf{E}(\log(\sigma(i)))$ is $\leq e^{-d\lambda} \cdot n$, and we may estimate $\sum_{2 \leq t \leq 2\lambda+1} C_t$ roughly by $O(\lambda^2)$.

Summing up, we obtain the following result. (The estimate for $\varepsilon > 0$ is derived in a similar way.)

Theorem 1. *In Algorithm 1 we have, for $\lambda = n/r$ an integer:* $\sum_{0 \leq i < r} \mathbf{E}(\log(\sigma(i))) \leq n(\log e + e^{-\Omega(\lambda)}) + O(\lambda^2)$.

The compression scheme from [11] will give an expected space bound of $n(\log e + O((\log(\lambda)+1)/\lambda)) + O(\lambda^2)$ (the O-notation refers to growing λ; details in the full paper [1]). More sophisticated compression schemes for the index table T, which can reduce the effect of the fact that the code length must be an integer while $\log(\sigma(i))$ is not, will be able to better approximate the optimum $n \log e$.

Finally, using arguments similar to those from Section 3.1 one may also obtain a simpler result that explicitly shows the dependence of construction time on λ and ε. (Proof in the full paper [1].)

Theorem 2. *The algorithm "Hash, displace, and compress", with $m = (1+\varepsilon)n$, and $\lambda = n/r$ an integer, requires expected time $O(n \cdot (2^\lambda + (1/\varepsilon)^\lambda)$ and scratch space $O(\log(1/\varepsilon)n)$ (no dependence on λ).*

4 Experimental Results

The purpose of this section is to evaluate the performance of Algorithm 1, referred to as "compressed hash-and-displace", or CHD, from here on. We also compare CHD with the algorithm proposed by Botelho, Pagh and Ziviani [2],

Table 1. Characteristics of the key sets used for the experiments

Key Set	n	Shortest Key	Largest Key	Average Key Length	Key Set Size (MB)
AllTheWeb	5,000,000	2	31	17.46	91
URL	20,000,000	8	496	58.77	2,150

which is the main practical (linear time) perfect hashing algorithm we found in the literature and will be referred to as BPZ algorithm from now on. The details of the implementation used for the experiments are described in the full paper [1]. The algorithms were implemented in the C language and are available at `http://cmph.sf.net` under the GNU Lesser General Public License (LGPL). The experiments were carried out on a computer running the Linux operating system, version 2.6, with a 1.86 gigahertz Intel Core 2 processor with a 4 megabyte L2 cache and 4 gigabyte of main memory.

To compare the algorithms we used the following metrics: (i) The amount of time to generate PHFs or MPHFs. (ii) The space requirement for storing the resulting PHFs or MPHFs. (iii) The amount of time required by a PHF or an MPHF for each retrieval. All results are averages on 50 trials and were statistically validated with a confidence level of 95%.

In our experiments we used two key sets: (i) a key set of unique query terms extracted from the AllTheWeb[2] query log, referred to as AllTheWeb key set; (ii) a key set of unique URLs collected from the Brazilian Web by the TodoBr[3] search engine, referred to as URL key set. Table 1 shows the main characteristics of each key set.

4.1 Comparing the CHD and BPZ Algorithms

In this section we show that the CHD algorithm is very competitive in practice. It generates in linear time the most compact PHFs and MPHFs we know of and those functions can also be computed in constant time. We have experimented with four different values for the load factor: $\alpha = 0.81, 0.90, 0.99$ and 1.00. MPHFs are generated when $\alpha = 1.00$. For each α we vary the average number of keys per bucket (λ) in order to obtain a tradeoff between generation time and representation size.

In our implementation we have replaced the family of indexed fully random hash functions with a pair (f, h) of simpler hash functions, and have used a linear combination of the two hash functions in order to simulate a family of indexed random hash functions. The search for suitable displacement values for buckets makes repeated access to the list of buckets sorted by size and to the list of keys in each bucket. By reimplementing the buckets and key lists using arrays instead of linked lists to exploit locality of reference, we have obtained a speedup of about a factor of 2 in construction time. Finally, for placing buckets we have used a greedy heuristic for finding the respective displacement values.

[2] AllTheWeb (`www.alltheweb.com`) is a registered trademark.

[3] TodoBr (www.todobr.com.br) is a registered trademark.

It improves space by about 5% at the expense of a slightly longer construction time (see the full paper for a detailed description [1]).

Figure 1 also shows a comparison with the BPZ algorithm [2]. We remark that the BPZ algorithm can generate PHFs only with $\alpha = 0.81$ without degrading its performance at evaluation time, whereas the CHD algorithm can achieve much higher load factors ($\alpha = 0.99$), see Figure 1(a),(b),(c). Moreover, the CHD algorithm is more flexible by offering a full trade-off between space usage and load factors obtaining 1.4 bits per key for $\alpha = 0.81$ against 1.98 bits per key for $\alpha = 0.99$. To obtain MPHFs with $\alpha = 1.00$ both algorithms need to use a succinct data structure that supports rank and select operations, and this requires a few more bits as shown in Figure 1(d). This figure also shows that for some parameter settings like $\lambda = 3$ the CHD algorithm outperforms the BPZ algorithm for both generation time and description size of the resulting functions. Also, with the CHD algorithm it is possible to obtain PHFs and MPHFs that are only a factor of 1.43 away from the information theoretic lower bound, at the expense of spending more time to generate those functions. There is no other algorithm in the perfect hashing literature that can get so close to the information theoretic lower bound and even so run in linear time.

We now compare the CHD and BPZ algorithms regarding the evaluation time for 5×10^6 keys of the AllTheWeb collection and 2×10^7 keys of the URL

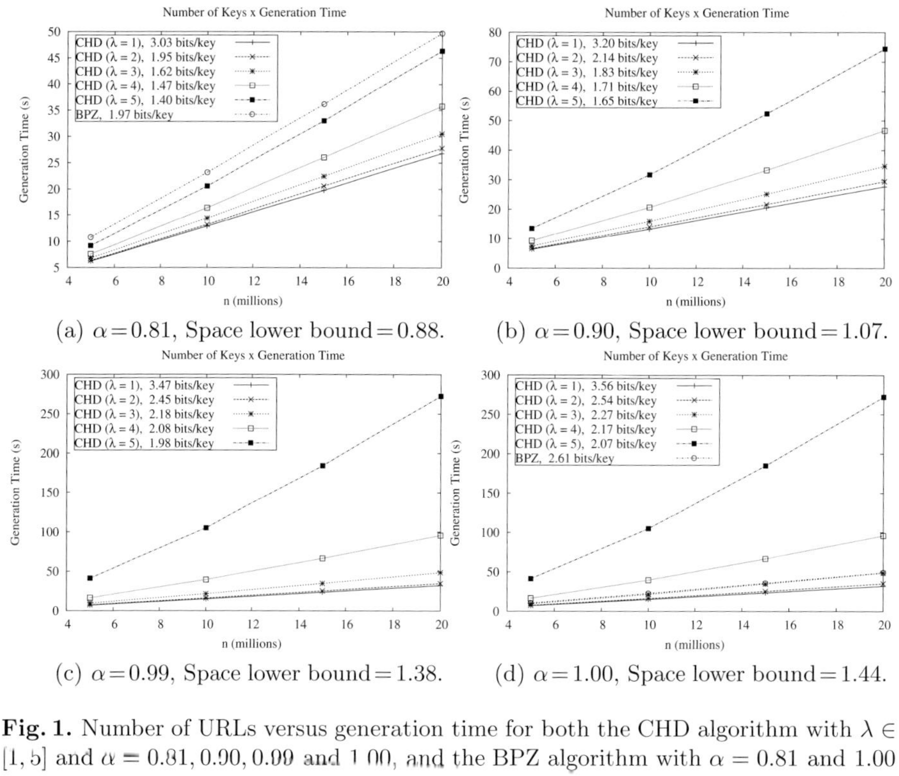

Fig. 1. Number of URLs versus generation time for both the CHD algorithm with $\lambda \in [1, 5]$ and $\alpha = 0.81, 0.90, 0.99$ and 1.00, and the BPZ algorithm with $\alpha = 0.81$ and 1.00

Table 2. Comparing the CHD and BPZ algorithms considering the time to evaluate 5×10^6 keys of the AllTheWeb collection and 2×10^7 keys of the URL collection

Algorithms	λ	AllTheWeb ($n = 5 \times 10^6$) Evaluation Time (sec)			URL ($n = 2 \times 10^7$) Evaluation Time (sec)		
		$\alpha = 0.81$	$\alpha = 0.99$	$\alpha = 1.0$	$\alpha = 0.81$	$\alpha = 0.99$	$\alpha = 1.0$
CHD	1	3.53	3.59	4.14	24.06	24.24	26.99
	2	3.41	3.46	4.01	23.24	23.70	26.26
	3	3.40	3.49	4.04	22.71	23.47	26.05
	4	3.42	3.44	4.01	22.58	23.06	25.65
	5	3.43	3.45	4.02	22.41	22.98	25.59
BPZ	1	2.80	–	3.19	19.76	–	22.12

collection. Table 2 shows that the functions generated by the BPZ algorithm are slightly faster than the ones generated by the CHD algorithm. However, all functions require less than 0.8 microseconds when their description fits in cache, which is the case for the AllTheWeb collection, and less than 1.4 microseconds when their description does not fit in cache, which is the case for the URL collection. We remark that functions with a larger description size are slightly slower due to cache effects (more cache misses). The MPHFs are also slightly slower because of the extra expense for the rank and select operations.

The evaluation time of the PHFs and MPHFs generated by the CHD algorithm depends on the compression technique used. For instance, it is possible to generate faster functions using the Elias-Fano scheme [21] instead of the one we used for the experiments [11] at the cost of generating functions with a slightly larger description size (for example, we obtained PHFs that require 2.08 bits per key instead of 1.98 bits per key for $\alpha = 0.99$ and $\lambda = 5$).

4.2 Results for k-Perfect Hashing

In this section we present experimental results showing that the CHD algorithm can generate very compact k-perfect hash functions. Such functions can be used as an index stored in fast memory for data stored in slow memory. Usually, memory blocks can contain more than one element. In the case where the number of elements per block is a small constant k, we can employ a k-perfect hash function, which uses less space than a perfect hash function. Note that such functioncs require only a single random access to the slow memory in the worst case, unlike other schemes like the linear hashing method proposed by Litwin [13] or bucketed cuckoo hashing [7,8].

Table 3. Generation time and description size of k-perfect hash functions with $\alpha = 0.99$

k	AllTheWeb ($n = 5 \times 10^6$) Generation Time(sec) Avg. Bucket Size (γ)			AllTheWeb Space (bits/key) Avg. Bucket Size (γ)			URL ($n = 2 \times 10^7$) Generation Time(sec) Avg. Bucket Size (γ)			URL Space (bits/key) Avg. Bucket Size (γ)		
	2	4	8	2	4	8	2	4	8	2	4	8
4	4.59	5.30	57.63	1.70	1.20	1.03	26.69	32.72	455.18	1.70	1.20	1.03
8	4.28	4.48	14.51	1.50	0.98	0.77	24.74	26.50	91.56	1.50	0.98	0.77
16	4.14	4.18	6.65	1.37	0.83	0.60	23.79	24.28	37.70	1.37	0.83	0.60

We have only experimented with load factor $\alpha = 0.99$ as this case is closest to the situation for which we know at least some lower bounds (Section 1). Table 3 presents the results for generation time and space usage. We omitted the evaluation times as these are very similar to the ones presented in Table 2.

5 Conclusions

We presented a novel way of constructing (minimal) perfect hash functions: "hash, displace, and compress", modifying Pagh's method "hash-and-displace" so as to obtain space bounds of $O(n)$ bits. A partial analysis of the space requirements of the functions is given, leaving open the exact analysis of the loss created by compression schemes. In experiments, for realistic set sizes the CHD algorithm generates the most compact PHFs we know of in $O(n)$ expected time. The time required to evaluate the generated functions is constant in the worst case (in practice less than 1.4 microseconds). The expected storage space of the resulting PHFs and MPHFs exceed the information theoretic lower bound by a constant factor smaller than 1.5. For certain parameter settings the CHD algorithm constructs more compact MPHFs faster than the BPZ algorithm [2] (the fastest algorithm available in the literature so far). The most impressive characteristic is that CHD has the ability, in principle, to approximate the information theoretic lower bound while being practical.

Acknowledgements. The third author wishes to thank Peter Sanders for sharing with him the idea of using random functions for the buckets in Pagh's hash-and-displace, which was subsequently explored in Lars Dietzel's M.Sc. thesis [4] at TU Ilmenau, which led to a solution with superlinear space. The authors are grateful to the reviewers for their careful reading of the manuscript and their helpful remarks.

References

1. Belazzougui, D., Botelho, F.C., Dietzfelbinger, M.: Hash, displace, and compress. Computer Research Repository (2009)
2. Botelho, F.C., Pagh, R., Ziviani, N.: Simple and space-efficient minimal perfect hash functions. In: Dehne, F., Sack, J.-R., Zeh, N. (eds.) WADS 2007. LNCS, vol. 4619, pp. 139–150. Springer, Heidelberg (2007)
3. Chazelle, B., Kilian, J., Rubinfeld, R., Tal, A.: The bloomier filter: An efficient data structure for static support lookup tables. In: Proc. of the 15th Annual ACM-SIAM Symposium on Discrete Algorithms, pp. 30–39. SIAM, Philadelphia (2004)
4. Dietzel, L.: Speicherplatzeffiziente perfekte Hashfunktionen. Master's thesis, Technische Universität Ilmenau (Novmber 2005) (in German)
5. Dietzfelbinger, M.: Design strategies for minimal perfect hash functions. In: Hromkovič, J., Královič, R., Nunkesser, M., Widmayer, P. (eds.) SAGA 2007. LNCS, vol. 4665, pp. 2–17. Springer, Heidelberg (2007)
6. Dietzfelbinger, M., Rink, M.: Applications of a splitting trick. In: Proc. of 36th International Colloquium on Automata, Languages and Programming. Springer, Heidelberg (to appear, 2009)

7. Dietzfelbinger, M., Weidling, C.: Balanced allocation and dictionaries with tightly packed constant size bins. Theoretical Computer Science 380(1-2), 47–68 (2007)
8. Erlingsson, U., Manasse, M., McSherry, F.: A cool and practical alternative to traditional hash tables. In: Proc. of the 7th Workshop on Distributed Data and Structures, pp. 1–6 (2006)
9. Ferragina, P., Venturini, R.: A simple storage scheme for strings achieving entropy bounds. Theoretical Computer Science 372(1), 115–121 (2007)
10. Fredman, M.L., Komlós, J.: On the size of separating systems and families of perfect hashing functions. SIAM Journal on Algebraic and Discrete Methods 5, 61–68 (1984)
11. Fredriksson, K., Nikitin, F.: Simple compression code supporting random access and fast string matching. In: Demetrescu, C. (ed.) WEA 2007. LNCS, vol. 4525, pp. 203–216. Springer, Heidelberg (2007)
12. Hagerup, T., Tholey, T.: Efficient minimal perfect hashing in nearly minimal space. In: Ferreira, A., Reichel, H. (eds.) STACS 2001. LNCS, vol. 2010, pp. 317–326. Springer, Heidelberg (2001)
13. Litwin, W.: Linear hashing: a new tool for file and table addressing. In: Proc. of the 6th International Conference on Very Large Data Bases, pp. 212–223. VLDB Endowment (1980)
14. Mehlhorn, K.: Data Structures and Algorithms 1: Sorting and Searching. Springer, Heidelberg (1984)
15. Mitzenmacher, M., Upfal, E.: Probability and Computing. Cambridge University Press, Cambridge (2005)
16. Pagh, R.: Hash and displace: Efficient evaluation of minimal perfect hash functions. In: Dehne, F., Gupta, A., Sack, J.-R., Tamassia, R. (eds.) WADS 1999. LNCS, vol. 1663, pp. 49–54. Springer, Heidelberg (1999)
17. Radhakrishnan, J.: Improved bounds for covering complete uniform hypergraphs. Information Processing Letters 41, 203–207 (1992)
18. Sanders, P.: Personal communication (2004)
19. Schmidt, J.P., Siegel, A.: The spatial complexity of oblivious k-probe hash functions. SIAM Journal on Computing 19(5), 775–786 (1990)
20. Tarjan, R.E., Yao, A.C.C.: Storing a sparse table. Communications of the ACM 22(11), 606–611 (1979)
21. Vigna, S.: Broadword implementation of rank/select queries. In: Proc. of the 7th International Workshop on Efficient and Experimental Algorithms, pp. 154–168. ACM Press, New York (2008)

Solving Dominating Set in Larger Classes of Graphs: FPT Algorithms and Polynomial Kernels

Geevarghese Philip, Venkatesh Raman, and Somnath Sikdar

The Institute of Mathematical Sciences, Chennai, India
{gphilip,vraman,somnath}@imsc.res.in

Abstract. We show that the k-DOMINATING SET problem is fixed parameter tractable (FPT) and has a polynomial kernel for any class of graphs that exclude $K_{i,j}$ as a subgraph, for any fixed $i, j \geq 1$. This strictly includes every class of graphs for which this problem has been previously shown to have FPT algorithms and/or polynomial kernels. In particular, our result implies that the problem restricted to bounded-degenerate graphs has a polynomial kernel, solving an open problem posed by Alon and Gutner in [3].

1 Introduction

The k-DOMINATING SET problem asks, for a graph $G = (V, E)$ and a positive integer k given as inputs, whether there is a vertex-subset $S \subseteq V$ of size at most k such that every vertex in $V \setminus S$ is adjacent to some vertex in S. Such a vertex-subset is called a *dominating set* of G. This problem is known to be NP-hard even in very restricted graph classes, such as the class of planar graphs with maximum degree 3 [14]. In the world of parameterized complexity, this is one of the most important hard problems: the problem parameterized by k is the canonical $W[2]$-hard problem [9]. The problem remains $W[2]$-hard even in many restricted classes of graphs — for example, it is $W[2]$-hard in classes of graphs with bounded average degree [15]. This latter fact implies that it is unlikely that the problem has a fixed-parameter-tractable (FPT) algorithm on graphs with a bounded average degree, that is, an algorithm that runs in time $f(k) \cdot n^c$ for *some* computable function $f(k)$ independent of the input size n, and some constant c independent of k.[1]

The problem has an FPT algorithm on certain restricted families of graphs, such as planar graphs [12], graphs of bounded genus [10], K_h-topological-minor-free graphs, and graphs of bounded degeneracy [2]; these last being, to the best of our knowledge, the most general graph class previously known to have an FPT algorithm for this problem. In the current paper, we show that the problem has an FPT algorithm in a class of graphs that encompasses, and is strictly larger than, all these classes — namely, the class of $K_{i,j}$-free graphs.

[1] To know more about the notions of FPT and W-hardness and to see why it is considered unlikely that a $W[2]$-hard problem will have an FPT algorithm, see [9].

A. Fiat and P. Sanders (Eds.): ESA 2009, LNCS 5757, pp. 694–705, 2009.

Closely related to the notion of an FPT algorithm is the concept of a *kernel* for a parameterized problem. For the k-Dominating Set problem parameterized by k, a kernelization algorithm is a polynomial-time algorithm that takes (G, k) as input and outputs a graph G' and a nonnegative integer k' such that the size of G' is bounded by some function $g(k)$ of k alone, $k' \leq h(k)$ for some function $h(k)$, and G has a dominating set of size at most k if and only if G' has a dominating set of size at most k'. The resulting instance G' is called a kernel for the problem. A parameterized problem has a kernelization algorithm if and only if it has an FPT algorithm [9], and so it is unlikely that the k-Dominating Set problem on general graphs or on graphs having a bounded average degree has a kernelization algorithm. For the same reason, this problem has a kernelization algorithm when restricted to those graph classes for which it has an FPT algorithm. But the size of the kernel obtained from such an algorithm could be exponential in k, and finding if the kernel size can be made smaller — in particular, whether it can be made polynomial in k — is an important problem in parameterized complexity.

Proving polynomial bounds on the size of the kernel for different parameterized problems has been a significant practical aspect in the study of the parameterized complexity of NP-hard problems, and many positive results are known. See [16] for a survey of kernelization results.

For the k-Dominating Set problem, the first polynomial kernel result was obtained by Alber et al. [1] in 2004: they showed that the problem has a *linear* kernel of at most $335k$ vertices in planar graphs. This bound for planar graphs was later improved by Chen et al. [5] to $67k$. Fomin and Thilikos [11] showed in 2004 that the same reduction rules as used by Alber et al. give a linear kernel (linear in $k + g$) for the problem in graphs of genus g. The next advances in kernelizing this problem were made by Alon and Gutner in 2008 [3]. They showed that the problem has a linear kernel in $K_{3,h}$-topological-minor-free graph classes (which include, for example, planar graphs), and a polynomial kernel (where the exponent depends on h) for K_h-topological-minor-free graph classes, these last being the most general class of graphs for which the problem has been previously shown to have a polynomial kernel. In the meantime, the same authors had shown in 2007 that the problem is FPT in (the strictly larger class of) graphs of bounded degeneracy [2], but had left open the question whether the problem has a polynomial kernel in such graph classes. In this paper, we answer this question in the affirmative, and show, in fact, that strictly larger classes of graphs — the $K_{i,j}$-free graph classes — have polynomial kernels for this problem.

See Table 1 for a summary of some FPT and kernelization results for the k-Dominating Set problem on various classes of graphs.

Our Results. We show that for any fixed $i, j \geq 1$, the k-Dominating Set problem has a polynomial kernel on graphs that do not have $K_{i,j}$ (a complete bipartite graph with the two parts having i and j vertices) as a subgraph. For input graph G and parameter k, the size of the kernel is bounded by k^c where c is a constant that depends only on i and j. A graph G is said to be *d-degenerate* if every subgraph of G has a vertex of degree at most d. Since a d-degenerate graph does not have $K_{d+1,d+1}$ as a subgraph, it follows that the k-Dominating

Table 1. Some FPT and kernelization results for k-DOMINATING SET. Results proved in this paper are marked with a †.

Graph Class	*FPT Algorithm Running Time*	*Kernel Size*
Planar	$O(k^4 + 2^{15.13\sqrt{k}}k + n^3)$ [12]	$O(k)$ [1,5]
Genus-g	$O((24g^2 + 24g + 1)^k n^2)$ [10]	$O(k + g)$ [11]
K_h-minor-free	$2^{O(\sqrt{k})}n^c$ [6],$O((\log h))^{hk/2} \cdot n$ [2]	$O(k^c)$ [3]
K_h-topological-minor-free	$(O(h))^{hk} \cdot n$ [2]	$O(k^c)$ [3]
d-degenerate	$k^{O(dk)}n$ [2]	$k^{O(dk)}$ [2], $O(k^{2(d+1)^2})$†
$K_{i,j}$-free	$O(n^{i+O(1)})$†	$O(k^{2i^2})$†

SET problem has a polynomial kernel on graphs of bounded degeneracy. This settles a question posed by Alon and Gutner in [3]. We also provide a slightly simpler and a smaller kernel for the version where we want the k-DOMINATING SET to be independent as well.

Note that except for d-degenerate and $K_{i,j}$-free graphs, all the other graph classes in Table 1 are *minor-closed* (See e.g., [7], Chapter 12, for the definition of a minor-closed graph class.). This seems to be indicative of the state of the art — the only other previous FPT or kernelization result for the k-DOMINATING SET problem on a non-minor-closed class of graphs that we know of is the $O(k^3)$ kernel and the resulting FPT algorithm for graphs that exclude triangles and 4-cycles [17]. In fact, this result can be modified to obtain similar bounds on graphs with just no 4-cycles (allowing triangles). Since a 4-cycle is just $K_{2,2}$, this result follows from the main result of this paper by setting $i = j = 2$.

Since a K_h-topological-minor-free graph has bounded degeneracy [3, Proposition 3.1] (for a constant h), the class of $K_{i,j}$-free graphs is more general than the class of K_h-topological-minor-free graphs. Thus we extend the class of graphs for which the k-DOMINATING SET problem is known to have (1) FPT algorithms and (2) polynomial kernels, to the class of $K_{i,j}$-free graphs.

Organization of the rest of the paper. In Section 2, we develop our main algorithm that runs in $O(n^{i+O(1)})$ time and constructs a kernel of size $O((j+1)^{2(i+1)}k^{2i^2})$ for k-DOMINATING SET on $K_{i,j}$-free graphs, for fixed $j \geq i \geq 2$. As a corollary we obtain, in Section 3, a polynomial kernel for d-degenerate graphs, with running time $O(n^{O(d)})$ and kernel size $O((d+2)^{2(d+2)}k^{2(d+1)^2})$. In Section 3.1 we describe an improvement to the above algorithm that applies to d-degenerate input graphs, yields a kernel of the same size as above, and runs in time $O(2^d dn^2)$. In Section 4 we describe a modification of the algorithm in Section 2 that constructs a polynomial kernel for the k-INDEPENDENT DOMINATING SET problem on $K_{i,j}$-free graphs. This kernel has $O(jk^i)$ vertices, and so implies a kernel of size $O((d+1)k^{d+1})$ for this problem on d-degenerate graphs. In Section 5 we state our conclusions and list some open problems.

Notation. All the graphs in this paper are finite, undirected and simple. In general we follow the graph terminology of [7]. We let $V(G)$ and $E(G)$ denote, respectively, the vertex and edge sets of a graph G. The *open-neighborhood* of a vertex v in a graph G, denoted $N(v)$, is the set of all vertices that are adjacent to v in G. A *k-dominating set* of graph G is a vertex-subset S of size at most k such that for each $u \in V(G) \setminus S$ there exists $v \in S$ such that $\{u, v\} \in E(G)$. Given a graph G and $A, B \subseteq V(G)$, we say that A dominates B if every vertex in $B \setminus A$ is adjacent in G to some vertex in A.

2 A Polynomial Kernel for $K_{i,j}$-Free Graphs

In this section we consider the parameterized k-DOMINATING SET problem on graphs that do not have $K_{i,j}$ as a subgraph, for fixed $j \geq i \geq 1$. It is easy to see that the problem has a linear kernel when $i = 1, j \geq i$, so we consider the cases $j \geq i \geq 2$. We solve a more general problem, namely the RWB-DOMINATING SET problem, defined as follows: Given a graph G whose vertex set V is partitioned into R_G, W_G, and B_G (colored red, white, and black, respectively) and a non-negative integer parameter k, is there a subset $S \subseteq V$ of size at most k such that $R_G \subseteq S$ and S dominates B_G? We call such an S an *rwb-dominating* set of G, and such a graph an *rwb-graph*.

Intuitively, the vertices colored red are those that will be picked up by the reduction rules in the k-dominating set D that we are trying to construct. In particular, if there is a k-dominating set in the graph, there will be one that contains all the red vertices. White vertices are those that have been already dominated. Clearly all neighbors of red vertices are white, but our reduction rules color some vertices white even if they have no red neighbors (at that point). These are vertices that will be dominated by one of some constant number of vertices identified by the reduction rules. See reduction rule 2 for more details. Black vertices are those that are yet to be dominated. It is easy to see that if we start with a general graph G and color all its vertices black to obtain an rwb-graph G', then G has a dominating set of size at most k if and only if G' has an rwb-dominating set of size at most k.

We describe an algorithm that takes as input an rwb-graph G on n vertices and a positive number k, and runs in $O(n^{i+O(1)})$ time. The algorithm either finds that G does not have any rwb-dominating set of size at most k, or it constructs an instance (G', k') on $O((j+1)^{i+1}k^{i^2})$ vertices such that G has an rwb-dominating set of size at most k if and only if G' has an rwb-dominating set of size at most k'.

The algorithm applies a sequence of reduction rules in a specified order. The input and output of each reduction rule are rwb-graphs.

Definition 1. *We say that graph G is* **reduced** *with respect to a reduction rule if an application of the rule to G does not change G.*

Each reduction rule satisfies the following correctness condition and preserves the invariants stated below:

Definition 2. (Correctness) *A reduction rule R is said to be* **correct** *if the following condition holds: if (G', k') is the instance obtained from (G, k) by one application of rule R then G' has an rwb-dominating set D' of size k' if and only if G has an rwb-dominating set D of size k.*[2]

Invariants:

1. None of the reduction rules introduces a $K_{i,j}$ into a graph.
2. In the rwb-graphs constructed by the algorithm, red vertices have all neighbors white.
3. Let R be any reduction rule, and let R' be a rule that precedes R in the given order. If G is a graph that is reduced with respect to R' and G' is a graph obtained by applying R to G, then G' is reduced with respect to R'.

2.1 The Reduction Rules and the Kernelization Algorithm

The kernelization algorithm assumes that the input graph is an rwb-graph. It applies the following rules exhaustively in the *given order*. Each rule is repeatedly applied till it causes no changes to the graph and then the next rule is applied.

To make it easier to present the reduction rules and the arguments of their correctness, we use a couple of notational conventions in this section. For each rule below, G denotes the graph on which the rule is applied, and G' the graph that results. Further, D and D' are as in Definition 2: D is an rwb-dominating set of size k of G, and D' an rwb-dominating of G' of size k'. [2]

Rule 1. Color all isolated black vertices of G red.

Rule 1 is correct as the only way to dominate isolated black vertices is by picking them in the proposed rwb-dominating set.

Rule 2. For $p = 1, 2, \ldots, i-2$, in this order, apply Rule 2.p repeatedly till it no longer causes any changes in the graph.

Rule 2.p Let $b = jk$ if $p = 1$, and $b = jk^p + k^{p-1} + k^{p-2} \cdots + k$ if $2 \le p \le i-2$. If a set of $(i-p)$ vertices $U = \{u_1, u_2, \ldots, u_{i-p}\}$, none of which is red, has more than b common black neighbors, then let B be this set of neighbors.

1. Color all the vertices in B white.
2. Add to the graph $(i-p)$ new (gadget) vertices $X = \{x_1, x_2, \ldots, x_{i-p}\}$ and all the edges $\{u, x\}; u \in U, x \in X$, as in Figure 1.
3. Color all the vertices in X black.

[2] Note, however, that none of our reduction rules changes the value of k, and so $k' = k$ for every one of these rules.

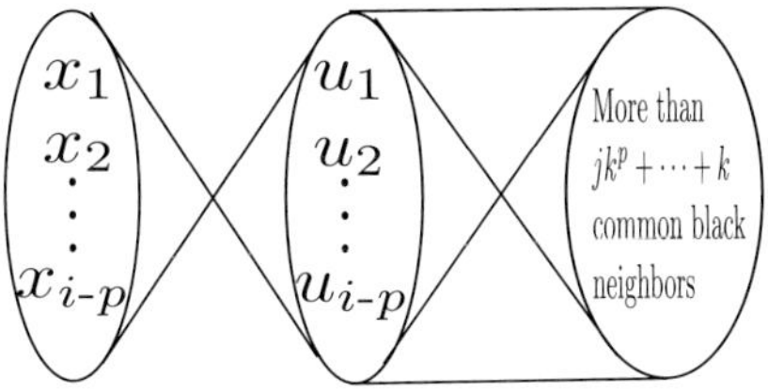

Fig. 1. Rule 2

Claim 1. Consider the application of Rule 2.$p, 1 \leq p \leq i-2$. If U is a set of vertices of G that satisfies the condition in Rule 2.p, then at least one vertex in U must be in any subset of $V(G)$ of size at most k that dominates B.

Proof. For $p = 1$, suppose that there is a rwb-dominating set D of G of size at most k that does not contain any vertex of U. Since U has more than $b = jk$ common black neighbors, there is a vertex in D that dominates at least $j+1$ common black neighbors of U (possibly including itself). That vertex along with U forms a $K_{i,j}$ in G, contradicting either the property of the input graph or the first invariant for the rules.

A similar argument works for $1 < p \leq i-2$, using the fact that Rule 2.p is applied to a graph that is reduced with respect to Rule 2.$(p-1)$; the reasoning is nearly the same as the one in the proof of Claim 2 below.

Lemma 1. *Rule* 2.p *is correct for* $1 \leq p \leq i-2$.

Proof. If G has an rwb-dominating set D of size k, then $D \cap U \neq \emptyset$ by Claim 1. So $D' := D$ is an rwb-dominating set of G', since $D \cap U$ dominates X. For the other direction, assume that D' exists. If $D' \cap U = \emptyset$ then since D' dominates X and X is independent, $X \subseteq D'$, and so set $D := D' \setminus X \cup U$. If $D' \cap X = \emptyset$ then since D' dominates X, $D' \cap U \neq \emptyset$, and so set $D := D'$. If $D' \cap U \neq \emptyset$ and $D' \cap X \neq \emptyset$ then pick an arbitrary vertex $b \in B$ and set $D := D' \setminus X \cup \{b\}$. □

Rule 3. If a black or white vertex u has more than $jk^{i-1} + k^{i-2} + \cdots + k^2 + k$ black neighbors, then color u red and color all the black neighbors of u white.

Claim 2. Let G be reduced with respect to Rule 1 and Rules 2.1 to 2.$(i-2)$. If a black or white vertex u of G has more than $h = jk^{i-1} + k^{i-2} + \cdots + k^2 + k$ black neighbors (let this set of neighbors be B), then u must be in any subset of $V(G)$ of size at most k that dominates B.

Proof. Let $S \subseteq V(G)$ be a set of size at most k that dominates B. If S does not contain u, then there is a $v \in S$ that dominates at least $(h/k)+1$ of the vertices in B. The vertex v is not red (because of the second invariant), and u, v have $h/k > jk^{i-2} + k^{i-3} + \cdots + 1$ common black neighbors, a contradiction to the fact that G is reduced with respect to Rule 2.$(i-2)$. □

This proves the correctness of Rule 3 on graphs reduced with respect to Rule 1 and Rules 2.1 to 2.$(i-2)$.

Rule 4. If a white vertex u is adjacent to at most one black vertex, then delete u and apply Rule 1.

It is easy to see that Rule 4 is correct, since if u has no black neighbor in G then u has no role in dominating B_G; if u has a single black neighbor v then we can replace u with v in D'.

Rule 5. If there is a white vertex u and a white or black vertex v such that $N(u) \cap B_G \subseteq N(v) \cap B_G$, then delete u and apply Rule 1.

The correctness of this rule follows from the fact that if D chooses u, then we can choose v in D'.

Rule 6. If $|R_G| > k$ or $|B_G| > jk^i + k^{i-1} + k^{i-2} + \cdots + k^2$ then output "NO".

The correctness of the rule when $|R_G| > k$ is obvious as the proposed dominating set we construct should contain all of R_G. Note that in a graph G reduced with respect to Rules 1 to 5, no white or black vertex has more than $jk^{i-1} + k^{i-2} + \cdots + k$ black neighbors, or else Rule 3 would have applied, contradicting the third invariant. Hence k of these vertices can dominate at most $jk^i + k^{i-1} + k^{i-2} + \cdots + k^2$ black vertices and hence if $|B_G| > jk^i + k^{i-1} + k^{i-2} + \cdots + k^2$, the algorithm is correct in saying "NO".

2.2 Algorithm Correctness and Kernel Size

The following claim giving the correctness of the kernelization algorithm follows from the correctness of the reduction rules.

Claim 3. Let G be the input rwb-graph and H the rwb-graph constructed by the algorithm after applying all the reduction rules. Then G has an rwb-dominating set of size at most k if and only if there is an rwb-dominating set of size at most k in H.

Now we move on to prove a polynomial bound on the size of the reduced instance.

Lemma 2. *Starting with a $K_{i,j}$-free rwb-graph G as input, if the kernelization algorithm says "NO" then G does not have an rwb-dominating set of size at most k. Otherwise, if the algorithm outputs the rwb-graph H, then $|V(H)| = O((j+1)^{i+1}k^{i^2})$.*

Proof. The correctness of Rule 6 establishes the claim if the algorithm says "NO". Now suppose the algorithm outputs H without saying "NO". The same rule establishes that $|R_H| \leq k$ and $b = |B_H| \leq jk^i + k^{i-1} + \cdots + k \leq (j+1)k^i$. Now we bound $|W_H|$. Note that no two white vertices have identical black neighborhoods, or else Rule 5 would have applied. Also each white vertex has at least two black neighbors, or else Rule 4 would have applied. Hence the number of white vertices that have less than i black neighbors is at most $\binom{b}{2} + \binom{b}{3} + \cdots + \binom{b}{i-1} \leq 2b^{i-1}$. No set of i black vertices has more than $(j-1)$ common white neighbors, or else these form a $K_{i,j}$. Hence the number of white vertices that have i or more black neighbors is at most $\binom{b}{i}(j-1) \leq (j-1)b^i$. The bound in the lemma follows. □

The algorithm can be implemented in $O(n^{i+O(1)})$ time, as the main Rule 2 can be applied by running through various subsets of $V(G)$ of size p for p ranging from 1 to $i-2$. Thus, we have

Lemma 3. *For any fixed $j \geq i \geq 1$, the* RWB-DOMINATING SET *problem (with parameter k) on $K_{i,j}$-free graphs has a polynomial kernel with $O((j+1)^{i+1}k^{i^2})$ vertices.*

To obtain a polynomial kernel for the k-DOMINATING SET problem on $K_{i,j}$-free graphs, we first color all the vertices black and use Lemma 3 on this RWB-DOMINATING SET problem instance. To transform the reduced colored instance H to an instance of (the uncolored) k-DOMINATING SET, we can start by deleting all the red vertices, since they have no black neighbors. But we need to capture the fact that the white vertices need not be dominated. This can be done by, for example, adding a new vertex v_x adjacent to every vertex x in W_H of the reduced graph H, and attaching $k+|W_H|+1$ separate pendant vertices to each of the vertices v_x. It is easy to see that the new graph does not have a $K_{i,j}$, $j \geq i \geq 2$, if H does not have one and that H has at most k black or white vertices dominating B_H if and only if the resulting (uncolored) graph has a dominating set of size at most $|W_H|+k$. Thus after reducing to the uncolored version, k becomes $k+|W_H|$ and the number of vertices increases by $(k+|W_H|+2)\cdot|W_H|$. Hence by Lemma 3, we have

Theorem 1. *For any fixed $j \geq i \geq 1$, the k-*DOMINATING SET *problem on $K_{i,j}$-free graphs has a polynomial kernel with $O((j+1)^{2(i+1)}k^{2i^2})$ vertices.*

3 A Polynomial Kernel for d-Degenerate Graphs

A d-degenerate graph does not contain $K_{d+1,d+1}$ as a subgraph, and so the kernelization algorithm of the previous section can be applied to a d-degenerate graph, setting $i=j=d+1$. The algorithm runs in time $O((d+1)^2 n^{d+O(1)})$ and constructs a kernel with $O((d+2)^{2(d+2)} \cdot k^{2(d+1)^2})$ vertices. Since a d-degenerate graph on v vertices has at most dv edges, we have:

Corollary 1. *The k-*DOMINATING SET *problem on d-degenerate graphs has a kernel on $O((d+2)^{2(d+2)} \cdot k^{2(d+1)^2})$ vertices and edges.*

Corollary 1 settles an open problem posed by Alon and Gutner in [3].

3.1 Improving the Running Time

We describe a modification of our algorithm to d-degenerate graphs that makes use of the following well known property of d-degenerate graphs, to reduce the running time to $O(2^d \cdot dn^2)$; the bound on the kernel size remains the same.

Algorithm 1. Faster implementation of Rule 2.p in d-degenerate graphs.

```
for l := 1 to n
do
  if v_l is black and its degree in G[v_{l+1}, ..., v_n] is at least d − p + 1
  then
    Find the neighborhood N of v_l in G[v_{l+1}, ..., v_n]
    for each (d − p + 1)-subset S of N
    do
      if S has more than (d + 1)k^p + k^{p−1} + ··· + k
      common black neighbors in G
      then
        Apply the three steps of Rule 2.p, taking S as U
      endif
    done
  endif
done
```

Fact 1. [13, Theorem 2.10] Let G be a d-degenerate graph on n vertices. Then one can compute, in $O(dn)$ time, an ordering $v_1, v_2, \ldots, v_n$ of the vertices of G such that for $1 \leq i \leq n$, v_i has at most d neighbors in the subgraph of G induced on $\{v_{i+1}, \ldots, v_n\}$.

The modification to the algorithm pertains to the way rules 2.1 to 2.$(d-1)$ are implemented: the rest of the algorithm remains the same.

In implementing Rule 2.p, $1 \leq p \leq (d-1)$, instead of checking each $(d-p+1)$-subset of vertices in the graph to see if it satisfies the condition in the rule, we make use of Fact 1 to quickly find such a set of vertices, if it exists. Let G be the graph instance on n vertices on which Rule 2.p is to be applied. First we delete, temporarily, all the red vertices in G. We then find an ordering $v_1, v_2, \ldots, v_n$ of the kind described in the above fact, of all the remaining vertices in G. Let U and B be as defined in the rule. The first vertex v_l in $U \cup B$ that appears in the ordering has to be from B, since each vertex in U has degree greater than d. The vertex v_l will then have a neighborhood of size $d-p+1$ that in turn has B as its common neighborhood. We use this fact to look for such a pair (U, B) and exhaustively apply Rule 2.p to G. See Algorithm 1 for pseudocode of the algorithm. We then add back the red vertices that we deleted prior to this step, along with all their edges to the rest of the graph.

As $|N| \leq d$, the inner *for* loop is executed at most $\binom{d}{p-1}$ times for each iteration of the outer loop. Each of the individual steps in the algorithm can be done in $O(dn)$ time, and so Rule 2.p can be applied in $O(dn\sum_{l=1}^{n}\binom{d}{p-1})$ time. All the rules 2.p can therefore be applied in $O(dn\sum_{l=1}^{n}\sum_{p=1}^{d-1}\binom{d}{p-1}) = O(2^d \cdot dn^2)$ time. Thus we have:

Theorem 2. *For any fixed $d \geq 1$, the k-*Dominating Set *problem on d-degenerate graphs has a kernel on $O((d+2)^{2(d+2)} \cdot k^{(2(d+1))^2})$ vertices and edges, and this kernel can be found in $O(2^d \cdot dn^2)$ time for an input graph on n vertices.*

4 A Polynomial Kernel for Independent Dominating Set on $K_{i,j}$-Free Graphs

The k-INDEPENDENT DOMINATING SET problem asks, for a graph G and a positive integer k given as inputs, whether G has a dominating set S of size at most k such that S is an independent set (i.e. no two vertices in S are adjacent). This problem is known to be NP-hard for general graphs [14], and the problem parameterized by k is $W[2]$-complete [9]. Using a modified version of the set of reduction rules in Section 2 we show that the k-INDEPENDENT DOMINATING SET has a polynomial kernel in K_{ij}-free graphs for $j \geq i \geq 1$. For $i = 1, j \geq 1$ we can easily obtain trivial kernels as before, and for $i = 2, j \geq 2$ a simplified version of the following algorithm gives a kernel of size $O(j^3k^4)$.

4.1 The Reduction Rules

Rule 1 is the same as for the DOMINATING SET kernel for K_{ij}-free graphs (Section 2.1). Rules 2.1 to 2.$(i-2)$ and Rule 3 are modified to make use of the fact that we are looking for a dominating set that is independent. A vertex u that is made white will never be part of the independent dominating set D that is sought to be constructed by the algorithm, since u is adjacent to some vertex $v \in D$. So a vertex can be deleted as soon as it is made white. Also, rules $1, 2.1 \ldots 2.(i-2)$ and 3 are the only rules. Rules 4 and 5 from that section do not apply, because of the same reason as above. The modified rules ensure that no vertex is colored white, and so they work on *rb-graphs*: graphs whose vertex set is partitioned into red and black vertices. Using these modified rules, the bounds of $|R_H|$ and $|B_H|$ in the proof of Lemma 2, and the fact that there are no white vertices, we have

Theorem 3. *For any fixed $j \geq i \geq 1$, the k-INDEPENDENT DOMINATING SET problem on $K_{i,j}$-free graphs has a polynomial kernel with $O(jk^i)$ vertices.*

For d-degenerate graphs, we have $i = j = d + 1$, and therefore we have:

Corollary 2. *For any fixed $d \geq 1$, the k-INDEPENDENT DOMINATING SET problem on d-degenerate graphs has a polynomial kernel with $O((d+1)k^{(d+1)})$ vertices.*

5 Conclusions and Future Work

In this paper, we presented a polynomial kernel for the k-DOMINATING SET problem on graphs that do not have $K_{i,j}$ as a subgraph, for any fixed $j \geq i \geq 1$. We used this to show that the k-DOMINATING SET problem has a polynomial kernel of size $O((d+2)^{2(d+2)} \cdot k^{2(d+1)^2})$ on graphs of bounded degeneracy, thereby settling an open problem from [3]. Our algorithm also yielded a slightly simpler and a smaller kernel for the k-INDEPENDENT DOMINATING SET problem on $K_{i,j}$-free and d-degenerate graphs. These algorithms are based on simple reduction rules that look at the common neighborhoods of sets of vertices.

Dom et al. [8] have shown, by extending the kernel lower-bound techniques of Bodlaender et al. [4], that the k-Dominating Set problem on d-degenerate graphs does not have a kernel of size polynomial in *both* d and k unless the Polynomial Hierarchy collapses to the third level. This shows that the kernel size that we have obtained for this class of graphs cannot possibly be significantly improved.

Many interesting classes of graphs are of bounded degeneracy. These include all nontrivial minor-closed families of graphs such as planar graphs, graphs of bounded genus, graphs of bounded treewidth, and graphs excluding a fixed minor, and some non-minor-closed families such as graphs of bounded degree. Graphs of degeneracy d are $K_{d+1,d+1}$-free. Since any $K_{i,j}; j \geq i \geq 2$ contains a 4-cycle, every graph of girth 5 is $K_{i,j}$-free. From [18, Theorem 1], there exist graphs of girth 5 and arbitrarily large degeneracy. Hence $K_{i,j}$-free graphs are strictly more general than graphs of bounded degeneracy. To the best of our knowledge, $K_{i,j}$-free graphs form the largest class of graphs for which FPT algorithms and polynomial kernels are known for the dominating set problem variants discussed in this paper.

One interesting direction of future work is to try to demonstrate kernels of size $f(d) \cdot k^c$ for the k-Dominating Set problem on d-degenerate graphs, where c is independent of d. Note that the result of Dom et al. mentioned above does *not* suggest that such kernels are unlikely. Another challenge is to improve the running times of the kernelization algorithms: to remove the exponential dependence on d of the running time for d-degenerate graphs, and to get a running time of the form $O(n^c)$ for $K_{i,j}$-free graphs where c is independent of i and j.

Acknowledgments. We thank Aravind Natarajan for pointing out the connection between $K_{i,j}$-free and d-degenerate graphs, Saket Saurabh and the other authors of [8] for sharing with us the lower-bound result mentioned in their paper, and Shai Gutner for his comments on an earlier draft of this paper.

References

1. Alber, J., Fellows, M.R., Niedermeier, R.: Polynomial-time data reduction for Dominating Set. Journal of the ACM 51(3), 363–384 (2004)
2. Alon, N., Gutner, S.: Linear time algorithms for finding a dominating set of fixed size in degenerated graphs. In: Lin, G. (ed.) COCOON 2007. LNCS, vol. 4598, pp. 394–405. Springer, Heidelberg (2007)
3. Alon, N., Gutner, S.: Kernels for the dominating set problem on graphs with an excluded minor. Technical Report TR08-066, The Electronic Colloquium on Computational Complexity (ECCC) (2008)
4. Bodlaender, H.L., Downey, R.G., Fellows, M.R., Hermelin, D.: On problems without polynomial kernels (Extended abstract). In: Aceto, L., Damgård, I., Goldberg, L.A., Halldórsson, M.M., Ingólfsdóttir, A., Walukiewicz, I. (eds.) ICALP 2008, Part I. LNCS, vol. 5125, pp. 563–574. Springer, Heidelberg (2008)
5. Chen, J., Fernau, H., Kanj, I.A., Xia, G.: Parametric duality and kernelization: Lower bounds and upper bounds on kernel size. SIAM Journal of Computing 37(4), 1077–1106 (2007)

6. Demaine, E.D., Fomin, F.V., Hajiaghayi, M., Thilikos, D.M.: Subexponential parameterized algorithms on bounded-genus graphs and H-minor-free graphs. Journal of the ACM 52(6), 866–893 (2005)
7. Diestel, R.: Graph theory, 3rd edn. Springer, Heidelberg (2005)
8. Dom, M., Lokshtanov, D., Saurabh, S.: Incompressibility through Colors and IDs. Accepted at ICALP 2009 (2009)
9. Downey, R.G., Fellows, M.R.: Parameterized Complexity. Springer, Heidelberg (1999)
10. Ellis, J.A., Fan, H., Fellows, M.R.: The Dominating Set problem is fixed parameter tractable for graphs of bounded genus. Journal of Algorithms 52(2), 152–168 (2004)
11. Fomin, F.V., Thilikos, D.M.: Fast parameterized algorithms for graphs on surfaces: Linear kernel and exponential speed-up. In: Díaz, J., Karhumäki, J., Lepistö, A., Sannella, D. (eds.) ICALP 2004. LNCS, vol. 3142, pp. 581–592. Springer, Heidelberg (2004)
12. Fomin, F.V., Thilikos, D.M.: Dominating sets in planar graphs: Branch-width and exponential speed-up. SIAM Journal of Computing 36(2), 281–309 (2006)
13. Franceschini, G., Luccio, F., Pagli, L.: Dense trees: a new look at degenerate graphs. Journal of Discrete Algorithms 4, 455–474 (2006)
14. Garey, M.R., Johnson, D.S.: Computers and Intractability: A Guide to the Theory of NP–Completeness. Freeman, San Francisco (1979)
15. Golovach, P.A., Villanger, Y.: Parameterized complexity for domination problems on degenerate graphs. In: Broersma, H., Erlebach, T., Friedetzky, T., Paulusma, D. (eds.) WG 2008. LNCS, vol. 5344, Springer, Heidelberg (2008)
16. Guo, J., Niedermeier, R.: Invitation to data reduction and problem kernelization. SIGACT News 38(1), 31–45 (2007)
17. Raman, V., Saurabh, S.: Short cycles make W-hard problems hard: FPT algorithms for W-hard problems in graphs with no short cycles. Algorithmica 52(2), 203–225 (2008)
18. Sachs, H.: Regular graphs with given girth and restricted circuits. Journal of the London Mathematical Society s1-38(1), 423–429 (1963)

Contraction Bidimensionality: The Accurate Picture*

Fedor V. Fomin[1], Petr Golovach[1], and Dimitrios M. Thilikos[2]

[1] Department of Informatics, University of Bergen, N-5020 Bergen, Norway
{fomin,Peter.Golovach}@ii.uib.no
[2] Department of Mathematics, University of Athens, Panepistimioupolis, GR15784 Athens, Greece
sedthilk@math.uoa.gr

Abstract. We provide new combinatorial theorems on the structure of graphs that are contained as contractions in graphs of large treewidth. As a consequence of our combinatorial results we unify and significantly simplify contraction bidimensionality theory—the meta algorithmic framework to design efficient parameterized and approximation algorithms for contraction closed parameters.

1 Introduction

The proof of Wagner's conjecture was the principal goal of the Graph Minors project of Robertson and Seymour. On the way to this goal, Robertson and Seymour have developed a powerful theory, which apparently became one of the most influential achievements in the modern Combinatorics. There are several very important consequences of Graph Minors theory for Algorithms Theory as well. In particular, obstruction theorems from Graph Minors give rise to powerful tools in the design of algorithms. Roughly speaking, such theorems say that either some width parameter (like pathwidth, treewidth, branchwidth, rank-width, etc.) of a graph is small (in which case one can proceed with dynamic programming techniques), or the graph contains some big pattern graph as a minor and such a graph can certify an answer to the problem. The most celebrated theorem of this type is the *Excluding Grid-minor Theorem* [15,13]: *"there is a function f, such that every graph G of treewidth at least $f(k)$ contains a $(k \times k)$-grid as a minor"*. Building on this theorem, Robertson and Seymour obtained their well known polynomial time algorithm for the disjoint path problems [14]. The bound in the obstruction theorem was refined by Robertson et al. [18] to $f(k) = 20^{2k^5}$. It is conjectured by Robertson et al. [18] that this bound is polynomial. For some classes of graphs like planar, graphs of bounded genus, and more generally, H-minor-free graphs, it is possible to prove that $f(k) = O(k)$ [4,5,18] and the

* The research of the first two authors was supported by the Norwegian Research Council. The research of the third author was Supported by the project "Kapodistrias" (AΠ 02839/28.07.2008) of the National and Kapodistrian University of Athens (project code. 70/4/8757).

A. Fiat and P. Sanders (Eds.): ESA 2009, LNCS 5757, pp. 706–717, 2009.

bidimensionality theory was built on the top of these results. Bidimensionality provides a meta-algorithmic framework for the design of *subexponential* parameterized algorithms (i.e. the algorithms of running time $2^{o(k)}n^{O(1)}$, where k is the parameter and n the length of the input) for wide families of combinatorial problems. There is also an evidence that such subexponential algorithms are optimal even for sparse structures such as planar graphs (see surveys [2,9] on bidimensionality). The applications of this theory are well understood and developed for minor closed graph optimization problems, i.e. the problems which only decrease under edge deletions/contractions.

However, many graph optimization problems are closed under edge contractions but *not* under edge deletions. Examples include most variants and extensions of the dominating set problem, connected dominating set, travelling salesman, or various distance modification problems. One of the challenges in Graph Theory and Algorithms is a possible extension of these results from graph minors to graph contractions (see [2,7]). The proofs from graph minors are very nontrivial, and such an extension is a difficult and sometimes an impossible task. For example, Demaine et al. in [7] disproved the analogue of Wagner' s Conjecture for graph contractions. Despite of that, it is possible to extend some algorithmic graph-minor results to graph contractions. In particular, it is possible to adapt obstruction theorems from graph minors to graph contractions. The extension of bidimensionality theory and meta theorems for contraction parameters was obtained in [3,4,5]. It is based on a modification of the grid-minor theorem for apex-minor-free graphs (graph class is *apex-minor-free* if it does not contain a graph with some fixed apex graph as a minor). An $(k \times k)$-augmented grid of span s is an $(k \times k)$-grid with some extra edges such that each vertex is attached to at most s non-boundary vertices of the grid. Fig. 1 provides an example of an augmented (6×6)-grid of span 5. The obstruction theorem for contractions [3,5] states that for every apex graph H there is a universal constant c_H such that every H-minor-free graph of treewidth at least $c_H \cdot k$ can be contracted into a $(k \times k)$-augmented grid of span c_H. This combinatorial theorem is the cornerstone of the meta algorithmic framework in the design of subexponential parameterized algorithms for contraction-closed parameters on different classes of sparse graphs. This framework, called *contraction bidimensionality*, was developed by Demaine et al. [3] (see also surveys [2,9]). (We postpone the definitions related to contraction bidimensionality untill Section 5, where we revise it.) Unfortunately, there is a drawback in the bidimensionality framework which is inherited by the "excluding-grid" theorem for contractions. The problem is that the number of augmented grids is huge. Even the number of planar augmented grids, i.e. graphs obtained by triangulating some faces of an $(k \times k)$-grid, is at least $2^{(k-1)^2}$. As a result, to verify if a parameter is apex-contraction bidimensional, one has to estimate its value on a graph family of exponential size. In this paper, we eliminate this main inefficiency of the meta algorithmic framework by redefining the notion of apex-contraction bidimensionality in simple and unifying way. With the new notion, we reduce the verification apex-contraction bidimensionality to the estimation of the value of the parameter in *one* specific

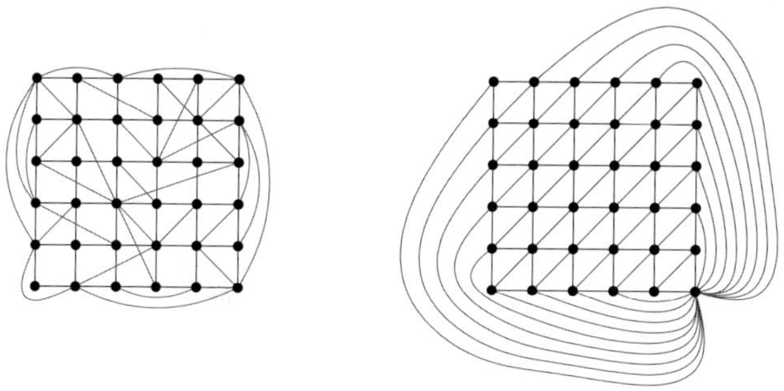

Fig. 1. An augmented (6×6)-grid of span 5 (on the left) and the graph Γ_6 (on the right)

triangulation of the $(k \times k)$-grid. The proof that this simple criterion holds for apex-contraction bidimensionality is highly nontrivial and requires the identification of a *single* pattern graph to which large-treewidth apex-minor free graphs can be contracted. This result has its own combinatorial merit as the *contraction counterpart* of the Excluding-grid Minor Theorem.

Let Γ_k be the graph obtained from the $(k \times k)$-grid by triangulating internal faces of $(k \times k)$-grid such that all internal vertices become of degree 6, all non-corner external vertices are of degree 4, and then one corner of degree two is joined by edges with all vertices of the external face. Graph Γ_6 is shown in Fig. 1. The main combinatorial contribution of our work is the following theorem.

Theorem 1. *Let H be an apex graph. There is a constant c_H such that every connected graph G excluding H as a minor and of treewidth at least $c_H \cdot k$, contains Γ_k as a contraction.*

In the conclusion of their survey [2], Demaine and Hajiaghayi mentioned that contraction bidimensionality is so far undefined for H-minor-free graphs (or general graphs). They also wrote that it would be quite interesting to explore an analogous theory of graph contractions paralleling the Graph Minor Theory. In this direction, we make a step by extending Theorem 1 for more general graph classes. We prove that for H-minor free graphs, big treewidth implies the existence of one two pattern graphs as contractions. In what follows, $\mathbf{tw}(G)$ denotes the treewidth of G.

Theorem 2. *Let G be a connected graph excluding a graph H as a minor. Then there exists some constant c_H such that if $\mathbf{tw}(G) \geq c_H \cdot k^2$, then G contains as a contraction either Γ_k or Π_k, where Π_k is the graph obtained from Γ_k by adding a vertex adjacent to all vertices of Γ_k.*

We stress that one can (artificially) define contraction-closed parameters whose value is non-trivial for Π_k and, for them, bidimensionality could be defined even for H-minor free graphs. However, it is interesting to notice that *for all* known contraction-closed parameters the value of Π_k is independent from k and this indicates that the classic Win/win approach for contraction closed parameters is *structurally confined* to apex-minor free graphs.

Finally, we show the following analogue of the Excluding Grid-minor Theorem. It follows that, instead of a single grid, there are three pattern graphs appearing as contractions in graphs with big treewidth (for minors, the only such pattern is the grid).

Theorem 3. *For any positive integer k, there is a constant c_k such that every connected graph G where* $\mathbf{tw}(G) > c_k$, *contains one of K_k, Γ_k, Π_k as a contraction, where K_k is a complete graph on k vertices.*

2 Basic Definitions

We consider finite undirected graphs without loops or multiple edges. The vertex set of a graph G is denoted by $V(G)$ and its edge set by $E(G)$.

Let G be a graph. For a vertex v, we denote by $N_G(v)$ its *(open) neighborhood*, i.e. the set of vertices which are adjacent to v. The *closed neighborhood* of v, i.e. the set $N_G(v) \cup \{v\}$, is denoted by $N_G[v]$. For $U \subseteq V(G)$, we define $N_G[U] = \bigcup_{v \in U} N_G[v]$ (we may omit index if the graph under consideration is clear from the context). If $U \subseteq V(G)$ (resp. $u \in V(G)$ or $E \subset E(G)$ or $e \in E(G)$) then $G - U$ (resp. $G - u$ or $G - E$ or $G - e$) is the graph obtained from G by the removal of vertices of U (resp. of vertex u or edges of E or of the edge e).

Surfaces. A *surface* Σ is a compact 2-manifold without boundary (we always consider connected surfaces). Whenever we refer to a *Σ-embedded graph G* we consider a 2-cell embedding of G in Σ. To simplify notations, we do not distinguish between a vertex of G and the point of Σ used in the drawing to represent the vertex or between an edge and the line representing it. We also consider a graph G embedded in Σ as the union of the points corresponding to its vertices and edges. That way, a subgraph H of G can be seen as a graph H, where $H \subseteq G$. Recall that $\Delta \subseteq \Sigma$ is an open (resp. closed) disc if it is homeomorphic to $\{(x, y) : x^2 + y^2 < 1\}$ (resp. $\{(x, y) : x^2 + y^2 \leq 1\}$). The *Euler genus* of a non-orientable surface Σ is equal to the non-orientable genus $\tilde{g}(\Sigma)$ (or the crosscap number). The *Euler genus* of an orientable surface Σ is $2g(\Sigma)$, where $g(\Sigma)$ is the orientable genus of Σ. We refer to the book of Mohar and Thomassen [12] for more details on graphs embeddings.

Contractions and minors. Given an edge $e = \{x, y\}$ of a graph G, the graph G/e is obtained from G by contracting the edge e, i.e. the endpoints x and y are replaced by a new vertex v_{xy} which is adjacent to the old neighbors of x and y (except from x and y). A graph H obtained by a sequence of edge-contractions is said to be a *contraction* of G. In this work we use contraction with different topological properties, and for this purpose it is convenient to give an alternative definition of contraction.

Let G and H be graphs and let $\phi : V(G) \to V(H)$ be a surjective mapping such that

1. for every vertex $v \in V(H)$, its codomain $\phi^{-1}(v)$ induces connected graph $G[\phi^{-1}(v)]$;
2. for every edge $\{v, u\} \in E(H)$, the graph $G[\phi^{-1}(v) \cup \phi^{-1}(u)]$ is connected;
3. for every $\{v, u\} \in E(G)$, either $\phi(v) = \phi(u)$, or $\{\phi(v), \phi(u)\} \in E(H)$.
We say that H *is a contraction of* G *via* ϕ, and denote it as $H \leq_c^{\phi} G$. Let us observe that H is a contraction of G if $H \leq_c^{\phi} G$ for some $\phi : V(G) \to V(H)$. In this case we simply write $H \leq_c G$. If $H \leq_c^{\phi} G$ and $v \in V(H)$ then we call the codomain $\phi^{-1}(v)$ by the *model of* v in G.

Let G be a graph embedded in some surface Σ and let H be a contraction of G via function ϕ. We say that H is a *surface contraction* of G if for each vertex $v \in V(H)$, $G[\phi^{-1}(v)]$ is embedded in some open disk in Σ.

Let G_0 be a graph embedded in some surface Σ of Euler genus γ and let G^+ be another graph that might share common vertices with G_0. We set $G = G_0 \cup G^+$. Let also H be some graph and let $v \in V(H)$. We say that G *contains a graph* H *as a* v*-smooth contraction* if $H \leq_c^{\phi} G$ for some $\phi : V(G) \to V(H)$ and there exists an closed disk D in Σ such that all the vertices of G that are outside D are exactly the model of v, i.e. $\phi^{-1}(v) = V(G) \setminus (V(G) \cap D)$.

A graph H is a *minor* of a graph G if H is the contraction of some subgraph of G and we denote it $H \leq_m G$. It is said that H is a *surface minor* of a graph G embedded in some surface Σ if H is the surface contraction of some subgraph of G. It can be easily noted that if H is a surface minor of a graph G embedded in a surface Σ then it can be assumed that H is embedded in a surface Σ' homeomorphic to Σ. For simplicity, we assume in such cases that Σ' and Σ are the same surface.

We say that a graph G is H*-minor-free* when it does not contain H as a minor. We also say that a graph class $\mathcal{G}$ is H*-minor-free* (or, excludes H as a minor) when all its members are H-minor-free. An *apex graph* is a graph obtained from a planar graph G by adding a vertex and making it adjacent to some of the vertices of G. A graph class $\mathcal{G}$ is *apex-minor-free* if $\mathcal{G}$ excludes a fixed apex graph H as a minor.

Grids and their triangulations. Let k and r be positive integers where $k, r \geq 2$. The $(k \times r)$*-grid* is the Cartesian product of two paths of lengths $k - 1$ and $r - 1$ respectively. A vertex of a $(k \times r)$-grid is a *corner* if it has degree 2. Thus each $(k \times r)$-grid has 4 corners. A vertex of a $(k \times r)$-grid is called *internal* if it has degree 4, otherwise it is called *external*.

A *partial triangulation* of a $(k \times r)$-grid is a planar graph obtained from a $(k \times r)$-grid (we call it the *underlying grid*) by adding edges. Let us note that there are many non-isomorphic partial triangulations of on underlying grid. For each partial triangulation of a $(k \times r)$-grid we use the terms *corner*, *internal* and *external* referring to the corners, the internal and the external vertices of the underlying grid.

Let us remind that we define Γ_k as the following (unique, up to isomorphism) triangulation of a plane embedding of the $(k \times k)$-grid. Let Γ be a plane embedding of the $(k \times k)$-grid such that all external vertices are on the boundary of the external face. We triangulate internal faces of the $(k \times k)$-grid such that all the internal vertices have degree 6 in the obtained graph and all non-corner external vertices have degree 4, and then one corner of degree two is joined by edges with all vertices of the external face (we call this corner *loaded*). We also use notation Γ_k^* for the graph obtained from Γ_k if we remove all edges incident to its loaded vertex that do not exist in its underlying grid. We define the graph Π_k as the graph obtained if we add a new vertex in Γ_k and connect it with all vertices of it. Let K be a clique of size 3 in Γ_k^*. Notice that exactly two of the edges of $\Gamma_k[K]$ are also edges of the underlying $(k \times k)$-grid of Γ_k. We call the unique vertex of K that is incident to both these two edges *rectangular* vertex of K.

Let G be a partial triangulation of a $(k \times k)$-grid and let m be a positive integer. Then by $\mathbf{P}_m(G)$ we denote the collection of m^2 vertex disjoint induced subgraphs of G where all of them are isomorphic to a $(\lfloor k/m \rfloor \times \lfloor k/m \rfloor)$-grid and where the union of their vertices induce a graph containing $(\lfloor k/m \rfloor \cdot m \times \lfloor k/m \rfloor \cdot m)$-grid as a spanning subgraph.

Suppose that G is a connected graph which contains as an induced subgraph a partially triangulated $((k+2) \times (k+2))$-grid Γ in such a way that internal vertices of Γ are not adjacent to vertices of $V(G) \setminus V(\Gamma)$. We define the *boundary contraction* of G to Γ as the partially triangulated $(k \times k)$-grid $\mathbf{bc}(G, \Gamma)$ obtained as follows: let v be a corner of the subgrid of Γ induced by the internal vertices which has the minimum degree (in this graph), all external vertices of Γ are contracted to v, and then all vertices of $V(G) \setminus V(\Gamma)$ are contracted to v. Note that if Γ is embedded in a disk of some surface Σ then $\mathbf{bc}(G, \Gamma)$ is a v-smooth contraction of G.

Treewidth and pathwidth. A *tree decomposition* of a graph G is a pair $(\mathcal{X}, T)$ where T is a tree and $\mathcal{X} = \{X_i \mid i \in V(T)\}$ is a collection of subsets of $V(G)$ such that:

1. $\bigcup_{i \in V(T)} X_i = V(G)$,
2. for each edge $\{x, y\} \in E(G)$, $\{x, y\} \subseteq X_i$ for some $i \in V(T)$; and
3. for each $x \in V(G)$ the set $\{i \mid x \in X_i\}$ induces a connected subtree of T.

The *width* of a tree decomposition $(\{X_i \mid i \in V(T)\}, T)$ is $\max_{i \in V(T)} \{|X_i| - 1\}$. The *treewidth* of a graph G is the minimum width over all tree decompositions of G. If, in the above definitions, we restrict the tree T to be a path then we define the notions of *path decomposition* and *pathwidth*. We write $\mathbf{tw}(G)$ and $\mathbf{pw}(G)$, respectively, for the treewidth and the pathwidth of a graph G.

Graph minor theorem. The proof of our results is using the Excluded Minor Theorem from the Graph Minor theory. Before we state it, we need some definitions.

Definition 1 (Clique-Sums). *Let G_1 and G_2 be two disjoint graphs, and $k \geq 0$ an integer. For $i = 1, 2$, let $W_i \subseteq V(G_i)$, form a clique of size h and let G_i' be the graph obtained from G_i by removing a set of edges (possibly empty) from the clique $G_i[W_i]$. Let $F : W_1 \to W_2$ be a bijection between W_1 and W_2. We define the* h-clique-sum *of G_1 and G_2, denoted by $G_1 \oplus_{h,F} G_2$, or simply $G_1 \oplus G_2$ if* *there is no confusion, as the graph obtained by taking the union of G_1' and G_2' by identifying $w \in W_1$ with $F(w) \in W_2$, and by removing all the multiple edges. The image of the vertices of W_1 and W_2 in $G_1 \oplus G_2$ is called the* join *of the sum.*

Note that some edges of G_1 and G_2 are not edges of G, since it is possible that they had edges which were removed by clique-sum operation. Such edges are called *virtual* edges of G. We remark that $\oplus$ is not well defined; different choices of G_i' and the bijection F could give different clique-sums. A sequence of h-clique-sums, not necessarily unique, which result in a graph G, is called a *clique-sum decomposition* of G.

Definition 2 (*h-nearly embeddable graphs*). *Let Σ be a surface with cycles $C_1, \ldots, C_h$, such that each cycle C_i is the border of an open disc Δ_i in Σ. A graph G is h-*nearly embeddable *in Σ, if G has a subset X of size at most h, called* apices, *so that there are (possibly empty) subgraphs $G_0, \ldots, G_h$ of $G \setminus X$ where*

i) $G' = G \setminus X = G_0 \cup G_1 \cup \cdots \cup G_h$ (we denote $G^+ = G_1 \cup \ldots \cup G_h$),
ii) G_0 is embeddable in Σ such that $G_0 \cap \bigcup_{i=1,\ldots,h} \Delta_h = \emptyset$, we fix an embedding of G_0,
iii) graphs $G_1, \ldots, G_h$ (called vortices*) are pairwise disjoint,*
iv) for $1 \leq i \leq h$, let $U_i := \{u_1^i, \ldots, u_{m_i}^i\} = V(G_0) \cap V(G_i)$, G_i has a path decomposition $\mathcal{B}_i = (B_j^i)_{1 \leq j \leq m_i}$, of width at most h such that
   a) for $1 \leq j \leq m_i$ we have $u_j^i \in B_j^i$,
   b) the vertices (we call them bases *of G_i) $u_1^i, \ldots, u_{m_i}^i$ appear on C_i in this order (either if we walk clockwise or anti-clockwise).*

The following proposition is known as the Excluded Minor Theorem [17] and is the cornerstone of Robertson and Seymour's Graph Minors theory.

Theorem 4 ([17]). *For every non-planar graph H, there exists an integer c_H, depending only on H, such that every graph excluding H as a minor can be obtained by c_H-clique-sums from graphs that can be c_H-nearly embedded in a surface Σ in which H cannot be embedded. Moreover, while applying each of the clique sums, at most three vertices from each summand other than apices and vertices in vortices are identified.*

Let us remark that by the result of Demaine et al. [6] such a clique-sum decomposition can be obtained in time $O(n^c)$ for some constant c which depends only from H (see also [1]).

Lemmata on treewidth. We need the following two well known results about treewidth.

Lemma 1. *If G_1 and G_2 are graphs, then $\mathbf{tw}(G_1 \oplus G_2) \leq \max\{\mathbf{tw}(G_1), \mathbf{tw}(G_2)\}$.*

Lemma 2. *If G is a graph and $X \subseteq V(G)$, then $\mathbf{tw}(G - X) \geq \mathbf{tw}(G) - |X|$.*

The following lemma is implicit in the proofs from [5,4].

Lemma 3. *Let G be a h-nearly embeddable graph without apices (i.e. where $X = \emptyset$). Then $\mathbf{tw}(G) \leq (h+1) \cdot (\mathbf{tw}(G_0) + 1) - 1$.*

3 Lemmata on Grids and Their Triangulations

In this section we give a sequence of auxiliary lemmata used to prove Lemma 11, the most important technical tool in the proofs of Theorems 1 and 2.

It is implicit in the proofs in [4, Theorem 4.12] and [16, (5.1)] that if the treewidth of a graph embedded in a surface with Euler genus $\mathbf{eg}(G)$ is large enough then this graph contains $(r \times r)$-grid as a surface minor. We state this with the following lemmata.

Lemma 4. *Let G be a graph embedded in a surface Σ of Euler genus γ. If the treewidth of G is more than $12r(\gamma+1)$, then G has the $(r \times r)$-grid as a surface minor.*

Lemma 5. *Let H be a partial triangulation of a $((2k+1) \times (2k+1))$-grid. Then H contains Γ_k as a contraction in a way that all external vertices of H belong to the model of the loaded corner of Γ_k.*

A basic ingredient of our proofs is a result roughly stating that if a graph G with a big grid as a minor is embedded on a surface Σ of small genus, then there is a disc in Σ containing a big enough part of the grid of G. This result is implicit in the work of Robertson and Seymour and there are simpler alternative proofs by Mohar and Thomassen [11,19] (see also [4, Lemma 3.3] and [8, Lemma 4.7]). By using a variant of this result from Geelen et al. [10] together with Lemmata 4 and 5 we prove the following.

Lemma 6. *Let G be a graph embedded in a surface of Euler genus γ and let k be a positive integer. If the treewidth of G is more than $12 \cdot (\gamma+1)^{3/2} \cdot (2k+4)$, then G contains Γ_k as a v-smooth contraction with v being one of the corners of Γ_k.*

The following is based on Lemma 3 and Lemma 6

Lemma 7. *There is a constant c such that if G is a graph h-nearly embedded in a surface of Euler genus γ without apices, where $\mathbf{tw}(G) \geq c \cdot \gamma^{3/2} \cdot h^{3/2} \cdot k$, then G contains as a v-smooth contraction the graph Γ_k with the loaded corner v.*

Let $\mathcal{C} = \{K_1, \ldots, K_r\}$ be a sequence of (not necessary different) cliques in a graph G and let $E \subseteq E(G[\cup_{i=1,\ldots,r} K_i])$. We define the $\mathbf{cl}(G, \mathcal{C}, E)$ to be the graph constructed from $G - E$ by adding for each non-empty K_i a new vertex $z_{\text{new}}^{(i)}$ and making it adjacent to all vertices in K_i.

The proof of the following lemma is based on Lemma 5.

Lemma 8. *Let G_0 be a graph embedded in surface Σ of Euler genus γ and let G^+ be another graph that might share common vertices with G_0. We set $G' = G_0 \cup G^+$. Let $\mathcal{C} = \{K_1, \ldots, K_r\}$ be a collection of cliques in G' such that each of them shares at most 3 vertices with G_0. Let $E \subseteq E(G'[\cup_{i=1,\ldots,r} K_i])$ and let $\hat{G}' = \mathbf{cl}(G', \mathcal{C}, E)$. Then, if G' contains Γ_{2k+5} with the loaded corner v as a v-smooth contraction, then $\hat{G}'$ contains Γ_k as a contraction.*

We also need the following two lemmata.

Lemma 9. *Let G be a graph and let $\mathcal{C} = \{K_1, \ldots, K_r\}$ be a sequence of cliques in G, let $E \subseteq E(G[\cup_{i=1,\ldots,r} K_i])$ and let $\hat{G} = \mathbf{cl}(G, \mathcal{C}, E)$. Let also $G' = G - X$ for some $X \subseteq V(G)$, where $|X| \leq h$. We set $\mathcal{C}' = \{K_1 \setminus X, \ldots, K_r \setminus X\}$, E' be the edges of E without endpoints in X and let $\hat{G}' = \mathbf{cl}(G', \mathcal{C}', E')$. Then if $\hat{G}'$ can be contracted to Γ_k, then $\hat{G}$ can be contracted to a graph H containing a vertex subset Y, $|Y| \leq h$, where $H - Y = \Gamma_k$.*

Lemma 10. *Let G be a connected graph that obtained from $\Gamma_{2^r k + 4(2^r - 1)}$ by adding $r \geq 1$ new vertices and an arbitrary number of edges incident to these vertices. Then G can be contracted to an apex graph which contains Γ_k and at most one additional vertex which is adjacent to some vertices of Γ_k.*

The following lemma is the most crucial technical result used in the proofs of Theorems 1 and 2.

Lemma 11. *Let G be a connected graph excluding a graph H as a minor. Then there exists some constant c_H such that if $\mathbf{tw}(G) \geq c_H \cdot k$, then G contains as a contraction a graph where the removal of at most one of its vertices results to Γ_k.*

Proof. Let G be a connected H-minor-free graph. If H is a planar graph then G has bounded treewidth [13] and the claim of the theorem is trivial. Assume that H is not planar. By Theorem 4, G can be represented as h-clique-sum $G = G_1 \oplus \cdots \oplus G_m$ such that each graph G_i can be h-nearly-embedded in a surface Σ (on which H cannot be embedded) where h is a constant which depends only on H. Let $F = G_i$ such that $\mathbf{tw}(F) = \max_{j=1,\ldots,m} \mathbf{tw}(G_j)$. By Lemma 1,

$$\mathbf{tw}(G) \leq \mathbf{tw}(F). \tag{1}$$

Assume that F is h-nearly-embedded in Σ and denote by X the set of apices of F. Recall that $|X| \leq h$. Let $F' = F - X$. By Lemma 2,

$$\mathbf{tw}(F) - |X| \leq \mathbf{tw}(F'). \tag{2}$$

Observe that F' is h-nearly embedded in Σ without apices. Using Lemma 7 and combining it's claim with inequalities 1 and 2, we note that there is a constant c_H which depends only on H such that if $\mathbf{tw}(G) \geq c_H \cdot k$ then F' contains as a v-smooth contraction the graph Γ_r where v is the loaded corner of Γ_r and $r = 2^{r+1} \cdot k + 8(2^r - 1) + 5$.

Denote by $S_1, \ldots, S_t$ components of the graph $G - V(F)$. For each S_i let K_i be the set of vertices of F which are adjacent to some vertex of S_i, and let $\mathcal{C} = \{K_1, \ldots, K_t\}$. By the definition of h-clique-sum each K_i is a clique of F. Denote by E the set of virtual edges of F. We assume that for any virtual edge $\{u, v\}$, there is a clique $K_i \in \mathcal{C}$ such that $u, v \in K_i$ (otherwise it is easy to redefine h-clique-sums in the representation of G and exclude such an edge). For every component S_i, all vertices of it are contracted into single vertex $z^{(i)}_{\text{new}}$. Denote by $\hat{F}$ obtained from G by these contractions. It can be easily seen that $\hat{F}$ is the graph $\mathbf{cl}(F, \mathcal{C}, E)$. We set $\mathcal{C}' = \{K_1 \setminus X, \ldots, K_t \setminus X\}$, E' be the edges of E without endpoints in X and let $\hat{F}' = \mathbf{cl}(F', \mathcal{C}', E')$. Since F' can be contracted to Γ_r, it follows immediately from Lemma 8 that $\hat{F}'$ contains Γ_s as a contraction for $s = (r-6)/2 = 2^r \cdot k + 4(2^r - 1)$. Then by Lemma 9, $\hat{F}$ (and consequently the graph G) can be contracted to a graph R containing a vertex subset $Y, |Y| \leq h$ such that $R - Y = \Gamma_s$. It remains to use Lemma 10 and note that R can be contracted to an apex graph which consists of Γ_k and at most one apex vertex which is adjacent to some vertices of Γ_k. The graph R is a contraction of G, so G contains as a contraction a graph which after the removal of at most one of its vertices results to Γ_r.

4 Proofs of Theorems

Proof (of Theorem 1). Let H be an apex graph. It was shown by Robertson et al. [13], that every planar graph on $\lceil h/7 \rceil$ vertices is a minor of an $(h \times h)$-grid, and without loss of generality, we can assume that H is a graph constructed from a $(h \times h)$-grid by adding one apex vertex adjacent to all vertices of the grid. By Lemma 11, if $\mathbf{tw}(G) \geq c_H \cdot k$, for some constant c_H, then G contains as a contraction a graph F such that the removal of at most one of its vertices results in $\Gamma = \Gamma_{h \cdot (k+2)}$. If $F = \Gamma$, then the theorem follows trivially. Thus we assume that F has an additional vertex u adjacent to some vertices of Γ. We consider the collection $\mathbf{P}_h(\Gamma)$ of h^2 vertex disjoint induced subgraphs of Γ. We claim that there is a subgraph in $\mathbf{P}_h(\Gamma)$ such that none of its vertices is adjacent to u. Indeed, if each subgraph in $\mathbf{P}_h(\Gamma)$ contains a vertex adjacent to u, then Γ contain an $(h \times h)$-grid as a minor such that the nodes of this grid are the neighbors of u. But this contradicts the assumption that G is H-minor-free.

Thus there is a subgraph Γ^*_{k+2} in $\mathbf{P}_h(\Gamma)$ such that none of the vertices of Γ^*_{k+2} is adjacent to u. The graph Γ^*_{k+2} can be seen as a graph obtained from Γ_{k+2} by removal all edges adjacent to the loaded corner of Γ_{k+2} that are not the edges of the underlying grid. Therefore, after applying the boundary contraction of F to Γ^*_{k+2}, the resulting graph $\mathbf{bc}(F, \Gamma')$ is Γ_k.

Proof (of Theorem 2). Let us assume that $\mathbf{tw}(G) \geq c_H \cdot k^2$, where c_H is the constant from Lemma 11. By the same lemma, G can be contracted to a graph H such that by the removal of at most one vertex of H the result is isomorphic to Γ_{k^2}. If H is itself isomorphic to Γ_{k^2} then we are done as Γ_{k^2} contains Γ_k as a contraction. Suppose then that G has an additional vertex x and let $S = N_G(x)$. Let $\mathcal{P}$ be a collection of k disjoint copies of Γ^*_k in Γ_{k^2}. In case for some $A \in \mathcal{P}$, $V(A) \cap S = \emptyset$, we contract all edges with both endpoints in $\cup_{H \in \mathcal{P} \setminus \{A\}} V(H)$. The obtained graph is Γ^*_k with one more vertex adjacent to all its external vertices and this graph can be further contracted to Γ_k. Suppose now that each graph in $\mathcal{P}$ intersects some neighbor of x. Then contract all edges of all graphs in $\mathcal{P}$ and the resulting graph is P_k.

Proof (of Theorem 3). Suppose that G does not contain $H = K_k$ as a contraction. Then G is an H-minor-free graph. By Theorem 2, there exists some constant c_H such that if $\mathbf{tw}(G) \geq c_H \cdot k^2$, then G contains as a contraction either Γ_k or Π_k. We put $c_k = c_H \cdot k^2$, which concludes the proof of the theorem.

5 Contraction Bidimensionality Revised

The theory of bidimensionality is a meta algorithmic framework for designing efficient fixed-parameter algorithms and approximation algorithms for a broad range of graph problems. Roughly speaking, graph problem is bidimensional if (a) the solution of the problem on the $(k \times k)$-grid (or some modification of the grid) is proportional to $\Omega(k^2)$ (b) the problem is closed under taking minor/contraction, which means that the solution value can only decrease with contracting or removing edges in the graph. Many problems are bidimensional. Classic examples are vertex cover, dominating set, and feedback vertex set.

A parameter P is a function mapping graphs to nonnegative integers. The decision problem associated with P asks, for a given graph G and nonnegative integer k, whether $P(G) \leq k$. Intuitively, a parameter is bidimensional if its value depends on the area of a grid and not on its width or height.

For minor-closed parameters, the definition of bidimensionality is easy [3]. A parameter P is *minor bidimensional* if (a) P is closed under taking of minors and (b) for the $(k \times k)$-grid Γ, $P(\Gamma) = \Omega(k^2)$. Examples of minor bidimensional parameters are sizes of a vertex cover, a feedback vertex set, or a minimum maximal matching in a graph.

For contraction-closed parameters, the definition of bidimensionality is much more complicated and depends on the class of graphs it is used for. Demaine et al. [3,4,2,8] defined parameter P as *contraction bidimensional* if the following hold: (**a**) P is closed under taking of contractions and (**b**) for a "$(k \times k)$-grid-like graph" Γ, $P(\Gamma) = \Omega(k^2)$. Here the property of being "grid-like graph" is different for different graph classes and is defined as follows.

- **b1**) For planar graphs and single-crossing-minor-free graphs, a "$(k \times k)$-grid-like graph" is a partially triangulated $(k \times k)$-grid;
- **b2**) For graphs of Euler genus γ, this is a partially triangulated $(k \times k)$-grid with up to γ additional handles;
- **b3**) For apex-minor-free graphs, this is $(k \times k)$-augmented grid, i.e. partially triangulated grid augmented with additional edges such that each vertex is incident to $O(1)$ edges to non-boundary vertices of the grid.

Typical examples of contraction bidimensional parameters are sizes of a dominating, clique-transversal, or edge domination sets.

The main contribution of Theorem 1 to contraction bidimensionality is that the notions of "grid-like" graphs (**b1**), (**b2**), and (**b3**) can be replaced by the following one

b$'$) $P(\Gamma_k) = \Omega(k^2)$.

This is justified by the following theorem, which is the main (meta) algorithmic contribution of this paper.

Theorem 5. *Let P be a graph parameter which satisfies conditions* (**a**) *and* (**b**$'$). *Let G be a n-vertex graph excluding an apex graph H as a minor. Then if P is computable in time $2^{O(\mathbf{tw}(G))} \cdot n^{O(1)}$, then deciding $P(G) \leq k$ can be done in time $2^{O(\sqrt{k})} \cdot n^{O(1)}$.*

References

1. Dawar, A., Grohe, M., Kreutzer, S.: Locally excluding a minor. In: LICS 2007, pp. 270–279. IEEE Computer Society, Los Alamitos (2007)
2. Demaine, E., Hajiaghayi, M.: The bidimensionality theory and its algorithmic applications. The Computer Journal 51, 292–302 (2007)
3. Demaine, E.D., Fomin, F.V., Hajiaghayi, M., Thilikos, D.M.: Bidimensional parameters and local treewidth. SIAM J. Discrete Math. 18, 501–511 (2004/2005)
4. Demaine, E.D., Fomin, F.V., Hajiaghayi, M., Thilikos, D.M.: Subexponential parameterized algorithms on graphs of bounded genus and H-minor-free graphs. J. ACM 52, 866–893 (2005)
5. Demaine, E.D., Hajiaghayi, M.: Linearity of grid minors in treewidth with applications through bidimensionality. Combinatorica 28, 19–36 (2008)
6. Demaine, E.D., Hajiaghayi, M., ichi Kawarabayashi, K.: Algorithmic graph minor theory: Decomposition, approximation, and coloring. In: FOCS 2005, pp. 637–646. IEEE Computer Society, Los Alamitos (2005)
7. Demaine, E.D., Hajiaghayi, M., ichi Kawarabayashi, K.: Algorithmic graph minor theory: Improved grid minor bounds and Wagner's contraction. Algorithmica (to appear, 2009)
8. Demaine, E.D., Hajiaghayi, M., Thilikos, D.M.: The bidimensional theory of bounded-genus graphs. SIAM J. Discrete Math. 20, 357–371 (2006) (electronic)
9. Dorn, F., Fomin, F.V., Thilikos, D.M.: Subexponential parameterized algorithms. Comp. Sci. Rev. 2, 29–39 (2008)
10. Geelen, J.F., Richter, R.B., Salazar, G.: Embedding grids in surfaces. European J. Combin. 25, 785–792 (2004)
11. Mohar, B.: Combinatorial local planarity and the width of graph embeddings. Canad. J. Math. 44, 1272–1288 (1992)
12. Mohar, B., Thomassen, C.: Graphs on surfaces. Johns Hopkins University Press, Baltimore (2001)
13. Robertson, N., Seymour, P., Thomas, R.: Quickly excluding a planar graph. J. Combin. Theory Ser. B 62, 323–348 (1994)
14. Robertson, N., Seymour, P.D.: Disjoint paths—a survey. SIAM J. Algebraic Discrete Methods 6, 300–305 (1985)
15. Robertson, N., Seymour, P.D.: Graph minors. V. Excluding a planar graph. J. Comb. Theory Series B 41, 92–114 (1986)
16. Robertson, N., Seymour, P.D.: Graph minors. X. Obstructions to tree-decomposition. J. Combin. Theory Ser. B 52, 153–190 (1991)
17. Robertson, N., Seymour, P.D.: Graph minors. XVI. Excluding a non-planar graph. J. Combin. Theory Ser. B 89, 43–76 (2003)
18. Robertson, N., Seymour, P.D., Thomas, R.: Quickly excluding a planar graph. J. Combin. Theory Ser. B 62, 323–348 (1994)
19. Thomassen, C.: A simpler proof of the excluded minor theorem for higher surfaces. J. Combin. Theory Ser. B 70, 306–311 (1997)

Minimizing Movement: Fixed-Parameter Tractability

Erik D. Demaine[1], MohammadTaghi Hajiaghayi[2], and Dániel Marx[3,⋆]

[1] MIT Computer Science and Artificial Intelligence Laboratory, 32 Vassar St., Cambridge, MA 02139, USA
edemaine@mit.edu

[2] AT&T Labs — Research, 180 Park Ave., Florham Park, NJ 07932, USA
hajiagha@research.att.com

[3] Department of Computer Science and Information Theory, Budapest University of Technology and Economics, Budapest H-1521, Hungary
dmarx@cs.bme.hu

Abstract. We study an extensive class of movement minimization problems which arise from many practical scenarios but so far have little theoretical study. In general, these problems involve planning the coordinated motion of a collection of agents (representing robots, people, map labels, network messages, etc.) to achieve a global property in the network while minimizing the maximum or average movement (expended energy). The only previous theoretical results about this class of problems are about approximation, and mainly negative: many movement problems of interest have polynomial inapproximability. Given that the number of mobile agents is typically much smaller than the complexity of the environment, we turn to fixed-parameter tractability. We characterize the boundary between tractable and intractable movement problems in a very general set up: it turns out the complexity of the problem fundamentally depends on the treewidth of the minimal configurations. Thus the complexity of a particular problem can be determined by answering a purely combinatorial question. Using our general tools, we determine the complexity of several concrete problems and fortunately show that many movement problems of interest can be solved efficiently.

1 Introduction

In many applications, we have a relatively small number of mobile agents (e.g., a team of autonomous robots or people) moving cooperatively in a vast terrain or complex building to achieve some task. The number of cooperative agents is often small because of their expense: only small groups of people (e.g., emergency response or SWAT teams) can effectively cooperate, and autonomous mobile robots are currently quite expensive (in contrast to, e.g., immobile sensors). Nonetheless, an accurate model of the immense/intricate environment they traverse, and

⋆ Supported by the Magyary Zoltán Felsőoktatási Közalapítvány and the Hungarian National Research Fund (OTKA grant 67651).

A. Fiat and P. Sanders (Eds.): ESA 2009, LNCS 5757, pp. 718–729, 2009.

their ability to communicate or otherwise interact (say, by limited-range wireless radios or walkie-talkies), is complicated and results in a large problem input. Thus, to compute the most energy-efficient motion in such a scenario, we allow the running time to be relatively large (exponential) in the number of agents, but it must be small (polynomial or even linear) in the complexity of the environment. This set up motivates the study of *fixed-parameter tractability* (FPT) [10] for minimizing movement, with running time $f(k) \cdot n^{O(1)}$ for some function f, parameterized by the number k of mobile agents.

A movement minimization problem is defined by a class of target configurations that we wish the mobile agents to form and a movement objective function. For example, we may wish to move the agents to form a connected network (for communication), an independent set (either dispersing robots or placing map labels), or another topology. See Section 5 for more formal examples of problems and how our theory applies to them.

In the general formulation of the movement problem, we are given an arbitrary metric defining feasible motion, a graph defining "connectivity" (possibly according to the infinite Euclidean plane), and a desired property of the connectivity among the agents defined by a class $\mathcal{G}$ of graphs. We view the agents as "pebbles" located at vertices of the connectivity graph (and we use the two terms interchangeably). Our goal is to move the agents so that they induce a subgraph of the connectivity graph that possesses the desired property, that is, belongs to the class $\mathcal{G}$. There are three natural measures of agent motion that we might want to minimize: the total amount of motion, the maximum motion of any agent, and the number of moved agents. To obtain further generality and to model a wider range of problems, we augment this model with additional features: the agents have types, desired solutions can require certain types of agents, multiple agents can be located at the same vertex, and the cost of the movement can be different (even nonmetric) for the different agents.

To what level of generality can we solve these movement problems? Several versions have been studied from an approximation algorithms perspective in SODA 2007 [7] and FOCS 2008 [8], in addition to various specific problems considered less formally in practical scenarios [2,4,5,9,12,13,14]. Unfortunately, most forms of the movement problem are NP-complete, and furthermore are often hard to approximate even within polynomial factors [7]. Nonetheless, the problems are of significant practical interest, and the motion must be kept small in order to minimize energy consumption. Fortunately, as motivated above, the number of mobile agents is often small. Thus we have a natural context for considering fixed-parameter algorithms, i.e., algorithms with running time $f(k) \cdot n^{O(1)}$, where parameter k is the number of mobile agents.

2 Main Results

We develop general efficient fixed-parameter algorithms for a broad family of movement problems. Furthermore, we show our results are tight by characterizing, in a very general setting, the line between fixed-parameter tractability and intractability. It turns out that the notion of treewidth plays an important role in defining this boundary line. Specifically we show that, for problems closed under edge addition (i.e., adding an edge to the connectivity graph cannot destroy a solution), the complexity of the problem depends solely on whether the edge-deletion minimal graphs of the property have bounded treewidth. If they all have bounded treewidth, we show how to solve a very general formulation of the problem with an efficient fixed-parameter algorithm. If they have unbounded treewidth, we show that even very simple questions are W[1]-hard, meaning there is no efficient fixed-parameter algorithm under the standard parameterized complexity assumption FPT $\neq$ W[1]. In Section 5, we use these results to characterize the complexity of several concrete problems.

Our results apply to a more general model of agents, which in particular lets us capture facility-location types of problems where the number of facilities can be arbitrary large (not a fixed parameter). Such problems arise, e.g., in organizing a small team within a large infrastructure of wired network hubs or mobile satellites. The general model we consider divides the agents into three types—client, facility, and obnoxious agents—and the parameter is just the number of clients, which can be much smaller than the total number of agents. The clients can require collocated or nearby *facility agents*, among a potentially large set of facility agents, which themselves are mobile. Intuitively, facilities provide some service needed by clients. Clients can also require at most a certain number (e.g., zero) of collocated *obnoxious agents* (again among a potentially large, mobile set), which can represent dangerous or undesirable resources. In other words, adding facility agents or removing obnoxious agents does not affect a solution. More generally, there can be many different subtypes of client, facility, and obnoxious agents, and we may require a particular pattern of these types.

Formally, our results are as follows. A movement problem specifies a *multicolored graph property*: an (infinite) set $\mathcal{G}$ of desired configurations, each specifying a desired subgraph and how that subgraph should be populated by different types of agents (a *multicolored graph*). In this way, we can specify different types of client agents that need to interact in a particular way, or need particular types of nearby facility agents. The goal of the *movement problem* is to move the agents into a configuration containing at most ℓ vertices that contain all k client agents and induce a "good" target pattern: either the induced multicolored graph is in the set $\mathcal{G}$ or it is better than some multicolored graph $G \in \mathcal{G}$, i.e., contains more facility agents and fewer obnoxious agents at each vertex.

A mild technical condition that we require is that the multicolored graph property $\mathcal{G}$ is *regular*: for every fixed numbers k and ℓ, there are only finitely many graphs in $\mathcal{G}$ with at most ℓ vertices and at most k client agents (as we do not bound the number of obnoxious and facility agents here, this is a nontrivial restriction). In other words, there should be only finitely many minimal ways to satisfy a bounded number of clients in a bounded subgraph. For example, the propery requiring that the number of facility agents is not less than the number of obnoxious agents is *not* a regular property. Note that this restriction does not say that there is only a finite number of good configurations: as mentioned in the previous paragraph, we allow configurations having any number of extra facility vertices. Furthermore, our main algorithmic result considers properties that are closed under edge addition; this is certainly true for properties that model some notion of connectivity.

Theorem 1. *If $\mathcal{G}$ is a regular multicolored graph property that is closed under edge addition, and if the edge-deletion minimal graphs in $\mathcal{G}$ have bounded treewidth, then the movement problem can be solved in $f(k,\ell) \cdot n^{O(1)}$ time, assuming that the movement cost function is the same on any two agents of the same obnoxious type that are initially located on the same vertex.*

Here the *movement cost function* is an arbitrary (polynomially computable) function for each agent specifying the nonnegative integer cost of moving that agent to each vertex in the graph. This definition allows nonmetric terrains, agents of different speeds, immobile agents, regions impassable by certain agents, etc. In the movement problem, we are given an initial configuration (a multicolored graph), and we wish to minimize the total cost of all movement subject to reaching one of the desired target configurations in $\mathcal{G}$ with at most ℓ vertices, where both ℓ and the number k of client agents are parameters. This problem in particular captures the variations of minimizing the maximum movement and minimizing the number of moved agents. For the latter, we simply specify a movement cost function for each agent of 0 to remain stationary and 1 to make any move. For the former, we can binary search on the maximum movement cost τ, and modify the movement cost function to jump to ∞ whenever exceeding τ.

Our main algorithm uses several tools from fixed-parameter tractability, color coding, and graph structure theory, in particular treewidth. This combination of techniques seems interesting in its own right.

We prove a matching hardness result for Theorem 1: if the edge-deletion minimal graphs in $\mathcal{G}$ have unbounded treewidth, then it is hard to answer even some very simple questions. Thus treewidth plays an essential role in the complexity of the problem, which is not apparent at first sight.

Theorem 2. *If $\mathcal{G}$ is any (possibly regular) multicolored graph property that is closed under edge addition, and for every $w \geq 1$, there is an edge-deletion minimal graph $G_w \in \mathcal{G}$ with treewidth at least w and at least one client agent on each vertex (but no other type of agent), then the movement problem is* W[1]*-hard with the combined parameter (k,ℓ), already in the special case where each agent is allowed to move at most one step.*

3 Further Results

In addition to our general classification, we present many additional fixed-parameter results. These results capture situations where the general classification cannot be applied directly, or the general results apply but problem-specific approaches enable more efficient algorithms. Specifically, we consider situations where the graphs are more specific (e.g., almost planar), the property is not closed under edge addition, or the number of client agents is not bounded. Our aim is to demonstrate that there are many problem variants that can be explored and that there is a vast array of algorithmic techniques that become relevant when studying movement problems. In particular, the fast set convolution algorithm of Björklund et al., results from algorithmic graph minor theory, Courcelle's Theorem, bidimensionality, Canny's Roadmap Algorithm, and a result of Khot and Raman all find uses in this framework.

Planar graphs and H-minor-free graphs. Our general characterization makes no assumptions on the connectivity structure: it is an arbitrary graph. However, significantly stronger results can be achieved if we have some restriction on the connectivity graph. For example, many road networks, fiber networks, and building floorplans can be accurately represented by planar graphs. We show that, for planar graphs, the fixed-parameter algorithms of Theorem 1 work even if we remove the requirement that $\mathcal{G}$ is closed under edge addition.

In many cases, approximation and fixed-parameter tractability results for planar graphs generalize to arbitrary surfaces, to bounded local treewidth graphs, and to H-minor-free graph classes. These generalizations are made possible by the algorithmic consequences of the Graph Minor Theorem [6]. To obtain maximum generality, we state the result on planar graphs generalized to arbitrary H-minor-free classes:

Theorem 3. *If $\mathcal{G}$ is a regular multicolored graph property, then for every fixed graph H, the movement problem can be solved on H-minor-free graphs in $f(k,\ell)\cdot n^{O(1)}$ time, assuming that the movement cost function is the same on any two agents of the same obnoxious type that are initially located on the same vertex.*

One possible application scenario where these generalizations of planar graphs play a role is the following. The terrain is a multi-level building, where the connectivity graph is planar on each level, and there are at most d connections between two adjacent levels. Now the graph is K_{d+1}-free for $d \geq 4$ (as a K_{d+1} minor would be contained on one level). Thus, for every fixed value of d, Theorem 3 applies for such connectivity graphs.

We also consider two specific problems in the context of planar graphs.

Bidimensionality. We consider parameterizing by the sum of all movement, instead of the number of pebbles, for the problem of DISPERSION (see Section 5). This parameterization is likely hard in general, but we show that it becomes fixed-parameter tractable in planar graphs, even in linear time (for every fixed maximum sum k). The proof uses a combination of bidimensionality theory, parameter-treewidth bounds, grid-minor theorems, Courcelle's Theorem, and monadic second-order logic.

Planar STEINER CONNECTIVITY. In the STEINER CONNECTIVITY problem (see Section 5), the goal is to connect one type of agents ("terminals") using another type of agents ("connectors"). Our general characterization shows that this problem is fixed-parameter tractable if the numbers of both types of agents are bounded. The problem becomes W[1]-hard if only the number of connector agents is bounded and the number of terminal pebbles is unbounded. On the other hand, we show that this version of the problem is fixed-parameter tractable for planar graphs, using problem-specific techniques.

Geometric graphs. In some of the applications, the environment can be naturally modeled by the infinite geometric graph defined by Euclidean space, where vertices correspond to points and edges connect two vertices that are within a fixed distance of each other, say 1. In this case, we develop efficient algorithms in a very general setting, even though the graph is infinite:

Theorem 4. *If $\mathcal{G}$ is any regular graph property, then the movement problem can be solved in Euclidean d-space up to multiplicative error $1+\varepsilon$ in $f(k,d) \cdot n^{O(1)} \lg(1/\varepsilon)$ time, where k is the total number of agents (including facility and obnoxious agents).*

The main tool for proving this theorem is Canny's Roadmap Algorithm for motion planning in Euclidean space [3], which lets us manipulate bounded-size semi-algebraic sets.

Hereditary properties. In addition to properties closed under edge addition, we investigate another general class of properties, *hereditary properties*, where if some $G \in \mathcal{G}$, then every induced subgraph of G is also in the property $\mathcal{G}$. For example, independence (having no edges) is such a property. We prove another general hardness result for hereditary properties:

Theorem 5. *Let $\mathcal{G}$ be a hereditary property where each vertex has exactly one client pebble and there are no other type of pebbles. If the maximum clique size is bounded in $\mathcal{G}$, then the movement problem is* W[1]*-hard with the combined parameter (k, ℓ), already in the special case where each agent is allowed to move at most one step in the graph.*

The proof of Theorem 5 uses a hardness result by Khot and Raman [11] on the parameterized complexity of finding induced subgraphs with hereditary properties. The theorem in particular establishes W[1]-hardness of DISPERSION (moving to an independent set); see Section 5.

Improving CONNECTIVITY with fast subset convolution. Finally, we optimize one particularly practical problem, CONNECTIVITY: moving the agents so that they form a connected subgraph. Our general characterization implies that this problem is fixed-parameter tractable. Using the recent algorithm of Björklund et al. [1] for fast subset convolution in the min-sum semiring, we design a more efficient algorithm for this problem: the exponential factor of the running time is only $O(2^k)$.

In summary, our results form a systematic study of the movement problem, using powerful tools to classify the complexity of the different variants. Our algorithms are general, so may not be optimal for any specific version of the problem, but they nonetheless characterize which problems are tractable, and lead the way for future investigation into more efficient algorithms for practical special cases.

4 Model and Definitions

In this section, we make precise the model described in the Introduction and introduce some additional notation.

Definition 1. *We fix three finite sets of colors: C_m (main colors), C_f (facility colors), C_o (obnoxious colors).*

Definition 2. *A* multicolored graph *is a graph with a multiset of colored pebbles assigned to each vertex (a vertex can be assigned multiple pebbles with the same color). We denote by $n_G(c, v)$ the number of pebbles with color c at vertex v in G. A* multicolored graph property *is a (possibly infinite) recursively enumerable set $\mathcal{G}$ of multicolored graphs. A graph property $\mathcal{G}$ is* regular *if for every fixed k, ℓ there is only a finite number of graphs in $\mathcal{G}$ with at most ℓ vertices and at most k main pebbles and there is an algorithm that, given k and ℓ, enumerates these graphs. A graph property $\mathcal{G}$ is* hereditary *if, for every $G \in \mathcal{G}$, every induced subgraph of G is also in $\mathcal{G}$. A graph property $\mathcal{G}$ is* closed under edge addition *if whenever G is in $\mathcal{G}$ and G' is obtained from G by connecting two nonadjacent vertices, then G' is also in $\mathcal{G}$. A graph $G \in \mathcal{G}$ is* edge-deletion minimal *if there is no graph $G' \in \mathcal{G}$ that can obtained from G by edge deletions.*

Definition 3. *Let G_1 and G_2 be two multicolored graphs whose underlying graphs are isomorphic. G_2* dominates *G_1 if there is an isomorphism $\phi : V(G_1) \rightarrow V(G_2)$ such that, for every $v \in V(G_1)$,*

1. *for every $c \in C_m$, vertices v and $\phi(v)$ have the same number of pebbles with color c;*
2. *for every $c \in C_f$, vertex $\phi(v)$ has at least as many pebbles with color c as v; and*
3. *for every $c \in C_o$, vertex $\phi(v)$ has at most as many pebbles with color c as vertex v.*

Definition 4. *For every set $\mathcal{G}$ of multicolored graphs, the* movement problem *has the following inputs:*

1. *a multicolored graph $G(V, E)$, P is the set of pebbles, k is the number of main pebbles;*
2. *a movement cost function $c_p : V \rightarrow Z^+$ for each pebble $p \in P$;*
3. *integer ℓ, the maximum solution size; and*
4. *integer C, the maximum cost.*

The task is to find a movement plan $m : P \rightarrow V$ such that

1. *the total cost $\sum_{p \in P} c_p(m(p))$ of the moves is at most C; and*
2. *after the movements, there is a set S of at most ℓ vertices such that S contains all the main pebbles and the multicolored graph $G[S]$ dominates some graph in $\mathcal{G}$.*

By using different movement cost functions, we can express various goals:

1. if $c_p(v)$ is the distance of p from v, then we have to minimize the sum of movements,
2. if $c_p(v) = 0$ if v is at distance at most d from p and ∞ otherwise, then we have to find a solution where p moves at most d steps,
3. if $c_p(v) = 0$ if v is the initial location of p and $c_p(v) = 1$ for every other vertex, then we have to minimize the number of pebbles that move.

Of course, we can express combinations of these goals or the different pebbles can have different movement graphs, etc. The formulation is very flexible.

5 Sample Problems of Interest

To illustrate the generality of our model and characterization, we define several specific movement problems similar to those mentioned informally in the Introduction, and determine their fixed-parameter tractability using Theorems 1 and 2. Using these tools, if a movement problem can be modeled with colored pebbles and the target patterns are closed under adding edges, then the complexity of the problem can be determined by solving the (sometimes nontrivial) combinatorial question of whether the minimal configurations have bounded treewidth. The minimal configurations are those pebbled graphs that are acceptable solutions, but removing any edge makes them unacceptable.

Example: CONNECTIVITY. Move the pebbles (agents) so that they are connected and on distinct vertices. The parameter is the number k of pebbles. Now there is only one, main color of pebbles, and $\mathcal{G}$ contains all connected graphs with exactly one pebble on each vertex. Clearly, $\mathcal{G}$ is closed under edge addition and the edge-deletion minimal graphs are trees. Trees have treewidth 1, hence by Theorem 1, this movement problem is fixed-parameter tractable for any movement cost function. The variant of the problem where it is not required that the pebbles are on distinct vertices is also FPT: in this case, $\mathcal{G}$ contains all connected graphs with *at least* one pebble on each vertex. □

Example: GRID. Move the k pebbles so that they form a $\lfloor\sqrt{k}\rfloor \times \lfloor\sqrt{k}\rfloor$ square grid. The parameter is the number k of pebbles. Again there is only one, main color of pebbles, and $\mathcal{G}$ contains all graphs containing a spanning square grid subgraph with exactly one pebble on each vertex. Clearly, $\mathcal{G}$ is closed under edge addition and the edge-deletion minimal graphs are grids, which have arbitrarily large treewidth. Thus Theorem 2 implies that it is W[1]-hard, parameterized by (k, ℓ), to decide whether there is a solution where each pebble moves at most one step. □

Example: s-t CONNECTIVITY (few pebbles). Move the pebbles to form a path of pebbled vertices between fixed vertices s and t. The parameter is the number k of pebbles. Now there are two main colors of pebbles, call them red and blue, and $\mathcal{G}$ consists of all graphs containing exactly two red pebbles and a path between them using only vertices with blue pebbles. We reduce s-t CONNECTIVITY to this movement problem by putting red pebbles at s and t, and giving them an infinite movement cost to any other vertices. Clearly, $\mathcal{G}$ is closed under edge addition and the edge-deletion minimal graphs are paths. Paths have treewidth 1, so by Theorem 1, this problem is fixed-parameter tractable. □

In the next example, we show that a much more general version of s-t CONNECTIVITY is FPT: instead of parameterizing by the number k of pebbles, we can parameterize by the maximum length L of the path. Thus we can have arbitrarily many pebbles that might form the path, as long as the formed path itself is small.

Example: s-t CONNECTIVITY (bounded length). Move the pebbles to form a path of pebbled vertices of length at most L between fixed vertices s and t. The parameter is the length L. Now we define one main color of pebbles, red, and one facility color of pebbles, blue, and we define $\mathcal{G}$ as in the previous example. Again by Theorem 1, this problem is fixed-parameter tractable in the combined parameter (k, ℓ); in the example, we have $k = 2$ and $\ell = L + 1$. □

Example: STEINER CONNECTIVITY. Connect the red pebbles (representing terminals) by moving the blue pebbles to form a Steiner tree. The parameter is the number of red pebbles plus the number of blue pebbles in the *solution* Steiner tree. This is simply a generalization of s-t CONNECTIVITY to more than two red pebbles. Again by Theorem 1 the problem is fixed-parameter tractable (the edge-deletion minimal graphs are trees), even when the number of blue pebbles is very large. □

Example: 2-CONNECTIVITY. Move the pebbles so that they induce a 2-connected graph and the pebbles are on distinct vertices. The parameter is the number k of pebbles. Now $\mathcal{G}$ contains all 2-connected graphs and clearly $\mathcal{G}$ is closed under edge addition. The edge-deletion minimal graphs have unbounded treewidth: subdividing every edge of a clique gives an edge-deletion-minimal 2-connected graph. Thus by Theorem 2, it is W[1]-hard to decide whether there is a solution where each pebble moves at most one step. □

Example: s-t d-CONNECTIVITY. Move the pebbles so that there are d vertex-disjoint paths using pebbled vertices between two fixed vertices s and t. The parameter is the total length L of the solution paths. Now we use one main color, red, and one facility color, blue, and $\mathcal{G}_d$ consists of all graphs containing two vertices with a red pebble on each, and having d vertex-disjoint paths between these two vertices, with blue pebbles on each path vertex. In the input instance, there are red pebbles on s and t, and the cost of moving them is infinite. Clearly, $\mathcal{G}_d$ is closed under edge addition and the edge-deletion minimal graphs are series-parallel (as they consist of d internally vertex disjoint paths connecting two vertices), which have treewidth 2. Hence, by Theorem 1, this movement problem is fixed-parameter tractable with respect to L, for every fixed d. Again the number of blue pebbles can be arbitrarily large. □

The previous example shows that s-t d-CONNECTIVITY is FPT for every fixed value of d. Furthermore, we can show that the problem remains FPT even if d appears as part of the input.

Example: s-t d-CONNECTIVITY (unbounded version). Move the pebbles so that there are d vertex-disjoint paths using pebbled vertices between two fixed vertices s and t, where d is a number given in the input. The parameter is the total length L of the solution paths. First, if d is larger than the bound on the total length of the paths, then there is no solution. Otherwise, we can assume d is a fixed parameter. Now we use two main colors, red and green, and one facility color, blue. A graph G is in $\mathcal{G}$ if the blue pebbles form d vertex-disjoint paths between two vertices containing red pebbles, where d is the number of green pebbles in G. Thus we use green pebbles to "label" a graph G in $\mathcal{G}$ according to what level of connectivity it attains. Again $\mathcal{G}$ is closed under edge addition and the edge-deletion minimal graphs are series-parallel, which have treewidth 2, so by Theorem 1, the movement problem is fixed-parameter tractable with respect to $k := 2$ and $\ell := L$. In the initial configuration, we put red pebbles on s and t with infinite movement cost, and we place d green pebbles arbitrarily in the graph. The target configuration we obtain will have exactly d green pebbles, and thus d vertex-disjoint paths, because these are main pebbles. □

We can also consider the edge-disjoint version of s-t connectivity. We need the following combinatorial lemma to characterize the minimal graphs:

Lemma 6. *Let G be a connected graph and assume that there are d edge-disjoint paths between vertices s and t in G, but for any edge $e \in E(G)$, there are at most $d-1$ edge-disjoint paths between s and t in $G \setminus e$. Then the treewidth of G is at most $O(d^2)$.*

Example: s-t d-EDGE-CONNECTIVITY. Move the pebbles so that there are d edge-disjoint paths of pebbled vertices between s and t. The parameter is the total length L of the paths. Now we use one main color, red, and one facility color, blue, and $\mathcal{G}_d$ contains all graphs containing two vertices with a red pebble on each and having d edge-disjoint paths between these two vertices, with blue pebbles on each path vertex. By Lemma 6, the edge-deletion minimal graphs have treewidth $O(d^2)$. Hence, by Theorem 1, the movement problem is fixed-parameter tractable with respect to L. □

The previous example shows that s-t d-EDGE-CONNECTIVITY is FPT for every fixed value of d. Somewhat surprisingly, unlike in the vertex-disjoint case, the problem becomes hard if d is part of the input:

Example: s-t d-EDGE-CONNECTIVITY (unbounded version). Move the pebbles so that there are d edge-disjoint paths of pebbled vertices between s and t, where d is a number given in the input. We use three main colors: red, green, and blue. A graph G is in $\mathcal{G}$ if the blue pebbles form d edge-disjoint paths between two vertices containing red pebbles, where d is the number of green pebbles in G. We show that $\mathcal{G}$ contains edge-deletion minimal graphs of arbitrary large treewidth, so by Theorem 2, it is W[1]-hard to decide whether there is a solution where each of the k pebbles move at most one step each. Assume d is even and let G be a graph consisting of vertices s, t, and d vertex-disjoint paths between s and t such that vertices $p_{i,1}, \ldots, p_{i,d}$ are the internal vertices of the ith path. Now for every odd i and odd j, identify vertices $p_{i,j}$ and $p_{i+1,j}$, and for every even $i < d$ and even j, identify $p_{i,j}$ and $p_{i+1,j}$. There are d edge-disjoint s-t paths in this graph, but there are at most $d-1$ such paths after the deletion of every edge. (It is easy to see that every edge is in an s-t cut of exactly d edges.) Thus G is an edge-deletion minimal member of $\mathcal{G}$. Furthermore, if for every odd i and odd j, we contract the edge $p_{i,j}p_{i,j+1}$, then we get a $d/2 \times d/2$ grid, so the treewidth is $\Omega(d)$. □

Example: FACILITY LOCATION (collocation version). Move client and facility pebbles so that each client pebble is collocated with at least one facility pebble and the client pebbles are at distinct locations. The parameter is the number of client pebbles. We use one main color, red, for the clients, and one facility color, blue, for the facilities, and $\mathcal{G}$ contains all graphs in which every vertex contains exactly one red and one blue pebble. The edge-deletion minimal graphs in $\mathcal{G}$ have no edges, so have treewidth 0. By Theorem 1, the movement problem is fixed-parameter tractable parameterized by the number of main pebbles, i.e., the number of clients. The number of facilities can be unbounded, which is useful, e.g., to organize a small team within a large infrastructure of wired network hubs or mobile satellites. □

Example: FACILITY LOCATION (distance-d version). Move client and facility pebbles so that each client pebble is within distance at most d from at least one facility pebble and the client pebbles are at distinct locations. Now we use two main colors, red and green, and one facility color, blue. Let $\mathcal{G}$ contain all graphs that contain some number d of green pebbles and each red pebble is at distance at most d from some blue pebble. Given a graph with k client (red) pebbles and some number of facility (blue) pebbles, we add d dummy green pebbles and ask whether there is a solution on $\ell := k(d+1)+d$ vertices. If we move the pebbles so that each red pebble is at distance d from some blue pebble, then there are $k(d+1)+d$ vertices that contain all d of the green pebbles and induce a graph in $\mathcal{G}$. We claim that the edge-deletion minimal graphs in $\mathcal{G}$ are forests, and hence have treewidth 1. Consider an edge-deletion minimal graph $G \in \mathcal{G}$, and for each vertex v without a blue pebble, select an edge uv that goes to a neighbor u that is closer to some blue pebble than v. If an edge is not selected in this process, then it can be removed (it does not change the distance to the blue pebbles), so by the minimality of G, every edge is selected. Each connected component contains at least one blue pebble. This means that, in each connected component, the number of selected edges is strictly smaller than the number of vertices, i.e., each component is a tree. Thus, by Theorem 1, the movement problem is FPT. □

On the other hand, FACILITY LOCATION becomes W[2]-hard if the parameter is the number of facilities, while the number of clients can be unbounded. We cannot obtain this result using Theorem 2 because, in this setting, the parameter is the number of facility pebbles.

Theorem 7. *For every fixed $d \geq 0$, FACILITY LOCATION (distance d version) is* W[2]*-hard parameterized by the number of facilities, even if each pebble is allowed to move at most one step in the graph.*

Example: MATCHING. Move the pebbles so that the pebbles are on distinct vertices and there is a perfect matching in the graph induced by the pebbles. The parameter is the number of pebbles. Now there is just one, main pebble color, and $\mathcal{G}$ contains all graphs that have a perfect matching. The edge-deletion minimal graphs are perfect matchings, so they have treewidth 1. By Theorem 1, the movement problem is FPT. □

Example: SEPARATION. Move client pebbles (say, representing population) and/or obnoxious pebbles (say, representing power plants) so that each client pebble is collocated with at most o obnoxious pebbles. The parameter is the number of client pebbles. Here $\mathcal{G}$ contains all graphs with the desired bounds, so the edge-deletion minimal graphs have no edges, which have treewidth 0. By Theorem 1, the movement problem is fixed-parameter tractable. As in previous examples, we can make o an input to the problem. □

Example: DISPERSION. Move the pebbles to distinct vertices and such that no two pebbles are adjacent. The parameter is the number k of pebbles. Here $\mathcal{G}$ contains all independent sets with exactly one pebble on each vertex. Because $\mathcal{G}$ is hereditary and the maximum clique size is 1, Theorem 5 implies that the movement problem is W[1]-hard, even in the case when each pebble is allowed to move at most one step. □

References

1. Björklund, A., Husfeldt, T., Kaski, P., Koivisto, M.: Fourier meets Möbius: fast subset convolution. In: STOC 2007, pp. 67–74 (2007)
2. Bredin, J.L., Demaine, E.D., Hajiaghayi, M., Rus, D.: Deploying sensor networks with guaranteed capacity and fault tolerance. In: MOBIHOC 2005, pp. 309–319 (2005)
3. Canny, J.F.: The Complexity of Robot Motion Planning. MIT Press, Cambridge (1987)
4. Corke, P., Hrabar, S., Peterson, R., Rus, D., Saripalli, S., Sukhatme, G.: Autonomous deployment of a sensor network using an unmanned aerial vehicle. In: ICRA 2004, New Orleans, USA (2004)
5. Corke, P., Hrabar, S., Peterson, R., Rus, D., Saripalli, S., Sukhatme, G.: Deployment and connectivity repair of a sensor net with a flying robot. In: ISER 2004, Singapore (2004)
6. Demaine, E.D., Hajiaghayi, M., Kawarabayashi, K.: Algorithmic graph minor theory: Decomposition, approximation, and coloring. In: FOCS 2005, pp. 637–646 (2005)
7. Demaine, E.D., Hajiaghayi, M., Mahini, H., Sayedi-Roshkhar, A.S., Oveisgharan, S., Zadimoghaddam, M.: Minimizing movement. ACM Trans. Algorithms
8. Friggstad, Z., Salavatipour, M.R.: Minimizing movement in mobile facility location problems. In: FOCS 2008, pp. 357–366 (2008)
9. Hsiang, T.-R., Arkin, E.M., Bender, M.A., Fekete, S.P., Mitchell, J.S.B.: Algorithms for rapidly dispersing robot swarms in unknown environments. In: WAFR 2003, pp. 77–94 (2003)
10. Hüffner, F., Niedermeier, R., Wernicke, S.: Techniques for practical fixed-parameter algorithms. Comput. J. 51(1), 7–25 (2008)
11. Khot, S., Raman, V.: Parameterized complexity of finding subgraphs with hereditary properties. Theoret. Comput. Sci. 289(2), 997–1008 (2002)
12. LaValle, S.M.: Planning Algorithms. Cambridge University Press, Cambridge (2006)
13. Reif, J.H., Wang, H.: Social potential fields: a distributed behavioral control for autonomous robots. In: WAFR 2005, pp. 331–345 (2005)
14. Schultz, A.C., Parker, L.E., Schneider, F.E. (eds.): Multi-Robot Systems: From Swarms to Intelligent Automata. Springer, Heidelberg (2003)

Storing a Compressed Function with Constant Time Access

Jóhannes B. Hreinsson, Morten Krøyer, and Rasmus Pagh*

IT University of Copenhagen, 2300 København S, Denmark
{johre06,kroyer,pagh}@itu.dk

Abstract. We consider the problem of representing, in a space-efficient way, a function $f : S \rightarrow \Sigma$ such that any function value can be computed in constant time on a RAM. Specifically, our aim is to achieve space usage close to the 0th order entropy of the sequence of function values. Our technique works for any set S of machine words, without storing S, which is crucial for applications.

Our contribution consists of two new techniques, of independent interest, that we use in combination with an existing result of Dietzfelbinger and Pagh (ICALP 2008). First of all, we introduce a way to support more space efficient approximate membership queries (Bloom filter functionality) with arbitrary false positive rate. Second, we present a variation of Huffman coding using approximate membership, providing an alternative that improves the classical bounds of Gallager (IEEE Trans. Information Theory, 1978) in some cases. The end result is an entropy-compressed function supporting constant time random access to values associated with a given set S. This improves both space and time compared to a recent result by Talbot and Talbot (ANALCO 2008).

1 Introduction

Compression is an important technique in modern computing systems. Often, some kind of random access is required, such that a given data item can be decompressed without decompressing all data. The standard way of dealing with this requirement is to split the data into blocks that are compressed and decompressed individually. The end result is a trade-off between compression and access time: Larger blocks lead to better compression, but slows down decompression. A recent theoretical breakthrough of Pătraşcu [17] shows how to combine optimal (in a certain sense) compression of an array with logarithmic decompression time. In this paper we are concerned with compression of *functions* that allows random access. Like most other compression methods we deal with the static case, and do not consider efficient updates.

The problem of storing a function f with certain specified values (referred to as the *retrieval* problem) has recently received renewed interest [7,9,18]. The

* This work was supported by the Danish National Research Foundation, as part of the project "Scalable Query Evaluation in Relational Database Systems."

A. Fiat and P. Sanders (Eds.): ESA 2009, LNCS 5757, pp. 730–741, 2009.

main finding is that there exist simple methods that store f in space close to the space that would be needed to store the (uncompressed) function values, and provide constant access time. For example, if f maps city names to weather, and there are two kinds of weather, f can be stored in close to 1 bit per value, with fast random access to the value for a particular city name. This is of course much less than the space usage required for also storing the set of city names. The price paid compared to standard dictionary representations is that the data structure does not detect when it is used on an input where no function value has been specified—in this case it will simply return some arbitrary value.

The retrieval problem is relevant in situations where the amount of data associated with each key is small, and it is either known that queries will only be asked on keys in a fixed set S, or where the query answers for keys not in S are insignificant. For example, suppose that we have ranked a collection of web pages. Then a retrieval data structure would be able to return the ranking of a given URL, without storing the URL itself. This might allow the ranking information to be stored entirely in RAM (e.g., a browser could show highly ranked links more prominently). Several applications of retrieval structures as building blocks in other data structures are described in [1,2].

Our starting point is that data values are often skewed (e.g., in some parts of the world it is sunny much more often than it is rainy), and we may like to extend the set of possible values (e.g., add more rare weather phenomena such as thunderstorms) without increasing the number of bits required to represent each function value. Therefore, it is desirable to have representations where the space usage is dependent on the *entropy* of the data values rather than on the number of possible values.

1.1 Our Results

Let $S = \{x_1, \ldots, x_n\}$ denote the domain of the function, and let Σ be the set of possible (or actual) function values. For simplicity we assume that $\Sigma = \{1, \ldots, \sigma\}$ — the general case can be handled by a separate data structure implementing a bijection between Σ and $\{1, \ldots, \sigma\}$ (e.g. a minimal perfect hash function [19,12,4] for Σ plus an array of size σ). We describe a new data structure that represents a function $f : S \rightarrow \Sigma$ in space that is close to the (empirical, 0th order) entropy H_0 of the sequence $f(x_1), \ldots, f(x_n)$. If $p_1, p_2, \ldots$ are the frequencies of different characters in the sequence, $H_0 = \sum_i p_i \log_2(1/p_i)$ is a lower bound on the number of bits needed per function value, assuming that the function values are independent of the corresponding input values.

We present our results in the Word RAM model of computation [11] with word size w. To simplify the presentation we assume that elements of S as well as values in Σ can be represented in a single word, and specifically that $w \geq 2 + \log \sigma$. We show the following:

Theorem 1. *Let n, w, and σ be positive integers, where $w^3 < n < 2^w$ and $\sigma \leq 2^w$, let $\delta > 0$ be a constant, and let $S = \{x_1, \ldots, x_n\} \subseteq \{0,1\}^w$ be a set of size n. Given a function $f : S \rightarrow \{1, \ldots, \sigma\}$, let H_0 denote the empirical (0th* *order) entropy of the sequence $f(x_1), \ldots, f(x_n)$, and let p_1 denote the frequency of the most common function value. If n is larger than some constant (depending on δ) there exists a retrieval data structure for f using space*

$$(1+\delta)H_0 + \min(p_1 + 0.086, 1.82(1-p_1)) \text{ bits}$$

per function value, plus $o(\sigma)$ bits to store information about the distribution, such that a function value can be computed in $O(1)$ time on a Word RAM with word size w.

Discussion. The term $o(\sigma)$ is negligible in most cases that are interesting from a compression point of view, i.e., in cases where the number σ of possible values is not much larger than the number of values n. Ignoring this term, the number of bits per character is at most H_0 (a lower bound) times $1+\delta$, plus a small additive term. This is similar to the space that would be obtained by Huffman coding the sequence of function values (with no random access). If p_1 is close to 1 the space usage becomes much better than using Huffman coding — in fact, the space per value can get arbitrarily close to 0, while Huffman coding uses at least 1 bit per value. An example where this is important is storage of functions with many undefined/NULL values.

While the algorithm used to construct our data structure is somewhat complex, the algorithm for evaluating f is extremely simple. It consists of looking up $O(1)$ w-bit strings, performing a bitwise exclusive or, and applying a constant time decoding procedure similar to Huffman decoding. This is illustrated in Figure 1. Thus, we show how to extend the simplicity of existing retrieval data structures (with no compression) to the compressed case.

Techniques. Technically, we first show how to extend existing retrieval data structures to support variable length bit strings as values. The method works under the condition that the set of possible values is prefix-free, and involves a nontrivial load balancing idea using slightly correlated hash functions. Combining this with a variation of (length-limited) Huffman coding, and showing how the decoding can be done in constant time, yields a result that is close to Theorem 1, but missing the second part of the minimum in the additive term. To strengthen the result in the important case where the majority of function values are identical we show how to use an *approximate membership* data structure (i.e., a data structure with the same functionality as a Bloom filter [3]) to decrease the space usage.

To reduce the redundancy as much as possible, we describe a general reduction that can be used to obtain space-efficient approximate membership data structures for any false positive rate $\varepsilon > 0$. Previous space-efficient methods (see [9] for an overview) provided false positive rates that are negative powers of two. This means that one needs to "pay" (with extra space usage) for a false positive rate that is up to 50% smaller than desired – our method reduces this to less than 6%.

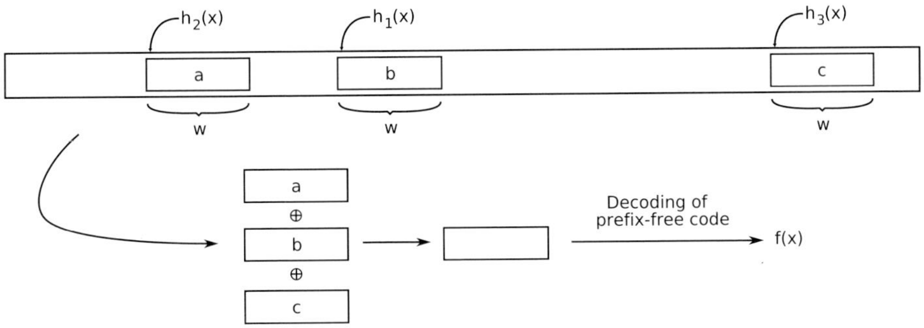

Fig. 1. Illustration of how a value $f(x)$ is computed from our data structure

1.2 Related Work

Several previous papers described data structures having two independent functionalities: They support retrieval queries (given $x \in S$, return $f(x)$), and approximate membership queries (given x, return "true" if $x \in S$, and return "false" with probability at least $1 - \varepsilon$ if $x \notin S$). A data structure with this interface is sometimes referred to as a *bloomier filter* [8]. In this paper we consider only the retrieval problem, but note that approximate membership queries can be added (without loss) by a separate approximate membership data structure.

A faster approach to space-efficient retrieval is through space-efficient minimal perfect hashing [19,12,4]. For $\Sigma = \{0, 1\}^r$, where $r \leq w$ is a positive integer, this gives a space usage of $nr + O(n)$ bits and $O(1)$ query time. However, this approach is rather complicated, and it seems especially complicated to generalize it to variable-length function values (which would be required for compression). Also, even without compression the use of minimal perfect hashing is known to yield a redundancy of at least $\log_2 e \approx 1.44$ bits per element in S, which is more than we achieve.

Rather than building on minimal perfect hashing, we use as a starting point the recent retrieval structure described by Dietzfelbinger and Pagh [9], which gives a very simple query algorithm for function values in $\Sigma = \{0, 1\}^r$ (with no compression). Similar techniques have recently been studied in two other papers [9,18]. The result of Porat [18] is that a space usage of $nr + o(n)$ bits is possible with $O(1)$ query time. While this is asymptotically superior to [9], it relies heavily on tabulation and does not seem to admit a similarly efficient generalization to the compressed case. We refer to the discussion in [9] for a history of related data structures.

The first paper to consider compressed retrieval was recently published by Talbot and Talbot [20]. The authors consider a relaxation of the retrieval problem, where a fraction ε of the function values are allowed to be *incorrect* ("misassignments"). Arguably, this makes their data structure unusable for some applications, but we still find it instructive to compare their result to ours (even assuming ε is close to 1). Talbot and Talbot show how to obtain two different trade-offs between compression and query time: Space $1.44H_0 + 1 + \log \log \sigma$ per value is possible with query time $O(\log^2 \sigma)$, or alternatively space $2.88H_0 + 2$ per value is possible with query time $O(\log \sigma)$. It is also stated (without proof) that the multiplicative constants can be improved to 1.23 and 2.46, respectively. The technique used in [20] is essentially adaptive decoding of Huffman encoded values (one bit at a time). Our data structure performs $O(1)$ non-adaptive memory accesses, does not have misassignments, and improves time as well as space of both trade-offs. In addition, the query algorithm is considerably simpler, and more likely to be of practical use.

Recently, another retrieval data structure that is able to take advantage of skew in the value distributions, "two-step MWHC", was described in [2]. While this method has $O(1)$ query time, its space usage is larger than ours, and cannot be bounded in terms of the 0th order entropy.

1.3 Preliminaries

Each element in $S = \{x_1, x_2, \ldots, x_n\} \subseteq D$ maps to the value $f(x_i)$ from an the alphabet $\Sigma = \{1, \ldots, \sigma\}$. Let a_i denote the ith most frequent value in Σ, breaking ties arbitrarily. For uniformly random $x \in S$ we let $p_i = \Pr\{f(x) = a_i\}$, for $i = 1, 2, \ldots, \sigma$.

Since we build on the data structure of Dietzfelbinger and Pagh [9], we now briefly describe it. Suppose that $k \geq 3$ hash functions $h_1, \ldots, h_k$ are given such that for any $x_i \in S$ the set $\{h_1(x_i), \ldots, h_k(x_i)\}$ is a random set of size k (no collisions), and the sets associated with different x_i are independent[1]. The idea is to set up a vector A of r-bits values, such that $f(x_i)$ can be calculated as a bitwise exclusive-or of k vector entries.

$$f(x_i) = \bigoplus_{j \in \{1,\ldots,k\}} A_{h_j(x_i)}. \tag{1}$$

On inputs $x \in D \setminus S$, the query algorithm returns an arbitrary value. The existence of a suitable vector A depends on the hash function values of the keys in S. It is shown in [9], using results from [5], that for any $\delta > 0$ there exists a constant $k = O(\log(1/\delta))$, such that array size $m = (1 + \delta)n$ suffices to ensure the existence of A with high probability, given that n exceeds some sufficiently large constant. Specifically, the probability of failure for $k \geq 4$ is $O(n^{-5/7})$ [9]. We observe that a failure is not a serious problem, since we may simply choose k new hash functions and try again until we succeed. Even a small value of k allows a very space-efficient data structure, e.g., the redundancy δ for $k = 3, 4, 5$ is around 12%, 3%, and 1%, respectively.

Computing the entries of vector A requires solving a system of linear equations over GF(2). At first glance this seems to require Gaussian elimination in time $O(n^3)$, but it is shown in [9] how to set up equation systems that are easier to solve, and require time $O(n^{1+\epsilon})$, for any $\epsilon > 0$, or even linear time (increasing

[1] The assumption that fully independent hash functions are available can be justified, in the sense that there is a simulation of full randomness that makes everything work. See [9] for details.

k by $O(1)$). The first technique is based on splitting the set S into buckets of size $n^{\epsilon/2}$, and this also applies to our setting in a straightforward fashion. This means we can achieve construction time $O(n^{1+\epsilon})$, for any $\epsilon > 0$. Since the details are very similar to [9] we do not further describe this.

Overview of paper. In section 2 we present an efficient data structure for the retrieval problem with variable length values from a prefix-free set of bit strings. Section 3 describes how to combine this with a variation of Huffman coding that can be decoded in constant time. Finally, in section 5 we show how to improve the result in the case where some function value is much more frequent than others, using a result on approximate membership described in section 4.

2 Retrieval with Variable-Length Values

We will use a prefix-free code for values in Σ, so that we get a function $f : S \to \{0,1\}^*$, where the keys are mapped to codewords of various length. We assume that the longest codeword has length at most w, which will be the case in our application. Our data structure will represent a similar function, $\hat{f} : S \to \{0,1\}^w$, so that for each key $x_i \in S$, $f(x_i)$ is a prefix of $\hat{f}(x_i)$. Since the values of f are prefixes of the values of $\hat{f}$, and the code is prefix-free, we can use a code tree to determine the value of $f(x_i)$ from $\hat{f}(x_i)$, so in effect this gives a representation of f. Section 3 will describe how to do the decoding efficiently.

The data structure and query algorithm. Our data structure is simply a *bit* array A of length m, where $m = (1 + \delta/3)\sum_i |f(x_i)| + O(w)$, i.e., m is essentially a factor $1+\delta/3$ larger than the total length of all strings in $f(S)$. We round up m to the nearest multiple of w, and implement "wrap-around" reads by duplicating the initial word of A after the last word. This means that when we look up w consecutive bits, the data structure behaves like a cyclic array of m bits.

Let $k \geq 3$ be an integer constant, and let $h_1, \ldots, h_k$ be hash functions mapping each key to k bit positions in A (details below). We arrange the array so that for any key $x \in S$, we can compute $\hat{f}(x)$ as the bitwise exclusive-or of the k words starting at the bit positions $h_1(x), \ldots, h_k(x)$. (Note that these words are not necessarily aligned with the Word RAM machine words.)

Hash functions and construction algorithm. Let $h'_1, \ldots, h'_k : S \to [m/w]$ be hash functions such that for any x, the set $\{h'_1(x), \ldots, h'_k(x)\}$ contains k (distinct) values, and is uniformly distributed over all such sets. (See [9] for a discussion on how to construct such hash functions in an efficient way.) Also, let $q : S \to [w]$ be a fully random hash function. We use the k hash functions defined by:

$$h_i(x) = h'_i(x)w + q(x) \ . \tag{2}$$

That is, for any x the $\log w$ least significant bits of the hash function values $h_1(x), \ldots, h_k(x)$ are identical, while the $\lceil \log(m/w) \rceil$ most significant bits are distinct.

In order to construct A we must solve the following system of linear equations

$$f(x_i)_d = \bigoplus_{j=1}^{k} A_{h'_j(x_i)w+q(x_i)+d}, \text{ for} \qquad (3)$$
$$i = 1, \ldots, n, \; d = 1, \ldots, |f(x_i)|$$

where $f(x_i)_d$ is the dth bit in the codeword associated with the key x_i. We observe that all equations involve only bit positions in A that have the same residue modulo w. This means that there are in effect w systems of equations that may be solved individually. In each such system, an equation for $f(x)_d$ involves the k variables at positions $\{h'_1(x), \ldots, h'_k(x)\}$ within the system. This is exactly the setting of [9], which means that the same construction algorithm and analysis applies to each system.

2.1 Analysis

Let n_d denote the number of equations in the linear system corresponding to bits in positions with residue d modulo w. Note that $n_d = \sum_i X_{d,i}$, where $X_{d,i}$ is an indicator random variable that is 1 if and only if some equation involving x_i involves bits in position with residue d modulo w. Note that $X_{d,i}$ depends entirely on $q(x_i)$. For every d, the random variables $X_{d,i}$, $i = 1, \ldots, n$, are independent, so the sum n_d is tightly concentrated around the expectation of $\sum_i |f(x_i)|/w$. By Chernoff bounds (e.g. [16, Theorem 4.1]), and using that $n > w^3$, we have that $\Pr[n_d > (1 + \delta/2)n/w] < 1/(2w)$ when n is sufficiently large. This means that with probability at least 1/2 (over the choice of q) the w equation systems are "well balanced" in the sense that $n_d \leq (1 + \delta/2)n/w$ for all d.

Conditioned on $n_d \leq (1 + \delta/2)n/w$ we can use Calkin's bounds [5,9], which imply that each equation system is solvable with probability $1-O((n/w)^{-5/7}) \geq 1 - o(1/w)$, where the inequality uses that $n \geq w^3$. By the union bound this means that with probability $1 - o(1)$ all equation systems are simultaneously solvable, and hence a suitable bit array A exists. In conclusion, the probability that the randomly chosen hash functions are suitable is bounded away from 0, so a constant number of trials suffices in expectation to find good hash functions for a given set S.

3 Constant Time Huffman-Like Decoding

Without loss of generality assume that $\delta < 1$ in the statement of Theorem 1. We wish to apply the data structure of section 2 with a near-optimal prefix-free code. One possibility would be a Huffman code [13], but since we are willing to sacrifice a factor $1 + \delta/3$ in space usage it is possible to do better, both in terms of decoding time and in terms of the size of the representation of the code tree.

The first step is to identify the set of values Σ' for which the number of occurrences is above $n/\sigma^{1/(1+\delta/3)}$. Observe that $|\Sigma'| \leq \sigma^{1/(1+\delta/3)}$. Conceptually, we substitute all function values in $\Sigma \backslash \Sigma'$ with a new value $\bot$, and then consider a code tree for the values of the resulting function f^*. Length-limited Huffman codes [14,15] provide codewords of maximum length $\ell = \log|\Sigma'| + O(1)$, whose redundancy is within an additive constant of Huffman codes. In fact, the additive constant can be made arbitrarily small by increasing the $O(1)$ term. Decoding of length-limited Huffman codes can be done in constant time using a table indexed by all bit strings of length ℓ, where an entry contains the value in Σ corresponding to its prefix. The size of the table is $2^\ell \log \sigma = O(|\Sigma'| \log \sigma) = o(\sigma)$ bits.

Whenever the value $\perp$ is observed in the lookup table, we fall back on a trivial encoding of the appropriate symbol. We store the codeword for $\perp$, and use the next $\lceil \log \sigma \rceil$ bits to encode the value. Since the frequency of each symbol encoded in this way is at most $\sigma^{-1/(1+\delta/3)}$, the total length of the resulting n codewords is at most a factor $1/(1 + \delta/3)$ from the length if an optimal code was used.

In conclusion, we have described a variation of Huffman coding that can be decoded in constant time, using a table of $o(\sigma)$ bits, at the expense of increasing the length of codewords by a factor arbitrarily close to (but bounded away from) 1. Using this with section 2 we get a result that is very close to Theorem 1, but where the additive term in the space usage does not decrease with p_1.

4 Approximate Membership with Arbitrary Error

As a building block to be used in the next section, we now consider the approximate membership problem. Using traditional theory on approximate membership (AM), the theoretical lower limit for the space usage of AM structures is $n \log_2(1/\varepsilon)$ bits [6]. In known optimal constructions, the limit can only be reached (or reached within a factor $1 + \delta$) when ε is a negative power of 2. We now describe a way to circumvent this limitation that is more efficient than simply choosing a lower false positive probability.

We describe an AM structure with near optimal space complexity and with an arbitrary false positive error rate, $\varepsilon = c\,2^{-i}$, where $c \in [1/2; 1]$, and i is a positive integer. Let g be a random hash function, $g : D \to [0; 1]$ mapping the keys in S uniformly and independently to $[0; 1]$ (a discrete approximation would be needed in an implementation, but the analysis would be essentially the same as in the idealized scenario). Let $\gamma = 2(1 - c) \in [0; 1]$, and divide S into two subsets, S_1 and S_2, mapping to values smaller and larger than γ respectively. Formally

$$S_1 = S \cap g^{-1}([0; \gamma])$$
$$S_2 = S \cap g^{-1}((\gamma; 1]).$$

Note that the expected number of keys in S_1 is a fraction $2(1-c)$ of all the keys in S.

The AM structure is made up of two ordinary AM structures, each with a false positive rate that is a negative power of 2: one for S, with false positive rate $\varepsilon_0 = 2^{-i}$, and one for S_1, with $\varepsilon_1 = 1/2$. A query on a key x is performed by first consulting the larger structure. In case of a negative answer, we know the key is not present. In case of a positive answer, we calculate $g(x)$. If $g(x)$ indicates that the key is in S_2, we return a positive answer. Otherwise, we consult the smaller AM structure, and return the answer obtained.

It is easy to analyze the expected false positive rate of the AM structure described above. Consider the false positives of the structure for S. For a fraction $2(1-c)$ of those, the hash function g will point to S_1 and we therefore consult the AM structure for S_1. This will in turn result in a negative result for half of the queries, since $\varepsilon_1 = 1/2$. That is, a fraction $1-c$ of the false positives of the structure for S has been eliminated. Since the structure for S has an error rate $\varepsilon_0 = 2^{-i}$, the error rate of the whole AM structure is $\varepsilon = c2^{-i}$.

The expected space usage of the AM structure is $i+2(1-c)$ bits per key, which is close to the optimal $\log(1/\varepsilon) = i - \log c$ bits per key. In fact, the expected space usage is at most $1 - \log_2 e + \log_2 \log_2 e \approx 0.086$ bits per key from the optimal value.

5 Improvement for Skewed Distributions

We now provide the last part needed to show Theorem 1. In particular, we focus on the setting where the probability p_1 of the most frequent value is large. The idea is to use one or more approximate membership data structures as "filters" that determine a large fraction of values at little cost.

The redundancy of a prefix-free code is the difference between the expected cost of coding and the theoretical lower limit cost. The lower limit for coding a sequence of symbols related to n keys (assuming that values and corresponding keys are independent) is nH_0, where H_0 is the 0th order entropy of the value distribution. The redundancy per key is therefore $r = E(|u|) - H_0$.

Gallager [10] established that the redundancy of a Huffman code could be bounded by

$$r \leq p_1 + \rho \tag{4}$$

where p_1 is the probability of the most frequent symbol a_1 and $\rho = 1 - \log_2 e + \log_2 \log_2 e \approx 0.086$. For $p_1 \geq 0.5$ the bound further reduces to $r \leq p_1$. Together with the analysis in section 3 this immediately implies Theorem 1 whenever $p_1 < 1/2$.

In the following we examine the case where $p_1 \geq 0.5$, and determine a constant bound lower than 1, by adding a filter for handling the most frequent symbol. More specifically, we create an approximate membership structure, using the approach from the previous section. The structure supports $O(1)$ time membership queries on n keys, in space

$$(1+\delta)n(\log_2(1/\varepsilon) + \rho) \text{ bits},$$

where $\rho \approx 0.086$ and ε is the false positive rate. Recall that the data structure returns "true" for any key in the filter, and "false" with probability at least $1-\varepsilon$ for other keys.

We build an AM structure for the subset S^* that contains all keys in S except those mapping to a_1. All queries start by consulting this AM structure. A negative result for a key $x \in S$ means that the key with certainty maps to a_1. A positive result, however, means that the key is either in S^* or is a false positive (in which case it maps to a_1). The remaining task is therefore to store the restriction of f to the set S_{AM} of keys for which the AM structure returned "true". This can be done using the method previously described (base case), or recursively using the same method — we analyze the former case, where only one filter is used. The expected number of keys in S_{AM} is $(1 - p_1 + \varepsilon p_1)n$.

In terms of our original code tree, this approach corresponds to adding a top level node, with the old root and a new a_1 leaf as children. The new code is one that allows two codewords for one symbol; one represented by the new leaf, and another in a modified version of the old tree. The advantage lies in the the low cost of coding the first bit using an AM structure, analyzed in the following.

5.1 Space Analysis

We now examine the space cost incurred by adding a filter as described. To simplify the calculations we pretend that Huffman coding is used rather than the method described in section 3. This means that all space bounds should be multiplied by a factor $1 + \delta/3$. The size of the approximate membership data structure with false positive rate ε is

$$n(1 - p_1)(\log_2(1/\varepsilon) + \rho) \text{ bits.} \tag{5}$$

If we let $\alpha = (1 - p_1 + \varepsilon p_1)$ and assume that a_2 is the second most frequent symbol, with probability p_2, Gallager [10] tells us that the space for encoding the remaining values using a Huffman code can be bounded by

$$\alpha n(H_0' + \frac{p_2}{\alpha} + \rho), \tag{6}$$

bits per value, where H_0' is the 0th order entropy of the distribution of values over keys in S_{AM}. Note that a_2 may be identical to a_1 if the same value remains the most frequent.

In order to determine the redundancy of the combined structure, we express H_0' in terms of the original entropy H_0, p_1, and ε,

$$H_0' = \sum_{i=2}^{\sigma} \left(\frac{n_i}{\alpha n} \log \frac{\alpha n}{n_i} \right) + \frac{\varepsilon n_1}{\alpha n} \log \frac{\alpha n}{\varepsilon n_1}, \tag{7}$$

By using

$$H_0 = \sum_{i=2}^{\sigma} \left(\frac{n_i}{n} \log \frac{n}{n_i} \right) + \frac{n_1}{n} \log \frac{n}{n_1}, \tag{8}$$

we may conclude from (7) that

$$\alpha H_0' = H_0 + p_1 \log p_1 + (1 - p_1) \log \alpha + \varepsilon p_1 \log \left(\frac{\alpha}{\varepsilon p_1} \right). \tag{9}$$

Summing the space usage from (5) and (6) and subtracting the lower bound H_0 we can bound the redundancy per value r by a function of p_1, p_2, and ε:

$$r < \log p_1 + \alpha \log\left(\frac{\alpha}{\varepsilon p_1}\right) + p_2 + (1 - p_1 + \alpha)\rho. \tag{10}$$

If we choose $\varepsilon = 1 - p_1$, which is a near-optimal choice, we get $\alpha = 1 - p_1^2$, and using $p_2 \leq 1 - p_1$ the redundancy is bounded by

$$\hat{r}(p_1) = \log p_1 + (1 - p_1^2)\log\left(\frac{1 + p_1}{p_1}\right) + (1 - p_1) + (2 - p_1 - p_1^2)\rho. \tag{11}$$

The function $\hat{r}$ is convex, $\hat{r}(1) = 0$, and $\frac{du}{dt}\hat{r}(1) > -1.82$. Therefore $\hat{r}(p_1) \leq 1.82(1 - p_1)$. At the same time, Gallager's bound means that we should not use filtering whenever the resulting redundancy is above p_1 bits per element (assuming $p_1 > 0.5$). The crossover happens around $p_1 = 0.63$, meaning that this is the maximum redundancy.

We have shown the following upper bound for the redundancy of our prefix-free code:

$$r < \min(p_1 + 0.086, 0.63, 1.82(1 - p_1)) \tag{12}$$

providing an alternative to Gallager's [10] bound for Huffman codes. Our bound is an improvement when there is a significant imbalance in the distribution of symbols, in the way that one symbol dominates with $p_1 > 0.63$. In the statement of Theorem 1 we have omitted the middle term, which is only a small improvement for a narrow range of p_1 values.

6 Conclusion

We have described a data structure for space efficiently representing a function with skewed function values. The representation uses space close to the entropy of the function values, and is independent of the size of the domain of the function.

Our adaptation of [9] to handle variable-length strings is nontrivial in the sense that the most straightforward generalizations do not seem to work, while our method of choosing hash functions that are slightly correlated does. In addition, we have introduced several techniques that may be of independent interest: A general reduction that gives approximate membership data structures with arbitrary error probability, the use of filtering for efficient compression, and constant time decoding of a Huffman-like code.

Acknowledgement. We thank the anonymous reviewers for their thorough comments.

References

1. Belazzougui, D., Boldi, P., Pagh, R., Vigna, S.: Monotone minimal perfect hashing: Searching a sorted table with $O(1)$ accesses. In: Proceedings of the 20th Annual ACM-SIAM Symposium on Discrete Algorithms (SODA 2009). ACM Press, New York (2009)

2. Belazzougui, D., Boldi, P., Pagh, R., Vigna, S.: Theory and practise of monotone minimal perfect hashing. In: Finocchi, I., Hershberger, J. (eds.) ALENEX, pp. 132–144. SIAM, Philadelphia (2009)
3. Bloom, B.H.: Space/time trade-offs in hash coding with allowable errors. Communications of the ACM 13(7), 422–426 (1970)
4. Botelho, F.C., Pagh, R., Ziviani, N.: Simple and space-efficient minimal perfect hash functions. In: Dehne, F., Sack, J.-R., Zeh, N. (eds.) WADS 2007. LNCS, vol. 4619, pp. 139–150. Springer, Heidelberg (2007)
5. Calkin, N.J.: Dependent sets of constant weight binary vectors. Combinatorics, Probability & Computing 6(3), 263–271 (1997)
6. Carter, L., Floyd, R., Gill, J., Markowsky, G., Wegman, M.: Exact and approximate membership testers. In: Proceedings of the 10th Annual ACM Symposium on Theory of Computing (STOC 1978), pp. 59–65. ACM Press, New York (1978)
7. Charles, D., Chellapilla, K.: Bloomier filters: A second look. In: Halperin, D., Mehlhorn, K. (eds.) ESA 2008. LNCS, vol. 5193, pp. 259–270. Springer, Heidelberg (2008)
8. Chazelle, B., Kilian, J., Rubinfeld, R., Tal, A.: The Bloomier filter: An efficient data structure for static support lookup tables. In: Proceedings of the 15th Annual ACM-SIAM Symposium on Discrete Algorithms (SODA 2004), pp. 30–39. ACM Press, New York (2004)
9. Dietzfelbinger, M., Pagh, R.: Succinct data structures for retrieval and approximate membership (Extended abstract). In: Aceto, L., Damgård, I., Goldberg, L.A., Halldórsson, M.M., Ingólfsdóttir, A., Walukiewicz, I. (eds.) ICALP 2008, Part I. LNCS, vol. 5125, pp. 385–396. Springer, Heidelberg (2008)
10. Gallager, R.: Variations on a theme by Huffman. IEEE Transactions on Information Theory 24(6), 668–674 (1978)
11. Hagerup, T.: Sorting and searching on the word RAM. In: Meinel, C., Morvan, M. (eds.) STACS 1998. LNCS, vol. 1373, pp. 366–398. Springer, Heidelberg (1998)
12. Hagerup, T., Tholey, T.: Efficient minimal perfect hashing in nearly minimal space. In: Ferreira, A., Reichel, H. (eds.) STACS 2001. LNCS, vol. 2010, pp. 317–326. Springer, Heidelberg (2001)
13. Huffman, D.A.: A method for the construction of minimum-redundancy codes. Proceedings of the Institute of Radio Engineers 40(9), 1098–1101 (1952)
14. Larmore, L.L., Hirschberg, D.S.: A fast algorithm for optimal length-limited Huffman codes. Journal of the ACM 37(3), 464–473 (1990)
15. Milidiú, R.L., Laber, E.S.: Bounding the inefficiency of length-restricted prefix codes. Algorithmica 31(4), 513–529 (2001)
16. Motwani, R., Raghavan, P.: Randomized algorithms. Cambridge University Press, Cambridge (1995)
17. Pătraşcu, M.: Succincter. In: Proceedings of the 49th Annual Symposium on Foundations of Computer Science (FOCS 2008), pp. 305–313 (2008)
18. Porat, E.: An optimal Bloom filter replacement based on matrix solving. CoRR, abs/0804.1845 (2008)
19. Schmidt, J.P., Siegel, A.: The spatial complexity of oblivious k-probe hash functions. SIAM J. Comput. 19(5), 775–786 (1990)
20. Talbot, D., Talbot, J.M.: Bloom maps. In: Proceedings of the Fourth Workshop on Analytic Algorithmics and Combinatorics (ANALCO). IEEE, Los Alamitos (2008)

Experimental Variations of a Theoretically Good Retrieval Data Structure*

Martin Aumüller, Martin Dietzfelbinger, and Michael Rink

Technische Universität Ilmenau, 98693 Ilmenau, Germany
{martin.aumueller,martin.dietzfelbinger,michael.rink}@tu-ilmenau.de

Abstract. A retrieval data structure implements a mapping from a set S of n keys to range $R = \{0,1\}^r$, e.g. given by a list of key-value pairs $(x, v) \in S \times R$, but an element outside S may be mapped to any value. Asymptotically, minimal perfect hashing allows to build such a data structure that needs $n \log_2 e + nr + o(n)$ bits of memory and has constant evaluation time. Recently, data structures based on other approaches have been proposed that have linear construction time, constant evaluation time and space consumption $O(nr)$ bits or even $(1+\varepsilon)nr$ bits for arbitrary $\varepsilon > 0$. This paper explores the practicability of one such theoretically very good proposal, bridging a gap between theory and real data structures.

1 Introduction

The retrieval problem is the following elementary task: Let a "universe" U of keys and a "range" R be given (we always assume $R = \{0,1\}^r$ for some $r \geq 1$). Given a mapping $f: S \to R$, for a subset $S \subseteq U$ of size n, provide a data structure that on input $x \in U$ will calculate $f(x)$ if $x \in S$ and an arbitrary value from R if $x \in U - S$. For a discussion of the problem and its applications we refer the reader to [3]. A common way to obtain a retrieval data structure is to use a hash table with a minimal perfect hash function. Hagerup and Tholey [4] showed how to obtain such a function with near minimal space needs and constant evaluation time, leading to a retrieval data structure with asymptotic space usage of about $1.44n + nr$ bits. However, that construction is rather of theoretical interest because it works only for extremely large n (see [1]). A simpler and more practical approach to construct a minimal perfect hash functions, given by Botelho, Pagh and Ziviani [1], leeds to a retrieval data structure with constant evaluation time and space usage of about $2.62n + nr$ bits.

In this experimental paper we study one algorithm, described in [3], that follows a different approach, based primarily on known bounds of the probability that almost square, sparse random binary matrices have full row rank. A retrieval data structure built by this algorithm will—in theory—have the following properties: The space needed is $(1 + \varepsilon) \cdot nr$ bits, the evaluation time is $O(\log(1/\varepsilon))$, and the construction time is expected linear. The data structure consists of two

* Research supported by DFG, Grant Di 412/10-1.

A. Fiat and P. Sanders (Eds.): ESA 2009, LNCS 5757, pp. 742–751, 2009.

parts and is constructed as follows. We have an algorithm A from [1] that builds a retrieval data structure with constant evaluation time and space usage $1.23nr$ bits in expected linear time. Furthermore we have an algorithm B from [3] that succeeds with high probability and if it does not fail constructs a basic retrieval data structure that has constant evaluation time and space usage $(1+\varepsilon)\cdot nr$ bits. The runtime of algorithm B is at least quadratic, since the construction involves solving a sparse system of linear equations. In [3] it is shown how to combine algorithms A and B to get a retrieval data structure with expected linear construction time and asymptotically optimal space usage. For that purpose the key set S is split by a hash function into subsets of expected size $n' = \frac{1}{2}\sqrt{\log n}$, and algorithm B is applied to each of the subsets separately, using the fact that the linear systems are so small that preprocessed lookup tables can be used for the linear algebra. For many of the subsets this will be successful. A vanishing fraction of the keys will be left; they can be accommodated in a "backup data structure" that is built by algorithm A. A paper with related (theoretical) results is [5], which also gives very space efficient linear-time constructions for data structures with the retrieval functionality, where the data structures used there utilize table lookup also for queries.

In this paper we report on experiments to turn the approach from [3] into an operational data structure with very small space usage, where we try to uncover the impact of the constants hidden in the asymptotic notation of the space and time consumption. Although the space usage of about $(1+\varepsilon)\cdot nr$ is asymptotically not better than the solution obtained by perfect hashing, it can be better at least for very small r and it is interesting in itself to see how far the approach from [3] carries. The best known practical retrieval data structure for our problem if that from [1]. Not only is it elegant but also very easily implemented and it runs in linear time. It needs space about $1.23nr$ bits. Thus, this space bound is the one to beat. Our experiments led to the following results:

(i) A naive implementation of the tabulation method for solving the linear equation systems hits space limits during construction time already for very small n' (around 5).

(ii) Employing tricks for compressing the tables, filtering out redundant information (matrices that differ only in a permutation of the rows), raises the n' that can be treated to 6. With better compression tricks and very fast and large machines, a value of $n' = 7$ may be achievable, but not beyond. For the "$o(n)$" estimate for the space requirements for the tables to take over, n must be extremely large.

(iii) With $n' \leq 7$, the space needed for the main data structure and the backup data structure is far above $1.23nr$, apart from the space needed in the construction. This does not contradict the theoretical results, since those assume that n' is sufficiently large. The consequence is that the construction from [3] cannot be used as described there for a very space-efficient data structure for realistic values of n.

(iv) A realizable construction that makes it possible to compress the space below $1.1nr$ resulted from using the bucketing idea of the theoretical construction, solving linear systems of, say, n' a few hundred equations without using precomputed tables (in expected linear time, for constant n'). Reducing the space further is possible, but it will increase the constant factor in the construction time.

2 The Basic Retrieval Data Structure

We briefly recall the basic retrieval data structure from [3] that realizes a function $f : S \to \{0,1\}^r$ for a key set $S = \{x_0, x_1, \ldots, x_{n-1}\}$. Assume we have a function ξ that maps each $x \in S$ to a set A_x, where A_x is a subset of $[m] = \{0,1,\ldots,m-1\}$ of size k. Let $\boldsymbol{v}_x \in \{0,1\}^m$ be the characteristic vector of A_x, that is, the number of 1-bits of $\boldsymbol{v}_x$ is exactly k. We build an $n \times m$ matrix $\boldsymbol{M} = (\boldsymbol{v}_{x_i}^\top)_{i\in[n]}$, using the vectors $\boldsymbol{v}_x$ as rows, and a vector $\boldsymbol{u} = (u_0, u_1, \ldots, u_{n-1})^\top \in (\{0,1\}^r)^n$ of the corresponding function values $u_i = f(x_i)$. Consider the field $\mathbb{F}_{2^r}$ of cardinality 2^r and suppose that $\text{rank}(\boldsymbol{M}) = n$. Then there is a solution $\boldsymbol{a} = (a_0, a_1, \ldots, a_{m-1})^\top \in (\{0,1\}^r)^m$ of the system of linear equations

$$\boldsymbol{M} \cdot \boldsymbol{a} = \boldsymbol{u} \ . \tag{1}$$

After determining such a solution we can simply retrieve $f(x)$ for each $x \in S$ via:

$$f(x) = \bigoplus_{i \in A_x} a_i, \quad \xi(x) = A_x \subseteq [m], \quad |A_x| = k \ , \tag{2}$$

where $\oplus$ denotes bitwise XOR. In [2] it is shown that if the mapping ξ is fully random on S, $k \geq 3$, n sufficiently large as well as $m \geq (1+\varepsilon) \cdot n$ then $\boldsymbol{M}$ has full row rank with high probability. More precisely we must have $\frac{1}{1+\varepsilon} < \beta(k)$ for some threshold $\beta(k)$ with $\beta(k) - (1 - e^{-k}/\ln 2) \to 0$ as $k \to \infty$, exponentially fast. Hence to get a retrieval data structure for an arbitrary function $f : S \to \{0,1\}^r$ with asymptotically optimal space usage $(1+\varepsilon) \cdot nr$ and evaluation time $O(k) = O(\log(1/\varepsilon))$, we need to construct a matrix $\boldsymbol{M} \in \mathbb{F}_2^{n\times m}$, $(1+\varepsilon) \cdot n = m$, solve the linear system (1), which costs time $\Omega(n^2)$, and store the solution vector $\boldsymbol{a}$ as well as the function $\xi = \xi(k)$.

3 Strategies for Solving the Linear System

Although there exist several methods for calculating the vector $\boldsymbol{a}$ we mainly concentrate on the well known Gaussian elimination. In Section 6 we give some comments on using other methods that utilize the fact that (1) is a sparse linear equation system.

3.1 Pseudoinverse

For a given key set S we map each key $x \in S$ to A_x via k hash functions $h_i : U \to [m-i], i \in [k]$, where the i-th element of A_x equals the $h_i(x)$-th number of $[m] - \{h_0(x), h_1(x), \ldots, h_{i-1}(x)\}$. We build the matrix $\boldsymbol{M} \in \mathbb{F}_2^{n\times m}$ from the index sets A_x and check if $\boldsymbol{M}$ has full row rank using Gaussian elimination. Since there is a positive probability that $\operatorname{rank}(\boldsymbol{M}) < n$, we provide a constant number of sets of hash functions $H^j = \{h_0^j, h_1^j, \ldots, h_{k-1}^j\}, j \in [l]$, and if $\boldsymbol{M}$ has not full row rank we simply choose a new set of functions and rebuild $\boldsymbol{M}$. Note: We need $\lceil \log l \rceil$ bits to write down the index j that is used. The probability that we do not end up with a matrix of full row rank can be made as small as $O(n^{-c})$ for an arbitrary constant $c > 0$. If $\operatorname{rank}(\boldsymbol{M}) = n$, Gaussian elimination yields also a matrix $\boldsymbol{C} \in \mathbb{F}_2^{n\times n}$ coding elementary row transformations (no row exchanges), such that the n unit vectors are columns of $\boldsymbol{C} \cdot \boldsymbol{M}$. Let $b_i \in [m]$ be the column number of the i-th unit vector and let $\boldsymbol{u}' = \boldsymbol{C} \cdot \boldsymbol{u}$ with $\boldsymbol{u}' = (u_0', u_1', \ldots, u_{n-1}')^\top$. Then $\boldsymbol{a} = (a_i)_{i\in[m]}$ with $a_{b_i} = u_i'$ and $a_i = 0$ for $i \notin \{b_0, b_1, \ldots, b_{n-1}\}$ is a solution of the system

$$(\boldsymbol{C} \cdot \boldsymbol{M}) \cdot \boldsymbol{a} = \boldsymbol{C} \cdot \boldsymbol{u} = \boldsymbol{u}' \ , \tag{3}$$

and hence a solution of (1). Since, for constant ε, Gaussian elimination has worst case runtime $T(n) = O(n^3)$, in [3] this approach is modified in order to handle large key sets.

3.2 Splitting into Tiny Subsets

For large n it is impractical to solve (1) directly. Hence in [3] it is proposed to reduce the problem of solving a large linear system for S by splitting the key set in n/n' subsets S_i of size about n' and solve a smaller system for each S_i. The split is done via a hash function $h' : U \to [n/n']$ by defining $S_i := \{x \mid h'(x) = i\}$. For each S_i we obtain the related $|S_i| \times m'$ matrix $\boldsymbol{M}_i = (\boldsymbol{v}_x^\top)_{x\in S_i}$ and calculate the pseudoinverse $\boldsymbol{C}_i$ for the corresponding system

$$\boldsymbol{M}_i \cdot \boldsymbol{a}_i = \boldsymbol{u}_i \ . \tag{4}$$

For possible rebuilds of matrices $\boldsymbol{M}_i$, l sets of k hash functions are provided for one S_i. (We can share these functions among all subsets, so no space problem is caused.) If a subset S_i is too large, that is, $|S_i| > (1+\varepsilon)\cdot n' = m'$, then no matrix $\boldsymbol{M}_i$ is created and the elements of S_i are classified as "special". Similarly, if all l attempts at finding a pseudoinverse $\boldsymbol{C}_i$ for $\boldsymbol{M}_i$ fail, the keys in S_i are declared "special". For all special keys together we use the data structure from [1], in the version of a retrieval structure, as backup data structure. This takes extra space of about $\alpha \cdot 1.23nr$, where α denotes the fraction of special keys. The (expected) runtime for the construction, excluding the time for the backup data structure, can be simply bounded by $O(n/n' \cdot T(n'))$, where $T(n')$ depends on the way the $\boldsymbol{C}_i$'s are obtained.

In [3] it was suggested that n' is chosen as $\frac{1}{2}\sqrt{\log n}$ and that precomputed tables of pseudoinverses for the matrices $\boldsymbol{M}_i$ are used, because in this way linear construction time can be obtained while the size of the tables and the number of special elements are upper bounded by $o(n)$. To anticipate some results of our experiments, it turned out that the approach with precomputed pseudoinverses is realizable only for very small n'. Even when using mild compression tricks it was not practical to choose n' larger than 6. Note: If $n' = \frac{1}{2}\sqrt{\log n} = 6$, then we have $n = 2^{144}$. This experience prompted us to try another solution strategy, using larger n' and refraining from precomputation.

4 Implementation Considerations

In this section we describe the framework for our study of the space-time-tradeoff for the construction of the retrieval data structure with and without using precomputed solutions. Our primary goal was to explore what orders of magnitude of n could be treated in reasonable time to beat the $1.23nr$ space bound of the retrieval data structure from [1].

4.1 Small Matrices — Precomputed Solutions

We studied how practical it is to precompute and store (on disc) all pseudoinverses for different parameters n', m' and k, such that during the construction of the retrieval data structure we are able to obtain the $\boldsymbol{C}_i$'s by simple lookups, that is in time $O(1)$. (Note that the whole matrix fits in $O(1)$ words, if $n' = \frac{1}{2}\sqrt{\log n}$.) After creating and storing the tables for the pseudoinverses one can fix some suitable m' and k and load tables for parameters $n'_1, n'_2, \ldots, n'_t$ with $n'_1 < n'_2 < \ldots < n'_t \leq m'$. Subsets S_i of size $n'' \leq n'_t$, $n'' \notin \{n'_1, \ldots, n'_t\}$, can be filled with dummy elements up to the next larger n'_j, $j \in [t]$. For subsets of size n'_j pseudoinverses are obtained by a lookup if they exist. Subsets for which none of the hash function sets works or that are larger than n'_t are treated using the backup data structure. Since the number of $n' \times m'$ binary matrices with k ones per row is $\binom{m'}{k}^{n'}$ it seems clear that this approach is worthwhile only for fairly small matrices, although only a fraction of them has full row rank. Furthermore a simple estimation of the size of the subsets S_i using Poisson distribution (for small mean n', under the assumption that the splitting hash function is fully random) shows that a substantial fraction of the keys will be inserted into the backup retrieval data structure because they sit in a S_i with $|S_i| > n'_t$.

4.2 Large Matrices — No Precomputation

The problem of using precomputed solutions is that even for small n' the costs in space (and time) will be too large, such that optimizing ε will be impractical. So we consider computing the pseudoinverses during the construction of the retrieval data structure. Using this approach we try to solve the linear systems for subsets S_i up to size m'. Storing already computed solutions to reuse them for other subsets later in the construction phase does not make sense, since the probability that the same matrix appears twice is vanishingly small. The special elements (from overfull subsets and matrices with no pseudoinverse) are handled like above using the backup data structure. The main problem remains in finding suitable parameter configurations that optimize the space and time requirements.

4.3 Basic Conditions

For the experiments we use fairly common but weak hash functions for obtaining the index sets A_x as well as for splitting the key set S. Each hash function $h : U \to R$ is defined as

$$h(x) = ((a \cdot x + b) \bmod p) \bmod |R| \ , \tag{5}$$

for a prime number $p > |U|$ and $a \in [p] - \{0\}, b \in [p]$ randomly chosen via a standard pseudo random number generator. We provide one hash function to split the key set S, as well as l sets of k hash functions $H^j = \{h_0, h_1, \ldots, h_{k-1}\}, j \in [l]$, to build the $\boldsymbol{M}_i$'s, that is, for each subset S_i one tries to find a solution using consecutively $H^0, H^1, \ldots,$ until one succeeds or H^{l-1} has been tried without success. For each S_i the "successful index" needs to be stored. In order to obtain a parameter configuration that optimizes the space usage, the fraction α of elements that have to be inserted into the backup data structure is crucial. We can estimate the space needed for the data structure by the following simple formula:

$$S(n) = n/n' \cdot (m'r + \lceil \log l \rceil) + \alpha \cdot 1.23nr \ , \tag{6}$$

ignoring $O(l \cdot k \cdot \log |U|)$ bits for the hash functions and other additional space depending on implementation and used architecture.

4.4 Experimental Setup

The experiments were implemented in C++ (g++ version 4.3.2 with optimization flags: -O3 -march=nocona -ffast-math -funsafe-loop-optimizations) on an Intel(R) Xeon(TM) machine with 3.20GHz CPU and 4 GB RAM under Linux (kernel 2.6.27), and ran on a single core. Except noted otherwise we used the universe $U = [2^{28}]$, and chose the n elements of S as well as the coefficients of the hash functions via *rand*(), a standard utility function for generating pseudo random numbers using the trinomial $x^{31} + x^3 + 1$. Since we considered the solution of (1) via a pseudoinverse, that is only when $\boldsymbol{M}_i$ has full row rank, the concrete retrieval function f is irrelevant. For testing purposes we used the identity function. Gaussian elimination and matrix multiplication were realized utilizing blockwise calculation.

5 Results

5.1 Small Matrices — Precomputed Solutions

We started our tests by calculating the number π of binary $n' \times m'$ matrices with k ones per row that have a pseudoinverse, for different parameters (n', m', k), as shown in Table 1. Since π becomes quite large even for small n' we used a simple compression method to reduce the amount of space for the lookup tables. Each $n' \times m'$ matrix $\boldsymbol{M}$ was normalized by sorting the rows lexicographically in descending order. This reduces the number of distinguishable matrices from

Table 1. $\mu \hat{=}$ # (n', m', k)-matrices, $\pi \hat{=}$ # matrices with pseudoinverses,$\mu_{\text{norm}} \hat{=}$ # normalized matrices, $\mu_{\text{sav}} \hat{=}$ # saved normalized matrices, $S_{\text{LT}} \hat{=}$ lower bound for the size of the lookup tables (in bytes)

n'	m'	k	μ	π	π/μ	μ_{norm}	μ_{sav}	S_{LT} [B]
5	6	3	$3.20 \cdot 10^6$	$1.54 \cdot 10^6$	0.482	$4.25 \cdot 10^4$	$1.29 \cdot 10^4$	$1.03 \cdot 10^5$
5	6	4	$7.59 \cdot 10^5$	$1.81 \cdot 10^5$	0.239	$1.16 \cdot 10^4$	$1.51 \cdot 10^3$	$1.21 \cdot 10^4$
5	7	3	$5.25 \cdot 10^7$	$3.58 \cdot 10^7$	0.682	$5.76 \cdot 10^5$	$2.99 \cdot 10^5$	$2.69 \cdot 10^6$
5	7	4	$5.25 \cdot 10^7$	$2.93 \cdot 10^7$	0.557	$5.76 \cdot 10^5$	$2.44 \cdot 10^5$	$2.20 \cdot 10^6$
6	7	3	$1.84 \cdot 10^9$	$8.78 \cdot 10^8$	0.478	$3.84 \cdot 10^6$	$1.22 \cdot 10^6$	$1.34 \cdot 10^7$
6	7	4	$1.84 \cdot 10^9$	$4.89 \cdot 10^8$	0.266	$3.84 \cdot 10^6$	$6.79 \cdot 10^5$	$7.47 \cdot 10^6$
6	8	3	$3.08 \cdot 10^{10}$	$2.04 \cdot 10^{10}$	0.663	$5.55 \cdot 10^7$	$2.84 \cdot 10^7$	$3.12 \cdot 10^8$
6	8	4	$1.18 \cdot 10^{11}$	$6.54 \cdot 10^{10}$	0.556	$2.01 \cdot 10^8$	$9.09 \cdot 10^7$	$9.99 \cdot 10^8$

$\mu = t^{n'}$, $t = \binom{m'}{k}$, to $\mu_{\text{norm}} = \binom{t-1+n'}{n'}$, the number of normalized matrices. If a normalized matrix $\boldsymbol{M}_{\text{norm}}$ had a pseudoinverse $\boldsymbol{C}$ then the tuple $(\boldsymbol{M}_{\text{norm}}, \boldsymbol{C})$ was saved into a file. For the construction of the retrieval data structure the file was loaded and a lookup hash table (with expected constant access time) was built using the saved tuples as key-value pairs. So each lookup needs some pre- and postprocessing, since we have to normalize the key $\boldsymbol{M}$ and do the inverse transformation on the value $\boldsymbol{C}$. The space usage S_{LT} for the lookup hash table is at least $(\lceil n' \cdot m'/8 \rceil + \lceil n' \cdot n'/8 \rceil) \cdot \mu_{\text{sav}}$ bytes, where $\mu_{\text{sav}} = \pi/n'!$ denotes the number of saved normalized $n' \times m'$ matrices and $n' \times n'$ pseudoinverses, respectively.

n	$\bar{T}(n')$
$1 \cdot 10^6$	14.2 s
$5 \cdot 10^6$	60.1 s
$10 \cdot 10^6$	125 s
$20 \cdot 10^6$	252 s

Table 2. Average (15 runs) construction time $\bar{T}(n')$, using precomputed tables of pseudoinverses for parameters $n' = 5$, $m' = 7$, $k = 3$, $l = 4$.

One observes that μ_{sav} becomes quite large even for small n', such that it is not practical to go beyond $n' = 6$. (With better compression a value of $n' = 7$ may be achievable.) Furthermore the probability that a random (n', m', k)-matrix $\boldsymbol{M}$ has a pseudoinverse, varies strongly. This has to be compensated using an adequate number l of sets of hash functions for rebuilding $\boldsymbol{M}$, see Section 4.3. We started to determine the best parameter configuration according to the average space usage $\bar{S}(n)$ using only one lookup table, that is all matrices $\boldsymbol{M}$ have the same number of rows n'. This has the advantage that each solution vector $\boldsymbol{a}_i = (a_0, a_1, \ldots, a_{m'-1})$ has a fixed number of $m' - n'$ zero-entries whose positions can be encoded with $\lceil \log(\binom{m'}{n'}) \rceil$ bits. Hence formula (6) has to be modified as follows:

$$\bar{S}(n) = n/n' \cdot (n'r + \lceil \log l \rceil + \lceil \log(\binom{m'}{n'}) \rceil) + \bar{\alpha} \cdot 1.23nr \ , \tag{7}$$

to evaluate the average space usage $\bar{S}(n)$, where $\bar{\alpha}$ is the mean fraction of elements which must be inserted into the backup data structure. It turned out that on the one hand $\frac{\pi}{\mu}$ is a good approximation for the probability that a constructed

(m', n', k)-matrix has full row rank and on the other hand the Poisson distribution gives a good approximation for the average fraction of overfull subsets for $n \geq 10^5$. Hence one can use the following rule of thumb to estimate $\bar{\alpha}$:

$$\bar{\alpha} \approx 1 - e^{-n'} \cdot \left(\sum_{i=0}^{n'-1} \tfrac{(n')^i}{i!}\right) \cdot \left(1 - \left(1 - \tfrac{\pi}{\mu}\right)^l\right) . \tag{8}$$

According to (7) parameters $(n', m', k, l) = (5, 7, 3, 4)$ are a good choice. The fraction of subsets with size greater than 5 is about 56 percent and setting $l = 4$ the probability that we get a solution for a subset S_i of size ≤ 5 is about 99 percent. Since encoding the index j of the set of hash functions H^j takes 2 and encoding the zero-entries in the solution vector takes $\lceil \log(\binom{7}{2}) \rceil = 5$ bits, we need space about $\bar{S}(n) = 1.69nr + 7/5n$ bits. One can reduce the space needs slightly by using a lookup table T' for the $(6, 7, 3)$-matrices in addition to the table T for the $(5, 7, 3)$-matrices to handle subsets S_i of size up to 5 with table T and subsets of size 6 with table T', see Section 4.1. This reduces the fraction of too large subsets to about 38 percent and $\bar{\alpha}$ to about 40 percent. Solution vectors for matrices with 5 and 6 rows are handled equally by encoding only one zero-entry. Therefore we get a space usage of $\bar{S}(n) = n/5 \cdot (6 \cdot r + 5) + \bar{\alpha} \cdot 1.23nr = 1.69nr + n$ bits. Only slightly less space would be needed using parameters $(6, 8, 3)$, but this lookup table has already a size of more than 300 MB. And even with $n' = 7$ one could not beat the $1.23nr$ space bound. Considering the rapidly growing number of pseudoinverses one concludes that the approach using precomputed solutions is limited in its practicability. For completeness we verified that the construction time for $\boldsymbol{a}$ is linear as expected using parameters $(5, 7, 3, 4)$, see Table 2.

5.2 Large Matrices — No Precomputation

In our experiments with computing pseudoinverses during the construction we concentrated on parameters n' from 10 up to 1000 and tried to minimize the space usage via local linear search in the parameter space. Table 3 shows the optimal values for m' according to (6). The results are obtained for parameters $n' \in \{10, 20, 35, 50, 100, 200, 350, 500, 750, 1000\}$, $k = 5$ and $l = 2$ via inserting $n = 10^7$ keys 7 times. With $n' \geq 100$ one beats the $1.23nr$ bound from [1] with high probability. Growing n' makes it possible to reduce the space overhead $\varepsilon = m'/m' - 1$ in the primary structure and simultaneously the mean fraction $\bar{\alpha}$ of special elements.

Remark 1. We verified the results from Table 3 with a more "natural" set of keys using all article names of the English wikipedia from 24/05/2008 ($n = 5.249.107$ different keys with an average length of 20 bytes).

For 10^7 keys and $n' \leq 250$ the construction time is below 20 min, for our reference value $n' = 100$ even below 4 min. But as expected the construction time is $O(n \cdot (n')^2)$, see Figure 1, and therefore reducing the overall space usage with growing n' becomes soon expensive. For completeness we verified that the construction time for fixed bin sizes is linear like expected, see Figure 2.

Table 3. values of m' optimizing average space usage $\bar{S}$, $\bar{\alpha}\hat{=}$ mean, $s(\alpha)\hat{=}$ standard deviation

n'	m'	$\bar{\alpha}$	$s(\alpha)$	$\bar{S}/(nr)$
10	14	0.178	0.017	1.62
20	26	0.153	0.029	1.49
35	45	0.075	0.021	1.38
50	61	0.082	0.043	1.32
100	119	0.023	0.014	1.22

n'	m'	$\bar{\alpha}$	$s(\alpha)$	$\bar{S}/(nr)$
200	230	0.012	0.011	1.16
350	379	0.027	0.017	1.11
500	540	0.017	0.014	1.10
750	796	0.017	0.014	1.08
1000	1057	0.011	0.014	1.07

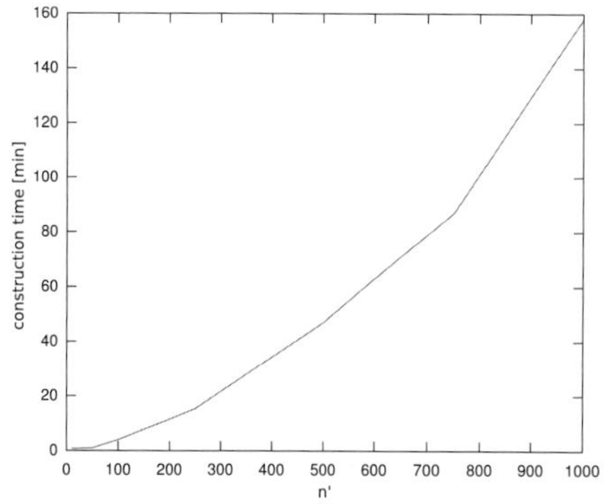

Fig. 1. average construction time (7 runs) for parameters $n = 10^7$ and n', m' from Table 3

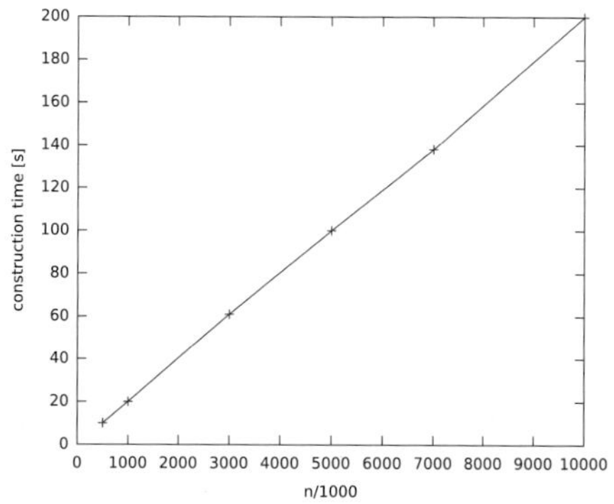

Fig. 2. average construction time (7 runs) for fixed $n' = 100, m' = 119$ and $n \in \{5 \cdot 10^5, 10^6, 3 \cdot 10^6, 5 \cdot 10^6, 7 \cdot 10^6, 10^7\}$

6 Conclusion and Outlook

In this paper we have studied how to transfer the theoretical good retrieval data structure from [3] into an operational "real world" data structure and beat the $1.23nr$ space bound from [1]. We showed that the construction from [3] using precomputed pseudoinverses cannot be used for a very space-efficient data structure since relevant parts of the actual complexity are hidden in the asymptotic estimates, and real input data and real machines are not large enough to make these estimates valid. In consequence we modified the theoretical construction in a straightforward way, using the same idea of "splitting the key set", but solving the linear equation systems of a few hundred equations one after the other without precomputation. This modification makes it possible to actually build retrieval structures with fewer than, say, $1.1nr$ bits of space, and expected linear construction time (for constant n') for realistic key set sizes.

Since our experiments showed that we can beat the $1.23nr$ space bound only via solving relatively large linear equation systems, it seems worthwhile to consider solution methods that utilize the sparse structure of the $\boldsymbol{M}_i$'s better than Gaussian elimination. We started tests with the conjugate gradient method and

Lanczos algorithm which can be applied to the system $\boldsymbol{M}_i^\top \cdot \boldsymbol{M}_i \cdot \boldsymbol{a}_i = \boldsymbol{M}_i^\top \cdot \boldsymbol{u}_i$ and have runtime $O(n' \cdot \Delta + (n')^2)$, where Δ denotes the number of non-zero entries of $\boldsymbol{M}_i$. The problem is that over finite fields these algorithms fail (even if a solution exists) when they encounter a non-zero vector which is self-orthogonal or self-orthogonal with respect to the inner product defined by $\boldsymbol{M}_i^\top \cdot \boldsymbol{M}_i$. First tests, using parameters n' and m' from Table 3 (starting with $n' = 200$), showed that this the case for more than half of the matrices, such that we find an existing solution in only 2/3 of the cases if we use $l = 2$ trials per subset S_i. Possibly techniques from [6] can be used to overcome this problem, but we haven't yet checked this. Furthermore we started investigating the suitability of Wiedemann's method [7]. Our first simple implementation using the deterministic variant of the algorithm (based on column elimination), which must be called $O(\varepsilon n')$ times in the worst case, beats Gaussian elimination at least for fairly large matrices with small ε, e.g. $n' = 1500, m' = 1575$, $\varepsilon = 0.05$. All in all it seems promising that one can achieve shorter runtime starting from the range $n' = 1000$ using sparse linear system solvers, which would make the approach "large matrices and no precomputation" even more practicable.

References

1. Botelho, F.C., Pagh, R., Ziviani, N.: Simple and space-efficient minimal perfect hash functions. In: Dehne, F., Sack, J.-R., Zeh, N. (eds.) WADS 2007. LNCS, vol. 4619, pp. 139–150. Springer, Heidelberg (2007)
2. Calkin, N.J.: Dependent sets of constant weight binary vectors. Combinatorics, Probability & Computing 6(3), 263–271 (1997)
3. Dietzfelbinger, M., Pagh, R.: Succinct data structures for retrieval and approximate membership (extended abstract). In: Aceto, L., Damgård, I., Goldberg, L.A., Halldórsson, M.M., Ingólfsdóttir, A., Walukiewicz, I. (eds.) ICALP 2008, Part I. LNCS, vol. 5125, pp. 385–396. Springer, Heidelberg (2008)
4. Hagerup, T., Tholey, T.: Efficient minimal perfect hashing in nearly minimal space. In: Ferreira, A., Reichel, H. (eds.) STACS 2001. LNCS, vol. 2010, pp. 317–326. Springer, Heidelberg (2001)
5. Porat, E.: An optimal Bloom filter replacement based on matrix solving. CoRR, abs/0804.1845 (2008)
6. Teitelbaum, J.: Euclid's algorithm and the Lanczos method over finite fields. Math. Comput. 67(224), 1665–1678 (1998)
7. Wiedemann, D.H.: Solving sparse linear equations over finite fields. IEEE Trans. Inf. Theor. 32(1), 54–62 (1986)

Short Labels for Lowest Common Ancestors in Trees

Johannes Fischer

Universität Tübingen, Center for Bioinformatics (ZBIT), Sand 14, D-72076 Tübingen
fischer@informatik.uni-tuebingen.de

Abstract. We evaluate the practical performance of labeling schemes for lowest common ancestors in trees. We develop different variants for encoding the labels. We then perform a thorough experimental evaluation of these schemes on a variety of tree shapes and sizes. We find that in general non-prefix-free codes lead to shorter labels than those that are prefix-free, while having roughly the same query time.

1 Introduction

In distributed environments, it is often desirable to compute certain functions on data without consulting *global* lookup tables. This may be, for example, because not all peers in a network can perform simultaneous reads in a global structure, and because it would be too costly (in terms of time, bandwidth, and/or space) to distribute these tables to all peers.

One particularly important instance of such distributed data structures are *labeling schemes* for trees or graphs with n nodes, where the aim is to assign a short (bit-)label to every node such that certain queries (aka functions) on the nodes can be computed *solely by looking at the labels* of the query nodes. Examples of such queries include node adjacency [1], ancestry [2,3], and distances [4, 5, 6]. In all of these examples, it has been shown that $O(\log n)$-size bit-labels suffice to compute the functions in $O(1)$ time.

Computing *nearest* or *lowest common ancestors* (LCAs) of two nodes in a tree is arguably more difficult than the examples mentioned above, in the sense that this problem is non-trivial already in a non-distributed setting [7]. The LCA of nodes u and v is defined as the deepest node ℓ that is an ancestor of both u and v. This definition shows a further complication for this type of query: the answer is not simply binary (e.g., yes/no to an "is-ancestor" query) or a natural number (e.g., a distance), but a node *label* itself (the one of node ℓ). Given these complications, it is perhaps surprising that Alstrup et al. have shown that $O(\log n)$-bit-labels suffice to answer LCA-queries in constant time [8]. Such a scheme *must* exploit the freedom to choose the node identifiers, for Peleg shows that if one further requires the answer ℓ to be a *predetermined* identifier of size $O(\log n)$ bits, a lower bound of $O(\log^2 n)$ bits exists on the label lengths [9]. In the same article, Peleg also gives a labeling scheme that achieves this bound. For the significance of LCA-queries in distributed environments, we refer to the excellent survey by Alstrup et al. [8].

A. Fiat and P. Sanders (Eds.): ESA 2009, LNCS 5757, pp. 752–763, 2009.

Despite of the significance of labeling schemes for LCAs, not much research has been done on their practical performance, although there is a thorough experimental study of labeling schemes for ancestor queries [3]. In last year's ESA, Caminiti et al. [10] compared the performance of labelling schemes for LCAs in trees. However, they only focused on the case where we require the query to return a predetermined node identifier (Peleg's scenario above), and hence considered only $O(\log^2 n)$ labeling schemes. Given this, Caminiti et al. experimented mainly on different options for *decomposing the tree into disjoint paths*[1], a key technique for most labeling schemes: decomposing by large child, maximum child, and by rank (see Sect. 2 for more details on tree decomposition). To summarize their contribution, tree-decomposition based on the *largest child* is shown to be always best in practice.

The aim of this article is to evaluate the practical performance of the $O(\log n)$-bit labeling scheme for LCAs due to Alstrup et a. [8], by experimenting on different *codes* used for constructing the node labels. Alstrup et al. themselves [8] give a code that achieves $O(\log n)$ label lengths, but there is certainly room for experiments. And improvements, as we shall see! Here, we rely on the experiments conducted before [10], and take it as granted that the largest-child decomposition is always best in practice. Interestingly, this is also the decomposition proposed in the original algorithm [8].

2 Labeling Trees for Lowest Common Ancestors

We describe a simplified version of the labeling scheme due to Alstrup et al. [8] for LCA-queries in a rooted tree T. Let $n = |T|$ denote the number of nodes in T, $\textsc{Parent}(v)$ the parent node of node v, and $\textsc{Children}(v)$ the set of v's children. Let T_v denote T's subtree that is rooted at v. We classify the nodes in T into *heavy* and *light* nodes as follows: The root of T is called light. For every internal node v, exactly one child $u \in \textsc{Children}(v)$ with $|T_u| = \max\{|T_w| : w \in \textsc{Children}(v)\}$ called *heavy*, whereas the other children are called *light*. The heavy nodes divide T into disjoint *heavy paths*; a heavy path $\langle v \rangle_k = \langle v_1, v_2, \ldots, v_k \rangle$ is defined recursively by following the heavy children $v_2, \ldots, v_k$, starting at a light node v_1. Node v_1 is called the *apex* of all nodes v_i on $\langle v \rangle_k$, denoted by $\textsc{Apex}(v_i)$ (note that $\textsc{Apex}(v_1) = v_1$ for light nodes v_1). For example, in the tree on the left side of Fig. 1, we have $\textsc{Apex}(A) = \textsc{Apex}(C) = \cdots = \textsc{Apex}(L) = A$.

We now assign a *heavy label* $hl(v)$ to every node v. The constraint on the heavy labels is that for a heavy path $\langle v \rangle_k$ and two given nodes v_i and v_j on $\langle v \rangle_k$, $1 \leq i, j \leq k$, we can decide from $hl(v_i)$ and $hl(v_j)$ alone which of the nodes v_i and v_j is an ancestor of the other. In other words, we need to decide whether $i < j$ solely by looking at $hl(v_i)$ and $hl(v_j)$.

We also assign a *light label* $ll(v)$ to each light node v, subject to the constraint that light nodes with the same parent get assigned different labels. The root gets assigned the empty string as a light label. Now, the label of node v is given by

[1] The authors [10] also consider a variation of Peleg's original scheme [9] called *CFP*, but it also uses $O(\log^2 n)$-bit-labels.

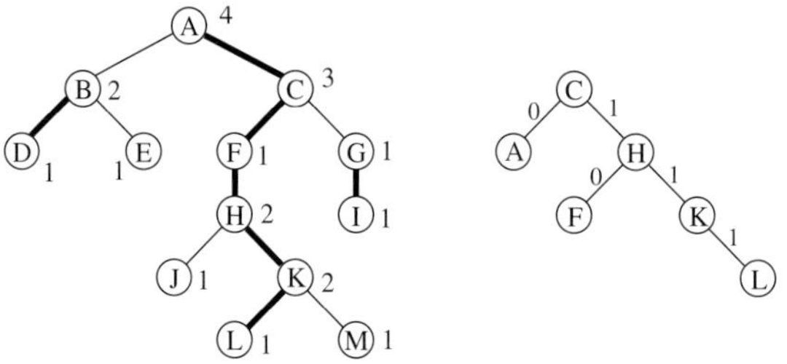

	codes $\langle c \rangle_k$	
node	actual	stored
A	0	1
C	ϵ	0
F	10	11
H	1	00
K	11	000
L	111	0000

Fig. 1. Left: A tree T with the light size drawn next to each node. Heavy paths are depicted by thicker edges. Middle: FLBST B for the heavy path $\langle v \rangle_k = \langle A, C, F, H, K, L \rangle$ in T, with light sizes $\langle y \rangle_k = \langle 4, 3, 1, 2, 2, 1 \rangle$. Right: Assigned heavy labels to nodes before (left) and after (right) adding the artificial '1' (needed to avoid the empty label).

$l(v) = l(\text{PARENT}(\text{APEX}(v))) \cdot ll(\text{APEX}(v)) \cdot hl(v)$, where "$\cdot$" denotes concatenation of strings. It follows from this definition that $l(v)$ is a concatenation of alternating heavy and light labels, $l(v) = h_0 \cdot l_1 \cdot h_1 \cdots l_t \cdot h_t$.

To see how these labels help for computing the label of $\ell = \text{LCA}(u, v)$ [8, Lemma 5], first assume that $l(u) = h_0 \cdot l_1 \cdot h_1 \cdots h_{i-1} \cdot l_i \cdot \alpha$, and $l(v) = h_0 \cdot l_1 \cdot h_1 \cdots h_{i-1} \cdot l'_i \cdot \alpha'$, with $l_i \neq l'_i$. Then u and v are in different subtrees T_x and T_y for two different light children x and y of ℓ; hence, $l(\ell) = h'_0 \cdot l'_1 \cdot h'_1 \cdots h'_{i-1}$. The same is true if $l_i \cdot \alpha$ or $l'_i \cdot \alpha'$ is empty, in which case one of u and v is an ancestor of the other.

If, on the other hand, $l(u) = h_0 \cdot l_1 \cdot h_1 \cdots l_i \cdot h_i \cdots$, and $l(v) = h_0 \cdot l_1 \cdot h_1 \cdots l_i \cdot h'_i \cdots$, with $h_i \neq h'_i$, then ℓ must be on the deepest heavy path $\langle v \rangle_k$ that is common to all paths from the root down to u and v. Heavy label h_i is a heavy label of some node v_x on $\langle v \rangle_k$, and h'_i is a heavy label of some node v_y on $\langle v \rangle_k$. Thus, if $x < y$, $l(\ell) = h'_0 \cdot l'_1 \cdot h'_1 \cdots l_i \cdot h_i$, due to the constraints on heavy labels. Otherwise, $l(\ell) = h'_0 \cdot l'_1 \cdot h'_1 \cdots l_i \cdot h'_i$.

Apart from finding longest common prefixes, the query algorithm needs to (1) find the beginning and end of a heavy/light label h_i/l_i in $l(v)$, given a position j within h_i/l_i, (2) determine whether position j occurs in a heavy or light label in $l(u)$, and (3) decide whether $i < j$ by looking at $hl(v_i)$ and $hl(v_j)$ for nodes v_i, v_j on a heavy path $\langle v \rangle_k$.

The third point above will be discussed separately in Sect. 3 for each different coding algorithm. Concerning the first two points, one possibility is to use an additional bitstring $k(v)$ of length $|l(v)|$, which we call the *helper label* of v, where $k(v)$ has a '1' at position j iff j within a heavy label, and a '0' otherwise. Then point (2) above becomes trivial, and point (1) amounts to finding the preceding/succeeding occurrence of a '0' or '1' bit before/after position j. This can be done with standard bit-manipulating techniques.

The label $l(v)$ of a heavy node v will be repeated in all labels $l(w)$ for w being a node in a subtree rooted at one of v's *light* children. In order to ensure that the label lengths are all within $O(\log n)$, we require the heavy labels $\langle hl(v) \rangle_k$ of a heavy path $\langle v \rangle_k$ satisfy $|c_i| \leq \log \sum_{j \neq i} y_j + O(1)$. Here, $y_i = |T_{v_i}| - |T_{v_{i+1}}|$ is called v_i's *light size* for $1 \leq i < k$ (define $y_k = 1$).

3 Coding Algorithms

In this section, we review different coding algorithms used in our experiments. Let $\langle v \rangle_k$ be the sequence of nodes that we want to encode into a sequence $\langle c \rangle_k$, and $\langle y \rangle_k$ be the sequence of light sizes of the node sequence $\langle v \rangle_k$ (if needed).

3.1 Non-Self-Delimiting Codes

We first discuss codes that need a *helper label* (as discussed in Sect. 2) for extracting an individual label in a juxtaposition of alternating heavy and light labels. Remember that helper labels double the total label size.

Canonical Coding. The *canonical code* is given by setting c_i to the i'th entry in the canonical ordering $0, 1, 00, 01, 10, 11, 000, 001, \ldots$ of all bit-strings [11]. Given c_i and c_j, checking whether $i < j$ is done by first checking whether $|c_i| < |c_j|$, and, in case of equality, checking if $c_i < c_j$ (numerically).

Hu-Tucker Coding. The problem of constructing an *optimal alphabetic search tree* [12] for $\langle y \rangle_k$ is to construct a binary tree B, where the i'th leftmost leaf is labeled with v_i, such that $\sum y_i d_i$ is minimal. Here, d_i denotes the depth of leaf i. Hu and Tucker [13] showed how to construct such a tree in $O(k \log k)$ time. The general idea of this algorithm is to first construct a Huffman-like binary tree, where the leaves are labeled with the v_i's, but not necessarily increasingly, and then rearrange this tree to get an increasing labeling. Although an exact linear time algorithm for our setting exists [14], we implemented an $O(k \log k)$-time approximation algorithm [15].

The code c_i for v_i is obtained by following the path from B's root down to the i'th leaf, reading a left child as a '0', and a right child as a '1'. This gives a prefix-free code (no code is a prefix of other codes), although this property would not be necessary for our scheme. Checking whether $i < j$ for given c_i and c_j is done by a simple lexicographic comparison: first align c_i and c_j by shifting the shorter to the left, and then do a standard integer comparison on the resulting words. In case of equality, check if $|c_i| < |c_j|$.

Alstrup et al.'s Coding. Alstrup et al. themselves [8, p. OF9] propose an interesting code by dividing the interval $[0, \sum_{i \leq k} y_i)$ into k sub-intervals I_i of length y_i, and then locating an integer z_i in I_i which has $\lfloor \log y_i \rfloor$ trailing 0s. The code c_i is obtained from z_i by chopping off these trailing 0s, i.e., taking only the $\lceil \log \sum_{i \leq k} y_i \rceil - \lfloor \log y_i \rfloor$ most significant bits of z_i.

Checking whether $i < j$ for given c_i and c_j is done by lexicographic comparison, as in the Hu-Tucker coding.

Mehlhorn's Coding. Alstrup et al. [8, p. OF10] already hint at the fact that Mehlhorn's linear-time method [15] for approximating *optimal binary search trees* [16] can be used to construct *prefix-free* codes (what they call *alphabetic codes*). However, we found that Mehlhorn's algorithm can also be used to construct much shorter (non-prefix-free) codes that are suitable for our setting, as explained next.

We define a *fully labeled binary search tree* (FLBST) for $\langle v \rangle_k$ as a binary search tree with k nodes, such that the i'th node in symmetric order (i.e., in-order) is labeled by v_i. In accordance with alphabetic search trees (see above), an optimum FLBST is one where $\sum y_i d_i$ is minimal (d_i is the depth of the node with in-order i). Computing an optimum FLBST could be done with Knuth's $O(k^2)$-time-and-space-algorithm for optimum binary search trees [16], but this would be too costly. Instead, we use Mehlhorn's linear-time algorithm [15] for *approximating* an optimum FLBST B. It is based on recursively *bisecting* the interval $[0, \sum_{i \leq k} y_i)$ such that the "weights" of the subtrees are roughly equalized, and shown to be very close to the optimum solution. We implemented this scheme in $O(k \log k)$ time instead of optimal linear time.

The code c_i for v_i is obtained by following the path from B's root down to the node with inorder i, again reading a left child as a '0', and a right child as a '1'. A technical detail is how to avoid the *empty* label at the root, because empty labels are not permitted in our setting. We overcome this by artificially adding 1 (arithmetically!) to each of the c_i's, increasing the length of the label in case of an overflow. These steps need to be undone when decoding the label. See Fig. 1 for examples.

Deciding whether $i < j$ from c_i and c_j alone is more complicated in this case. We first align c_i and c_j by shifting the shorter one to the left, and then set the lower $\max(|c_i|, |c_j|) - \min(|c_i|, |c_j|)$ bits in the shorter label to the complement of the corresponding bits in the longer label; let the results be c_i' and c_j', respectively. We then compute d as the leftmost (most significant) position in which c_i' and c_j' disagree. Note that the bit-sequence $(c_i')_1 \dots (c_i')_{d-1}$ is a code itself and corresponds to a node v in the FLBST B (in fact their LCA!). If the d'th bit is 0 in c_i', and 1 in c_j', we know that label c_i is found in v's left subtree (or at v itself if c_i is a prefix of c_j), whereas c_j is found in v's right subtree. Hence, $i < j$ in this case. The other case is symmetric.

As an example, look at the codes $c_i = 10$ and $c_j = 111$ in Fig. 1, corresponding to nodes F and L, respectively, in the FLBST B in the middle. Aligning and complementing yields $c_i' = 100$ and $c_j' = 111$, and the leftmost bit d where they differ is the middle one. We have that $(c_i')_1 \dots (c_i')_{d-1} = 1$, corresponding to node H in B. Because c_i' continues with a 0 and c_j' with a 1, we see that F is in H's left subtree, whereas L is in the right one. We conclude that F is above L in the original tree T.

3.2 Self-delimiting Codes

We call a code *self-delimiting* if in a juxtaposition C of m encoded words, the following operations can be computed in constant time (assuming C fits into a constant number of computer words):

- LOCATE$_C(j)$: locate the j'th encoded word for $1 \leq j \leq m$.
- COUNT$_C(j)$: compute the number of encoded words before position $j \leq |C|$.

As a simple example, encoding an integer $x \geq 0$ in *unary* as $0^x 1$ yields a self-delimiting code, as the locate- end count-operations above reduce to the well-known rank- and select-queries on binary sequences [17]. Because the unary code is obviously too long for our purposes, below we introduce two other self-delimiting codes.

Self-delimiting codes are relevant for our labeling scheme for the following reason. Assume we encode heavy and light labels with the *same* self-delimiting code, and concatenate the resulting bit-strings to obtain a node label $l(v)$ for node v. Then given a position j in $l(v)$, we can decide in $O(1)$ time if j occurs in a heavy or light label by deciding whether $x = \textsc{Count}_{l(v)}(j)$ is even or odd. Finding the beginning and end of the heavy/light label can be done by a subsequent $\textsc{Locate}_{l(v)}(x)$. Hence, a *helper label is not necessary* in this case.

Fibonacci Coding. Every y_i can be written uniquely as a sum $y_i = F_{k_1} + F_{k_2} + \cdots + F_{k_r}$ with $k_{j+1} < k_j + 1$ for all $1 < j < r$ and $F_{k_r} > 0$, where F_j is the j'th *Fibonacci number*, defined here $F_1 = 1, F_2 = 2$, and $F_{j+1} = F_j + F_{j-1}$ for $j > 2$. Hence, v_i can be encoded by a bit-vector c_i of length k_1, where a '1' at position j in c_i indicates that F_j is part of the sum ($k_x = j$ for some x). The property $k_{j+1} < k_j + 1$ guarantees that no '11'-pair is present in c_i; thus, prepending a '1' to c_i implies that the only '11'-pair appears at the beginning of an encoded word. To support the $\textsc{Count}_{l(v)}(j)$-operation, simply count the number of '11'-pairs up to position j in the sequence $l(v)$ of encoded labels. For the $\textsc{Locate}_{l(v)}(j)$-operation, find the j'th and the $(j-1)$'th '11'-pair in $l(v)$. We use a combination of bit-tricks ("broadword-computing") and table-lookups to support these two operations [18].

Deciding whether $i < j$ from c_i and c_j is simple: first check if $|c_i| < |c_j|$, and, in case of equality, perform a standard integer comparison on c_i and c_j.

Ternary Coding. Writing each y_i in *ternary*, then substituting a '0' by binary '00', a '1' by '01', and a '2' by '10', and finally prepending a '11'-pair to the resulting sequence, gives codewords c_i that are asymptotically even shorter than the Fibonacci-codes [19]. Now the two operations $\textsc{Count}(\cdot)$ and $\textsc{Locate}(\cdot)$ reduce to counting/locating the '11'-pairs at *even* positions, which can be done again by bit-tricks and table-lookups [18].

Deciding whether $i < j$ is similar to Fibonacci Coding.

3.3 Best Possible Heuristic for Light Labels

In the presence of a helper label, we can use different codings for the heavy and light labels. Thus, when using a helper label, it always makes sense to use the *Canonical Coding* for the *light* labels, because it is obviously the one that produces shortest codes. Since every light node v is the apex of a heavy path, $ll(v)$ will be repeated for all nodes w in T_v. Therefore, we *assign shorter labels to nodes v with larger* $|T_v|$. This rule produces the best possible light labels.

Table 1. Implemented and tested schemes

name	heavy labels	light labels	$O(\log n)$-bit-labels?	helper label?
H	Hu-Tucker [20]	Canonical	yes	yes
A	Alstrup et al. [8]	Canonical	yes	yes
M	Mehlhorn [15]	Canonical	yes	yes
C	Canonical	Canonical	no	yes
F	Fibonacci	Fibonacci	no	no
T	Ternary	Ternary	no	no

3.4 Implemented Labeling Schemes

The implemented labeling schemes are summarized in Tbl. 1. Because this is a practical paper, we also look at labelings that do *not* guarantee that the theoretical label length is $O(\log n)$.

4 Experiments

All schemes from Tbl. 1 were implemented in C++ and are available at `http://www-ab.informatik.uni-tuebingen.de/people/fischer/lca-labeling.zip`. All tests were run on a 64-bit machine with one Athlon DualCore 2.2GHz processor and with 16GB of RAM, using ScientificLinux 4.7, and GCC 4.1.2 with the same compiler options (-O3 -fomit-frame-pointer -funroll-loops) for all methods. All figures shown are averages over 10 different executions of the algorithms.

4.1 Performance Indicators

Average label length. As the average label length is directly correlated to the total size of the scheme, this is probably the most important measure.

Maximum label length. The length of the longest label plays a central role in the worst-case performance of the query algorithm.

Construction time. Construction time of the labels should not be over-estimated as a performance indicator (because it is done only once), but is certainly an interesting measure.

Query time. The average query time is especially important in a distributed setting, where new queries keep arriving at a high rate. More complicated codes could result in a higher query time.

4.2 Test Data

Random trees. We generated random trees with the method due to Rémy [21]. It is originally devised for binary trees, but using the natural isomorphism between binary and arbitrary ordered trees, it can also be used to produce random ordered trees. The advantage of this approach is that it does not produce "balanced" trees of height $O(\log n)$, but rather of height $O(\sqrt{n})$.

Skewed trees: path. A simple path of length n. This tree constitutes a test on the worst-case performance of the coder used for the heavy labels.

Skewed trees: star. A root node with $n-1$ leaves attached to it. This constitutes a test on the worst-case behavior of the construction of light labels.

Perfect binary trees. A tree of height $O(\log n)$, where each internal node has exactly two children.

4.3 Experimental Results: Label Length

Average Label Length. The results for the average label lengths can be seen in Fig. 2. Let us first look at (a), where random trees are examined. We can see that coding scheme M is always best. The other coding schemes are roughly equal, with one exception: scheme C is almost competitive to M for small trees (up to 10^4), but becomes the worst scheme of all for large trees (starting roughly at 10^6). Also, for small inputs, the schemes that do not need a helper label (F & T) are better than H & A, but this changes for growing tree size.

Regarding the skewed trees (b and c), we see that schemes F & T are always best, with slight advantages for T. For single paths (b), we further see that M & C are better than H & A. It is interesting to note that M & C are equal; the reason for

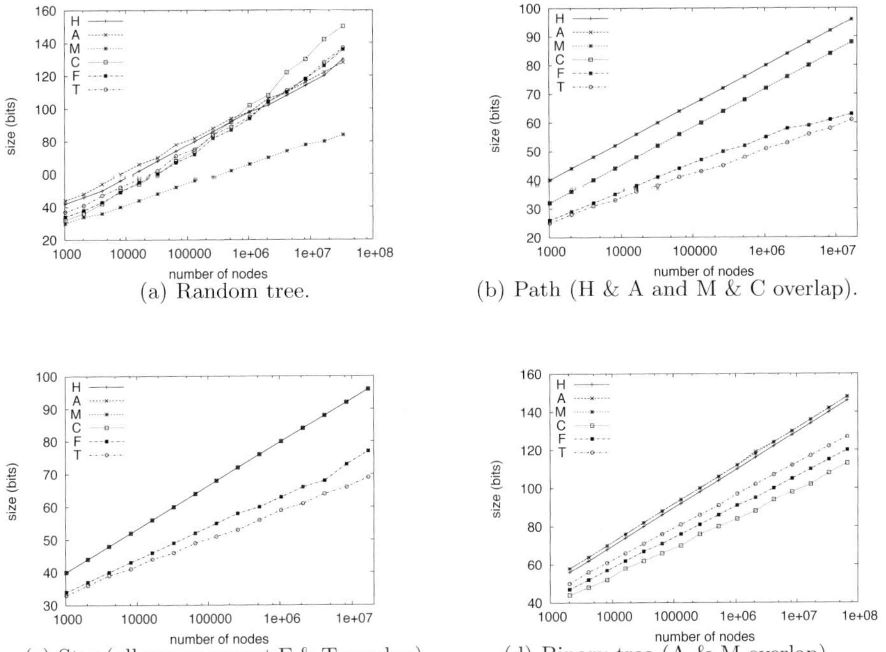

(a) Random tree.

(b) Path (H & A and M & C overlap).

(c) Star (all curves except F & T overlap).

(d) Binary tree (A & M overlap).

Fig. 2. Average label lengths for different coding methods and varying tree sizes

this "artifact" can be seen as follows. On a single path, all nodes have a light size of 0; hence, the "symbols" to be encoded with Mehlhorn's method are all equally likely. In this case, Mehlhorn's method just constructs the shortest (non-prefix-free) binary strings, which coincide with Canonical Codes in this case.

In (d), where complete binary trees are examined, we see a somehow reverse ordering compared with (a). This is interesting, because the label lengths for complete binary trees can be seen as the "overhead" of our coding scheme: the aim of the tree decomposition into heavy paths is to create "flat" trees of height $O(\log n)$; this is pointless in the case of binary trees, which already have the desired height. In this light, it is not surprising that scheme C is best (no overhead due to shortest possible codes), whereas the other schemes have an overhead that is inversely proportional to the performance of the scheme in the "average" case.

Maximum Label Length. Let us now turn our attention to the maximum label lengths. Again, Fig. 3 shows the results for varying tree shapes and sizes. For random trees (a), we have roughly the same picture as for the average label length, with the difference that the gap between the theoretically non-optimal coding methods (C, F, and T) and the optimal ones (H, A, and M) becomes larger. This underlines the fact that the whole machinery for $O(\log n)$-labeling schemes is primarily divised for avoiding the worst case.

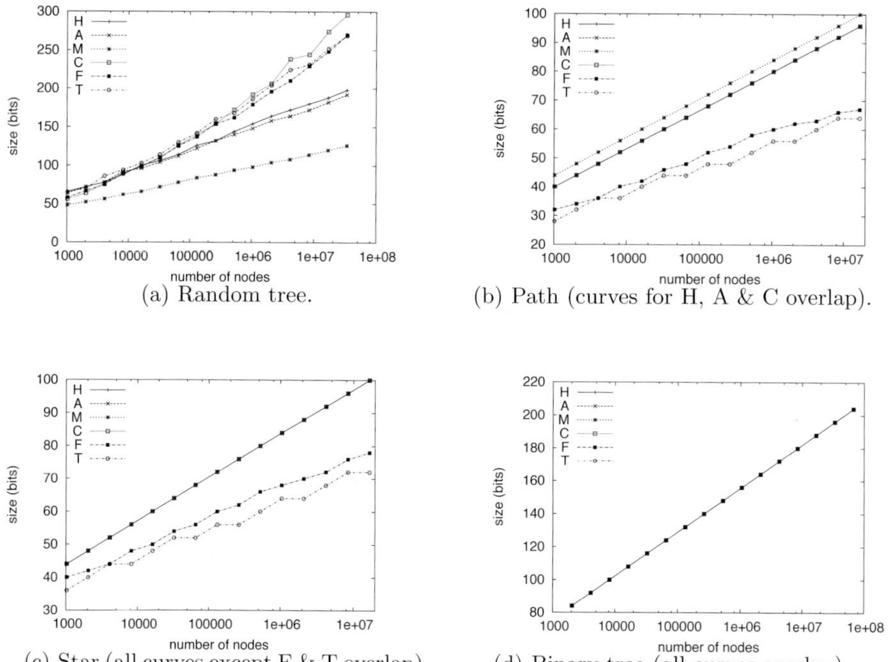

(a) Random tree.

(b) Path (curves for H, A & C overlap).

(c) Star (all curves except F & T overlap).

(d) Binary tree (all curves overlap).

Fig. 3. Maximum label lengths for different coding methods and varying tree sizes

Table 2. Construction times of the different coding schemes

# nodes	H	A	M	C	F	T
262,144	0.174	0.164	0.18	0.111	0.0943	0.0943
524,288	0.361	0.334	0.369	0.229	0.194	0.191
1,048,576	0.744	0.693	0.759	0.476	0.403	0.397
2,097,152	1.56	1.43	1.57	1.02	0.859	0.851
4,194,304	3.2	2.95	3.25	2.15	1.81	1.82
8,388,608	6.52	6.01	6.62	4.48	3.77	3.75
16,777,216	13.2	12.2	13.4	9.16	7.72	7.66
33,554,432	24.5	24.5	26.5	18.5	15.5	15.5

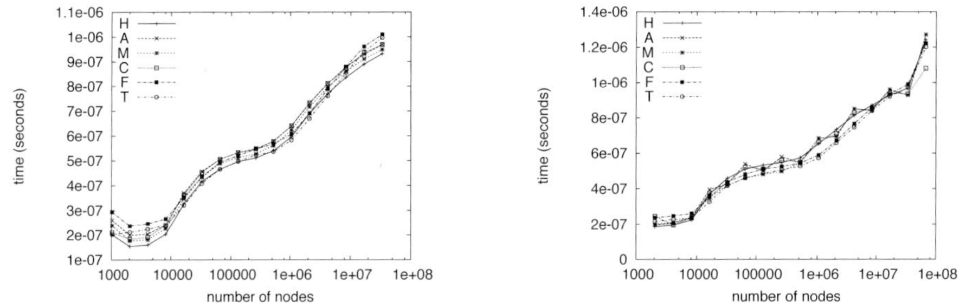

(a) Random tree. (b) Complete binary tree.

Fig. 4. Query times — all data points are averages over 1,000,000 queries

Also, for the pathological trees (b and c), the results do not deviate too much from the ones for average label lengths. Note, however, that in the case of a simple path (b), all three scheme H, A, and C produce equal-sized maximum labels.

4.4 Experimental Results: Construction Time

The construction times for random trees are shown in Tbl. 2 (including times for finding heavy paths). The times for the other tree types (path, star, and complete) do not deviate much from the random tree, hence we do not display them here. The first thing to note is that all coding schemes exhibit a nice linear-time construction, even those which are implemented in $O(n \log n)$ time.

It is not surprising that the non-optimal coding schemes C, F, and T are faster than the optimal ones (H, A, and M), by a factor of roughly 2. Among the optimal ones, little if no differences can be observed, maybe a slight advantage for A, and a slight disadvantage for M, mirroring their different degree of complexity.

4.5 Experimental Results: Query Time

Finally, we look at query times (including times for the retrieval of labels). We posed 1,000,000 random queries to each combination of tree type, size, and coding method.

The results for the different tree types show little variation; we therefore just show the graphs for random trees (Fig. 4(a)) and complete binary trees (b).

First, note that none of the methods exhibits its theoretical $O(1)$ worst-case query time in practice. This is not surprising, as the label lengths increase with growing tree size, implying that longer bit-vectors have to be handled by the query algorithm for larger trees.

We do not have an explanation for the "saddle-points" between 10^5 and 10^6 in both (a) and (b), but we stress that a similar phenomenon has been observed in practical tests of range minimum queries [22].

Last, we comment on the difference in query time between the coding schemes. The most interesting point is that no big differences can be observed at all; although we have seen earlier that the label sizes differ significantly, and that the decoding algorithms have a different degree of complexity. From this, we draw the conclusion that the most time-consuming part of the query procedures is to perform the bit-manipulations needed to extract the heavy/light labels, which is common to all methods. The time for the arithmetic operations needed to compare labels seems to be negligible.

Acknowledgments. We wish to thank Rossano Venturini for bringing our attention to the Hu-Tucker algorithm. Further thanks go to Simon Gog for his broadword-implementation of the Fibonacci- and Ternary Coding.

References

1. Kannan, S., Naor, M., Rudich, S.: Implicit representation of graphs. SIAM J. Discrete Math. 5(4), 596–603 (1992)
2. Abiteboul, S., Alstrup, S., Kaplan, H., Milo, T., Rauhe, T.: Compact labeling scheme for ancestor queries. SIAM J. Comput. 35(6), 1295–1309 (2006)
3. Kaplan, H., Milo, T., Shabo, R.: Compact labeling scheme for xml ancestor queries. Theory Comput. Syst. 40(1), 55–99 (2007)
4. Alstrup, S., Bille, P., Rauhe, T.: Labeling schemes for small distances in trees. SIAM J. Discrete Math. 19(2), 448–462 (2005)
5. Gavoille, C., Peleg, D., Pérennes, S., Raz, R.: Distance labeling in graphs. J. Algorithms 53(1), 85–112 (2004)
6. Cohen, E., Halperin, E., Kaplan, H., Zwick, U.: Reachability and distance queries via 2-hop labels. SIAM J. Comput. 32(5), 1338–1355 (2003)
7. Harel, D., Tarjan, R.E.: Fast algorithms for finding nearest common ancestors. SIAM J. Comput. 13(2), 338–355 (1984)
8. Alstrup, S., Gavoille, C., Kaplan, H., Rauhe, T.: Nearest common ancestors: A survey and a new algorithm for a distributed environment. Theory Comput. Syst. 37, 441–456 (2004)
9. Peleg, D.: Informative labeling schemes for graphs. Theor. Comput. Sci. 340(3), 577–593 (2005)
10. Caminiti, S., Finocchi, I., Petreschi, R.: Engineering tree labeling schemes: A case study on least common ancestors. In: Halperin, D., Mehlhorn, K. (eds.) ESA 2008. LNCS, vol. 5193, pp. 234–245. Springer, Heidelberg (2008)
11. Ferragina, P., Venturini, R.: A simple storage scheme for strings achieving entropy bounds. Theor. Comput. Sci. 372(1), 115–121 (2007)

12. Gilbert, E.N., Moore, E.F.: Variable length binary encodings. Bell System Tech. J. 38, 933–968 (1959)
13. Hu, T.C., Tucker, A.C.: Optimum computer search trees. SIAM J. Appl. Math. 21, 514–532 (1971)
14. Hu, T.C., Larmore, L.L., Morgenthaler, J.D.: Optimal integer alphabetic trees in linear time. In: Brodal, G.S., Leonardi, S. (eds.) ESA 2005. LNCS, vol. 3669, pp. 226–237. Springer, Heidelberg (2005)
15. Mehlhorn, K.: The best possible bound for the weighted path length of binary search trees. SIAM J. Comput. 6(2), 235–239 (1977)
16. Knuth, D.E.: Optimum binary search trees. Act Informatica 1, 14–25 (1971)
17. Munro, J.I.: Tables. In: Chandru, V., Vinay, V. (eds.) FSTTCS 1996. LNCS, vol. 1180, pp. 37–42. Springer, Heidelberg (1996)
18. Gog, S.: Broadword computing and Fibonacci code speed up compressed suffix arrays. In: Vahrenhold, J. (ed.) SEA 2009. LNCS, vol. 5526, pp. 161–172. Springer, Heidelberg (2009)
19. Witten, I.H., Moffat, A., Bell, T.C.: Managing Gigabytes: Compressing and Indexing Documents and Images, 2nd edn. Morgan Kaufmann, San Francisco (1999)
20. Garsia, A.M., Wachs, M.L.: A new algorithm for minimum cost binary trees. SIAM J. Comput. 6(4), 622–642 (1977)
21. Rémy, J.L.: Un procédé itératif de dénombrement d'arbres binaires et son application a leur génération aléatoire. Informatique Théorique et Applications 19(2), 179–195 (1985)
22. Fischer, J., Heun, V.: Theoretical and practical improvements on the RMQ-problem, with applications to LCA and LCE. In: Lewenstein, M., Valiente, G. (eds.) CPM 2006. LNCS, vol. 4009, pp. 36–48. Springer, Heidelberg (2006)

Disproof of the Neighborhood Conjecture with Implications to SAT

Heidi Gebauer*

Institute of Theoretical Computer Science, ETH Zurich, CH-8092 Switzerland
gebauerh@inf.ethz.ch

Abstract. We study a special class of binary trees. Our results have implications on Maker/Breaker games and SAT: We disprove a conjecture of Beck on positional games and construct an unsatisfiable k-CNF formula with few occurrences per variable, thereby improving a previous result by Hoory and Szeider and showing that the bound obtained from the Lovász Local Lemma is tight up to a constant factor.

A (k, s)-*CNF formula* is a boolean formula in conjunctive normal form where every clause contains exactly k literals and every variable occurs in at most s clauses. The (k, s)-SAT problem is the satisfiability problem restricted to (k, s)-CNF formulas. Kratochvíl, Savický and Tuza showed that for every $k \geq 3$ there is an integer $f(k)$ such that every $(k, f(k))$-formula is satisfiable, but $(k, f(k) + 1)$-SAT is already NP-complete (it is not known whether $f(k)$ is computable). Kratochvíl, Savický and Tuza also gave the best known lower bound $f(k) = \Omega\left(\frac{2^k}{k}\right)$, which is a consequence of the Lovász Local Lemma. We prove that, in fact, $f(k) = \Theta\left(\frac{2^k}{k}\right)$, improving upon the best known upper bound $O\left((\log k) \cdot \frac{2^k}{k}\right)$ by Hoory and Szeider.

Finally we establish a connection between the class of trees we consider and a certain family of positional games. The Maker/Breaker game we study is as follows. Maker and Breaker take turns in choosing vertices from a given n-uniform hypergraph $\mathcal{F}$, with Maker going first. Maker's goal is to completely occupy a hyperedge and Breaker tries to prevent this. Beck conjectures that if the maximum neighborhood size of $\mathcal{F}$ is smaller than $2^{n-1} - 1$ then Breaker has a winning strategy. We disprove this conjecture by establishing an n-uniform hypergraph with maximum neighborhood size $3 \cdot 2^{n-3}$ where Maker has a winning strategy. Moreover, we show how to construct an n-uniform hypergraph with maximum degree $\frac{2^{n-1}}{n}$ where Maker has a winning strategy.

In addition we show that each n-uniform hypergraph with maximum degree at most $\frac{2^{n-2}}{en}$ has a proper halving 2-coloring, which solves another open problem posed by Beck related to the Neighborhood Conjecture.

* Research is supported by the SNF Grant 200021-118001/1.

A. Fiat and P. Sanders (Eds.): ESA 2009, LNCS 5757, pp. 764–775, 2009.

1 Introduction

1.1 Trees

We first consider a special class of trees, which connect both to Maker/Breaker games and SAT. We regard binary trees where every node has either two or no children. In such a binary tree we say that a leaf v is l-*close* to a node w if w is an ancestor of v, at distance at most l from v. For positive integers k and d, we call a binary tree T a (k,d)-*tree* if (i) every leaf has depth at least $k-1$ and (ii) for every node u of T there are at most d leaves $(k-1)$-close to u. Clearly, every binary tree with all leaves at depth at least $k-1$ is a $(k, 2^{k-1})$-tree. The following lemma will be the main ingredient in proving some new results on Maker/Breaker games and SAT.

Lemma 1. *(i) A* $(k, \lfloor \frac{2^k}{k} \rfloor)$*-tree exists for every sufficiently large* k.
(ii) If k *is a sufficiently large power of 2 then a* $(k, \frac{2^{k-1}}{k})$*-tree exists.*
(iii) Let $c = \frac{63}{64}$*. For every sufficiently large* k *with* $\frac{k}{c}$ *being a power of 2 there is a* $(k, c\frac{2^{k-1}}{k})$*-tree.*

1.2 SAT

Lemma 1 has implications on SAT. Following the standard notation we denote by (k,s)-*CNF* the set of boolean formulas $\mathcal{F}$ in conjunctive normal form where every clause of $\mathcal{F}$ has exactly k distinct literals and each variable occurs in at most s clauses of $\mathcal{F}$. Moreover, we denote by (k,s)-SAT the satisfiability problem restricted to formulas in (k,s)-CNF. Tovey [17] proved that every (3,3)-CNF formula is satisfiable. For k a positive integer let $f(k)$ be defined as the largest integer, so that every $(k, f(k))$-CNF formula is satisfiable. Tovey also showed that $f(3) = 3$ and that, moreover, $(3,4)$-SAT is NP-complete. Kratochvíl, Savický and Tuza [11] generalized this result by showing that for every $k \geq 3$ $(k, f(k)+1)$-SAT is already NP-complete. This phenomenon, that $(k, f(k))$-SAT is trivial while $(k, f(k)+1)$-SAT is already NP-hard, is often referred to as *complexity jump*.

The best known lower bound for $f(k)$, a consequence of Lovász Local Lemma, is due to Kratochvíl, Savický and Tuza [11].

Theorem 1. (Kratochvíl, Savický and Tuza [11]) $f(k) \geq \lfloor \frac{2^k}{ek} \rfloor$

From the other side Savický and Sgall [14] showed that $f(k) = O(k^{(1-\alpha)} \cdot \frac{2^k}{k})$ where $\alpha = \log_3 4 - 1 \approx 0.26$. This was improved by Hoory and Szeider [8] who proved that $f(k) = O((\log k) \cdot \frac{2^k}{k})$, which is the best known upper bound. We close the gap between upper and lower bound up to a constant factor by showing that $f(k) = \Theta(\frac{2^k}{k})$.

Theorem 2. *For a large enough integer* k,

$$f(k) < \frac{2^{k+1}}{k}.$$

If k is a large enough power of 2 we have $f(k) < \frac{2^k}{k}$. This bound is not tight since $f(k) < \frac{63}{64}\frac{2^k}{k}$ for infinitely many k.

Hence the lower bound in Theorem 1 is best possible up to a factor slightly less than e.

Recently Moser [13] showed that for $s \leq \frac{2^{k-6}}{k}$ not only every (k,s)-CNF has a satisfying assignment but there is also an algorithm computing such an assignment efficiently. Theorem 2 proves that this bound is asymptotically tight. Indeed, for some $(k, \frac{2^k}{k})$-CNF formulas we can not find a satisfying assignment efficiently, simply because there is none.

A special class of unsatisfiable formulas. The class MU(1) of minimal unsatisfiable CNF-formulas $\mathcal{F}$ where $m(\mathcal{F}) - n(\mathcal{F}) = 1$ with $m(\mathcal{F})$ denoting the number of clauses of $\mathcal{F}$ and $n(\mathcal{F})$ denoting the number of variables of $\mathcal{F}$ has been widely studied (see, e.g., [1], [5], [10], [12], [16]). While it is not known whether $f(k)$ is computable Hoory and Szeider investigated $f_1(k)$, the largest integer such that every $(k, f_1(k))$-CNF in MU(1) is satisfiable. They showed that $f_1(k)$ is computable. With $f(k) \leq f_1(k)$ this allowed them to derive the best known upper bounds for $f(k)$ for small k: $f(5) \leq 7$, $f(6) \leq 11$, $f(7) \leq 17$, $f(8) \leq 29$, $f(9) \leq 51$.

While the derivation of the previous bound of $f(k) = O((\log k) \cdot \frac{2^k}{k})$ by Hoory and Szeider did not go via an MU(1) formula the constructions for proving Theorem 2 reside in MU(1).

Corollary 1. *For large enough k we have $f_1(k) < 2 \cdot \frac{2^k}{k}$, implying that $f(k) = \Theta(\frac{2^k}{k})$ and $f_1(k) = \Theta(\frac{2^k}{k})$. Moreover, for infinitely many k we have $f_1(k) < \frac{2^k}{k}$.*

It is an open question whether $f(k) = f_1(k)$, i.e., whether the unsatisfiable CNF-formulas with the smallest possible number of occurrences per variable (i.e. the unsatisfiable $(k, f(k)+1)$-CNF formulas) are members of MU(1). Scheder [15] showed that for almost disjoint k-CNF formulas (i.e. CNF-formulas where any two clauses have at most one variable in common) this is not true, i.e., no almost disjoint unsatisfiable $(k, \tilde{f}(k)+1)$-CNF formula is in MU(1), with $\tilde{f}(k)$ denoting the maximum s such that every almost disjoint (k,s)-CNF formula is satisfiable.

Bounded neighborhood size. The *neighborhood* $\Gamma(C)$ of a clause C in a CNF formula $\mathcal{F}$ is the set of clauses in $\mathcal{F}$ that share variables with C. Analogously to $f(k)$ let $l(k)$ denote the largest integer d such that every k-CNF formula for which $|\Gamma(C)| \leq d$ for every clause C of $\mathcal{F}$ is satisfiable. In fact, the proof of Theorem 1 shows that every k-CNF formula $\mathcal{F}$ with $|\Gamma(C)| \leq \frac{2^k}{e} - 1$ for all clauses C of $\mathcal{F}$ is satisfiable. (Note that Theorem 1 is a direct consequence of this; for, if in a k-CNF formula every variable occurs at most $\lfloor \frac{2^k}{ek} \rfloor$ times then no clause can collect more than $k(\lfloor \frac{2^k}{ek} \rfloor - 1) \leq \frac{2^k}{e} - 1$ neighbors.) Thus $l(k) \geq \frac{2^k}{e} - 1$. This bound is asymptotically tight. The most simple reason is that the complete formula consisting of all 2^k clauses of size k over k variables is clearly unsatisfiable and has neighborhoods of size $2^k - 1$ at each clause. The constructions for proving Theorem 2 further tighten the constant gap between the known lower and upper bounds.

Theorem 3. *(Lower bound in [11]) Let $c = \frac{63}{64}$. For infinitely many k,*

$$\lfloor \frac{2^k}{e} \rfloor - 1 \leq l(k) < c \cdot 2^{k-1}$$

Connection to (k,d)-trees. Theorem 2, Corollary 1 and the upper bound in Theorem 3 are a direct consequence of the following lemma, which establishes a connection between SAT and the (k,d)-trees described above.

Lemma 2. *Let T be a (k,d)-tree, k and d positive integers. Then there is an unsatisfiable k-CNF formula $\mathcal{F} = \mathcal{F}(T)$ with the following properties.*
(a) *Every literal occurs in at most d clauses of $\mathcal{F}$.*
(b) *For every two distinct clauses C, D having a variable in common there is a variable that appears in C and D with opposite signs.*
(c) *If T is minimum with respect to the number of leaves then $\mathcal{F}$ belongs to* MU(1).
(d) *$|\Gamma(C)| \leq kd$ for all clauses C in $\mathcal{F}$.*

In particular, $f(k) \leq 2d - 1$ and $l(k) \leq kd - 1$.

Note that Theorem 2, Corollary 1 and the upper bound in Theorem 3 follow directly from Lemma 2 and Lemma 1.

Implications on (k,d)-trees. By Theorem 3 and Lemma 2 we obtain the following.

Observation 1. *There is no (k,d)-tree for $d \leq \frac{2^k}{ek} - 1$.*

1.3 Maker/Breaker Games

A *hypergraph* is a pair (V, E), where V is a finite set whose elements are called *vertices* and E is a family of subsets of V, called *hyperedges*. A hypergraph is *n-uniform* if every hyperedge contains exactly n vertices. We study the following Maker/Breaker game. Maker and Breaker take turns in claiming one previously unclaimed vertex of a given n-uniform hypergraph $\mathcal{F}$, with Maker going first. Maker wins if he claims all vertices of some hyperedge of $\mathcal{F}$, otherwise Breaker wins. We say that Maker uses a *pairing strategy* if, after claiming his first vertex, he divides all but at most one of the remaining vertices of $\mathcal{F}$ into pairs and whenever Breaker claims one vertex of a pair he takes the other one.

Let $\mathcal{F}$ be an n-uniform hypergraph. The *degree* $d(v)$ of a vertex v is the number of hyperedges containing v and the *maximum degree* $\Delta(\mathcal{F})$ of a hypergraph $\mathcal{F}$ is the maximum degree of its vertices. The *neighborhood* $N(e)$ of a hyperedge e is the set of hyperedges of $\mathcal{F}$ which intersect e, excluding e itself, and the *maximum neighborhood size* of $\mathcal{F}$ is the maximum of $|N(e)|$ where e runs over all hyperedges of $\mathcal{F}$.

The famous Erdős-Selfridge Theorem [6] states that for each n-uniform hypergraph $\mathcal{F}$ with less than 2^{n-1} hyperedges Breaker has a winning strategy. This upper bound on the number of hyperedges is best possible as the following example shows. Let T be a rooted binary tree with n levels and let $\mathcal{G}$ be the hypergraph whose hyperedges are exactly the sets $\{v_0, \ldots v_{n-1}\}$ such that $v_0, v_1, \ldots, v_{n-1}$ is a path from the root to a leaf. Note that the number of hyperedges of $\mathcal{G}$ is 2^{n-1}. To win the game on $\mathcal{G}$, Maker can use the following strategy. In his first move he claims the root m_1 of T. Let b_1 denote the vertex occupied by Breaker in his subsequent move. In his second move Maker claims the child m_2 of m_1 such that m_2 lies in the subtree of m_1 not containing b_1. More generally, in his ith move Maker selects the child m_i of his previously occupied node m_{i-1} such that the subtree rooted at m_i contains no Breaker's node. Note that such a child m_i always exists since the vertex previously claimed by Breaker is either in the left or in the right subtree of m_{i-1} (but not in both!). Using this strategy Maker can achieve to own some set $\{v_0, \ldots, v_{n-1}\}$ of vertices such that $v_0, v_1, \ldots, v_{n-1}$ is a path from the root to a leaf, which corresponds to some hyperedge of $\mathcal{G}$. Hence Maker has a winning strategy on $\mathcal{G}$.

In this game Maker even has a winning *pairing* strategy. Indeed, by first claiming the root and then pairing every vertex with its sibling (i.e. the vertex having the same parent) he can achieve a path from the root to a leaf, which contains a hyperedge.

Note that the maximum degree of $\mathcal{G}$ is 2^{n-1}, thus equally large as the number of hyperedges of $\mathcal{G}$. This provides some evidence that in order to be a Maker's win a hypergraph must have largely overlapping hyperedges. Moreover, Beck [3] conjectured that the main criterion for whether a hypergraph is a Breaker's win is not the cardinality of the hyperedge set but rather the maximum neighborhood size, i.e. the actual reason why each hypergraph $\mathcal{H}$ with less than 2^{n-1} edges is a Breaker's win is that the maximum neighborhood size of $\mathcal{H}$ is smaller than $2^{n-1} - 1$.

Neighborhood Conjecture. (Open Problem 9.1(a), [3]) Assume that $\mathcal{F}$ is an n-uniform hypergraph, and its maximum neighborhood size is smaller than $2^{n-1} - 1$. Is it true that by playing on $\mathcal{F}$ Breaker has a winning strategy?

Further motivation for the Neighborhood Conjecture is the well-known Erdős-Lovász 2-coloring Theorem – a direct consequence of the famous Lovász Local Lemma – which states that every n-uniform hypergraph with maximum neighborhood size at most $\frac{2^{n-1}}{e} - 1$ has a proper 2-coloring. An interesting feature of this theorem is that the board size does not matter. We can prove by also applying the Lovász Local Lemma that every n-uniform hypergraph with maximum degree at most $\frac{2^{n-2}}{en}$ has a so called proper *halving* 2-coloring, i.e., a proper 2-coloring in which the number of red vertices and the number of blue vertices differ by at most 1 (see Theorem 6 for details). This guarantees the *existence* of a course of the game such that at the end Breaker owns at least one vertex of each hyperedge and thus is the winner. Hence it is a priori not completely impossible that Breaker has a winning strategy.

In our first theorem we prove that the Neighborhood Conjecture, in this strongest of its forms, is not true, even if we require Maker to use a pairing strategy.

Theorem 4. *There is an n-uniform hypergraph $\mathcal{H}$ with maximum neighborhood size $2^{n-2} + 2^{n-3}$ where Maker has a winning pairing strategy.*

In his book [3] Beck also poses the following weakening of the Neighborhood Conjecture.

Open Problem 1. *(Open Problem 9.1(b), [3]) If the Neighborhood Conjecture is too difficult (or false) then how about if the upper bound on the maximum neighborhood size is replaced by an upper bound $\frac{2^{n-c}}{n}$ on the maximum degree where c is a sufficiently large constant?*

In the hypergraph $\mathcal{H}$ we will construct to prove Theorem 4 one vertex has degree $2^{n-2}+1$, which is still high. However, the existence of vertices with high degree is not crucial. We also establish a hypergraph with maximum degree $\frac{2^{n-1}}{n}$ on which Maker has a winning strategy.

Theorem 5. *If n is a sufficiently large power of 2 there is an n-uniform hypergraph with maximum degree $\frac{2^{n-1}}{n}$ where Maker has a winning pairing strategy.*

The hypergraph of Theorem 5 has maximum neighborhood size at most $2^{n-1} - n$, which is weaker than Theorem 4 but also disproves the Neighborhood Conjecture.

In his book [3] Beck also poses several further weakenings of the Neighborhood Conjecture. The last one is as follows.

Open Problem 2. *(Open Problem 9.1(f), [3]) How about if we just want a proper halving 2-coloring?*

It is already known that the answer to Open Problem 2 is positive if the maximum degree is at most $\left(\frac{3}{2} - o(1)\right)^n$. According to Beck [3] the real question is whether or not $\frac{3}{2}$ can be replaced by 2. We prove that the answer is yes.

Theorem 6. *For every n-uniform hypergraph $\mathcal{F}$ with maximum degree at most $\frac{2^{n-2}}{en}$ there is a proper halving 2-coloring.*

Connection to trees. Let T be a binary tree where every leaf has depth at least $n-1$. Then we define $\mathcal{H}_T = \mathcal{H}_T(n)$ as the n-uniform hypergraph whose hyperedges are the paths of length $n-1$ in T ending at a leaf.

Lemma 3. *Let T be a binary tree where every leaf has depth at least $n-1$. Then Maker has a winning pairing strategy on $\mathcal{H}_T$.*

So in order to prove Theorem 4 it suffices to show the next lemma.

Lemma 4. *There is a binary tree T where every leaf has depth at least $n-1$ such that $\mathcal{H}_T$ has maximum neighborhood size $2^{n-2}+2^{n-3}$.*

By the construction of $\mathcal{H}_T$ and Lemma 3 we can immediately connect (k,d)-trees to the game we study.

Observation 2. *Let T be an (n,d)-tree. Then (i) Maker has a winning pairing strategy on $\mathcal{H}_T$ and (ii) every vertex of $\mathcal{H}_T$ occurs in at most d hyperedges.*

Theorem 5 is then a direct consequence of Observation 2 and Lemma 1.

Notation. Ceiling and floor signs are routinely omitted whenever they are not crucial for clarity. Throughout this paper log stands for the binary logarithm. Let T be a rooted binary tree. A *path* of T is a sequence of vertices $v_1, v_2, \ldots, v_j$ of T where v_k is a child of v_{k-1} for every $k = 2, \ldots, j$. Depending on the context we consider a hyperedge e of a hypergraph $\mathcal{H}_T$ either as a set or as a path in T. So we will sometimes speak of the start or end node of a hyperedge.

Organization of this paper. In Sect. 2 we establish a connection between the trees we study and SAT by proving Lemma 2. In Sect. 3 we show Lemma 3, which connects trees to the Maker/Breaker game we consider, and prove Lemma 4 refuting the Neighborhood Conjecture in the strongest of its forms.

In Sect. 4 we finally construct suitable (k,d)-trees, show a slightly weaker version of Lemma 1 and give a proof sketch of Lemma 1.(i)-(ii).

The proofs of Theorem 6 and Lemma 1 will appear in the full version.

2 Constructing Unsatisfiable k-CNF Formulas with Small Neighborhood

Proof of Lemma 2. Let m denote the number of leaves of T. We move to a binary tree $\widehat{T}$ by attaching the roots of two copies of T as the two children of a new root r. This yields a (k,d)-tree (actually, with all leaves of depth at least k). It has $2m$ leaves and $4m-1$ nodes altogether. Two nodes are called *siblings*, if they share the same parent. The $4m-2$ non-root nodes of $\widehat{T}$ can be partitioned into $2m-1$ sibling pairs.

For some set V of $2m-1$ boolean variables, we label the nodes of $\widehat{T}$ other than the root by literals in $V \cup \overline{V}$ so that every literal appears exactly once and siblings get complementary literals. With every leaf v we associate a clause C_v by walking along a path of length $k-1$ from v towards the root and collecting all labels encountered on this path (i.e. the labels of all nodes to which v is $(k-1)$-close). The set of clauses C_v, over all leaves v of $\widehat{T}$, constitutes $\mathcal{F}$.

$\mathcal{F}$ is unsatisfiable, for if an assignment α over V is given, it defines a path from the root to a leaf, say v, by always proceeding to the unique child whose label is mapped to 0 by α; thus C_v is violated by α.

The defining property of (k,d)-trees guarantees that no label is collected more than d times (hence (a)). We now settle (b). Let C_u, C_v be two clauses sharing at least one variable and let w denote the lowest common ancestor of u and v (i.e. the node of maximum depth that appears on both paths from u and v, respectively, to the root). Then one child of w occurs in C_u whereas the other child occurs in C_v. Since siblings have complementary literals (b) is shown.

Next we prove (c). Davydov, Davydova, and Kleine Büning [5] established the following characterization for MU(1)-formulas. ($\mathsf{vbl}(F)$ denotes the set of variables which occur in the formula F.)

Lemma 5. (Davydov, Davydova, and Kleine Büning [5]) $F \in \mathrm{MU}(1)$ *if and only if either* $F = \{\emptyset\}$ *or* F *is the disjoint union of formulas* F_1', F_2' *such that for a variable* x *we have*

- $\mathsf{vbl}(F_1') \cap \mathsf{vbl}(F_2') = \{x\}$ *and* $\{x, \bar{x}\} \subseteq \bigcup_{C \in F} C$*;*
- $F_1 := \{C \backslash \{x\} : C \in F_1'\} \in \mathrm{MU}(1)$*;*
- $F_2 := \{C \backslash \{\bar{x}\} : C \in F_2'\} \in \mathrm{MU}(1)$*.*

Note that due to our choice of T, every node u of T has one leaf descendant at distance at most $k-1$. Indeed, if some node u has distance at least k to all its leaf descendants then the subtree of T rooted at a child of u is a (k, d)-tree with fewer leaves than T. So by construction $\mathcal{F}$ has the properties stated in Lemma 5.

(d) follows from (a) and (b): Indeed, if we define $\mathrm{occ}(u)$ as the number of clauses of $\mathcal{F}$ containing a literal u, then (b) allows us to write $|\Gamma(C)|$ as $\sum_{u:\, C \text{ contains } u} \mathrm{occ}(\overline{u})$, which is at most kd for every clause C of $\mathcal{F}$. □

3 Counterexample to the Neighborhood Conjecture

Proof of Lemma 3. The set of non-root nodes of T can be divided into pairs of siblings. By first claiming the root of T and then pairing every node with its sibling Maker can finally achieve a path from the root to a leaf, which by assumption contains a hyperedge. □

Proof of Lemma 4. Let T' be a full binary tree with $n-1$ levels. For each leaf u of T' we proceed as follows: We add two children v, w to u and let v be a leaf. Then we attach a full binary tree S with $n-2$ levels to w (such that w is the root of S). For each leaf u' of S we add two children v', w' to u' and let v' be a leaf. Note that the hyperedge ending at v' starts at u. Finally, we attach a full binary S' with $n-1$ levels to w' (such that w' is the root of S'), see Fig. 1. Let T denote the resulting tree.

Clearly, every leaf of T has depth at least $n-1$. It remains to show that the maximum neighborhood of $\mathcal{H}_T$ is at most $2^{n-2} + 2^{n-3}$.

Claim. *Every hyperedge* e *of* $\mathcal{H}_T$ *intersects at most* $2^{n-2} + 2^{n-3}$ *other hyperedges.*

In order to prove this claim, we fix six vertices u, u', v, v', w, w' according to the above description, i.e., u is a node on level $n-2$ whose children are v and w, u' is

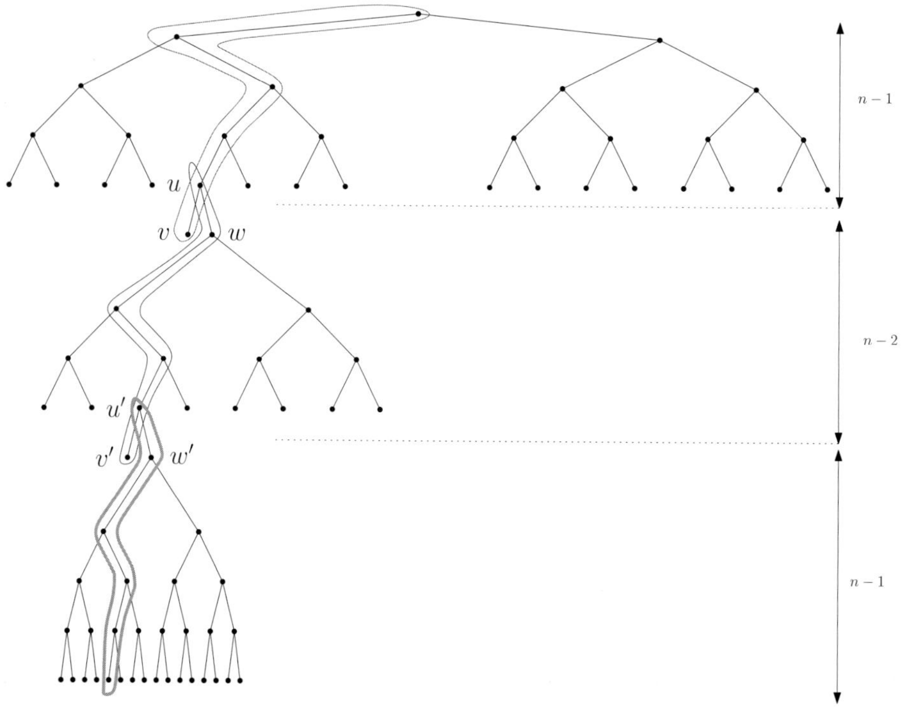

Fig. 1. An illustration of $\mathcal{H}_T$. The marked paths represent exemplary hyperedges.

a descendant of w on level $2n-4$ whose children are v' and w'. Let e be a hyperedge of $\mathcal{H}_T$. Note that the start node of e is either the root r of T, a node on the same level as u or a node on the same level as u'. We now distinguish these cases.

(a) The start node of e is r. By symmetry we assume that e ends at v. According to the construction of T the hyperedge e intersects the $2^{n-2}-1$ other hyperedges starting at r and the 2^{n-3} hyperedges starting at u. So altogether e intersects $2^{n-2}+2^{n-3}-1$ hyperedges, as claimed.

(b) The start node of e is on the same level as u. By symmetry we suppose that e starts at u and ends at v'. The hyperedges intersecting e can be divided into the following three categories.

- The hyperedge starting at r and ending at v,
- the $2^{n-3}-1$ hyperedges different from e starting at u, and
- the 2^{n-2} hyperedges starting at u',

implying that e intersects at most $2^{n-2}+2^{n-3}$ hyperedges in total.

(c) The start node of e is on the same level as u'. By symmetry we assume that e starts at u'. Then e intersects the $2^{n-2}-1$ other hyperedges starting at u' and the hyperedge starting at u and ending at v', thus 2^{n-2} hyperedges altogether. □

4 Constructing Suitable (k, d)-Trees

We need some notation first. Let T be a binary tree (not necessarily with all leaves having depth at least $k-1$) and let v be a vertex of T. In the following we denote by the *degree* $d(v)$ of v the number of leaf descendants which have distance at most $k-1$ from v. The proof of Lemma 1 is tedious and too long for this extended abstract. In order to show one of the main ideas of our proof we first prove a weaker claim.

Proposition 1. *A $(k, \lfloor \frac{2^{k+2}}{k} \rfloor)$-tree exists for every sufficiently large k.*

Proof. Let $s = \frac{2^{k+1}}{2^{\lfloor \log k \rfloor}}$ and note that $s \le \frac{2^{k+2}}{k}$. Let T' be a full binary tree of height $k-1$. We subdivide its leaves into intervals of length $\frac{2^{\lfloor \log k \rfloor}}{2}$. Let $\{v_0, \ldots, v_{\frac{2^{\lfloor \log k \rfloor}}{2}-1}\}$ be such an interval. Then we attach a full binary subtree of height i to v_i. Let T denote the resulting tree. It suffices to prove the following.

Proposition 2. *Let v be a vertex of T. Then $d(v) \le s$.*

Proof. We apply induction on the depth i of v. For $i = 0$ the claim is clearly true. Indeed, the degree of the root is $\frac{2^{k-1}}{\frac{2^{\lfloor \log k \rfloor}}{2}} = \frac{2^k}{2^{\lfloor \log k \rfloor}} = \frac{s}{2}$. Now suppose that v has depth $i \in \{1, \ldots, \frac{2^{\lfloor \log k \rfloor}}{2} - 1\}$. Note that the set of descendants of v on level $k-1$ can be subdivided into $\frac{2^{k-1-i}}{\frac{2^{\lfloor \log k \rfloor}}{2}} \ge 1$ intervals. Let v' denote the parent of v. By construction the number of leaf descendants which have distance at most $k-2$ from v equals $\frac{d(v')}{2}$. Moreover, every interval $\{v_0, \ldots, v_{\frac{2^{\lfloor \log k \rfloor}}{2}-1}\}$ gives rise to 2^i leaves on level $k-1+i$, implying that the number of leaf descendants of v which have distance exactly $k-1$ from v equals $\frac{2^{k-1-i}}{\frac{2^{\lfloor \log k \rfloor}}{2}} \cdot 2^i = \frac{2^k}{2^{\lfloor \log k \rfloor}} = \frac{s}{2}$. So altogether $d(v) \le \frac{d(v')}{2} + \frac{s}{2} \le s$. It remains to consider the case where v has depth at least $\frac{2^{\lfloor \log k \rfloor}}{2}$. By construction no leaf of T has depth larger than $\frac{2^{\lfloor \log k \rfloor}}{2} + k - 2$, implying that the degree of v is at most the degree of its parent. □

4.1 Proof Sketch of Lemma 1.(i) and 1.(ii):

Let $s = \frac{2^{k-1}}{2^{\lfloor \log k \rfloor}}$. It suffices to show that there is a (k, s)-tree. We need some notation first. To every node w of a binary tree T we assign a *distance-sequence* $D_w = (x_0, x_1, \ldots, x_{k-1})$ where $x_i \cdot \frac{s}{2^{i+1}}$ is the number of leaf descendants of w which have distance $k-1-i$ from w. This notation encodes the degree of w in a weighted fashion, which allows us to describe our most frequent operations in a more compact way. Note that $d(w) = \sum_{i=0}^{k-1} x_i \cdot \frac{s}{2^{i+1}}$. If, for a sequence $(y_0, y_1, \ldots, y_{k-1})$, we have that $x_i \le y_i$ for every i, $i = 1, \ldots, k-1$ then we write $D_w \le (y_0, y_1, \ldots, y_{k-1})$.

Observation 3. *We have*

(i) *Let T, T' be binary trees whose roots have distance sequence $(x_0, \ldots, x_{k-1})$ and $(x'_0, \ldots, x'_{k-1})$, respectively. Let v be a vertex with left subtree T and right subtree T'. Then*
$D_v = (\frac{x_1+x'_1}{2}, \ldots, \frac{x_{k-1}+x'_{k-1}}{2}, 0)$.

(ii) *Let T' be a binary tree whose root has distance sequence $(x_0, \ldots, x_{k-1})$ and let T be a full binary tree of height $h \leq k-1$. By attaching a copy of T' to every leaf l of T (such that l is the root of T') we obtain $D_v = (x_h, \ldots, x_{k-1}, 0, \ldots, 0)$ for the root v of T.*

We need some more notation. Let $x_0, \ldots, x_{k-1} \in \mathbb{N}$. An $(x_0, x_1, \ldots, x_{k-1})$-*tree* is a nonempty binary tree where (i) every node has degree at most s and (ii) for the root r, $D_r \leq (x_0, x_1, \ldots, x_{k-1})$ and (iii) $\sum_{i=0}^{k-1} x_i \cdot \frac{s}{2^{i+1}} \leq s$. To prove Lemma 1.(i) and (ii) it suffices to show the following.

Lemma 6. *There is an $(x_0, 0, 0, \ldots, 0)$-tree for some $x_0 \geq 0$.*

Lemma 6 guarantees that there is a nonempty binary tree where every vertex has degree at most s and every leaf has depth at least $k-1$, which implies Lemma 1.(i) and (ii). □

Proof sketch of Lemma 6. We divide the proof of Lemma 6 into three propositions. Let $r = \lfloor \frac{\log s}{2} \rfloor - 1$.

Proposition 3. *There is a* $(0, \underbrace{2, \ldots, 2}_{\lceil \frac{r}{2} \rceil}, 0, \underbrace{4, \ldots, 4}_{\lfloor \frac{r}{2} \rfloor}, 0, \ldots 0)$-*tree.*

Proposition 4. *Let $j \leq \lfloor \frac{r}{2} \rfloor - 1$.*
If there is a $(0, \underbrace{2, \ldots, 2}_{r-j-1}, 0, \underbrace{4, \ldots, 4}_{j+1}, 0, \ldots 0)$-*tree then*
there is a $(0, \underbrace{2, \ldots, 2}_{r-j}, 0, \underbrace{4, \ldots, 4}_{j}, 0, \ldots 0)$-*tree.*

Proposition 5. *Let $i \leq r-1$.*
If there is a $(0, \underbrace{2, 2, \ldots, 2}_{i+1}, 0, \ldots, 0)$-*tree then there is a* $(0, \underbrace{2, \ldots, 2}_{i}, 0, \ldots, 0)$-*tree.*

Note that Proposition 3 - 5 together imply Lemma 6 (with $x_0 = 0$). The proofs of Proposition 3 - 5 will appear in the full version.

Acknowledgment. We would like to thank Tibor Szabó for the intensive support and the many fruitful discussions, which significantly simplified the exposition of the proofs.

We are grateful to Emo Welzl for the numerous helpful remarks both on the content and the structure of this paper, in particular for the formulation of Lemma 2.

We are also indebted to Eoin P. Long for pointing out an inaccuracy in an earlier version of this paper.

References

1. Aharoni, R., Linial, N.: Minimal non-two-colorable hypergraphs and minimal unsatisfiable formulas. J. Combin. Theory Ser. A 43, 196–204 (1986)
2. Alon, N., Spencer, J.H.: The Probabilistic Method. John Wiley & Sons, Chichester (2002)
3. Beck, J.: Combinatorial Games: Tic Tac Toe Theory. Encyclopedia of Mathematics and Its Applications 114 (2008)
4. Beck, J.: Remarks on positional games. Acta Math. Acad. Sci. Hungar. 40, 65–71 (1982)
5. Davydov, G., Davydova, I., Kleine Büning, H.: An efficient algorithm for the minimal unsatisfiability problem for a subclass of CNF. Artif. Intell. 23, 229–245 (1998)
6. Erdős, P., Selfridge, J.L.: On a combinatorial game. J. Combinatorial Theory Ser. A 14, 298–301 (1973)
7. Erdős, P., Spencer, J.: Lopsided Lovász local lemma and Latin transversals. Discrete Appl. Math. 30, 151–154 (1991)
8. Hoory, S., Szeider, S.: A note on unsatisfiable k-CNF formulas with few occurrences per variable. SIAM J. Discrete Math. 20(2), 523–528 (2006)
9. Hoory, S., Szeider, S.: Computing Unsatisfiable k-SAT Instances with Few Occurrences per Variable. Theoretical Computer Science 337(1-3), 347–359 (2005)
10. Kleine Büning, H., Zhao, X.: On the structure of some classes of minimal unsatisfiable formulas. Discr. Appl. Math. 130(2), 185–207 (2003)
11. Kratochvíl, J., Savický, P., Tuza, Z.: One more occurrence of variables makes satisfiability jump from trivial to NP-complete. SIAM J. Comput. 22(1), 203–210 (1993)
12. Kullmann, O.: An application of matroid theory to the SAT problem. In: Fifteenth Annual IEEE Conference on Computational Complexity, pp. 116–124 (2000)
13. Moser, R.: A constructive proof of the Lovasz Local Lemma, Eprint, arXiv:0810.4812v2 (2008)
14. Savický, P., Sgall, J.: DNF tautologies with a limited number of occurrences of every variable. Theoret. Comput. Sci. 238(1-2), 495–498 (2000)
15. Scheder, D.: Existence, Size, and Resolution Complexity of Almost Disjoint CNF Formulas (submitted)
16. Szeider, S.: Homomorphisms of conjunctive normal forms. Discr. Appl. Math. 130(2), 351–365 (2003)
17. Tovey, C.A.: A simplified NP-complete satisfiability problem. Discr. Appl. Math. 8(1), 85–89 (1984)

Reconstructing 3-Colored Grids from Horizontal and Vertical Projections Is NP-hard

Christoph Dürr[1], Flavio Guiñez[2], and Martín Matamala[3]

[1] CNRS, LIX (UMR 7161), Ecole Polytechnique, 91128 Palaiseau, France
[2] DIM, Universidad de Chile, Casilla 170-3, Correo 3, Santiago, Chile
[3] DIM and CMM (UMI 2807, CNRS), Universidad de Chile, Casilla 170-3, Correo 3, Santiago, Chile

Abstract. We consider the problem of coloring a grid using k colors with the restriction that in each row and each column has an specific number of cells of each color. In an already classical result, Ryser obtained a necessary and sufficient condition for the existence of such a coloring when two colors are considered. This characterization yields a linear time algorithm for constructing such a coloring when it exists. Gardner *et al.* showed that for $k \geq 7$ the problem is NP-hard. Afterward Chrobak and Dürr improved this result, by proving that it remains NP-hard for $k \geq 4$. We solve the gap by showing that for 3 colors the problem is already NP-hard. Besides we also give some results on tiling tomography.

1 Introduction

Tomography consists of reconstructing spatial objects from lower dimensional projections, and has medical applications as well as non-destructive quality control. In the discrete variant, the objects to be reconstructed are discrete, as for example atoms in a crystaline structure, see [1]. One of the first studied problem in discrete tomography involves the coloring of a grid using a fixed number of colors with the requirement that each row and each column has a specific total number of entries of each color. More formally we are given a set of colors $\mathcal{C}$, and an $m \times n$ matrix M, whose items are elements of $\mathcal{C}$. The projection of M is a sequence of vectors $r^c \in \mathbb{N}^m, s^c \in \mathbb{N}^n$, for $c \in \mathcal{C}$, where $r_i^c = |\{j : M_{ij} = c\}|$ and $s_j^c|\{i : M_{ij} = c\}|$. In the reconstruction problem, we are given only a sequence of vectors satisfying: (1) for $1 \leq i \leq m$, $1 \leq j \leq n$, $c \in \mathcal{C}$, $\sum_c r_i^c = n, \sum_c s_j^c = m, \sum_i r_i^c = \sum_j s_j^c$. The goal is to compute *a* matrix M that has the given projections. If there are $k = |\mathcal{C}|$ colors, we call it the k-COLOR TOMOGRAPHY PROBLEM.[1] It was known since long time, that for 2 colors, the problem can be solved in polynomial time [8]. Ten years ago it was shown that the problem is NP-hard for 7 colors [5]. By NP-hardness, we mean that the decision variant — deciding whether a given instance is *feasible*, i.e. admits a solution — is NP-hard. Shortly after this proof was improved to show

[1] As the projection of one of the colors is redundant by (1), some earlier papers [2,5] refered to this problem as the k'-ATOMS CONSISTENCY PROBLEM for $k' = k - 1$.

A. Fiat and P. Sanders (Eds.): ESA 2009, LNCS 5757, pp. 776–787, 2009.

NP-hardness for 4 colors, leaving open the case when $|\mathcal{C}| = 3$ [2]. This paper closes the gap, by showing that for 3 colors already the problem is NP-hard. Just to fix the notation, for $|\mathcal{C}| = 2$ we denote the colors as *black* and *white*, and use symbols B, W. For $|\mathcal{C}| = 3$ we denote the colors as *red, green* and *yellow* and use symbols R, G, Y. Notice that we can think white and yellow as ground colors in the 2 and 3–color problem, respectively. Thus when we denote the instance of the tomography problem, we sometimes omit the white or yellow projections as they are redundant. In addition for a 2-color instance (r^B, s^B) we omit the superscript when the context permits it. First we recall some well known facts about the 2-COLOR TOMOGRAPHY PROBLEM.

Lemma 1 ([8]). *Let (r, s) be a feasible instance of the 2-color tomography problem. Let I be some set of rows, and J be some set of columns. If $\sum_{i\in I} r_i - \sum_{j\notin J} s_j = |I \times J|$, then any solution to the instance will be all black in $I \times J$ and all white in $\overline{I} \times \overline{J}$.*

Proof: The sets I,J divide the grid into four parts, $A = I \times J$, $B = I \times \overline{J}$, $C = \overline{I} \times J$ and $D = \overline{I} \times \overline{J}$. The value $\sum_{i\in I} r_i$ equals the number of black cells in A and B, and $\sum_{j\notin J} s_j$ the number of black cells in B and D. So the difference is the number of black cells in A minus the number of black cells in D. So when $\sum_{i\in I} r_i - \sum_{j\notin J} s_j = |A|$, A must be all black and D all white. □

Before stating the next lemma, we need to introduce some notation about vectors. The *conjugate* of a vector $s \in \{0, 1, \ldots, m\}^n$ is defined as the vector $s^* \in \{0, 1, \ldots, n\}^m$ where $s^*_i = |\{j : s_j \geq i\}|$. There is a very simple graphical interpretation of this. Let be an $m \times n$ matrix M, such in column j, the first s_j cells are colored black and the others are colored white. Then the conjugate of s is just the row projection of M. Note that s^* is always a non-increasing vector. If in addition s is non-increasing we have that $(s^*)^* = s$ since in this case $s^*_i = \max\{j : s_j \geq i\}$ and $s^*_i \geq j$ if and only if $s_j \geq i$. For every $s, t \in \mathbb{N}^n$ we say that s dominates t, denoted $s \succeq t$, if $\sum_{j=1}^{\ell} s_j \geq \sum_{j=1}^{\ell} t_j$ for every $1 \leq \ell \leq n$. For any $0 \leq k \leq n$ we define the set $\mathcal{X}_{n,k} := \{x \in \{0, 1\}^n : \sum x_i = k\}$. Clearly $\succeq$ defines a partial order on $\mathcal{X}_{n,k}$, and we show now that it has a *small depth.*

Lemma 2 ([2]). *Let n, k be two integers with $0 \leq k \leq n$. Let $b^0 \prec b^1 \prec \ldots \prec b^q$, be a strictly increasing sequence of vectors from $\mathcal{X}_{n,k}$. Then $q \leq k(n - k)$.*

Proof: For each vector $\alpha \in \mathcal{X}_{n,k}$ we associate the number $\varphi(\alpha)$ defined by $\varphi(\alpha) = \sum_{\ell=1}^{n} \sum_{i=1}^{\ell} \alpha_i$. If $\alpha \prec \beta$ then $\sum_{j=1}^{\ell} \alpha_j \leq \sum_{j=1}^{\ell} \beta_j$ for every $1 \leq \ell \leq n$ and the inequality is strict for at least one ℓ. We conclude that $\alpha \prec \beta$ implies $\varphi(\alpha) < \varphi(\beta)$. Then the vectors with extreme values for φ are $\alpha = (0, \ldots, 0, 1, \ldots, 1)$ and $\beta = (1, \ldots, 1, 0, \ldots, 0)$. Since $\varphi(\alpha) = k(k-1)/2$ and $\varphi(\beta) = k(k-1)/2+k(n-k)$, this concludes the proof. □

A well-known characterization of the feasible instances of the 2-COLOR TOMOGRAPHY PROBLEM can be expressed using dominance.

Lemma 3 ([8]). *Let (r, s) be an instance of the 2-color tomography problem, such that r is non-increasing. Then (r, s) is feasible if and only if $r \preceq s^*$.*

Moreover if $r = s^*$, then there is a single solution, namely the realization having the first s_j cells of column j colored black, and the others white.

There is a very simple graphical interpretation of this. Again let M be a matrix where in column j the first s_j cells are colored black and the remaining cells white. Then the row projection of M is s^*, and if $s^* = r$ we are done. Now if $s^* \neq r$, then some of the black cells in M have to be exchanged with some white cells in the same column but a lower row. These operations transform the matrix in such a way, that the new row projection is dominated by s^*. So if s^* does not dominate r, then there is no solution to the instance.

2 The Gadget

The gadget depends on some integers n, k, u, v with $1 \leq k, u, v \leq n$ and $u \neq v$ as well as on two vectors $\alpha, \beta \in \mathcal{X}_{n,k}$. It is defined as the instance of n rows, and $2n+2$ columns with the following projections for $1 \leq i, j \leq n$. If $i \in \{u, v\}$, then $r_i^R = i+1$ and $r_i^G = i$. Otherwise, $r_i^R = i$ and $r_i^G = i+1$. $s_j^R = n - j + \alpha_j$, $s_{n+1}^R = 1$, $s_{n+2}^R = n-k+1$ and $s_{n+2+j}^R = 0$, and $s_j^G = 0$, $s_{n+1}^G = n-1$, $s_{n+2}^G = k-1$, $s_{n+2+j}^G = n-j+1-\beta_j$.

Lemma 4. *If the instance above is feasible then $\alpha \preceq \beta$. Moreover, if $\alpha = \beta$ then the instance is feasible if and only if $\alpha_u + \alpha_v \geq 1$.*

Proof: Assume the instance is feasible, we will show that this implies $\alpha \preceq \beta$. Consider the yellow projection vectors $r^Y = 2n+2-r^R-r^G$ and $s^Y = n-s^R-s^G$. We have that $r_i^Y = 2(n-i)+1$ for $1 \leq i \leq n$. Note that r^Y is a non-increasing vector. Similarly, we obtain that $s_j^Y = j - \alpha_j$ and $s_{n+2+j}^Y = j - 1 + \beta_j$, for $1 \leq j \leq n$, and $s_{n+1}^Y = s_{n+2}^Y = 0$. The conjugate of the column yellow projections is a vector $(s^Y)^*$ with $(s^Y)_i^* = 2(n-i)+1-\alpha_i+\beta_i$. Then clearly $r^Y \preceq (s^Y)^*$ if and only if $\alpha \preceq \beta$. By assumption the 3-color instance $(r^R, r^G, r^Y, s^R, s^G, s^Y)$ is feasible, therefore the 2-color instance (r^Y, s^Y) is feasible as well — where yellow is renamed as black — which by Lemma 3 implies $r^Y \preceq (s^Y)^*$ and therefore also $\alpha \preceq \beta$. This shows the first part of the lemma. Now assume that the instance has a solution, and $\alpha = \beta$. The $n \times (2n+2)$ grid is divided into 3 parts (see figure 1): into an $n \times n$ block (called *RY-block*), a $n \times 2$ rectangle (called *2-column translator*) and another $n \times n$ block (called *GY-block*). Again every block is subdivided into an upper triangle, a diagonal and a lower triangle. Since $\alpha = \beta$, we have $r^Y = (s^Y)^*$. So by Lemma 3 any solution must color in yellow the s_j^Y first cells in every column j, and no other cell. In particular it means that the lower triangle of the RY-block must be red, the lower triangle of the GY-block must be green, and both upper triangles have to be yellow. Also on the first diagonal, the cell (i, i) has to be red if $\alpha_i = 1$ and yellow otherwise. On the second diagonal, the cell $(n+2+i, i)$ must be yellow if $\alpha_i = 1$ and green otherwise. What can we say about the colors of the translator? If $\alpha_u = \alpha_v = 0$, then the cells $(n+1, u), (n+2, u), (n+1, v), (n+2, v)$ have to be all red to satisfy the row projections. This contradicts the column projection $s_{n+1}^R = 1$, and hence the

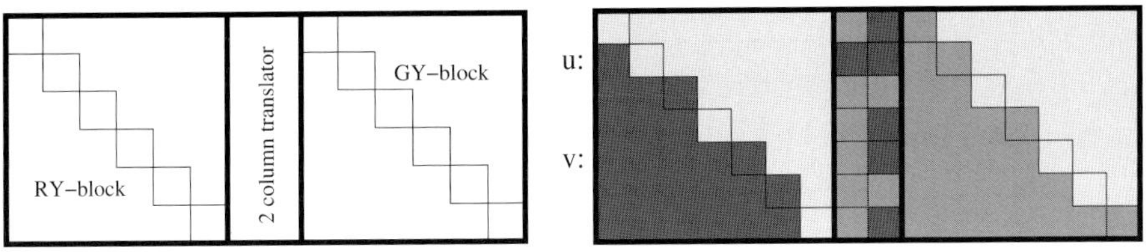

Fig. 1. The structure of the gadget (left) and a realization (right) for $n = 7$, $k = 3$, $u = 2$, $v = 5$ and $\alpha = \beta = (0, 0, 1, 0, 1, 1, 0)$

instance is not feasible. Conversely, assume $\alpha_u + \alpha_v \geq 1$. We will color the cells of the translator in a manner that respects the required projections. If $i \notin \{u, v\}$ and $\alpha_i = 1$ — that is (i, i) is red — we color the cells $(n + 1, i), (n + 2, i)$ in green. If $i \notin \{u, v\}$ and $\alpha_i = 0$, we color the cell $(n + 1, i)$ in green and $(n + 2, i)$ in red. Without loss of generality assume that $\alpha_u = 1$. Hence (u, u) is red and we color $(n + 1, u)$ in green and $(n + 2, u)$ in red. We color $(n + 1, v)$ in red. In addition we color $(n + 2, v)$ in red if $\alpha_v = 0$ and in green otherwise. It can be verified that the coloring defined above is a solution to the instance, which concludes the proof of the lemma. □

3 The Reduction

In this section we will construct a reduction from Vertex Cover to 3-color tomography. We basically use the same approach than in [2], but with a different gadget. Vertex Cover is a well known intractable problem, indeed one of the first 21 problems shown to be NP-hard by Karp [6]. Its input is a graph $G = (V, E)$ and an integer k and its output is a set $S \subseteq V$ of size $|S| = k$ such that $\forall (u, v) \in E$, $u \in S$ or $v \in S$.

Given an instance (G, k) of Vertex Cover, we construct an instance of the 3-color tomography problem which is feasible if and only if the former instance has a solution. Without loss of generality we assume that $k \leq n - 2$. Let be $n = |V|, m = |E|$, and $N = k(n - k)(m - 1) + 1$. We denote the m edges as $E = \{e_0, e_1, \ldots, e_{m-1}\}$, and the n vertices as $V = \{1, 2, \ldots, n\}$. We define an instance with $N(n + 1) + 1$ rows and $N(n + 2) + n$ columns. For row $p = 1, \ldots, N(n+1)+1$, let $x = \lfloor (p-1)/(n+1) \rfloor$ and $i = (p-1) \bmod (n+1)$. We think the set of rows as divided into N blocks of $n+1$ rows each, and a last block with a single row. We have x as the block index and i the row index relative to the block, with $0 \leq i \leq n$. Let $t = x \bmod m$ and consider the edge $e_t = (u, v)$. We define the projections $r^R_p = x(n+2) + z^R_p$ and $r^G_p = (N - x - 1)(n+2) + z^G_p$ where z^R and z^G are vectors defined by

$$z^R_p = \begin{cases} n - k & \text{if } x < N \text{ and } i = 0 \\ 0 & \text{if } x = N \text{ and } i = 0 \\ i + 1 & \text{if } i \in \{u, v\} \\ i & \text{if } i \in \{1, \ldots, n\} \setminus \{u, v\} \end{cases} \qquad z^G_p = \begin{cases} n + 2 & \text{if } x = 0 \text{ and } i = 0 \\ n + 2 + k & \text{if } x > 0 \text{ and } i = 0 \\ i & \text{if } i \in \{u, v\} \\ i + 1 & \text{if } i \in \{1, \ldots, n\} \setminus \{u, v\}. \end{cases}$$

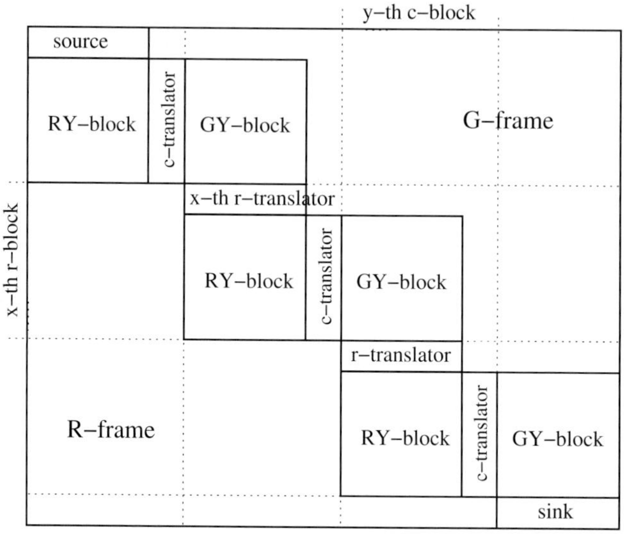

Fig. 2. The general structure of our reduction

In the same manner, for column $q = 1, \ldots, N(n+2)+n$, let $y = \lfloor (q-1)/(n+2) \rfloor$ and $j = ((q-1) \bmod (n+2)) + 1$. The reason for defining j this way, is that if cell (p, q) is part of an RY-block or an GY-block, then (i, j) will be the relative position inside the block with ranges $1 \leq i, j \leq n$. Similarly as for the rows, we think the set of columns as divided into N blocks with $n+2$ columns each and a last block with only n columns.

Again, we have y as the block index, and j as the column index relative to a block with $1 \leq j \leq n+2$. For $y = N$ we set $s^R_q = 0$, for each $j = 1, \ldots, n$. Similarly, for $y = 0$ the green column projections are: $s^G_q = 0$ if $j \in \{1, \ldots, n\}$, $s^G_q = n$ if $j = n+1$ and $s^G_q = k$ if $j = n+2$. For block $1 \leq y \leq N$ we define the red column projections as $s^R_q = (y-1)(n+1)+1+w^R_q$ and $s^G_q = (y-1)(n+1)+1+w^G_q$, where w^R and w^G are given by:

$$w^R_q = \begin{cases} n-j+1 & \text{if } j \in \{1, \ldots, n\} \\ 1 & \text{if } j = n+1 \\ n-k+1 & \text{if } j = n+2, \end{cases} \qquad w^G_q = \begin{cases} j & \text{if } j \in \{1, \ldots, n\} \\ n-1 & \text{if } j = n+1 \\ k-1 & \text{if } j = n+2. \end{cases}$$

As this is a polynomial time reduction, it only remains to show the following.

Theorem 1. *The 3-color tomography instance is feasible if and only if the vertex cover instance is feasible.*

Proof: For one direction of the statement, assume that the vertex cover instance is feasible, and let $b \in \mathcal{X}_{n,k}$ be the characteristic vector of a vertex cover of size k, i.e. $b_i = 1$ if and only if i belongs to the vertex cover. We construct now a solution to the tomography instance. Consider the partitioning of the grid, as in figure 2. For convenience we refer to the source also as the 0-th row translator and to the sink as the $(N+1)$-th row translator. The j-th cell of the x-th row translator is defined as $(x(n+1)+1, x(n+2)+j)$. We color the R-frame in red and the G-frame in green. Let be any $1 \le j \le n$. We color the j-th cell of the source in yellow if $b_j = 1$ and in red otherwise. For $x = 1, \ldots, N-1$ we color the j-cell of the x-th row translator in green if $b_j = 1$ and in red otherwise. In the sink we color the j-th cell in green if $b_j = 1$ and in yellow otherwise. Now for block $x = 0, \ldots, N-1$, consider the instance to the gadget defined by $\alpha = \beta = b$, and u, v such that $(u,v) = e_{x \bmod m}$. By Lemma 4 it is feasible, since b is a vertex cover and hence $b_u + b_v \ge 1$. Then we color the $(n+1) \times (2n+2)$ cells starting at $((x-1)(n+2)+1, (x-1)(n+1)+1)$ exactly as in the solution to the gadget. It is straightforward to check that this grid satisfies the required projections, and therefore the tomography instance is feasible. For the converse, assume that the tomography instance has a solution. For every $x = 1, \ldots, N$ we apply Lemma 1 for the red color and intervals $I = [x(n+1)+1, N(n+1)+1]$ and $J = [1, x(n+2)]$. We deduce that in the solution the R-frame must be all red, and all GY-blocks (and also the G-frame) must be free of any red. Similarly, we show that the G-frame must be all green, and all RY-blocks must be free of any green. This implies that in the source, k cells are yellow, and $n-k$ are red, in the row translators k cells are green and $n-k$ red, and in the sink k cells are green and $n-k$ yellow. We define the vectors $b^0, b^1, \ldots, b^N \in \mathcal{X}_{n,k}$, such that for all $1 \le j \le n$ we have (i) $b^0_j = 1$ iff the j-th cell in the source is yellow, (ii) $b^x_j = 1$ iff the j-th cell in the x-th row translator is green, for all $1 \le x \le N$. For $x = 0, \ldots, N$, consider the part P of the solution that is the intersection of rows $[x(n+1)+2, x(n+1)+n+1]$ and columns $[x(n+2)+1, x(n+2)+2n+2]$. We number the rows of P from 1 to n and the columns from 1 to $2n+2$. Let $(u,v) = e_{x \bmod m}$. By subtracting from the row projections the number of red and green cells in the frames, we deduce that row $1 \le i \le n$ in P contains $i+1$ red cells and i green cells if $i \in \{u,v\}$ and i red cells and $i+1$ green cells if $i \notin \{u,v\}$. We proceed similarly for the columns $n+1$ and $n+2$. By subtracting from the column projections the quantities that are in the frames, we deduce that column $n+1$ of P contains one red cell, and $n-1$ green cells, and column $n+2$ contains $n-k+1$ red cells and $k-1$ green cells. Column $x(n+2)+j$ for $1 \le j \le n$ contains $n-j+1$ red cells that are not in the R-frame. Since GY-blocks are free of red, these cells must either be in the x-th row translator or in column j of P. Note that the j-cell of the x-th row translator is red iff $b^x_j = 0$. Then column j of P contains $n-j+b^x_j$ red cells and no green cell. Similarly column $n+2+j$ of P contains $n-j+1-b^{x+1}_j$ green cells and no red cell. This implies that P is the solution to the gadget defined by u, v, α, β with $\alpha = b^x$ and $\beta = b^{x+1}$. Then by Lemma 4 we obtain that $b^x \preceq b^{x+1}$ and in general $b^0 \preceq b^1 \preceq \ldots \preceq b^N$. By the choice of N and Lemma 2 there exists an ℓ such that $b^\ell = b^{\ell+1} = \ldots = b^{\ell+m}$. By Lemma 4, we have $b^\ell_u + b^\ell_v \ge 1$ for all $(u,v) \in \{e_\ell, e_{\ell+1 \bmod m}, \ldots, e_{\ell+m-1 \bmod m}\} = E$. Then b^ℓ encodes a vertex cover of size k, and this completes the proof. □

4 Related Problems

4.1 Edge-Colored Graphs with Prescribed Degrees

In the Edge-decomposition with Prescribed degrees problem (EPD), a set of two colors $\{R, G\}$, a vertex set V and prescribed degrees $d^R, d^G : V \to \mathbb{N}$ are given, and we have to find two disjoint edge sets $E^R, E^G \subseteq V^2$ such that the graph $G(V, E^R \cup E^G)$ has the required degrees, i.e. for all $v \in V$, $d^R(v) = |\{u : (u,v) \in E^R\}|$ and $d^G(v) = |\{u : (u,v) \in E^G\}|$. Finding an uncolored graph with given degree sequences can be solved in polynomial time, see for example [7]. In constrast, we can reduce the 3-COLOR TOMOGRAPHY PROBLEM to EPD.

Lemma 5. *The problem EPD is NP-hard.*

Proof: We reduce from the 3-color tomography problem. Let (r^R, r^G, s^R, s^G) be an $m \times n$-instance of the 3-color tomography problem. We set $k = n + m$, $V = \{1, \ldots, k\}$, and the following degrees, for $1 \le i \le n$ and $1 \le j \le m$, $d^R(i) = r_i^R + n - 1$, $d^G(i) = r_i^G$, $d^R(n + j) = s_j^R$, and $d^G(n + j) = s_j^G + m - 1$. Now we show that the instance (r^R, r^G, s^R, s^G) is feasible if and only if the instance (d^R, d^G) is feasible. For one direction, assume that there is a solution M to the 3-color tomography instance. We construct a solution E^R, E^G to the graph problem as follows. For any $1 \le i \le n$ and $1 \le j \le m$, if $M_{ij} = R$, then $(i, n + j) \in E^R$, if $M_{ij} = G$, then $(i, n + j) \in E^G$. Also for any $1 \le i < i' \le n$, we have $(i, i') \in E^R$ and for any $1 \le j < j' \le m$, we have $(n + j, n + j') \in E^G$. Now clearly E^R, E^G satisfy the required degrees. For the converse, we define the quantity $\Phi = \sum_{i=1}^n d^R(i) - \sum_{j=1}^m d^R(n + j)$. By assumption (1) this value is $n(n-1)$. Since this value equals also $|E^R \cap \{1, \ldots, n\}^2| - |E^R \cap \{n+1, \ldots, n+m\}^2|$, there is a red edge between every pair of vertices (i, i') with $1 \le i < i' < n$, and no edge between every pair of vertices $(n + j, n + j')$ with $1 \le j < j' \le m$. Similarly we can show that there is a green edge between every pair of vertices $(n + j, n + j')$ with $1 \le j < j' \le m$. Now let M be the $m \times n$ grid, with cell (i, j) colored in red if $(i, n + j) \in E^R$, and in green if $(i, n + j) \in E^G$. By the degree requirements, M is a solution to the 3-color tomography instance. □

4.2 Tiling Tomography

Tiling tomography was introduced in [3], and it consists of constructing a tiling that satisfies some given row and column projections for each type of tiles we admit. Formally a tile is a finite set T of cells of the grid $\mathbb{N} \times \mathbb{N}$, that are 4-connected, in the sense that the graph $G(T; E)$ is connected for $E = \{((i,j), (i',j')) : |i - i'| + |j - j'| = 1\}$. By $T + (i', j') = \{(i + i', j + j') : (i, j) \in T\}$ we denote a copy of T that is shifted i' units down and j' units to the right. We say that a set of tiles is *feasible* if they do not intersect. In addition we say that it tiles the $m \times n$ grid if its (disjoint) union equals the set of all grid cells, and we refer it as a *tiling*. In the Tiling Tomography Problem, denoted by TTP in the sequel, we are given a finite set of tiles $\mathcal{T} = \{T_1, \ldots, T_k\}$, and vectors $r^d \in \mathbb{N}^m$, $s^d \in \mathbb{N}^n$ for $1 \le d \le k$. The goal is to compute a matrix $M \in \{0, 1, \ldots, k\}$ such that the set $\{T_{M_{ij}} + (i, j) : 1 \leq i \leq n, 1 \leq j \leq m\}$ is a tiling of the $m \times n$ grid, with the projections $r_i^d = |\{j : M_{ij} = d\}|$ and $s_j^d = |\{i : M_{ij} = d\}|$. By *width* and *height* of a tile T we understand the size of the smallest intervals I, J such that $T \subseteq I \times J$. This definition extends to set of tiles. A tile T is said to be *rectangle-like* if for every (i', j') such that $\{T, T + (i', j')\}$ is feasible, we have that the width of $\{T, T + (i', j')\}$ is at least twice the width of T or the height of the set is at least twice the height of T. It was conjectured in [3], that for T_1 being a single cell and T_2 a tile which is not rectangle-like, the $\{T_1, T_2\}$-tiling tomography problem is NP-hard. This question is still open and intriguing.

4.3 Rectangular Tiles

Consider two rectangular tiles, T_1 being a $p_1 \times q_1$ rectangle and T_2 a $p_2 \times q_2$ rectangle, i.e. $T_c = \{0, \ldots, p_c - 1\} \times \{0, \ldots, q_c - 1\}$, for $c \in \{1, 2\}$. What can be said about the complexity of the $\{T_1, T_2\}$-TTP? If $\gcd(p_1, p_2) = d > 1$, then clearly any solution $\bar{M} \in \{0, 1, 2\}^{m \times n}$ to a $\{T_1, T_2\}$-tiling tomography instance (r^c, s^c), must satisfy that if $\bar{M}_{ij} \neq 0$, then $i \bmod d = 1$. Then the $\{T_1, T_2\}$-TTP can be reduced to the $\{T_1', T_2'\}$-TTP, with T_1' being a $(p_1/d) \times q_1$ rectangle, and T_2' a $(p_2/d) \times q_2$ rectangle. We omit the formal reduction, which is straightforward. From now on suppose that $\gcd(p_1, p_2) = \gcd(q_1, q_2) = 1$. We distinguish the following cases, up to row-column symmetry: If $p_1 = p_2 = 1$, that is the tiles are two horizontal bars of length q_1 and q_2, then the problem can be solved in polynomial time (Theorem 2). We use an idea already present in [4], where it is proven for $q_1 = 1$; If $p_1 = q_2 = 1$ and $p_2 = q_1 = 2$, then the tiles are called dominoes, and again the problem can be solved in polynomial time, although with a more involved algorithm [9]; If $p_1 = q_2 = 1$, $p_2 \geq 2$ and $q_1 \geq 3$ then the problem is open. The first author conjectures that the problem is NP-hard, while the other two conjecture that it could be solved in polynomial time with a similar approach as in [9]; If $p_1, q_1 \geq 2$, then the problem is NP-hard (Theorem 3). In [3] the special case $p_1 = q_1 = 2$, $p_2 = q_2 = 1$ was related to the 3-color tomography problem, and it is therefore also NP-hard. We generalize this reduction in section 4.6; If there is a third rectangular tile T_3, then for the tile set $\{T_1, T_2, T_3\}$ the problem is NP-hard, see section 4.7.

4.4 An Algorithm for Vertical Bars

Theorem 2. *The TTP can be solved in polynomial time for two rectangular tiles of dimensions* $p_1 \times 1$ *and* $p_2 \times 1$.

Proof: The algorithm is the simple greedy algorithm, as the one used in [4]. It iteratively *stacks* bars in the matrix. Formally the algorithm is defined like this. We construct a matrix $A \in \{0, 1, 2\}^{m \times n}$ with the required projections. Initially A is all 0. We maintain a vector v such that v_j is the minimal i such that $A_{i,j} \neq 0$, and $v_j = m + 1$ if column j of A is all zero. Initially $v_j = m + 1$ for all $1 \leq j \leq n$. We also maintain vectors $\bar{r}^1, \bar{r}^2, \bar{s}^1, \bar{s}^2$, which represent the remaining projections. Initially they equal the given projections of the instance.

The vectors $(v, \bar{r}^1, \bar{r}^2, \bar{s}^1, \bar{s}^2)$ define a more general tiling problem, where in every column j, only the first $v_j - 1$ cells have to be tiled.

The algorithm: Let $i = \max v_j$. If $i = 1$ we are done, and return A, if all vectors $\bar{r}^1, \bar{r}^2, \bar{s}^1, \bar{s}^2$ are zero, and return "no solution" otherwise.
If $i > 1$, let $i_1 = i - p_1$ and $i_2 = i - p_2$. If $\bar{r}^1_{i_1} = \bar{r}^2_{i_2} = 0$, abort and return "no solution". Otherwise let $c \in \{1, 2\}$ such that $\bar{r}^c_{i_c} > 0$. Let j be a column with $v_j = i$ that maximizes $\bar{s}^c_j$. Then *drop* the bar $p_c \times 1$ in column j, i.e. set $A_{i_c,j} = c$, and decrease $\bar{r}^c_{i_c}$ and $\bar{s}^c_j$. Repeat the whole step.

Clearly, if this algorithm produces a matrix, then it defines a valid tiling with the required projections. We have to show that if the instance has a solution, then the algorithm will actually find one. For this purpose, suppose that some intermediate instance $\mathcal{I} := (v, \bar{r}^1, \bar{r}^2, \bar{s}^1, \bar{s}^2)$ is feasible. We show that an iteration of the algorithm preserves feasibility. Let $i = \max v_j$. If $i = 1$, then $\bar{r}^1, \ldots, \bar{s}^2$ are all zero, since the instance is feasible. Let $i_1 = i - p_1$ and $i_2 = i - p_2$. We have that either $M_{i_1,j} = 1$ or $M_{i_2,j} = 2$ for every column j satisfying $v_j = i$, since M is a valid tiling. Then some of $\bar{r}^1_{i_1}, \bar{r}^2_{i_2}$ must be non zero. Let c, j be the values the algorithm chooses. Let $\mathcal{I}'$ be the instance obtained after the iteration of the algorithm, that is $\bar{r}^c_{i_c}, \bar{s}^c_j$ are decreased by 1 and v_j by p_c. If $M_{i_c,j} = c$, then M' which equals M except for $M_{i_c,j} = 0$ is a solution to $\mathcal{I}'$. If $M_{i_c,j} \neq c$, then by the projections, there must be a another column k with $v_k = i$ and $M_{i_c,k} = c$. We will now transform M such that $M_{i_c,j} = c$. Then we are in the case above and done. By the choice of the algorithm we have $\bar{s}^c_k \leq \bar{s}^c_j$. By this inequality, there exists i_0 such that the total number of c's below the row i_0 is the same in both column j and column k. Take i_0 being the largest one satisfying that. By the choice of i_0 we have that $M_{i_0,k} = c$ and $M_{i_0,j} \neq c$. Since M is a valid tiling, then the restriction to cells below i_0 in column k is also a tiling and then $M_{i_0,j} \neq 0$. We conclude that between i_0 and i the number of 1's and 2's in column j is the same as in column k. Then exchanging the parts of columns j and k in M between i_0 and i, does not change the projections of M, and we obtain the required property $M_{i_c,j} = c$. □

4.5 A General NP-Hardness Proof Structure

In the next section we will reduce the 3-color tomography problem to the TTP for some fixed set of tiles $\mathcal{T}$. The proof uses a particular structure that we explain now. Let $(r^R, r^G, r^Y, s^R, s^G, s^Y)$ be an instance to the 3-color tomography problem for an $m \times n$ grid. In the reduction we will choose constant size grid $\ell \times k$ — that we call a *block* — and three $\mathcal{T}$-tilings of it, that we denote $\bar{M}^R, \bar{M}^G, \bar{M}^Y$. Let $\bar{r}^{c,d}, \bar{s}^{c,d}$ be the T_d-projections of the tiling $\bar{M}^c$ for $c \in \{R, G, Y\}$ and $d \in \{1, 2\}$. There will be two requirements: **The first requirement** is that the vectors $\{\bar{r}^{R,1}, \bar{r}^{G,1}, \bar{r}^{Y,1}\}$ are affine linear independent. The same requirement holds for the column projections $\{\bar{s}^{R,1}, \bar{s}^{G,1}, \bar{s}^{Y,1}\}$. This implies that every vector r spanned by $\bar{r}^{R,1}, \bar{r}^{G,1}, \bar{r}^{Y,1}$, has a unique decomposition into $r = n_R\bar{r}^R + n_G\bar{r}^G + n_Y\bar{r}^Y$ for $n_R + n_G + n_Y = n$. The reduction, consists of an $m\ell \times nk$ grid, and the projections $1 \leq i \leq \ell, 1 \leq j \leq k, 1 \leq x \leq m,$

$1 \le y \le n$, $d \in \{1,2\}$, $r^d_{x\ell-\ell+i} = \sum_c r^c_x \cdot \bar{r}^{c,d}_i$ and $s^d_{yk-k+j} = \sum_c s^c_y \cdot \bar{s}^{c,d}_j$. The idea is that the $m\ell \times nk$ is partitioned into mn blocks of dimension $\ell \times k$. **The second requirement** is that in every solution $\bar{M}$ to the tiling instance, all blocks of $\bar{M}$, are either $\bar{M}^R, \bar{M}^G, \bar{M}^Y$ or blocks that have equivalent projections.

Lemma 6. *The instance to the $\mathcal{T}$-tiling problem has a solution if and only if the instance to the 3-color tomography problem has a solution.*

Proof: Let $M \in \{R,G,Y\}^{m\times n}$ be a solution to the 3-color tomography problem. We transform it into a matrix $\bar{M} \in \{0,1,2\}^{m\ell\times nk}$ by replacing each cell (i,j) of M by the $\ell \times k$ matrix $\bar{M}^c$ for $c = M_{ij}$. By construction, this is a solution to the tiling problem. For the converse, suppose that there is a solution $\bar{M}$ to the tiling problem. By the second requirement, every block of $\bar{M}$ can be associated to one of the colors $\{R,G,Y\}$. We construct a matrix $M \in \{R,G,Y\}^{m\times n}$ such that $M_{xy} = c$ if the block (x,y) of $\bar{M}$ is $\bar{M}^c$, or something projection equivalent. Fix some arbitrary $1 \le x \le m$. By the first requirement, the projections of the rows $x\ell - \ell + 1, \ldots, x\ell$ have a unique decomposition into $n_R \bar{r}^{R,1} + n_G \bar{r}^{G,1} + n_Y \bar{r}^{Y,1}$ with $n_R + n_G + n_Y = n$. By the definitions of the projections $n_R = r^R_x, n_G = r^G_x, n_Y = r^Y_x$, and then row x of M has the required projections. We proceed in the same manner for the columns and show that M is a solution to the 3-color tomography instance. □

4.6 An NP-Hardness Proof for Two Rectangular Tiles

Theorem 3. *The TTP is NP-hard for two rectangular tiles of dimensions $p_1 \times q_1$ and $p_2 \times q_2$ with $\gcd(p_1,p_2) = \gcd(q_1,q_2) = 1$ and $p_1, q_1 \ge 2$.*

Proof: We apply Lemma 6 for $\ell = 2p_1p_2$ and $k = 2q_1q_2$. The 3 tilings of the $\ell \times k$ grid are defined formally as follows. Let us denote by $||a,b||$ the set of integers $\{a, a+1, \ldots, b\}$. The rows $I = \{1,\ldots,\ell\}$ and the columns $J = \{1,\ldots,k\}$ are partitioned into sets I_1, I_2, I_3, I_4 and J_1, J_2, J_3, J_4 defined as $I_1 = ||1,p_2||$, $J_1 = ||1,q_2||$, $I_2 = ||p_2+1, p_1p_2||$, $J_2 = ||q_2+1, q_1q_2||$, $I_3 = ||p_1p_2+1, p_1p_2+p_2||$, $J_3 = ||q_1q_2+1, q_1q_2+q_2||$, $I_4 = ||p_1p_2+p_2+1, 2p_1p_2||$, $J_4 = ||q_1q_2+q_2+1, 2q_1q_2||$. Then $\bar{M}^R$ is defined as the block tiling that covers $(I_1 \cup I_4) \times (J_3 \cup J_4)$ with T_2 and the rest with T_1, $\bar{M}^G$ is defined as the block tiling that covers $(I_3 \cup I_4) \times (J_1 \cup J_4)$ with T_2 and the rest with T_1, while $\bar{M}^Y$ is defined as a tiling using only T_1. These tilings are uniquely defined. Clearly the row T_1-projections of the 3 tilings are affine linear independent, so the first requirement of the construction is satisfied. The second requirement follows from a sequence of observations. Let $\bar{M}$ be the solution to the tiling instance, obtained by reduction from a 3-color instance $(r^R, r^G, r^Y, s^R, s^G, s^Y)$. First note that in the tilings $\bar{M}^R, \bar{M}^G, \bar{M}^Y$, every tile is completely contained in the $\ell \times k$ block. Then the tiling instance has zero projections for T_1 at rows x with $(x-1) \bmod \ell > \ell - p_1 + 2$. A similar observation holds for tile T_2 and for the column projections. As a result in $\bar{M}$ every tile is completely contained in some $\ell \times k$ block, and in other words every block of $\bar{M}$ is $\{T_1, T_2\}$-tiled. What can we say about the possible tilings? Again note that in the tilings $\bar{M}^R, \bar{M}^G, \bar{M}^Y$, every row in I_2 is completely covered by T_1-tiles. Then by the projections, this holds also for every block in $\bar{M}$. The same observation can be done about columns in J_2. Note that if $ap_1+bp_2 = 2p_1p_2$, then $(a, b) \in \{(0, 2p_1), (2p_2, 0), (p_2, p_1)\}$. This is simply because by $\gcd(p_1, p_2) = 1$, in any solution to $ap_1 = p_2(2p_1 - b)$, a must be a multiple of p_2. Together with the previous observation, this implies that every column of a block is either covered completely by T_1-tiles or covered half by T_1-tiles and half by T_2-tiles. The same observation holds for the rows. The trickiest observation of this proof is that in every block of $\bar{M}$, the region $I_1 \times J_1$ is covered by T_1. For a proof by contradiction, suppose it is covered by T_2, in fact by a single tile T_2 since $|I_1 \times J_1| = |T_2|$. But since $I_2 \times J$ is covered with T_2, and by $\gcd(q_1, q_2) = 1$, it must be that the cell (p_2+1, q_2+1) is covered by a tile $T_2 + (p_2+1, j)$ for some column $j \le q_2$. By the same argument, the cell $(p_2 + 1, q_2 + 1)$ is *also* covered by a tile $T_2 + (q_2 + 1, i)$ for some row $i \le p_2$. Then these two tiles overlap in $(p_2 + 1, q_2 + 1)$, which contradicts that M is a (valid) tiling. Now fix a block of $\bar{M}$. If row 1 is partly covered by T_2, then T_2-tiles must cover the half columns in J. Hence in the row 1 they cover exactly the columns in $J_3 \cup J_4$. The same argument shows that every column $j \in J_3 \cup J_4$ is then half covered by T_2-tiles. Previous observation state that $I_2 \times \{j\}$ is covered by T_1. But the length of I_2 is a not a multiple of p_1. Then $(p_1p_2 + 1, j)$ must then also be covered by T_1 and hence $(I_2 \cup I_3) \cup \{j\}$ is covered by T_1-tiles. Then $(I_1 \cup I_4) \times \{j\}$ is covered by T_2. The choice of j was arbitrary, and therefore the block-tiling is exactly $\bar{M}^R$. Similarly we deduce that if column 1 is covered partly by T_2, then the block-tiling is exactly $\bar{M}^G$. Now if row 1 and column 1 are completely covered by T_1, then $(I_1 \cup I_2) \times J$ and $I \times (J_1 \cup J_2)$ are completely covered by T_2-tiles. As a result the block-tiling only contains in $(I_3 \cup I_4) \times (J_3 \cup J_4)$ either T_1-tiles or T_2-tiles, that correspond with the $\bar{M}^Y$ tiling and another we call the bad tiling, respectively. We will show that no bad tiling appears in $\bar{M}$. Let N_R be the number of blocks in $\bar{M}$ that are $\bar{M}^R$. Similarly, let N_B be number of bad block-tilings in $\bar{M}$. Note that the row projection of a bad tiling equal the row projections of $\bar{M}^G$ and that the column projections equal the projections of $\bar{M}^R$. Then by the projections we have the equalities $N_R = \sum_i r_i^R$ and $N_R + N_B = \sum_j s_j^R$. Since by assumption $\sum_i r_i^R = \sum_j s_j^R$, we have $N_B = 0$. This shows the second requirement of our construction, and by Lemma 6 completes the proof. □

4.7 An NP-Hardness Proof for Three Rectangular Tiles

Theorem 4. *The TTP is NP-hard for any 3 rectangular tiles.*

Proof:[sketch] The idea of the construction is that we apply the general proof scheme from section 4.5 with 3 tilings $\bar{M}^R, \bar{M}^G, \bar{M}^Y$, such that $\bar{M}^R$ contains tile T_1 in position $(0, 0)$, $\bar{M}^G$ contains T_2 and $\bar{M}^Y$ contains T_3 in position $(0, 0)$. Moreover each of the 3 tiling minimizes lexicographically n_1, n_2, n_3, where n_c is the number of tiles T_c in the tiling. □

Acknowledgement

We thank Christophe Picouleau and Dominique de Werra for correcting an earlier version of this manuscript. This work is partially supported by the FONDAP and BASAL-CMM projects.

References

1. Alpers, A., Rodek, L., Poulsen, H.F., Knudsen, E., Herman, G.T.: Discrete Tomography for Generating Grain Maps of Polycrystals. In: Advances in Discrete Tomography and Its Applications, pp. 271–301. Birkhäuser, Basel (2007)
2. Chrobak, M., Dürr, C.: Reconstructing polyatomic structures from discrete X-rays: NP-completeness proof for three atoms. Theoretical Computer Science 259, 81–98 (2001)
3. Chrobak, M., Couperus, P., Dürr, C., Woeginger, G.: On tiling under tomographic constraints. Theoret. Comput. Sci. 290(3), 2125–2136 (2003)
4. Dürr, C., Goles, E., Rapaport, I., Rémila, E.: Tiling with bars under tomographic constraints. Theoret. Comput. Sci. 290(3), 1317–1329 (2003)
5. Gardner, R., Gritzmann, P., Prangenberg, D.: On the computational complexity of determining polyatomic structures by X-rays. Theoretical Computer Science 233, 91–106 (2000)
6. Garey, M.R., Johnson, D.S.: Computers and Intractability: A Guide to the Theory of NP-Completeness. W.H.Freeman and Co., New York (1979)
7. Kuba, A., Herman, G.T.: Discrete tomography: A Historical Overview. In: Discrete tomography: Foundations, Algorithms and Applications. Birkhäuser, Basel (1999)
8. Ryser, H.J.: Matrices of zeros and ones. Bull. Am. Math. Soc. 66, 442–464 (1960)
9. Thiant, N.: Constructions et reconstructions de pavages de dominos. PhD thesis, Université Paris 6 (2006)

Author Index